WIND EFFECTS ON CABLE-SUPPORTED BRIDGES

Erratum

The publisher would like to correct an error in the Table of Contents and the Foreword.

Please note that the first Foreword is written by Professor Ahsan Kareem only and his affiliations is as below:

Ahsan Kareem, NAE, Dist. Mem. ASCE

Robert M Moran Professor of Engineering

NatHaz Modeling Laboratory

University of Notre Dame, Indiana

The publisher apologizes for any inconvenience caused to Professor Kareem, Professor Xu and the readers. The amendment has been corrected in the online version as well as future reprints.

WIND EFFECTS ON CABLE-SUPPORTED BRIDGES

You-Lin Xu
The Hong Kong Polytechnic University, Hong Kong, P. R. China

WILEY

This edition first published 2013

Registered office

John Wiley & Sons Singapore Pte. Ltd., 1 Fusionopolis Walk, #07-01 Solaris South Tower, Singapore 138628

For details of our global editorial offices, for customer services and for information about how to apply for permission to reuse the copyright material in this book please see our website at www.wiley.com.

Library of Congress Cataloging-in-Publication Data

Xu, You-Lin, 1952-
Wind effects on cable-supported bridges / You-Lin Xu.
p. cm.
Includes bibliographical references and index.
ISBN 978-1-118-18828-6 (cloth)
1. Cable-stayed bridges. 2. Wind-pressure. I. Title.
TG405.X885 2013
624.2′52–dc23
2012038933

ISBN: 9781118188286

Set in 9/11 pt Times by Thomson Digital, Noida, India
Printed and bound in Singapore by Markono Print Media Pte Ltd

To

Wei-Jian, my wife; Anna, my daughter

–Y.L. Xu

Contents

Foreword

by Ahsan Kareem and Robert M Moran

Economic growth and modernization has resulted in greater demand for cable-supported bridges that are designed to carry large volumes of vehicle traffic and railways over a long span. At the end of year 2010, there were at least ten suspension bridges with a main span over 1,200 m and ten cable-stayed bridges with a main span over 700 m.

Ambitious cable-supported bridge projects are often constructed along coastal areas and in or near to cities that are vulnerable to typhoons and high winds; therefore, wind effects on cable-supported bridges are key issues in their design, construction, operation and maintenance. However, a good knowledge and understanding of bridge aerodynamics is not widespread amongst university students and practicing engineers, mainly because of the multi-disciplinary nature of the subject. Another reason is that there is no book that addresses this topic comprehensively.

In anticipation that more cable-supported bridges and some super-long-span cable-supported bridges will be designed and built in the 21 century throughout the world, the demand for such a book will increase as this subject shall be taught at major universities and colleges at least at the postgraduate level. A comprehensive book like this one on this subject, covering not only the fundamental knowledge but also state-of-the-art developments, will certainly facilitate learning and preparation of students to face the challenges posed by the bridges of tomorrow.

Dr. You-Lin Xu has conducted teaching, research and consultancy work in the field of wind engineering and bridge engineering for almost 30 years. He was engaged in wind tunnel studies and wind-induced vibration control in Australia from 1989 to 1995. Together with his students and colleagues, he has worked extensively on wind loading and effects on the Tsing Ma suspension bridge in Hong Kong since 1995 and on the Stonecutters cable-stayed bridge in Hong Kong since 2003. In recognition of his contribution, he was awarded the Robert H. Scanlan Medal by the American Society of Civil Engineers in 2012. This medal is awarded to those who make extraordinary contributions to mechanics and its applications, with special reference to bridge aeroelasticity. He has taught the subject "Wind Engineering" to MSc students at The Hong Kong Polytechnic University since 1998.

Essentially, the fundamentals presented in the book are drawn from his lecture notes, and most of the state-of-the-art developments presented in the book have resulted from their relevant publications in international journals and conference proceedings. Dr. Xu, in my opinion, is at the right stage of his career, with the distinguished academic and profession background, to synthesize this interdisciplinary work into a comprehensive and exhaustive book. I am confident that it will be very well received, both in academia and in design practice.

The history and latest developments of bridge wind engineering both demonstrate that any progress made in this subject evolves from the synergy between research and practice. This book does bridge the gap between the theoretical research and practical application. Covering a comprehensive range of topics and the most up to date information on the subject, it will also inspire researchers and academics to pursue new methodologies and innovative technologies for the design, construction, operation and maintenance of wind-excited cable-supported bridges.

Ahsan Kareem, NAE, Dist. Mem. ASCE
Robert M Moran, Professor of Engineering
NatHaz Modeling Laboratory
University of Notre Dame, Indiana

Foreword

by Hai-Fan Xiang

To meet the social and economic needs for efficient transportation systems, many cable-supported bridges have been built throughout the world. The Nanpu cable-stayed bridge, with a main span of 423 m, which was successfully built in Shanghai in 1991, marks a milestone in the history of the construction of cable-supported bridges in China. Since then, a high tide of construction of long-span cable-supported bridges has emerged in China. As of 2012, among the ten longest suspension bridges of a main span over 1200 m in the world, five are in China. Similarly, there are five cable-stayed bridges in China among the world's ten longest cable-stayed bridges with a main span over 700 m.

With accumulated experience and advanced technology, the construction of super-long-span cable-supported bridges to cross straits has also been planned around the world, such as Messina Strait in Italy, Qiongzhou Strait in China, Sunda Strait in Indonesia and Tsugaru Strait in Japan. However, as span length increases, cable-supported bridges are becoming lighter in weight, more slender in stiffness, lower in damping and more sensitive to wind-induced vibration. The requirements of functionality, safety, and sustainability of the bridges against wind hazards have presented new challenges to our wind engineering community. A comprehensive book like this one on this subject, covering not only the fundamental knowledge but also state-of-the-art developments, will definitely help the learning and preparation of our students and engineers who face these challenges.

Dr You-Lin Xu graduated from Tongji University, where I have been working for about 60 years. Dr Xu and his research team at The Hong Kong Polytechnic University have worked extensively on wind loading and effect on the Tsing Ma suspension bridge in Hong Kong since 1995 and on the Stonecutters cable-stayed bridge in Hong Kong since 2003. This book is structured to systemically move from introductory areas through to advanced topics with real-world examples. It should serve well to advance the research and practice in the field of wind engineering in general, and wind effects on cable-supported bridges in particular.

This book is actually a summary of the work they have done in the past 17 years. I would give my warm congratulation to Dr Xu for this excellent work.

Hai-Fan Xiang, Professor Emeritus,
MCAE, Advisory Dean
College of Civil Engineering
Tongji University, Shanghai, China

Preface

The well-known collapse of the original Tacoma Narrows Bridge in 1940 identified the importance of understanding wind effects on long-span cable-supported bridges. Extensive research and practice have been carried out since then. The pioneering work of Professors Robert Scanlan and Alan Davenport, among others in the 1960s and 1970s, laid down a foundation for the subject of bridge wind engineering. The advanced theories and modern technologies developed in the past 40 years have made it possible to construct suspension bridges with a main span over 1990 m and cable-stayed bridges with a main span over 1000 m.

The need to construct super-long-span cable-supported bridges to cross straits and to bring people together to live and work has become more obvious in this century. However, the great increase in destructive wind storms due to global climate change has affected many parts of the world, and the vulnerability of long-span cable-supported bridges to strong winds has been increased significantly. The demand imposed by this on the functionality, safety and sustainability of super-long-span cable-supported bridges against wind hazard has presented new challenges to our wind engineering community. This was the original incentive of the author in writing this book: to provide the fundamental knowledge from which modern bridge wind engineering has evolved for our graduate students, and to present the state-of-the-art development from the past 40 years in the field to both graduate students and practicing engineers, so that we are better prepared for new challenges in bridge aerodynamics.

Chapter 1 of this book presents the background materials, including basic notions of meteorology, basic types of wind storms, basic types of cable-supported bridges, wind damage to cable-supported bridges and history of bridge aerodynamics. Chapter 2 introduces wind characteristics in atmospheric boundary layer. Chapters 3 to 6 respectively describe mean wind load and aerostatic instability of bridges, wind-induced vibration and aerodynamic instability of bridges, wind-induced vibration of stay cables and wind-vehicle-bridge interaction. These contents cover the fundamentals of bridge aerodynamics, which are suitable as an elective subject for final-year undergraduate students. As three important tools in the studies of bridge aerodynamics in addition to theoretical analysis discussed in the previous chapters, Chapters 7 to 9 respectively introduce wind tunnel technique, computational wind engineering simulation and wind/structural health monitoring technology.

The materials presented in the first nine chapters are appropriate for graduate student courses. Special topics, such as buffeting response to skew winds, multiple loading-induced fatigue analysis, wind-induced vibration control and typhoon wind field simulation, are introduced in Chapters 10 to 13, respectively. These chapters are rather independent of the others and can be used individually. In Chapter 14, reliability analysis of wind-excited bridges is described, laying down a foundation for the probabilistic wind-resistant design of long-span cable-supported bridges. As a frontier in this field, Chapter 15 presents the preliminary study results on non-stationary and nonlinear buffeting responses. Finally, challenges and prospects of bridge aerodynamics as a scientific but practical subject are highlighted in Chapter 16.

The history and latest developments in bridge wind engineering both demonstrate that any progress made in this subject stems from the synergy between research and practice. Case studies on real long-span cable-supported bridges are provided within almost every chapter of this book.

I embarked on the field of wind engineering in 1983 as a master student at Tongji University, China, and continued my study in this field as a PhD student at University of Sydney in 1989 in Australia. I have, fortunately, been involved in long-term collaborative research and practice with the Hong Kong Highways Department since 1995 for wind effects on, and structural health monitoring of, the Tsing Ma suspension bridge and the Stonecutters cable-stayed bridge in Hong Kong. I have taught the subject "Wind Engineering" to Master of Science students since 1998 at The Hong Kong Polytechnic University. Most importantly, I have been inspired by the work of many outstanding scholars and engineers in the past 30 years, and I would like to dedicate this book to them.

In writing the book, I am always reminded that it mainly serves as a textbook for graduate students and practicing engineers to understand bridge aerodynamics and straddle the gap between theoretical research and practical application. Its readers are assumed to have some background in structural analysis, structural dynamics, probability theory, and random vibration.

I would be very happy to receive constructive comments and suggestions from readers.

You-Lin Xu
The Hong Kong Polytechnic University
ceylxu@polyu.edu.hk
August 2012

Acknowledgements

The writing of this book has been a challenging and time-consuming task that could not have been completed without the help of many individuals. I am grateful to the many people who helped in the preparation of this book.

A few PhD students, former and present, at The Hong Kong Polytechnic University participated in some research works presented in this book: Prof. Le-Dong Zhu, Prof. Wen-Hua Guo, Prof. Wen-Shou Zhang, Dr. Zhi-Wei Chen, Dr. Wen-Feng Huang, Dr. Wai-Shan Chan, Dr. Ling-Yun Wang, Dr. Kei-Man Shum, Dr. Chi-Lun Ng, Mr. Qing Zhu, and Mr. Bin Wang.

Several colleagues and research staff, former and present, at The Hong Kong Polytechnic University made contributions to some research works described in this book: Prof. He Xia, Prof. Jia-Hao Lin, Prof. Jun Chen, Prof. Yong-Le Li, Prof. Nan Zhang, Dr. Yong Xia, Dr. Quan-Shun Ding, Dr. Liang Hu, Dr. Hai-Jun Zhou, Dr. Guo Liu, Dr. Lin Huang, Dr. Ting-Ting Liu, Dr. Zhi-Gi Li, and Mr. Shan Zhan.

In particular, Mr. Cao-Dong Zhang was involved in the writing of Chapter 1; Dr. Wen-Feng Huang was involved in the writing of Chapters 2, 12 and 13; Mr. Qing Zhu participated in the writing of Chapters 3, 4, and 7; Miss Yi-Xin Peng drew some figures in Chapter 7; Mr. Bin Wang participated in the writing of Chapter 8; Dr. Yong Xia was engaged in the writing of Chapter 9; Dr. Zhi-Wei Chen prepared the first draft of Chapter 14; Dr Liang Hu prepared the first draft of Chapter 15; Mr Xiao-Liang Meng proof-read the book and prepared the index; and Mr Jian-Fu Lin examined all the figures and tables, for which I am most grateful.

I wish to thank Dr. Chi-Ho Hui and Dr. Kai-Yuen Wong, the chief engineer and the senior engineer respectively, from the Hong Kong Highways Department for their strong support when we were working on the relevant projects of the Stonecutters Bridge and the Tsing Ma Bridge in Hong Kong.

I wish to express my appreciation to my former professors who provided a foundation of knowledge on which I could build. I would also like to acknowledge the anonymous reviewers whose comments on the proposal of the book led to a much better presentation of the materials. I have been influenced by the work of many outstanding scholars and researchers in this field, and most of their names will be found in the reference list attached to each chapter.

My research work in this area has been largely supported by the Hong Kong Research Grants Council, The Hong Kong Polytechnic University, The Natural Science Foundation of China, and the Hong Kong Highways Department over many years. All the support is gratefully acknowledged.

A vote of thanks must go to Mr. James Murphy, Publisher at John Wiley & Sons, for his patience and encouragement from the beginning and during the preparation of this book, and to Ms. Clarissa Lim and Ms. Shelley Chow, Project Editors at John Wiley & Sons, for their patience and scrutiny in the editing of this book.

Finally, I am grateful to my family for their help, encouragement and endurance.

Acknowledgements

1

Wind Storms and Cable-Supported Bridges

1.1 Preview

Because of their competence for long spans, many cable-supported bridges, including both cable-stayed bridges and suspension bridges, have been built throughout the world. These long-span cable-supported bridges are often remarkably flexible, low in damping and light in weight. Therefore, they can also be susceptible to the action of wind. For example, the Tacoma Narrows suspension bridge, which had a main span of 853 m and was built to link the Olympic Peninsula with the rest of the state of Washington, oscillated through large displacements at a wind speed of about 19 m/s and collapsed on November 7, 1940, only four months and six days after the bridge was opened to the public.

Modern long-span cable-supported bridges carry a large volume of vehicles and may experience considerable vibration due both to moving vehicles and to turbulent winds. The considerable vibration of the bridge and the crosswinds may, in turn, affect the running safety of vehicles. Therefore, adequate treatment of wind effects in design is essential to the safety and functionality of both long-span cable-supported bridges and vehicles running on the bridges.

In this introductory chapter, the meteorology of wind storms is first introduced to provide information on the basic features of wind storms. The basic configuration, structural systems, and aerodynamic characteristics of both cable-stayed and suspension bridges are then described to facilitate understanding of aerodynamic phenomena and performance of the bridges discussed in the subsequent chapters. Wind-induced excessive vibration and damage to long-span cable-supported bridges are discussed, focusing on the lessons learned from them by the engineering profession. Finally, the history of bridge wind engineering, particularly for cable-supported bridges is outlined to look back to the past and look forward to the future.

1.2 Basic Notions of Meteorology

1.2.1 Global Wind Circulations

Wind, or the motion of air with respect to the surface of the earth, is fundamentally due to differences in the amount of solar heat received by the atmosphere over various areas of the earth's surface because of the shape of the earth and its position relative to the sun [1]. The differences in solar radiation between

Wind Effects on Cable-Supported Bridges, First Edition. You-Lin Xu.

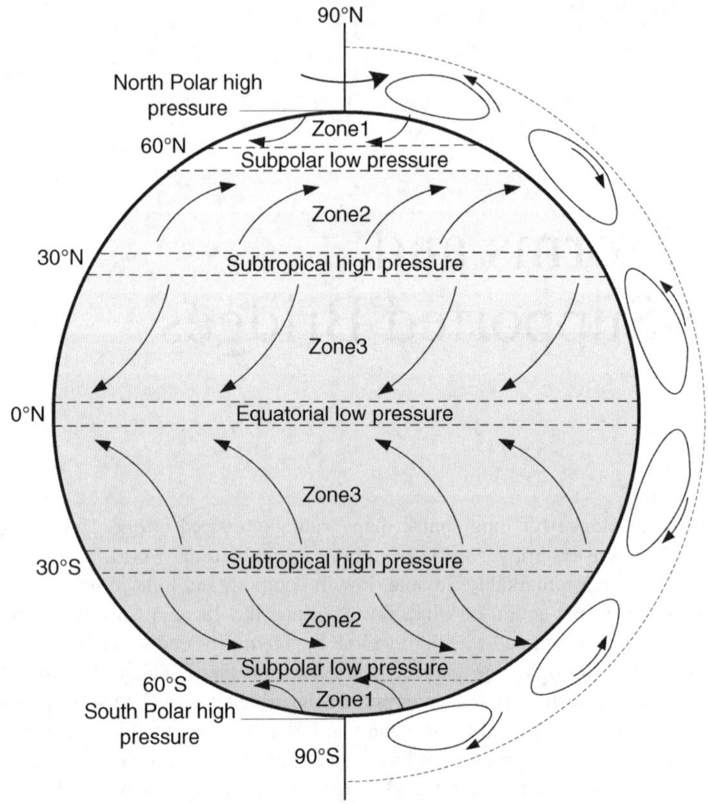

Figure 1.1 Idealized global circulation.

the poles and the equator produce temperature and pressure differences. These, together with the effects of the earth's rotation, the curvature of the path of motion and friction at the earth's surface, initiate and break down the air movement into six district circulations, as shown in Figure 1.1.

There are three in each hemisphere: easterly trade wind, westerly wind in the temperate zone, and polar easterly wind [2]:

- **Easterly trade wind:** the permanent subtropical high-pressure zone in the Northern Hemisphere initiates a flow towards the equator with low pressure at ground level. The rotation of the earth creates a virtual force which is known as the Coriolis force. The Coriolis force is perpendicular to the wind direction and to the right in the Northern Hemisphere. Thus, the flow is bent westwards and forms a trade wind. This is called the easterly trade wind, as it comes from the east. The subtropical high-pressure zone in the Northern Hemisphere is at approximately 30° latitude, because the flow away from the equator at high altitude cannot penetrate further north due to the Coriolis force.
- **Westerly wind in temperate zone:** in the layers near the ground of the temperate zone, flow is directed towards the north by the subtropical high-pressure zone and the polar front low-pressure zone. Due to the Coriolis force, it is then diverted eastwards and becomes a westerly wind in the temperate zone. The polar front is a surface where a warm westerly wind flows to the south of the polar front and a cold easterly wind flows to the north of the front. The

equilibrium at this front is very sensitive to changes in the temperature, velocity and humidity of the two air masses.

- **Easterly polar wind:** from the high pressure at the North Pole, the air flows to the south at low altitudes. The flow is then diverted to the west and becomes the cold easterly polar wind. The two main flows – the west wind in the temperate zone and the easterly polar wind – pass along each other at 50°–60° northern latitude, thereby forming the low pressure polar front.

1.2.2 Pressure Gradient Force

The most important forces acting on a particle of air are pressure gradient force, Coriolis force and frictional force [3]. The first two of these are particularly important to the upper level air, where the effect of frictional force is insignificant. Nevertheless, the frictional force must be taken into account for air near the earth's surface.

If there is a pressure gradient $\partial p/\partial x$ at a point in air in a given direction x, there is a resulting force called the pressure gradient force. The pressure gradient force per unit mass is given by Equation 1.1:

$$P = -\frac{1}{\rho_a}\frac{\partial p}{\partial x} \tag{1.1}$$

where ρ_a is the density of air.

1.2.3 Coriolis Force

Under the pressure gradient force, a particle of air may not flow absolutely in the direction of this force from a high pressure zone to a low pressure zone, but is deflected to some extent by the Coriolis force due to the earth's rotation [3]. The Coriolis force causes a moving particle on the surface of the earth to veer to the right in the Northern Hemisphere, or to the left in the Southern Hemisphere.

The magnitude and direction of the Coriolis force can be calculated by:

$$\boldsymbol{F}_c = 2m(\boldsymbol{v} \times \boldsymbol{\omega}) \tag{1.2}$$

where:

m is the mass of the particle;
$\boldsymbol{\omega}$ is the angular velocity vector of the earth;
$\boldsymbol{v}$ is the velocity of the air particle relative to a coordinate system rotating with the earth;
the symbol $\times$ represents the cross product operator;
$\boldsymbol{F}_c$ is the Coriolis force, which is perpendicular to $\boldsymbol{\omega}$ and $\boldsymbol{v}$, is directed according to the vector multiplication (right-hand) rule and has the magnitude $2m|\boldsymbol{\omega}||\boldsymbol{v}|\sin\alpha$, where α is the angle between $\boldsymbol{\omega}$ and $\boldsymbol{v}$.

The term $2\omega\sin\phi$ is commonly denoted as the Coriolis parameter f_c, where ϕ is the latitude of the air particle. It follows that the Coriolis force per unit mass acting in a plane parallel to the surface of the earth on the air particle moving in such a plane with velocity v relative to the earth will have a magnitude of $f_c v$. The Coriolis force is zero at the equator and is negligible in magnitude in the equatorial region, which is within about five degrees either side of the equator. This explains why the typhoons and other cyclonic storms will not form in the equatorial regions. The Coriolis effect is responsible for the rotation of cyclonic storms, which will be discussed later.

1.2.4 Geostrophic Wind

At sufficiently great height, wind speed and direction depend on only the horizontal pressure gradient force and the Coriolis force. The pressure gradient towards a low-pressure zone causes a particle of air

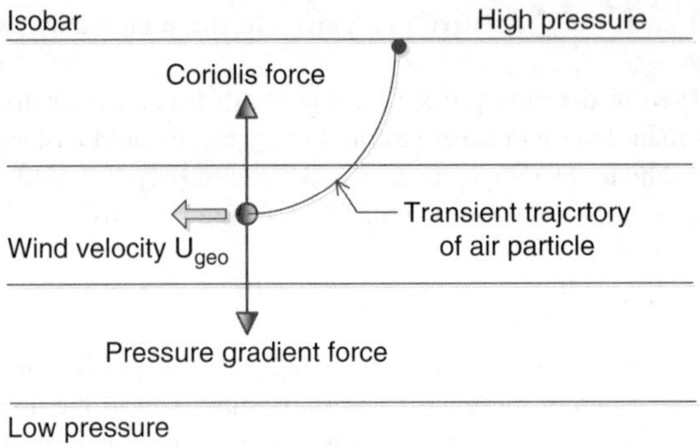

Figure 1.2 Frictionless wind balances in geostrophic flow (northern hemisphere).

to accelerate along a curve until a state of equilibrium is reached. In this case, the pressure gradient force (***P***) and Coriolis force ($\boldsymbol{F_c}$) are of equal magnitude, but in opposite direction, both of which are perpendicular to the wind direction, which is parallel to the isobars. Figure 1.2 shows the state of equilibrium in the Northern Hemisphere.

This kind of wind is called geostrophic wind U_{geo} [3]. By equating the pressure gradient force (see Equation 1.1) to the Coriolis force $f_c v$, the geostrophic wind speed can be expressed by:

$$U_{geo} = -\left(\frac{1}{\rho_a f_c}\right)\frac{\partial p}{\partial x} \tag{1.3}$$

Clearly, the geostrophic wind speed is proportional to the magnitude of the pressure gradient $\partial p/\partial x$.

1.2.5 Gradient Wind

Geostrophic wind occurs in cases where the radii of curvature of the isobars are so large that the centrifugal force is negligible. When the isobars have significant curvature, wind speed and direction not only depend on the pressure gradient force and the Coriolis force, but also on the centrifugal force. The value of the centrifugal force for per unit air mass (C) is U_{gr}^2/r, where U_{gr} is the resultant wind velocity and r is the radius of curvature of the isobars, as shown in Figure 1.3.

The direction of the centrifugal force is always normal to the isobars and away from the center of curvature of the isobars. The resultant wind velocity is called the gradient wind velocity [3] and can be found by the following equation in a polar coordinate system:

$$f_c U_{gr} \pm \frac{U_{gr}^2}{r} = \frac{1}{\rho_a}\frac{dp}{dr} \tag{1.4}$$

where, if the air mass is in the Northern Hemisphere, the positive or the negative sign is used as the circulation is cyclonic (around a low-pressure center) or anticyclonic (around a high-pressure center) respectively. The gradient wind velocity is equal to the geostrophic wind velocity in the particular case in which the curvature of the isobars is zero. If the radius of curvature is finite, in the Northern Hemisphere:

$$U_{gr} = -\frac{f_c r}{2} + \sqrt{\left(\frac{f_c r}{2}\right)^2 + \frac{r}{\rho_a}\frac{dp}{dr}} \tag{1.5}$$

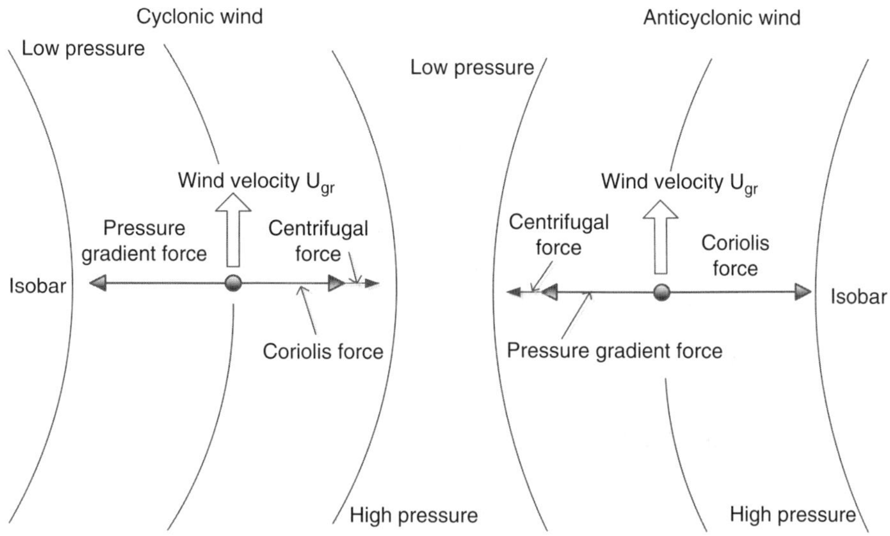

Figure 1.3 Frictionless wind balances in cyclonic and anticyclonic flow (northern hemisphere).

for cyclonic winds, and

$$U_{gr} = +\frac{f_c r}{2} - \sqrt{\left(\frac{f_c r}{2}\right)^2 - \frac{r}{\rho_a}\frac{dp}{dr}} \tag{1.6}$$

for anticyclonic winds.

By examining Equation 1.6, one may see that a maximum value, $f_c r/2$, of U_{gr} is obtained when the term under the square root is zero. Thus, there is an upper limit for anticyclonic winds, and anticyclones are therefore associated with low wind velocities. By contrast, cyclonic winds have no limits for their magnitude and give counterclockwise flow in the Northern Hemisphere. The geostrophic wind velocity can also be expressed in terms of the gradient wind velocity:

$$U_{geo} = U_{gr}\left[1 + \frac{U_{gr}}{f_c r}\right] \tag{1.7}$$

1.2.6 Frictional Effects

Frictional effects cannot be ignored as the earth's surface is approached. Air flow will be slowed down by the horizontal resistance of the earth's surface (frictional forces), and this resistance decreases as the height from the ground level increases, until it reaches a certain height, above which the friction effects can be ignored. The bottom part of the earth's atmosphere, affected by frictional forces, is called the "atmospheric boundary layer". The air above this is termed "free atmosphere", and the height of boundary between the two is termed as "atmospheric boundary layer height". Atmospheric boundary layer height varies with the weather conditions, terrain and surface roughness changes. Figure 1.4 shows the new balance of the four types of forces in the boundary layer [4]. Since long-span cable-supported bridges are often built in the atmospheric boundary layer, the wind conditions in the boundary layer – including the boundary layer thickness and the change of wind speed and wind direction with height – are those most relevant to bridge wind engineering.

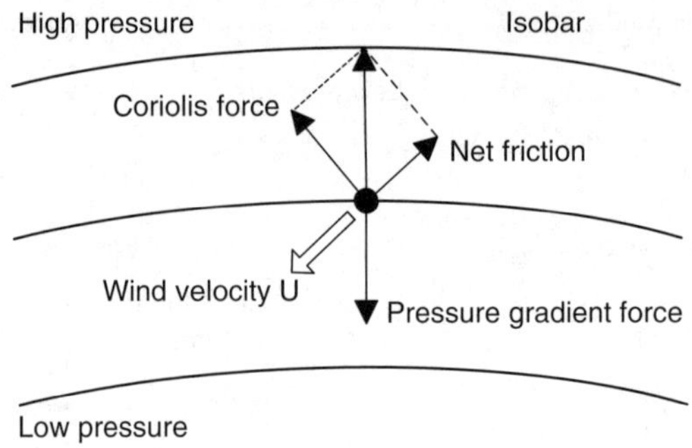

Figure 1.4 Balance of forces in the atmospheric boundary layer (northern hemisphere).

1.3 Basic Types of Wind Storms

Air movement caused by atmospheric difference at different areas of the earth generates global wind circulations, which are the most large-scale air movements on the earth. However, on more local scales, atmospheric difference may be caused by many other factors, such as sun elevation angle, atmospheric transparency, altitude, topography, earth surface physical properties and moisture. All of these can be the main reasons for local air movement.

Air movement usually includes combinations of different temporal and spatial scales. Spatial scale ranges from fractions of a meter to several thousand kilometers, while timescale can range from several seconds to several years. In meteorology, atmospheric motion is usually divided into several different scales, including microscale (less than 1 km), mesoscale (less than 10^2 km), synoptic scale (less than 10^3 km) and macroscale (less than 10^4 km).

1.3.1 Gales from Large Depressions

In the mid-latitudes, from about 40° to 60°, the strongest winds are gales generated by large and deep depressions (extra-tropical cyclones) [4]. These can also be significant contributors to strong winds in lower latitudes. The gales are usually large of synoptic scale in horizontal dimension. They can extend for more than 1000 km and can take several days to pass several countries, in the case of Europe. The winds tend to be quite turbulent near the ground, as the flow has adjusted to the frictional effects of the earth's surface over hundreds of kilometers. The direction of the winds remains quite constant over many hours.

1.3.2 Monsoons

Since the trade wind caused by the low and high pressure centers is related to the solar heating of the earth's surface, the inclination of the earth's rotation axis to the ecliptic causes a seasonal oscillation of the trade wind. These seasonal trade winds are called monsoons, and they are caused by the larger amplitude of the seasonal cycle of land temperature compared to that of nearby oceans.

In summer, the air over the land warms faster and reaches a higher temperature than the air over the ocean. The hot air over the land tends to rise, creating an area of low pressure and producing wind from ocean to land. In winter, the land cools off quickly, but the ocean retains heat longer. The cold air over the land creates a high pressure area which produces wind from land to ocean. Owing to the vast land mass of the Asian continent, monsoon effects are developed most strongly in Asia, where they have a considerable influence on the seasonal changes of weather patterns.

1.3.3 Tropical Cyclones (Hurricanes or Typhoons)

Tropical cyclones are intense cyclonic storms that occur over the tropical oceans, mainly in late summer and autumn [4]. They derive all their energy from the latent heat of the oceans and require a minimum sea temperature of about 26 °C to sustain them. They will not form within about 5° of the equator, because of the very small Coriolis force. They are usually at full strength when they are located between 20° and 30° latitude, but they can travel to higher latitudes if there are warm ocean currents to sustain them. They rapidly degenerate when they move over land or into cooler waters.

Tropical cyclones, hurricanes and typhoons are different names for the same type of severe storms occurring in different geographical regions. These encountered in the Far East such (e.g. Hong Kong, Taiwan, Mainland China, Japan, and the Philippines) are called typhoons. Those occurring in the United States, including Hawaii, are called hurricanes. Those affecting Australia are called tropical cyclones.

Full-scale tropical cyclones usually develop initially from cyclone eddies that usually last for several days and no more than a few weeks. Most of these eddies decay and fade away, while only a few intensify and develop into tropical cyclones. The triggering action that turns a cyclone eddy into a tropical cyclone is complicated and not well understood. After an eddy matures into a tropical cyclone, it sucks up from the ocean large quantities of water vapor, which condenses at higher altitudes. This latent heat of condensation is the prime source of energy supply, which intensifies the cyclone as it moves across the ocean. If this energy source is being cut off, the tropical cyclone will decay rapidly. Therefore, cyclone winds are strong only over the ocean and in adjacent coastal areas (within approximately 100 km of coastlines). The life span of a cyclone is of the order of one to three weeks. As a tropical cyclone is a large body of rotating air, in the Northern Hemisphere they always rotate in the counter-clockwise direction due to the Coriolis force generated by the earth's rotation. In contrast, tropical cyclones in the Southern Hemisphere rotate clockwise.

A tropical cyclone is a large funnel-shaped storm of a three-dimensional vortex structure with a wide top of the order of 1000 km in diameter and a narrow bottom of the order of 300–500 km in diameter (see Figure 1.5). The height of the cyclone is of the order of 10–15 km. The diameter of a cyclone, encompassing the region of relatively strong wind, is of the order of 500 km. The center part of a cyclone, which has a diameter of the order of 30 km, is called the eye; the boundary of the eye is called the wall. The eye is a region of clear to partly cloudy skies, absent of rain and strong winds. The wall is a region packed with high winds and intense rain. While rain falls in the inner region of the wall, warm, humid air rises in the outer part of the wall to supply energy to the cyclone.

While a tropical cyclone rotates around its center, the cyclone also moves forward as a whole. The translational speed of the cyclone is the speed at which its center moves. The translational speed can be anywhere between 0 and 100 km/h. Normally it is between 10–50 km/h.

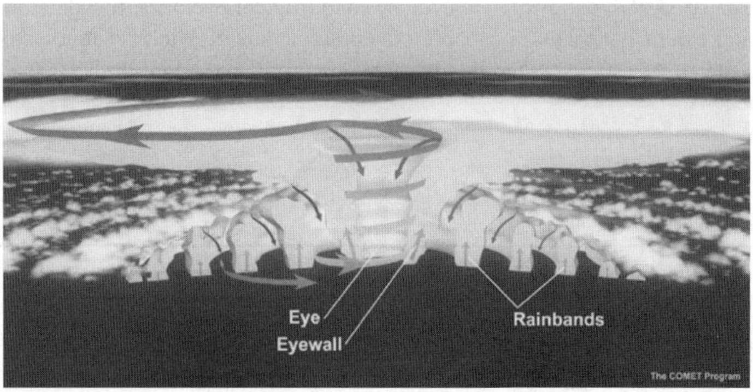

Figure 1.5 Cross-section of a mature tropical cyclone (Source: Wikipedia (http://zh.wikipedia.org/wiki/File:Hurricane_structure_graphic.jpg#file), Original upload by a staff of NOAA).

Outside of the eye of a tropical cyclone, the wind speed at upper level decays with the radial distance from the storm center. This wind speed can be determined by combining Equation 1.5 with a function for the pressure gradient, such as one suggested by Holland [5].

$$\frac{p - p_o}{p_n - p_o} = \exp\left(\frac{-A}{r^B}\right) \tag{1.8}$$

where:

p_o is the central pressure of the tropical cyclone;
p_n is the atmospheric pressure at the edge of the storm;
A and B are scaling parameters;
r is the radius from the storm center.

Differentiating Equation 1.8 and substituting it in Equation 1.5 yields:

$$U_{gr} = -\frac{f_c r}{2} + \sqrt{\left(\frac{f_c r}{2}\right)^2 + \frac{\Delta p AB}{\rho_a r^B}\exp\left(-\frac{A}{r^B}\right)} \tag{1.9}$$

where $\Delta p = p_n - p_o$ is the pressure difference, which is an indication of the strength of the storm.

The exponent B is found to be in the range 1.0–2.5 and to reduce with increasing central pressure [5]. The parameter $A^{1/B}$ is, to a good approximation, the radius of maximum winds in the cyclone [4].

An intensity scale for hurricanes has been proposed by Saffir and Simpson and is reproduced here in Table 1.1. It should be noted that the estimated wind speeds in hurricanes by using Table 1.1 are usually obtained from upper level aircraft readings.

1.3.4 *Thunderstorms*

A thunderstorm is a natural weather phenomenon producing lightning and thunder. It is usually accompanied by strong winds, heavy rain and, sometimes, hail. Those that cause hail to fall are called hailstorms. Thunderstorms can generally form and develop in any geographic location. In subtropical and temperate mid-latitudes, they usually occur in summer, and sometimes in winter owing to the impact of a cold front. Compared with extra-tropical cyclones and tropical cyclones, thunderstorms are of small size in horizontal extent but they are capable of producing severe winds. Thunderstorms contribute significantly to the strongest gusts recorded in many countries, including the United States, Australia and South Africa [4].

Table 1.1 Saffir–Simpson hurricane scale

Category	Wind speed range (3s gust, m/s)	Central pressure (mbar)	Damage
1	42~54	>980	Minimal
2	55~62	965~979	Moderate
3	63~74	945~964	Extensive
4	75~88	920~944	Extreme
5	>88	<920	Catastrophic

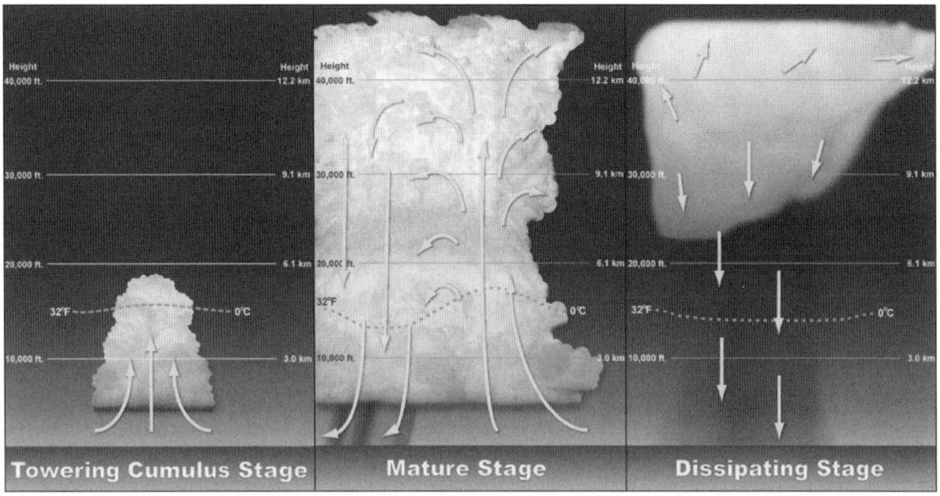

Figure 1.6 Stages of a thunderstorm's life (Source: Wikipedia (http://en.wikipedia.org/wiki/File:Thunderstorm_formation.jpg), Diagram from NOAA National Weather Service training materials).

Thunderstorms also derive their energy from heat. Warm, moist air is convected upwards to mix with the drier upper air. With evaporation, rapid cooling occurs and the air mass loses its buoyancy and starts to sink. Severe thunderstorms develop under three main conditions:

1. the humidity of the earth's low atmosphere is very high;
2. a negative temperature gradient with height greater than the adiabatic rate of the neutral atmosphere; and
3. a lifting mechanism produces the initial rapid convection due to a cold front or a mountain range.

The life cycle of a thunderstorm, regardless of type, mainly go through three stages (Figure 1.6): the cumulus stage, the mature stage, and the dissipation stage. These three stages take an average of 30 minutes to go through, but more powerful thunderstorms may last for several hours.

1.3.5 Downbursts

A particular type of thunderstorm wind is called a downburst or a thunderstorm downburst. It is generated by a falling mass of evaporative and cooled air in the parent thunderstorm [6]. As this falling air mass impinges on ground, it spreads out horizontally in all directions and produce strong winds for a short period of time – about 5–10 minutes (see Figure 1.7). The horizontal wind speed in a thunderstorm downburst with respect to the moving storm is similar to that in a jet of fluid impinging on a plain surface. It varies approximately linearly from the center of impact to a radius where the wind speed reaches its maximum, and then decays with increasing radius.

Fujita [7] classified downbursts into two groups: microburst and macroburst. A microburst has a small horizontal scale, of the order of a few hundred meters. On the other hand, a macroburst covers a larger area, of the order of 1–5 km. The forward velocity of the moving downburst can be a significant component of the total wind speed produced at ground level, and it must be added as a vector component to that produced by a jet [4].

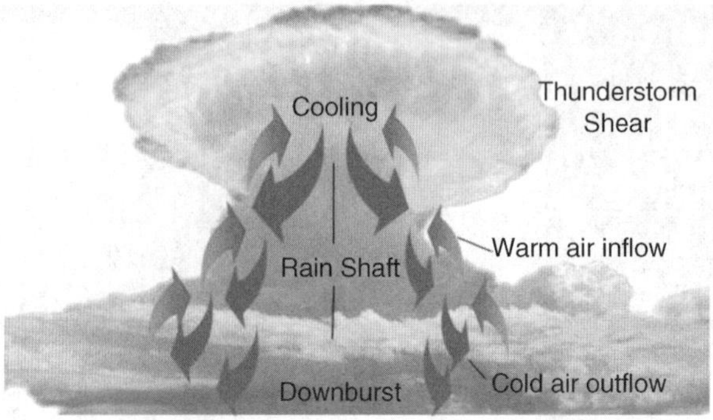

Figure 1.7 Cross section of a downburst.

1.3.6 Tornadoes

Both severe thunderstorms and tropical cyclones can generate tornadoes, with the former being the more frequent cause. A tornado is a vertical, funnel-shaped vortex created in thunderclouds, and it is the most destructive type of wind storm (see Figure 1.8). Most tornadoes have a diameter smaller than 400 m, but they can travel for quite long distances (up to 50 km) at an average translational speed of about 50–60 km/h before dissipating, producing a long, narrow path of destruction.

In contrast to tropical cyclones, the majority of tornadoes last no more than 30 minutes, but the strongest tornadoes may have a lifespan longer than an hour. Tornadoes are sometimes confused with downbursts; however, tornadoes can be identified by the appearance of the characteristic funnel vortex. Tornadoes have been observed in all parts of the world, with the United States being the country most frequently plagued by tornadoes. The existing field measurement data of tornadoes is quite sparse, because tornadoes have very rarely passed over weather recording stations, due to their small size. An intensity scale for tornadoes was proposed by Fujita in 1971 [8].

Figure 1.8 Tornado near Anadarko, Oklahoma (Source: Wikipedia (http://en.wikipedia.org/wiki/File:Dszpics1.jpg), Author: Daphne Zaras).

1.3.7 *Downslope Winds*

A foehn wind is a type of dry downslope wind that occurs in the lee (downwind side) of a mountain range. It is a rain shadow wind that results from the subsequent adiabatic warming of air that has dropped most of its moisture on windward slopes. As a consequence of the different adiabatic lapse rates of moist and dry air, the air on the leeward slopes becomes warmer than equivalent elevations on the windward slopes. Foehn winds can raise temperatures by as much as 30 °C (54 °F) in just a matter of hours. Downslope winds have been observed in certain regions, such as those near the Rocky Mountains in the United States, and also in Switzerland.

1.4 Basic Types of Cable-Supported Bridges

1.4.1 *Main Features of Cable-Supported Bridges*

Bridge engineering has been developed over centuries. According to the structural configuration, bridges can be mainly categorized into five types: beam bridges; cantilever bridges; arch bridges; truss bridges; and cable-supported bridges.

Cable-supported bridges are more suitable for longer main spans than any other types of bridges. The longest cable-stayed bridge in the world is currently the Russky Bridge, located in Vladivostok, Russia, which opened to the public in 2012 and which has a main span of 1104 m. The longest suspension bridge in the world, as of 2012, is the Akashi-Kaikyo Bridge in Kobe, Japan, completed in 1998, which has a main span of 1991 m. The suspension bridge is best suited for extending main span lengths even farther.

The structural system of a cable-supported bridge normally consists of four main components [9]:

1. stiffening girder (deck);
2. cable system;
3. pylons (towers);
4. anchors (anchorages).

The stiffening girder in the bridge deck is a major structural component, carrying most of the external loadings applied to the bridge. The cable system is used to support the stiffening girder and transfer the loadings from the stiffening girder to the pylons. The cable anchors in a cable-stayed bridge connect stay cables to the pylons and the girder. The anchorages in a suspension bridge are fundamental to the stability of the bridge.

Three kinds of forces operate on a cable-supported bridge, as with any other type of bridge: the dead load, the live load, and the dynamic load:

- "Dead load" refers to the weight of the bridge itself. It is possible for any kind of bridge to collapse simply because of the gravitational forces acting on the materials of which the bridge is made.
- "Live load" refers to the traffic (highway and/or railway) that moves across the bridge, as well as normal environmental factors such as changes in temperature and precipitation.
- "Dynamic load" refers to environmental factors that go beyond normal weather conditions, factors such as sudden gusts of wind and earthquakes.

All three factors must be taken into consideration when building a cable-supported bridge.

Cable-supported bridges are distinctly characterized by the configuration of cable system: they can be suspension bridges, cable-stayed bridges, or hybrid cable-supported bridges.

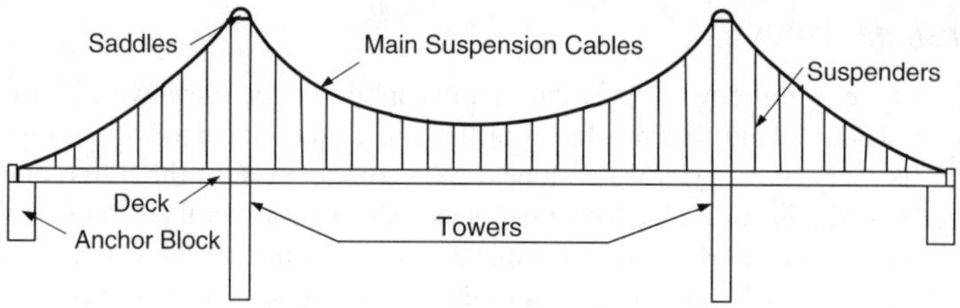

Figure 1.9 Main components of a suspension bridge.

1.4.2 Suspension Bridges

A suspension bridge is a type of cable-supported bridge in which the bridge deck (stiffening girder) is hung below suspension (main) cables on suspenders (hangers) that carry the weight of the deck, upon which the traffic crosses (see Figure 1.9). Stiffening girders may be I-girders, trusses or box girders. In long-span suspension bridges, trusses or box girders are typically adopted, but I-girders is not suitable because of their low torsional rigidity. There are both advantages and disadvantages to trusses and box girders, involving trade-off in bridge aerodynamics, construction and maintenance [10].

Suspenders, or hangers, connect the stiffening girder to the suspension cables, and they can be either vertical or diagonal. Generally, suspenders of most suspension bridges are vertical. Diagonal hangers have been used to increase the damping of the suspended girder. Occasionally, vertical and diagonal hangers are combined for more stiffness. Suspenders might be steel bars, steel rods, stranded wire ropes, parallel wire strands or other types. Stranded wire rope is most often used in modern suspension bridges. In early suspension bridges, chains, eye-bar chains or other materials were used for the main cables. In modern long-span suspension bridges, cold-drawn and galvanized steel wires have been used as parallel wire strands and stranded wire ropes that are bundled into a circle to form main cables.

The connections between the main cables and the towers are usually made through saddles, which support the main cables as they cross over the towers. Saddles are commonly made from fabricated steel or castings. A cover plate is provided for protection against corrosion and the whole unit is bolted down to resist movement. The main cables continue beyond the towers to deck-level supports and must be anchored at each end of the bridge in the ground, since any load applied to the bridge is transformed into a tension in these main cables.

The anchorage is fundamental to the stability of a suspension bridge. In general, anchorage structure includes the foundation, anchor block, bent block, cable anchor frames and protective housing. Inside the anchorages, the cables are spread over a large area to distribute the load evenly and to prevent the cables from breaking free. Anchorages are classified into gravity or tunnel anchorage system. Gravity anchorage relies on the mass of the anchorage itself to resist the tension of the main cables. This type is commonplace in many suspension bridges. Tunnel anchorage takes the tension of the main cables directly into the ground. Adequate geotechnical conditions are required.

Towers are classified into portal or diagonally braced types, and tower shafts can be either vertical or inclined. Typically, the center axis of inclined shafts coincides with the center line of the cable at the top of the tower. Careful examination of the tower configuration is important, in that towers dominate the bridge aesthetics. The force on the tower needs to be carefully balanced so that the force pulling inward is equal to the force pulling outward. As a result, the deck weight pulls directly down to the base of the towers. The towers can be fairly thin, since they are subjected mainly to axial compressive forces. The deck can also be thin, since it is being supported by a series of hangers. Thus, suspension bridges are more competitive for longer spans than any other type of bridge.

Suspension bridges are classified into single-span, two-span or three-span suspension bridges with two towers, and multi-span suspension bridges which have three or more towers. Three-span

Table 1.2 The top 10 longest suspension bridges in the world (as of September, 2012)

NO	Name	Main span/m	Country/Region	Year	Main girder	Control Measures
1	Akashi Kaikyō Bridge	1991	Japan	1998	Trussed girder	Slotted deck/Stabilizer
2	Xihoumen Bridge	1650	China	2009	Box girder	Slotted deck
3	Great Belt Bridge	1624	Denmark	1998	Box girder	Guide vanes
4	Yi Sun-sin bridge	1535	South Korea	2012	Twin-box girder	
5	Runyang Bridge	1490	China	2005	Box girder	Central stabilizer
6	Humber Bridge	1410	England	1981	Box girder	Nil
7	Jiangyin Bridge	1385	China	1997	Box girder	Nil
8	Tsing Ma Bridge	1377	Hong Kong	1997	Trussed girder	Truss outsourcing
9	Verrazano-Narrows Bridge	1298	USA	1964	Trussed girder	Nil
10	Golden Gate Bridge	1280	USA	1937	Trussed girder	Nil

suspension bridges are the most commonly used. In multi-span suspension bridges, the horizontal displacement of the tower tops might increase, due to the load conditions.

Suspension bridges are typically ranked by the length of their main span. The top ten longest suspension bridges in the world are listed in Table 1.2.

1.4.3 Cable-Stayed Bridges

A cable-stayed bridge is a type of cable-supported bridge that consists of one or more pylons (towers) with stay cables supporting the stiffening girder (bridge deck), as shown in Figure 1.10. A multiple-tower cable-stayed bridge may appear similar to a suspension bridge, but in fact it is very different in principle. A cable-stayed bridge carries mainly vertical loads acting on the girder. The stay cables provide intermediate supports for the girder, so that it can span a long distance. The basic structural form of a cable-stayed bridge is a series of overlapping triangles comprising the pylons, the cables, and the girder. All of these components are under predominately axial forces, with the cables under tension and both the pylons and the girder under compression. Axially loaded members are generally more efficient than flexural members. Moreover, for an almost symmetrical cable-stayed bridge, the horizontal forces in the girder balance and large ground anchorages are not required. All of these factors contribute to the

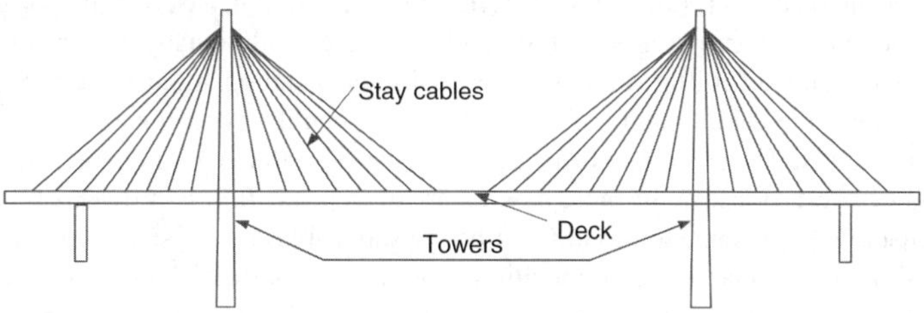

Figure 1.10 Main components of a cable-stayed bridge.

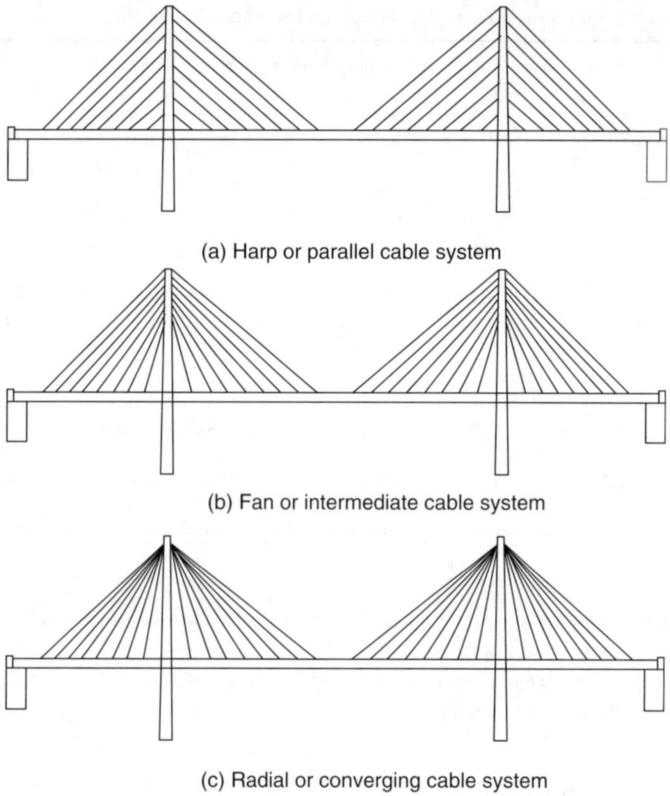

Figure 1.11 Major stay cable systems for cable-stayed bridges.

economy of a cable-stayed bridge. For spans up to about 1000 m, cable-stayed bridges are more economical than suspension bridges [10].

There are three major cable configurations, known as harp, fan and radial systems (see Figure 1.11). A harp-type cable arrangement offers a very clean and delicate appearance, because an array of parallel cables will always appear parallel, irrespective of the viewing angle. It also allows an earlier start of girder construction, because the cable anchors in the pylon begin at a lower elevation. A fan-type cable arrangement can also be very attractive, especially for a single-plane cable system. Because the cable slopes are steeper, the axial force in the girder, which is an accumulation of all horizontal components of cable forces, is smaller. This feature is advantageous for longer-span bridges, where compression in the girder may control the design. A radial arrangement of cables with all cables anchored at a common point at the pylon is quite efficient, but a good detail is difficult to achieve. Unless it is well treated, it may look clumsy.

The recently adopted design requires that all cables could be individually de-tensioned, dismantled and replaced under reduced traffic loading. The additional bending moment in the girder will not increase excessively if the cable spacing is small. The most popular type of cable employed nowadays uses seven-wire strands. These strands offer good workability and economy, and they can either be shop-fabricated or site-fabricated. In most cases, corrosion protection is provided by a high-density polyethylene pipe filled with cement grout. Usually, the stay cables are tensioned to about 40% of their ultimate strength under permanent load condition. The most important element in a stay cable is the anchor, which is the weakest point with respect to capacity and fatigue behavior. Strand cables with bonded sockets have performed very well in this aspect.

Table 1.3 The top 10 longest cable-stayed bridges in the world (as of September, 2012)

NO.	Name	Main span/m	Country/Region	Year	Control Measures
1	Russky Bridge	1104	Russia	2012	
2	Sutong Bridge	1088	China	2008	Scallops/Damper
3	Stonecutters Bridge	1018	Hong Kong	2009	Scallops/Damper
4	Edong Bridge	926	China	2010	Helical strakes /Damper
5	Tatara Bridge	890	Japan	1999	Scallops /Damper
6	Pont de Normandie	856	France	1995	Helical strakes /Damper
7	Jingyue Bridge	816	China	2010	Helical strakes /Damper
8	Incheon Bridge	800	South Korea	2009	Scallops /Damper
9	Zolotoy Rog Bridge	737	Russia	2012	
10	Shanghai Yangtze River Bridge	730	China	2009	Helical strakes /Damper

A properly designed and fabricated orthotropic deck is a good solution for a cable-stayed bridge. However, with increasing labor costs, the orthotropic deck becomes less commercially attractive except for very long spans. The composite deck with a concrete slab on a steel frame can be a competitive alternative. Furthermore, many concrete cable-stayed bridges have been completed.

In general, there have been two major developments: cast-in-place construction and precast construction. Several hybrid structures, with concrete side spans and steel main span, have been completed. There are two main reasons for the hybrid combination: to have heavier, shorter side spans to balance the longer main span; or to build the side spans the same way as the connecting approaches. The transition, however, must be carefully detailed to avoid problems.

The pylons are the most visible elements of a cable-stayed bridge. Free-standing pylons look very elegant. H-shaped pylons are the most logical shape structurally for a two-plane cable-stayed bridge. The A shape, the inverted Y and the diamond shape, however, are excellent choices for long-span cable-stayed bridges with very flexible decks. Cables are anchored at the upper part of the pylon, and criss-crossing the cables at the pylon is a good idea in a technical sense, being safe, simple, and economical. The difficulty is in the geometry. More details on stay cable anchorages at the pylon can be found in the literature [10].

Cable-stayed bridges are also ranked by the length of their main span. The top ten longest cable-stayed bridges in the world are listed in Table 1.3.

1.4.4 Hybrid Cable-Supported Bridges

A hybrid cable-supported bridge is a combination of suspension and cable-stayed bridge (also known as a cable-stayed-suspension bridge or a suspension-cable-stayed bridge). Figure 1.12 shows a

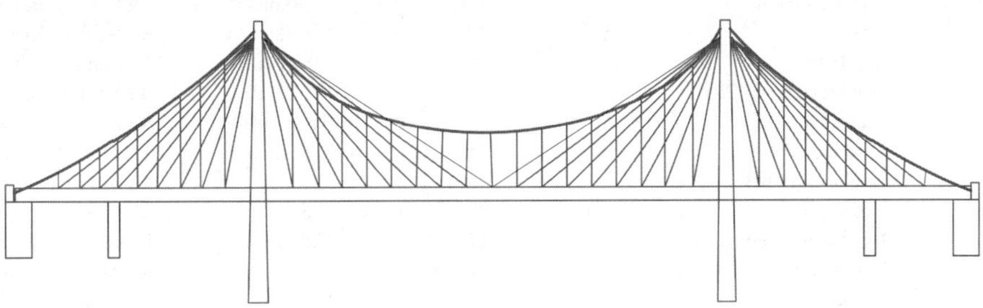

Figure 1.12 Hybrid cable-supported bridge.

preliminary design of a hybrid cable-supported bridge carrying both highway and railway with a 1500 m main span across the great Belt in Denmark [9]. The suspenders are arranged all along the bridge deck, with no difference from the general suspension bridge, and a number of stay cables are added in side spans and main span.

Hybrid cable-supported bridges have some unique features:

1. Compared with the same span suspension bridge, this hybrid system can lower tension forces in the main cables and reduce the cost of main cables and anchorages in particular, which makes the construction of a suspension bridge in soft ground possible.
2. Compared with the same span cable-stayed bridge, the compressive forces in the stiffening girder and the tension forces in the stay cables are greatly reduced and the optimal height of the pylons can be used, which makes longer spans possible in the hybrid system than in cable-stayed bridges.

However, the hybrid cable-supported system is not as widely applied as other two types. This is mainly because the cable system in this hybrid bridge is a highly indeterminate system which is difficult to analyze and construct.

1.5 Wind Damage to Cable-Supported Bridges

1.5.1 Suspension Bridges

Compared with other types of bridges, long-span cable-supported bridges are remarkably flexible, low in damping, light in weight, and therefore susceptible to the action of wind. From 1818 to 1889, windstorms worldwide caused ten suspension bridges to collapse or suffer major damage, including three in the United States (see Table 1.4).

In the half century that followed, such collapses decreased because the importance of making girders sufficiently stiff was recognized. However, the original Tacoma Narrows suspension bridge, linking the Olympic Peninsula with the rest of the state of Washington, USA, which had a main span of 853 m, oscillated through large displacements at a wind speed of about 19 m/s and collapsed on November 7, 1940, only four months and six days after the bridge was opened to the public (see Figure 1.13). The deck of the original Tacoma Narrows Bridge was stiffened with I-girders formed from built-up plates. The I-girders had low rigidity in torsion, resulting in aerodynamic instability of the bridge. In the three years immediately preceding this event, five newly completed bridges exhibited sensitivity to winds with significant oscillations. As a result, engineering awareness heightened about the potential

Table 1.4 List of the suspension bridges destroyed by winds

Year	Name	Main span(m)	Country	Designer
1818	Dyburgh Abbey	79	Scotland	Jorn & Willian Smith
1821	Union	139	Scotland	Sir Samuel Brown
1834	Scotland	75	German	Lossen & Wolf
1836	Brighton Chain Pier	78	Scotland	Sir Samuel Brown
1838	Montrose	132	Scotland	Sir Samuel Brown
1839	Menai Strait	155	Welsh	Tomas Telford
1852	Roche-Beruard	195	France	Le Blance
1854	Wheeling	308	U.S.A	Charles Ellet
1864	Lewiston-Queenston	317	U.S.A	Edward Serrell
1879	Tay bridge	75×4	Scotland	Sir Thomas Bouch
1889	Nigara-Clifton	384	U.S.A	Samuelm Keefer
1940	Tacoma Narrows Bridge	853	U.S.A	Leon Moisseiff

Figure 1.13 Collapse of the original Tacoma Narrows Bridge (Source: Wikipedia (http://en.wikipedia.org/wiki/File:Image-Tacoma_Narrows_Bridge1.gif)).

sensitivity of flexible bridges to winds. Wind tunnel tests for stiffening girders became routine in the investigation of aerodynamic stability. Truss-type stiffening girders and streamlined box girders, which give sufficient rigidity and aerodynamic stability, were adopted [10].

1.5.2 Cable-Stayed Bridges

Aerodynamic stability of cable-stayed bridges was a major concern for many bridge engineers in the early years. This was probably because cable-stayed bridges are extremely slender, and lessons learned from aerodynamic problems in suspension bridges led engineers to worry about cable-stayed bridges. Although cable-stayed bridges have been found surprisingly stable aerodynamically, several bridges have required special treatment against wind action.

The Kessock cable-stayed bridge in UK was planned during the 1970 s as part of the upgrading of the A9 road during the first North Sea oil boom [11]. The bridge was constructed in the late 1970 s and early 1980 s and opened on 19 July 1982. The bridge has an overall length of 1052 m and it is supported on twelve supporting piers. The main span has a length of 240 m and is supported by two planes of cables. The bridge is of steel construction with a steel bridge deck and two steel beams on the two sides. The overall width of the bridge deck is about 22 m and the depth of the side beams is about 3.3 m. The bridge is of an open section construction, such that the cross section is of an inverted "U" shape. Other than the two side girders, there are no longitudinal beams. The lengthwise stiffness of the bridge is all derived from the overall inverted U section. Figure 1.14 shows the overall bridge.

During the design of the bridge, wind tunnel tests were carried out and the results showed that the bridge would suffer large amplitude response due to vortex shedding for winds in the region of 20 m/s [11]. However, this behavior was not considered sufficiently important by the designers, based on the assumption that the response of prototype bridges is often significantly less than that predicted in the wind tunnel. However, just before closure of the main span, the structure displayed significant response to an east wind of 12.5 m/s. Again, in October 1982, soon after opening, peak-to-peak movements in excess of 300 mm were noted at mid-span. As a result, steps were taken to find a method of reducing future vibrations. Following further wind tunnel tests, it was proposed to install an array of vanes down each side of the bridge to prevent the formation of vortices. However, the designers preferred to install tuned mass dampers to reduce the level of vibration.

In general, the following analyses and wind tunnel tests will be conducted nowadays during the design and/or the construction of long-span cable-supported bridges:

1. Flutter analysis and section model tests: flutter is the most critical phenomenon in considering the dynamic stability of long-span cable-supported bridges because of the possibility of collapse.

Figure 1.14 Photo of Kessock Bridge (Source: Wikipedia (http://en.wikipedia.org/wiki/File:KessockNight_tgr.jpg), this document under the terms of the GNU Free Documentation License).

2. Buffeting analysis and section model/full aeroelastic model tests: buffeting is a forced vibration caused by randomly fluctuating wind loads present at all wind speeds.
3. Vortex-induced vibration analysis and section model tests: vortex-induced vibration is a forced vibration induced by vortex shedding in non-streamlined deck sections.

1.5.3 Stay Cables

Stay cables in cable-stayed bridges are laterally flexible structural members with very low fundamental frequency. Because of the range of different cable lengths, the collection of stay cables on a cable-stayed bridge has a practical continuum of fundamental and higher-mode frequencies. Thus, any excitation mechanism with any arbitrary frequency is likely to find one or more cables with either a fundamental or higher-mode frequency sympathetic to the excitation. Cables also have very little inherent damping and are therefore not able to dissipate much of the excitation energy, making them susceptible to large amplitude build-up. For this reason, stay cables can be somewhat lively by nature and have been known to be susceptible to excitations, especially during construction, wind and rain/wind conditions [12].

High-amplitude vibrations in stay cables have been observed under moderate rain combined with moderate wind conditions, and hence they are referred to as rain and wind-induced vibrations [13]. Peak-to-peak amplitudes of up to 2 m have been reported, with typical values of around 60 cm (Figure 1.15). Vibrations have been observed primarily in the lower cable modes of vibration, with frequencies ranging approximately from 1–3 Hz. Early reports described the vibrations simply as transverse in the vertical plane, but detailed observations suggest more complicated elliptical loci.

At several bridges in Japan, the observed vibrations were restricted to a wind velocity range of 6 to 17 m/s [14]. The stay cables of the Brotonne Bridge in France were observed to vibrate only when the wind direction was 20–30° relative to the bridge longitudinal axis. On the Meiko-Nishi Bridge in Japan, vibrations were observed with wind direction greater than 45° from the deck, and only on cables that declined in the direction of the wind [13]. Recognition of this susceptibility of stay cables has led to the incorporation of some mitigation measures on several of the earlier structures. These included cable cross-ties that effectively reduce the free length of cables, external dampers that increase cable damping and aerodynamic modifications of cable surface.

Figure 1.15 Rain-wind-induced vibration of stay cable (Source: Wind Induced Vibration of Cable Stay Bridges Workshop April 25–27, 2006 St Louis, MO FREYSSINET LLC (http://www.modot.org/csb/documents/24-New DevelopmentsinCableStayVibrationSuppression.pdf)).

1.5.4 Road Vehicles Running on Bridge

Wind-induced accidents involving road vehicles of various types have become a topic of increasing concern in recent years. This is because not only have vehicle numbers dramatically increased, but also vehicle weights are significantly reduced, owing to the use of more efficient structural design and lighter materials. When large numbers of road vehicles run on a long-span cable-supported bridge built in a wind prone region, the bridge will experience considerable vibration due both to moving vehicles and to turbulent winds [15]. The considerable vibration of the bridge may, in turn, affect the running safety of road vehicles. Moreover, road vehicles may be briefly shielded from the wind by the bridge tower or other road vehicles but, when they pass out of such shelter, they enter a sharp-edged crosswind gust and there is an obvious danger of the vehicles turning over.

On August 11, 2004, seven high-sided road vehicles were overturned by high winds when they ran on the Humen suspension bridge in China, just before a strong typhoon (see Figure 1.16a). A similar accident happened with road vehicles running on the Minjiang cable-stayed bridge in China in 2005 (see Figure 1.16b). Vehicle accidents caused injury, loss of lives, transportation interruptions and economic loss. Therefore, the safety of road vehicles subjected to crosswind gust when running on an oscillating cable-supported bridge, and the decision on the threshold of wind speed above which the bridge should be closed or the vehicle should be slowed down, are important social and economic issues.

1.6 History of Bridge Aerodynamics

Two excellent reviews of bridge aerodynamics were made by Miyata [16] and Xiang *et al.* [17]. The origins of cable-supported bridges go back a long way in history. Primitive suspension bridges were constructed with iron chain cables over 2000 years ago. However, the age of the fully developed suspended span with a horizontal traffic path began in the nineteenth century, with the adoption of the stiffening girder, which gave rigidity to the bridge in order to distribute the load through the hanger ropes and thus prevent excessive deformation of the cable. Although some suspension bridges collapsed under wind loads in the nineteenth century, the awareness of the necessity of aerodynamic

(a) Humen Bridge, China, 2004 (b) Minjiang Bridge, China, 2005

Figure 1.16 Road vehicle accidents on long-span cable-supported bridges. (a) (Source: Chinese news website (http://news.southcn.com/gdnews/gdpic/200408120258.htm)) (b) (Source: Chinese news website (http://www.fj.xinhuanet.com/news/2005-09/02/cntent_5035207.htm)).

investigations did not come until the original Tacoma Narrows suspension bridge was destroyed by a relatively low (19 m/s) wind in 1940.

The collapse of the original Tacoma Narrows Bridge had a determinant influence on the establishment of bridge aerodynamics as a scientific subject. Farquharson *et al.* conducted a series of wind tunnel tests on a model of the Tacoma Narrows Bridge [18]. Bleich, in 1948 [19], performed analytical studies to explore the possibility of torsional flutter of the bridge using the motion-dependent forces described by Theodorsen for a thin airfoil in the field of aeronautics [20]. Late on, Pugsley made comments on Bleich's work and pointed out that flutter derivatives of a bridge deck could be measured through wind tunnel tests [21]. In 1971, Scanlan and Tomko [22] jointly published an important paper on airfoil and bridge deck flutter derivatives, laying down a foundation for bridge flutter theory and providing a correct explanation for the cause of collapse of the Tacoma Narrows Bridge.

It is worth to note that, in the early 1960s, Davenport established the Alan G. Davenport Wind Loading Chain (see Figure 1.17) and stated clearly that wind resistant design of structures should be performed through five links: wind climate; terrain effects; aerodynamic effects; dynamic structural responses; and structural design criteria [23]. Davenport also pioneered the application of statistics and stochastic vibration theory to the buffeting analysis of long-span bridges in the frequency domain [24].

The quasi-static linear theory was employed in Davenport's theory to establish the buffeting forces and aeroelastic damping, and the buffeting response was analyzed mode by mode, based on the strip theory of aerodynamics. The concept of joint acceptance function was also introduced to consider the effectiveness of the temporal and spatial cross-correlation of buffeting loading, which was assumed to be the same as that of wind turbulence. Meanwhile, the concept of aerodynamic admittance was adopted to take into account the effects of unsteadiness and spatial variation of wind turbulence surrounding the cross-section of the bridge deck. However, the aerodynamic stiffness and aerodynamic coupling effects were not considered.

Scanlan and his co-workers believed that the self-excited (aeroelastic) forces would exert an important influence on the buffeting response of bridges. In the late 1970s, they proposed their buffeting

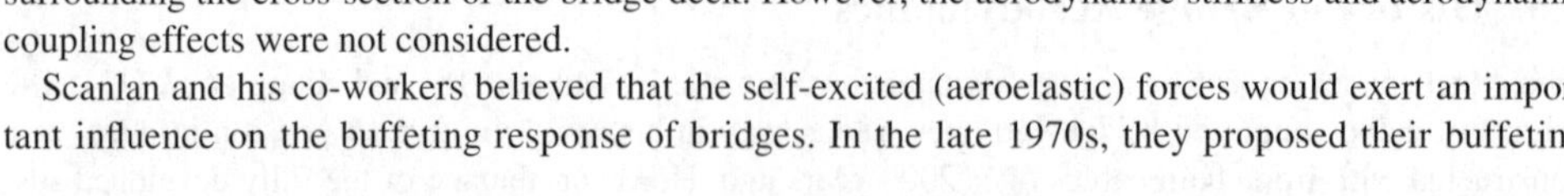

Figure 1.17 The Alan G. Davenport wind loading chain.

response analysis method by simultaneously considering both self-excited forces due to bridge deck motion and quasi-static aerodynamic forces caused by wind turbulence [25,26]. While the frequency domain methods were established for buffeting analysis, Lin and his co-workers presented the time-domain method for predicting buffeting response to turbulent winds using Ito's stochastic differential equations [27]. The effects of turbulence on stability of the bridge motion were included in their theory. In 1983, Lin and Yang [28] further proposed a general framework to handle multi-mode buffeting analysis of long-span bridges. With the rapid enhancement of computer capacity and speed, three-dimensional flutter and buffeting analyses, including multi-mode and inter-mode effects, can be performed nowadays for long-span cable-supported bridges, either in the time-domain [29,30] or in the frequency domain [31,33].

Wind flow against a bridge deck may form a stream of alternating vortices that create an alternative force in a direction normal to the wind flow. This alternative force is called the vortex shedding excitation. Vortex shedding excitation can induce significant, but limited, amplitude of vibration of a long-span cable-supported bridge in low wind speed and low turbulence conditions.

Much effort has been made to find an expression for forces resulting from vortex shedding [34]. However, since the interaction between the wind and bridge deck is very complex, no completely successful model has yet been developed for bridge decks. Nevertheless, the Scanlan model can be used for calculating the vortex-shedding force at "lock-in" when the vortex shedding frequency matches one of the natural frequencies of the bridge [3].

Advances in bridge aerodynamics during the past four decades make it possible to build even longer-span cable-supported bridges. The Akashi Kaikyo suspension bridge, with a main span of 1991 m, was built in 1998 in Japan, and is the longest suspension bridge in the world at present. The Russky cable-stayed bridge in Russia, which has a main span of 1104 m was open to the public in 2012 and became the longest cable-stayed bridge in the world. In Hong Kong, the Tsing Ma suspension bridge, with a main span of 1377 m, was completed in 1997 and is the longest suspension bridge carrying both highways and railways in the world. The Stonecutters cable-stayed bridge in Hong Kong, with a main span of 1018 m, was also open to the public in 2009.

On the other hand, the design and construction of long-span cable-supported bridges in the past three decades have brought many challenging issues to the bridge engineering community. These include wind-induced or rain-wind-induced vibration of stay cables, buffeting response to skew winds, wind-vehicle-bridge interaction, wind-induced fatigue, wind-induced vibration control, typhoon wind field simulation, non-stationary and non-linear flutter and buffeting, probabilistic analysis and reliability assessment. These challenging issues motivate not only the advancement of both theoretical analyses and wind tunnel test techniques, but also the development of new technologies, in which computational fluid dynamic (CFD) simulation and wind/structural health monitoring technology are two examples. All of these challenging issues and new technologies will be discussed in this book in detail, in addition to the fundamentals of bridge aerodynamics.

1.7 Organization of this Book

Background materials, including basic notions of meteorology, basic types of wind storms, basic types of cable-supported bridges, wind damage to cable-supported bridges, and history of bridge aerodynamics, have been provided in this chapter. Chapter 2 introduces wind characteristics in the atmospheric boundary layer, which is an extension of wind climate as discussed in Chapter 1 and covers the link "terrain effects" in the Alan G. Davenport wind loading chain. Chapters 3 to 6 describe, respectively, mean wind load and aerostatic instability of bridges, wind-induced vibration and aerodynamic instability of bridges, wind-induced vibration of stay cables, and wind-vehicle-bridge interaction. These contents cover the fundamentals of bridge aerodynamics and are associated with the two links "aerodynamic effects" and "dynamic structural responses" in the Alan G. Davenport wind loading chain.

As three important tools in the studies of bridge aerodynamics, in addition to theoretical analysis discussed in the previous chapters, Chapters 7 to 9 respectively introduce wind tunnel technique, computational wind engineering simulation and wind/structural health monitoring technology. Special topics such as buffeting response to skew winds, multiple loading-induced fatigue analysis, wind-induced vibration control, and typhoon wind field simulation are introduced in Chapters 10 to 13, respectively. These chapters are rather independent of the others and can be used individually.

In Chapter 14, reliability analysis of wind-excited bridges is described, laying down a foundation for the probabilistic wind resistant design of long-span cable-supported bridges and covering the last link, "design criteria", in the Alan G. Davenport wind loading chain. As a frontier in the field, Chapter 15 presents the preliminary research results on non-stationary and non-linear buffeting responses. Finally, challenges and prospects of bridge aerodynamics as a scientific but practical subject are highlighted in Chapter 16.

1.8 Notations

ρ_a	Density of air
p	Air pressure
F_c	Coriolis force
m	Mass of air particle
ω	Angular velocity vector of the earth
v	Velocity of air particle
α	The angle between ω and v
f_c	Coriolis parameter
ϕ	The latitude of air particle
U_{geo}	Geostrophic wind velocity
U_{gr}	Gradient wind velocity
r	Radius of curvature of isobars and radius from storm center
p_0	Central pressure of a tropical cyclone
p_n	Atmospheric pressure at the edge of the storm
A, B	Scaling parameters
P	Pressure gradient force per unit mass

References

1. ASCE (1987) *Wind Loading and Wind-Induced Structural Response*, American Society of Civil Engineers.
2. Dyrbye, C. and Hansen, S.O. (1996) *Wind Loads on Structures*, John Wiley & Sons Ltd, Chichester.
3. Simiu, E. and Scanlan, R.H. (1996) *Wind Effects on Structures*, 3rd edn, John Wiley & Sons, Inc., New York.
4. Holmes, J.D. (2007) *Wind Loading of Structures*, 2nd edn, Taylor & Francis, Abingdon.
5. Holland, G.J. (1980) An analytical model of the wind and pressure profile in hurricanes. *Monthly Weather Review*, **108**(8), 1212–1218.
6. Liu, H. (1991) *Wind Engineering-A Handbook for Structural Engineers*, Prentice-Hall, Inc., New Jersey.
7. Fujita, T.T. (1985) The downburst *SMRP Research Report No. 210*, University of Chicago, Chicago.
8. Fujita, T.T. (1971) Proposed characterization of tornadoes and hurricanes by area and intensity *SMRP Research Report No. 91*, University of Chicago, Chicago.
9. Gimsing, N.J. and Georgakis, C.T. (2011) *Cable Supported Bridges*, 3rd edn, John Wiley & Sons, Chichester.
10. Chen, W.F. and Duan, L. (2000) *Bridge Engineering Handbook*, CRC Press.
11. Owen, J.S., Vann, A.M., Davies, J.P., and Blakeborough, A. (1996) The prototype testing of Kessock Bridge: response to vortex shedding. *Journal of Wind Engineering and Industrial Aerodynamics*, **60**, 91–108.
12. FHWA (2007) *Wind-Induced Vibration of Stay Cables*, Report of Federal Highway Administration, FHWA-HRT-05–083.
13. Hikami, Y. and Shiraishi, N. (1988) Rain/wind-induced vibrations of cables in cable stayed bridges. *Journal of Wind Engineering and Industrial Aerodynamics*, **29**, 409–418.

14. Matsumoto, M., Saitoh, T., Kitazawa, M. *et al.* (1995) Response characteristics of rain-wind induced vibration of stay cables of cable-stayed bridges. *Journal of Wind Engineering and Industrial Aerodynamics*, **57**(2–3), 323–333.
15. Xu, Y.L. and Guo, W.H. (2003) Dynamic analysis of coupled road vehicle and cable-stayed bridge systems under turbulent wind. *Engineering Structures*, **25**, 473–486.
16. Miyata, T. (2003) Historical view of long-span bridge aerodynamics. *Journal of Wind Engineering and Industrial Aerodynamics*, **91**, 1393–1410.
17. Xiang, H.F. (2005) *Modern Theory and Practice on Bridge Wind Resistance*, China Communications Press, Beijing (in Chinese).
18. Farquharson, F.B., Smith, F.C., and Vincent, G.S. ((1949)–1954) Aerodynamic stability of suspension bridges. *Report Parts 1–5*, Structural Research Laboratory, University of Washington, USA.
19. Bleich, F. (1948) Dynamic instability of truss-stiffened suspension bridges under wind action. *American Society of Civil Engineers*, **74** (8), 1269–1314.
20. Theodorson, T. (1934) General theory of aerodynamic instability and the mechanism of flutter. *NACA, R496*, National Aeronautics and Space Administration.
21. Scruton, C. (1952). Experimental investigation of aerodynamic stability of suspension bridges with special reference to proposed Severn Bridge, *Proceedings of the Institute of Civil Engineers*, **2**, 189–222.
22. Scanlan, R.H. and Tomko, J.J. (1971) Airfoil and bridge deck flutter derivatives. *American Society of Civil Engineers*, **104** (EM4), 1717–1737.
23. Davenport, A.G. (1961) The application of statistical concepts to the wind loading of structures. Proceedings of the Institute of Civil Engineers, London, UK, 19 pp. 449–472.
24. Davenport, A.G. (1962) The response of slender, line-like structures to a gusty wind. Proceedings of the Institute of Civil Engineers, London, UK, 23 pp. 389–407.
25. Scanlan, R.H. and Gade, R.H. (1977) Motion of suspension bridge spans under gusty wind. *Journal of Structural Division, ASCE*, **103**(ST9), 1867–1883.
26. Scanlan, R.H. (1978) The action of flexible bridge under wind: II-buffeting theory. *Journal of Sound and Vibration*, **60**(2), 201–211.
27. Lin, Y.K. (1979) Motion of suspension bridges in turbulent winds. *Journal of Engineering Mechanics Division, ASCE*, **105** (EM6), 921–932.
28. Lin, Y.K. and Yang, J.N. (1983) Multimode bridge response to wind excitation. *Journal of Engineering Mechanics, ASCE*, **109** (2), 586–603.
29. Diana, G., Bruni, S., Collina, A., and Zasso, A. (1998) Aerodynamic challenges in super long bridge design, in Larsen, A. and Esdahl, S. (eds) *Bridge Aerodynamics*. Balkema, Rotterdam.
30. Chen, X.Z., Matsumoto, M., and Kareem, A. (2000) Time domain flutter and buffeting response analysis of bridges. *Journal of Engineering Mechanics, ASCE*, **126**(1), 7–16.
31. Jain, A., Jones, N.P., and Scanlan, R.H. (1996) Coupled flutter and buffeting analysis of long-span bridge. *Journal of Structural Engineering, ASCE*, **122**(7), 716–725.
32. Xu, Y.L., Sun, D.K., Ko, J.M., and Lin, J.H. (2000) Fully coupled buffeting analysis of Tsing Ma suspension bridge. *Journal of Wind Engineering and Industrial Aerodynamics*, **85**(1), 97–117.
33. Chen, X.Z., Matsumoto, M., and Kareem, A. (2000) Aerodynamic coupled effects on flutter and buffeting of bridges. *Journal of Engineering Mechanics, ASCE*, **126**(1), 17–26.
34. Ehsan, F. and Scanlan, R.H. (1990) Vortex-induced vibration of flexible bridges. *Journal of Engineering Mechanics, ASCE*, **116**, 1392–1411.

2

Wind Characteristics in Atmospheric Boundary Layer

2.1 Preview

As discussed in Chapter 1, frictional forces play an important role in the balance of forces on the moving air when it approaches the earth's surface. Frictional forces caused by the ground surface roughness will reduce wind speed until it reaches zero on the ground and, at the same time, will introduce turbulence near the ground surface. The wind region affected by the ground surface roughness is referred as the "atmospheric boundary layer". The depth of the atmospheric boundary layer may range from a few hundred meters to several kilometers, depending upon wind type, wind intensity, surface roughness and angle of latitude.

Since long-span cable-supported bridges are all placed on the ground, it is important to know turbulent wind characteristics in the atmospheric boundary layer. Therefore, this chapter will focus on turbulent wind characteristics near the ground surface. Turbulent winds are highly irregular and, accordingly, they are treated statistically rather than deterministically in this chapter. Turbulent wind is first defined based on the observation of measurement wind data in the field. Mean wind speeds and profiles are then introduced, followed by the characterization of turbulent winds in terms of theory of statistics and probability. For the design purpose of bridges, extreme wind speeds are discussed, the way of predicting design wind speed is provided and wind directional effects are pointed out. By taking the Tsing Ma suspension bridge in Hong Kong as an example, wind characteristics on the bridge site are finally demonstrated through extensive analyses of wind data recorded on the site.

2.2 Turbulent Winds in Atmospheric Boundary Layer

There are many different instruments used to measure winds, and they are collectively known as anemometers. The most common anemometer is the cup anemometer, which is simply three or four cup-shaped objects mounted on the arms of a center shaft, as shown in Figure 2.1a. The cups catch the wind and are spun around as a result. The speed at which the cups rotate is used to determine wind speed. Another common mechanical velocity anemometer is the windmill anemometer or propeller anemometer, in which a propeller is mounted on a nacelle which allows the windmill to rotate into the wind direction (Figure 2.1b). The wind speed is determined by how fast the propeller is spinning. The other forms of anemometers include ultrasonic anemometers (Figure 2.1c), hot-wire anemometers and laser

Wind Effects on Cable-Supported Bridges, First Edition. You-Lin Xu.

(a) Cup anemometer (b) Propeller anemometer (c) Ultrasonic anemometer

Figure 2.1 Three types of anemometers (a) Cup anemometer (Source: (http://en.wikipedia.org/wiki/File:Wea00920.jpg), Original uploader was Dhaluza at en.wikipedia) (b) Propeller anemometer (Source: (http://en.wikipedia.org/wiki/File:Wind_speed_and_direction_instrument_-_NOAA.jp), Original from NOAA Photo Library) (c) Ultrasonic anemometer (Source: (http://en.wikipedia.org/wiki/File:WindMaster.jpg), Original uploader was Gillinstruments at en.wikipedia).

Doppler anemometers. Further information on the measurement of wind and structures can be found in Chapter 9 of this book.

Figure 2.2 shows wind speeds recorded at three heights on a tall mast during a period of strong wind produced by gales and measured by sensitive cup anemometers. The records show the main characteristics of fully-developed boundary layer wind [1,2]:

1. the increase of the average wind speed as the height increases;
2. the gusty or turbulent nature of the wind at all heights;
3. the broad range of frequencies in the turbulent wind; and
4. some similarity in the patterns of gusts at all heights.

Within the atmospheric boundary layer and over a flat homogeneous terrain, the wind speed can normally be decomposed into a mean wind speed in the mean wind direction and three perpendicular turbulence components in a sufficiently long averaging time – say, ten minutes. With the x-axis in the

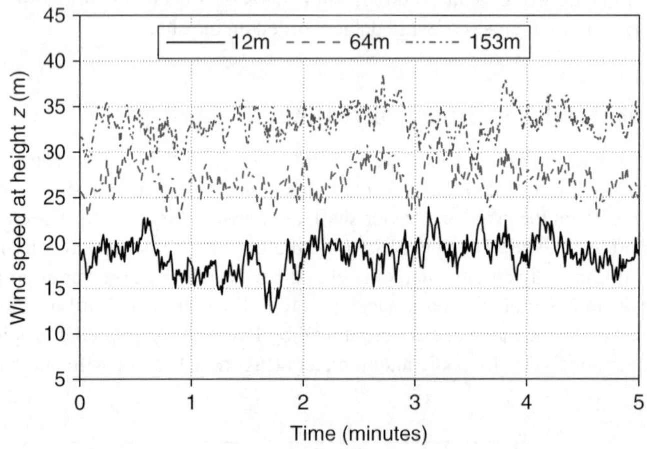

Figure 2.2 Wind speeds at three heights during gales.

direction of the mean wind speed, the y-axis being horizontal and the z-axis being vertical with positive upwards, the wind velocities at a given point (x, y, z) and a given time t can be expressed as:

$$\overline{U}(z) + u(x, y, z, t) \quad \text{(in the longitudinal direction)} \tag{2.1}$$

$$v(x, y, z, t) \quad \text{(in the lateral direction)} \tag{2.2}$$

$$w(x, y, z, t) \quad \text{(in the vertical direction)} \tag{2.3}$$

where:

$\overline{U}(z)$ is the mean wind speed depending on the height z above the ground;
u, v and w are the fluctuating parts of the wind in the x-, y- and z-directions, in which u is the turbulence component in the wind direction or the longitudinal direction;
v is the horizontal turbulence component perpendicular to the wind direction;
w is the vertical turbulence component perpendicular to u and v.

The above decomposition implies that the direction of the wind does not change along the height above the ground. Actually, though, the mean wind speed may change in direction slightly with height, which is known as the *Ekman Spiral* [2]. Furthermore, the thermal state of the atmosphere may influence the actual mean wind speed and turbulent components. However, the thermal effects can be ignored if the mean wind speed is large than 10 m/s. At higher wind speeds, mechanically generated wind conditions are normally far more important than thermal effects [3].

Define $U(x, y, z, t)$ as the instantaneous wind speed time history recorded from an anemometer in the mean wind direction and T is the averaging time. The mean wind speed $\overline{U}(z)$, shown in Equation 2.1, can be calculated by:

$$\overline{U}(z) = \frac{1}{T}\int_0^T U(x, y, z, t)dt \tag{2.4}$$

The turbulence component in the longitudinal direction can be calculated by:

$$u(x, y, z, t) = U(x, y, z, t) - \overline{U}(z) \tag{2.5}$$

Obviously, the mean wind speed depends on the averaging time T. The commonly-used averaging time for mean wind speed is either ten minutes or one hour. Figure 2.3 shows the relations among the instantaneous wind speed, the mean wind speed and the turbulence wind component in the longitudinal direction. The turbulence component can be treated mathematically as a stationary random process with a zero mean value for winds over a flat homogeneous terrain. Thus, the wind in the atmospheric boundary layer may be represented by the wind profile for the mean wind speed, and the additional turbulence components as stationary random processes.

2.3 Mean Wind Speed Profiles

For different types of strong winds, mean wind speed profiles are different. This section will introduce mean wind speed profiles mainly for monsoons, extra-tropical cyclones and tropical cyclones. Since downbursts may produce severe winds for short periods and are transient in nature, it is meaningless to define a mean wind speed for this type of event. For monsoons and extra-tropical cyclones, the mean

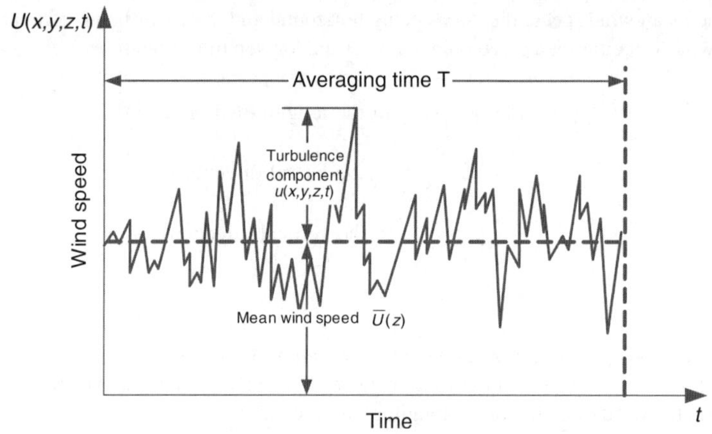

Figure 2.3 Instantaneous, mean and turbulence wind in longitudinal direction.

wind speed profile within the atmospheric boundary layer can be represented by either the so-called "logarithmic law" or the "power law".

2.3.1 The "Logarithmic Law"

In strong wind conditions, the most accurate mathematical expression for the mean wind profile due to monsoons and extra-tropical cyclones is the "logarithmic law". The logarithmic law was originally derived for the turbulent boundary layer on a flat plate, but it has been found to be valid in an unmodified form for strong winds in the atmospheric boundary layer near the ground. In consideration of wind shear in strong winds above the ground, a non-dimensional wind shear $\frac{d\overline{U}}{dz}\frac{z}{u_*}$ can be considered, in which $u_* = \sqrt{\tau_o/\rho_a}$ is known as the friction velocity, τ_0 is the surface shear stress denoting the retarding force per unit area exerted by the ground surface on the flow, and ρ_a is the air density. The non-dimensional wind shear is a constant.

$$\frac{d\overline{U}}{dz}\frac{z}{u_*} = \frac{1}{k} \tag{2.6}$$

The integration of Equation 2.6 then yields the usual form of the logarithmic law for the mean wind speed $\overline{U}(z)$ at any height z:

$$\overline{U}(z) = \frac{1}{k}u_* \ln\left(\frac{z}{z_0}\right) \tag{2.7}$$

where:

z_0 is the surface roughness length, which is an effective height of surface roughness elements;
k is von Karman's constant and has been found experimentally to have a value about 0.4.

Clearly, the mean wind speed varies with height. The mean wind speed is zero at the surface and it increases with height above the ground in the atmospheric boundary layer. The magnitude at the top of the boundary layer is referred to as the gradient speed. Outside the boundary layer, i.e. in the free

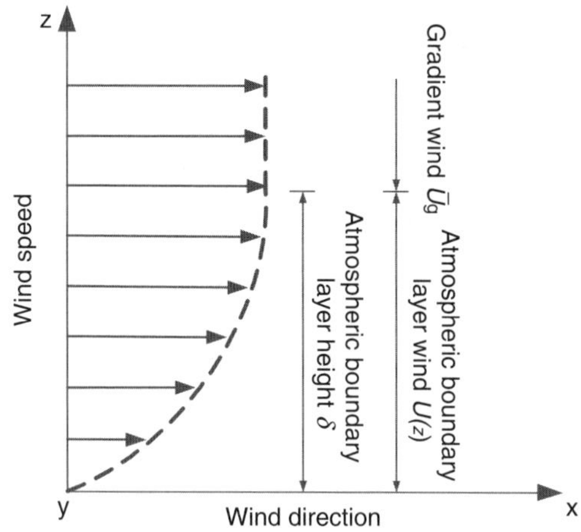

Figure 2.4 Mean wind speed profile in atmospheric boundary layer.

atmosphere, the wind flows approximately with the gradient speed without turbulence. Figure 2.4 shows a typical mean wind speed profile in the atmospheric boundary layer and above.

Applying Equation 2.7 to the mean wind speeds at two different heights, z_1 and z_2, gives the relationship of the two mean wind speeds:

$$\frac{\overline{U}(z_2)}{\overline{U}(z_1)} = \frac{\ln[z_1/z_0]}{\ln[z_2/z_0]} \tag{2.8}$$

It is normally accepted that the logarithmic law described by Equation 2.7 is not a good model for mean wind speed profile when the height z is over 100 m. Therefore, a revised logarithmic mean wind speed profile is proposed by Deaves and Harris [4], which can be expressed as:

$$\overline{U}(z) = \frac{u_*}{k}\left[\ln\frac{z-z_h}{z_0} + 5.75\left(\frac{z-z_h}{H_G}\right) - 1.88\left(\frac{z-z_h}{H_G}\right)^2 - 1.33\left(\frac{z-z_h}{H_G}\right)^3 + 0.25\left(\frac{z-z_h}{H_G}\right)^4\right] \tag{2.9}$$

where:

z_h is the zero-plane displacement for urban areas and it is zero for a flat terrain;
H_G is the equilibrium boundary layer height given by:

$$H_G = \frac{u_*}{6f_c} \tag{2.10}$$

where f_c is the Coriolis parameter.

It can be seen from Equations 2.7 and 2.9 that surface roughness has a profound effect on the mean wind speed in the atmospheric boundary layer. The rougher a terrain is, the more it retards the wind in the atmospheric boundary layer. The value of surface roughness length z_0 is between 1/20 and 1/30 of the average height of the roughness elements on the ground. The surface roughness can also be measured by the surface drag coefficient κ, which is a non-dimensional surface shear stress, defined as:

$$\kappa = \frac{\tau_0}{\rho\overline{U}_{10}^2} = \frac{u_*^2}{\overline{U}_{10}^2} \tag{2.11}$$

where $\overline{U}_{10}$ is the mean wind speed at 10 m height.

Table 2.1 Terrain types and roughness-related parameters

Terrain type	Surface roughness length z_0 (m)	Surface drag coefficient κ	Power law exponent α	Boundary layer height δ (m)
Very flat terrain (snow, desert)	0.001–0.005	0.002–0.003	0.12	300
Open terrain (grassland, few trees)	0.01–0.05	0.003–0.006	0.16	350
Suburban terrain (buildings 3–5 m)	0.1–0.5	0.0075–0.02	0.22	400
Dense urban (buildings 10–30 m)	1–5	0.03–0.3	0.30	450

By applying Equations 2.7 and 2.11 for z equal to 10 m, a relationship between the surface drag coefficient κ and the surface roughness length z_0 can be determined and expressed as:

$$\kappa = \left[\frac{k}{\ln(10/z_0)}\right]^2 \tag{2.12}$$

The approximate value of surface roughness length z_0 and surface drag coefficient κ for various terrain types is given in Table 2.1.

2.3.2 The "Power Law"

Although the logarithmic law has a sound theoretical basis, it cannot be used to evaluate mean wind speeds at a height z below the zero-plane displacement z_h and it is less easy to integrate. To avoid some of these problems, wind engineers have often preferred to use the power law. The power law is an empirical formula for mean wind speed profile. For mean wind speed $\overline{U}(z)$ at any height z, the power law can be written as:

$$\overline{U}(z) = \overline{U}_1\left(\frac{z}{z_1}\right)^{\alpha} \tag{2.13}$$

where:

$\overline{U}_1$ is the wind speed at any reference height z_1;
α is the power law exponent, which depends on surface roughness and other considerations. In practice, the power law exponent is often regarded as equivalent to the surface roughness length z_0.

Since the power law is valid for any value of z within the atmospheric boundary layer δ, we can set $\overline{U}_1 = \overline{U}_g$ at $z_1 = \delta$, which results in:

$$\overline{U}(z) = \overline{U}_g\left(\frac{z}{\delta}\right)^{\alpha} \tag{2.14}$$

where δ is the height of the boundary layer above the ground; and $\overline{U}_g$ is the mean wind speed at height δ above the ground. In general, the rougher a surface is, the higher the values of α and δ and the smaller the velocity $\overline{U}(z)$ is at any given height z. Table 2.1 also provides the values of α and δ for monsoons and extra-tropical cyclones.

2.3.3 Mean Wind Speed Profile Over Ocean

Since higher winds over the ocean create higher waves and, therefore, higher surface drag coefficients, the surface drag coefficients are dependent on mean wind speed. Using dimensional argument,

Charnock proposed a mean wind speed profile over the ocean [5], which implies that the roughness length z_0 should be given by [2]:

$$z_o = \frac{a_0}{g}\left(\frac{k\overline{U}_{10}}{\ln(10/z_o)}\right)^2 \tag{2.15}$$

where a_0 is an empirical constant lying between 0.01 and 0.02.

2.3.4 Mean Wind Speed Profile in Tropical Cyclone

Extreme winds experienced in tropical cyclones (typhoons and hurricanes) are of significant concern to bridge engineers, since these winds are notorious for their damaging effects on bridges.

The mean wind profiles in a typhoon may be different from those in a monsoon or extra-tropical cyclone in the atmospheric boundary layer, and this is still not well understood. A number of mean wind speed profiles have been obtained during typhoons by using the GPS dropwindsonde, an instrument which was developed by the National Center for Atmospheric Research in the middle 1990s in a joint effort with NOAA and the German Aerospace Research Establishment [6]. The results indicate a steep logarithmic-type profile up to a certain height (60–200 m), followed by a layer of strong convection, with nearly constant mean wind speed [2]. More investigations on the mean wind speed profile in tropical cyclones are required.

2.4 Wind Turbulence

In order to describe a turbulent flow, statistical methods must be applied. The three turbulence components are treated as stationary random processes mathematically and are described by means of their standard deviations, time scales and integral length scales, power spectral density functions that define the frequency distribution, and normalized co-spectra that specify the wind spatial correlation.

2.4.1 Standard Deviations

Mathematically, the standard deviations σ_u, σ_v and σ_w for the turbulence components in the longitudinal direction, lateral horizontal direction, and vertical direction can be written as:

$$\sigma_u = \sqrt{\frac{1}{T}\int_0^T u(t)^2 dt} \tag{2.16}$$

$$\sigma_v = \sqrt{\frac{1}{T}\int_0^T v(t)^2 dt} \tag{2.17}$$

$$\sigma_w = \sqrt{\frac{1}{T}\int_0^T w(t)^2 dt} \tag{2.18}$$

where $u(t)$, $v(t)$, and $w(t)$ are the turbulence components in the longitudinal direction, lateral horizontal direction and vertical direction respectively. Since the mean value of each turbulence component is zero, the above standard deviation is equal to its root-mean-square (RMS) value.

2.4.2 Turbulence Intensities

The turbulence intensity is often used to describe the intensity of turbulence. It is defined as the ratio of the standard deviation of each turbulence component to the mean wind speed of the same averaging time. Thus, the turbulence intensities $I_u(z)$, $I_v(z)$ and $I_w(z)$ for turbulence component in the longitudinal direction, lateral direction and vertical direction can be expressed as:

$$I_u(z) = \frac{\sigma_u(z)}{\overline{U}(z)} \tag{2.19}$$

$$I_v(z) = \frac{\sigma_v(z)}{\overline{U}(z)} \tag{2.20}$$

$$I_w(z) = \frac{\sigma_w(z)}{\overline{U}(z)} \tag{2.21}$$

Normally, the turbulence intensity increases with surface roughness and decreases with height. It also varies with the duration (averaging time) used in determining the mean wind speed $\overline{U}(z)$ and the standard deviation.

For gales (extra-tropical cyclones) near the ground, field measurements have found that the standard deviation of longitudinal turbulence component σ_u is roughly equal to $2.5u_*$. For the lateral and vertical turbulence components, the standard deviation of lateral and vertical component is approximately equal to $2.20u_*$ and $1.4u_*$ respectively. As a result, the turbulence intensity I_u, I_v and I_w can be given by the following equations [2].

$$I_u = \frac{2.5u_*}{(u_*/0.4)\ln(z/z_0)} = \frac{1}{\ln(z/z_0)} \tag{2.22}$$

$$I_v = \frac{2.2u_*}{(u_*/0.4)\ln(z/z_0)} = \frac{0.88}{\ln(z/z_0)} \tag{2.23}$$

$$I_w = \frac{1.4u_*}{(u_*/0.4)\ln(z/z_0)} = \frac{0.56}{\ln(z/z_0)} \tag{2.24}$$

Thus, the turbulence intensity is related to the surface roughness, as measured by the surface roughness length z_0. For a rural terrain, with a roughness length of 0.04 m, the longitudinal, lateral and vertical turbulence intensities for various height above the ground are given in Table 2.2. It can be seen that the turbulence intensity decreases with height above the ground. The turbulence intensities in tropical

Table 2.2 Longitudinal, lateral and vertical turbulence intensities for rural terrain ($z_0 = 0.04$ m)

Height z (m)	I_u	I_v	I_w
2	0.26	0.23	0.14
5	0.21	0.19	0.12
10	0.18	0.16	0.10
20	0.16	0.14	0.09
50	0.14	0.12	0.08
100	0.13	0.11	0.07

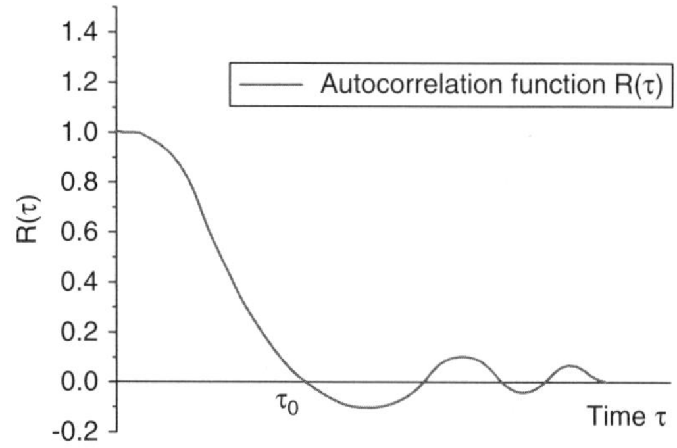

Figure 2.5 Typical variation of autocorrelation $R(\tau)$.

cyclones (typhoons and hurricanes) are generally believed to be higher than those in gales in temperate regions due to convective "squall-like" turbulence [2].

2.4.3 Time Scales and Integral Length Scales

The autocorrelation function $R(z, \tau)$ of the longitudinal turbulence component u is defined as the normalized mean value of the product of the turbulence components u at time t and at time $t+\tau$:

$$R(z,\tau) = \frac{1}{T}\int_0^T \{u(z,t)\cdot u(z,t+\tau)\}dt/\sigma_u^2(z) \tag{2.25}$$

The autocorrelation function gives the correlation of the turbulence component u with itself at a time τ later. Figure 2.5 shows a typical shape of $R(z, \tau)$. Note that the autocorrelation function may oscillate about the τ-axis after a time τ_0 that corresponds to the time when $R(z, \tau)$ has first reached zero.

The autocorrelation function depends on height z above the ground and on time difference τ. The turbulence component u may be said to have an average period of large eddies in flow, which is called the time scale $T_u(z)$. The formal definition of time scale $T_u(z)$ is:

$$T_u(z) = \int_0^\infty R(z,\tau)d\tau \tag{2.26}$$

The cross-correlation function between the turbulence component u at the two points separated longitudinally by a distance r_x and measured simultaneously is defined as:

$$f(z,r_x) = \frac{1}{T}\int_0^T \{u(z,x,t)\cdot u(z,x+r_x,t)\}dt/\sigma_u^2(z) \tag{2.27}$$

The cross-correlation gives the correlation of u with itself at a distance $x=r_x$ apart.

If all the eddies in a turbulent flow are carried downstream by a constant mean wind speed $\overline{U}$ (Taylor's hypothesis), it can be proven that $\partial/\partial t = \overline{U}\partial/\partial x$, and $R(z, \tau)=f(z, r_x)$ for $r_x = \overline{U}\tau$. This indicates that a statistical description of temporal turbulence variations could be based on spatial wind velocity field characteristics, and vice versa.

Corresponding to the time scale, there is an integral length scale. Integral length scales are a measure of the sizes of the vortices in the wind or, in other words, the average size of a gust in a given direction. As an example, L_u^x is the integral length scale for the longitudinal turbulence component u measured in the longitudinal direction x, and it is formally defined as:

$$L_u^x = \int_0^\infty f(z,r_x)dr_x \tag{2.28}$$

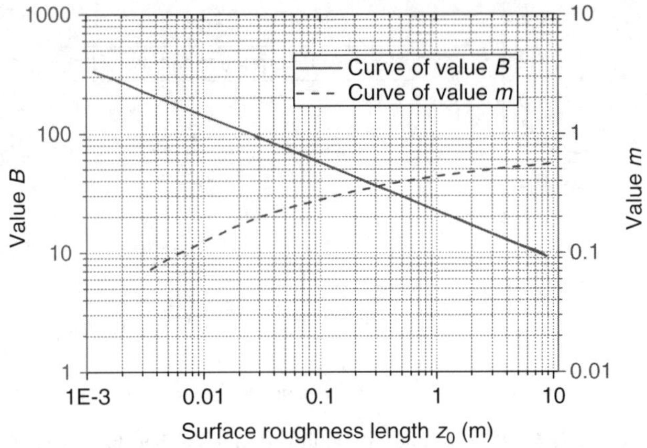

Figure 2.6 Values of B and m as functions of surface roughness.

Thus, the longitudinal integral length scale is equal to the time scale multiplied by the mean wind speed:

$$L_u^x(z) = \overline{U}(z)T_u(z) \tag{2.29}$$

Normally, full-scale measurements are used to estimate integral length scales. The integral length scales depend on the height z above the ground and the surface roughness, i.e. surface roughness length z_0. The mean wind speed may also influence the integral length scales on site.

Counihan has suggested the following purely empirical expression for the longitudinal integral length scale at height z in the range of 10–240 m [7].

$$L_u^x = Bz^m \tag{2.30}$$

where B and m depend on the surface roughness length z_0, as shown in Figure 2.6. z and L_u^x are stated in meters. According to Counihan [7], integral length scales decrease with increasing surface roughness.

In total, nine integral length scales can be defined similarly as shown in Equation 2.28. The remaining integral length scales are often expressed as a function of the longitudinal integral length scale L_u^x. The integral length scales L_u^y and L_u^z for the longitudinal turbulence component u in the lateral direction y and vertical direction z are approximately 1/3 to 1/2 of the integral length scale L_u^x.

2.4.4 Probability Density Functions

It is known that turbulence winds in the atmospheric boundary layer are generally random in nature, and therefore they must be described by the statistical method. Measurements have shown that the turbulence winds in the atmospheric boundary layer due to monsoons and extra-tropical cyclones follow closely the Gaussian distribution, which can be described as:

$$f(u) = \frac{1}{\sqrt{2\pi}\sigma_u} \exp\left[-\frac{1}{2}\left(\frac{u-\overline{U}}{\sigma_u}\right)^2\right] \tag{2.31}$$

where $f(u)$ is the probability density function of the longitudinal turbulence component u. This function has the characteristic bell shape as shown in Figure 2.7. It is defined by only the mean value $\overline{U}$ and standard deviation σ_u. The probability density functions of the lateral and vertical turbulence components in monsoons and extra-tropical cyclones can be similarly defined.

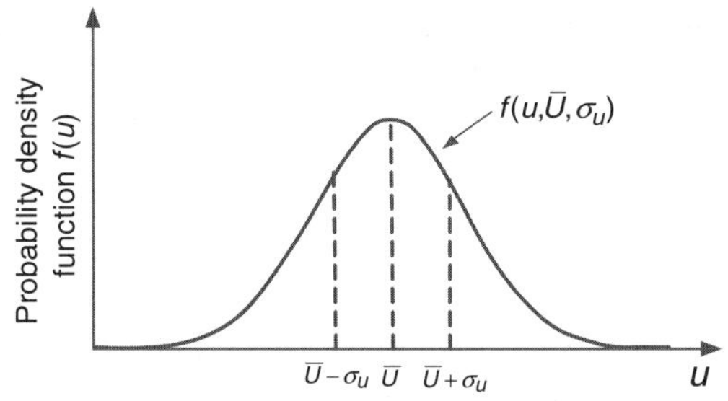

Figure 2.7 Gaussian probability density function.

2.4.5 Power Spectral Density Functions

Turbulence winds consist of a large number of eddies or velocity waves which have different amplitudes and frequencies. The power spectral density function, usually abbreviated to wind spectrum, is used to describe the distribution of turbulence with frequency,

The spectrum of turbulence winds can be measured by processing the electronic signal from an anemometer. The signal is first stripped of its mean value, i.e. the direct current (DC) component. Only the fluctuating part – the alternating current (AC) component – is kept and analyzed for determining the wind spectrum. The spectrum can be obtained with a spectrometer, which connects the signal (the AC component) to a set of electronic band pass filters, each having a narrow band width and a different filter frequency. The signal that has passed through each filter is measured by an RMS-meter for its intensity. Finally, the spectrometer plots the spectrum as a function of filter frequency n.

An alternate method to obtain a wind spectrum is to perform a Fourier analysis of signals on a digital computer. Most modern computers, including personal computers, can perform such an analysis and can plot the resulting spectrum using special software.

The wind spectrum shown in Figure 2.8 illustrates certain special features of all turbulence spectra. Due to the DC cut-off in turbulence measurements, all spectra pass through the origin of their coordinates. They all increase rapidly to reach a peak at a certain frequency, corresponding to the frequency of the strongest eddy in the wind. After passing the peak, spectral curves decrease gradually, approaching

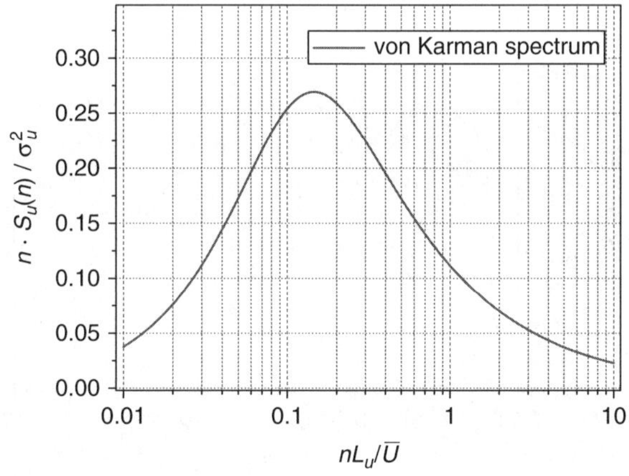

Figure 2.8 Normalized von Karman spectrum of longitudinal turbulence component.

zero value when the frequency approaches infinity. The spectral peak for the turbulence in natural winds is usually at rather low frequencies, of the order of 1 cycle per minute.

There are many mathematical forms that have been used for wind spectrum in meteorology and wind engineering. The most common and mathematically correct wind spectrum for longitudinal turbulence component is the von Karman-Harris form [8,9], which can be expressed in a non-dimensional form as:

$$\frac{nS_u(n)}{\sigma_u^2} = \frac{4x}{(1+70.8x^2)^{5/6}} \qquad x = \frac{nL_u^x(z)}{\overline{U}(z)} \tag{2.32}$$

where $S_u(n)$ is the longitudinal turbulence wind spectrum. The von Karman spectrum for longitudinal turbulence component is shown in Figure 2.8. It can be seen that the curve of the non-dimensional wind spectrum has a peak, and the value of integral length scale L_u^x determines the value of $n/\overline{U}$ at which the peak occurs. The integral length scale varies with both terrain roughness and height above the ground.

The other orthogonal components of wind turbulence have spectral densities with somewhat different characteristics [2]. For long-span cable-supported bridges, the wind spectrum of vertical turbulence component is the most important. A commonly-used wind spectrum of vertical turbulence component was proposed by Busch and Panofsky in the form [10]:

$$\frac{nS_w(n)}{\sigma_w^2} = \frac{2.15f}{\left[1+11.16f^{5/3}\right]} \qquad f = \frac{nz}{\overline{U}(z)} \tag{2.33}$$

where $S_w(n)$ is the vertical turbulence wind spectrum. The Busch and Panofsky spectrum for vertical turbulence component is shown in Figure 2.9.

2.4.6 *Covariance and Correlation*

Covariance and correlation are two important properties of turbulence wind, reflecting the statistical relation of fluctuating wind velocities between two points in space.

For example, consider the wind speed at two different heights on a bridge tower. The covariance between the turbulence winds at two different heights, z_1 and z_2, is defined as follows:

$$\overline{u(z_1)u(z_2)} = \frac{1}{T}\int_0^T \left[U(z_1,t) - \overline{U}(z_1)\right]\left[U(z_2,t) - \overline{U}(z_2)\right]dt \tag{2.34}$$

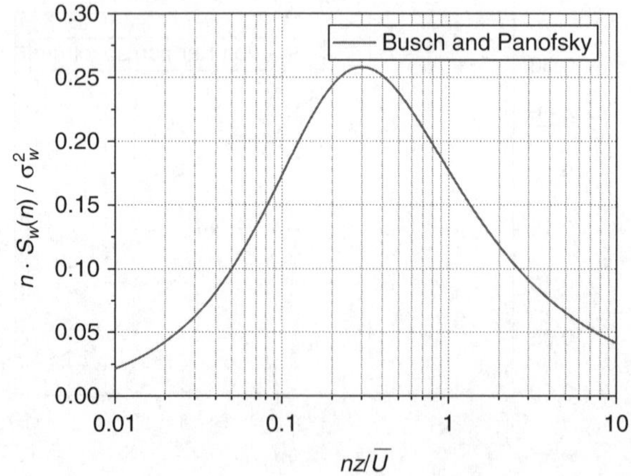

Figure 2.9 Normalized wind spectrum of vertical turbulence component.

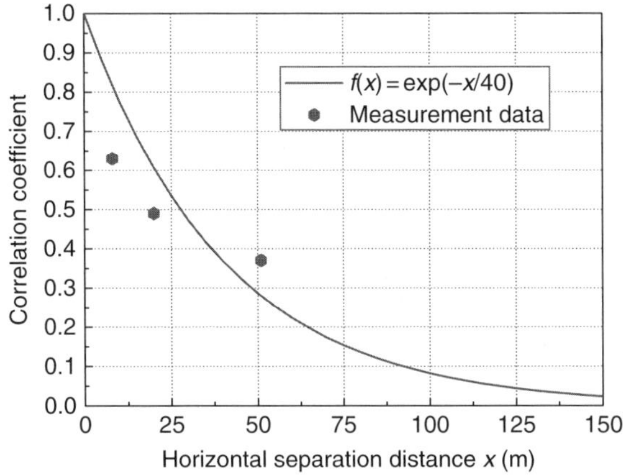

Figure 2.10 Correlation coefficient of longitudinal turbulence component.

Thus, the covariance is the product of the turbulence winds at two different heights averaged over time. Note that in the special case when z_1 is equal to z_2, the right hand side is then equal to the variance of the turbulence wind at the single height.

The correlation coefficient ρ is then defined by the following equation:

$$\rho = \frac{\overline{u(z_1)u(z_2)}}{\sigma_u(z_1) \cdot \sigma_u(z_2)} \tag{2.35}$$

When z_1 is equal to z_2, the value of ρ is $+1$. It can be shown that ρ must lie between -1 and $+1$. A value of 0 indicates no correlation and this usually occurs when heights z_1 and z_2 are widely separated.

A mathematical function which is useful for describing the correlation ρ is the exponential decay function:

$$\rho = \exp[-C|z_1 - z_2|] \tag{2.36}$$

This function is equal to $+1$ when z_1 is equal to z_2, and it tends to zero when $|z_1 - z_2|$ becomes very large. Figure 2.10 shows the correlation coefficient with C equal to $(1/40)\,\mathrm{m}^{-1}$ and some measurement results recorded at a height of 13.5 m [11].

2.4.7 *Cross-Spectrum and Coherence*

The normalized cross-spectrum describes the statistical dependence of the turbulence components at two points at a given frequency n. This dependence is due to the spatial dimension of the vortices in the wind field. It is important for considering the resonant response of bridges to wind.

The spatial distribution of the longitudinal turbulence component is often described by the normalized dimensionless cross-spectrum S_N, which can be expressed as:

$$S_N = \frac{S_{uu}(P_1, P_2, n)}{\sqrt{S_u(P_1, n)S_u(P_2, n)}} \tag{2.37}$$

where $S_{uu}(P_1, P_2, n)$ is the cross-spectrum of the two longitudinal turbulence components at points P_1 and P_2, respectively, and it is a complex quantity. S_u is the power (auto) spectrum of longitudinal turbulence component in the point specified by the argument of P_1 or P_2.

The normalized cross-spectrum is a complex variable, with real and imaginary components. The real part of the normalized cross-spectrum is called the normalized co-spectrum, ψ_u, and the imaginary part is called the normalized quad-spectrum, φ_u. The root-coherence function is defined as the absolute value of the normalized cross-spectrum, $\sqrt{\text{Coh}} = |\text{S}_\text{N}|$. The normalized co-spectrum may be regarded as a frequency-dependent covariance, whereas the root-coherence is approximately equivalent to a frequency-dependent correlation coefficient.

The normalized co-spectrum and the root-coherence function are identical when the phase spectrum (quad-spectrum) is zero. The normalized co-spectrum ψ_u decreases with the increasing distance r between the two points considered. This decrease depends on the size of the vortices, and a measure of size is the ratio of the mean wind speed to the frequency $\overline{U}/n$.

On a purely empirical basis, Davenport suggested an exponential expression as the normalized co-spectrum with a phase-spectrum of zero [12,13]:

$$\psi_u(r,n) = \exp\left(-Cr\frac{n}{\overline{U}}\right) \tag{2.38}$$

where:

C is a non-dimensional decay constant that determines the spatial extent of the correlation in the turbulence, used to fit measured data, and a typical range of values for atmospheric turbulence is 10 to 20;
r is the spatial distance between two points.

Equation 2.38 does not allow negative values – a theoretical problem [2].

For two points with transverse separation (r_y, r_z), Davenport extended the expression to allow for different normalized co-spectrum decays horizontally and vertically, respectively [14]:

$$\psi_u(r_y, r_z, n) = \exp\left(-\frac{n}{\overline{U}}\sqrt{(C_y r_y)^2 + (C_z r_z)^2}\right) \tag{2.39}$$

where:

$\overline{U}$ is the mean wind speed at the two points considered, equal to $\frac{1}{2}\left[\overline{U}(z_1) + \overline{U}(z_2)\right]$;
r_y and r_z are the lateral and vertical spatial distance;
C_y and C_z are non-dimensional decay constants.

When choosing numerical values for the decay constants, due consideration must be given to the fact that surface pressures are better correlated than the turbulence in the undisturbed wind field. Typical values are $C_y = 10$ and $C_z = 10$.

Thus, the cross-spectrum of two longitudinal turbulence components can be expressed as:

$$S_{uu}(P_1, P_2, n) = \sqrt{S_u(P_1, n) S_u(P_2, n)} \cdot \exp\left(-Cr\frac{n}{\overline{U}}\right) \tag{2.40}$$

$$S_{uu}(P_1, P_2, n) = \sqrt{S_u(P_1, n) S_u(P_2, n)} \cdot \exp\left(-\frac{n}{\overline{U}}\sqrt{(C_y r_y)^2 + (C_z r_z)^2}\right) \tag{2.41}$$

2.4.8 Gust Wind Speed and Gust Factor

Gust is the rapid fluctuation or instantaneous velocity of wind. Long-span cable-supported bridges are sensitive to peak gusts of a duration of the order of 2–3 seconds. Therefore, the use of any mean wind speed that has a much longer duration than 2–3 seconds without taking into account gust effect is inadequate for bridge design. One must design bridges to withstand the gust wind speed rather than the mean wind speed. Gust wind speed can be related to the mean wind speed in terms of the gust factor, which is defined as:

$$G_v = \frac{U_G}{\overline{U}} \tag{2.42}$$

where:

G_v is the gust factor;
U_G is the gust wind speed;
$\overline{U}$ is the mean wind speed.

Both the gust wind speed and the gust factor are the functions of gust duration t and the averaging time T for the mean wind speed. On the other hand, if the longitudinal wind velocity has a Gaussian probability distribution, the expected peak gust wind speed is given approximately by:

$$U_G = \overline{U} + g\sigma_u \tag{2.43}$$

where g is the peak factor equal to about 3.5.

Based on measurements in typhoons and Equation 2.43, the following expression for gust factor G has been proposed [15]:

$$G_v(t) = 1 + 0.5I_u \ln(T/t) \tag{2.44}$$

where:

I_u is the longitudinal turbulence intensity;
T is the averaging time for the mean wind speed;
t is the gust duration.

Thus, knowing the turbulence intensity of the site, the gust wind speed for any gust duration can be obtained by multiplying the mean wind speed with the gust factor G_v calculated from Equation 2.44.

Since gusts are the results of the effects of eddies and vortices in the atmospheric turbulence, it can be visualized that a certain gust has a certain size of influence; the smaller the size of the gust, the shorter will be its duration and the higher will be the gust wind speed. A larger sized gust has a longer duration and, hence, a lower average gust wind speed. A smaller gust has a higher gust wind speed but, because of its small physical dimensions, it can only produce high wind pressure on a small area of a bridge. In other words, a small gust can only create high wind loading on a small local area of the bridge, so it would be wrong to design the whole bridge based on this gust wind speed. The whole bridge should be designed to take into account the speed of a gust which is just big enough to affect the whole bridge simultaneously. Thus, the size of a bridge is a factor affecting its own wind loading.

There are two different ways in which the effect of gusts is being dealt with in the different codes of practice. The first is to use the mean wind speed as the basis, then multiply by the gust factor to obtain the design gust wind speed. The second method is to use the peak gust recording of the meteorology stations as the basis. The period of this peak gust depends on the anemometer type and, in most cases, a response time of 2–3 seconds is thought to be appropriate. For the design of bridges, which takes a gust of longer duration to encompass, this 2–3 second gust wind speed has to be multiplied by a reduction factor to give the appropriate loading.

2.5 Terrain and Topographic Effects

2.5.1 Change of Surface Roughness

All of the equations given above are for fully developed mean wind speed profiles that are for wind which has blown over a considerable distance of terrain with the same surface roughness. When winds in a fully developed boundary layer encounter a change of surface roughness, a process of adjustment in the boundary layer flow properties develops. For example, consider the change of terrain from smooth to rough. Due to the greater surface roughness at location 2 than at location 1, winds must slow down from 1 to 2, resulting in a rise in streamlines (see Figure 2.11).

This expansion of streamlines causes an increase in the boundary-layer thickness from δ_1 to δ_2 and an increase in the power-law exponent from α_1 to α_2. The gradient wind speed $\overline{U}_g$, however, remains unchanged. That is:

$$\overline{U}_1 = \overline{U}_g \left(\frac{z_1}{\delta_1}\right)^{\alpha_1} \quad \overline{U}_2 = \overline{U}_g \left(\frac{z_2}{\delta_2}\right)^{\alpha_2} \tag{2.45}$$

from which one has:

$$\frac{\overline{U}_2}{\overline{U}_1} = \left(\frac{z_2}{\delta_2}\right)^{\alpha_2} \cdot \left(\frac{\delta_1}{z_1}\right)^{\alpha_1} \tag{2.46}$$

In the transitional area between the two profiles, the relationship between the development height and the fetch distance is complicated. Wind tunnel tests, computational simulations and field measurements may be required to investigate such cases.

2.5.2 Amplification of Wind by Hills

Besides surface roughness of the terrain, the wind speed at a given site is also affected by natural and man-made local topography in the form of ridges, cliffs, and hills. This is particularly true when long-span cable-supported bridges are built to cross mountains. This local effect can sometimes be very strong, such that the wind speed distribution deviates far away from the wind speed profile expected for the type of terrain. In general, this problem is complicated and necessitates wind tunnel tests, computational simulations and field measurements.

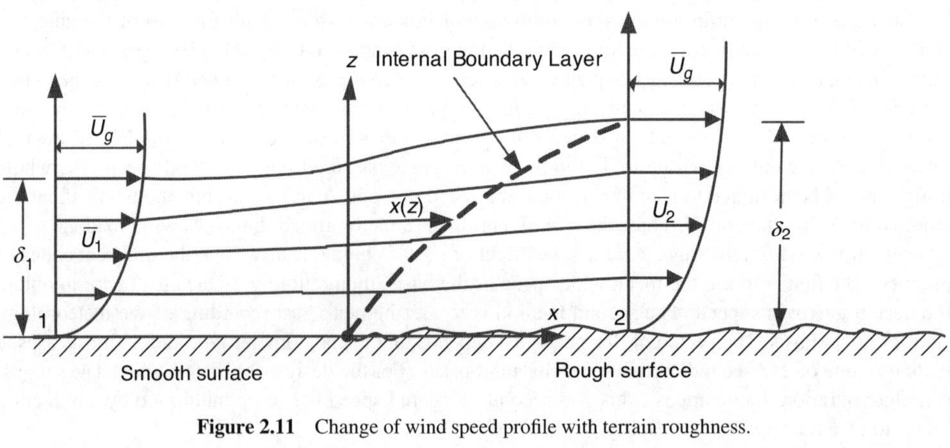

Figure 2.11 Change of wind speed profile with terrain roughness.

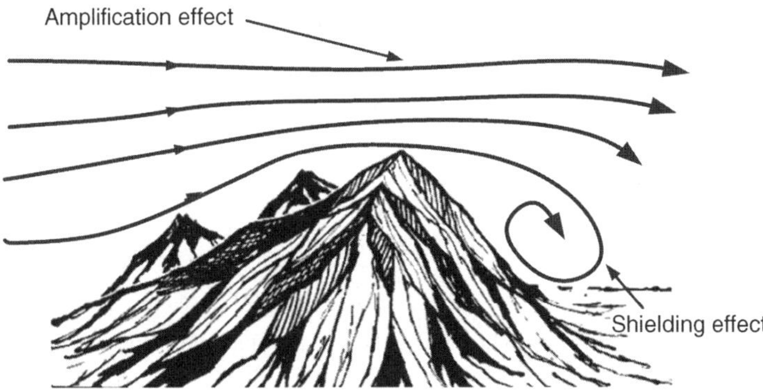

Figure 2.12 Effects of hill on wind field.

Ordinarily, the wind speed increases on the windward slope of a hill or a mountain peak, reaching a maximum at or near the summit (see Figure 2.12). Two major factors contribute to this amplification of wind. First, mountains and hills restrict the passage of wind, causing the streamline of the wind to converge on the windward slope. Since wind speed is inversely proportional to the spacing between streamlines of a two-dimensional flow, the wind speed continuously accelerates as the wind approaches a mountain (hill) top. This effect, due to the compression of streamlines, is most profound for two-dimensional mountains (hills) with winds perpendicular to ridgelines. This explains why two-dimensional mountains (hills) generally encounter greater amplification of wind than three-dimensional mountains (hills). Second, the height of mountains (hills) brings the gradient wind closer to surface, resulting in a reduction of the gradient height, which causes an increase in the wind speed within the boundary layer.

Generally, the wind speed above hills increases rapidly with height in a region very close to the surface (see Figure 2.13). Outside this small surface layer, the increase is at a rate less than for wind above flat surfaces, resulting in a decrease in the α value of the power law. In some cases, the wind speed above a hill may first increase with height until it reaches a maximum at certain height. Above this height, wind speed decreases with height instead of increasing. Such a profile cannot be represented adequately by the power law or the logarithmic velocity profiles, which are for winds above flat areas.

From wind tunnel studies and field measurements, it is known that for a two-dimensional hill with perpendicular to ridgeline, maximum amplification of wind occurs when the windward slope is approximately 1 : 3.5 (vertical : horizontal) and when the surface of the windward slope is smooth. Such a hill can cause surface wind speed to double at the hill top. This points to the great importance of considering the topographic effect of wind when designing bridges on mountains or hills.

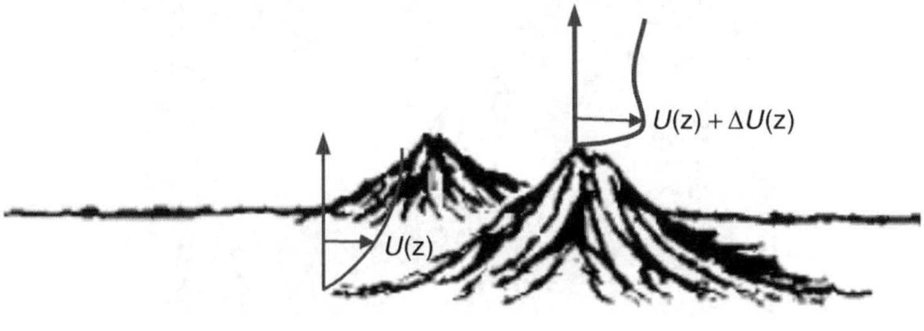

Figure 2.13 Velocity profile over the crest of a hill.

2.5.3 *Amplification Factor and Speed-up Ratio*

Amplification factor or speed-up ratio is used in the wind codes of many countries. The amplification of winds by hills, mountains and so on can be quantified by using the amplification factor A, defined as:

$$A = \frac{U'(z)}{U(z)} \tag{2.47}$$

where:

$U'(z)$ is the amplified wind speed at height z above the surface of a hill or slope;
$U(z)$ is the speed of the approaching wind at the same height z above the ground.

A value of $A = 2$ means the wind speed is doubled by the hill.
Furthermore, the speed-up ratio or fractional speed-up ratio β is defined as follows:

$$\beta = \frac{U'(z) - U(z)}{U(z)} = \frac{\Delta U(z)}{U(z)} \tag{2.48}$$

Obviously,

$$A = \beta + 1 \tag{2.49}$$

The range of β is normally between 0 and 1.0, and the range of A is between 1.0 and 2.0.

2.5.4 *Funneling Effect*

Another type of topography that will induce speed-up effect is the narrow pass or gap. As the wind is squeezed through the narrower flow section, the funneling effect produces a higher wind speed (see Figure 2.14). The speed-up for wind in narrow passes is more difficult to evaluate than flow over slopes.

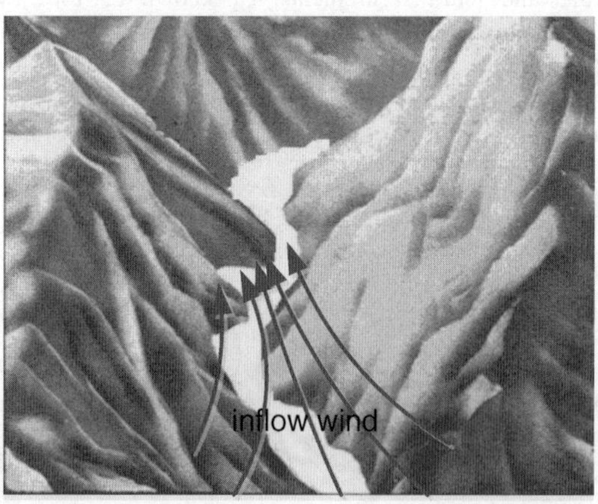

Figure 2.14 Speed-up of wind through canyons.

Parameters such as longitudinal gradient of the pass, slopes and the heights of the two sides will all affect the wind speed profile.

The effect of topographic features on wind near the earth's surface in tropical cyclones is much less clearly understood than those in the well-developed boundary layers of large-scale synoptic systems [2]. Some discussions on this topic can be found in Chapter 13 (Typhoon Wind Field Simulation).

2.6 Design Wind Speeds

The determination of design wind speeds is an important step of the calculation of design wind loads for a bridge. However, wind is a random process, and wind speed fluctuates in a random manner. Thus, the determination of design wind speeds is the most uncertain part of the calculation of design wind loads, which should be determined based on the concept of probability in conjunction with the recorded historical wind speed data, as a result of the work of Davenport [12–14]. This section focuses on the prediction of extreme wind speeds for the design of bridges in terms of extreme value analysis.

2.6.1 Exceedance Probability and Return Period

The basic probability of maximum winds needed for bridge design is the exceedance probability P_E, which is the probability that a given wind speed will be exceeded within a one-year period. The reciprocal of exceedance probability is called the return period, namely:

$$T_R = \frac{1}{P_E} \tag{2.50}$$

where P_E and T_R are the exceedance probability and the return period, respectively.

If the annual maximum is being considered, the return period is measured in years. Thus, a 50-year return period wind speed has a probability of exceedance of 0.02 in any one year. Normally, the exceedance probability P_E can be determined based on annual maximum wind data. Table 2.3 lists the annual maximum hourly mean wind speeds recorded at the Hong Kong Royal Observatory Station from 1970 to 1999. It can be observed that the values differ very much from year to year for a specified site.

To estimate the exceedance probability P_E, one can arrange these values into descending order and a rank number i is given to each value, such that $i = 1$ for the highest value data. The exceedance probability P_E can then be written as

$$P_E = \frac{i}{N+1} \tag{2.51}$$

Table 2.3 Annual maximum hourly mean wind speeds recorded at the Hong Kong Royal Observatory Station

Year	Mean wind speed (m/s)	Year	Mean wind speed (m/s)	Year	Mean wind speed (m/s)
1970	20.5	1980	21.5	1990	22.0
1971	36.0	1981	20.0	1991	29.0
1972	19.5	1982	20.0	1992	26.0
1973	24.5	1983	44.0	1993	30.0
1974	22.5	1984	24.5	1994	20.6
1975	32.5	1985	22.0	1995	28.8
1976	26.0	1986	29.0	1996	27.0
1977	23.0	1987	23.5	1997	30.5
1978	29.5	1988	19.0	1998	22.5
1979	38.5	1989	26.0	1999	42.0

Table 2.4 Exceedance probabilities of hourly mean wind speed for wind date in Table 2.3

Rank	Year	Mean wind speed	Exceedance probability	Rank	Year	Mean wind speed	Exceedance probability
1	1983	44.0	0.0323	16	1973	24.5	0.5484
2	1999	42.0	0.0645	17	1984	24.5	0.5484
3	1979	38.5	0.0968	18	1987	23.5	0.5806
4	1971	36.0	0.1290	19	1977	23.0	0.6129
5	1975	32.5	0.1613	20	1974	22.5	0.6774
6	1997	30.5	0.1935	21	1998	22.5	0.6774
7	1993	30.0	0.2258	22	1985	22.0	0.7419
8	1978	29.5	0.2581	23	1990	22.0	0.7419
9	1986	29.0	0.3226	24	1980	21.5	0.7742
10	1991	29.0	0.3226	25	1994	20.6	0.8065
11	1995	28.8	0.3548	26	1970	20.5	0.8387
12	1996	27.0	0.3871	27	1981	20.0	0.9032
13	1976	26.0	0.4839	28	1982	20.0	0.9032
14	1989	26.0	0.4839	29	1972	19.5	0.9355
15	1992	26.0	0.4839	30	1988	19.0	0.9677

*For the years with same mean wind speed, the larger exceedance probability was used.

where:

i is the rank number;
N is the total number of years.

Note that $i/(N+1)$ instead of i/N is used, to avoid $P_E = 1$ (certainty) for $i = N$. The exceedance probabilities of the hourly mean wind speed for the wind date listed in Table 2.3 are calculated using Equation 2.51 and are listed in Table 2.4.

2.6.2 *Probability Distribution Function*

In most cases, maximum wind data are often limited, so the predication of exceedance probability based on the measured wind data may not be reliable. Knowing the correct probability law (probability density functions and probability distribution functions) of high winds makes prediction more realistic, especially for large values of wind speeds.

The probability density function $f(U)$ of wind speed U is the probability per unit wind speed. From this definition, the probability for wind speed between any value U and an infinitesimally larger value $U + \mathrm{d}U$ is $f(U)\mathrm{d}U$.

The probability for wind speed to be less than U, designated as the probability distribution function $F(U)$, is:

$$F(U) = \int_0^U f(U)dU \tag{2.52}$$

from which:

$$f(U) = \frac{dF(U)}{dU} \tag{2.53}$$

Thus, the exceedance probability P_E is related to the probability distribution function $F(U)$ or probability density function $f(U)$, and it can be written as:

$$P_E(U) = 1 - F(U) = 1 - \int_0^U f(U)dU \tag{2.54}$$

This equation shows that the exceedance probability can be calculated easily from either the probability distribution function or the probability density function.

2.6.3 Generalized Extreme Value Distribution

There are three asymptotic extreme value distributions identified by Fisher and Tippett [16]. These distributions are asymptotic in the sense that they are the correct distributions for the largest of an infinite population of independent random variables of known probability distribution. However, there will be a finite number in a population in reality due to lack of measured wind data. Physical reasoning has sometimes been used to justify the use of one or other of the three asymptotic extreme value distributions [2]. The generalized extreme value (GEV) distribution introduced by Jenkinson [17] combines the three extreme value distributions into a single mathematical form:

$$F(U) = \exp\left\{-[1 - \xi(U-\mu)/a]^{1/\xi}\right\} \tag{2.55}$$

where:

a is the scale factor;
μ is the location parameter;
ξ is the shape factor.

When ξ tends to 0, the generalized extreme value distribution is known as Type I Extreme Value Distribution, or the Gumbel Distribution; when $\xi < 0$, the generalized extreme value distribution is called Type II Extreme Value Distribution, or the Frechet Distribution; and when $\xi > 0$, the generalized extreme value distribution is known as Type III Extreme Value Distribution, which is one form of the Weibull Distribution.

2.6.4 Extreme Wind Estimation by the Gumbel Distribution

Gumbel [18] gave an easily usable methodology for fitting recorded annual maximum wind speeds to the Type I extreme value distribution. The Type I extreme value distribution or the Gumbel distribution takes the form of Equation 2.56 for the cumulative distribution function $F(U)$:

$$F(U) = \exp\{-\exp[-\eta(U-\mu)]\} \tag{2.56}$$

where U is the annual maximum wind speed; μ is the mode of the distribution (the location parameter); and η is the dispersion of the distribution (equals to $1/a$).

Furthermore, the Type I extreme value distribution can be rewritten into a form of:

$$F(U) = \exp[-\exp(-y)] \tag{2.57}$$

$$y = \eta(U - \mu) \tag{2.58}$$

where y is the reduced variant. The probability density function of this distribution can be obtained by the differentiation of the probability distribution function:

$$f(U) = \exp(-y)\exp[-\exp(-y)] \tag{2.59}$$

The two parameters, the mode μ and the dispersion η in the Type I extreme value distribution, can be determined using the recorded annual maximum wind speed data. That is, combining Equations 2.51, 2.54 and 2.59 leads to:

$$1 - \frac{i}{N+1} = \exp[-\exp(-y_i)] \tag{2.60}$$

From Equation 2.60, the reduced variate y_i for the ith highest wind speed U_i can be calculated. It is noted that the relationship between y_i and U_i should be linear:

$$y_i = \eta(U_i - \mu) \tag{2.61}$$

Then plot U_i against y_i to obtain a set of data. Finally, by using a curve-fitting technique, the mode μ and the dispersion η can be obtained and the Type I extreme value distribution expressed by Equation 2.56 can be determined. Furthermore, substituting Equation 2.56 to Equation 2.54 and then to Equation 2.50 yields the design wind speed U, corresponding to a return period T_R (in years):

$$U = \mu + \frac{1}{\eta}\left\{-\ln\left[-\ln\left(1 - \frac{1}{T_R}\right)\right]\right\} \tag{2.62}$$

For large values of return period, T_R, Equation 2.62 can be written as:

$$U \approx \mu + \frac{1}{\eta}\ln T_R \tag{2.63}$$

2.6.5 Extreme Wind Estimation by the Method of Moments

By using Equation 2.56, the mean value and variance of the Type I extreme value distribution can be expressed as:

$$\overline{U} = \int_{-\infty}^{\infty} U dF(U) = \frac{0.5772}{\eta} + \mu \tag{2.64}$$

$$\sigma = \sqrt{\int_{-\infty}^{\infty} [U - \overline{U}]^2 dF(U)} = \frac{\pi}{\eta\sqrt{6}} \tag{2.65}$$

where:

$\overline{U}$ is the mean value of U;
σ is the standard deviation of U.

Then, by combining Equations 2.64 and 2.65, the mode μ and the dispersion η can be obtained as:

$$\eta = \frac{1.28255}{\sigma} \tag{2.66}$$

$$\mu = \overline{U} - 0.45005\sigma \tag{2.67}$$

Finally, by substituting Equations 2.66 and 2.67 into Equations 2.56, 2.54 and 2.50, it can be proved that the design wind speed U corresponding to a return period T_R (in years) can be expressed as:

$$U = \overline{U} + 0.78[\ln T_R - 0.577]\sigma \tag{2.68}$$

As a result, Equations 2.66, 2.67 and 2.68 can be used to determine the dispersion η, the mode μ and the extreme wind speed U corresponding to a return period T_R.

2.6.6 Design Lifespan and Risk

Knowing the extreme distribution of the annual maximum wind speed is only the first step in solving our design wind speed problem. The second step is to select a wind speed with a return period appropriate to our design problem. To begin with, let us say that the design life of our bridge is N years. The probability that the wind speed will not exceed U_0 in any one single year is $F(U_0)$. By assuming the probabilities are statistically independent of each other in N years, then the probability that this same value U_0 will not be exceeded in these N years is:

$$[F(U_0)]^N = \left[1 - \frac{1}{T_R}\right]^N \tag{2.69}$$

and the chance or risk that it will be exceeded in the N years will be:

$$R = 1 - [F(U_0)]^N = 1 - \left[1 - \frac{1}{T_R}\right]^N \tag{2.70}$$

where R is the associated risk of having a wind speed higher than U_0 in the N years.

Let us assume that a return period of 50 years is used. Thus, for a bridge with an expected life of 50 years, the risk we are taking is 63%. This shows that the use of a 50-year return period wind speed to design bridges results in a high probability of the design wind load being exceeded within the lifetime of the bridges. This, however, need not worry the reader, since having a wind higher than the design wind speed does not necessarily mean that the bridge will collapse or even suffer damage. Due to the use of required safety factors such as load factors in bridge design, a bridge is usually able to resist wind speeds higher than the design wind speed. Furthermore, designs are normally made by assuming that the wind is perpendicular to the bridge – the worst case possible. In reality, even when a bridge is hit by a wind higher than the design value, the wind may not be from the worst direction. Therefore, the probability that a bridge designed using a 50-year return-period wind will be destroyed or seriously damaged by wind within 50 years is much less than 63%.

On the other hand, if the life span of a bridge and the accepted risk are both fixed, then we have to use other values for the return period and its corresponding wind speed. For example, if the expected life of a bridge is 50 years and the accepted risk is limited to 40%, the return period T_R can be calculated as follows:

$$T_R = \frac{1}{1-(1-R)^{1/N}} = 98.4 \text{ years} \tag{2.71}$$

Then, by using Equations 2.62 or 2.68, the corresponding design wind speed can be determined.

2.6.7 *Parent Wind Distribution*

For some design applications, it is necessary to have information on the distribution of the complete population of wind speeds at a site. An example is the estimation of fatigue damage for which account must be taken of damage accumulation over a range of wind storms [2]. The complete population of wind speeds is usually fitted to the parent wind distribution, which is one form of the Weibull type:

$$f(U) = \frac{kU^{k-1}}{c^k} \exp\left[-\left(\frac{U}{c}\right)^k\right] \tag{2.72}$$

where:

c is the scale factor;
k is the shape factor.

The probability of exceedance of any given wind speed is given by the following equation:

$$1 - F(U) = \exp\left[-\left(\frac{U}{c}\right)^k\right] \tag{2.73}$$

Typical values of c are 3–10 m/s, and k is usually falls within the range of 1.3–2.0.

2.7 Directional Preference of High Winds

With increased knowledge of bridge aerodynamics, the variation of bridge response is known as a function of wind direction as well as wind speed. In addition, extreme wind speeds reported by most weather stations have preferred wind directions. The wind speed in a particular directional range for a given return period will be different from those in the other directional ranges.

To take wind directional effects into account, several methods have been proposed [19]. A simple method is presented in this section, which may be applied to any type of structure, including bridges subjected to aerodynamic amplification of aeroelastic effects. This method is to find the extreme value distribution of wind speed in each directional sector.

Assuming that there is no correlation between extreme wind speeds at any directional sectors [20], the cumulative probability distribution of the largest annual wind effect will be equal to the product of the cumulative probabilities of the equivalent wind speeds in each of the directional sectors [21]. The joint cumulative probability of all wind directional wind speeds can be expressed as:

$$\begin{aligned} P_E(v < U) &= P_E(v_1 < U, v_2 < U, \ldots, v_N < U) \\ &= P_E(v_1 < U) P_E(v_2 < U) \ldots P_E(v_N < U) \end{aligned} \tag{2.74}$$

where:

$P_E(v < U)$ and $P_E(v_1 < U, v_2 < U, \ldots, v_N < U)$ is the probability that a wind speed U is not exceeded for all wind directions;

$P_E(v_i < U)$ is the probability that a wind speed U is not exceeded for wind direction i, $i = 1, 2, 3, \ldots, N$.

The following tentative conclusions from related researches can be drawn as the directional preference of high winds.

1. Knowing the preferred wind direction at any location facilitates the development of effective strategies for mitigating wind damage.
2. Knowing the preferred wind direction, a bridge can be designed with its longitudinal axis parallel to that direction. This results in reduced wind loads on the bridge.
3. Because the high winds at any location may come from different directions, the probability of a certain wind speed being exceeded in a given direction is always smaller than the probability of exceeding the same wind speed without considering wind direction. This means that the common practice of disregarding wind direction in the determination of exceedance probability results in a conservative design.

2.8 Case Study: Tsing Ma Bridge Site

The Tsing Ma Bridge in Hong Kong is the longest suspension bridge in the world carrying both highway and railway, with an overall length of 2132 m and a main span of 1377 m between the Tsing Yi tower in the east and the Ma Wan tower in the west (see Figure 2.15). The Tsing Ma Bridge is also located in one of the most active typhoon-prone regions in the world. A comprehensive Wind And Structural Health Monitoring System (WASHMS) and a Global Positioning System-On-Structure Instrumentation System (GPS-OSIS) were installed in the Tsing Ma Bridge by the Hong Kong Highways Department in 1997 and 2000, respectively [22,23].

Hong Kong is situated at latitude N22.2° and longitude E114.1° and it is on the southeastern coast of China, facing the South China Sea. The weather system of Hong Kong is influenced by the land mass to its north as well as by the ocean to its south and east. Two types of wind conditions dominate Hong Kong: monsoon wind, prevailing in the months from November to April, and typhoon wind predominating in the summer. The local topography surrounding the bridge is quite unique and complex as it includes sea, islands and mountains from 69–500 m high (see Figure 2.16). The alignment of the bridge deck deviates for 17° counterclockwise from the east-west axis (see Figure 2.17). The complex topography makes wind characteristics at the bridge site very complicated.

The WASHMS installed in the bridge makes it possible to investigate wind characteristics at the bridge site and to gain a better understanding of wind loading on the bridge. Wind data recorded by the WASHMS have been analyzed, and wind characteristics such as mean wind speed, mean wind direction, turbulence components, turbulence intensities, integral scales and wind spectra have been obtained for both monsoon winds and typhoon winds. It is impossible to cover all the research activities

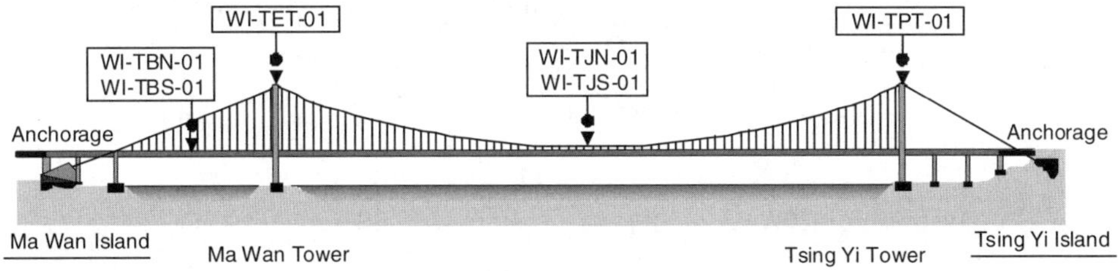

Figure 2.15 Tsing Ma suspension bridge and distribution of anemometers (from [25]) (Reproduced with permission from Taylor & Francis).

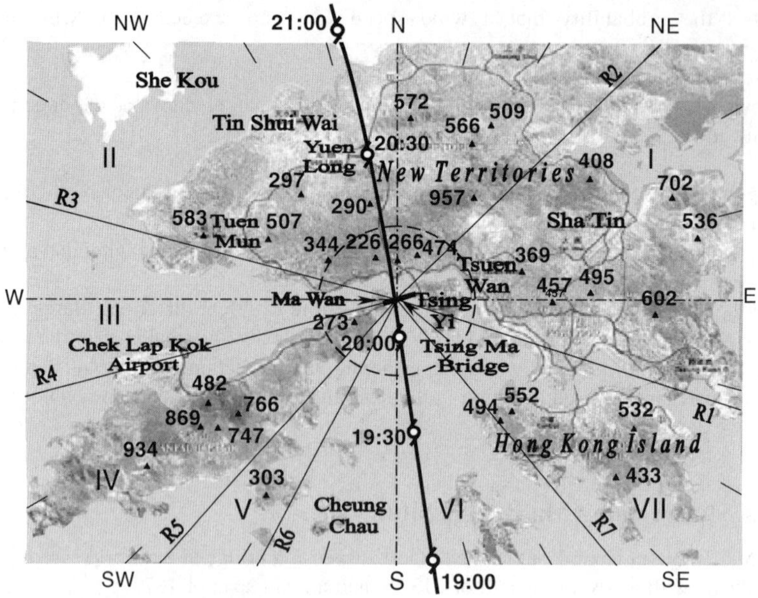

Figure 2.16 Local topography of Hong Kong (from [25]) (Reproduced with permission from Taylor & Francis).

and present all the research results here. Only wind characteristics of typhoons and the joint probability density function for monsoons at the bridge site are briefly introduced in this section.

2.8.1 Anemometers in WASHMS

The WASHMS of the bridge includes a total of six anemometers, with two at the middle of the main span, two at the middle of the Ma Wan side span and one of each on the Tsing Yi tower and Ma Wan

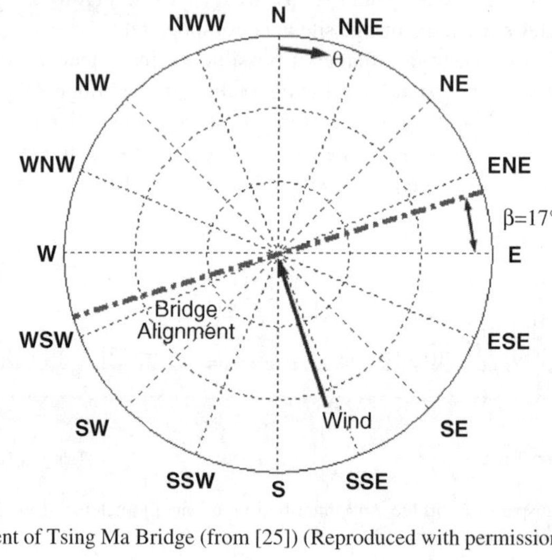

Figure 2.17 Alignment of Tsing Ma Bridge (from [25]) (Reproduced with permission from Taylor & Francis).

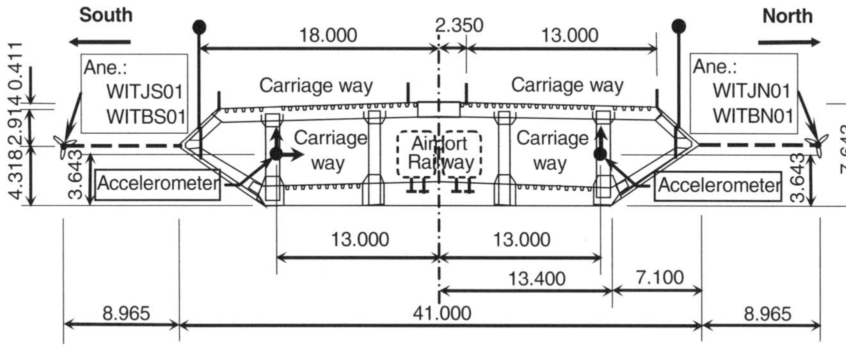

Figure 2.18 Deck cross-section and sensor positions (from [25]) (Reproduced with permission from Taylor & Francis).

tower (see Figure 2.15). To prevent disturbance from the bridge deck, the anemometers at the deck level were respectively installed on the north side and south side of the bridge deck via a 8.965 m long boom from the leading edge of the deck (see Figure 2.18).

The anemometers installed on the north side and south side of the bridge deck at the middle of main span, respectively specified as WI-TJN-01 and WI-TJS-01, are the digital type Gill Wind Master ultrasonic anemometers. Those located at the two sides of the bridge deck near the middle of the Ma Wan approach span, specified as WI-TBN-01 on the north side and WI-TBS-01 on the south side, are analog mechanical anemometers. Each analog anemometer consists of a horizontal component (RM Young 05106) with two channels, giving the horizontal resultant wind speed and its azimuth, and a vertical component (RM Young 27106) with one channel, providing the vertical wind speed. Another two analog mechanical anemometers which have a horizontal component only are arranged at 11 m above the top of each bridge tower. They are specified as WI-TPT-01 for the Tsing Yi tower and WI-TET-01 for the Ma Wan tower. The sampling frequency of measurement of wind speeds was set as 2.56 Hz.

2.8.2 Typhoon Wind Characteristics

To understand typhoon wind characteristics at the bridge site, typhoons with signal No. 3 and above hoisted by the Hong Kong Observatory (HKO) during the period from July 1997 to September 2005 were targeted. A total of 247 hourly typhoon data records were correspondingly retrieved. Four major steps were then taken for the data pre-processing of original data records. After the data pre-processing, a total of 147 hourly typhoon wind records were of acceptable quality for subsequent statistical analysis.

To understand the mean wind speed and mean wind direction of typhoon events experienced at the Tsing Ma bridge site, the wind records were further split into the four groups in terms of mean wind speed. These wind speed groups included:

1. less than 10 m/s;
2. between 10 and 18 m/s;
3. between 18 and 45.8 m/s;
4. greater than 45.8 m/s.

The latter three groups represent the stages 1, 2 and 3 specified in the high wind management system for the bridge. Figure 2.19 shows the polar plot of the ten-minute mean wind direction for typhoon events. It can be seen that almost 23% of the records are taken from the north direction. This observation is consistent with that made by the HKO.

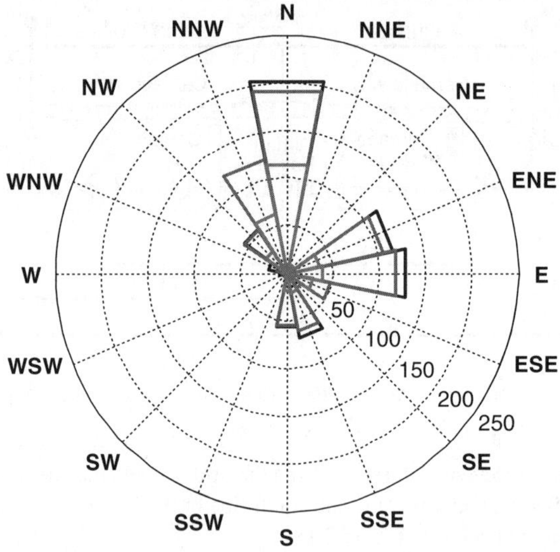

Figure 2.19 Polar plot of mean wind direction.

Figure 2.20 displays the polar plot of the ten-minute mean wind speed for typhoon events. It can be observed that wind speeds within the range from 10–18 m/s are dominant for typhoon events. This indicates that stage 1 of the high wind management system was hoisted during most of the typhoons at the bridge during the period concerned. Among the 16 direction sectors, the southwest direction shows the maximum ten-minute mean wind speed of 22.67 m/s.

The mean wind incidence is defined as the angle between the mean wind velocity and the horizontal plane. The positive mean wind incidence means the wind is blowing upward.

Figure 2.21 displays the polar plot of the ten-minute mean wind incidence recorded at the deck level of the bridge for typhoon events. All the wind incidences are within $\pm 10^{\circ}$ at the 95% upper and lower limits

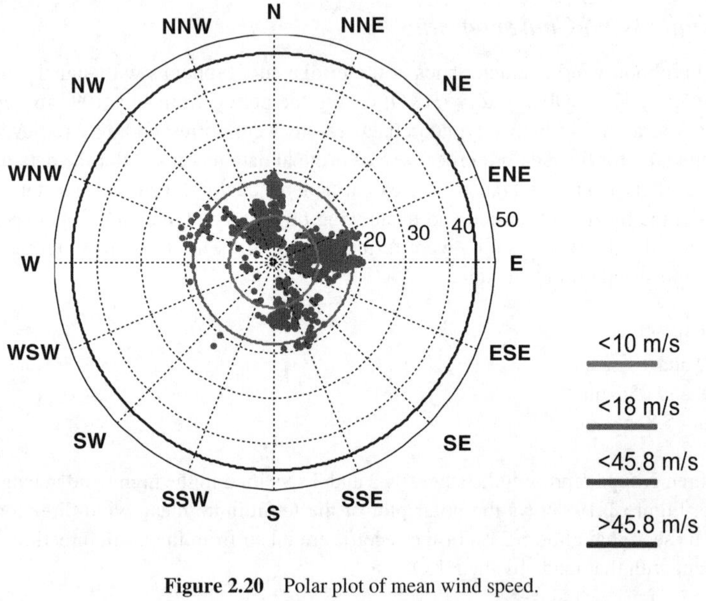

Figure 2.20 Polar plot of mean wind speed.

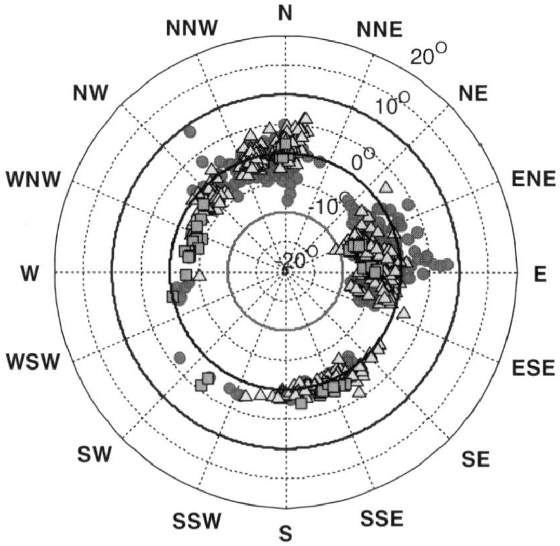

Figure 2.21 Polar plot of mean wind incidence.

of wind incidences under a mean wind speed of 20 m/s. However, the wind incidences measured in the easterly directions are much more scattered than those in other directions, which may be because easterly wind directions are almost parallel to the longitudinal direction of the bridge deck. There is a possibility that the flow of air could be disrupted by the bridge deck. It can also be seen that wind incidences tend to be approximately zero for the open-sea area. The mean values of ten-minute mean wind incidences were recorded as 0.33°, 0.09° and 0.52° in the southeast, south-southeast and south directions, respectively.

Figure 2.22 shows the polar plot of the ten-minute turbulence intensity in the longitudinal direction at the deck level of the bridge for typhoon events. It can be seen that the turbulence intensities measured

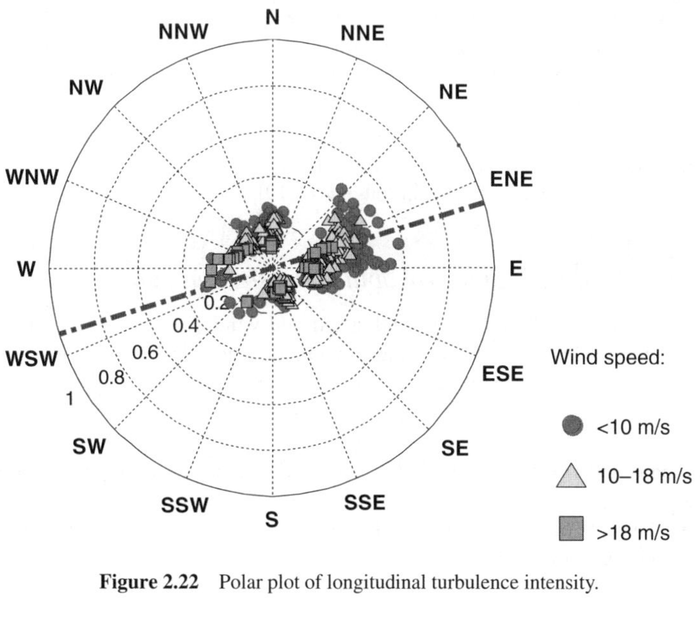

Figure 2.22 Polar plot of longitudinal turbulence intensity.

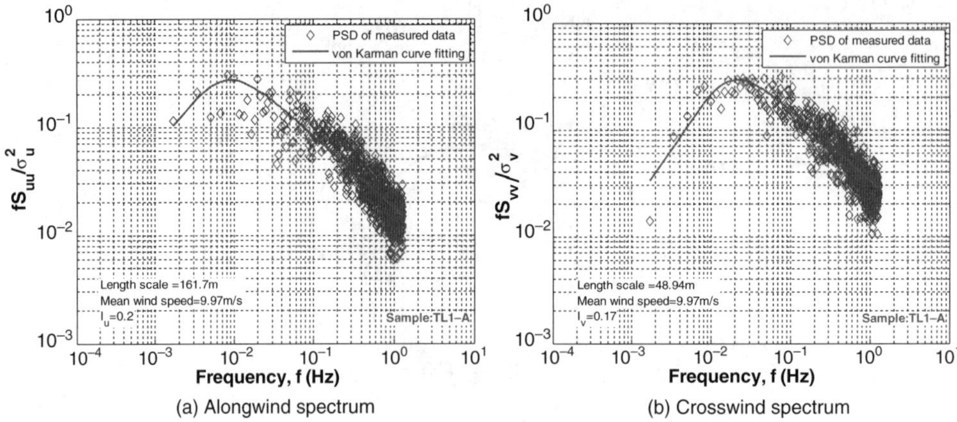

Figure 2.23 Wind spectra of typhoon sample 1 from over-land fetch.

in the northeasterly and easterly directions vary within a larger range than those measured in other directions. The northeasterly direction shows the most turbulent winds for typhoon events. The mean values of the longitudinal, lateral and vertical turbulence intensities in this direction are 38.6, 36.2 and 26.1%, respectively. In contrast, the least turbulent winds are in the southerly direction, which has mean longitudinal, lateral and vertical turbulence intensities of 8.6, 8.4 and 5.3%, respectively. The average ratio of the lateral to longitudinal turbulence intensities is 0.903 and the ratio of the vertical to longitudinal turbulence intensities is 0.703.

The alongwind, crosswind and upwind power spectra and the integral length scales derived by the curve-fitting method from four selected typhoon samples were investigated in detail. Figure 2.23 shows the alongwind and crosswind power spectra for one selected typhoon sample. The measured integral length scales for the overland exposure vary between 94.81 m and 188.52 m (the average value being 136.75 m) for L_u^x, between 31.34 m and 54.55 m (the average value being 44.96 m) for L_v^x, and between 33.18 m and 43.04 m (the average value being 37.93 m) for L_w^x.

In contrast, for the open-sea fetch the measured integral length scales range between 110.66 m and 539.66 m (the average value being 280.76 m) for L_u^x, between 43.74 m and 149.18 m (the average value being 83.85 m) for L_v^x, and between 29.15 m and 60.04 m (the average value being 40.91 m) for L_w^x. It seems that the integral length scales of L_u^x, L_v^x and L_w^x from the open-sea fetch are larger than those from the overland fetch. This observation appears to be consistent with a comment in the literature that the length scale is a decreasing function of terrain roughness [19].

2.8.3 *Monsoon Wind and Joint Probability Density Function*

The joint probability density function of wind speed and wind direction is essential when assessing wind-induced fatigue damage to the bridge. A practical joint probability distribution function has been adopted for a complete population of wind speed and wind direction, based on two assumptions:

1. The distribution of the component of wind speed for any given wind direction follows the Weibull distribution.
2. The interdependence of wind distribution in different wind directions can be reflected by the relative frequency of occurrence of wind.

$$P_{u,\theta}(U,\theta) = P_\theta(\theta)\left(1 - \exp\left[-\left(\frac{U}{c(\theta)}\right)^{k(\theta)}\right]\right) = \iint f_\theta(\theta) f_{u,\theta}(U,k(\theta),c(\theta))du\,d\theta \tag{2.75}$$

$$f_{u,\theta}(U,k(\theta),c(\theta)) = \frac{k(\theta)}{c(\theta)}\left(\frac{U}{c(\theta)}\right)^{k(\theta)-1}\exp\left[-\left(\frac{U}{c(\theta)}\right)^{k(\theta)}\right] \tag{2.76}$$

$$P_\theta(\theta) = \int_0^\theta f_\theta(\theta)d\theta \tag{2.77}$$

where:

$0 \le \theta < 2\pi$;
$P_\theta(\theta)$ is the relative frequency of occurrence of wind in wind direction θ.

The occurrence frequency $P_\theta(\theta)$, as well as the distribution parameters, $k(\theta)$ and $c(\theta)$, can be estimated using wind data recorded at the bridge site. Wind records of hourly mean wind speed and direction within the period between 1 January 2000 and 31 December 2005, from the anemometer installed

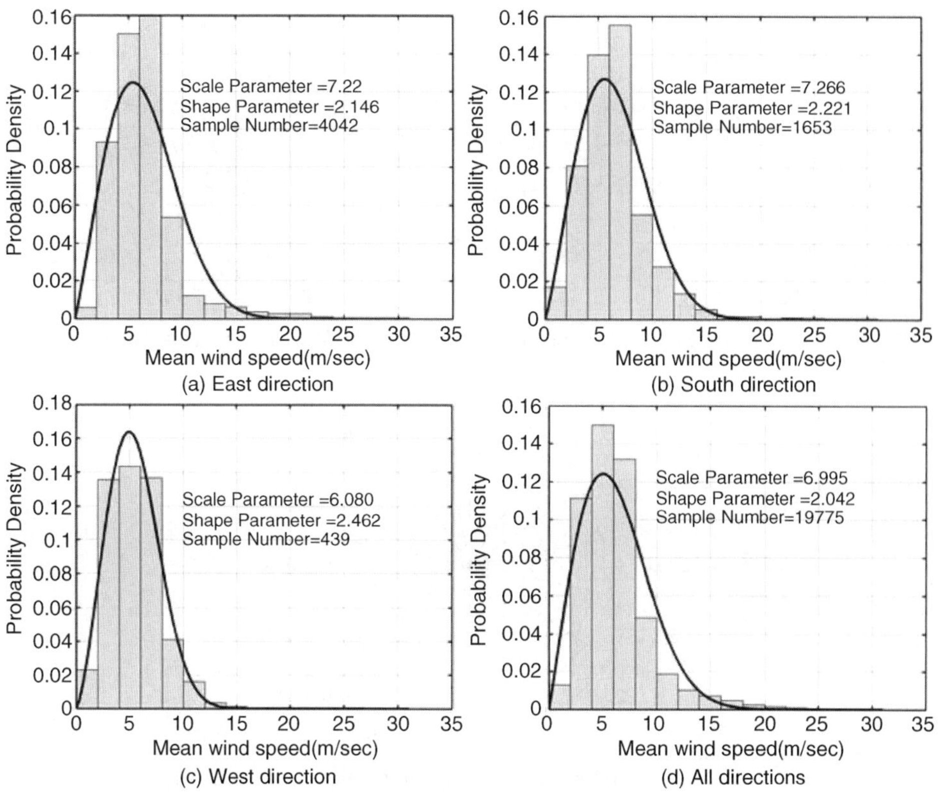

Figure 2.24 Weibull distribution of hourly mean wind speed (from [24]) (Reproduced with permission from Multi-Science Publishing Co Ltd).

on the top of the Ma Wan tower, were used to ascertain the joint probability density function of hourly mean wind speed and direction. The height of the anemometer is 214 m above the sea level. Wind records having an hourly mean wind speed lower than 1 m/sec were removed in order to avoid any adverse effect on the statistics. As a result, 19 775 hourly monsoon records were available for calculation of the joint probability density function of wind speed and direction. The number of hourly typhoon wind records during this period was so small that the corresponding joint probability density function could not be obtained at present.

All the monsoon records were classified into 16 sectors of the compass with an interval of $\Delta\theta = 22.5^\circ$ according to the hourly mean wind direction (see Figure 2.24). In each sector, mean wind speed was further divided into 16 ranges from zero to 32 m/sec, with an interval of $\Delta U = 2$ m/sec. This led to a total of 256 cells, and the relative frequency of hourly mean wind speed and wind direction in each cell was calculated. Based on the relative frequencies of wind speed and wind direction calculated, the theoretical expression of joint probability density function was deduced, based on Equation 2.76. The Weibull function was used to fit the histogram of hourly mean wind speed for each wind direction, and the typical results in the east, south and west directions are depicted in Figure 2.24a–c, respectively.

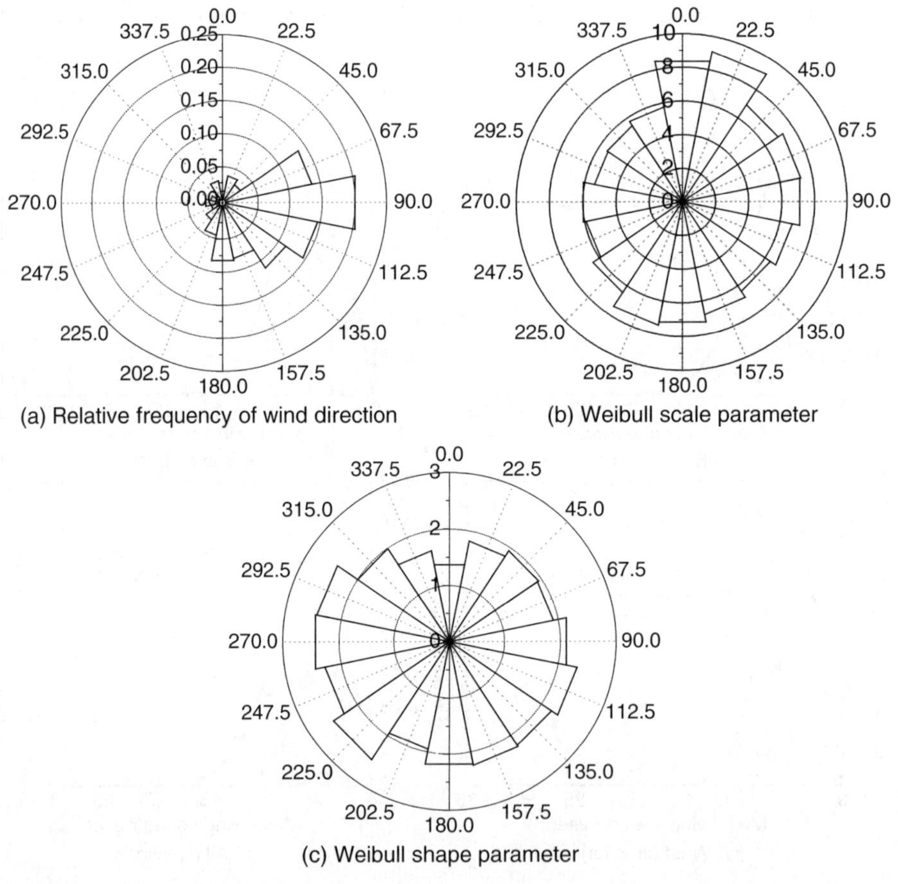

Figure 2.25 Relative frequency of wind direction and Weibull scale and shape parameters (from [24]) (Reproduced with permission from Multi-Science Publishing Co Ltd).

The Weibull function was also applied to the complete wind records without considering wind direction, as shown in Figure 2.24d. The results show that the Weibull function also fits the complete wind data adequately. The relative frequency of wind direction and the scale and shape parameters of the Weibull function obtained are given in polar plot in Figure 2.25a–c, respectively. It can be seen that the dominant monsoon direction is the east and that the scale and shape parameters do not vary significantly with wind direction. Further details can be found in the literature [24,25].

2.9 Notations

A	Amplification factor
B	Constant value depend on the roughness length z_0
C	Non-dimensional decay constant
A	Scale factor
a_0	Empirical constant
C	Scale factor for Weibull type
K	Shape factor for Weibull type
M	Constant value depend on the roughness length z_0
N	Frequency
u_*	Shear velocity or friction velocity
ρ_a	Air density
ρ	Correlation coefficient
α	Power-law exponent
β	Fractional speed-up ratio
σ	Standard deviations for turbulence components
μ	Location parameter
ξ	Shape factor
η	Dispersion
ω_{up}	Upper cutoff frequency
ψ_u	Normalized co-spectrum
φ_u	Normalized quad-spectrum
κ	Surface drag coefficient
δ	Atmospheric boundary layer height
θ	Wind direction
τ_0	Surface shear stress
k	Von Karman's constant
f_c	Coriolis parameter
g	Peak factor
u	Turbulence component in longitudinal direction
r	Spatial distance
v	Horizontal turbulence component perpendicular to the wind direction
w	Vertical turbulence component perpendicular to u and v
y	Reduced variate
z_0	Surface roughness length
z_h	Zero-plane displacement
G_v	Gust factor
H_G	Equilibrium boundary layer height
L_u	Integral length scale for the longitudinal turbulence component u
N	Total number of years

R	Accepted risk
T	Averaging time
T_R	Return period
U_G	Gust speed
$f(z, r_x)$	Cross correlation function in two points separated by a distance r_x
$f(V)$	Probability density function
$F(V)$	Probability distribution function
$I(z)$	Turbulence intensity
P_E	Exceedance probability
$k(\theta)$	Distribution parameter
$c(\theta)$	Distribution parameter
$P_\theta(\theta)$	Relative frequency of occurrence of wind in wind direction θ
$\overline{U}(z)$	Mean wind speed
$U(x, y, z, t)$	Measured instantaneous wind velocity time history
$S(n)$	Turbulence wind spectrum
S_N	Dimensionless normalized cross-spectrum
$S(P_1, P_2, n)$	Cross-spectrum at points P_1 and P_2
$R(z, \tau)$	Autocorrelation function
$T_u(z)$	Integral time scale

References

1. Deacon, E.L. (1955) Gust variation with height up to 150 meters. *Quarterly Journal of the Royal Meteorological Society*, **81**, 562–573.
2. Holmes, J.D. (2007) *Wind Loading of Structures*, 2nd edn, Taylor & Francis, Abingdon.
3. Dyrbye, C. and Hansen, S.O. (1996) *Wind Loads on Structures*, John Wiley & Sons Ltd, Chichester.
4. Deaves, D.M. and Harris, R.I. (1978) A Mathematical Model of the Structure of Strong Winds. Report 76, Construction Industry Research and Information Association, London, 49pp.
5. Charnock, H. (1955) Wind stress on a water surface. *Quarterly Journal of the Royal Meteorological Society*, **81**, 639–640.
6. Franklin, J.L., Black, M.L., and Valde, K. (2003) GPS dropwindsonde wind profiles in hurricanes and their operational implications. *Weather and Forecasting*, **18**, 32–44.
7. Counihan, J. (1975) Adiabatic atmospheric boundary layers: a review and analysis of data from the period from 1880–1972. *Atmospheric Environment*, **9**, 871–905.
8. von Karman, T. (1948) Progress in the statistical theory of turbulence. *Proceedings of the National Academy of Sciences of the United States of America*, **34**(11), 530–539.
9. Harris, R.I. (1968) On the Spectrum and Auto-correlation Function of Gustiness in High Winds. Report 5273, Electrical Research Association.
10. Busch, N. and Panofsky, H. (1968) Recent spectra of atmospheric turbulence. *Quarterly Journal of the Royal Meteorological Society*, **94**, 132–148.
11. Holmes, J.D. (1973) Wind Pressure Fluctuations on a Large Building. PhD Thesis, Monash University, Australia.
12. Davenport, A.G. (1961) The spectrum of horizontal gustiness near the ground in high winds. *Quarterly Journal of the Royal Meteorological Society*, **87**, 194–211.
13. Davenport, A.G. (1962) The response of slender line-like structures to a gusty wind. *Proceedings of the Institution of Civil Engineers*, **23**, 389–408.
14. Davenport, A.G. (1968) The dependence of wind load upon meteorological parameters. Proceedings of the International Research Seminar on Wind Effects on Buildings and Structures, University of Toronto Press, Toronto, pp. 19–82.
15. Ishizaki, H. (1983) Wind profiles, turbulence intensities and gust factors for design in typhoon-prone regions. *Journal of Wind Engineering and Industrial Aerodynamics*, **13**, 55–66.
16. Fisher, R.A. and Tippett, L.H.C. (1928) Limiting forms of the frequency distribution of the largest or smallest member of a sample. *Proceedings of the Cambridge Philosophical Society, Part 2*, **24**, 180–190.
17. Jenkinson, A.F. (1955) The frequency distribution of the annual maximum (or minimum) values of meteorological elements. *Quarterly Journal of the Royal Meteorological Society*, **81**, 158–171.
18. Gumbel, E.J. (1954) *Statistical Theory of Extreme Values and Some Practical Applications*, Applied Mathematics Series 33, National Bureau of Standards, Washington, DC.

19. Simiu, E. and Scanlan, R.H. (1996) *Wind Effects on Structures: Fundamentals and Applications to Design*, 3rd edn, John Wiley & Sons., Inc. New York.
20. Cook, N.J. (1982) Towards better estimation of extreme wind. *Journal of Wind Engineering and Industrial Aerodynamics*, **9**, 295–323.
21. Simiu, E., Hendrickson, E.M., Nolan, W.A. *et al.* (1985) Multi-variate distributions of directional wind speeds. *Journal of Structural Engineering, ASCE*, **111**, 939–943.
22. Wong, K.Y., Man, K.L. and Chan, W.Y.K. (2001) Monitoring of wind load and response for cable-supported bridges in Hong Kong. *Proceedings of SPIE 6th International Symposium on Health Monitoring and Management of Civil Infrastructure Systems* (Chase and Aktan, eds.), Newport Beach, California, **4337**, 292–303.
23. Wong, K.Y., Man, K.L., Chan, W.Y.K. (2001) Application of global positioning system to structural health monitoring of cable-supported bridges. *Proceedings of SPIE 6th International Symposium on Health Monitoring and Management of Civil Infrastructure Systems* (Chase and Aktan, eds.), Newport Beach, California, **4337**, 390–401.
24. Xu, Y.L., Chen, J., Ng, C.L., and Zhou, H.J. (2008) Occurrence probability of wind-rain-induced stay cable vibration. *Advances in Structural Engineering-An International Journal*, **11**, 53–69.
25. Xu, Y.L. and Xia, Y. (2012) *Structural Health Monitoring of Long-span Suspension Bridges*, Spon Press.

3

Mean Wind Load and Aerostatic Instability

3.1 Preview

As discussed in Chapter 2, wind velocity can normally be decomposed into a mean wind speed and three perpendicular turbulence components. The theory of random vibration manifests that the mean value of a structural response to an excitation in the form of a stationary random process is constant and proportional to the mean value of the excitation process. Therefore, to facilitate wind-resistant design of a long-span bridge, it is often assumed that the total bridge response is the sum of a mean response and a random response with a zero mean. The mean response depends on the mean wind load which, in turn, counts on the mean wind speed.

The mean wind load is an important load that should be considered in the design of long-span bridges. Because a long-span bridge is usually flexible, the mean wind loads on the bridge deck, towers and cables can cause considerable displacements. Moreover, aerostatic instability due to the mean wind load may occur, which can cause the bridge to collapse and should be prevented in the design stage.

This chapter first introduces the formulation of mean wind load and the concept of wind force coefficients. The torsional divergence and its critical wind speed of a long-span bridge are then discussed, based on a 2-D bridge deck section model. The 3-D non-linear aerostatic instability analysis, considering lateral-torsional divergence, follows, based on the finite element method (FEM). The finite element method-based mean wind response analysis is finally discussed and applied to the Stonecutters cable-stayed bridge as a case study.

In wind-resistant design of a long-span bridge, an accurate finite element model of the bridge is essential to facilitate effective analyses of various wind load effects on the bridge. The methods for analyzing both aerostatic instability and mean wind response, which will be introduced in this chapter, are also established on the finite element method. Therefore, the finite element modeling technique for long-span bridges is discussed in this chapter to some extent.

3.2 Mean Wind Load and Force Coefficients

3.2.1 Bernoulli's Equation and Wind Pressure

The region outside the boundary layer in the case of the airfoil, and the outer region of the bluff-body flow are regions of inviscid (zero viscosity) and irrotational (zero vorticity) flow [1]. In these regions,

Wind Effects on Cable-Supported Bridges, First Edition. You-Lin Xu.

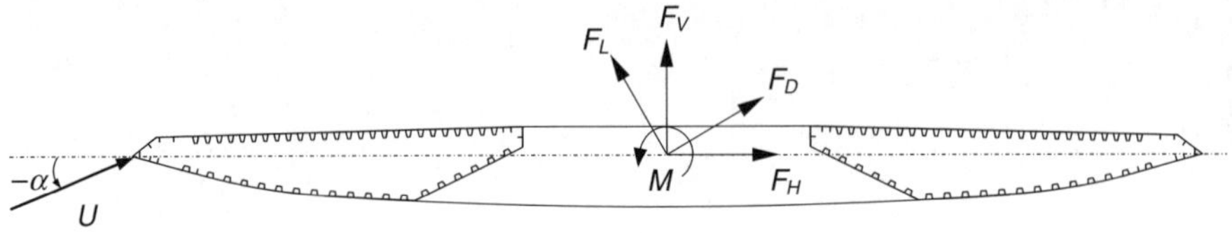

Figure 3.1 Mean wind load in wind coordinate system and structural coordinate system.

the pressure p and velocity U_b in the fluid are related by Bernoulli's equation:

$$\frac{1}{2}\rho U_b^2 + p = \text{const.} \tag{3.1}$$

where:

ρ is the air density;
U_b is the wind speed at a point on the streamline;
p is the instantaneous wind pressure at the chosen point.

Bernoulli's equation states that for an inviscid and irrotational flow, an increase in the speed of the fluid occurs simultaneously with a decrease in pressure.

By using Equation 3.1 and considering the atmospheric pressure p_n and velocity U in the region outside the influence of the bluff-body, one may have:

$$p - p_n = \frac{1}{2}\rho\left(U^2 - U_b^2\right) \tag{3.2}$$

For a bridge deck section immersed in a wind field (see Figure 3.1), some points on the deck surface are the stagnation points where U_b is zero. If the atmospheric (ambient) pressure is regarded as zero pressure, wind pressure at the stagnation pint is equal to $\frac{1}{2}\rho U^2$, which is known as the dynamic pressure. However, most of the points on the deck surface are not stagnation points and, consequently, the distribution of wind pressure on the deck surfaces is non-uniform.

3.2.2 Mean Wind Load

Conventionally, the mean wind load is expressed with respect to the wind coordinate system, as shown in Figure 3.1. In a typical 2-D analysis, the mean wind load can be split into three parts: the drag force F_D in the mean wind direction, the lift force F_L perpendicular to the mean wind direction, and the moment M with respect to the centroid of the section. The drag force equals the integral of wind pressures in the alongwind direction (the mean wind direction); the lift force equals the integral of wind pressures on the section in the acrosswind direction perpendicular to the alongwind direction; and the moment is the torsion, which is equal to the total resultant wind force times a moment distance with respect to the centroid of the section.

The mean wind load can also be expressed with respect to the structural coordinate system, as shown in Figure 3.1. In this system, the mean wind load can be split into the vertical force, F_V, the horizontal force, F_H, and the moment, M. The vertical force equals the integral of wind pressures in the vertical direction; the horizontal force equals the integral of wind pressures in the horizontal direction; and the moment is the same as is expressed in the wind coordinate system.

Wind forces can be transformed from the wind coordinate system to the structural coordinate system using the following equation:

$$\begin{pmatrix} F_V \\ F_H \end{pmatrix} = \begin{pmatrix} \cos(\alpha) & \sin(\alpha) \\ -\sin(\alpha) & \cos(\alpha) \end{pmatrix} \begin{pmatrix} F_L \\ F_D \end{pmatrix} \tag{3.3}$$

where α is the angle of attack of the incoming flow.

3.2.3 Wind Force Coefficients

The determination of wind forces on a bluff-body is quite difficult because they are dependent on a number of variables related to the geometry of the body and to the upwind flow characteristics. Therefore, geometrically scaled models are often used in practice to obtain pressure (or force) coefficients through wind tunnel tests, and these force coefficients are then applied to full-scale prototype structures. The pressure or force coefficients are non-dimensional quantities.

For a bridge deck section, three non-dimensional mean wind force coefficients, which are seen as the functions of an angle α of attack, are usually defined as follows:

$$F_D(\alpha) = \frac{1}{2}\rho U^2 B C_D(\alpha) \tag{3.4a}$$

$$F_L(\alpha) = \frac{1}{2}\rho U^2 B C_L(\alpha) \tag{3.4b}$$

$$M(\alpha) = \frac{1}{2}\rho U^2 B^2 C_M(\alpha) \tag{3.4c}$$

where:

$C_D(\alpha)$, $C_L(\alpha)$ and $C_M(\alpha)$ are the drag, lift and moment coefficients, respectively;
U is the incoming wind flow velocity;
α is the wind angle of attack;
B is the characteristic dimension of the structural section. In the case of a bridge deck, B is usually taken as the width of deck section.

Figure 3.2 shows wind force coefficients of a box girder section obtained from wind tunnel tests [2]. With these coefficients, the mean wind forces on the bridge girder can be determined according to its section size. Details on wind tunnel tests can be found in Chapter 7 of this book.

The utility of mean wind force coefficients of a bridge girder section is not limited in the determination of mean wind load. The coefficients and their derivatives with respect to the angle of attack are also used to determine buffeting forces and galloping stability [3,4]. Mean wind force coefficients are also called aerodynamic coefficients.

3.3 Torsional Divergence

Torsional divergence of long-span bridges was first observed in a full bridge wind tunnel test in 1967 [5]. The phenomenon of torsional divergence is characterized by torsional instability, a monotonically increasing rotation until failure at a critical wind speed. This phenomenon is non-oscillatory and can be described in a quasi-steady manner in terms of aerodynamic coefficients. Like most instabilities, torsional divergence occurs abruptly at the critical wind velocity and can cause a bridge to collapse.

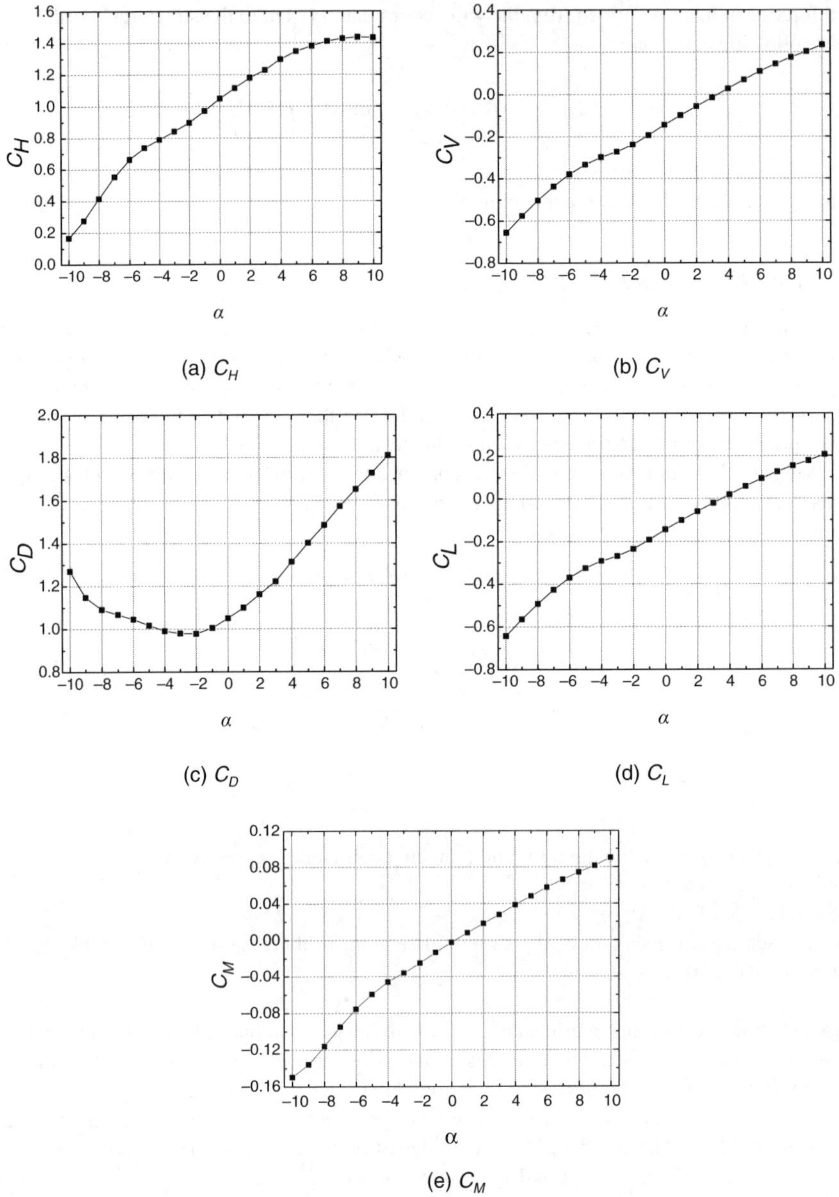

Figure 3.2 Aerodynamic coefficients of a box girder section.

To clearly demonstrate how the non-linear coupling between wind forces and bridge deck motion leads to torsional divergence, a simple one-dimensional (1-D) torsional divergence analysis is introduced in this section. The solution for the critical wind velocity of 1-D torsional divergence can be given straightforwardly in terms of aerodynamic moment coefficient as a function of wind angle of attack.

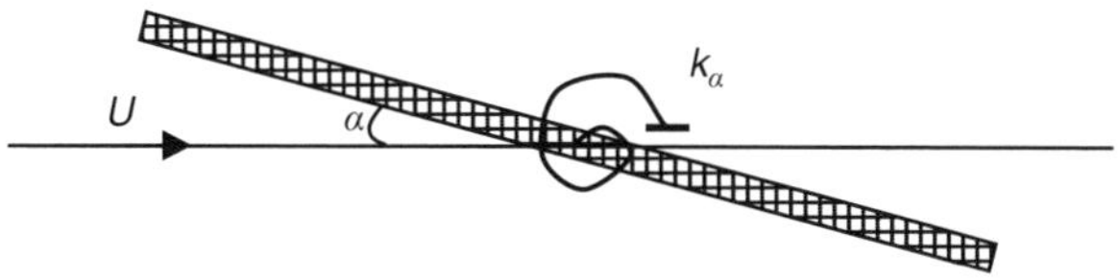

Figure 3.3 Calculation model for torsional divergence.

Let us consider a single degree of freedom (1-D) model, as shown in Figure 3.3. The torsional equilibrium equation of the bridge girder section can be written as:

$$K_\alpha \alpha = \frac{1}{2}\rho U^2 B^2 C_M(\alpha) \tag{3.5}$$

where:

K_α is the torsional stiffness of the bridge girder;
C_M is the aerodynamic moment coefficient.

The substitution of a linear expansion of $C_M(\alpha)$ as $C_M(\alpha) = C_M(0) + C'_M(0) \cdot \alpha$ into Equation 3.5 yields:

$$\left(K_\alpha - \frac{1}{2}\rho U^2 B^2 C'_M(0)\right)\alpha = \frac{1}{2}\rho U^2 B^2 C_M(0) \tag{3.6}$$

The term $-\rho U^2 B^2 C'_M(0)/2$ represents the negative torsional stiffness due to the interaction between the wind moment and the bridge rotation. It is obvious that the negative torsional stiffness increases with wind speed. If the total torsional stiffness becomes zero or negative, then the torsional displacement of the bridge girder becomes divergent.

The critical wind speed for torsional divergence is therefore the wind speed that makes the total stiffness equal to zero:

$$K_\alpha - \frac{1}{2}\rho U^2 B^2 C'_M(0) = 0 \tag{3.7}$$

As a result, the critical wind speed for torsional divergence can be given as:

$$U_{cr} = \sqrt{\frac{2K_\alpha}{\rho B^2 C'_M(0)}} \tag{3.8}$$

The torsional stiffness of the bridge girder can be expressed as:

$$K_\alpha = \omega_\alpha^2 \cdot I_m = (2\pi f_\alpha)^2 \cdot mr^2 \tag{3.9}$$

where:

ω_α represents the torsional circular frequency of the bridge girder;
I_m represents the mass moment of inertia of the bridge girder;

f_a represents the torsional frequency of the bridge girder;
m is the mass of the bridge girder;
r represents the mass radius of the bridge girder.

As a result, the critical wind speed for torsional divergence can be given by [6]:

$$U_{cr} = K_{cr} f_{\alpha} B \tag{3.10}$$

$$K_{cr} = \sqrt{\frac{\pi^3}{2} \cdot \mu \cdot \left(\frac{r}{b}\right)^2 \cdot \frac{1}{C'_M(0)}} \tag{3.11}$$

where $b = 0.5B$ and $\mu = m/(\pi \rho b^2)$.

It should be noted that real torsional divergence of a long-span bridge is usually coupled with its lateral displacement, because of large drag force acting on the bridge. Moreover, geometric non-linearity cannot be neglected in aerostatic divergence analysis of long-span bridges, because of large displacements. All of these issues have to be, and can be, addressed through 3-D aerostatic instability analysis.

3.4 3-D Aerostatic Instability Analysis

Aerostatic instability of long-span bridges usually takes place in the pattern of lateral-torsional divergence. Aerostatic instability was investigated by Boonyapinyo *et al.* [7] and Nagai *et al.* [8] for long-span cable-stayed bridges and by Cheng *et al.* [9] and Zhang *et al.* [10] for long-span suspension bridges. The results from these researches show that the incorporation of displacement-dependent wind forces and the geometric non-linearity in the analysis results in a significant reduction in the critical wind velocity for non-linear aerostatic instability.

An iterative method for 3-D non-linear aerostatic instability analysis, based on the finite element model of a long-span bridge, is introduced in this section. Both geometric non-linearity and non-uniformly distributed displacement-dependent wind forces are considered in this method.

In general, aerostatic instability analysis is all about forming and solving the following equation:

$$[K(\delta)]\{\delta\} = \{F(\alpha, U)\} \tag{3.12}$$

where:

$[K(\delta)]$ is the total structural stiffness matrix in which both linear stiffness and geometric stiffness are included;
$\{\delta\}$ is the displacement vector of the bridge structure;
$\{F(\alpha, U)\}$ is the wind force vector acting on the bridge structure, which is usually expressed as equivalent nodal forces as they are functions of both mean wind speed and wind angle of attack.

The iteration form of Equation (3.12) can be expressed as:

$$\left([K_e] + \left[K_g(\delta_{j-1})\right]\right) \cdot \{\Delta\delta_j\} = \left\{F_j(\alpha_j, U_k)\right\} - \left\{F_{j-1}(\alpha_{j-1}, U_k)\right\} \tag{3.13}$$

where:

$[Ke]$ and $[K_g(\delta_{j-1})]$ are the elastic-linear stiffness matrix and the geometric stiffness matrix at the $(j-1)$ step, respectively;

$\Delta\delta_j$ is the displacement increment at the j step;

$\{F_j(\alpha_j, U_k)\}$ and $\{F_{j-1}(\alpha_{j-1}, U_k)\}$ are the wind force vectors at the j step and the $(j-1)$ step, which correspond to the wind angle of attack at the j step and the $(j-1)$ step, respectively, but the mean wind speed at the k step.

The Euclidean norm of the aerostatic coefficients of lift, drag and pitch moment is taken as convergence criterion, which can be expressed as:

$$\left\{\frac{\sum_{j=1}^{N}\left[C_k(\alpha_j) - C_k(\alpha_{j-1})\right]^2}{\sum_{j=1}^{N}\left[C_k(\alpha_{j-1})\right]^2}\right\}^{1/2} \leq \varepsilon_k \quad (k = D, L, M) \tag{3.14}$$

where:

N is the total node number;

C_k $(k = D, L, M)$ are the aerodynamic force coefficients;

ε_k is the allowable tolerance.

The procedure of a 3-D aerostatic instability analysis can be summarized as follows:

1. Assume an initial wind velocity U_0, an initial wind angle of attack α_0, an initial displacement vector $\{\delta_0\}$, and decide a wind speed increment ΔU.
2. Calculate the total stiffness matrix using the initial displacement vector $\{\delta\}$ and decide wind forces on the bridge using the initial wind speed U with the initial wind angle of attack α.
3. Solve the global equilibrium equation (3.12) to obtain the displacement vector $\{\delta\}$ using a numerical method such as the Newton-Raphson method.
4. Obtain the torsional angles of bridge deck elements from the displacement vector $\{\delta\}$.
5. Calculate the effective angle of attack α for each element.
6. Check if the Euclidean norm of Equation 3.14 is less than the prescribed tolerance.
7. If Equation 3.14 is not satisfied, then go to step (2) with the new effective angle of attack and the new displacement vector.
8. If Equation 3.14 is satisfied, increase wind speed by ΔU and go to step (2).
9. Keep the iteration of steps (2) to (8) until the solution of Equation 3.12 at step (3) becomes divergent.
10. Return to step (8) using a slightly small wind speed increment ΔU.
11. Keep the iteration of steps (2) to (10) until the difference of wind speed increment ΔU between the previous and current steps is less than the preset threshold. The final wind speed is then the critical wind speed for 3-D aerostatic instability of a long-span bridge.

3.5 Finite Element Modeling of Long-Span Cable-Supported Bridges

The finite element model of a long-span bridge is essential to facilitate aerostatic instability analysis. It is also an essential model for mean wind response analysis which will be presented in the next section, for vortex-induced vibration analysis, buffeting analysis, and flutter analysis in Chapter 4, and for wind-vehicle-bridge coupling analysis in other chapters. In this connection, this section provides a concise introduction to finite element modeling technique for long-span cable-supported bridges. The further details on this topic can be found in many books [e.g. 11,12].

3.5.1 Theoretical Background

By the finite element approach, the bridge structure is assumed to be divided into a system of discrete elements which are interconnected only at a finite number of nodal points. The mass and stiffness matrices of the entire bridge are then formed by evaluating the properties of the individual finite elements and superimposing them appropriately [11].

Within an element, each node holds specific degrees of freedom (DOFs), depending on the problem described. Under external loads, the deflected shape of an element follows specific displacement function which satisfies nodal and internal continuity requirements. Based on the load-displacement relation, the element stiffness matrix can be established.

For mass property, the simplest method is to assume that the mass is concentrated at the nodes, at which the translational displacements are defined and rotational inertia is null. This is referred as lumped mass, and the matrix has a diagonal form. Following a similar method and the same displacement function in deriving the element stiffness matrix, the consistent mass matrix can also be calculated.

The concentrated load acting on the nodes can be directly applied. Other loads, such as distributed forces, can be evaluated in a similar manner as deriving the consistent mass matrix with the same displacement function.

In engineering practice, the mass and stiffness matrices and load vector are automatically computed by structural analysis computer software. However, it is utmost important for users to understand the theories, assumptions and limitations of numerical modeling using the finite element method, as well as the limitations of the computation algorithms.

There have been numerous studies on the finite element modeling of long-span cable-supported bridges to facilitate static and dynamic analyses [12]. Most of the studies are based on a simplified spine beam model of equivalent sectional properties to the actual structural components [13]. Such a simplified model is effective in capturing the dynamic characteristics and global structural behavior of the bridge without heavy computational effort. However, local structural behavior, such as stress and strain concentration at joints which is prone to cause local damage in static and/or dynamic loading conditions, cannot be estimated directly. In this regard, a delicate finite element model, with finer details in highlighting local behaviors of the bridge components, is needed.

The rapid development of information technology and improvements in speed and memory capacity of personnel computers (PC) have made it possible to establish a more detailed finite element model for a long-span cable-supported bridge. However, a finer finite element model will cost more computational resources including computational time and storage memory.

In general, a simplified spine beam finite element model of a bridge helps understanding of the global behaviors of the structure, preliminary design, aerodynamic analysis and so on. A fine model with solid and/or shell elements can be used for stress analysis, so that the computed stresses can be directly compared with the measured ones [14,15]. To trade off the computational efficiency and capability, one can establish a multi-scale model to cater for the objectives.

3.5.2 Spine Beam Model

In a spine beam model, components of a long-span cable-supported bridge are modeled by line elements including beam elements, truss elements, and rigid links. Pylons and piers are usually modeled with beam elements based on their geometric properties. Cables or suspenders are often modeled by truss elements, and the geometric non-linearity due to cable tension is taken into consideration. In a suspension bridge, the static equilibrium profile of the main cable needs to be calculated iteratively, based on the static horizontal tensions and the unit weight of both the cable and deck given in the design. In a cable-stayed bridge, the static equilibrium profile of the bridge also needs to be calculated iteratively, based on the tension forces in the stay cables and the weight of the bridge deck. The bridge deck, however, is more challenging to model. Usually, there are two approaches that can be employed to simplify the complicated bridge deck into a series of beam elements.

In the first approach, the deck is modeled as a central beam (the spine beam). The equivalent cross-sectional area of the deck is calculated by summing up all cross-sectional areas. In the case of a composite section, the areas should be converted to that of one single material, according to the modular ratio. The neutral axes and moments of inertia about the vertical and transverse axes are also determined in a similar way. The calculation of the torsional stiffness of the deck section should consider both pure and warping torsional constants. The mass moment of inertia of the deck should include those of all members, according to their distances from the centroid of the section.

For a more complicated deck, with a number of structural members, especially with diagonally placed braces and irregular members, an alternative to the equivalent sectional property method is to use the equivalent displacement method. As an example, Figure 3.4(a) shows a detailed finite element model of a typical 18 m long segment of the deck of the Tsing Ma Bridge. The segment is then simplified to an equivalent beam element with 12 DOFs, as shown in Figure 3.4(b), each corresponding to a generalized displacement and a generalized force. The resulting element stiffness matrix has a size of 12×12. A positive unit displacement is imposed on one DOF of the segment, while all other possible displacements in the three-dimensional model are prevented. The resultant generalized forces of the segment, representing the stiffness coefficients, are then calculated. Note that, in the section model, beam elements, shell elements and solid elements can be employed, although Figure 3.4(a) has beam elements and shell elements only.

Constraints, usually spring elements, rigid links or direct coupling of nodal displacements, are necessary to connect different parts of the model together and to enforce certain types of rigid-body features. For example, the nodes of the deck, bearings and towers do not coincide with each other. Rigid links are usually used to restrain their motions in different directions, depending on the bearing types and boundary conditions. Rigid links are also used to connect the spine beam with cables, as shown in Figure 3.5 in the spin beam model. Figure 3.6 displays a complete spin beam model of the Tsing Ma suspension bridge in Hong Kong. Further details on how to build the spin beam model of the Tsing Ma Bridge can be found in Xu *et al.* (1997) [13].

3.5.3 Multi-Scale Model

To capture the local responses of some critical members and/or to obtain more accurate results, the entire bridge can be modeled with different types of elements. The ever-increasing capacity of computers makes this multi-scale modeling more and more popular. In such a multi-scale model [16], the components of interest can be modeled with shell elements or solid elements and other components still with line elements. Special care must be taken at the interface between a solid element, a shell element and a beam element, because they have different DOFs. Multi-point constraints (MPC) are needed to connect the different elements in order to avoid displacement incompatibility among the nodes. Various MPC equations for beam to shell, shell to solid and beam to solid connections have been developed in terms of total energy principle and implemented in the commercial software [17].

Figure 3.7(a) shows a typical 18 m long segment of the Tsing Ma suspension bridge deck. The deck of the Tsing Ma Bridge is quite complicated in geometry and consists of deck plates and cross-frames. Figure 3.7(b) displays a multi-scale model of the same 18 m long segment. The plate is modeled with shell elements and the cross-frames with beam type. Since the nodes of the beam elements modeling cross-frames do not coincide with the nodes of the shell elements located in the middle plane of the plate, MPCs are needed to connect them to avoid displacement incompatibility among the nodes. There are totally 130 nodes, with 188 beam elements, 24 shell elements and 50 MPCs in this typical multi-scale model.

The towers, piers, cables and suspenders of the Tsing Ma Bridge are relatively simple in geometry, and thus beam elements may be sufficient to model them. Integrating the bridge components with the

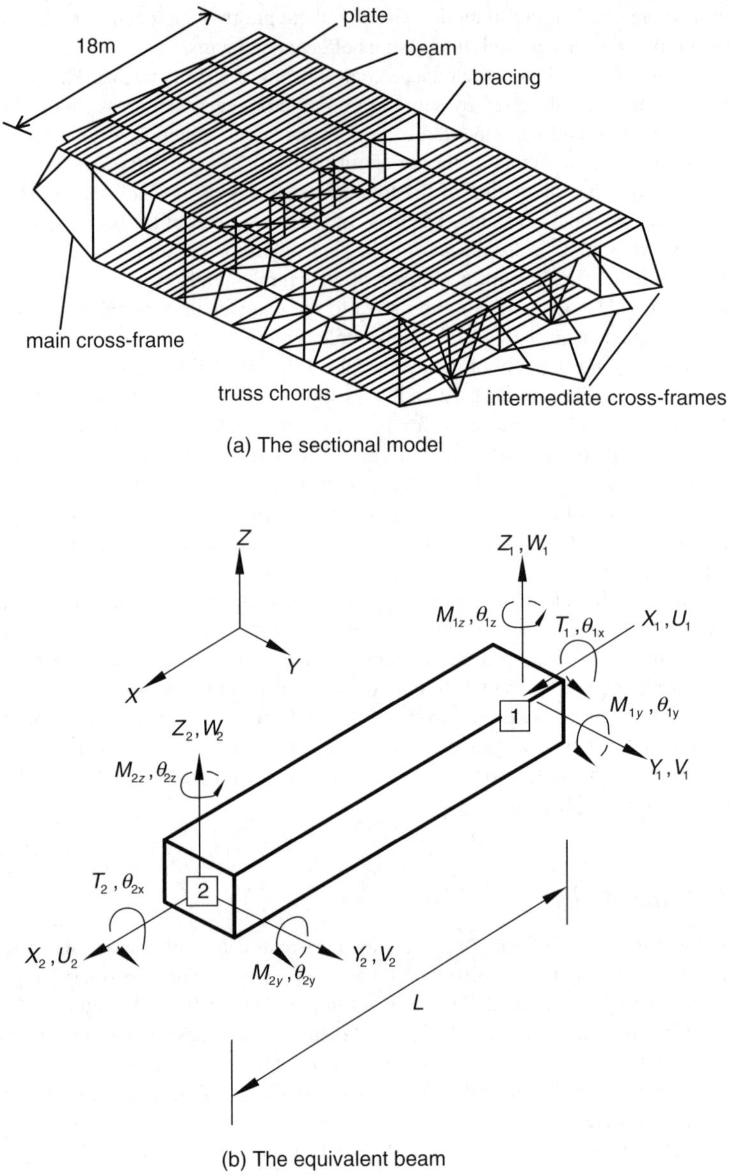

Figure 3.4 One segment of the deck system of Tsing Ma Bridge (from [12]) (Reproduced with permission from Taylor & Francis).

proper modeling of the connections and boundary conditions enables the global multi-scale model of the Tsing Ma Bridge to be established, as shown in Figure 3.8. The establishment of this multi-scale bridge model involves 12 898 nodes, 21 946 elements (2 906 shell elements and 19 040 beam elements) and 4788 MPCs. Further details on the establishment of the multi-scale model of the Tsing Ma Bridge can be found in Chapter 11.

In the establishment of the Tsing Ma global bridge model, the coordinates of the bridge structure are taken from as-built drawings. Therefore, the configuration of the bridge model is the target one. The

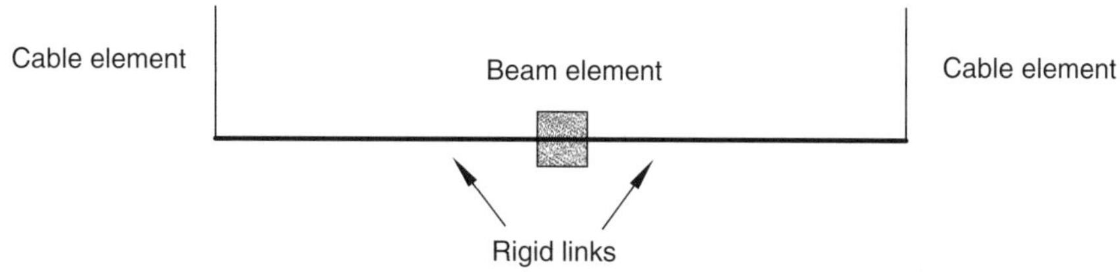

Figure 3.5 Connection between the spine beam element and the suspender cable element (from [12]) (Reproduced with permission from Taylor & Francis).

process for finding the target configuration of the bridge in the equilibrium state under dead loads is referred to as "shape-finding". This task is accordingly performed through iteration to form the final global bridge model for the subsequent model updating, static analysis and dynamic analysis. Two types of system identification techniques, the input-output methods and output-only methods, and the vibration-based FE model updating using the identified dynamic characteristics of the bridge, can be found in Xu *et al.* (2012) [12]. The subsequent static analysis can be found in Section 3.6 of this chapter and the subsequent dynamic analysis can be found in Chapter 4 and other chapters.

3.5.4 Modeling of Cables

Special attention should be paid to modeling of cables in considering the sag effect and the initial tension force under full dead loads. The sag effect of a cable is considered using the following equivalent elastic modulus for the straight truss element to replace the actual modulus of the cable:

$$E_{\mathrm{eq}} = \frac{E}{1 + \frac{(\rho_c A g \bar{l})^2 AE}{12T^3}} \tag{3.15}$$

where:

E_{eq} is the equivalent modulus of elasticity;
E is the effective modulus of elasticity of cable;

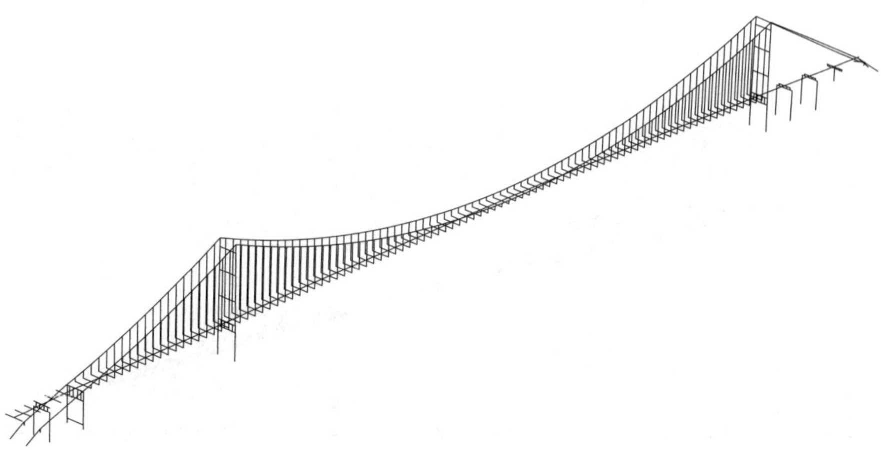

Figure 3.6 Spin beam finite element model of the Tsing Ma Bridge.

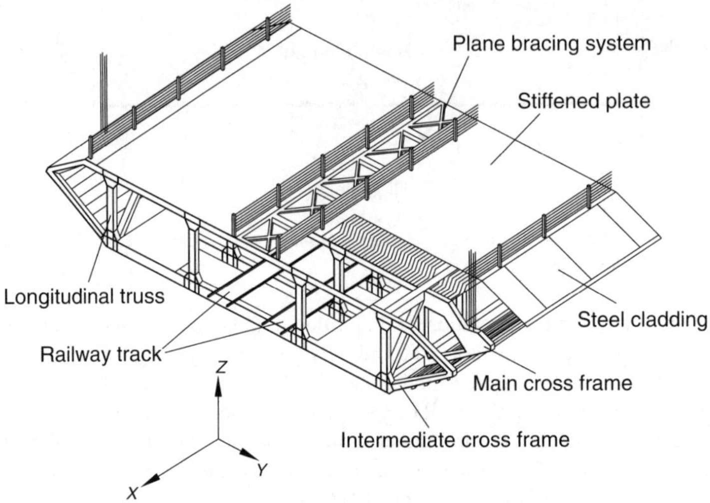

(a) An isometric view of the deck segment

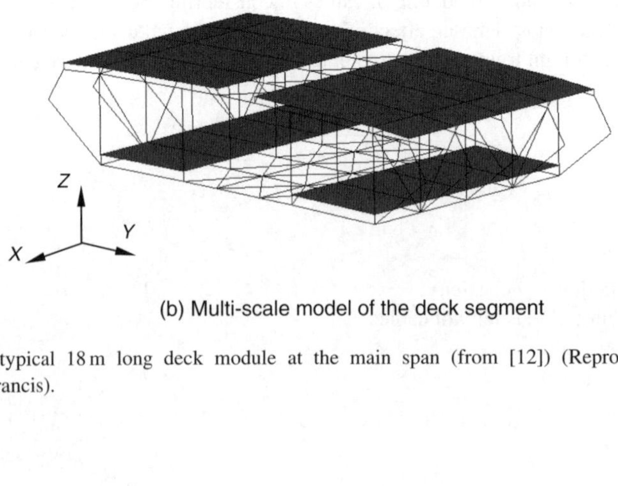

(b) Multi-scale model of the deck segment

Figure 3.7 A typical 18 m long deck module at the main span (from [12]) (Reproduced with permission from Taylor & Francis).

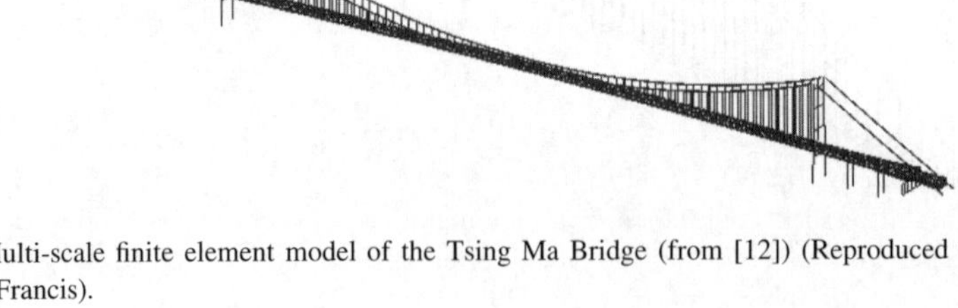

Figure 3.8 Multi-scale finite element model of the Tsing Ma Bridge (from [12]) (Reproduced with permission from Taylor & Francis).

ρ_c is the effective material density of the cable;
g the gravity acceleration;
$\bar{l}$ is the horizontal projected length of the element;
A the effective cross-sectional area;
T is the mean cable tension.

The elastic stiffness matrix of the cable element in the three-dimensional local coordinate is given by:

$$[K_{\mathrm{e}}] = \frac{AE_{\mathrm{eq}}}{L_{\mathrm{c}}} \begin{bmatrix} 1 & 0 & 0 & -1 & 0 & 0 \\ 0 & 0 & 0 & 0 & 0 & 0 \\ 0 & 0 & 0 & 0 & 0 & 0 \\ -1 & 0 & 0 & 1 & 0 & 0 \\ 0 & 0 & 0 & 0 & 0 & 0 \\ 0 & 0 & 0 & 0 & 0 & 0 \end{bmatrix} \tag{3.16}$$

where L_c is the chord length of the cable. Furthermore, a geometric stiffness matrix of the cable element should be included for consideration of the cable tension force:

$$[K_{\mathrm{g}}] = \frac{T}{L_c} \begin{bmatrix} 0 & 0 & 0 & 0 & 0 & 0 \\ 0 & 1 & 0 & 0 & -1 & 0 \\ 0 & 0 & 1 & 0 & 0 & -1 \\ 0 & 0 & 0 & 0 & 0 & 0 \\ 0 & 1 & 0 & 0 & -1 & 0 \\ 0 & 0 & 1 & 0 & 0 & -1 \end{bmatrix} \tag{3.17}$$

The tangent stiffness matrix of the cable element in the local coordinate is then obtained by:

$$[K_t] = [K_e] + [K_g] \tag{3.18}$$

The tangent stiffness matrix of the cable system of the bridge in the global coordinate can be assembled from the element tangent stiffness matrices.

3.6 Mean Wind Response Analysis

3.6.1 Determination of Reference Position

Before the mean wind response of a long-span cable-supported bridge is determined, a geometrically non-linear static analysis of the bridge, in which only the gravity forces of bridge components and the initial tension forces of cables are included, needs to be performed to determine a reference position of the bridge at its complete stage. The mean wind response is then computed with respect to this reference position, which is also required in the 3-D aerostatic instability analysis.

Geometrically non-linear phenomena in the static analysis of the bridge mainly results from stress-stiffening effect. The stress-stiffening effect is introduced in Section 3.5.4, where the tangent stiffness matrix of the cable system of the bridge in the global coordinate is formed. This effect is also important in the bridge deck and towers, especially for a long-span cable-stayed bridge, due to high compressive forces caused by tension forces in cables and dead loads.

3.6.2 Mean Wind Response Analysis

Under the action of incident wind, there are three major components of the wind forces acting on the bridge: the mean wind force due to mean wind; buffeting force due to turbulent wind; and self-excited force due to aeroelastic interaction between bridge motion and wind velocity. The response of the bridge is traditionally divided into response to mean wind (mean wind response) and response to buffeting force and self-excited force (buffeting response).

The mean wind response can be determined through a 3-D finite element-based static analysis, while the buffeting response can be determined using a 3-D finite element-based dynamic analysis by taking the deformed position of the bridge under the mean wind load as a reference position. The theoretical background of the 3-D finite element-based coupled buffeting analysis of a long-span cable-supported bridge is given in Chapter 4.

The mean wind load on a bridge component can be determined by the aerodynamic coefficient and mean wind speed at the component according to Equation 3.4, and then converted to the relevant nodes of the bridge model. The mean wind load on a stay cable can be converted to its two ends. Since long-span cable-supported bridges are often very flexible, their displacements under strong winds can be so large that small displacement assumption is no longer valid and that mean wind load is displacement-dependent, because aerodynamic coefficients vary with the effective angle of attack.

The situation can be more complicated because the torsional displacements of deck segments vary along the bridge deck. Therefore, the 3-D finite element-based mean wind response analysis is actually a non-linear static analysis with iterations, which is similar to the 3-D aerostatic instability analysis, but wind speed concerned in the mean wind response analysis is relatively lower than that in the instability analysis.

3.7 Case Study: Stonecutters Bridge

3.7.1 Main Features of Stonecutters Bridge

The Stonecutters Bridge, stretching from the Tsing Yi Island to the Stonecutters Island, is a cable-stayed bridge with a total length of 1596 m and a main span of 1018 m. Both the east and west side spans consist of four sub-spans of 69.25 m, 70 m, 70 m and 79.75 m, respectively. The configuration of the bridge is shown in Figure 3.9.

Each bridge tower consists of a reinforced concrete structure from the base to level +175 m and then a composite steel and concrete structure. The stay cables are anchored in a steel box inside the tower within the height from level +175 m to level +293 m. The top 5 m of the towers, from level +293 m to level +298 m, is primarily designed as an architectural lighting feature. The towers are founded on piled foundations composed of bored piles of 2.8 m diameter.

The bridge deck is made of steel in the main span. The steel deck consists of two streamlined twin girders supported by stay cables. At the location of the stay cables in the main span, the two longitudinal girders are interconnected by a series of cross-girders.

The two side spans are generally in concrete with the transition from steel deck to concrete deck located 49.75 m into the side spans. The concrete deck also consists of twin girders connected by concrete cross-girders.

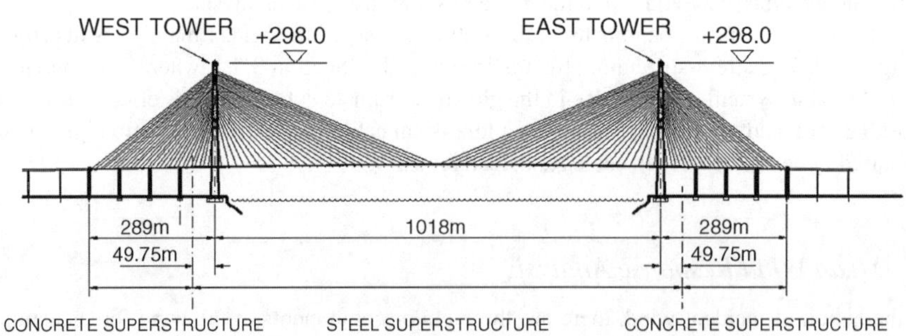

Figure 3.9 Configuration of Stonecutters Bridge.

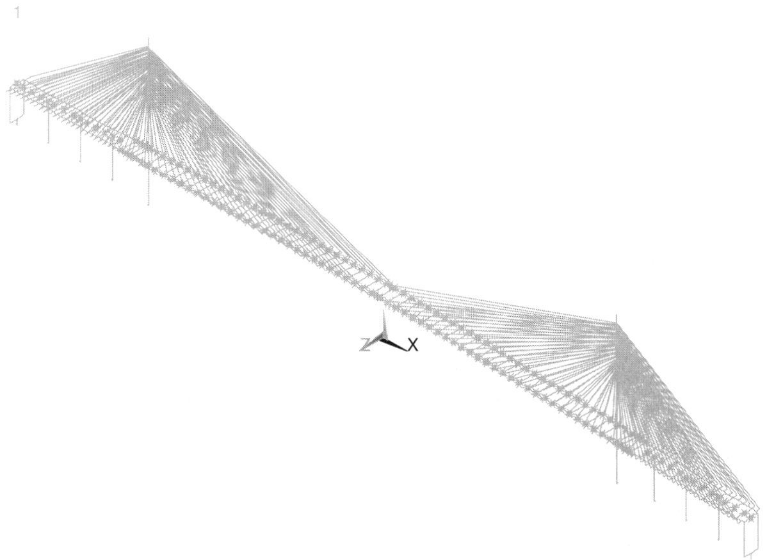

Figure 3.10 Isometric view of Stonecutters bridge model and global coordinate system.

The pier shafts in the two side spans are all monolithically connected to the bridge deck through cross-heads. The three intermediate pier shafts are of single rectangular tapered shape, while the pier at each end of the bridge has two rectangular shafts to increase stiffness of the side spans.

The stay cables are of the parallel wire strand type, made up of 7 mm wires. The length of the longest stay cable is about 540 m. More detailed information on the configuration of the Stonecutters Bridge can be found in [18].

3.7.2 Finite Element Modeling of Stonecutters Bridge

The global coordinate system used in the modeling of the entire bridge is selected as a right-hand coordinate system as shown in Figure 3.10. The spin beam finite element model is used for the bridge.

A typical cross-section of the steel deck is shown in Figure 3.11, while a typical cross-section of the concrete deck is shown in Figure 3.12. In the 3-D finite element model of the bridge, two longitudinal beams are used to represent the two longitudinal girders of the bridge deck. The centre of each longitudinal beam is located at the gravity centre of the corresponding longitudinal girder. Beam elements with six degrees of freedom at each node are used to model the longitudinal girders. Section properties of the beam elements are calculated based on the cross-sections of the longitudinal girders. Minor local elements, such as additional stiffeners near stay anchorages, are not

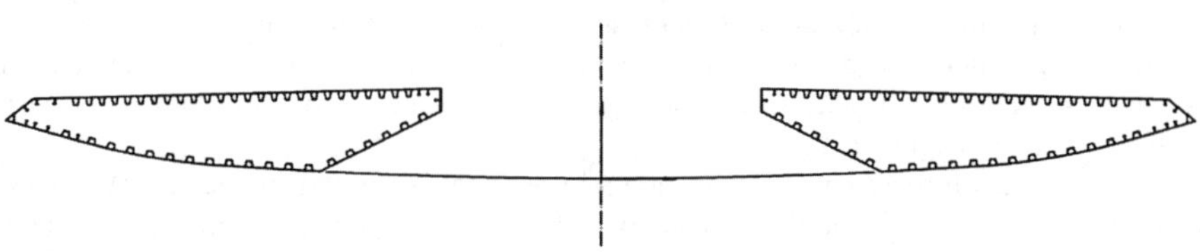

Figure 3.11 A typical cross-section of longitudinal steel girders.

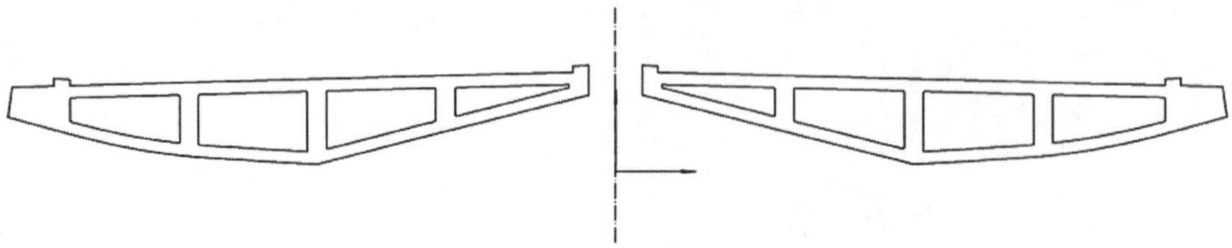

Figure 3.12 A typical cross-section of longitudinal concrete girders.

included in the calculation of section properties. Transverse elements in the longitudinal girders are also not included in the global finite element model, but their masses are taken into consideration as additional lumped masses at nodes.

Each cross-girder is modeled as a cross-beam, with the same kind of beam elements used to model the longitudinal girders. The end of each cross-beam is connected to the longitudinal beams through rigid arms. The effects of superimposed dead loads (e.g. the weight of road surfacing, services loads etc.), and the masses of transverse elements in the bridge components on bridge dynamic characteristics, are considered in terms of additional masses and mass moments of inertia lumped and distributed at the nodes of the longitudinal steel girders for steel part of the bridge deck and at the middle nodes of the concrete cross-girders for the concrete part of the bridge deck. Section layouts of both steel and concrete decks in the finite element model are shown in Figure. 3.13.

Each bridge tower is represented by a three-dimensional cantilever beam, which is divided into 49 beam elements. There are a total of nine elements, from level 3.2 m to level 77.75 m. The beam element properties are calculated based on a super-elliptic cross-section with a constant wall thickness of 2 m. From level 77.75 m to level 175.0 m, there are a total of ten elements, the properties of which are computed based on a circular cross-section. The wall thickness of the circular cross-section varies from 2.0 m at level 77.75 m to 1.4 m at level 175.0 m.

The material of the two towers below level 175.0 m is concrete. The mass and the mass moment of inertia per unit length are calculated based on the material property and the middle cross-sections of each element. From level 175.0 m to level 293.0 m, the tower is basically a circular composite structure with an inner concrete ring, the latter being covered with a 20 mm stainless steel skin plate. From level 195.65 m to level 280.5 m, a cable anchor is present inside the tower and can be seen as a hollow steel box. Figure 3.14 shows a typical composite cross-section of the upper tower for calculation of equivalent section properties.

In the calculation of equivalent section properties of a composite cross-section with concrete ring, stainless steel skin and anchor steel box, the plane assumption is adopted and the concrete material properties are taken as the reference properties. The equivalent material density is used for the elements in this range, which is equal to the total mass of the upper tower divided by the total volume of the upper tower. The additional masses from the stay anchorages are also taken into consideration. From level 293.0 m to level 298.0 m, the tower is of circular cross-section and made of steel, with 20 mm thickness. Only one element is used to model this part of the tower. The section properties are calculated based on the cross-section at level 293.0 m. Local elements in the towers, such as openings, lift shafts and staircases, are not taken into consideration because the global effects will be insignificant.

All supporting piers in the two side spans are reinforced concrete structures. Hollow boxes with internal walls are used to represent the concrete pier shafts. Each pier shaft is modeled by nine three-dimensional beam elements in order to consider the sloping outer surface of the pier shaft. Thus, for the end pier with two shafts, there are a total of 18 beam elements.

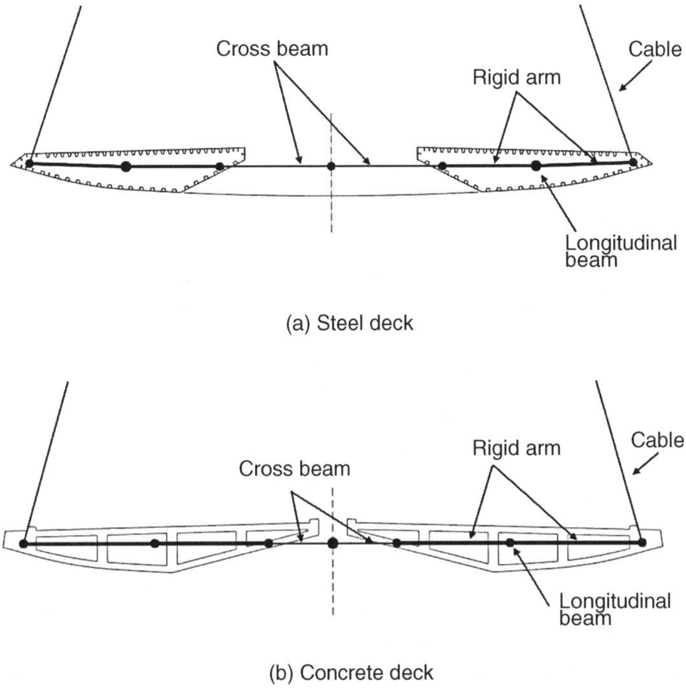

Figure 3.13 Connection between the cross-girder and longitudinal beams of the deck.

In the dynamic characteristics analysis, the stay cables connected to the towers and the longitudinal girders are modeled using spatial truss elements with three degrees of freedom at each end of the element. In consideration of the complexity of the 3-D finite element-based buffeting analysis of the entire bridge, and to focus on the global buffeting behavior of the bridge, each stay cable may have to be modeled as single truss element. Nevertheless, such a simple modeling of the cable should not affect the global static and dynamic characteristics of the bridge significantly. The equivalent modulus of a stay cable should be used to consider the effect of cable sag and tension load approximately. The equivalent elasticity modulus of the cable is calculated using Equation 3.15. In order to avoid local cable modes in the dynamic characteristics analysis, the mass of each stay cable is lumped at the end nodes of the cable.

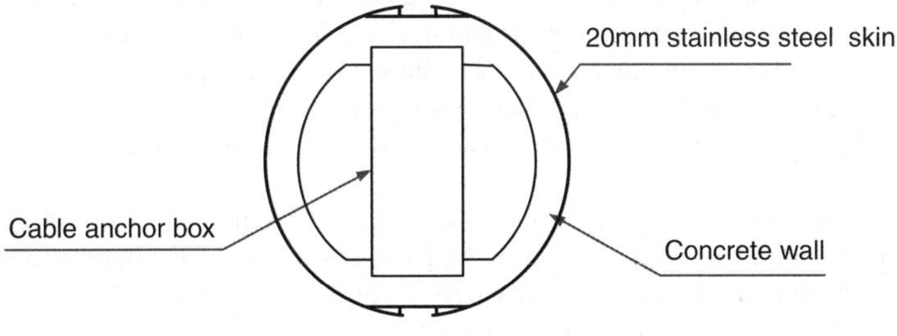

Figure 3.14 A typical composite cross-section of the upper tower.

For intermediate piers, the ends of each cross-girder (cross-head) are rigidly connected to the two longitudinal girders. The middle point of the cross-girder is then rigidly connected to the top of the pier shaft. The additional mass is placed at the middle point of the cross-girder to take account of the complicated geometry of the connection. For the end piers with two pier shafts, the ends of each cross-girder are rigidly connected to the two longitudinal girders. The top end of each pier shaft is then rigidly connected to the longitudinal girder. The mass property at the connection is adjusted in terms of additional mass at the middle point of the cross-girder.

The bridge deck is connected to the two towers through longitudinal hydraulic buffers and lateral pressure bearings. The functionality of the hydraulic buffers relies on the speed of the piston. For fast longitudinal movements of the bridge deck, the buffers will restrain the movements. Therefore, the hydraulic buffers are modeled as rigid bars, each with two pin-ends connecting the bridge deck to the tower. As the lateral pressure bearing is of contact pressure type, which functions non-linearly in a real situation, the lateral pressure bearings at the tower are simplified into master-slave couplings between the tower and the deck in the lateral direction only. The end nodes of stay cables are connected to the towers and to the longitudinal girders of the bridge deck using rigid connections. More detail information on the finite element modeling of the Stonecutters Bridge can be found in [18].

3.7.3 Aerodynamic Coefficients of Bridge Components

The aerodynamic coefficients of the Stonecutters bridge deck at the complete stage, without traffic, are provided by the Hong Kong Highways Department, based on the section model tests in a wind tunnel for the entire bridge deck in turbulent flow. The drag, lift and moment coefficients for the cross-section of the entire bridge deck at zero degrees of incidence are $C_D = 0.073$, $C_L = -0.155$, and $C_M = -0.018$. The derivatives of the drag, lift, and moment aerodynamic coefficients with wind angle of attack at zero degree are $dC_D/d\alpha = 0.069$, $dC_L/d\alpha = 2.510$ and $dC_M/d\alpha = 0.539$.

The drag coefficient of the bridge tower is taken as $C_D = 0.9$ along the entire height of the tower and normalized by the actual width of the tower perpendicular to the wind direction. The drag coefficient of the piers is taken as $C_D = 1.1$ for wind perpendicular to the bridge longitudinal axis and normalized by the actual width of the piers perpendicular to the wind direction. The drag coefficient of the stay cables is taken as $C_D = 0.8$ for wind perpendicular to the bridge longitudinal axis and normalized by the actual diameter of the stay cables. The aerodynamic lift force and moment on towers and stay cables are neglected in mean wind response analysis.

3.7.4 Mean Wind Response Analysis

The mean wind response can be determined through a 3-D finite element-based static analysis. For brevity, only the response caused by mean wind from the SW (southwest) direction is presented in this section.

The mean wind velocity is considered to be perpendicular to the bridge longitudinal axis, with zero angle of attack from the southwest direction. The bridge is in its complete stage without traffic.

The mean wind speed at the middle point of the main span of the bridge deck (+86 m) is taken as the reference mean wind velocity V_{dref}. The wind responses of the bridge are computed from 10 m/s to 100 m/s at an interval of reference mean wind speed of 10 m/s plus the design mean wind speed at the reference point.

Before the mean wind response of the bridge is determined, a geometrically non-linear static analysis of the bridge, in which only the gravity forces of bridge components and the initial forces of cables are included, is performed to determine a reference position of the bridge at its complete stage. The mean wind response is then computed with respect to the reference position.

Table 3.1 Mean wind displacement responses of the bridge at key locations (SW direction, $V_{dref} = 55$ m/s)

Location	Longitudinal (m)	Vertical (m)	Lateral (m)	Around x-axis (rad)	Around y-axis (rad)	Around z-axis (rad)
D1-windward	−2E-04	−0.297	−0.592	−0.002	−3E-06	3E-06
D1-leeward	−8E-05	−0.369	−0.592	−0.002	−3E-06	−7E-05
D2-windward	0.019	−0.089	−0.388	−0.003	0.001	−8E-04
D2-leeward	−0.017	−0.189	−0.387	−0.003	0.001	−8E-04
D5-windward	0.008	0.016	−0.042	−0.001	2E-04	2E-04
D5-leeward	7E-04	−0.015	−0.042	−0.001	2E-04	−8E-05
T1-Tower-top	0.065	−0.006	−1.071	−0.007	6E-05	−3E-04

The mean wind load on a bridge component can be determined by the aerodynamic coefficient and mean wind speed at the component and then converted to the relevant nodes of the bridge model. The mean wind load on a stay cable is converted to its two ends. The mean wind loads on bridge components may be affected by deformation of the bridge components. For instance, the rotation of the bridge deck may affect the mean wind loads on the bridge deck. Such non-linearity of mean wind loads is also considered in the determination of mean wind response of the bridge in this study.

The basic value of the ten-minute reference wind velocity at 10 m height above ground is 37 m/s for wind from the SW direction. The basic value of the ten-minute reference wind velocity at the deck level V_{dref} is thus 55 m/s, according to the mean wind speed profile specified in the design memorandum. The mean wind displacement responses of the complete bridge are computed, and those at the following key locations are listed in Table 3.1 for the basic mean wind velocity of 55 m/s:

(i) at middle point of main span (point D1 in Figure 3.15);
(ii) at quarter-point of main span (point D2 in Figure 3.15);
(iii) at middle point of side-spans (point D5 in Figure 3.15); and
(iv) at tower-top (point T1 in Figure 3.15).

The mean lateral, vertical and torsional displacement responses of the windward deck at the middle point of main span are −0.592 m, −0.297 m and −0.002 rad, respectively. The mean lateral, vertical and torsional displacements of the leeward deck at the middle point of main span are −0.592 m, −0.369 m and −0.002 rad, respectively. There is a difference in the vertical displacement response between the two girders because of the existence of the torsional displacement response.

The lateral and torsional displacement responses of the two girders are almost the same. The mean lateral displacement at the top of the tower reaches −1.071 m. The signs of the displacement responses comply with the global coordinate of the bridge as shown in Figure 3.10. More detailed information on the mean wind response analysis of the Stonecutters Bridge can be found in [19].

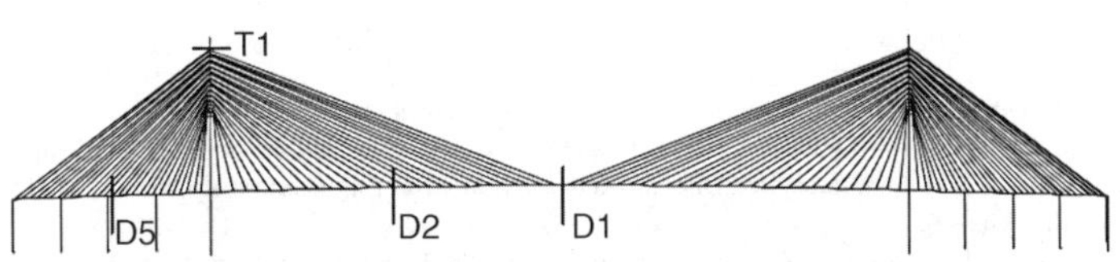

Figure 3.15 Key locations of Stonecutters Bridge in the computation.

3.8 Notations

A	Effective cross-sectional area
B	Deck section width
b	Half deck width
C_D	Drag force coefficient
C_L	Lift force coefficient
C_M	Moment coefficient
E	Effective modulus of elasticity of cable
E_{eq}	Equivalent modulus of elasticity
F_D	Drag force
F_L	Lift force
F_H	Horizontal wind force
F_V	Vertical wind force
$\{F(\alpha, U)\}$	Wind force vector
f_a	Torsional frequency
g	Gravity acceleration
I_m	Mass moment of inertia
$[K_t]$	Tangent stiffness matrix
$[K_e]$	Elastic stiffness matrix
$[K_g]$	Geometric stiffness matrix
$[K(\delta)]$	Total structure stiffness matrix
K_α	Torsional stiffness
$\bar{l}$	Horizontal projected length of cable element
L_c	Chord length of the cable
M	Aerodynamic moment
N	Total node number
p	Instantaneous wind pressure
R	Mass radius
T	Mean cable tension
U	Mean wind velocity
ΔU	Wind velocity increment
U_{cr}	Critical wind speed
V_{dref}	Reference mean wind velocity of bridge deck
α	Incident angle
$\{\delta\}$	Displacement vector of structure nodes
ε_k	Allowable tolerance
ρ	(i) Air density
	(ii) Effective density of cable
ω_α	Torsional circular frequency

References

1. Holmes, J.D. (2007) *Wind Loading of Structures*, 2nd edn, Taylor & Francis, Abingdon.
2. Zhu, L.D. and Guo, Z.S. (2010) Wind Tunnel Test and CFD Analysis on Wind Barriers of Ningbo Xiangshan Harbor Bridge, Research Report of State Key Laboratory for Disaster Reduction in Civil Engineering of China, No. SLDRCE WT201003, Tongji University (in Chinese).
3. Simiu, E. and Scanlan, R.H. (1996) *Wind Effects on Structures*, 3rd edn, John Wiley & Sons, USA.
4. Davenport, A.G. (1962) Buffeting of a suspension bridge by storm winds. *Journal of Structural Division, ASCE*, **88**(3), 233–270.

5. Hirai, A., Okauchi, I., Ito, M., and Miyata, T. (1967) Studies on the critical wind velocity for suspension bridges. Proceedings of International Research Seminar on Wind Effects on Buildings and Structures, University of Toronto Press, Ontario, Canada, pp. 81–103.
6. Xiang, H.F. (2005) *Modern Theory and Practice on Bridge Wind Resistance*, China Communications Press, Beijing (in Chinese).
7. Boonyapinyo, V., Yamada, H., and Miyata, T. (1994) Wind-induced non-linear lateral-torsional buckling of cable-stayed bridges. *Journal of Structural Engineering, ASCE*, **120**(2), 486–506.
8. Nagai, M., Ishida, T., Xie, X. *et al.* (1998) Static and dynamic instability behaviors of long-span cable-stayed bridges under wind load. *Proceedings of Sixth East Asia-Pacific Conference on Structural Engineering & Construction, Taiwan*, **3**, 2281–2286.
9. Cheng, J., Jiang, J.J., Xiao, R.C., and Xiang, H.F. (2002) Nonlinear aerostatic stability analysis of Jiang Yin suspension bridge. *Engineering Structures*, **24**(6), 773–781.
10. Zhang, X., Xiang, H.F., and Sun, B. (2002) Nonlinear aerostatic and aerodynamic analysis of long-span suspension bridges considering wind-structure interactions. *Journal of Wind Engineering and Industrial Aerodynamics*, **90**(9), 1065–1080.
11. Clough, R.W. and Penzien, J. (2003) *Dynamics of Structures*, 3rd edn, Computers and Structures, Berkeley, USA.
12. Xu, Y.L. and Xia, Y. (2012) *Structural Health Monitoring of Long-span Suspension Bridges*, London: Spon Press.
13. Xu, Y.L., Ko, J.M., and Zhang, W.S. (1997) Vibration studies of Tsing Ma suspension bridge. *Journal of Bridge Engineering*, **2**(4), 149–156.
14. Fei, Q.G., Xu, Y.L., Ng, C.L. *et al.* (2007) Structural health monitoring oriented finite element model of Tsing Ma bridge tower. *International Journal of Structural Stability and Dynamics*, **7**(4), 647–668.
15. Duan, Y.F., Xu, Y.L., Fei, Q.G. *et al.* (2011) Advanced finite element model of Tsing Ma bridge for structural health monitoring. *International Journal of Structural Stability and Dynamics*, **11**(2), 313–344.
16. Li, Z.X., Zhou, T.Q., Chan, T.H.T., and Yu, Y. (2007) Multi-scale numerical analysis on dynamic response and local damage in long-span bridges. *Engineering Structures*, **29**(7), 1507–1524.
17. McCune, R.W., Armstrong, C.G., and Robinson, D.J. (2000) Mixed-dimensional coupling in finite element models. *International Journal for Numerical Methods in Engineering*, **49**(6), 725–750.
18. The Hong Kong Polytechnic University (2004) Stonecutters Bridge: 3-D Finite Element Based Buffeting Analysis: Static and Dynamic Characteristics Study, Report No. 2 to Highways Department, Hong Kong.
19. The Hong Kong Polytechnic University (2004) Stonecutters Bridge: 3-D Finite Element Based Buffeting Analysis: Buffeting Response Analysis, Report No. 5 to Highways Department, Hong Kong.

4

Wind-Induced Vibration and Aerodynamic Instability

4.1 Preview

Besides mean wind load and aerostatic instability, as discussed in Chapter 3, there are several mechanisms, in various wind speed ranges, that can excite dynamic response and aerodynamic instability of long-span cable-supported bridges. Wind-induced vibration is an important source of loads on bridge structures, whereas wind-induced aerodynamic instability is very dangerous to bridge structures and may cause the bridge to collapse.

This chapter is going to introduce four types of wind-induced vibration and aerodynamic instability problems that all need to be considered in the design of a long-span cable-supported bridge:

- Vortex shedding excitation usually occurs at low wind speeds and low turbulence conditions, but it can cause considerable vibration of the bridge deck. The interaction of the bridge with vortex flow may result in the so-called "lock-in" phenomenon that leads to excessive bridge vibration. Vortex-induced response of the bridge should be controlled to a certain limit to ensure normal operation and to avoid fatigue damage to the bridge.
- Galloping instability is caused by self-excited forces, and it occurs in vertical modes of the bridge deck. Galloping happens when the bridge deck continuously absorbs energy from the incoming wind flow and the vibration becomes divergent. It happens abruptly and can cause the bridge to collapse, so it should be avoided in the design of the bridge.
- Flutter instabilities of several types occur at very high wind speeds for bridge decks, as a result of self-excited aerodynamic forces. Flutter always involves torsional motions and may also involve vertical bending motions so it, too, should be avoided in the design of bridges. It is flutter that caused the collapse of the original Tacoma Narrows Bridge.
- Buffeting excitation is caused by fluctuating forces induced by turbulence. It occurs over a wide range of wind speeds and normally increases monotonically with increasing wind speed. Excessive buffeting may cause fatigue problem in bridge components and affect functionality of the bridge. The buffeting should be considered in the design stage.

Wind Effects on Cable-Supported Bridges, First Edition. You-Lin Xu.

4.2 Vortex-Induced Vibration

4.2.1 Reynolds Number and Vortex Shedding

Vortex-induced vibrations are motions induced on a structure which is interacting with an external flow and are produced by vortex shedding of the flow [1]. The phenomenon that vortices are shed alternately from opposite sides of a circular cylinder is one of the classical open-flow problems in fluid mechanics. At very low Reynolds numbers, the streamlines of the resulting flow around the cylinder are perfectly symmetric, as expected from the potential theory [2]. However, as the Reynolds number increases, the flow becomes asymmetric and the so-called Kármán vortex street occurs, as shown in Figure 4.1.

The flow pattern of incompressible flow around a long circular cylinder perpendicular to the flow depends on the Reynolds number. The Reynolds number is an important parameter in all branches of fluid mechanics and is defined as the ratio of fluid inertia forces to viscous forces:

$$Re = \frac{\rho V D}{\mu} \tag{4.1}$$

where:

Re is the Reynolds number;
ρ is the density of the fluid;
V is the velocity of the fluid relative to the cylinder;
D is the cylinder diameter;
μ is the dynamic viscosity of the fluid.

The flow pattern around a cylinder in different ranges of Reynolds number is shown in Figure 4.2 [3]. For a Reynolds number greater than approximately 30, but less than about 5000, regular shedding of vortices from the two sides of the cylinder occurs, forming a Kármán vortex street or vortex trail downstream from the cylinder, as shown in Figure 4.2a. The flow in this case is unsteady but laminar. As the Reynolds number exceeds approximately 5000, the wake downstream of the cylinder becomes turbulent, whereas the flow around the cylinder upstream of the wake remains laminar. The wake becomes more and more turbulent as the Reynolds number increases.

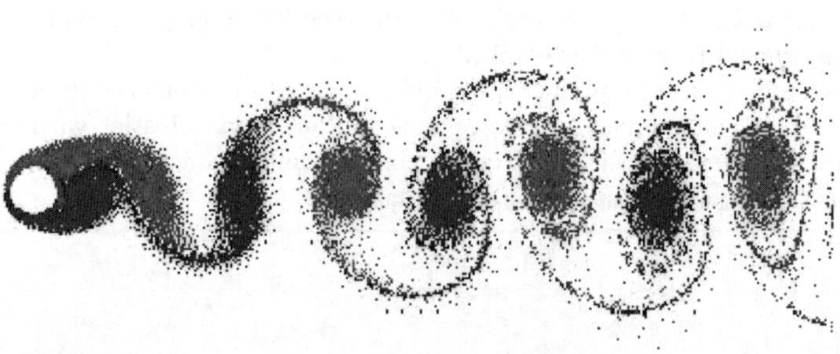

Figure 4.1 Kármán vortex street (Source: Wikipedia (http://en.wikipedia.org/wiki/File:Vortex-street-animation.gif), Reproduced with permission from original author: Cesareo de La Rosa Siqueira).

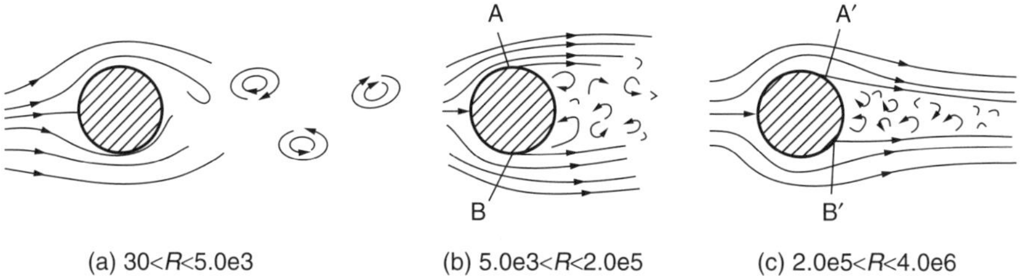

Figure 4.2 Change of flow pattern with Reynolds number.

Before reaching the critical Reynolds number (2×10^5), flow separation occurs on the two sides of the cylinder at points A and B, as shown in Figure 4.2b. The width of the wake is rather wider than the cylinder diameter, and vortex shedding is rather regular. As the Reynolds number exceeds 2×10^5, the flow separation points suddenly shift downstream from A and B to A′ and B′, respectively, as shown in Figure 4.2c. This causes a narrower wake and a sudden decrease in drag. The vortex shedding during this stage becomes rather random. Finally, when the Reynolds number exceeds approximately 4×10^6, the vortex shedding restores some regularity.

From the foregoing discussion, it can be seen that there are three distinctly different ranges of Reynolds number with distinctly different characteristics in vortex shedding, as listed in Table 4.1. Note that in the literature of vortex shedding, the supercritical range is sometimes referred to as critical range, and the hypercritical range is sometimes referred to as transcritical range. The terms used in this chapter are consistent with the use of subsonic, supersonic and hypersonic in studying compressible flow.

4.2.2 *Strouhal Number and Lock-In*

The Strouhal number, which is named after a Czech scientist, represents the vortex shedding characteristic of a structure and relates the frequency of shedding to the velocity of the flow and a characteristic dimension of the structure (e.g. diameter in the case of a cylinder). Using the Strouhal number, the frequency of vortex shedding can be determined by:

$$f_{st} = S_t \cdot \frac{U}{D} \tag{4.2}$$

Table 4.1 Ranges of Reynolds number and shedding characteristics

Range	Reynolds number	Shedding characteristics
Subcritical	$30 - 2 \times 10^5$	Regular (constant frequency)
Supercritical	$2 \times 10^5 - 4 \times 10^6$	Random (variable frequency)
Hypercritical	$>4 \times 10^6$	Regular (constant frequency)

where:

f_{st} is the vortex shedding frequency of a structure at rest;
D is the characteristic dimension, usually the across wind width, of the structure;
U is the velocity of the ambient wind flow;
S_t is the Strouhal number.

For a cylinder, the Strouhal number is 0.2 over a wide range of flow velocities.

Assuming that U is slowly increasing from zero, then f_{st} will increase according to Equation 4.2. As each vortex is shed from a structure, a strong crosswind force is induced towards the side of the shed vortex. In this way, the alternate shedding of vortices induces a nearly harmonic cross-wind force. Thus, resonance will first occur when f_{st} becomes equal to the lowest natural frequency of the structure with respect to vibrations in the acrosswind direction or torsion. Further increase of U will cause resonance to occur when f_{st} is equal to the next natural frequency, and so on.

Theoretically, resonance can occur at any natural frequency of the structure [4]. When resonance occurs, the vibration of the structure can be sufficiently large so that the structural vibration can control the frequency of vortex shedding, resulting in a phenomenon known as "lock-in", as shown in Figure 4.3. The lock-in means that the resonance can sustain through a certain range of wind velocity [5]. Finally, when wind speed is increased to above that causing lock-in, the frequency of shedding will be controlled by the wind again, rather than the structural frequency. The wind speed referring to lock-in is sometimes called the critical wind speed.

4.2.3 Vortex-Induced Vibration

In a long-span cable-supported bridge, the phenomenon of lock-in happens when vortex shedding frequency becomes close to one of the natural frequencies of a bridge component, such as a bridge deck or cable. When it happens, large vibration occurs and becomes an important source of fatigue damage. Therefore, the amplitude of vortex-induced vibrations should be restrained to certain limit in bridge design, and the general mechanism of vortex-induced bridge response should be explored.

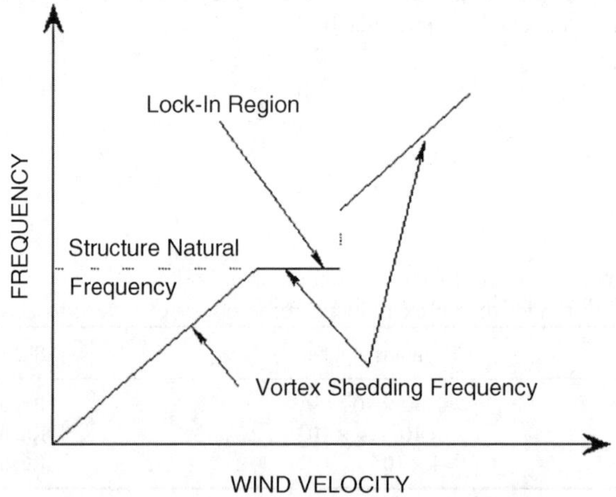

Figure 4.3 Lock-in phenomenon in vortex-induced vibration.

However, because of complex geometrical shapes of structures, the flow patterns around the structures and their relevant responses are also complex. Consequently, vortex-induced vibration has been investigated for simplified structural sections only, such as 2-D rectangular and circular cylinders, and mainly in wind tunnels and in smooth flows.

The governing equation for a structure under vortex-induced load is:

$$\mathbf{M}\ddot{X} + \mathbf{C}\dot{\mathbf{X}} + \mathbf{K}\mathbf{X} = \mathbf{L}_{VS} \tag{4.3}$$

where:

$\mathbf{M}$, $\mathbf{C}$, and $\mathbf{K}$ are the structural mass, damping, and stiffness matrix, respectively;
$\mathbf{X}$, $\dot{\mathbf{X}}$, $\ddot{\mathbf{X}}$ are the nodal displacement, velocity, and acceleration response vector; respectively;
$\mathbf{L}_{vs}$ is the vortex-induced lift force.

At the very beginning, due to the complexity of fluid-structure interaction, simplified analytical models are used to simulate vortex-induced force. Vortex-induced force is assumed to be simple harmonic, as the phenomenon is very similar to a simple harmonic vibration. As a result, the governing equation of vortex induced vibration of a SDOF structure can be written as:

$$m\left(\ddot{y} + 2\xi\omega_n\dot{y} + \omega_n^2 y\right) = \frac{1}{2}\rho\overline{U}^2 DC_L \sin(\omega_s t + \phi) \tag{4.4}$$

where:

m is the mass of the structure;
y is the vertical displacement;
ξ is the structural damping ratio;
ω_n is the structural natural frequency;
C_L is the lift force coefficient;
ω_s is the vortex shedding frequency;
ϕ is the phase angle.

When $\omega_s = \omega_n$, the solution of Equation 4.4 gives the lock-in response of the structure as:

$$y_{\max} = \frac{DC_L\rho\overline{U}^2}{4m\xi\omega_n^2} = \frac{DC_L}{16\pi^2 S_c \cdot S_t^2} \tag{4.5}$$

where $S_c = \dfrac{m\xi}{\rho D^2}$ is the Scruton number, which is an important non-dimensional structural mass-damping parameter for analysis of wind-induced structural response. Generally speaking, vortex-induced vibrations tend to be mitigated by increasing the Scruton number.

The simple harmonic model of vortex-induced force does not take into account motion-induced forces. One method to consider motion-induced force in modeling vortex-induced force is to assume that the structure is a Van der Pol oscillator. Therefore, the lift force coefficient C_L is time-dependent and should satisfy the following equation:

$$\ddot{C}_L + a_1\dot{C}_L + a_2\dot{C}_L^3 + a_3C_L = a_4\dot{y} \tag{4.6}$$

where a_i $(i = 1 - 4)$ are the parameters to be determined by experiments.

A more engineering-oriented model was proposed by Simiu and Scanlan in 1986 [6], in which vortex-induced force can be expressed as:

$$L_{VS}(t) = \frac{1}{2}\rho\overline{U}^2 D\left[Y_1(K)\frac{\dot{y}}{\overline{U}} + Y_2(K)\frac{y}{D} + C_L(K)\sin(\omega_n t + \phi)\right] \tag{4.7}$$

where:

$K = \omega B/\overline{U}$ is the reduced frequency;
$Y_1(K)$, $Y_2(K)$ and $C_L(K)$ are all the functions of the reduced frequency K at lock-in, to be determined by experiments.

A simple harmonic part of the vortex-induced force is included in Equation 4.7 so that the maximum vibration amplitude, which is usually the most important for engineering applications, can be easily calculated. Besides, in the Simiu and Scanlan model, motion-induced stiffness and damping are considered linear.

In 1990, Ehsan and Scanlan [7] revised the above model by adding a non-linear aeroelastic damping coefficient. The revised model can be expressed as:

$$L_{VS}(t) = \frac{1}{2}\rho\overline{U}^2 D\left[Y_1(K)\left(1 - \varepsilon\frac{y^2}{D^2}\right)\frac{\dot{y}}{\overline{U}} + Y_2(K)\frac{y}{D} + C_L(K)\sin(\omega_n t + \phi)\right] \tag{4.8}$$

where ε is the non-linear aeroelastic damping parameter.

The aeroelastic parameters, Y_2 and C_L, are usually ignored, since they have negligible effects on the response of lock-in. Y_1 and ε can be extracted from wind tunnel test results of steady-state amplitudes of the model at lock-in based on the following equation:

$$\frac{y_0}{D} = 2\left[\frac{Y_1 - 8\pi \cdot S_c \cdot S_t}{\varepsilon \cdot Y_1}\right]^{1/2} \tag{4.9}$$

where:

y_0/D is the reduced amplitude;
y_0 is the vibration amplitude of the vertical displacement;
$S_c = m\xi/\rho D^2$ is the Scruton number;
$S_t = f_{st}D/\overline{U}$ is the Strouhal number;
$f_n = f_{st}$ is the vortex-shedding frequency or the natural frequency.

Once these parameters are obtained, the vortex shedding force acting on the bridge deck per unit length can be obtained by Equation 4.8. The vortex shedding analysis of the bridge can then be performed and vortex shedding-induced response of the bridge can be determined by using Equation 4.3, considering the spatial correlation of vortex shedding-induced forces along the bridge deck and employing the mode superposition method [8].

4.3 Galloping Instability

4.3.1 Galloping Mechanism

"Galloping" is the term used to describe large amplitude vibrations occurring in a direction normal to the mean wind at frequencies much lower than those of vortex shedding from the structure [9].

Galloping is a common instability mode for transmission lines that have non-circular cross-sections due to particular reasons. It can happen to some forms of bridges, usually those with light weight, but it is not a typical instability mode for long-span cable-supported bridges.

Galloping usually occurs at very low reduced frequencies, $B\omega/\overline{U}$, where B is the typical chord or deck width dimension, ω is the frequency of vibration and $\overline{U}$ is the mean free stream velocity. Because the reduced frequency is low, the aerodynamic pressure or force on the bridge deck can be assumed to vary with the incoming velocity in the same manner as for steady flow (the quasi-steady assumption). Therefore, mean or average static aerodynamic data (lift and drag coefficients) can be used to assess the susceptibility of a particular bridge deck to the galloping phenomena.

When a steady wind flow passes an oscillating structure, the effective angle of attack changes with the motion of the structure, even if the incoming flow has a fixed angle of attack. The changing effective angle of attack results in the change of aerodynamic forces and leads to self-excited forces. Consider a 2-D steady flow passing a structure section, as shown in Figure 4.4.

Although the incoming flow velocity U is horizontal, the actual wind velocity acting on the structure is U_α, with an effective angle of attack α, because of the motion of the structure in the y-direction. Based on the quasi-steady assumption, the drag and lift forces can be expressed as:

$$D(\alpha) = \frac{1}{2}\rho U_\alpha^2 B C_D(\alpha) \tag{4.10a}$$

$$L(\alpha) = \frac{1}{2}\rho U_\alpha^2 B C_L(\alpha) \tag{4.10b}$$

where:

$D(\alpha)$ and $L(\alpha)$ are the drag and lift forces on the structure section, respectively;
U_α is the wind velocity with effective angle of attack α;
$B = 2b$ is the bridge deck width;
$C_D(\alpha)$ and $C_L(\alpha)$ are the drag and lift coefficients of the structure section, respectively.

The vertical force in the y-direction can then be calculated:

$$\begin{aligned} F_y(\alpha) &= \frac{1}{2}\rho U_\alpha^2 B(C_D(\alpha)\sin\alpha + C_L(\alpha)\cos\alpha) \\ &= \frac{1}{2}\rho U^2 B(C_D(\alpha)\tan\alpha + C_L(\alpha)\sec\alpha) \end{aligned} \tag{4.11}$$

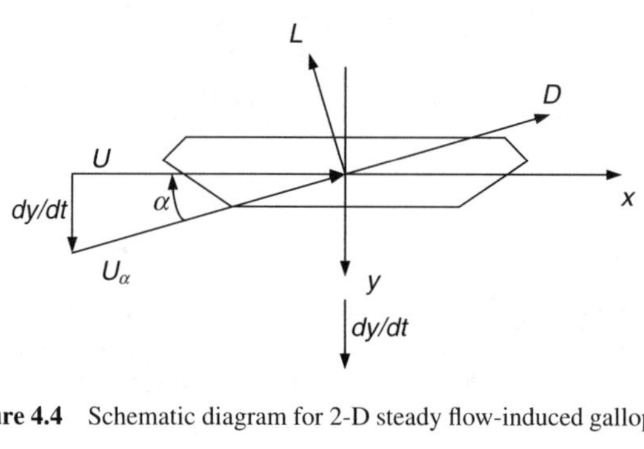

Figure 4.4 Schematic diagram for 2-D steady flow-induced galloping.

Assume that the oscillation is small, thus $\alpha \approx \frac{\dot{y}}{U} \to 0$. Applying Taylor's expansion to Equation 4.11 and considering only the first two terms yields:

$$F_y(\alpha) = F_y(0) + \frac{\partial F_y}{\partial \alpha}\bigg|_{\alpha=0} \cdot \alpha = \frac{1}{2}\rho U^2 B C_L + \frac{1}{2}\rho U^2 B\left(\frac{dC_L}{d\alpha} + C_D\right)\bigg|_{\alpha=0} \cdot \frac{\dot{y}}{U} \tag{4.12}$$

Neglecting the static components yields:

$$F_y(\alpha) = \frac{1}{2}\rho U^2 B\left(\frac{dC_L}{d\alpha} + C_D\right)\bigg|_{\alpha=0} \cdot \frac{\dot{y}}{U} \tag{4.13}$$

which is a quasi-steady expression of aerodynamic force acting on the structure section in the y-direction.

4.3.2 Criterion for Galloping Instability

The governing equation of vertical vibration of the SDOF structure can be written as:

$$m\ddot{y} + c\dot{y} + ky = \frac{1}{2}\rho U^2 B\left(\frac{dC_L}{d\alpha} + C_D\right)\bigg|_{\alpha=0} \cdot \frac{\dot{y}}{U} \tag{4.14}$$

where:

c is the damping coefficient of the structure;
k is the stiffness coefficient.

The right side of the above equation represents the aerodynamic damping force. Galloping happens when the negative aerodynamic damping force exceeds the structural damping force. It is obvious that if galloping instability is to occur, at least the following equation should be satisfied:

$$\frac{dC_L}{d\alpha} + C_D < 0 \tag{4.15}$$

The above equation can be used to judge the galloping instability of a flexible structure. Clearly, circular cylinders cannot gallop. This equation is called the Glauert-Hartog criterion. A more exact condition (sufficient condition) of galloping is:

$$\frac{dC_L}{d\alpha} + C_D < -\frac{2c}{\rho U B} \tag{4.16}$$

4.3.3 Wake Galloping

Wake galloping is a phenomenon in which oscillations of a downstream cylinder are induced by the flow in the turbulent wake of an upstream cylinder. Consider two cylinders separated at a few diameters away from each other with one cylinder in the wake of the other, as shown in Figure 4.5. Due to the circulation of the flow inside the wake (clockwise for the upper half of the wake in Figure 4.5 and

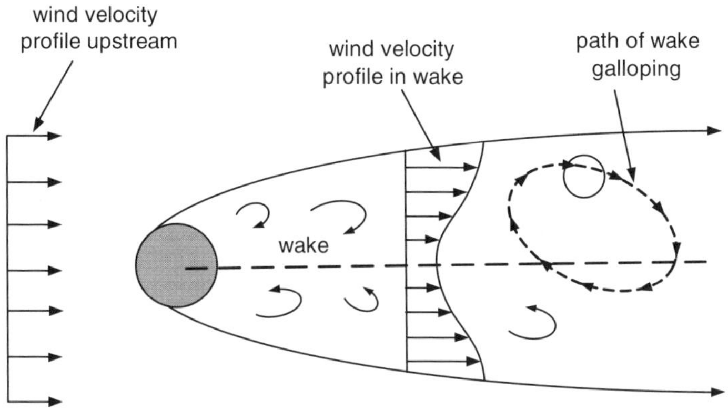

Figure 4.5 Wake galloping (from [3]).

counterclockwise of the lower half), the cylinder located in the upper half of the wake, if allowed to move, will oscillate in a clockwise, elliptic path as shown in Figure 4.5. Likewise, a cylinder free to move in the lower half would oscillate in the counterclockwise direction. Such oscillation or vibration is wake galloping [3].

Wake galloping can happen to stay cables in a cable-stayed bridge, or to the hangers of a suspension bridge. It occurs at high wind speeds and leads to large amplitude oscillations. These oscillations have been found to cause fatigue of the outer strands of hangers at end clamps on suspension bridges, and similar fatigue problems are a theoretical possibility on cable-stayed bridges. Nevertheless, wake galloping occurs only under conditions where the frequency of vibration of the downstream cylinder is low compared to the vortex-shedding frequencies of both the downstream and upstream cylinders. In such cases, when the downstream cylinder is located a few diameters behind the upstream cylinder, it enters a region of galloping instability. Researches show that the instability region is about $8D$–$20D$, where D is the diameter of the upstream cylinder.

4.4 Flutter Analysis

4.4.1 Introduction

Flutter is a self-feeding and potentially destructive vibration to a long-span cable-supported bridge where aerodynamic forces on the bridge deck couple with its motion. If the energy input by the aerodynamic forces due to strong winds in a cycle is larger than that dissipated by the damping in the bridge structure system, the amplitude of vibration of the bridge deck will increase. This increasing vibration will then amplify the aerodynamic forces, resulting in self-excited forces and self-exciting oscillations. The vibration amplitude of the bridge deck can build up until it results in the collapse of the bridge. One famous example of flutter phenomena was the collapse of the original Tacoma Narrows Bridge in 1940.

The term of classical flutter is originally applied to thin airfoils in the aircraft industry. It means an aeroelastic phenomenon in which two degrees of freedom (DOFs) of a structure, torsional and vertical, couple together in a flow-driven, unstable oscillation. It is also called 2-D flutter. Single degree of freedom (1-D) flutter may manifest itself in a vertical or torsional mode of vibration of a structure, but torsional action is more serious by far. The celebrated failure of the original Tacoma Narrows Bridge exhibited two forms of 1-D flutter – initially a non-catastrophic vertical flutter and then, at a higher wind speed, a large-amplitude torsional flutter [9]. For modern long-span cable-supported bridges, flutter instability may involve multiple modes of vibration.

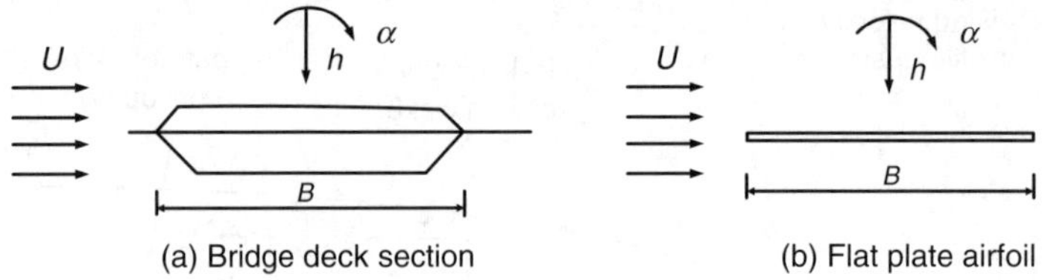

Figure 4.6 2-D structures for flutter analysis.

4.4.2 Self-Excited Forces and Aerodynamic Derivatives

Equation 4.13 gives the self-excited force in a 1-D case, in which only the first two terms are considered in Taylor's expansion and the static component is not considered. In a more general way, by considering only the first two terms in Taylor's expansion and by ignoring the static components, the self-excited forces on a 2-D structure involving both vertical and torsional vibrations (see Figure 4.6a) can be written as:

$$L_{se} = \frac{\partial L_{se}}{\partial \dot{h}}\dot{h} + \frac{\partial L_{se}}{\partial \dot{\alpha}}\dot{\alpha} + \frac{\partial L_{se}}{\partial h}h + \frac{\partial L_{se}}{\partial \alpha}\alpha \tag{4.17a}$$

$$M_{se} = \frac{\partial M_{se}}{\partial \dot{\alpha}}\dot{\alpha} + \frac{\partial M_{se}}{\partial \dot{h}}\dot{h} + \frac{\partial M_{se}}{\partial \alpha}\alpha + \frac{\partial M_{se}}{\partial h}h \tag{4.17b}$$

where:

L_{se} and M_{se} are the self-excited lift and moment on the structure respectively;
h and α are the vertical and torsional displacements of the structure, respectively.

The self-excited lift and moment are non-linear functions of the vertical and torsional displacements and their derivatives. However, the incipient flutter condition, which separates the stable and unstable regimes, may be treated as having only small amplitude. With such assumptions, Scanlan and Tomko [10] introduced aerodynamic derivatives to express the self-excited forces as:

$$L_{se} = \frac{1}{2}\rho\overline{U}^2 B\left[KH_1^*\frac{\dot{h}}{\overline{U}} + KH_2^*\frac{B\dot{\alpha}}{\overline{U}} + K^2H_3^*\alpha + K^2H_4^*\frac{h}{B}\right] \tag{4.18a}$$

$$M_{se} = \frac{1}{2}\rho\overline{U}^2 B^2\left[KA_1^*\frac{\dot{h}}{\overline{U}} + KA_2^*\frac{B\dot{\alpha}}{\overline{U}} + K^2A_3^*\alpha + K^2A_4^*\frac{h}{B}\right] \tag{4.18b}$$

where:

$\overline{U}$ is the mean wind velocity;
H_i^* and A_i^* ($i = 1$–4) are the aerodynamic derivatives.

The aerodynamic derivatives can be obtained from either wind tunnel tests or computational fluid dynamics (CFD) simulation, which will be introduced in Chapters 7 and 8, respectively.

4.4.3 Theodorsen Circulatory Function

The theoretical expressions for self-excited lift and moment on a flat plate airfoil subject to sinusoidal motions (see Figure 4.6b) are given by Theodorsen [11]:

$$L_{se} = \pi\rho b\left\{-b\ddot{h} - 2UC(k)\dot{h} - [1 + C(k)]Ub\dot{\alpha} - 2U^2C(k)\alpha\right\} \tag{4.19a}$$

$$M_{Se} = \pi\rho b^2\left\{UC(k)\dot{h} - \frac{b^2\ddot{\alpha}}{8} + \left[-\frac{1}{2} + \frac{1}{2}C(k)\right]Ub\dot{\alpha} + U^2C(k)\alpha\right\} \tag{4.19b}$$

where:

b is the half of the plate width;
$C(k)$ is the Theodorsen cyclical function, which can be expressed as:

$$C(k) = F(k) + iG(k) \tag{4.20}$$

The functions $F(K)$ and $G(K)$ are the real and imaginary parts of the function $C(k)$, as shown in Figure 4.7, which can be expressed mathematically in terms of the Bessel function of the first and second kind.

In consideration of sinusoidal displacements of the flat plate airfoil in Equation 4.19 and then by comparing this equation with Equation 4.18, the aerodynamic derivatives of the flat plate airfoil can be found as:

$$H_1^*(K) = -\frac{\pi F(k)}{k} \quad A_1^*(K) = -\frac{\pi F(k)}{4k} \tag{4.21a}$$

$$H_2^*(K) = \frac{\pi}{4k}\left[1 + F(k) + \frac{2G(k)}{k}\right] \quad A_2^*(K) = -\frac{\pi}{16k}\left[1 - F(k) - \frac{2G(k)}{k}\right] \tag{4.21b}$$

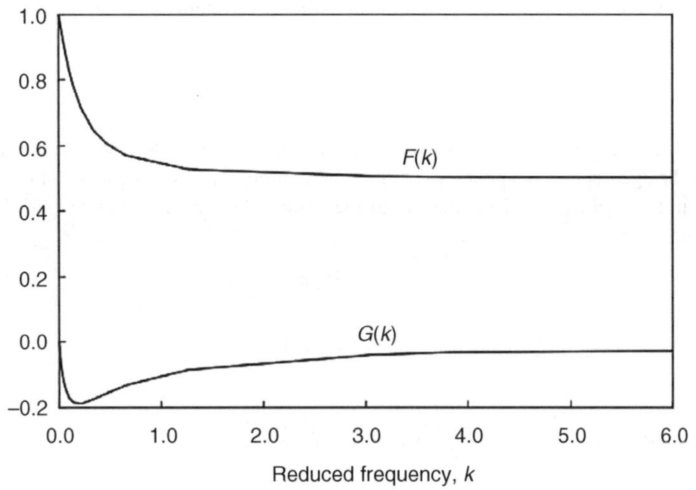

Figure 4.7 Real and imaginary parts of the Theodorsen function.

$$H_3^*(K) = \frac{\pi}{2k^2}\left[F(k) - \frac{kG(k)}{2}\right] \quad A_3^*(K) = -\frac{\pi}{8k^2}\left[F(k) - \frac{kG(k)}{2}\right] \tag{4.21c}$$

$$H_4^*(K) = \frac{\pi}{2}\left[1 + \frac{2G(k)}{k}\right] \quad A_4^*(K) = -\frac{\pi G(k)}{4k} \tag{4.21d}$$

where $k = K/2$.

4.4.4 1-D Flutter Analysis

The 1-D flutter analysis considered here is the pure torsional motion of the bridge deck. The governing equation of motion of the bridge deck for 1-D flutter analysis can be derived from Equation 4.18b as:

$$I(\ddot{\alpha} + 2\xi_{0\alpha}\omega_{0\alpha}\dot{\alpha} + \omega_{0\alpha}^2\alpha) = \frac{1}{2}\rho\overline{U}^2B^2\left[KA_2^*\frac{B\dot{\alpha}}{U} + K^2A_3^*\alpha\right] \tag{4.22}$$

where:

I represents the mass moment of inertia of the structure;
$\xi_{0\alpha}$ is the structural damping ratio in the torsional vibration;
$\omega_{0\alpha}$ is the structural circular natural frequency in the torsional vibration.

Thus, the total stiffness k_α for the wind-structure system is:

$$k_\alpha = I\omega_{0\alpha}^2 - \frac{1}{2}\rho\overline{U}^2B^2K^2A_3^*(K) \tag{4.23}$$

From the above equation, one can get the vibration frequency for the wind-structure system as:

$$\omega_\alpha = \omega_{\alpha 0}\sqrt{1 - \frac{\rho B^4}{2I}\frac{\omega_\alpha^2}{\omega_{\alpha 0}^2}A_3^*\left(\frac{\omega_\alpha B}{\overline{U}}\right)} \approx \omega_{\alpha 0}\sqrt{1 - \frac{\rho B^4}{2I}A_3^*\left(\frac{\omega_{\alpha 0}B}{\overline{U}}\right)} \tag{4.24}$$

In a similar way, the total damping ξ_α for the wind-structure system can be expressed as:

$$\xi_\alpha = 2I\xi_{0\alpha}\omega_{0\alpha} - \frac{1}{2}\rho\overline{U}B^3KA_2^*(K) \tag{4.25}$$

It is clear that if A_2^* is negative, the total damping is positive and the torsional vibration of the structure is stable. If bridge decks have positive A_2^*, the total damping may be negative. This causes flutter to occur and vibration amplitude will increase until the structure collapses. Therefore, the critical condition can be expressed as:

$$2I\xi_{0\alpha}\omega_{0\alpha} = \frac{1}{2}\rho U_{cr}B^3KA_2^*(K) \tag{4.26}$$

Because $K = \omega_\alpha B/\overline{U}$, the above equation can be rewritten as:

$$A_2^*\left(\frac{\omega_\alpha B}{U_{cr}}\right) = \frac{4I}{\rho B^4}\xi_{0\alpha}\frac{\omega_{0\alpha}}{\omega_\alpha} \tag{4.27}$$

Using Equations 4.27 and 4.24, the critical flutter wind speed for pure torsional motion can be calculated directly.

4.4.5 2-D Flutter Analysis

2-D flutter analysis aims to calculate the critical flutter wind speed for an ideal structure with two degrees of freedom – normally, one in the vertical direction and the other in the torsional direction.

Although a long-span cable-supported bridge is a complicated structure in which flutter may involve multiple modes of vibration, the first vertical mode of vibration and the first torsional mode of vibration are often considered to be most important. Moreover, in terms of the modal decomposition method, the first vertical mode and the first torsional mode can be simplified into a 2-D system. Therefore, a simplified 2-D flutter analysis can be performed for a real bridge by assuming that only the first vertical and torsional modes participate in the coupled flutter, although the critical wind speed derived from such a flutter analysis may be over-estimated due to the simplification.

It can be seen from Equation 4.18a that the self-excited forces are coupled in the vertical and torsional directions. Due to this coupling effect, a bridge can have flutter instability even if its girder section has a negative aerodynamic damping derivative A_2^* in the torsional direction.

The governing equation of motion of the structure for 2-D flutter analysis can be written as:

$$M(\ddot{h} + 2\xi_{0h}\omega_{0h}\dot{h} + \omega_{0h}^2 h) = \frac{1}{2}\rho\overline{U}^2 B\left[KH_1^*\frac{\dot{h}}{\overline{U}} + KH_2^*\frac{B\dot{\alpha}}{\overline{U}} + K^2H_3^*\alpha + K^2H_4^*\frac{h}{B}\right] \tag{4.28a}$$

$$I(\ddot{\alpha} + 2\xi_{0\alpha}\omega_{0\alpha}\dot{\alpha} + \omega_{0\alpha}^2 \alpha) = \frac{1}{2}\rho\overline{U}^2 B^2\left[KA_1^*\frac{\dot{h}}{\overline{U}} + KA_2^*\frac{B\dot{\alpha}}{\overline{U}} + K^2A_3^*\alpha + K^2A_4^*\frac{h}{B}\right] \tag{4.28b}$$

where:

M represents the structural mass;
ξ_{0h} denotes the structural damping ratio in the vertical direction;
ω_{0h} denotes the structural natural frequency in the vertical direction.

Let $X = [h\ \alpha]^{\mathrm{T}}$, for brevity. By considering the structure and air as an integrated system and transposing the terms on the right side of Equation 4.28 to the left side, Equation 4.28 can be rewritten as:

$$\ddot{X} + C^e\dot{X} + K^eX = 0 \tag{4.29}$$

where:

$$C^e = \begin{pmatrix} 2\xi_{0h}\omega_{0h} - \dfrac{\rho B^2 L}{M}\omega_h H_1^* & \dfrac{\rho B^3 L}{M}\omega_\alpha H_2^* \\ \dfrac{\rho B^3 L}{I}\omega_h A_1^* & 2\xi_{0\alpha}\omega_{0\alpha} - \dfrac{\rho B^4 L}{I}\omega_\alpha A_2^* \end{pmatrix} \tag{4.30a}$$

$$K^e = \begin{pmatrix} \omega_{0h}^2 - \dfrac{\rho B^2 L}{M}\omega_h^2 H_4^* & \dfrac{\rho B^3 L}{M}\omega_\alpha^2 H_3^* \\ \dfrac{\rho B^3 L}{I}\omega_h^2 A_4^* & \omega_{0\alpha}^2 - \dfrac{\rho B^4 L}{I}\omega_\alpha^2 A_3^* \end{pmatrix} \tag{4.30b}$$

C^e and K^e are the damping and stiffness matrix of the coupled wind-structure system respectively; ω_h and ω_α denote the frequency of the vibration system in the vertical and torsional direction respectively.

Letting $Y = [X\quad \dot{X}]^{\mathrm{T}}$, the governing equation can be written in the state space as:

$$\dot{Y} = AY \tag{4.31}$$

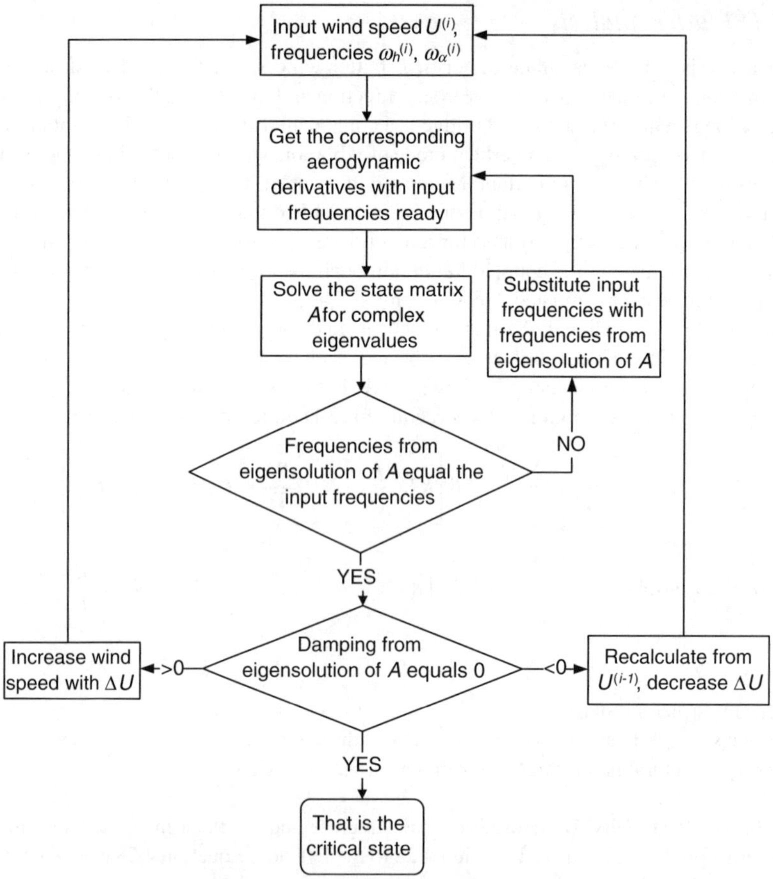

Figure 4.8 Flow chart of 2-D flutter analysis.

where:

$A = \begin{pmatrix} 0 & I \\ -K^e & -C^e \end{pmatrix}$ is the state matrix;

I denotes the unit matrix.

The real parts of the eigenvalues of the matrix A ($-\xi_i\omega_i$, $i = h, \alpha$) represent the modal damping of the coupled system, while the imaginary parts of the eigenvalues of A ($\sqrt{1-\xi_i^2}\omega_i$, $i = h, \alpha$) represent the modal frequencies of the coupled system. Through an iterative eigenvalue analysis of A, the critical flutter wind speed for a 2-D coupled system can be calculated. The procedure of a 2-D flutter analysis is shown in Figure 4.8.

4.4.6 3-D Flutter Analysis in the Frequency Domain

The main purpose of a flutter analysis is to determine the critical flutter wind speed for a specific bridge. It is usually conducted in the frequency domain for calculation efficiency. The theoretical backgrounds of coupled flutter analysis of long-span cable-supported bridges can be found in the literature [12–16] and are briefly introduced in this section. It is assumed that the buffeting forces have no influence on aerodynamic stability and are excluded in the flutter analysis. As a result, the governing equation of motion of a bridge deck under self-excited forces is given by:

$$\mathbf{M}\ddot{\mathbf{X}} + \mathbf{C}\dot{\mathbf{X}} + \mathbf{K}\mathbf{X} = \mathbf{F}_{se} \tag{4.32}$$

where:

$\mathbf{M}$, $\mathbf{C}$, and $\mathbf{K}$ are the structural mass, damping, and stiffness matrix, respectively;
$\mathbf{X}, \dot{\mathbf{X}}, \ddot{\mathbf{X}}$ are the nodal displacement, velocity, and acceleration vector, respectively;
$\mathbf{F}$ indicates the nodal equivalent force vector; the subscript *se* represents the self-excited force.

The self-excited vertical and lateral forces and self-excited moment acting on the bridge deck per unit length are expressed often in terms of Scanlan's format as follows:

$$L_{se}(t) = \frac{1}{2}\rho\overline{U}^2(2B)\left(KH_1^*\frac{\dot{h}}{\overline{U}} + KH_2^*\frac{B\dot{\alpha}}{\overline{U}} + K^2H_3^*\alpha + K^2H_4^*\frac{h}{B} + KH_5^*\frac{\dot{p}}{\overline{U}} + K^2H_6^*\frac{p}{B}\right) \tag{4.33a}$$

$$D_{se}(t) = \frac{1}{2}\rho\overline{U}^2(2B)\left(KP_1^*\frac{\dot{p}}{\overline{U}} + KP_2^*\frac{B\dot{\alpha}}{\overline{U}} + K^2P_3^*\alpha + K^2H_4^*\frac{h}{B} + KH_5^*\frac{\dot{p}}{\overline{U}} + K^2H_6^*\frac{p}{B}\right) \tag{4.33b}$$

$$M_{se}(t) = \frac{1}{2}\rho\overline{U}^2(2B^2)\left(KA_1^*\frac{\dot{h}}{\overline{U}} + KA_2^*\frac{B\dot{\alpha}}{\overline{U}} + K^2A_3^*\alpha + K^2A_4^*\frac{h}{B} + KA_5^*\frac{\dot{p}}{\overline{U}} + K^2A_6^*\frac{p}{B}\right) \tag{4.33c}$$

where:

h, p, and α are the vertical, lateral, and torsional displacements of the bridge deck, respectively;
the over-dot denotes the partial differentiation with respect to time t;
H_i^*, P_i^*, A_i^* $(i = 1-6)$ are the non-dimensional aerodynamic derivatives (flutter derivatives) which can be obtained from wind tunnel tests.

At present, a two-degrees-of-freedom section model of the bridge deck is widely used to identify the flutter derivatives H_i^* and A_i^* $(i = 1-4)$. The drag components associated with the lateral motion and some coupling terms are generally negligible. If these flutter derivatives are not available from the wind tunnel tests, the empirical expressions based on the quasi-steady theory can be used in the analysis:

$$P_1^* = -\frac{1}{K}C_D, \quad P_2^* = \frac{1}{2K}C_D', \quad P_3^* = \frac{1}{2K^2}C_D' \tag{4.34a}$$

$$P_5^* = \frac{1}{2K}C_D', \quad H_5^* = \frac{1}{K}C_L, \quad A_5^* = -\frac{1}{K}C_M \tag{4.34b}$$

$$P_4^* = P_6^* = H_6^* = A_6^* = 0 \tag{4.34c}$$

where:

C_L, C_D, and C_M are the aerodynamic lift, drag, and moment coefficients referring to the deck width B, respectively;
$C_D' = dC_D/d\alpha$.

By using a complex notation, the self-excited forces on the bridge deck per unit length may be expressed as:

$$L_{se}(t) = \omega^2 \rho B^2 (C_{Lh}h + C_{Lp}p + BC_{L\alpha}\alpha) \tag{4.35a}$$

$$D_{se}(t) = \omega^2 \rho B^2 (C_{Dh}h + C_{Dp}p + BC_{D\alpha}\alpha) \tag{4.35b}$$

$$M_{se}(t) = \omega^2 \rho B^2 (BC_{Mh}h + BC_{Mp}p + B^2 C_{M\alpha}\alpha) \tag{4.35c}$$

where C_{rs} ($r = D, L, M$; $s = h, p, \alpha$) are the complex flutter derivatives of self-excited forces.

The relationships between the real and complex flutter derivatives can be found as:

$$C_{Lh} = H_4^* + iH_1^*, \quad C_{Lp} = H_6^* + iH_5^*, \quad C_{L\alpha} = H_3^* + iH_2^* \tag{4.36a}$$

$$C_{Dh} = P_6^* + iP_5^*, \quad C_{Dp} = P_4^* + iP_1^*, \quad C_{D\alpha} = P_3^* + iP_2^* \tag{4.36b}$$

$$C_{Mh} = A_4^* + iA_1^*, \quad C_{Mp} = A_6^* + iA_5^*, \quad C_{M\alpha} = A_3^* + iA_2^* \tag{4.36c}$$

In a 3-D finite element-based flutter analysis, the distributed self-excited forces acting on an element of a bridge deck need to be converted into equivalent nodal loads at two ends of the element.

$$\mathbf{F}_{se}^e = \omega^2 \mathbf{A}_{se}^e \mathbf{X}^e \tag{4.37}$$

where:

the subscript e represents the local coordinates of the element;
$\mathbf{A}_{se}^e$ is a 12×12 aeroelastic matrix.

For an element with length L, the aeroelastic matrix is

$$\mathbf{A}_{se}^e = \begin{bmatrix} \mathbf{A}_1 & \mathbf{0} \\ \mathbf{0} & \mathbf{A}_1 \end{bmatrix} \tag{4.38}$$

where:

$$\mathbf{A}_1 = \frac{1}{2}\rho B^2 L \begin{bmatrix} 0 & 0 & 0 & 0 & 0 & 0 \\ 0 & C_{Lh} & C_{Lp} & BC_{L\alpha} & 0 & 0 \\ 0 & C_{Dh} & C_{Dp} & BC_{D\alpha} & 0 & 0 \\ 0 & BC_{Mh} & BC_{Mp} & B^2 C_{M\alpha} & 0 & 0 \\ 0 & 0 & 0 & 0 & 0 & 0 \\ 0 & 0 & 0 & 0 & 0 & 0 \end{bmatrix} \tag{4.39}$$

Since aeroelastic (self-excited) forces are non-conservative, the aeroelastic matrix of the element is generally unsymmetric and it is a function of the reduced frequency. When the aeroelastic matrix of the element is transformed into the global coordinate system and is assembled, one may have:

$$\mathbf{F}_{se} = \omega^2 \mathbf{A}_{se} \mathbf{X} \tag{4.40}$$

where $\mathbf{A}_{se}$ is the aeroelastic matrix of the bridge, and it is a complex matrix. The governing equation of motion of the bridge for flutter analysis can then be expressed as:

$$\mathbf{M}\ddot{\mathbf{X}} + \mathbf{C}\dot{\mathbf{X}} + \mathbf{K}\mathbf{X} = \omega^2 \mathbf{A}_{se}\mathbf{X} \tag{4.41}$$

Let $\mathbf{X}$ equal $\mathbf{R}\, e^{st}$, where $\mathbf{R}$ is the complex modal response amplitude vector of the system. Denote the complex frequency $s=(-\xi+i)\omega$ (where ξ and ω are the damping ratio and circular frequency of the complex mode of vibration, respectively, and $i^2=-1$). The governing equation of motion can be written as:

$$(s^2\mathbf{M} + s\mathbf{C} + \mathbf{K} - \omega^2\mathbf{A}_{se})\mathbf{R}e^{st} = \mathbf{0} \tag{4.42}$$

The complex mode response of the system can be given approximately by the first m modes of vibration:

$$\mathbf{R} = \mathbf{\Phi}\mathbf{q} \tag{4.43}$$

where:

$\mathbf{\Phi}$ is the matrix of mode shapes, obtained from the modal analysis of the structure;
$\mathbf{q}$ is the vector of generalized coordinates.

Applying the above linear transformation to Equation 4.41 yields:

$$[s^2\mathbf{I} - \omega^2\overline{\mathbf{A}}_{se} + s\overline{\mathbf{C}} + \mathbf{\Lambda}]\mathbf{q}e^{st} = \mathbf{0} \tag{4.44}$$

where $\mathbf{\Lambda}$ is the diagonal eigenvalue matrix obtained from modal analysis; and the matrix $\overline{\mathbf{A}}_{se} = \mathbf{\Phi}^{\mathrm{T}}\mathbf{A}_{se}\mathbf{\Phi}$ and $\overline{\mathbf{C}} = \mathbf{\Phi}^{\mathrm{T}}\mathbf{C}\mathbf{\Phi}$.

Considering the fact that the damping ratios of the system (positive or negative) are small, the approximate relation $\omega^2 = -s^2$ exists. As a result, one may have:

$$[s^2(\mathbf{I} + \overline{\mathbf{A}}_{se}) + s\overline{\mathbf{C}} + \mathbf{\Lambda}]\mathbf{q}e^{st} = \mathbf{0} \tag{4.45}$$

The above equation can be further expressed in the following state-space format:

$$(\mathbf{A} - s\mathbf{I})\mathbf{Y}e^{st} = \mathbf{0} \tag{4.46}$$

where:

$$\mathbf{Y} = \begin{Bmatrix} \mathbf{q} \\ s\mathbf{q} \end{Bmatrix}, \quad \mathbf{A} = \begin{bmatrix} \mathbf{0} & \mathbf{I} \\ -\overline{\mathbf{M}}\mathbf{\Lambda} & -\overline{\mathbf{M}}\overline{\mathbf{C}} \end{bmatrix} \tag{4.47a}$$

$$\overline{\mathbf{M}} = (\mathbf{I} + \overline{\mathbf{A}}_{se})^{-1} \tag{4.47b}$$

Thus, to have a non-trivial solution, the following equation must be satisfied, leading to a standard eigenvalue problem:

$$\mathbf{A}\mathbf{Y} = s\mathbf{Y} \tag{4.48}$$

where the characteristic matrix **A** is a 2 m × 2 m complex matrix and a function of reduced frequency K (or reduced velocity) only. Thus, the above equation can be solved for only two variables, s and K.

For a given K, standard linear eigensolvers are available to find the 2 m sets of eigenvalues s and corresponding eigenvectors **Y** from Equation 4.48:

$$s = (-\xi + i)\omega \tag{4.49a}$$

$$\mathbf{q} = \mathbf{a} + \mathbf{b}i \tag{4.49b}$$

The m eigenvalues with positive imaginary part are the complex frequencies of the system, and the upper half vector **q** in the corresponding eigenvector **Y** is the complex mode shape of the system. In a prescribed complex mode shape, the magnitude and phase of the kth natural mode are given as:

$$|q_k| = \sqrt{\mathrm{a}_k^2 + \mathrm{b}_k^2} \tag{4.50a}$$

$$\phi_k = \tan^{-1}(\mathrm{b}_k/\mathrm{a}_k) \tag{4.50b}$$

If the damping ratios of all complex modes are positive, the system is stable; if at least one damping ratio is equal to zero, the system is neutrally stable; if at least one damping ratio is negative, the system is unstable. Therefore, the flutter analysis described above is able to find the critical state through searching the reduced frequency K. The corresponding circular frequency is the flutter circular frequency ω_f and the critical wind speed U_{cr} is then equal to $B\omega_f/K$. At the critical wind speed, the generalized modal coordinate vector $\mathbf{q}(t)$ and the nodal displacement vector of the bridge can be expressed as:

$$\mathbf{q}(t) = \{|\mathbf{q}_i|\sin(\omega_f t + \phi_i)\} \tag{4.51}$$

$$\mathbf{X}(t) = \sum_{i=1}^{m} \boldsymbol{\varphi}_i|\mathbf{q}_i|\sin(\omega_f t + \phi_i) = \mathbf{X}_0 \sin(\omega_f t + \phi_i) \tag{4.52}$$

where:

$\boldsymbol{\varphi}_i$ is the ith natural mode shape;
ω_f is the flutter circular frequency;
X_0 and ϕ_i are the amplitude and phase of $X(t)$;
m is the number of participating modes.

It is clear that the coupled flutter motion is three-dimensional and that the phase shift exists among mode components. The total energy in the characteristic motion (flutter motion) of the bridge at the lowest critical wind speed is:

$$E = \frac{1}{2}\{\dot{\mathbf{X}}_{\max}\}^{\mathrm{T}}\mathbf{M}\{\dot{\mathbf{X}}_{\max}\} = \frac{1}{2}\omega_f^2\sum_{i=1}^{m}|\mathbf{q}_i|^2 \tag{4.53}$$

The energy in the ith natural mode of vibration of the bridge is expressed as:

$$E_i = \frac{1}{2}\omega_f^2|\mathbf{q}_i|^2 \tag{4.54}$$

The ratio of the ith modal energy over the total energy E_i/E is defined as the modal energy ratio e_i. Clearly, the modal energy ratio provides a uniform measurement of the contribution of a particular vibration mode to the flutter instability of the whole bridge.

4.4.7 Flutter Analysis in the Time Domain

Flutter analysis has been predominantly conducted in the frequency domain for computational efficiency particularly because self-excited forces are functions of reduced frequency.

However, the frequency domain approach is restricted to linear structures excited by the stationary wind loads without considering aerodynamic and/or structural non-linearities. Schemes for time-domain coupled multimode flutter analysis have been proposed by introducing the unsteady self-excited forces in terms of rational function approximations [17]. The advantages of the time-domain approach are that an iterative solution for determining flutter conditions is unnecessary because the equations of motion are independent of frequency, and aerodynamic non-linearities can be taken into account in flutter analysis. The time-domain modeling of self-excited forces is also used in the time-domain buffeting analysis, which will be introduced in Section 4.7 in detail.

4.5 Buffeting Analysis in the Frequency Domain

4.5.1 Background

When a long-span cable-supported bridge is immersed in a wind field, the bridge will be subjected to static and dynamic wind forces caused by mean and fluctuating wind speeds, respectively. Buffeting action on a long-span bridge is a random vibration caused by fluctuating winds that appear within a wide range of wind speeds. In wind resistance design of a long-span bridge, consideration of the buffeting responses is normally dominant in order to determine the size of structural members. In addition to buffeting action, the self-excited forces induced by wind-structure interaction is also important for predicting the buffeting response of long-span bridges, as the additional energy injected into the oscillating structure by self-excited forces increases the magnitude of vibrations. To model the action of buffeting wind load, the buffeting forces resulting from turbulent wind and the self-excited forces due to the wind-bridge interaction should be taken into account. The buffeting response prediction can be performed in both the frequency domain [18–21] and the time domain [17,22,23].

4.5.2 Buffeting Forces and Aerodynamic Admittances

Under the quasi-steady assumption, the transient aerodynamic forces on the bridge deck per unit length can be written as:

$$L_\alpha(t) = \frac{1}{2}\rho \overline{U}^2(t) C_L(\alpha_0 + \Delta\alpha) \cdot B \tag{4.55a}$$

$$D_\alpha(t) = \frac{1}{2}\rho \overline{U}^2(t) C_D(\alpha_0 + \Delta\alpha) \cdot B \tag{4.55b}$$

$$M_\alpha(t) = \frac{1}{2}\rho \overline{U}^2(t) C_M(\alpha_0 + \Delta\alpha) \cdot B^2 \tag{4.55c}$$

where:

the subscript α denotes that the force is expressed in the transient wind axis;
α_0 is the attack angle of mean wind speed $\overline{U}$;
$\Delta\alpha$ is the additional attack angle induced by fluctuation (see Figure 4.9).

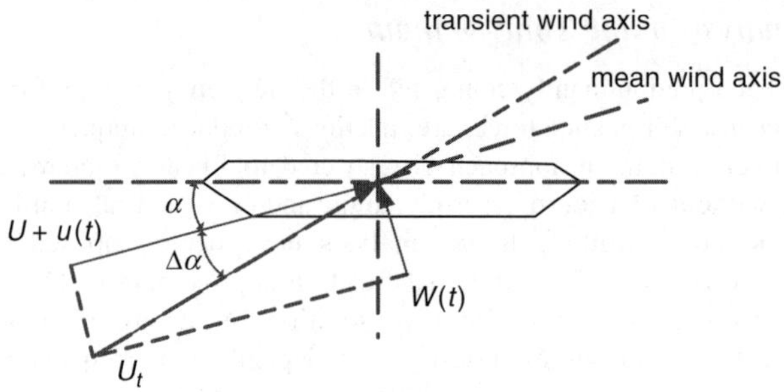

Figure 4.9 Wind and buffeting forces on bridge deck.

The aerodynamic coefficients can be expressed in terms of the Taylor's expansion to the first two terms:

$$C_L(\alpha_0 + \Delta\alpha) = C_L(\alpha_0) + C'_L(\alpha_0) \cdot \Delta\alpha \tag{4.56a}$$

$$C_D(\alpha_0 + \Delta\alpha) = C_D(\alpha_0) + C'_D(\alpha_0) \cdot \Delta\alpha \tag{4.56b}$$

$$C_M(\alpha_0 + \Delta\alpha) = C_M(\alpha_0) + C'_M(\alpha_0) \cdot \Delta\alpha \tag{4.56c}$$

Equation 4.55a can be transformed into the mean wind speed axis as:

$$L(t) = L_\alpha(t) \cdot \cos(\Delta\alpha) + D_\alpha(t) \cdot \sin(\Delta\alpha) \tag{4.57a}$$

$$D(t) = D_\alpha(t) \cdot \cos(\Delta\alpha) - L_\alpha(t) \cdot \sin(\Delta\alpha) \tag{4.57b}$$

$$M(t) = M_\alpha(t) \tag{4.57c}$$

Assuming that $\Delta\alpha$ is very small, then $\sin(\Delta\alpha) \approx \alpha$ and $\cos(\Delta\alpha) \approx 1 - \Delta\alpha^2/2$. Equation 4.57 becomes:

$$L(t) = \frac{1}{2}\rho\overline{U}^2 B\left[C_L(\alpha_0) \cdot \left(\frac{2u(t)}{\overline{U}}\right) + (C'_L(\alpha_0) + C_D(\alpha_0)) \cdot \frac{w(t)}{\overline{U}}\right] + \frac{1}{2}\rho\overline{U}^2 B C_L(\alpha_0) \tag{4.58a}$$

$$D(t) = \frac{1}{2}\rho\overline{U}^2 B\left[C_D(\alpha_0) \cdot \left(\frac{2u(t)}{\overline{U}}\right) + C'_D(\alpha_0) \cdot \frac{w(t)}{\overline{U}}\right] + \frac{1}{2}\rho\overline{U}^2 B C_D(\alpha_0) \tag{4.58b}$$

$$M(t) = \frac{1}{2}\rho\overline{U}^2 B\left[C_M(\alpha_0) \cdot \left(\frac{2u(t)}{\overline{U}}\right) + C'_M(\alpha_0) \cdot \frac{w(t)}{\overline{U}}\right] + \frac{1}{2}\rho\overline{U}^2 B^2 C_M(\alpha_0) \tag{4.58c}$$

It is obvious that the second terms on the right side of the above equation are static forces. Therefore, the buffeting forces can be written as:

$$L_b(t) = \frac{1}{2}\rho\overline{U}^2 B\left[C_L(\alpha_0) \cdot \left(\frac{2u(t)}{\overline{U}}\right) + (C'_L(\alpha_0) + C_D(\alpha_0)) \cdot \frac{w(t)}{\overline{U}}\right] \tag{4.59a}$$

$$D_b(t) = \frac{1}{2}\rho U^2 B\left[C_D(\alpha_0) \cdot \left(\frac{2u(t)}{\overline{U}}\right) + C'_D(\alpha_0) \cdot \frac{w(t)}{\overline{U}}\right] \tag{4.59b}$$

$$M_b(t) = \frac{1}{2}\rho U^2 B\left[C_M(\alpha_0) \cdot \left(\frac{2u(t)}{\overline{U}}\right) + C'_M(\alpha_0) \cdot \frac{w(t)}{\overline{U}}\right] \tag{4.59c}$$

The quasi-steady buffeting theory assumes that the wind loads are fully correlated along the bridge span. Such an assumption can only be satisfied when the turbulence has a larger scale than the bridge width, otherwise the incomplete span-wise correlation should be taken into account when modeling buffeting forces. Davenport used six aerodynamic admittance functions to represent the span-wise correlation of wind loads [18]. After the modification, the buffeting forces can be written as:

$$L_b = \frac{1}{2}\rho\overline{U}^2 B\left[2C_L\chi_{Lu}\frac{u}{\overline{U}} + (C_L' + C_D)\chi_{Lw}\frac{w}{\overline{U}}\right] \tag{4.60a}$$

$$D_b = \frac{1}{2}\rho\overline{U}^2 B\left[2C_D\chi_{Du}\frac{u}{\overline{U}} + C_D'\chi_{Dw}\frac{w}{\overline{U}}\right] \tag{4.60b}$$

$$M_b = \frac{1}{2}\rho\overline{U}^2 B^2\left[2C_M\chi_{Mu}\frac{u}{\overline{U}} + C_M'\chi_{Mw}\frac{w}{\overline{U}}\right] \tag{4.60c}$$

where:

$C_L' = dC_L/d\alpha$, $C_D' = dC_D/d\alpha$, and $C_M' = dC_M/d\alpha$;

χ_{Lu}, χ_{Lw}, χ_{Du}, χ_{Dw}, χ_{Mu}, χ_{Mw} are the aerodynamic admittance functions, which are functions of the reduced frequency and dependent on the geometrical configuration of the cross section of the bridge deck;

u and w are the longitudinal and vertical turbulences, respectively.

How to determine the aerodynamic admittance functions is a key issue in buffeting analysis. One conventional way is to compare measured wind spectra and buffeting force spectra and then obtain the empirical aerodynamic admittance functions by the following equations:

$$\chi_{Lu}(K) = \frac{4}{\rho^2\overline{U}^2B^2}\cdot\frac{S_{L_bL_b}(x,K)}{S_{uu}(x,K)} \quad \chi_{Lw}(K) = \frac{4}{\rho^2\overline{U}^2B^2}\cdot\frac{S_{L_bL_b}(x,K)}{S_{ww}(x,K)} \tag{4.61a}$$

$$\chi_{Du}(K) = \frac{4}{\rho^2\overline{U}^2B^2}\cdot\frac{S_{D_bD_b}(x,K)}{S_{uu}(x,K)} \quad \chi_{Du}(K) = \frac{4}{\rho^2\overline{U}^2B^2}\cdot\frac{S_{D_bD_b}(x,K)}{S_{ww}(x,K)} \tag{4.61b}$$

$$\chi_{Mu}(K) = \frac{4}{\rho^2\overline{U}^2B^4}\cdot\frac{S_{M_bM_b}(x,K)}{S_{uu}(x,K)} \quad \chi_{Mu}(K) = \frac{4}{\rho^2\overline{U}^2B^4}\cdot\frac{S_{M_bM_b}(x,K)}{S_{ww}(x,K)} \tag{4.61c}$$

where:

$S_{i_bi_b}(x,K)$ is the buffeting force spectrum ($i = L$, D, and M);
$S_{jj}(x,K)$ is the wind spectrum ($j = u$, w).

4.5.3 3-D Buffeting Analysis in the Frequency Domain

The main purpose of buffeting analysis is to calculate the dynamic response of a bridge under both self-excited and buffeting forces. It can be conducted in either frequency domain or time domain. This section will introduce buffeting analysis in the frequency domain, which offers better computational efficiency because the analysis can be carried in the selected mode. The governing equation of motion of a bridge excited by fluctuating winds with respect to the static equilibrium position can be given in a matrix form by:

$$\mathbf{M}\ddot{\mathbf{X}} + \mathbf{C}\dot{\mathbf{X}} + \mathbf{K}\mathbf{X} = \mathbf{F}_{se} + \mathbf{F}_b + \mathbf{F}_s \tag{4.62}$$

where:

F indicates the nodal equivalent force vector;
the subscript *se*, *b and s* represents the self-excited force, buffeting force and mean wind force components, respectively.

The mean deformation of the bridge caused by the mean wind forces can be readily determined using the mean wind speeds and the other parameters determined by wind tunnel tests as introduced in Chapter 3. The formulation of self-excited force $\boldsymbol{F}_{se}$ is already mentioned in the above section, as expressed by Equation 4.40. The vertical and lateral buffeting forces and buffeting moment acting on the bridge deck per unit length due to wind fluctuations are given in Equation 4.60. If the aerodynamic admittance functions are taken as units, the buffeting forces aforementioned can be expressed as follows:

$$\mathbf{F}_b = 0.5\rho\overline{U}(\mathbf{C}_{bu}u + \mathbf{C}_{bw}w) \tag{4.63}$$

where:

$$\mathbf{F}_b = \begin{Bmatrix} L_b \\ D_b \\ M_b \end{Bmatrix}, \quad \mathbf{C}_{bu} = B\begin{Bmatrix} 2C_L \\ 2C_D \\ 2BC_M \end{Bmatrix}, \quad \mathbf{C}_{bw} = B\begin{Bmatrix} C'_L + C_D \\ C'_D \\ BC'_M \end{Bmatrix} \tag{4.64}$$

When the element is small enough, it can be assumed that the longitudinal and vertical wind fluctuations are distributed linearly on the element:

$$u = \begin{bmatrix} 1 - \frac{x}{L} & \frac{x}{L} \end{bmatrix}\begin{Bmatrix} u_1 \\ u_2 \end{Bmatrix} = \mathbf{Au}^e \tag{4.65a}$$

$$w = \begin{bmatrix} 1 - \frac{x}{L} & \frac{x}{L} \end{bmatrix}\begin{Bmatrix} w_1 \\ w_2 \end{Bmatrix} = \mathbf{Aw}^e \tag{4.65b}$$

where:

x and L are the axial location and the length of the element, respectively;
the subscripts 1 and 2 indicate the two ends of the element.

The consistent buffeting forces at the element ends in the local coordinate system can be obtained by the following definite integral:

$$\mathbf{F}_b^e = \int_L \mathbf{B}^T\mathbf{F}_b dx = 0.5\rho\overline{U}\left(\int_L \mathbf{B}^T\mathbf{C}_{bu}\mathbf{A}dx\mathbf{u}^e + \int_L \mathbf{B}^T\mathbf{C}_{bw}\mathbf{A}dx\mathbf{w}^e\right) = 0.5\rho\overline{U}(\mathbf{A}_{bu}^e\mathbf{u}^e + \mathbf{A}_{bw}^e\mathbf{w}^e) \tag{4.66}$$

where:

$\mathbf{A}_{bu}^e$ and $\mathbf{A}_{bw}^e$ are the buffeting force matrices of the element corresponding to the longitudinal and vertical wind fluctuations, respectively;

B is the matrix of interpolated functions:

$$B = \begin{bmatrix} 0 & -N_1 & 0 & 0 & 0 & -N_3 & 0 & -N_2 & 0 & 0 & 0 & N_4 \\ 0 & 0 & -N_1 & 0 & N_3 & 0 & 0 & 0 & -N_2 & 0 & -N_4 & 0 \\ 0 & 0 & 0 & -N_5 & 0 & 0 & 0 & 0 & 0 & -N_6 & 0 & 0 \end{bmatrix} \tag{4.67}$$

where:

$$N_1 = 1 - 3\left(\frac{x}{L}\right)^2 + 2\left(\frac{x}{L}\right)^3; \quad N_2 = 3\left(\frac{x}{L}\right)^2 - 2\left(\frac{x}{L}\right)^3; \quad N_3 = x\left(1 - \frac{x}{L}\right)^2; \tag{4.68a}$$

$$N_4 = \frac{x^2}{L}\left(1 - \frac{x}{L}\right); \quad N_5 = 1 - \frac{x}{L}; \quad N_6 = \frac{x}{L} \tag{4.68b}$$

The matrices $\mathbf{A}^e_{bu}$ and $\mathbf{A}^e_{bw}$ can be derived as:

$$\mathbf{A}^e_{bu} = \frac{-BL}{30}\begin{bmatrix} 0 & 21C_L & 21C_D & 20BC_M & -3LC_D & 3LC_L \\ 0 & 9C_L & 9C_D & 10BC_M & -2LC_D & 2LC_L \\ 0 & 9C_L & 9C_D & 10BC_M & 2LC_D & -2LC_L \\ 0 & 21C_L & 21C_D & 20BC_M & 3LC_D & -3LC_L \end{bmatrix}^T \tag{4.69}$$

$$\mathbf{A}^e_{bw} = \frac{-BL}{60}\begin{bmatrix} 0 & 21(C'_L + C_D) & 21C'_D & 20BC'_M & -3LC'_D & 3L(C'_L + C_D) \\ 0 & 9(C'_L + C_D) & 9C'_D & 10BC'_M & -2LC'_D & 2L(C'_L + C_D) \\ 0 & 9(C'_L + C_D) & 9C'_D & 10BC'_M & 2LC'_D & -2L(C'_L + C_D) \\ 0 & 21(C'_L + C_D) & 21C'_D & 20BC'_M & 3LC'_D & -3L(C'_L + C_D) \end{bmatrix}^T \tag{4.70}$$

The local nodal buffeting forces can be converted into the global coordinate system using the coordinate transformation matrix. As a result, the global nodal buffeting force vector can be obtained as:

$$\mathbf{F}_b = 0.5\rho\overline{U}(\mathbf{A}_{bu}\mathbf{u} + \mathbf{A}_{bw}\mathbf{w}) \tag{4.71}$$

where:

$\mathbf{A}_{bu}$ and $\mathbf{A}_{bw}$ are the global buffeting force matrices;
u and **w** are the r-row nodal fluctuating wind vectors for the longitudinal and vertical components, respectively, where r is the number of nodes subjected to wind fluctuations.

Apart from the bridge deck, the buffeting forces also act on the bridge towers, the cables and other components. These forces can be determined using a similar way to the determination of the forces acting on the bridge deck. It is thus possible to have a buffeting analysis of the bridge as a whole, rather than the bridge deck only. Based on the preceding discussion, the governing equation of motion of the bridge as a whole can be written as:

$$\mathbf{M}\ddot{\mathbf{X}} + \mathbf{C}\dot{\mathbf{X}} + \mathbf{K}\mathbf{X} - \omega^2\mathbf{A}_{se}\mathbf{X} = \mathbf{F}_b \tag{4.72}$$

The buffeting response of the bridge is dominant by the first m modes of vibration and thus a linear transformation is introduced as:

$$\mathbf{X} = \mathbf{\Phi}\mathbf{q} \tag{4.73}$$

where:

$\mathbf{\Phi}$ is the matrix of mode shapes, which are obtained from the modal analysis;
$\mathbf{q}$ is the m vector of generalized coordinates.

The equation of motion of the bridge can be then expressed as:

$$\ddot{\mathbf{q}} + \overline{\mathbf{C}}\dot{\mathbf{q}} + \mathbf{\Lambda q} - \omega^2 \overline{\mathbf{A}}_{se}\mathbf{q} = \mathbf{Q}_b \tag{4.74}$$

where:

$\mathbf{\Lambda}$ is the diagonal eigenvalue matrix obtained from the dynamic characteristic analysis;
$\overline{\mathbf{C}} = \mathbf{\Phi}^{\mathrm{T}}\mathbf{C\Phi}$; $\overline{\mathbf{A}}_{se} = \mathbf{\Phi}^{\mathrm{T}}\mathbf{A}_{se}\mathbf{\Phi}$; $\mathbf{Q}_b = \overline{\mathbf{A}}_{bu}\mathbf{u} + \overline{\mathbf{A}}_{bw}\mathbf{w}$ ($\overline{\mathbf{A}}_{bu} = \mathbf{\Phi}^{\mathrm{T}}\mathbf{A}_{bu}$ and $\overline{\mathbf{A}}_{bw} = \mathbf{\Phi}^{\mathrm{T}}\mathbf{A}_{bw}$) are the generalized buffeting force vector.

According to the random vibration theory, the power spectral density (PSD) matrices of the vectors of generalized modal response $\mathbf{q}$ and nodal displacement $\mathbf{X}$ have the following relations:

$$\mathbf{S}_q(\omega) = \mathbf{H}^*(\omega)\mathbf{S}_{Q_b}(\omega)\mathbf{H}^{\mathrm{T}}(\omega) \tag{4.75}$$

$$\mathbf{S}_X(\omega) = \mathbf{\Phi}\mathbf{H}^*(\omega)\mathbf{S}_{Q_b}(\omega)\mathbf{H}^{\mathrm{T}}(\omega)\mathbf{\Phi}^{\mathrm{T}} \tag{4.76}$$

where $\mathbf{H}(\omega)$ is the transfer function matrix:

$$\mathbf{H}(\omega) = [-\omega^2(\mathbf{I} + \overline{\mathbf{A}}_{se}) + i\omega\overline{\mathbf{C}} + \mathbf{\Lambda}]^{-1} \tag{4.77}$$

in which the subscript $*$ and T denote the complex conjugate and transpose, respectively.

The PSD matrix of the generalized buffeting forces is given by:

$$\mathbf{S}_{Q_b}(\omega) = 0.25\rho^2 \overline{U}^2 (\overline{\mathbf{A}}_{bu}\mathbf{S}_{uu}\overline{\mathbf{A}}_{bu}^{\mathrm{T}} + \overline{\mathbf{A}}_{bw}\mathbf{S}_{ww}\overline{\mathbf{A}}_{bw}^{\mathrm{T}}) \tag{4.78}$$

where $\mathbf{S}_{uu}$ and $\mathbf{S}_{ww}$ are the PSD matrices of $\mathbf{u}$ and $\mathbf{w}$ components, respectively.

If the aerodynamic admittance functions are taken into consideration, the above PSD matrix includes the aerodynamic admittance functions. The power spectra of wind components $\mathbf{u}$ and $\mathbf{w}$ are functions of the circular frequency ω. The cross-spectral density functions of the wind component between two points can be expressed in a conventional form [24]:

$$S_{uu(ww)}(z_1, z_2, \omega) = \sqrt{S_{uu(ww)}(z_1, \omega)S_{uu(ww)}(z_2, \omega)e^{-\hat{f}_{u(w)}}} \tag{4.79}$$

where:

z_1 and z_2 denote the two points;
$S_{uu(ww)}(z, \omega)$ is the auto spectra identified from the wind characteristics study;
$e^{-f^{u(w)}}$ is the coherence function of fluctuating winds given by the wind characteristics study.

The components of the matrices S_q and S_x can be expressed as:

$$S_{q_{ij}}(\omega) = \sum_{k=1}^{m}\sum_{l=1}^{m} H_{ik}^{*}(\omega)S_{Qb_{kl}}(\omega)H_{jl}(\omega) \tag{4.80}$$

$$S_{X_i}(\omega) = \sum_{k=1}^{m}\sum_{l=1}^{m} \varphi_{ik}S_{q_{kl}}(\omega)\varphi_{il} \tag{4.81}$$

The variances of the *i*th generalized modal response and nodal displacement are thus given by:

$$\sigma_{q_{ii}}^2 = \int_0^\infty S_{q_{ii}}(\omega)d\omega \tag{4.82}$$

$$\sigma_{X_i}^2 = \int_0^\infty S_{X_i}(\omega)d\omega = \sum_{k=1}^{m}\sum_{l=1}^{m} \varphi_{ik}\left(\int_0^\infty S_{q_{kl}}(\omega)d\omega\right)\varphi_{il} \tag{4.83}$$

Figure 4.10 provides the flow chart of 3-D finite element based buffeting analysis for a long-span bridge.

4.6 Simulation of Stationary Wind Field

In order to carry out the buffeting analysis in the time domain, the stationary wind field of turbulent winds should be simulated, and this is generally represented by turbulence wind components $u(t)$, $v(t)$ and $w(t)$. Among various simulation methods, the spectral representation methods appear to be most widely used because they are fast and conceptually straightforward. In this section, the simulation of a stationary wind field mainly concerns the simulation of one-dimensional, multivariate, stationary stochastic process, and the algorithm of the spectral representation method proposed by Shinozuka and Jan [25] and Deodatis [26] is introduced.

Consider a set of n one-dimensional stationary stochastic process $\{u_j^0(t)\}(j = 1, 2, \ldots, n)$ with zero as the mean value, where the superscript 0 denotes the target function. The two-sided target cross-spectral density function $\mathbf{S}^0(\omega)$ of the stochastic process is given by:

$$\mathbf{S}^0(\omega) = \begin{bmatrix} S_{11}^0(\omega) & S_{12}^0(\omega) & \cdots & S_{1n}^0(\omega) \\ S_{21}^0(\omega) & S_{22}^0(\omega) & \cdots & S_{2n}^0(\omega) \\ \vdots & \vdots & \ddots & \vdots \\ S_{n1}^0(\omega) & S_{n2}^0(\omega) & \cdots & S_{nn}^0(\omega) \end{bmatrix} \tag{4.84}$$

According to Shinozuka and Jan [25] and Deodatis [26], the stochastic process $\{u_j^0(t)\}$ $(j = 1, 2, \ldots, n)$ can be simulated by the following series:

$$u_j(t) = 2\sqrt{\Delta\omega}\sum_{m=1}^{j}\sum_{l=1}^{N} \left|H_{jm}(\omega_{ml})\cos(\omega_{ml}t - \theta_{jm}(\omega_{ml}) + \phi_{ml})\right| \tag{4.85}$$

where:

N is a sufficiently large number;
$\Delta\omega = \omega_{up}/N$ is the frequency increment;

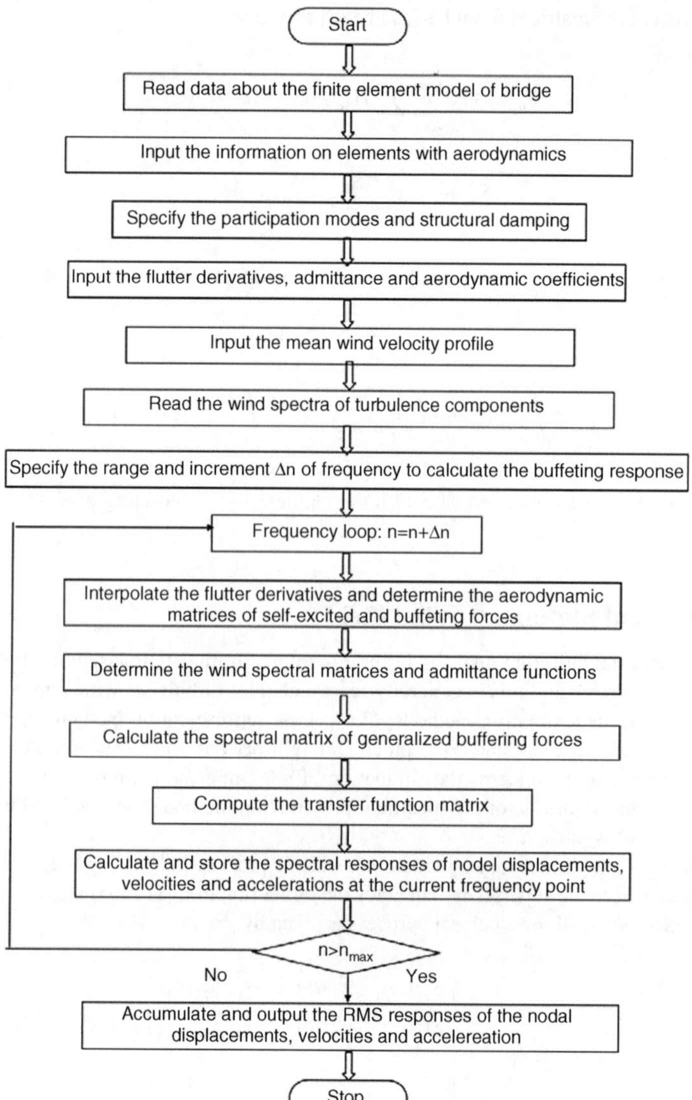

Figure 4.10 Flow chart of 3-D finite element based buffeting analysis for a long-span bridge.

ω_{up} is the upper cutoff frequency, with the condition that, when $\omega > \omega_{up}$, the value of $S^0(\omega)$ is trivial;
ϕ_{ml} is the sequence of independent random phase angles, uniformly distributed over the interval $[0, 2\pi]$;
ω_{ml} is the double-indexing frequency:

$$\omega_{ml} = (l-1)\Delta\omega + \frac{m}{n}\Delta\omega \quad l = 1, 2, \ldots, N \tag{4.86}$$

$H_{\mathrm{jm}}(\omega)$ is a typical element of the matrix $H(\omega)$, which is defined as Cholesky decomposition of cross-spectral density matrix $S^0(\omega)$:

$$S^0(\omega) = H(\omega)H^{T*}(\omega) \tag{4.87}$$

$$H(\omega) = \begin{bmatrix} H_{11}(\omega) & 0 & \cdots & 0 \\ H_{21}(\omega) & H_{22}(\omega) & \cdots & 0 \\ \cdots & \cdots & \cdots & \cdots \\ H_{n1}(\omega) & H_{n2}(\omega) & \cdots & H_{nn}(\omega) \end{bmatrix} \tag{4.88}$$

$\theta_{jm}(\omega)$ is the complex angle of $H_{jm}(\omega)$ and is given by:

$$\theta_{jm}(\omega) = \tan^{-1}\left\{\frac{\mathrm{Im}\left[H_{jm}(\omega)\right]}{\mathrm{Re}\left[H_{jm}(\omega)\right]}\right\} \tag{4.89}$$

where $\mathrm{Im}[H_{jm}(\omega)]$ and $\mathrm{Re}[H_{jm}(\omega)]$ are the imaginary and real parts of the complex function $H_{jm}(\omega)$, respectively.

It has been proved that both ensemble and temporal auto- and cross- correlation functions of any sample function obtained by Equation 4.85 will approach to the target auto- and cross-correlation functions as $\Delta\omega \to 0$ and $N \to \infty$.

In order to avoid aliasing, the time step Δt has to obey the condition:

$$\Delta t \leq \frac{2\pi}{2\omega_{up}} \tag{4.90}$$

The period of the sample functions expressed by Equation 4.85 is:

$$T_0 = \frac{2\pi n}{\Delta\omega} = \frac{2\pi N}{\omega_{up}} \tag{4.91}$$

The ergodicity of the results of Equation 4.85 has been proved by Deodatis [26]. One can be confident that the one-dimensional stationary, multivariate, Gaussian stochastic process can be simulated quite well by means of Equation 4.85, when the cross-spectral density matrix is given and the values of the parameters N, ω_{up} and $\Delta\omega$ are properly chosen.

However, the number of wind velocity processes to be simulated is often very large in the buffeting analysis of long-span cable-supported bridges, when the stochastic wind loads on all major structural components such as bridge deck, cables, and towers shall be taken into consideration. It is difficult to do the simulation using the traditional method above, so an improved algorithm of the spectral representation method, which is considered suitable for the simulation of wind velocity filed on long-span cable-supported bridges, has been proposed by Ding *et al.* [27]. An interpolation approximation is introduced to simplify the computation of the lower triangular matrix with the Cholesky decomposition of the cross-spectral density matrix. Fast Fourier transform technique is used to further enhance the efficiency of computation in the simulation of stationary wind field.

4.7 Buffeting Analysis in the Time Domain

The buffeting response of a bridge induced by both self-excited and buffeting forces can also be calculated in the time domain. Compared to the frequency domain approach, time domain analysis offers the benefit of capturing the effects of non-linearities of both structural and aerodynamic origins and also the influence of non-stationary features in the approaching wind in the analysis [17]. The theoretical background of time domain buffeting analysis is introduced in this section.

The self-excited forces per unit span can be expressed in terms of convolution integrals [28]:

$$L_{se}(t) = \frac{1}{2}\rho\bar{U}^2 \int_{-\infty}^{t} \left[I_{L_{seh}}(t-\tau)h(\tau) + I_{L_{sep}}(t-\tau)p(\tau) + I_{L_{se\alpha}}(t-\tau)\alpha(\tau)\right]d\tau \quad (4.92a)$$

$$D_{se}(t) = \frac{1}{2}\rho\bar{U}^2 \int_{-\infty}^{t} \left[I_{D_{seh}}(t-\tau)h(\tau) + I_{D_{sep}}(t-\tau)p(\tau) + I_{D_{se\alpha}}(t-\tau)\alpha(\tau)\right]d\tau \quad (4.92b)$$

$$M_{se}(t) = \frac{1}{2}\rho\bar{U}^2 \int_{-\infty}^{t} \left[I_{M_{seh}}(t-\tau)h(\tau) + I_{M_{sep}}(t-\tau)p(\tau) + I_{M_{se\alpha}}(t-\tau)\alpha(\tau)\right]d\tau \quad (4.92c)$$

where $I_{()}$ is the impulse function of the self-excited forces, in which the subscript represents the corresponding force component.

The impulse functions can be obtained using the flutter derivatives measured from wind tunnel tests and the rational function approximation approach [17].

The relationship between the impulse functions and flutter derivatives in the frequency domain can be obtained by taking the Fourier transform of Equation 4.92 and comparing with Equation 4.33:

$$\bar{I}_{L_{seh}} = 2k^2(H_4^* + iH_1^*);\quad \bar{I}_{L_{sep}} = 2k^2(H_6^* + iH_5^*);\quad \bar{I}_{L_{se\alpha}} = 2k^2b(H_3^* + iH_2^*) \quad (4.93a)$$

$$\bar{I}_{D_{seh}} = 2k^2(P_6^* + iP_5^*);\quad \bar{I}_{D_{sep}} = 2k^2(P_4^* + iP_1^*);\quad \bar{I}_{D_{se\alpha}} = 2k^2b(P_3^* + iP_2^*) \quad (4.93b)$$

$$\bar{I}_{M_{seh}} = 2k^2b(A_4^* + iA_1^*);\quad \bar{I}_{M_{sep}} = 2k^2b(A_6^* + iA_5^*);\quad \bar{I}_{M_{se\alpha}} = 2k^2b^2(A_3^* + iA_2^*) \quad (4.93c)$$

where the over-bar denotes the Fourier transform operator and $b = 0.5B$.

From the classical airfoil theory, the aerodynamic transfer functions in Equation 4.93 (the Fourier transform of the impulse functions) can be reasonably approximated by the rational functions. The frequency independent coefficients of the rational functions can then be determined by the non-linear least-squares method, using the flutter derivatives at different reduced frequencies. The impulse functions can be finally obtained by the inverse Laplace transform [17,29].

Because the aerodynamic derivatives are normally obtained from wind tunnel tests at discrete values of the reduced frequency k, some numerical methods are needed to extend these discrete values into continuous functions of the reduced frequency for time domain analysis.

Similarly, the buffeting forces per unit length can be expressed in terms of convolution integrals involving the aerodynamic impulse functions and fluctuating wind velocities:

$$L_b(t) = -\frac{1}{2}\rho\bar{U}^2 \int_{-\infty}^{t} \left[I_{L_{bu}}(t-\tau)\frac{u(\tau)}{\bar{U}} + I_{L_{bw}}(t-\tau)\frac{w(\tau)}{\bar{U}}\right] d\tau \quad (4.94a)$$

$$D_b(t) = -\frac{1}{2}\rho\bar{U}^2 \int_{-\infty}^{t} \left[I_{D_{bu}}(t-\tau)\frac{u(\tau)}{\bar{U}} + I_{D_{bw}}(t-\tau)\frac{w(\tau)}{\bar{U}}\right] d\tau \quad (4.94b)$$

$$M_b(t) = -\frac{1}{2}\rho\bar{U}^2 \int_{-\infty}^{t} \left[I_{M_{bu}}(t-\tau)\frac{u(\tau)}{\bar{U}} + I_{M_{bw}}(t-\tau)\frac{w(\tau)}{\bar{U}}\right] d\tau \quad (4.94c)$$

where $I_{()}$ is the impulse function of the buffeting force with the subscript representing the corresponding force component.

Similar to the self-excited forces, the impulse functions of the buffeting forces can be obtained by the inverse Laplace transform of the aerodynamic transfer functions, whereas the aerodynamic transfer functions of the buffeting forces can be related to aerodynamic coefficients and aerodynamic admittances as:

$$\bar{I}_{L_{bu}} = 4bC_L\chi_{L_{bu}}; \quad \bar{I}_{L_{bw}} = 2b(C'_L + C_D)\chi_{L_{bw}} \tag{4.95a}$$

$$\bar{I}_{D_{bu}} = 4bC_D\chi_{D_{bu}}; \quad \bar{I}_{D_{bw}} = 2bC'_D\chi_{D_{bw}} \tag{4.95b}$$

$$\bar{I}_{M_{bu}} = 8b^2C_M\chi_{M_{bu}}; \quad \bar{I}_{D_{bw}} = 4b^2C'_M\chi_{M_{bw}} \tag{4.95c}$$

Self-excited forces are usually assumed to be fully correlated. Thus, the self-excited force for an entire beam element can be expressed as:

$$L^e_{se}(t) = L^c_{se}(t)L \tag{4.96a}$$

$$D^e_{se}(t) = D^c_{se}(t)L \tag{4.96b}$$

$$M^e_{se}(t) = M^c_{se}(t)L \tag{4.96c}$$

where the superscript c indicates the center of the element.

The span-wise correlation of the buffeting forces in an element should be considered, and the buffeting forces for an entire beam element can be expressed as:

$$L^e_b(t) = L\int_0^t \left[J_{L_{bu}}(t-\tau)L^c_{bu}(\tau) + J_{L_{bw}}(t-\tau)L^c_{bw}(\tau)\right]d\tau \tag{4.97a}$$

$$D^e_b(t) = L\int_0^t \left[J_{D_{bu}}(t-\tau)D^c_{bu}(\tau) + J_{D_{bw}}(t-\tau)D^c_{bw}(\tau)\right]d\tau \tag{4.97b}$$

$$M^e_b(t) = L\int_0^t \left[J_{M_{bu}}(t-\tau)M^c_{bu}(\tau) + J_{M_{bw}}(t-\tau)M^c_{bw}(\tau)\right]d\tau \tag{4.97c}$$

where:

L^c_{bu} and L^c_{bw} are the first and second terms in Equation 4.94a at the center of the element;
$J_{()}$ is the impulse function, whose Fourier transform counterpart is referred to as the joint acceptance function:

$$\bar{J}^2_{L_{bu}} = \frac{1}{L^2}\int_0^L\int_0^L coh_{L_{bu}}(x_1, x_2, \omega)\mathrm{d}x_1\mathrm{d}x_2 \tag{4.98}$$

where:

$$coh_{L_{bu}}(x_1, x_2, \omega) = \frac{S_{L_{bu}}(x_1, x_2, \omega)}{S_{L^c_{bu}}(\omega)} \tag{4.99}$$

It is the span-wise coherence of the buffeting lift force component; $S_{L_{bu}}(x_1, x_2, \omega)$ is the cross-spectra between the buffeting forces at two different positions x_1 and x_2; and $S_{L^c_{bu}}(\omega)$ is the auto-spectral density of the buffeting force at the center of the element.

Once the local nodal self-excited forces and buffeting forces are obtained from Equation 4.96 and Equation 4.97, they can be converted into the global coordinate system using the coordinate transformation matrix. As a result, the global nodal buffeting force and self-excited force vectors $\mathbf{F}_{se}$ and $\mathbf{F}_b$ in Equation 4.62 can be obtained. The solution of the equation of motion (Equation 4.62) in the time domain can be obtained by the Newmark-Beta method. Because the self-excited forces are dependent on the motion, iteration is needed for each time-step until certain convergence criterion is satisfied [29].

4.8 Effective Static Loading Distributions

4.8.1 Gust Response Factor and Peak Factor

Gust response factor G is a commonly used term in wind engineering and may be defined as the ratio of the expected maximum response of the structure in a defined period (e.g. ten minutes or one hour) to the mean or time-averaged response in the same time period [30–32]. Clearly, this term really has meaning only in stationary or near-stationary winds. The expected maximum response of the structure can be buffeting-induced maximum bending moment, deflection, or stress. It can be written as:

$$\hat{X} = \overline{X} + g\sigma_x \tag{4.100}$$

where:

$\hat{X}$ and $\overline{X}$ are the maximum and mean response of the structure;
σ_x is the standard deviation of the structural response;
g is a peak factor which depends on the frequency range of the response and the time interval for which the maximum value is calculated.

From Equation 4.100, the gust response factor can be expressed as:

$$G = \frac{\hat{X}}{\overline{X}} = 1 + g\frac{\sigma_x}{\overline{X}} \tag{4.101}$$

For a stationary Gaussian response to wind, Davenport [33] derived the following expression for the peak factor g:

$$g = \sqrt{2\log_e(\upsilon T)} + \frac{0.577}{\sqrt{2\log_e(\upsilon T)}} \tag{4.102}$$

where:

υ is the cycling rate or effective frequency for the response (often conservatively taken as the natural frequency of the structure);
T is the time period over which the maximum value is required.

For building structures, alongwind-induced responses are often estimated in terms of the sum of three components: mean component, background component and resonant component. Accordingly, the gust response factor can be expressed as:

$$G = \frac{\hat{X}}{\overline{X}} = 1 + 2g\frac{\sigma_u}{\overline{U}}\sqrt{B+R} \tag{4.103}$$

where:

σ_u is the standard deviation of turbulent wind in along-wind direction;
B is the background factor representing the quasi-static response caused by gusts below the first natural frequency of the structure;
R is the resonant factor describing the resonant response caused by gusts near the first natural frequency of the structure.

Equations 4.101 and 4.103 are used in many codes and standards for wind loading and, particularly, for alongwind dynamic loading of building structures. This is an approximate approach which works reasonably well for some structures and load effects, such as the base bending moment of tall buildings. However, in other cases it gives significant errors and should be used with caution [32].

4.8.2 *Effective Static Loading Distributions*

Effective static loading distributions are those loadings that produce the correct expected values of peak load effects generated by the fluctuating wind loading. For building structures, effective static peak loading distributions can be derived from three components: mean components, background component and resonant component. The main advantage of the effective static loading distribution approach is that the distributions can be applied to a static structural analysis computer program for use in final structural design and can be used relatively easily by engineers and incorporated in design codes and standards. The following introduction on this topic refers to Holmes's book [32].

By taking alongwind (drag) response of a long-span bridge deck as an example, and regarding the bridge deck as a continuous beam, the mean wind force per unit length on the bridge deck can be written as:

$$\bar{f}(x) = \frac{1}{2}\rho\bar{U}^2(x)B(x)C_D(x) \tag{4.104}$$

The mean value of any load effect can be obtained by integrating the local force with the influence line along the length of the bridge deck. If the purpose is to derive an effective static loading, Equation 4.104 is already in the correct form.

The background wind loading is the quasi-static loading produced by fluctuations due to turbulence, but with frequencies too low to excite any resonant response. The load-response correlation formula derived by Kasperski and Niemann [34] can be used to derive the effective background fluctuating loading distribution. In the form of a continuous distribution, this term can be written as:

$$f_B(x) = g_B\rho(x)\sigma_p(x) \tag{4.105}$$

where:

g_B is the peak factor for the background response;
$\sigma_p(x)$ is the standard deviation of the fluctuating load at position x;
$\rho(x)$ is the correlation coefficient between the fluctuating load at position x and the load effect of interest [32]:

$$\rho(x) = \frac{\int_s^L \overline{f'(x)f'(x_1)}I_r(x_1)B(x_1)\mathrm{d}x_1}{\sqrt{\overline{f'^2}(x)}\cdot\sqrt{\int_s^L\int_s^L \overline{f'(x_1)f'(x_2)}I_r(x_1)I_r(x_2)B(x_1)B(x_2)\mathrm{d}x_1\mathrm{d}x_2}} \tag{4.106}$$

where:

$I_r(x)$ is the influence line for the load effect r, as a function of position x;
$f'(x)$ is the fluctuating alongwind force on the bridge deck;
$\overline{f'(x)f'(x_1)}$ is the covariance for the fluctuating alongwind forces at positions x and x_1.

The equivalent load distribution for the resonant response in the first mode of vibration can be represented as a distribution of inertial forces over the length of the bridge deck:

$$f_R(x) = g_R m(x)(2\pi n_1)^2 \sqrt{\overline{a'^2}}\,\phi_1(x) \tag{4.107}$$

where:

$f_R(x)$ is the equivalent load distribution for the resonant response at position x;
g_R is the peak factor for resonant response;
$m(x)$ is the mass per unit length at position x;
n_1 is the first mode natural frequency;
$\sqrt{\overline{a'^2}}$ is the standard deviation of modal coordinate (resonant contribution only);
$\phi_1(x)$ is the first mode shape.

Determining the standard deviation of modal coordinate due to the resonant contribution only requires knowledge of the spectral density of the fluctuating forces, the correlation of those forces at the natural frequency and the modal damping and stiffness. The total effective loading distribution can be obtained by:

$$f_c(x) = \bar{f}(x) + w_{back} f_B(x) + w_{res} f_R(x) \tag{4.108}$$

where w_{back} and w_{res} are the weight factors for background and resonant response, respectively.

Assuming that the fluctuating background and resonant components are uncorrelated with each other, the weight factors can be calculated by:

$$|w_{back}| = \frac{g_B \sigma_{r,B}}{\sqrt{g_B^2 \sigma_{r,B}^2 + g_R^2 \sigma_{r,R}^2}} \tag{4.109a}$$

$$|w_{res}| = \frac{g_R \sigma_{r,R}}{\sqrt{g_B^2 \sigma_{r,B}^2 + g_R^2 \sigma_{r,R}^2}} \tag{4.109b}$$

where:

$\sigma_{r,B}$ and $\sigma_{r,R}$ are the standard deviation of the background response and the resonant response, respectively;
the subscript r means the load effect consistent with the response.

The method of effective static loading distribution approach described above can be applied to common bridges in principle. However, for long-span cable-supported bridges, it is often the case that several modes of vibration must be considered. Holmes [35] proposed a method of combining inertial force distributions from more than one resonant mode of vibration. By neglecting the background response, the total effective loading distribution for the peak load effect, r, is [32]:

$$f_{eff}(x) = \bar{f}(x) + m(x)\sum_{j=1}^{N} W_j \phi_j(x) \tag{4.110}$$

where W_j is the peak inertial force in the mode j.

4.9 Case Study: Stonecutters Bridge

4.9.1 Dynamic and Aerodynamic Characteristics of Stonecutters Bridge

The finite element modeling technique for long-span cable-supported bridges and its application for developing a 3D finite element model of the Stonecutters Bridge have been introduced in Chapter 3. The frequency domain flutter and buffeting analyses of the Stonecutters Bridge are introduced in this section as a case study using that model [16,36].

The first 50 natural modes of the bridge were obtained from the modal analysis of the finite element model. The first 50 natural frequencies, as well as the corresponding mode shape descriptions, are listed in Table 4.2. It can be seen from this table that the first 50 natural frequencies of the bridge range from 0.145 Hz to 1.447 Hz. The lowest natural frequency of 0.145 Hz corresponds to the first horizontal mode of vibration, in which the motion of the bridge deck is almost in symmetry in the main span. The first vertical vibration mode dominated by the bridge deck is almost symmetric in the main span at a natural frequency of 0.201. The first torsional mode of vibration dominated by the bridge deck is almost symmetric in the main span at a natural frequency of 0.425 Hz.

The aerodynamic coefficients of the Stonecutters bridge deck without traffic were obtained from the section model tests in a wind tunnel. The drag, lift, and moment coefficients for the cross section of the entire bridge deck at zero degree of attack angle are $C_D = 0.073$, $C_L = -0.155$ and $C_M = -0.018$. The derivatives of the drag, lift, and moment aerodynamic coefficients with wind angle of attack at zero degree are $dC_D/d\alpha = 0.069$, $dC_L/d\alpha = 2.510$ and $dC_M/d\alpha = 0.539$. These aerodynamic coefficients are normalized by the overall deck width of 53.3 m.

The drag coefficient of the bridge tower is taken as $C_D = 0.9$ along the entire height of the tower and is normalized by the actual width of the tower perpendicular to the wind direction. The drag coefficient of the piers is taken as $C_D = 1.1$ for wind perpendicular to the bridge longitudinal axis and is normalized by the actual width of the piers perpendicular to the wind direction. The drag coefficient of the stay cables is taken as $C_D = 0.8$ for wind perpendicular to the bridge longitudinal axis and is normalized by the actual diameter of the stay cables. As the cross sections of the stay cables, towers and piers are symmetric, their aerodynamic lift and moment coefficients for wind perpendicular to the bridge longitudinal axis are taken as zero.

The flutter derivatives of the Stonecutters Bridge deck without traffic were also obtained from the section model tests in a wind tunnel. Only the flutter derivatives H_i^* and $A_i^*(i = 1-4)$ are available, and they are listed in Table 4.3 and plotted in Figure 4.11. Since the flutter derivatives related to the lateral motion of the bridge deck and the flutter derivatives H_5^*, H_6^*, A_5^*, A_6^* are not available, they are calculated based on the quasi-steady theory in terms of the aerodynamic coefficients:

$$P_1^* = -\frac{1}{K}C_D, \quad P_2^* = \frac{1}{2K}C_D', \quad P_3^* = \frac{1}{2K^2}C_D' \tag{4.111a}$$

$$P_5^* = \frac{1}{2K}C_D', \quad H_5^* = \frac{1}{K}C_L, \quad A_5^* = -\frac{1}{K}C_M \tag{4.111b}$$

$$P_4^* = P_6^* = H_6^* = A_6^* = 0 \tag{4.111c}$$

The flutter derivatives and the aerodynamic coefficients mentioned above are assumed to be uniform along the bridge deck in the both flutter and buffeting analysis.

4.9.2 Flutter Analysis of Stonecutters Bridge

The aerodynamically coupled flutter analysis is carried out in the complex frequency domain using in-house software packages with theoretical background, as introduced in Section 4.4. By using the modal coordinates of the bridge, the governing equation of the bridge for the flutter analysis is converted into a

Table 4.2 Dynamic characteristics of Stonecutters Bridge

Mode no.	Natural frequency (Hz)	Period (sec)	Mode shape description
1	0.145	6.875	1.symmetric horizontal, deck
2	0.190	5.260	1.asymmetric horizontal, towers
3	0.197	5.073	1.symmetric horizontal, towers
4	0.201	4.968	1.symmetric vertical, deck
5	0.246	4.072	1.asymmetric vertical, deck
6	0.302	3.311	2.asymmetric vertical, deck
7	0.311	3.217	2.symmetric vertical, deck
8	0.361	2.772	1.asymmetric horizontal, deck
9	0.376	2.660	3.asymmetric vertical, deck
10	0.422	2.372	3. symmetric vertical, deck
11	0.425	2.354	1.symmetric torsional, deck
12	0.444	2.250	Horizontal, deck and piers, west side span
13	0.452	2.214	Horizontal, deck and piers, east side span
14	0.490	2.040	4.asymmetric vertical, deck
15	0.525	1.903	2.symmetric horizontal, towers
16	0.535	1.869	4.symmetric vertical, deck
17	0.579	1.728	2.asymmetric horizontal, towers
18	0.591	1.693	1.asymmetric torsional, deck
19	0.592	1.688	5.symmetric vertical, deck
20	0.620	1.613	5.asymmetric vertical, deck
21	0.682	1.466	2.symmetric torsional, deck
22	0.694	1.440	6.symmetric vertical, deck
23	0.729	1.372	Symmetric horizontal and torsional, deck
24	0.768	1.303	6.asymmetric vertical, deck
25	0.827	1.209	2.asymmetric torsional, deck
26	0.842	1.188	7.symmetric vertical, deck
27	0.900	1.111	Horizontal, deck and piers, west & east side spans
28	0.907	1.103	Horizontal, deck and piers, west & east side spans
29	0.912	1.096	7.asymmetric vertical, deck
30	0.948	1.055	3.symmetric torsional, deck
31	0.969	1.032	Asymmetric horizontal, deck and tower
32	0.973	1.027	8.symmetric vertical, deck
33	1.003	0.998	1.asymmetric longitudinal, tower
34	1.034	0.967	Symmetric longitudinal, deck and tower
35	1.049	0.953	8.asymmetric vertical, deck
36	1.072	0.933	3.asymmetric torsional, deck
37	1.129	0.886	9.symmetric vertical, deck
38	1.132	0.884	Symmetric horizontal, deck
39	1.193	0.839	4.symmetric torsional, deck
40	1.216	0.822	9.asymmetric vertical, deck
41	1.245	0.803	Asymmetric horizontal, deck and tower
42	1.254	0.797	10. symmetric vertical, deck
43	1.265	0.791	10.asymmetric vertical, deck
44	1.298	0.771	4.asymmetric torsional, deck
45	1.314	0.761	Symmetric vertical, towers and deck
46	1.345	0.744	Asymmetric vertical, towers and deck
47	1.364	0.733	11.symmetric vertical, deck
48	1.379	0.725	5.symmetric torsional, deck
49	1.407	0.711	11.asymmetric vertical, deck
50	1.447	0.691	Symmetric horizontal, deck and tower

Table 4.3 Flutter derivatives of the Stonecutters Bridge deck

(a) Flutter derivatives H_1^*, $H_4^* A_1^*$, A_4^*

U/fB	H_1^*	H_4^*	A_1^*	A_4^*
0.00	0.000	0.000	0.000	0.000
2.35	−0.280	−0.020	0.040	0.044
3.50	−0.390	0.000	0.110	0.080
4.71	−0.550	−0.080	0.170	0.100
5.88	−0.760	−0.170	0.240	0.120
6.93	−1.020	−0.330	0.310	0.140
8.08	−1.310	−0.430	0.360	0.160
9.18	−1.570	−0.470	0.400	0.124
10.44	−1.830	−0.450	0.430	0.064
11.56	−2.040	−0.470	0.460	0.050
13.00	−2.370	−0.500	0.500	0.036

(b) Flutter derivatives H_2^*, $H_3^* A_2^*$, A_3^*

U/fB	H_2^*	H_3^*	A_2^*	A_3^*
0.00	0.000	0.000	0.000	0.000
2.02	−0.210	−0.320	−0.037	−0.014
2.88	−0.390	−0.430	−0.083	0.024
3.81	−0.440	−0.590	−0.119	0.069
5.00	−0.360	−0.900	−0.172	0.131
6.32	−0.180	−1.440	−0.237	0.204
7.41	0.120	−2.110	−0.339	0.308
9.52	0.380	−2.850	−0.403	0.415
10.93	0.580	−3.920	−0.419	0.569
13.44	−2.750	−6.600	−0.543	0.956

complex characteristic equation with only two variables. A single parameter-searching method is then used to find the lowest critical wind speed without choosing participating modes beforehand. The major participating modes of vibration causing the flutter instability and the phase angles between the participating modes of vibration can also be found.

The effects of the towers, pier shafts and stay cables on the flutter instability of the bridge are also taken into consideration. The flutter derivatives for these bridge components are calculated based on the quasi-steady theory. Since the cross-sections of the tower, pier shafts and stay cables are basically circular, only the drag coefficients are needed to calculate the flutter derivatives required. Since the method used for flutter analysis has no limit on the number of modes, the first 50 modes of vibration of the bridge are employed as the participating modes for the flutter analysis of the bridge.

The lowest critical wind speed for the bridge is computed as being 230.4 m/s at a reduced velocity of 13.08 and a flutter frequency of 0.331 Hz. Such a high wind speed will not appear in practice. Therefore, the computed critical wind speed is of theoretical interest only, and it indicates no flutter instability problem for the concerned bridge.

Corresponding to the lowest critical wind speed or the flutter frequency, a flutter eigenvector can be found. From this, one may observe the distribution of modal motion and modal energy over all the participating modes of vibration and the distribution of modal phase angle among all the participating modes of vibration. From these distributions, the dominant modes of vibration causing the flutter instability can then be identified.

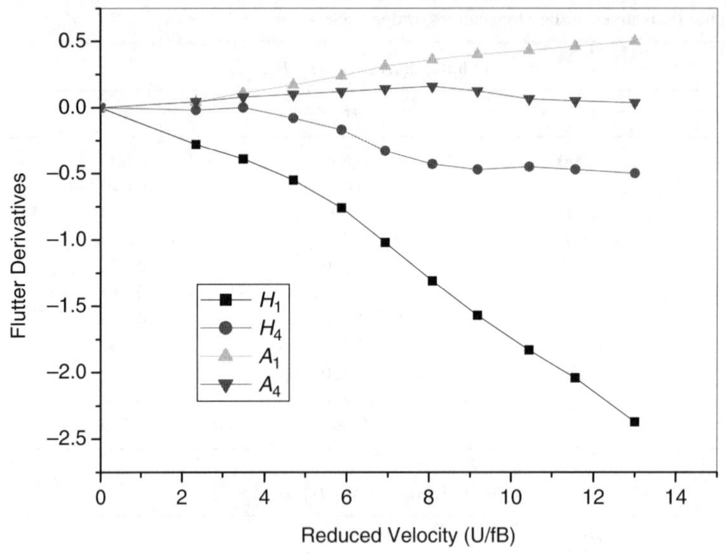

(a) Flutter derivatives H_1^* , H_4^* A_1^* , z_4^*

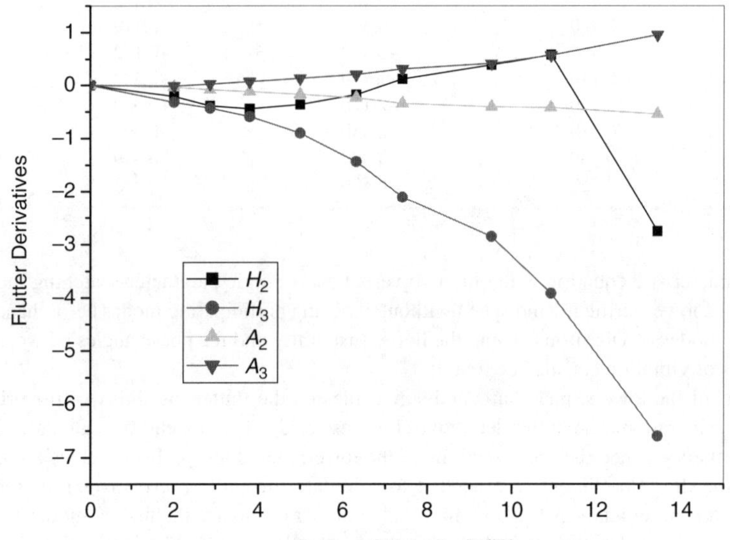

(b) Flutter derivatives H_2^* , H_3^* A_2^* , A_3^*

Figure 4.11 Flutter derivatives of Stonecutters bridge deck without traffic.

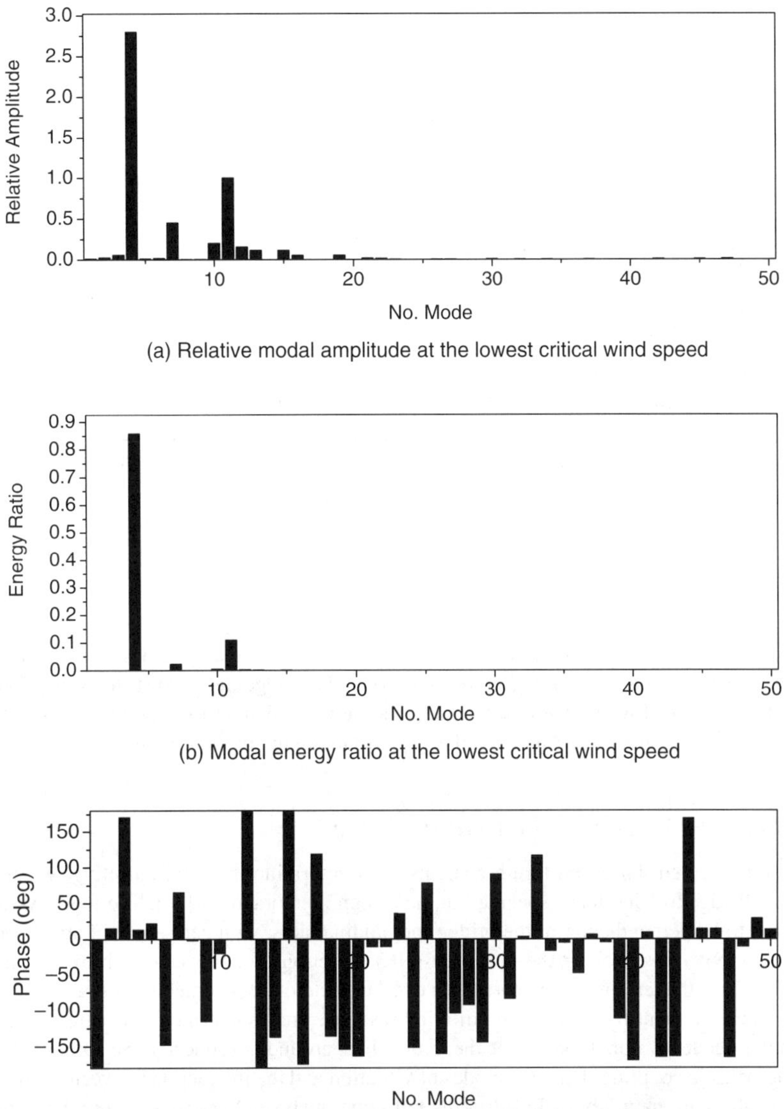

(a) Relative modal amplitude at the lowest critical wind speed

(b) Modal energy ratio at the lowest critical wind speed

(c) Modal phase angle at the lowest critical wind speed

Figure 4.12 Modal information of the bridge at the lowest critical wind speed.

Figure 4.12 shows the distribution of relative modal amplitude, modal energy ratio and modal phase angle over the 50 participating modes of vibration, respectively. It is noted that vibration modes 4 (first symmetric vertical mode) and 11 (first symmetric torsional mode) are the two major participating modes dominating the flutter instability of the bridge, with almost the same phase angle. Although the lateral vibration modes of the bridge deck also participate in the flutter motion, the degree of their participations is very small.

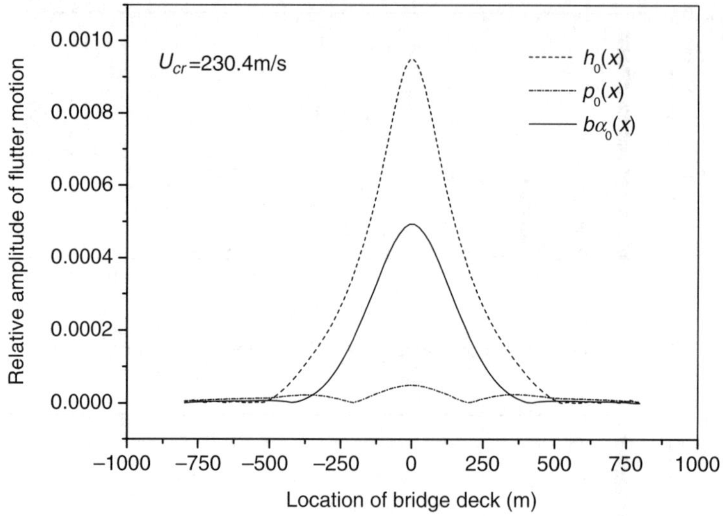

Figure 4.13 Relative amplitudes of deck motion along the bridge longitudinal axis.

Apart from the modal information, Figure 4.13 shows the relative amplitudes of the vertical, lateral and torsional displacement responses (flutter motion) of the bridge deck along the bridge longitudinal axis at the lowest critical wind speed. Note again that the vertical and torsional vibrations dominate the flutter motion of the bridge deck. More details can be found in the literature [16].

4.9.3 Buffeting Analysis of Stonecutters Bridge

The 3-D finite element-based buffeting analysis is to determine the total buffeting response of the Stonecutters Bridge to wind loads specified in the design memorandum [36]. The mean wind velocity is considered to be perpendicular to the bridge longitudinal axis, with zero angle of attack from either the southwest (SW) direction or the northeast (NE) direction. For the sake of brevity, only the NE direction, in which the terrain is considered as a overland fetch, is concerned in this case study.

The 3-D finite element-based coupled buffeting response analysis of the Stonecutters Bridge is carried out in the frequency domain, with the theoretical background introduced in Section 4.5. Both structural and aeroelastic couplings between modes of vibration and the interaction between the bridge deck, towers and cables are naturally included in the buffeting analysis. Varying structural properties along the bridge deck are taken into consideration. The structural damping ratio of the bridge is assumed to be 0.36% for each mode of vibration in the buffeting analysis, according to the design specification. This level of structural damping is considered a realistic value. The first 50 modes of vibration (excluding local cable modes) are included in the buffeting analysis.

The one-hour mean wind velocity profile $\overline{U}(z)$ in m/s for wind from the NE direction is taken as:

$$\overline{U}(z) = 0.66 \cdot 35 \cdot \left(\frac{z}{10}\right)^{0.29} \tag{4.112}$$

The ten-minute mean wind velocity profile $\overline{U}(z)$ in m/s for wind from the NE direction is taken as:

$$\overline{U}_{10}(z) = \overline{U}_{10}(10) \cdot \left(\frac{z}{10}\right)^{0.29} \tag{4.113}$$

The ten-minute mean wind velocity $\overline{U}_{10}(10)$ at 10 m height above ground is defined as the reference wind velocity, and its basic value is 24.5 m/s for wind from the NE direction. The longitudinal turbulent intensity I_u for wind from the NE direction is given by the following expression:

$$I_u = \frac{\sigma_u}{\overline{U}_{10}(z)} = 0.37 \cdot \left(\frac{10}{z}\right)^{0.29} \tag{4.114}$$

where σ_u is also assumed to be constant with height.

For lateral (v) and vertical (w) turbulence intensities, the following relationships are assumed, based on analysis of typhoon data recorded at the Tsing Ma Bridge for the overland fetch:

$$I_v = 1.00 \cdot I_u; \quad I_w = 0.75 \cdot I_u \tag{4.115}$$

The power spectra of the longitudinal (u), lateral (v) and vertical (w) turbulent wind components are defined according to the von Kármán spectral model.

For longitudinal turbulent wind component:

$$\frac{f \cdot S_{uu}(f)}{\sigma_u^2} = \frac{4\dfrac{L_u \cdot f}{\overline{U}_{10}(z)}}{\left[1 + 70.8 \cdot \left(\dfrac{L_u \cdot f}{\overline{U}_{10}(z)}\right)^2\right]^{5/6}} \tag{4.116}$$

For lateral and vertical turbulent wind components:

$$\frac{f \cdot S_{vv,ww}(f)}{\sigma_{v,w}^2} = \frac{4\dfrac{L_{v,w} \cdot f}{\overline{U}_{10}(z)}\left[1 + 755 \cdot \left(\dfrac{L_{v,w} \cdot f}{\overline{U}_{10}(z)}\right)^2\right]}{\left[1 + 283 \cdot \left(\dfrac{L_{v,w} \cdot f}{\overline{U}_{10}(z)}\right)^2\right]^{11/6}} \tag{4.117}$$

where:

$S_{ii}(f)$ is the wind energy spectrum for turbulent wind components i;
f is the frequency in Hz;
L_i is the integral length scale for turbulent wind component i.

Turbulence integral scale is specified in the design memorandum as:

$$L_i = L_{i50} \cdot \left(\frac{z}{50}\right)^{0.45} \tag{4.118}$$

where L_{i50} represents the length scale at 50 m height above ground for turbulent wind component i, and $L_{u50} = 160$ m, $L_{v50} = L_{u50}/3$, $L_{w50} = L_{u50}/6$ are specified for all wind directions.

The root coherence, which defines the statistical dependency between the turbulent wind components at two different points, is given by the following expression:

$$\sqrt{coh} = \exp\left(-C \cdot \frac{D \cdot f}{\overline{U}_{10}(z)}\right) \tag{4.119}$$

Table 4.4 Values of the decay factor C

	Lateral separation	Longitudinal separation	Vertical separation
Longitudinal turbulent wind component	8	2	8
Lateral turbulent wind component	4	4	8
Vertical turbulent wind component	8	4	4

where the parameter C is the decay factor; and the parameter D is the distance between the two points. The values of the decay factor C for all 9 coherences are listed in Table 4.4.

Since the basic value of the ten-minute mean wind speed at 10 m height above ground is 24.4 m/s for wind from the N-E direction, the corresponding basic value of the ten-minute mean wind speed at the middle point of the main span of the bridge deck is 46 m/s. The mean wind response of the complete bridge is then computed for the deck-reference mean wind velocity of 46 m/s.

The mean wind displacement responses of the bridge are computed, and those at the key locations are listed in Table 4.5. The positions of the key locations are the same as shown in Figure 3.14. It is noted that the mean lateral, vertical and torsional displacement responses of the windward deck at the middle point of main span are −0.449 m, −0.202 m and −0.002 rad, respectively. The mean lateral, vertical and torsional displacements of the leeward deck at the middle point of main span are −0.449 m, −0.268 m and −0.002 rad, respectively. The mean lateral and vertical displacement responses for wind from the NE direction are relatively smaller compared with those for wind from the SW direction (as introduced in Chapter 3), because of smaller design mean wind speed. The mean lateral displacement at the top of the tower reaches −0.869 m. For other mean wind speeds, the mean wind displacement responses of the bridge at the key locations are also computed. Figure 4.14 shows the mean displacement responses of the windward deck at the middle of main span versus NE mean wind speed.

The coupled buffeting responses of the bridge are computed using the mean wind speed at the middle point of the main span of the bridge deck as a reference mean wind speed of 46 m/s as well. By using the Sears function as admittance functions for the bridge deck, the buffeting peak responses of the bridge are computed and the peak displacement and acceleration responses at the key locations are listed in Table 4.6 and Table 4.7, respectively, for the basic mean wind velocity of 46 m/s.

The peak buffeting response refers to the standard deviation buffeting response multiplied by a peak factor. The peak factor of the buffeting response is taken as 3.5 for the purposes of this case study. The total wind response, then, refers to the mean wind response plus the peak buffeting response.

Table 4.5 Mean wind displacement responses of the bridge at key locations (NE direction, V_{dref} = 46 m/s)

Location	Longitudinal (m)	Vertical (m)	Lateral (m)	Around x-axis (rad)	Around y-axis (rad)	Around z-axis (rad)
D1-windward	−1E-04	−0.202	−0.449	−0.002	−2E-06	−1E-06
D1-leeward	−4E-05	−0.268	−0.449	−0.002	−2E-06	−7E-05
D2-windward	0.014	−0.054	−0.294	−0.003	1E-03	−6E-04
D2-leeward	−0.013	−0.138	−0.294	−0.002	1E-03	−6E-04
D5-windward	0.006	0.013	−0.033	−9E-04	2E-04	1E-04
D5-leeward	2E-04	−0.012	−0.033	−9E-04	2E-04	−7E-05
T1-tower-top	0.046	−0.004	−0.869	−0.006	5E-05	−2E-04

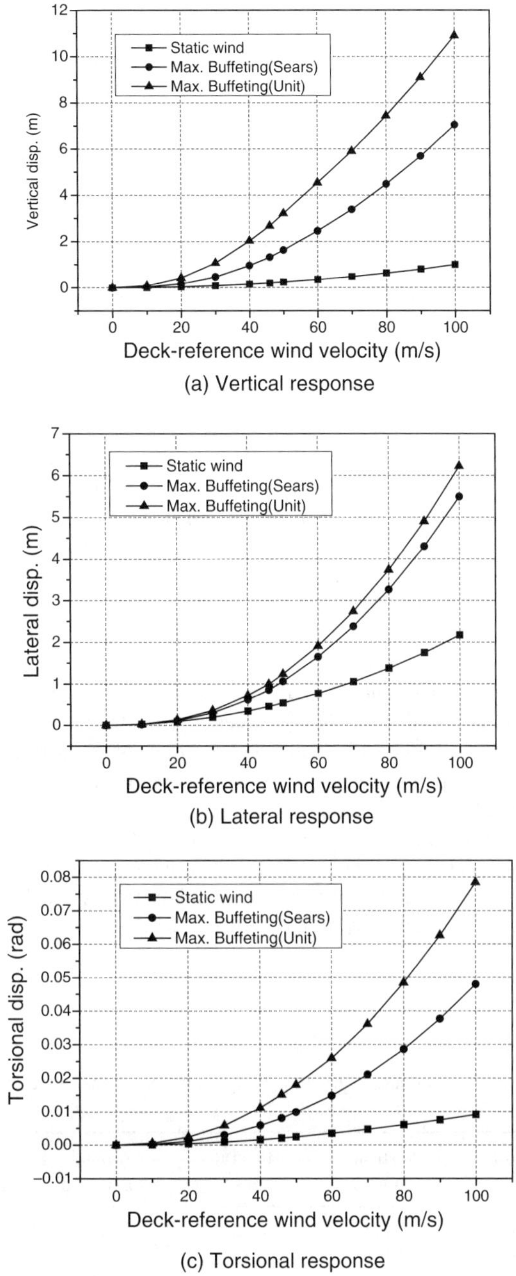

Figure 4.14 Displacement responses of the windward deck at the middle of main span.

Table 4.6 Buffeting peak displacement responses of the bridge at key locations (NE direction, V_{dref} = 46 m/s, Sears admittance)

Location	Longitudinal (m)	Vertical (m)	Lateral (m)	Around x-axis (rad)	Around y-axis (rad)	Around z-axis (rad)
D1-windward	0.051	1.322	0.848	0.008	7E-04	0.008
D1-leeward	0.051	1.219	0.848	0.008	7E-04	0.008
D2-windward	0.057	0.505	0.529	0.006	0.002	0.004
D2-leeward	0.057	0.444	0.529	0.007	0.002	0.004
D5-windward	0.053	0.029	0.088	0.002	3E-04	2E-04
D5-leeward	0.053	0.029	0.088	0.002	3E-04	2E-04
T1-tower-top	0.257	0.009	1.519	0.009	1E-04	0.002

Note that the lateral, vertical and torsional buffeting peak displacement responses of the windward deck at the middle point of main span are 0.848 m, 1.322 m and 0.008 rad, respectively. The lateral, vertical and torsional buffeting peak displacement responses of the leeward deck at the middle point of main span are 0.848 m, 1.219 m and 0.008 rad, respectively. The lateral buffeting peak displacement response at the top of the tower reaches 1.519 m. The lateral and vertical buffeting peak acceleration responses of the windward deck at the middle point of main span are 0.783 m/s^2 and 2.111 m/s^2, respectively, while the lateral buffeting peak acceleration response at the top of the tower reaches 2.146 m/s^2.

The buffeting peak displacement and acceleration responses of the bridge at the key locations are also computed for other wind speeds from the NE direction using the Sears function and the unit admittance, respectively. Figures 4.14 and 4.15 display the variations of the buffeting peak (maximum) displacement and acceleration responses, respectively, of the windward deck at the middle of main span, versus NE mean wind speed.

The total wind response of the complete bridge is equal to the sum of the absolute mean wind response and the buffeting peak response of the bridge. For the Sears function used as admittance functions for the bridge deck under the N-E wind, the total wind responses of the bridge are computed. The total displacement responses at the key locations are listed in Table 4.8 for the basic mean wind velocity of 46 m/s.

Table 4.7 Buffeting peak acceleration responses of the bridge at key locations (NE direction, V_{dref} = 46 m/s, Sears admittance)

Location	Longitudinal (m/s^2)	Vertical (m/s^2)	Lateral (m/s^2)	Around x-axis (rad)	Around y-axis (rad)	Around z-axis (rad)
D1-windward	0.186	2.111	0.783	0.034	0.002	0.028
D1-leeward	0.186	2.130	0.783	0.034	0.002	0.028
D2-windward	0.195	1.582	0.546	0.026	0.003	0.025
D2-leeward	0.195	1.609	0.546	0.026	0.003	0.025
D5-windward	0.221	0.048	0.180	0.003	0.001	0.002
D5-leeward	0.221	0.049	0.180	0.003	0.001	0.002
T1-tower-top	0.493	0.020	2.146	0.016	5E-04	0.008

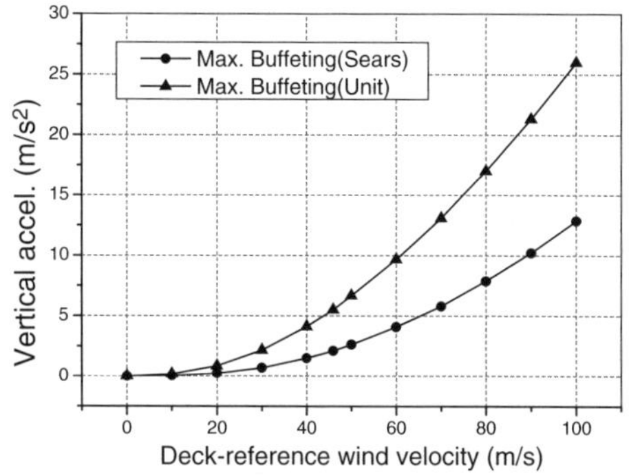

(a) Vertical response

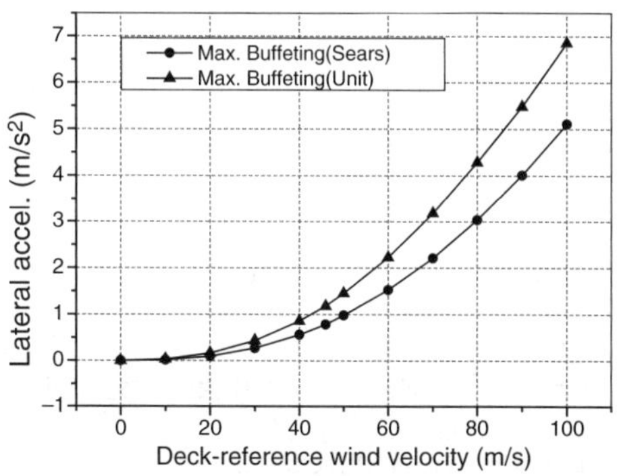

(b) Lateral response

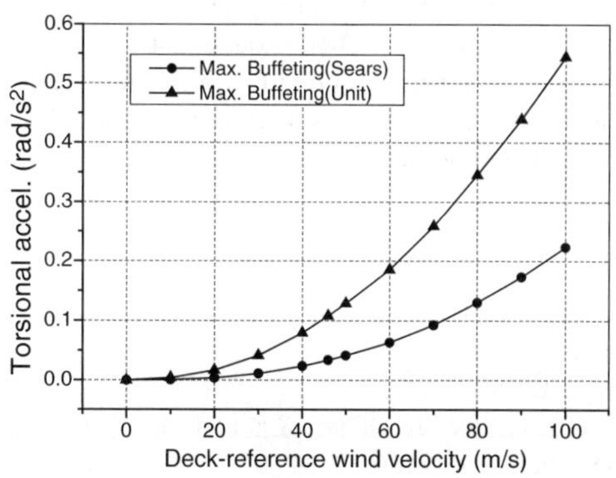

(c) Torsional response

Figure 4.15 Acceleration responses of the windward deck at the middle of main span versus N-E mean wind speed.

Table 4.8 Total displacement responses of the bridge at key locations (NE direction, $_{dref}$ = 46 m/s, Sears admittance)

Location	Longitudinal (m)	Vertical (m)	Lateral (m)	Around x-axis (rad)	Around y-axis (rad)	Around z-axis (rad)
D1-windward	0.051	1.523	1.297	0.010	7E-04	0.008
D1-leeward	0.051	1.487	1.297	0.010	7E-04	0.008
D2-windward	0.071	0.559	0.822	0.009	0.003	0.005
D2-leeward	0.070	0.583	0.822	0.009	0.003	0.005
D5-windward	0.059	0.042	0.122	0.003	5E-04	4E-04
D5-leeward	0.053	0.041	0.122	0.003	5E-04	3E-04
T1-tower-top	0.303	0.013	2.388	0.015	2E-04	0.003

4.10 Notations

a_i	Parameters to be determined by experiments
$\sqrt{a'^2}$	Standard deviation of modal coordinate
A	State matrix
$\mathbf{A}_{bu}$	Global buffeting force matrices corresponding to the longitudinal wind fluctuation
$\mathbf{A}_{bw}$	Global buffeting force matrices corresponding to the vertical wind fluctuation
A^e_{bu}	Buffeting force matrices of element corresponding to the longitudinal wind fluctuation
A^e_{bw}	Buffeting force matrices of element corresponding to the vertical wind fluctuation
A^*_i $(i = 1-6)$	Aerodynamic derivatives
A_{se}	Complex aeroelastic matrix of the bridge
b	Half of the plate width
B	(i) Typical chord or deck width dimension (ii) Matrix of interpolated function (Equation 4.66) (iii) Background factor representing the quasi-static response caused by gusts below the first natural frequency of the structure (Equation 4.103)
c	Damping coefficient of the structure
$\sqrt{coh}$	Root coherence
C	Decay factor
$\mathbf{C}$	Structural damping matrix
C_D	Drag force coefficient
C_L	Lift force coefficient
C_M	Moment coefficient
$C_L(K)$	Function of the reduced frequency K at lock-in
$C(k)$	Theodorsen cyclical function
C_{rs}	Complex flutter derivatives of self-excited forces
C^e	Damping matrix of the coupled wind-structure system
D	(i) Cylinder diameter (Equation 4.1) (ii) Characteristic dimension (Equation 4.2) (iii) Bridge deck width (Equation 4.10) (iv) Distance between the two points (Equation 4.119)
$D(\alpha)$	Aerodynamic drag force
E	Total energy in the characteristic motion of the bridge at the lowest critical wind speed
f	Frequency
$f_B(x)$	Load-response correlation formula

$f_c(x)$	Total effective loading distribution
f_{st}	Vortex shedding frequency
f_n, f_s	Vortex-shedding frequency or the natural frequency
$f_R(x)$	Equivalent load distribution for the resonant response at position x
$f(x)$	Mean wind force per unit length on the bridge deck
$f'(x)$	Fluctuating along-wind force on the bridge deck
$\overline{f'(x)f'(x_1)}$	Covariance for the fluctuating along-wind forces at positions x and x_1
$\mathbf{F}$	Nodal equivalent force vector
$\mathbf{F}_{se}$	Global nodal self-excited force vector
$\mathbf{F}_b$	Global nodal buffeting force vector
F_s	Static force
$F(K)$	Real part of Theodorsen cyclical function
$F_y(\alpha)$	Vertical force in the y-direction
g	Peak factor
g_B	Peak factor for background response
g_R	Peak factor for resonant response
G	Gust response factor
$G(K)$	Imaginary part of Theodorsen cyclical function
h	Vertical displacement of structure or bridge deck
$H_i^*(i=1-6)$	Aerodynamic derivatives
$H(\omega)$	(i) Transfer function matrix (Equation 4.77)
	(ii) Cholesky decomposition of cross-spectral density matrix $S^0(\omega)$ (Equation 4.86)
I	(i) Mass moment of inertia of the structure (Equation 4.22)
	(ii) Unit matrix (Equation 4.31)
$I_{()}$	Impulse function of the self-excited forces
$I_r(x)$	Influence line for the load effect r
I_u, I_v, I_w	Longitudinal, lateral and vertical turbulence intensity, respectively
$J_{()}$	Impulse function whose Fourier transform counterparts are referred to as the joint acceptance function
k	Stiffness coefficient
k_α	Total stiffness for the wind-structure system
$\mathbf{K}$	Structural stiffness matrix
K	Reduced frequency
K^e	Stiffness matrix of the coupled wind-structure system
$L(\alpha)$	Aerodynamic lift force
L_{se}	Self-excited lift force
$\mathbf{L}_{vs}$	Vortex-induced lift force
L_{i50}	Integral length scale at 50m height above ground for turbulent wind component i
L_u, L_v, L_w	Longitudinal, lateral and vertical turbulence integral scale, respectively
$\mathbf{M}$	Structural mass matrix
M	Aerodynamic pitching moment
M_{se}	Self-excited moment
m	(i) Mass of the structure
	(ii) Number of participating mode
N	Sufficiently large number
N_P	Number of points in the wind field
n_1	First mode natural frequency
p	Lateral displacement of the bridge deck
P_i^* $(i=1-6)$	Aerodynamic derivatives
q	Generalized modal response

$\mathbf{Q}_b$	Generalized buffeting force vector
Re	Reynolds number
R	(i) Complex modal response amplitude vector (Equation 4.41) (ii) Resonant factor describing the resonant response (Equation 4.103)
s	Complex frequency
S	(i) PSD matrix (Equation 4.78) (ii) Cross-spectra (Equation 4.79)
$\mathrm{S}^0(\omega)$	Cross-spectral density matrix
S_t	Strouhal number
S_c	Scruton number
$S_{ii}(f)$	Wind energy spectrum for turbulent wind components i
$\mathbf{S}^0(\omega)$	Two-side target cross-spectral density function
$S_{i_b i_b}(x, K)$	Buffeting force spectrum
$S_{jj}(x, K)$	Wind spectrum
$S_{L_{bu}}(x_1, x_2, \omega)$	Cross-spectra between the buffeting forces at two different positions x_1 and x_2
$S_{L^c_{bu}}(\omega)$	Auto-spectral density of the buffeting force at the center of the element
T	Time period over which the maximum value is required
T_0	Period of the sample function
U	Incoming wind velocity
$\overline{U}$	Mean wind velocity
U_α	Wind velocity with an effective angle of attack α
U_{cr}	Critical wind speed
$\overline{U}(z)$	One-hour mean wind velocity at height z
$u(t)$	Longitudinal fluctuating wind speed component
$u_j^0(t)$	One-dimensional stationary stochastic process
V	Velocity of the fluid relative to the cylinder
w_{back}	Weight factors for background response
w_{res}	Weight factors for resonant response
$w(t)$	Vertical fluctuating wind speed component
W_{j}	Peak inertial force in the mode j
X	Nodal displacement
$X(t)$	Displacement vector of the bridge
X_0	Amplitude of displacement vector of the bridge
$\hat{X}$	Maximum response of the structure
$\overline{X}$	Mean response of the structure
$\dot{X}$	Nodal velocity
$\ddot{X}$	Nodal acceleration
y	Vertical displacement
y_0	Vibration amplitude of the vertical displacement
y_{max}	Lock-in response of the structure
$\mathbf{Y}$	Eigenvector
$Y_1(K)$, $Y_2(K)$	Functions of the reduced frequency K at lock-in
z	Height
z_1, z_2	Two points
α	(i) Effective angle of attack (ii) Torsional displacements of the structure (Equation 4.17)
α_0	Attack angle of mean wind speed
$\Delta\alpha$	additional incident angle induced by fluctuation
ε	Non-linear aeroelastic damping coefficient
ξ	Structural damping ratio

ξ_α	Total damping for the wind-structure system
$\xi_{0\alpha}$	Structural damping coefficient in the torsional direction
Λ	Diagonal eigenvalue matrix obtained from modal analysis
υ	Cycling rate or effective frequency for the response
μ	Dynamic viscosity of the fluid
ρ	Air density
$\rho(x)$	Correlation coefficient between the fluctuating load at position x and the load effect of interest
σ_x	Standard deviation of the structural response
σ_u	Standard deviation of wind turbulence in the along-wind direction
$\sigma_p(x)$	Standard deviation of fluctuating load at position x
$\sigma_{r,B}$	Standard deviation of the background response
$\sigma_{r,R}$	Standard deviation of the resonant response
Φ	Matrix of modal shapes
ϕ	Phase angle
ϕ_{ml}	Sequence of independent random phase angles
$\boldsymbol{\varphi}_i$	ith natural mode shape
$\phi_1(x)$	First mode shape
$\theta_{jm}(\omega)$	Complex angle of $H_{jm}(\omega)$ for cholesky decomposition of cross-spectral density matrix $\mathrm{S}^0(\omega)$
χ_{Du}	Aerodynamic transfer function between the horizontal fluctuating wind velocity and aerodynamic drag force
χ_{Dw}	Aerodynamic transfer function between the vertical fluctuating wind velocity and aerodynamic drag force
χ_{Lu}	Aerodynamic transfer function between the horizontal fluctuating wind velocity and aerodynamic lift force
χ_{Lw}	Aerodynamic transfer function between the vertical fluctuating wind velocity and aerodynamic lift force
χ_{Mu}	Aerodynamic transfer function between the horizontal fluctuating wind velocity and aerodynamic moment
χ_{Mw}	Aerodynamic transfer function between the vertical fluctuating wind velocity and aerodynamic moment
ω	Circular frequency of vibration
ω_f	Flutter circular frequency
ω_{ml}	Double-indexing frequency
ω_n	Structural natural frequency
ω_h	Frequency of vibration in the vertical direction
ω_s	Vortex-shedding frequency
ω_α	Frequency of vibration in the torsional direction
ω_{0h}	Structural natural frequency in the vertical direction
$\omega_{0\alpha}$	Structural natural frequency in the torsional direction
ω_{up}	Upper cutoff frequency
$\Delta\omega$	Frequency increment
Δt	Time step

References

1. Bearman, P.W. (1984) Vortex shedding from oscillating bluff bodies. *Annual Review of Fluid Mechanics*, **16**, 195–222.
2. Williamson, C.H.K. and Govardhan, R. (2004) Vortex-induced vibrations. *Annual Review of Fluid Mechanics*, **36**, 413–455.
3. Liu, H. (1991) *Wind Engineering-A Handbook for Structural Engineers*, Prentice-Hall, Inc., New Jersey.

4. Sarpkaya, T. (1979) Vortex-induced oscillations: A selective review. *Journal of Applied Mechanics*, **46**(2), 241–258.
5. Sarpkaya, T. (2004) A critical review of the intrinsic nature of vortex-induced vibrations. *Journal of Fluids and Structures*, **19**(4), 389–447.
6. Simiu, E. and Scanlan, R.H. (1986) *Wind Effects on Structures*, 2nd edn, John Wiley & Sons.
7. Ehsan, F. and Scanlan, R.H. (1990) Vortex-induced vibration of flexible bridges. *Journal of Engineering Mechanics*, **116**(6), 1392–1411.
8. Simiu, E. and Scanlan, R.H. (1996) *Wind Effects on Structures*, 3rd edn, John Wiley & Sons.
9. ASCE (1987) *Wind Loading and Wind-Induced Structural Response*, American Society of Civil Engineers.
10. Scanlan, R.H. and Tomko, J.J. (1971) Airfoil and bridges deck flutter derivatives. *Journal of the Engineering Mechanics Division*, **97**(6), 1717–1737.
11. Theodorson, T. (1934) *General theory of aerodynamic instability and the mechanism of flutter*. NACA Report 496, National Aeronautics and Space Administration, Washington DC.
12. Agar, T.J.A. (1991) Dynamic instability of suspension bridges. *Computers & structures*, **41**(6), 1321–1328.
13. Namini, A., Albrecht, P., and Bosch, H. (1992) Finite element-based flutter analysis of cable-suspended bridges. *Journal of Structural Engineering*, **118**(6), 1509–1526.
14. Beith, J.G. (1998) A practical engineering method for the flutter analysis of long-span bridges. *Journal of Wind Engineering and Industrial Aerodynamics*, **77–78**, 357–366.
15. Ding, Q.S., Chen, A.R., and Xiang, H.F. (2002) A state space method for coupled flutter analysis of long span bridges. *Structural Engineering and Mechanics*, **14**(4), 491–504.
16. Ding, Q.S. and Xu, Y.L. (2004) 3-D Finite Element Based Flutter Analysis of Stonecutters Bridge, Report No. 03 to Hong Kong Highways Department, Department of Civil and Structural Engineering, The Hong Kong Polytechnic University, Hong Kong.
17. Chen, X.Z., Matsumoto, M., and Kareem, A. (2000) Time domain flutter and buffeting response analysis of bridges. *Journal of Engineering Mechanics, ASCE*, **126**(1), 7–16.
18. Davenport, A.G. (1962) Buffeting of a suspension bridge by storm winds. *Journal of Structural Division, ASCE*, **88**(3), 233–268.
19. Scanlan, R.H. (1978) The action of flexible bridge under wind, II: buffeting theory. *Journal of Sound and Vibration*, **60**(2), 201–211.
20. Jain, A., Jones, N.P., and Scanlan, R.H. (1996) Coupled flutter and buffeting analysis of long-span bridges. *Journal of Structural Engineering*, **122**(7), 716–725.
21. Xu, Y.L., Sun, D.K., Ko, J.M., and Lin, J.H. (2000) Fully coupled buffeting analysis of Tsing Ma suspension bridge. *Journal of Wind Engineering and Industrial Aerodynamics*, **85**(1), 97–117.
22. Bucher, G.C. and Lin, Y.K. (1988) Stochastic stability of bridges considering coupled modes. *Journal of Engineering Mechanics, ASCE*, **114**(12), 2055–2071.
23. Xiang, H.F., Liu, C.H., and Gu, M. (1995) Time-domain analysis for coupled buffeting response of long span bridge. Proceeding of the 9th ICWE, New Delhi, pp. 881–892.
24. Davenport, A.G. (1961) The application of statistical concepts to the wind loading of structures. *Proceedings, Institution of Civil Engineers*, **19**(4), 449–472.
25. Shinozuka, M. and Jan, C.M. (1972) Digital simulation of random processes and its applications. *Journal of Sound and Vibration*, **25**(1), 111–128.
26. Deodatis, G. (1996) Simulation of ergodic multivariate stochastic processes. *Journal of Engineering Mechanics, ASCE*, **122**(8), 778–787.
27. Ding, Q.S., Zhu, L.D., and Xiang, H.F. (2006) Simulation of stationary Gaussian stochastic wind velocity field. *Wind and Structures*, **9**(3), 231–243.
28. Lin, Y.K. and Yang, J.N. (1983) Multimode bridge response to wind excitations. *Journal of Engineering Mechanics, ASCE*, **109**(2), 586–603.
29. Xiang, H.F. (2005) *Modern Theory and Practice on Bridge Wind Resistance*, China Communications Press, Beijing (in Chinese).
30. Davenport, A.G. (1967) Gust loading factors. *Journal of the Structural Division, ASCE*, **93**, 11–34.
31. Vickery, B.J. (1966) On the assessment of wind effects on elastic structures. *Australian Civil Engineering Transactions*, **CE8**, 183–192.
32. Holmes, J.D. (2007) *Wind Loading of Structures*, 2nd edn, Taylor & Francis, Abingdon.
33. Davenport, A.G. (1964) Note on the distribution of the least value of a random function with application to gust loading. *Proceedings of the Institution of Civil Engineers*, **28**, 187–196.
34. Kasperski, M. and Niemann, H.J. (1992) The L.R.C. (load-response-correlation) - method a general method of estimating unfavorable wind load distributions for linear and non-linear structural behavior. *Journal of Wind Engineering and Industrial Aerodynamics*, **43**(1–3), 1753–1763.
35. Holmes, J.D. (1999) Equivalent static load distributions for resonant dynamic response of bridges. Proceedings of 10th International Conference on Wind Engineering, Copenhagen, Denmark, Balkema, Rotterdam, pp. 907–911.
36. Ding, Q.S. and Xu, Y.L. (2004) 3-D Finite Element Based Buffeting Analysis of Stonecutters Bridge, Report No. 05 to Hong Kong Highways Department, Department of Civil and Structural Engineering, The Hong Kong Polytechnic University, Hong Kong.

5

Wind-Induced Vibration of Stay Cables

5.1 Preview

With widespread popularity of cable-stayed bridges around the world, the bridges that have longer spans – over 1000 m – have been constructed by employing increasingly longer stay cables, some over 500 m long. The stay cables are laterally flexible structural members with very low fundamental frequency and very little inherent damping. For this reason, they have been found to be susceptible to various types of vibrations, including wind-induced vibration, rain-wind-induced vibration (RWIV) and deck-induced vibration. These vibrations, as well as other types that have been observed but not effectively categorized, are of significant concern to bridge engineers because they can reach large amplitude and cause unexpected stresses both in the stay cables and in the connections between the cables and other bridge components, potentially leading to strength or fatigue failure of stay cables and other structural members. Recognition of this susceptibility of stay cables leads to the use of various vibration mitigation measures aerodynamically or mechanically. In consideration of the scope of this book, only wind-induced vibration and RWIV of stay cables are discussed in this chapter, whereas vibration mitigation measures for stay cables will be discussed in Chapter 12, together with those for bridge decks and towers.

This chapter will first introduce fundamentals of cable dynamics to lay down a foundation for better understanding of wind-induced vibration and RWIV of stay cables. Wind-induced vibration mainly refers to buffeting by wind turbulence, vortex-induced vibration, galloping of dry inclined cables and wake galloping for groups of cables. A possible mechanism of RWIV will then be discussed and a simple analytical model for RWIV will be presented. The analytical model, together with the fundamentals of cable dynamics, will be applied to predict the RWIV of full-scale stay cables. In order to foresee whether RWIV will occur for a particular cable-stayed bridge at a particular bridge site, a framework for predicting occurrence probability of RWIV will be finally presented.

5.2 Fundamentals of Cable Dynamics

The study of cable dynamics has enjoyed a long and rich history and it is still a popular topic for contemporary research. This section reviews some of basic, but important, concepts in cable dynamics that are relevant to the problem of wind-induced vibration and RWIV of stay cables. It serves as the

Wind Effects on Cable-Supported Bridges, First Edition. You-Lin Xu.

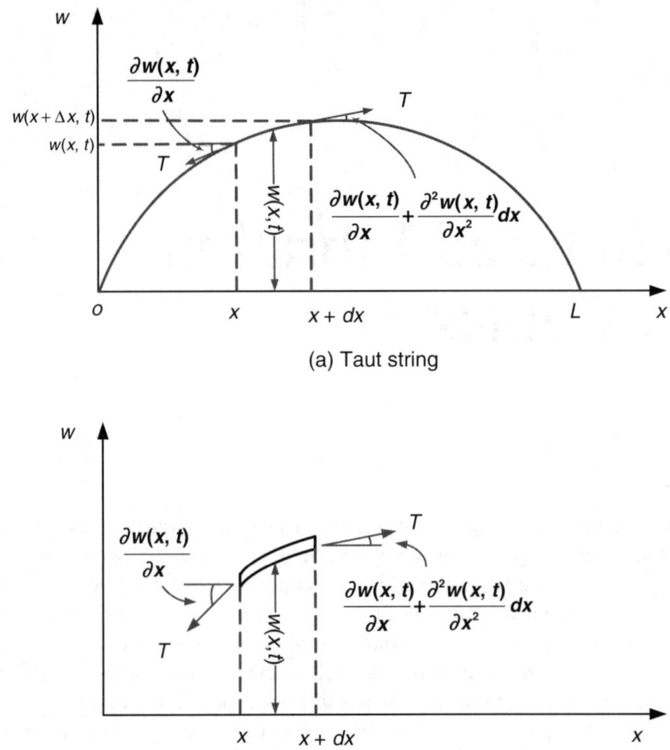

(b) Static equilibrium of element

Figure 5.1 A taut string and its element.

background for the discussions in the subsequent sections. The review on this topic is based on those presented in Meirovitch [1] and Zuo [2] for vibration of a taut string, and in Irvine [3] and Zuo [2] for vibration of an inclined cable with sag.

5.2.1 Vibration of a Taut String

Vibration of a taut string is the simplest model for stay cable vibration, which neglects the bending stiffness of the cable and the sag due to self weight. Figure 5.1a shows a taut string, while Figure 5.1b displays an element of the taut string in dynamic equilibrium.

If the displacement w is small, the dynamic equilibrium of the element requires:

$$m\frac{\partial^2 w}{\partial t^2} - T\frac{\partial^2 w}{\partial x^2} + c\frac{\partial w}{\partial t} + F(x,t) = 0 \tag{5.1}$$

with the boundary conditions $w(x, t) = w(L, t) = 0$.

In Equation 5.1: $w(x, t)$ is the in-plane transverse displacement normal to the cable axis at position x;
m is the cable mass per unit length;

T is the cable tension force with a constant value;
c is the internal cable damping coefficient;
$F(x, t)$ is the wind-induced force on the cable;
L is the length of the cable.

For an undamped free vibration of the taut string, Equation 5.1 can be reduced to:

$$m\frac{\partial^2 w}{\partial t^2} = T\frac{\partial^2 w}{\partial x^2} \tag{5.2}$$

Assuming that the displacement $w(x, t)$ is in the form of:

$$w(x, t) = W(x)f(t) \tag{5.3}$$

where $f(t)$ is a harmonic function of time with frequency ω, Equation 5.2 becomes:

$$T\frac{\partial^2 W(x)}{\partial x^2} = -\omega^2 mW(x) \tag{5.4}$$

The characteristic equation of Equation 5.4 with the given boundary conditions can be expressed as:

$$\sin\left(\omega L\sqrt{\frac{m}{T}}\right) = 0 \tag{5.5}$$

to which the solution is an infinite number of natural frequencies of the taut string in the form of:

$$\omega_n = \frac{n\pi}{L}\sqrt{\frac{T}{m}}, \quad n = 1, 2, \ldots \tag{5.6}$$

with corresponding mode shapes:

$$W_n(x) = A_n \sin\frac{n\pi x}{L}, \quad n = 1, 2, \ldots \tag{5.7}$$

where A_n is a constant. The vibration of the taut string can be composed in terms of these natural mode shapes. The undamped free vibration of the taut string can then be completely represented as:

$$w(x, t) = \sum_{i=1}^{n} A_n \sin\left(\frac{n\pi x}{L}\right)\sin\left(\frac{n\pi}{L}\sqrt{\frac{T}{m}}t\right) \tag{5.8}$$

According to Equation 5.8, the different modes of vibration of a taut string is uncoupled. Therefore, the individual modes can be treated as single degree of freedom (SDOF) oscillators.

5.2.2 *Vibration of an Inclined Cable with Sag*

When the self weight of a stay cable cannot be ignored, the cable develops continuous sag along its chord due to gravity, and thus can no longer be treated as a taut string. The existence of the sag makes the dynamic behavior of the cable in the gravitational plane different from that in the lateral plane and, depending on the magnitude of the sag, this can potentially introduce various kinds of non-linear

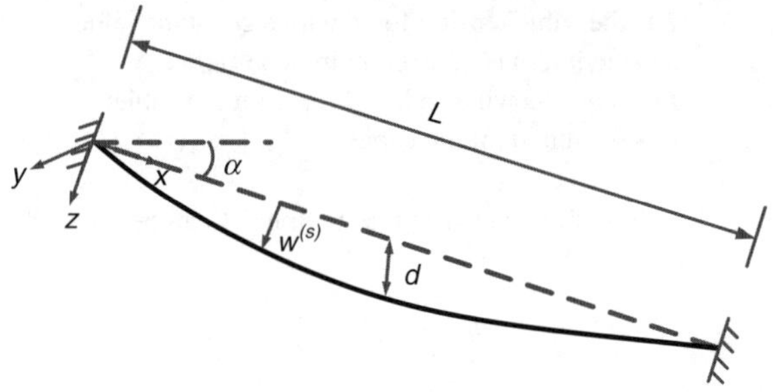

Figure 5.2 Static profile of an inclined, sagged cable.

phenomena [4]. For the stay cables in cable-stayed bridges, the sag is usually very small, with the sag-to-span ratio of the order of 0.001 [4]. As a result, the effect of sag is usually not very significant in stay cable vibrations except in some specific cases, when the vibration is induced by the large amplitude oscillation of the bridge deck [5]. Nevertheless, it is essential to investigate the effects of cable sag, because super long-span cable-stayed bridges with very long stay cables are emerging.

Figure 5.2 depicts the static profile of an inclined, sagged cable in the Cartesian coordinate (x, y, z). The mass of the cable is uniformly distributed with density m; the chord length is L; the sag and chord inclination angle are d and α, respectively. For a metallic cable, the weight to tension ratio and the static strain are relatively small. As a result, the static configuration of the cable can be assumed to be parabolic in the gravitational plane [3].

$$w^s(x) = \frac{1}{2}\left(mg\cos\alpha L^2/T\right)\left[(x/L) - (x/L)^2\right] \tag{5.9a}$$

$$u^s(x) = v^s(x) = 0 \tag{5.9b}$$

where:

u, v and w are the axial, out-of-plane transverse and in-plane transverse displacements of the cable, respectively;
g is the gravitational constant;
the superscript, s, denotes the static equilibrium state of the cable.

The equation of motion of the cable with sag can be expressed, with respect to its static equilibrium position, as [3]:

$$\frac{\partial}{\partial x}\left[(T+\tau)\frac{\partial u}{\partial x} + \tau\right] + F_x(x,t) = m\frac{\partial^2 u}{\partial t^2} + c_x\frac{\partial u}{\partial t} \tag{5.10a}$$

$$\frac{\partial}{\partial x}\left[(T+\tau)\frac{\partial v}{\partial x}\right] + F_y(x,t) = m\frac{\partial^2 v}{\partial t^2} + c_y\frac{\partial v}{\partial t} \tag{5.10b}$$

$$\frac{\partial}{\partial x}\left[(T+\tau)\frac{\partial w}{\partial x} + \tau\frac{dw^s}{dx}\right] + F_z(x,t) = m\frac{\partial^2 w}{\partial t^2} + c_z\frac{\partial w}{\partial t} \tag{5.10c}$$

where:

τ is the additional dynamic tension due to the cable vibration;
$F_x(x, t)$, $F_y(x, t)$, and $F_z(x, t)$ are the wind-induced forces in the x-,y- and z- directions, respectively;
c_x, c_y, and c_z are the internal cable damping coefficients in the x-, y- and z-directions, respectively.

Assume that the displacement of the cable is small and the stress is still in its elastic range. The dynamic tension can be expressed as:

$$\tau(x,t) = EA\varepsilon(x,t) \tag{5.11}$$

where:

E is the Young's modulus of the cable;
A is the area of the cross-section of the cable;
ε is the dynamic strain, which can be expressed as a function of the displacement as:

$$\varepsilon(x,t) = \frac{\partial u}{\partial x} + \frac{dw^s}{dx}\frac{\partial w}{\partial x} + \frac{1}{2}\left(\frac{\partial u}{\partial x}\right)^2 + \frac{1}{2}\left(\frac{\partial v}{\partial x}\right)^2 + \frac{1}{2}\left(\frac{\partial w}{\partial x}\right)^2 \tag{5.12}$$

If the displacement is small, the second-order terms can be dropped and Equation 5.12 can be reduced to:

$$\varepsilon(x,t) = \frac{\partial u}{\partial x} + \frac{dw^s}{dx}\frac{\partial w}{\partial x} \tag{5.13}$$

For a tightly-stressed cable such as the stay cables in cable-stayed bridges, the axial motion is usually considered unimportant. As a result, Equation 5.10a can be ignored. For the problem of undamped free vibration, the linearized forms of Equations 5.10b and 5.10c reduce to

$$T\frac{\partial^2 v}{\partial x^2} = m\frac{\partial^2 v}{\partial t^2} \tag{5.14a}$$

$$T\frac{\partial^2 w}{\partial x^2} + \tau\frac{d^2 w^s}{dx^2} = m\frac{\partial^2 w}{\partial t^2} \tag{5.14b}$$

Note that Equation 5.14a is essentially the same as Equation 5.2. This suggests that the dynamic behavior of an inclined, sagged cable in the out-of-plane vibration is the same as that of a taut string, which also means that the natural frequencies and associated mode shapes of the out-of-plane vibration are the same as those expressed by Equations 5.6 and 5.7. For the asymmetric modes of the in-plane vibration, the additional dynamic tension is equal to zero and Equation 5.14b can also be reduced to the same as Equation 5.2 or Equation 5.14a. This indicates that the natural frequencies and mode shapes associated with the asymmetric modes of the in-plane vibration are the same as those expressed by Equations 5.6 and 5.7.

In the case of symmetric in-plane modes, additional dynamic tension is not equal to zero, and therefore the natural frequencies and mode shapes of the symmetric modes are different from the corresponding out-of-plane modes. The natural frequencies of the symmetrical in-plane modes can be found by solving the transcendental equation as follows [3]:

$$\tan\frac{\varpi}{2} = \frac{\varpi}{2} - \frac{4}{\lambda^2}\left(\frac{\varpi}{2}\right)^3 \tag{5.15}$$

where the normalized circle frequency ϖ is:

$$\varpi = \omega L\sqrt{\frac{m}{T}} \tag{5.16}$$

and the sag parameters λ^2 is defined as the ratio of the elastic-to-centenary stiffness:

$$\lambda^2 = \left(\frac{mgL}{T}\right)^2 \frac{EA}{T} \tag{5.17}$$

Equation 5.15 is valid for cables of considerable sag-to-span ratio. For cables of small sag, such as the stay cables in cable-stayed bridges, Warnitchai *et al.* [6] suggests that the mode shapes of the symmetric in-plane modes can also be approximately treated as sinusoidal. Therefore, when the sag-to-span ratio and the displacement of the cable both are small, the effects of geometric and material non-linearities can be ignored and, as a result, the dynamic properties of an inclined, sagged cable are very similar to those of a taut string.

Note that the above analysis is based on assumptions that the sag-to-span ratio and the displacement of the cable are both small. Without these two assumptions, geometric and material non-linearities need to be considered, leading to various complicated vibration phenomena, as discussed by many researchers [7–10].

5.3 Wind-Induced Cable Vibrations

There are a number of mechanisms that can potentially lead to wind-induced cable vibrations. This section reviews several wind-induced cable vibrations that require careful consideration by bridge designers. These are: buffeting by wind turbulence; vortex-induced vibration; galloping of dry inclined cable; and wake galloping for groups of cables. The review on this topic follows that presented by FHWA [11].

5.3.1 Buffeting by Wind Turbulence

Flexible structures such as long stay cables undergo substantial motions in strong winds simply because of the random buffeting action of wind turbulence. Very long stay cables will have their lower modes of vibration excited by this effect, but it is not an aeroelastic instability; even very aerodynamically stable structures can be seen to move in strong winds if they are flexible. Buffeting motions are not typically a problem for stay cables in cable-stayed bridges. The buffeting motions increase gradually with wind speed, rather than in the sudden fashion associated with an instability [11].

5.3.2 Vortex-Induced Vibration

Vortex-induced vibration is a classical wind-induced vibration, characterized by limited-amplitude vibrations at relatively low wind speeds. Vortex-induced vibration of a stay cable is caused by the alternate shedding of vortices from the two sides of the cable when the wind is approximately perpendicular to the cable axis. The wind velocity at which the vortex excitation frequency matches the natural frequency of the cable can be found in terms of the Strouhal number:

$$U = \frac{fD}{S} \tag{5.18}$$

where:

U is the mean wind speed;
f is the natural frequency of the cable;
D is the cable diameter;
S is the Strouhal number.

For circular cross-section cables in the Reynolds number range, 10^4 to about 3×10^5, S is about 0.2.

The amplitude of the cable oscillations is inversely proportional to the Scruton number S_c. It can be seen from $S_c = m\zeta/\rho D^2$ that increasing the mass per unit length m and damping ratio ζ of the cable increases the Scruton number and therefore reduces oscillation amplitudes.

A realistic estimate of inherent cable damping ratios on in-service bridges is in the range from 0.001 to 0.005. For example, a cable consisting of steel strands grouted inside the cable pipe and with a damping ratio of 0.005 has a Scruton number of about 12, and the amplitude of oscillation is only about 0.5% of the cable diameter [11]. During construction and before grouting, the damping ratio of stay cables can be extremely low (e.g. 0.001), and the amplitude could conceivably increase to about 4% of the cable diameter, which is still small [11]. Therefore, vortex shedding from the cables is unlikely to be a major vibration problem for cable-stayed bridges. By adding a small amount of damping, vortex excitation will be suppressed effectively.

5.3.3 *Galloping of Dry Inclined Cables*

Single cables of circular cross-section do not gallop when they are aligned normal to the wind. However, when the wind velocity has a component that is not normal to the cable axis, an instability with the same characteristics as galloping has been observed; for a single inclined cable, the wind actually acts on an elliptical cross-section of cable [11].

Saito *et al.* [12] conducted a series of wind tunnel experiments on a section of bridge cable mounted on a spring suspension system. Their data suggest an instability criterion given approximately by the following:

$$\left(\frac{U}{fD}\right)_{crit} = 40\sqrt{S_c} \tag{5.19}$$

This formula was for cases where the angle between the cable axis and wind direction was 30° to 60°. The above criterion is a difficult condition to satisfy, particularly for the longer cables of cable-stayed bridges with a typical diameter of 150–200 mm. It was apparent that galloping of dry inclined cables presented the biggest concern and biggest unknown for wind-induced vibration. Further experimental research was necessary to confirm the results of Saito *et al.* [12]. Furthermore, all of their experiments used low levels of damping, so it was important to investigate whether galloping of an inclined cable is possible at damping ratios of 0.005 and higher.

New sets of wind tunnel tests were carried out and the results were reported by FHWA [11]. Testing was performed for various levels of structural damping, cable frequency ratios, surface roughness and at various angles of wind flow. The cable model orientation was changed against the mean wind flow direction for several configurations. Limited-amplitude oscillations were observed under a variety of conditions. The limited-amplitude vibrations occurred within a narrow wind speed range only, which is characteristic of vortex excitation of the high-speed type described by Matsumoto [13].

For the typical cable diameters and wind speeds of concern on cable-stayed bridges, the Reynolds number is in the critical range, i.e. where large changes in the airflow patterns around the cable occur for relatively small changes in the Reynolds number. The excitation mechanism is thus likely to be linked with these changes. The maximum amplitude of the response depended on the orientation angle of the cable.

The results of the new sets of tests obtained by FHWA [11] showed a deviation from the criterion presented by Saito *et al.* [12]. While significant oscillations of the cable occurred (double amplitude up to 1D), it is not conclusive that this was dry inclined cable galloping. In fact, as indicated above, the oscillations had similar characteristics to high speed vortex excitation as discovered by Matsumoto [13]. Divergent oscillations only occurred for one test setup at very low damping, and the vibration had to be suppressed since the setup only allowed for amplitude of 1D. Large vibrations were found only at the lowest damping ratios less than 0.001. Above a damping ratio of 0.003, no significant vibration larger than 10 mm was observed. It can thus be concluded that, if even a low amount of structural damping greater than 0.003 is provided, then vortex shedding and inclined cable galloping vibrations are not significant. This damping corresponds to a Scruton number of approximately 3 [11]. Therefore, dry cable instability should be suppressed by default if enough damping is provided.

5.3.4 Wake Galloping for Groups of Cables

Wake galloping is the elliptical movement caused by variations in drag and acrosswind forces for cables in the wake of other structural components, such as towers or other cables. This occurs at high wind speeds and leads to large amplitude oscillations. These oscillations have been found to cause fatigue of the outer strands of bridge hangers at end clamps on suspension and arch bridges. Similar fatigue problems are a theoretical possibility on cable-stayed bridges but, to date, none have been documented.

The Scruton number is an important parameter with regard to wake galloping effects. An approximate equation for the minimum wind velocity U_{crit}, above which instability can be expected due to wake galloping effects, has been proposed as follows [14,15]:

$$U_{crit} = cfD\sqrt{S_c} \tag{5.20}$$

For circular sections, the constant c has an approximate median value of 40. For cable-stayed bridges, this constant depends on the clear spacing between cables: $c = 25$ for closely spaced cables (2D to 6D spacing); and $c = 80$ for normally spaced cables (generally 10D and higher) [11]. Due to the level of uncertainty associated with practical applications, it is recommended that these values be applied conservatively, exercising engineering judgment.

The critical wind velocity may be low enough to occur commonly during the life of the bridge. Wake galloping, therefore, has the potential to cause serviceability problems. The equation for U_{crit} suggests several possibilities for mitigation. By increasing the Scruton number or natural frequency, the cables will be stable up to a higher wind velocity. However, increasing the frequency is far more effective in raising U_{crit}, due to the square root manifestation of S_c in Equation 5.20. The Scruton number increases with additional damping. The natural frequency may be increased by installing spacers or crossties along the cables to shorten the effective length of cable for the vibration mode of concern.

It should be noted that wake galloping is not a major design concern for normal, well-separated cable arrangements. For unusual cases, however, it is recommended that some attention be paid to the possibility of wake galloping.

5.4 Mechanism of Rain-Wind-Induced Cable Vibration

5.4.1 Background

The combination of rain and moderate wind speeds can cause high-amplitude cable vibrations at low frequencies. This phenomenon has been observed on many cable-stayed bridges and has been studied in detail [11].

Rain-wind-induced vibrations (RWIV) were first identified by Hikami and Shiraishi on the Meiko-Nishi cable-stayed bridge [16]. Since then, these vibrations have been observed on other cable-stayed bridges around the world [17,18]. Large amplitude cable vibrations under simultaneous occurrence of wind and rain may induce undue stresses in the cable in the vicinity of the anchorage, potentially leading to fatigue fractures of internal wire strands [19]. Sometimes they may also damage the steel tubes installed on the anchorage for protecting the cable [20].

Many studies have thus been conducted to reveal the mechanism and conditions of RWIV of stay cables. These studies include field measurements of RWIV of stay cables [16,21], wind-rain tunnel tests of sectional cable models [22–24] and analytical studies for exploring the mechanism of wind-rain-induced cable vibration [23,25–27]. It was found that these vibrations occurred typically when there was rain and moderate wind speeds (8–15 m/s) in the direction angled 20° to 60° to the cable plane, with the cable declined in the direction of the wind. The frequencies were low, typically less than 3 Hz. The peak amplitudes were very high, in the range of 0.25 m to 1.0 m – violent movements, resulting in the clashing of adjacent cables observed in several cases.

Wind-rain tunnel tests have shown that rivulets of water running down the upper and lower surfaces of the cable in rainy weather were the essential component of this aeroelastic instability [16]. The water rivulets changed the effective shape of the cable and moved as the cable oscillated, causing cyclical changes in the aerodynamic forces that led to the wind feeding energy into oscillations. The wind direction causing the excitation was approximately 45° to the cable plane. The particular range of wind velocities that caused the oscillations appears to be that which maintained the upper rivulet within a critical zone on the upper surface of the cable.

Since some bridges have been built without experiencing problems from RWIV of stay cables, it appears probable that, in some cases, the level of damping naturally present is sufficient to avoid the problem. The rig test data of Saito *et al.* [12], obtained using realistic cable mass and damping values, are useful in helping to define the boundary of instability for rain and wind oscillations. Based on their results, it appears that rain-wind oscillations can be reduced to a harmless level if the Scruton number is greater than 10 [15].

RWIV are highly problematic vibrations, with their large amplitudes and relatively frequent occurrence, and this is among the most significant considerations in the design of mitigation measures for stay cables. One of the primary research components is to acquire a more in-depth understanding of the underlying mechanics of rain-wind-induced stay cable vibration and to develop an analytical model with the capability to predict major characteristics of RWIV for an arbitrary stay cable. Such a model will enable a much more comprehensive treatment of the problem, both in terms of when such vibrations are likely to occur and what will be the effect of various mitigation approaches. An initial attempt at solving such a problem is presented in the following sections for reference [27–29].

5.4.2 *Analytical Model of SDOF*

Let us use a rigid and uniform inclined cylinder to represent a stay cable segment (see Figure 5.3). The inclination of the cylinder is denoted by angle α, and the yaw angle of the incident wind is designated by angle β. The cylinder is supposed to be supported by springs at its ends in the plane 1-5-7. The consideration of such a cylinder, rather than a real cable, is because many researchers used it in their rain-wind simulation tests [16,30], and some of the test results have to be used in this section for exploration of mechanism of RWIV.

As a preliminary analytical study, the upper rivulet is assumed to distribute uniformly along the longitudinal axis of the cylinder and vibrate circumferentially over the surface of the cylinder. Turbulent effect and axial flow effect are not considered. The static position of the upper rivulet due to the mean wind when the cylinder is stationary is defined by angle θ_0, and the dynamic angular displacement of the rivulet as the cylinder vibrates is designated by θ with reference to θ_0 (see Figure 5.4).

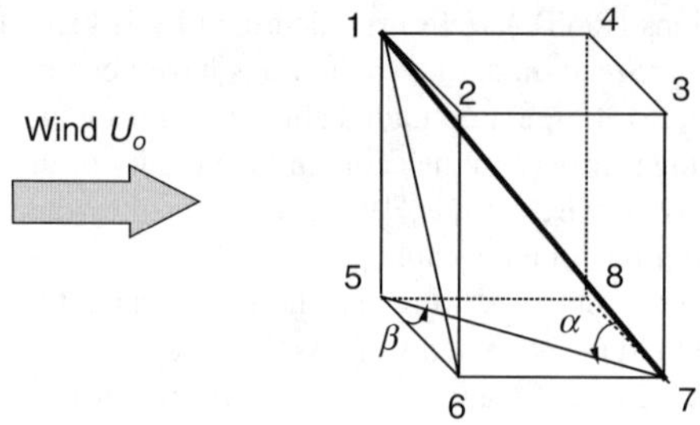

Figure 5.3 Orientation of inclined cylinder (from [27]) (Reproduced with permission from Elsevier).

Since the cylinder is not perpendicular to the direction of the mean wind speed U_0, one needs to find the component of mean wind speed perpendicular to the cylinder, U, using the following equation.

$$U = U_0\sqrt{\cos^2\beta + \sin^2\alpha\sin^2\beta} = U_0\sqrt{\sin^2\alpha + \cos^2\alpha\cos^2\beta} \tag{5.21}$$

The angle of attack of the mean wind speed component U is defined as γ (see Figure 5.4), which indicates the stagnation point of incident wind on the surface of cylinder. The position of the stagnation point depends on many factors, such as the cross-section of cylinder with rivulet, roughness of cylinder surface, wind turbulence and cable motion. No experimental results are available to the author at this stage with respect to the position of stagnation point on a stay cable with rivulet. Thus, the angle of attack in this study is selected as the ideal angle of attack multiplied by an influence factor ε.

$$\gamma = \varepsilon\sin^{-1}\left(\frac{\sin\alpha\sin\beta}{\sqrt{\cos^2\beta + \sin^2\alpha\sin^2\beta}}\right) \tag{5.22}$$

When ε is set as 1, γ represents the ideal angle of attack for the cylinder without rivulet [31]. When ε is selected as zero, it indicates that the position of the stagnation point is the same as that on the

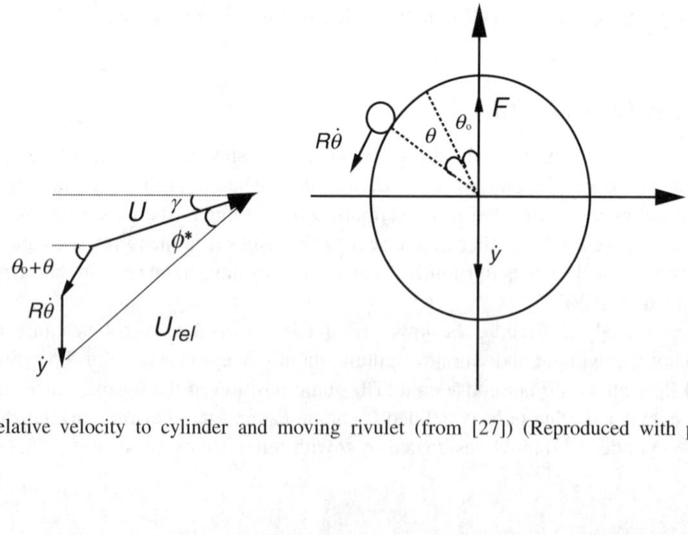

Figure 5.4 Relative velocity to cylinder and moving rivulet (from [27]) (Reproduced with permission from Elsevier).

cylinder without rivulet and yaw angle. The effects of the mean wind speed component along the cylinder axis and wind turbulence are not considered in this study.

In consideration of the transverse vibration of the cable of velocity $\dot{y}(t)$ and the angular vibration of the upper rivulet of velocity $\dot{\theta}(t)$, the relative velocity of mean wind to the cylinder with moving rivulet is, therefore:

$$U_{rel} = \sqrt{\left(U\cos\gamma + R\dot{\theta}\cos(\theta+\theta_0)\right)^2 + \left(U\sin\gamma + \dot{y} + R\dot{\theta}\sin(\theta+\theta_0)\right)^2} \tag{5.23}$$

The angle between the relative velocity U_{rel} and the horizontal axis in Figure 5.4 is defined as $\varphi^*(t)$, which can be expressed as:

$$\varphi^* = \tan^{-1}\frac{U\sin\gamma + \dot{y} + R\dot{\theta}\sin(\theta+\theta_0)}{U\cos\gamma + R\dot{\theta}\cos(\theta+\theta_0)} \tag{5.24}$$

where R is the radius of the cylinder and the size of the rivulet is considered small compared with the diameter of the cylinder. The field and laboratory observations indicate that $R\dot{\theta}$ is very small compared with the concerned mean wind speed U. Thus, if the angles γ and $\varphi^*(t)$ are limited a certain range, Equation 5.24 can be reduced as

$$\begin{aligned}\varphi^* &\approx \frac{U\sin\gamma + \dot{y} + R\dot{\theta}\sin(\theta+\theta_0)}{U\cos\gamma} \\ &\approx \frac{U\sin\gamma + \dot{y} + R\dot{\theta}\left[(\theta+\theta_0) - \frac{1}{6}(\theta+\theta_0)^3\right]}{U\cos\gamma}\end{aligned} \tag{5.25}$$

The net vertical force on the cylinder per unit length in the y direction is then:

$$F = \frac{\rho D U_{rel}^2}{2}\left[C_L(\varphi)\cos\varphi^* + C_d(\varphi)\sin\varphi^*\right] \tag{5.26}$$

where:

ρ is the density of the air;
D is the diameter of the cross-section of the cylinder;
C_d is the drag coefficient;
C_L is the lift coefficient.

The drag and lift coefficients of the cylinder with rigid rivulet measured from the wind tunnel tests are often expressed as the function of the angle φ defined in Figure 5.5 [16,32]. The relationship between the angles φ^* and φ is given by:

$$\varphi = \varphi^* - \theta - \theta_0 = \frac{U\sin\gamma + \dot{y} + R\dot{\theta}\left[(\theta+\theta_0) - \frac{1}{6}(\theta+\theta_0)^3\right]}{U\cos\gamma} - \theta - \theta_0 \tag{5.27}$$

In consideration that the cylinder structural damping is viscous and the cylinder mass is uniformly distributed, the equation of vertical motion of the cylinder can be written as:

$$\ddot{y} + \omega_c^2 y + 2\xi_c\omega_c\dot{y} + \frac{F}{m} = 0 \tag{5.28}$$

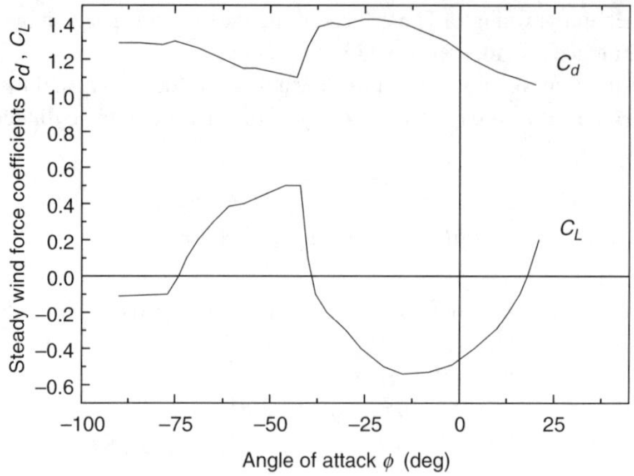

Figure 5.5 Aerodynamic coefficients vs. wind angle of attack (from [27]) (Reproduced with permission from Elsevier).

where:

ω_c is the circular natural frequency of the cylinder;
ξ_c is the structural damping ratio of the cylinder;
m is the mass of the cylinder per unit length.

Now fit the drag coefficient and lift coefficient in Equation 5.26 with respect to φ using the first three terms of Taylor's series:

$$C_L(\varphi) = A_o + A_1\varphi + \frac{A_2}{2}\varphi^2 + \frac{A_3}{6}\varphi^3 \tag{5.29a}$$

$$C_d(\varphi) = B_o + B_1\varphi + \frac{B_2}{2}\varphi^2 + \frac{B_3}{6}\varphi^3 \tag{5.29b}$$

The substitution of Equations 5.29a and 5.29b into Equation 5.26 and then the expansion of the sine and cosine functions in Equation 5.26 with respect to φ^* and the use of the first two terms of the power series yield the vertical force acting on the cylinder per unit length of the form:

$$F = \frac{\rho D U_{rel}^2}{2}\left[\left(A_0 + A_1\varphi + \frac{1}{2}A_2\varphi^2 + \frac{1}{6}A_3\varphi^3\right)\left(1 - \frac{\varphi^{*2}}{2}\right) + \left(B_0 + B_1\varphi + \frac{1}{2}B_2\varphi^2 + \frac{1}{6}B_3\varphi^3\right)\left(\varphi^* - \frac{\varphi^{*3}}{6}\right)\right] \tag{5.30}$$

The substitution of Equations 5.21, 5.23, 5.25 and 5.27 into Equation 5.30 and then the reservation of the linear terms of $\dot{y}$, θ and $\dot{\theta}$ only lead to

$$F = \frac{\rho D}{2}\left(\Gamma_1 R U_0 \dot{\theta} + \Gamma_2 U_0 \dot{y} + \Gamma_3 U_0^2 \theta\right) \tag{5.31}$$

where Γ_1, Γ_2, Γ_3 are the rain-wind aerodynamic functions. They are not only the function of the cable inclination, the wind yaw angle, the wind stagnation point and the mean wind speed via the static position of rivulet, but also the function of cable motion and rivulet motion, which may change the values of drag coefficient and lift coefficient.

$$
\begin{aligned}
\Gamma_1 = {} & \left[c_0c_4\left(1-\frac{c_5^2}{2}\right)+c_1c_5\left(1-\frac{c_5^2}{6}\right)\right]B_0+\left[c_0c_4c_5\left(1-\frac{c_5^2}{6}\right)+c_0c_4c_6\left(1-\frac{c_5^2}{2}\right)+c_1c_5c_6\left(1-\frac{c_5^2}{6}\right)\right]B_1 \\
& +\left[c_0c_4c_5c_6\left(1-\frac{c_5^2}{6}\right)+\frac{c_0c_4c_6^2}{2}\left(1-\frac{c_5^2}{2}\right)+\frac{c_1c_5c_6^2}{2}\left(1-\frac{c_5^2}{6}\right)\right]B_2 \\
& +\left[\frac{c_0c_4c_5c_6^2}{2}\left(1-\frac{c_5^2}{6}\right)+\frac{c_0c_4c_6^3}{6}\left(1-\frac{c_5^2}{2}\right)+\frac{c_1c_5c_6^3}{6}\left(1-\frac{c_5^2}{6}\right)\right]B_3 \\
& +\left[c_1\left(1-\frac{c_5^2}{2}\right)-c_0c_4c_5\right]A_0+\left[c_0c_4\left(1-\frac{c_5^2}{2}\right)+c_1c_6\left(1-\frac{c_5^2}{2}\right)-c_0c_4c_5c_6\right]A_1 \\
& +\left[c_0c_4c_6\left(1-\frac{c_5^2}{2}\right)-\frac{c_0c_4c_5c_6^2}{2}+\frac{c_1c_6^2}{2}\left(1-\frac{c_5^2}{2}\right)\right]A_2 \\
& +\left[\frac{c_0c_4c_6^2}{2}\left(1-\frac{c_5^2}{2}\right)-\frac{c_0c_4c_5c_6^3}{6}+\frac{c_1c_6^3}{6}\left(1-\frac{c_5^2}{2}\right)\right]A_3
\end{aligned} \tag{5.32a}
$$

$$
\begin{aligned}
\Gamma_2 = {} & \left[c_0c_3\left(1-\frac{c_5^2}{2}\right)+c_2c_5\left(1-\frac{c_5^2}{6}\right)\right]B_0+\left[c_0c_3c_5\left(1-\frac{c_5^2}{6}\right)+c_0c_3c_6\left(1-\frac{c_5^2}{2}\right)+c_2c_5c_6\left(1-\frac{c_5^2}{6}\right)\right]B_1 \\
& +\left[c_0c_3c_5c_6\left(1-\frac{c_5^2}{6}\right)+\frac{c_0c_3c_6^2}{2}\left(1-\frac{c_5^2}{2}\right)+\frac{c_2c_5c_6^2}{2}\left(1-\frac{c_5^2}{6}\right)\right]B_2 \\
& +\left[\frac{c_0c_3c_5c_6^2}{2}\left(1-\frac{c_5^2}{6}\right)+\frac{c_oc_3c_6^3}{6}\left(1-\frac{c_5^2}{2}\right)+\frac{c_2c_5c_6^3}{6}\left(1-\frac{c_5^2}{6}\right)\right]B_3 \\
& +\left[c_2\left(1-\frac{c_5^2}{2}\right)-c_0c_3c_5\right]A_0+\left[c_0c_3\left(1-\frac{c_5^2}{2}\right)+c_2c_6\left(1-\frac{c_5^2}{2}\right)-c_0c_3c_5c_6\right]A_1 \\
& +\left[c_0c_3c_6\left(1-\frac{c_5^2}{2}\right)-\frac{c_0c_3c_5c_6^2}{2}+\frac{c_2c_6^2}{2}\left(1-\frac{c_5^2}{2}\right)\right]A_2 \\
& +\left[\frac{c_0c_3c_6^2}{2}\left(1-\frac{c_5^2}{2}\right)-\frac{c_0c_3c_5c_6^3}{6}+\frac{c_2c_6^3}{6}\left(1-\frac{c_5^2}{2}\right)\right]A_3
\end{aligned} \tag{5.32b}
$$

$$
\Gamma_3 = -c_0c_5\left(1-\frac{c_5^2}{6}\right)\left(B_1+B_2c_6+\frac{1}{2}B_3c_6^2\right)-c_0\left(1-\frac{c_5^2}{2}\right)\left(A_1+A_2c_6+\frac{1}{2}A_3c_6^2\right) \tag{5.32c}
$$

in which:

$$c_0 = \cos^2\beta + \sin^2\alpha\sin^2\beta \tag{5.33a}$$

$$c_1 = \left[2-(\gamma-\theta_0)^2\right]\sqrt{c_0} \tag{5.33b}$$

$$c_2 = 2\sin\gamma\sqrt{c_0} \tag{5.33c}$$

$$c_3 = \frac{1}{\sqrt{c_0}\cos\gamma} \tag{5.33d}$$

$$c_4 = \frac{\theta_0 - \dfrac{\theta_0^3}{6}}{\sqrt{c_0}\cos\gamma} \tag{5.33e}$$

$$c_5 = \tan\gamma \tag{5.33f}$$

$$c_6 = c_5 - \theta_0 \tag{5.33g}$$

With the information on the orientation of a stay cable (α, β), the stagnation point influence factor (ε), the static position of upper rivulet related to the mean wind speed (θ_0) and the drag and lift coefficients of a cable segment with artificial rivulet measured from wind tunnel tests (A_i, B_i, $i=0, 1, 2, 3$), one can readily identify the aerodynamic functions Γ_1, Γ_2, Γ_3.

The combination of Equation 5.28 with Equation 5.31 results in:

$$\ddot{y} + {\omega_c}^2 y + \left(2\xi_c\omega_c + \frac{\rho D \Gamma_2 U_0}{2m}\right)\dot{y} = -\frac{\rho D}{2m}\left(\Gamma_1 R U_0 \dot{\theta} + \Gamma_3 U_0^2 \theta\right) \tag{5.34}$$

Based on the observations from either field measurements or simulated rain-wind tunnel tests [16], the motion of upper rivulet, θ, can be assumed to be harmonic as long as the steady-state vibration is concerned:

$$\theta = a \sin\widehat{\omega} t \tag{5.35}$$

The frequency of rivulet motion is almost the same as that of cable motion, as observed from rain-wind tunnel tests when the vibration is in steady state [16]. The amplitude of rivulet motion, a, is also taken from rain-wind tunnel tests. Thus, the motion of the rivulet is assumed as a known function of time in this section. Substituting Equation 5.35 into Equation 5.34 produces:

$$\ddot{y} + \omega_c^2 y + 2\tilde{\xi}\omega_c \dot{y} = -\frac{\rho D a}{2m}\left(\Gamma_1 R U_0 \widehat{\omega}\cos\widehat{\omega} t + \Gamma_3 U_0^2 \sin\widehat{\omega} t\right) = \tilde{F}(t) \tag{5.36}$$

in which:

$$\tilde{\xi} = \xi_c + \frac{\rho D \Gamma_2 U_0}{4m\omega} = \xi_c + \xi_a \tag{5.37}$$

where:

$\tilde{\xi}$ is the total damping ratio of the cylinder;
ξ_a is the rain-wind aerodynamic damping ratio;
$\tilde{F}(t)$ is the normalized force due to the motion of rivulet.

Since Γ_2 changes with the mean wind speed U_0 through the static position of rivulet, and depends on the motions of cable and rivulet, the aerodynamic damping ratio and the total damping ratio are a function of time.

5.4.3 *Horizontal Cylinder with Fixed Rivulet*

There are two approaches currently used to simulate rivulet on a cable section model in wind tunnel simulation tests: one is to spray water appropriately onto the surface of the cable model to form moving rivulet [33]; and the other is to stick artificial rivulet on the cable surface [32]. To start with the simplest case, this section investigates the dynamic behavior of horizontal cylinder with fixed rivulet using the developed model, then compares the results with the wind tunnel test results obtained by Gu *et al.* [32]. Note that this simplest case does not represent the RWIV of stay cables and it is only for exploring the mechanism of RWIV using the proposed analytical model.

When the rivulet is fixed, the dynamic motion $\theta(t)$ of the rivulet relative to the cylinder is equal to zero and the static position of the rivulet θ_0 is no longer the function of the mean wind speed. The equation of vertical motion of the horizontal cylinder becomes:

$$\ddot{y} + \omega_c^2 y + 2\tilde{\xi}\omega_c \dot{y} = 0 \tag{5.38}$$

Γ_2 in Equation 5.37 is also changed. For instance, Γ_2 can be expressed by the following equation for the horizontal cylinder with zero wind yaw angle β:

$$\Gamma_2 = B_0 - B_1\theta_0 + \frac{B_2}{2}\theta_0^2 - \frac{B_3}{6}\theta_0^3 + A_1 - A_2\theta_0 + \frac{A_3}{2}\theta_0^2 \tag{5.39}$$

For the cylinder with fixed rivulet, the coefficients A_i and B_i ($i=0$, 1, 2, 3) may change with the angle φ, which, in turn depends on the cylinder velocity $\dot{y}$ and the angle θ_0. Therefore, Equation 5.38 is a non-linear differential equation. The Runge-Kutta method, using the computer package MATLAB as a platform, is employed to find the solution of the equation of motion of the cylinder with rivulet.

Let us consider a horizontal cylinder of 120 mm diameter with a fixed upper rivulet [32]. The upper rivulet is modeled as a half ellipse, with a long axis of 14.5 mm and a short axis of 10 mm. The measured drag and lift coefficients are plotted in Figure 5.5. It can be seen that when φ is equal to $-43°$, the derivative of lift coefficient with angle of attack has a sudden change from a positive value to a negative value, whereas the derivative of the drag coefficient changes from a negative value to a positive value. To have the best fit of the measured aerodynamic coefficients using the first three terms of Taylor's series, the measured aerodynamic coefficients are divided into the two ranges, distinguished by the critical angle φ of $-43°$.

Figure 5.6a shows the time history of displacement response of the horizontal cylinder with fixed rivulet, obtained by using Equation 5.27 and Equations 5.37 to 5.39 with a small initial displacement. The cylinder has the following parameters: the mass per unit length of the cylinder m is 3.48 kg/m; the natural frequency f_n is 1.28 Hz; and the structural damping ratio ξ_s is 0.2%. The mean wind speed U_0 and yaw angle β are 10 m/s and zero, respectively. The air density is 1.225 kg/m^3 and the position of the fixed rivulet θ_0 is 32°.

The computed result shows that, after giving a small initial displacement, the vibration amplitude of the cylinder increases with time within the first 320 seconds, because the total damping ratio $\tilde{\xi}$ is negative and constant during this period (see Figure 5.6b). After the amplitude of the cylinder is increased to a certain level, the angle φ reaches the value of $-43°$. The aerodynamic damping ratio and the total damping ratio then change alternately between a negative value and a positive value, rather than remaining constant (Figure 5.6b). As a result of the alteration of aerodynamic damping, the motion of the cylinder becomes almost periodic, with a nearly constant peak-to peak-amplitude after 320 seconds.

To know the effect of position of the fixed rivulet θ_0 on cylinder vibration, the motion of the horizontal cylinder with a wind yaw angle of 30° is computed against a series of angles θ_0. The computed results are plotted in Figure 5.7a, together with the test results [32], in which A_{max} means the maximum peak to peak amplitude. A_{allow} means the allowable peak-to-peak amplitude, which was set as 530 mm in the simulated wind tunnel tests. The mean wind speed U_0 used in the computation is 18 m/s. $A_{max}/A_{allow} = 1$ indicates that the vibration amplitude is larger than the allowable amplitude.

It can be seen that the computed and measured results both demonstrate that, within a certain range of U_0, the cylinder has large amplitude vibration. However, out of this range, the cylinder has very small vibration, or it stops vibrating very quickly after a small initial displacement or disturbance. The

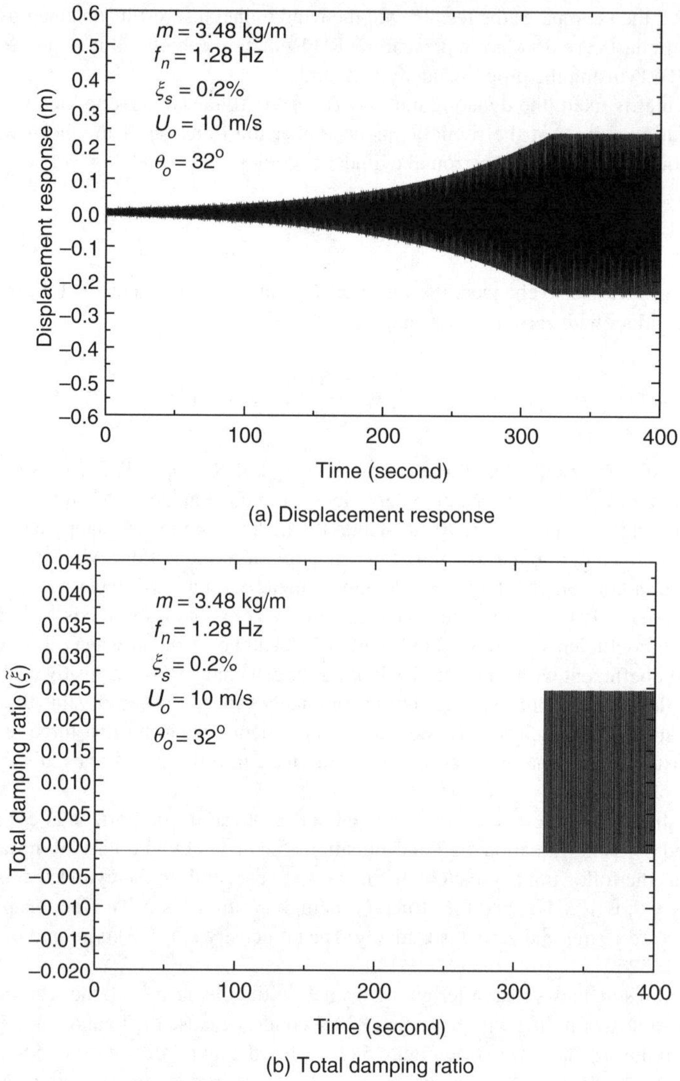

Figure 5.6 Time histories of horizontal cylinder vibration with fixed rivulet (from [27]) (Reproduced with permission from Elsevier).

cylinder starts its large amplitude vibration at 30°, measured by the wind tunnel test and also predicted by the analytical model. There is some difference in the upper bounds of the range between the computed and measured results.

Figure 5.7b shows the variation of the normalized vibration amplitude of the cylinder with mean wind speed for the case of 30° wind yaw angle and 35° static rivulet position. Again, the computed results are in good agreement with the test results. Both the wind tunnel test results and the computed results show a kind of amplitude-restricted vibration. The reason why the vibration amplitude is limited is that when the vibration amplitude of the cylinder reaches a certain level, the aerodynamic damping ratio and the total damping ratio change periodically between a positive peak value and a negative peak value.

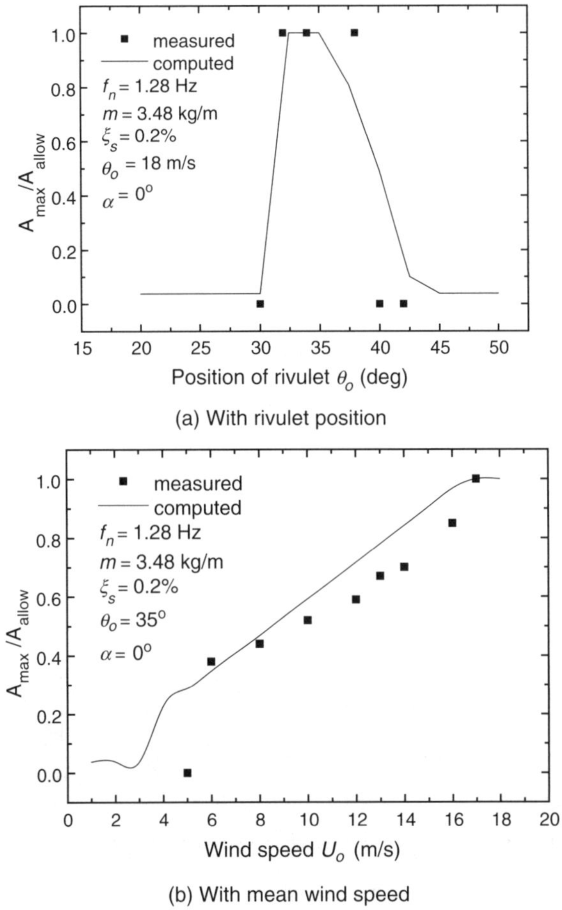

Figure 5.7 Variations of cylinder vibration amplitude (from [27]) (Reproduced with permission from Elsevier).

5.4.4 Inclined Cylinder with Moving Rivulet

To investigate the capability of the analytical model for predicting the motion of inclined cable with moving rivulet, the inclined cylinder tested in a simulated rain-wind tunnel by Hikami and Shiraishi [16] is selected in this study. The diameter of the cylinder is 140 mm; the mass per unit length is 10.2 kg/m; both the inclination and yaw angles are 45°; and the natural frequency is 1 Hz. The structural damping ratio and the stagnation influence factors are assumed to be 0.7% and 0.4, respectively. The frequency of steady-state upper rivulet motion is the same as the natural frequency of the cylinder [16]. The drag and lift coefficient curves of the cylinder with upper rivulet ($d/D = 0.1$) were reported by Yamaguchi [25] and are reproduced in Figure 5.8a.

It can be seen that there is a sudden change in the gradient of the curves at an angle of −55°. The curves are subsequently fitted to Taylor's series of the first three terms distinguished by the critical angle of −55°. For the moving rivulet, its static position θ_0 is the function of mean wind speed. The wind tunnel test results related to this position [16] are plotted in Figure 5.8b and fitted by a quadratic function.

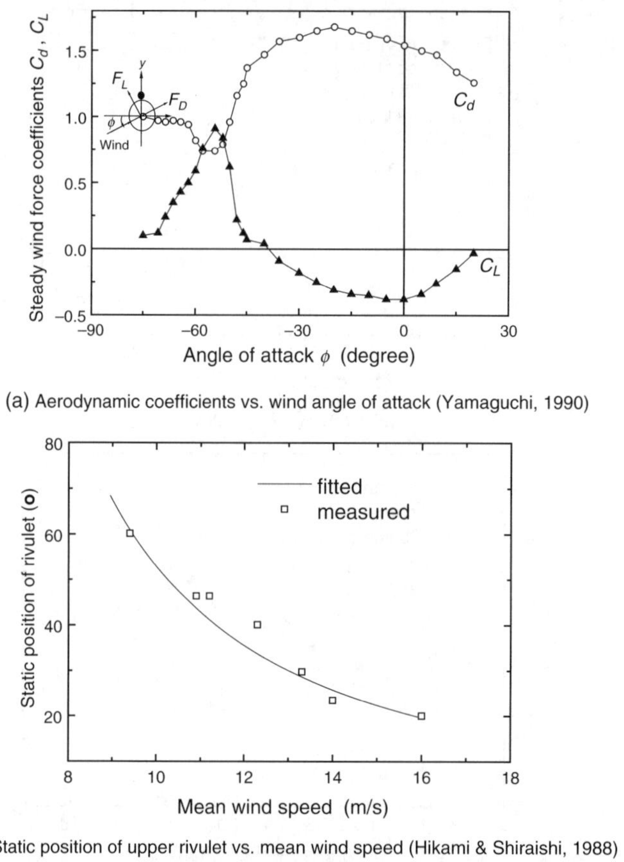

Figure 5.8 Aerodynamic properties of rain-wind-induced cylinder vibration (from [27]) (Reproduced with permission from Elsevier).

The results obtained from the analytical model are shown in Figure 5.9, together with the rain-wind tunnel test results. In this figure, the x-coordinate is the mean wind speed U_0 and the y-coordinate is the maximum displacement response amplitude of the cylinder. Both analytical and measured results show that the rain-wind-induced cylinder vibration only occurs within a certain range of mean wind speed. It is also a vibration of restricted amplitude rather than galloping. The predicted maximum vibration amplitude is moderately larger than that from the rain-wind tunnel tests but the onset mean wind speed at which the large cylinder vibration starts is slightly lower in the computation than in the rain-wind tunnel test. Such differences may be due to uncertainties in the selection of stagnation influence factor and structural damping ratio.

One particular mean wind speed corresponding to the global maximum displacement response, i.e. 9.8 m/s, is selected to examine cable vibration features. It is found that at this mean wind speed, the angle $\varphi(t)$ changes with time, and its negative amplitude is less than $-55°$. Thus, the aerodynamic damping ratio and the total damping ratio alternate with time. Because the structural damping ratio is relatively large, the total damping ratio is kept within the positive value range. However, the upper and lower bounds of the total damping ratio are quite small, but the peak-to-peak amplitude of excitation force due to the motion of rivulet is relatively larger. As a result, the cylinder exhibits a large amplitude vibration. One may thus conclude that the restricted large amplitude vibration of the cylinder at 9.8 m/s

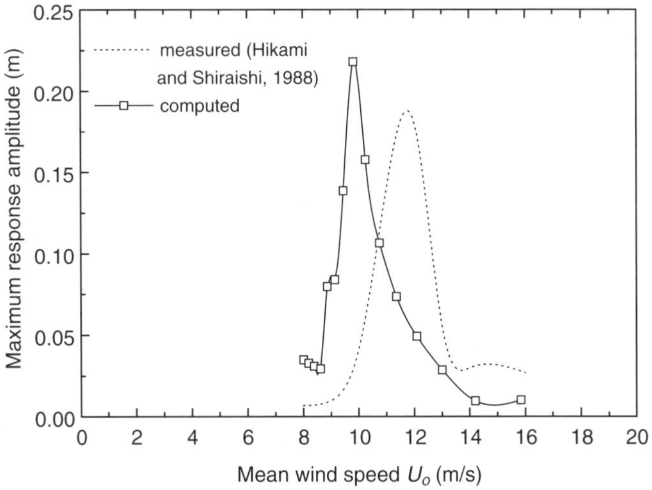

Figure 5.9 Maximum cylinder vibration amplitude vs. mean wind speed (from [27]) (Reproduced with permission from Elsevier).

wind speed is because of a very small total damping ratio and relatively large amplitude force due to the motion of rivulet.

The time-history of the displacement response of the cylinder is shown in Figure 5.10. During the first 50 seconds, the cylinder vibration is in a transient vibration period and the smallest amplitude of the angle $\varphi(t)$ is above $-55°$. This transient vibration period does not reflect the real situation because of the steady-state rivulet motion assumption used in this study. However, when the vibration amplitude of cylinder increases to a certain level, the smallest amplitude of the angle $\varphi(t)$ becomes less than $-55°$ and the aerodynamic damping ratio and the total damping ratio no longer remain constant. The

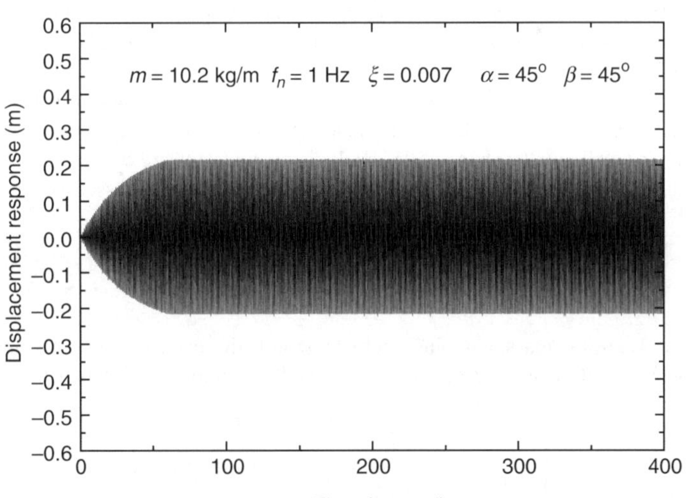

Figure 5.10 Time history of cylinder vibration with moving rivulet (from [27]) (Reproduced with permission from Elsevier).

normalized force also changes with time because of the interaction between incident wind, moving rivulet and cylinder motion. All of these features of rain-wind-induced cable vibration cause the cylinder to vibrate with almost constant amplitude.

5.4.5 Analytical Model of 2DOF

The discussions above show that the SDOF model could capture main features of rain-wind-induced cable vibration, such as velocity-restricted vibration and amplitude-restricted vibration. The occurrence of RWIV of the cylinder was mainly because of alternating aerodynamic damping and/or aerodynamic force due to the interaction between rivulet, cable and wind. However, the motion of rivulet was assumed to be a known harmonic motion in the SDOF model and, accordingly, only the steady-state RWIV could be predicted. In this section, the SDOF model is extended to a 2DOF model by including the equation of motion of the upper rivulet. In the 2DOF model, the rotating motion of the rivulet around the central axis of the cylinder is considered together with the transverse motion of the inclined cylinder. The interaction between the upper rivulet and the cylinder is described in terms of non-linear damping force, linear restoring force and inertia force [29].

For the upper rivulet, it can be assumed that the mean wind force, the supporting force from the cylinder and the gravity force on the rivulet all keep the rivulet in the static position θ_0. The dynamic equilibrium of the rivulet in the circumferential direction of the cylinder depends on the rivulet inertia force and the interacting force from the cylinder at the contacting surface with the rivulet. As the absolute acceleration of the rivulet is the superimposition of the cylinder vertical acceleration at the contacting point and the acceleration of the rivulet relative to the cylinder, the acceleration of the rivulet in the tangential direction at the contacting point with the cylinder is given by $\ddot{y}\,\sin(\theta+\theta_0)+R\ddot{\theta}$.

As well as the inertial force, the rivulet is subjected to turbulent wind force, aerodynamic damping force, restoring force due to water surface tension and friction force between water and cylinder surface. The restoring force and the friction force depend on many factors, such as the contacting surface roughness and the size and density of the rivulet. The turbulent wind force and the aerodynamic damping force are also difficult to be quantified. For the sake of simplification, the combined effect of all damping forces on the rivulet is approximately represented by $-c_r|\dot{x}|^{\alpha_r}\dot{x}$. Here, $\dot{x}$ is the relative velocity of the rivulet to the cylinder, which is equal to $R\dot{\theta}$; c_r is the damping coefficient of the rivulet; α_r is a predetermined exponent; $\alpha_r=0$ represents a viscous damping force; $\alpha_r=1$ indicates the damping force from an orifice of square law.

The restoring force due to water surface tension is assumed to be a linear function of the relative movement of the rivulet to the cylinder at this stage. The turbulence wind force is ignored, since the size of the rivulet is small and the shape of the rivulet is naturally toward the streamline. Within all the above considerations, the dynamic equation of motion of the rivulet is expressed as:

$$m_r\left(\ddot{\theta}+\frac{\ddot{y}\,\sin(\theta_0+\theta)}{R}\right)+c_r|\dot{\theta}|^{\alpha_r}\dot{\theta}+k_r\theta=0 \tag{5.40}$$

where m_r, c_r, and k_r are the mass, damping coefficient, and stiffness coefficient per unit length of the rivulet. Equation 5.36 and Equation 5.40 can be combined and simplified as follows:

$$\begin{cases} \ddot{y}+2\tilde{\xi}\omega_c\dot{y}+\omega_c^2 y=\tilde{F} \\ \ddot{\theta}+\dfrac{\ddot{y}\,\sin(\theta_0+\theta)}{R}+2\xi_r\omega_r|\dot{\theta}|^{\alpha_r}\dot{\theta}+\omega_r^2\theta=0 \end{cases} \tag{5.41}$$

Equation 5.41 is a strong non-linear equation. The 4th order Runge-Kutta method can be applied to find the solution of motion for both the cylinder and the rivulet. Figure 5.11 displays the maximum

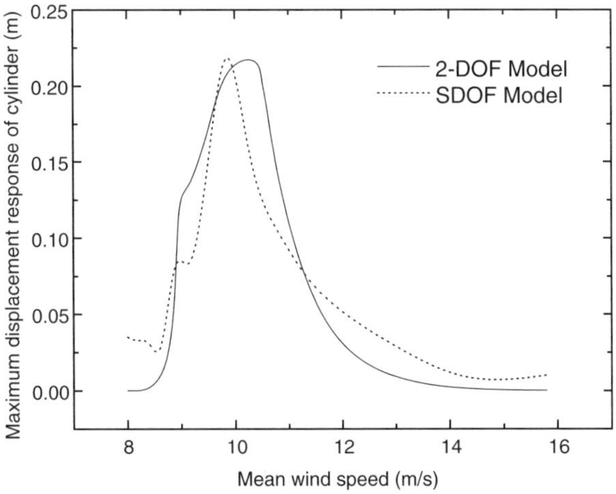

Figure 5.11 Comparison between 2DOF model and SDOF model (from [29]) (Reproduced with permission from Techno.Press).

displacement response of the cylinder against mean wind speed obtained by both the SDOF model and the 2DOF model. The parameters of the cylinder are the same for both the computed cases.

It can be seen that, at very low and high mean wind speeds, the vibration amplitude of the cylinder predicted by the SDOF model is slightly larger than that predicted by the 2DOF model. Within the range around the critical mean wind speed, the vibration amplitude of the cylinder predicted by the 2DOF model is slightly larger than that computed by the SDOF model. In comparison with the measured results shown in Figure 5.9, the computed curve of vibration amplitude of the cylinder by the 2DOF model is slightly closer to the measured results than that computed from the SDOF model. It seems that the two degrees of freedom model is a better model to predict and explain the phenomena and mechanism of rain-wind-induced cable vibration than the single degree of freedom model. Nevertheless, the single degree of freedom model is acceptable to obtain the global maximum vibration response of the cylinder in respect of its simplicity. More detailed information on the 2DOF model can be found in [29].

It should be pointed out that the above analytical models of rain-wind-induced vibration did not consider the effects of axial flow and wind turbulence, so their effects on RWIV need further investigation. Moreover, many new investigations are emerging using either wind tunnel experiments [20,34,35], full-scale measurements [36], analytical methods [37] or CFD numerical simulation methods [38]. The new results shed light on the mechanism of RWIV of stay cables.

5.5 Prediction of Rain-Wind-Induced Cable Vibration

Although analytical models could predict major features of RWIV observed from field measurements and rain-wind tunnel tests, they are based on cylinder models without considering the variation of mean wind speed along a stay cable and the effect of mode shapes of cable vibration. The comparison between the analytical and field measurement results cannot be performed satisfactorily, and rain-wind tunnel test results alone cannot be used for full-scale stay cables. Furthermore, damping devices such as viscous dampers have been installed on stay cables near anchorages to mitigate RWIV, even though the mechanism of RWIV has not been fully understood. There is almost no analytical model available to evaluate the effectiveness of viscous dampers for vibration control of full-scale stay cables under rain-wind excitation.

This section thus presents an analytical model for investigating RWIV of stay cables in a cable-stayed bridge. The single degree of freedom (SDOF) rain-wind excitation model developed in Section 5.4 is first applied to a taut stay cable, taking into consideration the variation of mean wind speed along the stay cable and the effect of mode shapes of cable vibration. The computed results from the analytical model are then compared with available field measurement results. After a satisfactory comparison, parameter studies are performed to explore the mechanism of RWIV of full-scale stay cables.

5.5.1 Analytical Model for Full-Scale Stay Cables

Most stay cables in a cable-stayed bridge are of small sag in the order of 1% sag-to-length ratio, but with a high tension-to-weight ratio [3]. As a preliminary analytical study, the effects of sag and bending stiffness on cable vibration are neglected and only a taut inclined cable is considered and shown in Figure 5.12.

The inclination of the taut cable is denoted by angle α, and the yaw angle of incident wind is designated by angle β. The in-plane transverse motion $w(x, t)$ of a taut cable under rain-wind excitation can be described by the following partial differential equation of motion, which is essentially the same as Equation 5.1:

$$m\frac{\partial^2 w}{\partial t^2} - T\frac{\partial^2 w}{\partial x^2} + c\frac{\partial w}{\partial t} + F(x,t) = 0 \tag{5.42}$$

with the boundary conditions $w(x, t) = w(L, t) = 0$.

In Equation 5.42: $w(x, t)$ is the in-plane transverse displacement normal to the cable axis at position x;
m is the cable mass per unit length;
T is the cable tension force;
c is the internal cable damping coefficient;
$F(x, t)$ is the rain-wind induced force on the cable;
L is the length of the cable.

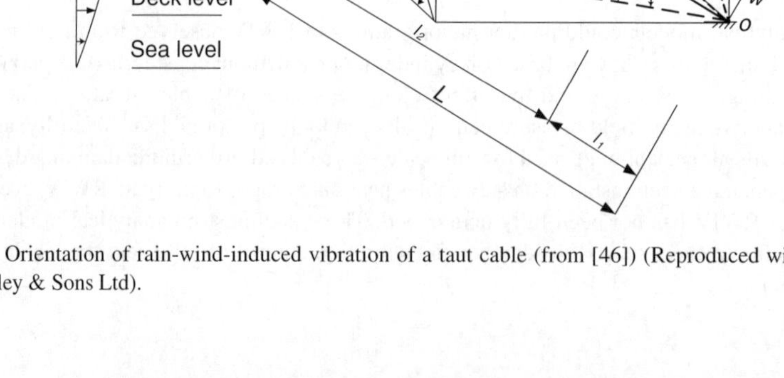

Figure 5.12 Orientation of rain-wind-induced vibration of a taut cable (from [46]) (Reproduced with permission from John Wiley & Sons Ltd).

By assuming that the upper rivulet vibrates circumferentially over the surface of the cable, rain-wind-induced force normal to the cable axis per unit length in the vertical direction, without considering turbulence and axis flow effects, can be expressed in terms of Equation 5.31:

$$F(x,t) = \frac{\rho D U_0(x)}{2}\left[R\dot{\theta}(x,t)\Gamma_1(x,t) + \dot{w}(x,t)\Gamma_2(x,t) + U_0(x)\theta(x,t)\Gamma_3(x,t)\right] \tag{5.43}$$

where:

R is the cable radius;
Γ_1, Γ_2, Γ_3 are the rain-wind aerodynamic functions.

The rain-wind aerodynamic functions not only include the function of cable inclination angle, wind yaw angle, wind stagnation point, and the mean wind speed via the static position of rivulet, but also the function of cable motion and rivulet motion, which may change the values of drag and lift coefficients as expressed by Equations 5.32 and 5.33.

In RWIV, the motion of upper rivulet is actually coupled with the motion of the cable. Therefore, the motion of upper rivulet should be taken as a variable and predicted together with the motion of cable in principle. This, however, requires a complicated analytical model for RWIV. Based on the observations from either field measurements or simulated rain-wind tunnel tests, the motion of upper rivulet, $\theta(x, t)$, is assumed to be harmonic and to follow the excited mode shape of the cable so far as a steady-state cable vibration is concerned.

$$\theta(x,t) = a \cdot W(x)\sin(\omega_r t) \tag{5.44}$$

where:

a is the maximum amplitude of rivulet motion, assumed to be a constant in this study;
$W(x)$ is the motion profile of rivulet along the cable axis, assumed to be the same as the excited mode shape of cable;
ω_r is the frequency of rivulet motion, assumed to be the same as the natural frequency of cable corresponding to the excited mode shape.

For a taut cable without damper, the motion profile and frequency of rivulet corresponding to the ith mode of cable vibration are given by:

$$W(x) = \sin\left(\frac{\pi i}{L}x\right) \tag{5.45}$$

$$\omega_i = i\omega_{o1} \tag{5.46}$$

where $\omega_{o1} = \frac{\pi}{L}\sqrt{\frac{T}{m}}$ is the first circular natural frequency of the taut cable.

For a stay cable in a cable-stayed bridge, mean wind speed actually varies along with cable axis. Mean wind speed is zero at the sea level, and it increases with height above the sea level in the atmospheric boundary layer as shown in Figure 5.12. The mean wind speed profile could be approximately expressed by a power law function:

$$U_0(h) = U_r\left(\frac{h}{h_r}\right)^{\alpha_1} \tag{5.47}$$

where:

$U_0(h)$ is the mean wind speed at height h above sea level;
U_r is the mean wind speed at a chosen reference height h_r;
α_1 is the exponent depending on the surface roughness.

Since the height h can be expressed as a function of x along the cable axis, Equation 5.47 can be rewritten as:

$$U_0(x) = U_0(h(x)) = U_r\left(\frac{h_0}{h_r} + \frac{x\sin\alpha}{h_r}\right)^{\alpha_1} \tag{5.48}$$

where h_0 is the height of lower anchorage of the cable above the sea level.

Equation 5.42 is a strongly non-linear equation due to physical interaction between cable, rivulet, and wind. The rain-wind-induced force expressed in terms of Equation 5.43 is also an implicit function of cable motion and rivulet motion. The Runge-Kutta-Fehlberg method is therefore employed to find the numerical solutions [39]. This method is adaptive and easily implemented for the concerned numerical problem, in which a stay cable is divided into a number of sections along the cable axis and the numerical solutions are found at these sections for a series of time intervals using the relevant shape functions.

When the Runge-Kutta-Fehlberg method is used, only six evaluations are required per step, and local truncation error can be kept below a prescribed value by varying the step length. Note that rain-wind-induced force depends on static rivulet position and cable motion, while the static rivulet position varies with wind speed. Rain-wind-induced force on the cable needs to be computed section by section, starting from the static rivulet position and cable motion at the previous time step.

5.5.2 *Prediction of Rain-Wind-Induced Vibration of Full-Scale Stay Cable*

In this section, the analytical model and the numerical method proposed above are applied to a stay cable in a real cable-stayed bridge to compute its rain-wind-induced response and the computed response is compared with the measured one.

Stay cable No.14 of the Meikonishi West Bridge, as investigated by Hikami and Shiraishi [16] has been selected for this purpose. The length of the cable is 75 m; its diameter is 0.14 m; the mass per unit length is 51 kg/m; the tension force in the cable is 1.147×10^6 N; the first structural damping ratio of the cable is estimated at 0.0011; the inclination of the cable is about 45°; and the height of lower anchorage of the cable above the sea level h_0 is 45 m.

The field measurement carried out by Hikami and Shiraishi [16] recorded that when cable No. 14 was down inclined along wind direction, with a wind yawed angle of about 45°, the simultaneous occurrence of wind and rain caused excessive vibration of the cable in the first mode of vibration, with a natural frequency about 1 Hz. They also conducted a series of wind tunnel tests, with rain conditions simulated, and reproduced rain-wind-induced cable vibration in their model tests. They observed from their model tests that the rivulet oscillated in a circumferential direction at the same period of the cable motion, and that the amplitude of rivulet motion was around 10°. The drag and lift coefficient curves of the cylinder with artificial upper rivulet, and the rivulet static position curve as the function of mean wind speed, were also given by Yamaguchi [25] and Hikami and Shiraishi [16], respectively. These are reproduced in Figures 5.8a and 5.8b respectively.

In addition to the information provided above, the power exponent α_1 in the wind profile is selected as 0.16 for an open fetch. The stagnation point influence factor ε is selected as 0.4. The wind reference height h_r is 47 m above the sea level and the mean wind speed at the reference height is taken as a variable. The cable is then divided into 50 sections along the cable axis. The basic parameters used in the simulation of rain-wind-induced cable vibration are listed in Table 5.1. The analytical model and the

Table 5.1 Basic parameters used in the simulation of rain-wind-induced cable vibration (from [46]) (Reproduced with permission from John Wiley & Sons Ltd)

Parameters	L (m)	D (m)	T ($\times 10^6$N)	m (kg/m)	ζ_s	α (°)	β (°)	h_0 (m)	h_r (m)	ε	α_1
Values	75	0.14	1.147	51	0.11%	45	45	45	47	0.4	0.16

numerical method proposed are finally applied to the cable, with proper initial conditions and an error tolerance of 10^{-4}, to compute rain-wind-induced displacement response of the cable mode by mode.

Rain-wind-induced maximum displacement responses of cable No. 14, computed for the first three modes of vibration, are plotted in Figure 5.13 against different mean wind speeds at the reference level. The measured maximum displacement responses of the cables in the field are also presented in Figure 5.13, which includes cable No. 14 with its first two modes of vibration and cable No. 15 with its second and third modes of vibration [16].

Although rain-wind-induced cable vibration is a very complicated air-fluid-solid interaction problem, the computed results based on the proposed rain-wind-induced force model could capture the main features observed in the field. Both the computed and measured results demonstrate that rain-wind-induced cable vibration only occurs within a certain range of mean wind speed. It is also a kind of vibration of restricted amplitude other than galloping.

The predicted global maximum response is, however, larger than the measured one. This is because there exist uncertainties in both the measurement data and the modeling of rain-wind-induced cable vibration. For instance, the so-called measured maximum response was actually calculated from the response measured at a point 2 m above the deck level by assuming a sinusoidal mode shapes of cable vibration [16], the stagnation point influence factor used is approximate, and the influence of turbulent intensity and axial flow is neglected. Well-planned and carefully implemented full-scale measurements and wind tunnel tests are needed in the future in order to improve the quality of comparison and to refine the proposed model if necessary.

The computed displacement time history of the cable at its middle point in the first mode of vibration is displayed in Figure 5.14. It can be seen that rain-wind-induced response increases very quickly at the beginning, but the increasing rate becomes smaller and smaller. After about 300 seconds, the response amplitude remains constant, with the amplitude being about 0.22 m. The computed rain-wind-induced maximum displacement responses depicted in Figure 5.13 also manifest that for cable No. 14, the maximum displacement response in the first mode of vibration is largest while the maximum

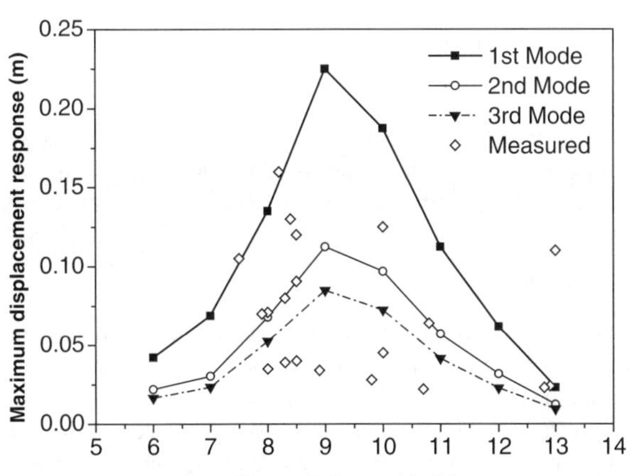

Figure 5.13 Computed and measured rain-wind-induced maximum displacement responses (from [46]) (Reproduced with permission from John Wiley & Sons Ltd).

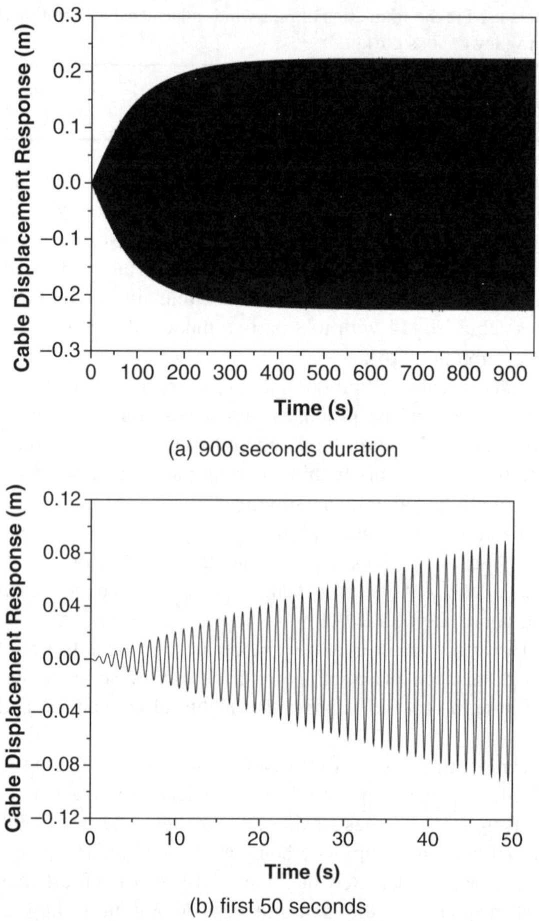

Figure 5.14 The global maximum displacement response time history of the cable in the first mode of vibration (from [46]) (Reproduced with permission from John Wiley & Sons Ltd).

displacement response in the third mode of vibration is at its smallest. However, the unfavorable mean wind speed range in which rain-wind-induced cable vibration occurs is almost the same for the first three modes of vibration.

5.5.3 Parameter Studies

To understand further the mechanisms of rain-wind-induced vibration of the stay cable, the effects of cable tension force, wind profile and cable internal damping are investigated. The investigation is performed by altering one parameter while keeping all other parameters unchanged.

Let us first consider the effect of cable tension on rain-wind-induced response of cable No. 14. While all the other parameters of the cable and the rivulet are kept the same as those used in the last section, the tension force of the cable is taken as a variable. The values of the tension force are selected as 0.765×10^6 (2/3 the original tension force T), 1.147×10^6 (the original tension force T) and $1.529 \times 10^6 N$ (4/3 the original tension force T).

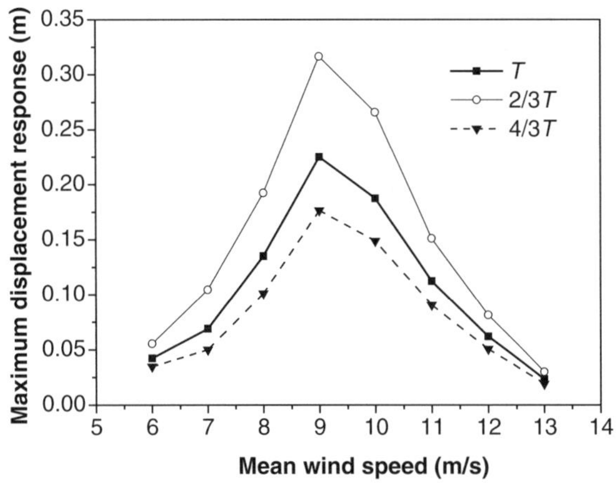

Figure 5.15 Effect of cable tension force on cable response in the first mode of vibration (from [46]) (Reproduced with permission from John Wiley & Sons Ltd).

Figure 5.15 shows rain-wind-induced maximum displacement response of the cable in the first mode of vibration against the mean wind speed at the reference level for three tension forces. It can be seen that the cable vibration remains velocity-restricted and amplitude-restricted features under three different tension forces. The tension force in the cable does not change the unfavorable mean wind speed range in which rain-wind-induced cable vibration occurs. The critical mean wind speed at which the global maximum displacement response of the cable occurs is almost the same under three different tension forces. However, as the tension force decreases (i.e. the cable becomes more flexible), the displacement responses of the cable increase. Similar results are found in the second and third modes of vibration, but with relatively small displacement responses.

Let us now consider the influence of mean wind speed profile on rain-wind-induced response of cable No. 14. The power exponent α_1 in the wind profile reflects the change of terrain. The increase in power exponent α_1 corresponds to the change of terrain from open sea to overland. While all the other parameters of the cable and the rivulet remain unchanged, including the reference level of mean wind speed, α_1 is taken as a variable and is selected as 0.16, 0.28 and 0.40 respectively.

Figure 5.16 shows the rain-wind-induced maximum displacement responses of the cable in the first and second modes of vibration against the mean wind speed at the reference level for three different power exponents. It can be seen that rain-wind-induced vibration of the cable remains velocity-restricted and amplitude-restricted for three different power exponents. However, as α_1 increases, the unfavorable wind speed range in which rain-wind-induced cable vibration occurs, and the critical mean wind speed at which the global maximum displacement response of the cable occurs, shift moderately to the lower wind speed side. This is because the change of power exponent of mean wind speed actually alters the mean wind speed which, in turn, changes the static position of upper rivulet and then the lift coefficient and, finally, the unfavorable wind speed range. On the other hand, the change in cable tension force does not affect the static position of rivulet and, therefore, it does not affect unfavorable wind speed range.

It can also be seen from Figure 5.16 that the global maximum displacement response of the cable at the critical mean wind speed also decreases as the power exponent increases. In addition to the effect of turbulence, one may say that rain-wind-induced cable vibration will be less if the terrain is overland.

The last parameter considered is the internal structural damping ratio in cable No. 14. While all other parameters of the cable and the rivulet are kept the same as before, the internal damping ratio is taken as a variable. For the first mode of vibration, the internal modal damping ratio is selected as 0.0011,

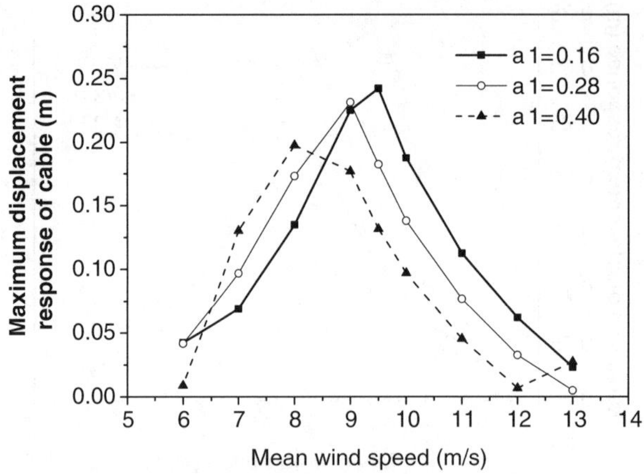

(a) the first mode of vibration

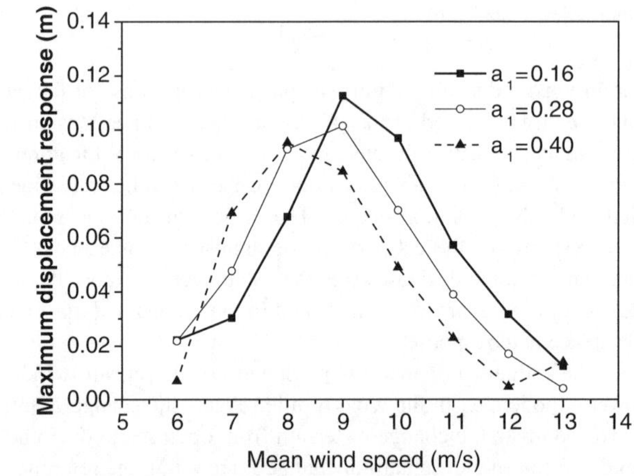

(b) the second mode of vibration

Figure 5.16 Effect of power exponent on cable response (from [46]) (Reproduced with permission from John Wiley & Sons Ltd).

0.0065 and 0.0111, and the computed rain-wind-induced maximum displacement responses of the cable are plotted in Figure 5.17 against the mean wind speed at the reference level. It can be seen that an increase of structural damping ratio to 0.65% can almost suppress rain-wind-induced cable vibration. Similar results are found for the second and third modes of vibration. It can thus be concluded that increasing cable damping is a very effective way to mitigate rain-wind-induced cable vibration. Rain-wind-induced vibration control of a stay cable will be discussed in Chapter 12.

5.6 Occurrence Probability of Rain-Wind-Induced Cable Vibration

To facilitate the engineering decision on whether or not RWIV will occur and anti-vibration measures of stay cables should be implemented, this section presents a framework for estimating the occurrence probability of RWIV in stay cables. The proposed framework is founded on the statistical analysis of wind speed, wind direction and rainfall intensity and the theoretical analysis of occurrence range of

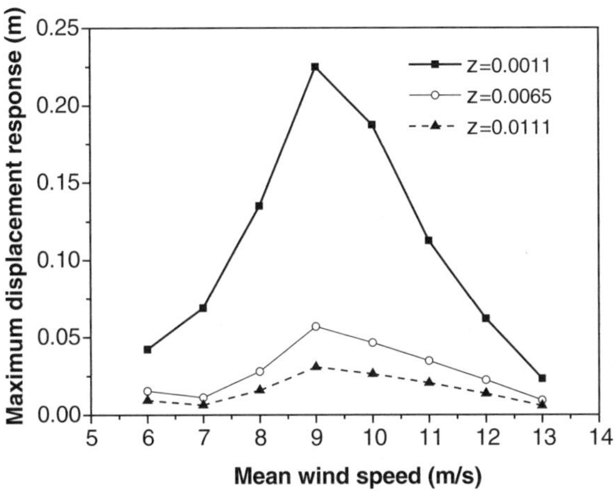

Figure 5.17 Effect of cable structural damping ratio on cable response in the first mode of vibration (from [46]) (Reproduced with permission from John Wiley & Sons Ltd).

RWIV in stay cables. The joint probability density function of wind speed and wind direction, and the probability density function of rainfall intensity are first obtained for the site of a bridge through statistical analysis. The occurrence range of wind speed and wind direction for RWIV in a stay cable greater than a vibration amplitude threshold value is then found based on a theoretical model described in Section 5.5. The occurrence probability and risk of RWIV are finally calculated according to the probability theory.

Figure 5.18 is a flow chart of the proposed framework, showing the estimation procedure for occurrence probability. The mathematical derivations and assumptions involved in each step are discussed in detail in subsequent sections. The practical use of the proposed framework is demonstrated in this section by taking a typical stay cable in a real cable-stayed bridge as a case study.

5.6.1 Joint Probability Density Function (JPDF) of Wind Speed and Direction

Excessive RWIV occurs mainly under low or moderate mean wind speeds. Therefore, a complete population of wind speed is of greater concern to this study than extreme values of wind speed. Various probability density functions (PDF) have been proposed to model a complete population of wind speed [40,41], but the convenience of the two-parameter Weibull distribution has encouraged its greater use than the other distributions [42]. With the lower limit being zero, the cumulative distribution function (CDF) and probability density function (PDF) of the Weibull form are, respectively:

$$P_u(U) = 1 - \exp\left[-\left(\frac{U}{c}\right)^k\right] \tag{5.49}$$

$$f_u(U) = \frac{k}{c}\left(\frac{U}{c}\right)^{k-1}\exp\left[-\left(\frac{U}{c}\right)^k\right] \tag{5.50}$$

where:

U is the wind speed;
$c(>0)$ is the scale parameter with the same unit as the wind speed;
$k(>0)$ is the shape parameter without dimension.

One possible weakness in the two-parameter Weibull distribution for wind speed is that it neglects the effect of wind direction [43,44]. Since excessive RWIV is closely related to wind direction, a joint

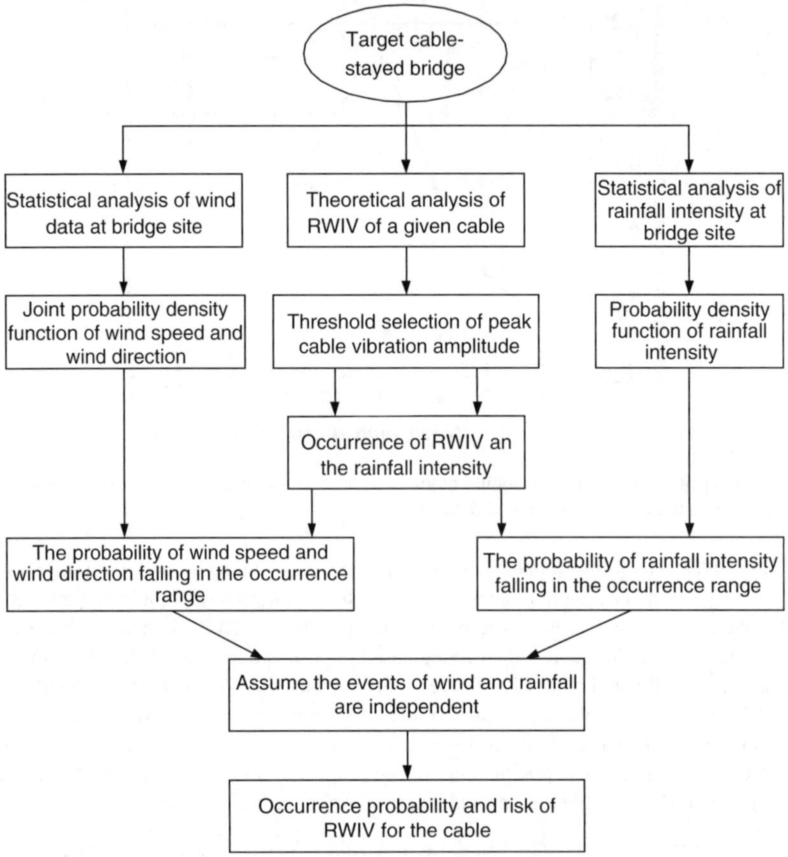

Figure 5.18 Flow chart of proposed probability-based framework (from [48]) (Reproduced with permission from Multi-Science Publishing Co Ltd).

probability distribution of wind speed and wind direction has to be used for the study. To this end, a practical joint probability distribution function (see Chapter 2) is adopted in this study for a complete population of wind speed and wind direction based on two assumptions:

1. the distribution of the component of wind speed for any given wind direction follows the Weibull distribution;
2. the interdependence of wind distribution in different wind directions can be reflected by the relative frequency of occurrence of wind.

$$P_{u,\theta}(U,\theta) = P_{\theta}(\theta)\left\{1 - \exp\left[-\left(\frac{U}{c(\theta)}\right)^{k(\theta)}\right]\right\} = \iint f_{\theta}(\theta) f_{u,\theta}(U,k(\theta),c(\theta))dud\theta \tag{5.51a}$$

$$f_{u,\theta}(U,k(\theta),c(\theta)) = \frac{k(\theta)}{c(\theta)}\left(\frac{U}{c(\theta)}\right)^{k(\theta)-1} \exp\left[-\left(\frac{U}{c(\theta)}\right)^{k(\theta)}\right] \tag{5.51b}$$

$$P_{\theta}(\theta) = \int_0^{\theta} f_{\theta}(\theta)d\theta \tag{5.51c}$$

where:

$0 \leq \theta \leq 2\pi$;
$P_\theta(\theta)$ is the relative frequency of occurrence of wind in wind direction θ.

The occurrence frequency $P_\theta(\theta)$, as well as the distribution parameters $k(\theta)$ and $c(\theta)$, can be estimated using wind data recorded at the bridge site.

5.6.2 *Probability Density Function of Rainfall Intensity*

Rainfall intensity is another crucial factor for the onset of rain-wind-induced cable vibration. Either very light or very heavy rainfall could not provoke the relatively stable rivulet to be formed on the upper surface of a stay cable [19]. Statistical analysis of rainfall data at the bridge site is thus necessary to determine the probability distribution function of rainfall intensity, by which the probability of occurrence range of rainfall intensity that would give rise to cable vibration can be identified.

Suppose A and B denote, respectively, the rainy event and the rainy event that might induce cable vibration. The probability of event B is clearly the probability that both events A and B occur. This can be expressed as:

$$P_r(B) = P_r(A \cap B) = P_r(A)P_r(B|A) \tag{5.52}$$

The probability of event B can be determined by two ways: one is to calculate $P_r(B)$ by directly considering all the rainy events that might induce cable vibration; the other is to calculate the probability of event A and the conditional probability $P_r(B|A)$ to obtain $P_r(B)$. In this study the latter method is used, in order to be consistent with the method used by climatologists. To calculate the conditional probability $P_r(B|A)$, the probability density function of rainfall intensity should be established.

There are various probability density functions (PDF) proposed to model rainfall intensity [45]. The Gamma and Weibull distributions are explored in this study. The mathematical expression of the Gamma and Weibull probability density functions are, respectively, given by:

$$f_r(R) = \frac{1}{b^a \Gamma(a)} R^{a-1} e^{-\frac{R}{b}} \tag{5.53}$$

$$f_r(R) = \frac{d}{c}\left(\frac{R}{c}\right)^{d-1} e^{-\left(\frac{R}{c}\right)^d} \tag{5.54}$$

where:

$\Gamma(a)$ is the Gamma function $= \int_0^\infty y^{a-1} e^{-y} dy$;
R is the rainfall intensity;
a and b are the shape parameter and the scale parameter respectively for the Gamma distribution;
c and d are the shape parameter and the scale parameter, respectively, for the Weibull distribution.

5.6.3 *Occurrence Range of Rain-Wind-Induced Cable Vibration*

The occurrence range of wind speed and wind direction in which large-amplitude cable vibration will occur can be computed based on a recently-developed theoretical model presented in Section 5.5 [46].

Due to physical interaction between cable motion, rivulet motion and wind, the equation of motion of a taut cable under rain-wind excitation is non-linear. The Runge-Kutta-Fehlberg method is employed

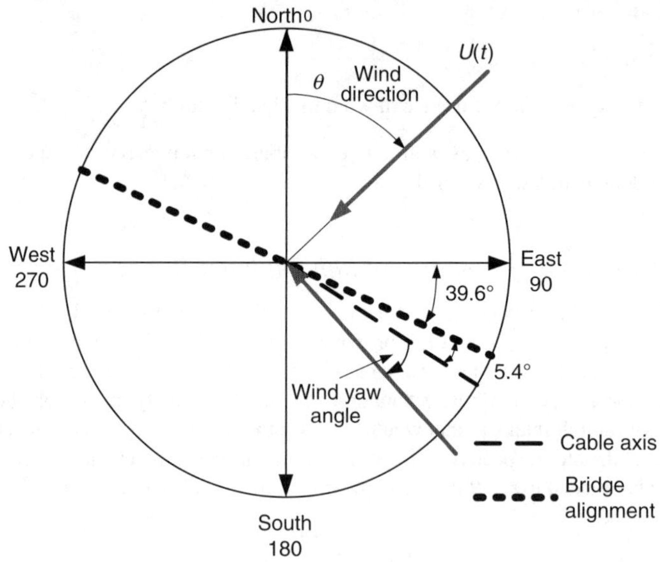

Figure 5.19 Definition of wind direction and wind yaw angle (from [48]) (Reproduced with permission from Multi-Science Publishing Co Ltd).

to find the numerical solution. The peak vibration amplitude, $A(U, \beta)$, of a given stay cable with a given inclination under rain-wind excitation is computed for a series of mean wind speed, U, and wind yaw angle, β. It should be pointed out that β is presented with reference to the cable axis while the mean wind direction, θ, is counted with respect to the north in clockwise, as shown in Figure 5.19.

Thus, an angle transformation is necessary to transform $A(U, \beta)$ to $A(U, \theta)$ for a given stay cable in a given cable-stayed bridge. Once $A(U, \theta)$ is obtained, a threshold value A_s, beyond which RWIV is regarded unacceptable, can be selected. The occurrence range of wind speed and wind direction in which cable vibration amplitude exceeds the threshold value A_s can be determined and denoted as:

$$\Omega_{u,\theta}^{A_s} = \{U \in [U_1, U_2] \cap \theta \in [\theta_1, \theta_2] | A(U, \theta) \geq A_s\} \tag{5.55}$$

where U_1, U_2 and θ_1, θ_2 are the lower and upper bound of occurrence range of wind speed and wind direction, respectively.

5.6.4 Occurrence Probability of Rain-Wind-Induced Cable Vibration

The probability of wind speed and wind direction falling in the occurrence range $\Omega_{u,\theta}^{A_s}$ can be calculated as:

$$P_{u,\theta}\left\{(U, \theta) \in \Omega_{u,\theta}^{A_s}\right\} = \iint\limits_{\Omega_{u,\theta}^{A_s}} f_\theta(\theta) f_{u,\theta}(U, k(\theta), c(\theta)) du d\theta \tag{5.56}$$

Suppose Ω_r denotes the occurrence range of rainfall intensity for rain-wind-induced cable vibration. The probability of rainfall intensity falling in the occurrence range Ω_r can be calculated as $P_r(B|A) = P_r(B \in \Omega_r) = \int_{\Omega_r} f_r(R) dR$.

The occurrence probability of rain-wind-induced cable vibration with cable vibration amplitude larger than the threshold value can be finally calculated by:

$$P\Big\{(U,\theta) \in \Omega_{u,\theta}^{A_s} \cap B \in \Omega_r\Big\} = P_r(A)P_r(B \in \Omega_r)P_{u,\theta}\Big((U,\theta) \in \Omega_{u,\theta}^{A_s} | B \in \Omega_r\Big) \tag{5.57}$$

Assume that the two events of rain and wind occurrence are independent at this stage. Equation 5.57 can be reduced to:

$$P\Big\{(U,\theta) \in \Omega_{u,\theta}^{A_s} \cap B \in \Omega_r\Big\} = P_r(A)P_r(B \in \Omega_r)P_{u,\theta}\Big((U,\theta) \in \Omega_{u,\theta}^{A_s}\Big) \tag{5.58}$$

This assumption may be reasonably valid for monsoon situations in Hong Kong, in consideration that RWIV occurs mainly at low and moderate wind speeds, but it is not necessarily applicable to typhoon situations or other climates. Nevertheless, the framework proposed here can be extended without any considerable difficult if the correlation between the two events can be quantified in the future.

In terms of probability density functions of wind speed, wind direction and rainfall intensity, Equation 5.58 can be further written as:

$$P\Big\{(U,\theta) \in \Omega_{u,\theta}^{A_s} \cap B \in \Omega_r\Big\} = P_r(A)\int_{\Omega_r} f_r(R)dR \iint_{\Omega_{u,\theta}^{A_s}} f_\theta(\theta)f_{u,\theta}(U,k(\theta),c(\theta))dud\theta \tag{5.59}$$

5.7 Case Study: Stonecutters Bridge

In this section, the proposed method is applied to the Stonecutters cable-stayed bridge to demonstrate its practical use. Hourly mean wind data and hourly rainfall data recorded near the bridge site are first analyzed to obtain the probabilistic models of both wind and rainfall intensity. By taking one typical cable as an example, the occurrence range of RWIV is then determined. The hourly occurrence probability and annual risk of rain-wind-induced vibration of the example cable are finally computed for different threshold values.

5.7.1 Statistical Analysis of Wind Data

There is a wind monitoring system consisting of four propeller anemometers and two ultrasonic anemometers set up nearby the concerned cable-stayed bridge. Wind records of hourly mean wind speed and wind direction within the period between 1 January 2000 to 31 December 2005 from the propeller anemometer are used in this study to find the joint probability density function of hourly mean wind speed and direction. The height of the propeller anemometer is 214 m above the sea level.

The wind records are carefully checked and abnormal records are eliminated first. Wind records having an hourly mean wind speed lower than 1 m/sec are then removed in order to avoid any adverse effect on the statistics. These qualified data are further divided into two categories – monsoon and typhoon – according to the records of the typhoon warning signal hoisted by the Hong Kong Observatory (HKO) during the period concerned. As a result, 19 775 hourly monsoon records are available for calculation of the joint probability density function of wind speed and direction.

All of the monsoon records are classified into 16 sectors of the compass, with an interval of $\Delta\theta = 22.5°$ according to the hourly mean wind direction. In each sector, mean wind speed is further divided into 16 ranges, from zero to 32 m/sec, with an interval of $\Delta U = 2$ m/sec. This leads to a total of 256 cells, and the relative frequency of hourly mean wind speed and wind direction in each cell is listed in Table 5.2. The last row in Table 5.2 gives the relative frequency of mean wind speed without

Table 5.2 Relative frequencies of hourly mean wind speed and wind direction (from [48]) (Reproduced with permission from Multi-Science Publishing Co Ltd)

Direction	0–2	2–4	4–6	6–8	8–10	10–12	12–14	14–16	16–18	18–20	20–22	22–24	24–26	26–28	28–30	30–32	SUM
N	0.081	0.465	0.126	0.152	0.081	0.071	0.076	0.106	0.071	0.035	0.020	0.005	0.010	0.000	0.005	0.000	1.305
NNE	0.101	0.753	0.647	0.647	0.440	0.324	0.314	0.197	0.142	0.066	0.046	0.015	0.000	0.000	0.000	0.000	3.692
NE	0.086	0.870	0.728	0.389	0.288	0.197	0.167	0.025	0.010	0.015	0.005	0.000	0.000	0.000	0.000	0.000	2.781
ENE	0.142	3.075	4.420	3.363	1.290	0.435	0.177	0.278	0.263	0.147	0.056	0.046	0.010	0.010	0.000	0.005	13.714
E	0.243	3.798	6.144	6.528	2.190	0.501	0.329	0.253	0.152	0.116	0.116	0.046	0.020	0.005	0.000	0.000	20.440
ESE	0.253	3.004	4.733	4.243	1.446	0.470	0.217	0.137	0.056	0.020	0.005	0.010	0.010	0.000	0.000	0.000	14.604
SE	0.293	2.908	3.965	2.655	0.860	0.430	0.142	0.051	0.040	0.025	0.005	0.010	0.000	0.000	0.000	0.005	11.388
SSE	0.172	1.760	2.528	2.215	0.753	0.354	0.121	0.091	0.051	0.010	0.010	0.005	0.005	0.000	0.000	0.000	8.076
S	0.283	1.350	2.336	2.599	0.925	0.465	0.228	0.086	0.025	0.025	0.005	0.015	0.010	0.005	0.000	0.000	8.359
SSW	0.187	0.910	1.163	1.077	0.531	0.142	0.167	0.137	0.106	0.025	0.000	0.005	0.000	0.000	0.000	0.000	4.450
SW	0.071	0.531	0.855	0.753	0.314	0.051	0.015	0.005	0.005	0.000	0.005	0.000	0.000	0.000	0.000	0.000	2.604
WSW	0.101	0.278	0.455	0.329	0.066	0.030	0.010	0.005	0.005	0.000	0.000	0.000	0.000	0.000	0.000	0.000	1.279
W	0.101	0.602	0.637	0.607	0.182	0.071	0.015	0.005	0.000	0.000	0.000	0.000	0.000	0.000	0.000	0.000	2.220
WNW	0.091	0.475	0.435	0.303	0.061	0.015	0.010	0.000	0.000	0.000	0.000	0.000	0.000	0.000	0.000	0.000	1.391
NW	0.051	0.379	0.142	0.147	0.051	0.035	0.000	0.000	0.000	0.005	0.000	0.000	0.000	0.000	0.000	0.000	0.809
NNW	0.308	1.102	0.622	0.389	0.223	0.131	0.025	0.035	0.030	0.005	0.010	0.000	0.005	0.000	0.000	0.000	2.887
SUM	2.564	22.260	29.937	26.397	9.699	3.722	2.013	1.411	0.956	0.496	0.283	0.157	0.071	0.020	0.005	0.010	100.0

considering wind direction, and the last column shows the relative frequency of mean wind direction without considering wind speed.

5.7.2 *Joint Probability Density Function of Wind Speed and Wind Direction*

Based on the relative frequencies of wind speed and wind direction calculated above, the theoretical expression of joint probability density function is deduced, based on Equation 5.51. The Weibull function is used to fit the histogram of hourly mean wind speed for each wind direction, and the typical results in the east, south and west directions are depicted in Figures 5.20a to 5.20c, respectively. The Weibull parameters identified for each wind direction are listed in Table 5.3, together with the relative frequency of mean wind direction. The coefficient of determination to measure the quality of fitting is also given in Table 5.3 for each wind direction.

It can be seen that, except for the north direction, the Weibull function fits wind data satisfactorily. The Weibull function is also applied to the complete wind records without considering wind direction, as shown in Figure 5.20d. The Weibull parameters identified for this case are

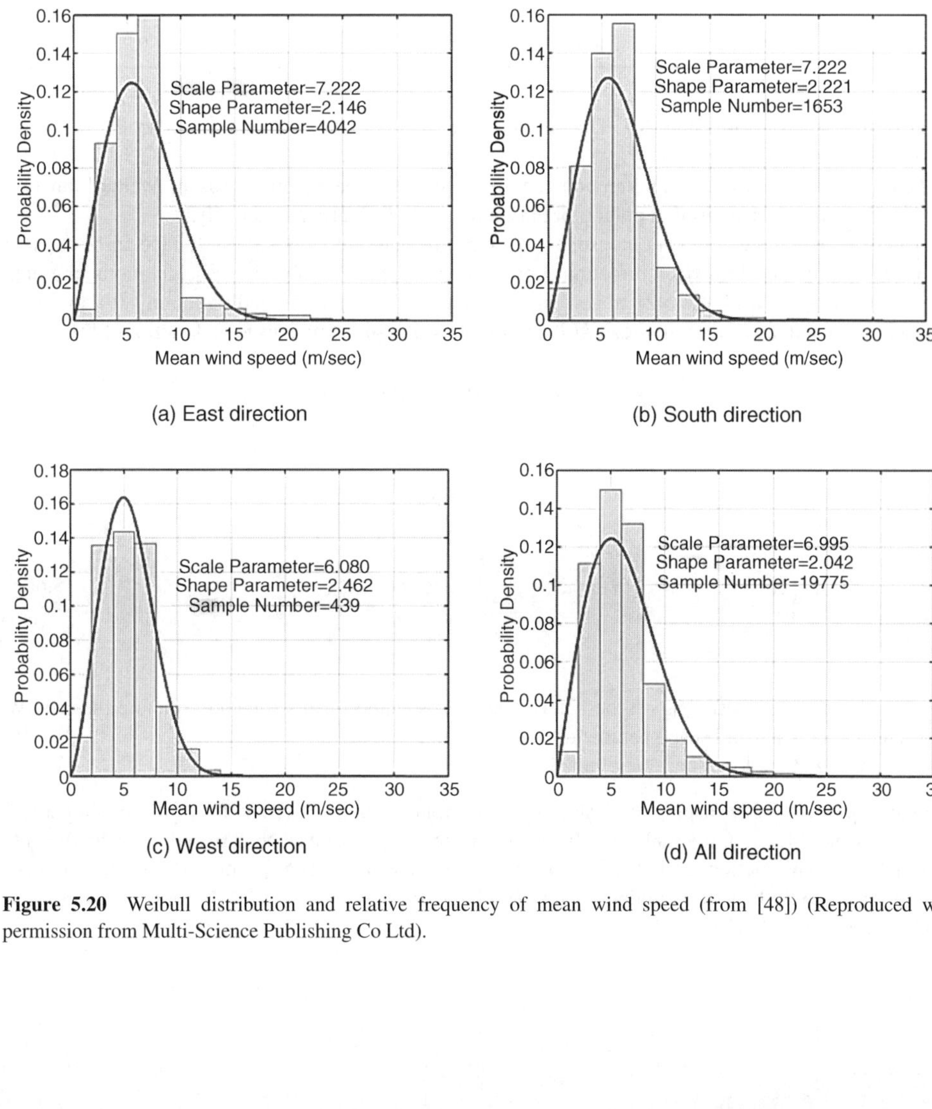

Figure 5.20 Weibull distribution and relative frequency of mean wind speed (from [48]) (Reproduced with permission from Multi-Science Publishing Co Ltd).

Table 5.3 Identified parameters in different wind direction sectors (from [48]) (Reproduced with permission from Multi-Science Publishing Co Ltd)

Direction	Record No.	$f_\theta(\theta)$	$c(\theta)$	$k(\theta)$	Coefficient of determination
N	258	0.013	8.501	1.394	0.6943
NNE	730	0.037	9.100	1.823	0.9165
NE	550	0.028	6.931	1.940	0.8914
ENE	2712	0.137	7.270	1.913	0.9826
E	4042	0.204	7.222	2.146	0.9840
ESE	2888	0.146	6.764	2.344	0.9929
SE	2252	0.114	6.384	2.215	0.9916
SSE	1597	0.081	6.819	2.242	0.9887
S	1653	0.084	7.266	2.221	0.9753
SSW	880	0.045	7.460	1.949	0.9723
SW	515	0.026	6.465	2.521	0.9987
WSW	253	0.013	5.904	2.295	0.9823
W	439	0.022	6.080	2.462	0.9771
WNW	275	0.014	5.392	2.438	0.9840
NW	160	0.008	5.454	1.992	0.8268
NNW	571	0.029	5.745	1.649	0.9248
Total	19 775	1.000	6.995	2.042	0.9869

given in the last row of Table 5.3. The results show that the Weibull function also fits the complete wind data adequately.

The relative frequency of wind direction and the scale and shape parameters of the Weibull function calculated above are also given in polar plot in Figures 5.21a to 5.21c, respectively.

It can be seen that the dominant monsoon direction is the east, and the scale and shape parameters do not vary significantly with wind direction. For the convenience of subsequent calculations, the data given in Table 5.3 and Figure 5.21, regarding the relative frequency of wind direction $f_\theta(\theta)$, the scale parameter $c(\theta)$ and the shape parameter $k(\theta)$, are fitted by the following harmonic functions [47]:

$$f_\theta(\theta) = a^f + \sum_{m=1}^{n_f} b_m^f \cos\left(m\theta - c_m^f\right) \tag{5.60a}$$

$$c(\theta) = a^c + \sum_{m=1}^{n_c} b_m^c \cos\left(m\theta - c_m^c\right) \tag{5.60b}$$

$$k(\theta) = a^k + \sum_{m=1}^{n_k} b_m^k \cos\left(m\theta - c_m^k\right) \tag{5.60c}$$

where:

a, b_m and c_m are the coefficients to be determined, whose superscripts f, c and k denote the relative frequency, the scale and shape parameters respectively;

n_f, n_c and n_k are the order of harmonic functions.

Figures 5.22a to 5.22c display the histograms and the fitted harmonic functions for the relative frequency of wind direction, the scale and shape parameters, respectively. The corresponding coefficients identified for each function are listed in Table 5.4, together with the coefficient of determination. Five terms are used to fit the shape parameter and the relative frequency. For the scale parameter, only four terms are needed.

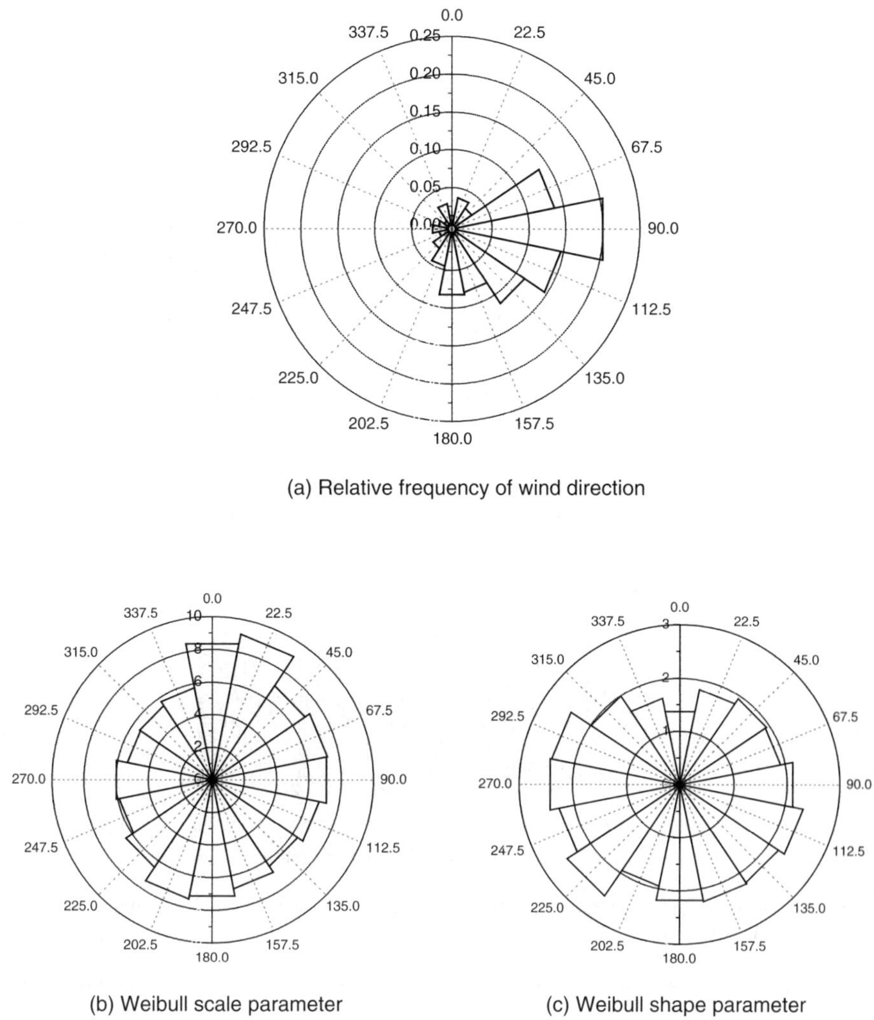

Figure 5.21 Relative frequency of wind direction and Weibull scale and shape parameters (from [48]) (Reproduced with permission from Multi-Science Publishing Co Ltd).

5.7.3 *Statistical Analysis of Rainfall Data*

The sources of rainfall data used in this case study were acquired from the Hong Kong Observatory (HKO) and comprise nine years' worth of hourly rainfall recordings from 1997 to 2005, recorded at the HKO station, which is fairly close to the concerned cable-stayed bridge. These statistics are considered to represent the general condition of rainfall in the vicinity of the bridge.

There are 78 888 available hourly rainfall records, which are accurate to 0.1 mm/hour. Any hourly rainfall less than 0.1 mm/hour is marked as trace rainfall in the collected records and is recorded as rainless in the analysis. As a result, the total of rainy records with hourly rainfall above 0.1 mm/hour comes to only 7385 over the nine years. Hourly rainfall is considered because a full development of rain-wind-induced cable vibration often needs more than half an hour, and this is also consistent with the hourly mean wind speed used in the statistical analysis of wind data.

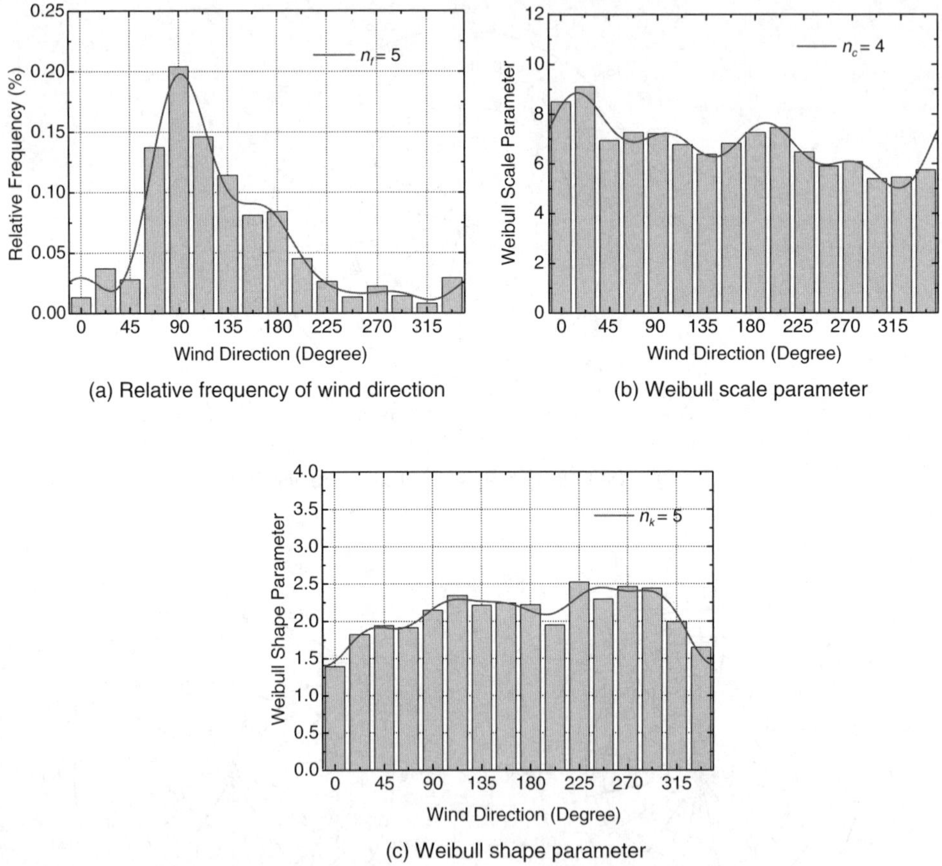

Figure 5.22 Histograms and fitted harmonic functions (from [48]) (Reproduced with permission from Multi-Science Publishing Co Ltd).

Table 5.4 Identified coefficients in harmonic functions (from [48]) (Reproduced with permission from Multi-Science Publishing Co Ltd)

Coefficients	$f_\theta(\theta)$ $n_f = 5$	$c(\theta)$ $n_c = 4$	$k(\theta)$ $n_k = 5$
a	0.06 256	6.797	2.097
b_1	0.06 882	0.7342	−0.2904
c_1	1.949	1.325	0.3654
b_2	−0.02 908	0.9061	−0.2432
c_2	0.42	7.064	37.82
b_3	−0.02 043	0.2764	0.09 093
c_3	1.491	0.6311	1.947
b_4	0.01 851	0.6069	0.07 955
c_4	−0.1218	0.6603	−10.22
b_5	0.00 645	—	0.07 557
c_5	1.212	—	1.996
Coefficient of determination	0.9733	0.9157	0.9077

Let M and N ($M < N$) represent the total rainy hours and the total hours in the considered nine years, respectively. The relative frequency (probability) of rainy event $P_r(A) = M/N = 0.0936$. Figure 5.23a shows the annual rainfall and rainy hours from 1997 to 2005, based on the 7385 rainy records. Figure 5.23b depicts the average monthly rainfall and rainy hours.

It can be seen that, although the average monthly rainy hours vary over each month, the annual rainy hours remains relatively stable over the years. It can be also noted that the annual or average monthly rainfall is fairly correlated with the annual or average monthly rainy hours. To obtain the probability density function of hourly rainfall, the 7385 rainy records are classified according to their rainfall intensity into rainfall cells, starting from $R = 0.1$ mm/hour with an increment of $\Delta R = 0.25$ mm/hour. The resulting relative frequency of hourly rainfall is shown in Figures 5.24a and 5.25a.

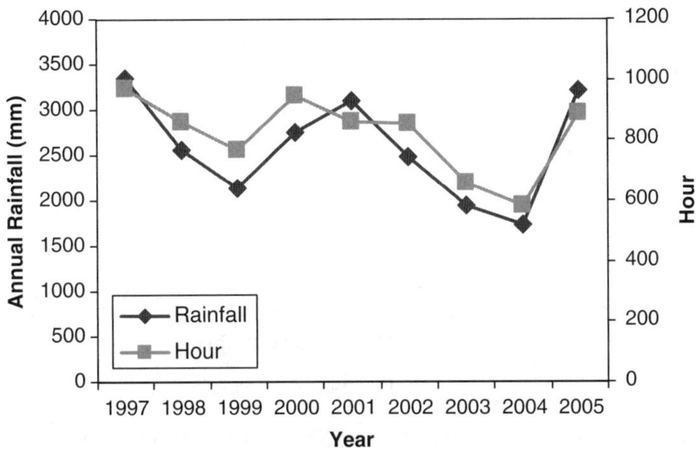

(a) Annual rainfall and rainy hours

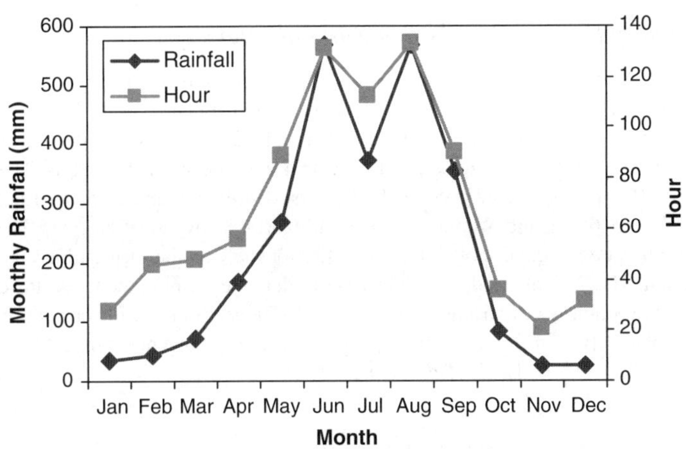

(b) Average monthly rainfall and rainy hours

Figure 5.23 Rainfall intensity and rainy hours (from [48]) (Reproduced with permission from Multi-Science Publishing Co Ltd).

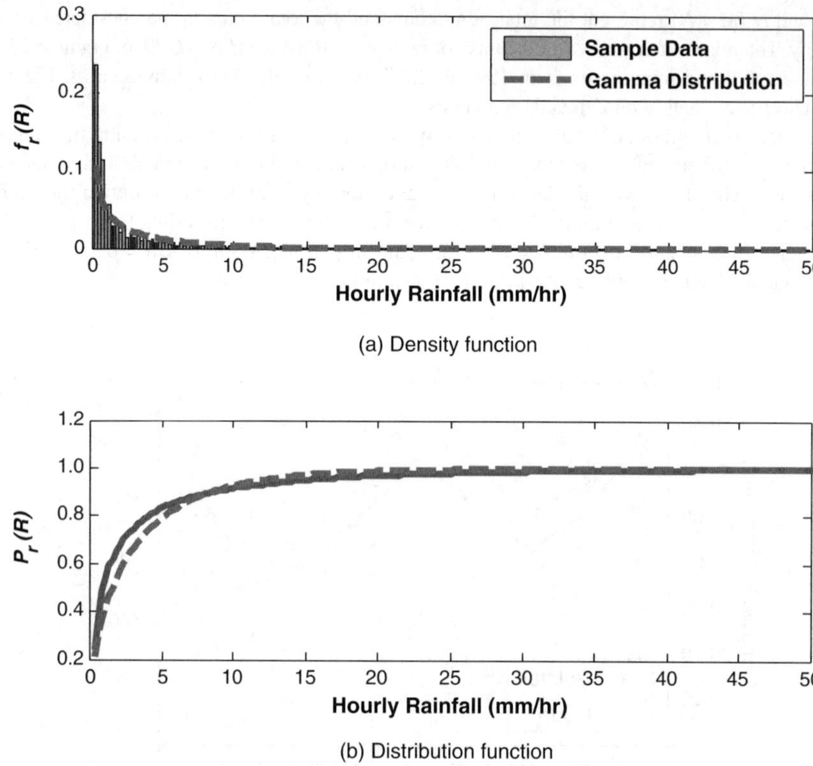

Figure 5.24 Gamma distribution for hourly rainfall (from [48]) (Reproduced with permission from Multi-Science Publishing Co Ltd).

5.7.4 Probability Density Function of Rainfall Intensity

Both Gamma and Weibull distributions are adopted in this study to model the rainfall intensity. Figures 5.24a and 5.24b show the Gamma density function and the Gamma distribution function, respectively, together with the statistical results. Figures 5.25a and 5.25b display the Weibull density function and the Weibull distribution function, respectively, together with the statistical results. The estimated two parameters are $a = 0.5415$ and $b = 6.0406$ for the Gamma distribution expressed by Equation 5.53, and $c = 2.0849$ and $d = 0.6365$ for the Weibull distribution expressed by Equation 5.54. The coefficient of determination to measure the quality of fitting is 0.7432 for the Gamma distribution and 0.8192 for the Weibull distribution. The Weibull model is slightly better than the Gamma model in this case study.

By taking 3–30 mm/hour as the occurrence range Ω_r of hourly rainfall for rain-wind-induced cable vibration, the probability of rainfall intensity falling in the occurrence range Ω_r is calculated as $P_r(B \in \Omega_r|A) = P_r(B \in \Omega_r) = \int_{\Omega_r} f_r(R)dR = 0.2391$.

5.7.5 Occurrence Range of Rain-Wind-Induced Cable Vibration

After wind and rainfall data have been analyzed, a typical stay cable (No. 314) of the Stonecutters cable-stayed bridge is selected as a representative. The total length L, diameter D, tension force T, mass per length m, structural damping ratio in the first mode of vibration ζ_s, cable inclination α, and the first natural frequency n_{01} $(=\omega_{01}/2\pi)$ of the cable are listed in Table 5.5.

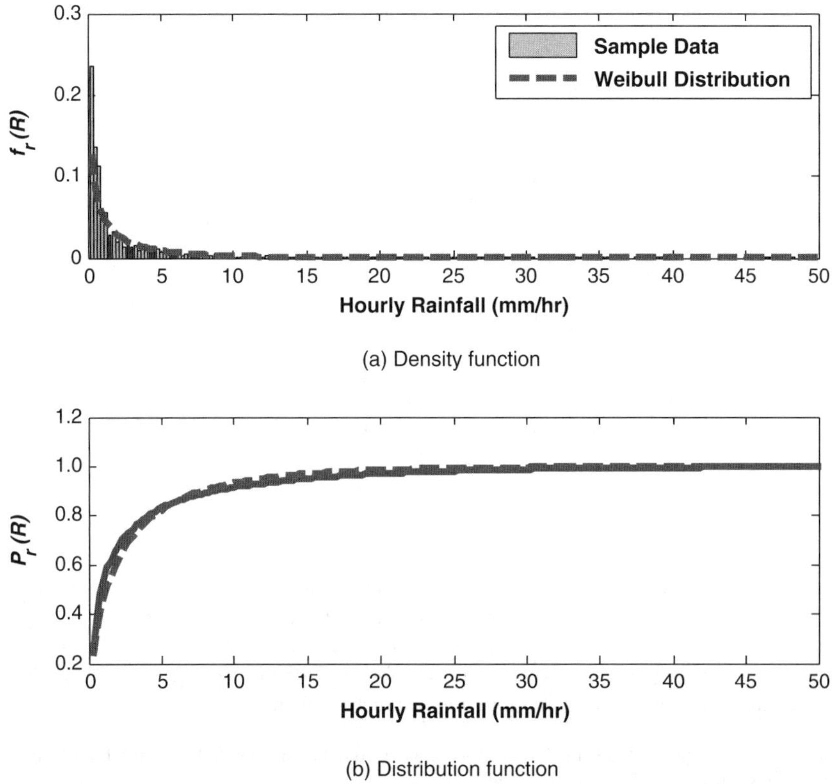

Figure 5.25 Weibull distribution for hourly rainfall (from [48]) (Reproduced with permission from Multi-Science Publishing Co Ltd).

The cable is inclined downwards from southeast to northwest, as shown in Figure 5.19. The yaw wind from east or south, as shown in Figure 5.19, is considered in the determination of occurrence range of rain-wind-induced cable vibration. The bridge deck or the low anchorage of the cable is about 86 m above sea level. Since the bridge faces mainly an open-sea fetch from the south-west direction and mainly an over-land fetch from the north-east direction, two different mean wind speed profiles of power law form are considered. The power law exponent α_1 is 0.19 for the wind from the sea and 0.29 for the wind from the land. Wind reference height h_r is selected at the bridge deck level, i.e. 86 m above the sea level. The static position of upper rivulet is assumed to have the following expression for different yaw angle β, which is derived from the experiment results [20]:

$$\theta_0 = \frac{5550 \cdot ln(0.045|\beta|)}{U^2} \qquad 25^\circ \leq |\beta| \leq 85^\circ \tag{5.61}$$

Figure 5.26 shows the comparison of the static position of rivulet calculated using Equation 5.61 with the experiment results available in the literature. The aerodynamic functions, Γ_1, Γ_2, Γ_3 and the wind stagnation point influence factor ε suggested in the literature [27] are chosen in this case study.

Table 5.5 Main parameters of example stay cable used in case study (from [48]) (Reproduced with permission from Multi-Science Publishing Co Ltd)

Cable no.	L (m)	D (m)	T (ton)	m (kg/m)	ζ_s	α (°)	n_{01} (Hz)
314	292.5	0.155	668.5	98.6	0.0015	30	0.44

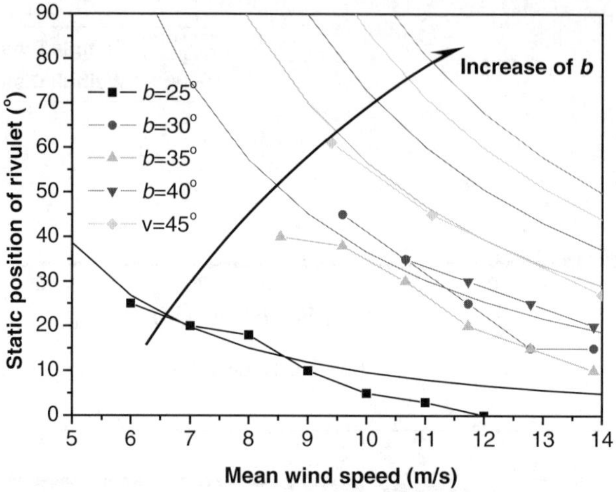

Figure 5.26 The static position of rivulet vs. yaw angle and wind speed (from [48]) (Reproduced with permission from Multi-Science Publishing Co Ltd).

The range of mean wind speed at the reference height is considered from 5 m/s to 14 m/s at 1 m/s interval in the calculation, because the reference height is actually at the low anchorage of the cable. Only the cable response amplitude in the first mode of vibration is reported in Table 5.6 in terms of cable diameter for the wind from the sea, and in Table 5.7 for the wind from the land. The cable response amplitudes in higher modes of vibration are much smaller.

Table 5.6 Cable vibration amplitude in diameter for the wind from the sea (from [48]) (Reproduced with permission from Multi-Science Publishing Co Ltd)

β (°)	Mean wind speed (m/s)									
	5	6	7	8	9	10	11	12	13	14
25	0.5	0	0	0	0	0	0	0	0	0
35	0	0.5	1.5	3	2.5	1	0.5	0	0	0
45	0	0	0	0.5	2	3.5	2	1	0.5	0
55	0	0	0	0	0.5	1.5	2	1	0.5	0.5
65	0	0	0	0	0	0.5	1	0.5	0.5	0
75	0	0	0	0	0	0	0.5	0	0	0
85	0	0	0	0	0	0	0	0	0	0

Table 5.7 Cable vibration amplitude in diameter for the wind from the land (from [48]) (Reproduced with permission from Multi-Science Publishing Co Ltd)

β (°)	Mean Wind Speed (m/s)									
	5	6	7	8	9	10	11	12	13	14
25	0.5	0	0	0	0	0	0	0	0	0
35	0	0	2	3.5	1.5	1	0.5	0	0	0
45	0	0	0	1.5	2.5	2.5	1	0.5	0	0
55	0	0	0	0	1	2	1.5	0.5	0.5	0
65	0	0	0	0	0	1	1	0.5	0	0
75	0	0	0	0	0	0	0.5	0	0	0
85	0	0	0	0	0	0	0	0	0	0

Table 5.8 Hourly occurrence probability and annual risk of rain-wind-induced cable vibration (from [48]) (Reproduced with permission from Multi-Science Publishing Co Ltd)

Threshold A_s	$0.5D$	$1D$	$2D$	$3D$
Probability ($\times 10^{-4}$)	18	11	6.17	1.98
Risk (%)	Almost 100%	Almost 100%	99.6	82.4

Rain-wind-induced cable vibration with mixed modes is not considered in this study. It can be seen from Tables 5.6 and 5.7 that the most severe cable vibration occurs when wind yaw angle β is between 35° and 55° and mean wind speed is between 7 m/s to 11 m/s, which are consistent with the field observations of rain-wind-induced vibration of stay cables in cable-stayed bridges.

5.7.6 *Hourly Occurrence Probability and Annual Risk*

Based on all the above results, the hourly occurrence probability of RWIV of the selected cable with cable vibration amplitude larger than a threshold value can be calculated using Equation 5.59. Four levels of vibration threshold A_s have been selected. They are $A_s \geq 0.5D$, $1D$, $2D$ and $3D$, where D is the diameter of the cable. The calculated hourly occurrence probability of RWIV of the cable for each threshold level is summarized in Table 5.8. It can be seen from this table that the hourly occurrence probability of RWIV of the cable is moderate.

For the selected cable, RWIV is about 16 hours, 10 hours, 5 hours and 2 hours in a year for the threshold $A_s = 0.5D$, $1D$, $2D$ and $3D$, respectively. The annual risk of RWIV of the cable is almost 100% for the threshold $A_s = 0.5D$ and $1D$, and 99.6% and 82.4% for the threshold $A_s = 2D$ and $3D$, respectively. It is noted that the cable selected is in the worst orientation for RWIV of all the stay cables.

The same procedure can be applied to other stay cables in the bridge, and the hourly occurrence probability and annual risk of all the stay cables can then be estimated. More detailed information on Sections 5.6 and 5.7 can be found in the literature [48].

5.8 Notations

a	(i) Amplitude of rivulet motion (Equation 5.35) (ii) Shape parameter for Gamma distribution (Equation 5.53) (iii) Coefficient (Equation 5.60)
b	(i) Scale parameter for Gamma distribution (Equation 5.53) (ii) Coefficient (Equation 5.60)
c	(i) Internal cable damping coefficient (Equation 5.1) (ii) Constant value (Equation 5.20) (iii) Scale parameter for Weibull distribution (Equation 5.50) (iv) Coefficient (Equation 5.60)
c_r	Damping coefficient of rivulet
d	(i) Sag inclination angle (ii) Scale parameter for Weibull distribution (Equation 5.54)
f	Natural frequency of cable
g	Gravitational constant
k	Shape parameter (Equation 5.35)
h_0	Height of lower anchorage of the cable above the sea level
k_r	Stiffness coefficient per unit length of the rivulet
m	Cable mass per unit length

m_r	Mass per unit length of rivulet
n	Order of harmonic function
s	Static equilibrium state of cable
u	Axial displacement
v	Out-of-plane transverse displacement
w	In-plane transverse displacement
$\dot{x}$	Relative velocity of the rivulet to the cylinder
ρ	Air density
α	Inclination angle
α_r	Predetermined exponent
α_1	Exponent depending on the surface roughness
ε	(i) Influence factor
	(ii) Dynamic strain
β	Yaw angle of incident wind
θ	(i) Motion of upper rivulet (Equation 5.35)
	(ii) Dynamic angular displacement of the rivulet as the cylinder vibrates
θ_0	Static position angle of upper rivulet due to mean wind
τ	Additional dynamic tension
γ	Ideal attack angle for the cylinder without rivulet
φ	Wind angle of attack
$\varphi*$	Angle between relative velocity and the horizontal axis
ω_c	Circular natural frequency of cylinder
ω_r	Frequency of rivulet motion
ω_{01}	First circular natural frequency of taut cable
Γ	Rain-wind aerodynamic function
ϖ	Normalized circle frequency
ζ	Damping ratio of cable
ζ_a	Rain-wind aerodynamic damping ratio
ξ_c	Structural damping ratio of cylinder
ζ_s	Structural damping ratio
$\tilde{\xi}$	Total damping ratio of cylinder
λ^2	Sag parameter
A	Cross-section area of cable
A_i	Coefficients changing with the angle φ
A_n	Constant value (Equation 5.7)
A_{max}	Maximum peak to peak amplitude
A_{allow}	Allowable peak to peak amplitude
B_i	Coefficients changing with the angle φ
C_d	Drag coefficient
C_L	Lift coefficient
D	Cable diameter
E	Young's modulus of cable
L	Cable length
R	(i) Radius (Equation 5.24)
	(ii) Rainfall intensity (Equation 5.54)
S	Strouhal number
S_c	Scruton number
T	Cable tension force
U	Mean wind speed
$U_0(h)$	Mean wind speed at height h above the sea level

U_r	Mean wind speed at a chosen reference height h_r
U_{crit}	Minimum wind velocity for wake galloping instability
U_{rel}	Relative velocity of mean wind to the cylinder with moving rivulet
$f(t)$	Harmonic function of time with frequency ω
$f_u(U)$	Probability density function of Weibull distribution
$f_r(R)$	Mathematical expression of Gamma and Weibull probability density functions
$W(x)$	Motion profile of rivulet along the cable axis
$F(x, t)$	Wind-induced force on the cable
$\tilde{F}(t)$	Normalized force due to the motion of rivulet
$P_\theta(\theta)$	Relative frequency of occurrence of wind in wind direction θ
$f_\theta(\theta)$	Relative frequency of wind direction
$P_u(U)$	Cumulative distribution function of Weibull distribution
$\Gamma(a)$	Gamma function (Equation 5.54)
$\Omega_{u,\theta}^{A_s}$	Occurrence range of wind speed and wind direction

References

1. Meirovitch, L. (1967) *Analytical Methods in Vibrations*, Macmillan.
2. Zuo, D.L. (2005) Understanding Wind- and Rain-Wind-Induced Stay Cable Vibrations, PhD Thesis, The Johns Hopkins University, Maryland, USA.
3. Irvine, H.M. (1981) *Cable Structures*, MIT Press.
4. Yamaguchi, H. and Fujino, Y. (1998) Stayed cable dynamics and its vibration control, in International Symposium on Advances in Bridge Aerodynamics, Balkema, Rotterdam, Netherlands, pp. 235–253.
5. Pinto da Costa, A., Martins, J.A.C., Branco, F., and Lilien, J.L. (1996) Oscillations of bridge stay cables induced by periodic motions of deck and/or towers. *Journal of Engineering Mechanics, ASCE*, **122**, 613–622.
6. Warnitchai, P., Fujino, Y., and Susumpow, T. (1995) A non-linear dynamic model for cables and its application to a cable-structure system. *Journal of Sound and Vibration*, **187**(4), 695–712.
7. Takahashi, K. and Konishi, Y. (1987) Non-linear vibrations of cables in three dimensions, Part II: out-of-plane vibrations under in-plane sinusoidally time-varying load. *Journal of Sound and Vibration*, **118**(1), 85–97.
8. Benedettini, F., Rega, G., and Alaggio, R. (1995) Non-linear oscillations of a four-degree-of-freedom model of a suspended cable under multiple internal resonance conditions. *Journal of Sound and Vibration*, **182**(5), 775–798.
9. Lee, C. and Perkins, N.C. (1995) Three-dimensional oscillations of suspended cables involving simultaneous internal resonances. *International Journal of Non-linear Dynamics*, **8**, 45–63.
10. Nayfeh, A.H. and Mook, D.T. (1995) *Non-linear Oscillations*, John Wiley & Sons, Inc., New York, USA.
11. FHWA (2007) Wind-Induced Vibration of Stay Cables, Report of Federal Highway Administration, FHWA-HRT-05-083.
12. Saito, T., Matsumoto, M., and Kitazawa, M. (1994) Rain-wind excitation of cables on cable- stayed Higashi-Kobe Bridge and cable vibration control. Proceedings of the International Conference on Cable-Stayed and Suspension Bridges (AFPC), Deauville, France, **2**, pp. 507–514.
13. Matsumoto, M. (1998) Observed behavior of prototype cable vibration and its generation mechanism, Bridge Aerodynamics. Proceedings of the International Symposium on Advances in Bridge Aerodynamics, Copenhagen, Denmark, pp. 189–211.
14. Irwin, P.A. (1997) Wind vibrations of cables on cable-stayed bridges. Proceedings of Structural Congress XV, Portland, OR, pp. 383–387.
15. PTI Guide Specification (2001) *Recommendations for Stay Cable Design, Testing and Installation*, 4th edn, Post-Tensioning Institute Committee on Cable-Stayed, Bridges.
16. Hikami, Y. and Shiraishi, N. (1988) Rain-wind induced vibrations of cables in cables stayed bridges. *Journal of Wind Engineering and Industrial Aerodynamics*, **29**(1–3), 409–418.
17. Matsumoto, M., Saitoh, T., Kitazawa, M. *et al.* (1995) Response characteristics of rain-wind induced vibration of stay-cables of cable-stayed bridges. *Journal of Wind Engineering and Industrial Aerodynamics*, **57**(2–3), 323–333.
18. Main, J.A. and Jones, N.P. (1999) Full-scale measurements of stay cable vibration. Proceedings of 10th International Conference on Wind Engineering, Copenhagen, Denmark, pp. 963–970.
19. Main, J.A. (2002) Modeling the Vibrations of a Stay Cable with Attached Damper, PhD Thesis, The Johns Hopkins University, USA.
20. Gu, M. and Du, X. (2005) Experimental investigation of rain-wind-induced vibration of cables in cable-stayed bridges and its mitigation. *Journal of Wind Engineering and Industrial Aerodynamics*, **93**(1), 79–95.
21. Main, J.A. and Jones, N.P. (2001) Evaluation of viscous dampers for stay-cable vibration mitigation. *Journal of Bridge Engineering, ASCE*, **6**(6), 385–397.

22. Matsumoto, M., Shiraishi, N., Kitazawa, M. *et al.* (1990) Aerodynamic behavior of inclined circular cylinders-cable aerodynamics. *Journal of Wind Engineering and Industrial Aerodynamics*, **33**(1–2), 63–72.
23. Cosentino, N., Flamand, O., and Ceccoli, C. (2003) Rain-wind induced vibration of inclined stay cables. Part II: Mechanical modeling and parameter characterization. *Wind and Structures*, **6**(6), 485–498.
24. Xu, Y.L., Li, Y.L., Shum, K.M. *et al.* (2006) Aerodynamic coefficients of inclined circular cylinders with artificial rivulet in smooth flow. *Advances in Structural Engineering-An International Journal*, **9**(2), 265–278.
25. Yamaguchi, H. (1990) Analytical study on growth mechanism of rain vibration of cables. *Journal of Wind Engineering and Industrial Aerodynamics*, **33**(1–2), 73–80.
26. Peil, U. and Nahrath, N. (2003) Modeling of rain-wind induced vibrations. *Wind and Structures*, **6**(1), 41–52.
27. Xu, Y.L. and Wang, L.Y. (2003) Analytical study of wind-rain-induced cable vibration: SDOF model. *Journal of Wind Engineering and Industrial Aerodynamics*, **91**(1–2), 27–40.
28. Xu, Y.L. and Wang, L.Y. (2001) Analytical study of wind-rain-induced cable vibration. Proceedings of the Fifth Asia-Pacific Conference on Wind Engineering, Kyoto, Japan, pp. 109–112.
29. Wang, L.Y. and Xu, Y.L. (2003) Analytical study of wind-rain-induced cable vibration: 2 DOF model. *Wind & Structures*, **6** (4), 291–306.
30. Matsumoto, M., Shirashi, N., and Shirato, H. (1992) Rain-wind induced vibration of cables of cable-stayed bridges. *Journal of Wind Engineering and Industrial Aerodynamics*, **43**(1–3), 2011–2022.
31. Matsumoto, M., Yamagishi, M., Aoki, J., and Shiraishi, N. (1995) Various mechanism of inclined cable aerodynamics. Proceedings of 9th International Conference on Wind Engineering, Wiley Eastern, New Delhi, **2**, pp. 759–769.
32. Gu, M., Xu, Y.L., Liu, C.J., and Xiang, H.F. (2000) Wind tunnel study of response characteristics of cables with artificial rivulet, in Proceedings of International Conference on Advances in Structural Dynamics, **2**, Elsevier Science, pp. 1479–1486.
33. Flamand, O. (1995) Rain-wind induced vibration of cables. *Journal of Wind Engineering and Industrial Aerodynamics*, **57** (2–3), 353–362.
34. Cosentino, N., Flamand, O., and Ceccoli, C. (2003) Rain-wind induced vibration of inclined stay cables. Part I: experimental investigation and physical explanation. *Wind and Structures*, **6** (6), 471–484.
35. Zhan, S., Xu, Y.L., Zhou, H.J., and Shum, K.M. (2008) Experimental study of wind-rain-induced cable vibration using a new model setup scheme. *Journal of Wind Engineering and Industrial Aerodynamics*, **96**(12), 2438–2451.
36. Zuo, D., Jones, N.P., and Main, J.A. (2008) Field observation of vortex- and rain-wind- induced stay-cable vibrations in a three-dimensional environment. *Journal of Wind Engineering and Industrial Aerodynamics*, **96**(6–7), 1124–1133.
37. Gu, M. and Huang, L. (2008) Theoretical and experimental studies on the aerodynamic instability of a two-dimensional circular cylinder with a moving attachment. *Journal of Fluids and Structures*, **24**(2), 200–211.
38. Robertson, A.C., Taylor, I.J., Wilson, S.K. *et al.* (2010) Numerical simulation of rivulet evolution on a horizontal cable subject to an external aerodynamic field. *Journal of Fluids and Structures*, **26**(1), 50–73.
39. Butcher, J.C. (1987) *The Numerical Analysis of Ordinary Differential Equations: Runge-Kutta and General Linear Methods*, John Wiley & Sons, Inc., NY, USA.
40. Corotis, R.B., Sigl, A.B., and Klein, J. (1978) Probability models of wind velocity magnitude and persistence. *Solar Energy*, **20**(6), 483–493.
41. Hennessy, J.P. (1978) A comparison of the Weibull and Rayleigh distributions for estimating wind power potential. *Wind Engineering*, **2** (3), 156–164.
42. Morgan, V.T. (1995) Statistical distributions of wind parameters at Sydney, *Australia. Journal of Renewable Energy*, **6** (1), 39–47.
43. McWilliams, B., Newmann, M.M., and Sprevak, D. (1979) The probability distribution of wind velocity and direction. *Wind Engineering*, **3**(4), 269–273.
44. Dixon, J.C. and Swift, R.H. (1984) The directional variation of wind probability and Weibull speed parameters. *Atmospheric Environment*, **18** (10), 2041–2047.
45. Mishra, A.K. and Desai, V.R. (2005) Drought forecasting using stochastic models. *Journal of Stochastic Environmental Research and Risk Assessment*, **19**(5), 326–339.
46. Zhou, H.J. and Xu, Y.L. (2007) Wind-rain-induced vibration and control of stay cables in a cable-stayed bridge. *Journal of Structural Control and Health Monitoring*, **14**(7), 1013–1033.
47. Coles, S.G. and Walshaw, D. (1994) Directional modeling of extreme wind speeds, *Applied Statistics-Journal of the Royal Statistical Society Series C*, **43**(1), 139–157.
48. Xu, Y.L., Chen, J., Ng, C.L. and Zhou, H.J. (2008) occurrence probability of wind-rain-induced stay cable vibration, *Advances in Structural Engineering-An International Journal*, **11**(1), 53–69.

6

Wind-Vehicle-Bridge Interaction

6.1 Preview

Wind-induced accidents involving road vehicles of various types have become a topic of increasing concern in recent years. This is because not only traffic volumes have increased dramatically, but also vehicle weights have become significantly lower, due to the use of more efficient structural design and lighter materials. When large numbers of road vehicles run on a long-span cable-supported bridge built in a wind-prone region, the bridge will experience considerable vibration due to both moving vehicles and turbulent winds. This vibration may, in turn, affect the running safety of road vehicles. Moreover, road vehicles will be shielded briefly from the wind by the bridge towers or other road vehicles but, when they pass out of this shelter, they will enter a sharp-edged crosswind gust and there is an obvious danger of the vehicles turning over. Therefore, the safety of road vehicles subjected to crosswind gust and running on an oscillating cable-supported bridge, and the decision on the threshold of wind speed above which the bridge should be closed or vehicles should be slowed down, are important social and economic issues.

Dynamic response of railway vehicles running on a track has also been a subject of great interest to vehicle designers and maintenance engineers, as well as track designers, for many years. This interest is motivated by the desire to improve ride quality, to reduce wear to vehicle and track components, to prevent vehicle hunting and, most important of all, to ensure operational safety. With ever-increasing trailing tonnage and higher running speeds, it becomes more important to improve further the performance of railway vehicles and their suspension system. Under strong crosswinds, the aerodynamic forces on a moving railway vehicle may be sufficiently large to overturn the vehicle. To be able to guarantee comfort and safety for a railway vehicle in crosswind, one thus has to understand the effects of crosswinds on dynamic interaction between the vehicle subsystem and the track subsystem.

Furthermore, more and more long-span cable-supported bridges carrying both highway and railway have been built throughout the world in recent years. Heavy trains moving on a long-span cable-stayed bridge may significantly change the dynamic behavior and affect the fatigue life of the bridge. The vibration of the bridge will, in turn, affect the running safety of trains and the comfort of the passengers. If such a long-span cable-stayed bridge is built in wind-prone area, there will be a complicated dynamic interaction problem between the bridge, the trains and crosswinds. No sophisticated way can be followed at present to make a rational decision on the threshold of wind speed above which the bridge should be closed to trains.

Wind Effects on Cable-Supported Bridges, First Edition. You-Lin Xu.

In this respect, this chapter first discusses wind-road vehicle interaction, in which dynamic behavior and possible accidents of road vehicles running on the ground and entering a sharp-edged crosswind gust are explored, including effects of road surface roughness and vehicle suspension system. The framework for performing dynamic interaction analysis between road vehicles and cable-supported bridges under crosswind is then presented, in which the ride comfort and safety of road vehicles running on the bridge under crosswind are evaluated. In a similar way, this chapter also presents a framework for simulating railway vehicle and track interaction under crosswind, by which the safety and comfort performance of the moving train in crosswind can be assessed. Finally, the framework is extended to the dynamic interaction problem of a long-span cable-supported bridge with running trains subjected to crosswinds by using the most up to date information in the area of wind-bridge interaction, bridge-train interaction and wind-train interaction.

6.2 Wind-Road Vehicle Interaction

6.2.1 Wind-Induced Vehicle Accidents

Road vehicles are susceptible to a sharp-edged crosswind gust, which may cause vehicle accidents such as overturning, excessive sideslip, or exaggerated rotation. Baker and Reynolds carried out a post-disaster investigation on wind-induced vehicle accidents that occurred in the United Kingdom during the major storm of 25 January 1990 [1]. They found that among 400 wind-induced vehicle accidents in that event, overturning accidents were the most common type, accounting for 47% of the total. Course deviation accidents made up 19% of the total and 66% of the accidents involved high-sided lorries or vans, while only 27% involved cars.

Baker developed a theoretical model that described the dynamics of vehicles in crosswind [2]. Coleman and Baker performed a series of wind tunnel tests to determine the aerodynamic forces and moments on road vehicles of certain types [3]. Baker then quantified accident wind speeds of road vehicles for overturning, sideslip and rotation accidents and estimated accident risk [4]. A wind-induced vehicle accident was said to occur if, within 0.5 seconds of the vehicle entering a sharp-edged crosswind gust, one of the tire reactions fell to zero (an overturning accident), or the lateral displacement exceeded 0.5 m (a sideslip accident), or the rotational displacement exceeded 0.2 radians (a rotational accident). The accident vehicle speed at which any one of the three accident criteria is first exceeded can therefore be estimated.

It should be pointed out that, although the above definition will be used in the following study, it is conservative in several aspects. First, it was assumed that, within the 0.5 seconds, the driver of the vehicle would not react to correct any lateral or rotational displacement. Second, a sharp-edged crosswind gust was assumed. Third, the aerodynamic force coefficients were assumed to be constant within the 0.5 seconds.

6.2.2 Modeling of Road Vehicle

Take a high-sided road vehicle as an example. To investigate the possible course deviation of the vehicle caused by a sudden crosswind gust, not only the vertical vibration but also the lateral and rotational vibrations are considered [5]. A high-sided road vehicle is modeled as a combination of several rigid bodies connected by a series of springs, dampers and pivots (see Figure 6.1), so that the effects of vehicle suspension on vehicle accidents can be investigated. The rigid bodies are used to represent the vehicle bodies, the axles, the wheels or other components.

The centre of gravity of each rigid body is taken as a node, which has six degrees of freedom in general: three translational degrees and three rotational degrees with respect to the local co-ordinate originated at the node. The displacements and rotations of the vehicle body and axles are assumed to

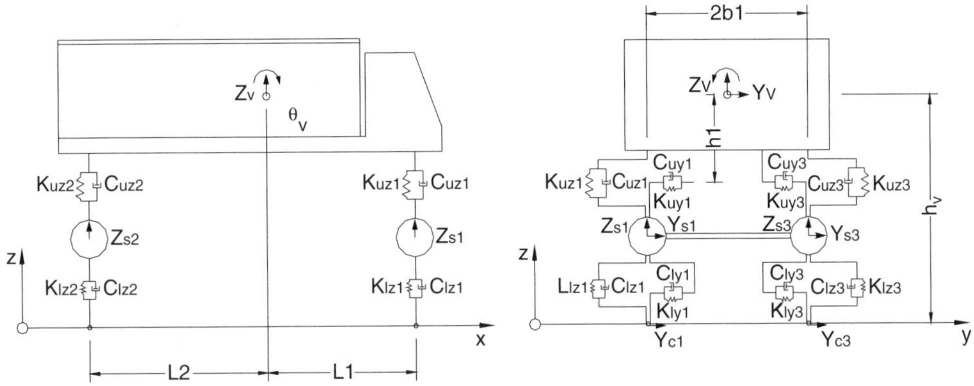

Figure 6.1 Vehicle model used in case study (from [5]) (Reproduced with permission from Techno.Press).

remain small throughout the analysis, so that the sines of the angles of rotation may be taken equal to angles themselves and the cosines of the angles of rotation may be taken as unity, leading to linear equations of motion for the vehicle itself [5]. The mass and/or the mass moments of inertia of each rigid body are calculated from the weight distribution and dimension of the body with respect to its local coordinate.

A vehicle tire is assumed to be a point, so the contact between the road surface and the vehicle tire is thus a point contact. The road surface is not so rough as to make the vehicle tire jump or leave the riding surface. The tires of the vehicle therefore remain in contact with road surface at all times, except in the case of the vehicle overturning, in which the contact force between the road surface and any one of the vehicle tires becomes zero. As a result, the vertical displacement of the tire is not an independent degree of freedom and can thus be determined by the vertical road surface profile and its relative position. However, the lateral displacement of the tire should be taken as an independent degree of freedom, because of the vehicle sideslip relative to the road surface.

The springs can be used to model the suspension system, the flexibility of a tire or others. Each spring is assumed to be massless. Apart from the stiffness coefficient of each spring, the positions of the two ends of the spring connecting two rigid bodies, or connecting one rigid body and one contact point, are required as input data. The energy dissipation capacity of the suspension system and the tires can be modeled by dampers. If the damper is of a viscous type, the damping coefficient can be used as a sole parameter for the damping device. The pivots may be used to connect the trailer to the tractor, for which the constraint equations should be correspondingly developed.

In summary, the input data about road vehicles required by the computer program are the dynamic properties and positions of all the rigid bodies and the springs and the dampers, the positions of all the contact points and the constraint conditions for all the pivots. Based on these input data, the mass matrix, the damping matrix and the stiffness matrix of the vehicle and the force vectors due to road surface roughness can be automatically assembled, using a fully computerized approach developed by Guo and Xu [6].

6.2.3 Modeling of Road Surface Roughness

Many investigations have shown that the roughness of road surface is an important factor that affects the dynamic response of a vehicle [7]. It is thus envisaged that the road surface roughness also has an impact on vehicle accidents. The road surface roughness in the vertical direction may be described as a realization of a random process that can be described by a power spectral density (PSD) function. The following PSD functions were proposed by Dodds and Robson for road surface roughness of a highway in the vertical direction [8]:

$$S(\bar{\varphi}) = A_r\left(\frac{\bar{\varphi}}{\bar{\varphi}_0}\right)^{-w_1}, \quad \bar{\varphi} \leq \bar{\varphi}_0 \tag{6.1}$$

$$S(\bar{\varphi}) = A_r\left(\frac{\bar{\varphi}}{\bar{\varphi}_0}\right)^{-w_2}, \quad \bar{\varphi} \geq \bar{\varphi}_0 \tag{6.2}$$

where:

$S(\bar{\varphi})$ is the PSD function (m^3/cycle) for the road surface roughness in the vertical direction;
$\bar{\varphi}$ is the spatial frequency (cycle/m);
$\bar{\varphi}_0$ is the discontinuity frequency of $\frac{1}{2\pi}$ (cycle/m);
A_r is the roughness coefficient (m^3/cycle) depending on the road condition. The power exponents w_1 and w_2 vary from 1.36 to 2.28.

To simplify the description of the road surface roughness, Wang and Huang suggested the following PSD function [7]:

$$S(\bar{\varphi}) = A_r\left(\frac{\bar{\varphi}}{\bar{\varphi}_0}\right)^{-2} \tag{6.3}$$

The vertical road surface roughness is assumed to be a periodically modulated random process, and its time-history can be generated through an inverse Fourier transform:

$$r(x) = \sum_{k=1}^{N} \sqrt{2S(\bar{\varphi}_k)\Delta\bar{\varphi}}\,\cos(2\pi\bar{\varphi}_k x + \theta_k) \tag{6.4}$$

where θ_k is the random phase angle uniformly distributed from 0 to 2π. While the road surface roughness in the vertical direction can be modeled, the road surface roughness in the lateral direction cannot be considered at this stage, owing to a lack of relevant information.

6.2.4 *Aerodynamic Forces and Moments on Road Vehicle*

In this study, the wind velocity U_m is assumed to be perpendicular to the longitudinal axis of the road and the road vehicle runs at a constant velocity of U_v. The wind velocity relative to the vehicle, U_R, and its yaw angle, Ψ (see Figure 6.2a), can then be expressed as:

$$U_R = \sqrt{U_m^2 + U_v^2} \tag{6.5}$$

$$\psi = \arctan\left(\frac{U_m}{U_v}\right) \tag{6.6}$$

The aerodynamic forces and moments acting on the vehicle are determined based on the quasi-steady assumption and expressed by:

$$F_x = \frac{1}{2}\rho U_R^2 C_D(\psi) A_f \tag{6.7a}$$

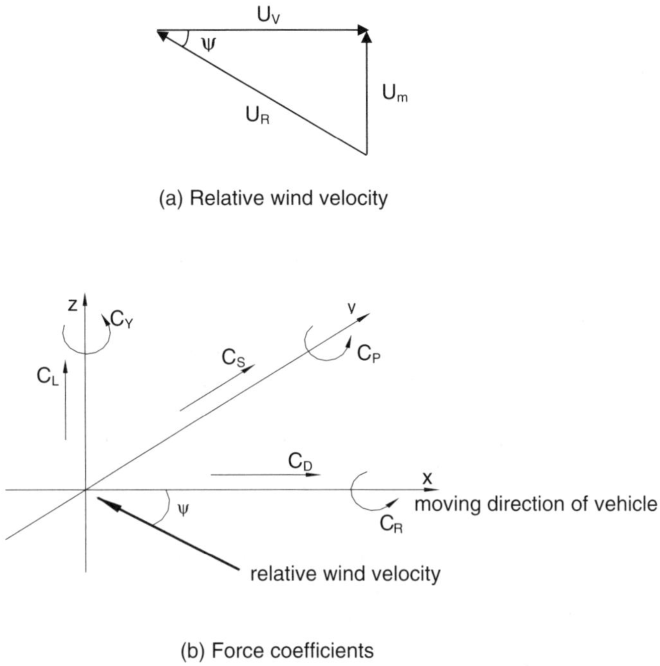

Figure 6.2 Sign convention for wind velocities and wind forces (from [5]) (Reproduced with permission from Techno.Press).

$$F_y = \frac{1}{2}\rho U_R^2 C_S(\psi) A_f \tag{6.7b}$$

$$F_z = \frac{1}{2}\rho U_R^2 C_L(\psi) A_f \tag{6.7c}$$

$$M_x = \frac{1}{2}\rho U_R^2 C_R(\psi) A_f h_v \tag{6.7d}$$

$$M_y = \frac{1}{2}\rho U_R^2 C_P(\psi) A_f h_v \tag{6.7e}$$

$$M_z = \frac{1}{2}\rho U_R^2 C_Y(\psi) A_f h_v \tag{6.7f}$$

where:

F_x, F_y, F_z, M_x, M_y, and M_z are the drag force, side force, lift force, rolling moment, pitching moment, and yawing moment acting on the vehicle respectively;

$C_D(\psi)$, $C_S(\psi)$, $C_L(\psi)$, $C_R(\psi)$, $C_P(\psi)$ and $C_Y(\psi)$ are the drag force coefficient, side force coefficient, lift force coefficient, rolling moment coefficient, pitching moment coefficient, and yawing moment coefficient, respectively, which are a function of yaw angle Ψ;

A_f is the reference area, which is normally taken as the frontal area of the vehicle;

h_v is the reference height, which is normally taken as the height of the vehicle centre of gravity above the ground [2].

The force and moment coefficients are usually obtained through wind tunnel tests and a sign convention for these coefficients is given in Figure 6.2b. A more sophisticated way of determining the aerodynamic forces and moments on a vehicle can be found in Section 6.5.5.

6.2.5 Governing Equations of Motion of Road Vehicle

A large truck shown schematically in Figure 6.1 is taken as an example to investigate the dynamic behaviour and possible accidents of high-sided road vehicles subject to a sudden crosswind gust [5]. The vehicle comprises nine rigid bodies: one for the vehicle body, two for the front axle set, two for the rear axle set and four for the tires (see Figure 6.1). It is assumed that the vehicle runs at a constant velocity on a straight road.

Thus, the position of the vehicle in the x-direction can be determined accurately at any given time as long as the initial position of the vehicle is known. In this connection, the degrees of freedom in the x-direction are not included. The vehicle body is assigned five degrees of freedom with respect to its gravity centre: the vertical displacement (Z_v), the lateral displacement (Y_v), the rotation about the y-axis (pitching angle θ_v), the rotation about the x-axis (rolling angle φ_v) and the rotation about the z-axis (yawing angle ϕ_v).

Each rigid body in either the front axle set or the rear axle set is assigned two degrees of freedom in the z-direction (Z_{si}) and the y-direction (Y_{si}). Since the vehicle tire is assumed to be a point and to have a sideslip, each rigid body for one tire is assigned one degree of freedom (Y_{ci}) in the y-direction. As a result, the vehicle concerned has a total of 17 degrees of freedom:

$$\{v_v\} = \{Z_v \quad Y_v \quad \theta_v \quad \varphi_v \quad \phi_v \quad Z_{s1} \quad Y_{s1} \quad Z_{s2} \quad Y_{s2} \quad Z_{s3} \quad Y_{s3} \quad Z_{s4} \quad Y_{s4} \quad Y_{c1} \quad Y_{c2} \quad Y_{c3} \quad Y_{c4}\} \tag{6.8}$$

In Equation 6.8, all the vertical displacements of the vehicle are measured from the position of static equilibrium. Each rigid body in either the front or rear axle set is connected to the vehicle body through two suspension units: one is the parallel combination of a linear elastic spring of stiffness K_{uzi} and a viscous damper of damping coefficient C_{uzi} in the z-direction; the other is the parallel combination of a linear elastic spring of stiffness K_{uyi} and a viscous damper of damping coefficient C_{uyi} in the y-direction.

The connection of each rigid body in either the front axle set or the rear axle set to the tire is realized through the two units representing the dynamic characteristics of the tire: one is the parallel combination of a linear elastic spring of stiffness K_{lzi} and a viscous damper of damping coefficient C_{lzi} in the z-direction, while the other is the parallel combination of a linear elastic spring of stiffness K_{lyi} and a viscous damper of damping coefficient C_{lyi} in the y-direction. The horizontal distance between the two rigid bodies in either the front axle set or the rear axle set is $2b_l$. The other major parameters of the vehicle are listed in Table 6.1.

The use of the fully computerized approach [6] can easily lead to the equations of motion of the vehicle under a sudden crosswind gust, established from the static equilibrium position of the vehicle.

The equation of motion of the vehicle body in the y-direction is:

$$\begin{aligned}
&M_v\ddot{Y}_v + C_{uy1}(\dot{Y}_v + h_1\dot{\varphi}_v + L_1\dot{\phi}_v - \dot{Y}_{s1}) + C_{uy2}(\dot{Y}_v + h_1\dot{\varphi}_v - L_2\dot{\phi}_v - \dot{Y}_{s2}) \\
&+ C_{uy3}(\dot{Y}_v + h_1\dot{\varphi}_v + L_1\dot{\phi}_v - \dot{Y}_{s3}) + C_{uy4}(\dot{Y}_v + h_1\dot{\varphi}_v - L_2\dot{\phi}_v - \dot{Y}_{s4}) \\
&+ K_{uy1}(Y_v + h_1\varphi_v + L_1\phi_v - Y_{s1}) + K_{uy2}(Y_v + h_1\varphi_v - L_2\phi_v - Y_{s2}) \\
&+ K_{uy3}(Y_v + h_1\varphi_v + L_1\phi_v - Y_{s3}) + K_{uy4}(Y_v + h_1\varphi_v - L_2\phi_v - Y_{s4}) = F_{vyw}
\end{aligned} \tag{6.9}$$

where F_{vyw} is the aerodynamic force acting on the vehicle body in the y-direction; other parameters can be found in Table 6.1.

Table 6.1 Major parameters of the road vehicle used in case study (from [5]) (Reproduced with permission from Techno.Press)

Parameter	Unit	Value
Full length of vehicle (L)	m	13.450
Total weight of vehicle (W_v)	kN	73.575
Mass of truck body (M_v)	kg	4480
Pitching moment of inertia of truck body (J_{yv})	kg·m^2	5516
Rolling moment of inertia of truck body (J_{xv})	kg·m^2	1349
Yawing moment of inertia of truck body (J_{zv})	kg·m^2	100 000
Mass of axle set ($M_{s1} = M_{s3}$)	kg	800
Mass of axle set ($M_{s2} = M_{s4}$)	kg	710
Mass of tires ($M_{c1} = M_{c2} = M_{c3} = M_{c4}$)	kg	0.0
Upper vertical spring stiffness ($K_{uz1} = K_{uz2} = K_{uz3} = K_{uz4}$)	kN/m	399
Upper lateral spring stiffness ($K_{uy1} = K_{uy2} = K_{uy3} = K_{uy4}$)	kN/m	299
Upper vertical damper damping coefficient ($C_{uz1} = C_{uz3}$)	kN·s/m	23.21
Upper lateral damper damping coefficient ($C_{uy1} = C_{uy3}$)	kN·s/m	23.21
Upper vertical damper damping coefficient ($C_{uz2} = C_{uz4}$)	kN·s/m	5.18
Upper lateral damper damping coefficient ($C_{uy2} = C_{uy4}$)	kN·s/m	5.18
Lower vertical spring stiffness ($K_{lz1} = K_{lz2} = K_{lz3} = K_{lz4}$)	kN/m	351
Lower lateral spring stiffness ($K_{ly1} = K_{ly2} = K_{ly3} = K_{ly4}$)	kN/m	121
Lower vertical damper damping coefficient ($C_{lz1} = C_{lz2} = C_{lz3} = C_{lz4}$)	kN·s/m	0.80
Lower lateral damper damping coefficient ($C_{ly1} = C_{ly2} = C_{ly3} = C_{ly4}$)	kN·s/m	0.80
Reference area (A_f)	m^2	10.50
Reference height (h_v)	m	1.50
Distance (L_1)	m	3.00
Distance (L_2)	m	5.00
Distance (b_1)	m	1.10
Distance (h_1)	m	0.80

The equation of motion of the vehicle body in the z-direction is:

$$\begin{aligned} &M_v\ddot{Z}_v + C_{uz1}(\dot{Z}_v - L_1\dot{\theta}_v - b_1\dot{\varphi}_v - \dot{Z}_{s1}) + C_{uz2}(\dot{Z}_v + L_2\dot{\theta}_v - b_1\dot{\varphi}_v - \dot{Z}_{s2}) \\ &+ C_{uz3}(\dot{Z}_v - L_1\dot{\theta}_v + b_1\dot{\varphi}_v - \dot{Z}_{s3}) + C_{uz4}(\dot{Z}_v + L_2\dot{\theta}_v + b_1\dot{\varphi}_v - \dot{Z}_{s4}) + K_{uz1}(Z_v - L_1\theta_v - b_1\varphi_v - Z_{s1}) \\ &+ K_{uz2}(Z_v + L_2\theta_v - b_1\varphi_v - Z_{s2}) + K_{uz3}(Z_v - L_1\theta_v + b_1\varphi_v - Z_{s3}) + K_{uz4}(Z_v + L_2\theta_v + b_1\varphi_v - Z_{s4}) = F_{vzw} \end{aligned} \tag{6.10}$$

where F_{vzw} is the aerodynamic force acting on the vehicle body in the z-direction.

The dynamic equilibrium condition of the vehicle body about the x-axis leads to:

$$\begin{aligned} &J_{xv}\ddot{\varphi}_v - C_{uz1}(\dot{Z}_v - L_1\dot{\theta}_v - b_1\dot{\varphi}_v - \dot{Z}_{s1})b_1 - C_{uz2}(\dot{Z}_v + L_2\dot{\theta}_v - b_1\dot{\varphi}_v - \dot{Z}_{s2})b_1 \\ &+ C_{uz3}(\dot{Z}_v - L_1\dot{\theta}_v + b_1\dot{\varphi}_v - \dot{Z}_{s3})b_1 + C_{uz4}(\dot{Z}_v + L_2\dot{\theta}_v + b_1\dot{\varphi}_v - \dot{Z}_{s4})b_1 \\ &- K_{uz1}(Z_v - L_1\theta_v - b_1\varphi_v - Z_{s1})b_1 - K_{uz2}(Z_v + L_2\theta_v - b_1\varphi_v - Z_{s2})b_1 \\ &+ K_{uz3}(Z_v - L_1\theta_v + b_1\varphi_v - Z_{s3})b_1 + K_{uz4}(Z_v + L_2\theta_v + b_1\varphi_v - Z_{s4})b_1 \\ &+ C_{uy1}(\dot{Y}_v + h_1\dot{\varphi}_v + L_1\dot{\phi}_v - \dot{Y}_{s1})h_1 + C_{uy2}(\dot{Y}_v + h_1\dot{\varphi}_v - L_2\dot{\phi}_v - \dot{Y}_{s2})h_1 \\ &+ C_{uy4}(\dot{Y}_v + h_1\dot{\varphi}_v - L_2\dot{\phi}_v - \dot{Y}_{s4})h_1 + K_{uy1}(Y_v + h_1\varphi_v + L_1\phi_v - Y_{s1})h_1 \\ &+ K_{uy2}(Y_v + h_1\varphi_v - L_2\phi_v - Y_{s2})h_1 + C_{uy3}(\dot{Y}_v + h_1\dot{\varphi}_v + L_1\dot{\phi}_v - \dot{Y}_{s3})h_1 \\ &+ K_{uy3}(Y_v + h_1\varphi_v + L_1\phi_v - Y_{s3})h_1 + K_{uy4}(Y_v + h_1\varphi_v - L_2\phi_v - Y_{s4})h_1 = M_{vxw} \end{aligned} \tag{6.11}$$

where M_{vxw} is the aerodynamic moment acting on the vehicle body about the x-axis.

The dynamic equilibrium condition of the vehicle body about the y-axis yields:

$$\begin{aligned}&J_{yv}\ddot{\theta}_v - C_{uz1}(\dot{Z}_v - L_1\dot{\theta}_v - b_1\dot{\varphi}_v - \dot{Z}_{s1})L_1 + C_{uz2}(\dot{Z}_v + L_2\dot{\theta}_v - b_1\dot{\varphi}_v - \dot{Z}_{s2})L_2 - C_{uz3}(\dot{Z}_v - L_1\dot{\theta}_v + b_1\dot{\varphi}_v - \dot{Z}_{s3})L_1\\&+ C_{uz4}(\dot{Z}_v + L_2\dot{\theta}_v + b_1\dot{\varphi}_v - \dot{Z}_{s4})L_2 - K_{uz1}(Z_v - L_1\theta_v - b_1\varphi_v - Z_{s1})L_1 + K_{uz2}(Z_v + L_2\theta_v - b_1\varphi_v - Z_{s2})L_2\\&- K_{uz3}(Z_v - L_1\theta_v + b_1\varphi_v - Z_{s3})L_1 + K_{uz4}(Z_v + L_2\theta_v + b_1\varphi_v - Z_{s4})L_2 = M_{vyw}\end{aligned} \tag{6.12}$$

where M_{vyw} is the aerodynamic moment acting on the vehicle body about the y-axis.

The dynamic equilibrium condition of the vehicle body about the z-axis results in:

$$\begin{aligned}&J_{zv}\ddot{\phi}_v + C_{uy1}(\dot{Y}_v + h_1\dot{\varphi}_v + L_1\dot{\phi}_v - \dot{Y}_{s1})L_1 - C_{uy2}(\dot{Y}_v + h_1\dot{\varphi}_v - L_2\dot{\phi}_v - \dot{Y}_{s2})L_2\\&+ C_{uy3}(\dot{Y}_v + h_1\dot{\varphi}_v + L_1\dot{\phi}_v - \dot{Y}_{s3})L_1 - C_{uy4}(\dot{Y}_v + h_1\dot{\varphi}_v - L_2\dot{\phi}_v - \dot{Y}_{s4})L_2\\&+ K_{uy1}(Y_v + h_1\varphi_v + L_1\phi_v - Y_{s1})L_1 - K_{uy2}(Y_v + h_1\varphi_v - L_2\phi_v - Y_{s2})L_2\\&+ K_{uy3}(Y_v + h_1\varphi_v + L_1\phi_v - Y_{s3})L_1 - K_{uy4}(Y_v + h_1\varphi_v - L_2\phi_v - Y_{s4})L_2 = M_{vzw}\end{aligned} \tag{6.13}$$

where M_{vzw} is the aerodynamic moment acting on the vehicle body about the z-axis.

The equations of motion of the left rigid body in the front axle set in the y- and z-directions can be expressed as:

$$\begin{aligned}&M_{s1}\ddot{Y}_{s1} - C_{uy1}(\dot{Y}_v + h_1\dot{\varphi}_v + L_1\dot{\phi}_v - \dot{Y}_{s1}) - K_{uy1}(Y_v + h_1\varphi_v + L_1\phi_v - Y_{s1})\\&+ C_{ly1}(\dot{Y}_{s1} - \dot{Y}_{c1}) + K_{ly1}(Y_{s1} - Y_{c1}) = 0\end{aligned} \tag{6.14}$$

$$\begin{aligned}&M_{s1}\ddot{Z}_{s1} - C_{uz1}(\dot{Z}_v - L_1\dot{\theta}_v - b_1\dot{\varphi}_v - \dot{Z}_{s1}) - K_{uz1}(Z_v - L_1\theta_v - b_1\varphi_v - Z_{s1})\\&+ C_{lz1}(\dot{Z}_{s1} - \dot{Z}_{c1}) + K_{lz1}(Z_{s1} - Z_{c1}) = 0\end{aligned} \tag{6.15}$$

The equations of motion of the left rigid body in the rear axle set in the y- and z-directions can be expressed as:

$$\begin{aligned}&M_{s2}\ddot{Y}_{s2} - C_{uy2}(\dot{Y}_v + h_1\dot{\varphi}_v - L_2\dot{\phi}_v - \dot{Y}_{s2}) - K_{uy2}(Y_v + h_1\varphi_v - L_2\phi_v - Y_{s2})\\&+ C_{ly2}(\dot{Y}_{s2} - \dot{Y}_{c2}) + K_{ly2}(Y_{s2} - Y_{c2}) = 0\end{aligned} \tag{6.16}$$

$$\begin{aligned}&M_{s2}\ddot{Z}_{s2} - C_{uz2}(\dot{Z}_v + L_2\dot{\theta}_v - b_1\dot{\varphi}_v - \dot{Z}_{s2}) - K_{uz2}(Z_v + L_2\theta_v - b_1\varphi_v - Z_{s2})\\&+ C_{lz2}(\dot{Z}_{s2} - \dot{Z}_{c2}) + K_{lz2}(Z_{s2} - Z_{c2}) = 0\end{aligned} \tag{6.17}$$

The equations of motion of the right rigid body in the front axle set in the y- and z-directions can be written as:

$$\begin{aligned}&M_{s3}\ddot{Y}_{s3} - C_{uy3}(\dot{Y}_v + h_1\dot{\varphi}_v + L_1\dot{\phi}_v - \dot{Y}_{s3}) - K_{uy3}(Y_v + h_1\varphi_v + L_1\phi_v - Y_{s3})\\&+ C_{ly3}(\dot{Y}_{s3} - \dot{Y}_{c3}) + K_{ly3}(Y_{s3} - Y_{c3}) = 0\end{aligned} \tag{6.18}$$

$$\begin{aligned}&M_{s3}\ddot{Z}_{s3} - C_{uz3}(\dot{Z}_v - L_1\dot{\theta}_v + b_1\dot{\varphi}_v - \dot{Z}_{s3}) - K_{uz3}(Z_v - L_1\theta_v + b_1\varphi_v - Z_{s3})\\&+ C_{lz3}(\dot{Z}_{s3} - \dot{Z}_{c3}) + K_{lz3}(Z_{s3} - Z_{c3}) = 0\end{aligned} \tag{6.19}$$

The equations of motion of the right rigid body in the rear axle set in the y- and z-directions can be written as:

$$
\begin{aligned}
& M_{s4}\ddot{Y}_{s4} - C_{uy4}(\dot{Y}_v + h_1\dot{\varphi}_v - L_2\dot{\phi}_v - \dot{Y}_{s4}) - K_{uy4}(Y_v + h_1\varphi_v - L_2\phi_v - Y_{s4}) \\
& \quad + C_{ly4}(\dot{Y}_{s4} - \dot{Y}_{c4}) + K_{ly4}(Y_{s4} - Y_{c4}) = 0
\end{aligned}
\tag{6.20}
$$

$$
\begin{aligned}
& M_{s4}\ddot{Z}_{s4} - C_{uz4}(\dot{Z}_v + L_2\dot{\theta}_v + b_1\dot{\varphi}_v - \dot{Z}_{s4}) - K_{uz4}(Z_v + L_2\theta_v + b_1\varphi_v - Z_{s4}) \\
& \quad + C_{lz4}(\dot{Z}_{s4} - \dot{Z}_{c4}) + K_{lz4}(Z_{s4} - Z_{c4}) = 0
\end{aligned}
\tag{6.21}
$$

In Equations 6.15, 6.17, 6.19 and 6.21, Z_{ci} ($i = 1, 2, \ldots, 4$) denotes the vertical displacement of the ith tire. Assuming that the vertical road surface profile is not so rough as to make the vehicle jump or leave the riding surface, the tire of the vehicle is assumed to be a point and remains in contact with the road surface at all times except in the occurrence of an overturning accident. As a result, the vertical displacement, velocity and acceleration of each contact point can be expressed in terms of the road surface profile:

$$Z_{ci} = r_{ci}(x) \tag{6.22}$$

$$\dot{Z}_{ci} = \frac{\partial r_{ci}(x)}{\partial x} U_v \tag{6.23}$$

$$\ddot{Z}_{ci} = \frac{\partial^2 r_{ci}(x)}{\partial x^2} U_v^2 \tag{6.24}$$

where $r_{ci}(x)$ is the road surface roughness under the ith contact point. Clearly, Z_{ci} ($i = 1, 2, \ldots, 4$) and its derivatives are known quantities, so they do not appear as independent degrees of freedom.

The consideration of dynamic equilibrium condition of each tire in the y-direction yields:

$$M_{ci}\ddot{Y}_{ci} + C_{lyi}(\dot{Y}_{ci} - \dot{Y}_{si}) + K_{lyi}(Y_{ci} - Y_{si}) = F_{hi} \quad (i = 1, 2, \ldots, 4) \tag{6.25}$$

where F_{hi} ($i = 1, 2, \ldots, 4$) is the lateral contact force between the ith tire and the road surface, namely the tire sideslip force.

The tire sideslip forces can be related, very approximately, to the vertical reactions by equations of the form [2]:

$$F_{hi} = -m\left(\frac{\dot{Y}_{ci}}{U_v} + \delta\right)F_{vi} \quad (i = 1, 3) \tag{6.26}$$

$$F_{hi} = -m\frac{\dot{Y}_{ci}}{U_v}F_{vi} \quad (i = 2, 4) \tag{6.27}$$

where:

m is a coefficient of sideslip friction, and the negative sign ensures that the sideslip force resists the lateral motion of the tire relative to the road surface;

F_{vi} is the vertical contact force between the ith tire and the road surface;

δ is a steering angle, that is the angle of the front wheels to the vehicle axis.

The introduction of the steering angle of the front wheels is to consider driver behaviour for course correction, but it is not considered here. Note that Equations 6.26 and 6.27 control the sideslip of the vehicle and introduce the non-linear terms to the governing equations of motion of the vehicle. Substituting Equations 6.26 and 6.27 into Equation 6.25 then yields:

$$M_{ci}\ddot{Y}_{ci} + \left(C_{lyi} + m\frac{F_{vi}}{U_v}\right)\dot{Y}_{ci} - C_{lyi}\dot{Y}_{si} + K_{lyi}(Y_{ci} - Y_{si}) = -\delta m F_{vi} \quad (i = 1, 3) \tag{6.28}$$

$$M_{ci}\ddot{Y}_{ci} + \left(C_{lyi} + m\frac{F_{vi}}{U_v}\right)\dot{Y}_{ci} - C_{lyi}\dot{Y}_{si} + K_{lyi}(Y_{ci} - Y_{si}) = 0 \quad (i = 2, 4) \tag{6.29}$$

The vertical contact forces are given by:

$$F_{vi} = M_{ci}\ddot{Z}_{ci} + C_{lzi}(\dot{Z}_{ci} - \dot{Z}_{si}) + K_{lzi}(Z_{ci} - Z_{si}) + F_{Gi} \quad (i = 1, 2, \ldots, 4) \tag{6.30}$$

where F_{Gi} $(i = 1, 2, \ldots, 4)$ is the force on the ith tire due to the gravity of the vehicle, which can be calculated by:

$$F_{Gi} = M_v g \frac{L_2}{2(L_1 + L_2)} + (M_{si} + M_{ci})g \quad (i = 1, 3) \tag{6.31}$$

$$F_{Gi} = M_v g \frac{L_1}{2(L_1 + L_2)} + (M_{si} + M_{ci})g \quad (i = 2, 4) \tag{6.32}$$

where g is the acceleration due to gravity.

Equations 6.9 to 6.21 and Equations 6.28 and 6.29 are regarded as the governing equations of motion of a high-sided road vehicle running on the road at a constant velocity and subjected to a sudden crosswind gust. However, these are non-linear equations, because the time-varying vertical contact forces expressed by Equation 6.30 are functions of the motions of the front axle set and the rear axle set in the z-direction (Z_{si}), which are coupled with the motions of the front axle set and the rear axle set in the y-direction (Y_{si}). Thus, iterations have to be used at each time step in order to find the numerical solution of the governing equations of motion of the vehicle.

The two rigid bodies in either the front axle set or the rear axle set are connected by a massless rigid rod in this numerical study. The following two restriction equations are used to reduce the number of governing equations of motion from 17 to 15:

$$Y_{s1} = Y_{s3} \tag{6.33}$$

$$Y_{s2} = Y_{s4} \tag{6.34}$$

6.2.6 *Case Study*

A computer program for determining the dynamic response and the accident vehicle speed of a moving road vehicle subject to a sudden crosswind gust was developed according to the proposed framework. The equations of motion assembled by the computer program were a set of coupled second-order non-linear differential equations. The Wilson-θ method was used in this study to find the solutions [9]. The θ value and the time interval used in the computation were 1.4 and 0.005 seconds, respectively. Within each time step, the iterations were performed in consideration of the sideslip forces between the tire and the road surface.

The value of roughness coefficient A_r in Equation 6.3 is taken as 80×10^{-6}, 20×10^{-6} and 5×10^{-6} m^3/cycle for the average, good and very road surface, respectively. A total of 16 384 (2^{14}) data points are generated within the sample length of 2048 m. The vertical road surface profile averaged from five simulations with the first 512 m is shown in Figures 6.3a, 6.3b and 6.3c, respectively, for very good, good, and average road conditions.

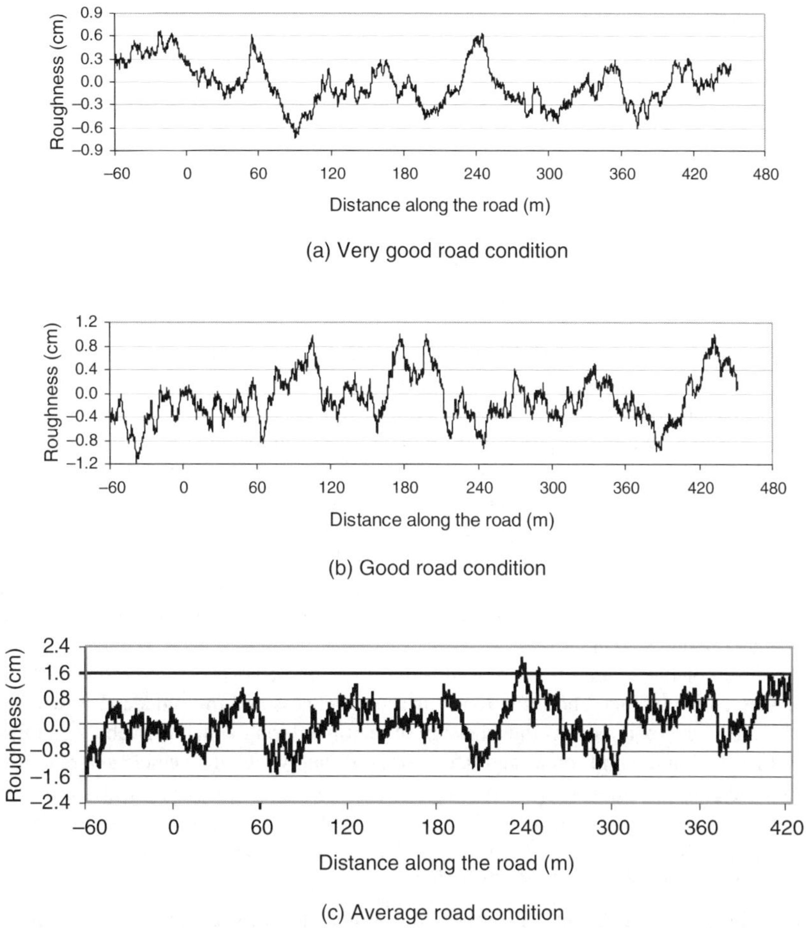

Figure 6.3 Vertical road surface profiles (from [5]) (Reproduced with permission from Techno.Press).

The selected vehicle in this study is similar to that investigated by Coleman and Baker [3] for ground vehicles under crosswinds. The following formulae were suggested for determining the aerodynamic force and moment coefficients of the vehicle as a function of the yaw angle:

$$C_s(\psi) = 5.2(\psi)^{0.382} \tag{6.35a}$$

$$C_L(\psi) = 0.93(1 + \sin 3\psi) \tag{6.35b}$$

$$C_D(\psi) = 0.5(1 + 2\sin 3\psi) \tag{6.35c}$$

$$C_Y(\psi) = -2.0(\psi)^{1.77} \tag{6.35d}$$

$$C_P(\psi) = -2.0(\psi)^{1.32} \tag{6.35e}$$

$$C_R(\psi) = 7.3(\Psi)^{0.294} \tag{6.35f}$$

These force and moment coefficients were obtained with respect to the gravity centre of the vehicle body.

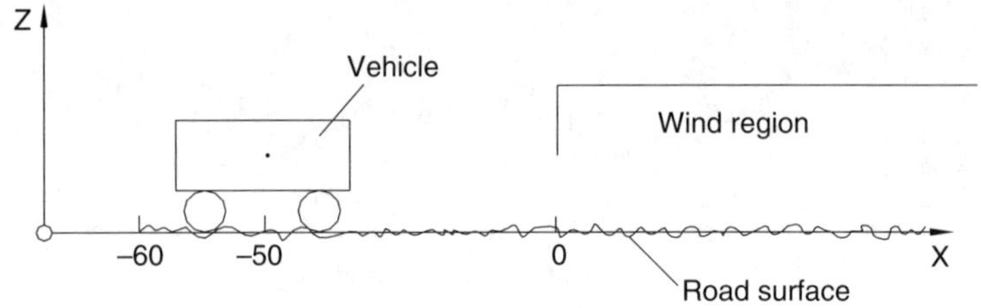

Figure 6.4 The x-coordinates used for a moving road vehicle (from [5]) (Reproduced with permission from Techno.Press).

To properly simulate the case in which a moving road vehicle was subjected to a sudden crosswind gust, the x-axis was set along the road with the coordinate of −60 m for the starting point of the road surface profile, as shown in Figure 6.4.

The x-coordinate of the initial position of the centroid of the vehicle body was taken as −50 m, where the vehicle started to run, with all the initial conditions being zero except for the vehicle speed in the x-direction. When the vehicle had run to the place with the zero x-coordinate for its centroid, a sudden crosswind gust was imposed on the vehicle. When the centroid of the vehicle body was of negative x-coordinate, there was no crosswind and the relative wind velocity became the same as the vehicle velocity in the x-direction. Since the corresponding wind lift force due to the relative wind velocity (i.e. the vehicle velocity) was very small, it was neglected and only the road surface roughness was considered as the vibration source.

As the centroid of the vehicle body moved to the location of positive x-coordinate, the vibration sources of the vehicle included both the road surface roughness and the sudden crosswind gust. The total computation time required was thus the sum of the time during which the vehicle ran through the first 50 m of road without wind forces and 0.5 seconds, during which the vehicle entered into a sharp-edged crosswind gust.

6.2.7 Effects of Road Surface Roughness

Displayed in Figure 6.5a are the time histories of the vertical displacement responses (Z_v) of the vehicle at its centroid for four road surface conditions in the vertical direction: no roughness, very good, good, and average road conditions. The vehicle speed used in the computation is constant at 40 km/h, and the sudden crosswind gust is 20 m/s. It thus takes 4.5 seconds for the vehicle to run through a 50 m distance without wind forces and then to enter a sudden crosswind gust for 0.5 seconds.

It can be seen from Figure 6.5a that, within the first 4.5 seconds, the vertical displacement response remains zero for the vehicle running on the road of no roughness. When the vehicle runs on the rough road, however, the vehicle vibrates vertically with a dominant frequency around the first natural frequency of the vehicle in the vertical direction.

The vertical displacement response of the vehicle depends on the road surface profile. When the vehicle enters a sudden crosswind gust at 4.5 seconds, the vehicle experiences a vertical vibration even when it runs on the road without roughness. The peak response of the vehicle, however, depends on the road surface and the initial condition of the vehicle when it enters the sudden crosswind gust.

The time histories of the lateral and rotational displacement responses of the vehicle are plotted in Figure 6.5b and Figure 6.5c, respectively. During the first 4.5 seconds, when there is no wind force acting on the vehicle, the lateral and rotational displacement responses of the vehicle remain zero. Afterwards, the lateral and rotational displacement responses increase rapidly due to suddenly applied wind forces. For example, the lateral displacement response of the vehicle at its centroid, Y_v, is zero at 4.5 seconds, but it reaches 0.256 m at 5.0 seconds. Such a lateral displacement response is actually the

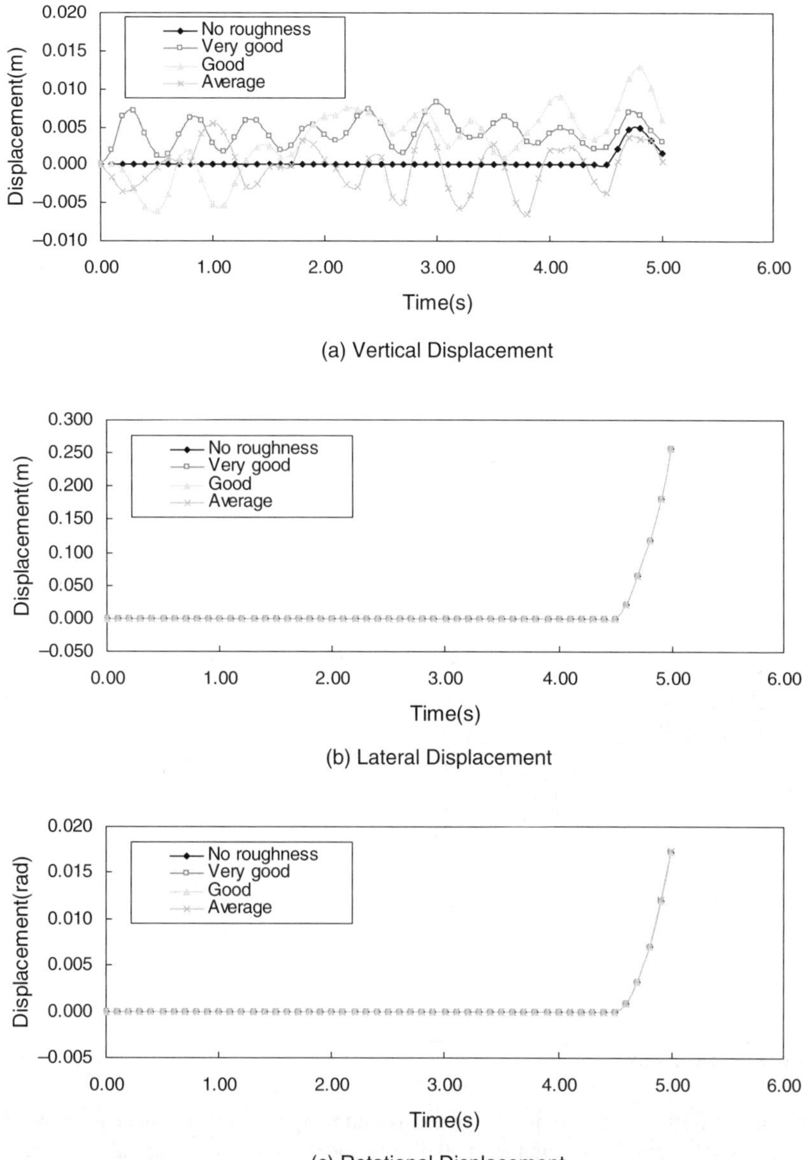

Figure 6.5 Effects of road roughness on dynamic displacement responses ($U_v = 40\,\text{km/h}$, $U_m = 20\,\text{m/s}$) (from [5]) (Reproduced with permission from Techno.Press)

algebraic summation of two parts – the motion of the vehicle as a whole due to the sideslip, and the vibration of the vehicle itself.

In this case, the sideslip of the vehicle at the front left tire, Y_{c1}, is computed as 0.246 m. The total lateral displacement of the front axle set, Y_{s1}, is 0.274 m. As a result, the relative lateral displacement of the vehicle at its centroid to its front axle set is 0.018 m only, and the relative lateral displacement of the front axle set to the front left tire is 0.028 m only. These results show that, even though the total

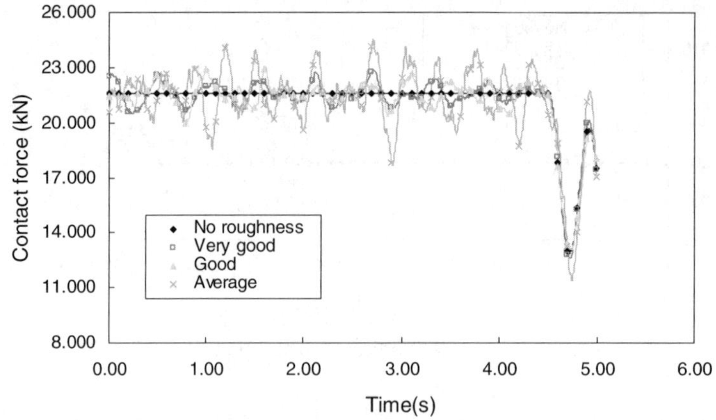

(a) Contact force on 1st tyre

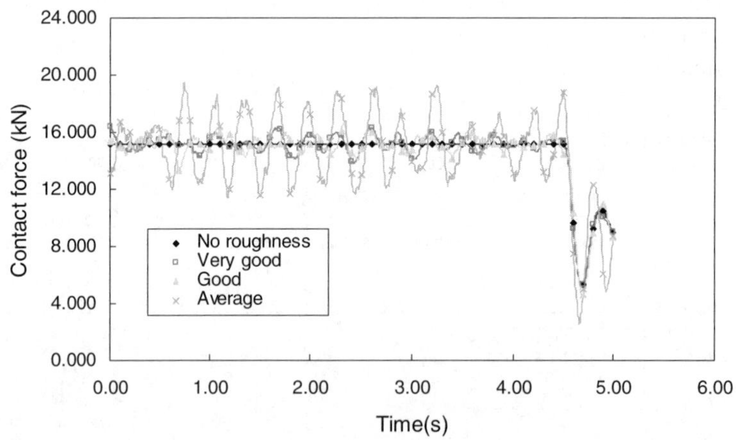

(b) Contact force on 2nd tyre

Figure 6.6 Effects of road roughness on contact forces ($U_v = 40\,\text{km/h}$, $U_m = 20\,\text{m/s}$) (from [5]) (Reproduced with permission from Techno.Press).

lateral displacement of the vehicle at its centroid is quite large, its relative displacement within the vehicle is quite small, complying with the small displacement assumption. It can also be seen that both the lateral and rotational displacement responses of the vehicle keep the same for all the road conditions concerned. This indicates that the vertical road surface condition does not affect the lateral and rotational displacement responses of the vehicle.

Figures 6.6a to 6.6d show the time histories of vertical contact forces acting on the first, second, third and fourth tires, respectively. During the first 4.5 seconds, the vertical contact forces acting on the first, second, third and fourth tires remain constant at 21.58, 15.21, 21.58, and 15.21 kN, respectively. These constant contact forces on the tires are solely determined by the gravity force of the vehicle, as there are no wind forces and no road roughness. Furthermore, because of the symmetry of the vehicle with respect to the x-axis, the contact force on the first tire is the same as that on the third tire, while the contact force on the second tire is the same as that on the fourth tire.

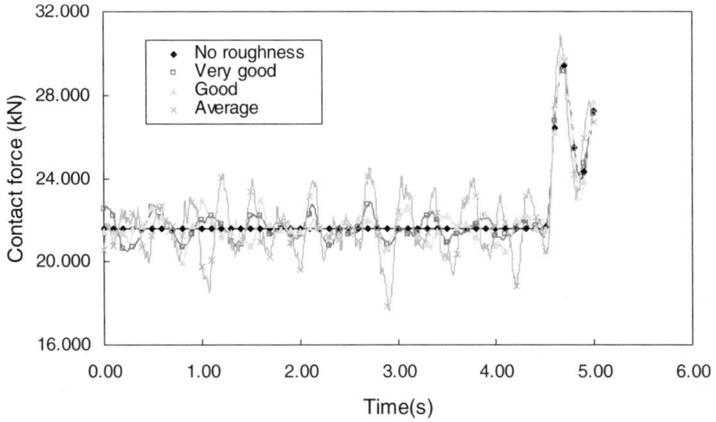

(c) Contact force on 3rd tyre

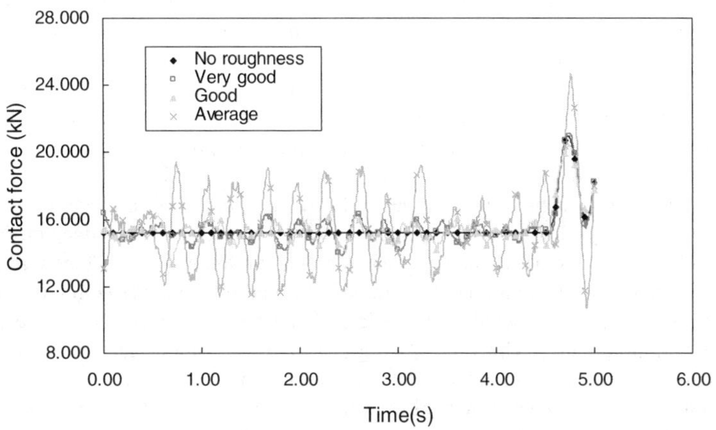

(d) Contact force on 4th tyre

Figure 6.6 (*Continued*)

When the vehicle runs on the rough road, the contact forces fluctuate around the constant contact forces caused by the gravity force of the vehicle. It is clear that within the first 4.5 seconds, road conditions are better, and the peak contact force for any one of the four tires is smaller. When the vehicle enters into a sudden crosswind gust, the contact forces on the first and second tires, on the windward side, significantly decrease, whereas the contact forces on third and fourth tires, on the leeward side, considerably increase, compared with the constant contact forces in the first 4.5 seconds. The decrease of peak contact forces on the windward tires and the increase of peak contact forces on the leeward tires are larger for rougher road conditions.

The minimum peak contact force on the second tire are 2.52, 4.50, 5.12, and 5.22 kN for the average, good, very good and no roughness road conditions, respectively. These results indicate that the overturning accident is controlled by the second tire; the rougher the road condition, the smaller the contact force on the second tire. From this, one may conclude that the road surface condition in the vertical direction does affect the vertical displacement response of the vehicle and the contact forces on the vehicle tires. They thus may, in turn, affect the accident vehicle speed for a given gust wind speed.

6.2.8 Effects of Vehicle Suspension System

To investigate the effects of vehicle suspension system on the dynamic performance of the vehicle, all of the spring stiffness coefficients and damper damping coefficients of the vehicle, as listed in Table 6.1, are multiplied by the same value of 100, 10, 2, 1 (normal), and 0.5, respectively, to form the suspension systems of No. 1, 2, 3, 4, and 5 correspondingly. The multiplication factor over 10 is not realistic, but it implies the rigid suspension assumption used in the previous study [2] so that the validity of this assumption can be assessed.

Other parameters of the vehicle, such as its geometric dimensions, mass, and the mass moments of inertia, remain unchanged. As a result, the first natural frequency of the vehicle is 9.80, 3.10, 1,39, 0.98, and 0.69 Hz in the lateral direction, and 18.60, 5.71, 2.55, 1.81, and 1.28 Hz in the vertical direction, corresponding to the suspension systems of Nos. 1, 2, 3, 4 and 5, respectively. The discussion in the last section was for the No. 4 suspension system (normal system). To achieve a reasonable comparison between different vehicle suspension systems, the dynamic analyses of the vehicle are carried out for the vehicle running on the road of no roughness.

Figure 6.7a shows the time histories of vertical displacement responses of the vehicle for different suspension systems at a vehicle speed of 40 km/h and under a sudden crosswind gust of 20 m/s. During the first 4.5 seconds, when the vehicle runs on the smooth road without wind forces, the vertical displacement response of the vehicle remains zero for all types of suspension systems. When the vehicle enters into a sudden crosswind gust, it experiences vertical vibration. The peak vertical displacement response of the vehicle increases from zero to 0.0, 0.5, 2.6, 5.3 and 10.7 mm, respectively, corresponding to the suspension systems of No. 1, 2, 3, 4, and 5. Clearly, the softer the spring stiffness and the lighter the damper damping, the larger the peak displacement response of the vehicle when it is subjected to a sudden crosswind gust.

Figures 6.7b and 6.7c show the time histories of lateral and rotational displacements of the vehicle for different suspension systems at a vehicle speed of 40 km/h and under a sudden crosswind gust of 20 m/s. Again, during the first 4.5 seconds, all the lateral and rotational displacement responses of the vehicle remain zero because there are no wind forces and road roughness affecting the vehicle. When the vehicle enters a sudden crosswind gust, the lateral and rotational displacement responses of the vehicle increase monotonously with time. At a given time, the lateral and rotational displacement responses increase with the increase in the number of the suspension system.

The maximum lateral displacement responses of the vehicle occur at 5.0 seconds and are 0.22, 0.23, 0.24, 0.26, and 0.29 m for the suspension systems of No. 1, 2, 3, 4, and 5, respectively. The maximum rotational displacement responses of the vehicle also occur at 5.0 seconds and are 0.0155, 0.0156, 0.0163, 0.0172, and 0.0184 radians for the suspension systems of No. 1, 2, 3, 4, and 5, respectively. Clearly, the softer the spring stiffness and the lighter the damper damping, the larger the lateral and rotational displacement responses.

Plotted in Figures 6.8a to 6.8d are the time histories of vertical contact forces on the first, second, third and fourth tires, respectively. During the first 4.5 seconds, when the vehicle runs on the road of no roughness and without wind forces, the vertical contact forces on the first and third tires remain unchanged at 21.58 kN, whereas the contact forces on the second and fourth tires remain at 15.21 kN. These vertical contact forces are caused by the vehicle weight only. When the vehicle enters a sudden crosswind gust, the vertical contact forces on the windward first and second tires have a sudden decrease, whereas those on the leeward third and fourth tires have a sudden increase. Afterwards, the contact force on each tire fluctuates at the natural frequency of the vehicle and around a new dynamic equilibrium position.

The contact forces on each tire for different suspension systems finally reach a common value, which is determined by both the static wind forces and the weight of the vehicle. Again, the minimum contact force occurs on the second tire of 7.94, 6.36, 5.49, 5.22 and 5.02 kN for the suspension systems of No. 1, 2, 3, 4 and 5, respectively. Therefore, one may conclude that the vehicle suspension does affect the dynamic

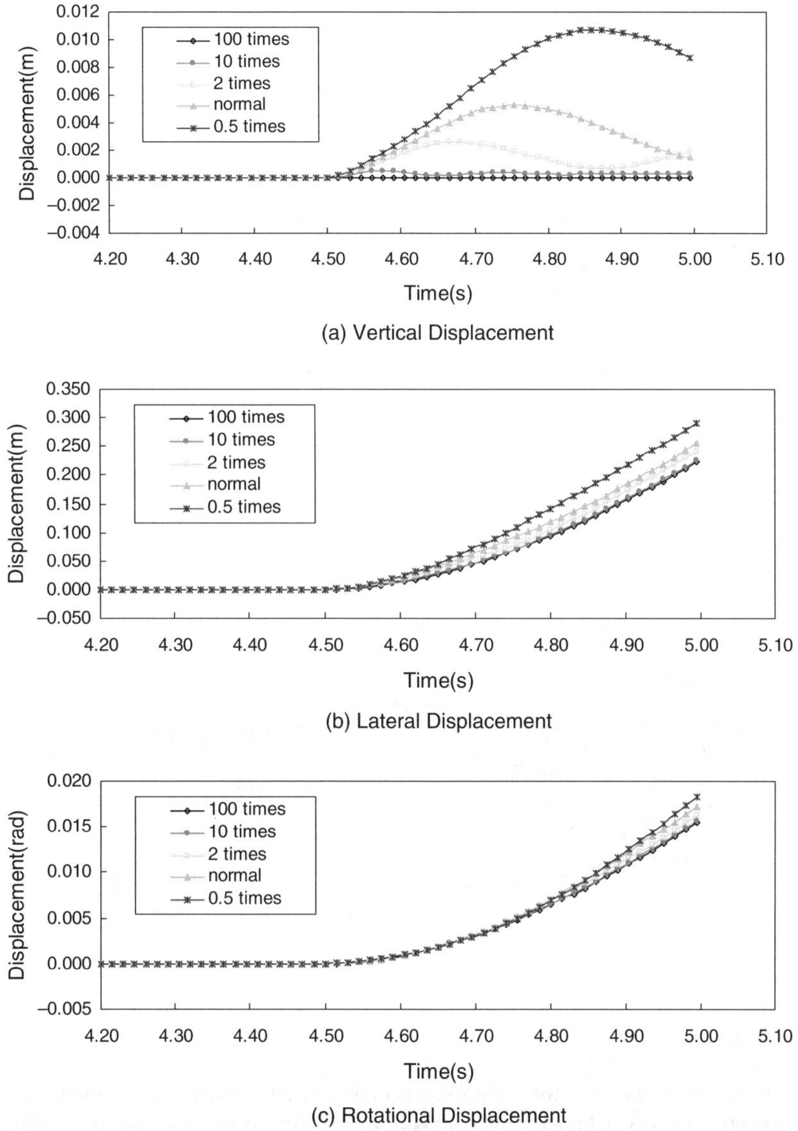

Figure 6.7 Effects of vehicle suspension on dynamic displacement response ($U_v = 40\,\text{km/h}$, $U_m = 20\,\text{m/s}$) (from [5]) (Reproduced with permission from Techno.Press).

responses and contact forces of the vehicle subject to a sudden crosswind gust. The suspension systems of softer spring and lighter damper may lead to a lower accident vehicle speed for a given wind speed.

6.2.9 Accident Vehicle Speed

To investigate the effects of road surface roughness on the accident vehicle speed of the high-sided road vehicle selected, the suspension system of the vehicle is taken as the normal case (No. 4) and four road conditions are considered: no roughness, very good, good and average road surfaces. The computation

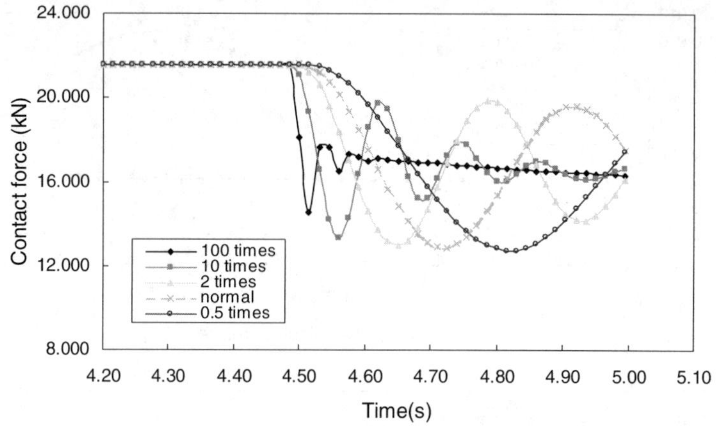

(a) Contact force on 1st tyre

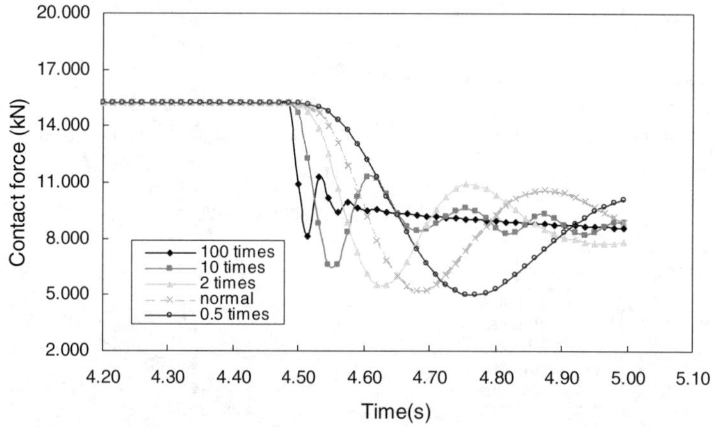

(b) Contact force on 2nd tyre

Figure 6.8 Effects of vehicle suspension on contact forces ($U_v = 40\,\text{km/h}$, $U_m = 20\,\text{m/s}$) (from [5]) (Reproduced with permission from Techno.Press).

is carried out in such a way that, for a given road condition and a given sudden crosswind speed, the dynamic responses and contact forces of the moving vehicle are computed at a series of vehicle speeds in an ascending order. The increment of the vehicle speed is taken as 2.5 km/h. If the dynamic responses and contact forces computed indicate that wind-induced vehicle accident does not occur, a higher vehicle speed that equals the current vehicle speed plus an increment of 2.5 km/h is adopted for the next step computation, until the computation results show that at least one type of vehicle accident occurs. Correspondingly, the final vehicle speed is called the "accident vehicle speed" for the wind speed specified.

The computed accident vehicle speeds for different road surface conditions are plotted in Figure 6.9 and listed in Table 6.2, together with the type of vehicle accidents. It is seen that with increasing gust wind speed, the accident vehicle speed decreases. For a given crosswind speed, a better road condition leads to a relatively higher accident vehicle speed. For a given vehicle speed, a better road condition also gives a relatively higher accident wind speed. The safety of the concerned high-sided road vehicle

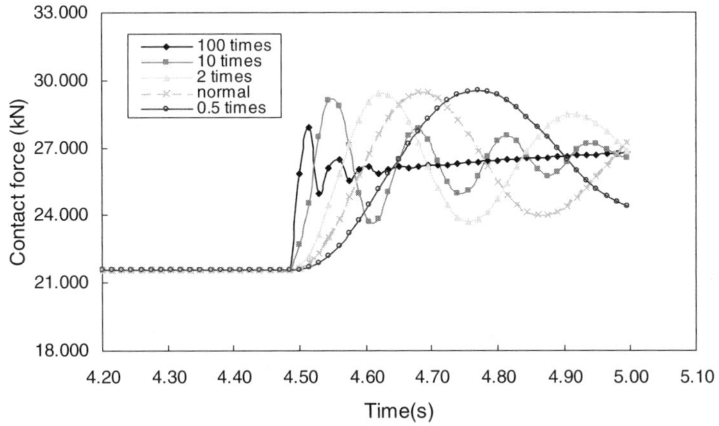

(c) Contact force on 3rd tyre

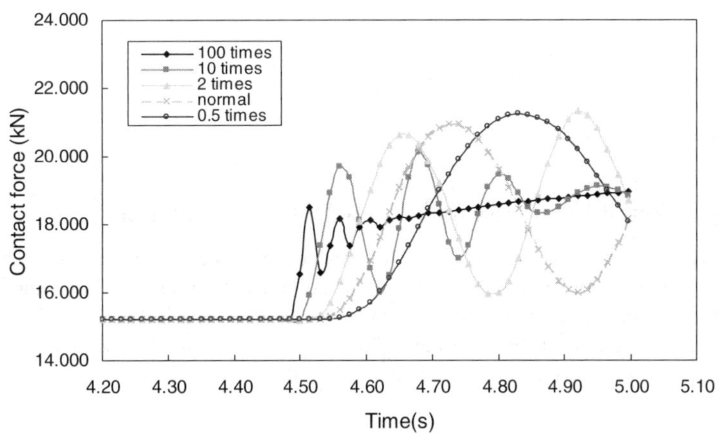

(d) Contact force on 4th tyre

Figure 6.8 *(Continued)*

subject to a sudden crosswind gust mainly results in an overturning accident; there is just one exception to this, in which the crosswind speed is 10 m/s and the road has no roughness, resulting in a sideslip accident at an accident vehicle speed of 135 km/h.

To investigate the effects of vehicle suspension on the accident vehicle speed of the high-sided road vehicle selected, the suspension systems of the vehicle from No. 1 to No. 5 are considered, but only one road condition (i.e. no roughness) is selected for the computation. The computation is carried out in such a way that, for a given suspension system and a given sudden crosswind speed, the dynamic responses and contact forces of the moving vehicle are computed at a series of vehicle speeds in an ascending order of increment of 2.5 km/h until the computation results show that at least one type of vehicle accident occurs. The obtained accident vehicle speed for different suspension systems are plotted in Figure 6.10 and listed in Table 6.3.

It can be seen that with the increase in the gust wind speed, the accident vehicle speed decreases. For a given crosswind speed, the softer suspension system leads to a lower accident vehicle speed. For a

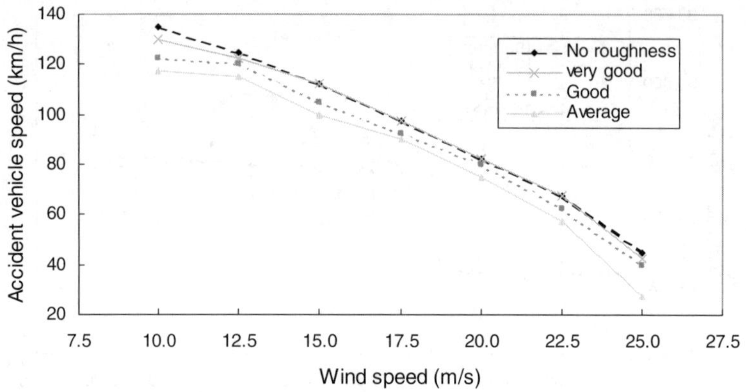

Figure 6.9 Effects of road roughness on accident vehicle speed (from [5]) (Reproduced with permission from Techno.Press).

given vehicle speed, the softer suspension system yields a lower accident wind speed. It is interesting to see that for the suspension system of the highest spring stiffness and the heaviest damper damping (No. 1), all the vehicle accidents are of the sideslip type (see Table 6.3). For the suspension system of smaller spring stiffness and lighter damper damping, the dominant accident type is the overturning accident. These results indicate that the rigid suspension assumption is not suitable for estimating accident vehicle speed and type.

6.3 Formulation of Wind-Road Vehicle-Bridge Interaction

Coupled road vehicle and long-span bridge system under crosswind presents a complicated dynamic interaction problem, involving wind-road vehicle interaction, wind-bridge interaction and road vehicle-bridge interaction. Wind-road vehicle interaction has been discussed in Section 6.2 of this chapter, while wind-bridge interaction has been introduced in Chapter 4 in detail. This section will explore road vehicle-bridge interaction and then wind-road vehicle-bridge interaction.

6.3.1 Equations of Motion of Coupled Road Vehicle-Bridge System

Dynamic vehicle-bridge interaction problems have been studied by many investigators since the middle of the nineteenth century [10–14]. Because of the limitations of computation capacity in

Table 6.2 Accident vehicle speeds for different road conditions (normal suspension system No. 4) (from [5]) (Reproduced with permission from Techno.Press)

Mean wind speed (m/s)	Average (km/h)	Good (km/h)	Very good (km/h)	No roughness (km/h)
10.0	117.5 (o)	122.5 (o)	130.0 (o)	135.0 (s)
12.5	115.0 (o)	120.0 (o)	122.5 (o)	125.0 (o)
15.0	100.0 (o)	105.0 (o)	112.5 (o)	112.5 (o)
17.5	90.0 (o)	92.5 (o)	97.5 (o)	97.5 (o)
20.0	75.0 (o)	80.0 (o)	82.5 (o)	82.5 (o)
22.5	57.5 (o)	62.5 (o)	67.5 (o)	67.5 (o)
25.0	27.5 (o)	40.0 (o)	42.5 (o)	45.0 (o)

Note: (o): Overturning accident; (s): Sideslip accident.

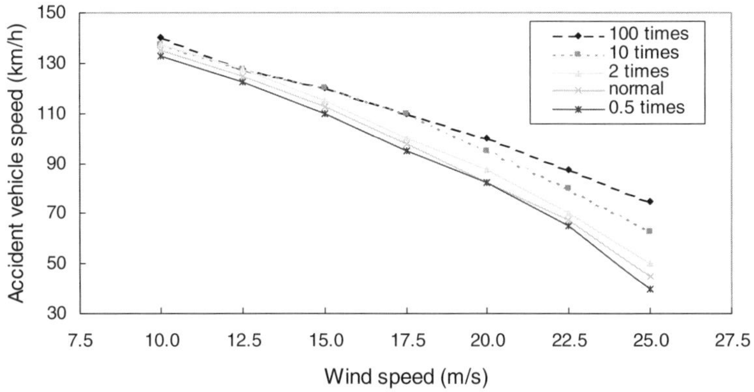

Figure 6.10 Effects of vehicle suspension on accident vehicle speed (from [5]) (Reproduced with permission from Techno.Press).

the past, only simplified models of vehicle-bridge systems could be considered. For instance, a moving vehicle was modeled as a moving load without considering the effect of inertia force [15], and later a moving-mass model was used instead of a moving load to include inertia force effects [10]. Nowadays, the volume of traffic and the speed of vehicles have increased considerably, and the configurations of vehicles have also changed dramatically. More sophisticated and rational models and computerized approaches are required.

However, most of the existing methods are not fully computerized to form the equations of motion of a bridge with different types of vehicles running over it. Thus, the types of vehicles on the bridge should be decided before the formulation. The cases in which a vehicle enters onto the bridge or leaves the bridge are also difficult to handle. Some methods also cannot efficiently compute the dynamic responses of both bridge and vehicles simultaneously.

In this respect, a fully computerized approach for assembling equations of motion of any types of coupled vehicle-bridge systems has been proposed by Guo and Xu [6]. Heavy road vehicles are idealized as a combination of a number of rigid bodies, connected by a series of springs and dampers, while the bridge is modeled using the conventional finite element method. The mass matrix, stiffness matrix, damping matrix and force vector of coupled vehicle-bridge systems are automatically assembled using the fully computerized approach and taking into account road surface roughness. This approach can easily lead to the following equations of motion of the coupled vehicle and bridge system, established from the static equilibrium

Table 6.3 Accident vehicle speeds for different suspension systems (no roughness) (from [5]) (Reproduced with permission from Techno.Press)

Mean wind speed (m/s)	100 times (km/h)	10 times (km/h)	2 times (km/h)	Normal (km/h)	0.5 times (km/h)
10.0	140.0 (s)	137.5 (s)	137.5 (s)	135.0 (s)	132.5 (s)
12.5	127.5 (s)	127.5 (s)	127.5 (s)	125.0 (o)	122.5 (o)
15.0	120.0 (s)	120.0 (s)	115.0 (o)	112.5 (o)	110.0 (o)
17.5	110.0 (s)	110.0 (o)	100.0 (o)	97.5 (o)	95.0 (o)
20.0	100.0 (s)	95.0 (o)	87.5 (o)	82.5 (o)	82.5 (o)
22.5	87.5 (s)	80.0 (o)	70.0 (o)	67.5 (o)	65.0 (o)
25.0	75.0 (s)	62.5 (o)	50.0 (o)	45.0 (o)	40.0 (o)

Note: (o): Overturning accident; (s): Sideslip accident.

position of the system, though it is not necessary for these to be written out explicitly in the computer implementation.

$$\begin{bmatrix} M_b + M_{bbv} & 0 \\ 0 & M_v \end{bmatrix} \begin{Bmatrix} \ddot{v}_b \\ \ddot{v}_v \end{Bmatrix} + \begin{bmatrix} C_b + C_{bbv1} & C_{bv1} \\ C_{vb1} & C_v + C_{v1} \end{bmatrix} \begin{Bmatrix} \dot{v}_b \\ \dot{v}_v \end{Bmatrix} + \begin{bmatrix} K_b + K_{bbv1} & K_{bv1} \\ K_{vb1} & K_v + K_{v1} \end{bmatrix} \begin{Bmatrix} v_b \\ v_v \end{Bmatrix}$$
$$= \begin{Bmatrix} P_{bvg} + P_{bvr1} + P_{bvr2} + P_{bvr3} \\ P_{vvr2} + P_{vvr3} \end{Bmatrix} \quad (6.36)$$

where:

v_b, $\dot{v}_b$, $\ddot{v}_b$ are the nodal displacement, velocity, and acceleration vectors of the bridge, respectively;
v_v, $\dot{v}_v$, $\ddot{v}_v$ are the displacement, velocity, and acceleration vectors of all the vehicles running on the bridge, respectively;
M_b, C_b and K_b are the mass matrix, damping matrix, and stiffness matrix of the bridge, respectively, obtained by the conventional finite element method;
M_{bbv} is related to the inertia forces from the masses of all the tires of the vehicles at the contact points due to the bridge accelerations in the vertical direction;
M_v corresponds to the inertia forces of all the rigid bodies of the vehicles, excluding the masses of the tires at the contact points.

For the dampers whose relative velocities are a function of the degrees of freedom of the vehicles only, the damper forces lead to the matrix C_v. For the dampers connected to the contact points, their relative velocities depend not only on the degrees of freedom of the vehicles but also on the degrees of freedom of the bridge and the deck surface roughness. As a result, the coupled damping matrices C_{bv1} and C_{vb1}, the additional damping matrix C_{bbv1} to the bridge damping matrix C_b, the additional damping matrix C_{v1} to the vehicle damping matrix C_v, the additional force vector on the bridge P_{bvr2} due to the deck surface roughness, and the additional force vector on the vehicles P_{vvr2} due to the deck surface roughness, are generated.

Similarly, for the springs whose relative displacements are the function of the degrees of freedom of the vehicles only, the spring forces lead to the matrix K_v. For the springs connected to the contact points, the stiffness matrices K_{bv1}, K_{vb1}, K_{bbv1} and K_{v1}, and the additional force vectors due to deck surface roughness P_{bvr3} and P_{vvr3}, are constituted. The inertia forces of all the masses of the vehicle at the contact points due to the road surface roughness constitute the force vector P_{bvr1}.The external forces on the bridge due to the gravity forces of the vehicles are denoted by the force vector P_{bvg}.

6.3.2 Equations of Motion of Coupled Wind-Road Vehicle-Bridge System

Based on wind-bridge interaction introduced in Chapter 4, wind-road vehicle interaction discussed in Section 6.2 of this chapter and road vehicle-bridge interaction presented in Section 6.3.1 of this chapter, wind-road vehicle-bridge interaction can be studied in this section [15]. A three-dimensional finite element model, which takes into account the geometric non-linearity, is used to model a long-span cable-stayed bridge. Road vehicles are modeled as a combination of several rigid bodies connected by a series of springs and damping devices. The random roughness of the bridge road surface is included to consider the interaction between the bridge and road vehicles in the vertical direction.

The wind forces acting on the bridge, which comprise mean wind forces, buffeting forces and self-excited forces, are simulated in the time domain. The mean and fluctuating wind forces acting on the road vehicles are also simulated in the time domain as the function of incident wind velocity, vehicle velocity and steady state aerodynamic parameters. The equations of motion of coupled road vehicle-bridge systems under turbulent wind are established using a fully computerized approach [6]. The detailed investigation on the safety of road vehicles running on an oscillating cable-stayed bridge subjected to crosswinds will be presented in Section 6.4.

In the study of wind-road vehicle-bridge interaction, the lateral displacement of a tire of the vehicle should also be taken as an independent degree of freedom to consider the course deviation of the vehicle caused by a sudden crosswind gust when the vehicle runs over the bridge. Since the lateral sideslip force between the tire and the bridge deck depends on the motion of the bridge, this problem should be handled carefully, compared with the vehicle subjected to crosswinds but running on the ground.

Furthermore, the turbulent wind velocity used in the calculation of wind forces on the vehicle should maintain compatibility with those used in the calculation of wind forces on the bridge. That is, the turbulent wind speed used to determine wind forces on the vehicle at a given time and a given position should be the same as the turbulent wind speed at the corresponding position of the bridge deck. Thus, the turbulent wind speed impacting the vehicle is not only a function of time but also of position. The turbulence wind speeds, both in the horizontal and vertical directions, along with the bridge deck, are generated at the bridge nodes only. Therefore, if the position of the vehicle is located between the two nodes, the wind speed used to determine wind forces on the vehicle is determined by a linear interpolation.

The use of the fully computerized approach [6] can easily lead to the equations of motion of coupled road vehicle and stayed-cable bridge system under crosswinds in the following form, established from the static equilibrium position of the system:

$$\begin{aligned}\begin{bmatrix} M_b + M_{bbv} & 0 \\ 0 & M_v \end{bmatrix}\begin{Bmatrix} \ddot{v}_b \\ \ddot{v}_v \end{Bmatrix} &+ \begin{bmatrix} C_b + C_{bbv1} + C_{bbv2} & C_{bv1} + C_{bv2} \\ C_{vb1} + C_{vb2} & C_v + C_{v1} + C_{v2} \end{bmatrix}\begin{Bmatrix} \dot{v}_b \\ \dot{v}_v \end{Bmatrix} \\ &+ \begin{bmatrix} K_b + K_{bbv1} & K_{bv1} \\ K_{vb1} & K_v + K_{v1} \end{bmatrix}\begin{Bmatrix} v_b \\ v_v \end{Bmatrix} \\ &= \begin{Bmatrix} P_{bvg} + P_{bvr1} + P_{bvr2} + P_{bvr3} + P_{bw} \\ P_{vvr2} + P_{vvr3} + P_{vw} \end{Bmatrix}\end{aligned} \tag{6.37}$$

where:

P_{bw} is the total wind force vector on the bridge deck and is the sum of the mean wind force, the buffeting force and the self-excited forces that are functions of the bridge motion;

P_{vw} is the wind force vector on the vehicles only.

The presence of the damping matrices C_{bv2}, C_{vb2} and C_{v2}, which are functions of the vertical contact forces between the tires and the bridge deck (and are, in turn functions of the motions of both the vehicle and bridge in the vertical direction), results from the sideslip forces between the tires and the bridge deck in the lateral direction. Thus, the equation of motion of the system is non-linear and iterations in the numerical computation are inevitable. All other terms can be found in Equation 6.36.

Although the governing equation of motion established is a set of coupled second-order differential equations with time-varying coefficients, not all of the sub-matrices of the system are time-dependent. It is necessary to distinguish the feature of each sub-matrix and external force vector in the computer program so that the computation efforts can be reduced significantly. The sub-matrices and external force vectors in Equation 6.37 can be divided into three categories:

- The sub-matrix of the system remains unchanged with time. This means the sub-matrix is independent on the relative position of the vehicles to the bridge deck. For instance, the dynamic matrices M_b, C_b and K_b of the bridge remain unchanged with time. For the vehicles, the sub-matrices M_v, C_v, C_{v1}, K_v and K_{v1}, which correspond to the degrees of freedom of the vehicles only, remain unchanged. This is because the geometric information and dynamic properties of the vehicles and the sign convention all remain unchanged. As a result, the above eight sub-matrices can be assembled and stored in the computer program once, without any change during the computation. In the computation implementation, the eight sub-matrices are put into the proper positions in the system matrices as the first step for the establishment of equation of motion of the entire system.
- The sub-matrix and the external force vector vary with time because the relative positions of the vehicles to the bridge deck vary with time. The sub-matrices M_{bbv}, C_{bbv1}, C_{bv1}, C_{vb1}, K_{bbv1}, K_{bv1}, K_{vb1}, and the external force vectors P_{bvg}, P_{bvr1}, P_{bvr2}, P_{bvr3}, P_{vvr2}, P_{bvr2}, belong to this category. They should be assembled and put into the proper positions of the system matrices at each time step but without iterations. The external force vector P_{bbu} should also be computed at each time step due to the time variation of turbulent winds. The same thing is applied to the external force vector P_{vw}, because of the time variation of both the position of vehicle and turbulent winds.
- The sub-matrix and the external force vector are dependent on the motion of the system. The sub-matrices C_{bbv2}, C_{bv2}, C_{vb2}, C_{v2} and the external force vector P_{bse} can be classified into this category and should be determined through iterations within each time step.

The above classifications of the sub-matrices and the loading vectors of the system make the computation of the wind-vehicle-bridge interaction possible and efficient.

6.4 Safety Analysis of Road Vehicles on Ting Kau Bridge under Crosswind

This section will perform the safety analysis of a high-sided road vehicle running on a long-span cable-stayed bridge when the road vehicle enters a sharp-edged crosswind gust while the bridge is oscillating under fluctuating winds [16]. Road vehicle accidents, including overturning, excessive sideslip, and exaggerated rotation, as discussed in Section 6.2, are applied. The equations of motion of coupled road vehicle-bridge systems under crosswind established in Section 6.3.2 are used, which include road surface roughness, vehicle suspension and the sideslip of the vehicle tire relative to the bridge deck in the lateral direction. The associated computer programs are developed. A case study is conducted, using a real long-span cable-stayed bridge and a high-sided road vehicle. Extensive computation work is performed to obtain a series of accident vehicle speeds against mean crosswind speed. The obtained accident vehicle speeds are also compared with those for the same vehicle running on the ground.

6.4.1 Ting Kau Bridge

The triple-tower cable-stayed bridge selected for the case study is the Ting Kau Bridge in Hong Kong, which has an overall length of 1177 m,with the two main spans measured at 475 m and 448 m and two side spans of 127 m each (see Figure 6.11).

The bridge deck is separated into two carriageway structures, and each carriageway structure is formed by two longitudinal steel plate girders with steel beams spanning transversely between them at 4.5 m centers (see Figure 6.12). A three-dimensional dynamic finite element model is established for the triple-tower cable-stayed bridge. Three-dimensional Timoshenko beam elements are used to model the three bridge towers. The stay cables and stabilizing cables are modeled by cable elements accounting for geometric non-linearity due to cable tension. Each carriageway structure is represented by a

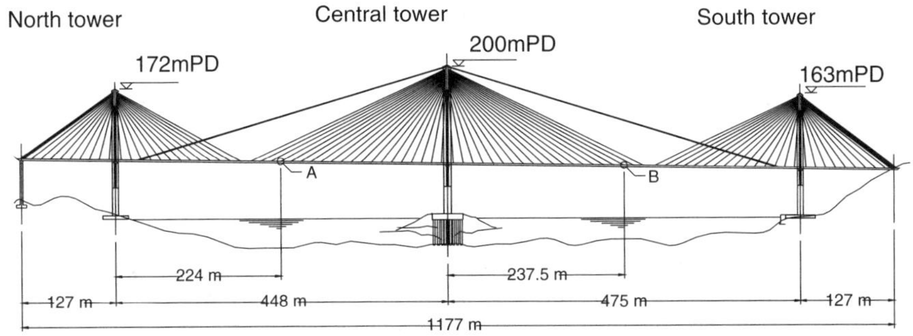

Figure 6.11 Configuration of Ting Kau Bridge used in case study (from [16]) (Reproduced with permission from ASCE).

three-girder model consisting of one central girder and two side girders, connected by transverse links. All of the girders are modeled by the three-dimensional Timoshenko beam elements. The connections between the bridge components and the supports of the bridge are properly modeled. The fundamental frequencies in the lateral, vertical, and torsional directions are 0.216 Hz, 0.189 Hz and 0.510 Hz, respectively. The damping ratios of the bridge are taken as 1%.

6.4.2 Wind Forces on Bridge

In the simulation of horizontal and vertical fluctuating wind speed time-histories with prescribed spectral characteristics along the bridge beck, the following horizontal and vertical wind auto-spectra are adopted for the concerned bridge [16]:

$$R_{u,N}(n) = \frac{nS_u(n)}{\sigma_u^2} = 4\left(\frac{nL_x^u}{U_m}\right)\left[1 + 70.8\left(\frac{nL_x^u}{U_m}\right)^2\right]^{-5/6} \tag{6.38a}$$

$$R_{w,N}(n) = \frac{nS_w(n)}{\sigma_w^2} = 4\left(\frac{nL_x^u}{U_m}\right)\left[1 + 755\left(\frac{nL_x^u}{U_m}\right)^2\right]\left[1 + 283\left(\frac{nL_x^u}{U_m}\right)^2\right]^{-11/6} \tag{6.38b}$$

where:

$R_{u,N}(n)$ and $R_{w,N}(n)$ are the non-dimensional PSD functions for wind turbulence components in the horizontal and vertical direction, respectively;
$n = \omega/2\pi$ is the frequency in Hz;

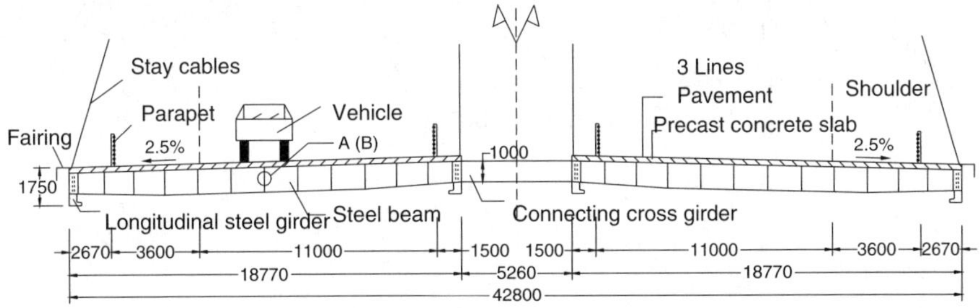

Figure 6.12 Typical deck cross section of bridge (from [16]) (Reproduced with permission from ASCE).

σ_u and σ_w are the standard deviation of wind turbulence in the horizontal and vertical direction respectively;
L_x^u is the integral scale of wind turbulence in the horizontal direction;
U_m is the mean wind speed at the deck level.

In this case study, the turbulence intensity used is 0.105 and the ratio of the turbulent intensity in the vertical direction to the turbulence intensity in the longitudinal direction is 0.825. The integral scale of wind turbulence in the horizontal direction is assumed to remain constant at 612.4 m. The exponential decay coefficient λ, used in the coherence function of wind turbulence along the bridge deck, is taken as 16. The sampling frequency and duration used in the simulation of wind speeds are, respectively, 100 Hz and 140 seconds. The frequency interval and the total number of frequency intervals used in the simulation are, respectively, 0.002 Hz and 1000.

6.4.3 Scenario for Extreme Case Study

The most severe condition that a vehicle accident may occur is when the vehicle encounters a sharp-edged crosswind gust. The safety of a high-sided road vehicle running on the ground and subjected to a sudden crosswind gust was investigated in Section 6.2. Since the motion of the bridge deck at the bridge tower is very small compared with that at the midpoint of the main span of the bridge, the bridge motion will have a little effect on the safety of the vehicle when the vehicle passes through the bridge towers. One may thus envisage that the accident vehicle speeds due to crosswind for the vehicle passing through the bridge tower and then entering a sudden crosswind gust are close to those for the same vehicle running on the ground.

Therefore, the scenario for an extreme case investigated in this study is to let a high-sided vehicle run on the oscillating cable-stayed bridge and be shielded from crosswind by other vehicles that run parallel to the high-sided vehicle concerned. It is then assumed that the high-sided vehicle overtakes the other vehicles at the midpoint of the left main span of the bridge and enters a sudden crosswind gust. To this end, the x-axis is set along the longitudinal direction of the cable-stayed bridge and the left end of the bridge deck is assigned as the zero x-coordinate (see Figure 6.13).

The x-coordinate of initial position of the centroid of the vehicle body is taken as 151 m, where the vehicle starts to run with the initial condition being zero except for the vehicle speed in the x-direction. The start point of vertical road surface profile is taken as x-coordinate of 141 m to

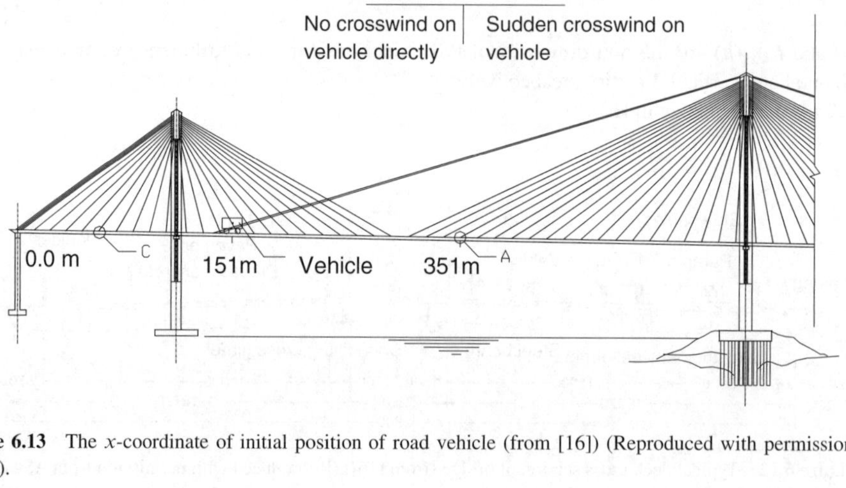

Figure 6.13 The x-coordinate of initial position of road vehicle (from [16]) (Reproduced with permission from ASCE).

ensure the whole vehicle runs on the bridge deck with road roughness. It is assumed that no crosswind acts on the vehicle when it runs from 151 m to 351 m but it then enters a sudden crosswind gust at the midpoint of the main span of the bridge. However, the buffeting forces and the self-excited forces due to crosswind acting on the bridge start when the vehicle is at its initial position, that is, at 151 m of the x-coordinate. The Wilson-θ method is used to solve the equation of motion numerically, and the value and the time interval used in the computation are 1.4 and 0.01 seconds, respectively. The total computation time required is thus the sum of the time during which the vehicle runs through the 200 m long bridge deck without crosswind and the additional 0.5 seconds during which the vehicle enters into a sharp-edged crosswind gust.

6.4.4 Dynamic Response of High-Sided Road Vehicle

To demonstrate the proposed approach and to understand how to assess the safety of a high-sided road vehicle, two mean crosswind speeds of 15.0 and 22.5 m/s, respectively, are selected to compute dynamic responses of the vehicle. A good road surface condition is selected and the vehicle speed is taken as 60 km/h. If any one type of wind-induced vehicle accident occurs, the computation will be terminated automatically and the type of vehicle accident will be recorded.

Depicted in Figure 6.14 are the time histories of the lateral and rotational displacements of the vehicle at two different mean wind speeds. It takes 12.0 seconds for the vehicle to run to point A, the midpoint of the main span of the bridge, without crosswind, and then to enter a sudden crosswind gust

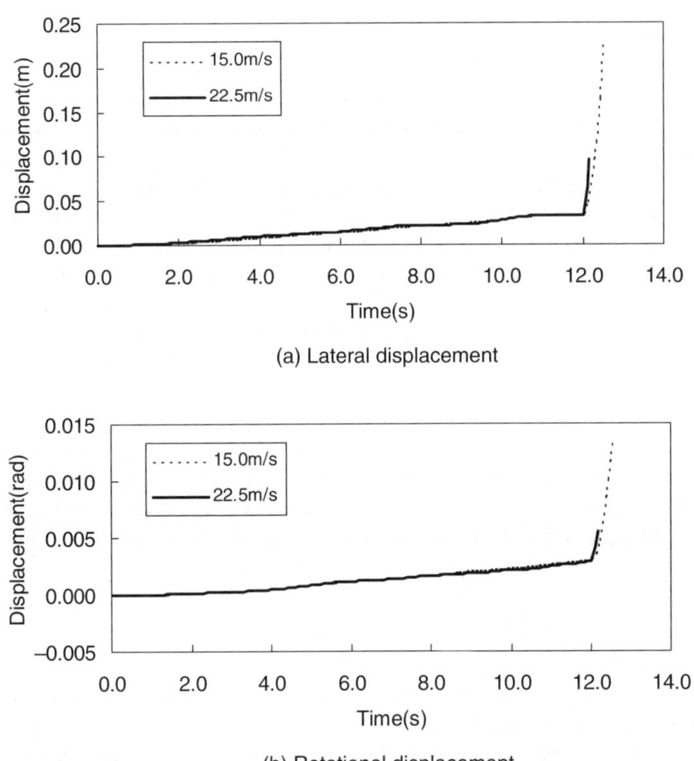

Figure 6.14 Dynamic displacement responses of high-sided vehicle (from [16]) (Reproduced with permission from ASCE).

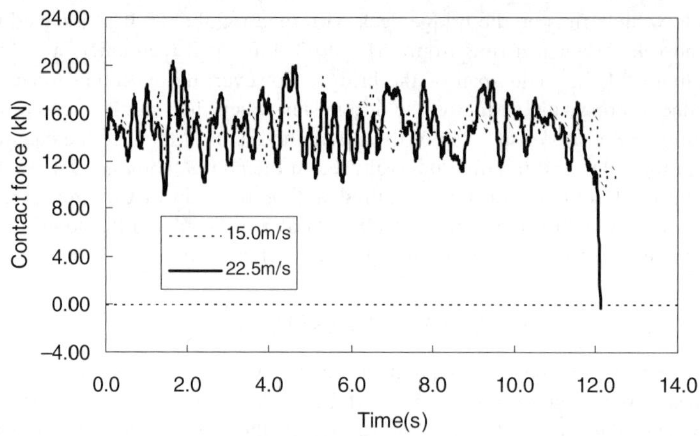

Figure 6.15 Dynamic vertical contact force on second tire of high-sided vehicle (from [16]) (Reproduced with permission from ASCE).

for 0.5 seconds. The total computation time required is thus 12.5 seconds, except for the case of 22.5 m/s mean wind speed, in which the vertical contact force on the second tire of the vehicle falls to zero at 12.13 seconds and the computation is thus terminated automatically. During the first 12.0 seconds, when the vehicle is shielded from crosswind, the lateral and rotational displacement responses of the vehicle are small and less affected by the bridge motion. Afterwards, the lateral and rotational displacement responses increase rapidly and monotonously due to suddenly applied wind forces. However, the maximum lateral and rotational displacements of the vehicle at 12.5 seconds for 15.0 m/s wind speed and at 12.13 seconds for 22.5 m/s wind speed are smaller than the allowable values of 0.5 m and 0.2 radians respectively. Thus, one may say that no sideslip or rotational accident occurs.

Figure 6.15 shows the time histories of vertical contact forces acting on the second tire of the vehicle under two different mean wind speeds. During the first 12.0 seconds, when the vehicle runs on the bridge deck and bears only static wind lift, the maximum vertical contact force at the mean wind speed of 22.5 m/s is greater than that at the mean wind speed of 15 m/s. This indicates that the wind-induced vibration of the bridge can affect the vertical contact forces considerably. When the vehicle enters into a sudden crosswind gust, the contact force on the second tire on the windward side significantly decreases, compared with the vertical contact force in the first 12.0 seconds. Because the minimum value of the contact force on the second tire becomes almost zero at 12.13 seconds under the mean wind velocity of 22.5 m/s, an overturning accident of the vehicle is said to occur. However, under the mean wind speed of 15 m/s, the minimum peak contact force on the second tire at 12.5 seconds is 8.80 kN and the maximum lateral and rotational displacements of the vehicle are 0.224 m and 0.013 rad, respectively. Thus, it may be said that no vehicle accident occurs at 15 m/s mean wind speed.

6.4.5 Accident Vehicle Speed

To investigate the accident vehicle speed of the selected high-sided road vehicle for various mean wind speeds, four road conditions are considered: no roughness, very good, good, and average road surfaces. The computation is carried out in such a way that for a given road condition and a given sudden crosswind speed, the dynamic responses and contact forces of the moving vehicle are computed at a series of vehicle speeds in an ascending order, at increments of 2.5 km/h. If the dynamic responses and contact forces computed indicate that no type of wind-induced vehicle accident occurs, a higher vehicle speed that equals the current vehicle speed plus the vehicle speed increment of 2.5 km/h is adopted for the

Table 6.4 Accident vehicle speeds for different road conditions and wind speeds (from [16]) (Reproduced with permission from ASCE)

U_m (m/s)	Average (km/h)	Good (km/h)	Very good (km/h)	No roughness (km/h)
10.0	115.0 (o)	122.5 (o)	130.0 (o)	135.0 (s)
12.5	110.0 (o)	115.0 (o)	120.0 (o)	122.5 (o)
15.0	95.0 (o)	97.5 (o)	107.5 (o)	107.5 (o)
17.5	82.5 (o)	87.5 (o)	90.0 (o)	92.5 (o)
20.0	65.0 (o)	70.0 (o)	75.0 (o)	77.5 (o)
22.5	42.5 (o)	47.5 (o)	52.5 (o)	57.5 (o)
25.0	15.0 (o)	27.5 (o)	30.0 (o)	32.5 (o)

Note: (o): Overturning accident; (s): Sideslip accident.

next step computation, until the computation results show that at least one type of vehicle accident occurs. Correspondingly, the final vehicle speed is called the "accident vehicle speed" for the mean wind velocity specified.

The computed accident vehicle speeds of the selected high-sided vehicle are listed in Table 6.4 for different road surface conditions, together with the type of vehicle accidents. It is seen that as mean wind speed increases, accident vehicle speed decreases. For a given mean wind speed, a better road condition leads to a higher accident vehicle speed. In all cases but one, the type of accident that occurs is an overturning accident. The exception is the case in which the mean cross-wind speed is 10 m/s and the road has no roughness, where a sideslip accident occurs with the accident vehicle speed of 135 km/h.

6.4.6 Comparison of Safety of Road Vehicle Running on Bridge and Ground

To study the effects of bridge motion on the safety of the road vehicle under a sudden crosswind gust, the dynamic response of the road vehicle running on the cable-stayed bridge subjected to a sudden crosswind gust is compared with that of the same road vehicle running on the ground [16]. Displayed in Figure 6.16 are the time histories of vertical contact force acting on the second tire of the vehicle running on the bridge deck and the ground at a speed of 45 km/h, respectively.

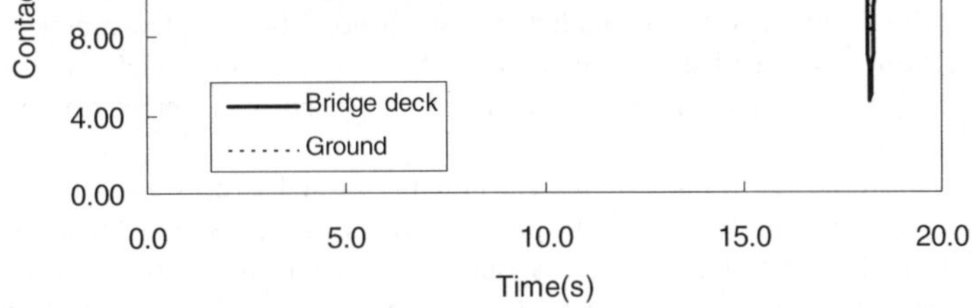

Figure 6.16 Comparison of vertical contact force on second tire of high-sided vehicle (from [16]) (Reproduced with permission from ASCE).

Table 6.5 Comparison of accident vehicle speeds for vehicle on bridge and ground (from [16]) (Reproduced with permission from ASCE)

U_m (m/s)	Average (km/h)		Good (km/h)		Very good (km/h)		No roughness (km/h)	
	Bridge	Ground	Bridge	Ground	Bridge	Ground	Bridge	Ground
10.0	115.0 (o)	117.5 (o)	122.5 (o)	122.5 (o)	130.0 (o)	130.0 (o)	135.0 (s)	135.0 (s)
12.5	110.0 (o)	115.0 (o)	115.0 (o)	120.0 (o)	120.0 (o)	122.5 (o)	122.5 (o)	125.0 (o)
15.0	95.0 (o)	100.0 (o)	97.5 (o)	105.0 (o)	107.5 (o)	112.5 (o)	107.5 (o)	112.5 (o)
17.5	82.5 (o)	90.0 (o)	87.5 (o)	92.5 (o)	90.0 (o)	97.5 (o)	92.5 (o)	97.5 (o)
20.0	65.0 (o)	75.0 (o)	70.0 (o)	80.0 (o)	75.0 (o)	82.5 (o)	77.5 (o)	82.5 (o)
22.5	42.5 (o)	55.0 (o)	47.5 (o)	62.5 (o)	52.5 (o)	65.0 (o)	57.5 (o)	67.5 (o)
25.0	15.0 (o)	27.5 (o)	27.5 (o)	40.0 (o)	30.0 (o)	42.5 (o)	32.5 (o)	45.0 (o)

Note: (o): Overturning accident; (s): Sideslip accident.

It can be seen that within the first 18.0 seconds, when the vehicle is shielded from crosswind, the amplitude of the vertical contact force of the vehicle running on the bridge is much greater than that of the vehicle running on the ground. This indicates that the oscillating cable-stayed bridge can greatly affect the vertical contact forces. Furthermore, the minimum vertical contact force acting on the second tire of the vehicle running on the bridge is lower than that of the vehicle running on the ground. One may thus conclude that the oscillating cable-stayed bridge can affect the accident vehicle speed for a given road condition and a given mean wind velocity.

The computed accident vehicle speeds of the selected high-sided vehicle running on the cable-stayed bridge and the ground for different road conditions are listed in Table 6.5. It can be seen that when a sudden crosswind speed is lower than 12.5 m/s, accident vehicle speeds of the vehicle running on the bridge are slightly less than or equal to those of the same vehicle running on the ground for a given road condition. When mean crosswind speed reaches 25 m/s, the accident vehicle speeds of the vehicle running on the bridge are considerably smaller than those of the same vehicle running on the ground. This highlights again that the larger vibration of the bridge caused by higher crosswind can affect the accident vehicle speed considerably.

6.5 Formulation of Wind-Railway Vehicle Interaction

Early studies [17–19] on vibration of a railway track under moving vehicles did not take into consideration the coupling effects between vehicles and railway track. Later, several models were developed to consider the coupling of vehicles and railway track [20–23]. For instance, Cai and Raymond [20] reported a track dynamic model with one bogie to examine the effect of various wheel and rail defects on dynamic responses. Zhai and Sun [23] presented a more detailed coupled model, in which a wagon with two bogies was represented by two multi-body systems and the track was modeled as an infinite Euler beam supported on a discrete elastic foundation consisting of three layers, with sleeper and ballast included. Furthermore, the coupled vehicle-track model in the vertical direction was extended to include interactive vibration of vehicle and track in the lateral direction [24,25]. A review report in modeling vehicle and track interaction can be found in Knothe and Grassie [26].

The aerodynamic forces on a railway vehicle moving through a crosswind may be sufficiently large to overturn the vehicle. To be able to guarantee comfort and safety for a vehicle in crosswind, one has to understand effects of crosswinds on dynamic interaction between the vehicle subsystem and the track subsystem. Balzer [27] developed a theory to estimate aerodynamic forces on a moving vehicle, in which Taylor's hypothesis of "frozen turbulence" was employed. For engineering applications, Cooper [28] proposed the power spectral density, square-root coherence function, phase-lag function and aerodynamic admittance function for unsteady side forces on a moving vehicle, and laid down a foundation for investigating wind effects on a moving vehicle in the frequency domain. Baker [29,30] further investigated both steady and unsteady aerodynamic forces on a variety of vehicles and carried out extensive studies on the interaction between aerodynamic forces and moving vehicles.

This section focuses on the simulation of dynamic response of railway vehicles running on a track in crosswind in the time domain [30]. Each four-axle vehicle in a train is modeled by a dynamic system with 27 degrees of freedom. Two parallel rails of a track are modeled as two continuous beams supported by a discrete elastic foundation of three layers, with sleepers and ballasts included. The vehicle subsystem and the track subsystem are coupled through contacts between wheels and rails, based on the contact theory. Vertical and lateral rail irregularities, simulated using an inverse Fourier transform, are also taken into consideration. The steady and unsteady aerodynamic forces on a moving railway vehicle in crosswind are derived and simulated in the time domain. The Hilber-Hughes-Taylor α-method is employed to solve the non-linear equations of motion of coupled vehicle and track systems in crosswind. The proposed framework is finally applied to a railway vehicle running on a straight track substructure in crosswind. The safety and comfort performance of the moving vehicle in crosswind are discussed.

6.5.1 Modeling of Vehicle Subsystem

A four-axle railway vehicle with two suspension systems, which is a common railway vehicle used in China, is taken as an example to demonstrate the modeling of vehicle subsystem (see Figure 6.17).

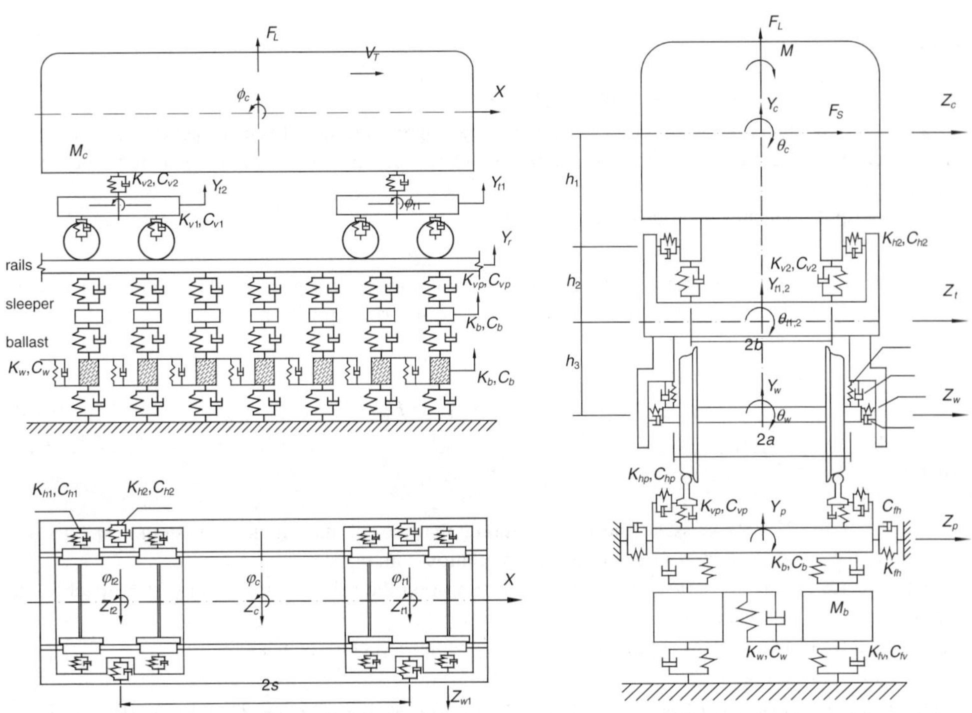

Figure 6.17 Modeling of vehicle and track interaction in crosswind (from [30]) (Reproduced with permission from Elsevier).

The vehicle consists of a car body, two bogies, four wheel-sets, and the connections between the three components. To simplify the analysis while retaining sufficient accuracy, the following assumptions are used in the modeling of the vehicle subsystem:

1. the vehicle is running on a straight railway track at a constant velocity V_T;
2. the car body, bogies and wheel-sets are regarded as rigid components, neglecting their elastic deformation during vibration;
3. the connections between a bogie and a wheel-set are characterized by two linear springs and two viscous dashpots of the same properties in either the horizontal direction or the vertical direction, named the first suspension system;
4. the connections between the car body and a bogie are represented by two linear springs and two viscous dashpots of the same properties in either the horizontal direction or the vertical direction, named the secondary suspension system.

The car body or each bogie is assigned five degrees of freedom: the vertical displacement Y, the lateral displacement Z; the roll displacement θ; the yaw displacement φ; and the pitch displacement ϕ with respect to its mass centre. Each wheel-set has three degrees of freedom: the vertical displacement Y_w; the lateral displacement Z_w; and the roll displacement θ_w with respect to its mass centre. As a result, the total degrees of freedom of the vehicle number 27. The nodal displacement vector of the vehicle can be written as:

$$X_v^{\mathrm{T}} = \{X_c^{\mathrm{T}}, X_{t1}^{\mathrm{T}}, X_{t2}^{\mathrm{T}}, X_{w1}^{\mathrm{T}}, X_{w2}^{\mathrm{T}} X_{w3}^{\mathrm{T}}, X_{w4}^{\mathrm{T}}\} \tag{6.39}$$

where:

X_v is the 27 × 1 displacement vector of the vehicle;
X_c, X_t, X_w are the displacement vectors of the car body, bogies, and wheel-sets, respectively;
the subscript 1 or 2 indicates the first bogie (or wheel-set) and the second bogie (or wheel-set), respectively;
the superscript T indicates the transpose operation.

By assuming that displacement responses of vehicle components are small, the equation of motion of the vehicle subsystem with respect to the static equilibrium position can be derived using the Lagrangian approach, as follows:

$$M_v \ddot{X}_v + C_v \dot{X}_v + K_v X_v = F_v^c + F_v^w \tag{6.40}$$

where:

M_v, C_v, K_v are the 27 × 27 mass, stiffness and damping matrices of the vehicle subsystem, respectively;
each dot in the vector X_v denotes the partial differentiation with respect to time t;
F_v^c is the vector of forces exerted by the track subsystem on the vehicle subsystem with respect to the mass centre of the wheel-set;
F_v^w is the vector of wind forces acting on the car body with respect to its mass centre.

The detailed derivation of Equation 6.40 can be referred to the literature [31], while the two force vectors will be discussed in the subsequent sections.

6.5.2 Modeling of Track Subsystem

The track subsystem consists of rails, sleepers and ballasts. The two parallel rails of the track are modeled as two continuous beams supported on a discrete elastic foundation consisting of three layers, with sleepers and ballasts included (see Figure 6.17). If it is considered that the frequency range of interest for the coupled vehicle-track system in crosswind is below 30 Hz, sleepers and ballasts can be regarded as rigid bodies.

The connections between the rails and the sleepers are represented by linear springs and viscous dashpots of the same properties in either the horizontal direction or the vertical direction. The connections between the sleepers and the ballasts are represented by linear springs and viscous dashpots of the same properties in the vertical direction. The horizontal stiffness and damping of both the sleeper and the ballast are modeled by linear springs and viscous dashpots horizontally installed at the ends of the sleeper against the ground. In order to account for shearing continuity of the particles between the adjacent ballasts, linear springs and viscous dashpots are introduced between adjacent ballasts to model shear coupling effects. Moreover, the ballasts are connected to the ground through linear springs and viscous dashpots in the vertical direction.

In the modeling, the rail between two adjacent sleepers is taken as one beam element. The degrees of freedom of the beam element in the longitudinal (x-) and torsional directions are not considered. Each sleeper has three degrees of freedom and each ballast block has one degree of freedom in the vertical direction only. As a result, the nodal displacement vector of the track subsystem at the ith sleeper can be written as

$$X_{si}^{\mathrm{T}} = \{Y_{rl}, Z_{rl}, \theta_{rl}^{Y}, \theta_{rl}^{Z}, Y_{rr}, Z_{rr}, \theta_{rr}^{Y}, \theta_{rr}^{Z}, Y_{p}, Z_{p}, \theta_{p}, Y_{bl}, Y_{br}\} \tag{6.41}$$

where:

the first subscripts r, p, and b, stand for the rail, sleeper, and ballast, respectively;
the second subscripts l and r stand for left and right side, respectively;
the superscripts Y and Z stand for the axis around which the beam rotates.

In terms of the general procedure of finite element method, the equation of motion of the track subsystem can be assembled as:

$$M_s\ddot{X}_s + C_s\dot{X}_s + K_sX_s = F_s^c \tag{6.42}$$

where:

M_s, C_s, K_s is the mass, stiffness and damping matrix of the track subsystem, respectively;
X_s is the total nodal displacement vector of the track subsystem;
F_s^c is the vector of the contact forces transmitted from the wheels to the rails at all contact points, which will be discussed in the subsequent section.

6.5.3 Wheel and Rail Interaction

Wheel and rail interaction is an essential element that couples the vehicle subsystem with the track subsystem. The interaction between a wheel and a rail involves two basic issues – the geometric relationship and the contact forces between the wheel and the rail.

As mentioned before, each wheel-set has three degrees of freedom: the vertical, lateral, and rolling motions with respect to its mass centre. In this study, the vertical motion of the wheel-set is assumed to be independent of its lateral and rolling motions. The rolling displacement of the wheel-set consists of two parts: one is due to non-uniform configurations of the right and left rails; the other is the rolling angle induced by the lateral displacement of the wheel-set relative to the rails due to the profiles of the wheel and rail cross sections.

Before simulating the interaction of vehicle and track in crosswind, a geometric analysis should be carried out to find the geometric contact information as a function of the lateral displacement of the wheel-set. The geometric contact information includes, but it is not limited to: the relative rolling angle of the wheel-set to the rails; the position of the contact point between the wheel and rail; the contact angle at contact point between the wheel and rail; and the radius of curvature at contact point for either wheel or rail. Under strong crosswind, the lateral displacement of the wheel-set relative to the rails may be large, which may cause a strong non-linearity in the geometric relationship.

Based on the Kalker creepage theory [32], the creeping forces between the wheel and rail can be calculated as

$$T_x = -f_{11}\xi_x \tag{6.43a}$$

$$T_z = -f_{22}\xi_z - f_{23}\xi_{sp} \tag{6.43b}$$

$$M_{sp} = -f_{23}\xi_z - f_{33}\xi_{sp} \tag{6.43c}$$

where:

T_x, T_z and M_{sp} are the longitudinal creeping force, lateral creeping force and spin creeping moment, respectively;

f_{11}, f_{22}, f_{23} and f_{33} are the creepage coefficients;

ξ_x, ξ_z, and ξ_{sp} are the creepage ratios in the longitudinal, lateral, and spin directions respectively, which can be expressed as follows:

$$\xi_x = \frac{V_{wx} - V_{rx}}{V_T}, \quad \xi_z = \frac{V_{wz} - V_{rz}}{V_T}, \quad \xi_{sp} = \frac{\Omega_w - \Omega_r}{V_T} \tag{6.44}$$

where:

V_T is the nominal travelling speed of the wheel-set;

V_{wx} and V_{wz} are the velocities of the wheel at contact point in the longitudinal and lateral direction, respectively;

V_{rx} and V_{rz} are the velocities of the rail at contact point in the longitudinal and lateral direction, respectively;

Ω_w and Ω_r are the rotational velocities of spin motions of the wheel and rail at contact point, respectively.

For large creepage ratios, Kalker's linear creepage theory may cause some errors in calculating creeping forces. Johnson's non-linear creepage law can be used to modify Kalker's linear creepage theory [32].

The vehicle subsystem and the track subsystem are coupled through contacts between the wheels and rails. As shown in Figure 6.18, the contact forces transmitted from the wheels to the left and right rails in the y- and z- directions can be expressed as

$$F'_{yl} = N_l \cos\delta_l - T_{zl}\sin\delta_l, \quad F'_{zl} = N_l \sin\delta_l + T_{zl}\cos\delta_l \tag{6.45a}$$

$$F'_{yr} = N_r \cos\delta_r + T_{zr}\sin\delta_r, \quad F'_{zr} = -N_r \sin\delta_r + T_{zr}\cos\delta_r \tag{6.45b}$$

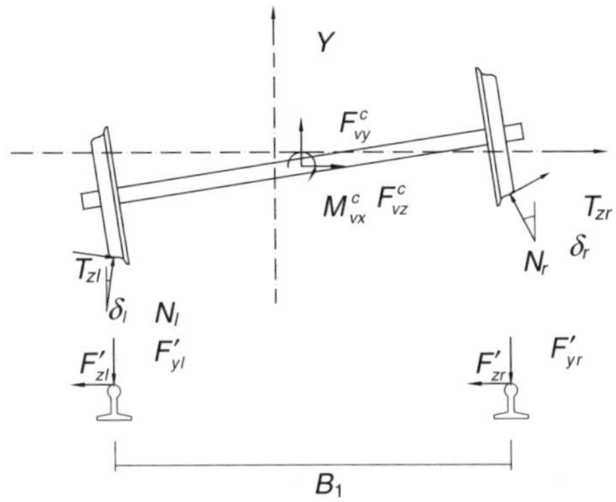

Figure 6.18 Contact forces on wheel-set and rails (from [30]) (Reproduced with permission from Elsevier).

where:

N and T_z are the normal contact force and the lateral creeping force between the wheel and rail, respectively;
δ is the position angle calculated based on the contact angle and the relative rolling angle of the wheel-set;
the subscripts l and r stand for the left and right sides, respectively;
the contact forces F'_{yl}, F'_{yr}, F'_{zl}, F'_{zr} at all contact points on the rails constitute the force vector of F^c_s in Equation 6.42.

In Equation 6.40, F^c_v is the vector of forces exerted by the track subsystem on the vehicle subsystem with respect to the mass centre of the wheel-sets. Figure 6.18 shows that, with the assumption of small displacement, the forces transmitted from the rails to the wheel-set at its mass centre can be expressed as:

$$F^c_{vy} = F'_{yl} + F'_{yr}, \quad F^c_{vz} = F'_{zl} + F'_{zr} \tag{6.46a}$$

$$M^c_{vx} = (F'_{yl} - F'_{yr}) * B_1/2 - (F'_{zl} + F'_{zr}) * R_0 \tag{6.46b}$$

where:

B_1 is the distance between the left and right contact points;
R_0 is the nominal radius of the wheel.

The forces F^c_{vy}, F^c_{vz}, and M^c_{vx} acting on all the wheel-sets of the vehicle constitute the force vector F^c_v in Equation 6.40.

6.5.4 *Rail Irregularity*

Rail irregularities provide self-excitation in a coupled vehicle-track system. Rail irregularities are, however, of random nature, and their statistical characteristics are influenced by many factors. For engineering applications, rail irregularities can be approximately regarded as stationary stochastic processes which can be simulated by numerical methods. Wheel hunting is usually omitted in the vehicle-track analysis, due to its weak effect.

In this study, the lateral, vertical, and rotational irregularities are all assumed to be zero-mean stationary Gaussian random processes. The rail irregularity profile $r(x)$ can then be generated using a simple inverse Fourier transform:

$$r(x) = \sum_{k=1}^{N} \sqrt{2S(f_k)\Delta f}\cos(2\pi f_k x + \theta_k) \tag{6.47a}$$

$$f_k = (k-1)\Delta f + \frac{\Delta f}{2}, \quad k = 1, 2, \ldots, N \tag{6.47b}$$

where:

$S(f_k)$ is the power spectral density (PSD) function (m^3/cycle) of the rail irregularity;
f_k is the spatial frequency (cycle/m);
Δf is the increment of spatial frequency;
θ_k is the random phase angle uniformly distributed from 0 to 2π.

6.5.5 Wind Forces on Ground Railway Vehicles

Wind forces acting on a ground railway vehicle in crosswind can be divided into two parts: the steady and unsteady aerodynamic forces. The steady wind forces are due to the mean wind speed component, while the unsteady wind forces are caused by the fluctuating wind speed components of natural wind. Since a train is often much longer than a road vehicle, unsteady wind forces acting on a train must consider spatial correlation like a bridge deck. As a result, a more sophisticated way is used in this section to determine aerodynamic forces on a vehicle.

The mean wind speed is assumed to be horizontal and normal to the direction of motion of the vehicle in this study. Only wind forces acting on the car body of the vehicle are taken into account. Wind forces acting on the car body of the vehicle refer mainly to drag, lift and moment, as shown in Figure 6.19. If the vehicle considered is not the first or last vehicle in a long train, the conventional strip theory and quasi-static theory for bridge decks can be applied to the aerodynamics of the vehicle [33,34].

As shown in Figure 6.19, the instantaneous wind velocity V and its angle of incidence α can be given by:

$$V^2 = (\bar{u} + u)^2 + w^2, \; \alpha = \arctan\left(\frac{w}{\bar{u} + u}\right) \tag{6.48}$$

where:

$\bar{u}$ is the mean wind speed component;
u and w are the longitudinal and vertical fluctuating wind speed components, respectively.

Since the vehicle runs along the track at a constant velocity V_T, the wind velocity V_R relative to the vehicle and its yaw angle φ can be derived as:

$$V_R^2 = V_T^2 + V^2 = V_T^2 + (\bar{u} + u)^2 + w^2 \tag{6.49a}$$

$$\tan\varphi = \sqrt{(\bar{u} + u)^2 + w^2}/V_T \tag{6.49b}$$

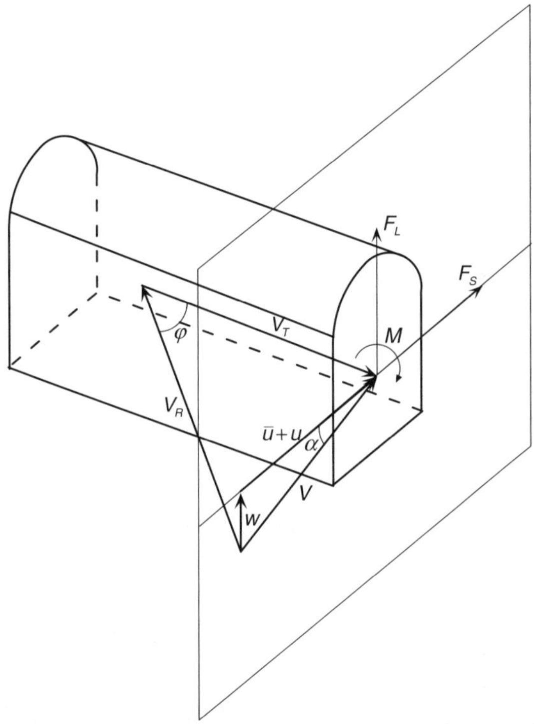

Figure 6.19 Wind forces on a moving vehicle (from [30]) (Reproduced with permission from Elsevier).

In most locations, the horizontal fluctuation u and the vertical fluctuation w are considerably smaller than the mean wind speed $\bar{u}$. The higher order fluctuations in Equations 6.48 and 6.49, such as u^2, w^2, uw, can be neglected, which leads to:

$$V^2 = \bar{u}^2 + 2\bar{u}u, \quad \alpha \approx \arctan\left(\frac{w}{\bar{u}}\right) \tag{6.50}$$

$$V_R^2 = V_T^2 + \bar{u}^2 + 2\bar{u}u, \quad \varphi \approx \arctan\left(\frac{\bar{u}}{V_T}\right) \tag{6.51}$$

Based on quasi-steady theory, the aerodynamic forces on a moving railway vehicle can be expressed as follows:

$$F_S = \frac{1}{2}\rho A V_R^2 C_{F_S}(\alpha, \varphi) \tag{6.52a}$$

$$F_L = \frac{1}{2}\rho A V_R^2 C_{F_L}(\alpha, \varphi) \tag{6.52b}$$

$$M = \frac{1}{2}\rho A H V_R^2 C_M(\alpha, \varphi) \tag{6.52c}$$

where:

F_S, F_L and M are the side force, vertical force and rolling moment with respect to the mass centre of the car body in the car body coordinate system, respectively;

ρ is the air density;
A is the reference area;
H is the reference height, which is taken as the height of the car body;
C_{FS}, C_{FL} and C_M are the aerodynamic force coefficients, which are the function of incidence angle α and yaw angle φ.

Aerodynamic pitch and yawing moments on a long train are small and are neglected in this study. The aerodynamic coefficients can be expanded in the Taylor's series form at $\alpha = 0$ as:

$$C_i(\alpha, \varphi) = C_i(\varphi) + C_i'(\varphi)\alpha \approx C_i(\varphi) + C_i'(\varphi)\frac{w}{\bar{u}} \tag{6.53}$$

where $C_i(\varphi)$ and $C_i'(\varphi)(i = F_S, F_L, M)$ are the side force, lift force and moment coefficients and their derivatives at $\alpha = 0$.

By substituting Equations 6.50, 6.51 and 6.53 into Equation 6.52, and after some manipulations, wind forces on the car body can be obtained:

$$F_S = \frac{1}{2}\rho A\bar{V}_R^2 C_{F_S}(\varphi) + \frac{1}{2}\rho A\bar{V}_R^2\left[C_{F_S}(\varphi)\frac{2\bar{u}u}{\bar{V}_R^2} + C_{F_S}'(\varphi)\frac{w}{\bar{u}}\right] \tag{6.54a}$$

$$F_L = \frac{1}{2}\rho A\bar{V}_R^2 C_{F_L}(\varphi) + \frac{1}{2}\rho A\bar{V}_R^2\left[C_{F_L}(\varphi)\frac{2\bar{u}u}{\bar{V}_R^2} + C_{F_L}'(\varphi)\frac{w}{\bar{u}}\right] \tag{6.54b}$$

$$M = \frac{1}{2}\rho AH\bar{V}_R^2 C_M(\varphi) + \frac{1}{2}\rho AH\bar{V}_R^2\left[C_M(\varphi)\frac{2\bar{u}u}{\bar{V}_R^2} + C_M'(\varphi)\frac{w}{\bar{u}}\right] \tag{6.54c}$$

where $\bar{V}_R^2 = V_T^2 + \bar{u}^2$.

In the above equations, the first term in each case is the steady aerodynamic force and the last two terms are called the unsteady aerodynamic or buffeting forces. The aerodynamic admittance functions are often invoked to reduce the errors involved in the quasi-steady theory for the unsteady aerodynamic forces as follows:

$$F_S^{bu} = \frac{1}{2}\rho A\bar{V}_R^2\left[\chi_{F_S u}(n)C_{F_S}(\varphi)\frac{2\bar{u}u}{\bar{V}_R^2} + \chi_{F_S w}(n)C_{F_S}'(\varphi)\frac{w}{\bar{u}}\right] \tag{6.55a}$$

$$F_L^{bu} = \frac{1}{2}\rho A\bar{V}_R^2\left[\chi_{F_L u}(n)C_{F_S}(\varphi)\frac{2\bar{u}u}{\bar{V}_R^2} + \chi_{F_L w}(n)C_{F_S}'(\varphi)\frac{w}{\bar{u}}\right] \tag{6.55b}$$

$$M^{bu} = \frac{1}{2}\rho AH\bar{V}_R^2\left[\chi_{Mu}(n)C_M(\varphi)\frac{2\bar{u}u}{\bar{V}_R^2} + \chi_{Mw}(n)C_M'(\varphi)\frac{w}{\bar{u}}\right] \tag{6.55c}$$

where:

$\chi_{F_s u}(n)$, $\chi_{F_L u}(n)$, $\chi_{F_S w}(n)$, $\chi_{F_L w}(n)$, $\chi_{Mu}(n)$ and $\chi_{Mw}(n)$ are the aerodynamic transfer functions between the fluctuating wind velocities and aerodynamic forces;
n is the frequency in Hz. The absolute magnitudes of these transfer functions are called the aerodynamic admittance functions.

Obviously, the fluctuation components of turbulence, u and w, should be given to determine the unsteady aerodynamic forces. In this study, the turbulent wind speeds, u and w, are simulated at a series of points along a horizontal line passing through the mass centre of the vehicle with wind turbulence coherence included. The two-side cross-spectral density matrix $S^0(\omega)$ of each fluctuation component is given by:

$$S^0(\omega) = \begin{bmatrix} S^0_{11}(\omega) & S^0_{12}(\omega) & \cdots & S^0_{1n}(\omega) \\ S^0_{21}(\omega) & S^0_{22}(\omega) & \cdots & S^0_{2n}(\omega) \\ \vdots & \vdots & \ddots & \vdots \\ S^0_{n1}(\omega) & S^0_{n2}(\omega) & \cdots & S^0_{nn}(\omega) \end{bmatrix} \tag{6.56}$$

where n is the number of points where the fluctuation component is simulated. The simulation is performed using the following formula:

$$v_j(t) = 2\sqrt{\Delta\omega}\sum_{m=1}^{j}\sum_{l=1}^{N}\left|H_{jm}(\omega_l)\right|\cos(\omega_l t - \theta_{jm}(\omega_l) + \phi_{ml}) \tag{6.57a}$$

$$\omega_l = (l-1)\Delta\omega + \Delta\omega/2, \quad l = 1, 2, \ldots, N \tag{6.57b}$$

where:

v stands for either u or w;
N is a sufficiently large number; $\Delta\omega = \omega_{up}/N$ is the frequency increment;
ω_{up} is the upper cut-off frequency with the condition that, when $\omega > \omega_{up}$, the value of $S^0(\omega)$ is trivial;
ϕ_{ml} is the sequence of independent random phase angles uniformly distributed over the interval $[0, 2\pi]$;
$H_{jm}(\omega)$ is a typical element of matrix $H(\omega)$, which is defined by the Cholesky decomposition of cross-spectral density matrix $S^0(\omega)$; and $\theta_{jm}(\omega)$ is the complex angle of $H_{jm}(\omega)$.

The time histories of the unsteady aerodynamic forces at n points are computed using Equation 6.55, in which fluctuating wind speeds are interpolated based on Taylor's frozen turbulence hypothesis in predicting dynamic response of the coupled vehicle-track system in crosswind. Furthermore, when the aerodynamic admittance should be taken into consideration, equivalent wind spectra, including the admittance functions, can be used to simulate the equivalent fluctuating wind speeds.

6.5.6 Numerical Solution

The dynamic response of coupled vehicle-track system in crosswind is predicted in the time domain in this study. Since the coupled system is a non-linear system due to non-linear contacts between wheels and rails, the Hilber-Hughes-Taylor α-method [35], which is regarded as a modified Newmark method, is used to find numerical solution to avoid spurious high-frequency oscillations in non-linear contact problems.

In the traditional Newmark method [36], the velocity $\dot{X}_{t+\Delta t}$ and displacement $X_{t+\Delta t}$ at time $t+\Delta t$ are calculated in terms of the acceleration $\ddot{X}_{t+\Delta t}$ at time $t+\Delta t$ using the following algorithm:

$$\dot{X}_{t+\Delta t} = \dot{X}_t + (1-\gamma)\ddot{X}_t\Delta t + \gamma\ddot{X}_{t+\Delta t}\Delta t \tag{6.58a}$$

$$X_{t+\Delta t} = X_t + \dot{X}_t\Delta t + \left[\left(\frac{1}{2} - \beta\right)\ddot{X}_t + \beta\ddot{X}_{t+\Delta t}\right]\Delta t^2 \tag{6.58b}$$

where γ and β are the two weighting parameters. Let us denote:

$$c_0 = \frac{1}{\beta\Delta t^2}, \quad c_1 = \frac{\gamma}{\beta\Delta t}, \quad c_2 = \frac{1}{\beta\Delta t}, \quad c_3 = \frac{1}{2\beta} - 1 \tag{6.59a}$$

$$c_4 = \frac{\gamma}{\beta} - 1, \quad c_5 = \frac{\Delta t}{2}\left(\frac{\gamma}{\beta} - 2\right), \quad c_6 = \Delta t(1-\gamma), \quad c_7 = \Delta t\gamma \tag{6.59b}$$

Equation 6.58 can be rewritten as:

$$\ddot{X}_{t+\Delta t} = c_0(X_{t+\Delta t} - X_t) - c_2\dot{X}_t\Delta t - c_3\ddot{X}_t \tag{6.60a}$$

$$\dot{X}_{t+\Delta t} = \dot{X}_t + c_6\ddot{X}_t + c_7\ddot{X}_{t+\Delta t} \tag{6.60b}$$

To use the α-method, the equation of motion of either the vehicle subsystem or the track subsystem is modified as:

$$M\ddot{X}_{t+\Delta t} + (1+\alpha)C\dot{X}_{t+\Delta t} - \alpha C\dot{X}_t + (1+\alpha)KX_{t+\Delta t} - \alpha KX_t = (1+\alpha)F_{t+\Delta t} - \alpha F_t \tag{6.61}$$

where $-1/3 \leq \alpha \leq 0$. Inserting Equation 6.60 into Equation 6.61 then yields:

$$\hat{K}X_{t+\Delta t} = \hat{F}_{t+\Delta t} \tag{6.62}$$

where:

$$\hat{K} = c_0M + (1+\alpha)K + c_1(1+\alpha)C \tag{6.63a}$$

$$\begin{aligned}\hat{F}_{t+\Delta t} = {} & (1+\alpha)F_{t+\Delta t} + M(c_0X_t + c_2\dot{X}_t + c_3\ddot{X}_t) \\ & + (1+\alpha)C(c_1X_t + c_4\dot{X}_t + c_5\ddot{X}_t) - \alpha(F_t - C\dot{X}_t - KX_t)\end{aligned} \tag{6.63b}$$

By using Equation 6.62, the displacement of either subsystem can be solved at time step $t+\Delta t$, and the corresponding velocity and acceleration of the subsystem can then be obtained using Equation 6.60. Clearly, if $\alpha = 0$, the α-method is reduced to the traditional Newmark method. It is shown that when the parameters are selected as:

$$\gamma = (1-2\alpha)/2, \quad \beta = (1-\alpha)^2/4 \tag{6.64}$$

the α-method results in unconditional stability and second-order accuracy and improves convergence in non-linear contact problems.

The main procedure for the numerical integration of the equations of motion of the coupled vehicle and track system in crosswind can be summarized as follows:

(a) Estimate the motion of wheel-sets based on the motion of two rails and rail irregularities.
(b) Estimate the contact forces on the wheel-set based on the contact theory through iteration.
(c) Compute wind forces and solve Equation 6.40 to find the first approximation of motion of vehicle subsystem.
(d) Because of the non-linear nature of contact forces, an internal iteration is required until the solution of Equation 6.40 converges.

(e) The contact forces computed from the vehicle subsystem are then applied to the track subsystem, and Equation 6.42 is integrated to find the new approximation of motion of the track subsystem.
(f) Steps (a) to (e) are repeated until convergence is reached for both Equation 6.40 and Equation 6.42. At this point, a new time step can be started from (a).

6.6 Safety and Ride Comfort of Ground Railway Vehicle under Crosswind

A computer program is written based on the framework discussed in Section 6.5 and is used to perform a case study, in which dynamic responses of a railway vehicle running on a straight track subsystem in crosswind are predicted, and the safety and comfort performance of the moving vehicle in crosswind are assessed.

6.6.1 Vehicle and Track Models

The railway vehicle model consists of seven rigid bodies: one car body, two bogies and four wheel-sets. The seven rigid bodies are connected with springs and dashpots, forming a vehicle subsystem with 27 degrees of freedom. The main parameters of the railway vehicle used in the case study are listed in Table 6.6. The height and length of the car body are 3.2 m and 22.5 m respectively. The average static axle load of the vehicle is 10,150 kg. The fundamental frequency of the railway vehicle is 0.49 Hz in the lateral direction and 1.06 Hz in the vertical direction.

The railway track model includes two rails and a series of sleepers and ballasts, which are connected to each other using springs and dashpots. The main parameters of the railway track subsystem are listed

Table 6.6 Main parameters of the railway vehicle model used in the case study (from [30]) (Reproduced with permission from Elsevier)

Parameter	Units	Value
Half distance of two wheel-sets (q)	M	1.200
Half distance of bogie (s)	M	9.0
Half span of the first suspension system (a)	M	0.978
Half span of the second suspension system (b)	M	0.978
Mass of wheel-set (M_w)	Kg	1900
Mass moment of inertia of wheel-set around x-axis (I_w)	kg m^2	1067
Distance between car body and second suspension system (h_1)	M	1.415
Distance between second suspension system and bogie (h_2)	M	−0.081
Distance between bogie and wheel-set (h_3)	M	0.14
Radius of wheel (R_w)	M	0.4575
Mass of bogie (M_b)	Kg	1700
Mass moment of inertia of bogie around x-axis (I_{bX})	kg m^2	1600
Mass moment of inertia of bogie around y-axis (I_{bY})	kg m^2	1700
Mass moment of inertia of bogie around z-axis (I_{bZ})	kg m^2	1700
Mass of car body (M_c)	Kg	29 600
Mass moment of inertia of car body around x-axis (I_{cX})	kg m^2	58 020
Mass moment of inertia of car body around y-axis (I_{cY})	kg m^2	2 139 000
Mass moment of inertia of car body around z-axis (I_{cZ})	kg m^2	2 139 000
Lateral damping of first suspension system (per side) (C_{h1})	N s/m	25 000
Lateral damping of second suspension system (per side) (C_{h2})	N s/m	0
Lateral stiffness of first suspension system (per side) (K_{h1})	N/m	5 100 000
Lateral stiffness of second suspension system (per side) (K_{h2})	N/m	300 000
Vertical damping of first suspension system (per side) (C_{v1})	N s/m	30 000
Vertical damping of second suspension system (per side) (C_{v2})	N s/m	108 700
Vertical stiffness of first suspension system (per side) (K_{v1})	N/m	873 000
Vertical stiffness of second suspension system (per side) (K_{v2})	N/m	410 000

Table 6.7 Main parameters of the track model used in the case study (from [30]) (Reproduced with permission from Elsevier)

Parameter	Unit	Value
Distance between the centers of two rails	m	1.435
Sleeper spacing	m	0.545
Mass of steel rail per unit length	kg/m	60.64
Mass of sleeper	kg	237
Lumped mass of ballast	kg	739
Vertical bending stiffness of rail	$MN\,m^2$	6.62
Lateral bending stiffness of rail	$MN\,m^2$	1.079
Thickness of ballasts	m	0.45
Density of ballasts	kg/m^3	1800
Elastic modulus of ballasts	MPa/m	1.1×10^8
Vertical stiffness of pads and fasteners	MN/m	120
Vertical damping of pads and fasteners	kN s/m	75
Elastic modulus of roadbeds	MPa/m	8.0×10^7
Vertical stiffness of roadbeds	MN/m	65
Vertical damping of roadbeds	kN s/m	31

in Table 6.7. The total length of the track subsystem considered in this case study is 1090 m. The spacing distance between two adjacent sleepers is 0.545 m. Thus, there are a total of 2001 sleepers. The distance between the centers of two rails is 1.435 m. Both the vehicle and track subsystems used in this study represent a conventional railway line in China [37].

The cross-section profiles of both rails and wheels are important in considering vehicle and track interaction. The TB60 rail and worn wheel, commonly used in China for analysis of vehicle and track interaction, are adopted in this study. Then, the relationships of the relative rolling angle, the contact angle, and others with the lateral displacement of wheel-set are found [37]. Figure 6.20 shows the relative rolling angle and the contact angle as functions of the lateral displacement of the wheel-set at a zero yaw angle. The sign of the angles complies with the x-y-z coordinate system shown in Figure 6.17. Clearly, their relationships are strongly non-linear.

6.6.2 Wind Forces on Railway Vehicle

For simulation of the railway vehicle and track interaction in crosswind, aerodynamic data for the moving vehicle, including both steady force coefficients and turbulence characteristics, are required. Several researchers have acquired some valuable results in this aspect using either an experimental or a numerical approach.

The steady coefficients of the side force, vertical force and rolling moment on the vehicle used in this case study are depicted in Figure 6.21 as a function of the wind yaw angle. These coefficients are quite similar to those reported in [38]. Also given in Figure 6.21 is the derivative of steady side-force coefficient with respect to wind inclination at zero angle. The derivatives of other force (moment) coefficients are not available. Note that the steady coefficients of side force, vertical force and rolling moment at the yaw angle of 90° are 1.1, 0.5 and 0.8, respectively.

In the simulation of unsteady aerodynamic forces on the vehicle, the following von Kármán longitudinal and vertical wind auto-spectra are adopted [39]:

$$\frac{nS_{uu}(n)}{\sigma_u^2} = \frac{4\dfrac{L_u n}{\bar{u}}}{\left[1 + 70.8\left(\dfrac{L_u n}{\bar{u}}\right)^2\right]^{5/6}} \tag{6.65a}$$

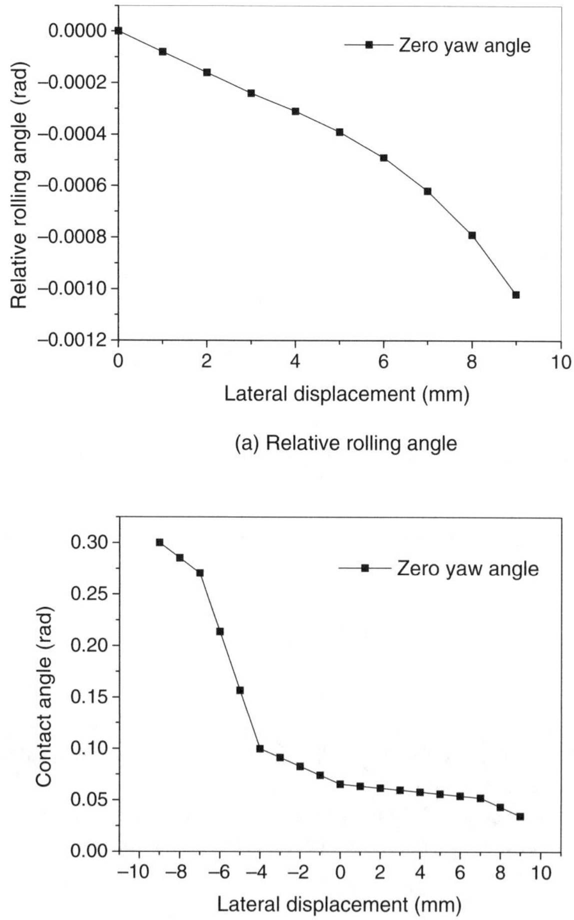

Figure 6.20 Relationship of relative rolling angle and contact angle with lateral displacement of wheel-set (from [30]) (Reproduced with permission from Elsevier).

$$\frac{nS_{ww}(n)}{\sigma_w^2} = \frac{4\dfrac{L_w n}{\bar{u}}\left[1 + 755\left(\dfrac{L_w n}{\bar{u}}\right)^2\right]}{\left[1 + 283\left(\dfrac{L_w n}{\bar{u}}\right)^2\right]^{11/6}} \tag{6.65b}$$

where:

σ_u and σ_w are the standard deviations of fluctuating wind in the longitudinal and vertical direction, which are taken as $0.15\bar{u}$ and $0.075\bar{u}$, respectively, in this study;

n is the frequency in Hz;

L_u and L_w are the integral length scales of fluctuating wind in the longitudinal and lateral directions, which are set to 45 m and 1 m, respectively, in this study.

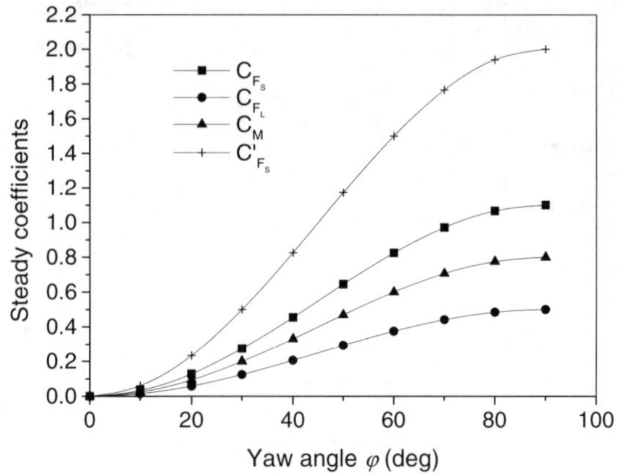

Figure 6.21 Steady aerodynamic coefficients for the vehicle used in the case study (from [30]) (Reproduced with permission from Elsevier).

The turbulence coherence which defines the statistical dependency between the turbulence components at two different points is given by

$$\sqrt{\text{coh}} = \exp\left(-\frac{CDn}{\bar{u}}\right) \tag{6.66}$$

where:

C is the decay factor, selected as 10 for longitudinal turbulence and 8 for vertical turbulence;
D is the distance between the two points.
Due to lack of information, the aerodynamic admittance functions for the unsteady aerodynamic forces are assumed to be equal to unit, and the co-spectra between the longitudinal and vertical turbulence components are set to zero.

Based on the wind spectra and coherence functions, time histories of longitudinal and vertical fluctuating wind speeds are simulated along the line of track subsystem at the level of vehicle mass center and at a distance interval of 5 m. The total number of the time histories is 219 in either the longitudinal direction or the vertical direction. The duration of each time history is 164 seconds, and the sampling frequency is 50 Hz. The unsteady aerodynamic forces on the moving vehicle can then be computed using Equation 6.55a and applied to the vehicle based on Taylor's frozen turbulence hypothesis, with a proper interpolation at a given time.

6.6.3 *Rail Irregularity*

In this case study, vertical and lateral irregularities are considered for both the right and left rails of the track subsystem. The rail irregularity in railway engineering is often represented by a one-sided power spectral density (PSD) function. The PSD functions of rail irregularities developed by the Research Institute of China Railway Administration [37] are used in this case study. All of rail irregularities are expressed using a unified rational formula as:

$$S(f) = \frac{A(f^2 + Bf + C)}{f^4 + Df^3 + Ef^2 + Ff + G}\ \text{mm}^2/\text{m}^{-1} \tag{6.67}$$

where:

$f = 1/\lambda$ is the spatial frequency in cycle/m (λ is the wavelength);
A to G are the specific parameters, which are different for vertical and lateral rail irregularities. The values of these parameters can be found in [37].

Figure 6.22 shows simulated vertical and lateral rail irregularities of the right and left rails. In the simulation, the length of the track subsystem is taken as 1090 m and the sampling points are 2001, which is the same as the number of sleepers.

6.6.4 *Response of Coupled Vehicle-Track System in Crosswind*

Let us consider that the vehicle runs at a constant velocity of 160 km/hr in crosswind. The wind is normal to the motion of the vehicle and of 20 m/s at the level of the vehicle mass center. The relative wind velocity V_R is, therefore, 48.7 m/s and the wind yaw angle, φ, is 24°.

Figure 6.23 depicts the time histories of vertical and lateral displacement responses of the car body at its center, with and without wind forces. Figure 6.24 displays the time history of lateral displacement response of the first wheel-set of the vehicle, with and without wind forces. It can be seen that both steady and unsteady aerodynamic forces have significant influence on the vertical and lateral displacement responses of the moving vehicle, in particular in the lateral direction, where the lateral displacement of the car body is very small without wind forces but increases significantly under wind forces.

Figures 6.25 and 6.26 show the time histories of vertical and lateral acceleration responses of the car body at its center without and with wind forces, respectively. Compared with the displacement responses of the car body, effects of unsteady aerodynamic forces on acceleration responses of the car body are relatively smaller.

Figures 6.27 and 6.28 show the horizontal and vertical contact forces between the first wheel-set and rails on both windward and leeward sides, without and with wind forces, respectively. It can be seen that, without wind forces, the horizontal contact forces on the windward rail and the leeward rail are similar in magnitude but opposite in direction. The vertical contact forces on the windward rail and the leeward rail are similar in both magnitude and direction. With wind forces, both the horizontal and vertical contact forces on the leeward side are much larger than those on the windward side. The horizontal and vertical contact forces on the leeward side are also much larger with wind forces than without wind forces. Wind forces thus certainly affect the safety and comfort performance of the moving railway vehicle.

To know if the elasticity of the track subsystem will affect the coupled vehicle-track system in crosswind, the responses of the coupled vehicle-track system in crosswind are also computed using a rigid track subsystem, in which the stiffness of all springs in the original track subsystem are assumed to be infinitely large. The displacement responses of the car body running on the elastic and rigid track subsystems are plotted in Figure 6.29 for both vertical and lateral directions.

It can be observed that both the lateral and vertical displacement responses of the car body running on the elastic track subsystem are almost the same as those running on the rigid track subsystem. This is because the stiffness of the track subsystem used in this study is much higher than that of the railway vehicle. However, if the stiffness of the track subsystem is comparable with that of the vehicle, the elastic track subsystem, rather than the rigid track subsystem, should be used.

6.6.5 *Safety and Ride Comfort Performance*

The safety of a railway vehicle concerns mainly the risk of derailment. There are two important factors that should be considered in the evaluation of the safety of a railway vehicle. One is the derail factor, defined as the ratio of lateral force Q acting on the wheel to the total vertical force P acting on the same

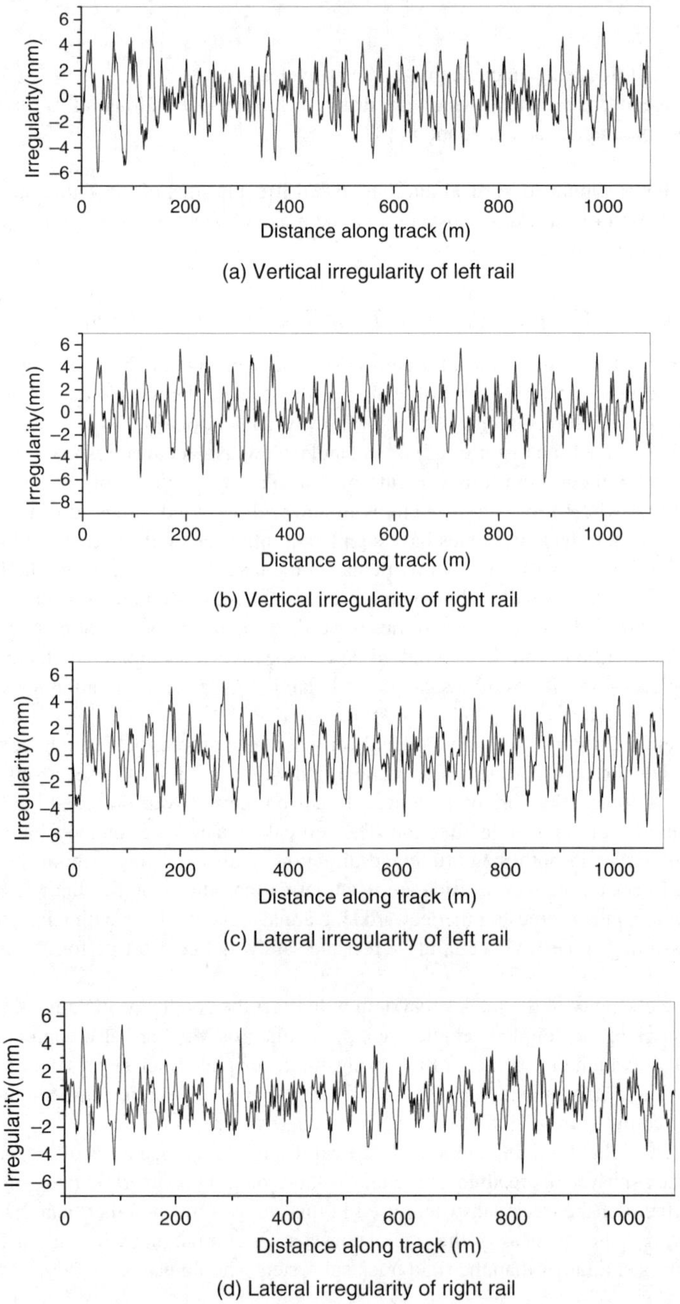

Figure 6.22 Vertical and lateral rail irregularities (from [30]) (Reproduced with permission from Elsevier).

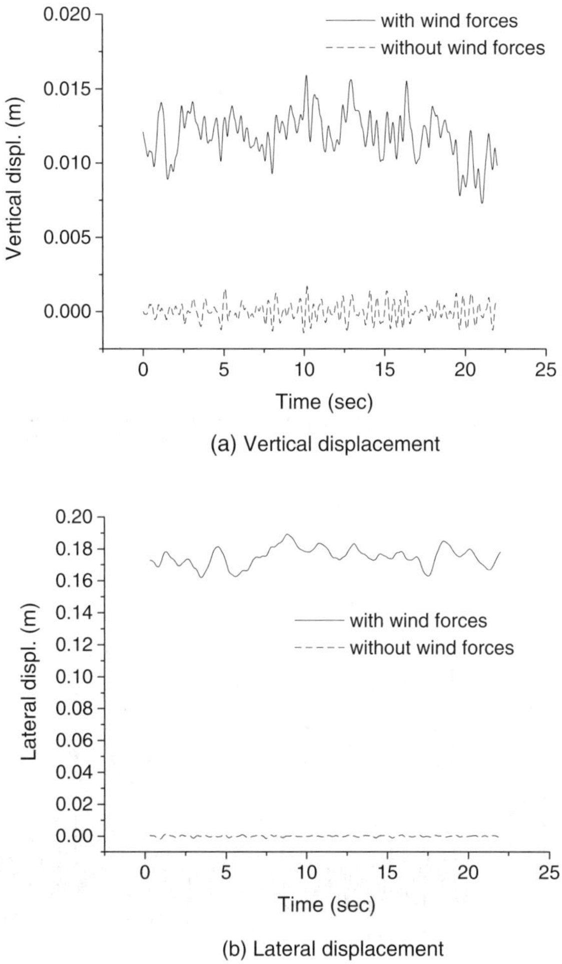

Figure 6.23 Displacement responses of the car body at its center ($\bar{u} = 20\,\text{m/s}$, $V_T = 160\,\text{km/hr}$) (from [30]) (Reproduced with permission from Elsevier).

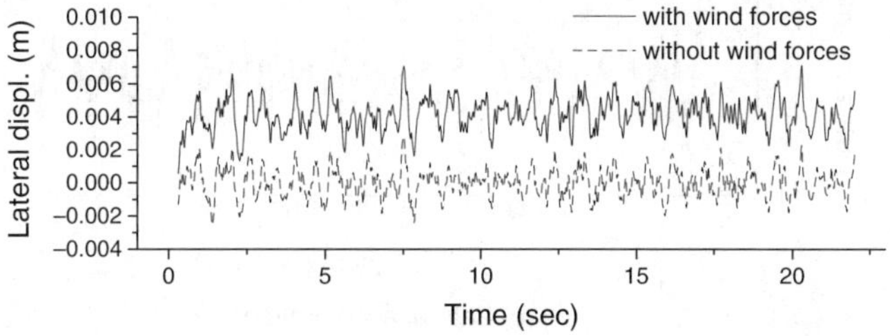

Figure 6.24 Lateral displacement responses of the first wheel-set ($\bar{u} = 20\,\text{m/s}$, $V_T = 160\,\text{km/hr}$) (from [30]) (Reproduced with permission from Elsevier).

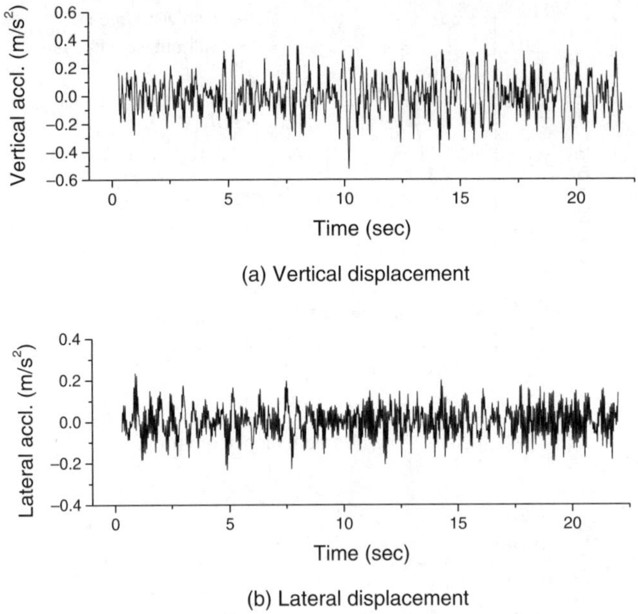

(a) Vertical displacement

(b) Lateral displacement

Figure 6.25 Acceleration responses of the car body at its center without wind forces ($V_T = 160$ km/hr) (from [30]) (Reproduced with permission from Elsevier).

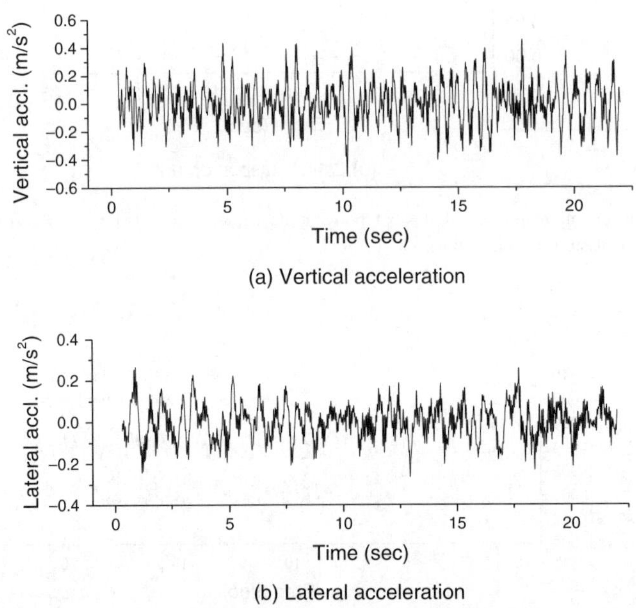

(a) Vertical acceleration

(b) Lateral acceleration

Figure 6.26 Acceleration responses of the car body at its center with wind forces ($\bar{u} = 20$ m/s, $V_T = 160$ km/hr) (from [30]) (Reproduced with permission from Elsevier).

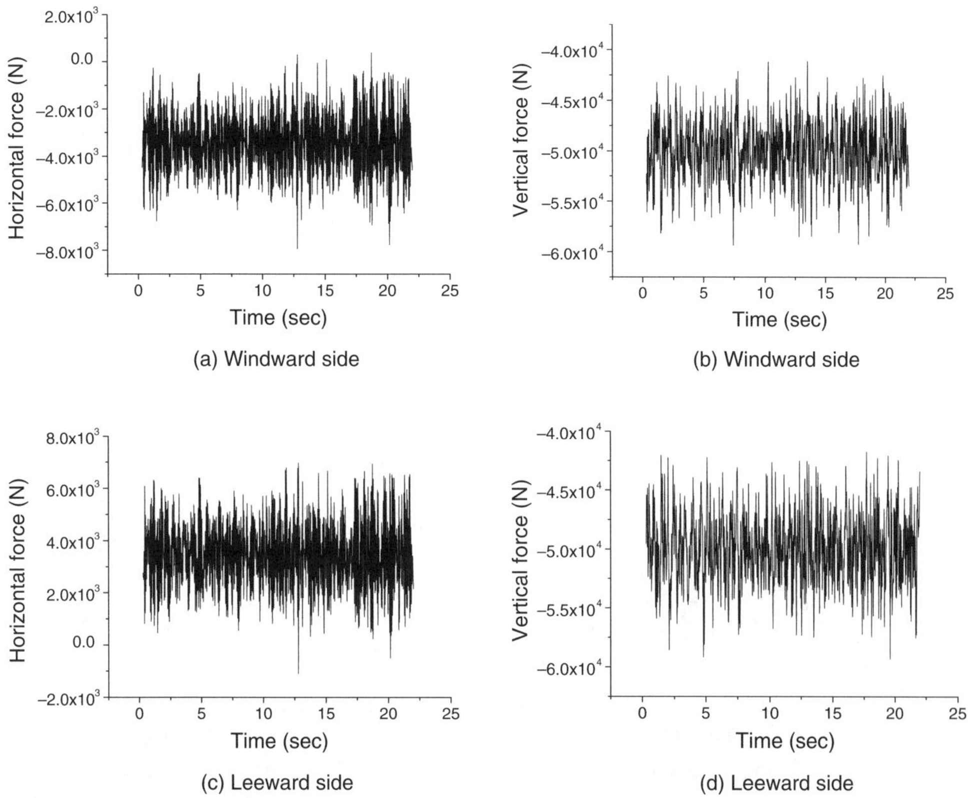

Figure 6.27 Horizontal and vertical contact forces between the first wheel-set and rails without wind forces (V_T= 160 km/hr) (from [30]) (Reproduced with permission from Elsevier).

wheel. The total vertical force is the sum of the self-weight of the vehicle per wheel and the dynamic vertical forces on the wheel. The other factor is the load reduction factor, defined as the ratio of the reduction in the vertical force to the static wheel load:

$$\Delta P/P_0 = (P - P_0)/P_0 \tag{6.68}$$

in which P_0 is the static wheel load and ΔP is the reduction in the wheel load with respect to P_0. The allowable derail factor (Q/P) and load reduction factor specified in the Chinese design guideline are 1.0 and 0.6, respectively [37]. Therefore, if the conditions:

$$Q/P \leq 1.0, \quad \Delta P/P_0 \leq 0.6 \tag{6.69}$$

are satisfied at the same time, the railway vehicle is said to be safe.

The ride comfort of the passenger coach in a running train can be assessed using the Sperling comfort index, is defined as:

$$W = 0.896\left(\frac{a^3}{f}F(f)\right)^{\frac{1}{10}} \tag{6.70}$$

where:

a is the acceleration of the car body in cm/s^2;

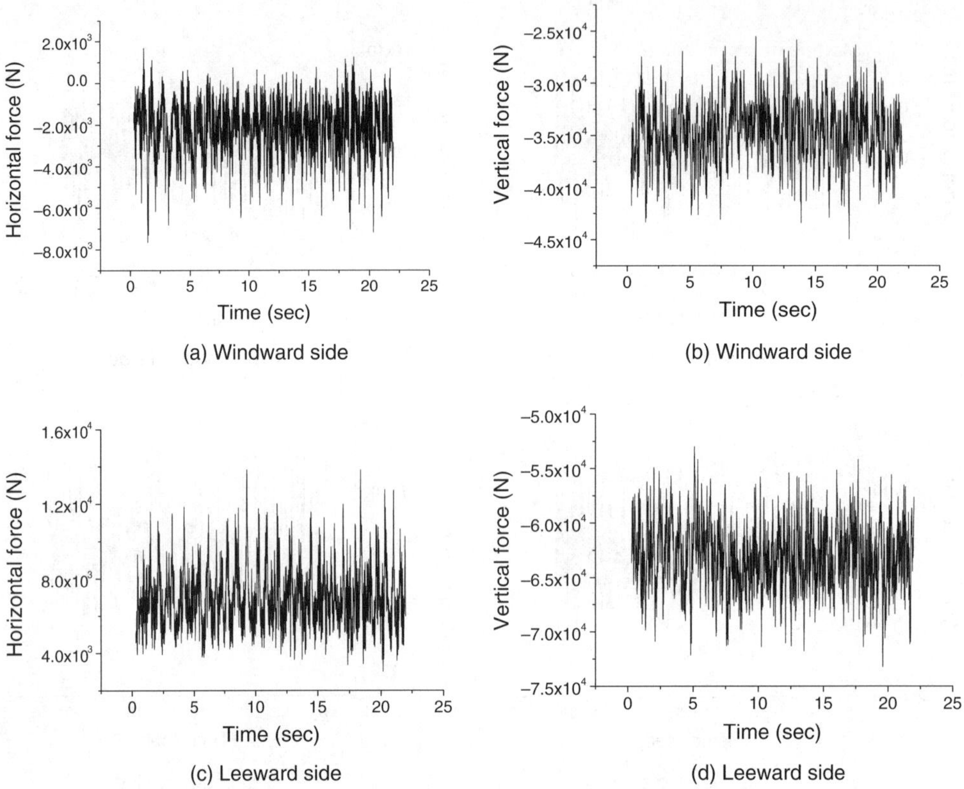

Figure 6.28 Horizontal and vertical contact forces between the first wheel-set and rails with wind forces ($\bar{u} = 20\,\text{m/s}$, $V_T = 160\,\text{km/hr}$) (Reproduced with permission from Elsevier).

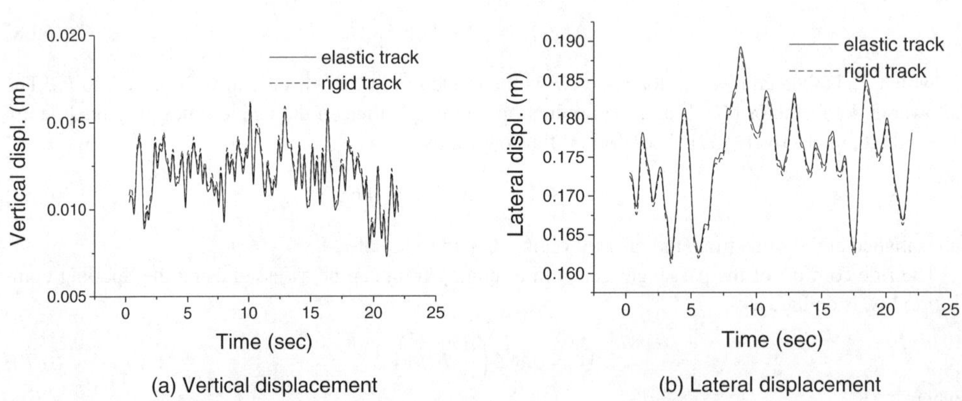

Figure 6.29 Comparison of displacement responses of car body running on the elastic and rigid track subsystems ($\bar{u} = 20\,\text{m/s}$, $V_T = 160\,\text{km/hr}$) (from [30]) (Reproduced with permission from Elsevier).

f is the frequency in Hz;
$F(f)$ is the modification coefficient of frequency.

When vertical vibration is concerned:

$$F(f) = \begin{cases} 0.325f^2 & (f = 0.5 \sim 5.9\,\text{Hz}) \\ 400/f^2 & (f = 5.9 \sim 20\,\text{Hz}) \\ 1 & (f > 20\,\text{Hz}). \end{cases} \tag{6.71a}$$

When lateral vibration is concerned:

$$F(f) = \begin{cases} 0.8f^2 & (f = 0.5 \sim 5.4\,\text{Hz}) \\ 650/f^2 & (f = 5.4 \sim 26\,\text{Hz}) \\ 1 & (f > 26\,\text{Hz}). \end{cases} \tag{6.71b}$$

The acceleration response of the car body is random due to random wind forces and rail irregularities, and it contains a wide range of vibration frequencies. Thus, the Sperling comfort index is calculated for a series of frequencies based on the Fourier spectrum of the acceleration response time history. Its mean value is then taken for the assessment of vehicle comfort. The allowable value of vehicle comfort specified in the Chinese design guideline is 3.0 for both lateral and vertical vibrations.

To investigate the effects of wind speed on the safety and ride comfort of the railway vehicle concerned in this study, both displacement and acceleration responses of the coupled vehicle and track system are computed for mean wind speed ranging from zero to 30 m/s at a 5 m/s interval. The running speed of the vehicle remains at 160 km/hr. The computed results relating to the safety and comfort indexes of the vehicle are listed in Table 6.8. The variations of the derail factor and the load reduction factor with wind speed are plotted in Figure 6.30. It can be seen that the safety of the vehicle is controlled by the load reduction factor rather than the derail factor. Based on the allowable value of the load reduction factor, $\Delta P/P_0 = 0.6$, the critical wind speed should be about 21 m/s. Note also that, even at the critical wind speed, the lateral Sperling index is less than 2, indicating that the ride comfort is satisfactory.

Generally speaking, the safety and comfort performance of the vehicle also vary with vehicle speed. However, wind forces on the moving vehicle concerned in this study are dominated by wind speed rather than vehicle speed. It is noted from the analysis that the effects of vehicle speed on the safety and comfort performance are relatively small. Nevertheless, further studies are required to investigate effects of vehicle speed on the safety and comfort performance when the required experimental data are available.

Table 6.8 Safety and comfort performance of the railway vehicle used in the case study (from [30]) (Reproduced with permission from Elsevier)

Wind velocity (m/s)	Derailment factor (Q/P)	$\Delta P/P_0$	Lateral Sperling index (W)	$\ddot{Z}_{max}$ $(m/s^2)^1$
0.0	0.074	0.285	1.413	0.234
5.0	0.077	0.293	1.413	0.234
10.0	0.087	0.350	1.414	0.236
15.0	0.101	0.446	1.420	0.237
20.0	0.129	0.580	1.537	0.327
25.0	0.156	0.753	1.703	0.437
30.0	0.216	0.969	2.016	0.678

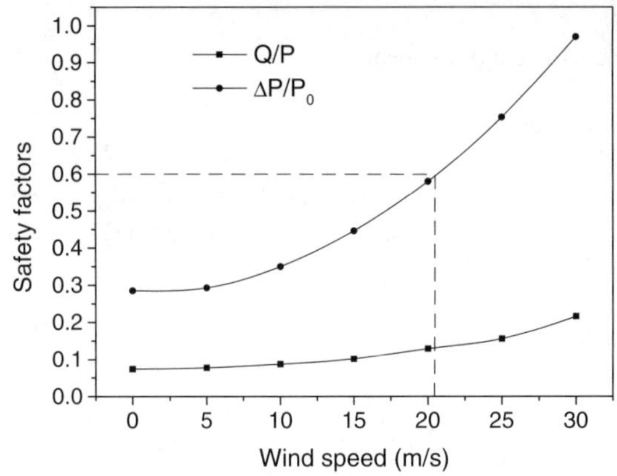

Figure 6.30 Variations of safety factors of the moving vehicle with wind speed (from [30]) (Reproduced with permission from Elsevier).

6.7 Wind-Railway Vehicle-Bridge Interaction

A framework for dynamics of coupled wind-railway vehicle-bridge systems in the time domain presented in [40] is introduced in this section. The framework is established based on wind-railway vehicle interaction introduced in Section 6.5, wind-bridge interaction discussed in Chapter 4, and railway vehicle-bridge interaction, which is similar to the railway vehicle-track subsystem interaction presented in Section 6.5.

A three-dimensional finite element model is used to represent a long-span cable-stayed bridge carrying two railway tracks laid on the bridge deck. Two parallel rails of a track are modeled as two continuous beams supported by the bridge deck directly, without sleepers and ballasts being included. Wind forces acting on the bridge, including both buffeting and self-excited forces, are generated in the time domain using a fast spectral representation method, measured aerodynamic coefficients and flutter derivatives.

Each vehicle in the train is modeled as a mass-spring-damper system. Wind forces acting on the train, including both steady and unsteady aerodynamic forces, are simulated in the time domain, taking into account the effects of vehicle speed and the spatial correlation with wind forces on the bridge. In consideration of the difficulties in measuring the aerodynamic coefficients of a moving vehicle over a bridge deck, the cosine rule is adopted to consider yaw angle effects on wind forces on the train. The dynamic interaction between the bridge and train is realized through the contact forces between the wheels and track, with track irregularities and wheel hunting included. The proposed framework is finally applied to a real long-span cable-stayed bridge carrying high-speed railway vehicles in crosswinds.

6.7.1 *Formulation of Wind-Railway Vehicle-Bridge Interaction*

In the formulation of wind-railway vehicle-bridge interaction by Li *et al.* [40], the bridge and railway vehicles are regarded as two subsystems. The equations of motion of the coupled wind-railway vehicle-bridge system can be expressed in the following forms:

$$\begin{bmatrix} M_b & 0 \\ 0 & M_v \end{bmatrix}\begin{Bmatrix} \ddot{u}_b \\ \ddot{u}_v \end{Bmatrix} + \begin{bmatrix} C_b & 0 \\ 0 & C_v \end{bmatrix}\begin{Bmatrix} \dot{u}_b \\ \dot{u}_v \end{Bmatrix} + \begin{bmatrix} K_b & 0 \\ 0 & K_v \end{bmatrix}\begin{Bmatrix} u_b \\ u_v \end{Bmatrix} = \begin{Bmatrix} F_{stb} + F_{bub} + F_{seb} + F_{vb} \\ F_{stv} + F_{buv} + F_{bv} \end{Bmatrix} \tag{6.72}$$

where the subscript b and v indicate bridge and vehicle, respectively.

The coupling of the two subsystems depends on interaction forces and relative geometries between wheels and rail. Each subsystem can be solved independently, but an iterative process needs to be applied to make sure that the displacement and force compatibility conditions at the wheel-rail contact points are satisfied at each time step. An iterative solution can be found through the following steps:

(a) Combine rail irregularities and deck movements at the last time step to obtain equivalent rail irregularities for the current time step.
(b) Calculate the forces acting on vehicles at the current time step induced by the equivalent rail irregularities and cross winds.
(c) The equation of motion for the vehicle subsystem can be independently solved with the Newmark integration method to obtain the initial vehicle response $(\ddot{u}_v^t, \dot{u}_v^t, u_v^t)$ at the current time step.
(d) Based on the computed vehicle responses, the wheel-rail interaction forces at the current time step can be determined, and the static and buffeting forces on the bridge structure at the current time step can also be calculated.
(e) The bridge responses $(\ddot{u}_b^{t-1}, \dot{u}_b^{t-1}, u_b^{t-1})$ at the last time step are taken as the initial values, and the self-excited forces on the bridge at the current time step can be obtained.
(f) The load matrix of the bridge can be formed, and the equation of motion for the bridge subsystem can be independently solved by the Newmark integration method. The bridge responses $\ddot{u}_b^t, \dot{u}_b^t, u_b^t$ at the current time step can be obtained.
(g) The bridge responses $\ddot{u}_b^t, \dot{u}_b^t, u_b^t$ at the current time step are then taken as the initial value in step (e), and steps (e) to (g) are repeated until the error of the self-excited force is less than the allowable value.
(h) The deck movements obtained at step (g) are adopted for step (a). Steps (a) to (h) are repeated until the geometric and force compatibility conditions at the wheel-rail contact points are satisfied, and the calculations are then continued for the next time step.

6.7.2 Engineering Approach for Determining Wind Forces on Moving Vehicle

The aerodynamic coefficients in the equation of wind loads for a moving vehicle are described as the function of both vehicle velocity and wind yaw angle. For a ground vehicle, the aerodynamic coefficients of a moving vehicle can be measured by wind tunnel tests with a certain yaw angle to reflect the effect of vehicle movement [41,42]. However, in a vehicle-bridge system, only the vehicle moves, while the bridge deck is still. As a result, it is difficult to measure aerodynamic coefficients of the moving vehicle on the bridge deck if stationary models of both vehicle and bridge deck are still being used. It may be necessary to use a moving vehicle model on a bridge deck model for the measurement of aerodynamic coefficients, and setting up tests with moving vehicle models is very difficult [41,43].

Therefore, some approximations have to be introduced in this study from an engineering viewpoint. For line-like structures under crosswinds at a certain yaw angle, such as long-span bridge decks and long trains, the cosine rule can be used to determinate wind loads approximately [44,45]. Experiments on an idealized train model suggested that the pressure distribution over the train surface is essentially two-dimensional at locations away from the nose, and without the front-end and back-end effect, the pressure distribution around the train at yaw angle φ could be predicted by simply scaling the prediction for 0° yaw angle by a factor of $\cos^2\varphi$ [45]. The following equation can thus be used to find the aerodynamic coefficients of a moving vehicle, according to the cosine rule and the vehicle velocity:

$$C_{F_S}(\alpha, \varphi) = C_{F_S}(\alpha)\cos^2\varphi \tag{6.73a}$$

$$C_{F_L}(\alpha, \varphi) = C_{F_L}(\alpha)\cos^2\varphi \tag{6.73b}$$

$$C_M(\alpha, \varphi) = C_M(\alpha)\cos^2\varphi \tag{6.73c}$$

where $C_i(\alpha)$, $i = F_S, F_L, F_M$, are the drag, lift and moment coefficients at 0° yaw angle, which can be easily measured through wind tunnel tests.

With the approximation above, wind loads on the vehicle are independent of vehicle velocity, and the aerodynamic coefficients can be obtained through wind tunnel model tests. Note, however, that the aerodynamic effect of the existing deck on wind loads of the vehicle should be considered in the determination of aerodynamic coefficients of the vehicle in wind tunnel tests.

6.7.3 Case Study

A high-speed railway in China has a length about 1300 km. Its design speed for railway vehicles is about 300 km/h and its infrastructure is designed to support a maximum speed of 350 km/h. The bridge within this railway is a three-pylons cable-stayed bridge across the Yangtze River. The overall length of the bridge is 1464 m, two main spans are 488 m, and three pylons are about 170 m in height (see Figure 6.31). The bridge deck is composed of a pre-stressed concrete (PC) box and a steel truss as shown in Figure 6.32.

Beam elements are used to model the steel truss, and the concrete box is modeled with three dimensional spine beams. The fundamental frequencies in the vertical, lateral and torsional directions are 0.197, 0.200 and 0.793 Hz, respectively. The design wind speed of the bridge is 38.4 m/s at the deck level, and the mean wind velocity profile, which is used in the simulation of the wind speed fields along pylons, is expressed as:

$$\frac{\bar{U}_Z}{\bar{U}_d} = \left(\frac{Z}{Z_d}\right)^{\alpha} \tag{6.74}$$

where $\bar{U}_z$ is the mean wind velocity at the height of Z, $\bar{U}_d$ is the design wind velocity at the deck height of Z_d, and $\alpha = 0.12$ is the dimensionless power exponent related to ground surface roughness.

The wind velocity field of the bridge is simplified into eight independent one-dimensional multivariate processes, as listed in Table 6.9. The distribution of simulation point is shown in Figure 6.31.

92 points are uniformly distributed along the bridge deck longitudinal axis, with an interval distance of 16 m. 17 points are uniformly distributed, with a 10 m interval, along each of pylons. The horizontal and lateral wind spectra are adopted in Kaimal's form [46], while the vertical spectrum is in the form presented by Lumley and Panofsky [47]. The coherence function adopted is in Davenport's form [48].

Horizontal wind spectrum:

$$\frac{nS_u(f)}{u_*^2} = \frac{200f}{(1+50f)^{5/3}} \tag{6.75}$$

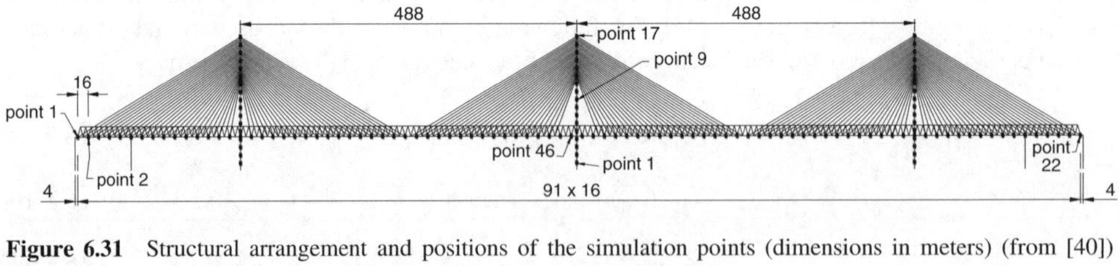

Figure 6.31 Structural arrangement and positions of the simulation points (dimensions in meters) (from [40]) (Reproduced with permission from Elsevier).

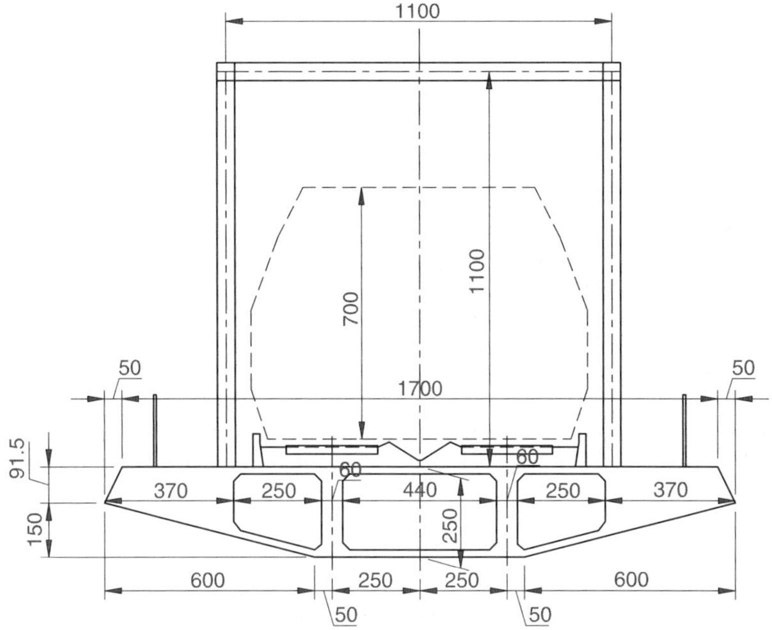

Figure 6.32 Cross section of bridge deck (dimensions in mm) (from [40]) (Reproduced with permission from Elsevier).

Lateral wind spectrum:

$$\frac{nS_v(f)}{u_*^2} = \frac{15f}{(1 + 9.5f)^{5/3}} \tag{6.76}$$

Vertical wind spectrum:

$$\frac{nS_w(f)}{u_*^2} = \frac{3.36f}{(1 + 10f)^{5/3}} \tag{6.77}$$

Table 6.9 One-dimensional wind fields (from [40]) (Reproduced with permission from Elsevier)

No.	Position	Direction	Number of points
1	Left pylon	Lateral[a]	17
2	Left pylon	Longitudianal[b]	17
3	Middle pylon	Lateral	17
4	Middle pylon	Longitudinal	17
5	Right pylon	Lateral	17
6	Right pylon	Longitudinal	17
7	Bridge deck	Lateral	92
8	Bridge deck	Vertical	92

[a]Lateral direction is normal to the deck longitudinal axis.
[b]Longitudinal direction is along the deck axis.

Coherence function:

$$Coh_{jm}(\omega) = \exp\left(-\lambda \frac{\omega r_{jm}}{2\pi \bar{U}(z)}\right) \tag{6.78}$$

where:

f is the dimensionless normalized frequency;
n and ω are the frequency in Hz and rad/s respectively;
u_* is the shear velocity of the flow;
r_{jm} is the distance between point j and point m;
$\bar{U}(z)$ is the average of the mean wind speeds at point j and point m;
$\lambda = 7$ is the coefficient relating to wind correlation.

The train used for the high speed railway is composed of two locomotives and 14 coaches. The design vehicle velocity is up to 350 km/h and the German rail irregularity spectra for high speed railway are adopted as follows:

Alignment irregularity:

$$S_a(\Omega) = \frac{A_a \Omega_c^2}{(\Omega^2 + \Omega_r^2)(\Omega^2 + \Omega_c^2)} \tag{6.79a}$$

Vertical-profile irregularity:

$$S_v(\Omega) = \frac{A_v \Omega_c^2}{(\Omega^2 + \Omega_r^2)(\Omega^2 + \Omega_c^2)} \tag{6.79b}$$

Cross-level irregularity:

$$S_c(\Omega) = \frac{(A_v/b^2)\Omega_c^2 \Omega^2}{(\Omega^2 + \Omega_r^2)(\Omega^2 + \Omega_c^2)(\Omega^2 + \Omega_s^2)} \tag{6.79c}$$

where:

b is half of the horizontal distance between two wheels;
Ω is the space frequency;
$A_a = 2.19 \times 10^{-7}$ m-rad, $A_v = 4.032 \times 10^{-7}$ m-rad are the roughness parameters;
$\Omega_c = 0.8246$ rad/m, $\Omega_r = 0.0206$ rad/m, $\Omega_s = 0.4380$ rad/m are the break frequencies.

The direct time integration is adopted using the Newmark method, with parameters $\alpha = 0.25$, $\beta = 0.5$ and time interval = 0.00288 seconds. The section model tests were executed in the wind tunnel to obtain the wind aerodynamic coefficients of the deck and vehicles with a special cross-slot system to separate the aerodynamic loads of the vehicle and the bridge.

The static force coefficients of both the deck and vehicles can be found in the literature [40]. Eight vertical and torsional flutter derivatives of the bridge deck were identified by the weighting ensemble least-square method [49]. The 4 flutter derivatives of the bridge are displayed in Figure 6.33.

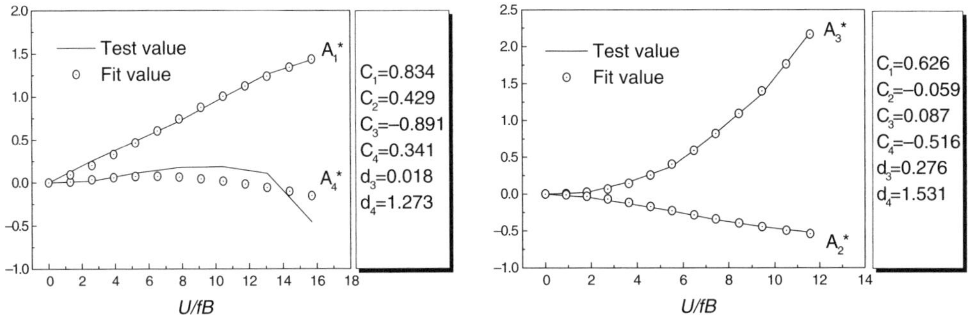

Figure 6.33 Flutter derivatives of bridge deck (from [40]) (Reproduced with permission from Elsevier).

Based on the quasi-steady theory, the first three lateral flutter derivatives are approximately estimated by the following equation:

$$P_1^* = -\frac{2C_D}{K}, \quad P_2^* = \frac{C_D'}{K}, \quad P_3^* = \frac{C_D'}{K^2} \tag{6.80}$$

As far as pylons and cables concerned, only drag coefficients are taken into consideration with 2.0 and 0.7 respectively. Since both pylons and bridge deck are blunt, aerodynamic admittances can be adopted as 1.0 conservatively.

To demonstrate the feasibility of the proposed framework, two cases are calculated using self-developed computer software. In case 1, wind speed is 0 m/s and vehicle speed is 250 km/h; in case 2, wind speed is 25 m/s and vehicle speed is 250 km/h. The responses of both the bridge and vehicle are computed and compared in Figures 6.34 to 6.36.

In Figure 6.34, the bridge deck has a static lateral displacement due to static wind loads, and the bridge deck vibration in case 2 is larger than that in case 1. The lateral and vertical accelerations of the vehicle body increase because of wind action, as shown in Figure 6.35. In case 2, the vehicle wheel-sets are subject to additional moments induced by the lateral wind loads acting on vehicle bodies. Thus, vertical loads acting on the wheels on one side of the wheel-sets increase, but those acting on wheels on the other side of the wheel-sets decrease.

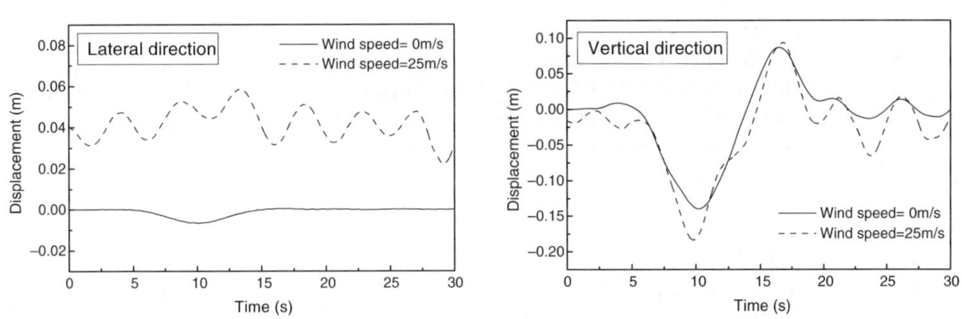

Figure 6.34 Deck displacements at the middle of left main span (from [40]) (Reproduced with permission from Elsevier).

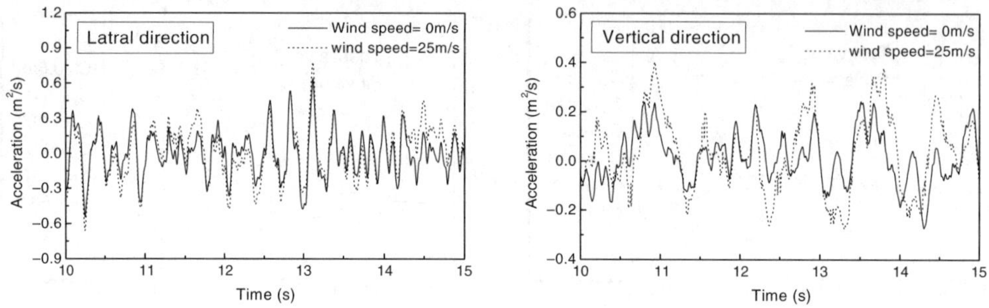

Figure 6.35 Accelerations of the first coach body (from [40]) (Reproduced with permission from Elsevier).

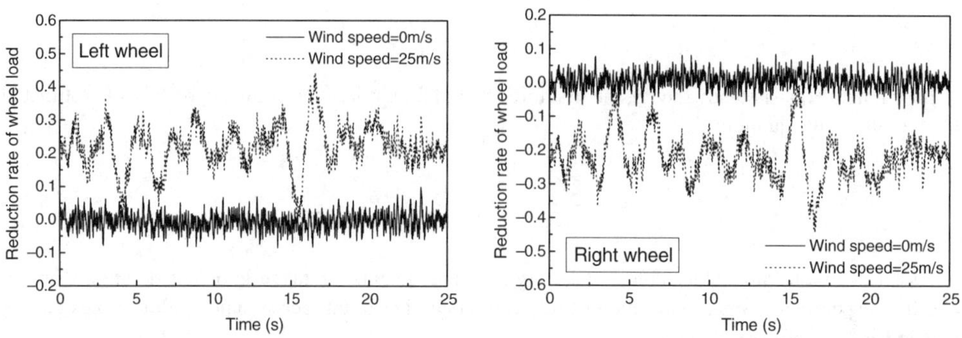

Figure 6.36 Reduction rate of wheel load at the first wheel-set of the first coach (from [40]) (Reproduced with permission from Elsevier).

The reduction rate of vertical wheel loads is the ratio of the fluctuating vertical wheel load on one side to the mean vertical wheel load on both sides. It can be found that reduction rates of vertical wheel loads on one side are obviously increased, while those on the other side are remarkably decreased, indicating that the safety factors of vehicles are reduced under wind action.

6.8 Notations

A, A_f	Reference area
A_r	Roughness coefficient
a	Acceleration of the car body
B_1	Distance between the left and right contact points
b	Half of the horizontal distance between two wheels
b_1	Half of the horizontal distance between the two rigid bodies
C	Decay factor
$C_D(\psi)$	Drag force coefficient
$C_S(\psi)$	Side force coefficient
$C_L(\psi)$	Lift force coefficient
$C_R(\psi)$	Rolling moment coefficient
$C_P(\psi)$	Pitching moment coefficient
$C_Y(\psi)$	Yawing moment coefficient
C_{uz}	Viscous damper of damping coefficient in the z-direction to the vehicle body

C_{uy}	Viscous damper of damping coefficient in the *y*-direction to the vehicle body
C_{lz}	Viscous damper of damping coefficient in the *z*-direction to the tire
C_{ly}	Viscous damper of damping coefficient in the *y*-direction to the tire
C_b	Damping matrix of the bridge
C_{bbv1}	Damping on the bridge from the tires
C_{vb1}	Coupled damping on bridge
C_{bv1}	Coupled damping on vehicle
C_v	Damping on the vehicle related only to the DOFs of the vehicle
C_{v1}	Additional damping on the vehicles due to the deck surface roughness
C_{bbv2}	Damping on the bridge from the tires at the contact points due to the sideslip forces between the tires and the bridge deck in the lateral direction
C_{vb2}	Coupled damping on bridge due to the sideslip forces between the tires and the bridge deck in the lateral direction
C_{bv2}	Coupled damping on vehicle due to the sideslip forces between the tires and the bridge deck in the lateral direction
C_{v2}	Additional damping on the vehicles due to the sideslip forces between the tires and the bridge deck in the lateral direction
$C_{Fs}(\alpha)$	Aerodynamic drag coefficient
$C_{Fl}(\alpha)$	Aerodynamic lift coefficient
$C_{Fm}(\alpha)$	Aerodynamic moment coefficient
C_{s}:	Damping matrix of the track subsystem
D	Distance between two measured points
F_x	Drag force
F_y	Side force
F_z	Lift force
F_{Gi}	Force on the tire due to the gravity of the vehicle
F_{hi}	Lateral contact force between the tire and the road surface
F_{vi}	Vertical contact force between the tire and the road surface
F_S	Side force
F_L	Lift force
F_v^c	Forces exerted by the track subsystem on the vehicle subsystem
F_v^w	Wind forces acting on the car body
F_s^c	Contact forces transmitted from the wheels to the rails
F_{vyw}	Aerodynamic force acting on the vehicle body in the *y*-direction
F_{vzw}	Aerodynamic force acting on the vehicle body in the *z*-direction
F'_{yl}	Contact force transmitted from the wheels to the left rail in the *y*-direction
F'_{zl}	Contact force transmitted from the wheels to the left rail in the *z*-direction
F'_{yr}	Contact forces transmitted from the wheels to the right rail in the *y*-direction
F'_{zr}	Contact forces transmitted from the wheels to the right rail in the *z*-direction
$F(f)$	Modification coefficient of frequency
f	Frequency
$f_{11}, f_{22}, f_{23}, f_{33}$	Creepage coefficients
f_k:	Spatial frequency (cycle/m)
Δf	Increment of spatial frequency
g	Acceleration due to gravity
H	Reference height
$H(\omega)$	Cholesky decomposition of cross-spectral density matrix $S_0(\omega)$
$H_{jm}(\omega)$	A typical element of matrix $H(\omega)$
h_v	Reference height of the vehicle
K_s:	Stiffness matrix of the track subsystem

K_{uy}	Linear elastic spring of stiffness in the y-direction to the vehicle body
K_{uz}	Linear elastic spring of stiffness in the z-direction to the vehicle body
K_{ly}	Linear elastic spring of stiffness in the y-direction to the tire
K_{lz}	Linear elastic spring of stiffness in the z-direction to the tire
K_b	Stiffness matrix of the bridge
K_{bbv1}	Stiffness of the bridge from the tires at the contact points
K_{vb1}	Coupled stiffness on bridge
K_{bv1}	Coupled stiffness on vehicle
K_v	Stiffness on the vehicle related only to the DOFs of the vehicle
K_{v1}	Additional stiffness on the vehicles due to the deck surface roughness
L_u	Integral length scales of fluctuating wind in the longitudinal direction
L_w	Integral length scales of fluctuating wind in the vertical direction
L_x^u	Integral scale of wind turbulence in the horizontal direction
M, M_x	Rolling moment acting on the vehicle
M_y	Pitching moment acting on the vehicle
M_z	Yawing moment acting on the vehicle
M_{vxw}	Aerodynamic moment acting on the vehicle body about the x-axis
M_{vyw}	Aerodynamic moment acting on the vehicle body about the y-axis
M_{vzw}	Aerodynamic moment acting on the vehicle body about the z-axis
M_b	Mass matrix of the bridge
M_{bbv}	Masses of all the tires of the vehicles
M_v	Masses of all the rigid bodies of the vehicles, excluding the masses of the tires
M_s	Mass matrix of the track subsystem
M_{sp}	Spin creeping moment
m	Coefficient of sideslip friction
N	Normal contact force between the wheel and rail
P	Vertical force acting on the wheel
P_O	Static wheel load
ΔP	Reduction in the wheel load with respect to P_O
P_{bvg}	External forces on the bridge due to the gravity forces of the vehicles
P_{bvr1}	Inertia force of all the masses of the vehicle due to the road surface roughness
P_{bvr2}	Additional force on the bridge due to the deck surface roughness
P_{vvr2}	Additional force on the vehicle due to the deck surface roughness
P_{bw}	Total wind force on the bridge deck
P_{vw}	Wind force on the vehicle
Q	Lateral force on the wheel
R_O	The nominal radius of the wheel
$R_{u,N}(n)$	Non-dimensional PSD functions for wind turbulence components in the horizontal direction
$R_{w,N}(n)$	Non-dimensional PSD functions for wind turbulence components in the vertical direction
$r(x)$	The vertical road surface roughness
$r_{ci}(x)$	Road surface roughness under the ith contact point
r_{jm}	Distance between point j and point m
$S(\bar{\varphi})$	PSD function for the road surface roughness in the vertical direction
$S(f_k)$	PSD function of the rail irregularity
$S_a(\Omega)$	Alignment irregularity
$S_c(\Omega)$	Cross-level irregularity
$S_v(\Omega)$	Vertical-profile irregularity
$S^0(\omega)$	Two-side cross-spectral density matrix

T_x	Longitudinal creeping force
T_z	Lateral creeping force
U_m	Mean wind velocity
U_v	Road vehicle velocity
U_R	The wind velocity relative to the vehicle
$\bar{U}_z$	Mean wind velocity at the height Z
$\bar{U}_d$	Design wind velocity at the deck height Z_d
u	Longitudinal fluctuating wind speed components
$\bar{u}$	Mean wind speed component
u^*	Shear velocity of the flow
u_v^t	Vehicle displacement response at time t
u_b^t	Bridge displacement response at time t
V_T	Nominal traveling speed of the wheel-set
V_R	Wind velocity relative to the vehicle
V_{wx}	Velocity of the wheel in the longitudinal direction
V_{wz}	Velocity of the wheel in the lateral direction
V_{rx}	Velocity of the rail in the longitudinal direction
V_{rz}	Velocity of the rail in the lateral direction
W	Sperling comfort index
w	Vertical fluctuating wind speed components
$X_{s:}$	Nodal displacement of the track subsystem
Y, Y_v	Lateral displacement of the vehicle
Y_{bl}	Lateral displacement of the left ballast
Y_{br}	Lateral displacement of the right ballast
$Y_{w:}$	Lateral displacement of the wheel-sets
Y_{si}	Motions of the front axle set and the rear axle set in the y-direction
Y_{rl}	Displacement of the left rail in horizontal direction
Y_{rr}	Displacement of the right rail in horizontal direction
$Y_{p:}$	Displacement of the sleeper in horizontal direction
Z, Z_v	Vertical displacement of the vehicle
Z_{ci}	Vertical displacement of the ith tire
$Z_{w:}$	Vertical displacement of the wheel-sets
Z_{si}	Motions of the front axle set and the rear axle set in the z-direction
Z_{rl}	Displacement of the left rail in vertical direction
Z_{rr}	Displacement of the right rail in vertical direction
$Z_{p:}$	Displacement of the sleeper in vertical direction
α	Incident angle
δ	Angle of the wheel-sets to the vehicle axis
θ	Roll displacement response of the vehicle
$\theta_{jm}(\omega)$	Complex angle of $H_{jm}(\omega)$
θ_k	Random phase angle uniformly distributed from 0 to 2π
θ_v	Pitching angle
θ_w	Roll displacement of the wheel-sets
θ_{rl}^Y	Displacement of the left rail in rolling direction around the y-axis
θ_{rl}^Z	Displacement of the left rail in rolling direction around the z-axis
θ_{rr}^Y	Displacement of the right rail in rolling direction around the y-axis
θ_{rr}^Z	Displacement of the right rail in rolling direction around the z-axis
θ_P	Displacement of the sleeper in the rolling direction
λ	Exponential decay coefficient used in the coherence function of wind turbulence along the bridge deck

ξ_x	Creepage ratio in the longitudinal direction
ξ_z	Creepage ratio in the vertical direction
ξ_{sp}	Creepage ratio in the spin direction
ρ	Air density
σ_u	Standard deviation of wind turbulence in the horizontal direction
σ_w	Standard deviation of wind turbulence in the vertical direction
ϕ_{ml}	Sequence of independent random phase angles uniformly distributed from 0 to 2π
ϕ_v	Yaw angle of the vehicle
φ	Yaw angle of the wind velocity relative to the vehicle
φ_v	Roll angle of the vehicle
$\bar{\varphi}$	Spatial frequency
$\bar{\varphi}_0$	Discontinuity frequency
$\chi_{Fs,u}(n)$	Aerodynamic transfer functions between the horizontal fluctuating wind velocity and aerodynamic drag force
$\chi_{Fs,w}(n)$	Aerodynamic transfer functions between the vertical fluctuating wind velocity and aerodynamic drag force
$\chi_{Fl,u}(n)$	Aerodynamic transfer functions between the horizontal fluctuating wind velocity and aerodynamic lift force
$\chi_{Fl,w}(n)$	Aerodynamic transfer functions between the vertical fluctuating wind velocity and aerodynamic lift force
$\chi_{M,u}(n)$	Aerodynamic transfer functions between the horizontal fluctuating wind velocity and aerodynamic moment
$\chi_{M,w}(n)$	Aerodynamic transfer functions between the vertical fluctuating wind velocity and aerodynamic moment
Ψ	Yaw angle of the wind velocity relative to the vehicle
Ω	Space frequency
Ω_w	Rotational velocities of spin motions of the wheel
Ω_r	Rotational velocities of spin motions of the rail
ω_{up}	Upper cut-off frequency
$\Delta\omega$	Frequency increment

References

1. Baker, C.J. and Reynolds, S. (1992) Wind induced accidents of road vehicles. *Accident Analysis and Prevention*, **24**(6), 559–575.
2. Baker, C.J. (1986) A simplified analysis of various types of wind-induced road vehicle accidents. *Journal of Wind Engineering and Industrial Aerodynamics*, **22**, 69–85.
3. Coleman, S.A. and Baker, C.J. (1990) High-sided road vehicle in cross winds. *Journal of Wind Engineering and Industrial Aerodynamic*, **36**, 1383–1392.
4. Baker, C.J. (1987) Measures to control vehicle movement at exposed sites during windy periods. *Journal of Wind Engineering and Industrial Aerodynamics*, **25**, 151–161.
5. Xu, Y.L. and Guo, W.H. (2003) Dynamic behaviour of high-sided road vehicles subject to a sudden crosswind gust. *Wind & Structures*, **6**(5), 325–346.
6. Guo, W.H. and Xu, Y.L. (2001) Fully computerized approach to study cable-stayed bridge-vehicle interaction. *Journal of Sound and Vibration*, **248**, 745–761.
7. Wang, T.L. and Huang, D.Z. (1992) Cable-stayed bridge vibration due to road surface roughness. *Journal of Structural Engineering, ASCE*, **118**, 1354–1374.
8. Dodds, C.J. and Robson, J.D. (1973) The description of road surface roughness. *Journal of Sound and Vibration*, **31**, 175–183.
9. Chopra, A.G. (1995) *Dynamics of Structures*, Prentice-Hall, Upper Saddle River, New Jersey.
10. Blejwas, T.E., Feng, C.C., and Ayre, R.S. (1979) Dynamic interaction of moving vehicles and structures. *Journal of Sound and Vibration*, **67**, 513–521.
11. Olsson, M. (1985) Finite element modal co-ordinate analysis of structures subjected to moving loads. *Journal of Sound and Vibration*, **99**, 1–12.

12. Green, M.F., Cebon, D., and Cole, D.J. (1995) Effects of vehicle suspension design on dynamics of highway bridges. *Journal of Structural Engineering ASCE*, **121**, 272–282.
13. Wang, T.L., Huang, D.Z., and Shahawy, M. (1996) Dynamic behaviour of continuous and cantilever thin-walled box girder bridges. *Journal of Bridge Engineering, ASCE*, **1**, 67–75.
14. Cheung, Y.K., Au, F.T.K., Zheng, D.Y., and Cheng, Y.S. (1999) Vibration of multi-span non-uniform bridges under moving vehicles and trains by using modified beam vibration functions. *Journal of Sound and Vibration*, **228**, 611–628.
15. Xu, Y.L. and Guo, W.H. (2003) Dynamic analysis of coupled road vehicle and cable stayed bridge systems under turbulent wind. *Engineering Structures*, **25**, 473–486.
16. Guo, W.H. and Xu, Y.L. (2006) Safety analysis of moving road vehicles on a long bridge under crosswind. *Journal of Engineering Mechanics, ASCE*, **132** (4), 438–446.
17. Jenkins, H.H., Stephenson, J.E., Clayton, G.A. *et al.* (1974) The effect of track and vehicle parameters on wheel/rail vertical dynamic forces. *Railway Engineering Journal*, **3**(1), 2–16.
18. Newton, S.G. and Clark, R.A. (1979) An investigation into the dynamic effects on the track of wheel flats on railway vehicles. *Journal of Mechanical Engineering Society*, **21**(4), 287–297.
19. Grassie, S.L., Gregory, R.W., Harrison, D., and Johnson, K.L. (1982) The dynamic response of railway track to high frequency vertical excitation. *Journal of Mechanical Engineering Society*, **24**, 77–90.
20. Cai, Z. and Raymond, G.P. (1992) Theoretical model for dynamic wheel/rail and track interaction. Proceedings of 10th International Wheelset Congress, Sydney, Australia, pp. 127–131.
21. Dahlberg, T. (1995) Vertical dynamic train/track interaction verifying a theoretical model by full-scale experiments. *Vehicle System Dynamics*, **24** (Suppl.) 45–57.
22. Nielsen, J.C.O. (1993) *Train/track interaction: coupling of moving and stationary dynamic system. Dissertation*, Chalmers University of Technology, Goteborg, Sweden.
23. Zhai, W. and Sun, X. (1994) A detailed model for investigating interaction between railway vehicle and track. *Vehicle System Dynamics*, **23** (Suppl.) 603–614.
24. Zhai, W.M., Cai, C.B., and Guo, S.Z. (1996) Coupling model of vertical and lateral vehicle/track interactions. *Vehicle System Dynamics*, **26**(1), 61–79.
25. Diana, G., Cheli, F., Bruni, S., and Collina, A. (1994) Interaction between railroad superstructure and railway vehicles. *Vehicle System Dynamics*, **23** (Suppl.) 75–86.
26. Knothe, K.L. and Grassie, S.L. (1993) Modeling of railway track and vehicle/track interaction at high frequencies. *Vehicle System Dynamics*, **22**, 209–262.
27. Balzer, Z.A. (1977) Atmospheric turbulence encountered by high-speed ground transport vehicles. *Journal of Mechanical Engineering Society*, **19**, 227–235.
28. Cooper, R.K. (1984) Atmospheric turbulence with respect to moving ground vehicles. *Journal of Wind Engineering and Industrial Aerodynamics*, **17**, 215–238.
29. Baker, C.J. (1991) Ground vehicles in high cross wind, Part I: steady aerodynamic forces. *Journal of Fluids and Structures*, **5**, 69–90.
30. Xu, Y.L. and Ding, Q.S. (2006). Interaction of railway vehicles with track in crosswinds. *Journal of Fluids and Structures*, **22**(3), 295–315.
31. Xu, Y.L., Zhang, N., and Xia, H. (2004) Vibration of coupled train and cable-stayed bridge systems in cross-wind. *Engineering Structures*, **26**, 1389–1406.
32. Kalker, J.J. (1990) *Three-dimensional Elastic Bodies in Rolling Contact*, Kluwer Academic Publishers, Dordrecht.
33. Scanlan, R.H. and Gade, R.H. (1977) Motion of suspension bridge spans under gusty wind. *Journal of the Structural Division, ASCE*, **103** (ST9), 1867–1883.
34. Lin, Y.K. and Yang, J.N. (1983) Multimode bridge response to wind excitations. *Journal of Engineering Mechanics, ASCE*, **109**(2), 586–603.
35. Hughes, T.J.R. (1987) *The Finite Element Method: Linear Static and Dynamic Finite Element Analysis*, Prentice-Hall, Inc., New Jersey.
36. Chaudhary, A.B. and Bathe, K.J. (1986) A solution method for static and dynamic analysis of three-dimensional contact problems with friction. *Computers and Structures*, **24**, 855–873.
37. Zhai, W.M. (2002) *Vehicle-track Coupling Dynamics*, 2nd edn, Chinese Railway Publishing House, Beijing, (in Chinese).
38. Baker, C.J., Jones, J., Lopez-Calleja, F., and Numday, J. (2003) Measurements of the cross wind forces on trains. Proceedings of 11th International Conference on Wind Engineering, Texas Tech University, Lubbock, Texas, USA, pp. 845–852.
39. Simiu, P.D. and Scanlan, R.H. (1996) *Wind Effects on Structures*, John Wiley & Sons, New York.
40. Li, Y.L., Qiang, S.Z., Liao, H.L., and Xu, Y.L. (2005) Dynamics of wind-rail vehicle-bridge systems. *Journal of Wind Engineering and Industrial Aerodynamics*, **93**, 483–507.
41. Cooper, R.K. (1981) The effect of cross-winds on trains. *Journal of Fluids Engineering*, **103**, 170–178.
42. Suzuki, M., Tanemoto, K., and Maeda, T. (2003) Aerodynamic characteristics of train/vehicles under cross winds. *Journal of Wind Engineering and Industrial Aerodynamics.*, **91**, 209–218.
43. Baker, C.J. (1986) Train aerodynamic forces and moments from moving model experiments. *Journal of Wind Engineering and Industrial Aerodynamics*, **24**, 227–251.

44. Tanaka, H. and Davenport, A.G. (1982) Response of taut strip models to turbulent wind. *Journal of Engineering Mechanics, ASCE*, **108**, 33–49.
45. Chiu, T.W. and Squire, L.C. (1992) An experimental study of the flow over a train in a cross-wind at large yaw angles up to 90°. *Journal of Wind Engineering and Industrial Aerodynamics*, **45**, 47–74.
46. Kaimal, J.C., Wyngaard, J.C., Izumi, Y., and Cote, O.R. (1972) Spectral characteristics of surface-layer turbulence. *Journal of Royal Meteorological Society*, **98**, 563–589.
47. Lumley, J.L. and Panofsky, H.A. (1964) *The Structure of Atmospheric Turbulence*, Wiley, New York.
48. Davenport, A.G. (1968) The dependence of wind load upon meteorological parameters. Proceedings of International Research Seminar on Wind Effects on Buildings and Structures, University of Toronto Press, Toronto, Canada, pp. 19–82.
49. Li, Y.L., Liao, H.L., and Qiang, S.Z. (2003) Weighting ensemble least-square method for flutter derivatives of bridge decks. *Journal of Wind Engineering and Industrial Aerodynamics*, **91**, 713–721.

7

Wind Tunnel Studies

7.1 Preview

As can be seen from earlier chapters, wind-induced response analyses of long-span cable-supported bridges or coupled vehicle-bridge systems require some basic aerodynamic parameters or functions such as force coefficients, flutter derivatives and aerodynamic admittance functions. At present, these parameters and functions are obtained mainly from wind tunnel section model tests. To ensure the functionality and safety of long-span cable-supported bridges under strong wind conditions, full aeroelastic model tests are often carried out in wind tunnels to observe and quantify their aeroelastic behavior. Wind tunnel studies are also performed to explore some new aerodynamic or aeroelastic phenomena, such as rain-wind-induced cable vibration. There are also many situations in which the wind load and the wind-induced responses of bridges and vehicles cannot be predicted with sufficient accuracy, either to assure functionality and safety or to avoid using uneconomically large safety factors. In such situations, it may be desirable to conduct wind tunnel tests of structural models. This is particularly true when seeking aerodynamic measures to mitigate vortex shedding-induced vibration or increase critical flutter velocity, as will be discussed in Chapter 12.

Techniques for the modeling of wind effects on bridge structures have improved considerably in the past with the advent of several large wind tunnels designed to produce turbulent boundary layer models of the natural wind. New measurement techniques used in wind tunnel tests have also been developed to significantly improve the measurement accuracy.

This chapter first introduces the common types of boundary-layer wind tunnels used for bridge wind engineering, followed by a general discussion for model scaling requirements and boundary wind simulation. The most common types of wind tunnel tests for bridge wind engineering are then presented, and the methods for identifying aerodynamic parameters and functions are introduced. Special wind tunnel tests for rain-wind-induced cable vibration and wind-vehicle-bridge interaction are finally described. Since not all of the model scaling requirements and boundary wind simulation requirement can be satisfied in most wind tunnel tests, the validation of wind tunnel tests by field measurements is sometimes necessary.

7.2 Boundary-Layer Wind Tunnels

A boundary-layer wind tunnel must have capabilities for creating flows that simulate the basic characteristics of natural wind at a site in order to obtain wind-effect data representative of full-scale conditions [1]. It must have a test section that is sufficiently long to generate a thick

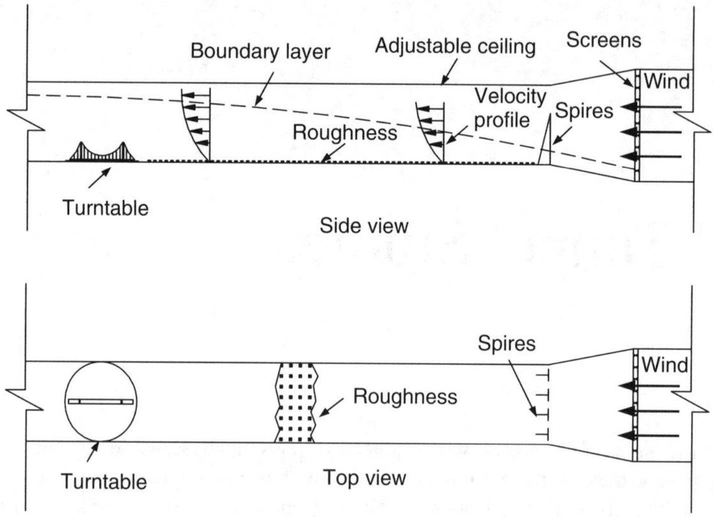

Figure 7.1 Components of boundary-layer wind tunnel test section.

vertical boundary layer, be sufficiently high so that the boundary layer generated will not touch the tunnel ceiling, and be sufficiently wide so that neighboring structures and topographical features can be incorporated into the model. Furthermore, the blockage ratio (the ratio of the cross-sectional area of the model blocking the flow and the cross-sectional area of the tunnel test section) must be less than approximately one-twentieth. These requirements necessitate rather large tunnels for bridge wind engineering.

To facilitate the rapid growth of a vertical boundary layer along the tunnel test section, not only must roughness elements be placed on the tunnel floor, but also additional devices such as spires must be installed upstream (see Figure 7.1). The roughness and the spires (or other vortex-generating devices such as barrier and grid) must be designed to produce a type of velocity profile and the type of turbulence that is similar to that encountered by the prototype structure.

Typically, the model tested in a wind tunnel is placed on a turntable so that it can be studied for winds from different directions. To simulate the wind field correctly, the model should include not only the particular structure to be tested, but also all neighboring structures and terrain features – all constructed to the same scale ratio.

To maintain a constant pressure (zero gradient of pressure) along tunnel test sections, the cross-sectional area of a tunnel test section must be slightly increased in the direction of the wind. The increasing cross-section causes a decreasing wind velocity and an increasing pressure in the wind direction, compensating for the pressure decrease caused by friction. Normally, the area increase is accomplished by using an adjustable ceiling for the tunnel test section. The ceiling slope is adjusted to produce a zero pressure gradient in the wind direction.

Boundary-layer wind tunnels used in bridge wind engineering are low-speed wind tunnels. There are mainly two types of wind tunnel according to flow circuit: open-circuit wind tunnel, and closed-circuit wind tunnel. The following is a brief description of each type of wind tunnel.

7.2.1 Open-Circuit Wind Tunnel

An open-circuit wind tunnel is normally a straight structure. Air is drawn into the tunnel from a funnel-shaped intake at one end, and the air exits the tunnel through a funnel-shaped outlet at the other end.

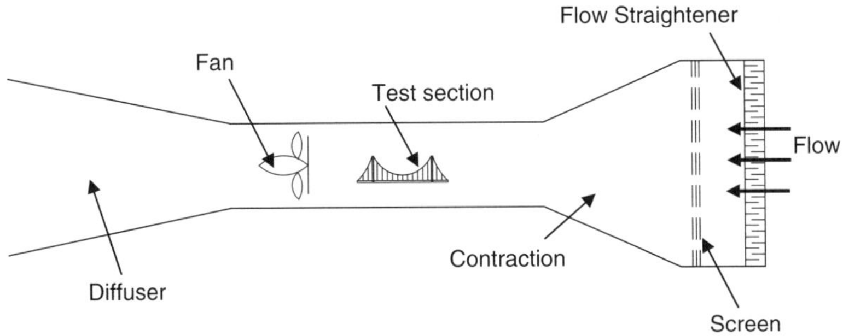

Figure 7.2 Layout of a sucking-type open-circuit wind tunnel.

Figure 7.2 shows the main components in an open-circuit wind tunnel with an axial-flow fan downstream of the test section [2]. There is a contraction section at the intake, usually with a flow straightener and fine mesh screens, to smooth mean flow variations and reduce turbulence in the test section. There is a diffuser at the outlet to conserve power by reducing the amount of kinetic energy that is lost with the discharging air. The tunnel shown in Figure 7.2 is a "sucking" type. An alternative is a "blowing" arrangement, in which the test section is downstream of the fan. For modeling atmospheric boundary layer flows, it may not be essential to include a contraction section and a diffuser, but it would be at the expense of high electricity consumption.

7.2.2 Closed-Circuit Wind Tunnel

The closed-circuit wind tunnel has a re-circulating loop, as shown in Figure 7.3, in which air is circulated during tests [3]. It may occupy a large space if the loop is horizontal.

Consequently, indoor closed-circuit wind tunnels are sometimes arranged in a vertical loop to save space. Some large tunnels of the closed-circuit type utilize the enlarged return section as an additional test section for low-speed tests. The advantages of this arrangement are as follows: (a) it does not cause undesirable wind in laboratories housing the wind tunnel; (b) it is generally less noisy than the open-circuit type; (c) it is usually more efficient; and (d) more than one test section with different characteristics can be incorporated. However, this type of wind tunnel has a higher capital cost and the air heats up over a long period of operation before reaching a steady-state temperature.

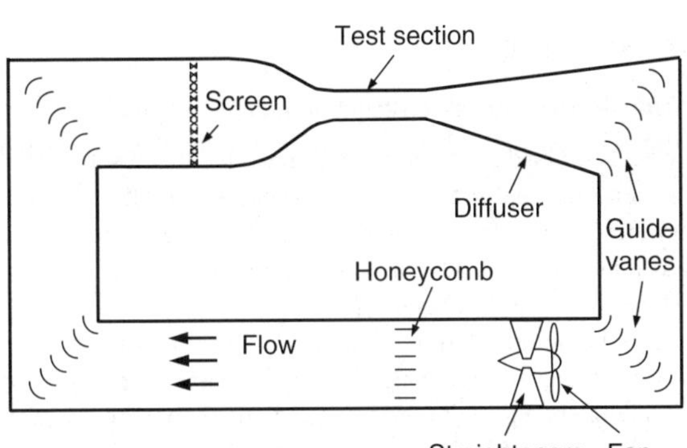

Figure 7.3 Layout of a closed-circuit wind tunnel.

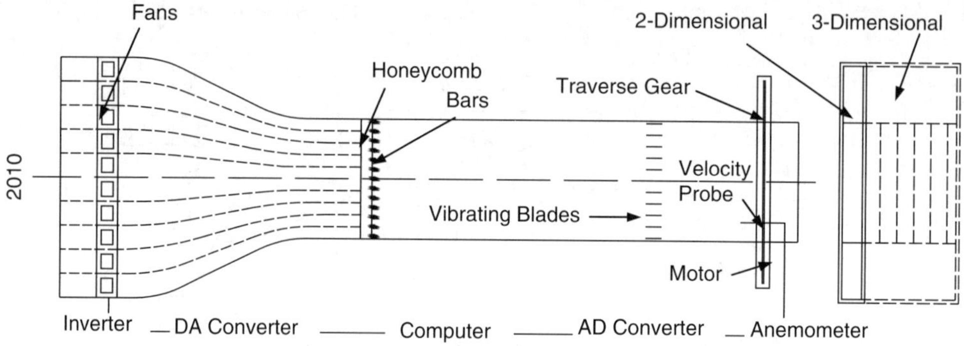

Figure 7.4 Schematic diagram of two- and three-dimensional actively-controlled wind tunnels (unit: mm).

7.2.3 Actively Controlled Wind Tunnel

In conventional open- or closed-circuit wind tunnels, turbulence characteristics are commonly produced by using spires (or other vortex generating devices) and roughness blocks set on the floor of the wind tunnel. It is difficult for these conventional tunnels to produce complex turbulence characteristics such as those which are often encountered by long-span cable-supported bridges surrounded by a complex topography. Consequently, new types of wind tunnels, with actively controlled multiple fans, have been proposed and constructed [4–6]. Figure 7.4 is a schematic diagram of two- and three-dimensional actively-controlled wind tunnels with multiple fans [6].

The two-dimensional multiple-fan wind tunnel has 11 fans arranged vertically. The three-dimensional one has six rows of two-dimensional wind tunnels with a total of 66 fans. The test section dimensions (length × width × height) are $3.8 \times 0.18 \times 1.0\,\text{m}$ for the two-dimensional wind tunnel and $5 \times 1.0 \times 1.0\,\text{m}$ for the three-dimensional wind tunnel, respectively. The lengths of the actively-controlled wind tunnels are very short compared to open or closed-circuit tunnels. The fans installed in the two-dimensional multiple-fan wind tunnel are driven by high-performance AC servo-motors which are, in turn, controlled by a computer. By using proper control algorithms, wind statistical parameters such as mean wind velocity, turbulence intensity, turbulence scale and power spectrum in a given boundary layer can be satisfactorily reproduced in the actively-controlled wind tunnel. Nevertheless, such tunnels have not been widely used in practice.

7.3 Model Scaling Requirements

7.3.1 General Model Scaling Requirements

The modeling of wind effects on bridge structures first requires that a satisfactory model of the natural wind be produced, and then that a similarly scaled structural model be tested in this modeled wind environment. Criteria and techniques for achieving a model of the natural wind in a wind tunnel have been discussed in various depths [7–11]. The principles and requirements of modeling of buildings and structures have been discussed by many scholars [11–14].

One rational approach to the determination of the overall scaling or similarity requirements for any model test is to assemble all the criteria, through either dimensional analysis or similarity arguments, and then to form a physical understanding of each criterion to enable the criteria to be pruned to what is essential for a given model test [15,16]. To this end, a short list of variables on which the modeling of wind effects on buildings and structures depends is suggested as follows:

$$\overline{U}_g, \overline{U}_z, u, v, w, L, n, S_u(n), S_v(n), S_w(n), \rho, \mu, p, L_s, n_s, \rho_s, E, g, \delta. \tag{7.1}$$

where:

$\overline{U}_g$ is the mean wind speed at the gradient height;
$\overline{U}_z$ is the mean wind speed at height z;
u, v, w are the fluctuating velocity components in the longitudinal, lateral and vertical directions, respectively;
L is the length associated with the external air flow (i.e. natural wind boundary layer);
n is the frequency associated with the external air flow;
$S_u(n)$, $S_v(n)$, $S_w(n)$ are the power spectral density function of the respective velocity components;
ρ is the air density;
μ is the air viscosity;
p is the air pressure;
L_s is the length associated with the structure;
n_s is the frequency associated with the structure;
ρ_s is the density of the structure;
E is the elastic modulus of the structure;
g is the acceleration due to gravity;
δ is the damping for the structure.

Other combinations used are: σ_u is the standard deviation of the fluctuating component; I_u is the turbulence intensity; and $\zeta = \delta/(2\pi)$ is the critical damping ratio.

By using either dimensional analysis or similarity arguments, it can be shown that dynamic similarity with respect to these nineteen variables can be achieved if the sixteen non-dimensional groups listed in Table 7.1 are maintained constant between model and full scale. The list of variables and the consequent list of non-dimensional parameters have been deliberately basic. There are others used regularly in this type of modeling which are either related to, or similar to, those given or are only required for special situations. For example: force and moment coefficients which express the same force ratio as the pressure coefficient could be used alternatively; Strouhal number related to the structure could have replaced the density ratio or the Cauchy number; structural stresses could be used to replace stiffness and is often used as a measured variable; shear stress distribution in a boundary layer is sometimes used.

7.3.2 *Notes on Model Scaling Requirements*

To achieve dynamic similarity, it is necessary to maintain constant ratios between all the forces affecting the phenomena. In practice, it is rarely possible to satisfy all these requirements. Therefore, some physical understanding of the phenomena is required to facilitate an assessment of the relative importance of the various forces, so that scaling of the less significant forces may be neglected. It may also be necessary to permit the distortion of some criteria to allow a practicable model to be constructed, and this again requires some analytical knowledge of the phenomena.

Full-scale Reynolds numbers can rarely be achieved, and the problems thus created are common to all of the tests and deserve special attention. The Reynolds number expresses the ratio of fluid inertia to viscous forces. In general, exact maintenance of this ratio is only required when the viscous forces become of the same order of magnitude as the inertia forces. For wind effects on structures, this is only achieved in a boundary layer, and it is most significant where boundary layer separation occurs. Fortunately, most structures are sharp-edged and, hence, separations occur at these edges. In these cases, for high Reynolds numbers, the flow field and hence pressure distributions seem to be relatively independent of Reynolds number, so errors due to incorrect scaling are small.

The reverse situation occurs when the structure has a curved surface on which a flow separation could occur. Model tests of these configurations must be treated very carefully, even if separation is forced with trips. The best example, of course, is the drag of a circular cylinder,which is very Reynolds number-dependent.

Table 7.1 Non-dimensional groups

Non-dimensional group	Name	Physical meaning
$\dfrac{\overline{U}_z}{\overline{U}_g}$	Velocity profile	Velocity ratio which defines the vertical velocity profile
$\dfrac{\sigma_u}{\overline{U}_z}, \dfrac{\sigma_v}{\overline{U}_z}, \dfrac{\sigma_w}{\overline{U}_z}$	Turbulence intensity	Expression relating total energy of the fluctuating components
$\dfrac{nS_u(n)}{\sigma_u^2}, \dfrac{nS_v(n)}{\sigma_v^2}, \dfrac{nS_w(n)}{\sigma_w^2}$	Normalized power spectral density	Expression giving turbulent energy distribution with respect to frequency
$\dfrac{nL}{\overline{U}}$ (L is any related length)	Strouhal number or reduced frequency (or inverse of reduced velocity)	Time scale
$\dfrac{\rho\overline{U}L}{\mu}$ (L is any related length)	Reynolds number	$\dfrac{\text{Inertia force (fluid)}}{\text{Viscous force}}$
$\dfrac{p}{1/2\rho\overline{U}^2}$	Pressure coefficient (or force and moment coefficient)	$\dfrac{\text{Pressure force (fluid)}}{\text{Inertia force (fluid)}}$
$\dfrac{L}{L_s}$	Length ratio	Ratio of lengths in boundary layer and structure
$\dfrac{n}{n_s}$	Frequency ratio	Ratio of frequency or time in boundary layer and structure
$\dfrac{\rho}{\rho_s}$	Density ratio	$\dfrac{\text{inertia force (fluid)}}{\text{inertia force (structure)}}$
$\dfrac{\rho\overline{U}^2}{E}$	Cauchy number	$\dfrac{\text{Inertia force}}{\text{Elastic force}}$
$\dfrac{\overline{U}}{\sqrt{Bg}}$	Froude number	$\dfrac{\text{Inertia force}}{\text{Gravity force}}$
δ or ζ	Logarithmic damping decrement or critical damping ratio	$\dfrac{\text{Energy dissipated/cycle}}{\text{Total energy of oscillation}}$

One other situation which can occur, even with relatively sharp edged bodies, is the formation of a leading edge separation bubble and subsequent re-attachment, which can cause very high suction peaks. This phenomenon is Reynolds number- and turbulence-dependent and results can be misleading, particularly when relatively low test Reynolds numbers are used.

7.3.3 Blockage Consideration

Another aspect of wind tunnel model tests which deserves attention is blockage corrections. A body placed in a wind tunnel will partially obstruct the passage of the air, causing the flow to accelerate. This effect is referred to as "blockage". If the blockage is substantial, the flow around the model and the aerodynamic behavior of the model are no longer of prototype conditions [17].

The blockage corrections depend on the body shape, the nature of the aerodynamic effect of concern, the characteristics of the wind tunnel flow and the relative body/wind tunnel dimensions. The magnitude of blockage correction varies with the blockage ratio, which is the ratio of the model projected area normal to wind flow over the wind tunnel cross-sectional area. For example, for a blockage ratio of 2%, the blockage corrections for measured pressures and forces are likely to be within 3–5% [18]. In general, blockage corrections are required if the blockage ratio exceeds 5%. Tests involving a blockage ratio of around 20% or higher are virtually worthless.

In the case of drag, the following approximate relation may be used for the great majority of model configurations:

$$C_{D_c} = \frac{C_D}{1 + KS/C} \tag{7.2}$$

where:

C_{D_c} is the corrected drag coefficient;
C_D is the drag coefficient measured in the wind tunnel;
S is the reference area for the drag coefficients C_{D_c} and C_D;
C is the wind tunnel cross-sectional area;
K is the coefficient.

The ratio of S over C is defined as blockage ratio.

7.4 Boundary Wind Simulation

The paramount requirement for physical modeling of wind engineering phenomena is the proper simulation of natural wind characteristics. Normally, the minimum requirements for modeling of the natural wind are similarity of velocity profile, $\overline{U}_z/\overline{U}_G$, longitudinal turbulence intensity, $\sigma_u/\overline{U}_z$, and power spectral density of the longitudinal component of turbulence, $nS_u(n)/\sigma_u^2$. For bridge wind engineering, the similarity of vertical turbulence intensity, $\sigma_w/\overline{U}_z$, and power spectral density of the vertical component of turbulence, $nS_w(n)/\sigma_w^2$, are also important. Four methods – the natural growth method, the augmented method, actively-controlled spires or grids, and actively-controlled multiple-fans – have been used to simulate the turbulent boundary layer flow of the natural wind in wind tunnels.

7.4.1 *Natural Growth Method*

The natural growth method requires a wind tunnel with a very long fetch of roughness elements, over which a rough wall boundary layer is developed, and a few special wind tunnels have been designed and built to meet this requirement. The roughness elements for the natural growth method should be of a size and density characteristic of the terrain surrounding the site and should extend sufficiently far downstream for the required turbulent shear layer to become fully established. This may give a good representation of the real conditions, but very long test sections are required [19].

7.4.2 *Augmented Method*

Most of the boundary-layer wind tunnels have smaller and shorter test sections than required for the natural growth method. Thus, it is necessary to accelerate the growth of the boundary layer thickness artificially. This can be done by introducing special devices at the entrance of the test section that modify the momentum distribution across the boundary layer, generating the same losses as those that can be achieved by a longer region of roughness elements.

The augmented method, which uses passive and/or active devices and requires only a relatively short fetch length, has been successfully used in practice. Large-scale turbulence is generated by installing tripping devices (e.g. grids, barriers, spires, screens and air-injection systems) upstream of a reduced

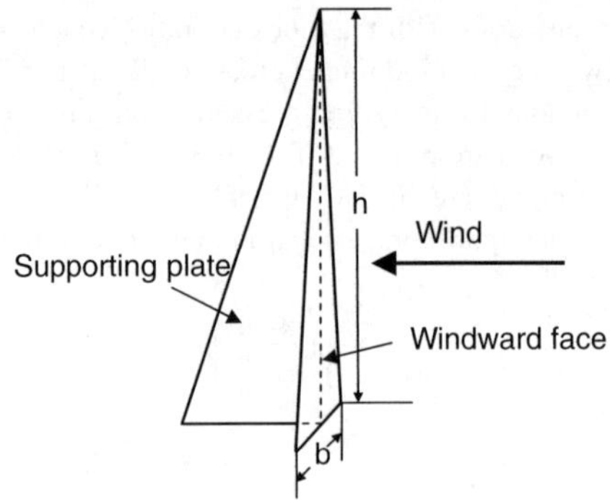

Figure 7.5 A proposed spire configuration.

fetch length covered with roughness elements. The size and spacing of various devices are varied to generate different wind models.

For example, the following procedure was proposed for the design of spires with the configuration shown in Figure 7.5 [20]:

(a) select the desired boundary-layer depth δ;
(b) select the desired shape of mean velocity profile defined by the power law exponent α;
(c) obtain the height h of the spires from the relation $h = \dfrac{1.39\delta}{1+\alpha/2}$;
(d) obtain the width b of the spire base from Equation 7.3, in which H is the height of the tunnel test section.

$$\frac{b}{h} = \frac{\phi(H/\delta)}{2(1+\phi)}(1+\alpha/2) \tag{7.3}$$

$$\phi = \beta\{[2/(1+2\alpha)] + \beta - [1.13\alpha/(1+\alpha)(1+\alpha/2)]\}/(1-\beta)^2, \quad \beta = (\delta/H)\alpha/(1+\alpha) \tag{7.4}$$

The desired mean wind profile occurs at a distance $6h$ downstream from the spires. The wind tunnel floor downwind of the spires should be covered with roughness elements, e.g. cubes with height k such that [20]:

$$\frac{k}{\delta} = \exp\left\{\left(\frac{2}{3}\right)\ln\left(\frac{D}{\delta}\right) - 0.1161\left[\left(\frac{2}{C_f}\right) + 2.05\right]^{1/2}\right\} \tag{7.5}$$

$$C_f = 0.136\left[\frac{\alpha}{1+\alpha}\right]^2 \tag{7.6}$$

where D is the spacing of the roughness elements. Equation 7.5 is valid in the range $30 < \delta D^2/k^3 < 2000$.

The boundary-layer wind tunnel at the University of Sydney, as shown in Figure 7.6, is an open-circuit wind tunnel with a 2.4×2 m working cross-section. The augmented method is used, where large scale turbulence fluctuations, generated by spires or trip-boards, are superimposed on a rough boundary layer growing over carpet or roughness elements. The size of the tripping devices and the spacing of the roughness elements are varied to develop wind models of open country, suburban and city centre terrains, of different scales. The flow characteristics, namely the mean velocity profile, longitudinal turbulence intensity profile and power spectral density function of the longitudinal component of turbulence, are given in Figures 7.7 to 7.9 for the 1 : 400 scale wind models.

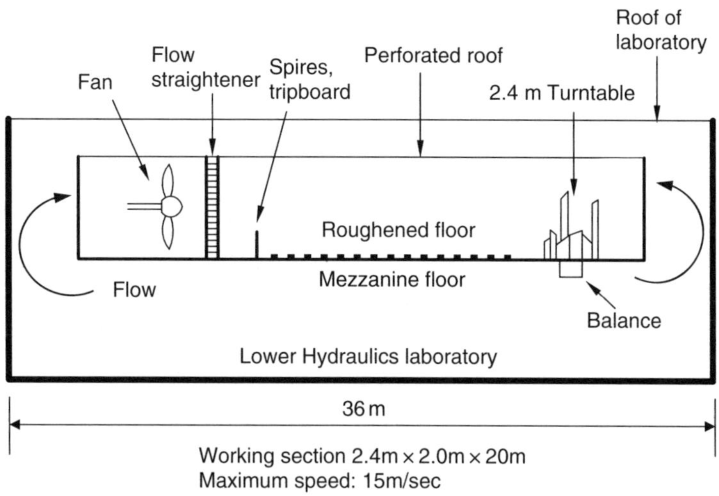

Figure 7.6 Arrangement of the boundary layer wind tunnel at the University of Sydney.

7.4.3 *Actively Controlled Grids and Spires*

Physical modeling for some applications requires flows with larger scales of turbulence than can be developed in a wind tunnel by passive devices such as stationary grids and spires. An example of this is the investigation of aerodynamic stability for long-span bridge decks by means of a section model in

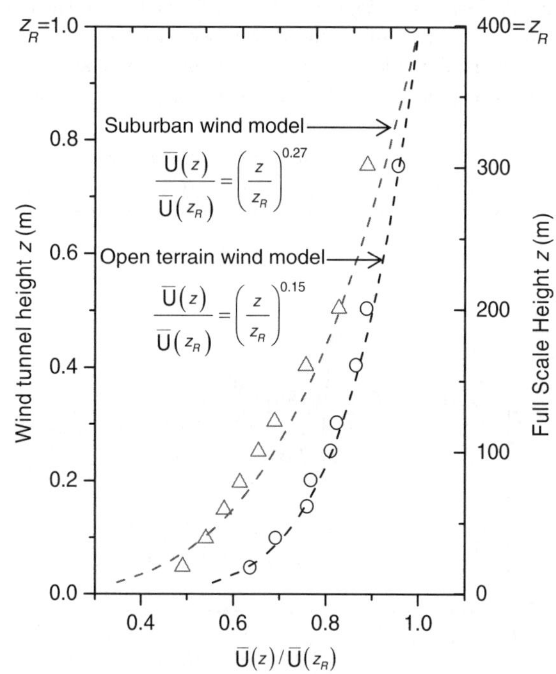

Figure 7.7 Mean wind velocity profiles (1 : 400 scale).

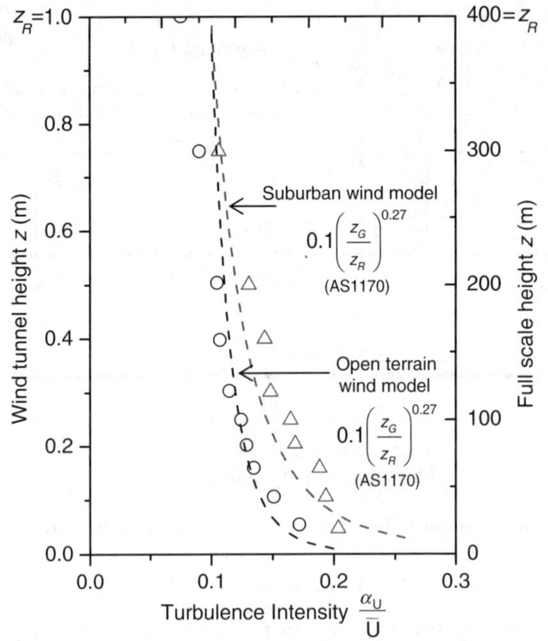

Figure 7.8 Longitudinal turbulence intensity profile (1 : 400 scale).

which the ratios of bridge-deck width to length scales of horizontal and vertical turbulence components corresponding to full-scale values are confronted by the passive wind tunnels.

A logical way to overcome the limitation of a passive grid is to devise an actively-controlled mechanical grid by which higher turbulent intensity and larger integral scale can be achieved owing to its 'active'

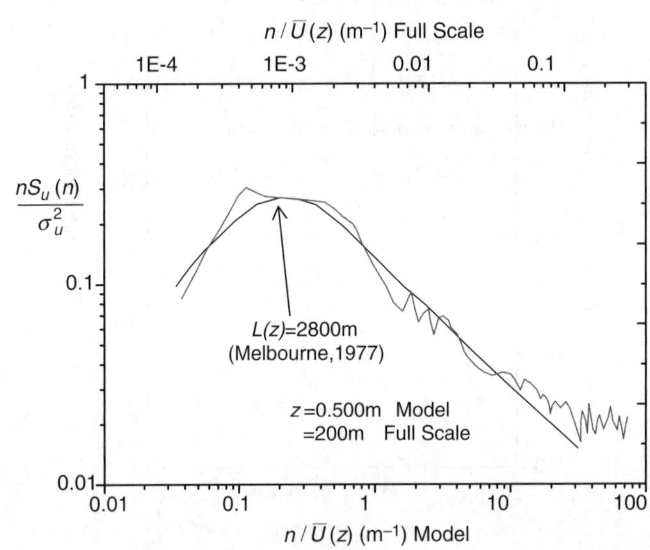

Figure 7.9 Longitudinal turbulence spectrum (1 : 400 scale open terrain wind model).

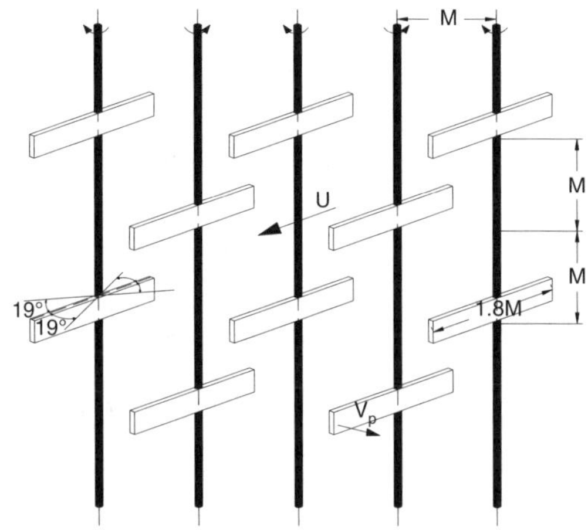

Figure 7.10 Actively controlled mechanical grid.

method of generating the flow. Figure 7.10 shows an actively-controlled mechanical grid and Figure 7.11 displays a flow chart of actively controlled boundary layer wind simulation [21].

Using the actively-controlled grids, the downstream turbulence intensity can easily reach up to 10% and the integral length scale of turbulence wind can reach up to 1 m. Inspired by this research, several actively-controlled wind tunnels have been developed. For example: Teunissen [22] developed a wind tunnel with jets to simulate the boundary layer; Cermak [23] developed a wind tunnel with active cascades of moving airfoils to examine the effects of large-scale turbulence over bridge deck models; Cogotti [24] used five actively-controlled spires with flaps made of carbon fiber to generate turbulent wind field with an integral length scale of turbulence up to 1.0–2.0 m.

7.4.4 Actively Controlled Multiple Fans

The power spectrum modification method is used to produce the required turbulence in actively-controlled wind tunnels with multiple-fans [6]. The target spectrum is selected first, with the assumed

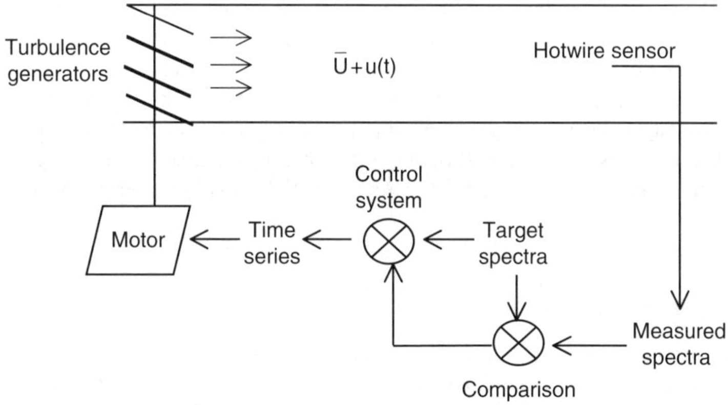

Figure 7.11 Flow chart of actively controlled boundary layer wind simulation.

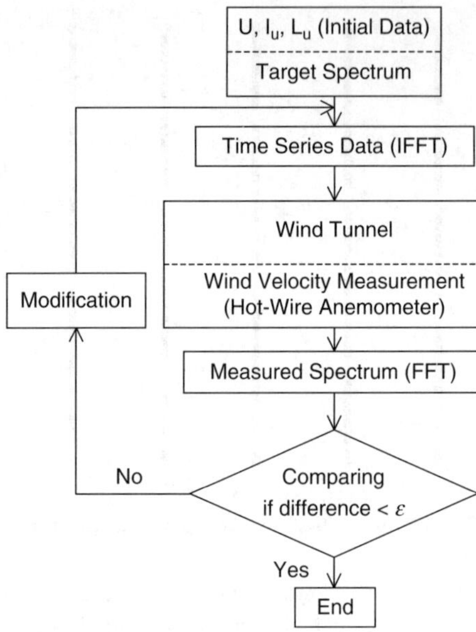

Figure 7.12 Block diagram for actively controlled turbulence simulation.

parameters such as U, I_u and L_u, and is then analyzed by making use of the inverse fast Fourier transform program to produce a time history of turbulent wind. The time history is given to the fan motor and the turbulent wind speed generated by the fans is measured by a hot-wire anemometer. The measured data are analyzed by the FFT method and the obtained spectrum is compared to the target one. Based on the deviated part of the measured spectrum from the target one, a new spectrum is estimated and given to the fan again, and the procedure is repeated until the satisfactory targeted spectrum is achieved. This is called the "spectrum modification method". The flow chart of the iterative procedure to simulate the targeted turbulence is shown in Figure 7.12.

There is a certain time lag in the above actively controlled wind simulation. When a fluctuating input signal is given to the fan motor, the same fluctuating velocity could be measured at the test section with a certain time lag. This time lag is mainly attributed to the operation of fans and the flow time and it depends on the inertia of rotating parts of the fans and motors and the aerodynamic delay caused by the acceleration of air mass in the tunnel path. This was assumed to be a first-order lag by the experiments. The time lag constant can be estimated and the time-lag modification can be implemented, i.e. the input data are modified by using the estimated time constant before the first run and the modified data are then given to the fans. After one or two iterations, a satisfactory time series of turbulence can be produced.

7.4.5 Topographic Models

Information on the characteristics of the full-scale wind applicable to a bridge site may not be available if the bridge is situated in complex topography or terrain. In such situations, small-scale topographic models (see Figure 7.13), constructed at scales in the range of 1 : 1000 to 1 : 5000, can be effective for estimating the full-scale mean wind flow [1]. Such data can form the basis for the subsequent modeling of the wind at a larger scale, as required for studying wind effects on bridge structures.

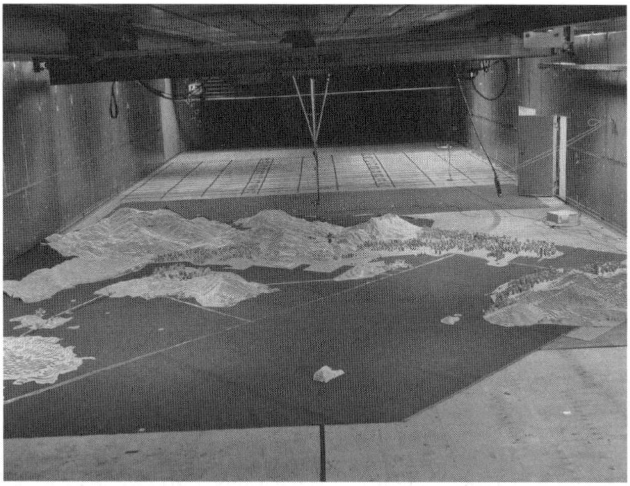

Figure 7.13 Topographic model for Stonecutters Bridge (scale 1 : 1500). (Reproduced with permission from Dr. Michael Hui).

7.4.6 *Instrumentation for Wind Measurement in Wind Tunnel*

The Pitot tube is the basic instrument used for measuring wind speed in a wind tunnel. It is based on the principle of conversion of kinetic energy to pressure at a stagnation point. The pressure differential sensed by the tube is proportional to the square of the velocity. This instrument is accurate, reliable, convenient and economical, but it is inaccurate at low speeds and unsuitable for measuring turbulence.

Hot-wire and hot-film anemometers can be used in wind tunnels to measure both mean wind speed and turbulence (see Figure 7.14). The sensing element of a hot-wire anemometer is a wire finer than a human hair. The turbulence in the wind causes changes of heat transfer from the wire, which in turn causes the resistance of the wire to fluctuate. The electronic circuit automatically adjusts the current going through the wire to keep the wire at a constant temperature. Consequently, the velocity fluctuations can be determined from the fluctuations of the current through the wire. The sensing element of a hot-film anemometer is a coated metal film laid over

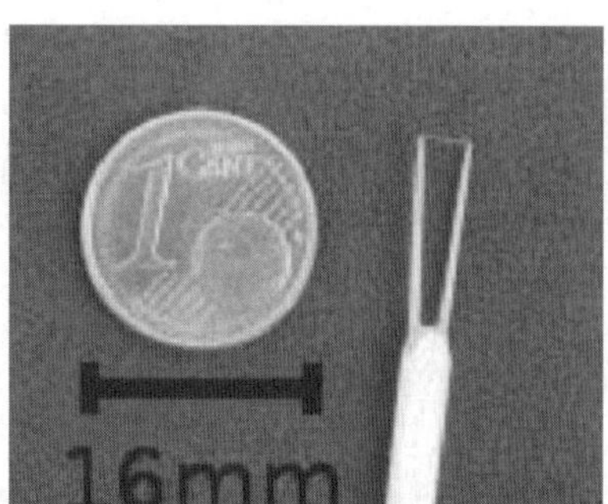

Figure 7.14 Hot-wire anemometer (Source: (http://en.wikipedia.org/wiki/File:Hd_sonde.jpg), Permission released under the GNU Free Documentation License. Original text : 28.06.2007 Bergmann/Kaiser).

a tiny glass wire. The rest is the same as for hot wires. Both hot-wire and hot-film anemometers can measure rapid changes of velocities with frequency response higher than 1 kHz.

There are many other types of instruments available for measuring various quantities of interest in wind tunnel model studies. Some of these are highly complex and include online data acquisition capabilities. There is also a growing level of automation, with computers used to control wind tunnel tests and to analyze data. However, it is not the intent of this chapter to list all the various instruments and associated data acquisition systems and data analysis systems. Some instruments will be introduced in the following sections, when we discuss particular wind tunnel tests.

7.5 Section Model Tests

7.5.1 Models and Scaling

For long-span bridges, both full bridge and partial bridge models have been tested in wind tunnels for different purposes [1]. The most basic of the partial bridge models has been the section model, which represents a section of the deck. Typical model-to-prototype scale ratios include λ_L (geometric length), λ_ρ (density) and λ_U (velocity). The density ratio is usually near unity, while the geometric length ratio and velocity ratio are fixed mainly by the size of the model and available wind tunnel speeds. If turbulence is introduced into the wind tunnel flow, it is important that its intensities correspond to those in full scale, and that typical turbulence lengths are scaled in accordance with the geometric length ratio. Full scaling of the turbulence is impractical at the geometric scales used in section model tests. In such cases, partial simulation of turbulence up to scales of turbulent length comparable to that of the deck width is desirable. For full models of suspension bridges, Froude number similarity is required in situations in which gravity forces substantially contribute to the stiffness. Froude scaling implies that:

$$\frac{\lambda_L}{\lambda_t^2} = \lambda_L \lambda_f^2 = 1 \tag{7.7}$$

where λ_L, λ_t and λ_f are the length, time and frequency scales, respectively. As a result, if the geometric scale is decided the frequency scale and velocity scale, λ_U can be determined by:

$$\lambda_f = \frac{1}{\sqrt{\lambda_L}}, \quad \lambda_U = \sqrt{\lambda_L} \tag{7.8}$$

For full bridge models of cable-stayed bridges, it is usually possible to avoid Froude number scaling if pre-tensioning of the stay cables to the design dead-load tension is provided to avoid the slackening of the model stays during a test.

For dynamic models in free vibration, the damping ratio shall be the same in the model as in the prototype. In all dynamic modeling cases, the scaling of the reduced velocity is required.

$$\left(\frac{U}{fB}\right)_m = \left(\frac{U}{fB}\right)_p \tag{7.9}$$

where:

f is the frequency of interest, either of the fluid or the structure;
B is the characteristic length, usually deck width;
subscripts m and p mean model and prototype, respectively.

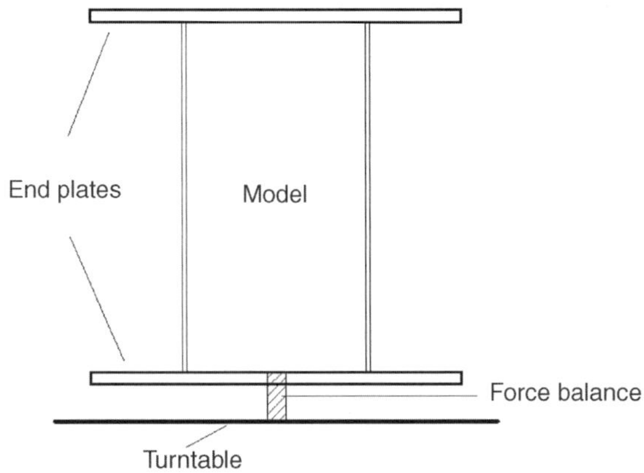

Figure 7.15 Set-up of section model tests for force coefficients.

7.5.2 Section Model Tests for Force Coefficients

As discussed in Chapters 3 and 4, force coefficients of a bridge deck are required in aerostatic and aerodynamic analyses. The section model test on a stationary representative section of a bridge deck is a common way to obtain the force coefficients, the first level aerodynamic characteristics of the deck cross-section. The section model for such a test is usually rigid. Only geometric similarity needs to be respected.

The common set-up of section model tests for force coefficients is shown in Figure 7.15. A rigid model representing a spanwise portion of a bridge deck is mounted on a force balance for measurement of wind forces on the model. The force balance must be designed properly according to the mass of the model and the frequency of interest. The force balance, of which the sensing element is often strain gauges, is usually fixed to a turntable so that wind forces on the model can be measured at different attack angles. If it is sensitive enough, the force balance can measure both mean and dynamic forces. End plates are usually used to eliminate end effects.

Section model tests for force coefficients are usually conducted under smooth wind flow, and the force coefficients can be obtained from the measured wind forces with the following equations:

$$
\begin{aligned}
C_D(\alpha) &= \frac{F_D(\alpha)}{\frac{1}{2}\rho\overline{U}^2 B_m L_m} \\
C_L(\alpha) &= \frac{F_L(\alpha)}{\frac{1}{2}\rho\overline{U}^2 B_m L_m} \\
C_M(\alpha) &= \frac{M(\alpha)}{\frac{1}{2}\rho\overline{U}^2 B_m^2 L_m}
\end{aligned}
\tag{7.10}
$$

where:

C_D, C_L and C_M are the drag, lift and moment coefficients, respectively;
F_D, F_L and M are the drag, lift and moment forces, respectively;
B_m and L_m are the width and length of the model, respectively.

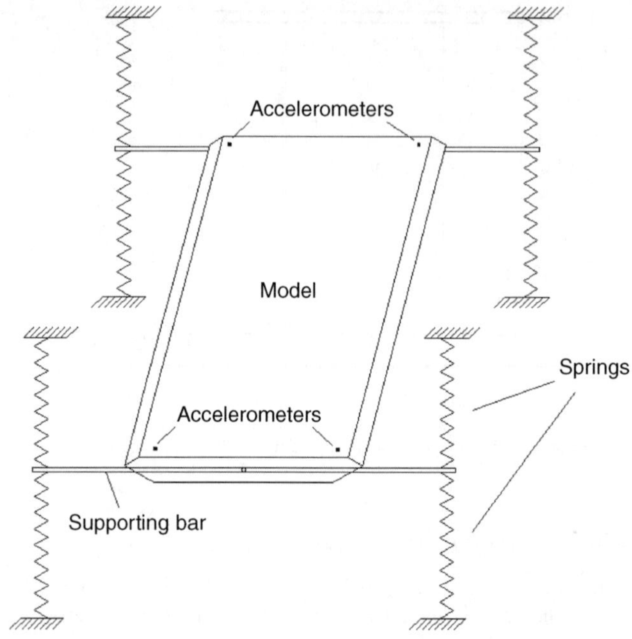

Figure 7.16 Set-up of free vibration tests of section model.

7.5.3 Section Model Tests for Flutter Derivatives and Vortex-Induced Vibration

A common technique to confirm the aerodynamic stability of the decks of long-span cable-supported bridges is the section model test in either free vibration or forced vibration. The flutter (aerodynamic) derivatives of the deck cross-section can be identified from the measured results. Free vibration tests can also be used to detect signs of vortex-induced vibrations. Section models are relatively inexpensive and can be constructed to scales of the order of 1/50 to 1/25, so that the discrepancies between full-scale and model Reynolds number are smaller than in the case of full-bridge tests.

A sectional model for free vibration test is made of a representative spanwise section of the bridge deck, constructed to the scale and spring-supported at the two ends to allow both vertical and torsional motions. Figure 7.16 shows a common set-up of a free vibration test of the section model. By suitable adjustment of the springs, the model frequencies in rotation and vertical translation can be arranged to have the same ratio as those for the primary bending and torsional modes of the prototype bridge. In order to achieve similarity between model and prototype, the reduced frequencies should be kept equal.

The model is also required to satisfy the density scaling requirement that the density ratio should be the same in model and full scale. Table 7.2 lists the scaling requirements for free vibration model tests. Usually, the effect of turbulence is to reduce or mark the vortex shedding-induced amplitude peaks. For this reason, section model tests are conducted in flow conditions with low turbulence intensity, in order to estimate the amplitudes of motion conservatively. Various edge treatments can be investigated to improve the response to vortex shedding. In addition, to ensure that the section is aerodynamically stable up to an acceptably high prototype wind speed, the model is also tested with the same low turbulence flows, to provide conservative estimates of the flutter derivatives of the section. The response is often investigated at various structural damping ratios.

Table 7.2 Scaling requirements for free vibration section model

Parameters	Scale ratio
Length	$1/\lambda_L$
Wind velocity	$1/\lambda_U$
Mass per unit length	$1/\lambda_L^2$
Mass inertial per unit length	$1/\lambda_L^4$
Structural frequency	λ_L/λ_U
Damping ratio	1

A section model for a forced vibration test consists of a representative spanwise section of the deck constructed, to the required scale and attached to the force balances at the ends. The models are driven to oscillate by driving devices through the force balances, as shown in Figure 7.17.

Total forces acting on the model can be measured by the force balances. Time-histories of displacement and acceleration responses of the section should also be measured in the tests, using displacement transducers and/or accelerometers. Section models for forced vibration are rigid models, so only geometric similarity needs to be respected. Ukeguchi first used forced vibration test technique to identify flutter derivatives in 1966 [25]. Since then, this technique has been further developed by many researchers [26].

The forced vibration method is somewhat expensive, compared with the free vibration method, since it involves sizeable and more complicated test equipment. Furthermore, the frequency and amplitude of model vibration in a forced vibration test is pre-set manually and cannot represent fully the aerodynamic characteristics of the bridge deck in a natural wind. The free vibration method seems to be more tractable than forced vibration testing. However, at high reduced wind speeds, the vertical bending motion of the bridge will decay rapidly, due to the effect of positive vertical bending aerodynamic damping, and thus the length of time history available for system identifications will decrease in free vibration, thereby adding more difficulties to the system identification [1].

7.5.4 Section Model Tests with Pressure Measurements

The relationship between incident velocity fluctuations and resulting wind loads on the bridge deck is an important source of uncertainty in bridge aerodynamics. The analysis of the distribution of deck fluctuating pressures will allow some insights into the mechanism of wind excitation on the bridge.

Compared with the measurement of static and dynamic forces on a stationary section model with force balances, pressure measurement tests have the advantages of obtaining the mean and fluctuating

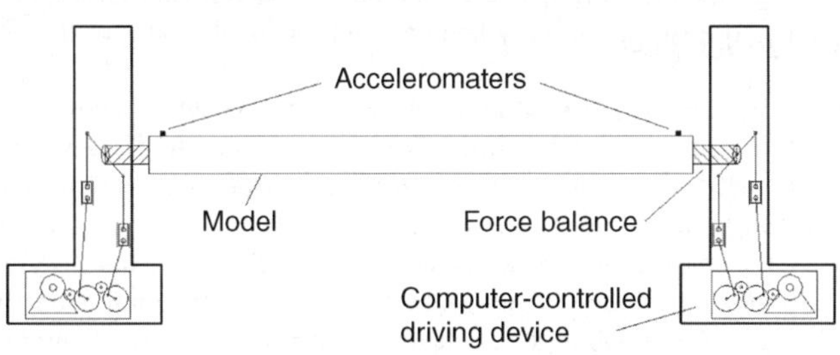

Figure 7.17 Set-up of forced vibration tests of section model.

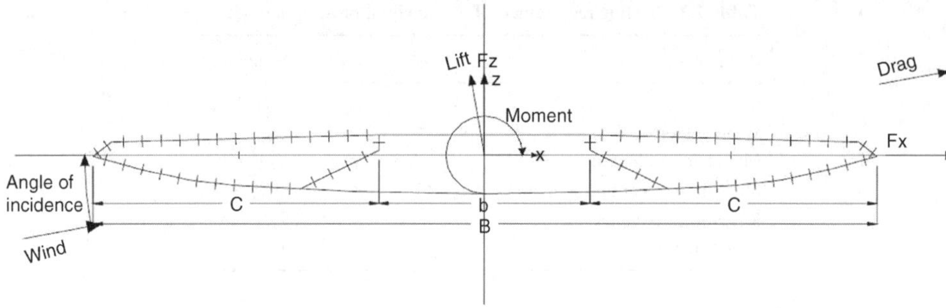

Figure 7.18 Section model tests with pressure measurement.

pressure distribution over the bridge deck section and the spanwise coherence of the fluctuating forces [35]. Because high-speed pressure scanning systems allow for simultaneous pressure measurement at a sufficiently high rate, pressure measurement tests usually give more accurate frequency domain information of fluctuating section forces than direct force measurement tests [27].

For a section model test with pressure measurements, pressure taps are embedded in the sectional model to measure pressure distribution over the external surface, as shown in Figure 7.18. Pressure should be measured at a sufficient number of locations, so that no significant aerodynamic events are missed. Theoretically, measurements should be taken in all areas of significance, with particular attention to areas of high local pressure or potential aerodynamic "hot spots". In practice, however, it is difficult to install pressure taps at edges, or auxiliary components such as guard rails in a section model. Therefore, it is likely that some aerodynamic events are missed in such pressure tests.

7.5.5 Section Model Tests for Aerodynamic Admittance

The aerodynamic admittances can be estimated by direct measurement of the aerodynamic forces of a stationary section model, or can be extracted from the steady random responses of the aeroelastic model of the bridge deck by system identification techniques [28–33]. The force measurement methods can be further classified into two categories: force measurement using force balances, and surface pressure integral method with pressure measurement tests. The former requires a rather higher inherent frequency for the model-balance system (more than 80 Hz) and a deck section model rigid enough not to deform or vibrate in the test. These requirements generally bring considerable difficulty for wind tunnel tests. The surface pressure integral technique is only applicable for closed box girder bridges, which somewhat limits its popularity in aerodynamic admittance estimation.

7.6 Taut Strip Model Tests

The taut strip model is a three-dimensional partial-bridge model which is between an aeroelastic section model and a full aeroelastic model [34,35]. A supporting structure, consisting of taut wires or tubes, supports the geometrically simulated deck structural form, as shown in Figure 7.19. Usually, fundamental vertical and torsion deck modes are simulated by the model, while bridge components such as towers, piers and cables are neglected. A taut strip model is less expensive than a full aeroelastic model and it can simulate the three-dimensional effects of wind-structural interaction.

Froude number ($Bg/\overline{U}^2$) similitude may be relaxed when both vortex shedding-induced oscillations at low wind speeds and aerodynamic instability at high wind speed are to be studied. A taut strip model can be considered as an extended section model of a bridge deck. Model scaling requirements for taut

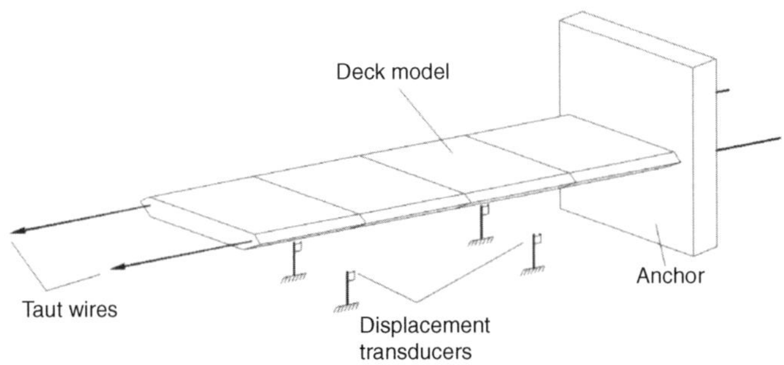

Figure 7.19 Taut strip model.

strip models are similar to those for free-vibration section models, except that more modes of the bridge deck may be considered.

The stiffness of a taut strip model is mainly provided by the taut wires, and therefore the fundamental bending and torsional frequencies of a taut strip model can be calculated by:

$$f_h = \frac{1}{2L}\sqrt{\frac{T}{m}}$$
$$f_\alpha = \frac{1}{2L}\sqrt{\frac{1}{I} \times \frac{B^2 T}{4}} \tag{7.11}$$

where:

f_h and f_α are the vertical and torsional frequency of the model, respectively;
L is the length of the model;
T is the tension force of the wires;
m is the mass per unit length;
I is the mass moment of inertia per unit length.

The main objective of taut strip model tests is to estimate the structural responses of a bridge deck under wind action. This technique can also be used to identify aerodynamic derivatives of a bridge deck. Nevertheless, to transfer taut-strip model results to the prototype invariably requires an analytical interpretation [1].

7.7 Full Aeroelastic Model Tests

In addition to being geometrically similar to the full bridge, the full aeroelastic model must satisfy similarity requirements pertaining to mass distribution, reduced frequency, mechanical damping and shapes of vibration modes. Modeling requirements for similarity between the model and prototype bridge components are listed in Table 7.3. Froude number similitude is generally respected for suspension bridges, where the geometrical stiffness of the main cables is the dominant stiffness.

To minimize Reynolds number effects, the geometric scale should be made as large as feasible. The geometric scale of 1 : 100 has been realized for full bridge models [36]. However, large-scale models require comparably large wind tunnel facilities, and the usual scale of such models is of the order of 1 : 200 [37–39]. Because of the complexity of full aeroelastic models, finite element methods are usually used in the design and calibration of such models.

Table 7.3 Scaling requirements for full aeroelastic model

Components	Geometric similarity	Stiffness similarity			Aerodynamic force similarity
		Longitudinal	Bending	Torsional	
Deck	✓		✓	✓	
Cables, suspenders	✓	✓			✓
Tower	✓	✓	✓		✓
Piers	✓				

Figure 7.20 Full aeroelastic model test of Stonecutters Bridge (Reproduced with permission from Dr. Michael Hui).

The construction of a full aeroelastic model is elaborate and its cost is relatively high. Thus, a full aeroelastic model test is usually used as a final design check for the aerodynamic characteristics of long-span cable-supported bridges.

The general practice in full aeroelastic model tests is to estimate the bridge response to smooth and turbulent flow in terms of mean response, RMS displacement and acceleration responses, spectrum response, total damping modifications due to wind effects, and flutter speed limits. Full aeroelastic model tests can also be used to identify aeroelastic parameters [39,40].

Full aeroelastic model tests can offer extensive exploration of prototype conditions, such as the effects of winds at various approach angles, velocity profiles, turbulence levels. Such models are also highly effective in studies of bridge performance during different stages of construction. A full aeroelastic model of a long-span cable-stayed bridge is shown in Figure 7.20.

7.8 Identification of Flutter Derivatives

7.8.1 Free Vibration Test of Section Model

Scanlan proposed the free vibration test method for flutter derivatives of a bridge deck in the 1970s, and he also proposed the semi-experimental and semi-analytical approach for estimating critical flutter wind speed [40,41]. In this technique, a spring-suspended sectional model is tested, and free decay vibration signals are recorded to extract the flutter derivatives.

A great advantage of the free vibration technique is its simplicity, but the original method needs three groups of tests. Torsional and vertical bending motions have to be constrained, respectively in order to obtain the so-called direct derivatives. Furthermore, the vertical and torsional motions of the model must have the same frequency at all wind velocities in order to obtain cross derivatives.

Yamada *et al.* [42] introduced the extended Kalman filter (EKF) method into the identification procedure of these derivatives based on the coupled vibration time histories. In this method, the time histories of the displacement and velocity, as well as information on the initial condition, are simultaneously recorded. Jokobsen and Hansen [43] proposed a method for the determination of the flutter derivatives, employing a conversion of buffeting response data to response covariance functions. Poulsen *et al.* [44] used a method which combines control theory with system identification techniques to extract flutter derivatives from section model tests for the Great Belt East Bridge. Juang and Pappa presented an eigensystem realization algorithm (ERA) for modal parameter identification and model reduction [45]. In 1994, Sarkar and Scanlan [46] developed the modified Ibrahim time-domain (MITD) method to extract all the direct and cross-derivatives from the coupled free vibration data of the 2-DOF model.

Gu *et al.* [47] proposed an identification method based on unifying least-squares theory. In this method, a unified error function which is linearly composed of two error components of vertical and torsional motions is defined as the objective function to optimize the flutter derivatives. The initial value in the iteration procedure for the optimization of the flutter derivatives is provided by the MITD method. The unifying least-squares method was refined by introducing two weighting factors to the error functions and one scaling factor to the iteration in order to enhance the accuracy and convergence rate of the identification results [48].

The above identification methods are based on free vibration of the section model with given initial conditions in smooth flow. The accuracy of identified flutter derivatives will be affected at high reduced velocities due to the nature of free vibration. Therefore, the identification of flutter derivatives of the section model under turbulence wind excitation is also explored, and Gu *et al.* [33] proposed a stochastic subspace identification (SSI) method that can directly identify flutter derivatives and aerodynamic admittances together.

In this section, the unifying least-squares method proposed by Gu *et al.* [47] is introduced as an example. As already introduced in Chapter 4, the governing equation of motion of the wind-structure system for 2-D flutter analysis can be written as:

$$\ddot{\mathbf{X}} + \mathbf{C}^{\mathbf{e}}\dot{\mathbf{X}} + \mathbf{K}^{\mathbf{e}}\mathbf{X} = \mathbf{0} \tag{7.12}$$

where:

$$\mathbf{X} = [h, \alpha]^{\mathrm{T}} \tag{7.13a}$$

$$\mathbf{C}^{\mathbf{e}} = \begin{pmatrix} 2\xi_{0h}\omega_{0h} - \dfrac{\rho B^2}{m}\omega_h H_1^* & \dfrac{\rho B^3}{m}\omega_\alpha H_2^* \\ \dfrac{\rho B^3}{I}\omega_h A_1^* & 2\xi_{0\alpha}\omega_{0\alpha} - \dfrac{\rho B^4}{I}\omega_\alpha A_2^* \end{pmatrix} \tag{7.13b}$$

$$\mathbf{K}^{\mathbf{e}} = \begin{pmatrix} \omega_{0h}^2 - \dfrac{\rho B^2}{m}\omega_h^2 H_4^* & \dfrac{\rho B^3}{m}\omega_\alpha^2 H_3^* \\ \dfrac{\rho B^3}{I}\omega_h^2 A_4^* & \omega_{0\alpha}^2 - \dfrac{\rho B^4}{I}\omega_\alpha^2 A_3^* \end{pmatrix} \tag{7.13c}$$

where:

h and α are the vertical and torsional displacements of the section model, respectively;
ξ_{0h} denotes the structural damping ratio in the vertical direction;
ω_{0h} denotes the structural natural frequency in the vertical direction;
$\xi_{0\alpha}$ is the structural damping ratio in the torsional vibration;
$\omega_{0\alpha}$ is the structural circular natural frequency in the torsional vibration;
ρ is the density of the fluid;

B is the section model width;
m represents the model mass per unit length;
I represents the mass moment of inertia of the model per unit length;
ω_h and ω_α denote the frequency of the vibration system in the vertical and torsional direction respectively;
H_i^* and A_i^* $(i = 1\sim4)$ are the flutter derivatives;
$\mathbf{C}^{\mathrm{e}}$ and $\mathbf{K}^{\mathrm{e}}$ are the total damping and stiffness matrix of the coupled wind-structure system respectively.

The total damping and stiffness matrices of the coupled vibration system are first identified from the displacement and acceleration time-histories recorded during the free vibration test. High-frequency noise in the original time-history data should be removed before identification. Aerodynamic damping and stiffness matrices, $\mathbf{C}^{\mathrm{ad}}$ and $\mathbf{K}^{\mathrm{ad}}$, can then be obtained by deducting structural damping and stiffness matrices $\mathbf{C}^0$ and $\mathbf{K}^0$ from the total damping and stiffness matrices:

$$\begin{aligned} \mathbf{C}^{\mathrm{ad}} &= \mathbf{C}^{\mathrm{e}} - \mathbf{C}^0 \\ \mathbf{K}^{\mathrm{ad}} &= \mathbf{K}^{\mathrm{e}} - \mathbf{K}^0 \end{aligned} \tag{7.14}$$

Eight flutter derivatives can be identified from $\mathbf{C}^{\mathrm{ad}}$ and $\mathbf{K}^{\mathrm{ad}}$ using the following relationship:

$$\begin{aligned} H_1^* &= \frac{-m_h}{\rho B^2 \omega_h} c_{11}^{ad} & A_1^* &= \frac{-I_\alpha}{\rho B^3 \omega_h} c_{21}^{ad} \\ H_2^* &= \frac{-m_h}{\rho B^3 \omega_\alpha} c_{12}^{ad} & A_2^* &= \frac{-I_\alpha}{\rho B^4 \omega_\alpha} c_{22}^{ad} \\ H_3^* &= \frac{-m_h}{\rho B^3 \omega_\alpha^2} k_{12}^{ad} & A_3^* &= \frac{-I_\alpha}{\rho B^4 \omega_\alpha^2} k_{22}^{ad} \\ H_4^* &= \frac{-m_h}{\rho B^2 \omega_h^2} k_{11}^{ad} & A_4^* &= \frac{-I_\alpha}{\rho B^3 \omega_h^2} k_{21}^{ad} \end{aligned} \tag{7.15}$$

where c_{ij}^{ad} and k_{ij}^{ad} $(i, j = 1, 2)$ are the elements of the aerodynamic damping and stiffness matrices $\mathbf{C}^{\mathrm{ad}}$ and $\mathbf{K}^{\mathrm{ad}}$, respectively.

It can be seen that the most important step in this identification process is to identify $\mathbf{C}^{\mathrm{e}}$ and $\mathbf{K}^{\mathrm{e}}$ from the coupled vibration time-histories of the section model. The unifying least-squares method is introduced below.

Equation 7.12 can be written in the state space with the matrix $\mathbf{A} = \begin{pmatrix} \mathbf{0} & \mathbf{I} \\ -\mathbf{K}^{\mathrm{e}} & -\mathbf{C}^{\mathrm{e}} \end{pmatrix}$, being the state matrix of the coupled wind-structure system. The real parts of the eigenvalues of $\mathbf{A}$ represent the damping of the system, while the imaginary parts of the eigenvalues of $\mathbf{A}$ represent the frequency of the system.

For the 2-DOF section model, the complex eigenvalues λ_r and eigenvectors $\boldsymbol{\Psi}_r$ of $\mathbf{A}$ can be written as:

$$\lambda_r = \sigma_r + i\beta_r,\ \lambda_r^* = \sigma_r - i\beta_r \quad (r = 1,\ 2) \tag{7.16a}$$

$$\boldsymbol{\Psi}_r = \begin{Bmatrix} \boldsymbol{\phi}_r \\ \lambda_r \boldsymbol{\phi}_r \end{Bmatrix},\ \boldsymbol{\phi}_r = \begin{Bmatrix} \phi_{hr} \\ \phi_{\alpha r} \end{Bmatrix} = \begin{Bmatrix} U_{hr} + iV_{hr} \\ U_{\alpha r} + iV_{\alpha r} \end{Bmatrix} \quad (r = 1,\ 2) \tag{7.16b}$$

where:

σ_r and β_r are the real and imaginary parts of the rth eigenvalue;
U_{hr}, V_{hr}, $U_{\alpha r}$ and $U_{\alpha r}$ are the coefficients of the rth eigenvector.

Finally, the total damping and stiffness matrices of the coupled wind-structure system can be expressed in terms of the complex dynamic characteristics as follows:

$$[\mathbf{K}^e \quad \mathbf{C}^e] = -[\boldsymbol{\phi}\boldsymbol{\Lambda} \quad \boldsymbol{\phi}^*\boldsymbol{\Lambda}^{*2}]\begin{bmatrix} \boldsymbol{\phi} & \boldsymbol{\phi}^* \\ \boldsymbol{\phi}\boldsymbol{\Lambda} & \boldsymbol{\phi}^*\boldsymbol{\Lambda}^* \end{bmatrix}^{-1} \tag{7.17a}$$

$$\boldsymbol{\phi} = [c_1\boldsymbol{\phi}_1 \quad c_2\boldsymbol{\phi}_2]; \quad \boldsymbol{\Lambda} = \begin{bmatrix} \lambda_1 & \\ & \lambda_2 \end{bmatrix} \tag{7.17b}$$

where c_1 and c_2 are the two constants to be determined by the initial conditions of vibration of the section model.

Furthermore, the theoretical displacement responses of the 2-DOF section model can be expressed as:

$$\begin{aligned} \hat{h}_m &= \sum_{r=1}^{2} e^{\sigma_r m\Delta t}[U_{hr}\cos(\beta_r m\Delta t) + V_{hr}\sin(\beta_r m\Delta t)] \\ \hat{\alpha}_m &= \sum_{r=1}^{2} e^{\sigma_r m\Delta t}[U_{\alpha r}\cos(\beta_r m\Delta t) + V_{\alpha r}\sin(\beta_r m\Delta t)] \end{aligned} \tag{7.18}$$

where:

m is the number of measured time points;
Δt is the time increment.

The errors between the theoretical and measured response time-histories can be written as:

$$\begin{aligned} \mathbf{e}_h^T &= \left\{h_1 - \hat{h}_1, h_2 - \hat{h}_2, \ldots, h_m - \hat{h}_m\right\} \\ \mathbf{e}_\alpha^T &= \{\alpha_1 - \hat{\alpha}_1, \alpha_2 - \hat{\alpha}_2, \ldots, \alpha_m - \hat{\alpha}_m\} \end{aligned} \tag{7.19}$$

Because of the possible large difference between the absolute values of vertical and torsional displacements, a unifying error function is introduced here:

$$J = \mathbf{e}_h^T\mathbf{e}_h + \mathbf{e}_\alpha^T\mathbf{e}_\alpha \tag{7.20}$$

Iterative steps of the unifying least-squares method are required to obtain the optimal value of σ_r and β_r, which makes J value smaller than an allowable value. The initial value of σ_r and β_r can be obtained by the modified Ibrahim time-domain method [46].

7.8.2 Forced Vibration Test of Section Model

In a forced vibration test, the section model is driven by a specially designed device to perform a simple harmonic vibration with designated frequency and amplitude. Self-excited forces, displacement and acceleration responses of the section model are directly measured. Flutter derivatives can then be identified, based on the measured data in either the frequency domain or the time domain. Compared with the free vibration method, the forced vibration method has advantages of simplicity in the identification algorithm and accuracy in identification results, particularly in high reduced velocity, but it requires special testing devices. The testing devices commonly used include the SDOF device for vertical vibration or torsional vibration and the 2DOF device for coupled vertical and torsional vibration. Only the frequency domain method for identifying torsional flutter derivatives from forced vibration tests with a torsional vibration device is introduced in this subsection [25]. Other flutter derivatives can be identified with a similar procedure.

The section model is forced to have a torsional vibration as:

$$\alpha(t) = \alpha_0 e^{i(\omega_\alpha t + \varphi_\alpha)} \tag{7.21}$$

where:

$\alpha(t)$ is the time history of torsional displacement;
α_0 is the vibration amplitude;
φ_a is the phase angle.

The total dynamic forces acting on the section model include several parts:

$$\begin{aligned} L_{dy}(\alpha, \dot{\alpha}, \ddot{\alpha}) &= L_{se}(\alpha, \dot{\alpha}) + L_{no} \\ D_{dy}(\alpha, \dot{\alpha}, \ddot{\alpha}) &= D_{se}(\alpha, \dot{\alpha}) + D_{no} \\ M_{dy}(\alpha, \dot{\alpha}, \ddot{\alpha}) &= M_{se}(\alpha, \dot{\alpha}) + M_{in}(\ddot{\alpha}) + M_{no} \end{aligned} \tag{7.22}$$

where:

the subscript *dy* denotes the total dynamic force;
the subscript *se* denotes the self-excited force;
the subscript *in* denotes the inertial force;
the subscript *no* denotes noise.

The self-excited force can be obtained by the total dynamic force minus the total dynamic force identified at zero wind speed:

$$\begin{aligned} L_{se}(f_\alpha) &= L_{dy}(f_\alpha) - L_{dy,0}(f_\alpha) \\ D_{se}(f_\alpha) &= D_{dy}(f_\alpha) - D_{dy,0}(f_\alpha) \\ M_{se}(f_\alpha) &= M_{dy}(f_\alpha) - M_{dy,0}(f_\alpha) \end{aligned} \tag{7.23}$$

where:

the subscript *dy,0* denotes the total dynamic force identified at zero wind speed;
f_α is the torsional vibration frequency.

Equation 7.23 can be rewritten in complex form as:

$$\begin{aligned} \mathrm{Re}[L_{se}(f_\alpha)] &= \mathrm{Re}[L_{dy}(f_\alpha)] - \mathrm{Re}[L_{dy,0}(f_\alpha)] \\ \mathrm{Im}[L_{se}(f_\alpha)] &= \mathrm{Im}[L_{dy}(f_\alpha)] - \mathrm{Im}[L_{dy,0}(f_\alpha)] \\ \mathrm{Re}[D_{se}(f_\alpha)] &= \mathrm{Re}[D_{dy}(f_\alpha)] - \mathrm{Re}[D_{dy,0}(f_\alpha)] \\ \mathrm{Im}[D_{se}(f_\alpha)] &= \mathrm{Im}[D_{dy}(f_\alpha)] - \mathrm{Im}[D_{dy,0}(f_\alpha)] \\ \mathrm{Re}[M_{se}(f_\alpha)] &= \mathrm{Re}[M_{dy}(f_\alpha)] - \mathrm{Re}[M_{dy,0}(f_\alpha)] \\ \mathrm{Im}[M_{se}(f_\alpha)] &= \mathrm{Im}[M_{dy}(f_\alpha)] - \mathrm{Im}[M_{dy,0}(f_\alpha)] \end{aligned} \tag{7.24}$$

where Re and Im refer to the real and imaginary parts of the force, respectively.

The corresponding three-dimensional aeroelastic governing equations of the self-excited forces can be expressed in complex form as:

$$
\begin{aligned}
\mathrm{Re}[L_{se}(f_\alpha)] &= \rho\overline{U}^2 B[K_\alpha^2 H_3^*(K_\alpha)\alpha_o] \\
\mathrm{Im}[L_{se}(f_\alpha)] &= \rho\overline{\overline{U}}^2 B[K_\alpha^2 H_2^*(K_\alpha)\alpha_o] \\
\mathrm{Re}[D_{se}(f_\alpha)] &= \rho\overline{U}^2 B[K_\alpha^2 P_3^*(K_\alpha)\alpha_o] \\
\mathrm{Im}[D_{se}(f_\alpha)] &= \rho\overline{U}^2 B[K_\alpha^2 P_2^*(K_\alpha)\alpha_o] \\
\mathrm{Re}[M_{se}(f_\alpha)] &= \rho\overline{U}^2 B[K_\alpha^2 A_3^*(K_\alpha)\alpha_o] \\
\mathrm{Im}[L_{se}(f_\alpha)] &= \rho\overline{U}^2 B[K_\alpha^2 A_2^*(K_\alpha)\alpha_o]
\end{aligned}
\tag{7.25}
$$

where:

H_i^*, $P_i^*$and $A_i^*(i=2, 3)$ are the flutter derivatives related to torsional vibration;
$K_\alpha = B\omega_\alpha/\overline{U}$ is the reduced frequency.

Comparing Equation 7.24 with Equation 7.25 yields the following equations for torsional flutter derivatives:

$$
\begin{aligned}
H_2^*(K_\alpha) &= \frac{1}{\rho B^3\alpha_0\omega_\alpha^2}\{\mathrm{Im}[L_{dy}(f_\alpha)] - \mathrm{Im}[L_{dy,0}(f_\alpha)]\} \\
H_3^*(K_\alpha) &= \frac{1}{\rho B^3\alpha_0\omega_\alpha^2}\{\mathrm{Re}[L_{dy}(f_\alpha)] - \mathrm{Re}[L_{dy,0}(f_\alpha)]\} \\
P_2^*(K_\alpha) &= \frac{1}{\rho B^3\alpha_0\omega_\alpha^2}\{\mathrm{Im}[D_{dy}(f_\alpha)] - \mathrm{Im}[D_{dy,0}(f_\alpha)]\} \\
P_3^*(K_\alpha) &= \frac{1}{\rho B^3\alpha_0\omega_\alpha^2}\{\mathrm{Re}[D_{dy}(f_\alpha)] - \mathrm{Re}[D_{dy,0}(f_\alpha)]\} \\
A_2^*(K_\alpha) &= \frac{1}{\rho B^4\alpha_0\omega_\alpha^2}\{\mathrm{Im}[M_{dy}(f_\alpha)] - \mathrm{Im}[M_{dy,0}(f_\alpha)]\} \\
A_3^*(K_\alpha) &= \frac{1}{\rho B^4\alpha_0\omega_\alpha^2}\{\mathrm{Re}[M_{dy}(f_\alpha)] - \mathrm{Re}[M_{dy,0}(f_\alpha)]\}
\end{aligned}
\tag{7.26}
$$

7.8.3 Free Vibration Test of Taut Strip Model and Full Aeroelastic Model

Unlike the identification methods for section models, the identification of flutter derivatives with a taut strip model or a full aeroelastic model involves bridge vibration of multiple modes and three-dimensional characteristics of wind-bridge interaction.

The generalized self-excited force of the *i*th mode due to the motion of the *j*th mode can be expressed in the frequency domain as:

$$
\overline{F}_{ij}(x) = \int_0^L F_{ij}(x)\phi_i(x)dx \tag{7.27}
$$

where:

subscripts i and j denotes the ith and jth mode respectively;
F_{ij} is the self-excited force of the ith mode due to the motion of the jth mode;
ϕ_i is the ith modal shape;
L is the length of the bridge deck.

It can be demonstrated that Equation 7.27 can be expanded to [48]:

$$\begin{aligned}\overline{F}_{ij}(x) &= \int_0^L \left(c_{ij}^{ad}\dot{q}_j + k_{ij}^{ad}q_j\right)\phi_j(x)\phi_i(x)dx \\ &= \left(c_{ij}^{ad}\dot{q}_j + k_{ij}^{ad}q_j\right)\int_0^L \phi_j(x)\phi_i(x)dx\end{aligned} \tag{7.28}$$

where c_{ij}^{ad} and k_{ij}^{ad} are the aerodynamic damping and stiffness coefficients, respectively, corresponding to the ith mode due to the motion of jth mode ($i, j = \alpha, h, p$).

Similar to free vibration section model tests, the system damping and stiffness matrices, $\mathbf{C}^{\mathrm{e}}$ and $\mathbf{K}^{\mathrm{e}}$, are obtained first using modal identification methods such as ERA. The aerodynamic damping and stiffness matrices, $\mathbf{C}^{\mathrm{ad}}$ and $\mathbf{K}^{\mathrm{ad}}$, can then be obtained by deducting the structural damping and stiffness matrices, $\mathbf{C}^0$ and $\mathbf{K}^0$, from the total damping and stiffness matrices, $\mathbf{C}^{\mathrm{e}}$ and $\mathbf{K}^{\mathrm{e}}$. As a result, the aerodynamic damping and stiffness coefficients can be obtained as:

$$c_{ij}^{ad} = \frac{c_{ij}^e - c_{ij}^0}{\int_0^L \phi_j(x)\phi_i(x)dx} \tag{7.29}$$

$$k_{ij}^{ad} = \frac{k_{ij}^e - k_{ij}^0}{\int_0^L \phi_j(x)\phi_i(x)dx} \tag{7.30}$$

where:

c_{ij}^e and c_{ij}^0 are the system and structural damping coefficient, respectively;
k_{ij}^e and k_{ij}^0 are the system and structural stiffness coefficients, respectively.

Once the aerodynamic damping and stiffness coefficients are obtained, a total of 18 equations, which are similar to Equation 7.15, can be used to find the 18 flutter derivatives.

7.9 Identification of Aerodynamic Admittance

As mentioned in subsection 7.5.5, aerodynamic admittance can be estimated by directly measuring the aerodynamic forces of a stationary section model [28–30], or can be extracted from the steady random responses of the aeroelastic model of the bridge deck by system identification techniques [31–33].

As discussed in Chapter 4, the buffeting forces can be expressed as:

$$
\begin{aligned}
L_b &= \frac{1}{2}\rho\overline{U}^2 B\left[2C_L\chi_{Lu}\frac{u}{\overline{U}} + (C_L' + C_D)\chi_{Lw}\frac{w}{\overline{U}}\right] \\
D_b &= \frac{1}{2}\rho\overline{U}^2 B\left[2C_D\chi_{Du}\frac{u}{\overline{U}} + C_D'\chi_{Dw}\frac{w}{\overline{U}}\right] \\
M_b &= \frac{1}{2}\rho\overline{U}^2 B^2\left[2C_M\chi_{Mu}\frac{u}{\overline{U}} + C_M'\chi_{Mw}\frac{w}{\overline{U}}\right]
\end{aligned}
\tag{7.31}
$$

where:

C_L, C_D and C_M are the lift, drag, and moment coefficients referred to the deck width B, respectively;
$C_L' = dC_L/d\alpha$, $C_D' = dC_D/d\alpha$, and $C_M' = dC_M/d\alpha$; χ_{Lu}, χ_{Lw}, χ_{Du}, χ_{Dw}, χ_{Mu}, χ_{Mw} are the aerodynamic admittance functions, which are functions of the reduced frequency and dependent on the geometrical configuration of the cross-section of the bridge deck;
and u and w are the longitudinal and vertical fluctuating winds, respectively.

By neglecting the cross-spectra of the fluctuating winds, Equation 7.31 can be expressed in the frequency domain as:

$$
\begin{aligned}
S_{LL}(\omega) &= \rho^2 U^2 B^2\left[C_L^2 S_{uu}(\omega)|\chi_{Lu}(\omega)|^2 + \frac{1}{4}(C_L' + C_D)^2 S_{ww}(\omega)|\chi_{Lw}(\omega)|^2\right] \\
S_{DD}(\omega) &= \rho^2 U^2 B^2\left[C_D^2 S_{uu}(\omega)|\chi_{Du}(\omega)|^2 + \frac{1}{4}C_D'^2 S_{ww}(\omega)|\chi_{Dw}(\omega)|^2\right] \\
S_{MM}(\omega) &= \rho^2 U^2 B^4\left[C_M^2 S_{uu}(\omega)|\chi_{Mu}(\omega)|^2 + \frac{1}{4}C_M'^2 S_{ww}(\omega)|\chi_{Mw}(\omega)|^2\right]
\end{aligned}
\tag{7.32}
$$

where:

S_{LL}, S_{DD}, and S_{MM} are the power spectrum density (PSD) functions of the vertical, longitudinal and torsional buffeting forces respectively;
S_{uu} and S_{ww} are the PSD functions of the longitudinal and vertical wind fluctuations, respectively.

There are six aerodynamic admittance functions in Equation 7.32, but there are only three equations. Some assumptions have to be made, and the following is a common one.

$$
\begin{aligned}
\chi_{Lu}(\omega) &= \chi_{Lw}(\omega) = \chi_L(\omega) \\
\chi_{Du}(\omega) &= \chi_{Dw}(\omega) = \chi_D(\omega) \\
\chi_{Mu}(\omega) &= \chi_{Mw}(\omega) = \chi_M(\omega)
\end{aligned}
\tag{7.33}
$$

where $\chi_L(\omega)$, $\chi_D(\omega)$ and$\chi_M(\omega)$ are called the equivalent aerodynamic admittance functions in the respective direction.

As a result, the equivalent aerodynamic admittance functions can be estimated by the following equations:

$$
\begin{aligned}
\chi_L^2(\omega) &= \frac{4S_{LL}(\omega)}{(\rho U^2 B)^2\left[4C_L^2(S_{uu}(\omega)/U^2) + (C_L' + C_D)^2(S_{ww}(\omega)/U^2)\right]} \\
\chi_D^2(\omega) &= \frac{4S_{DD}(\omega)}{(\rho U^2 B^2)^2\left[4C_D^2(S_{uu}(\omega)/U^2) + C_D'^2(S_{ww}(\omega)/U^2)\right]} \\
\chi_M^2(\omega) &= \frac{4S_{MM}(\omega)}{(\rho U^2 B^2)^2\left[4C_M^2(S_{uu}(\omega)/U^2) + C_M'^2(S_{ww}(\omega)/U^2)\right]}
\end{aligned}
\tag{7.34}
$$

For aerodynamic admittance identification by force balance technique, buffeting forces can be directly measured and the force spectra can then be obtained. Together with the measured wind spectra and force coefficients, the equivalent aerodynamic admittance functions can be achieved. For aerodynamic admittance identification by pressure measurements, buffeting forces can be obtained by integration of the pressures over the section model surface.

For aerodynamic admittance identification with aeroelastic section model, buffeting forces cannot be obtained directly. In this case, the system stiffness matrix $\mathbf{K}^e$ and the system damping matrix $\mathbf{C}^e$ need to be identified first; the PSD functions of the aerodynamic forces $S_{ff}(\omega)$ can then be obtained by the following equation:

$$S_{ff}(\omega) = Z(\omega)S_{yy}(\omega)Z^{\mathrm{T}}(\omega) \tag{7.35}$$

where:

$Z(\omega) = K^e + j\omega C^e - \omega^2 M$ is the impedance matrix of the system;
S_{yy} is the PSD function of the model response.

Again, an important task here is to identify the system stiffness matrix $\mathbf{K}^e$ and the system damping matrix $\mathbf{C}^e$. Buffeting forces are random functions of time. As a result, unlike identifying $\mathbf{K}^e$ and $\mathbf{C}^e$ in smooth wind flow, the identification of system characteristic matrices from model response data in turbulent flow can be seen as a typical inverse problem in the theory of random vibration, and it can be solved by stochastic system identification techniques [45,49–51]. Among all these methods, the stochastic subspace identification (SSI) technique [49–51] has proven appropriate for structural engineering application.

7.10 Cable Model Tests

As discussed in Chapter 5, stay cables in cable-stayed bridges are susceptible to vibration due to their long length and their low stiffness and damping. Excessive vibration of stay cables has been observed, particularly in the combined environment of wind and rain. Many studies have thus been carried out to explore the mechanism of rain-wind-induced cable vibration, such as field measurements [52–54], wind tunnel tests [55–58] and analytical studies [59–62].

It was concluded that excessive rain-wind-induced cable vibration could occur on the down-inclined stay cables along wind direction, with a certain wind yaw angle at a mean wind speed ranging from 10 m/s to 15 m/s. The main reason is that a rivulet is formed along the upper surface of a cable in the rain-wind condition. The original cable cross-section is therefore changed, leading to a different aerodynamic characteristic for the cable.

The common methods of studying rain-wind-induced cable vibration can be classified into two types. The first is to measure the aerodynamic forces acting on the cable in consideration of the effect of rivulet, and then to calculate the dynamic responses of the cables through an analytical model. In this way, the rivulet is usually modeled with some solid material called "artificial rivulet" in order to consider the influence of rivulet on the aerodynamic forces of the cable. This way is also called the dry cable test. The second method is to observe rain-wind-induced cable vibration directly after modeling the rainfall in a wind tunnel. In this section, the two methods are introduced, based on the work presented in [63] and [64] respectively.

7.10.1 Inclined Dry Cable Tests

The dry cable test involves obtaining the aerodynamic forces or force coefficients of a stay cable in consideration of the effect of rivulet. To ensure Reynolds number similarity and the same surface

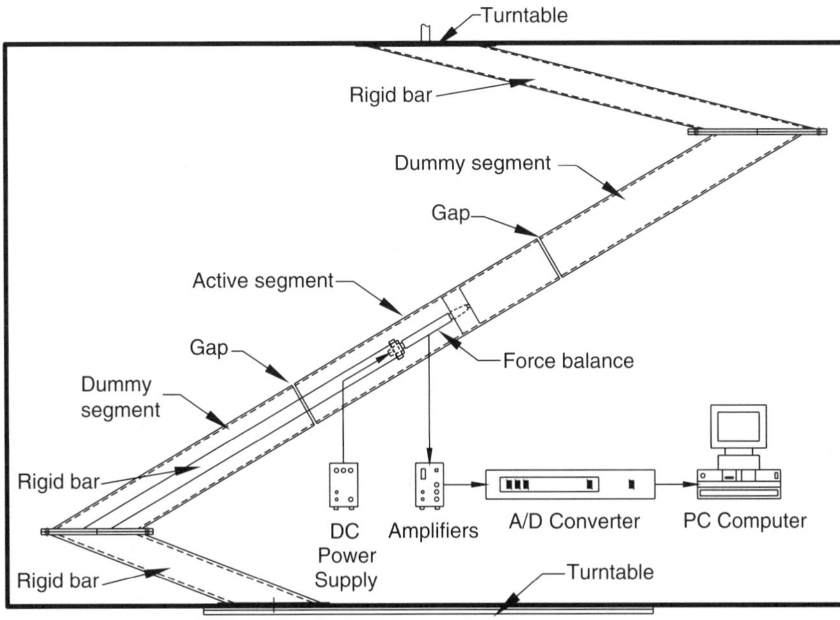

Figure 7.21 Cable test model and experimental set-up (from [63]) (Reproduced with permission from Multi-Science Publishing Co Ltd).

conditions as the prototype stay cables of a real bridge, the 1 : 1 cable models were manufactured as stainless steel cylinders covered with high-density polyethylene tube. Different cylinders with smooth surface configuration were manufactured with different diameters for the real bridge. One of the cable models is shown in Figure 7.21.

The total length of the cylinder (cable model) was about 3 m, and it was divided into three segments: the active segment was located in the middle, with a length of 1 m, and two dummy segments were distributed at the two ends. In the tests, the rivulet was simplified by using solid materials. The artificial rivulet was made of light wood and stuck on the smooth surface of the active segment along the cylinder axis (see Figure 7.22).

A special test apparatus (see Figure 7.21) was designed to take into consideration the effects of artificial rivulet and different space attitude of the cable. The bottom part of the test apparatus was first installed on the turntable of a wind tunnel, using high-strength bolts. The lower dummy cylinder segment of the cable model was then mounted on the test apparatus. The active cylinder

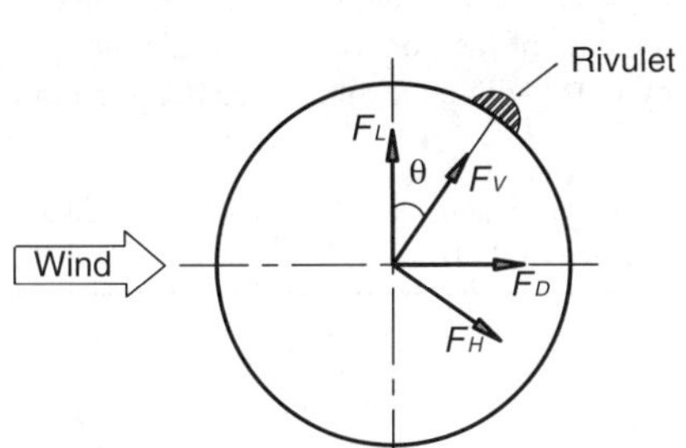

Figure 7.22 Position of rivulet and aerodynamic forces (from [63]) (Reproduced with permission from Multi-Science Publishing Co Ltd).

segment was mounted next to the lower dummy segment through a force balance, which was of high stiffness and sensitivity.

The aerodynamic forces acting on the active cylinder segment (see Figure 7.22) were therefore measured through the force balance, one end of which was connected to the active segment, while the other was supported by an inclined rigid bar of the test apparatus. A gear mechanism was installed at the top end of the inclined rigid bar, so that the active cylinder segment and the force balance could rotate synchronically around the cylinder axis for 360° with respect to the dummy segments, in order to facilitate adjustment of the position of the artificial rivulet. Finally, the upper dummy cylinder segment was installed next to the active cylinder segment, through a rigid steel tube which was mounted on the ceiling of the wind tunnel.

There was a small gap between the active segment and the lower segment and between the active segment and the upper segment, to ensure that only wind forces acting on the active segment were measured. The lower dummy segment and the upper dummy segment had the same diameter and types of surface and material as the active segment, to ensure a proper flow would be simulated at both the ends of the active segment. The ratio of length to diameter of the dummy segment was more than 5, so as to render the end effects unimportant.

The test apparatus was mounted in the high-speed working section of the CLP Wind Tunnel at the Hong Kong University of Science and Technology. The cross-section of the high speed working section is 3 m wide and 2 m high. Similar to the conventional measurement of aerodynamic coefficients of a bridge deck, smooth flow condition is considered. The two net screens were installed in the working section before the turntable to generate a uniform flow with turbulence intensity less than 1%. The test mean wind speed ranged from 10–16 m/s.

The instrumentation for measuring aerodynamic forces acting on the active cylinder segment was comprised of a force balance of six components, a DC power supply, a DC voltage amplifier of six channels, a NI PCI-6052E data acquisition board, a LabVIEW Full Development System software package and a PC computer working in the *Windows 2000* operating system. The force balance was a strain balance and its electric bridge was powered by a DC power supply and connected to a six-channel DC voltage amplifier. A temperature compensation system was built in the electric bridge to eliminate the temperature effects automatically. The voltage signals from the amplifier were transmitted to the PC computer through the data acquisition board. The signals were sampled at a frequency of 1000 Hz and duration of 30 seconds.

The drag and lift coefficients of test cylinders with and without an artificial rivulet were measured and compared against each other to see the effect of artificial rivulet on aerodynamic forces. Four cases were selected and these are listed in Table 7.4 for comparison. The orientation of the test cylinder is shown in Figure 7.23. The cylinder diameter and inclination selected were 160 mm and 30°, respectively. The yaw angle was selected as 0° and 35°: case 1 and case 2 with zero yaw angle, and case 3 and case 4 with a yaw angle of 35°. The artificial rivulet was of a half-elliptic cross-section, and the rivulet angle θ varied from −90° to 20°. The test wind speed for the four cases was the same, namely 14 m/s.

The aerodynamic force coefficients of the non-yawed cylinder are plotted in Figure 7.24a for case 1 and case 2, while those of the cylinder with a yaw angle are depicted in Figure 7.24b for cases 3 and 4.

Table 7.4 Four test cases selected for investigating the effect of artificial rivulet (from [63]) (Reproduced with permission from Multi-Science Publishing Co Ltd)

Case No.	α (deg)	β (deg)	Θ (deg)	U (m/s)	D (mm)	Rivulet
1	30	0	−90~20	14	160	Without
2	30	0	−90~20	14	160	With
3	30	35	−90~20	14	160	Without
4	30	35	−90~20	14	160	With

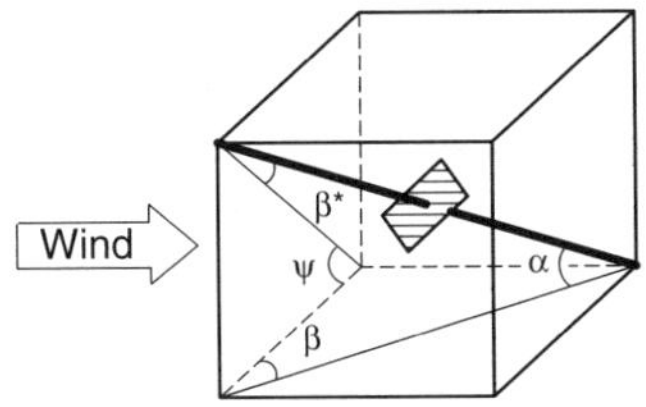

Figure 7.23 Orientation of test stay cable (from [63]) (Reproduced with permission from Multi-Science Publishing Co Ltd).

The drag and lift coefficients of the cylinder without artificial rivulet remain constant and are not a function of angle θ, in theory. The drag coefficient of the non-yawed cylinder without artificial rivulet is close to 0.95, but the lift coefficient is close to zero. The drag and lift coefficients of the yawed cylinder without artificial rivulet are close to 0.80 and 0.65, respectively.

With the artificial rivulet, the drag and lift coefficients of the cylinder no longer remain constant and vary with the position of the rivulet. For the non-yawed cylinder, the maximum and minimum drag

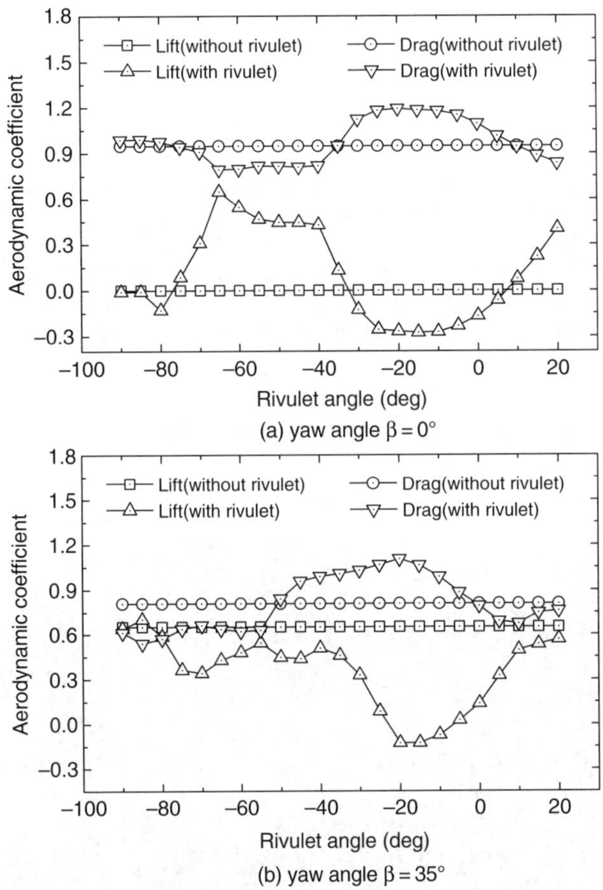

Figure 7.24 Comparison of aerodynamic coefficients of cylinder with and without rivulet (from [63]) (Reproduced with permission from Multi-Science Publishing Co Ltd).

coefficients are 1.20 and 0.78, respectively, while the maximum and minimum lift coefficients are 0.68 and −0.30, respectively. For the yawed cylinder, the maximum and minimum drag coefficients are 1.10 and 0.52, respectively, while the maximum and minimum lift coefficients are 0.70 and −0.15, respectively.

Note in particular that the negative slopes of the lift coefficient appear for the non-yawed cylinder when the rivulet angle θ ranges from −85° to −80°, and then from −65° to −15°, and for the yawed cylinder when the rivulet angle θ ranges from −85° to −70°, then from −55° to −45°, and finally from −40° to −20°. The occurrence of the negative slope of lift coefficient indicates the potential wind-rain-induced cable vibration. Clearly, the existence of artificial rivulet changes aerodynamic coefficients significantly. More results and discussion on this topic can be found in [63].

7.10.2 Rain-Wind Simulation of Inclined Stay Cable

Different from the dry cable test, the dynamics of a stay cable should be modeled in a rain-wind tunnel test. Figure 7.25 shows the model of a section of the prototype stay cable.

The cable section model was designed and made of 1 : 1 scale stainless steel circular pipe, covered with high-density polyethylene. It was then pulled at its two ends by two pre-tensioned springs. The far end of each spring was connected to a steel frame via a steel arc track. The tension forces in the springs could be changed by adjusting the spring lengths and monitored by a load cell. Thus, the targeted natural frequency of the prototype cable could be obtained by adjusting the tension force, and the restoring force of the cable model actually came from the tension force, which was similar to the prototype cable in principle. Clearly, such an arrangement could simulate the vibration of cable model in both vertical and horizontal directions. The inclination angle, α, of the cable model could also be changed easily by moving the sliding blocks to which the springs were connected along the arc tracks. The steel frame alone was capable of rotation around the vertical axis at its middle section, so as to change wind yaw angle, β, of the cable model.

The tests were carried out in the low-speed working section of the CLP Wind Tunnel at the Hong Kong University of Science and Technology. A new contraction section was designed and installed to convert the original cross-section of $5.0 \times 4.0\,\text{m}^2$ (W × H) into a working cross-section of $2.0 \times 3.0\,\text{m}^2$

Figure 7.25 Setup of rain-wind-induced cable vibration (from [64]) (Reproduced with permission from Elsevier).

(W × H), so as to increase wind speed and form an open jet to facilitate wind/rain-wind-induced cable vibration tests. Mean wind speeds and turbulence intensities were measured at the cross-section of 2 m downstream from the end of the contraction section. The maximum turbulence intensity in this section was less than 7%, and the average mean wind speed ranged from 0 to 16 m/s. A water-spraying system was used to simulate rainfall (see Figure 7.25). A water-adding device was attached to the upper end of the cable model to simulate the upstream rivulet running down along the cable. The water flow rate was measured by a flow meter and could be adjusted by a control valve. The rainfall direction could also be changed by rotating the orientation of the nozzles of the water-spraying system. The added water was set at 0.31 L/min for wind-rain induced cable vibration tests, and the rainfall intensity was set at 0.2 to 1.8 L/min/m^2 to investigate the effects of rainfall intensity on wind-rain-induced cable vibration.

Two triaxial piezoelectric accelerometers (BK4321), protected against water, were installed, one on each end of the cable model, to measure the cable responses in the vertical (y) and horizontal (x) directions. The signals from the sensors were amplified and converted to cable displacement by the signal conditioners BK2692 and transmitted to the computer data acquisition system DEWESoft 6 for data analysis. A load cell was installed on the top end of the upper spring to measure the tension force in the spring. A video photographer was installed to capture the vibration of cable model. A digital camera was set near the lower end of the model to capture the upper rivulet movements on the cable model. A blue grid was marked on the lower end of the cable model to record approximately the positions of the rivulet.

The cable model was tested with an inclination α of 30° and a yaw angle of 35° with simulated rain. The added water was set at 0.31 L/min and the rainfall was set at 0.16 L/min. The area of rainfall was taken as 0.8 m^2, and the rainfall intensity was set as 0.2 L/min/m^2. Figure 7.26 shows the maximum vertical amplitude of the cable model plotted against the mean wind speed, ranging from 9.0 to 14.0 m/s. The corresponding horizontal amplitude of the cable model is also depicted in Figure 7.26.

It can be seen that wind-rain-induced cable vibration is a kind of velocity-restricted and amplitude-restricted vibration. The horizontal amplitude of the cable model is much smaller than the vertical amplitude. Figure 7.27a shows the first 110 seconds of time history of vertical displacement of the cable model at a mean wind speed of 9.59 m/s. It can be seen that the vibration amplitude increases with time, and the vibration exhibits a sine wave of increasing amplitude. By selecting several time points with different vibration amplitudes within the time history, a few oscillating curves can be obtained and the total damping ratio and aerodynamic damping ratio can be estimated. The calculated aerodynamic

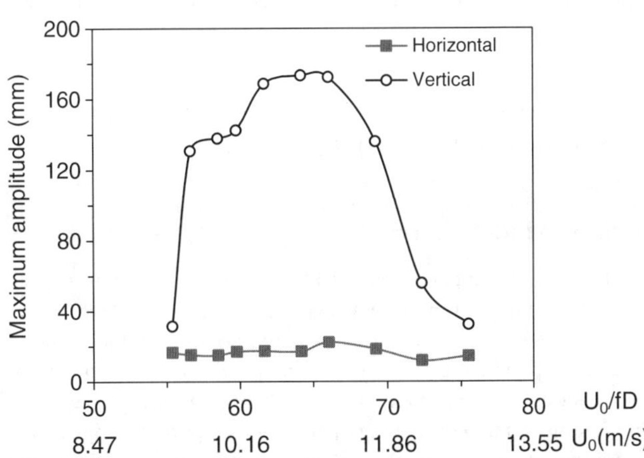

Figure 7.26 Rain-wind-induced cable vibration ($\alpha = 30°$, $\beta = 35°$) (from [64]) (Reproduced with permission from Elsevier).

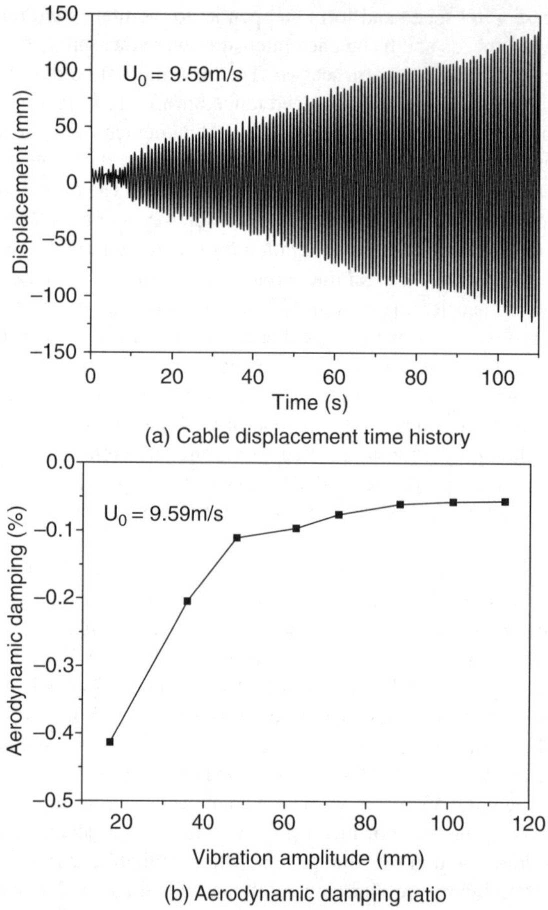

Figure 7.27 Displacement time history and aerodynamic damping ratio in vertical plane (from [64]) (Reproduced with permission from Elsevier).

damping ratio is shown in Figure 7.27b; it is negative and ranges from −0.45% to −0.075%. More test cases and result discussions can be found in [64].

7.11 Vehicle-Bridge Model Tests

Vehicles may become unsafe in wind conditions. Vehicle safety under winds, therefore, is an important topic in wind engineering. A typical report about wind-induced vehicle accidents is the post-disaster investigation after a major storm in the UK [65], where about 400 wind-induced vehicle accidents occurred. Later on, Baker and his colleagues conducted a series of wind tunnel experiments to study the aerodynamic forces acting on the ground vehicle [66–69]. With a growing requirement for high-speed transportation systems, a number of long-span cable-supported bridges have been built worldwide. Some of these bridges are located in areas prone to high wind conditions, and the safety of road vehicles running on long span bridges is thus of increasing concern to the public [70,71]. To carry out safety analysis of vehicles running on long-span bridges in high winds, knowledge of aerodynamic forces and moments on the vehicles is required [72,73], and this could be obtained through wind tunnel investigations. In this section, the aerodynamic force measurements of stationary vehicle on the ground and on the bridge deck, as well as the moving vehicle on the bridge deck, are briefly introduced.

7.11.1 Vehicles on Ground

In consideration of currently-used vehicles in Hong Kong and mainland China, a double-decker bus, an articulated lorry, a medium truck and a light car were selected as four major types of vehicles to be investigated [74]. The geometric scale of the four vehicle models was set as 1 : 25 after a careful consideration of many factors, which included the size of the bridge deck model, the size of working section of the wind tunnel and the sensitivity and capacity of the force balance. The dimensions of the four vehicle models were then determined, and they are illustrated in Figure 7.28.

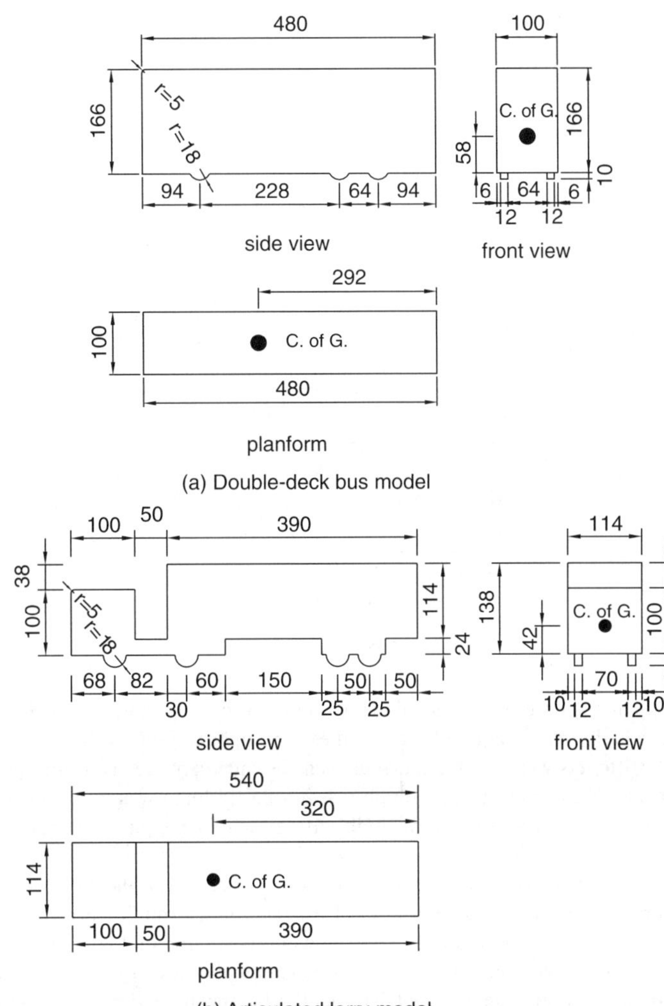

Figure 7.28 Four vehicle models of 1 : 25 scale (unit: mm) (from [74]) (Reproduced with permission from Elsevier).

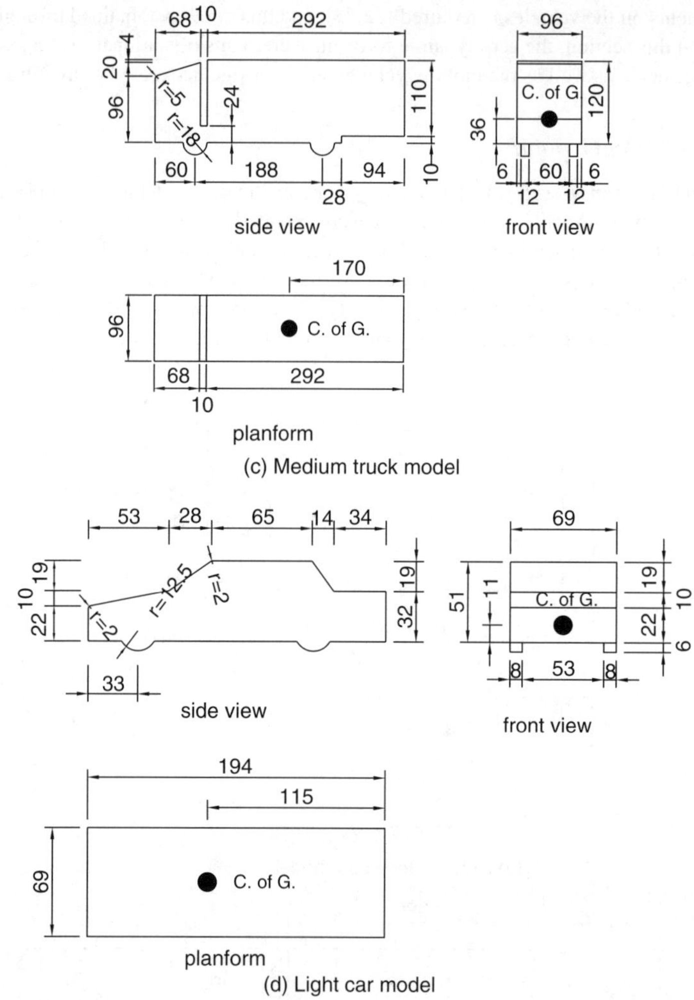

Figure 7.28 (*Continued*)

Since the force balance measurement technique was employed to measure aerodynamic forces on a vehicle model, the vehicle model required high stiffness and light weight. In this regard, high-quality wooden plates with stiffeners were used to make the vehicle undercarriage, as well as one wheel set to which the force balance was connected, and high-density plastic foam was used to make the vehicle body and other wheel sets, so as to achieve high stiffness, light weight and proper aerodynamic configuration.

Wind tunnel tests of vehicle models on the ground were carried out in the TJ-1 wind tunnel of the State Key Laboratory for Disaster Reduction in Civil Engineering at Tongji University in China. This wind tunnel is a boundary layer wind tunnel, having a working section of 1.8 m wide, 1.8 m high and 14 m long. The size of this wind tunnel is suitable for economically measuring aerodynamic forces on vehicle models without the bridge deck model. The achievable mean wind speed in the tunnel ranged from 0.5–30 m/s, and the wind speed could be adjusted continuously. The turbulence intensity in smooth flow tests was measured less than 1.0%.

(a) Vehicle model on a circular clapboard

(b) Force balance horizontally mounted in TJ-1 Wind Tunnel

Figure 7.29 Vehicle model and force balance mounted (from [74]) (Reproduced with permission from Elsevier).

To eliminate the boundary layer effect of the wind tunnel floor on the aerodynamic forces of a vehicle model, a circular-clapboard made of Perspex plate was mounted on the turntable at 0.4 m above the wind tunnel floor (see Figure 7.29a). Four steel posts and seven small steel bars were used to support the Perspex clapboard. A high-frequency force balance of six components was used to measure aerodynamic forces and moments on the vehicle model. The force balance was positioned horizontally underneath the clapboard, and it was connected to the turntable through a high-strength steel post to provide the sufficient stiffness to the balance (see Figure 7.29b). A special connecting part, of high stiffness and light weight, was designed to connect the wheel set of the vehicle model firmly to the balance.

A high-frequency six-components balance was used to measure the aerodynamic forces acting on the vehicle model. The electric bridge of the strain force balance was powered by a high-accuracy DC power supply and connected to an eight-channel DC voltage amplifier. A multi-channel data acquisition unit and a computer were used to digitize and record the measurement data. Calibration of the measurement system was performed before the test to obtain the calibration matrix of the force balance, reflecting the relationship between the applied forces/moments and the normalized output voltages.

Figure 7.30 shows the aerodynamic coefficients of the articulated lorry on the ground obtained from the present tests and Baker's tests [66]. Baker's tests were carried out in three test conditions:

- The first test (Case 1) was carried out with a low turbulence flow, a length scale of 1/25 and a Reynolds number of 2.4×10^5.
- The second test (Case 2) was conducted with a low turbulence flow, a length scale of 1/50 and a Reynolds number of 0.85×10^5.
- The third test condition (Case 3) included a normal turbulence flow of 10.6% turbulent intensity (I_u), a length scale of 1/50 and a Reynolds number of 0.88×10^5.

It can be seen from Figure 7.30 that the aerodynamic coefficients obtained from the present test and Baker's tests have similar variation pattern with wind yaw angle, except for the pitching moment coefficient (C_{MP}) from Case 1, which is different from the others when the yaw angle is greater than 50°. The side force coefficient (C_S) obtained from the present tests is very close to those gained from all three Baker's tests, in spite of different test conditions.

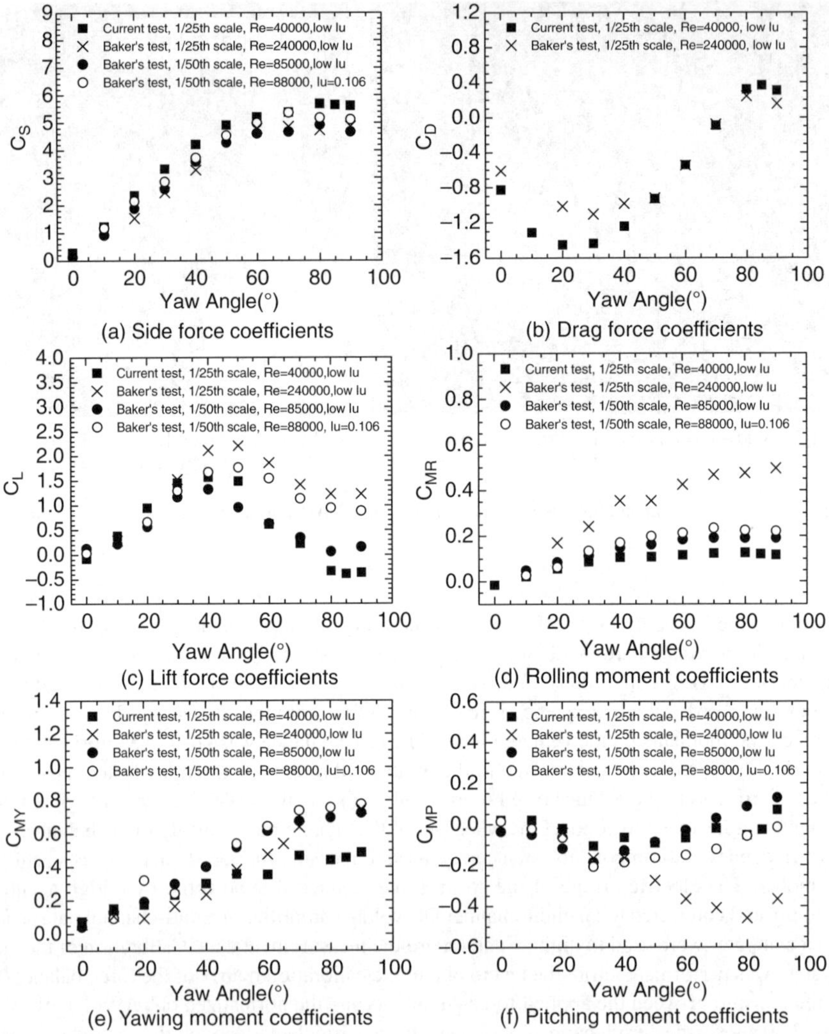

Figure 7.30 Comparison of aerodynamic coefficients of articulated lorry on ground (from [74]) (Reproduced with permission from Elsevier).

The side force coefficient (C_S) from the present test is slightly higher than those from the Baker's tests, especially after the yaw angle reaches 80°. This may be because the Reynolds number in the present test is smaller than those in the Baker tests. The drag coefficient (C_D) obtained from the present test is also consistent with the Baker test result of Case 1 for wind yaw angles between 50° and 90°, but it is more negative than the Baker result for the yaw angles below 50°. This may be due to the differences in the Reynolds number and the testing set-up.

The lift coefficient (C_L) and the yawing moment coefficient (C_{MY}) obtained from the present test are also close to those obtained from Baker's tests when the yaw angle is below 50°. However, they are smaller than the Baker results when the yaw angle is larger than 50°. The rolling moment coefficient (C_{MR}) and the pitching moment coefficient (C_{MP}) measured in the present test are well consistent with the results obtained from Baker's tests in Case 2 and Case 3, with a length scale of 1/50. However, they

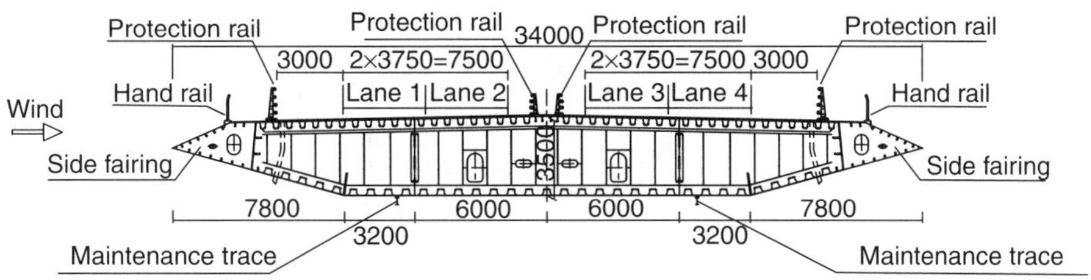

Figure 7.31 Cross-section of bridge deck with flat box girder (unit: mm) (from [74]) (Reproduced with permission from Elsevier).

deviate significantly from the Baker results for Case 1 with a length scale of 1/25, particularly when the yaw angle is larger than 50°. Further information can be found in [74].

7.11.2 Stationary Vehicle on Bridge Deck

The models of vehicles are the same ones described in subsection 7.11.1. Figure 7.31 shows a 1 : 25 deck model of a typical long-span highway bridge. The geometric scale of the deck section model was set as the same as the scale of the vehicle models. The prototype typical deck section is 34.0 m wide and 3.5 m high, carrying a dual two-lane highway on the upper surface. The deck section model was designed with a width of 1.36 m and a length of 5 m.

Since this investigation concerned aerodynamic forces of road vehicles on a bridge deck, the details of the bridge deck needed to be considered. In this regard, two sets of side fairing were utilized to enhance the aerodynamic performance of the bridge deck, and two lines of hand rail and four lines of protection rail, mounted on the bridge deck, were included in the wind tunnel investigation. Two lines of I-shape maintenance trace, with a height of 0.42 m and a width of 0.122 m, installed under the bottom surface of the deck, were also modeled. The deck's exterior appearance, including the side fairings, hand rails, protection rails and maintenance traces, was modeled in the light of the principle of geometric similarity.

Figure 7.32a shows the vehicle model and bridge deck section model with the details. To mount a vehicle model on the bridge deck via a force balance, the deck section model was designed as three segments along it lengthwise. Two end segments were made by using Perspex plates, while in the middle segment, a steel cross frame was introduced as the strong support to the force balance to make sure that the balance could perform well.

Wind tunnel tests of vehicle models on the bridge deck model were carried out in the TJ-3 wind tunnel of the State Key Laboratory for Disaster Reduction in Civil Engineering at Tongji University in China. The TJ-3 is a closed-type boundary layer wind tunnel with a working section of 15 × 2 × 14 m. The achievable mean wind speed in the tunnel ranged from 1.0 to 17.6 m/s. The bridge deck model was installed on the turntable of the wind tunnel at 1.14 m above the tunnel floor, and the deck model was supported by ten steel posts. Those steel posts also provided high horizontal stiffness for the deck model to ensure the measurement accuracy of aerodynamic forces on vehicle models.

As shown in Figure 7.32b, a high frequency force balance of six components was also used to measure aerodynamic forces and moments on the vehicle models for the deck case. It was mounted horizontally in the box of the deck model and connected firmly to the steel cross frame of the middle segment. The balance could be moved to pre-set positions for the concerned lanes (see Figure 7.32b) to measure the aerodynamic forces and moments on the vehicle in different lanes. The vehicle model was mounted on the perceiving end of the balance by using a Γ-shaped link and a ⊔-shaped link made of aluminum (Figure 7.32c). After the force balance with the Γ-shaped link was mounted, dismountable

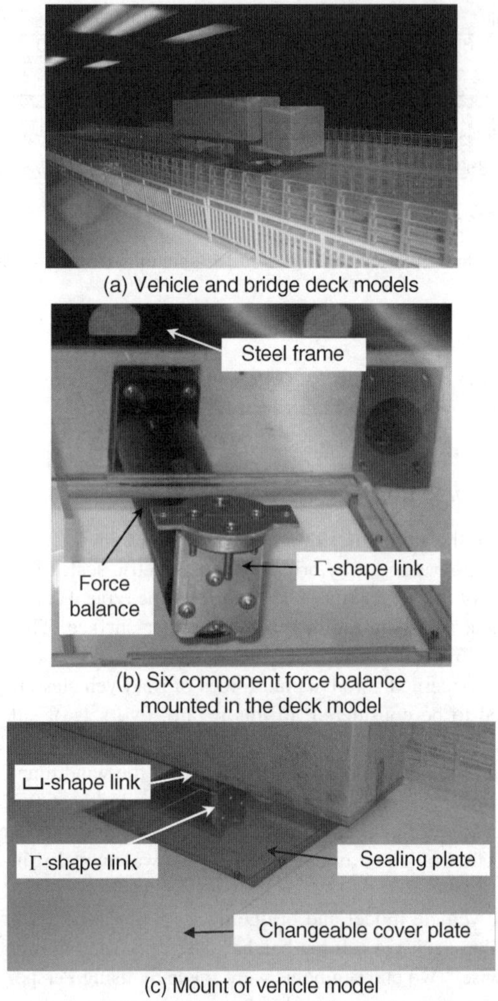

Figure 7.32 Vehicle and bridge deck models connected by force balance (from [74]) (Reproduced with permission from Elsevier).

plastic plates, including a changeable cover plate and a sealing plate, were used to enclose the steel frame, forming the complete middle segment of the bridge deck model. This arrangement avoided significant flow leakages between the bridge deck and the metal parts connecting the vehicle model to the force balance. The same measurement instruments were used as those presented in subsection 7.11.1.

Compared to the ground condition, the effects of bridge deck on the aerodynamic forces of vehicles come from two major aspects:

(a) The existence of the bridge deck changes the flow pattern around a vehicle.
(b) The existence of either handrails or protection parapets changes the windward flow of a vehicle.

Both effects could lower the mean wind velocity approaching to the vehicles and reduce side forces on the vehicles accordingly.

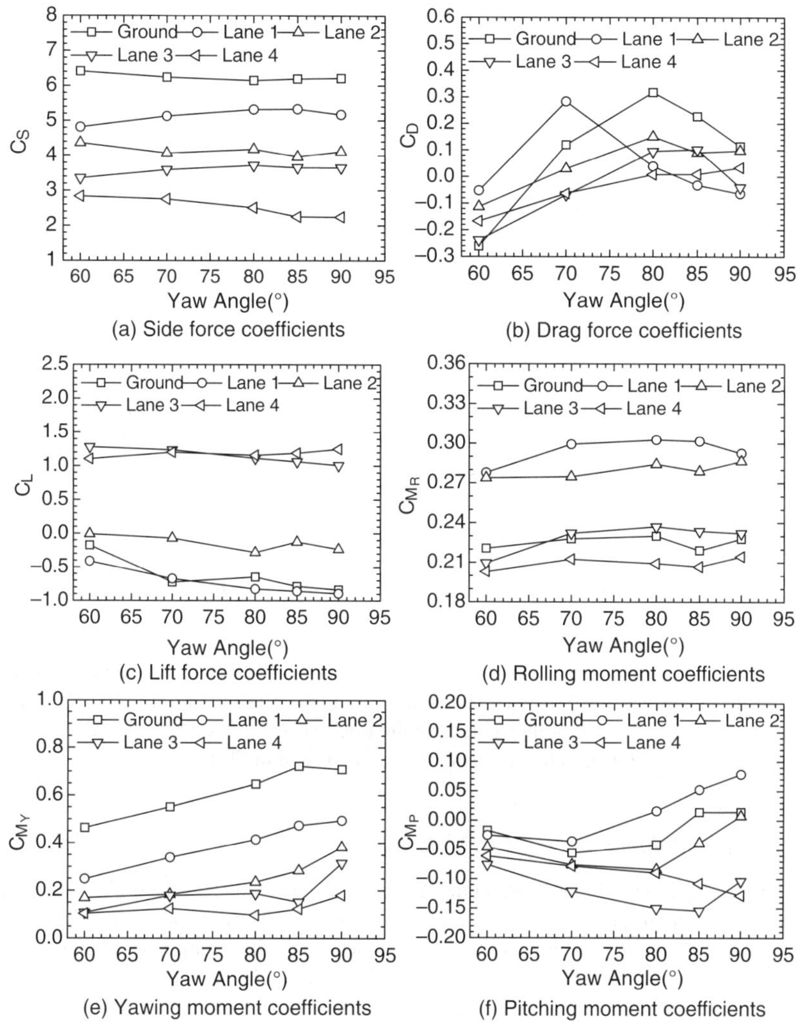

Figure 7.33 Aerodynamic coefficients of double-decker bus on deck (from [74]) (Reproduced with permission from Elsevier).

The six aerodynamic coefficients of the double-decker bus on each of four lanes of the bridge deck are depicted in Figure 7.33 for wind yaw angles ranging from 60° to 90°. The aerodynamic coefficient of the vehicle on the ground is also shown in the figure for comparison.

It can be seen that the side force coefficients of the vehicle on the bridge deck concerned are significantly smaller than those on the ground, as a result of the reduction effects of the bridge deck, rail and parapets. Moreover, the side force coefficients of the double-decker bus on the bridge deck decrease as expected when they are moved from a windward lane to a leeward lane across the bridge deck. The existence of the bridge deck also affects the drag force coefficients of the double-decker bus, but this is not so significant, because the vehicle is 4.4 m in height and the influence of the bridge deck, including rails and parapets, on the flows around the front and rear parts of the vehicle is relatively small. The lift force coefficients of the double-decker bus on the bridge deck are larger than those on the ground in general.

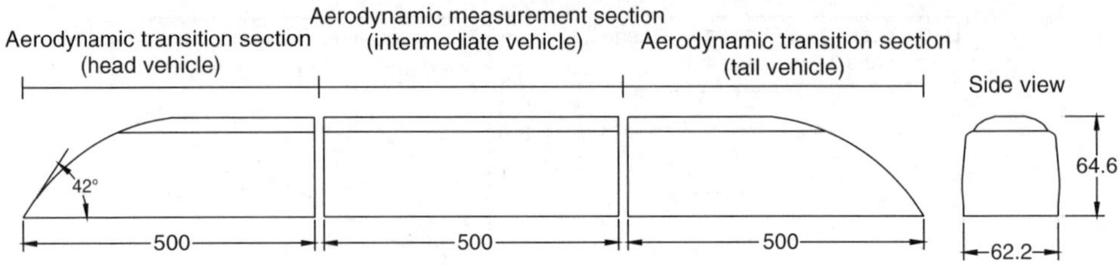

Figure 7.34 Schematic diagram of vehicle model (dimensions in mm).

The lift force coefficients of the vehicle on windward lanes (lane 1 and lane 2) are different from those on leeward lanes (lane 3 and lane 4). The lift force coefficients of the vehicle on leeward lanes are positive, while those on windward lanes may be negative. This may be due to the reduction of upward lift force acting on the vehicle chassis when the vehicle moves to lane 3 and lane 4.

The rolling moment coefficients of the vehicle on the windward lanes (lane 1 and lane 2) are larger than those on the ground. One of the major reasons for this phenomenon may be the significant reduction of mean wind speeds within the lower part of the vehicle, which may raise the action point of the resultant side force on the windward side of the vehicle. However, the changes of the rolling moment coefficients of the vehicle on the leeward lanes (lane 3 and lane 4) are relatively insignificant compared with those on the ground.

The rolling moment coefficients of the double-decker bus are positive, whether it is on the ground or on the bridge deck. The existence of the bridge deck obviously reduces the yawing moment coefficients of the double-decker bus on all four lanes, compared with those on the ground. The change pattern in the pitching moment coefficients of the vehicle on the bridge deck is not clear, and it depends on which lane the vehicle runs on. More detailed information can be found in [74].

7.11.3 Moving Vehicle on Bridge Deck

It is a challenging task to carry out wind tunnel tests on a moving vehicle on a bridge deck. Because of the size of a wind tunnel and the moving time of a vehicle, both vehicle and bridge models must be scaled down. Figure 7.34 shows a three-vehicle model, designed with a length of 0.5 m for each vehicle, to model rail transit vehicles such as trains and light rail vehicles [75].

The intermediate vehicle model was used to simulate the middle vehicle of a train, and the two vehicle models at the ends were used to simulate the head vehicle and tail vehicle of a train. As the head vehicle and tail vehicle correspond to the transition section of aerodynamic forces, the aerodynamic forces on the middle vehicle could keep relatively stable because the effect of three-dimensional ambient flow around the two ends of the model are weakened. For road vehicle simulation such as cars and trucks, only the middle vehicle model was used to reflect the three-dimensional aerodynamic characteristics. The vehicle model was made from hollow light wood to reduce the mass. The total length of the scaled bridge model (see Figure 7.35) was 13.07 m, with a constant cross-section. A measured deck section of 0.5 m length was set for measuring the aerodynamic forces of bridge deck. Vehicle model and bridge model were separated from each other without the model of the wheels (see Figure 7.36).

The experiment was conducted in the XNJD-3 wind tunnel of the Southwest Jiaotong University in China. The test section has a width of 22.5 m, a length of 36 m and a height of 4.5 m. The test wind speed can be adjusted from 1.0 m/s to 17 m/s. The model system (see Figure 7.35) was comprised of four parts: guide rail device, sliding device, motor driving device and buffer device. The total length of guide rail, with a continuous smooth chute, was 18 m. The guide rail was made rigid to decrease the deformation and to ensure the running stability of the vehicle. The sliding device was composed of two

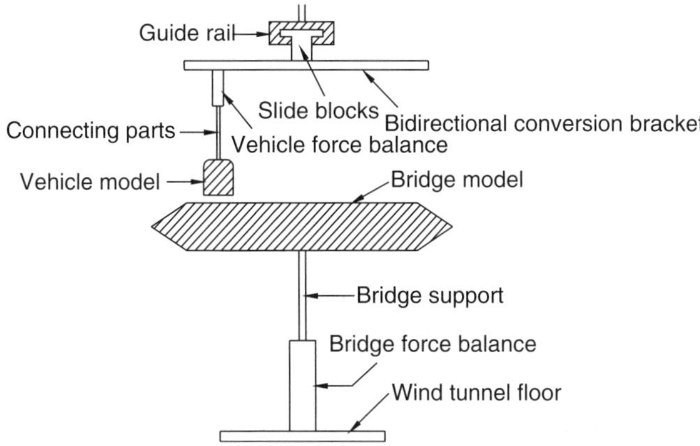

Figure 7.35 Arrangement of vehicle and bridge.

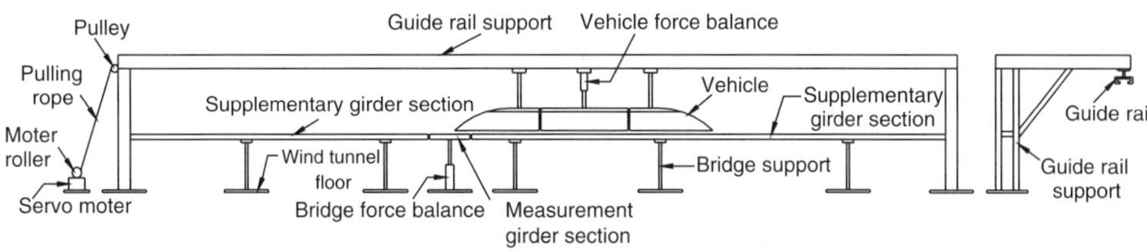

Figure 7.36 Schematic diagram of testing system.

slide blocks and a bidirectional conversion bracket. The two slide blocks were set on the head and the tail ends of a bidirectional conversion bracket.

The vehicle model was connected to the bidirectional conversion bracket through the connecting parts above the vehicle model. The vehicle model could run along the guide rail smoothly and a servo motor was adopted as a motor driving device, with its roller linked to the sliding device through a pulling rope. The motor speed could be accelerated to a given speed and kept constant after the acceleration.

The measurement instrumentation included both force balance devices and a data acquisition system. Two force balances were used to measure the aerodynamic forces on the middle vehicle and on the deck section of 0.5 m long. A six-component balance was joined to the middle vehicle model. Another five-component balance connected the selected bridge deck section to the wind tunnel floor. The data acquisition system of 11 channels simultaneously collected the signals from the two force balances. The test results can be found in [75].

7.12 Notations

A	State matrix of the coupled wind-structure system
A_i^* $(i=1\sim6)$	Aerodynamic derivatives
B	Deck width
C	Damping coefficient; Wind tunnel cross-sectional area
C_D	Drag force coefficient
C_{D_c}	Corrected drag coefficient
C_L	Lift force coefficient

C_M	Moment coefficient
$\mathbf{C^0}$	Structural damping matrix
$\mathbf{C^{ad}}$	Aerodynamic damping matrix
$\mathbf{C^e}$	Total damping matrix of the coupled wind-structure system
c_{ij}^0	Structural damping coefficients
c_{ij}^{ad}	Aerodynamic damping coefficients
c_{ij}^e	System damping coefficients
D	Spacing of the roughness elements
D_b:	Buffeting drag force
D_{dy}:	Total dynamic force in along-wind direction
D_{no}:	Noise in along-wind dynamic force
D_{se}:	Self-excited drag force
E	Elastic modulus of the structure
$e_{h,\alpha}$:	Error functions between the theoretical and measured response time-histories
$F_{ij}(x)$	Self-excited force of the *i*th mode due to the motion of the *j*th mode
F	Frequency of turbulence
f_h	Vertical bending frequency
f_α	Torsional frequency
G	Gravity acceleration
H	Height of the tunnel test section
H_i^* $(i=1{\sim}6)$	Vertical aerodynamic derivatives
H	Height of spires; vertical displacement
I	Model mass moment of inertial per unit length
$I_{u,w}$	Turbulence intensity
K	Height of cube; Stiffness of the spring
$\mathbf{K^0}$	Structural stiffness matrix
$\mathbf{K^{ad}}$	Aerodynamic stiffness matrix
$\mathbf{K^e}$	Total stiffness matrix of the coupled wind-structure system
k_{ij}^0	Structural stiffness coefficients
k_{ij}^{ad}	Aerodynamic stiffness coefficients
k_{ij}^e	System stiffness coefficients
L	Length associated with the external air flow; Length of model
L_o	Initial length
L_b:	Buffeting lift force
L_{dy}:	Total dynamic force in vertical direction
L_{no}:	Noise in vertical dynamic force
L_s	Length associated with the structure; Length of the spring
L_{se}:	Self-excited lift force
$L_{u,w}$	Integral length scale in longitudinal or vertical direction
L	Original length of spring
M	Total mass of the cable model
M_b:	Buffeting moment
M_{dy}:	Total torsional dynamic force
M_{in}	Torsional inertia force
M_{no}:	Noise in torsional dynamic force
M_{se}:	Self-excited pitching moment
M	Model mass per unit length; Number of measured time points
N	Frequency associated with the external air flow
n_s	Frequency associated with the structure
P_i^* $(i=1{\sim}6)$	Aerodynamic derivatives

P	Air pressure
S	Reference area for the drag coefficients; PSD matrix; Cross spectra
$S(n)$	Power spectral density of velocity
T	Tension force of the wires
T_0	Initial tension force
T_s	Tension forces of the spring
T_{s1}, T_{s2}	Tension forces of springs at the ends of the cable
Δt	Time increment
$\overline{U}$	Mean wind velocity
$\overline{U}_g$	Mean freestream longitudinal velocity or velocity at the gradient height
$\overline{U}_z$	Mean longitudinal velocity at height z
U	Fluctuating velocity components in the longitudinal direction
V	Fluctuating velocity components in the lateral direction
W	Fluctuating velocity components in the vertical direction
$y_m(t+\Delta t)$	Modified velocity
Δy	Increment of output within the interval of Δt
Z	Height
A	Wind incidence angle; torsional displacement
α_0	Torsional vibration amplitude
ζ	Critical damping ratio
Λ	Model-to-prototype scale ratio
$\lambda_{r:}$	Eigen values of state matrix **A**
M	Air viscosity
$\xi_{0h,0\alpha}$:	Structural damping ratio
$\xi_{h,\alpha}$	Damping ratio of the vibration system
ρ_s	Density of the structure
σ_u	Root mean square of the fluctuating component
ϕ_i	Modal shape of the *i*th mode
φ_α	Torsional vibration phase angle
χ_{Du}	Aerodynamic transfer functions between the horizontal fluctuating wind velocity and aerodynamic drag force
χ_{Dw}	Aerodynamic transfer functions between the vertical fluctuating wind velocity and aerodynamic drag force
χ_{Lu}	Aerodynamic transfer functions between the horizontal fluctuating wind velocity and aerodynamic lift force
χ_{Lw}	Aerodynamic transfer functions between the vertical fluctuating wind velocity and aerodynamic lift force
χ_{Mu}	Aerodynamic transfer functions between the horizontal fluctuating wind velocity and aerodynamic moment
χ_{Mw}	Aerodynamic transfer functions between the vertical fluctuating wind velocity and aerodynamic moment
$\mathbf{\Psi}_r$	Eigen vectors of state matrix **A**
$\omega_{0h,0\alpha}$	Structural natural frequency
$\omega_{h,\alpha}$	Frequency of the vibration system

References

1. ASCE Manuals and Reports on Engineering Practice No. 67 (1999) *Wind Tunnel Studies of Buildings and Structures*, American Society of Civil Engineers.
2. Holmes, J.D. (2007) *Wind Loading of Structures*, 2nd edn, Taylor and Francis, London and New York.

3. Liu, H. (1991) *Wind Engineering: A Handbook for Structural Engineers*, Prentice Hall, Englewood Cliffs, New Jersey.
4. Nishi, A., Miyagi, H., and Higuchi, K. (1993) A computer-controlled wind tunnel. *Journal of Wind Engineering and Industrial Aerodynamics*, **46, 47**, 837–846.
5. Kikitsu, H., Kanda, J., and Iwasaki, R. (1999) Flow simulation by wind tunnel with computer-controlled multiple fans. *Journal of Wind Engineering and Industrial Aerodynamics*, **83**, 421–429.
6. Nish, A., Kikugawa, H., Matsuda, Y., and Tashiro, D. (1999) Active control of turbulence for an atmospheric boundary layer model in a wind tunnel. *Journal of Wind Engineering and Industrial Aerodynamics*, **83**, 409–419.
7. Davenport, A.G. and Isyumov, N. (1967) The application of the boundary layer wind tunnel to the prediction of wind loading. Proceedings of International Seminar on Wind Effects on Building and Structure, Ottawa.
8. Cermak, J.E. (1965) *Simulation of Atmospheric Motion by Wind Tunnel Flow*, Technical Report CER66, Colorado State University, USA.
9. Mcvehil, G.E., Ludwjg, G.R., and Sundaram, T.R. (1967) *On the Feasibility of Modeling Small Scale Atmospheric Motions*, Report 2B-232B, Cornell Aero Laboratory, USA.
10. Nemoto, S. (1968) Similarity between natural local wind in the atmosphere and model wind in a wind tunnel. *Meteorology and Geophysics*, **19**, 131–230.
11. Melbourne, W.H. (1972) Modeling of structures to measure wind effects. Proceedings of Structural Models Conference, University of Sydney, Australia.
12. Scruton, C. (1963) On the wind-excited oscillations of stacks, towers and masts. Proceedings of the First International Conference on Wind Effects on Buildings and Structures, Teddington, Middlesex, UK, pp. 798–819.
13. Whitbread, R.E. (1963) Model simulation of wind effects on structures. Proceedings of the First International Conference on Wind Effects on Buildings and Structures, Teddington, Middlesex, UK, pp. 284–302.
14. Vickery, B.J. (1972) Lift or across-wind response to tapered stacks. *Journal of the Structural Division*, **98**(1), 1–20.
15. Pankhurst, R.C. (1964) *Dimensional Analysis and Scale Factors*, Chapman & Hall.
16. Langhaar, H. (1951) *Dimensional Analysis and Theory of Models*, John Wiley, New York.
17. Simiu, E. and Scanlan, R.H. (1996) *Wind Effects on Structures–Fundamentals and Applications to Design*, 3rd edn, John Wiley & Sons, Inc.
18. Melbourne, W.H. (1982) Wind tunnel blockage effects and correlations, in *Wind Tunnel Modeling for Civil Engineering Applications* (ed. T.A. Reinhold), Cambridge University Press, Cambridge, pp. 197–216.
19. Cermak F J.E. (1982) Physical modelling of the atmospheric boundary layer (ABL) in long boundary layer wind tunnels (BLWT), in Proceedings of the International Workshop on Wind Tunnel Modelling Criteria and Techniques in Civil Engineering Applications, Cambridge University Press, Gaithersburg, Maryland, USA.
20. Irwin, P.A. (1981) The design of spires for wind simulation. *Journal of Wind Engineering and Industrial Aerodynamics*, **7**, 361–366.
21. Ling, S.C. and Wan, C.A. (1972) Decay of isotropic turbulence generated by a mechanically agitated grid. *The Physics of Fluids*, **15**(8), 1363–1369.
22. Teunissen, H.W. (1975) Simulation of the planetary boundary layer in a multiple-jet wind tunnel. *Atmospheric Environment*, **9** (2), 145–174.
23. Cermak, J.E. and Cochran, L.S. (1992) Physical modeling of the atmospheric surface layer. *Journal of Wind Engineering and Industrial Aerodynamics*, **41–44**, 935–946.
24. Cogotti, A. (2008) Evolution of performance of an automotive wind tunnel. *Journal of Wind Engineering and Industrial Aerodynamics*, **96**, 667–700.
25. Ukeguchi., N., Saketa., H., and Nishitani, H. (1966) An investigation of aeroelastic instability of suspension bridges. Proceedings of the International Symposium on Suspension Bridges, Lisbon, Spain, pp. 273–284.
26. Falco, M., Curami, A., and Zaiia, A. (1992) Nonlinear effects in sectional model aeroelastic parameter identification. *Journal of Wind Engineering and Industrial Aerodynamics*, **42**, 1321–1332.
27. Ricciardelli, F. and Hangan, H. (2001) Pressure distribution and aerodynamic forces on stationary box bridge sections. *Wind and Structures*, **4**(5), 399–412.
28. Sankaran, R. and Jancanskas, E.D. (1992) Direct measurement of the aerodynamic admittance of two-dimensional rectangular cylinders in smooth and turbulent flows. *Journal of Wind Engineering and Industrial Aerodynamics*, **41**(1–3), 601–611.
29. Diana, G., Bruni, S., Cigada, A., and Zappa, E. (2002) Complex aerodynamic admittance function role in buffeting response of a bridge deck. *Journal of Wind Engineering and Industrial Aerodynamics*, **90**(12–15), 2057–2072.
30. Hatanaka, A. and Tanaka, H. (2002) New estimation method of aerodynamic admittance function. *Journal of Wind Engineering and Industrial Aerodynamics*, **90**(12–15), 2073–2086.
31. Kawatani, M. and Kim, H. (1992) Evaluation of aerodynamic admittance for buffeting analysis. *Journal of Wind Engineering and Industrial Aerodynamics*, **41**, 613–624.
32. Larose, G.L. (1997) The Dynamic Action of Gusty Winds on Long Span Bridges, PhD Thesis, Technical University of Denmark, Denmark.
33. Gu, M. and Qin, X.R. (2004) Direct identification of flutter derivatives and aerodynamic admittances of bridge decks. *Engineering Structures*, **26**, 2161–2172.

34. Davenport, A.G., King, J.P.C., and Larose, G.L. (1992) Taut strip model tests. Proceedings of the First International Symposium on Aerodynamics of Large Bridges, Copenhargen, Denmark, pp. 113–124.
35. Larose, G.L. (1992) The Response of a Suspension Bridge Deck to Turbulent Wind: the Taut Strip Model Approach, MSc. Thesis, The University of Western Ontario, Canada.
36. Miyata, T., Yokoyama, K., Yasuda, M., and Hikami, Y. (1992) Akashi Kaikyo Bridge: Wind effects and full model wind tunnel tests. Proceedings of the First International Symposium on Aerodynamics of Large Bridges, Copenhargen, Denmark, pp. 19–21.
37. Davenport, A.G. and King, J.P.C. (1984) Dynamic wind forces on long span bridges. Proceedings of the 12th IABSE Congress, Vancouver.
38. Reinhold, T.A., Brinch, M., and Damsgaard, A. (1992) Wind tunnel tests for the Great Belt Link. Proceedings of the First International Symposium on Aerodynamics of Large Bridges, Copenhargen, Denmark, pp. 255–267.
39. Irwin, P.A. (1992) Full aeroelastic model tests. Proceedings of the First International Symposium on Aerodynamics of Large Bridges, Copenhargen, Denmark, pp. 125–135.
40. Zasso, A., Cigada, A., and Negri, S. (1996) Flutter derivatives identification through full bridge aerodynamic model transfer function analysis. *Journal of Wind Engineering and Industrial Aerodynamics*, **60**, 17–33.
41. Scanlan, R.H. and Tomko, J.J. (1971) Airfoil and bridges deck flutter derivatives. *Journal of Engineering Mechanics Division*, **97** (6), 1717–1737.
42. Yamada, H. and Ichikawa, H. (1992) Measurement of aerodynamic parameters by extended Kalman Filter algorithm. *Journal of Wind Engineering and Industrial Aerodynamics*, **42**, 1255–1263.
43. Jokobsen, J.B. and Hansen, E. (1995) Determination of the aerodynamic derivatives by a system identification method. *Journal of Wind Engineering and Industrial Aerodynamics*, **57**, 295–305.
44. Poulsen, N.K., Damsgaard, A., and Reinhold, T.A. (1992) Determination of flutter derivatives for the Great Belt Bridge. *Journal of Wind Engineering and Industrial Aerodynamics*, **41**, 153–164.
45. Juang, J.N. and Pappa, R.S. (1985) An eigensystem realization algorithm for modal parameter identification and model reduction. *Journal of Guidance, Control, and Dynamics*, **8**(5), 620–627.
46. Sarkar, P.P., Jones, N.P., and Scanlan, R.H. (1994) Identification of aeroelastic parameters of flexible bridges. *Journal of Engineering Mechanics*, **120**(8), 1718–1741.
47. Gu, M., Zhang, R.X., and Xiang, H.F. (2000) Identification of flutter derivatives of bridge decks. *Journal of Wind Engineering and Industrial Aerodynamics*, **84**, 151–162.
48. Xiang, H.F. (2005) *Modern Theory and Practice on Bridge Wind Resistance*, China Communications Press.
49. Gu, M. and Qin, X.R. (2004) Direct identification of flutter derivatives and aerodynamic admittance of bridge decks. *Engineering Structures*, **26**(14), 2161–2172.
50. Overschee, P.V. (1991) Subspace algorithms for the stochastic identification problem. Proceedings of the 30th Conference on Decision and Control, Brighton, England, pp. 1321–1326.
51. Peters, B. (1999) Reference-based stochastic subspace identification for out-put only modal analysis. *Mechanical Systems and Signal Processing*, **13**(6), 855–878.
52. Hikami, Y. and Shiraishi, N. (1988) Rain-wind induced vibration of cables in cable stayed bridges. *Journal of Wind Engineering and Industrial Aerodynamics*, **29**, 409–418.
53. Main, J.A. and Jones, N.P. (1999) Full-scale measurements of stay cable vibration. Proceeding of 10th International Conference on Wind Engineering, Copenhagen, Denmark, June, 2, pp. 963–970.
54. Matsumoto, M. (2002) Cable aerodynamics of cable stayed bridges. Proceeding of 2nd International Symposium On Wind and Structures, Busan, Korea, August, pp. 21–42.
55. Matsumoto, M., Shiraishi, N., and Shirato, H. (1992) Rain-wind induced vibration of cables of cable-stayed bridges. *Journal of Wind Engineering and Industrial Aerodynamics*, **41–44**, 2011–2022.
56. Flamand, O. (1995) Rain-wind induced vibration of cables. *Journal of Wind Engineering and Industrial Aerodynamics*, **57**, 353–362.
57. Bosdogianni, A. and Olivari, D. (1996) Wind- and rain-induced oscillations of cables of stayed bridges. *Journal of Wind Engineering and Industrial Aerodynamics*, **64**, 171–185.
58. Cosentino, N., Flamand, O., and Ceccoli, C. (2003) Rain-wind induced vibration of inclined stay cables-Part I: Experimental investigation and physical explanation. *Wind and Structures*, **6**, 471–484.
59. Cosentino, N., Flamand, O., and Ceccoli, C. (2003) Rain-wind induced vibration of inclined stay cables-Part II: Mechanical modeling and parameter characterization. *Wind and Structures*, **6**, 485–498.
60. Yamaguchi, H. (1990) Analytical study on growth mechanism of rain vibration of cables. *Journal of Wind Engineering and Industrial Aerodynamics*, **33**, 73–80.
61. Xu, Y.L. and Wang, L.Y. (2003) Analytical study of wind-rain-induced cable vibration: SDOF model. *Journal of Wind Engineering and Industrial Aerodynamics*, **91**, 27–40.
62. Peil, U. and Nahrath, N. (2003) Modeling of rain-wind induced vibrations. *Wind and Structures*, **6**, 41–52.
63. Xu, Y.L., Li, Y.L., Shum, K.M. *et al.* (2006) Aerodynamic coefficients of inclined circular cylinders with artificial rivulet in smooth flow. *Advances in Structural Engineering*, **9**(2), 265–278.
64. Zhan, S., Xu, Y.L., Zhou, H.J., and Shum, K.M. (2008) Experimental study of wind-rain-induced cable vibration using a new model setup scheme. *Journal of Wind Engineering and Industrial Aerodynamics*, **96**, 2438–2451.

65. Baker, C.J. and Reynolds, S. (1992) Wind induced accidents of road vehicles. *Accident Analysis and Prevention*, **24**(6), 559–575.
66. Baker, C.J. (1991) Ground vehicles in high cross winds Part 1: Steady aerodynamic forces. *Journal of Fluids and Structures*, **5**, 69–90.
67. Coleman, S.A. and Baker, C.J. (1990) High sided road vehicles in cross winds. *Journal of Wind Engineering and Industrial Aerodynamics*, **36**, 1383–1392.
68. Coleman, S. and Baker, C. (1994) An experimental study of the aerodynamic behaviour of high sided lorries in cross winds. *Journal of Wind Engineering and Industrial Aerodynamics*, **53**(3), 401–429.
69. Humphreys, N.D. and Baker, C.J. (1992) Forces on vehicles in cross winds from moving model tests. *Journal of Wind Engineering and Industrial Aerodynamics*, **41–44**, 2673–2684.
70. Xu, Y.L. and Guo, W.H. (2003) Dynamic analysis of coupled road vehicle and cable-stayed bridge systems under turbulent wind. *Engineering Structures*, **25**, 473–486.
71. Guo, W.H. and Xu, Y.L. (2006) Safety analysis of moving road vehicles on a long bridge under crosswind. *Journal of Engineering Mechanics*, **132**(4), 438–446.
72. Xu, Y.L. and Guo, W.H. (2004) Effects of bridge motion and crosswind on ride comfort of road vehicles. *Journal of Wind Engineering and Industrial Aerodynamics*, **92**(7–8), 641–662.
73. Cai, C.S. and Chen, S.R. (2004) Framework of vehicle-bridge-wind dynamic analysis. *Journal of Wind Engineering and Industrial Aerodynamics*, **92**, 579–607.
74. Zhu, L.D., Li, L., Xu, Y.L., and Zhu, Q. (2012) Wind tunnel investigations of aerodynamic coefficients of road vehicles on bridge deck. *Journal of Fluids and Structures*, **30**, 35–50.
75. Li, Y.L., Hu, P., Xu, Y.L. *et al.* (2011) Aerodynamic characteristics of vehicle-bridge system under cross wind by wind tunnel tests with moving vehicle model. Proceedings of 13th International Conference on Wind Engineering, July 10–15, Amsterdam, Netherlands.

8

Computational Wind Engineering

8.1 Preview

Computational wind engineering (CWE) is a technique for solving wind engineering problems using a computational fluid dynamics (CFD) method. The CWE method is sometimes called "numerical wind tunnels" and, with the increased availability of high-end computing capability and user-friendly commercial CFD codes, it is gradually being adopted as an attractive tool to solve wind engineering problems. Compared with analytical methods (Chapters 3–6), wind tunnel test technique (Chapter 7), and field measurements (Chapter 9), CWE is especially useful for providing detailed flow visualization around, and wind pressure distribution over, the surface of a structure. CFD codes can also produce extremely large volumes of results at almost no added expense, and thus they provide an ideal tool for parametric studies. Nevertheless, there are some challenging issues remaining in CWE and uncertainties in CWE results. The accuracy of the computational results from CWE needs to be validated against the results from either analytical methods, wind tunnel tests or field measurements.

This chapter first introduces the fundamentals of CFD, including the governing equations, turbulence models, numerical methods, grid generations, boundary conditions and computational algorithms. With wind effects on long-span cable-supported bridges as a topic in this book, the applications of CFD to obtain aerodynamic (force) coefficients of a bridge deck, a vehicle on the ground and a vehicle on the bridge deck are illustrated, and the obtained results are validated through comparisons with wind tunnel results. The use of CFD to estimate flutter derivatives of a bridge deck is then presented, and the nonlinear aerodynamic forces on a bridge deck with large oscillation amplitude is finally explored using CFD.

8.2 Governing Equations of Fluid Flow

The governing equations of fluid flow are based on the universal laws of conservation: mass conservation, momentum conservation and energy conservation [1,2].

8.2.1 Mass Conservation

The mass conservation law applied to a fluid passing through an infinitesimal volume yields the following equation of mass conservation in a Cartesian coordinate system.

$$\frac{\partial \rho}{\partial t} + \frac{\partial(\rho u)}{\partial x} + \frac{\partial(\rho v)}{\partial y} + \frac{\partial(\rho w)}{\partial z} = 0 \tag{8.1}$$

Wind Effects on Cable-Supported Bridges, First Edition. You-Lin Xu.

where:

ρ is the density of fluid;
t is the time;
u, v, w represent the x, y, z components of the velocity vector.

The equation of mass conservation is also called the continuity equation. The first term on the left hand side of this equation represents the rate of change in time of the density (mass per unit volume). The other three terms describe the net flow of mass out of the volume across its boundaries and is called the convective term.

For an incompressible fluid, the density is constant and the continuity equation becomes:

$$\frac{\partial u}{\partial x}+\frac{\partial v}{\partial y}+\frac{\partial w}{\partial z}=0 \tag{8.2}$$

The above equation can be written in a vector form as:

$$\mathrm{div}(\mathbf{u})=0 \tag{8.3}$$

where:

div represents the divergence operation;
u is the vector $u\mathbf{i}+v\mathbf{j}+w\mathbf{k}$, in which **i**, **j**, **k** is the unit vector in the x, y, z directions respectively.

8.2.2 *Momentum Conservation*

Newton's Second Law, applied to a fluid passing through an infinitesimal volume, yields the following momentum equations in a Cartesian coordinate system:

$$\frac{D(\rho u)}{Dt}=\rho f_x+\frac{\partial \sigma_{xx}}{\partial x}+\frac{\partial \sigma_{xy}}{\partial y}+\frac{\partial \sigma_{xz}}{\partial z} \tag{8.4a}$$

$$\frac{D(\rho v)}{Dt}=\rho f_y+\frac{\partial \sigma_{yx}}{\partial x}+\frac{\partial \sigma_{yy}}{\partial y}+\frac{\partial \sigma_{yz}}{\partial z} \tag{8.4b}$$

$$\frac{D(\rho w)}{Dt}=\rho f_z+\frac{\partial \sigma_{zx}}{\partial x}+\frac{\partial \sigma_{zy}}{\partial y}+\frac{\partial \sigma_{zz}}{\partial z} \tag{8.4c}$$

where:

σ_{ij} $(i=x, y, z; j=x, y, z)$ is the stress tensor;
f_i is the body force per unit mass;
$\frac{D()}{Dt}$ represents the total derivative.

The term on the left hand side of the above equation represents the rate of increase of momentum per unit volume in the respective direction. The first term on the right hand side is the body force per unit volume. The rest of the terms on the right hand side are the surface forces per unit volume. The most common body force is the gravitational force, and in this case f_z is equal to the acceleration of gravity g and $f_x=f_y=0$.

For an incompressible fluid, the density is constant and the momentum equations become:

$$\rho \frac{Du}{Dt} = \rho f_x + \frac{\partial \sigma_{xx}}{\partial x} + \frac{\partial \sigma_{xy}}{\partial y} + \frac{\partial \sigma_{xz}}{\partial z} \tag{8.5a}$$

$$\rho \frac{Dv}{Dt} = \rho f_y + \frac{\partial \sigma_{yx}}{\partial x} + \frac{\partial \sigma_{yy}}{\partial y} + \frac{\partial \sigma_{yz}}{\partial z} \tag{8.5b}$$

$$\rho \frac{Dw}{Dt} = \rho f_z + \frac{\partial \sigma_{zx}}{\partial x} + \frac{\partial \sigma_{zy}}{\partial y} + \frac{\partial \sigma_{zz}}{\partial z} \tag{8.5c}$$

8.2.3 Energy Conservation and Newtonian Flow

The First Law of Thermodynamics, applied to a fluid passing through an infinitesimal volume, can yield the energy equation. The energy equation states that the rate of change of energy of a fluid particle is equal to the rate of heat addition to the fluid particle, plus the rate of work done on the particle. The details on the derivation of the energy equation can be found in the literature [1,2]. For a compressive fluid, the linkage between the energy equation and the mass conservation and momentum equations arises through the possibility of density variations as a result of pressure and temperature variations in the flow field. For a non-compressive fluid, without density variations there is no linkage between the energy equation and the mass conservation and momentum equations.

The flow field can often be solved by considering mass conservation and momentum equations only. The energy equation needs to be solved alongside the others if the problem involves heat transfer. For bridge wind engineering, wind is regarded as a non-compressive flow and, in most cases, only the mass conservation and momentum equations need to be considered.

From Equation 8.4 or 8.5, one may see that the momentum equations contain the nine unknown stress components σ_{ij}. The most useful way is to introduce a suitable model to relate the stresses to the local deformation rate or strain rate, which is called a constitutive relation. In bridge wind engineering, the flow is considered as Newtonian flow, which means that the flow is isotropic and the stress is proportional to the strain in a linear way. The constitutive relation of Newtonian flow can be expressed as:

$$\sigma_{xx} = -p + 2\mu\left[S_{xx} - \frac{1}{3}div(\mathbf{u})\right];\ \sigma_{yy} = -p + 2\mu\left[S_{yy} - \frac{1}{3}div(\mathbf{u})\right];\ \sigma_{zz} = -p + 2\mu\left[S_{zz} - \frac{1}{3}div(\mathbf{u})\right] \tag{8.6a}$$

$$\sigma_{xy} = \sigma_{yx} = 2\mu S_{xy};\ \sigma_{xz} = \sigma_{zx} = 2\mu S_{zy};\ \sigma_{yz} = \sigma_{zy} = 2\mu S_{yz} \tag{8.6b}$$

$$S_{xx} = \frac{\partial u}{\partial x};\ S_{yy} = \frac{\partial v}{\partial y};\ S_{zz} = \frac{\partial w}{\partial z} \tag{8.7a}$$

$$S_{xy} = S_{yx} = \frac{1}{2}\left(\frac{\partial u}{\partial y} + \frac{\partial v}{\partial x}\right);\ S_{xz} = S_{zx} = \frac{1}{2}\left(\frac{\partial u}{\partial z} + \frac{\partial w}{\partial x}\right);\ S_{yz} = S_{zy} = \frac{1}{2}\left(\frac{\partial v}{\partial z} + \frac{\partial w}{\partial y}\right) \tag{8.7b}$$

where:

μ is the dynamic viscosity coefficient of fluid;
S_{ij} is the rate of strain;
p is the pressure.

For an incompressible fluid:

$$\text{div}(\mathbf{u}) = \frac{\partial u}{\partial x} + \frac{\partial v}{\partial y} + \frac{\partial w}{\partial z} = 0 \tag{8.8}$$

Therefore, while other equations remain unchanged Equation 8.6a becomes:

$$\sigma_{xx} = -p + 2\mu S_{xx};\ \sigma_{yy} = -p + 2\mu S_{yy};\ \sigma_{zz} = -p + 2\mu S_{zz} \tag{8.9}$$

8.2.4 Navier-Stokes Equations

By submitting Equation 8.6 into Equation 8.4, the famous Navier-Stokes (N-S) equations can be derived as:

$$\frac{D(\rho u)}{Dt} = \rho f_x - \frac{\partial p}{\partial x} + \mu\left(\frac{\partial^2 u}{\partial x^2} + \frac{\partial^2 u}{\partial y^2} + \frac{\partial^2 u}{\partial z^2}\right) + \left(1 - \frac{2}{3}\mu\right)\frac{\partial}{\partial x}\left(\frac{\partial u}{\partial x} + \frac{\partial v}{\partial y} + \frac{\partial w}{\partial z}\right) \tag{8.10a}$$

$$\frac{D(\rho v)}{Dt} = \rho f_y - \frac{\partial p}{\partial y} + \mu\left(\frac{\partial^2 v}{\partial x^2} + \frac{\partial^2 v}{\partial y^2} + \frac{\partial^2 v}{\partial z^2}\right) + \left(1 - \frac{2}{3}\mu\right)\frac{\partial}{\partial y}\left(\frac{\partial u}{\partial x} + \frac{\partial v}{\partial y} + \frac{\partial w}{\partial z}\right) \tag{8.10b}$$

$$\frac{D(\rho w)}{Dt} = \rho f_z - \frac{\partial p}{\partial z} + \mu\left(\frac{\partial^2 w}{\partial x^2} + \frac{\partial^2 w}{\partial y^2} + \frac{\partial^2 w}{\partial z^2}\right) + \left(1 - \frac{2}{3}\mu\right)\frac{\partial}{\partial z}\left(\frac{\partial u}{\partial x} + \frac{\partial v}{\partial y} + \frac{\partial w}{\partial z}\right) \tag{8.10c}$$

Submitting Equation 8.2 into Equation 8.10 provides the N-S equations for incompressible fluid:

$$\rho\frac{D(u)}{Dt} = \rho f_x - \frac{\partial p}{\partial x} + \mu\left(\frac{\partial^2 u}{\partial x^2} + \frac{\partial^2 u}{\partial y^2} + \frac{\partial^2 u}{\partial z^2}\right) \tag{8.11a}$$

$$\rho\frac{D(v)}{Dt} = \rho f_y - \frac{\partial p}{\partial y} + \mu\left(\frac{\partial^2 v}{\partial x^2} + \frac{\partial^2 v}{\partial y^2} + \frac{\partial^2 v}{\partial z^2}\right) \tag{8.11b}$$

$$\rho\frac{D(w)}{Dt} = \rho f_z - \frac{\partial p}{\partial z} + \mu\left(\frac{\partial^2 w}{\partial x^2} + \frac{\partial^2 w}{\partial y^2} + \frac{\partial^2 w}{\partial z^2}\right) \tag{8.11c}$$

In bridge wind engineering, the moving velocity of air is relatively low and wind is always considered as being an incompressible fluid. Additionally, the body forces acting on the unit mass of air is neglected ($f_i = 0$). As a result, the N-S equations in bridge wind engineering are simplified as follows:

$$\rho\frac{D(u)}{Dt} = -\frac{\partial p}{\partial x} + \mu\left(\frac{\partial^2 u}{\partial x^2} + \frac{\partial^2 u}{\partial y^2} + \frac{\partial^2 u}{\partial z^2}\right) \tag{8.12a}$$

$$\rho\frac{D(v)}{Dt} = -\frac{\partial p}{\partial y} + \mu\left(\frac{\partial^2 v}{\partial x^2} + \frac{\partial^2 v}{\partial y^2} + \frac{\partial^2 v}{\partial z^2}\right) \tag{8.12b}$$

$$\rho\frac{D(w)}{Dt} = -\frac{\partial p}{\partial z} + \mu\left(\frac{\partial^2 w}{\partial x^2} + \frac{\partial^2 w}{\partial y^2} + \frac{\partial^2 w}{\partial z^2}\right) \tag{8.12c}$$

8.2.5 Governing Equations of Wind Flow

The governing equations of compressible fluid flow comprise the continuity equation (Equation 8.1), the momentum equation (Equation 8.10) and the energy equation. The governing equations of incompressible fluid flow comprise the continuity equation (Equation 8.2), the momentum equation (Equation 8.11) and the energy equation. Strictly speaking, the term "Navier-Stokes equations" refers to the components of the momentum equations (Equation 8.10 or 8.11). However, it is common practice that the governing equations are also called as the Navier-Stokes equations. Finally, the governing equations of wind flow or the Navier-Stokes equations for wind flow comprise of the continuity equation (Equation 8.2) and the momentum equation (Equation 8.12).

8.2.6 Vorticity Description of Navier-Stokes Equations

In the governing equations above, the physical quantities for describing the flow are the flow velocity and pressure. As an alternative, the vorticity can be used to describe the flow. The vorticity components are defined as:

$$\omega_x = \frac{\partial w}{\partial y} - \frac{\partial v}{\partial z}; \omega_y = \frac{\partial u}{\partial z} - \frac{\partial w}{\partial x}; \omega_z = \frac{\partial v}{\partial x} - \frac{\partial u}{\partial y} \tag{8.13}$$

The partial differentiation of Equation 8.12c to y minus the partial differentiation of Equation 8.12b to z leads to:

$$\rho\frac{D\omega_x}{Dt} = \rho\left(\omega_x\frac{\partial u}{\partial x} + \omega_y\frac{\partial u}{\partial y} + \omega_z\frac{\partial u}{\partial z}\right) + \mu\left(\frac{\partial \omega_x}{\partial x^2} + \frac{\partial \omega_x}{\partial y^2} + \frac{\partial \omega_x}{\partial z^2}\right) \tag{8.14a}$$

In a similar way, one may have other two equations as:

$$\rho\frac{D\omega_y}{Dt} = \rho\left(\omega_x\frac{\partial v}{\partial x} + \omega_y\frac{\partial v}{\partial y} + \omega_z\frac{\partial v}{\partial z}\right) + \mu\left(\frac{\partial \omega_y}{\partial x^2} + \frac{\partial \omega_y}{\partial y^2} + \frac{\partial \omega_y}{\partial z^2}\right) \tag{8.14b}$$

$$\rho\frac{D\omega_z}{Dt} = \rho\left(\omega_x\frac{\partial w}{\partial x} + \omega_y\frac{\partial w}{\partial y} + \omega_z\frac{\partial w}{\partial z}\right) + \mu\left(\frac{\partial \omega_z}{\partial x^2} + \frac{\partial \omega_z}{\partial y^2} + \frac{\partial \omega_z}{\partial z^2}\right) \tag{8.14c}$$

Equations 8.14 are the Navier-Stokes equations for wind flow in terms of vorticity.

8.3 Turbulence and its Modeling

As discussed in Chapter 2, within the atmospheric boundary layer and over a flat, homogeneous terrain, the wind velocity can normally be decomposed into a mean wind speed in the mean wind direction and three perpendicular turbulence components in a sufficiently long averaging time. When turbulent winds interact with a structure, turbulence structures around the surface of the structure become more complex and very geometric-dependent. Thus, wind flow calculations must include sufficiently accurate and general descriptions of the turbulence that capture not only the turbulence of incoming wind, but also that caused by the interactions between incoming flow and the structure.

However, this represents a very difficult problem. All presently known turbulence models have limitations: the ultimate turbulence model has yet to be developed. Our expectations in turbulence modeling are reduced from "seeking the ultimate" to "seeking models that have reasonable accuracy over a limited range of flow conditions". The purpose of this section is to introduce the methodology commonly used in turbulence modeling. It is important to remember that turbulence models must be verified through comparison with wind tunnel test or field measurement results.

8.3.1 Direct Numerical Simulation

Turbulent winds have a random characteristics, with a wide range of length and time scales. The ratios of the smallest scale in length η and time τ, to the largest scale, are approximated as:

$$\text{Length-scale ratio}: \frac{\eta}{L} = \text{Re}^{-3/4} \tag{8.15}$$

$$\text{Time-scale ratio}: \frac{\tau}{L/U} = \text{Re}^{-3/4} \tag{8.16}$$

where:

U is the mean wind speed;
L is the characteristic length of a structure;
Re is the Reynolds number.

The governing equations for wind flow are Equations 8.2 and 8.12. There are a total of four equations with four variables, indicating that the equations can be solved directly with initial and boundary conditions. The direct solving method is named as "direct numerical simulation" (DNS), and its aim is to compute the mean flow and all turbulent fluctuations at the same time. The unsteady Navier-Stokes equations are solved on spatial grids that are so fine that they can resolve the length scales at which energy dissipation takes place, and with time steps sufficiently small to resolve the period of the fastest fluctuations. The potential benefits of DNS include providing precise details of turbulence and instantaneous results.

However, in bridge wind engineering, the Reynolds number (Re) is high. For example, if a bridge deck is 10 m in width, Re is then 6.8×10^6 under a mean wind of 10 m/s. According to Equation 8.15, the number of numerical discrete cells has a magnitude about $(L/\eta)^3 = \text{Re}^{9/4}$ in three-dimensional flows, or $(L/\eta)^2 = \text{Re}^{3/2}$ in two-dimensional flows. The time step should be as small as $\text{Re}^{-3/4}(L/U)$, according to Equation 8.16.

For the deck mentioned above, the quantity of numerical cells is approximately 1.8×10^{10} and 2.4×10^{15} for two-dimensional and three-dimensional flows respectively. The corresponding time step is about 7.5×10^{-6} seconds, and thus a huge number of time steps are needed in practice. It is therefore difficult to apply DNS to bridge wind engineering at present, although it is likely to play an increasingly important role in turbulence research in the near future as the computer performance becomes further enhanced.

8.3.2 Reynolds Averaged Method

For some problems in bridge wind engineering, the time-averaged properties, such as mean wind speed and mean wind pressure, are good enough for the design purpose, and it is unnecessary to resolve the details of turbulent fluctuations. In this regard, the Navier-Stokes equations are time averaged, leading to the so-called Reynolds averaged Navier-Stokes (RANS) equations or the Reynolds equations of motion. Time averaging the Navier-Stokes equations gives rise to new terms, due to the interactions between various turbulent fluctuations and the effects of turbulence on mean flow properties. These new terms must be related to the mean flow variables through turbulence models. This process needs to introduce further assumptions and approximations in order to close the system of equations.

In the conventional Reynolds decomposition, the randomly changing flow variables are replaced by time averages plus fluctuations about the average:

$$\phi(t) = \overline{\phi(t)} + \phi'(t) \tag{8.17}$$

$$\overline{\phi(t)} = \frac{1}{\Delta t}\int_0^{\Delta t} \phi(t)dt \tag{8.18}$$

where:

$\phi(t)$ represents a variable such as wind speed u or wind pressure p;
$\overline{\phi(t)}$ and $\phi'(t)$ are the mean and fluctuation parts of $\phi(t)$;
Δt is the time interval.

In theory, the time interval will approach infinity but Equation 8.18 can provide meaningful time averages if Δt is larger than the time scale associated with the slowest variations (due to the largest eddies) of $\phi(t)$.

The time averaged mean value and the fluctuation value have the following properties:

$$\overline{\phi'(t)} = 0; \ \overline{\phi_1(t) + \phi_2(t)} = \overline{\phi_1(t)} + \overline{\phi_2(t)}; \ \overline{\left(\frac{\partial \phi}{\partial q}\right)} = \frac{\partial \overline{\phi}}{\partial q} \tag{8.19}$$

Replace the flow variables u, v, w and p in the governing equations (Equations 8.2 and 8.12) by the sum of its mean and fluctuation parts, as expressed by Equation 8.17, and then time-average the governing equations. The famous Reynolds averaged Navier-Stokes (RANS) equations can be derived for the incompressible fluid as follows:

$$\frac{\partial \bar{u}}{\partial x} + \frac{\partial \bar{v}}{\partial y} + \frac{\partial \bar{w}}{\partial z} = 0 \tag{8.20}$$

$$\rho\frac{\partial \bar{u}}{\partial t} + \rho\bar{u}\frac{\partial \bar{u}}{\partial x} + \rho\bar{v}\frac{\partial \bar{u}}{\partial y} + \rho\bar{w}\frac{\partial \bar{u}}{\partial z} = -\frac{\partial \bar{p}}{\partial x} + \mu\left(\frac{\partial^2 \bar{u}}{\partial x^2} + \frac{\partial^2 \bar{u}}{\partial y^2} + \frac{\partial^2 \bar{u}}{\partial z^2}\right) - \rho\frac{\partial \overline{u'u'}}{\partial x} - \rho\frac{\partial \overline{u'v'}}{\partial y} - \rho\frac{\partial \overline{u'w'}}{\partial z} \tag{8.21a}$$

$$\rho\frac{\partial \bar{v}}{\partial t} + \rho\bar{u}\frac{\partial \bar{v}}{\partial x} + \rho\bar{v}\frac{\partial \bar{v}}{\partial y} + \rho\bar{w}\frac{\partial \bar{v}}{\partial z} = -\frac{\partial \bar{p}}{\partial y} + \mu\left(\frac{\partial^2 \bar{v}}{\partial x^2} + \frac{\partial^2 \bar{v}}{\partial y^2} + \frac{\partial^2 \bar{v}}{\partial z^2}\right) - \rho\frac{\partial \overline{v'u'}}{\partial x} - \rho\frac{\partial \overline{v'v'}}{\partial y} - \rho\frac{\partial \overline{v'w'}}{\partial z} \tag{8.21b}$$

$$\rho\frac{\partial \bar{w}}{\partial t} + \rho\bar{u}\frac{\partial \bar{w}}{\partial x} + \rho\bar{v}\frac{\partial \bar{w}}{\partial y} + \rho\bar{w}\frac{\partial \bar{w}}{\partial z} = -\frac{\partial \bar{p}}{\partial z} + \mu\left(\frac{\partial^2 \bar{w}}{\partial x^2} + \frac{\partial^2 \bar{w}}{\partial y^2} + \frac{\partial^2 \bar{w}}{\partial z^2}\right) - \rho\frac{\partial \overline{w'u'}}{\partial x} - \rho\frac{\partial \overline{w'v'}}{\partial y} - \rho\frac{\partial \overline{w'w'}}{\partial z} \tag{8.21c}$$

There are ten unknowns – $\bar{p}$, $\bar{u}$, $\bar{v}$, $\bar{w}$,$\overline{u'u'}$, $\overline{v'v'}$, $\overline{w'w'}$, $\overline{u'v'}$, $\overline{u'w'}$, and $\overline{v'w'}$ – in the above RANS equations. They cannot be solved by the four RANS equations directly. Of these unknowns, $-\rho\overline{u'u'}$, $-\rho\overline{v'v'}$, $-\rho\overline{w'w'}$, $-\rho\overline{u'v'}$, $-\rho\overline{u'w'}$, $-\rho\overline{v'w'}$ are the so-called Reynolds stresses. In addition to these, other four unknowns exist with their time-averaged values. The Reynolds stresses represent the contribution of fluctuation flow to the mean flow. To solve for the mean flow, the Reynolds stresses should be related to the time-averaged variables through turbulence models.

Many turbulence models have been proposed, mostly based on Boussinesq's assumption. Boussinesq suggested, more than 100 year ago, that the apparent turbulent shearing stresses might be related to the rate of mean strain through an apparent scalar turbulent or eddy viscosity. For the general Reynolds stresses, the Boussinesq assumption gives:

$$-\rho\overline{u'u'} = 2\mu_t\overline{S_{xx}} - \frac{2}{3}\rho k; \ -\rho\overline{v'v'} = 2\mu_t\overline{S_{yy}} - \frac{2}{3}\rho k; \ -\rho\overline{v'v'} = 2\mu_t\overline{S_{zz}} - \frac{2}{3}\rho k \tag{8.22a}$$

$$-\rho\overline{u'v'} = 2\mu_t\overline{S_{xy}}; \ -\rho\overline{u'v'} = 2\mu_t\overline{S_{xy}}; \ -\rho\overline{v'w'} = 2\mu_t\overline{S_{yz}} \tag{8.22b}$$

where μ_t is called the eddy viscosity to be determined; k is the kinetic energy of turbulence per unit mass; and ε is the dissipation per unit mass that is associated with k.

$$k = \frac{1}{2}(\overline{u'u'} + \overline{v'v'} + \overline{w'w'}) \tag{8.23a}$$

$$\varepsilon = \frac{\mu}{\rho}\left(\overline{\frac{\partial u'}{\partial x}\frac{\partial u'}{\partial x}} + \overline{\frac{\partial v'}{\partial y}\frac{\partial v'}{\partial y}} + \overline{\frac{\partial w'}{\partial z}\frac{\partial w'}{\partial z}}\right) \tag{8.23b}$$

Turbulence models based on the Boussinesq assumption are called the viscosity models. The commonly used viscosity models are the Spalart-Allmara model, the standard k-ε model, the RNG k-ε model, the Wilcox k-ω model and the SST k-ω model.

(a) Spalart-Allmara Model

The Spalart-Allmara model involves one equation only for kinematic eddy viscosity parameter, $\tilde{\nu}$, and a specification of a length scale by means of an algebraic function, f_{v1}, to obtain the eddy viscosity in Equation 8.22, where the turbulence kinetic energy term is ignored. This model provides economical computation of boundary layers in external aerodynamics [3]. The eddy viscosity in this model is related to $\tilde{\nu}$ by:

$$\mu_t = \rho\tilde{\nu}f_{v1} \tag{8.24}$$

$\tilde{\nu}$ is solved from the following transport equation:

$$\rho\frac{\partial\tilde{\nu}}{\partial t} + \rho\bar{u}\frac{\partial\tilde{\nu}}{\partial x} + \rho\bar{v}\frac{\partial\tilde{\nu}}{\partial y} + \rho\bar{w}\frac{\partial\tilde{\nu}}{\partial z} = \frac{1}{\sigma}\frac{\partial}{\partial x}\left[(\mu+\rho\tilde{\nu})\frac{\partial\tilde{\nu}}{\partial x}\right] + \frac{1}{\sigma}\frac{\partial}{\partial y}\left[(\mu+\rho\tilde{\nu})\frac{\partial\tilde{\nu}}{\partial y}\right] + \frac{1}{\sigma}\frac{\partial}{\partial z}\left[(\mu+\rho\tilde{\nu})\frac{\partial\tilde{\nu}}{\partial z}\right]$$
$$+c_{b1}(1-f_{v2})\rho\tilde{S}\tilde{\nu} - c_{w1}f_w\rho\left(\frac{\tilde{\nu}}{d}\right)^2 + \rho\frac{c_{b2}}{\sigma}\left(\frac{\partial^2\tilde{\nu}}{\partial x^2} + \frac{\partial^2\tilde{\nu}}{\partial y^2} + \frac{\partial^2\tilde{\nu}}{\partial z^2}\right) \tag{8.25}$$

where the closure coefficients and functions are:

$$f_{v1} = \frac{\chi^3}{\chi^3 + c_{v1}^3};\ f_{v2} = 1 - \frac{\chi}{1+\chi f_{v1}};\ f_w = g\left(\frac{1+c_{w3}^6}{g^6 + c_{w3}^6}\right);\ \chi = \frac{\tilde{\nu}}{\nu} \tag{8.26}$$

$$g = r + c_{w2}(r^6 - r);\ r = \frac{\tilde{\nu}}{\tilde{S}\kappa^2 d^2};\ \tilde{S} = \Omega + \frac{\tilde{\nu}}{\kappa^2 d^2}f_{v2} \tag{8.27}$$

$$\Omega = 2\sqrt{(\Omega_{xy}\Omega_{xy} + \Omega_{xz}\Omega_{xz} + \Omega_{yz}\Omega_{yz})};\ \Omega_{xy} = \frac{1}{2}\left(\frac{\partial\bar{u}}{\partial y} - \frac{\partial\bar{v}}{\partial x}\right) \tag{8.28}$$

$$\Omega_{xz} = \frac{1}{2}\left(\frac{\partial\bar{u}}{\partial z} - \frac{\partial\bar{w}}{\partial x}\right);\ \Omega_{yz} = \frac{1}{2}\left(\frac{\partial\bar{v}}{\partial z} - \frac{\partial\bar{w}}{\partial y}\right);\ c_{w1} = \frac{c_{b1}}{\kappa^2} + \frac{1+c_{b2}}{\sigma};\ c_{b1} = 0.1355 \tag{8.29}$$

$$c_{b2} = 0.622;\ c_{v1} = 7.1;\ \sigma = 2/3;\ c_{w2} = 0.3;\ c_{w3} = 2;\ \kappa = 0.41 \tag{8.30}$$

where:

$\Omega_{ij}$is the rotation tensor;
d is the distance from the closest surface.

(b) Standard k-ε Model

The standard k-ε model expresses the eddy viscosity in Equation 8.22a as the ratio of turbulent kinetic energy k and its dissipation rate ε [4].

$$\mu_t = \rho C_\mu k^2/\varepsilon \tag{8.31}$$

k and ε are determined from the following transport equation:

$$\rho\frac{\partial k}{\partial t}+\rho\bar{u}\frac{\partial k}{\partial x}+\rho\bar{v}\frac{\partial k}{\partial y}+\rho\bar{w}\frac{\partial k}{\partial z}=\frac{\partial}{\partial x}\left[\left(\mu+\frac{\mu_t}{\sigma_k}\right)\frac{\partial k}{\partial x}\right]+\frac{\partial}{\partial y}\left[\left(\mu+\frac{\mu_t}{\sigma_k}\right)\frac{\partial k}{\partial y}\right]+\frac{\partial}{\partial z}\left[\left(\mu+\frac{\mu_t}{\sigma_k}\right)\frac{\partial k}{\partial z}\right]$$
$$+\left(2\mu_t\overline{S_{xx}}-\frac{2}{3}\rho k\right)\frac{\partial\bar{u}}{\partial x}+2\mu_t\overline{S_{xy}}\frac{\partial\bar{u}}{\partial y}+2\mu_t\overline{S_{xz}}\frac{\partial\bar{u}}{\partial z}+\left(2\mu_t\overline{S_{yy}}-\frac{2}{3}\rho k\right)\frac{\partial\bar{v}}{\partial y}+2\mu_t\overline{S_{yx}}\frac{\partial\bar{v}}{\partial x}$$
$$+2\mu_t\overline{S_{yz}}\frac{\partial\bar{v}}{\partial z}+\left(2\mu_t\overline{S_{zz}}-\frac{2}{3}\rho k\right)\frac{\partial\bar{w}}{\partial z}+2\mu_t\overline{S_{zx}}\frac{\partial\bar{w}}{\partial x}+2\mu_t\overline{S_{zy}}\frac{\partial\bar{w}}{\partial y}-\rho\varepsilon \tag{8.32}$$

$$\rho\frac{\partial \varepsilon}{\partial t}+\rho\bar{u}\frac{\partial \varepsilon}{\partial x}+\rho\bar{v}\frac{\partial \varepsilon}{\partial y}+\rho\bar{w}\frac{\partial \varepsilon}{\partial z}=\frac{\partial}{\partial x}\left[\left(\mu+\frac{\mu_t}{\sigma_\varepsilon}\right)\frac{\partial \varepsilon}{\partial x}\right]+\frac{\partial}{\partial y}\left[\left(\mu+\frac{\mu_t}{\sigma_\varepsilon}\right)\frac{\partial \varepsilon}{\partial y}\right]+\frac{\partial}{\partial z}\left[\left(\mu+\frac{\mu_t}{\sigma_\varepsilon}\right)\frac{\partial \varepsilon}{\partial z}\right]$$
$$+C_{\varepsilon 1}\frac{\varepsilon}{k}\left[\left(2\mu_t\overline{S_{xx}}-\frac{2}{3}\rho k\right)\frac{\partial\bar{u}}{\partial x}+2\mu_t\overline{S_{xy}}\frac{\partial\bar{u}}{\partial y}+2\mu_t\overline{S_{xz}}\frac{\partial\bar{u}}{\partial z}+\left(2\mu_t\overline{S_{yy}}-\frac{2}{3}\rho k\right)\frac{\partial\bar{v}}{\partial y}+2\mu_t\overline{S_{yx}}\frac{\partial\bar{v}}{\partial x}\right.$$
$$\left.+2\mu_t\overline{S_{yz}}\frac{\partial\bar{v}}{\partial z}+\left(2\mu_t\overline{S_{zz}}-\frac{2}{3}\rho k\right)\frac{\partial\bar{w}}{\partial z}+2\mu_t\overline{S_{zx}}\frac{\partial\bar{w}}{\partial x}+2\mu_t\overline{S_{zy}}\frac{\partial\bar{w}}{\partial y}\right]-C_{\varepsilon 2}\rho\frac{\varepsilon^2}{\kappa} \tag{8.33}$$

The model constants for the standard k-ε model are:

$$C_\mu=0.09,\ C_{\varepsilon 1}=1.44,\ C_{\varepsilon 2}=1.92,\ \sigma_k=1.0,\ \sigma_\varepsilon=1.3 \tag{8.34}$$

The standard k-ε model is a two-equation model. Until the last decade of the twentieth century this model was the most popular two-equation model by far.

(c) RNG k-ε Model

Using the techniques from the renormalization group (RNG) theory, Yakhot and Orszag have developed what is known as the RNG k-ε model [5]. The eddy viscosity equation (Equation 8.31) remains unchanged, but the solution equations for k and ε are:

$$\rho\frac{\partial k}{\partial t}+\rho\bar{u}\frac{\partial k}{\partial x}+\rho\bar{v}\frac{\partial k}{\partial y}+\rho\bar{w}\frac{\partial k}{\partial z}=\frac{\partial}{\partial x}\left[\alpha_k(\mu+\mu_t)\frac{\partial k}{\partial x}\right]+\frac{\partial}{\partial y}\left[(\alpha_k(\mu+\mu_t)\frac{\partial k}{\partial y}\right]+\frac{\partial}{\partial z}\left[\alpha_k(\mu+\mu_t)\frac{\partial k}{\partial z}\right]$$
$$+\left(2\mu_t\overline{S_{xx}}-\frac{2}{3}\rho k\right)\frac{\partial\bar{u}}{\partial x}+2\mu_t\overline{S_{xy}}\frac{\partial\bar{u}}{\partial y}+2\mu_t\overline{S_{xz}}\frac{\partial\bar{u}}{\partial z}+\left(2\mu_t\overline{S_{yy}}-\frac{2}{3}\rho k\right)\frac{\partial\bar{v}}{\partial y}+2\mu_t\overline{S_{yx}}\frac{\partial\bar{v}}{\partial x}$$
$$+2\mu_t\overline{S_{yz}}\frac{\partial\bar{v}}{\partial z}+\left(2\mu_t\overline{S_{zz}}-\frac{2}{3}\rho k\right)\frac{\partial\bar{w}}{\partial z}+2\mu_t\overline{S_{zx}}\frac{\partial\bar{w}}{\partial x}+2\mu_t\overline{S_{zy}}\frac{\partial\bar{w}}{\partial y}-\rho\varepsilon \tag{8.35}$$

$$\rho\frac{\partial \varepsilon}{\partial t}+\rho\bar{u}\frac{\partial \varepsilon}{\partial x}+\rho\bar{v}\frac{\partial \varepsilon}{\partial y}+\rho\bar{w}\frac{\partial \varepsilon}{\partial z}=\frac{\partial}{\partial x}\left[\alpha_\varepsilon(\mu+\mu_t)\frac{\partial \varepsilon}{\partial x}\right]+\frac{\partial}{\partial y}\left[\alpha_\varepsilon(\mu+\mu_t)\frac{\partial \varepsilon}{\partial y}\right]+\frac{\partial}{\partial z}\left[\alpha_\varepsilon(\mu+\mu_t)\frac{\partial \varepsilon}{\partial z}\right]$$
$$+C^*_{\varepsilon 1}\frac{\varepsilon}{k}\left[\left(2\mu_t\overline{S_{xx}}-\frac{2}{3}\rho k\right)\frac{\partial\bar{u}}{\partial x}+\overline{S_{xy}}\frac{\partial\bar{u}}{\partial y}+2\mu_t\overline{S_{xz}}\frac{\partial\bar{u}}{\partial z}+\left(2\mu_t\overline{S_{yy}}-\frac{2}{3}\rho k\right)\frac{\partial\bar{v}}{\partial y}+2\mu_t\overline{S_{yx}}\frac{\partial\bar{v}}{\partial x}+2\mu_t\overline{S_{yz}}\frac{\partial\bar{v}}{\partial z}\right.$$
$$\left.+\left(2\mu_t\overline{S_{zz}}-\frac{2}{3}\rho k\right)\frac{\partial\bar{w}}{\partial z}+2\mu_t\overline{S_{zx}}\frac{\partial\bar{w}}{\partial x}+2\mu_t\overline{S_{zy}}\frac{\partial\bar{w}}{\partial y}\right]-C^*_{\varepsilon 2}\rho\frac{\varepsilon^2}{\kappa} \tag{8.36}$$

$$C_\mu = 0.0845,\ C_{\varepsilon 1} = 1.42,\ C_{\varepsilon 2}^{*} = 1.68 \tag{8.37}$$

$$\alpha_k = 1.39,\ \alpha_\varepsilon = 1.39 \tag{8.38}$$

$$C_{\varepsilon 1}^{*} = C_{\varepsilon 1} - \frac{\eta(1-\eta/\eta_0)}{1+\beta\eta^3},\ \eta = \frac{k}{\varepsilon}S,\ \eta_0 = 4.377,\ \beta = 0.012 \tag{8.39}$$

$$S = \sqrt{2(\overline{S_{xx}}\,\overline{S_{xx}} + \overline{S_{yy}}\,\overline{S_{yy}} + \overline{S_{zz}}\,\overline{S_{zz}} + \overline{S_{xy}}\,\overline{S_{xy}} + \overline{S_{xz}}\,\overline{S_{xz}} + \overline{S_{yz}}\,\overline{S_{yz}} + \overline{S_{yx}}\,\overline{S_{yx}} + \overline{S_{zx}}\,\overline{S_{zx}} + \overline{S_{zy}}\,\overline{S_{zy}})} \tag{8.40}$$

Note that the RNG k-ε model contains a strain-dependent correction term in the constant $C_{\varepsilon 1}$ of the production term. This model removes the small scales of motion from the governing equations by expressing their effects in terms of larger scale motions and a modified viscosity.

(d) Wilcox k-ω Model

One of the most prominent two-equation models is the Wilcox k-ω model [6]. In this model, the turbulence frequency$\omega = \varepsilon/k$ is used as the second variable in addition to k. The eddy viscosity in Equation 8.22 is expressed as the product of the turbulent kinetic energy k and the turbulence frequency ω as follows:

$$\mu_t = \rho k/\omega \tag{8.41}$$

The equations for k and ω are:

$$\begin{aligned}\rho\frac{\partial k}{\partial t} + \rho\bar{u}\frac{\partial k}{\partial x} + \rho\bar{v}\frac{\partial k}{\partial y} + \rho\bar{w}\frac{\partial k}{\partial z} &= \frac{\partial}{\partial x}\left[\left(\mu + \frac{\mu_t}{\sigma_k}\right)\frac{\partial k}{\partial x}\right] + \frac{\partial}{\partial y}\left[\left(\mu + \frac{\mu_t}{\sigma_k}\right)\frac{\partial k}{\partial y}\right] + \frac{\partial}{\partial z}\left[\left(\mu + \frac{\mu_t}{\sigma_k}\right)\frac{\partial k}{\partial z}\right]\\
&+ \left(2\mu_t\overline{S_{xx}} - \frac{2}{3}\rho k\right)\frac{\partial \bar{u}}{\partial x} + 2\mu_t\overline{S_{xy}}\frac{\partial \bar{u}}{\partial y} + 2\mu_t\overline{S_{xz}}\frac{\partial \bar{u}}{\partial z} + \left(2\mu_t\overline{S_{yy}} - \frac{2}{3}\rho k\right)\frac{\partial \bar{v}}{\partial y} + 2\mu_t\overline{S_{yx}}\frac{\partial \bar{v}}{\partial x} + 2\mu_t\overline{S_{yz}}\frac{\partial \bar{v}}{\partial z}\\
&+ \left(2\mu_t\overline{S_{zz}} - \frac{2}{3}\rho k\right)\frac{\partial \bar{w}}{\partial z} + 2\mu_t\overline{S_{zx}}\frac{\partial \bar{w}}{\partial x} + 2\mu_t\overline{S_{zy}}\frac{\partial \bar{w}}{\partial y} - \beta^{*}\rho k\omega\end{aligned} \tag{8.42}$$

$$\begin{aligned}\rho\frac{\partial \omega}{\partial t} + \rho\bar{u}\frac{\partial \omega}{\partial x} + \rho\bar{v}\frac{\partial \omega}{\partial y} + \rho\bar{w}\frac{\partial \omega}{\partial z} &= \frac{\partial}{\partial x}\left[\left(\mu + \frac{\mu_t}{\sigma_\varepsilon}\right)\frac{\partial \omega}{\partial x}\right] + \frac{\partial}{\partial y}\left[\left(\mu + \frac{\mu_t}{\sigma_\varepsilon}\right)\frac{\partial \omega}{\partial y}\right] + \frac{\partial}{\partial z}\left[\left(\mu + \frac{\mu_t}{\sigma_\varepsilon}\right)\frac{\partial \omega}{\partial z}\right]\\
&+ \alpha_1\frac{\omega}{k}\left[\left(2\mu_t\overline{S_{xx}} - \frac{2}{3}\rho k\right)\frac{\partial \bar{u}}{\partial x} + 2\mu_t\overline{S_{xy}}\frac{\partial \bar{u}}{\partial y} + 2\mu_t\overline{S_{xz}}\frac{\partial \bar{u}}{\partial z} + \left(2\mu_t\overline{S_{yy}} - \frac{2}{3}\rho k\right)\frac{\partial \bar{v}}{\partial y} + 2\mu_t\overline{S_{yx}}\frac{\partial \bar{v}}{\partial x}\right.\\
&\left.+ 2\mu_t\overline{S_{yz}}\frac{\partial \bar{v}}{\partial z} + \left(2\mu_t\overline{S_{zz}} - \frac{2}{3}\rho k\right)\frac{\partial \bar{w}}{\partial z} + 2\mu_t\overline{S_{zx}}\frac{\partial \bar{w}}{\partial x} + 2\mu_t\overline{S_{zy}}\frac{\partial \bar{w}}{\partial y}\right] - \beta_1\rho\omega^2\end{aligned} \tag{8.43}$$

The constants in the above transport equations are:

$$\sigma_k = 2.0,\ \sigma_\omega = 2.0,\ \alpha_1 = 0.553,\ \beta_1 = 0.075,\ \beta^{*} = 0.09 \tag{8.44}$$

The Wilcox k-ω model initially attracted attention because integration to the wall does not require wall-damping functions in low Reynolds number applications.

(e) SST k-ω Model

The standard k-ε model is much less sensitive to the assumed values in the free stream, but its near-wall performance is unsatisfactory for boundary layers with adverse pressure gradients. On the other hand, the k-ω model is sensitive to the assumed free stream values but performs well in the boundary layers. Menter [7] thus proposed the shear-stress transport (SST) k-ω model to effectively blend the robust and accurate formulation of the *k-ω* model in the near-wall region with the k-ε model in the far field. The eddy viscosity in Equation 8.22 is expressed as the product of the turbulent kinetic energy k and the turbulence frequency ω as follows:

$$\mu_t = \frac{a_1 \rho k}{\max(a_1 \omega, SF_2)} \tag{8.45}$$

$$F_2 = \tanh(\Phi_2^2) \tag{8.46}$$

$$\Phi_2 = \max(\frac{\sqrt{k}}{0.045\omega y}, \frac{500\mu}{\rho y^2 \omega}) \tag{8.47}$$

$$\alpha_1 = 0.31 \tag{8.48}$$

where y is the distance to the near surface.

The equations for k and ω are:

$$\begin{aligned} &\rho\frac{\partial k}{\partial t} + \rho\bar{u}\frac{\partial k}{\partial x} + \rho\bar{v}\frac{\partial k}{\partial y} + \rho\bar{w}\frac{\partial k}{\partial z} = \frac{\partial}{\partial x}\left[\left(\mu + \frac{\mu_t}{\sigma_k}\right)\frac{\partial k}{\partial x}\right] + \frac{\partial}{\partial y}\left[\left(\mu + \frac{\mu_t}{\sigma_k}\right)\frac{\partial k}{\partial y}\right] + \frac{\partial}{\partial z}\left[\left(\mu + \frac{\mu_t}{\sigma_k}\right)\frac{\partial k}{\partial z}\right] \\ &+ \min\left[\left(2\mu_t\overline{S_{xx}} - \frac{2}{3}\rho k\right)\frac{\partial \bar{u}}{\partial x} + 2\mu_t\overline{S_{xy}}\frac{\partial \bar{u}}{\partial y} + 2\mu_t\overline{S_{xz}}\frac{\partial \bar{u}}{\partial z} + \left(2\mu_t\overline{S_{yy}} - \frac{2}{3}\rho k\right)\frac{\partial \bar{v}}{\partial y} + 2\mu_t\overline{S_{yx}}\frac{\partial \bar{v}}{\partial x}\right. \\ &\left. + 2\mu_t\overline{S_{yz}}\frac{\partial \bar{v}}{\partial z} + \left(2\mu_t\overline{S_{zz}} - \frac{2}{3}\rho k\right)\frac{\partial \bar{w}}{\partial z} + 2\mu_t\overline{S_{zx}}\frac{\partial \bar{w}}{\partial x} + 2\mu_t\overline{S_{zy}}\frac{\partial \bar{w}}{\partial y}, 10\rho\beta^* k\omega\right] - \beta^*\rho k\omega \end{aligned} \tag{8.49}$$

$$\begin{aligned} &\rho\frac{\partial \omega}{\partial t} + \rho\bar{u}\frac{\partial \omega}{\partial x} + \rho\bar{v}\frac{\partial \omega}{\partial y} + \rho\bar{w}\frac{\partial \omega}{\partial z} = \frac{\partial}{\partial x}\left[\left(\mu + \frac{\mu_t}{\sigma_\varepsilon}\right)\frac{\partial \omega}{\partial x}\right] + \frac{\partial}{\partial y}\left[\left(\mu + \frac{\mu_t}{\sigma_\varepsilon}\right)\frac{\partial \omega}{\partial y}\right] + \frac{\partial}{\partial z}\left[\left(\mu + \frac{\mu_t}{\sigma_\varepsilon}\right)\frac{\partial \omega}{\partial z}\right] \\ &+ \frac{\rho\alpha'}{\mu_t}\left[\left(2\mu_t\overline{S_{xx}} - \frac{2}{3}\rho k\right)\frac{\partial \bar{u}}{\partial x} + 2\mu_t\overline{S_{xy}}\frac{\partial \bar{u}}{\partial y} + 2\mu_t\overline{S_{xz}}\frac{\partial \bar{u}}{\partial z} + \left(2\mu_t\overline{S_{yy}} - \frac{2}{3}\rho k\right)\frac{\partial \bar{v}}{\partial y} + 2\mu_t\overline{S_{yx}}\frac{\partial \bar{v}}{\partial x}\right. \\ &\left. + 2\mu_t\overline{S_{yz}}\frac{\partial \bar{v}}{\partial z} + \left(2\mu_t\overline{S_{zz}} - \frac{2}{3}\rho k\right)\frac{\partial \bar{w}}{\partial z} + 2\mu_t\overline{S_{zx}}\frac{\partial \bar{w}}{\partial x} + 2\mu_t\overline{S_{zy}}\frac{\partial \bar{w}}{\partial y}\right] - \rho\beta'\omega^2 \\ &+ 2(1 - F_1)\rho\sigma_{\omega,2}\frac{1}{\omega}\left(\frac{\partial k}{\partial x}\frac{\partial \omega}{\partial x} + \frac{\partial k}{\partial y}\frac{\partial \omega}{\partial y} + \frac{\partial k}{\partial z}\frac{\partial \omega}{\partial z}\right) \end{aligned} \tag{8.50}$$

where:

$$\sigma_k = \frac{1}{F_1/\sigma_{k,1} + (1 - F_1)/\sigma_{k,2}} \tag{8.51}$$

$$\sigma_\omega = \frac{1}{F_1/\sigma_{\omega,1} + (1 - F_1)/\sigma_{\omega,2}} \tag{8.52}$$

$$F_1 = \tanh(\Phi_1^4) \tag{8.53}$$

$$\Phi_1 = \min\left[\max\left(\frac{\sqrt{k}}{0.09\omega y}, \frac{500\mu}{\rho y^2 \omega}\right), \frac{4\rho k}{\sigma_{\omega,2} D_\omega^+ y^2}\right] \tag{8.54}$$

$$D_\omega^+ = \max\left[2\rho \frac{1}{\sigma_{\omega,2}} \frac{1}{\omega} \frac{\partial k}{\partial x_j} \frac{\partial \omega}{\partial x_j}, 10^{-10}\right] \tag{8.55}$$

$$\sigma_{k,1} = 1.176, \ \sigma_{\omega,1} = 2.0 \tag{8.56}$$

$$\sigma_{k,2} = 1.0, \ \sigma_{\omega,2} = 1.168 \tag{8.57}$$

$$\beta^* = \beta^*_\infty \left(\frac{4/15 + (\mathrm{Re}_t/R_\beta)^4}{1 + (\mathrm{Re}_t/R_\beta)^4}\right) \tag{8.58}$$

$$\mathrm{Re}_t = \frac{\rho k}{\mu \omega} \tag{8.59}$$

$$\beta^*_\infty = 0.09, \ R_k = 6 \tag{8.60}$$

$$\beta' = F_1 \beta_{i,1} + (1 - F_1)\beta_{i,2} \tag{8.61}$$

$$\alpha' = \frac{\alpha'_\infty}{\alpha^*}\left(\frac{\alpha_0 + \mathrm{Re}_t/R_\omega}{1 + \mathrm{Re}_t/R_\omega}\right) \tag{8.62}$$

$$\alpha'_\infty = F_1 \alpha_{\infty,1} + (1 - F_1)\alpha_{\infty,2} \tag{8.63}$$

$$\alpha_{\infty,1} = \frac{\beta_{i,1}}{\beta^*_\infty} - \frac{\kappa^2}{\sigma_{\omega,1}\sqrt{\beta^*_\infty}} \tag{8.64}$$

$$\alpha_{\infty,2} = \frac{\beta_{i,2}}{\beta^*_\infty} - \frac{\kappa^2}{\sigma_{\omega,2}\sqrt{\beta^*_\infty}} \tag{8.65}$$

$$\beta_{i,1} = 0.075, \ \beta_{i,2} = 0.0828, \ \kappa = 0.41 \tag{8.66}$$

Menter *et al.* [8] summarized a series of modifications to optimize the performance of the SST k-ω model, based on experience with the model in general-purpose application.

8.3.3 Large Eddy Simulation

In the RANS method, many turbulence models have been developed to consider the effect of turbulence to the averaged flow. However, they are limited to special flow problems and cannot be used to solve all of the flow problems. The main reason for this limitation is that the RANS method ignores the scales effect of flow in space, and a single turbulence model is employed for all eddies with different scales. The turbulence actually has multi-scales in space; the difference of size between the smallest and the largest scales is huge in high Reynolds number flow. The smaller eddies have a universal behaviour and perform isotropically, while the larger eddies interact with the mean flow and are related to the geometry of the flow domain.

A different approach, the large eddy simulation (LES), has therefore been proposed [9,10]. In the LES approach, the larger eddies are computed directly for each problem with a time-dependent simulation, while the universal behavior of the smaller eddies is captured with a compact model. This is accomplished by using a spatial filtering operation on the governing equations of fluid flow to obtain a set of equations that separate the larger eddies from the small eddies, compared with the temporal average used in deriving the Reynolds equations.

The filtering of small scale eddies in the original turbulent flow $\phi(x', y', z', t)$in LES is completed via a filter function:

$$\bar{\phi}(x,y,z,t) = \int_{-\infty}^{\infty}\int_{-\infty}^{\infty}\int_{-\infty}^{\infty} G(x,y,z,x',y',z',\Delta)\phi(x',y',z',t)dx'dy'dz' \tag{8.67}$$

where:

$\bar{\phi}(x, y, z, t)$ is the filtered variable;
x', y', z' are the original spatial coordinates before filtering;
Δ is the cutoff-width of the filter as an indicative measure of the size of eddies that are retained in the computations and the eddies that are rejected;
G is the filter function;
and the overbar demotes the filter operation in space domain not in time domain.

The commonest forms of the filtering function in three-dimensional LES computation are the sharp cut-off filter, the top hat filter, and the Gaussian filter.

The sharp cut-off or spectral filter:

$$G(x,y,z,x',y',z',\Delta) = \frac{1}{\Delta^3}\frac{\sin[(x-x')/\Delta]}{(x-x')/\Delta}\frac{\sin[(y-y')/\Delta]}{(y-y')/\Delta}\frac{\sin[(z-z')/\Delta]}{(z-z')/\Delta} \tag{8.68}$$

The top hat or box filter:

$$G(x,y,z,x',y',z',\Delta) = \begin{cases} 1/\Delta^3 & if \quad r < \Delta/2 \\ 0 & if \quad r \geq \Delta/2 \end{cases} \tag{8.69a}$$

$$r = \sqrt{(x-x')^2 + (y-y')^2 + (z-z')^2} \tag{8.69b}$$

The Gaussian filter:

$$G(X,\Delta) = \left(\sqrt{\frac{\gamma}{\pi}\frac{1}{\Delta^2}}\right)^3 e^{-\gamma\frac{r^2}{\Delta^2}}, \ \gamma = 6 \tag{8.70}$$

The top-hat filter is used in finite volume implementations of LES, while the Gaussian filter was introduced for LES in finite differences. The sharp cut-off filter is preferred in the research literature. The most common selection of the cutoff width Δ is to take the same order as the grid size.

After filtering in space, the flow variables can be decomposed into the sum of large scale and small scale parts as follows:

$$\phi = \bar{\phi} + \hat{\phi} \tag{8.71}$$

where:

$\hat{\phi}$ is the small scale part of flow containing unresolved spatial variations at a length scale smaller than the filter cutoff width;
$\bar{\phi}$ is the filtered function with spatial variations that are larger than the cutoff width.

Let us filter the Navier-Stokes continuity and momentum equations for wind flow. The governing equations of LES can be derived as:

$$\frac{\partial \bar{u}}{\partial x}+\frac{\partial \bar{v}}{\partial y}+\frac{\partial \bar{w}}{\partial z}=0 \tag{8.72}$$

$$\rho\frac{\partial \bar{u}}{\partial t}+\rho\frac{\partial(\bar{u}\bar{u})}{\partial x}+\rho\frac{\partial(\bar{u}\bar{v})}{\partial y}+\rho w\frac{\partial(\bar{u}\bar{w})}{\partial z}=-\frac{\partial \bar{p}}{\partial x}+\mu\left(\frac{\partial^2 \bar{u}}{\partial x^2}+\frac{\partial^2 \bar{u}}{\partial y^2}+\frac{\partial^2 \bar{u}}{\partial z^2}\right)-\frac{\partial \tau_{xx}}{\partial x}-\frac{\partial \tau_{xy}}{\partial y}-\frac{\partial \tau_{xz}}{\partial z} \tag{8.73a}$$

$$\rho\frac{\partial \bar{v}}{\partial t}+\rho\frac{\partial(\bar{v}\bar{u})}{\partial x}+\rho\frac{\partial(\bar{v}\bar{v})}{\partial y}+\rho w\frac{\partial(\bar{v}\bar{w})}{\partial z}=-\frac{\partial \bar{p}}{\partial x}+\mu\left(\frac{\partial^2 \bar{v}}{\partial x^2}+\frac{\partial^2 \bar{v}}{\partial y^2}+\frac{\partial^2 \bar{v}}{\partial z^2}\right)-\frac{\partial \tau_{xy}}{\partial x}-\frac{\partial \tau_{yy}}{\partial y}-\frac{\partial \tau_{yz}}{\partial z} \tag{8.73b}$$

$$\rho\frac{\partial \bar{w}}{\partial t}+\rho\frac{\partial(\bar{w}\bar{u})}{\partial x}+\rho\frac{\partial(\bar{w}\bar{v})}{\partial y}+\rho w\frac{\partial(\bar{w}\bar{w})}{\partial z}=-\frac{\partial \bar{p}}{\partial x}+\mu\left(\frac{\partial^2 \bar{w}}{\partial x^2}+\frac{\partial^2 \bar{w}}{\partial y^2}+\frac{\partial^2 \bar{w}}{\partial z^2}\right)-\frac{\partial \tau_{xz}}{\partial x}-\frac{\partial \tau_{yz}}{\partial y}-\frac{\partial \tau_{zz}}{\partial z} \tag{8.73c}$$

$$\tau_{xx}=\rho\overline{u_x u_x}-\rho\overline{u_x}\,\overline{u_x};\ \tau_{yy}=\rho\overline{u_y u_y}-\rho\overline{u_y}\,\overline{u_y};\ \tau_{zz}=\rho\overline{u_z u_z}-\rho\overline{u_z}\,\overline{u_z} \tag{8.74a}$$

$$\tau_{xy}=\tau_{yx}=\rho\overline{u_x u_y}-\rho\overline{u_x}\,\overline{u_y};\ \tau_{yz}=\tau_{zy}=\rho\overline{u_y u_z}-\rho\overline{u_y}\,\overline{u_z};\ \tau_{xz}=\tau_{zx}=\rho\overline{u_x u_z}-\rho\overline{u_x}\,\overline{u_z} \tag{8.74b}$$

where τ_{ij} is the subgrid stress that represents the contributions of the smaller eddies to the larger eddies from three groups: Leonard stresses, cross-stresses, and LES Reynolds stresses [1].

Clearly, the filtering here is different from the time averaging. The subgrid stress can be modeled using the so-called subgrid stress (SGS) models. The typical SGS models are the Smagorinsky model and the dynamic SGS model.

(a) Smagorinsky Model

The Smagorinsky SGS model adopts the Boussinesq hypothesis and relates the subgrid stress to the strain rate of the resolved flow.

$$\tau_{xx}=-2\mu_{SGS}\overline{S_{xx}}+\frac{1}{3}\tau_0;\ \tau_{yy}=-2\mu_{SGS}\overline{S_{yy}}+\frac{1}{3}\tau_0;\ \tau_{zz}=-2\mu_{SGS}\overline{S_{zz}}+\frac{1}{3}\tau_0 \tag{8.75a}$$

$$\tau_{xy}=\tau_{yx}=-2\mu_{SGS}\overline{S_{xy}};\ \tau_{yz}=\tau_{zy}=-2\mu_{SGS}\overline{S_{yz}};\ \tau_{xz}=\tau_{zx}=-2\mu_{SGS}\overline{S_{xz}} \tag{8.75b}$$

$$\tau_0=\frac{1}{3}(\tau_{xx}+\tau_{yy}+\tau_{zz}) \tag{8.75c}$$

where μ_{SGS} is the subgrid eddy viscosity and can be approximated by:

$$\mu_{SGS}=\rho(C_s\Delta)^2|\bar{S}| \tag{8.76}$$

$$|\bar{S}|=\sqrt{2(\overline{S_{xx}}\,\overline{S_{xx}}+\overline{S_{yy}}\,\overline{S_{yy}}+\overline{S_{zz}}\,\overline{S_{zz}}+\overline{S_{xy}}\,\overline{S_{xy}}+\overline{S_{xz}}\,\overline{S_{xz}}+\overline{S_{yz}}\,\overline{S_{yz}}+\overline{S_{yx}}\,\overline{S_{yx}}+\overline{S_{zx}}\,\overline{S_{zx}}+\overline{S_{zy}}\,\overline{S_{zy}})} \tag{8.77}$$

where C_s is the Smagorinsky constant, within a range from 0.1–0.24, and is adjusted case-by-case.

(b) Dynamic SGS Model

The dynamic SGS model replaces the Smagorinsky constant (Equation 8.76) with a new parameter C_d:

$$\mu_{SGS} = \rho C_d \Delta^2 |\bar{S}| \tag{8.78}$$

The parameter C_d is obtained through the filtering of the governing equations 8.73a, once again with a filter size of $\hat{\Delta}$ ($\hat{\Delta} > \Delta$) [11].

8.3.4 *Detached Eddy Simulation*

To save computation time while, at the same time, retaining the computation accuracy, some hybrid models have been presented. Among these, the detached eddy simulation (DES) method is the most popular approach [12]. The DES method is a cost-effective procedure that treats the largest eddies through a conventional LES and handles boundary layers and thin shear layers with the conventional RANS method. The crucial ingredient in the DES method is how to distinguish the RANS and LES portions of computation – the so-called DES blending problem. The most acceptable way in which we distinguish the appropriate length scale and computational mode (LES or RANS) is that based on either the Spalart-Allmaras model or the SST k-ω model.

(a) DES Based on Spalart-Allmars Model

The distance from the wall d is taken from the original Spalart-Allmars model as a length scale to judge the change between the RANS and LES. The judgment parameter is defined as:

$$\hat{d} = \min(d, C_{des}\Delta) \tag{8.79}$$

where $C_{des} = 0.65$ is a closure coefficient whose value has been determined by the results of homogeneous turbulence computations. Equation 8.79 means that the Spalart-Allmars model will be used if the distance from the wall d is less than the value of $C_{des}\Delta$.

(b) DES Based on SST k-ω model

The length scale of turbulence in the SST k-ω model is expressed as:

$$l = \frac{\sqrt{k}}{\beta^* \omega} \tag{8.80}$$

A new length scale of turbulence is defined as:

$$\delta = \min(l, C_{DES}\Delta) \tag{8.81}$$

where $C_{DES} = 0.78$.

Based on this modification, the turbulent kinetic energy equation in the original SST k-ω model is changed to:

$$\begin{aligned}
\rho\frac{\partial k}{\partial t} + \rho\bar{u}\frac{\partial k}{\partial x} + \rho\bar{v}\frac{\partial k}{\partial y} + \rho\bar{w}\frac{\partial k}{\partial z} &= \frac{\partial}{\partial x}\left[\left(\mu + \frac{\mu_t}{\sigma_k}\right)\frac{\partial k}{\partial x}\right] + \frac{\partial}{\partial y}\left[\left(\mu + \frac{\mu_t}{\sigma_k}\right)\frac{\partial k}{\partial y}\right] + \frac{\partial}{\partial z}\left[\left(\mu + \frac{\mu_t}{\sigma_k}\right)\frac{\partial k}{\partial z}\right] \\
&+ \min\left[\left(2\mu_t\overline{S_{xx}} - \frac{2}{3}\rho k\right)\frac{\partial \bar{u}}{\partial x} + 2\mu_t\overline{S_{xy}}\frac{\partial \bar{u}}{\partial y} + 2\mu_t\overline{S_{xz}}\frac{\partial \bar{u}}{\partial z} + \left(2\mu_t\overline{S_{yy}} - \frac{2}{3}\rho k\right)\frac{\partial \bar{v}}{\partial y} + 2\mu_t\overline{S_{yx}}\frac{\partial \bar{v}}{\partial x}\right. \\
&\left. + 2\mu_t\overline{S_{yz}}\frac{\partial \bar{v}}{\partial z} + \left(2\mu_t\overline{S_{zz}} - \frac{2}{3}\rho k\right)\frac{\partial \bar{w}}{\partial z} + 2\mu_t\overline{S_{zx}}\frac{\partial \bar{w}}{\partial x} + 2\mu_t\overline{S_{zy}}\frac{\partial \bar{w}}{\partial y}, 10\rho\beta^* k\omega\right] - \beta^*\frac{k^{3/2}}{\delta}
\end{aligned} \tag{8.82}$$

If the eddy length scale δ is smaller than $C_{DES}\Delta$, the RANS method is adopted. Otherwise, the LES method is used.

8.3.5 *Discrete Vortex Method*

The flow around a body can also be simulated by solving Equations 8.14. The method to solve the vorticity is called the discrete vortex method. Larsen and Walther [13] and Taylor and Vezza [14] used this method to obtain the aerodynamic characteristics of a bridge deck section of two dimensions. In a two-dimensional plane x-y, $w=0$ and $\partial(\,)/\partial z = 0$, therefore $\omega_x = \omega_y = 0$. The governing equations become:

$$\omega_z = \frac{\partial v}{\partial x} - \frac{\partial u}{\partial y} \tag{8.83}$$

$$\rho \frac{D\omega_z}{Dt} = \mu \left(\frac{\partial \omega_z}{\partial x^2} + \frac{\partial \omega_z}{\partial y^2} \right) \tag{8.84}$$

The total vorticity field is composed of vorticity elements:

$$\omega_z(\mathbf{r}, t) = \sum_{j=1}^{N} \omega_{zj} \delta(\mathbf{r} - \mathbf{r}_j(t)) \tag{8.85}$$

where:

$\mathbf{r} = x\,\mathbf{i} + y\,\mathbf{j}$ is the location vector;
ω_{zj} and $\mathbf{r}_j$ are the vorticity and location of the j_{th} vorticity element;
N is the total number of vorticity elements;
$\delta()$ is the Dirac Delta function.

The locations of the j_{th} vorticity elements are updated during calculation:

$$\frac{\partial \mathbf{r}_j(t)}{\partial t} = \mathbf{u}(\mathbf{r}_j, t) \tag{8.86}$$

$\mathbf{u}$ is obtained based on the Biot-Savart formation:

$$\mathbf{u}(\mathbf{r}, t) = \mathbf{K} * \omega_z = \int \mathbf{K}(\mathbf{r} - \mathbf{r}')\omega_z(\mathbf{r}', t) d\mathbf{r}' + \mathbf{u}_\infty \tag{8.87}$$

$$\mathbf{K}(\mathbf{r} - \mathbf{r}') = \frac{-(y - y')\mathbf{i} + (x - x')\mathbf{j}}{2\pi |\mathbf{r} - \mathbf{r}'|^2} \tag{8.88}$$

where $\mathbf{u}_\infty$ is the flow velocity at the infinite location.

8.4 Numerical Considerations

The governing equations of fluid flow are partial differential equations (PDE) which are continuous in both space and time. In the numerical calculation, the continuous equations should be discretized in space and time domain with the formation of a system of algebraic equations for the variables at a set of discrete points. The algebraic equations can usually be solved once at a time.

Several considerations determine whether the solution will be a good approximation to the exact solution of the original PDE. These considerations are truncation error, consistency, and stability [2]. The commonly-used discretization methods for solving the governing equations of fluid flow numerically are the finite difference method, the finite element method and the finite volume method. In this section, the steady governing equation of momentum in the x-direction of two-dimensional flow is taken as an example to illustrate the three discretization methods. The expression of the concerned equation is:

$$\rho\left(\frac{\partial u}{\partial x}u+\frac{\partial u}{\partial y}u\right)=-\frac{\partial p}{\partial x}+\mu\left(\frac{\partial^2 u}{\partial x^2}+\frac{\partial^2 u}{\partial y^2}\right) \tag{8.89}$$

8.4.1 Finite Difference Method

The finite difference method (FDM) replaces the continuous problem domain by a finite difference mesh or grid [2]. Let us establish a grid on the domain using a uniform discretized nodes arranged with intervals of Δx and Δy, as shown in Figure 8.1.

The variables at each node are the velocity u, v and pressure p. Applying Equation 8.89 to this grid yields the following discrete equation:

$$\frac{\partial u(i,j)}{\partial x}\rho u(i,j)+\frac{\partial u(i,j)}{\partial y}\rho v(i,j)=-\frac{\partial p(i,j)}{\partial x}+\mu\left(\frac{\partial^2 u(i,j)}{\partial x^2}+\frac{\partial^2 u(i,j)}{\partial y^2}\right) \tag{8.90}$$

In the above equation, the derivatives $\frac{\partial u(i,j)}{\partial x}$, $\frac{\partial u(i,j)}{\partial y}$, $\frac{\partial p(i,j)}{\partial x}$, $\frac{\partial^2 u(i,j)}{\partial x^2}$, $\frac{\partial^2 u(i,j)}{\partial y^2}$ are not the basic variables to be solved, and they have to be approximated by the variables at nodes. If $u(x+\delta x,\ y)$ is expanded in a Taylor-series about $u(x,\ y)$:

$$u(x+\delta x,y)=u(x,y)+\delta x\frac{\partial u(x,y)}{\partial x}+\frac{\delta x^2}{2}\frac{\partial^2 u(x,y)}{\partial x^2}+\frac{\delta x^3}{3!}\frac{\partial^3 u(x,y)}{\partial x^3}+\cdots \tag{8.91}$$

we can have:

$$\frac{\partial u(x,y)}{\partial x}=\frac{u(x+\delta x,y)-u(x,y)}{\delta x}-\frac{\delta x}{2}\frac{\partial^2 u(x,y)}{\partial x^2}-\frac{\delta x^2}{3!}\frac{\partial^3 u(x,y)}{\partial x^3}-\cdots \tag{8.92}$$

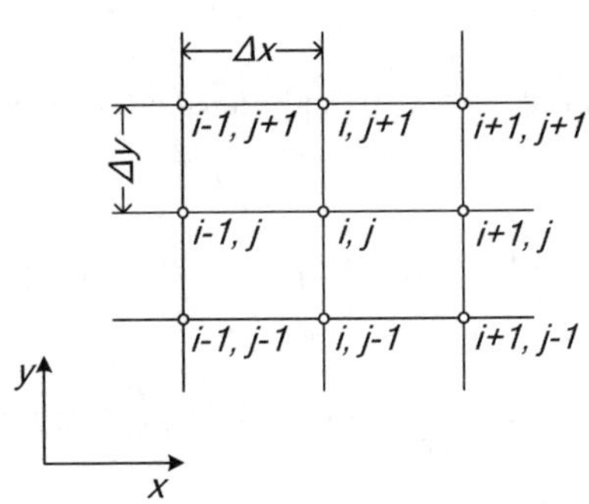

Figure 8.1 A typical finite difference grid.

Then, expanding u at the node (i, j) with Δx and $-\Delta x$ gives:

$$\frac{\partial u(i,j)}{\partial x} = \frac{u(i+1,j) - u(i,j)}{\Delta x} - \frac{\Delta x}{2}\frac{\partial^2 u(i,j)}{\partial x^2} - \frac{\Delta x^2}{3!}\frac{\partial^3 u(i,j)}{\partial x^3} - \cdots \tag{8.93}$$

$$\frac{\partial u(i,j)}{\partial x} = \frac{u(i,j) - u(i-1,j)}{\Delta x} + \frac{\Delta x}{2}\frac{\partial^2 u(i,j)}{\partial x^2} + \frac{\Delta x^2}{3!}\frac{\partial^3 u(i,j)}{\partial x^3} + \cdots \tag{8.94}$$

Subtracting Equation 8.94 from Equation 8.93 yields:

$$\frac{\partial^2 u(i,j)}{\partial x^2} = \frac{u(i+1,j) - 2u(i,j) + u_x(i-1,j)}{(\Delta x)^2} - \frac{2\Delta x^2}{3!}\frac{\partial^3 u(i,j)}{\partial x^3} - \cdots \tag{8.95}$$

Therefore, $\frac{\partial u(i,j)}{\partial x}$ and $\frac{\partial^2 u(i,j)}{\partial x^2}$ can be approximated as follows:

$$\frac{\partial u(i,j)}{\partial x} = \frac{u(i+1,j) - u(i,j)}{\Delta x} \tag{8.96}$$

$$\frac{\partial u(i,j)}{\partial x} = \frac{u(i,j) - u(i-1,j)}{\Delta x} \tag{8.97}$$

$$\frac{\partial^2 u(i,j)}{\partial x^2} = \frac{u(i+1,j) - 2u(i,j) + u(i-1,j)}{(\Delta x)^2} \tag{8.98}$$

Equation 8.96 is called the first-order forward difference, and Equation 8.97 is the first-order backward difference. Equation 8.98 is the central difference. In a similar way, the expressions for other derivatives can obtained:

$$\frac{\partial u(i,j)}{\partial y} = \frac{u(i,j+1) - u(i,j)}{\Delta y} \tag{8.99}$$

$$\frac{\partial p(i,j)}{\partial x} = \frac{p(i+1,j) - p(i,j)}{\Delta x} \tag{8.100}$$

$$\frac{\partial^2 u(i,j)}{\partial y^2} = \frac{u(i,j+1) - 2u(i,j) + u(i,j-1)}{(\Delta y)^2} \tag{8.101}$$

Substituting Equations 8.96 and 8.98–8.101 into Equation 8.90 leads to the following algebraic equation:

$$\begin{aligned}&\left[-\frac{\rho}{\Delta x}u(i,j) - \frac{\rho}{\Delta y}v(i,j) + \frac{2\mu}{(\Delta x)^2} + \frac{2\mu}{(\Delta y)^2}\right]u(i,j) - \frac{\mu}{(\Delta x)^2}u(i-1,j)\\ &+\left[\frac{\rho}{\Delta x}u(i,j) - \frac{\mu}{(\Delta x)^2}\right]u(i+1,j) - \frac{\mu}{(\Delta y)^2}u(i,j-1) + \left[\frac{\rho}{\Delta y}v(i,j) - \frac{\mu}{(\Delta y)^2}\right]u(i,j+1)\\ &= -\frac{1}{\Delta x}[p(i+1,j) - p(i,j)]\end{aligned} \tag{8.102}$$

A series of algebraic equations similar to Equation 8.102 can be constructed for all the nodes. The grids shown in Figure 8.1 are regular, with definite relation between the neighboring nodes. Grids with such a feature are called structured grids. For a structure of complex geometry in bridge wind engineering, it is difficult to use structured grids. Therefore, FDM is often used for the flow around a structure of simple geometry.

8.4.2 Finite Element Method

In the finite element method (FEM), some weight functions are used to bridge the values at the concerned positions with the values at the corners (nodes) of the element. The element can be of an arbitrary shape, so that FEM has the ability to solve the flow around a structure of complex geometry. There are two options for the formulation of finite element equations: variational methods or weighted residual methods. However, variational principles are not available in exact forms for non-linear governing equations in general. Thus, it is logical to seek the weighted residual methods [15]:

Figure 8.2 shows a typical quadrilateral finite element with four nodes, taken as an example to illustrate how the steady governing equations are discretized in FEM. Local interpolation functions $\mathbf{N_v^e}$ and $\mathbf{N_p^e}$ are formed to approximate the relationships of flow variables inside the element to those at the nodes. The approximate variables $\hat{u}^e$, $\hat{v}^e$ and $\hat{p}^e$ inside the element can be expressed as:

$$\hat{u}^e = \mathbf{N_v^e}\mathbf{u_N}^\mathbf{e} \quad \hat{v}^e = \mathbf{N_v^e}\mathbf{v_N}^\mathbf{e} \quad \hat{p}^e = \mathbf{N_p^e}\mathbf{p_N}^\mathbf{e} \tag{8.103}$$

$$\mathbf{u_N}^\mathbf{e} = \{u_1\ u_2\ u_3\ u_4\}^T;\ \mathbf{v_N}^\mathbf{e} = \{v_1\ v_2\ v_3\ v_4\}^T;\ \mathbf{p_N}^\mathbf{e} = \{p_1\ p_2\ p_3\ p_4\}^T \tag{8.104a}$$

$$\mathbf{N_v}^\mathbf{e} = \{N_{v1}^e\, N_{v2}^e\, N_{v3}^e\, N_{v4}^e\}^T; \mathbf{N_p}^\mathbf{e} = \left\{N_{p1}^e\, N_{p2}^e\, N_{p3}^e\, N_{p4}^e\right\}^T \tag{8.104b}$$

where the superscript e represents the element coordinate.

Replacing the variables in Equation 8.89 with the approximated values given by Equation 8.103 yields the residual of the equation as:

$$R^e = \frac{\partial \hat{u}^e}{\partial x}\rho\hat{u}^e + \frac{\partial \hat{u}^e}{\partial y}\rho\hat{v}^e + \frac{\partial \hat{p}^e}{\partial x} - \mu\left(\frac{\partial^2 \hat{u}^e}{\partial x^2} + \frac{\partial^2 \hat{u}^e}{\partial y^2}\right) \tag{8.105}$$

The basic idea of the weighted residual method is to construct a mathematical process in which the residual is minimized to zero. This can be done by forming a subspace spanned by the weighting functions and projecting the residual orthogonally onto this subspace. This process is called the inner product of the residual, which can be expressed as:

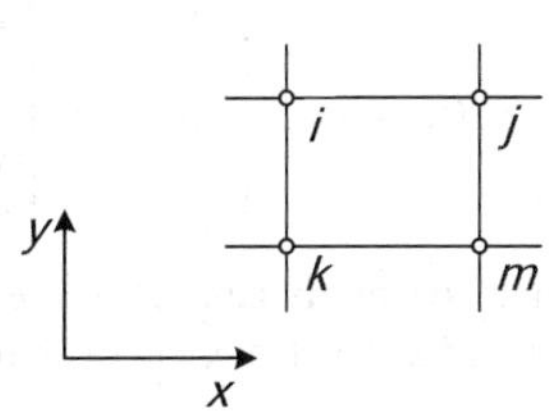

Figure 8.2 A typical quadrilateral finite element.

$$\iint_A WR^e dA = 0 \tag{8.106}$$

where:

W is the weighting function;
A is the area of the element.

If the weighting functions are replaced by the interpolation functions, the above equation is known as the Galerkin method. Therefore, the substitution of Equation 8.103 into Equation 8.105 and then into Equation 8.106 gives:

$$\begin{aligned}&\left\{\iint_A \left[N_{vi}{}^e\left(\rho \mathbf{N}_\mathbf{v}^\mathbf{e}\mathbf{u_N}{}^\mathbf{e}\frac{\partial \mathbf{N}_\mathbf{v}^\mathbf{e}}{\partial x} + \rho \mathbf{N}_\mathbf{v}^\mathbf{e}\mathbf{v_N}{}^\mathbf{e}\frac{\partial \mathbf{N}_\mathbf{v}^\mathbf{e}}{\partial y}\right) + \mu\left(\frac{\partial N_{vi}{}^e}{\partial x}\frac{\partial \mathbf{N}_\mathbf{v}^\mathbf{e}}{\partial x} + \frac{\partial N_{vi}{}^e}{\partial y}\frac{\partial \mathbf{N}_\mathbf{v}^\mathbf{e}}{\partial y}\right)\right] dA\right.\\ &\left. - \iint_\Gamma \mu N_{vi}{}^e\left(\frac{\partial \mathbf{N}_\mathbf{v}^\mathbf{e}}{\partial x} + \frac{\partial \mathbf{N}_\mathbf{v}^\mathbf{e}}{\partial y}\right) d\Gamma\right\}\mathbf{u_N}{}^\mathbf{e} = -\iint_A N_{vi}^e\left[\frac{\partial \mathbf{N}_\mathbf{p}^\mathbf{e}}{\partial x}\mathbf{p_N}{}^\mathbf{e}\right] dA\end{aligned} \tag{8.107}$$

where:

$N_{vi}^e = W_i$ (i=1, 2, 3, 4) is the ith interpolation function;
Γ is the boundary of the element shown in Figure 8.2.

For the sake of simplicity, Equation 8.107 can be shortened to:

$$A_i \mathbf{u_N}{}^\mathbf{e} = B_i; \quad i = 1,2,3,4 \tag{8.108}$$

where:

$$\begin{aligned}A_i = &\iint_A \left[N_{vi}{}^e\left(\rho \mathbf{N}_\mathbf{v}^\mathbf{e}\mathbf{u_N}{}^\mathbf{e}\frac{\partial \mathbf{N}_\mathbf{v}^\mathbf{e}}{\partial x} + \rho \mathbf{N}_\mathbf{v}^\mathbf{e}\mathbf{v_N}{}^\mathbf{e}\frac{\partial \mathbf{N}_\mathbf{v}^\mathbf{e}}{\partial y}\right) + \mu\left(\frac{\partial N_{vi}{}^e}{\partial x}\frac{\partial \mathbf{N}_\mathbf{v}^\mathbf{e}}{\partial x} + \frac{\partial N_{vi}{}^e}{\partial y}\frac{\partial \mathbf{N}_\mathbf{v}^\mathbf{e}}{\partial y}\right)\right] dA\\ &- \iint_\Gamma \mu N_{vi}{}^e\left(\frac{\partial \mathbf{N}_\mathbf{v}^\mathbf{e}}{\partial x} + \frac{\partial \mathbf{N}_\mathbf{v}^\mathbf{e}}{\partial y}\right) d\Gamma\end{aligned} \tag{8.109}$$

$$B_i = -\iint_A N_{vi}^e\left[\frac{\partial \mathbf{N}_\mathbf{p}^\mathbf{e}}{\partial x}\mathbf{p_N}{}^\mathbf{e}\right] dA \tag{8.110}$$

After considering all the four nodes of the element shown in Figure 8.2, the above equation can be expanded as:

$$\mathbf{A}\mathbf{u_N}{}^\mathbf{e} = \mathbf{B}; \ \mathbf{A} = \begin{bmatrix} A_1 \\ A_2 \\ A_3 \\ A_4 \end{bmatrix}; \ B = \begin{bmatrix} B_1 \\ B_2 \\ B_3 \\ B_4 \end{bmatrix} \tag{8.111}$$

Assuming that the element coordinate system is parallel to the global coordinate, as shown in Figure 8.2, the discretized governing equation for the element in the global coordinate is:

$$\mathbf{A}\mathbf{u_N} = \mathbf{B} \tag{8.112}$$

Equation 8.112 can be applied to all the elements in the flow region, leading to a series of algebraic equations for the steady governing equations.

8.4.3 Finite Volume Method

In the finite volume method (FVM), the solution domain is divided into a series of continuous control volumes and the steady governing equations are applied to each control volume [1]. The steady governing equations are employed in the integral form. Interpolation is used to express variable values at the control volume surface in terms of the values at the center of the control volume. As a result, an algebraic equation can be formed for each control volume. The FVM can be applied to any kind of grids, and therefore it is suitable for a structure of complex geometry.

In Figure 8.3, a solution domain is divided into a series of control volumes (In the 2-D plane, the length perpendicular to the plane can be viewed as unit). P is the center of the concerned control volume, marked with oblique lines. N, W, S and E are the centers of the other control volumes surrounding the concerned control volume P. n, w, s and e are the points on the corresponding interfaces between these control volumes, respectively. The integration of Equation 8.89 on the control volume V_P with P as its center gives:

$$\int_{V_P} \rho\left(\frac{\partial u}{\partial x}u + \frac{\partial u}{\partial y}v\right)dv = -\int_{V_P}\frac{\partial p}{\partial x}dv + \int_{V_P}\mu\left(\frac{\partial^2 u}{\partial x^2} + \frac{\partial^2 u}{\partial y^2}\right)dv \tag{8.113}$$

According to Gauss's divergence theorem, the above equation can be changed to the surface integration:

$$\int_{A_P} \rho u\mathbf{u}\cdot\mathbf{n}dA = -\int_{V_P}\frac{\partial p}{\partial x}dv + \int_{A_P}\mu\left(\frac{\partial u}{\partial x}\mathbf{i} + \frac{\partial u}{\partial y}\mathbf{j}\right)\cdot\mathbf{n}dA \tag{8.114}$$

where:

A_P is the surfaces enclosing the control volume V_P;
$\mathbf{n}$ is the direction normal to the surface.

Each term in Equation 8.111 can be approximated as:

$$\begin{aligned}\int_{A_P}\rho u\mathbf{u}\cdot\mathbf{n}dA &= \int_{A_w}\rho u\mathbf{u}\cdot\mathbf{n}dA + \int_{A_e}\rho u\mathbf{u}\cdot\mathbf{n}dA + \int_{A_n}\rho u\mathbf{u}\cdot\mathbf{n}dA + \int_{A_s}\rho u\mathbf{u}\cdot\mathbf{n}dA\\ &= \int_{A_e}\rho uudA - \int_{A_w}\rho uudA + \int_{A_n}\rho uvdA - \int_{A_s}\rho uvdA \\ &\approx \rho {u_e}^2\Delta y - \rho {u_w}^2\Delta y + \rho u_s u_s\Delta x - \rho u_s u_n\Delta x\end{aligned} \tag{8.115a}$$

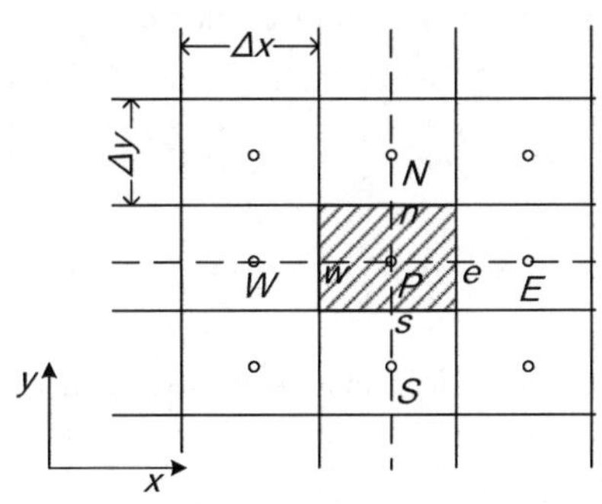

Figure 8.3 A typical finite volume grid.

$$\int_{V_P} \frac{\partial p}{\partial x} dv \approx \frac{\partial p_P}{\partial x} \Delta x \Delta y \tag{8.115b}$$

$$\begin{aligned}
&\int_{A_P} \mu\left(\frac{\partial u}{\partial x}\mathbf{i} + \frac{\partial v}{\partial y}\mathbf{j}\right) \cdot \mathbf{n} dA = \int_{A_w} \mu\left(\frac{\partial u}{\partial x}\mathbf{i} + \frac{\partial v}{\partial y}\mathbf{j}\right) \cdot \mathbf{n} dA + \int_{A_e} \mu\left(\frac{\partial u}{\partial x}\mathbf{i} + \frac{\partial v}{\partial y}\mathbf{j}\right) \cdot \mathbf{n} dA \\
&+ \int_{A_n} \mu\left(\frac{\partial u}{\partial x}\mathbf{i} + \frac{\partial v}{\partial y}\mathbf{j}\right) \cdot \mathbf{n} dA + \int_{A_s} \mu\left(\frac{\partial u}{\partial x}\mathbf{i} + \frac{\partial v}{\partial y}\mathbf{j}\right) \cdot \mathbf{n} dA \\
&= \int_{A_e} \mu \frac{\partial u}{\partial x} dA - \int_{A_w} \mu \frac{\partial u}{\partial x} dA + \int_{A_n} \mu \frac{\partial v}{\partial y} dA - \int_{A_s} \mu \frac{\partial v}{\partial y} dA \\
&\approx \mu \frac{\partial u_e}{\partial x} \Delta y - \mu \frac{\partial u_w}{\partial x} \Delta y + \mu \frac{\partial v_n}{\partial y} \Delta x - \mu \frac{\partial v_s}{\partial y} \Delta x
\end{aligned} \tag{8.115c}$$

where A_w,A_e, A_n and A_s are the interface between the concerned control volume P and the control volumes W, E, N and S, respectively.

Submitting Equation 8.115 into Equation 8.113 yields the following equation:

$$\begin{aligned}
&\rho u_e{}^2 \Delta y - \rho u_w{}^2 \Delta y + \rho u_n v_n \Delta x - \rho u_s v_s \Delta x \\
&= \mu \frac{\partial u_e}{\partial x} \Delta y - \mu \frac{\partial u_w}{\partial x} \Delta y + \mu \frac{\partial v_n}{\partial y} \Delta x - \mu \frac{\partial v_s}{\partial y} \Delta x + \frac{\partial p_P}{\partial x} \Delta x \Delta y
\end{aligned} \tag{8.116}$$

In FVM, the unknown variables are placed at the center of the control volume. Therefore, the values at the interfaces and the derivatives in Equation 8.116 need to be determined. Typical methods are the central difference, the upstream difference or the QUICK scheme. In the central difference method, the value at the interface is linear, interpolated from the values of the neighboring volumes:

$$u_w = \frac{1}{2}(u_W + u_P) \tag{8.117}$$

In the upstream method, the value at the interface is taken to be equal to the value at the upstream control volume:

$$u_w = u_W \ \textit{if}\ u_w > 0; \ \textit{or}\ u_w = u_P \ \textit{if}\ u_w < 0 \tag{8.118}$$

In the QUICK scheme, a three-point upstream-weighted quadratic interpolation for the interface value is employed.

$$u_e = \frac{6}{8} u_P + \frac{3}{8} u_E - \frac{1}{8} u_W \ \textit{if}\ u_e > 0 \tag{8.119}$$

$$u_w = \frac{6}{8} u_P + \frac{3}{8} u_W - \frac{1}{8} u_E \ \textit{if}\ u_w < 0 \tag{8.120}$$

Take the central difference scheme as an example to derive algebraic equations:

$$u_w = \frac{1}{2}(u_W + u_P);\ u_e = \frac{1}{2}(u_E + u_P);\ u_s = \frac{1}{2}(u_S + u_P) \tag{8.121}$$

$$u_n = \frac{1}{2}(u_N + u_P);\; u_s = \frac{1}{2}(u_S + u_P);\; v_n = \frac{1}{2}(v_N + v_P) \tag{8.122}$$

$$\frac{\partial u_w}{\partial x} = \frac{u_P - u_W}{\Delta x};\; \frac{\partial u_e}{\partial x} = \frac{u_E - u_P}{\Delta x};\; \frac{\partial v_s}{\partial y} = \frac{v_P - v_S}{\Delta y} \tag{8.123}$$

$$\frac{\partial v_n}{\partial y} = \frac{v_N - v_P}{\Delta y};\; \frac{\partial p_P}{\partial x} = \frac{p_E - p_W}{2\Delta x} \tag{8.124}$$

Submitting Equations 8.121 to 8.124 into Equation 8.116 gives a algebraic equation as follows:

$$\begin{aligned}&\left[2\rho u_E\Delta y - 2\rho u_W\Delta y + \rho(v_N + v_P)\Delta x - \rho(v_S + v_P)\Delta x + 8\mu\frac{\Delta y}{\Delta x} + 8\mu\frac{\Delta x}{\Delta y}\right]u_P\\&+\left(\rho u_E\Delta y - 4\mu\frac{\Delta y}{\Delta x}\right)u_E + \left(-\rho u_W\Delta y - 4\mu\frac{\Delta y}{\Delta x}\right)u_W + \left[\rho(v_N + v_P)\Delta x - 4\mu\frac{\Delta x}{\Delta y}\right]u_N\\&+\left[-\rho(v_S + v_P)\Delta x - 4\mu\frac{\Delta x}{\Delta y}\right]u_S = 2\Delta y(p_E - p_W)\end{aligned} \tag{8.125}$$

The above equation can be re-written as:

$$u_P = a_W u_W + a_E u_E + a_S u_S + a_N u_N + b \tag{8.126}$$

The coefficients in the above equation can be found easily through the comparison with Equation 8.125. In a similar way, a series of algebraic equations can be constructed for all the control volumes in the solution domain.

8.4.4 Solution Algorithms for Pressure-Velocity Coupling in Steady Flows

The governing equations are intricately coupled, because every velocity component appears in each momentum equation and in the continuity equation. Moreover, there are non-linear quantities in the momentum equations. These features also appear in the discretized algebraic equations for steady flows (e.g. Equation 8.102). For incompressible steady flows, the density is constant and hence, by definition, there is no pressure-velocity linkage in the continuity equation. Both the problems associated with the non-linearities in the momentum equations, and the lack of velocity-pressure linkage in the continuity equation, can be resolved by adopting an iterative solution strategy such as the SIMPLE algorithm [16], the SIMPLER algorithm [17] and the SIMPLEC algorithm [18].

(a) SIMPLE

SIMPLE is the short name for the Semi-Implicit Method for Pressure-Linked Equation [16]. It is essentially a guess-and-correct procedure for the calculation of pressure. The method is illustrated using the x-momentum equation in the two-dimensional steady flow with the FVM method. To initiate the SIMPLE calculation process, a pressure field p^* and velocity fields u^*, v^* are guessed. The discretized momentum equations (e.g. Equation 8.125) are solved using the guessing pressure field to yield velocity components.

Taking the discretized steady governing equation (Equation 8.125) as an example, the velocity component in the x-direction can be expressed as

$$u_P{}^* = \sum_{nb} a_{nb} u_{nb}{}^* + c_x(p_E{}^* - p_W{}^*) \tag{8.127}$$

where:

nb represents the neighboring control volumes (E, W, S, N) surrounding the volume P;
a_{nb}, c_x are the coefficients of the corresponding velocities.

Now we define the correction p' as the difference between the correct pressure field p and the guessed pressure field p^*, so that:

$$p = p^* + p\prime \tag{8.128}$$

Similarly, we define the correct velocity field as the sum of the guessed values $u*$ and the correction u':

$$u_P = u_P{}^* + u_P{}' \tag{8.129}$$

The correct velocity field and the correct pressure field also satisfy the discretized momentum equation:

$$u_P = \sum a_{nb} u_{nb} + c_x(p_E - p_W) \tag{8.130}$$

Subtraction of Equation 8.127 from Equation 8.130 yields:

$$u_P{}' = \sum a_{nb} u_{nb}{}' + c_x(p_E{}' - p_W{}') \tag{8.131}$$

At this point, an approximation is introduced:

$$\sum a_{nb} u_{nb}{}' = 0 \tag{8.132}$$

Omission of the above term is the main approximation of the SIMPLE algorithm. With this approximation, Equation 8.131 becomes:

$$u_P{}' = c_x(p_E{}' - p_W{}') \tag{8.133}$$

Therefore:

$$u_P = u_P{}^* + u_P{}' = u_P{}^* + c_x(p_E{}' - p_W{}') \tag{8.134}$$

In a similar way, we can have the equation in the y-direction:

$$v_P = v_P{}^* + v_P{}' = v_P{}^* + c_y(p_N{}' - p_S{}') \tag{8.135}$$

Integrating the continuity equation on the control volume as shown in Figure 8.3 gives:

$$u_e A_e - u_w A_w + v_n A_n - v_s A_s = 0 \tag{8.136}$$

Applying the centre difference scheme on the above equation yields:

$$\frac{u_E + u_P}{\Delta x} A_e - \frac{u_W + u_P}{\Delta x} A_w + \frac{v_N + v_P}{\Delta y} A_n - \frac{v_S + v_P}{\Delta y} A_s = 0 \tag{8.137}$$

In short, the above equation can be expressed as:

$$\sum_{i=P,E,W} b_{x,i} u_i + \sum_{i=P,N,S} b_{y,j} v_j = 0 \tag{8.138}$$

Substitution of Equation 8.134 and 8.135 into Equation 8.138 leads to a linear algebraic equation:

$$\sum_{i=P,E,W,N,S} d_i p_i' = \sum_{i=P,E,W} d_{x,i} u_i^* + \sum_{i=P,N,S} d_{y,j} \hat{v}_j^* \tag{8.139}$$

which means that the pressure difference can be solved through the guessed velocity fields.

The implementation of the SIMPLE algorithm is therefore:

1. Guess the pressure field and the velocity field.
2. Solve the momentum equation based on the guessed fields and update the guessed velocity fields.
3. Solve Equation 8.139 to find the pressure difference between the guessed pressure and the correct pressure.
4. Correct the pressure and velocity field using Equations 8.134 and 8.135.

(b) SIMPLER

The SIMPLER (SIMPLE Revised) is an improved version of SIMPLE [17]. Instead of the pressure correction equation in SIMPLE, the pressure field is solved directly. Express Equation 8.127 as follows:

$$u_P = \hat{u}_P + c_x(p_E - p_W) \tag{8.140}$$

where:

$$\hat{u}_P = \sum a_{nb} u_{nb} \tag{8.141}$$

Similarly, in the y direction:

$$u_P = \hat{u}_P + c_y(p_N - p_S) \tag{8.142}$$

By substituting Equations 8.140 and 8.142 into Equation 8.138, a linear algebraic equation can be obtained:

$$\sum_{i=P,E,W,N,S} \hat{d}_i p_i = \sum_{i=P,E,W} \hat{d}_{x,i} \hat{u}_i + \sum_{i=P,N,S} \hat{d}_{y,j} \hat{v}_j \tag{8.143}$$

which means that the pressure can be solved directly.

Subsequently, the SIMPLER procedure is completed by taking the pressure field obtained above as the guessed pressure field.

(c) SIMPLEC

The SIMPLEC (SIMPLE-Consistent) algorithm follows the same steps as SIMPLE, with differences in the correction equations for velocity [18]. The velocity correction equation (see Equation 8.133) is changed into the following formulation:

$$u_P{}' = \frac{c_x}{1 - \sum a_{nb}}(p_E{}' - p_W{}') \tag{8.144}$$

The sequence of operation of SIMPLEC is identical to that of SIMPLE.

8.4.5 Solution for Unsteady Flows

Having discussed the numerical methods for steady flows, we are now in a position to consider unsteady flows. The unsteady governing equation of momentum for a two-dimensional unsteady flow in the x-direction is taken as an example to illustrate the numerical methods for unsteady flows. The expression of the corresponding governing equation is:

$$\rho\left(\frac{\partial u}{\partial t} + \frac{\partial u}{\partial x}u + \frac{\partial u}{\partial y}v\right) = -\frac{\partial p}{\partial x} + \mu\left(\frac{\partial^2 u}{\partial x^2} + \frac{\partial^2 u}{\partial y^2}\right) \tag{8.145}$$

The first term in the above equation represents the rate of change with time and is zero for steady flows. To predict transient problems, we must retain this term in the discretization process. By discretizing Equation 8.145 with a time step Δt, using the first-order forward difference scheme, the discretized equation can be written as:

$$\rho(\frac{u^{(t+\Delta t)} - u^t}{\Delta t} + \frac{\partial u^t}{\partial x}u^t + \frac{\partial u^t}{\partial y}v^t) = -\frac{\partial p^t}{\partial x} + \mu\left(\frac{\partial^2 u^t}{\partial x^2} + \frac{\partial^2 u^t}{\partial y^2}\right) \tag{8.146}$$

After discretizing Equation 8.146 in the space, the variable at the location p at the time $t + \Delta t$ can be calculated as:

$$u_p{}^{(t+\Delta t)} = \alpha u_p{}^t + \sum_{nb} \alpha_{nb} u_{nb}{}^t + \beta \tag{8.147}$$

The second term on the right hand side is the sum of the neighboring nodes, or elements or volumes next to the location p. The right hand side of Equation 8.147 only contains values at the old time step, so that the variable of the left hand side at the new time can be calculated. It is called the explicit method.

Equation 8.147 can also be discretized on the time domain in another way:

$$\rho\left(\frac{u^{(t+\Delta t)} - u^t}{\Delta t} + \frac{\partial u^{(t+\Delta t)}}{\partial x}u_x{}^{(t+\Delta t)} + \frac{\partial u^{(t+\Delta t)}}{\partial y}v^{(t+\Delta t)}\right) = -\frac{\partial p^{(t+\Delta t)}}{\partial x} + \mu\left(\frac{\partial^2 u^{(t+\Delta t)}}{\partial x^2} + \frac{\partial^2 u^{(t+\Delta t)}}{\partial y^2}\right) \tag{8.148}$$

After discretizing Equation 8.148 in the space, the variable at the location p at the time $t + \Delta t$ can be calculated as:

$$u_p{}^{(t+\Delta t)} = \alpha' u_p{}^t + \sum_{nb} \alpha'_{nb} u_{nb}{}^{t+\Delta t} + \beta' \tag{8.149}$$

Thus, the variables at the time $t+\Delta t$ are solved requiring the values at both time t and time $t+\Delta t$. It is called the implicit method.

8.4.6 Boundary Conditions

All CFD problems are defined in terms of initial and boundary conditions. In transient problems, the initial values of all the flow variables need to be specified at all solution points in the flow domain. This involves no special measures other than initializing the appropriate data arrays in the CFD code [1]. Let us focus on the implementation of boundary conditions.

The common boundary conditions in bridge wind engineering are inlet, outlet, symmetry, periodicity and wall. Winds blow from the inlet boundaries. The magnitudes and angles of upcoming wind are set at the inlet boundaries. For RANS and DES method, the turbulence-related quantities such as k, ε, and ω are also defined at the inlet boundaries. Usually, the outlet boundaries are far from the target objects, and the flows there reach a fully developed state. The gradients of all flow variables (except the pressure) are assumed to be zero at the outlet boundaries. For the symmetry boundary, no flow crosses the boundary, and flow velocities normal to the boundary are zero. For the periodicity boundary, a pair of boundaries are coupled together with the same values as the flow variables. Wall boundaries are attached on the surfaces of the target objects directly. The relative velocity between the wall and the fluid at the surface is assumed to be zero, which is called the non-slip boundary condition.

In a wall boundary, the velocity parallel to the wall is expressed by a dimensionless velocity:

$$u^+ = u\sqrt{\frac{\tau_w}{\rho}} \tag{8.150}$$

It varies with the dimensionless height from the wall:

$$y^+ = \frac{\rho y}{\mu}\sqrt{\frac{\tau_w}{\rho}} \tag{8.151}$$

where:

y is the distance from the wall;
τ_w is the shear stress at the wall.

If $y^+ \leq 11.63$, the flow is located in a viscous sublayer and:

$$u^+ = y^+ \tag{8.152}$$

If $y^+ > 11.63$, the flow is assumed to be in a log-law region and:

$$u^+ = \frac{1}{\kappa}\ln(Ey^+) \tag{8.153}$$

where:

κ is von Karman's constant;
the value of E for smooth walls is 9.793.

8.4.7 Grid Generation

The first step to compute a numerical solution to the governing equations is the construction of a grid. The physical domain must be covered with a mesh, so that discrete points, elements or volumes are identified where the conservation laws can be applied. A well-constructed grid greatly improves the quality of the solution and, conversely, a poorly constructed grid is a major contributor to a poor result.

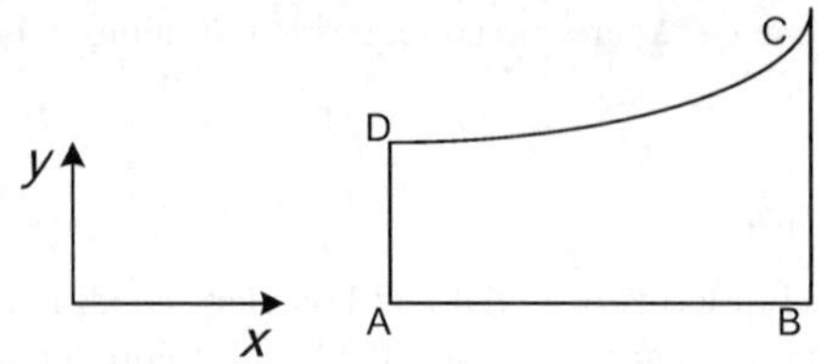

Figure 8.4 Nozzle geometry and physical domain.

Techniques for generating grids can be classified as structured and unstructured approaches [2]. Techniques for structured grid generation include complex variable methods, algebraic methods and differential equation methods, while those for unstructured grid generation are point insertion schemes, advancing front methods and domain decomposition methods. Unstructured methods are primarily based on using triangular or prismatic elements. Complex variable methods for the structured grid generation are restricted to two dimensions. Thus, only the algebraic method to generate the structured grids is explained here to illustrate the basic concept of so-called mapping.

Suppose a grid generation is needed in order to solve the flow in a diverging nozzle, as shown in Figure 8.4. The boundary lines of the physical domain of the nozzle are ABCD. The mathematic expressions of the boundaries are:

AB: $1 \leq x \leq 2, y = 0$;
BC: $x = 2, 0 \leq y \leq 4$;
CD: $1 \leq x \leq 2, y = x^2$;
DA: $x = 1, 0 \leq y \leq 1$.

To generate computational grids using the algebraic method, known functions are used to take arbitrarily shaped physical domain into rectangular computational domain, as shown in Figure 8.5.

Although the computational domain is not required to be rectangular, the usual procedure uses a rectangular domain for simplicity. The computational domain is a rectangle with the Cartesian coordinates (ξ, η). The boundary lines abcd are corresponding to the ABCD in the physical domain with the following expressions:

ab: $0 \leq \xi \leq 1, \eta = 0$;
bc: $\xi = 1, 0 \leq \eta \leq 1$;
cd: $0 \leq \xi \leq 1, \eta = 1$;
da: $\xi = 0, 0 \leq \eta \leq 1$.

A relationship between the computational domain and the physical domain has to be constructed for the boundary lines:

$$x = \xi + 1;\ y = \eta(\xi + 1)^2 \tag{8.154}$$

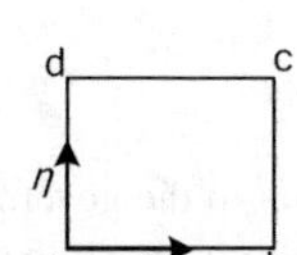

Figure 8.5 Computational domain.

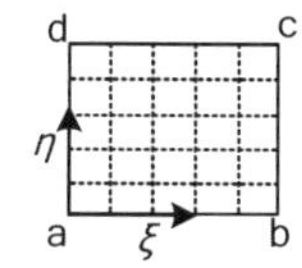

Figure 8.6 Meshing in computational domain.

Then a uniform mesh are generated in the computational domain as shown in Figure 8.6, with $\Delta\xi = 0.2$ and $\Delta\eta = 0.2$. The coordinates of each of the nodes in the computational domain can be obtained easily, and then the coordinates of the corresponding nodes in the physic domain can be calculated using Equation 8.154. The grid system in the physical domain generated is shown in Figure 8.7. It is clear that the grid points generated in the physical domain are mapped from the rectangular computational domain in a unique way.

8.4.8 Computing Techniques

A system of linear algebraic equations is generated through the discretization of the governing equations for fluid flow in the previous sections. The complexity and size of the set of algebraic equations depends on the dimensionality of the problem, the number of grid nodes and the discretization practice. There are two families of solution techniques for linear algebraic equations: direct methods and iterative methods. Iterative methods are commonly used in practice and are introduced in this section. Well-known iterative methods include Jacobi, Gauss-Seidel and relaxation iterations. In order to improve the convergence rate of iterative methods, the multi-grid method is also introduced here.

(a) Jacobi Iteration

A linear algebraic system of n equations and n unknowns have the following expression

$$\sum_{j=1}^{n} a_{ij}x_j = b_i \; (i = 1, 2, \ldots, n) \tag{8.155}$$

In all iterative methods, the expression is rearranged and the variables to be solved are located at the left:

$$x_i = \frac{b_i}{a_{ii}} - \sum_{j=1, j\neq i}^{n} \frac{a_{ij}}{a_{ii}} x_j \tag{8.156}$$

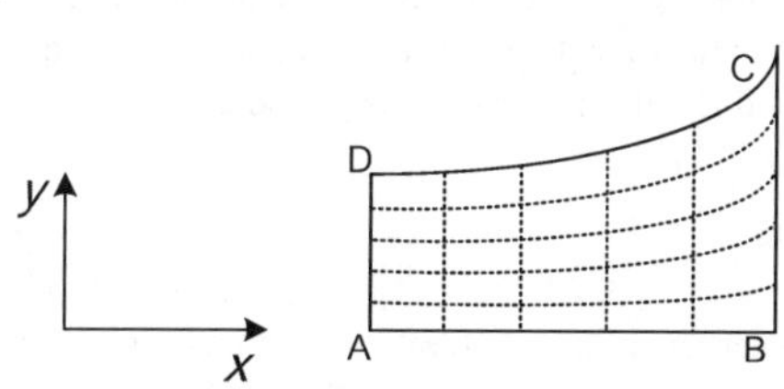

Figure 8.7 Meshing in physical domain.

In the Jacobi method, the values after the kth iteration are calculated based on the variable values from the last iteration step:

$$x_i^{(k)} = \frac{b_i}{a_{ij}} - \sum_{j=1, j\neq i}^{n} \frac{a_{ij}}{a_{ii}} x_j^{k-1} \tag{8.157}$$

This equation is the iteration equation for the Jacobi method in the form used for actual calculations.

(b) Gauss-Serdel Iteration

The kth iteration can be completed based on the values in the k-1 iteration and the values already updated in the kth iteration by the following equation:

$$x_i^{(k)} = \frac{b_i}{a_{ij}} - \sum_{j=1}^{i-1} \frac{a_{ij}}{a_{ii}} x_j^{k} - \sum_{j=i+1}^{n} \frac{a_{ij}}{a_{ii}} x_j^{k-1} \tag{8.158}$$

This is the Gauss-Serdel iteration.

(c) Relaxation Method

The convergence rate of the Jacobi and Gauss-Seidel methods depends on the properties of the iteration matrix. The iteration procedure of Jacobi method can also be written as:

$$x_i^{(k)} = x_i^{(k-1)} + \left(\frac{b_i}{a_{ij}} - \sum_{j=1}^{n} \frac{a_{ij}}{a_{ii}} x_j^{k-1} \right) \tag{8.159}$$

It can be improved by the introduction of a relaxation parameter α on the second term in the right hand of Equation 8.159:

$$x_i^{(k)} = x_i^{(k-1)} + \alpha \left(\frac{b_i}{a_{ij}} - \sum_{j=1}^{n} \frac{a_{ij}}{a_{ii}} x_j^{k-1} \right) \tag{8.160}$$

The relaxation parameter can also be introduced to the Gauss-Seidel iteration:

$$x_i^{(k)} = x_i^{(k-1)} + \alpha \left(\frac{b_i}{a_{ij}} - \sum_{j=1}^{i-1} \frac{a_{ij}}{a_{ii}} x_j^{k} - \sum_{j=i}^{n} \frac{a_{ij}}{a_{ii}} x_j^{k-1} \right) \tag{8.161}$$

Different values of parameter α will result in different iterative rates. If $0 < \alpha < 1$, the procedure is an under-relaxation method, whereas $\alpha > 1$ is called over-relaxation. Relaxation method may be advantageous if selecting an optimum value of α. Unfortunately, the optimum value of the relaxation parameter is meshing-dependent and it is difficult to give precise guidance.

(d) Multi-grid Method

It has been found that the convergence rate of iterative methods rapidly reduces as the mesh is refined. Multi-grid methods are therefore designed to improve this problem. In this method, iterations are completed on grids with different intervals in space.

Let us take a simple two-stage multi-grid procedure as an example. The target equation is $\mathbf{AX} = \mathbf{B}$. The iterations, based on the iterative method described above, are performed on the finer grid with interval h. The solution ($\mathbf{X}^h$) and the residual errors ($\mathbf{R}^h = \mathbf{B}^h - \mathbf{A}^h\mathbf{X}^h$) are obtained at the same time. The calculations then turn to the coarser grid, with interval mh ($m > 1$). The target equation becomes $\mathbf{A}^{mh}\mathbf{E}^{mh} = \mathbf{R}^{mh}$ ($\mathbf{A}^{mh}$ and $\mathbf{R}^{mh}$ are obtained from $\mathbf{A}^h$ and $\mathbf{R}^h$). After iterations, the error array $\mathbf{E}^{mh}$ is solved and transferred into $\mathbf{E}^h$ for the finer grids. Finally, the results can be combined as: $\mathbf{X}^h = \mathbf{X}^h + \mathbf{E}^h$. It has been proved that the iterations on finer grid can reduce the errors with long waves, while the short wave errors can be reduced on the coarser grids.

8.4.9 Verification and Validation

Errors and uncertainties are unavoidable in CFD simulation. It is necessary to develop rigorous methods to quantify the level of confidence in its results: verification and validation [1]. The process of verification involves quantification of the errors. For example, round-off error can be assessed by comparing CFD results obtained using different levels of machine accuracy. Iterative convergence error can be quantified by investigating the effects of systematic variation of the truncation criteria for all residuals on target quantities of interest. Discretization error is quantified by systematic refinement of the space and time meshes.

The process of validation involves quantification of the input uncertainty and physical model uncertainty. For example, input uncertainty can be estimated by means of sensitivity analysis or uncertainty analysis. The physical modeling uncertainty requires comparison of CFD results with high-quality experimental or field measurement results.

8.4.10 Applications in Bridge Wind Engineering

CFD technique has been applied to bridge wind engineering since 1990s, starting with the exploration of aerodynamic characteristics of a bridge deck. The static aerodynamic force coefficients are basic properties of a bridge deck, which have been computed using CFD and compared with the results from wind tunnel tests [13,14,19–26]. The flow features around different types of bridge decks, with and without ancillary, have been explored extensively [19,21,27,28]. The dynamic aerodynamic characteristics, including vortex-induced phenomenon [13,14,19] and flutter derivatives [13,24,25,27,29,30], have been computed and compared with wind tunnel test results. With the rapid development of computer hardware, CFD can also be employed to investigate more complex flow conditions, such as bridge deck vibration of large magnitude and the vehicle-bridge coupled system. The following sections present some of these applications.

8.5 CFD for Force Coefficients of Bridge Deck

In this section, the process of computing static aerodynamic forces on a bridge deck using the CFD tool is introduced, and the simulated aerodynamic force coefficients are validated through comparison with the results from wind tunnel tests.

8.5.1 Computational Domain

Figure 8.8 shows the cross-section of the box girder of a completed long-span bridge. The width and height of the deck are denoted as B and H, respectively. It is taken as an example to illustrate how to obtain the static aerodynamic forces, using CFD as a tool.

For the completed bridge deck, the side fairings are fixed to form a approximate streamlined shape. The protection rail, hand rail and maintenance trace are also equipped on the upper and bottom surfaces

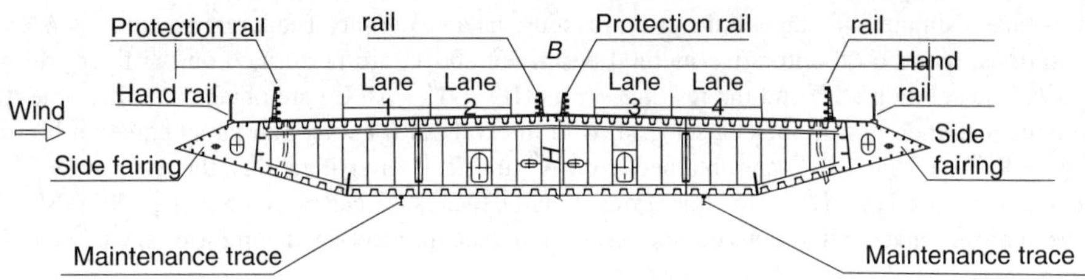

Figure 8.8 Cross-section of a completed bridge deck (from [35]) (Reproduced with permission from Elsevier).

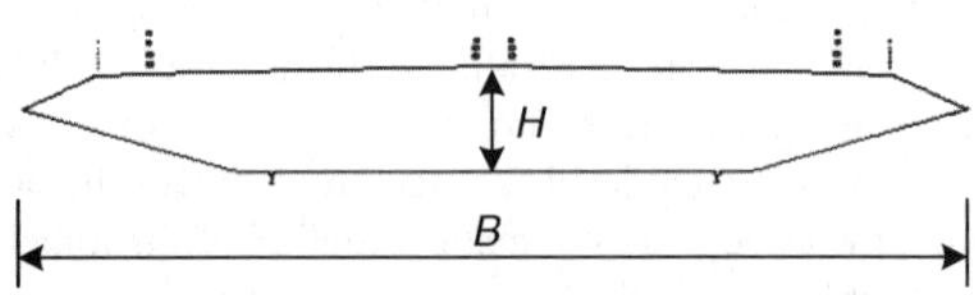

Figure 8.9 Cross-section of 2-D bridge deck model.

of the deck. The first step is to simplify the cross-section and generate a 2-D geometric model. In the flow field of air, only the outer boundaries of the deck interact with air, so the outer boundaries of the deck are taken as the basic geometric model.

Figure 8.9 shows the simplified geometric model for the deck. Unlike the prototype deck, the rails in the 2-D model do not have an identical cross-section to that of the prototype, because of vertical posts. The vertical posts are taken out, but their influence is taken into consideration through the horizontal bars, based on the principle that the porosity ratio of the simplified rails in the 2-D model is equal to that of the prototype deck.

In the real situation, the bridge deck is exposed to almost unbounded air. However, artificial boundaries have to be formed to limit the computational domain. As shown in Figure 8.10, four outer boundaries (B_In, B_Out, B_Up, B_Down) are provided. The computational domain is the area surrounded by the four outer boundaries. The size of the computational domain (the sizes of L_u, L_d, H_u, H_d in Figure 8.10) is obtained through a series of parameter optimizations, until the aerodynamic forces are not too sensitive to the size.

8.5.2 Meshing

After the computational domain is determined, the meshing needs to be performed. As shown in Figure 8.11, the mesh near the deck surfaces is dense because the gradients of flow variables are

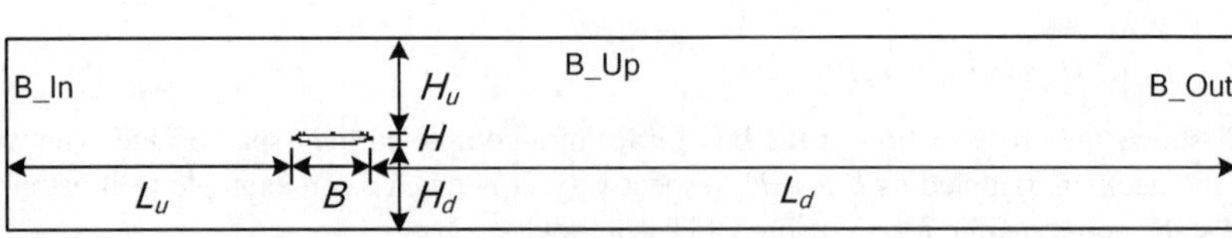

Figure 8.10 Computational domain of bridge deck model.

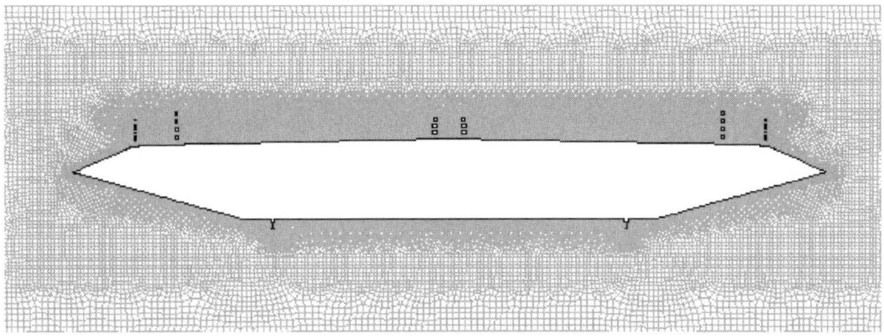

Figure 8.11 Meshing around bridge deck.

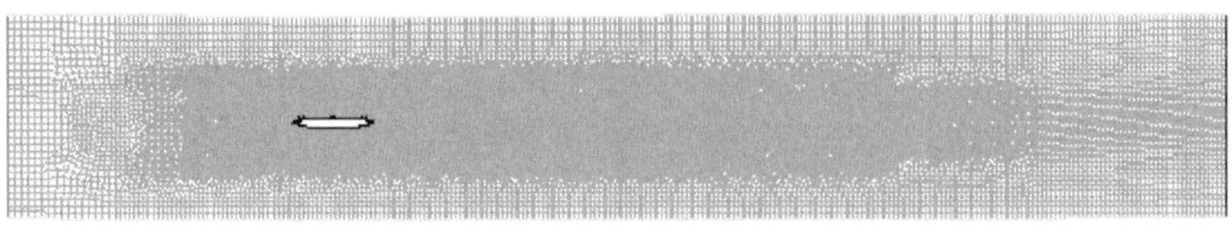

Figure 8.12 Meshing for entire computational domain.

large in these locations. The maximum y^+ of the first layer grids near the deck is less than 2. As shown in Figure 8.12, the mesh near the outer boundaries becomes sparse because the flow there is little affected by the bridge deck. In the transition region from the surface of the deck to the out boundaries, the mesh changes smoothly from dense status to sparse status.

8.5.3 Boundary Conditions and Numerical Method

For the surface of the bridge deck, a non-slip wall boundary condition is employed. The outer boundary B_In in Figure 8.10 is defined as velocity-inlet boundary, with a uniform wind velocity. The outer boundary B_Up and B_Down are set as symmetric boundary, which means that the gradients of flow variables perpendicular to the boundary are equal to zero. B_Out is defined as pressure-outlet boundary with a pressure value of zero. The steady RANS method is employed to solve the flow and the SST k-ω model is used to model the turbulence. The coupled pressure and velocity are computed using the SIMPLEC method. The pressure is discretized in second order, while other flow variables are discretized in the QUICK scheme.

8.5.4 Aerodynamic Force Coefficients and Flow Field

By slightly rotating the deck shown in Figure 8.10, the cases of different wind angles of attack to the bridge deck can be realized. The simulated aerodynamic force coefficients of the bridge deck in the wind coordinate system are computed and compared with the results from wind tunnel tests. The results are shown in Figure 8.13, which shows that the simulated aerodynamic forces agree well with the experimental results. The velocity contours around the deck at the attack angles of $-8°$, $-4°$, $0°$, $4°$ and $8°$ are shown in Figure 8.14; these cannot be obtained easily in wind tunnel tests.

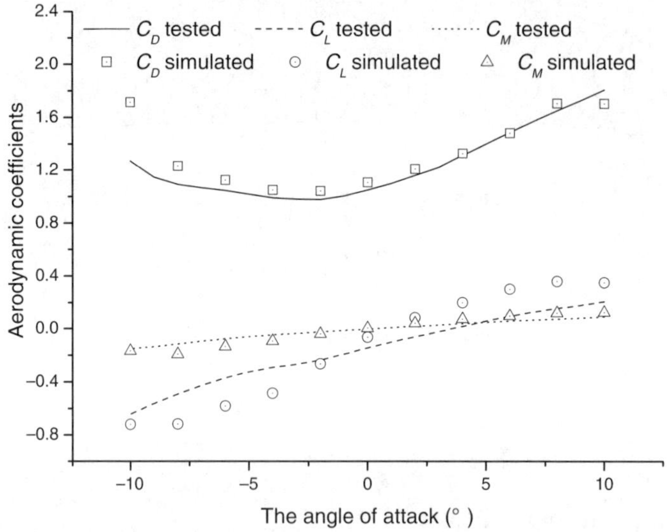

Figure 8.13 Static aerodynamic force coefficients.

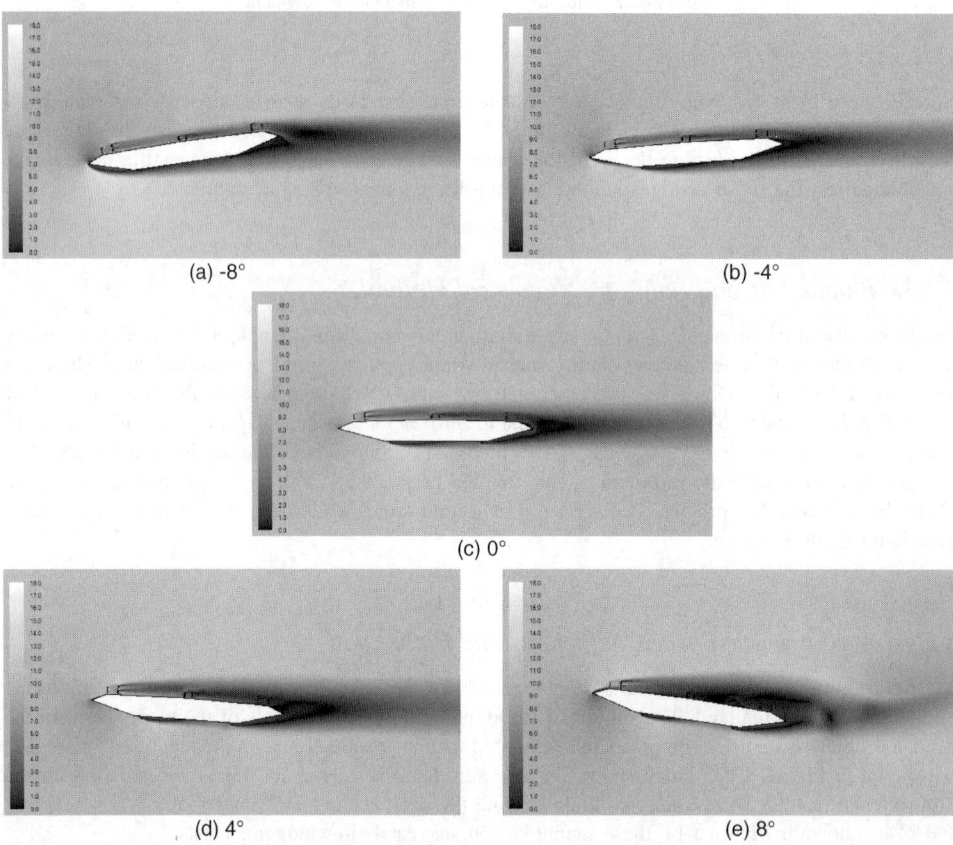

Figure 8.14 Velocity contour around the deck at different attack angle.

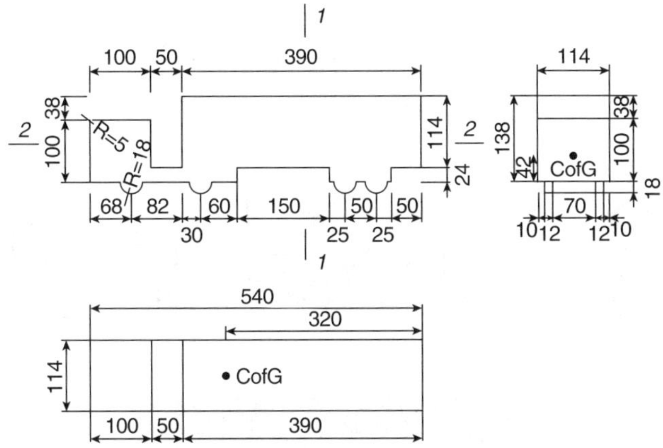

Figure 8.15 Dimensions of vehicle model (unit: mm).

8.6 CFD for Vehicle Aerodynamics

Road vehicles may potentially overturn or slide when exposed to a strong crosswind environment. The aerodynamic loads on a vehicle should be determined in order to ascertain the safety of the vehicle. In this section, the aerodynamic forces of a road vehicle on the ground are determined by CFD. The numerical model is generated and simulated in a 3-D domain.

8.6.1 Computational Domain

A high-sided articulated vehicle, a type which has been investigated by many scholars [31–34], is selected as a reference vehicle. The geometric model of the vehicle is scaled with a ratio of 1: 25 and is shown in Figure 8.15.

A 3-D box computational domain is formed around the vehicle. Figure. 8.16 shows the top view of the domain. The length and width of the domain are 15L and 10L, respectively, in which L is the length of the vehicle. The height of the domain (vertical to the plane of Figure 8.16) is 6.7 times the height of

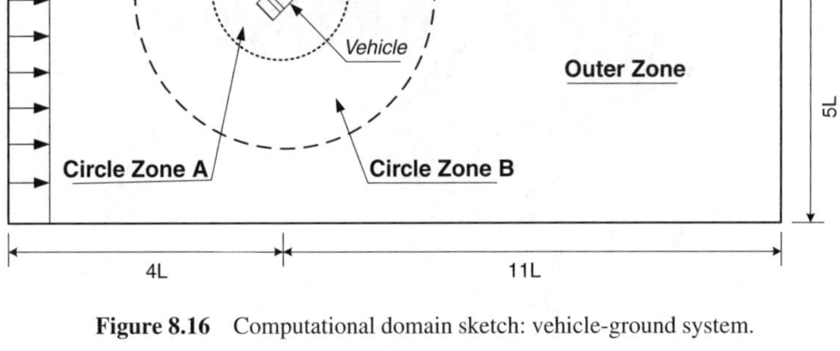

Figure 8.16 Computational domain sketch: vehicle-ground system.

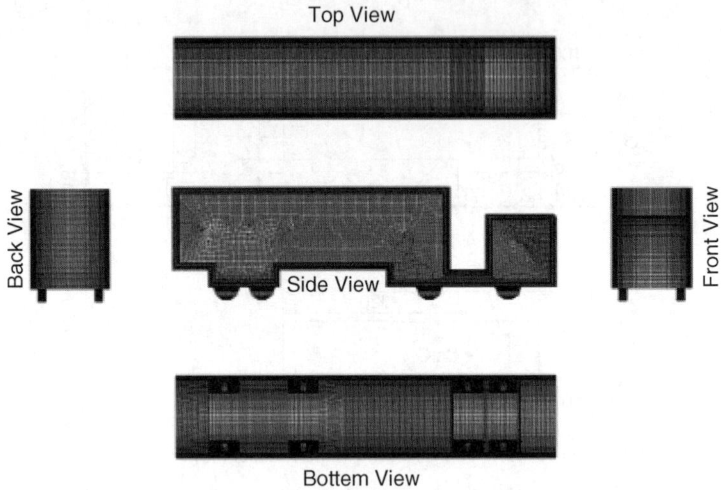

Figure 8.17 Meshing for vehicle.

vehicle. To take into consideration various yaw angles, α (the angle between inflow direction and the vehicle longitudinal axis, as shown in Figure 8.16), a rotating region is established for numerical simulation of the vehicle on the ground. The circle marked Zone A in which the vehicle model is located, is designed to be rotatable to achieve any designated yaw angle. Zone B is constructed as a transition region between Zone A and the outer zone to accommodate the change of grid size.

8.6.2 Meshing

Meshes change gradually from smaller sizes at the locations near the vehicle, to larger sizes near the outer boundaries of the computational domain. The height of the first layer grids over the surfaces of the vehicle and the ground underneath the vehicle is 1×10^{-5}m only. The maximum y^+ of the first layer grids near the vehicle and the ground is less than 1. A total of 5.4 million grids are created for the case of the vehicle on the ground (see Figures 8.17 and 8.18).

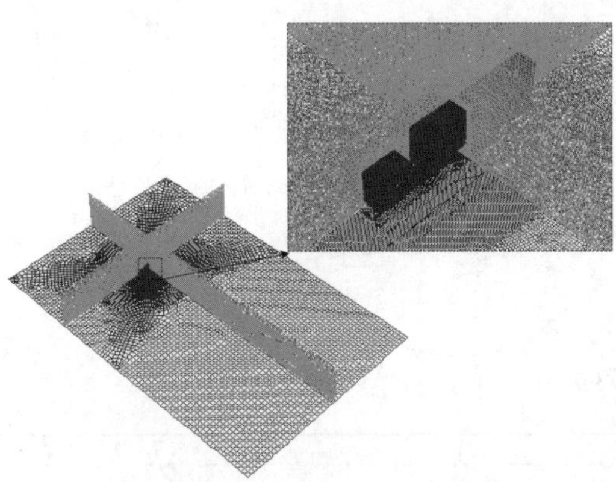

Figure 8.18 Meshing for vehicle on ground.

8.6.3 Boundary Conditions and Numerical Method

The surfaces of the vehicle are modeled as non-slip wall boundaries, and the entire computational domain is enclosed by six outer boundaries. Four outer boundaries from the top view can be seen from Figure 8.16. The left boundary is set as flow inlets with a uniform velocity of 10 m/s, which leads to a Reynolds number of 1.13×10^5 in terms of the height of the vehicle. The right boundary is specified as flow outlets with zero pressure. The other outer boundaries are all defined in such a way that the gradients of flow variables (including velocity and pressure) normal to those boundary faces are zero. The unsteady RANS method is used together with the SST k-ω turbulence model. The governing equations are discretized using the QUICK scheme, based on the finite volume method. The SIMPLEC algorithm is employed for the coupling of velocity and pressure. The time integration is performed using the second-order implicit method.

8.6.4 Simulation Results

There are six aerodynamic load components acting on the vehicle in the Cartesian coordinate system: lift force F_L, drag force F_D, side force F_S, pitching moment M_P, yawing moment M_Y and rolling moment M_R, as shown in Figure 8.19.

All of these components refer to the center of gravity of the unloaded vehicle and vary with wind yaw angle α. The non-dimensional aerodynamic force coefficients are defined by:

$$C_L = \frac{F_L}{qA};\ C_D = \frac{F_D}{qA};\ C_S = \frac{F_S}{qA} \tag{8.162}$$

$$C_P = \frac{M_P}{qAL};C_Y = \frac{M_Y}{qAL};C_R = \frac{M_R}{qAL} \tag{8.163}$$

$$q = 0.5\rho U^2 \tag{8.164}$$

where:

ρ is the air density;
U represents the mean wind speed at the inlet boundary of the computational domain;
A is the frontal project area of the vehicle without wheels and refers to the project area in the X-Y plane in this study, as shown in Figure 8.19;
L represents the maximum length of the vehicle in the Z-Y plane, as shown in Figure 8.19.

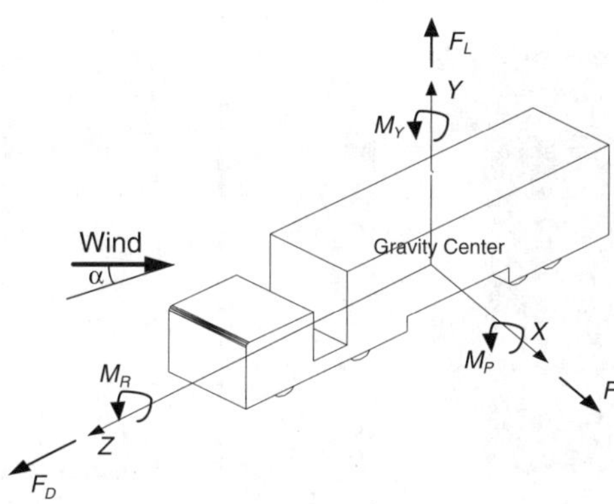

Figure 8.19 Definitions of aerodynamic forces.

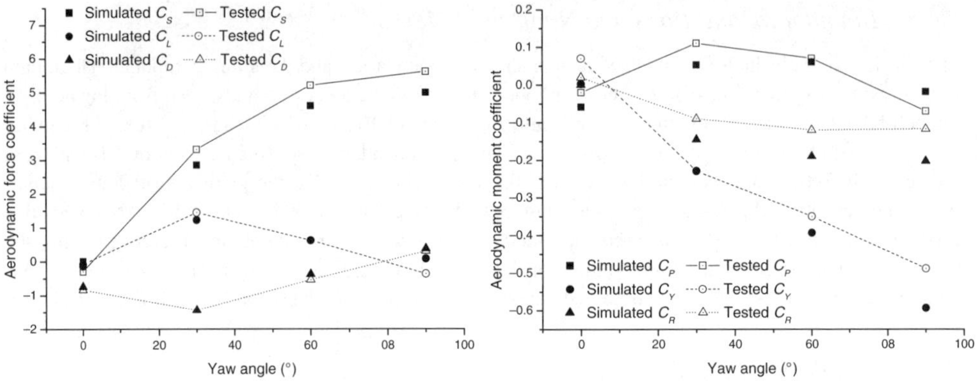

Figure 8.20 Aerodynamic force coefficients of vehicle.

Four cases, with wind yaw angles of 0°, 30°, 60° and 90°, are simulated. The comparisons of numerical and experimental aerodynamic coefficients [35] are performed and the results are displayed in Figure 8.20. It can be seen that the varying trend of numerical aerodynamic coefficients with the yaw angle is the same as that of the experimental results.

The non-dimensional pressure coefficients for pressures acting on the surface of the vehicle are defined by:

$$C_p = \frac{p}{q} \tag{8.165}$$

where p is the air pressure acting on the vehicle.

The distributions of mean pressure coefficients over the surfaces of the vehicle at 90° yaw angle are shown in Figure 8.21. The stagnation area can be observed clearly on the windward surface of the vehicle, in which high positive pressures occur. Over other surfaces of the vehicle, there are all negative wind pressures because of flow separations and wakes.

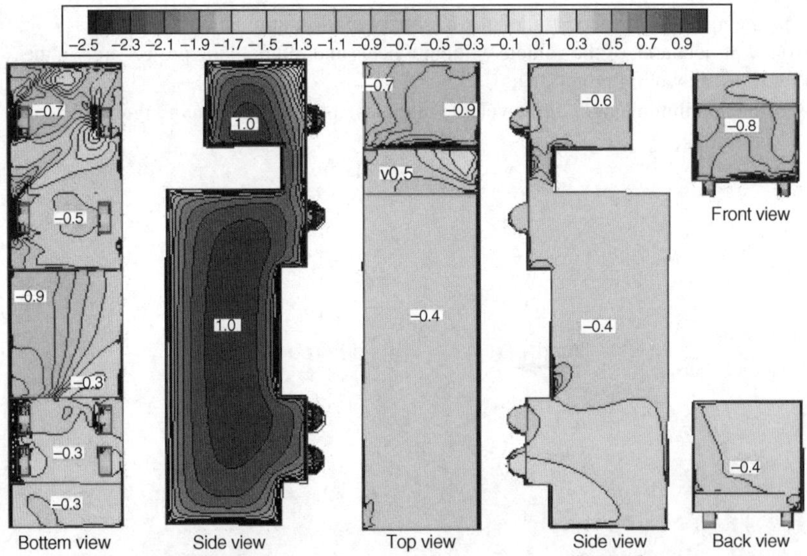

Figure 8.21 Mean pressure coefficient distributions over vehicle surfaces.

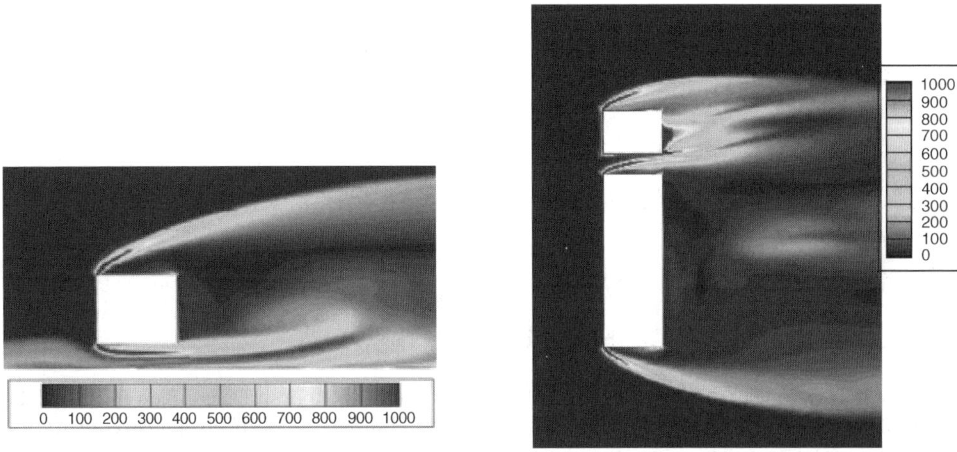

(a) Section 1-1 (b) Section 2-2

Figure 8.22 Contours of instantaneous vorticity magnitude for vehicle on ground (unit: sec^{-1}).

The instantaneous flow structures around the vehicle at the 90° yaw angle are illustrated in Figure 8.22 for the cross-sections located at the middle length of the trailer of the vehicle (section 1-1 shown in Figure 8.15) and the middle height of the vehicle (section 2-2 shown in Figure 8.15).

It can be seen clearly that vortexes are generated from the sharp edges between the windward surface and the top surface, the bottom surface and the side surfaces of the vehicle, forming separation zones. At the back of the vehicle – particularly the trailer – the flow is highly turbulent, forming wake regions. The mean flow structures are shown in Figure 8.23 in terms of projected streamlines and velocity contours for section 1-1 and section 2-2, respectively. It can be seen that after separations, the flow is

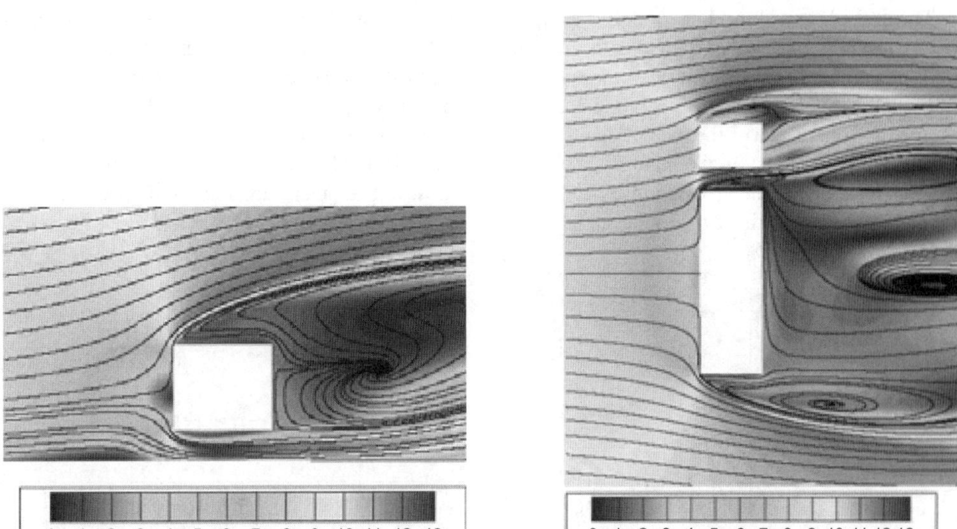

(a) Section 1-1 (b) Section 2-2

Figure 8.23 Projected streamlines and averaged velocity magnitude contours for vehicle on ground (unit: m/s).

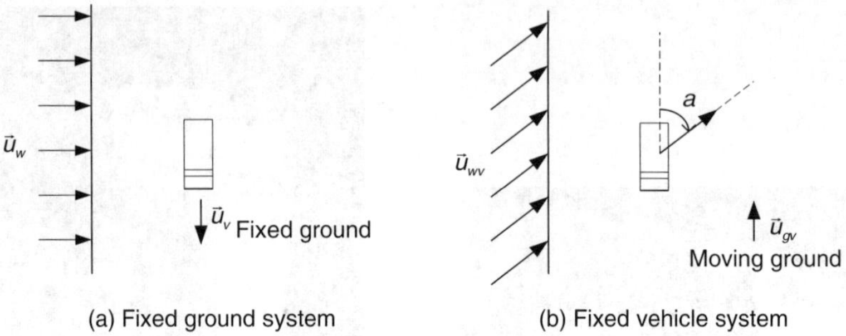

Figure 8.24 System transform.

accelerated. In particular, because of the small gaps between the bottom surface of the vehicle and the ground and between the tractor and the trailer, the mean wind speeds are relatively high over there. The mean wind speeds are relatively small in the wake regions.

8.6.5 Vehicle Moving on Ground

The previous discussion is for the vehicle on ground without movement. Now let us consider road vehicles moving on ground. As illustrated in Figure 8.24a, a vehicle moves with a velocity $\mathbf{u}_v$ and the winds blow perpendicular to the vehicle with a velocity of $\mathbf{u}_w$. If the reference coordinate system is fixed on the vehicle as in Figure 8.24b, the moving velocity of the ground $\mathbf{u}_{gv}$, the velocity of upcoming wind $\mathbf{u}_{wv}$ and its yaw angle α are calculated by:

$$\mathbf{u}_{gv} = -\mathbf{u}_v;\ \mathbf{u}_{wv} = \mathbf{u}_w - \mathbf{u}_v;\ \alpha = arctan\left(\frac{|\mathbf{u}_w|}{|\mathbf{u}_v|}\right) \tag{8.166}$$

The case presented in Figure 8.24b is actually replaced by that shown in Figure 8.25a in the computational simulation. The yaw angle in this case is caused by the movement of the vehicle.

As illustrated in the simulation of the vehicle on ground without movement, the yaw angle is realized through the rotation of the vehicle model, as shown in Figure 8.25b. What we want to know is whether there is any difference of force coefficients between the vehicle moving on ground and the static vehicle of the same yaw angle. In the numerical setting, the ground is fixed for the static vehicle. In the case of the moving vehicle, the ground moves with a velocity of $\mathbf{u}_{gv}$. The other numerical settings of the moving vehicle cases are the same as the static vehicle cases.

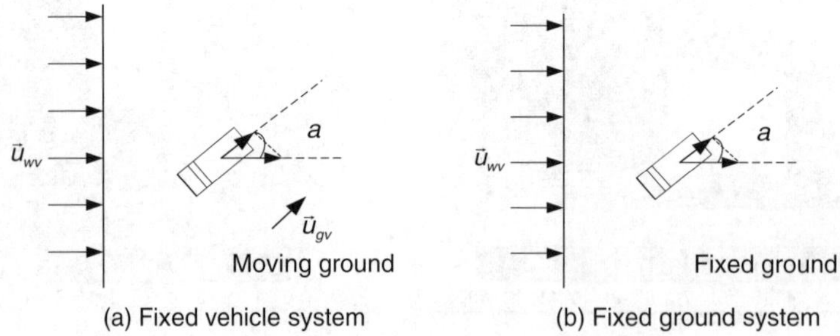

Figure 8.25 Modeling of moving vehicle.

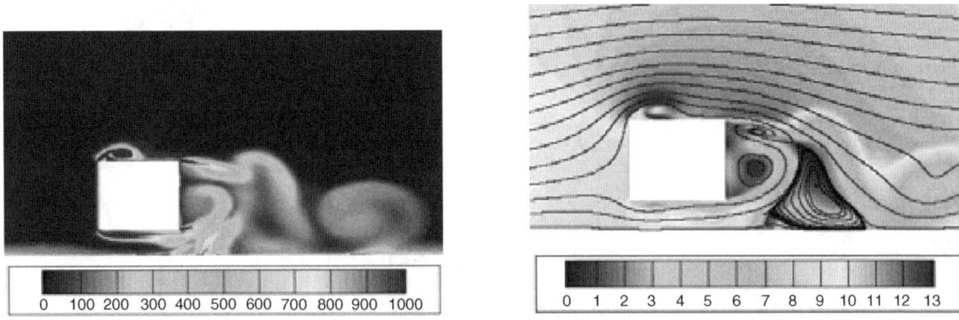

(a) Instantaneous vorticity magnitude (b) Averaging velocity contour and streamline

Figure 8.26 Flow characteristics in Section 1–1 at 30° yaw angle.

For the sake of illustration, the moving vehicle with a yaw angle of 30° is simulated. The flow characteristics of the moving vehicle in sections 1-1 and 2-2 are shown in Figures 8.26 and 8.27 respectively. Figure 8.28 shows the surface pressure coefficient distribution of the moving vehicle at a 30° yaw angle.

Figure 8.29 shows the aerodynamic coefficients of the vehicle in both the moving and static situations, with the yaw angles defined above. It can be seen that the differences of aerodynamic coefficients between the two conditions are small. The absolute aerodynamic coefficients of the moving vehicle seem slightly larger than those of the static vehicle.

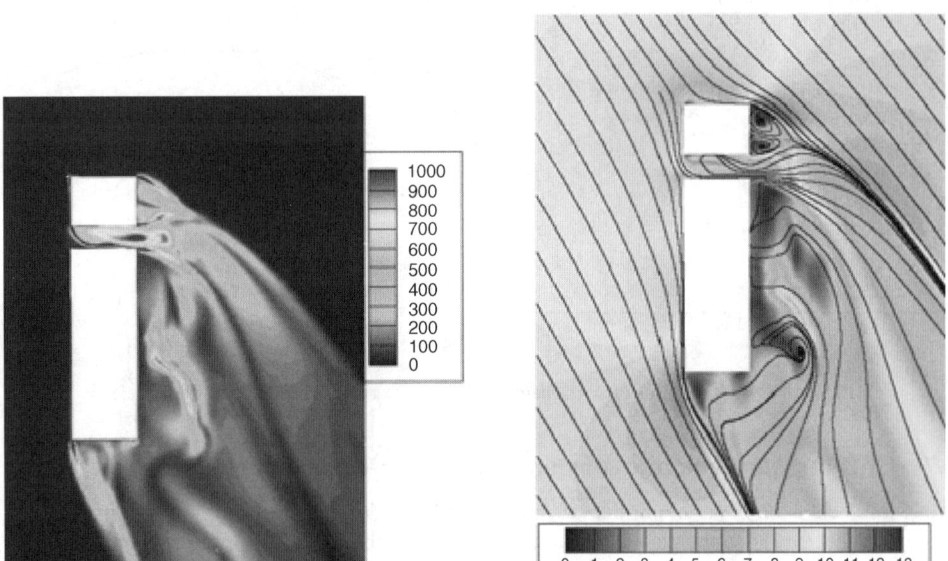

(a) Instantaneous vorticity magnitude (b) Averaging velocity contour and streamline

Figure 8.27 Flow characteristics in Section 2–2 at 30° yaw angle.

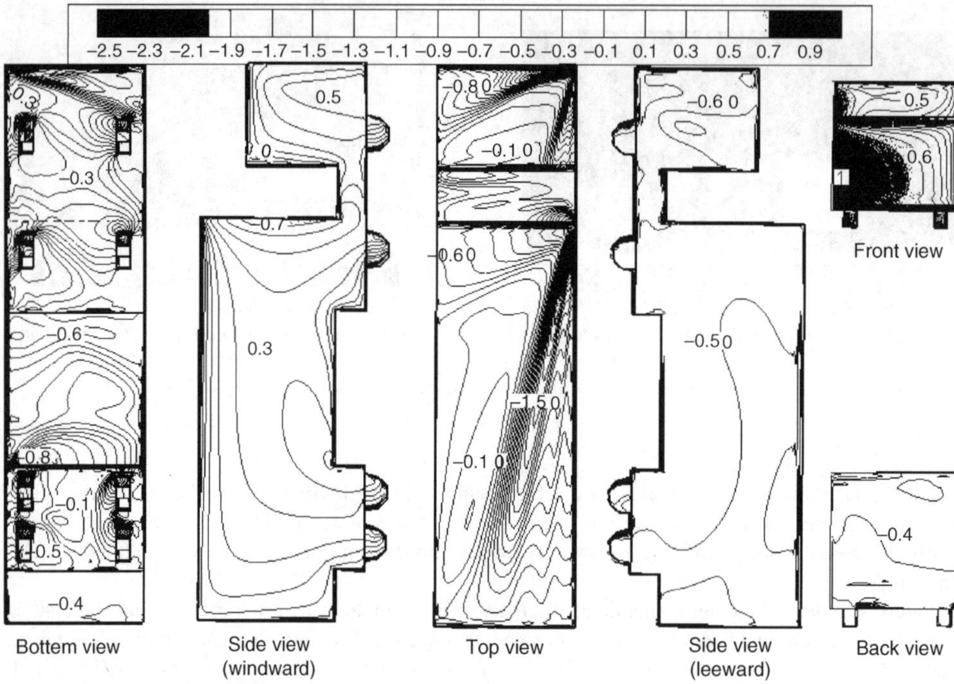

Figure 8.28 Mean pressure coefficients of moving vehicle at 30° yaw angle.

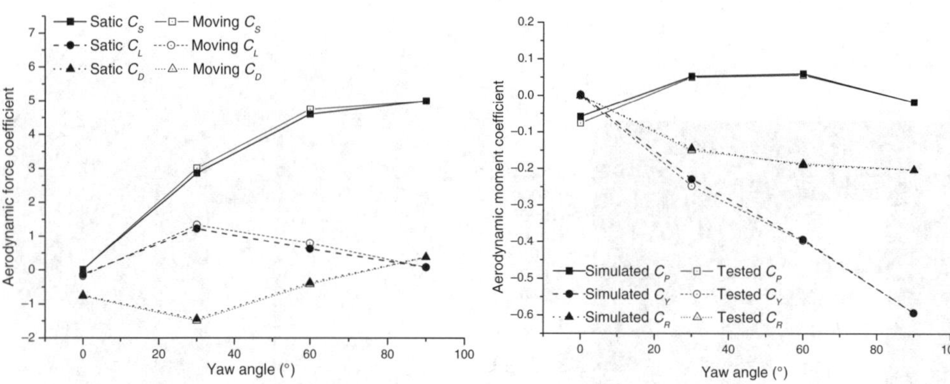

Figure 8.29 Comparison of aerodynamic coefficients between moving and static vehicle.

8.7 CFD for Aerodynamics of Coupled Vehicle-Bridge Deck System

When a vehicle is running on a bridge deck, a coupled vehicle-bridge deck system is formed. Compared with the vehicle on the ground, the aerodynamic characteristics of the vehicle-bridge system are much more complex because of the mutual influence on flow around the vehicle and the bridge deck. In this section, the vehicle on the bridge deck is simulated for a wind yaw angle of 90° (i.e. wind perpendicular to the bridge deck), and then the computational results are compared with the wind tunnel test results.

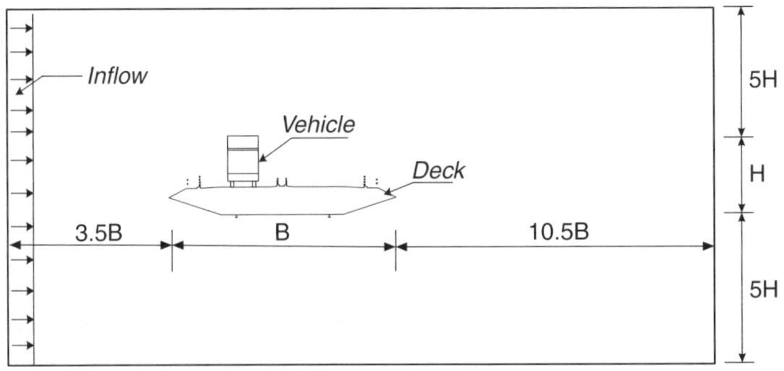

Figure 8.30 Computational domain sketch: vehicle-bridge system.

8.7.1 Computational Domain

The same vehicle model as used in the case of vehicle on ground is adopted, and the same deck model as used in the case for force coefficient simulation is chosen. Figure 8.30 shows the cross-section of the computational flow domain of the vehicle-bridge system. The vehicle is located on the first lane of the bridge deck. The length and height of the computational flow domain are $15B$ and $11H$, respectively, where B and H represent the width of the deck and the height of the vehicle-bridge system. The width of the computational domain, which is the length of the bridge deck model, is seven times the length of the vehicle. The vehicle is located at the middle of the width of the computational domain.

8.7.2 Meshing

As above, meshes change gradually form smaller sizes at locations near the vehicle and deck, to larger sizes near the outer boundaries of the computational domain. The height of the first layer grids, adjacent to the surfaces of the vehicle, is set as about 1×10^{-5} m, with $y^+ < 1$, while the height of the first layer grids, next to the surfaces of the bridge deck is about 1.0×10^{-4} m, with $y^+ < 7$. A total of 4.8 million grids are generated for the vehicle-bridge system (see Figure 8.31).

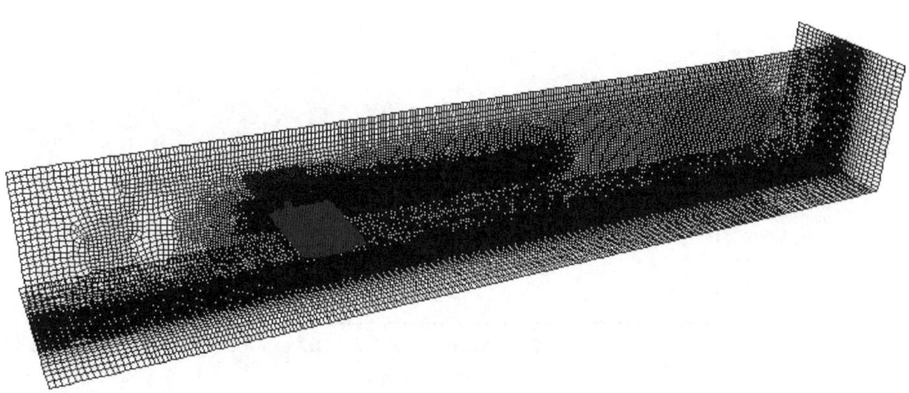

Figure 8.31 Meshing for vehicle-deck system.

Table 8.1 Aerodynamic coefficients of vehicle on bridge deck

	C_S	C_L	C_D	C_P	C_Y	C_R
Simulation	4.439	−0.003	0.254	−0.123	−0.627	−0.229
Wind tunnel	4.000	−0.641	0.287	−0.281	−0.602	−0.226

8.7.3 Boundary Conditions and Numerical Method

The surfaces of both the vehicle and the bridge deck are all modeled as non-slip wall boundaries. The entire computational domain is formed by six boundaries. The left boundary in Figure 8.30 is set as flow inlets with a uniform velocity of 10 m/s, which leads to a Reynolds number of 1.13×10^5 in terms of the height of the vehicle. The right boundary in Figure 8.30 is specified as flow outlets with zero pressure. All of the other boundaries are defined in such a way that the gradients of flow variables (including velocity and pressure) normal to those boundary faces are zero. The unsteady RANS method is used, together with the SST k-ω turbulence model. The governing equations are discretized using the QUICK scheme, based on the finite volume method. The SIMPLEC algorithm is employed for the coupling of velocity and pressure. The time integration is performed using the second-order implicit method.

8.7.4 Simulation Results

The aerodynamic force coefficients of the vehicle on the bridge deck are listed and compared with wind tunnel test results in Table 8.1. The simulation results and wind tunnel results are compatible in general, but there is a large difference in the lift force coefficient, which needs to be investigated further.

For the vehicle on the bridge deck, the simulated aerodynamic coefficients of the bridge deck along its length are presented and compared with the wind tunnel test results, and also the CFD results of the bridge deck only, in Figure 8.32. The vehicle is located from $z = 0$–0.54, with its head toward the

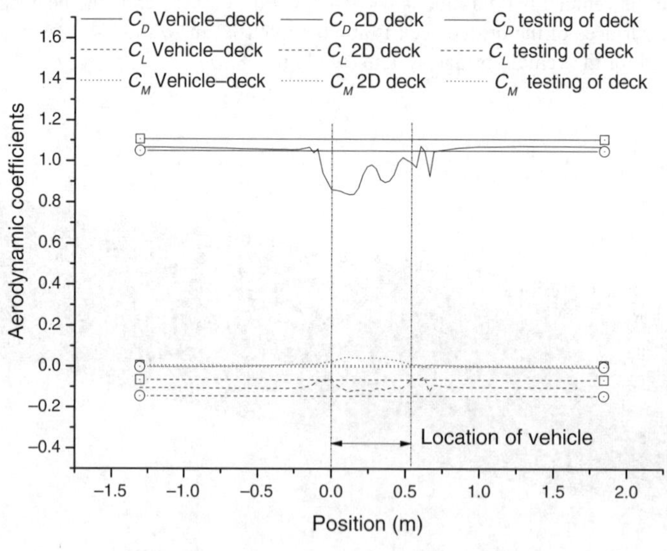

Figure 8.32 Aerodynamic coefficients of bridge deck.

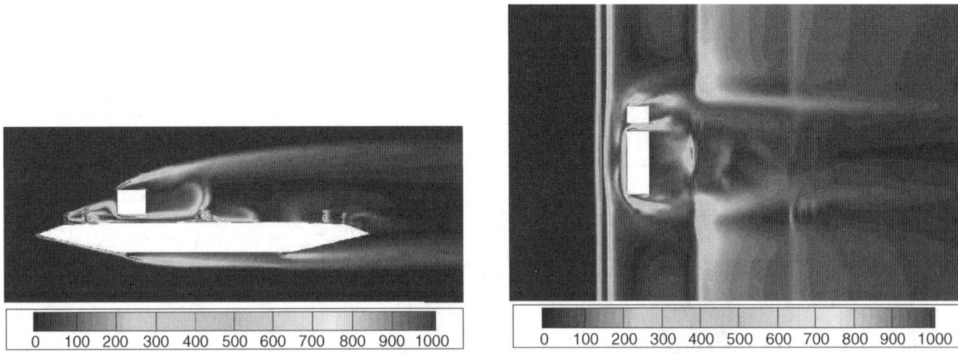

(a) Instantaneous verticity magnitude (b) Averaging velocity contour and streamline

Figure 8.33 Flow characteristics around vehicle-deck system.

positive direction of the z-axis. At locations away from the vehicle, the aerodynamic coefficients of the bridge deck are close to the wind tunnel test results. However, due to the existence of the vehicle, the aerodynamic coefficients of the bridge deck (from $z = 0$–0.54) undergo some changes compared with those of other deck sections without vehicle influence (e.g. $z = -1.0$). The drag force coefficient C_D of the bridge deck is reduced, while its moment coefficient C_M is increased slightly because of the vehicle. The lift force coefficient C_L of the bridge deck reaches a peak value at the head and tail positions of the vehicle.

The instantaneous contours of vorticity magnitude are shown in Figure 8.33. The vortex shedding from the vehicle moves downward and interacts with the vortex shedding from the deck. The projected streamlines and velocity contours of the vehicle-deck system are shown in Figure 8.34. The upcoming flow separates at the bottom surface of the bridge deck. However, due to the existence of the vehicle, the upcoming flow above the bridge deck directly pushes on the windward surface of the vehicle, then separates at the sharp lines and forms a wake region at the back of the vehicle. Moreover, the wind flow from the gap between the bottom surface of the vehicle and the top surface of the deck leads to a low speed area just above the top surface of the bridge deck.

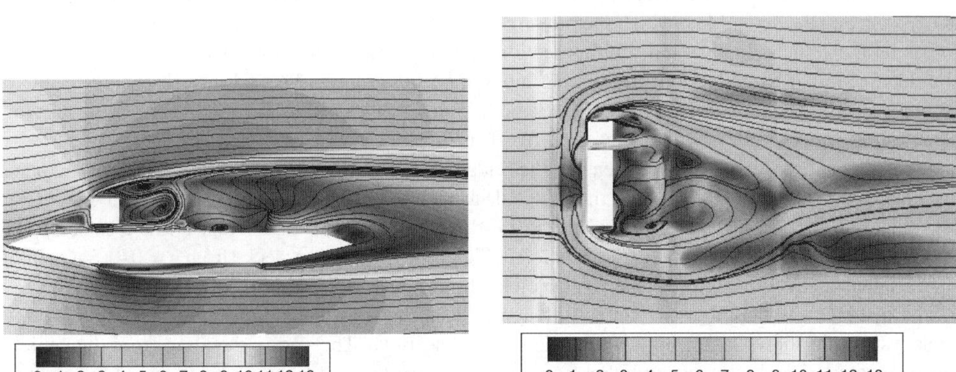

(a) Instantaneous verticity magnitude (b) Averaging velocity contour and streamline

Figure 8.34 Projected streamlines and velocity contours.

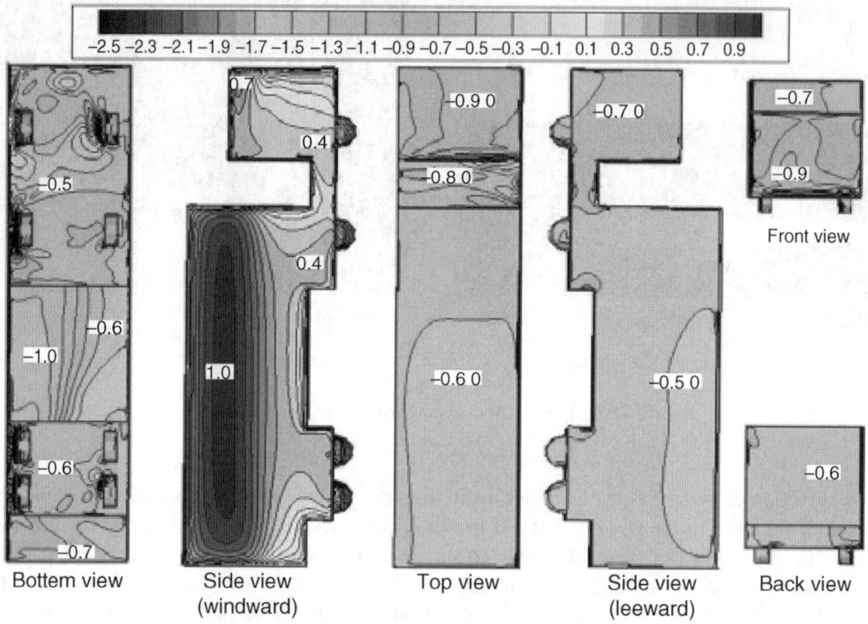

Figure 8.35 Mean pressure coefficients over vehicle.

The distributions of mean pressure coefficients over the surfaces of the vehicle on the deck are displayed in Figure 8.35. The stagnation area can be observed clearly on the windward surface of the vehicle, in which high positive pressures occur. Over other surfaces of the vehicle, all the wind pressures are negative because of flow separations and wakes.

8.7.5 Moving Vehicle on Bridge Deck

In a real situation, the vehicle actually moves over the bridge deck and there is relative motion between the vehicle and the deck. Meanwhile, the moving velocity of the vehicle contributes to the aerodynamic forces on the vehicle-deck system. In this section, the flow around the moving vehicle-deck system is simulated. Similar to the case of the vehicle moving on the ground, the bridge deck is forced with a velocity to simulate the moving effect of the vehicle-bridge deck system. The simulation of the moving vehicle-bridge deck system with a yaw angle of 30° is illustrated here.

The simulated aerodynamic coefficients of the bridge deck with the moving vehicle along its length are presented and compared with those of the bridge deck with fixed vehicle along its length in Figure 8.36. The vehicle is located from $z = 0$–0.54, with its head toward the positive direction of the z-axis. Due to the movement of the vehicle, the aerodynamic coefficients of the bridge deck (from $z = 0$–0.54) exhibit significant changes compared with the vehicle on the bridge deck without movement.

The instantaneous contours of vorticity magnitude of the flow around the moving vehicle-bridge deck system are shown in Figure 8.37 for the selected sections. The vortex shedding from the vehicle moves downward and interacts with that from the deck, forming a complex vortex street at the back of the vehicle-deck system. The projected streamlines and velocity contours of the vehicle-deck system are displayed in Figure 8.38. The upcoming flow separates from the deck and vehicle, and a wake region at the back of the vehicle is formed.

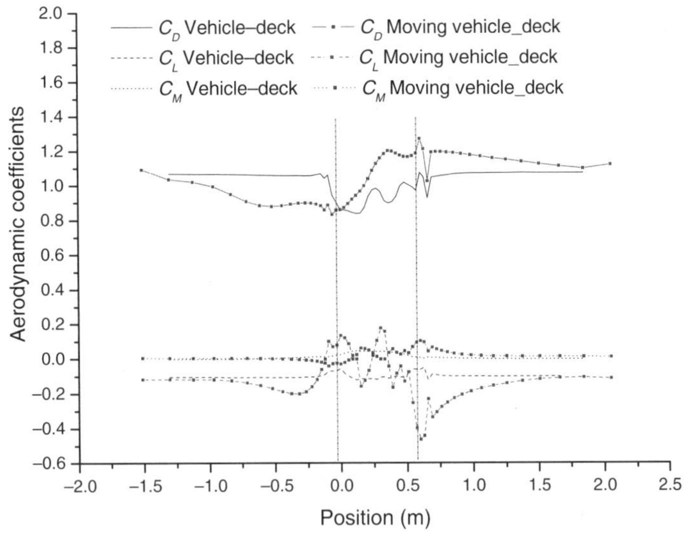

Figure 8.36 Aerodynamic coefficients of bridge deck.

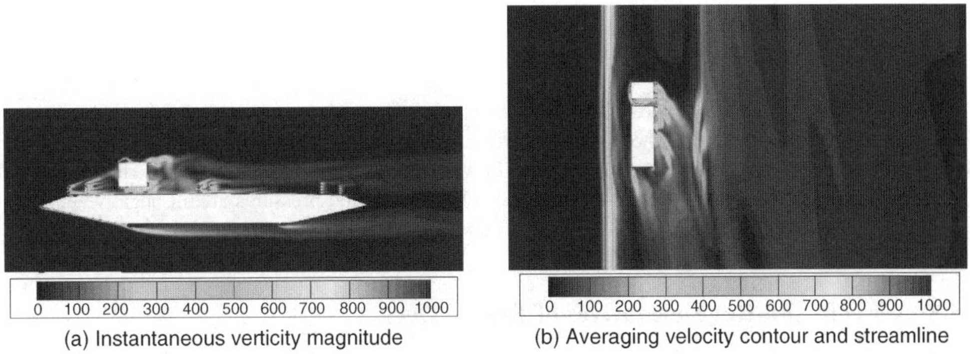

Figure 8.37 Flow characteristics around moving vehicle-deck system.

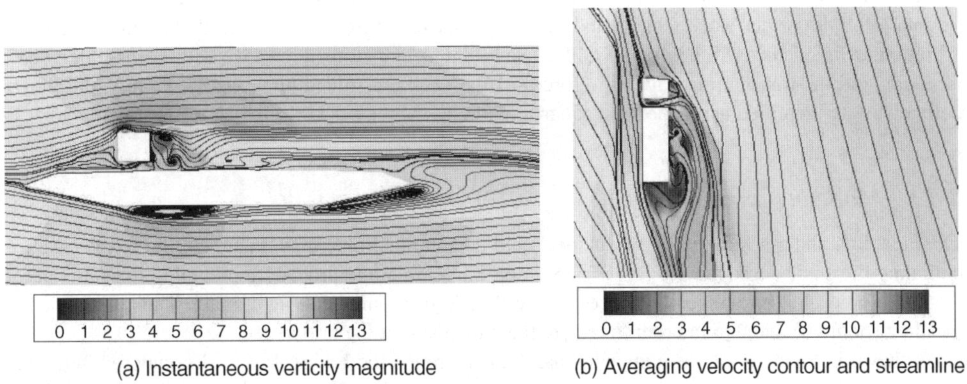

Figure 8.38 Projected streamlines and velocity contours of moving vehicle-deck system.

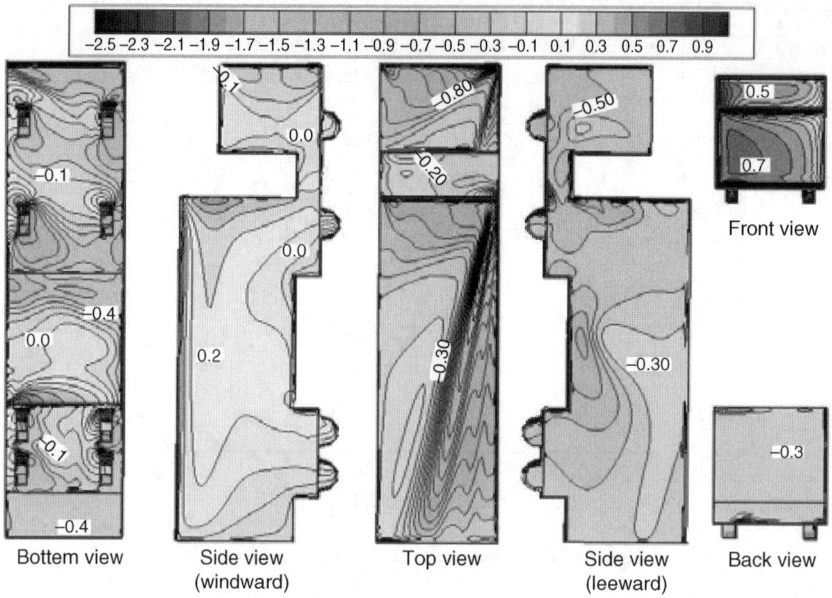

Figure 8.39 Mean pressure coefficients over moving vehicle.

The distributions of mean pressure coefficient over the surfaces of the moving vehicle on the bridge deck are shown in Figure 8.39. The stagnation area can be observed clearly on the head surface of the vehicle, in which high positive pressures occur. Lower positive wind pressures are distributed on the windward side surface. Over other surfaces of the vehicle, the wind pressures are all negative because of flow separations and wakes.

8.8 CFD for Flutter Derivatives of Bridge Deck

Flutter derivatives are the basic aerodynamic parameters for the buffeting and aeroelastic analyses of long-span cable-supported bridges. They represent the aeroelastic forces or self-excited forces acting on the bridge deck due to its motion. The forced vibration test technique is one of the common methods to obtain the flutter derivatives in a wind tunnel, as discussed in Chapter 7. In this section, the force vibration technique is used in terms of CFD. The bridge deck is forced to move with a single degree of freedom, and the transient aerodynamic forces of the deck are solved numerically. Finally, the flutter derivatives are identified using the least squares method.

8.8.1 Modeling and Meshing

The modeling and meshing of the bridge deck and its computational domain are similar to the case for obtaining the static aerodynamic forces in Section 8.3. Unlike the procedure obtaining the static aerodynamic forces, the bridge deck is forced to oscillate in the computational domain. Thus, the dynamic meshing should be adopted in order to realize the movement of the bridge deck.

As shown in Figure 8.40, the computational domain around the bridge deck is decomposed into three parts: the quadrilateral mesh region R_A, around the deck, the quadrilateral mesh region near the outer boundaries, R_B, and the triangle mesh region R_C, in the middle region. The grids in R_A move together

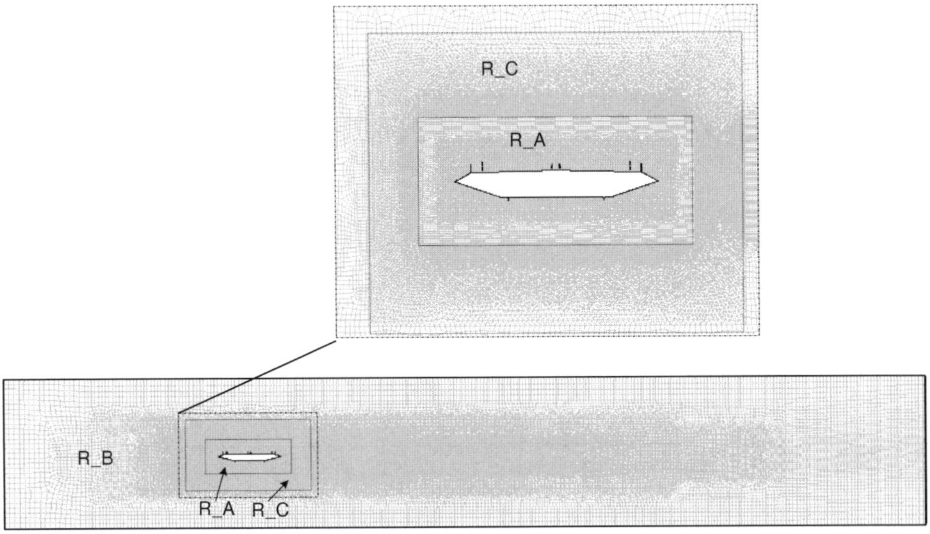

Figure 8.40 Modeling and meshing for flutter derivatives.

with the bridge deck, while the grids in R_B have no motion. As a result, the meshes in R_C vary with the motion of R_A and the bridge deck. The grids in the R_C region should be deformed every time step.

The motion of the deck can be divided into two types: vertical and rotational. For the single degree of freedom motion in the vertical direction, the center of the bridge deck is forced to have the following displacement:

$$y = y_{st}\sin(2\pi f_v t) \tag{8.167}$$

where:

y_{st} is the amplitude of vertical oscillation, selected as 0.065 m;
f_v is the frequency of the vertical vibration with 1.8 Hz.

For the single degree of freedom rotational motion, the angle displacement refers to the section center with the following form:

$$\theta = \theta_{st}\sin(2\pi f_r t) \tag{8.168}$$

where:

θ_{st} is the amplitude of the rotational vibration, selected as 2°;
f_r is the frequency of the rotational vibration with 3 Hz.

8.8.2 Numerical Method

In order to get the transient aerodynamic forces, the unsteady RANS equations are solved. The calculated time step is 0.002s. Second implicit time integration is adopted. The schemes of deformation of the dynamic mesh in the region R_C are "Smoothing" and "Remeshing" in the commercial software *FLUENT*. Different upcoming wind velocity conditions are set to obtain a range of reduced wind velocity. For vertical motion, the upcoming wind velocities are 3, 6, 9, 12, 15, 18 and 21 m/s, and the corresponding reduced wind velocities ($V_r = U/fB$) are 2.70, 5.39, 8.09, 10.79, 13.48, 16.18 and 18.88 m/s. For

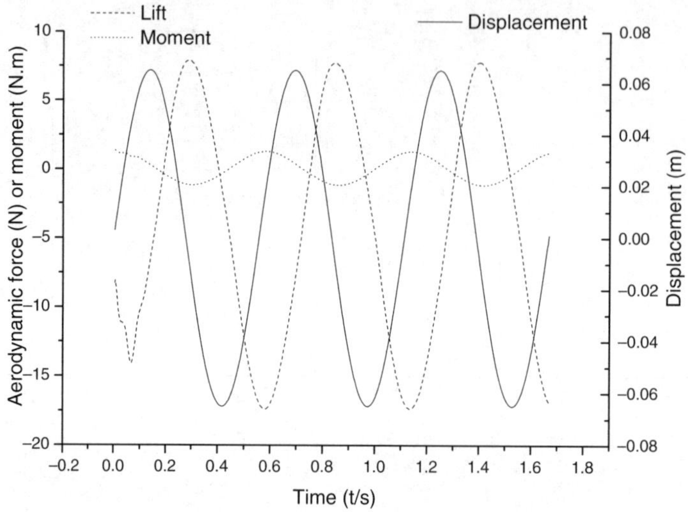

Figure 8.41 Displacement, aerodynamic lift and moment of bridge deck.

rotational motion, the upcoming wind velocities are 1.5, 3.0, 4.5, 6.0, 7.5, 9.0 and 10.5 m/s, and the corresponding reduced wind velocities are 0.81, 1.62, 2.43, 3.24, 4.05, 4.85 and 5.66 m/s. The other settings are the same as the solution for the static aerodynamic forces of the deck in Section 8.3.

8.8.3 Simulation Results

Using the CFD simulation, the transient aerodynamic lift and moment are obtained. Taking the vertical motion at an upcoming wind velocity of 12 m/s as an example, the displacement, transient aerodynamic lift and moment are computed and plotted in Figure 8.41.

From the expressions of self-excited forces in Chapter 4, the aerodynamic lift and moment are related to the displacement and the velocity of deck motion through the flutter derivatives (see Chapter 4). After the simulation of the vertical and the rotational vibration of the bridge deck at different reduced velocities, the flutter derivatives can be obtained by fitting and using the least squares method. Figure 8.42

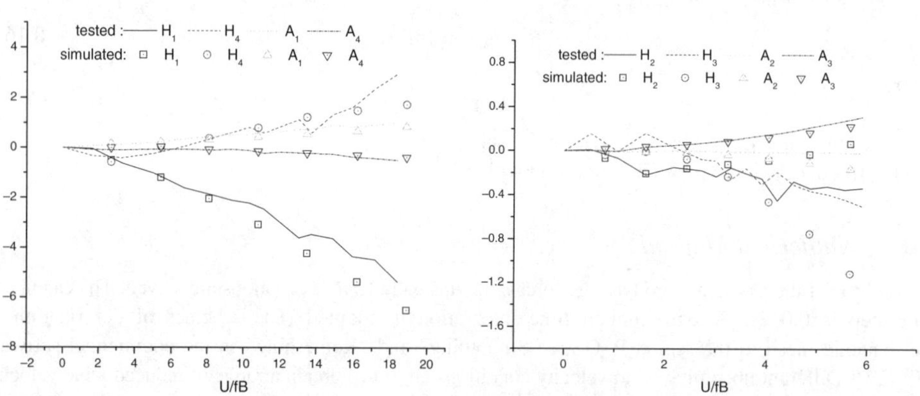

Figure 8.42 Comparison of flutter derivatives of bridge deck.

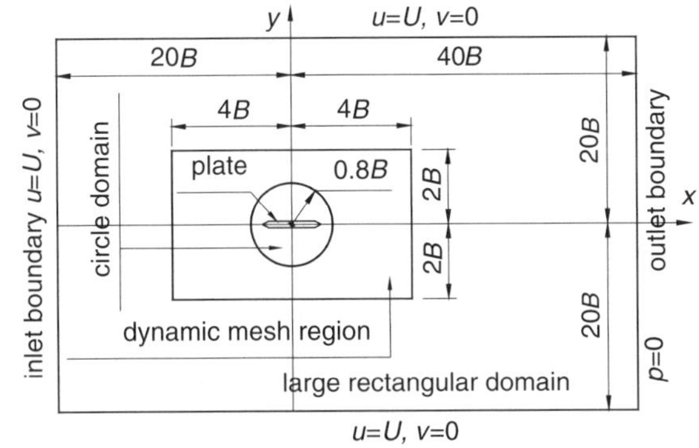

Figure 8.43 Sketch of domain decomposition method.

shows the comparison of flutter derivatives obtained from this simulation with wind tunnel test results. The comparative results are satisfactory.

8.9 CFD for Non-Linear Aerodynamic Forces on Bridge Deck

As a first step investigation, this section selects a thin flat plate as a bridge deck configuration to determine non-linear aerodynamic forces on the plate of large amplitude oscillation through integration of the CFD method and the forced asymptotic oscillation method. Similar work, but on a square section, can be found in the literature [36].

8.9.1 Modeling and Meshing

A two-dimensional thin flat plate of width $B = 0.7$ m and thickness $H = 3.5 \times 10^{-3}$ m is selected as a numerical example. The set-up of the numerical wind tunnel, using the domain decomposition algorithm, is shown in Figures 8.43 and 8.44. The height of body-fitted meshes is smaller than 2 mm to simulate properly the boundary layer around the plate. The initial 68 638 meshes in the computational domain are shown in Figure 8.44a, in which the circle domain (Figure 8.44b) has 20 543 quadrangular meshes and the dynamic mesh region (Figure 8.44c) has 34 564 triangular meshes.

The thin flat plate is forced to oscillate in either torsion or vertical direction, with a single frequency but asymptotic amplitude. The asymptotically oscillating displacement of the plate $A(t)$ is given by:

$$A(t) = An(t)\sin(2\pi f t) \tag{8.169a}$$

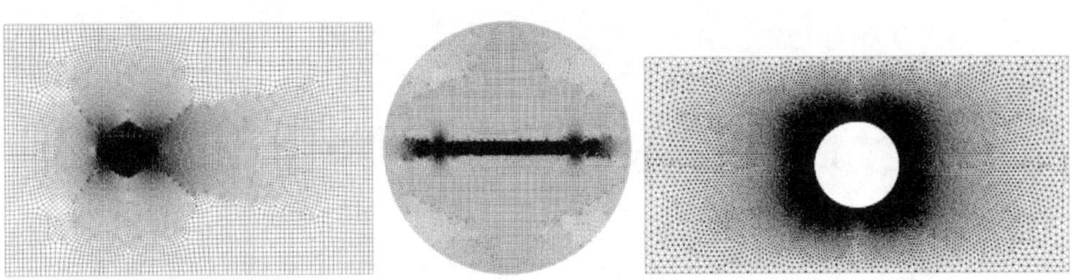

(a) Entire domain (b) Circle domain (c) Dynamic mesh region.

Figure 8.44 Meshing of computational domain.

$$A_n(t) = \begin{cases} A_0 & 0 \le t \le t_1 \\ A_0 \exp[2\pi\lambda f(t - t_1)] & t_1 \le t \le t_2 \\ A_0 \exp[2\pi\lambda f(t_2 - t_1)] & t \ge t_2 \end{cases} \tag{8.169b}$$

where:

f is the oscillation frequency in Hz;
$A_n(t)$ is called the instantaneous oscillation amplitude;
A_0 is a constant representing the initial amplitude of the plate oscillation;
λ is a constant of very small value, so that the amplitude of oscillation slowly increases with time as indicated in the second part of Equation 8.158b.

Note that for the vertical oscillation, the vertical displacement $A(t)$ is applied to the entire plate. For the torsional oscillation, the torsional displacement $A(t)$ is applied only to the rotation point (middle point) of the plate.

8.9.2 Numerical Method

The inflow is set as a uniform flow with a mean speed of $U = 15$ m/s and a turbulence intensity of 0.5%. The parameters involved in the forced oscillating displacement $A(t)$ expressed by Equation 8.169a are selected as $A_0 = 3°$ or 3 cm, $f = 3.571$ Hz, $\lambda = 0.005$. As a result, the reduced velocity $V_r = U/fB$ is 6.0 and the Reynolds number is $\mathrm{Re} = 7.2 \times 10^5$, based on the plate width. In the computation, the time interval gradually reduces as the amplitude increases, with an initial time interval of 3×10^{-4} seconds at time $t = 0$.

8.9.3 Simulation Results

Figure 8.45 shows the computed lift coefficient time history $C_L(t)$ and the computed moment coefficient time history $C_M(t)$ due to the forced asymptotically torsional oscillation.

After the wavelet transform of the coefficient time history is completed, the snake penalization method is used to find the ridges of the wavelet transform. For the concerned plate, three ridges are found at the three instantaneous frequencies (f_1, f_2 and f_3), being approximately 1, 3 and 5 times the forced oscillation frequency. Once the ridges are obtained, the instantaneous amplitudes $D_{Li}(t)$ and $D_{Mi}(t)$, corresponding to all the three instantaneous frequencies, can be determined. The instantaneous amplitudes $D_{Li}(t)$ and $D_{Mi}(t)$ are further normalized by the instantaneous amplitude $A_n(t)$ of the forced oscillation:

$$D_{LRi}(t) = D_{Li}(t)/A_n(t) D_{MRi}(t) = D_{Mi}(t)/A_n(t) \tag{8.170}$$

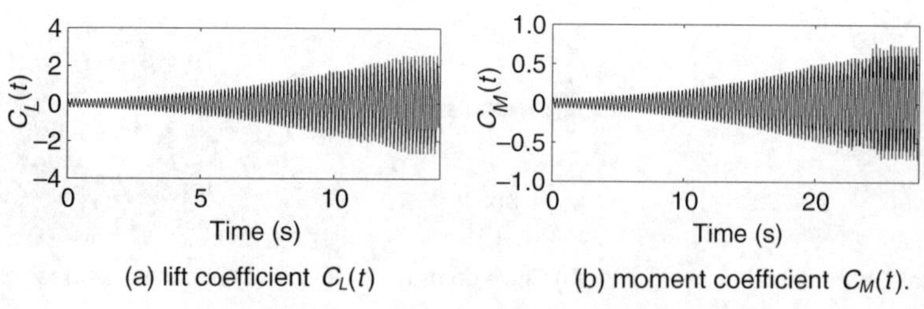

(a) lift coefficient $C_L(t)$ (b) moment coefficient $C_M(t)$.

Figure 8.45 Computed force coefficients.

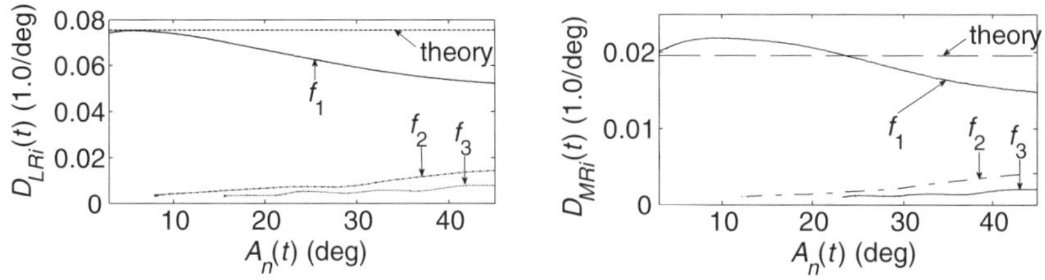

(a) lift force $D_{LRi}(t)$ (b) moment $D_{MRi}(t)$

Figure 8.46 Variation of instantaneous amplitude with torsional oscillation amplitude.

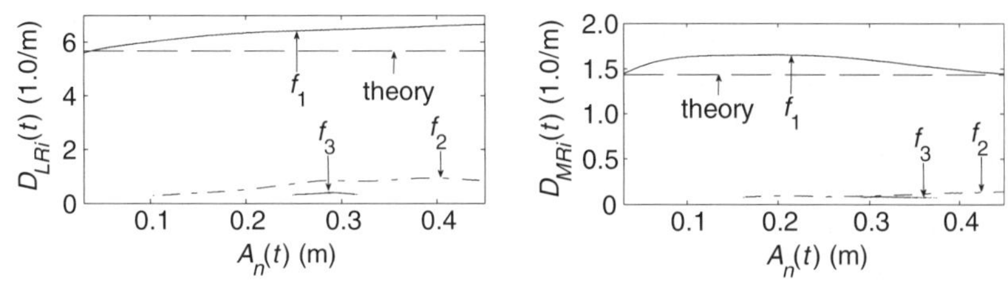

(a) lift force $D_{LRi}(t)$ (b) moment $D_{MRi}(t)$.

Figure 8.47 Variation of instantaneous amplitude with vertical oscillation amplitude.

Figures 8.46 and 8.47 show the normalized instantaneous amplitudes $D_{LRki}(t)$ and $D_{MRi}(t)$ for the three frequencies (f_1, f_2 and f_3) for the torsional and vertical oscillation, respectively. In these figures, the instantaneous amplitudes at f_2 and f_3 are not plotted when they are smaller than the 5% instantaneous amplitude at f_1 (equal to the forced frequency in this case). The line labeled with "theory" is directly computed from Scanlan's formulas [37], based on Theodorsen's circulation function [38] at the forced oscillation frequency.

From these figures, it can be easily observed that the instantaneous amplitude of non-linear aerodynamic forces on the plate is dominated by the forced oscillation frequency, but this also depends on the forced amplitude as well as the instantaneous amplitudes at the other two frequencies. These phenomena clearly demonstrate the non-linearities of aerodynamic forces on the plate when it oscillates in large amplitude.

8.10 Notations

A	Frontal project area of the vehicle
C_D	Drag force coefficient
C_L	Lift force coefficient
C_P	Pitching moment coefficient
C_R	Rolling moment coefficient
C_S	Side force coefficient
C_Y	Yawing moment coefficient
C_d	Constant in dynamic SGS model
C_p	Pressure coefficient
C_s	Smagorinsky constant

F_D	Drag force
F_L,	Lift force
F_S,	Side force
f_x	Body force per unit mass along x-coordinate axis
f_y	Body force per unit mass along y-coordinate axis
f_z	Body force per unit mass along z-coordinate axis
G	Filter function
H_u	Size parameter of computational domain
H_d	Size parameter of computational domain
$\boldsymbol{i}$	Unit vector in x-coordinate axle
$\boldsymbol{j}$	Unit vector in y-coordinate axle
$\boldsymbol{k}$	Unit vector in z-coordinate axle
k	Turbulence kinetic energy
L	Maximum length of the vehicle
L_d	Size parameter of computational domain
L_u	Size parameter of computational domain
M_R	Rolling moment
M_P	Pitching moment
M_Y	Yawing moment
$\boldsymbol{N}_{\mathbf{p}}^{\mathbf{e}}$	Local interpolation functions
$\boldsymbol{N}_{\mathbf{v}}^{\mathbf{e}}$	Local interpolation functions
$\boldsymbol{p}_{\mathbf{N}}^{\mathbf{e}}$	Pressure vector for nodes in element
p	Pressure
$\hat{p}^e$	Pressure in element coordinate system
p^*	Guessed pressure
p'	The difference between the correct and guessed pressure
R^e	Residual
$\boldsymbol{r}$	Location vector
Re	Reynolds number
S_{ij}	Deformation rate
$\overline{S_i}$	Deformation rate after filtering in space or average in time
t	Time
$\boldsymbol{u}$	Velocity vector
$\boldsymbol{u}_\infty$	Velocity vector at the infinite location
$u_{\mathbf{N}}^{\mathbf{e}}$	Vector of velocity component for nodes in element
U	Mean wind speed at the inlet boundary
u_∞	flow velocity at infinite positions
u	Velocity component along x-coordinate axis
$\bar{u}$	Mean part or large scale part of u after averaging or filtering in space
u'	Fluctuation part of u
$\hat{u}$	Small scale of u after filtering in space
$\hat{u}^e$	Velocity component in element coordinate system
u^*	Guessed velocity along x-direction
u'	The difference between the correct and guessed velocity along x-direction
V_{BV}	Velocity of deck related to vehicle
V_V	Moving speed of vehicle
V_W	Velocity of cross wind
V_{WV}	Velocity of wind related to vehicle
$v_{\mathbf{N}}^{\mathbf{e}}$	Vector of velocity component for nodes in element
v	Velocity component along y coordinate axis

$\bar{v}$	Mean part or large scale part of v after averaging or filtering in space
v'	Fluctuation part of v
$\hat{v}$	Small scale of v after filtering in space
$\hat{v}^e$	Velocity component in element coordinate system
v^*	Guessed velocity along y-direction
W	Weighting function
w	Velocity component along z-coordinate axis
$\bar{w}$	Mean part or large scale part of w after averaging or filtering in space
w'	Fluctuation part of w
$\hat{w}$	Small scale of w after filtering in space
$\hat{w}^e$	Velocity component in element coordinate system
x	Coordinate component in Cartesian coordinate system
y	Coordinate component in Cartesian coordinate system
z	Coordinate component in Cartesian coordinate system
α	Yaw angle
β	Angle between V_{WV} and the moving direction of vehicle
Δ	Filter width
Δt	Averaging time
δ	Dirac delta function
ε	Turbulence dissipation rate
η	Length scale of the smallest scale turbulence
μ	Dynamic viscosity coefficient of fluid
μt	Eddy viscosity
μ_{SGS}	subgrid eddy viscosity
ρ	Density of air
σ_{ij}	Stress tensor
τ	Time scale of the smallest scale turbulence
τ_{ij}	Subgrid stress
Π	Flow variable
$\bar{\phi}$	Mean part or large scale part of Π after averaging or filtering in space
Π'	Fluctuation part of $\Pi(t)$
$\hat{\phi}$	Small scale part of flow variable after filtering in space
ω	Turbulence frequency
ω_x	Vorticity in x-direction
ω_y	Vorticity in y-direction
ω_z	Vorticity in z-direction

References

1. Versteeg, H.K. and Malalasekera, W. (2007) *An Introduction to Computational Fluid Dynamics: The Finite Volume Method*, 2nd edn, Pearson Education Limited, England.
2. Tannehill, J.C., Anderson, D.A., and Pletcher, R.H. (1997) *Computational Fluid Mechanics and Heat Transfer*, 2nd edn, Taylor & Francis, London.
3. Spalart, P.R. and Allmaras, S.R. (1994) One-equation turbulence model for aerodynamic flows. *Recherche Aerospatiale*, **1**, 5–21.
4. Wilcox, D.C. (2006) *Turbulence Modeling for CFD*, 3rd edn, DCW Industries, California.
5. Yakhot, V. and Orszag, S.A. (1986) Renormalization group analysis of turbulence: basic theory. *Journal of Scientific Computing*, **1**, 3–11.
6. Wilcox, D.C. (1998) Reassessment of the scale-determining equation for advanced turbulence models. *AIAA Journal*, **26** (1), 1299–1310.
7. Menter, F.R. (1994) Two-equation eddy-viscosity turbulence models for engineering applications. *AIAA Journal*, **32** (8), 1598–1605.

8. Menter, F.R., Kuntz, M., and Langtry, R. (2003) Ten years of industrial experience with the SST turbulence model. Proceedings of the fourth International Symposium on Turbulence, Heat and Mass Transfer, Begell House, Redding, CT.
9. Ghosal, S., Lund, T.S., Moin, P., and Akselvoll, K. (1995) A dynamic localization model for large-eddy simulation of turbulent flow. *Journal of Fluid Mechanics*, **286**, 229–255.
10. Sagaut, P. and Germano, M. (2004) *Large Eddy Simulation for Incompressible Flows*, 2nd edn, Springer-Verlag, New York, NY.
11. Germano, M., Piomelli, U., Moin, P., and Cabot, W. (1991) A dynamic subgrid scale eddy viscosity model. *Physics of Fluids*, **3**, 1760–1765.
12. Spalart, P.R., Deck, S., Shur, M.L. *et al.* (2006) A new version of detached-eddy-simulation, resistant to ambiguous grid densities. *Theoretical and Computational Fluid Dynamics*, **20**, 181–195.
13. Larsen, A. and Walther, J.H. (1997) Aeroelastic analysis of bridge girder sections based on discrete vortex simulations. *Journal of Wind Engineering and Industrial Aerodynamics*, **67–68**253–265.
14. Taylor, I. and Vezza, M. (2001) Application of a discrete vortex method for the analysis of suspension bridge deck sections. *Wind and Structures*, **4** (4), 333–352.
15. Chung, T.J. (2002) *Computational Fluid Dynamics*, Cambridge University Press, UK.
16. Patankar, S.V. and Spalding, D.B. (1972) A calculation procedure for heat, mass and momentum transfer in three-dimensional parabolic flows. *International Journal of Heat Mass Transfer*, **15**, 1787.
17. Patankar, S.V. (1980) *Numerical Heat Transfer and Fluid Flow*, Hemisphere Publishing Corporation, Taylor& Francis, Group, New York.
18. Van Doormal, J.P. and Raithby, G.D. (1984) Enhancements of the SIMPLE method for predicting incompressible fluid flow. *Numerical Heat Transfer*, **7**, 147–163.
19. Fujiwara, A., Kataoka, H., and Ito, M. (1993) Numerical-simulation of flow-field around an oscillating bridge using finite-difference method. *Journal of Wind Engineering and Industrial Aerodynamics*, **46–47**567–575.
20. Onyemelukwe, O.U., Torkamani, M.A.M., and Bosch, H.R. (1997) Numerical simulation of wind-induced forces on bridge deck sections of long-span bridges. *Computers & Structures*, **62** (4), 667–679.
21. Kuroda, S. (1997) Numerical simulation of flow around a box girder of a long span suspension bridge. *Journal of Wind Engineering and Industrial Aerodynamics*, **67–68**239–252.
22. Selvam, R., Tarini, M., and Larsen, A. (1998) Computer modelling of flow around bridges using LES and FEM. *Journal of Wind Engineering and Industrial Aerodynamics*, **77–78**643–651.
23. Bruno, L., Khris, S., and Marcillat, J. (2001) Numerical simulation of the effect of section details and partial streamlining on the aerodynamics of bridge decks. *Wind and Structures*, **4** (4), 315–332.
24. Selvam, R., Govindaswamy, S., and Bosch, H. (2002) Aeroelastic analysis of bridges using FEM and moving grids. *Wind and Structures*, **5** (2–4), 257–266.
25. Frandsen, J.B. (2004) Numerical bridge deck studies using finite elements. Part I: flutter. *Journal of Fluids and Structures*, **19** (2), 171–191.
26. Fransos, D. and Bruno, L. (2010) Edge degree-of-sharpness and free-stream turbulence scale effects on the aerodynamics of a bridge deck. *Journal of Wind Engineering and Industrial Aerodynamics*, **98** (10–11), 661–671.
27. Amandolèse, X. and Crémona, C. (2005) Analyzing fluid loadings on moving bluff bodies using proper orthogonal decomposition. *Journal of Fluids and Structures*, **20** (4), 577–587.
28. Sarwar, M., Ishihara, T., Shimada, K. *et al.* (2008) Prediction of aerodynamic characteristics of a box girder bridge section using the LES turbulence model. *Journal of Wind Engineering and Industrial Aerodynamics*, **96** (10–11), 1895–1911.
29. Jeong, U.Y. and Kwon, S. (2003) Sequential numerical procedures for predicting flutter velocity of bridge sections. *Journal of Wind Engineering and Industrial Aerodynamics*, **91** (1–2), 291–305.
30. Zhu, Z., Gu, M., and Chen, Z. (2007) Wind tunnel and CFD study on identification of flutter derivatives of a long-span self-anchored suspension bridge. *Computer-Aided Civil and Infrastructure Engineering*, **22** (8), 541–554.
31. Coleman, S. and Baker, C. (1994) An experimental study of the aerodynamic behaviour of high sided lorries in cross winds. *Journal of Wind Engineering and Industrial Aerodynamics*, **53** (3), 401–429.
32. Xu, Y.L. and Guo, W.H. (2003) Dynamic analysis of coupled road vehicle and cable-stayed bridge systems under turbulent wind. *Engineering Structures*, **25**, 473–486.
33. Cai, C.S. and Chen, S.R. (2004) Framework of vehicle–bridge–wind dynamic analysis. *Journal of Wind Engineering and Industrial Aerodynamics*, **92**, 579–607.
34. Chen, S.R. and Wu, J. (2008) Performance enhancement of bridge infrastructure systems: Long-span bridge, moving trucks and wind with tuned mass dampers. *Engineering Structures*, **30**, 3316–3324.
35. Zhu, L.D., Li, L., Xu, Y.L., and Zhu, Q. (2011) Wind tunnel investigation of aerodynamic coefficients of road vehicles on bridge deck. *Journal of Fluids and Structures*, **30**, 35–50.
36. Huang, L., Liao, H.L. and Xu, Y.L. (2012) Nonlinear aerodynamic forces on square section: numerical study, Proceedings of the 7th International Colloquium on Bluff Body Aerodynamics and Applications (BBAA7), Shanghai, China, 2-6 September, 889–898.
37. Scanlan, R.H. and Tomko, J.J. (1971) Airfoil and bridge deck flutter derivatives. *Journal of the Engineering Mechanics Division, ASCE*, **97**, 1717–1737.
38. Theodorsen, T. (1935) General theory of aerodynamic instability and the mechanism of flutter. NACA Report No. 496, US National Advisory Committee for Aeronautics, Langley, VA, pp. 1–23.

9

Wind and Structural Health Monitoring

9.1 Preview

In recent decades, long-term wind and structural health monitoring systems (WASHMS) have been developed to measure the loading environment and responses of bridges in order to assess serviceability and safety while tracking the symptoms of operational incidents and potential damage. This advanced technology is based on a comprehensive sensory system and a sophisticated data processing system, which are implemented with advanced information technology and supported by cultivated computer algorithms. It stems from short-term field measurements, but covers much more than just field measurements. Using this technology, real loading conditions of a bridge can be monitored, real performance of the bridge under various service loads can be assessed, the design rules or assumptions employed can be verified or updated, damage and deterioration of the bridge may be identified, and bridge rating and maintenance can be more effective and efficient. This all helps to ensure that the bridge functions properly during a long service life and guards against catastrophic failure under extreme events.

The Tsing Ma Bridge in Hong Kong is one of the pioneering long-span suspension bridges that were installed with an advanced WASHMS. The author has carried out long-term collaborative researches on the wind and structural health monitoring of the Tsing Ma Bridge with the Hong Kong Highways Department since 1995. This chapter will first outline the design criteria of a WASHMS for a long-span cable-supported bridge. The commonly used types of sensors, data acquisition systems, basic signal processing techniques and data management systems will then be introduced. The WASHMS for the Tsing Ma Bridge will be introduced as an example. Finally, the monitoring results and the modal parameter identifications of the Tsing Ma Bridge during Typhoon Victor will be presented to demonstrate partially the functions of the WASHMS.

Since structural health monitoring technology is a mountainous topic, the emphasis of this chapter will be placed on the monitoring of wind and wind effects, rather than other types of loadings. Furthermore, the materials presented in this chapter provide practical and technical support to Chapters 2 to 6 for wind data pertinent to wind characteristics, wind-induced load and wind-induced vibration. The technology presented in this chapter also provides a most reliable way to verify wind tunnel technology, as introduced in Chapter 7, and computational simulation, as discussed in Chapter 8.

Wind Effects on Cable-Supported Bridges, First Edition. You-Lin Xu.

9.2 Design of Wind and Structural Health Monitoring Systems

Design of a WASHM system is a systematic work integrating various expertises [1–4], and the content introduced here on this topic is based on the work presented in the literature [5]. An online WASHM system generally consists of the following modules (see Figure 9.1):

- Sensory system (SS).
- Data acquisition and transmission system (DATS).
- Data processing and control system (DPCS).
- Data management system (DMS).
- Structural evaluation system (SES).

The first two of these systems are often embedded on the bridge, whereas the other three are usually located in the control office of the bridge administrative department. The sensory system is composed of various types of sensors that are distributed along the bridge to capture different signals of interest. The data acquisition and transmission system is responsible for collecting signals from the sensors and transmitting the signals to the central database server. The data processing and control system is designed to control the data acquisition and data transmission, process and store the data, and display the data. The data management system comprises the database system for temporal and spatial data management.

In accordance with monitoring objectives, the structural evaluation system may have different applications. It may include an online structural condition evaluation system and/or an offline structural health and safety assessment system. The former (online) is mainly to compare the measurement data with the design values, analysis results and pre-determined thresholds and patterns, in order to provide a prompt evaluation on the structural condition. The latter (offline) incorporates varieties of model-based and data-driven algorithms, e.g. loading identification, modal identification and model updating, bridge rating system, and damage diagnosis and prognosis.

Design of a WASHM system requires the designer to understand the needs of monitoring, characteristics of the bridge structure, environmental condition, hardware performance and economic considerations. Different bridges have different characteristics and demands. Designers of a WASHM system should work together with the designers of the bridge and know their main concerns. The monitoring items and the corresponding information should be identified, which may include [6]:

(a) the parameters to be monitored, such as wind, displacement and acceleration;
(b) the nominal value and expected ranges of the parameters;
(c) the spatial and temporal properties of the parameters, e.g. variation speed of the measurands, location and correlation of the measurands;
(d) the accuracy requirement;
(e) the environmental condition of the monitoring; and
(f) the duration of the monitoring.

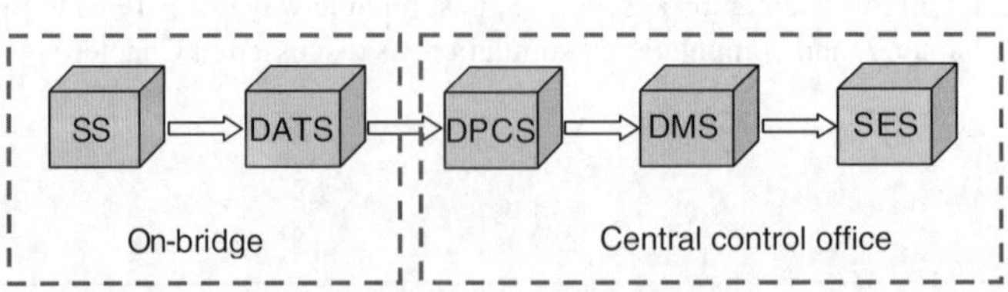

Figure 9.1 Architecture of a WASHM system (from [5]) (Reproduced with permission from Taylor & Francis).

After the monitoring parameters are identified, the number of sensors should be determined, according to the size and complexity of the bridge structure and the monitoring objectives. The types of sensors are then carefully selected, such that their performance can meet the requirement of the monitoring. Important sensor performance characteristics include measurement range, sampling rate, sensitivity, resolution, linearity, stability, accuracy, repeatability, frequency response, durability and so forth.

In addition, sensors must be compatible with the monitoring environment, such as temperature range, humidity range, size, packaging, isolation and thermal effect. The data acquisition units (DAUs) should be compatible with the sensors, too. Location and number of DAUs should be determined to trade-off the distance from the sensors to the DAUs and the number of channels of the DAUs. Sampling rate, resolution, accuracy and working environment should be taken into account for selection of hardware. The designer should also consider the budget of the project, availability of hardware, wiring, and the installation of and protection of the hardware. In practical monitoring projects, wires or cables are more easily damaged than the sensors. Special protection of sensors and wires is worth the effort. Maintenance is also a factor to be considered during the design stage. Important sensors and DAUs should be accessible for check and repair after installation.

9.3 Sensors and Sensing Technology

In a bridge monitoring system, the sensors are mainly employed for monitoring three types of parameters:

(a) loading sources such as wind, seismic, and traffic loading;
(b) structural responses such as strain, displacement, inclination, and acceleration; and
(c) environmental effects including temperature, humidity, rain, and corrosion.

This section will introduce the commonly used sensors for monitoring wind and wind effects on long-span cable-supported bridges.

9.3.1 Anemometers and Other Wind Measurement Sensors

Propeller and ultrasonic anemometers are the most commonly used instruments for measuring wind speed and direction on site. The propeller anemometer directly records wind speed and direction. It is convenient and relatively reliable, and is sustainable in harsh environments, but it is not sensitive enough to capture turbulent winds of higher frequencies. This is particularly true in situations when wind speed or direction changes rapidly. The ultrasonic anemometer measures wind velocity through its two or three orthogonal components. It is quite sensitive, but it is not sustainable in harsh environments.

The accuracy requirement for wind velocity measurement by the anemometers must be maintained under heavy rain, i.e. there should be no occurrence of spikes during heavy rainstorms. For a long-span cable-supported bridge, the anemometers are often installed at a few bridge deck sections on both sides and along the height of the towers, so that not only wind characteristics at key positions can be measured, but also the correlation of wind speed velocity in both horizontal and vertical directions can be determined. The positions of the anemometers must be selected so as to minimize the effect of the adjacent edges of the bridge deck and towers on the airflow towards them. To meet this requirement, anemometer booms or masts are often needed, so that the anemometer can be installed a few meters away from the bridge edges. The boom or mast must be equipped with a retrievable device to enable retracting of the anemometer in an unrestricted and safe manner for inspection and maintenance.

Occasionally, pressure transducers are installed to measure wind pressures and pressure distribution over a particular part of the bridge envelope. Wind pressure transducers sense differential pressure and convert this pressure difference to a proportional electrical output for either unidirectional or bidirectional pressure ranges. To measure the pressure difference accurately, the location of the reference pressure transducer needs to be selected appropriately in order to avoid possible disturbances from the surrounding environment.

As discussed in Chapter 2, mean wind speed profiles in the atmospheric boundary layer are very important for determining wind loads on the bridge, but they are difficult to measure. In recent years, Doppler radar [7], GPS drop-sonde [8] and Doppler sodar [9] have become powerful devices for measuring boundary layer wind profiles at high altitudes during typhoons.

9.3.2 Accelerometers

Accelerometers are widely used to measure wind-induced acceleration of a bridge. The acceleration responses of a bridge are closely related to the serviceability and functionality of the bridge.

Basically, an accelerometer is a mass-spring-damper system that produces electrical signals in proportion to the acceleration of the base where the sensor is mounted. Selection of accelerometers should consider the following parameters: usable frequency response, sensitivity, base strain sensitivity, dynamic range and thermal transient sensitivity. Installation of accelerometers and cables is also critical for a good vibration measurement. There are four main types of accelerometers available: piezoelectric type, piezoresistive type, capacitive type, and servo force balance type.

Piezoelectric type accelerometers are robust and stable in long-term use. However, their major drawback is that they are not capable of a true DC (0 Hz) response, which makes this type inappropriate for long-span cable-supported bridges of very low frequencies (e.g. around 0.1 Hz); the lower frequency limit of piezoelectric accelerometers is generally above 1 Hz. Piezoresistive and capacitive accelerometers are adequate for bridge structures, as they can measure accelerations from DC level. Capacitive type accelerometers are accurate and appropriate for low frequency and low-level vibration measurement such as micro-g (gravity acceleration = $9.80\,\text{m/sec}^2$). Force balance sensors are suitable for DC and low frequency measurement, providing milli-g measurement capability.

9.3.3 Displacement Transducers and Level Sensors

Displacement of a bridge structure serves as an effective indicator of its structural performance condition. Excessive displacements may affect the bridge's structural integrity, so displacement monitoring is therefore needed. Equipment to measure displacement includes linear variable differential transformers, level sensing stations, Global Positioning System (GPS), etc.

A linear variable differential transformer (LVDT) is a commonly used electro-mechanical facility for measuring relative displacements based on the principle of mutual inductance. It consists of a hollow metallic tube containing a primary and two secondary coils, with a separate movable ferromagnetic core. The coils produce an electrical signal that is in proportion to the position of the moving core. The frictionless movement of the core leads to a long mechanical life, high resolution, good zero repeatability and long-term stability. LVDTs are available in a wide range of linear strokes, ranging from micrometers to 0.5 m.

The measurement of vertical displacement by the level sensors is based, in principle, on the pressure difference. The system basically consists of two or more interconnected cells filled with a fluid (usually water). Relative vertical movement of the cells causes movement of water, and variation in the water level is measured. The conventional level sensing system can detect the elevation difference of about 0.5 mm.

9.3.4 Global Positioning Systems

The dynamic displacement response of the bridge can be measured using displacement transducers, or obtained by integrating the acceleration records twice with time. The absolute static or quasi-static displacement responses of the bridge deck and towers caused by the long-period component in the coming wind cannot be captured using these devices, but the total displacement (static plus dynamic) is imperative for monitoring the integrity and safety of the bridge. To circumvent this problem, Global Positioning System (GPS) technology is preferred.

GPS technology, developed by the U.S. Department of Defense in 1973, was originally designed to assist soldiers, military vehicles, planes, and ships [10]. It consists of three parts: the space segment, the control segment and the user segment. The space segment comprises 32 satellites in six orbital planes. The control segment consists of a master control station, an alternate master control station and shared ground antennas and monitor stations. The user segment is composed of thousands of military users of the secure precise positioning service, as well as millions of civil, commercial, and scientific users of the standard positioning service.

GPS provides a powerful ability to track dynamic, as well as static, displacements of long-span cable-supported bridges in high winds, in which the Real Time Kinematic (RTK) technique is often used on the basis of carrier phase measurements of the GPS, where a reference station provides the real-time corrections [11–17]. A RTK system usually consists of a base station receiver and a number of mobile units. The base station re-broadcasts the phase of the carrier that it measures, and the mobile units compare their own phase measurements with the ones received from the base station. This system can achieve a nominal accuracy of 1 cm $\pm$ 2 parts per million (ppm) horizontally and 2 cm $\pm$ 2 ppm vertically. This is likely to be improved further in the foreseeable future. Nevertheless, a few factors affect the accuracy of GPS measurement – in particular, the atmospheric conditions and multi-path effects. Inconsistencies of atmospheric conditions affect the speed of the GPS signals as they pass through the Earth's atmosphere. The signals are also reflected by surrounding obstacles, causing delay of signals.

9.3.5 Strain Gauges

Foil strain gauges, vibrating wire strain gauges and fiber optic strain gauges are commonly used sensors measuring strain in bridge structures.

Foil strain gauges are the most common type of strain gauge, consisting of a thin, insulting backing which supports a fine metallic foil. The gauge is attached to the object by a suitable adhesive. As the object is deformed, the foil is stretched or shortened, causing a change in its electrical resistance in proportion to the amount of strain, which is usually measured using a Wheatstone bridge. The physical size of most foil strain gauges is about a few millimeters to centimeters in length. Its full measurement range is about a few milli-strain. These gauges are economical and can measure dynamic strains, but their long-term performance (e.g. zero stability) is not as good as the alternatives, particularly in a harsh environment. For example, the presence of moisture may result in electrical noise in the measurement and zero drift.

A vibrating wire strain gauge consists of a thin steel wire held in tension between two end anchorages. When the distance between the anchorages changes, the tension of the wire changes and so does the natural frequency. The change in the vibration frequency of the wire is transferred into the change in strain. Consequently, the captured strain can be transmitted over a relatively long distance (a few hundred meters to kilometers) without much degradation. This is one advantage of vibrating wire gauges over foil gauges. Vibrating wire gauges are easy to install on the surface or embedded in concrete, and they are typically is about 100~200 mm long, with a measurement range of 3000 $\mu\varepsilon$ and a resolution of 1.0 $\mu\varepsilon$, which is suitable for monitoring of bridge structures. A main drawback of these gauges is that they can measure the static strain only, as it takes seconds to obtain the frequency of the vibrating wire.

9.3.6 Fiber Optic Sensors

Optical fibers can be used as sensors to measure strain, temperature, pressure and other quantities. The sensors modify a fiber so that the quantity to be measured modulates the intensity, phase, polarization and wavelength of light in the fiber. Accordingly, fiber optic sensors can be classified into four categories: intensity modulated sensors, interferometric sensors, polarimetric sensors and spectrometric sensors [18].

A significant advantage of fiber optic sensors is multiplexing – that is, several fiber optic sensors can be written at the same optical fiber and interrogated at the same time via one channel. In addition, fiber optic sensors are very small in size, compared with conventional strain gages, and are immune to electromagnetic interferences. They are also suitable for both static and dynamics measurements, with a frequency from hundreds to thousands Hertz. Their major drawback is the high cost of both sensors and the acquisition unit (or readout unit). In addition, the fibers are rather fragile and need to be handled very carefully in the field installation.

In bridge monitoring, Fiber Bragg grating (FBG) sensors are commonly used for strain measurement [19,20]. They work on the principle that the strain variation causes a shift in the central Bragg wavelength. Consequently, FBG strain sensors monitor changes in the wavelength of the light. Commercially available white light sources have a spectral width around 40 to 60 nm. An FBG sensor with the measurement range of 3000 $\mu\varepsilon$ takes a wavelength of 3 nm. Counting the spectral space between the sensors, one optical fiber can accommodate six to ten FBG sensors.

9.3.7 Laser Doppler Vibrometers

A laser Doppler vibrometer (LDV) is an instrument that is used to make non-contact vibration measurements of a surface. The LDV basically uses the Doppler principle to measure velocity at a point to which its laser beam is directed. The reflected laser light is compared with the incident light in an interferometer to give the Doppler-shifted wavelength, which provides information on surface velocity in the direction of the incident laser beam. Some advantages of an LDV over similar measurement devices such as an accelerometer include that the LDV can be directed at targets that are difficult to access, or which may be too small or too hot to attach a physical transducer. Also, the LDV makes the vibration measurement without mass-loading the target, which is especially important for tiny devices such as micro-electro-mechanical systems.

Abe *et al.* [21] and Kaito *et al.* [22] applied an LDV to measure vibration of bridge deck and stay cables. They found that when the measurement grid was pre-determined, the LDV automatically scanned 45 points at a high frequency, such that one LDV could measure the vibration of all points at the same time. With one reference, the modal properties could be extracted.

Other non-contact measurement techniques, the photogrammetric and videogrammetric techniques, have been developed with the advance of inexpensive and high-performance charge-coupled-device cameras and associated image techniques. Bales [23] applied a close-range photogrammetric technique to several bridges for estimation of crack sizes and deflection measurement. Li and Yuan [24] developed a 3D photogrammetric vision system consisting of video cameras and 3-D control points for measuring bridge deformation. Olaszek [25] incorporated the photogrammetric principle with the computer vision technique to investigate the dynamic characteristics of bridges. Others applications include [26]. Ji and Chang [27] and Zhou *et al.* [28] employed the techniques for cable vibration measurement.

9.3.8 Weather Stations

In some applications, it is desirable to measure environmental conditions such as ambient temperature, humidity, rainfall, air pressure and solar irradiation. A typical weather station usually integrates a few types of sensors and can measure the above-mentioned parameters as well as wind speed and direction. Solar irradiation intensity, air temperature and wind are important parameters for deriving the temperature distribution of structures. With temperature distribution, the thermal effect on the structural responses can be evaluated quantitatively.

9.3.9 Wireless Sensors

Advances in micro-electro-mechanical systems (MEMS) technology, wireless communications and digital electronics have enabled rapid development of wireless sensor technology since the late twentieth century [29–33]. A wireless sensor network can comprise all of the components in a wire-based WASHM system described previously, such as SS, DATS, DPCS, DMS, and SES, whereas it has its unique characteristics as compared with the wire-based WASHM systems. Although wireless sensors and networks have been developed rapidly, at present wireless monitoring is not ready to use for continuous health monitoring of long-span cable-supported bridges. Traditional wire-based systems still dominate practical WASHM projects, and wireless sensor nodes are mainly for research purposes or supplementary to the wired systems. Nevertheless, wireless sensors and networks might be a future direction for WASHM.

9.4 Data Acquisition and Transmission System (DATS)

9.4.1 Configuration of DATS

Sensors generate analog or digital signals that represent the physical parameters being monitored. Data acquisition is an intermediate device between the sensors and computers, which collects the signals generated by the sensors, converts them and transmits the signals to the computers for processing. For a small laboratory-based experiment, the above function can be achieved with a card-based data acquisition unit in a personal computer (PC). However, configuration of a data acquisition and transmission system (DATS) in a long-term bridge monitoring system is generally much more complicated. It usually consists of local cabling network, stand-alone data acquisition units (DAUs) or substations, and global cabling network, as illustrated in Figure 9.2. The local cabling network refers to the cables connecting the distributed sensors to the individual DAUs, while the global cabling network refers to the cables connecting the DAUs to central database servers.

For long-span cable-supported bridges, appropriate deployment of DAUs plays a significant role in assuring the quality and fidelity of acquired data. As the distributed sensors are far from the central control office, the length of the wires causes noise and significant loss in analog signals. It is also inefficient to wire all sensors to one central server. Therefore, DAUs are assigned at a few cross-sections of the bridge to collect the signals from surrounding sensors, condition the signals and transmit the digital data into the central database server.

Note that some proprietary sensors, such as GPS, video cameras, and fiber optic sensors, have specific data acquisition systems. These united systems capture corresponding information, transform the information into digital data and connect directly to the central data server for processing.

For such a system, as illustrated in Figure 9.2, it is desirable to have a uniform platform to assure the scalability, functionality and durability of the system.

9.4.2 Hardware of Data Acquisition Units

A data acquisition unit (DAU) generally comprises a number of electronic components, including signal conditioner, memory and data storage unit, microcontroller, communication device, uninterruptible

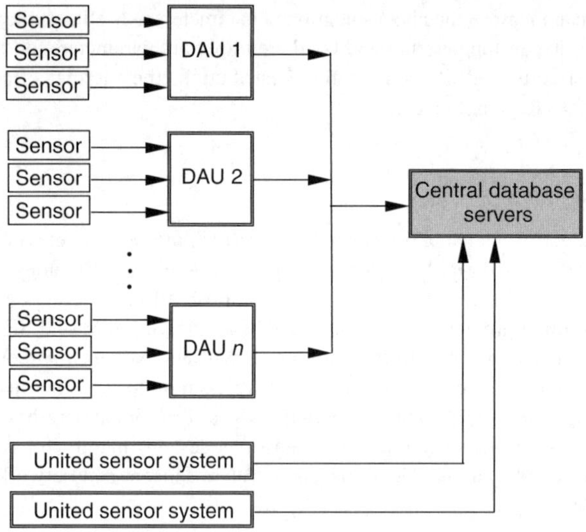

Figure 9.2 Configuration of a data acquisition and transmission system (from [5]) (Reproduced with permission from Taylor & Francis).

power supply, fan/air conditioner, lightning conductor and GPS time synchronizer. All of these components are integrated in a waterproof, rugged enclosure or cabinet for the long-term monitoring purpose.

It is common to include different types of sensors, with different output signals and different sampling rates, in one DAU. Consequently, a DAU can facilitate this flexibility and may have more than one signal conditioner. A signal conditioner manipulates an analog signal such that it meets the requirements of further processing. Signal conditioning usually includes amplification, filtering, analog-to-digital (A/D) conversion and isolation. Amplification serves to amplify the analog signal before A/D conversion to utilize the full range of the A/D converter, thus increasing the signal-to-noise ratio and resolution of the input signal. Filtering is used to remove the unwanted frequency components, and most signal conditioners employ low-pass filters. Isolation is used to isolate the possible ground loop and protect the hardware from damage.

As the sampling rate in the measurement data of bridges is usually low, a single A/D converter can perform A/D conversion by switching between several channels. This is much less expensive than having a separate A/D converter for each channel, and is thus adopted in most practical SHM systems. The internal memory serves as a temporary buffer of data for transmission, and is usually integrated with the microcontroller. The data storage unit can save measurement data for relatively longer, in case the global cabling network does not work appropriately. The data can be retrieved manually to the external storage devices, or automatically to the database server when the global cabling network recovers.

The microcontroller consists of internal electronic circuitry to execute commands sent by the users, and to control other hardware components. For example, the sampling rates and acquisition duration of the sensors can be changed by the users. The communication device is responsible for communication between the DAU and the computer. Usually, an Ethernet interface is employed.

The power supply provides power to the data acquisition system and to some sensors that require an external power source. An uninterruptible power supply provides instantaneous or near-instantaneous protection from unexpected power interruption or unstable input voltage. A fan or air conditioner is used to cool the temperature of the DAU. A lightning conductor can provide the DAU with protection from lightning damage. DAUs were previously synchronized through a synchronization signal sent from the central station regularly but, nowadays, GPS time synchronizers have become more popular. These can provide an easy way to keep the DAUs and united sensor systems accurately synchronized.

9.4.3 Network and Communication

In the DATS, a uniform network communication is crucial to assure the data can be transmitted over the entire system. Various communication network technologies, such as Ethernet, RS-232, RS-485 or IEEE-1394 can be employed for the common network.

In a WASHM system for a long-span bridge, the distance between the DAUs and the central control office may be as far as a few kilometers, and fiber optic cabling is desirable. There are basically two types of fiber optic cables: single mode and multi-mode fibers. Multi-mode fibers generally have a larger core diameter and are used for shorter distance communication. Single mode fibers are used for communication links longer than 600 meters, so are preferable for long-span bridges.

Wireless communication and networking have been rapidly developed and employed for data transmission, but transmission speed and accuracy are still not comparable with the cable-based network at the moment. Nevertheless, the wireless network shows promise for the near future. It has advantages in some situations, particularly for construction monitoring when the cable network is not ready.

9.4.4 Operation of Data Acquisition and Transmission

After the hardware has been installed, the DATS should be tested or verified through field tests – for example, controlled load tests, because the actual performance of the hardware is uncertain under long-term exposure to harsh conditions. Moreover, it is difficult in practice to identify, repair and change damaged facilities after the bridge is put into service. Field verification can thus help to identify problems in hardware, installation, cabling and software, such that these problems can be fixed before normal operation.

During normal operation, data acquisition and transmission are carried out in a systematic and organized manner. Depending on the nature of the monitored parameters, some sensors may work continuously (long-term mode), while others may work in the trigger mode (short-term mode), in which the signals are collected only when the parameters are above a certain threshold, due to some extreme event. These two modes can operate simultaneously in one data acquisition system.

Sampling rate (or sampling frequency) is an important factor affecting the data acquisition speed. It relies on the variation speed of the monitored parameters and can be programmed by users. If a parameter is not sampled quickly enough (under-sampling), the resulting digitized signal will not represent the actual signal accurately, and this error is called as aliasing error. To avoid this error, the Nyquist criterion requires that the sampling rate should be more than twice the highest frequency component of the original signal.

In bridge monitoring exercises, the sampling rate of most signals is usually not higher than 100 Hz, as the fundamental frequencies of long-span bridges are relatively low, unless some special measurements like acoustic methods or guided-wave methods are employed. For the low sampling rate, multiplexed sampling, rather than simultaneous sampling, can be employed. Multiplexed sampling allows different channels to share one A/D converter and be sampled sequentially. This can reduce cost compared with simultaneous sampling, in which each signal channel has an individual A/D converter.

After operation for a period in the adverse environment, DAUs are inevitably subject to error or malfunction, so it is preferable to carry out periodical calibrations. As mentioned previously, the DAUs should be accessible for maintenance.

9.5 Data Processing and Control System

Functions of the data processing and control system include:

1. control and display of the operation of the data acquisition system;
2. pre-processing of the raw signals received from the data acquisition system;
3. data archive into a database or storage media;
4. post-processing of the data; and
5. viewing the data.

9.5.1 Data Acquisition Control

A large-scale WASHM system comprises various types of data acquisition hardware, so centralized data acquisition control is preferable. As described above (Section 9.4.4), the signals can be collected in a long-term or short-term mode. Therefore, the data acquisition control unit should be flexible in handling both continuous monitoring mode and scheduled trigger modes. In practical WASHM systems, the centralized control unit is located in the central control office and operated by the users for carrying out communication with the local acquisition facilities via a graphical user interface.

The graphical user interface (GUI) program is an interface between the data acquisition hardware and the hardware driver software. It controls the operation of the DATS, including how and when the DATS collect data, and where to transmit. It provides users with an easy interface.

9.5.2 Signal Pre-Processing and Post-Processing

The collected raw signals are pre-processed prior to permanent storage. The data pre-processing has two primary functions:

1. Transforming the digital signals into the monitored physical data.
2. Removing abnormal or undesirable data.

Signal transforming is simply done by multiplying the corresponding calibration factor or sensitivities. There are several data-elimination criteria for the removal of typical abnormalities associated with various types of statistical data. The source of abnormal data is possibly derived from malfunction of the measurement instrument. There are a few criteria defining abnormal data; for example, extremely large or extremely small data may be regarded as abnormal if they do not have any physical meaning.

The pre-processed signal will be saved into a database system for future management, or on storage media such as hard disks and tapes after proper packaging and tagging. The stored data are post-processed for various uses. A few basic data processing techniques include data mining methods in the time domain (e.g. regression analysis, generic algorithm, artificial neural networks, support vector machine), spectral analyzes in the frequency-domain (e.g. Fourier transform, power spectrum, correlation analysis, frequency response function) and time-frequency analysis tools (e.g. short-time Fourier transform [34], wavelet transform [35], Hilbert-Huang transform [36]).

A WASHM system usually includes various types of sensors located in different spatial positions, and different types of sensors located in the same position may capture different signals. Spatially distributed sensors may also demonstrate different features of the bridge structure. Therefore, integration of data from different sensors and integration of the results made by different algorithms are important to a robust monitoring exercise [37]. In this regard, data fusion is an important data processing tool [38,39].

9.6 Data Management System

The collected data and processed data, or results in a WASHM system, should be stored and managed properly for display, query and further analysis. Relevant information on the WASHM system, computational models and design files also need to be documented. These tasks are completed by the DMS via a standard database management system such as MySQL or ORACLE.

9.6.1 Components and Functions of Data Management System

A standard database management system allows users to store and retrieve data in a structured way, so that later assessment is more efficient and reliable. In a long-term WASHM system for long-span cable-supported bridges, the data size is large in different types. Therefore, a large WASHM system usually consists of a few databases, including the device database, measurement data database, structural analysis data database, health evaluation data database and user data database. The DMS manages each database to fulfill their corresponding functions.

The device database records the information on all sensors and substations. For sensors, their ID, label, substation, location, specifications, manufacturer, installation time, initial values, sampling rate, thresholds, working condition and maintenance record should be recorded. For substations, their ID, label, sensors, location, specifications, manufacturer, installation time, working condition and maintenance files should be recorded. These data are necessary to examine the collected measurements in the long term. Various types of sensors should be labeled properly so that they can be easily identified by users. The DMS has the facility to insert and delete sensors and substations, and can monitor their conditions.

The measurement data database records all the data collected from the sensor system, including the loads, structural responses and environmental parameters. For efficiency, the database usually stores data for a limited period only, e.g. one year. Historical data beyond this period are archived in storage media, such as tapes and DVDs. For data safety, all measurement data should have a spare backup. The DMS has facilities for automatic retrieval and output of measurement data and data query from authorized users.

The structural analysis data database records the finite element model data, input parameters of the models and major output data, design drawings and basic design parameters. For long-span bridges, there might be more than one finite element model for cross-checking and for different applications. The models can be input into the corresponding analysis software, and output data are employed for comparison and evaluation.

The health evaluation data database records the evaluation time, parameters, objects, criteria, results and reporting. The structural health evaluation may be performed regularly for normal operation. Once an extreme event occurs (e.g. an earthquake), a specified evaluation should be carried out immediately. For each kind of evaluation, pre-determined criteria should be provided.

The user data database manages users' information, such as username, ID, user group, and personal data and contact information. Different users will have been authorized different rights by the DMS to log into the WASHM system.

A DMS should provide security management, which may include network security, data protection, database backup and user operation audit.

Finally a DMS should provide an alarming function. The alarming module can automatically generate warning messages when some pre-defined criteria are satisfied. Important alarms should be sent to relevant staff through email and short messages until countermeasures are taken.

9.6.2 Maintenance of Data Management System

A large WASHM system is operated by authorized persons who have received basic training in computer technology and structural engineering. When the databases start working, their functions and performance need to be tested. After a period of operation, increase in sizes of the databases may cause physical storage malfunctions and reduce the efficiency of the databases. Therefore maintenance of the DMS is necessary. The duties of the administrator of a DMS include backup and restoration of the databases, monitoring and improvement of the databases, reconstruction and reconfiguration of the databases.

9.7 Structural Health Monitoring System of Tsing Ma Bridge

9.7.1 Overview of WASHMS

A Wind And Structural Health Monitoring System (WASHMS) for the Tsing Ma Bridge has been devised, installed, and operated by the Highways Department of the Government of Hong Kong Special Administrative Region since 1997 [3]. The central control office is located in the Tsing Yi administration building. The system architecture of the WASHMS is comprised of six integrated modules: the SS, DATS, DPCS, structural health evaluation system (SHES), portable data acquisition system (PDAS) and portable inspection and maintenance system (PIMS).

The layout of the sensory system for the Tsing Ma Bridge is shown in Figure 9.3. The sensors are of seven major types: anemometers, temperature sensors, weigh-in-motion sensors, accelerometers, displacement transducers, level sensing stations and strain gauges, as listed in Table 9.1.

Tag numbers for all sensors except the GPS have the following format that is used in this system: AA-TBX-YYC. The letters and numbers have the following meanings: AA – sensor code as listed in Table 9.1, T – for Tsing Ma Bridge, B – section code as shown in Figure 9.3, X – location code ('A' for abutment, 'C' for main cable, 'N' for north side of deck, 'S' for south side of deck, 'T' for tower), YY – sequential number, and C – identifier. For some sensors, YYC is not shown here for brevity. In 2001, a GPS was installed in the Tsing Ma Bridge, which includes 14 rover stations on the bridge and two reference stations. Details of the various sensors are described in the following sections.

The DATS for the Tsing Ma Bridge has three DAUs, connected by a token ring fiber optic network. The positions of the DAUs are shown in Figure 9.3. One DAU controls 64 to 128 data channels. The DPCS comprises two UNIX-based 64-bit Alpha Servers and two 32-bit SGI Intel-based (Quad-CPU) Visual Workstations. The DPCS carries out the overall control of data acquisition, processing, transmission, filing, archiving, backup, display and operation. The application software systems for WASHMS are customized MATLAB, customized GPS software and a MATLAB data analysis suite.

The SHES comprises one UNIX-based 64-bit (Quad-CPU) Alpha server and one UNIX-based 64-bit Alpha workstation. The sever is used for structural health evaluation works, based on a customized bridge rating system, together with advanced finite element solvers such as MSC/NASTRAN, ANSYS/Mulitiphysics, ANSYS/LS-DYNA, ANSYS/FE-SAFE and MATLAB data analysis suite. The workstation is used for the preparation and display of graphical input and output files based on advanced graphical input/output tools such as MSC/PATRAN.

The PDAS comprises a 32-channel PC-based datalogger, 24 portable biaxial servo-type accelerometers, five portable uniaxial servo-type accelerometers and 16 cable drums. It is equipped with a customized LABVIEW software system for acquisition, processing, archiving, storage and display. It is used to measure the tensile forces in cables and to assist the fixed servo-type accelerometers in identifying the global dynamic characteristics of the bridge.

The PIMS comprises three portable notebook computers, which are used to carry out the inspection and maintenance work on the SS and DATS. Another major function of the PIMS is to facilitate the system inspection and maintenance by storing and updating all the system design information, including all drawings and all operation and maintenance manuals.

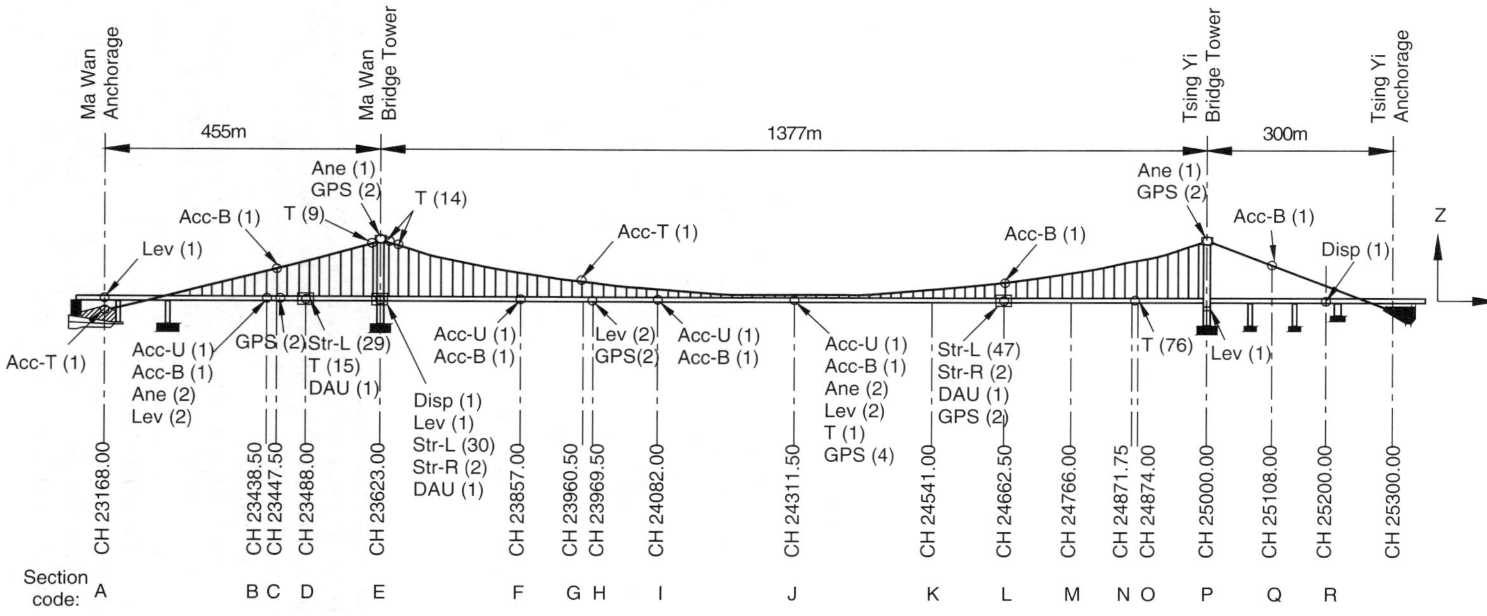

Figure 9.3 Layout of sensors and DAUs of the Tsing Ma Bridge (from [5]) (Reproduced with permission from Taylor & Francis).

Note: (1) Numbers in parentheses are the numbers of sensors. (2) Lev: Level sensing (9); Ane: Anemometer (6); Acc–U: Uniaxial Accelerometer (4); Acc–B: Biaxial Accelerometer (7); Acc–T: Triaxial Accelerometer (2); Str–L: Linear strain gage (106); Str–R: Rosette strain gage (4); T: Temperature sensor (115); Disp: Displacement transducer (2); DAU: Data acquisition unit (3). (3) Weigh-in-motion sensors are not located on the bridge and therefore are not shown in the figure.

Table 9.1 Sensors deployed on the Tsing Ma Bridge (from [5]) (Reproduced with permission from Taylor & Francis)

Monitoring item	Sensor type	Sensor code	No. of sensors	Position
Wind speed and direction	Anemometer	WI	6	2: deck of main span 2: deck of the Ma Wan side span 2: top of towers
Temperature	Thermometer	P1–P6, TC	115	6: ambient 86: deck section 23: main cables
Highway traffic	Weigh-in-motion station	WI	7	Approach to Lantau Toll Plaza
Displacement	Displacement transducer	DS	2	1: lowest portal beam of the Ma Wan tower (lateral) 1: deck at the Tsing Yi abutment (longitudinal)
	GPS station	TM	14	4: top of towers 2: middle of main cables 2: middle of the Ma Wan side span 6: $^1/_4$, $^1/_2$ and $^3/_4$ of main span
	Level sensing station	LV	9	1: abutment; 2: towers 2: deck of the Ma Wan side span 4: deck of main span
Acceleration	Accelerometer	AS, AB, AT	13	4: uniaxial, deck 7: biaxial, deck and main cables 2: triaxial, main cables and the Ma Wan abutment
Strain	Strain gage	SP, SR, SS	110	29: Ma Wan side span 32: cross frame at the Ma Wan tower 49: main span
Total	276			

9.7.2 Anemometers in WASHMS

The WASHMS of the bridge includes a total of six anemometers, with two at the middle of the main span, two at the middle of the Ma Wan side span and one of each on the Tsing Yi Tower and Ma Wan Tower (see Figure 9.4). To prevent disturbance from the bridge deck, the anemometers at the deck level were installed on the north side and south side of the bridge deck respectively, via a boom 8.965 m long from the leading edge of the deck (see Figure 9.5).

The anemometers installed on the north side and south side of the bridge deck at the middle of the main span, respectively specified as WI-TJN-01 and WI-TJS-01, are digital-type Gill Wind Master ultrasonic anemometers. The anemometers located at the two sides of the bridge deck near the middle of the Ma Wan approach span, specified as WI-TBN-01 on the north side and WI-TBS-01 on the south side, are analog mechanical (propeller) anemometers. Each analog anemometer consists of a horizontal component (RM Young 05106) with two channels, giving the horizontal resultant wind speed and its azimuth, and a vertical component (RM Young 27106) with one channel, providing the vertical wind speed. The other two anemometers, arranged at 11 m above the top of the bridge towers and specified as

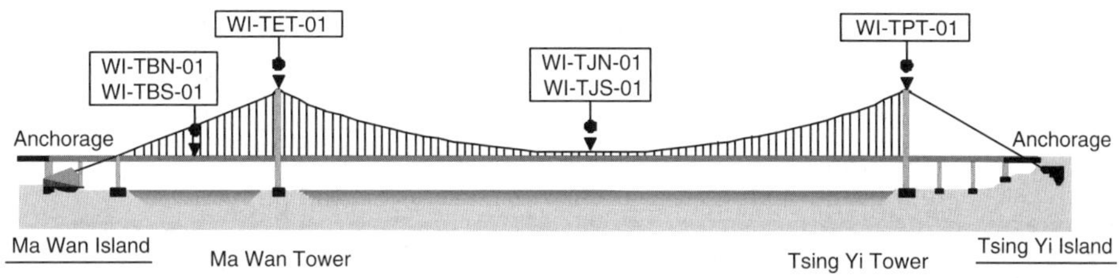

Figure 9.4 Distribution of anemometers in the Tsing Ma Bridge (from [5]) (Reproduced with permission from Taylor & Francis).

WI-TPT-01 for the Tsing Yi tower and WI-TET-01 for the Ma Wan tower, respectively, are analog mechanical anemometers of a horizontal component only. The sampling frequency of measurement of wind speeds was set as 2.56 Hz.

9.7.3 *Temperature Sensors in WASHMS*

The total number of temperature sensors installed on the bridge is 115, and their positions are shown in Figure 9.6. The collected temperature data from WASHMS can be grouped into three categories:

1. ambient temperature (P3, 6 in number);
2. section temperature (P1, P2, P4, and P6, 86 in number); and
3. cable temperature (TC, 23 in number).

The TC type is a thermocouple sensor and P1 to P6 are PT100 Platinum resistance temperature sensors.

One air temperature sensor (P3-TJS) is located approximately at the middle of the main span and attached on a sign gantry which stands on the upper deck. The other five (P3-TON and P3-TOS) measure the ambient temperature around the bridge deck section of the main span near the Tsing Yi tower. 23 thermocouples (TC-TEC) were embedded inside the main cables at three different locations to measure the cable temperature.

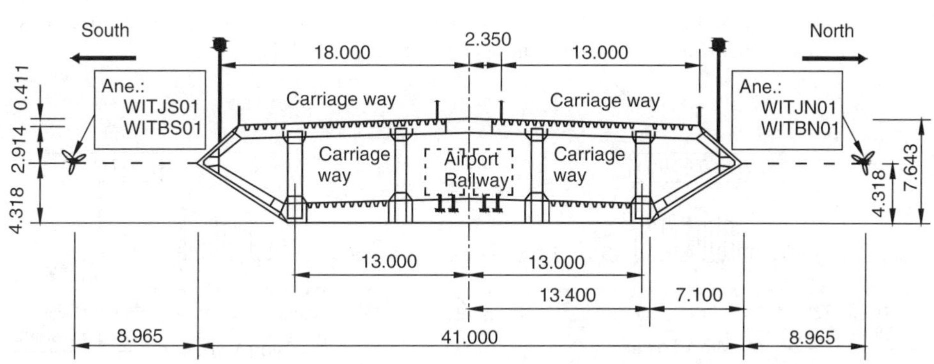

Figure 9.5 Deck cross-section and position of anemometers (unit: m) (from [5]) (Reproduced with permission from Taylor & Francis).

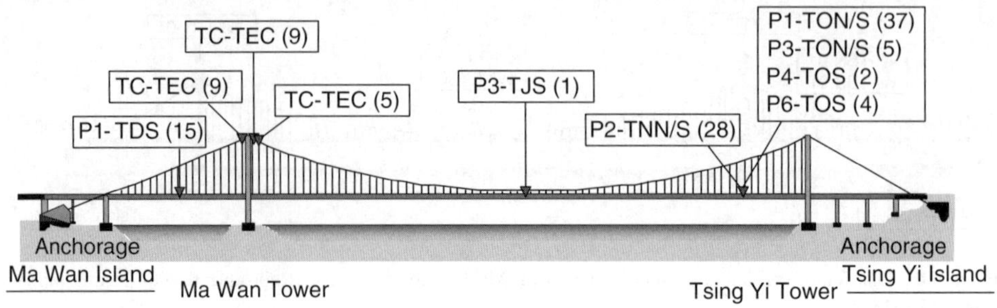

Figure 9.6 Distribution of temperature sensors in the Tsing Ma Bridge (from [5]) (Reproduced with permission from Taylor & Francis).

The deck section in the Ma Wan side span is equipped with 15 sensors (P1-TDS). Deck section 'O', close to the Tsing Yi tower, is equipped with 71 sensors – 43 sensors (P1-TON, P1-TOS, P4-TOS, and P6-TOS) installed on the cross-frame and 28 (P2-TNN and P2-TNS) mounted on the orthotropic deck plates, 2.25 m away from the section. The sampling frequency of all of the temperature sensors is 0.07 Hz.

9.7.4 *Displacement Transducers in WASHMS*

The displacement of the Tsing Ma Bridge in the three orthogonal directions – longitudinal (x-direction), lateral (y-direction) and vertical (z-direction) – is reflected in the measurement data recorded from displacement transducers, level sensing stations and GPS stations together.

The lateral and longitudinal movements of the bridge deck are measured by two displacement transducers at two locations, as shown in Figure 9.7. One displacement transducer (DS-TEN-01), which indicates the lateral motion with a positive value if the deck sways to the north side, is installed in the north side of a bearing frame that sits on the lowest portal beam of the Ma Wan tower with bearing connection. The transducer is connected between the bearing frame and the tower leg, and thus measures the relative lateral movement of the deck to the Ma Wan tower.

The longitudinal movement of the bridge deck at the Tsing Yi abutment is recorded by another displacement transducer (DS-TRA-01). This transducer, which is exactly underneath the expansion joint bearing, is attached between the Tsing Yi abutment and the bottom chord of a cross-frame next to the abutment in order to give the absolute displacement of the bridge deck. The measurements from this, in millimeters, have positive values if the deck moves toward the Tsing Yi Island.

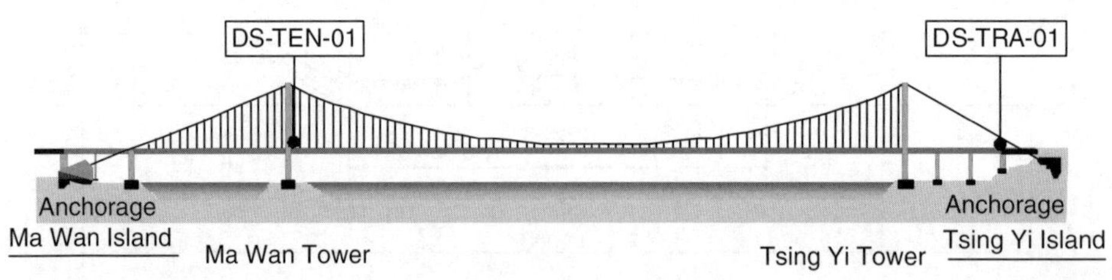

Figure 9.7 Distribution of displacement transducers in the Tsing Ma Bridge (from [5]) (Reproduced with permission from Taylor & Francis).

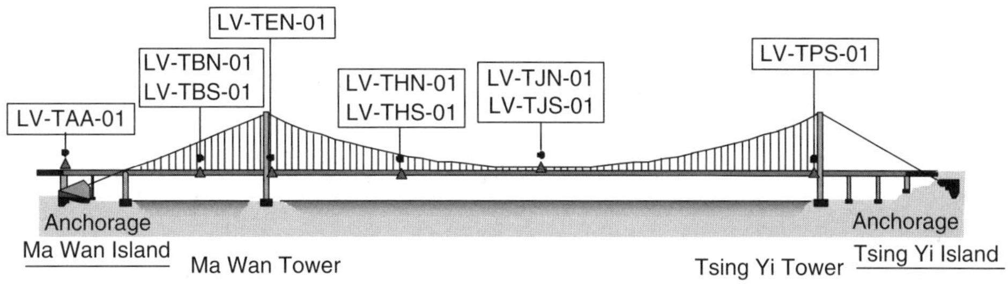

Figure 9.8 Distribution of level sensing stations in the Tsing Ma Bridge (from [5]) (Reproduced with permission from Taylor & Francis).

9.7.5 *Level Sensing Stations in WASHMS*

The vertical motion of the bridge deck is monitored by the level sensing stations which give positive values when the deck moves downward. The level sensing stations are mounted at six locations along the bridge deck as displayed in Figure 9.8.

There are, in total, three level sensing stations installed on one side of the bridge deck at the location of the Ma Wan abutment (LV-TAA-01), the Ma Wan tower (LV-TEN-01) and the Tsing Yi tower (LV-TPS-01). The vertical displacement at the Ma Wan approach span and the main span are monitored by one pair (LV-TBN/S-01) and two pairs (LV-THN/S-01 and LV-TJN/S-01) of level sensing stations, respectively. The installation positions of pairs of level sensing stations at the deck sections can be seen in Figure 9.9. The torsion of the deck sections is determined by the difference in vertical displacements between the two level sensing stations divided by the distance of separation [40].

The level sensing stations, at a sampling rate of 2.56 Hz and cut-off frequency of 1.28 Hz, provide real-time monitoring of displacements with a measurement accuracy of approximately 2 mm at typical deck sections, at vertical planes only.

9.7.6 *GPS in WASHMS*

Real-time measurement accuracy of GPS has been improved to centimeter-level, making it well suited to monitor three-dimensional displacement of bridges in response to wind, temperature, and traffic loads. The commissioning of the GPS in January 2001 brought an additional 14 rover stations into the existing WASHMS, for improving the efficiency and accuracy of the bridge health monitoring system.

The components of the bridge implemented with the GPS receivers include bridge towers, main cables and the bridge deck, as shown in Figure 9.10. Two base reference stations sit atop a storage

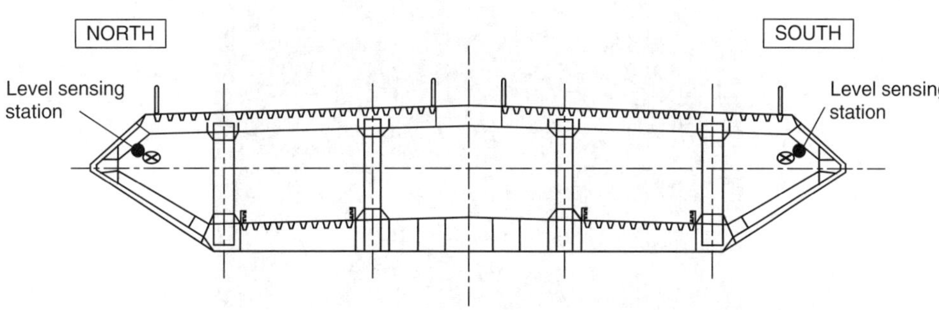

Figure 9.9 Mounting position of level sensing stations in the Tsing Ma Bridge (from [5]) (Reproduced with permission from Taylor & Francis).

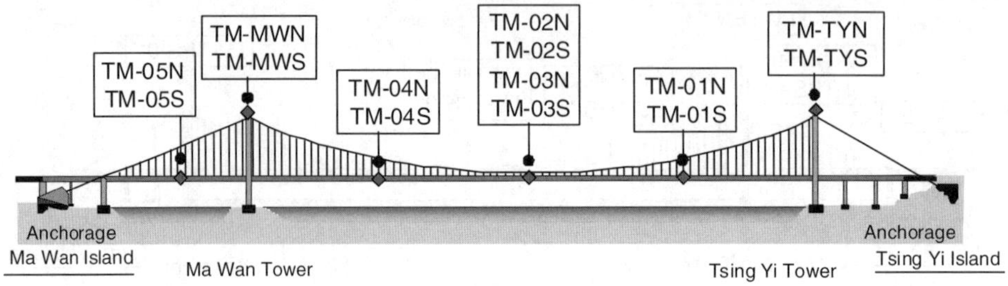

Figure 9.10 Distribution of GPS receivers in the Tsing Ma Bridge (from [5]) (Reproduced with permission from Taylor & Francis).

building adjacent to the bridge monitoring room (see Figure 9.11). The Ma Wan tower and the Tsing Yi tower are each equipped with a pair of GPS receivers, which they are mounted at the tops of saddles on each of the tower legs. The displacement of the main cables is monitored through a pair of GPS receivers at the mid-span. The mid-span of the Ma Wan side span is equipped with a pair of receivers. Three pairs of GPS receivers are positioned at one quarter, one half, and three quarters of the distance along the main span of the bridge deck.

The displacements monitored by these GPS receivers indicate positive movements in the corresponding three directions if the bridge component moves toward the Tsing Yi Island, sways to the north side or goes upward, respectively.

9.7.7 Strain Gauges in WASHMS

There are 110 strain gauges installed at three sections of the Tsing Ma Bridge, with a sampling rate of 51.2 Hz, as shown in Figure 9.12. The strain gauges have three types of configuration, i.e. 44 single gauges, 62 pairs and four rosettes.

Figure 9.11 A GPS reference station on the roof of a storage building (from [5]) (Reproduced with permission from Taylor & Francis).

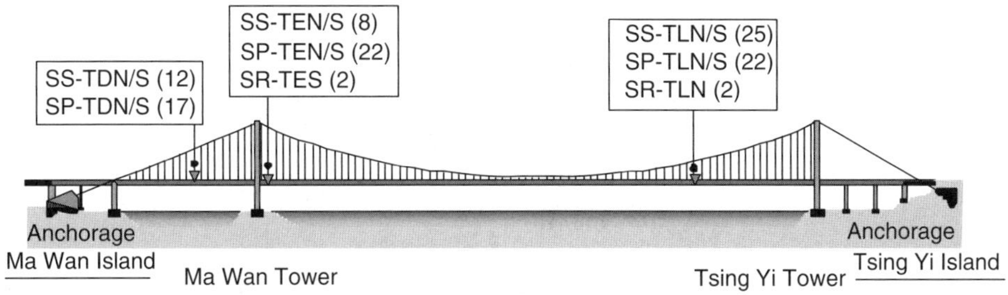

Figure 9.12 Distribution of strain gauges in the Tsing Ma Bridge (from [5]) (Reproduced with permission from Taylor & Francis).

9.7.8 *Accelerometers in WASHMS*

The WASHMS of the bridge includes 13 Honeywell Q-Flex QA700 accelerometers: four uni-axial and four bi-axial accelerometers at four sections of the bridge deck; three bi-axial and one tri-axial on the main cables; and one tri-axial at the Ma Wan abutment, as shown in Figure 9.13.

In each of the four deck sections, one bi-axial accelerometer (AB-TBS-01, AB-TFS-01, AB-TIS-01, and AB-TJS-01) is installed on the south side for measuring acceleration in the vertical and lateral directions, while one uni-axial accelerometer (AS-TBN-01, AS-TFN-01, AS-TIN-01, and AS-TJN-01) is installed on the north side to measure the vertical acceleration. Three bi-axial accelerometers (AB-TCC-01, AB-TGC-01, and AB-TQC-01) are installed on the main cables for measuring the accelerations in the vertical and horizontal directions. Tri-axial accelerometers AT-TGC-01 and AT-TAA-01 measure the acceleration of the main cable and the Ma Wan abutment respectively in the longitudinal, vertical and lateral directions. Hourly acceleration time histories are recorded on tapes, with a sampling frequency of 25.6 Hz before 2001 and 51.2 Hz from 2002 onward.

9.8 Monitoring Results of Tsing Ma Bridge during Typhoon Victor

9.8.1 *Typhoon Victor*

The Tsing Ma Bridge was opened to the public on 20 May 1997 [41]. On 31 July 1997, less than three months after the opening of the bridge, the tropical depression designated "Victor" formed in the middle of the South China Sea [42]. It first moved northwesterly for 12 hours, then made a sudden turn to near north and remained in almost the same direction during its passage over Hong Kong (see Figure 9.14).

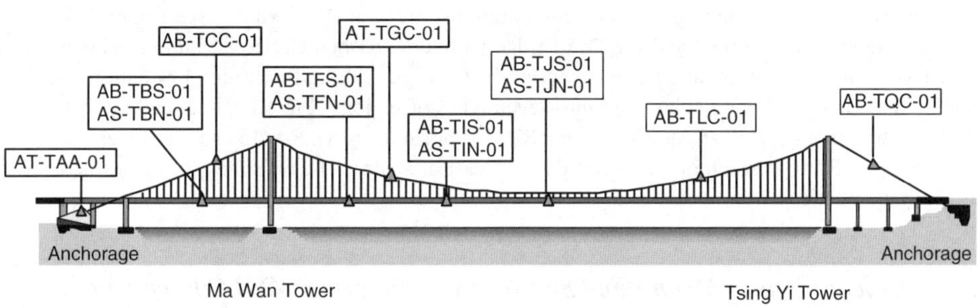

Figure 9.13 Distribution of accelerometers in the Tsing Ma Bridge (from [5]) (Reproduced with permission from Taylor & Francis).

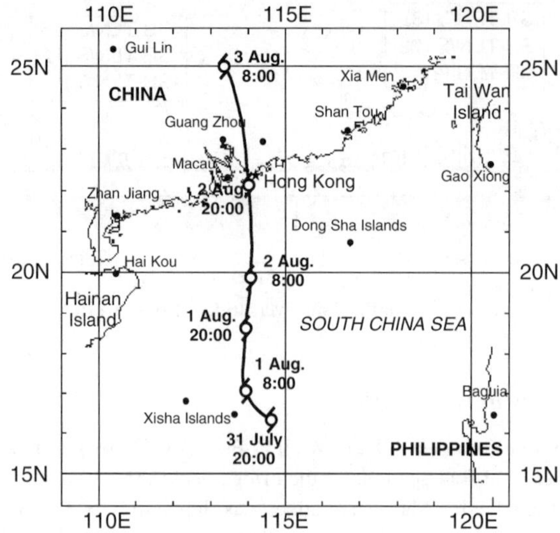

Figure 9.14 Moving path of Typhoon Victor (from [43]) (Reproduced with permission from Techno.Press).

Tropical depression Victor became a real typhoon when it entered the region of 250 km south of Hong Kong at 8:00 HKT (Hong Kong Time) on 2 August, 1997. At 19:00 HKT on 2 August, the centre of the typhoon victor moved into the region about 8 km east of Cheung Chau Island. The lowest air pressure measured on Cheung Chau Island at sea level was 972 hPa. Typhoon Victor then crossed over the Tsing Ma Bridge at 20:05 and made landfall over the western part of the New Territories. It crossed over the whole of Hong Kong within two hours at an average translational speed about 25 km per hour.

The WASHMS timely recorded time-histories of wind velocity and bridge responses. The wind and acceleration data recorded from 17:00 to 24:00 HKT of 7 hours duration [43] are used in this chapter as an example. Thus, the data number of each 7-hour time history is 64 512 for wind speed and 64 5120 for bridge acceleration response. Using MATLAB as a platform, some programs were developed to analyze the measured data to obtain wind characteristics and bridge acceleration response. The frequency resolution is 0.00175 Hz for wind spectral analysis.

9.8.2 Local Topography

Hong Kong is situated in the coastal area of South China. Not only there are many islands in Hong Kong, but there are also many mountains covering most areas of the territory (see Figure 9.15). The local topography surrounding the Tsing Ma Bridge within the dashed circle of 5 km in radius includes sea, islands, and mountains ranging between 69–500 m high. Taking the bridge as a centre, the surrounding area may be roughly classified into seven types of regions (I to VII), bounded by seven lines R1 (18° south of east), R2 (45° north of east), R3 (15° north of west), R4 (15° south of west), R5 (49° south of west), R6 (63° south of west) and R7 (40° east of south). The alignment of bridge deck deviates from the west-east axis for about 17° anticlockwise.

9.8.3 Calculations of Mean Wind Speed and Fluctuating Wind Components

Before wind characteristics can be analyzed, the mean wind and the three turbulence components of fluctuating wind should be extracted from the wind data recorded by different types of anemometers.

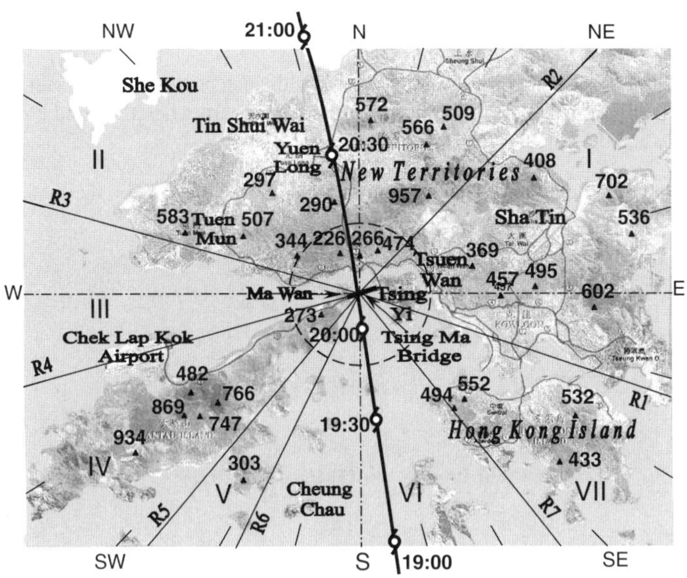

Figure 9.15 Schematic diagram of the topography of Hong Kong (from [43]) (Reproduced with permission from Techno.Press).

For this purpose, the wind coordinate system $X_S Y_E Z_U$ displayed in Figure 9.16 is selected, in which axis X_S points to the due south horizontally, axis Y_E points to the due east horizontally and axis Z_U is vertically upward.

$U_S(t)$, $U_E(t)$ and $U_U(t)$ are, respectively, denoted as the horizontally southward and eastward components of the transient wind velocity and the vertically upward component. For the ultrasonic anemometers used in this chapter, $U_S(t)$ is equal to the southward component of the anemometer whilst $U_E(t)$ and $U_U(t)$ take the minus values of the westward and downward components of the anemometer. For the mechanical anemometers, $U_U(t)$ takes the minus values of the downward component of the anemometer while $U_S(t)$ and $U_E(t)$ should be determined from the simultaneously recorded horizontal transient wind speed and azimuth by using the following equations.

$$U_S(t) = U_H(t) \times \cos \beta_H(t) \tag{9.1a}$$

$$U_E(t) = -U_H(t) \times \sin \beta_H(t) \tag{9.1b}$$

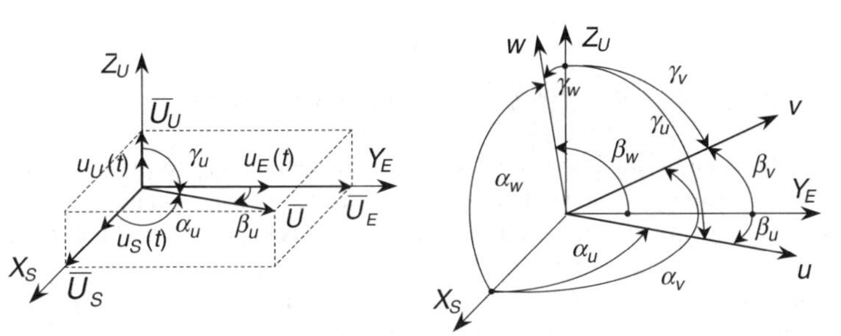

(a) Reference wind coordinate system (b) Directional angels of fluctuating wind

Figure 9.16 Definition of positive directions of mean wind and fluctuating wind (from [43]) (Reproduced with permission from Techno.Press).

where:

$U_H(t)$ is the horizontal resultant wind speed measured by the mechanical anemometer;
β_H indicates the azimuth of $U_H(t)$; the azimuth angle will be zero when the wind blows from the due north and 90° when the wind is from the due east.

By assuming the recorded wind velocity is an ergodic stochastic process, the mean values of $U_S(t)$, $U_E(t)$ and $U_U(t)$ can be obtained by the following formulae:

$$\overline{U}_S = \frac{1}{T}\int_0^T U_S(t)dt = \frac{1}{M}\sum_{i=1}^{M} U_{Si} \tag{9.2a}$$

$$\overline{U}_E = \frac{1}{T}\int_0^T U_E(t)dt = \frac{1}{M}\sum_{i=1}^{M} U_{Ei} \tag{9.2b}$$

$$\overline{U}_U = \frac{1}{T}\int_0^T U_U(t)dt = \frac{1}{M}\sum_{i=1}^{M} U_{Ui} \tag{9.2c}$$

where:

T is the duration of wind speed time history sample;
$M = T/\Delta t = Tf_s$ is the discrete data number of each sample;
Δt is the time interval of sampling;
f_s is the sampling frequency.

The three turbulence components of fluctuating wind along the axes X_S, Y_E and Z_U can be determined as follows:

$$u_S(t) = U_S(t) - \overline{U}_S \tag{9.3a}$$

$$u_E(t) = U_E(t) - \overline{U}_E \tag{9.3b}$$

$$u_U(t) = U_U(t) - \overline{U}_U \tag{9.3c}$$

The resultant mean speed and its directional cosine vector in the $X_SY_EZ_U$ coordinate system (see Figure 9.16a) can be determined by the following equations:

$$\overline{U} = \sqrt{\left(\overline{U}_S^2 + \overline{U}_E^2 + \overline{U}_U^2\right)} \tag{9.4a}$$

$$(\cos\alpha_u, \cos\beta_u, \cos\gamma_u) = \left(\overline{U}_S, \overline{U}_E, \overline{U}_U\right)/\overline{U} \tag{9.4b}$$

As the positive direction of longitudinal turbulence component $u(t)$ is normally defined to be in the same direction as the resultant mean wind, the positive direction of $u(t)$ is indicated by the unit vector of ($\cos\alpha_u$, $\cos\beta_u$, $\cos\gamma_u$) (see Figure 9.16b). Furthermore, it is stipulated that the lateral turbulence component $v(t)$ is horizontal and normal to the mean wind. The vertical turbulence component $w(t)$ is upward and perpendicular to both $u(t)$ and $v(t)$. As a result, the unit vector ($\cos\alpha_v$, $\cos\beta_v$, $\cos\gamma_v$), indicating the positive direction of $v(t)$, and the unit vector ($\cos\alpha_w$, $\cos\beta_w$, $\cos\gamma_w$), indicating the positive direction of $w(t)$, can be determined as follows:

$$(\cos\alpha_v, \cos\beta_v, \cos\gamma_v) = \frac{(-\cos\beta_u, \cos\alpha_u, 0)}{\sqrt{\cos^2\alpha_u + \cos^2\beta_u}} \tag{9.4c}$$

$$(\cos \alpha_w, \cos \beta_w, \cos \gamma_w) = \frac{(-\cos \alpha_u \cos \gamma_u, -\cos \beta_u \cos \gamma_u, \cos^2 \alpha_u + \cos^2 \beta_u)}{\sqrt{\cos^2 \alpha_u + \cos^2 \beta_u}} \tag{9.4d}$$

After having determined the above directional cosine vectors, the alongwind, lateral and upward components of wind turbulence can then be obtained with the following formulae:

$$\begin{aligned} u(t) &= u_S(t) \cos \alpha_u + u_E(t) \cos \beta_u + u_U(t) \cos \gamma_u \\ &= U_S(t) \cos \alpha_u + U_E(t) \cos \beta_u + U_U(t) \cos \gamma_u - \overline{U} \end{aligned} \tag{9.5a}$$

$$\begin{aligned} v(t) &= u_S(t) \cos \alpha_v + u_E(t) \cos \beta_v + u_U(t) \cos \gamma_v \\ &= U_S(t) \cos \alpha_v + U_E(t) \cos \beta_v + U_U(t) \cos \gamma_v \end{aligned} \tag{9.5b}$$

$$\begin{aligned} w(t) &= u_S(t) \cos \alpha_w + u_E(t) \cos \beta_w + u_U(t) \cos \gamma_w \\ &= U_S(t) \cos \alpha_w + U_E(t) \cos \beta_w + U_U(t) \cos \gamma_w \end{aligned} \tag{9.5c}$$

It should be noted that the positive direction of $w(t)$ is upward, but it may not to be in the vertical direction if the upward component of mean wind speed $\overline{U}_U$ is not zero. Figure 9.17 shows typical time histories of three fluctuating wind components measured at the middle point of the main span.

9.8.4 Mean Wind Speed and Direction

Ten-minute mean wind speed, mean wind direction and mean wind inclination are presented in this section. Displayed in Figures 9.18 and 9.19 are the variations of mean wind speed and mean wind direction, respectively, with time at the positions of the anemometers WITJS01, WITBS01, WITPT01, and WITET01.

Figure 9.19 shows that the mean wind to the bridge blew from the northeast in Region I between 17:00 and 19:50 HKT, from the southwest in Region V between 21:00 and 22:00 HKT, and from the southwest in Region VI between 22:00 and 24:00 HKT. There was a sudden change of wind direction from northeast to southwest between 19:50 and 20:10 HKT. During this period, mean wind speed was

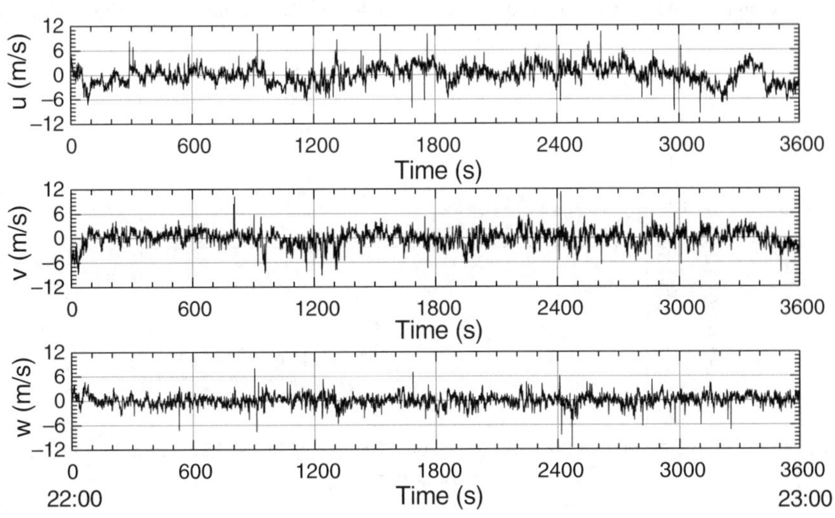

Figure 9.17 Typical time histories of fluctuating wind components (from [43]) (Reproduced with permission from Techno.Press).

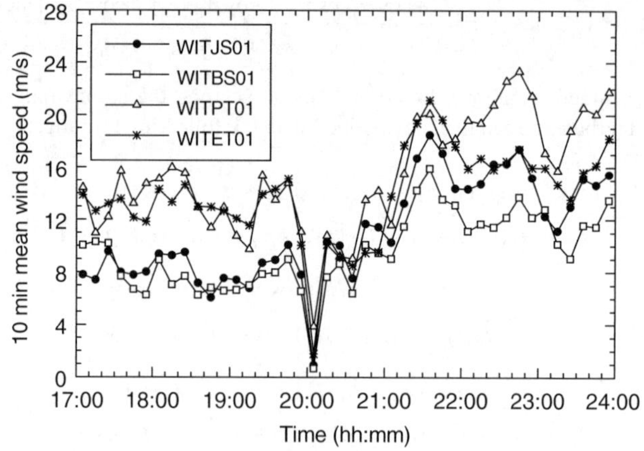

Figure 9.18 Variation of ten-minute mean wind speed (south side) (from [43]) (Reproduced with permission from Techno.Press).

very small, as shown in Figure 9.18. The reason for such a sudden change is that, during this period, Typhoon Victor's eye crossed over the bridge. In consideration of the bridge alignment, the mean wind yaw angles between the mean wind direction measured at the top of the Tsing Yi tower and the longitudinal axis of the bridge were about 20°, 34° and 52° when the wind blew in Regions I, V, and VI, respectively.

The maximum ten-minute mean wind speeds were measured as 12.9 m/s at the deck level and 16 m/s at the tower-top level before Typhoon Victor crossed the bridge. After the crossing, they became, respectively, 18.5 m/s and 21.1 m/s between 21:00 and 22:00 HKT, and 17.4 m/s and 23.3 m/s between 22:00 and 24:00 HKT. Clearly, the maximum ten-minute mean wind speed was larger after Typhoon Victor crossed the bridge than before the crossing.

For the horizontal distribution of mean wind speed, one may compare the mean wind speed measured from anemometer WITJS01 with that from anemometer WITBS01 (see Figure 9.18), because they were arranged on the same side of the bridge deck with a horizontal distance about 860 m. Although the

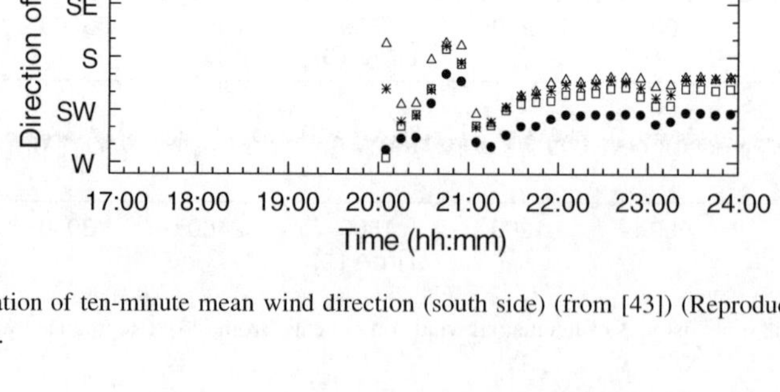

Figure 9.19 Variation of ten-minute mean wind direction (south side) (from [43]) (Reproduced with permission from Techno.Press).

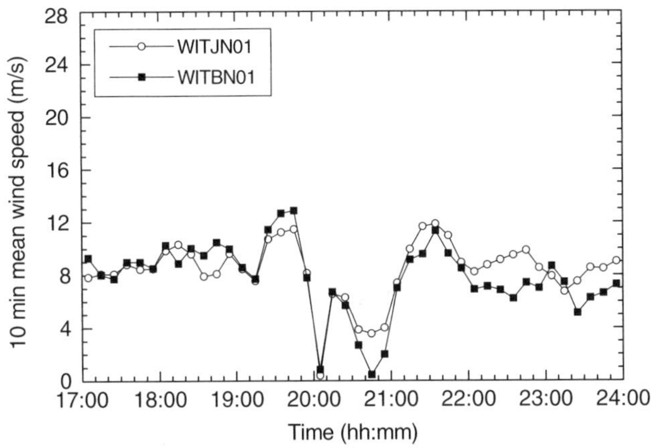

Figure 9.20 Variation of ten-minute mean wind speed (north side) (from [43]) (Reproduced with permission from Techno.Press).

patterns of variation of the mean wind speed with time were similar to each other, there were some differences in the values of mean wind speed, particularly after the crossing of Typhoon Victor. The same feature can be also found in Figure 9.20 for the mean speeds from anemometers WITJN01 and WITBN01, both of which are located on the north side of the bridge deck.

The mean wind speed profile during a typhoon is not yet well known. Also, it could not be exactly explored this time, as only two level wind speeds were available. However, by fitting two level mean wind speeds to the power law mean wind profile, it was found that the value of exponent for the power law varied with time. The mean value was about 0.324 when the wind blew from the northeast in Region I, and 0.199 when the wind blew from the southwest in Region V.

The variation of mean wind inclination with time is shown in Figure 9.21 for the anemometers near the middle of the main span of the bridge (WITJS01 and WITBN01). It is seen that the mean wind inclination, i.e. the transient angle between the mean wind direction and the horizontal plane, ranges from $+6^\circ$ to -6°.

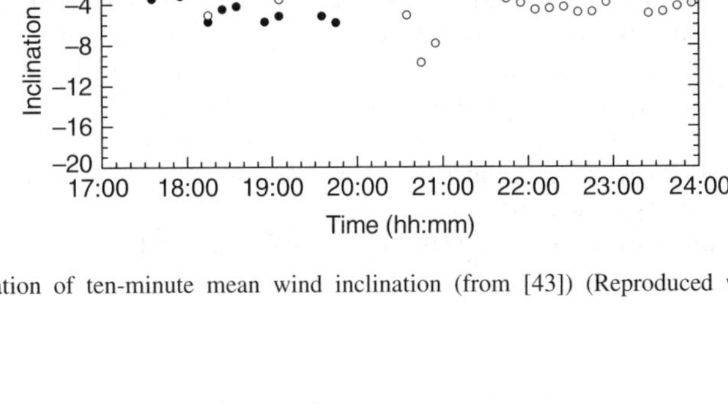

Figure 9.21 Variation of ten-minute mean wind inclination (from [43]) (Reproduced with permission from Techno.Press).

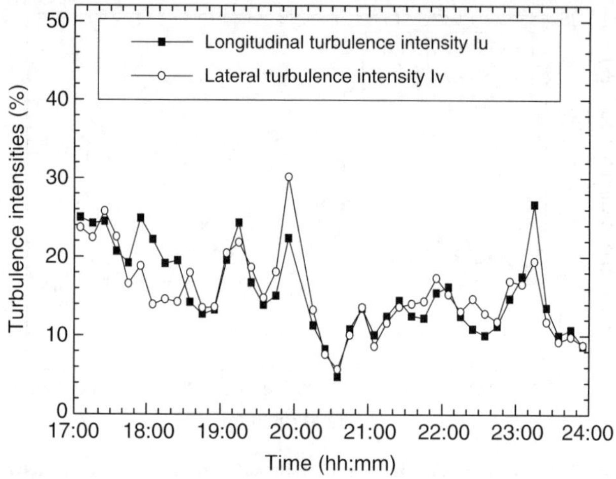

Figure 9.22 Variation of turbulence intensity at WITET01 (from [43]) (Reproduced with permission from Techno.Press).

9.8.5 *Turbulence Intensity and Integral Scale*

As discussed in Chapter 2, turbulence intensity is the ratio of the standard deviation $\sigma(z)$ of the fluctuating wind to the mean wind speed $\overline{U}(z)$at height z. It represents the intensity of the fluctuating wind and is expressed as:

$$I_i = \frac{\sigma_i(z)}{\overline{U}(z)} \quad i = u, v, w \tag{9.6}$$

Figures 9.22 and 9.23 display the longitudinal and lateral turbulence intensities at the top of the Ma Wan tower and the Tsing Yi tower (206 m above the sea level), respectively, using samples of ten minutes duration.

It was found that the mean longitudinal turbulence intensity at the top of the Ma Wan tower (anemometer WITET01) was 20.6% between 17:00 and 20:00 HKT for Region I, 14.3% between 21:00 and 22:00 HKT for Region V and 14.0% between 22:00 and 24:00 HKT for Region VI. Clearly, the longitudinal turbulence intensity varied with wind direction. The ratio of the mean lateral turbulence intensity to the mean longitudinal turbulence intensity at the top of the Ma Wan tower was 0.97 for Region I, 1.05 for Region V and 1.03 for Region VI. These ratios are much higher than those under the seasonal wind condition [44].

At the top of the Tsing Yi tower (anemometer WITPT01), the mean longitudinal turbulence intensity was 25.8% for Region I, 14.3% for Region V and 10.0% for Region VI. Compared with those at the top of the Ma Wan tower, it was found that turbulence intensity may not uniformly distribute along the longitudinal axis of the bridge under the typhoon condition. Furthermore, the mean longitudinal turbulence intensity at the middle main span of the deck was 30% for Region I, 15.8% for Region V and 12.8 % for Region VI. Compared with the same quantities at the top of the Ma Wan tower and the Tsing Yi tower, the mean longitudinal turbulence intensities at the deck level are larger at most times, but occasionally smaller under the typhoon condition.

The integral scale of turbulence represents the average size of the turbulence eddies in the flow (see Chapter 2). The integral scales of the turbulence component u, v, and w in the mean wind direction can be estimated by:

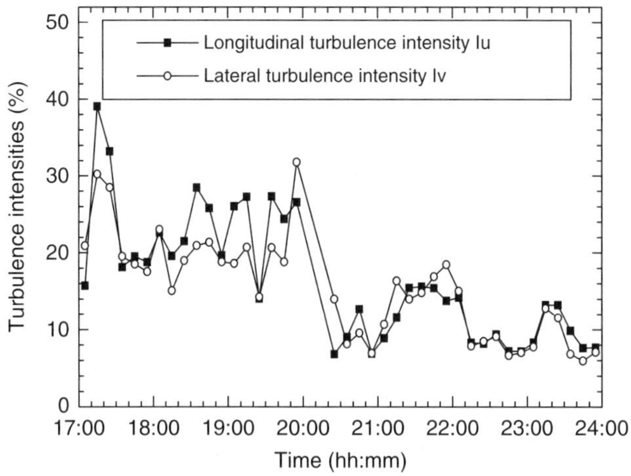

Figure 9.23 Variation of turbulence intensity at WITPT01 (from [43]) (Reproduced with permission from Techno .Press).

$$L_i^x = \overline{U} \int_0^{\infty} C_i(\tau) d\tau \quad i = u, v, w \tag{9.7}$$

where $C_i(\tau)$ is the auto-variance function normalized by the variance.

The mean integral scale of the turbulence component u at the top of the Tsing Yi tower was 210 m between 17:00 and 20:00 for Region I, 243 m between 21:00 and 22:00 for Region V, and 294 m between 22:00 and 24:00 for Region VI. The ratio of the mean integral scale of the turbulence component v to the turbulence component u at the top of the Tsing Yi tower was 0.54 for Region I, 0.78 for Region V and 0.46 for Region VI, respectively. Clearly, these ratios obtained under the typhoon condition were much higher than those obtained under the seasonal trade wind condition [44].

It was also found that the mean integral scales at the top of the Tsing Yi tower were different from those at the top of the Ma Wan tower. The mean integral scales of the turbulence component u and w at the deck level are usually smaller than those at the top of the towers.

9.8.6 Wind Spectra

The auto-spectrum of turbulence describes wind energy distribution over frequency n. For the turbulence component u, von Karman [45], Kaimal *et al.* [46] and Simiu and Scanlan [44] presented Equation 9.8, Equation 9.9 and Equation 9.10, respectively, as the normalized alongwind auto-spectrum:

$$\frac{nS_u(n)}{u_*^2} = \frac{4\sigma_u^2 n L_u^x}{u_*^2 \overline{U}[1 + 70.8(nL_u^x/\overline{U})^2]^{5/6}} \tag{9.8}$$

$$\frac{nS_u(n)}{u_*^2} = \frac{105f}{(1 + 33f)^{5/3}} \tag{9.9}$$

$$\frac{nS_u(n)}{u_*^2} = \frac{200f}{(1 + 50f)^{5/3}} \tag{9.10}$$

where $f = nz/\overline{U}$. Friction velocity u_* is estimated through the horizontal shear stress [47]. The mean friction velocities were found to be 1.23, 1.09 and 0.86 m/s for Regions I, V and VI, respectively, at the deck level.

For the turbulence component v, von Karman [45], Kaimal *et al.* [46] and Simiu and Scanlan [44] proposed Equation 9.11, Equation 9.12 and Equation 9.13, respectively, as the normalized lateral auto-spectrum:

$$\frac{nS_v(n)}{u_*^2} = \frac{4\sigma_v^2 nL_v^x[1 + 755.2(nL_v^x/\overline{U})^2]}{u_*^2\overline{U}[1 + 283.2(nL_v^x/\overline{U})^2]^{11/6}} \tag{9.11}$$

$$\frac{nS_v(n)}{u_*^2} = \frac{17f}{(1+9.5f)^{5/3}} \tag{9.12}$$

$$\frac{nS_v(n)}{u_*^2} = \frac{15f}{(1+9.5f)^{5/3}} \tag{9.13}$$

For the turbulence component w, von Karman [45], Kaimal *et al.* [46] and Simiu and Scanlan [44] recommended Equation 9.14, Equation 9.15 and Equation 9.16, respectively, as the normalized vertical auto-spectrum:

$$\frac{nS_w(n)}{u_*^2} = \frac{4\sigma_w^2 nL_w^x[1 + 755.2(nL_w^x/\overline{U})^2]}{u_*^2\overline{U}[1 + 283.2(nL_w^x/\overline{U})^2]^{11/6}} \tag{9.14}$$

$$\frac{nS_w(n)}{u_*^2} = \frac{2f}{(1+5.3f^{5/3})} \tag{9.15}$$

$$\frac{nS_w(n)}{u_*^2} = \frac{3.36f}{(1+10f^{5/3})} \tag{9.16}$$

The spectral analysis was carried out on the measured turbulence components at different locations and in different time periods, using samples of one hour duration, and the obtained auto-spectra were then compared with those expressed by Equations 9.8 to 9.16. It was found that the normalized auto-spectra of three components of fluctuating wind (nS_u/u_*^2, nS_v/u_*^2, nS_w/u_*^2) varied significantly with time due to the change of wind direction and upwind terrain, and also with the height of the anemometer position. It was also found that the von Karman spectra, using measured integral scales, fit the measured spectral much better than the Kaimal and Simiu spectra, especially in the low frequency region. Figures 9.24 to 9.26 display the longitudinal, lateral, and vertical wind spectra measured from WITJS01 during the period of 23:00 to 24:00 HKT.

The cross-spectra between three turbulence components at each position were also analyzed. The cross-spectrum is a complex quantity. Its real part is called the co-spectrum, which is an even function of frequency, and its imaginary part is named as the quadrature spectrum, which is an odd function of frequency. The cross-spectrum mainly describes the statistical dependence between the turbulence components at a given frequency. A typical co-spectrum is plotted in Figure 9.27; this was measured from WITJS01 during the period of 23:00 to 24:00 HKT.

The following empirical formula of the co-spectrum [46] is also plotted in Figure 9.27 for comparison:

$$\frac{nC_{uw}(n)}{u_*^2} = -\frac{7f}{(1+9.6f)^{2.4}} \tag{9.17}$$

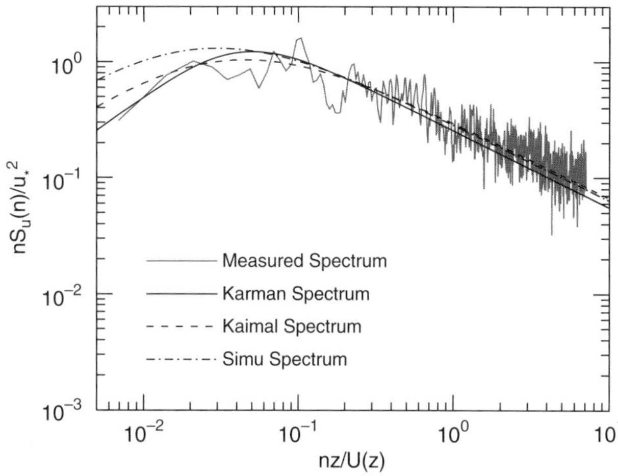

Figure 9.24 Longitudinal normalized spectra at WITJS01 (from [43]) (Reproduced with permission from Techno.Press).

9.8.7 Acceleration Response of Bridge Deck

Figure 9.28 illustrates the time histories of lateral, vertical and torsional acceleration responses of the bridge deck at the mid-main span during the period of 18:00 to 24:00 HKT. The lateral response was directly taken from the lateral accelerometer. The vertical response was obtained by averaging the signals from the two vertical accelerometers. The torsional acceleration response is expressed as the difference between the two vertical signals and divided by 2.

Figure 9.29 shows the variations of standard deviations of lateral, vertical and torsional accelerations with time in ten-minute intervals. The maximum lateral, vertical and torsional standard deviation acceleration responses were found to be $0.588\,\text{cm/sec}^2$, $3.082\,\text{cm/sec}^2$ and $0.0010\,\text{rad/sec}^2$, occurring at about

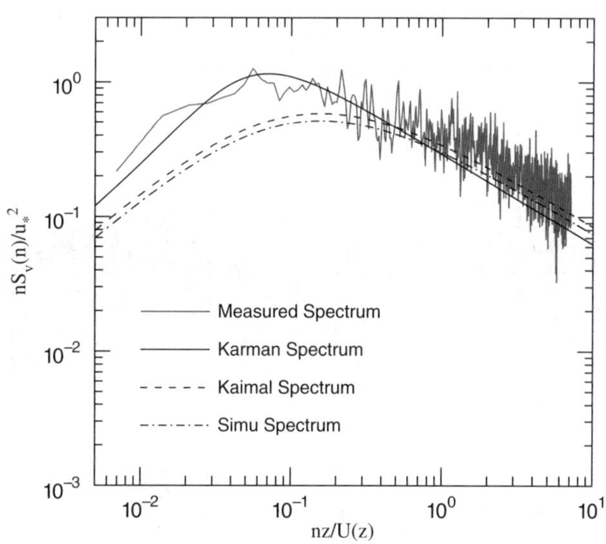

Figure 9.25 Lateral normalized spectra at WITJS01 (from [43]) (Reproduced with permission from Techno.Press).

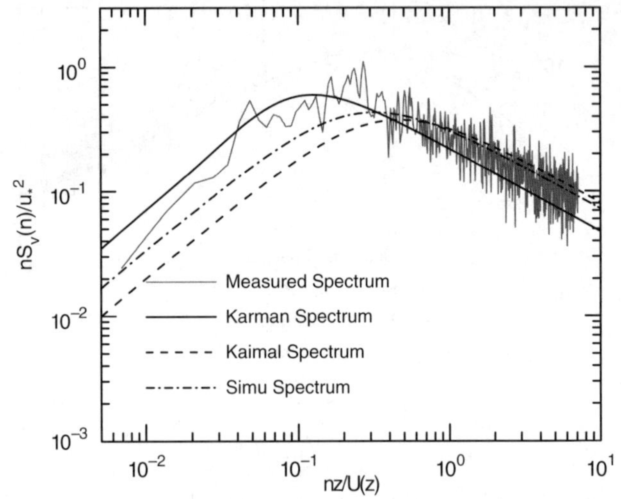

Figure 9.26 Vertical normalized spectra at WITJS01 (from [43]) (Reproduced with permission from Techno .Press).

21:35, 23:35 and 22:45 HKT, respectively. Furthermore, by plotting the lateral, vertical and torsional standard deviation acceleration responses against ten-minute mean wind speed (see Figure 9.30), it is seen that the vertical and torsional acceleration responses increase almost proportionally to the cube of mean wind speed. The lateral acceleration response increases almost proportionally to the square of mean wind speed. Clearly, the vertical acceleration is much larger than the lateral acceleration.

9.8.8 Acceleration Response of Bridge Cable

The acceleration responses of bridge cable at four positions were also analyzed. The maximum lateral and vertical standard deviation acceleration responses of the main cable at the accelerometer ABTLC

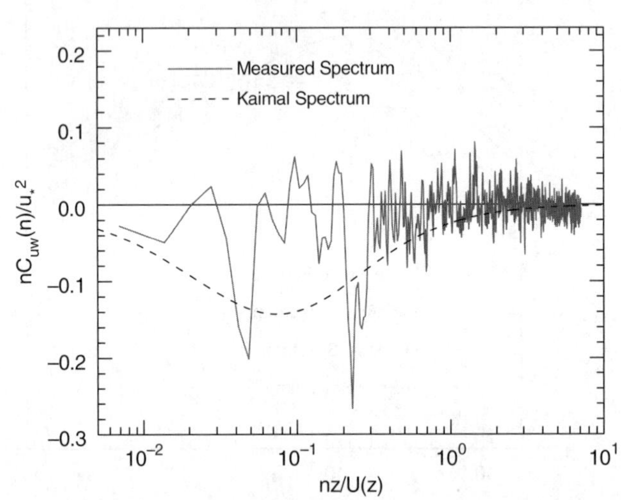

Figure 9.27 Normalized co-spectra between *u* and *w* at WITJS01 (from [43]) (Reproduced with permission from Techno.Press).

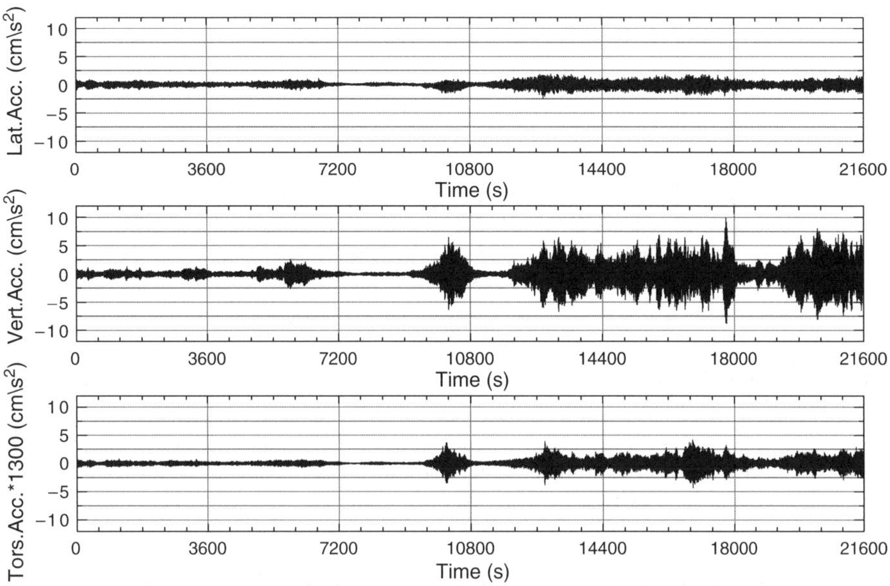

Figure 9.28 Time histories of lateral, vertical and torsional accelerations (18:00 to 24:00) (from [43]) (Reproduced with permission from Techno.Press).

were 1.574 m/sec^2 and 2.965 m/sec^2, occurring at about 21:30 and 22:40 HKT, respectively. For the cable in the Tsing Yi side span, the maximum lateral and vertical standard deviation acceleration responses measured by the accelerometer ABTQC were 4.296 m/sec^2 and 3.133 m/sec^2, respectively, at 23:50 HKT. Clearly, the lateral and vertical standard deviation acceleration responses of the bridge cable are sometime larger than those of the bridge deck.

9.8.9 Remarks

From the measured wind characteristics and bridge responses of the Tsing Ma Bridge during Typhoon Victor, it is clear that for a long-span suspension bridge located in a complex terrain and attacked by a typhoon, wind direction may not be perpendicular to the longitudinal axis of the bridge. The mean wind speed, mean wind inclination, turbulence intensity, auto-spectrum and other factors also vary with the

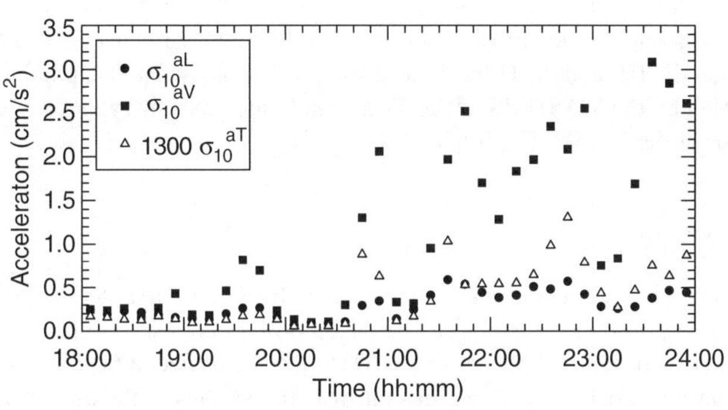

Figure 9.29 Variation of standard deviation of acceleration response (from [43]) (Reproduced with permission from Techno.Press).

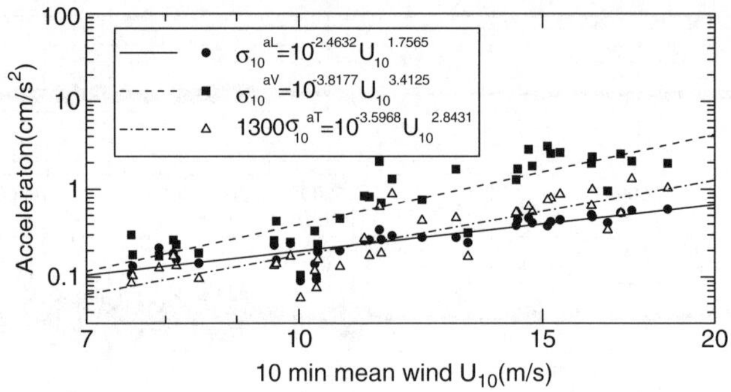

Figure 9.30 Standard deviation acceleration response vs. mean wind speed (from [43]) (Reproduced with permission from Techno.Press).

longitudinal axis of the bridge. The turbulence intensity and integral scales under the typhoon condition are usually higher than those used for the seasonal trade wind case.

9.9 System Identification of Tsing Ma Bridge during Typhoon Victor

9.9.1 Background

Natural frequencies, mode shapes and modal damping ratios are the most important modal parameters to be determined when designing a long-span cable-supported bridge. Knowledge of these properties is essential in order to understand and interpret with confidence the response of the bridge to strong winds. Therefore, it is always desirable to carry out field measurements on long-span bridges to provide such information in order to check the modal parameters used in the design, to understand the actual dynamic performance of the bridge under strong winds and to develop better design theories for future bridges.

For a long-span bridge, the most popular approach for identifying modal parameters is to carry out ambient vibration measurements and then to apply the Fast Fourier Transform (FFT) based method to the measured bridge response time histories. Though this approach can identify natural frequencies quite accurately, it often overestimates modal damping ratios, due to bias error involved in the spectral analysis. Also, this approach cannot provide information on how modal parameters (dynamic characteristics) vary with bridge vibration caused by winds. Furthermore, this approach may become more questionable when identifying the modal damping ratios from non-stationary bridge response time history.

Motivated by this problem, the (EMD + HT) method [48], which consists of mainly the empirical mode decomposition (EMD) and the Hilbert transform (HT), was used and applied to the field measurement results recorded by the WASHMS of the Tsing Ma Bridge during Typhoon Victor to demonstrate one of the functions of the WASHMS [49].

9.9.2 EMD + HT Method

The EMD + HT method is a two-step data analyzing method. The first step is the EMD, by which a measured structural response time history can be decomposed into a series of intrinsic mode functions (IMF) that admit well-behaved Hilbert transforms (HT) [36]. This decomposition is based on the direct extraction of the energy associated with various intrinsic time scales of the time history itself. To avoid mode mixing during the sifting process, the intermittency check is used to separate the waves of different periods into different modes, based on the period length, to obtain modal response time histories.

The random decrement technique (RDT) is then applied to each of the modal response time histories in order to obtain the free modal response time histories. The second step of the EMD + HT method is implemented by performing the HT to each of the free modal response time histories to find the corresponding natural frequency and modal damping ratio.

The EMD developed by Huang *et al.* [36] is a relatively new data processing method, which can decompose any data set into several IMFs by a procedure called the sifting process. Suppose $y(t)$ is the signal to be decomposed. The sifting process is conducted by first constructing the upper and lower envelope of $y(t)$ by connecting its local maxima and local minima through a cubic spline. The mean of the two envelopes is then computed and subtracted from the original time history. The difference between the original time history and the mean value is called the first IMF, c_1, if it satisfies the following two conditions:

1. within the data range, the number of extreme and the number of zero-crossings are equal or differ by 1 only; and
2. the envelope defined by the local maxima and the envelope defined by the local minima are symmetric with respect to the mean.

The difference between $y(t)$ and c_1 is then treated as a new time history and is subjected to the same sifting process, giving the second IMF. The EMD procedure continues until the residue becomes so small that it is less than a predetermined value of consequence, or the residue becomes a monotonic function. The original time history, $y(t)$, is finally expressed as the sum of the IMF components plus the final residue:

$$y(t) = \sum_{j=1}^{N} c_j(t) + r(t) \tag{9.18}$$

where:

N is the number of IMF components;
$r(t)$ is the final residue.

To extract the modal responses from the measured acceleration response time history of the bridge, an intermittency check [50] should be imposed during the sifting process in the EMD method in order to separate the waves of different periods into different modes based on the period length. The intermittency check is designed as the lower limit of the frequency that can be included in any given IMF component. It can be achieved by specifying a cut-off frequency, ω_c, for each IMF during its sifting process. Data having frequencies lower than ω_c will be removed from the resulting IMF. The cut-off frequencies can be decided according to the natural frequencies of the bridge estimated from the modal analysis, or the natural frequencies identified from the power spectrum of the response time history. As a result, the measured acceleration response of the bridge, $\ddot{x}(t)$, can be decomposed by the EMD as follows [4.48]:

$$\ddot{x}(t) = \sum_{j=1}^{k} \ddot{x}_j(t) + \sum_{i=1}^{N-k} c_i(t) + r_N(t) \tag{9.19}$$

where:

$\ddot{x}_j(t)$ is the jth modal acceleration response;
$c_i(t)$ is the ith IMF.

From each modal acceleration response $\ddot{x}_j(t)$, the random decrement technique [51] is then applied to obtain the free modal response time history. In this regard, a threshold level should be selected to obtain L segments. The jth free modal response, $\ddot{x}_j^f(i\Delta t)$, can then be obtained from the ensemble average of all segments as:

$$\ddot{x}_j^f(i\Delta t) = \frac{1}{L}\sum_{k=1}^{L} \ddot{x}_j(t_k + i\Delta t); \quad i = 1, 2, \ldots, M \tag{9.20}$$

where:

$\ddot{x}_j f(t)$ is the jth free modal response;
L is the number of segments related to the threshold level $\ddot{x}_s$;
t_k is the starting time for each segment;
Δt is the time interval of the modal response time history;
M is an integer, with $M\Delta t = T$ being the duration of each segment.

Having obtained the free modal response $\ddot{x}_j^f(t)$, it is possible to identify the natural frequency and modal damping ratio using the HT applicable to a single degree of freedom (SDOF) system.

The Hilbert transform (HT) has long been used to study linear and non-linear dynamic systems and to identify their modal parameters in the frequency-time domain [52]. For a linear SDOF system under the impulsive loading, the impulse displacement response function of the system $v(t) = 0$ for $t < 0$ and:

$$v(t) = A_0 e^{-\xi\omega_0 t} \sin \omega_d t, \quad t \geq 0 \tag{9.21}$$

where:

ω_0 is the natural circular frequency of the system;
ξ is the damping ratio;
ω_d is the damped natural circular frequency;
A_0 is the constant depending on the intensity of impulsive loading and the mass and frequency of the system.

For convenience of applying the HT, the impulse response function can be extended to the negative domain by considering its mirror image:

$$v(t) = A_0 e^{-\xi\omega_0 |t|} \sin \omega_d t, \quad -\infty < t < \infty \tag{9.22}$$

According to the HT theory [53], the analytical signal $z(t)$ associated with $v(t)$ is defined as:

$$z(t) = v(t) + i\tilde{v}(t) = A(t)e^{-i\theta(t)} \tag{9.23}$$

where $\tilde{v}(t)$ is the HT of $v(t)$.

For a special case in which ξ is small and ω_0 is large, the amplitude $A(t)$ and the phase angle $\theta(t)$ for the SDOF system can be obtained by:

$$A(t) = A_0 e^{-\xi\omega_0 t} \tag{9.24}$$

$$\theta(t) = \omega_d t - \pi/2 \tag{9.25}$$

By introducing the logarithmic and differential operators to Equations 9.24 and 9.25 respectively, one obtains:

$$\ln A(t) = -\xi\omega_0 t + \ln A_0 \tag{9.26}$$

$$\omega(t) = \frac{d\theta(t)}{dt} = \omega_d \tag{9.27}$$

Therefore, the damped natural circular frequency ω_d can be identified from the instantaneous frequency $\omega(t)$. With the identified ω_d and the slope $-\xi\omega_0$ of the straight line of the decaying amplitude $A(t)$ in a semi-logarithmic scale, the damping ratio ξ can be identified from the function $\omega_d = \omega_0\sqrt{1-\xi^2}$.

In consideration that the instantaneous frequency may fluctuate around its mean value due to the amplitude variation of the signal and that the requirement of small damping ratio may limit the application of the HT method, the following procedure suggested in [48] is used for the system identification of SDOF systems based on the HT method:

1. determine the damped frequency ω_d from the slope of the phase function $\theta(t)$by the linear least-squares fit technique;
2. determine the damping ratio ξ by applying the linear least-squares fit technique to the decaying amplitude $A(t)$ in a semi-logarithmic scale.

By applying the above-mentioned identification procedure to the *j*th free modal response, $\ddot{x}_j^f(t)$, the *j*th natural frequency and modal damping ratio of the bridge can be identified easily.

9.9.3 *Natural Frequencies and Modal Damping Ratios*

The EMD + HT method and the FFT method are applied to the measured acceleration response time histories of one hour duration of the Tsing Ma Bridge during Typhoon Victor for identifying the natural frequencies and modal damping ratios of Lat1, Lat2, Ver1, Ver2, and Tor1, which stand for the first and second lateral vibration modes, the first and second vertical vibration modes and the first torsional vibration mode, respectively [54].

The resulting natural frequencies from six hour-long records are listed in Table 9.2, and the resulting modal damping ratios are shown in Table 9.3, in which HKT means Hong Kong Time. Note that the damping identified here is the total damping, which consists of the net structural damping and the aerodynamic damping. The aerodynamic damping ratios in the lateral, vertical and torsional modes of vibration can be estimated in terms of aerodynamic coefficients. After removing the aerodynamic

Table 9.2 Comparison of identified natural frequencies by EMD-HT and FFT (ATTID) (from [49]) (Reproduced with permission from Elsevier)

HKT	Lat1		Lat2		Ver1		Ver2		Tor1	
	EMD	FFT	EMD	FFT	EMD	FFT	EMD	FFT	EMD	FFT
18:00–19:00	0.0686	0.0688	0.1602	0.1625	0.1137	0.1141	0.1365	0.1359	0.2652	0.2656
19:00–20:00	0.0677	0.0672	0.1608	0.1609	0.1138	0.1149	0.1373	0.1359	0.2653	0.2656
20:00–21:00	0.0683	0.0688	0.1610	0.1609	0.1139	0.1141	0.1387	0.1375	0.2655	0.2641
21:00–22:00	0.0677	0.0672	0.1598	0.1594	0.1133	0.1141	0.1362	0.1359	0.2641	0.2641
22:00–23:00	0.0681	0.0688	0.1587	0.1594	0.1141	0.1141	0.1376	0.1375	0.2645	0.2641
23:00–24:00	0.0677	0.0672	0.1601	0.1594	0.1145	0.1141	0.1373	0.1359	0.2645	0.2641

Table 9.3 Comparison of identified modal damping ratios by EMD-HT and FFT (ATTID) (from [49]) (Reproduced with permission from Elsevier)

HKT	Lat1		Lat2		Ver1		Ver2		Tor1	
	EMD	FFT	EMD	FFT	EMD	FFT	EMD	FFT	EMD	FFT
18:00–19:00	0.80%	2.54%	1.27%	1.39%	0.35%	1.24%	1.01%	1.68%	0.29%	0.45%
19:00–20:00	1.52%	2.09%	0.91%	1.02%	1.27%	1.61%	0.66%	0.98%	0.75%	0.81%
20:00–21:00	0.82%	1.97%	0.81%	1.27%	1.03%	0.97%	0.78%	0.93%	0.56%	0.54%
21:00–22:00	0.55%	1.67%	0.58%	1.09%	0.68%	1.09%	0.49%	0.92%	0.35%	0.48%
22:00–23:00	1.31%	2.22%	1.07%	1.61%	0.73%	1.14%	0.94%	1.62%	0.47%	0.62%
23:00–24:00	1.58%	2.20%	1.33%	1.24%	0.89%	1.55%	0.62%	0.89%	0.47%	0.54%

damping ratio from the identified total damping ratio for a given mode of vibration, the structural damping ratio can be obtained [4.49].

It is encouraging to see from Table 9.2 that, for a given mode of vibration, the natural frequencies identified by the EMD + HT method from the six records are very close to each other. It is also interesting to see that the natural frequencies identified by the EMD + HT method are very close to those obtained by the FFT-based method. The damping ratios identified by the EMD + HT method are smaller than those obtained from the FFT-based method (see Table 9.3). It is well known that the bias error in the FFT-based method always leads to an overestimation of damping because of the limitation on the frequency resolution, so hundreds of stationary samples should be used in the damping ratio estimation to reduce the random error. Obviously, in reality, during a typhoon it is almost impossible to obtain a stationary record long enough for damping estimation. Considering all these factors, and the adaptive nature of the EMD + HT method performed in the frequency-time domain, one may say that the EMD + HT method is superior to the FFT-based method for modal damping estimation.

The variations of natural frequency, total modal damping ratio, and structural modal damping ratio with vibration amplitude and mean wind speed were examined, and the variation of structural modal damping ratio with modal frequency was also investigated. The results demonstrated that the natural frequencies of the bridge decreased very slightly with the increase in either mean wind speed or vibration amplitude (see Figure 9.31).

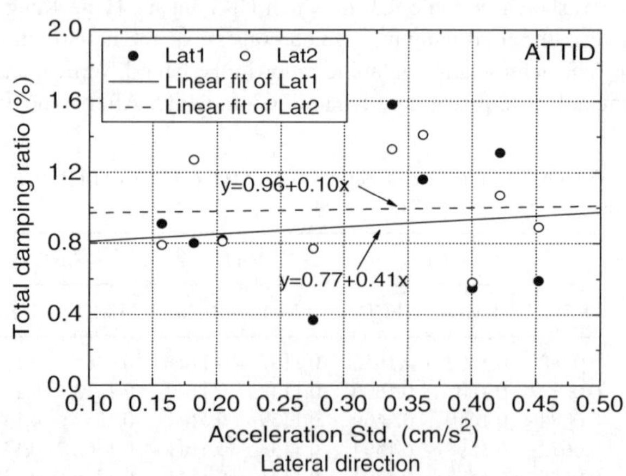

Figure 9.31 Variations of total damping ratio (from [49]) (Reproduced with permission from Elsevier).

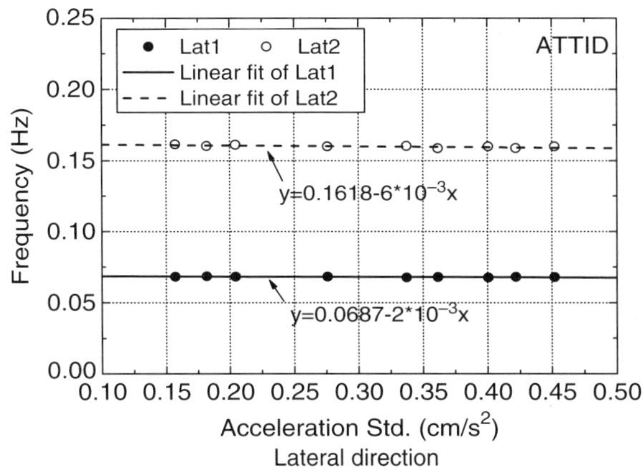

Figure 9.32 Variations of natural frequency (from [49]) (Reproduced with permission from Elsevier).

The total modal damping ratios and the structural damping ratios both exhibited an increasing trend with increasing vibration amplitude or increasing mean wind speed (see Figure 9.32). In consideration that most of the bridge response time histories recorded during Typhoon Victor were non-stationary, it may be concluded that the EMD + HT method is superior to the FFT-based method for damping identification of large civil structures under strong typhoon. Further discussions on non-stationary wind and wind-induced bridge responses can be found in Chapter 15.

9.10 Notations

$A(t)$	Decaying amplitude
$C_i(\tau)$	Auto-variance function normalized by the variance
$c_i(t)$	*i*th IMF
f_s	Sampling frequency
I	$i = u, v, w$, denotes turbulence component
I_i	Turbulence intensity of turbulence component i
L_i^x	Integral scale of turbulence component i
M	Discrete data number of each sample
N	Number of IMF components
n	Frequency
$r(t)$	Final residue
$S_i(n)$	Auto-spectrum of turbulence component i
T	Duration of wind speed time history sample
$\overline{U}$	Mean wind speed
u_*	Friction velocity
$u_E(t)$	Turbulence components of fluctuating wind along the axis Y_E
$U_E(t)$	Horizontally eastward component of transient wind velocity
$\overline{U}_E$	Mean value of $U_E(t)$
$U_H(t)$	Horizontal resultant wind speed measured by the mechanical anemometer
$u_S(t)$	Turbulence components of fluctuating wind along the axis X_S
$U_S(t)$	Horizontally southward of transient wind velocity
$\overline{U}_S$	Mean value of $U_S(t)$

$u_U(t)$	Turbulence components of fluctuating wind along the axis Z_U
$U_U(t)$	Vertically upward component of transient wind velocity
$\overline{U}_U$	Mean value of $U_U(t)$
$X_S Y_E Z_U$	Wind coordinate system
$\ddot{x}_s$	Threshold level
$\ddot{x}_j(t)$	jth modal acceleration response
$\ddot{x}_j^f(t)$	jth free modal acceleration response
β_H	Horizontal resultant wind azimuth
ω_0	Natural circular frequency
ω_d	Damped natural circular frequency
ξ	Damping ratio
Δt	Time interval of sampling
$\sigma_i(z)$	Standard deviation of turbulence component i
$\theta(t)$	Phase angle

References

1. Mufti, A. (2001) *Guidelines for Structural Health Monitoring*, Intelligent Sensing for Innovative Structures Canada, Winnipeg.
2. Sohn, H., Farrar, C.R., Hemez, F.M. *et al.* (2003) A Review of Structural Health Monitoring Literature: 1996–2001, Los Alamos National Laboratory Report, LA-13976-MS, Los Alamos National Laboratory, Los Alamos.
3. Wong, K.Y. (2004) Instrumentation and health monitoring of cable-supported bridges. *Structural Control and Health Monitoring*, **11**(2), 91–124.
4. Karbhari, V.M. and Ansari, F. (2009) *Structural Health Monitoring of Civil Infrastructure Systems*, Woodhead Publishing Limited, Cambridge.
5. Xu, Y.L. and Xia, Y. (2012) *Structural Health Monitoring of Long-Span Suspension Bridges*, Spon Press, London.
6. Aktan, A.E., Catbas, F.N., Grimmelsman, K.A., and Pervizpour, M. (2002) Development of a Model Health Monitoring Guide for Major Bridges, *Federal Highway Administration Research and Development Report* (No. DTFH61-01-P-00347), Drexel Intelligent Infrastructure and Transportation Safety Institute, Philadelphia.
7. Hayashida, H., Fukao, S., Kobayashi, T. *et al.* (1996) Remote sensing of wind velocity by boundary layer radar. *Journal of Wind Engineering*, **67**, 39–42.
8. Powell, M.D., Reinhold, T.A., and Marshall, R.D. (1999) GPS sonde insights on boundary layer wind structure in hurricanes, in *Wind Engineering into the 21st Century* (eds G.L. Larose, A. Larsen, and F.M. Livesey), Taylor & Francis, pp. 307–314.
9. Tamura F.Y., Suda, K., Sasaki, A. *et al.* (1999) Wind speed profiles measured over ground using Doppler sodars. *Journal of Wind Engineering and Industrial Aerodynamics*, **83**(1–3), 83–93.
10. Sahin, M., Tari, E., and Ince, C.D. (1999) Continuous earthquake monitoring with global positioning system (GPS), in *Proceedings of ITU-IAHS International Conference on Kocaeli Earthquake, A Scientific Assessment and Recommendations for Re-Building, Istanbul, Turkey* (eds M. Karaca and D.N. Ural), Istanbul Technical University, Istanbul, pp. 231–238.
11. Ashkenazi, V. and Roberts, G.W. (1997) Experimental monitoring of the Humber Bridge using GPS. *ICE Proceedings*, **120**(4), 177–182.
12. Fujino, Y., Murata, M., Okano, S., and Takeguchi, M. (2000) Monitoring system of the Akashi Kaikyo Bridge and displacement measurement using GPS, in *Proceedings of Nondestructive Evaluation of Highways, Utilities, and Pipelines IV*, vol. **3995** (eds A.E. Aktan and S.R. Gosselin), SPIE, pp. 229–236.
13. Nakamura, S. (2000) GPS measurement of wind-induced suspension bridge girder displacements. *Journal of Structural Engineering*, **126**(12), 1413–1419.
14. Miyata, T., Yamada, H., Katsuchi, H., and Kitagawa, M. (2002) Full-scale measurement of Akashi-Kaikyo Bridge during typhoon. *Journal of Wind Engineering and Industrial Aerodynamics*, **90**(12–15), 1517–1527.
15. Tamura, Y., Matsui, M., Luisa-Carlotta, P. *et al.* (2002) Measurement of wind-induced response of buildings using RTK-GPS. *Journal of Wind Engineering and Industrial Aerodynamics*, **90**(12–15), 1983–1973.
16. Chan, W.S., Xu, Y.L., Ding, X.L. *et al.* (2006) Assessment of dynamic measurement accuracy of GPS in three directions. *Journal of Surveying Engineering*, **132**(3), 108–117.
17. Kijewski-Correa, T. and Kochly, M. (2007) Monitoring the wind-induced response of tall buildings: GPS performance and the issue of multipath effects. *Journal of Wind Engineering and Industrial Aerodynamics*, **95**(9–11), 1176–1198.
18. Casas, J.R. and Cruz, P.J.S. (2003) Fiber optic sensors for bridge monitoring. *Journal of Bridge Engineering*, **8**(6), 362–373.
19. Seim, J.M., Udd, E., Schulz, W.L., and Laylor, H.M. (1999) Health monitoring of an Oregon historical bridge with fiber grating strain sensors, in *Proceedings of SPIE Smart Structures and Materials 1999: Smart Systems for Bridges, Structures, and Highways*, vol. **3671** (ed. S.C. Liu), SPIE, Bellingham, Washington, pp. 128–134.

20. Stewart, A., Carman, G., and Richards, L. (2005) Health monitoring technique for composite materials utilizing embedded thermal fiber optic sensors. *Journal of Composite Material*, **39**(3), 199–213.
21. Abe, M., Fujino, Y., and Kaito, K. (2001) Damage detection of civil concrete structures by laser Doppler vibrometry, in *Proceedings of the 19th International Modal Analysis Conference* (eds A.L. Wicks and R. Singhal), Society for Experimental Mechanics, Bethel, pp. 704–709.
22. Kaito, K., Abe, M., and Fujino, Y. (2001) An experimental modal analysis for RC bridge decks based on non-contact vibration measurement, in *Proceedings of the 19th International Modal Analysis Conference* (eds A.L. Wicks and R. Singhal), Society for Experimental Mechanics, Bethel, pp. 1561–1567.
23. Bales, F.B. (1985) Close-range photogrammetry for bridge measurement, in *Transportation Research Record (950)*, Transportation Research Board, Washington, D.C, pp. 39–44.
24. Li, J.C. and Yuan, B.Z. (1988) Using vision technique for bridge deformation detection. Proceedings of the International Conference on Acoustic, Speech and Signal Processing, Institute of Electrical and Electronics Engineers, Piscataway, New York, pp. 912–915.
25. Olaszek, P. (1999) Investigation of the dynamic characteristic of bridge structures using a computer vision method. *Measurement*, **25**(3), 227–236.
26. Patsias, S. and Staszewski, W.J. (2002) Damage detection using optical measurements and wavelets. *Structural Health Monitoring*, **1**(1), 7–22.
27. Ji, Y.F. and Chang, C.C. (2008) Non-target image-based technique for bridge cable vibration measurement. *Journal of Bridge Engineering*, **13**(1), 34–42.
28. Zhou, X.Q., Xia, Y., Deng, Z.K., and Zhu, H.P. (2010) Experimental study on videogrammetric technique for vibration displacement measurement, in *Proceedings of the 11th International Symposium on Structural Engineering* (eds J. Cui*et al.*), Science Press, Beijing, pp. 1059–1063.
29. Kahn, J.M., Katz, R.H., and Pister, K.S.J. (1999) Mobile networking for smart dust. ACM/IEEE International Conference on Mobile Computing and Networking, Seattle, WA.
30. Hollar, S. (2000) *COTS Dust*, Master Thesis, Department of Electronic Engineering, University of California at Berkeley.
31. Lynch, J.P., Law, K.H., Kiremidjian, A.S. *et al.* (2001) The design of a wireless sensing unit for structural health monitoring, in *Proceedings of the 3rd International Workshop on Structural Health Monitoring*, (ed R.A. Livingston), Stanford University, Stanford, CA.
32. Kling, R.M. (2003) Intel Mote: an enhanced sensor network node, in CD-ROM. Proceedings of International Workshop on Advanced Sensors, Structural Health Monitoring, and Smart Structures, Tokyo, Japan.
33. Rice, J.A. and Spencer, B.F. Jr (2008) Structural health monitoring sensor development for the Imote2 platform, in *Proceedings of SPIE Sensors and Smart Structures Technologies for Civil, Mechanical, and Aerospace Systems* (ed. M. Tomizuka), SPIE, Bellingham, Washington.
34. Oppenheim, A.V. and Schafer, R.W. (1989) *Discrete-Time Signal Processing*, Prentice-Hall, Englewood Cliffs.
35. Daubechies, I. (1992) *Ten Lectures on Wavelets*, Society for Industrial and Applied Mathematics, Philadelphia.
36. Huang, N.E., Shen, Z., Long, S.R. *et al.* (1998) The empirical mode decomposition and the Hilbert spectrum for nonlinear and non-stationary time series analysis. *Proceedings of Royal Society London A*, **454**(1971), 903–995.
37. Chan, W.S., Xu, Y.L., Ding, X.L., and Dai, W.J. (2006b) Integrated GPS-accelerometer data processing techniques for structural deformation monitoring. *Journal of Geodesy*, **80**(12), 705–719.
38. Hall, D.L. (1992) *Mathematical Techniques in Multisensor Data Fusion*, Artech House Inc., Norwood.
39. Hall, D.L. and Llinas, J. (1997) An introduction to multisensor data fusion. *Proceedings of IEEE*, **85**(1), 6–23.
40. Xu, Y.L., Guo, W.W., Chen, J. *et al.* (2007) Dynamic response of suspension bridge to typhoon and trains. I: Field measurement results. *Journal of Structural Engineering*, **133**(1), 3–11.
41. Lau, C.K., Wong, K.Y., and Chan, K.W.Y. (1998) Preliminary monitoring results of Tsing Ma Bridge. in The 14th National Conference on Bridge Engineering, (Vol. 2), Shanghai, pp. 730–740.
42. Li, P.W., Poon, H.T., and Lai, S.T. (1998) Observational study of Typhoon Victor (9712) during its passage over Hong Kong. in 12th Guangdong-Hong Kong-Macau Seminar on Hazardous Weather, Hong Kong, (in Chinese).
43. Xu, Y.L., Zhu, L.D., Wang, K.Y., and Chan, K.W.Y. (2000) Field measurement results of Tsing Ma suspension Bridge during typhoon Victor. *Structural Engineering and Mechanics*, **10**(6), 545–559.
44. Simiu, E. and Scanlan, R.H. (1996) *Wind Effects on Structures*, 3rd Edition, John Wiley & Sons, Inc., New York.
45. Morfiadakis, E.E., Glinou, G.L., and Koulouvari, M.J. (1996) The suitability of the von Karman spectrum for the structure of turbulence in a complex terrain wind farm. *Journal of Wind Engineering and Industrial Aerodynamics*, **62**(2–3), 237–257.
46. Kaimal, J.C., Wyngaard, J.C., Izumi, Y., and Cote, R. (1972) Spectral characteristics of surface-layer turbulence. *Quarterly Journal of the Royal Meteorological Society*, **98**(417), 132–148.
47. Tieleman, H.W. and Mullins, S.E. (1980) The structure of moderately strong winds at a Mid-Atlantic coastal site (below 75 m), *Proceedings of Fifth International Conference on Wind Engineering*, vol. **1**, Pergamon Press, Oxford, pp. 145–159.
48. Yang, J.N., Lei, Y., Lin, S.L., and Huang, N. (2004) Identification of natural frequencies and damping of *in situ* tall buildings using ambient wind vibration data. *Journal of Engineering Mechanics, ASCE*, **130**(5), 570–577.
49. Chen, J., Xu, Y.L., and Zhang, R.C. (2004) Modal parameter identification of Tsing Ma suspension bridge under Typhoon Victor: EMD-HT method. *Journal of Wind Engineering and Industrial Aerodynamics*, **92**, 805–827.

50. Huang, N.E., Shen, Z., and Long, S.R. (1999) A new view of nonlinear water waves: the Hilbert spectrum. *Annual Review of Fluid Mechanics*, **31**, 417–457.
51. Ibrahim, S.R. (1977) Random decrement technique for modal identification of structures. *Journal of Spacecraft*, **14**, 696–700.
52. Feldman, M. (1994) Non-linear system vibration analysis using the Hilbert transform –I: free vibration analysis method FREEVIB. *Mechanical System & Signal Process*, **8**, 309–318.
53. Bendat, J.S. and Piersol, A.G. (1986) *Random Data: Analysis and Measurement Procedures*, 2nd Edition, John Wiley & Sons, NY.
54. Xu, Y.L., Ko, J.M., and Zhang, W.S. (1997) Vibration studies of Tsing Ma suspension bridge. *Journal of Bridge Engineering, ASCE*, **2**, 149–156.

10

Buffeting Response to Skew Winds

10.1 Preview

As discussed in Chapter 4, buffeting analysis methods have been continuously refined as a result of the enhancement of computer technique and capacity, as well as the demand for more accurate prediction of the buffeting response of modern long-span cable-supported bridges. Nowadays, not only the effects of multi-modes, inter-mode coupling and aerodynamic coupling, but also the interaction between major bridge components can be included in either the time domain analysis or the frequency domain analysis.

However, most of the previous investigations take incident mean wind at a right angle to the longitudinal axis of the bridge. This may not always be the case when the bridge is located in a complex and heterogeneous topography and is attacked by a typhoon which has a three-dimensional vortex structure. The field measurements on the Tsing Ma suspension bridge during typhoons by the wind and structural health monitoring system (WASHMS), as discussed in Chapter 9, demonstrated that strong typhoon winds seldom attacked the bridge in a direction normal to the bridge longitudinal axis.

Although some efforts have been made to take into consideration skew winds in buffeting analysis of bridges, they are all based on the decomposition approach (cosine rule and sine rule). That is, the mean yaw wind is decomposed into two components: one normal and the other parallel to the bridge span. The contribution of the parallel mean wind component is then either ignored or is separately analyzed from that of the normal mean wind component.

If the parallel wind component is neglected, the buffeting response of a long-span cable-supported bridge due to yaw wind, obtained by taking the decomposition approach, would always be smaller than that due to the normal wind of the same wind speed. However, some wind tunnel tests revealed that buffeting response due to yaw wind would reach the same level as that due to the normal wind. This indicates that the decomposition approach may underestimate the buffeting response of bridges under yaw wind and may not truly reflect the effects of buffeting action due to skew wind on the bridge.

Therefore, the rational buffeting analysis method for a long-span cable-supported bridge under skew winds is needed for more accurate prediction and comparison with field measurement results, in order to verify various assumptions made in the analytical methods. Furthermore, the rational buffeting analysis method for skew winds is also required to enable a better evaluation of probability or risk analysis of the bridge exposed to local wind climate.

In this connection, this chapter first presents a finite element (FE)-based framework for buffeting analysis of long-span cable-supported bridges under skew winds in the frequency domain, based on the linear quasi-steady theory and the oblique strip theory, in conjunction with the pseudo-excitation

Wind Effects on Cable-Supported Bridges, First Edition. You-Lin Xu.

method. The frequency domain framework is then extended to the time domain in terms of the spectral representation method, the convolution integral and the Newmark numerical method. Fundamental aerodynamic information required in the frameworks, such as the aerodynamic coefficients and the flutter derivatives of the bridge deck under skew winds, are measured in wind tunnels by using specially designed test rigs and typical oblique strip models of a bridge deck. The proposed frameworks are finally applied to the Tsing Ma suspension bridge to compute its buffeting response caused by skew winds during Typhoon Sam, and to compare the computed responses with field measurement results.

10.2 Formulation in the Frequency Domain

10.2.1 Basic Assumptions

The development of the framework for buffeting analysis of long-span cable-supported bridges under skew winds requires some basic assumptions [1]. First, the incident wind is assumed to be stationary and its mean wind speed is sufficiently larger than any of three fluctuating wind components. Second, the mean wind speed used in buffeting analysis falls outside the range that may cause either aeroelastic instability or vortex shedding of the bridge. Third, the average scales of the turbulence are assumed to be sufficiently larger than the chord-wise dimension of the bridge deck, so that the quasi-steady theory is applicable. Finally, wind-induced bridge vibration is small, so that the linear approach can be accepted.

In the following, the expressions of buffeting forces on a long-span cable-supported bridge under skew winds are derived based on the quasi-steady theory and the aerodynamic strip theory, as used for the conventional buffeting analysis [2,3]. An oblique strip of a bridge deck along the mean wind direction is introduced. All of the six components of the aerodynamic force acting on the oblique strip due to skew winds are included. The buffeting forces are first formed for an element in the wind coordinate system, then are transferred to the structural coordinate system and finally used to assemble the global buffeting force vector. The transformations are conducted on the buffeting forces rather than on the mean and fluctuating winds, thus avoiding the difficulties involved in the traditional decomposition method [4–6].

10.2.2 Coordinate Systems and Transformation Matrices

Some Cartesian coordinate systems, obeying the right-handed rule, are introduced to facilitate the formulation of buffeting forces. As shown in Figure 10.1, *XYZ* is the global structural coordinate system

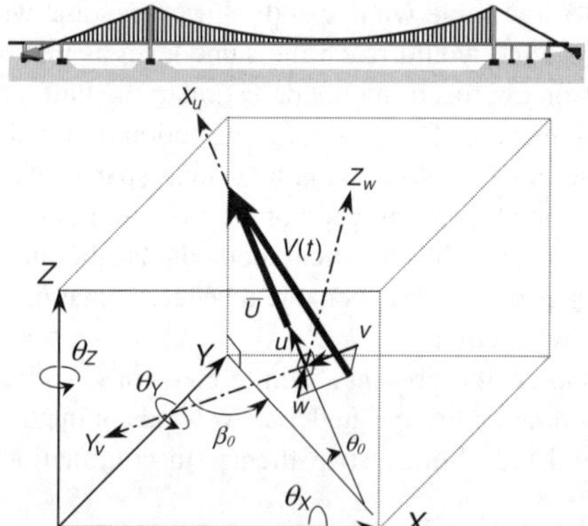

Figure 10.1 Global structural and wind coordinate systems (from [1]) (Reproduced with permission from Elsevier).

used to describe the bridge structural model and the overall dynamic equilibrium conditions. $X_uY_vZ_w$ is called the global wind coordinate system, required to define wind direction and fluctuating wind components. The axis X_u is along the direction of mean wind velocity $\overline{U}$, and the axis Y_v is located in a horizontal plane, while the axis Z_w is always positive upward. There are many possible combinations between the XYZ-system and the $X_uY_vZ_w$-system in the practice.

Figure 10.1 just shows a typical example of one of these combinations, where the axis Y_v is parallel to the X-Y plane. The mean wind direction in the global structural coordinate system can then be determined, with a global mean wind yaw angle (β_0) in conjunction with a global mean wind inclination angle (θ_0). As shown in Figure 10.1, β_0 is defined as the angle between the vertical plane normal to the bridge longitudinal axis and the vertical plane with the mean wind velocity included; θ_0 is interpreted as the angle between the mean wind velocity and the horizontal plane. Furthermore, β_0 is positive when mean wind comes from the right side of the Y-Z plane, and θ_0 is positive when the vertical component of the mean wind velocity is upward.

Figure 10.2 shows a typical combination of two local coordinate systems of an arbitrary element (strip), where xyz is called the local structural coordinate system used to present the elemental parameters, such as the matrices of elemental mass, stiffness, damping and loading, and qph is the local reference coordinate system introduced to define the wind direction with respect to the element. The axis q is along the longitudinal axis of the element. The axis p is located in either the deck plane or the tower plane. Its positive direction should be determined in such a way that the angle between the axis p and the mean wind direction is less than 90° and, when the mean wind is parallel to the qp plane and normal to the axis q, the axis p is along the mean wind direction.

The local mean wind yaw angle ($\overline{\beta}$) is defined as the angle between the ph-plane and the plane constituted of the mean wind $\overline{U}$ and the axis h. $\overline{\beta}$ is positive when the mean wind component $\overline{U}_q$ along the axis q is negative. The local mean wind inclination ($\overline{\theta}$) is defined as the angle between the mean wind velocity and the qp-plane. It takes a positive value when the mean wind component $\overline{U}_h$ along the axis h is positive. Figure 10.2 also shows the local yaw angle $\tilde{\beta}$ and the inclination angle $\tilde{\theta}$ of instantaneous wind, which together determine the direction of the instantaneous wind $V(t)$. The rules for defining the angles $\tilde{\beta}$ and $\tilde{\theta}$ are similar to those for the angles $\overline{\beta}$ and $\overline{\theta}$.

Figures 10.3 and 10.4 show the local mean wind coordinates, the $\overline{qph}$-system, and the local instantaneous wind coordinates, the $\tilde{q}\tilde{p}\tilde{h}$-system, respectively, which are used to derive six-component buffeting forces acting on an oblique strip. The $\overline{qph}$-system is attained by rotating the qph-system by an angle of $\overline{\beta}$ around the axis h first, and then by an angle of $\overline{\theta}$ around the axis $\overline{q}$. Similarly, the $\tilde{q}\tilde{p}\tilde{h}$-system is attained by rotating the qph-system by an angle of $\tilde{\beta}$ around the axis h first, and then by an angle of $\tilde{\theta}$

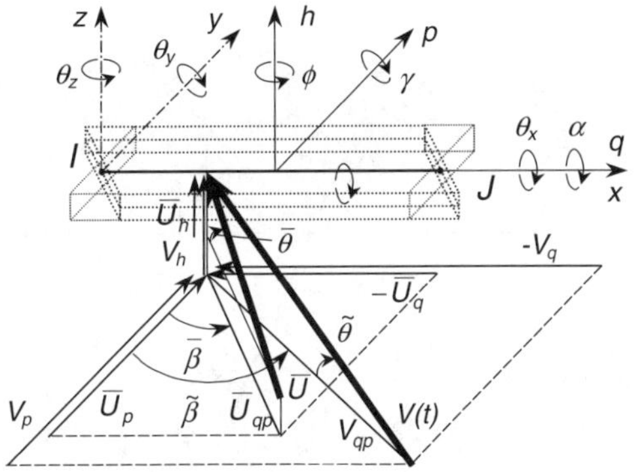

Figure 10.2 Local reference coordinate system (from [1]) (Reproduced with permission from Elsevier).

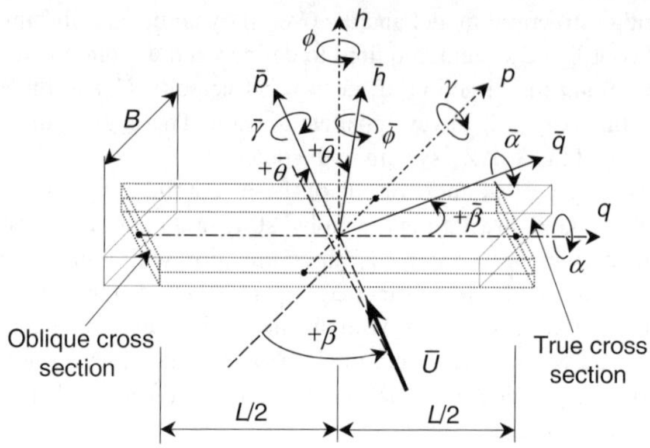

Figure 10.3 Local mean wind coordinate system (from [1]) (Reproduced with permission from Elsevier).

around the axis $\tilde{q}$. As a result, the axes $\overline{q}$ and $\tilde{q}$ are located in the qp-plane. The axis $\overline{p}$ is along the mean wind direction, while the axis $\tilde{p}$ is along the instantaneous wind direction.

Various coordinate transformation matrices are to be used during the formulation of buffeting forces. Denote by $\mathbf{T}_{S_1S_2}$ the 3×3 transformation matrix from the system S_2 to the system S_1, where the subscripts S_1 and S_2 ($S_1 \neq S_2$) can be any two of G_W, G_S, L_s, L_r, $L\overline{w}$ and $L\tilde{w}$ which represents, respectively, the global wind coordinate $X_uY_vZ_w$-system, the global structural coordinate XYZ-system, the local structural coordinate xyz-system, the local reference coordinate qph-system, the local mean wind coordinate $\overline{q}\overline{p}\overline{h}$-system and the local instantaneous wind coordinate $\tilde{q}\tilde{p}\tilde{h}$-system. If the size of S_3 is the same as those of S_1 and S_2, then:

$$\mathbf{T}_{S_1S_3} = \mathbf{T}_{S_1S_2}\mathbf{T}_{S_2S_3} \tag{10.1}$$

Furthermore, it can be proved that:

$$\mathbf{T}_{S_1S_2} = \mathbf{T}_{S_2S_1}^{-1} = \mathbf{T}_{S_2S_1}^{\mathrm{T}}, \ \mathbf{T}_{S_2S_1} = \mathbf{T}_{S_1S_2}^{-1} = \mathbf{T}_{S_1S_2}^{\mathrm{T}} \tag{10.2}$$

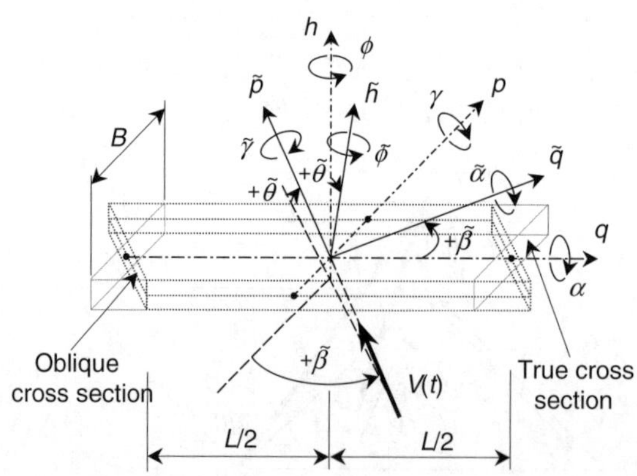

Figure 10.4 Local instantaneous wind coordinate system (from [1]) (Reproduced with permission from Elsevier).

For example, for the two global systems shown in Figure 10.1, the transformation matrix can be expressed as:

$$\mathbf{T}_{GsGw} = \mathbf{T}_{GwGs}^{\mathrm{T}} = \begin{bmatrix} -\cos\theta_0 \sin\beta_0 & -\cos\beta_0 & \sin\theta_0 \sin\beta_0 \\ \cos\theta_0 \cos\beta_0 & -\sin\beta_0 & -\sin\theta_0 \cos\beta_0 \\ \sin\theta_0 & 0 & \cos\theta_0 \end{bmatrix} \tag{10.3}$$

The transformation matrix between the global wind coordinate system and the local reference coordinate system can be found as:

$$\mathbf{T}_{LrGw} = \mathbf{T}_{LrLs}\mathbf{T}_{LsGs}\mathbf{T}_{GsGw} = \begin{bmatrix} t_{11} & t_{12} & t_{13} \\ t_{21} & t_{22} & t_{23} \\ t_{32} & t_{32} & t_{33} \end{bmatrix} \tag{10.4}$$

where t_{ij} $(i,j = 1,2,3)$ is the element of $\mathbf{T}_{LrGw}$ at the ith row and jth column, and it is the function of β_0 and θ_0 and satisfies the following relationships:

$$\sum_{k=1}^{3} t_{ik}t_{jk} = \begin{cases} 1, & i=j \\ 0, & i\neq j \end{cases}, \sum_{k=1}^{3} t_{ki}t_{kj} = \begin{cases} 1, & i=j \\ 0, & i\neq j \end{cases}, \quad (i,j = 1,2,3) \tag{10.5}$$

Moreover, one can derive the transformation matrix between the local mean wind coordinate system and the local reference coordinate system as:

$$\mathbf{T}_{Lr\overline{Lw}} = \begin{bmatrix} \cos\overline{\beta} & -\cos\overline{\theta}\sin\overline{\beta} & \sin\overline{\theta}\sin\overline{\beta} \\ \sin\overline{\beta} & \cos\overline{\theta}\cos\overline{\beta} & -\sin\overline{\theta}\cos\overline{\beta} \\ 0 & \sin\overline{\theta} & \cos\overline{\theta} \end{bmatrix} \tag{10.6}$$

The 6×6 transformation matrix from the local mean wind coordinate system to the local instantaneous wind coordinate system can be expressed as:

$$\overline{\mathbf{T}}_{L\overline{w}L\tilde{w}} = \mathbf{I} + \overline{\mathbf{T}}_v \frac{v}{\overline{U}} + \overline{\mathbf{T}}_w \frac{w}{\overline{U}} \tag{10.7}$$

where $\mathbf{I}$ is the 6×6 identity matrix, and

$$\overline{\mathbf{T}}_v = \begin{bmatrix} \mathbf{T}_v & \mathbf{0} \\ \mathbf{0} & \mathbf{T}_v \end{bmatrix}, \mathbf{T}_v = \begin{bmatrix} 0 & -s_1 & s_2 t_{31} \\ s_1 & 0 & -s_3 \\ -s_2 t_{31} & s_3 & 0 \end{bmatrix} \tag{10.8a}$$

$$\overline{\mathbf{T}}_w = \begin{bmatrix} \mathbf{T}_w & \mathbf{0} \\ \mathbf{0} & \mathbf{T}_w \end{bmatrix}, \mathbf{T}_w = \begin{bmatrix} 0 & -s_4 & s_5 t_{31} \\ s_4 & 0 & -s_6 \\ -s_5 t_{31} & s_6 & 0 \end{bmatrix} \tag{10.8b}$$

$$s_1 = (t_{11}t_{22} - t_{21}t_{12})/\sqrt{t_{11}^2 + t_{21}^2}, s_2 = (t_{11}t_{22} - t_{21}t_{12})/(t_{11}^2 + t_{21}^2) \tag{10.9a}$$

$$s_3 = t_{32}/\sqrt{t_{11}^2 + t_{21}^2}, s_4 = (t_{11}t_{23} - t_{21}t_{13})/\sqrt{t_{11}^2 + t_{21}^2} \tag{10.9b}$$

$$s_5 = (t_{11}t_{23} - t_{21}t_{13})/(t_{11}^2 + t_{21}^2), s_6 = t_{33}/\sqrt{t_{11}^2 + t_{21}^2} \tag{10.9c}$$

10.2.3 Wind Components and Directions

As shown in Figure 10.1, the alongwind, lateral and upward components of wind turbulence, $u(t)$, $v(t)$, and $w(t)$, are defined as the three velocity fluctuations along the axes X_u, Y_v and Z_w, respectively, and their positive directions are consistent with those of axes X_u, Y_v and Z_w. Thus, $u(t)$ is along the direction of mean wind ($\overline{U}$), $v(t)$ is horizontal and normal to the mean wind direction, and $w(t)$ is upward and normal to the mean wind direction. The resultant wind is therefore:

$$V(t) = \sqrt{[\overline{U} + u(t)]^2 + v^2(t) + w^2(t)} \tag{10.10}$$

The aerodynamic coefficients and flutter derivatives of a bridge deck are measured via wind tunnel tests and are expressed as the functions of mean wind yaw angle and inclination angle with respect to the test section model. The mean wind yaw angle and inclination angle refer to $\overline{\beta}$ and $\overline{\theta}$ of the section model, and they may be different from the global ones of β_0 and θ_0. Therefore, in order to use correctly the measured aerodynamic coefficients and flutter derivatives when determining wind-induced forces acting on an arbitrary element, $\overline{\beta}$ and $\overline{\theta}$ should be determined first.

Note that, though the vector of wind speed shows its different appearances in the global structural coordinate system and the local reference coordinate system, the absolute wind speed and direction are actually independent of the used coordinate systems. Therefore, the angles $\overline{\beta}$ and $\overline{\theta}$ for each element can be computed from β_0 and θ_0, with the following trigonometric functions derived via the coordinate transformation.

$$\cos\overline{\theta} = \sqrt{t_{11}^2 + t_{21}^2}, \sin\overline{\theta} = t_{31} \tag{10.11a}$$

$$\cos\overline{\beta} = t_{21}/\sqrt{t_{11}^2 + t_{21}^2}, \sin\overline{\beta} = -t_{11}/\sqrt{t_{11}^2 + t_{21}^2} \tag{10.11b}$$

where t_{ij} ($i,j = 1, 2, 3$) is the element of the ith row and jth column of the 3×3 transformation matrix ($\mathbf{T}_{LrGw}$) from the $X_uY_vZ_w$-system to the qph-system of the element concerned, and it is the function of θ_0 and β_0.

Furthermore, the angles $\tilde{\beta}$ and $\tilde{\theta}$ for three fluctuating wind components can be derived from $\overline{\beta}$ and $\overline{\theta}$. In consideration that the fluctuating wind components are much smaller than the mean wind speed, the increments of local wind yaw angle ($\Delta\beta = \tilde{\beta} - \overline{\beta}$) and local wind inclination angle ($\Delta\theta = \tilde{\theta} - \overline{\theta}$) can thus be expressed as the following linear functions of u, v, and w by using Taylor's expansion formula and ignoring all the non-linear items of u, v, and w:

$$\Delta\theta \approx \sin\Delta\theta = \frac{t_{32}}{\sqrt{t_{11}^2 + t_{21}^2}}\frac{v}{\overline{U}} + \frac{t_{33}}{\sqrt{t_{11}^2 + t_{21}^2}}\frac{w}{\overline{U}} \tag{10.12a}$$

$$\Delta\beta \approx \sin\Delta\beta = \frac{t_{11}t_{22} - t_{12}t_{21}}{t_{11}^2 + t_{21}^2}\frac{v}{\overline{U}} + \frac{t_{11}t_{23} - t_{13}t_{21}}{t_{11}^2 + t_{21}^2}\frac{w}{\overline{U}} \tag{10.12b}$$

10.2.4 Buffeting Forces and Spectra under Skew Winds

Figure 10.5 shows an oblique strip (element) parallel to the mean wind direction. The vector of the total aerodynamic wind forces acting on an element per unit length, $\tilde{\mathbf{f}}^{ad}(t)$, due to the instantaneous wind velocity $V(t)$, can be expressed as:

$$\tilde{\mathbf{f}}^{ad}(t) = \left(C_{\tilde{q}}^{ad}(t), D_{\tilde{p}}^{ad}(t), L_{\tilde{h}}^{ad}(t), M_{\tilde{\alpha}}^{ad}(t), M_{\tilde{\gamma}}^{ad}(t), M_{\tilde{\phi}}^{ad}(t)\right)^{\mathrm{T}} = \frac{1}{2}\rho V(t)^2\mathbf{B}\,\tilde{\mathbf{C}}(\tilde{\beta},\tilde{\theta}) \tag{10.13}$$

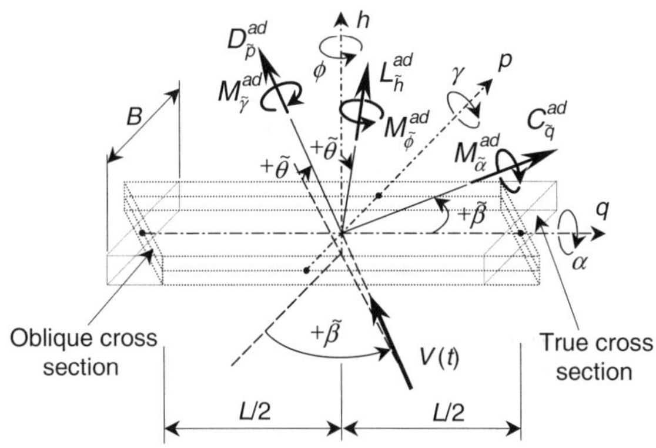

Figure 10.5 Aerodynamic forces in local instantaneous wind coordinate system (from [1]) (Reproduced with permission from Elsevier).

where:

$C^{ad}_{\tilde{q}}(t)$, $D^{ad}_{\tilde{p}}(t)$, $L^{ad}_{\tilde{h}}(t)$ are the total crosswind force, drag force and lift force, respectively, along the axis $\tilde{q}$, the axis $\tilde{p}$ and the axis $\tilde{h}$ of the $\tilde{q}\tilde{p}\tilde{h}$-system;

$M^{ad}_{\tilde{\alpha}}(t)$, $M^{ad}_{\tilde{\gamma}}(t)$ and $M^{ad}_{\tilde{\phi}}(t)$ are, respectively, the total pitching moment, rolling moment and yawing moment around the axis $\tilde{q}$, the axis $\tilde{p}$ and the axis $\tilde{h}$, respectively;

ρ is the air density; and

$$\mathbf{B} = \mathbf{Diag}(B, B, B, B^2, B^2, B^2) \tag{10.14}$$

in which B is the characteristic width of the elemental true cross section normal to the longitudinal axis of the element. $\tilde{\mathbf{C}}(\tilde{\beta}, \tilde{\theta})$ is the aerodynamic coefficient vector corresponding to $\tilde{\beta}$ and $\tilde{\theta}$. $\tilde{\mathbf{C}}(\tilde{\beta}, \tilde{\theta})$ can be linearized using Taylor's formula and ignoring all non-linear terms of u, v, and w:

$$\tilde{\mathbf{C}}(\tilde{\beta}, \tilde{\theta}) = \overline{\mathbf{C}}(\overline{\beta}, \overline{\theta}) + \overline{\mathbf{C}}'^{\beta}(\overline{\beta}, \overline{\theta})\Delta\beta + \overline{\mathbf{C}}'^{\theta}(\overline{\beta}, \overline{\theta})\Delta\theta \tag{10.15}$$

where:

$$\overline{\mathbf{C}}(\overline{\beta}, \overline{\theta}) = \left(C_{C_{\bar{q}}}, C_{D_{\bar{p}}}, C_{L_{\bar{h}}}, C_{M_{\bar{\alpha}}}, C_{M_{\bar{\gamma}}}, C_{M_{\bar{\phi}}}\right)^{\mathrm{T}}_{(\overline{\beta}, \overline{\theta})} \tag{10.16a}$$

is the aerodynamic coefficient vector corresponding to $\overline{\beta}$ and $\overline{\theta}$, and

$$\overline{\mathbf{C}}'^{\beta}(\overline{\beta}, \overline{\theta}) = \partial\overline{\mathbf{C}}(\overline{\beta}, \overline{\theta})/\partial\beta = \left(C'^{\beta}_{C_{\bar{q}}}, C'^{\beta}_{D_{\bar{p}}}, C'^{\beta}_{L_{\bar{h}}}, C'^{\beta}_{M_{\bar{\alpha}}}, C'^{\beta}_{M_{\bar{\gamma}}}, C'^{\beta}_{M_{\bar{\phi}}}\right)^{\mathrm{T}}_{(\overline{\beta}, \overline{\theta})} \tag{10.16b}$$

$$\overline{\mathbf{C}}'^{\theta}(\overline{\beta}, \overline{\theta}) = \partial\overline{\mathbf{C}}(\overline{\beta}, \overline{\theta})/\partial\theta = \left(C'^{\theta}_{C_{\bar{q}}}, C'^{\theta}_{D_{\bar{p}}}, C'^{\theta}_{L_{\bar{h}}}, C'^{\theta}_{M_{\bar{\alpha}}}, C'^{\theta}_{M_{\bar{\gamma}}}, C'^{\theta}_{M_{\bar{\phi}}}\right)^{\mathrm{T}}_{(\overline{\beta}, \overline{\theta})} \tag{10.16c}$$

in which:

$C_{C_{\bar{q}}}$, $C_{D_{\bar{p}}}$, $C_{L_{\bar{h}}}$, $C_{M_{\bar{\alpha}}}$, $C_{M_{\bar{\gamma}}}$ and $C_{M_{\bar{\phi}}}$ are, respectively, the aerodynamic coefficients of crosswind force, drag force, lift force, pitching moment, rolling moment and yawing moment with respect to the local mean wind coordinate $\overline{qph}$-system, which will be discussed in Section 10.4;

the subscript $(\overline{\beta}, \overline{\theta})$ means that the aerodynamic coefficients take the values at $\overline{\beta}$ and $\overline{\theta}$ based on the oblique strip;

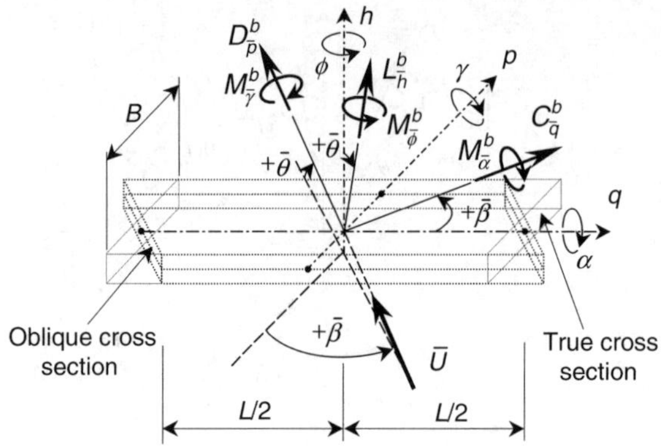

Figure 10.6 Buffeting forces in local mean wind coordinate system (from [1]) (Reproduced with permission from Elsevier).

$(\,)'^{\beta} = \partial(\,)/\partial\beta$ and $(\,)'^{\theta} = \partial(\,)/\partial\theta$ represent the partial derivatives with respect to either the local mean wind yaw angle or the local mean wind inclination.

As shown in Figure 10.6, $C^b_{\bar{q}}(t)$, $D^b_{\bar{p}}(t)$, $L^b_{\bar{h}}(t)$, $M^b_{\bar{\alpha}}(t)$, $M^b_{\bar{\gamma}}(t)$ and $M^b_{\bar{\phi}}(t)$ are the buffeting crosswind force, drag force, lift force, pitching moment, rolling moment and yawing moment due to the fluctuating wind, with respect to the local mean wind coordinate $\overline{qph}$-system, and the buffeting force vector can then be expressed as:

$$\begin{aligned}\bar{\mathbf{f}}^b(t) &= \left(C^b_{\bar{q}}(t),\, D^b_{\bar{p}}(t), L^b_{\bar{h}}(t), M^b_{\bar{\alpha}}(t), M^b_{\bar{\gamma}}(t), M^b_{\bar{\varphi}}(t)\right)^{\mathrm{T}} = \overline{\mathbf{T}}_{L\bar{w}L\tilde{w}}\tilde{\mathbf{f}}^{ad}(t) - \frac{1}{2}\rho\bar{U}^2\mathbf{B}\,\overline{\mathbf{C}}(\bar{\beta},\bar{\theta}) \\ &= \frac{1}{2}\rho V(t)^2\overline{\mathbf{T}}_{L\bar{w}L\tilde{w}}\mathbf{B}\left[\overline{\mathbf{C}}(\bar{\beta},\bar{\theta}) + \overline{\mathbf{C}}'^{\beta}(\bar{\beta},\bar{\theta})\Delta\beta + \overline{\mathbf{C}}'^{\theta}(\bar{\beta},\bar{\theta})\Delta\theta\right] - \frac{1}{2}\rho\bar{U}^2\mathbf{B}\,\overline{\mathbf{C}}(\bar{\beta},\bar{\theta})\end{aligned} \quad (10.17)$$

where:

$\overline{\mathbf{T}}_{L\bar{w}L\tilde{w}}$ is determined by Equation 10.7;
$\Delta\beta$ and $\Delta\theta$ are determined by Equation 10.12.

Then, by ignoring the non-linear terms of $u(t)$, $v(t)$, and $w(t)$, one obtains:

$$\bar{\mathbf{f}}^b(t) \approx \overline{\mathbf{A}}^b\,\bar{\mathbf{a}}(t) \quad (10.18)$$

where:

$$\bar{\mathbf{a}}(t) = [u(t), v(t), w(t)]^{\mathrm{T}} \quad (10.19a)$$

and $\overline{\mathbf{A}}^b$ is the 6×3 aerodynamic coefficient matrix of buffeting forces, i.e.

$$\overline{\mathbf{A}}^b = \left[\overline{\mathbf{A}}^u, \overline{\mathbf{A}}^v, \overline{\mathbf{A}}^w\right] \quad (10.19b)$$

$$\overline{\mathbf{A}}^u = \rho\bar{U}\mathbf{B}\boldsymbol{\chi}_u(\bar{\beta},\bar{\theta},K)\overline{\mathbf{C}}(\bar{\beta},\bar{\theta}) = \frac{1}{2}\rho\bar{U}\begin{Bmatrix} 2C_{C_{\bar{q}}}B\chi_{C_{\bar{q}}u} \\ 2C_{D_{\bar{p}}}B\chi_{D_{\bar{p}}u} \\ 2C_{L_{\bar{h}}}B\chi_{L_{\bar{h}}u} \\ 2C_{M_{\bar{\alpha}}}B^2\chi_{M_{\bar{\alpha}}u} \\ 2C_{M_{\bar{\gamma}}}B^2\chi_{M_{\bar{\gamma}}u} \\ 2C_{M_{\bar{\phi}}}B^2\chi_{M_{\bar{\phi}}u} \end{Bmatrix}_{(\bar{\beta},\bar{\theta})} \quad (10.20a)$$

$$\overline{\mathbf{A}}^v = \frac{1}{2}\rho\overline{U}\Big\{\mathbf{T}_v\mathbf{B}\boldsymbol{\chi}_v(\overline{\beta},\overline{\theta},K)\overline{\mathbf{C}}(\overline{\beta},\overline{\theta}) + s_2\mathbf{B}\boldsymbol{\chi}_v(\overline{\beta},\overline{\theta},K)\overline{\mathbf{C}}'^{\beta}(\overline{\beta},\overline{\theta}) + s_3\mathbf{B}\boldsymbol{\chi}_v(\overline{\beta},\overline{\theta},K)\overline{\mathbf{C}}'^{\theta}(\overline{\beta},\overline{\theta})\Big\}$$

$$= \frac{1}{2}\rho\overline{U}\left\{\begin{array}{c} [-s_1C_{D_{\overline{p}}} + s_2t_{31}C_{L_{\overline{h}}} + s_2C'^{\beta}_{C_{\overline{q}}} + s_3C'^{\theta}_{C_{\overline{q}}}]B\chi_{C_{\overline{q}}v} \\ [s_1C_{C_{\overline{q}}} - s_3C_{L_{\overline{h}}} + s_2C'^{\beta}_{D_{\overline{p}}} + s_3C'^{\theta}_{D_{\overline{p}}}]B\chi_{D_{\overline{p}}v} \\ [-s_2t_{31}C_{C_{\overline{q}}} + s_3C_{D_{\overline{p}}} + s_2C'^{\beta}_{L_{\overline{h}}} + s_3C'^{\theta}_{L_{\overline{h}}}]B\chi_{L_{\overline{h}}v} \\ [-s_1C_{M_{\overline{\gamma}}} + s_2t_{31}C_{M_{\overline{\varphi}}} + s_2C'^{\beta}_{M_{\overline{\alpha}}} + s_3C'^{\theta}_{M_{\overline{\alpha}}}]B^2\chi_{M_{\overline{\alpha}}v} \\ [s_1C_{M_{\overline{\alpha}}} - s_3C_{M_{\overline{\varphi}}} + s_2C'^{\beta}_{M_{\overline{\gamma}}} + s_3C'^{\theta}_{M_{\overline{\gamma}}}]B^2\chi_{M_{\overline{\gamma}}v} \\ [-s_2t_{31}C_{M_{\overline{\alpha}}} + s_3C_{M_{\overline{\gamma}}} + s_2C'^{\beta}_{M_{\overline{\varphi}}} + s_3C'^{\theta}_{M_{\overline{\varphi}}}]B^2\chi_{M_{\overline{\varphi}}v} \end{array}\right\}_{(\overline{\beta},\overline{\theta})} \tag{10.20b}$$

$$\overline{\mathbf{A}}^w = \frac{1}{2}\rho\overline{U}\Big(\mathbf{T}_w\mathbf{B}\boldsymbol{\chi}_w(\overline{\beta},\overline{\theta},K)\overline{\mathbf{C}}(\overline{\beta},\overline{\theta}) + s_5\mathbf{B}\boldsymbol{\chi}_w(\overline{\beta},\overline{\theta},K)\overline{\mathbf{C}}'^{\beta}(\overline{\beta},\overline{\theta}) + s_6\mathbf{B}\boldsymbol{\chi}_w(\overline{\beta},\overline{\theta},K)\overline{\mathbf{C}}'^{\theta}(\overline{\beta},\overline{\theta})\Big\}$$

$$= \frac{1}{2}\rho\overline{U}\left\{\begin{array}{c} [-s_4C_{D_{\overline{p}}} + s_5t_{31}C_{L_{\overline{h}}} + s_5C'^{\beta}_{C_{\overline{q}}} + s_6C'^{\theta}_{C_{\overline{q}}}]B\chi_{C_{\overline{q}}w} \\ [s_4C_{C_{\overline{q}}} - s_6C_{L_{\overline{h}}} + s_5C'^{\beta}_{D_{\overline{p}}} + s_6C'^{\theta}_{D_{\overline{p}}}]B\chi_{D_{\overline{p}}w} \\ [-s_5t_{31}C_{C_{\overline{q}}} + s_6C_{D_{\overline{p}}} + s_5C'^{\beta}_{L_{\overline{h}}} + s_6C'^{\theta}_{L_{\overline{h}}}]B\chi_{L_{\overline{h}}w} \\ [-s_4C_{M_{\overline{\gamma}}} + s_5t_{31}C_{M_{\overline{\varphi}}} + s_5C'^{\beta}_{M_{\overline{\alpha}}} + s_6C'^{\theta}_{M_{\overline{\alpha}}}]B^2\chi_{M_{\overline{\alpha}}w} \\ [s_4C_{M_{\overline{\alpha}}} - s_6C_{M_{\overline{\varphi}}} + s_5C'^{\beta}_{M_{\overline{\gamma}}} + s_6C'^{\theta}_{M_{\overline{\gamma}}}]B^2\chi_{M_{\overline{\gamma}}w} \\ [-s_5t_{31}C_{M_{\overline{\alpha}}} + s_6C_{M_{\overline{\gamma}}} + s_5C'^{\beta}_{M_{\overline{\varphi}}} + s_6C'^{\theta}_{M_{\overline{\varphi}}}]B^2\chi_{M_{\overline{\varphi}}w} \end{array}\right\}_{(\overline{\beta},\overline{\theta})} \tag{10.20c}$$

in which the coefficients s_i ($i = 1, \ldots, 6$) are determined by Equation 10.9 and depend on β_0 and θ_0:

$$\boldsymbol{\chi}_u(\overline{\beta},\overline{\theta},K) = \mathbf{Diag}(\chi_{C_{\overline{q}}u}, \chi_{D_{\overline{p}}u}, \chi_{L_{\overline{h}}u}, \chi_{M_{\overline{\alpha}}u}, \chi_{M_{\overline{\gamma}}u}, \chi_{M_{\overline{\varphi}}u}) \tag{10.21a}$$

$$\boldsymbol{\chi}_v(\overline{\beta},\overline{\theta},K) = \mathbf{Diag}(\chi_{C_{\overline{q}}v}, \chi_{D_{\overline{p}}v}, \chi_{L_{\overline{h}}v}, \chi_{M_{\overline{\alpha}}v}, \chi_{M_{\overline{\gamma}}v}, \chi_{M_{\overline{\varphi}}v}) \tag{10.21b}$$

$$\boldsymbol{\chi}_w(\overline{\beta},\overline{\theta},K) = \mathbf{Diag}(\chi_{C_{\overline{q}}w}, \chi_{D_{\overline{p}}w}, \chi_{L_{\overline{h}}w}, \chi_{M_{\overline{\alpha}}w}, \chi_{M_{\overline{\gamma}}w}, \chi_{M_{\overline{\phi}}w}) \tag{10.21c}$$

In the above equations, $K = \omega B/\overline{U}$ is the reduced frequency of the turbulence eddy with circular frequency ω; and $\chi_{\overline{f}a}(\overline{\beta},\overline{\theta},K)$ ($\overline{f} = C_{\overline{q}}$, $D_{\overline{p}}$, $L_{\overline{h}}$, $M_{\overline{\alpha}}$, $M_{\overline{\gamma}}$, $M_{\overline{\phi}}$; $a = u$, v, w) are the 18 aerodynamic admittance functions, considering the unsteadiness of wind turbulence and the partial coherence of the turbulence along the chord of the oblique cross-section in skew wind direction.

All of these admittance functions are functions of the reduced frequency and wind direction. The coherence of wind turbulence is higher for turbulence components with longer wavelength (lower frequency or higher velocity) than for those with shorter wavelength (higher frequency or lower velocity). As a result, the values of the aerodynamic admittance functions will drop with an increasing value of K.

Equations 10.18 to 10.21 are the universal expressions of the unit length quasi-steady buffeting forces acting on an arbitrary oblique element of the bridge deck under skew winds. The conventional cases can be deduced from these expressions. For instance, if $\overline{\beta} = \beta_0 = 0$ and $\overline{\theta} = \theta_0 = 0$, all the values of $C_{C_{\bar{q}}}$, $C_{M_{\bar{\gamma}}}$, $C_{M_{\bar{\phi}}}$ and $C_{\bar{f}}^{\prime\beta}$ ($\bar{f} = C_{\bar{q}}, D_{\bar{p}}, L_{\bar{h}}, M_{\bar{\alpha}}, M_{\bar{\gamma}}, M_{\bar{\phi}}$) are equal to zero for a straight element with constant cross-section. As a result, Equation 10.19b can be simplified as:

$$\overline{\mathbf{A}}^b = \frac{1}{2}\rho \bar{U} \begin{bmatrix} 0 & -C_{D_{\bar{p}}} B \chi_{C_{\bar{q}}v} & 0 \\ 2BC_{D_{\bar{p}}}\chi_{D_{\bar{p}}u} & 0 & \left(-C_{L_{\bar{h}}} + C_{D_{\bar{p}}}^{\prime\theta}\right) B\, \chi_{D_{\bar{p}}w} \\ 2BC_{L_{\bar{h}}}\chi_{L_{\bar{h}}u} & 0 & \left(C_{D_{\bar{p}}} + C_{L_{\bar{h}}}^{\prime\theta}\right) B\, \chi_{L_{\bar{h}}w} \\ 2B^2 C_{M_{\bar{\alpha}}}\chi_{M_{\bar{\alpha}}u} & 0 & C_{M_{\bar{\alpha}}}^{\prime\theta} B^2 \chi_{M_{\bar{\alpha}}w} \\ 0 & C_{M_{\bar{\alpha}}} B^2 \chi_{M_{\bar{\alpha}}v} & 0 \\ 0 & 0 & 0 \end{bmatrix}_{(0,0)} \tag{10.22}$$

Equation 10.22 is the same as that used in the traditional buffeting analysis if the mean wind is normal to the bridge deck and the lateral component of fluctuating wind along the bridge deck is ignored.

In the finite element model of a long-span cable-supported bridge, the length of some elements, such as cable and tower leg elements, may be quite long. To ensure the accuracy of buffeting analysis, these elements can be further divided into a number of segments so that a constant mean and fluctuating wind speeds, and a fully coherent turbulence wind, can be assumed within each segment. Suppose that n_k is the total number of segments in the kth element; $L_{i,k}$ ($i = 1, \ldots n_k$) is the length of ith segment, according to the principle of virtual work, the 12×1 vector of buffeting forces at the nodes of the kth element, $\mathbf{f}_{e,k}^{b}(t)$, in the xyz-system can be derived as:

$$\begin{aligned} \mathbf{f}_{e,k}^{b}(t) &= \left(F_{xI}^b, F_{yI}^b, F_{zI}^b, M_{\theta_{xI}}^b, M_{\theta_{yI}}^b, M_{\theta_{zI}}^b, F_{xJ}^b, F_{yJ}^b, F_{zJ}^b, M_{\theta_{xJ}}^b, M_{\theta_{yJ}}^b, M_{\theta_{zJ}}^b\right)_k^{\mathrm{T}} \\ &= L_k \int_0^1 \mathbf{N}_{\delta,k}^{\mathrm{T}}(\xi)\overline{\mathbf{T}}_{LsLr,k}\overline{\mathbf{T}}_{LrL\overline{w},k}\overline{\mathbf{A}}_k^b \bar{\mathbf{a}}_k(\xi, t)d\xi \\ &= L_k \sum_{i=1}^{n_k}\left(\left[\overline{\mathbf{T}}_{LsLr,k}^{\mathrm{T}} \int_{\xi_{i-1,k}}^{\xi_{i,k}} \mathbf{N}_{\delta,k}(\xi)d\xi\right]^{\mathrm{T}} \overline{\mathbf{T}}_{LrL\overline{w},k}\overline{\mathbf{A}}_{i,k}^b \bar{\mathbf{a}}_{i,k}(t)\right) \end{aligned} \tag{10.23}$$

where:

the subscript k indicates the kth element;
the subscript i indicates the ith segment;
I and J represent the left and right nodes of the element;
$\xi = x/L_k$ ($0 \le \xi \le 1$) is the reduced coordinate;
$\xi_{i,k} = x_{i,k}/L_k$; $L_k = \sum_{i=1}^{n_k} L_{i,k}$ is the total length of the kth element;
$x_{i,k}$ is the distance from the right end of the ith segment to the left node of the kth element with $x_{0,k} = 0$;
$\mathbf{N}_{\delta,k}(\xi)$ is the 6×12 matrix of displacement interpolation function of the kth element with respect to the local xyz-system, reflecting the relationship between the displacement at an arbitrary position within the kth element and the nodal displacements.

Designate $\mathbf{F}_{e,k}^{b}(t)$ the 12×1 nodal buffeting force vector of the kth element in XYZ-system. It can then be obtained through the coordinate transformation as follows:

$$\begin{aligned} \mathbf{F}_{e,k}^{b}(t) &= \left(F_{XI}^b, F_{YI}^b, F_{ZI}^b, M_{\theta_{XI}}^b, M_{\theta_{YI}}^b, M_{\theta_{ZI}}^b, F_{XJ}^b, F_{YJ}^b, F_{ZJ}^b, M_{\theta_{XJ}}^b, M_{\theta_{YJ}}^b, M_{\theta_{ZJ}}^b\right)_k^{\mathrm{T}} \\ &= \overline{\mathbf{T}}_{GsLs,k}\, \mathbf{f}_{e,k}^{b}(t) = \sum_{i=1}^{n_k} \mathbf{P}_{i,k}^{b} \bar{\mathbf{a}}_{i,k}(t) \end{aligned} \tag{10.24}$$

where:

$\overline{\mathbf{T}}_{GsLs,k}$ is the 12×12 transformation matrix of the kth element from the xyz-system to the XYZ-system;
$\mathbf{P}^b_{i,k}$ is the 12×3 coefficient matrix of nodal buffeting forces with respect to the XYZ-system for the ith segment of the kth element and is determined by:

$$\mathbf{P}^b_{i,k} = L_k \overline{\mathbf{T}}_{GsLs,k} \left[\overline{\mathbf{T}}^{\mathrm{T}}_{LsLr,k} \int_{\xi_{i-1,k}}^{\xi_{i,k}} \mathbf{N}_{\delta,k}(\xi) d\xi \right]^{\mathrm{T}} \overline{\mathbf{T}}_{LrL\overline{w},k} \overline{\mathbf{A}}^b_{i,k} \tag{10.25}$$

Designate $\mathbf{F}^b(t)$ the $6N \times 1$ global vector of nodal buffeting forces of the entire bridge assembled in the XYZ-system, where N is the number of nodes in the finite element model of the bridge. It can then be assembled from all $\mathbf{F}^b_{e,k}(t)$ as follows:

$$\mathbf{F}^b(t) = \mathbf{P}^b \mathbf{a}(t) \tag{10.26}$$

where $\mathbf{P}^b$ is the $6N \times 3m$ coefficient matrix of buffeting forces, assembled from all $\mathbf{P}^b_{i,k}$, and it is frequency-dependent and complex in nature if aerodynamic admittance functions are included in the computation. m is the total number of segments of the entire bridge and also the total point number of random excitations of fluctuating wind. $3m$ is the total number of the random excitations.

$$\mathbf{a}(t) = \left(\overline{\mathbf{a}}^{\mathrm{T}}_{1,1}, \cdots, \overline{\mathbf{a}}^{\mathrm{T}}_{n_1,1}, \cdots, \overline{\mathbf{a}}^{\mathrm{T}}_{n1,k}, \cdots, \overline{\mathbf{a}}^{\mathrm{T}}_{n_k,k}, \cdots, \overline{\mathbf{a}}^{\mathrm{T}}_{1,M}, \cdots, \overline{\mathbf{a}}^{\mathrm{T}}_{n_M,M} \right)^{\mathrm{T}} \tag{10.27}$$

is a $3m \times 1$ vector of the fluctuating wind components of the entire bridge.

The Fourier transformation of the buffeting force vector $\mathbf{F}^b(t)$ is:

$$\widehat{\mathbf{F}}^b(\omega) = \mathbf{P}^b(\omega) \widehat{\mathbf{a}}(\omega) \tag{10.28}$$

where $\widehat{\mathbf{a}}(\omega)$ is the Fourier transformation of fluctuating wind vector $\mathbf{a}(t)$. By using the following relationship:

$$\widehat{\mathbf{F}}^{b^*}(\omega) \widehat{\mathbf{F}}^{b^{\mathrm{T}}}(\omega) = \mathbf{P}^{b^*}(\omega)\, \widehat{\mathbf{a}}^*(\omega)\, \widehat{\mathbf{a}}^{\mathrm{T}}(\omega)\, \mathbf{P}^{b^{\mathrm{T}}}(\omega) \tag{10.29}$$

the spectral density function matrix of buffeting forces, $\mathbf{S}^b_{\mathrm{FF}}(\omega)$, can be obtained as:

$$\mathbf{S}^b_{\mathrm{FF}}(\omega) = \mathbf{P}^{b^*}(\omega) \mathbf{S}_{\mathrm{aa}}(\omega) \mathbf{P}^{b^{\mathrm{T}}}(\omega) \tag{10.30}$$

where:

the superscript asterisk denotes a complex conjugation operation;
$\mathbf{S}_{\mathrm{aa}}(\omega)$ is the spectral density function matrix of fluctuating wind components. It is a $3m \times 3m$ matrix and can be expressed as follows:

$$\mathbf{S}_{\mathrm{aa}}(\omega) = \begin{bmatrix} \mathbf{S}_{\bar{a}_{1,1}\bar{a}_{1,1}}(\omega) & \cdots & \mathbf{S}_{\bar{a}_{1,1}\bar{a}_{n_k,1}}(\omega) & \cdots & \mathbf{S}_{\bar{a}_{1,1}\bar{a}_{1,M}}(\omega) & \cdots & \mathbf{S}_{\bar{a}_{1,1}\bar{a}_{n_M,M}}(\omega) \\ \vdots & \ddots & \vdots & \ddots & \vdots & \ddots & \vdots \\ \mathbf{S}_{\bar{a}_{n_k,1}\bar{a}_{1,1}}(\omega) & \cdots & \mathbf{S}_{\bar{a}_{n_k,1}\bar{a}_{n_k,1}}(\omega) & \cdots & \mathbf{S}_{\bar{a}_{n_k,1}\bar{a}_{1,M}}(\omega) & \cdots & \mathbf{S}_{\bar{a}_{n_k,1}\bar{a}_{n_M,M}}(\omega) \\ \vdots & \ddots & \vdots & \ddots & \vdots & \ddots & \vdots \\ \mathbf{S}_{\bar{a}_{1,M}\bar{a}_{1,1}}(\omega) & \cdots & \mathbf{S}_{\bar{a}_{1,M}\bar{a}_{n_k,1}}(\omega) & \cdots & \mathbf{S}_{\bar{a}_{1,M}\bar{a}_{1,M}}(\omega) & \cdots & \mathbf{S}_{\bar{a}_{1,M}\bar{a}_{n_M,M}}(\omega) \\ \vdots & \ddots & \vdots & \ddots & \vdots & \ddots & \vdots \\ \mathbf{S}_{\bar{a}_{n_M,M}\bar{a}_{1,1}}(\omega) & \cdots & \mathbf{S}_{\bar{a}_{n_M,M}\bar{a}_{n_k,1}}(\omega) & \cdots & \mathbf{S}_{\bar{a}_{n_M,M}\bar{a}_{1,M}}(\omega) & \cdots & \mathbf{S}_{\bar{a}_{n_M,M}\bar{a}_{n_M,M}}(\omega) \end{bmatrix} \tag{10.31}$$

where $\mathbf{S}_{\bar{a}_{i,k}\bar{a}_{j,l}}(\omega)$ $(i=1, \ldots, n_k; j=1, \cdots, n_l; k=1, \ldots, M;, l=1, \ldots, M)$ is the 3×3 matrix of auto/cross-spectral densities between the three fluctuating wind components at the center of the ith segment of the kth element, as well as at the center of the jth segment of the lth element:

$$\mathbf{S}_{\bar{a}_{i,k}\bar{a}_{j,l}}(\omega) = \begin{bmatrix} S_{u_{i,k}u_{j,l}}(\omega) & S_{u_{i,k}v_{j,l}}(\omega) & S_{u_{i,k}w_{j,l}}(\omega) \\ S_{v_{i,k}u_{j,l}}(\omega) & S_{v_{i,k}v_{j,l}}(\omega) & S_{v_{i,k}w_{j,l}}(\omega) \\ S_{w_{i,k}u_{j,l}}(\omega) & S_{w_{i,k}v_{j,l}}(\omega) & S_{w_{i,k}w_{j,l}}(\omega) \end{bmatrix} \tag{10.32}$$

in which, for example, $S_{v_{i,k}u_{j,l}}$ means the cross-spectrum between v at the center of the ith segment of the kth element and u at the center of the jth segment of the lth element. The diagonal elements of $\mathbf{S}_{aa}(\omega)$ are the auto-spectra of fluctuating wind at a designated location and they are real quantities. All of the non-diagonal elements of $\mathbf{S}_{aa}(\omega)$ are cross-spectra and complex in general, with its real part being co-spectra of even function and its imaginary part being quadrature spectra of odd function. Thus, the two-sided spectra $(-\infty<\omega<+\infty)$ are to be adopted in this study. In addition, it can be proved that $\mathbf{S}_{aa}(\omega)$ is a non-negative definite Hermitian matrix.

The conventional expressions of cross-spectra between the fluctuating wind components at two different spatial points, P_1 and P_2, can be found frequently in the literature [7–10]. To carry out a fully coupled buffeting analysis of a three dimensional bridge, these conventional expressions are extended with the following forms, which at the same time take into account the turbulence coherence in all three directions along the axes of the global wind coordinate $X_uY_vZ_w$-system:

$$S_{a_{1P_1}a_{2P_2}}(\omega) = \sqrt{S_{a_{1P_1}a_{2P_1}}(\omega)S_{a_{1P_2}a_{2P_2}}(\omega)}\,R_{a_{1P_1}a_{2P_2}}(\omega) \tag{10.33}$$

$$R_{a_{1P_1}a_{2P_2}}(\omega) = (1-\hat{f}_{a_{1P_1}a_{2P_2}})\exp\{-\hat{f}_{a_{1P_1}a_{2P_2}}(\omega) + i\varphi_{a_{1P_1}a_{2P_2}}(\omega)\} \tag{10.34}$$

$$\hat{f}_{a_{1P_1}a_{2P_2}}(\omega) = [\hat{f}_{a_{1P_1}a_{1P_2}}(\omega) + \hat{f}_{a_{2P_1}a_{2P_2}}(\omega)]/2 \tag{10.35a}$$

$$\hat{f}_{a_{P_1}a_{P_2}}(\omega) = \frac{2n_{xa}\sqrt{\left[C^a_{X_u}\left(X_{uP_1}-X_{uP_2}\right)\right]^2+\left[C^a_{Y_v}\left(Y_{vP_1}-Y_{vP_2}\right)\right]^2+\left[C^a_{Z_w}\left(Z_{wP_1}-Z_{wP_2}\right)\right]^2}}{\overline{U}_{P_1}+\overline{U}_{P_2}} \tag{10.35b}$$

$$\varphi_{a_{1P_1}a_{2P_2}}(\omega) = [\varphi_{a_{1P_1}a_{1P_2}}(\omega) + \varphi_{a_{2P}a_{2P_2}}(\omega)]/2 \tag{10.35c}$$

where:

each of the subscripts a, a_1 and a_2 can be one of u, v and w;

$R_{a_{1P_1}a_{2P_2}}$ is the root-coherence function between the fluctuating wind components a_1 at point P_1 and a_2 at point P_2;

$i=\sqrt{-1}$;

X_{uP_j}, Y_{vP_j} and Z_{wP_j} are the coordinates of point P_j $(j=1, 2)$ in the global wind coordinate $X_uY_vZ_w$ -system;

$\overline{U}_{P_j}$ is the mean wind speed at point P_j, determined by the specified design wind speed and mean wind profile;

$S_{a_{1P_j}a_{2P_j}}$ is the cross-spectrum between a_1 and a_2 at point P_j and is a complex function of ω when $a_1 \neq a_2$;

$C^a_{X_u}$, $C^a_{Y_v}$ and $C^a_{Z_w}$ are the nine decay coefficients of turbulence coherence;

$\varphi_{a_1a_{2P_2}}$ is the phase spectrum between the turbulent component a_1 at point P_1 and the turbulent component a_2 at point P_2, and is traditionally set to zero in practice because there is little information about it;

and n_{xa} are the modified frequency determined by the following equation:

$$n_{xa} = \frac{\Gamma(5/6)}{2\sqrt{\pi}\Gamma(1/3)}\sqrt{1 + 70.78\left(\frac{nL_a^{x_u}}{\overline{U}(z)}\right)^2}\frac{\overline{U}(z)}{L_a^{x_u}} \approx \sqrt{n^2 + \frac{1}{70.78}\left(\frac{\overline{U}(z)}{L_a^{x_u}}\right)^2} \tag{10.36}$$

where:

$n = \omega/2\pi$;
$L_a^{x_u}$ is the length scale of turbulence component a in the alongwind direction;
Γ represents the Gamma function.

10.2.5 Aeroelastic Forces under Skew Winds

Figure 10.7 shows an arbitrary oblique segment of the bridge element submerged in a skew wind field with the mean wind speed $\overline{U}$ and the local mean wind yaw angle $\overline{\beta}$ and inclination $\overline{\theta}$. When the segment oscillates due to wind, there will be some motion-dependent aeroelastic forces and moments acting on the segment, caused by the interaction between the segment motion and the wind around the segment. These self-excited forces are often expressed in terms of Scanlan's flutter derivatives [11–14].

Since this investigation concerns skew winds, Scanlan's flutter derivatives are to be measured from the wind tunnel tests with an oblique sectional model under skew winds. Thus, they are not only the function of the reduced frequency but also the function of $\overline{\beta}$ and $\overline{\theta}$. Theoretically, there should be six components of the aeroelastic forces/moments, but only the three major components, i.e. the pitching moment M_α^{se}, the drag D_p^{se} and the lift L_h^{se}, as shown in Figure 10.7, are generally regarded as being significant to the buffeting response prediction of the bridge.

In this chapter, the M_α^{se}, D_p^{se} and L_h^{se} are defined with respect to the qph-system rather than the $\overline{qp}\overline{h}$-system, because they are induced by structural motions. The positive directions of M_α^{se}, D_p^{se} and L_h^{se} are determined by the qph-system and are independent of $\overline{\beta}$ and $\overline{\theta}$. Similar to that in [11–14], the following expressions are then adopted in this study for the self-excited aeroelastic forces acting on a unit length segment of a bridge component under skew winds:

$$\begin{aligned} M_\alpha^{se} = \frac{1}{2}\rho\overline{U}^2B^2\Big[&KA_1^*(\overline{\beta},\overline{\theta},K)\frac{\dot{\delta}_h}{\overline{U}} + KA_2^*(\overline{\beta},\overline{\theta},K)\frac{B\dot{\delta}_\alpha}{\overline{U}} + K^2A_3^*(\overline{\beta},\overline{\theta},K)\delta_\alpha \\ &+K^2A_4^*(\overline{\beta},\overline{\theta},K)\frac{\delta_h}{B} + KA_5^*(\overline{\beta},\overline{\theta},K)\frac{\dot{\delta}_p}{\overline{U}} + K^2A_6^*(\overline{\beta},\overline{\theta},K)\frac{\delta_p}{B}\Big] \end{aligned} \tag{10.37a}$$

$$\begin{aligned} D_p^{se} = \frac{1}{2}\rho\overline{U}^2B\Big[&KP_1^*(\overline{\beta},\overline{\theta},K)\frac{\dot{\delta}_p}{\overline{U}} + KP_2^*(\overline{\beta},\overline{\theta},K)\frac{B\dot{\delta}_\alpha}{\overline{U}} + K^2P_3^*(\overline{\beta},\overline{\theta},K)\delta_\alpha \\ &+ K^2P_4^*(\overline{\beta},\overline{\theta},K)\frac{\delta_p}{B} + KP_5^*(\overline{\beta},\overline{\theta},K)\frac{\dot{\delta}_h}{\overline{U}} + K^2P_6^*(\overline{\beta},\overline{\theta},K)\frac{\delta_h}{B}\Big] \end{aligned} \tag{10.37b}$$

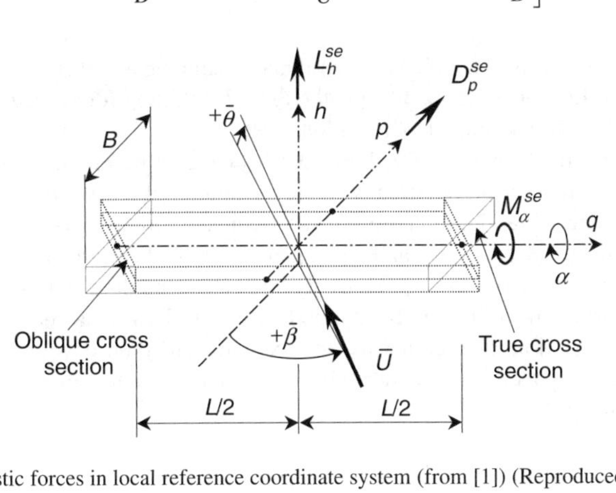

Figure 10.7 Aeroelastic forces in local reference coordinate system (from [1]) (Reproduced with permission from Elsevier).

$$L_h^{se} = \frac{1}{2}\rho\bar{U}^2 B\left[KH_1^*(\bar{\beta},\bar{\theta},K)\frac{\dot{\delta}_h}{\bar{U}} + KH_2^*(\bar{\beta},\bar{\theta},K)\frac{B\dot{\delta}_\alpha}{\bar{U}} + K^2H_3^*(\bar{\beta},\bar{\theta},K)\delta_\alpha \right.$$
$$\left. + K^2H_4^*(\bar{\beta},\bar{\theta},K)\frac{\delta_h}{B} + KH_5^*(\bar{\beta},\bar{\theta},K)\frac{\dot{\delta}_p}{\bar{U}} + K^2H_6^*(\bar{\beta},\bar{\theta},K)\frac{\delta_p}{B}\right] \tag{10.37c}$$

where:

$\delta_p(t)$ and $\delta_h(t)$ are the dynamic displacements along the axis p and the axis h and $\delta_\alpha(t)$ is the dynamic angular displacement about the axis q;

each over-dot denotes one order of partial differentiation with respect to time;

$P_i^*(\bar{\beta},\bar{\theta},K)$, $H_i^*(\bar{\beta},\bar{\theta},K)$ and $A_i^*(\bar{\beta},\bar{\theta},K)$ $(i = 1, \ldots, 6)$ are the flutter derivatives of the oblique cross section in the mean wind direction, and take the width of the true cross-section (B) as the reference characteristic width.

Finally, the global vector of self-excited aeroelastic forces of the whole bridge, $\mathbf{F}^{se}(t)$, can be expressed as:

$$\mathbf{F}^{se}(t) = -\mathbf{K}^{se}\boldsymbol{\Delta}(t) - \mathbf{C}^{se}\dot{\boldsymbol{\Delta}}(t) \tag{10.38}$$

where $\mathbf{K}^{se}$ and $\mathbf{C}^{se}$ are the $6N \times 6N$ global aerodynamic stiffness matrix and damping matrix with respect to the *XYZ*-system, which can easily be obtained based on Equation 10.37 according to the finite element technique. $\boldsymbol{\Delta}(t)$ is the $6N \times 1$ vector of the nodal displacement of the whole bridge referring to the *XYZ*-system.

10.2.6 Governing Equation and Solution in the Frequency Domain

Under the framework of finite element approach, the governing equation for buffeting analysis of a long-span cable-supported bridge under skew winds in the frequency domain can be expressed as:

$$\mathbf{M}\ddot{\boldsymbol{\Delta}}(t) + \mathbf{C}\dot{\boldsymbol{\Delta}}(t) + \mathbf{K}\boldsymbol{\Delta}(t) = \mathbf{F}^b(t) \tag{10.39}$$

where:

$$\mathbf{M} = \mathbf{M}^s, \mathbf{K} = \mathbf{K}^s + \mathbf{K}^{se}, \mathbf{C} = \mathbf{C}^s + \mathbf{C}^{se} \tag{10.40}$$

$\mathbf{M}_s$, $\mathbf{C}_s$ and $\mathbf{K}_s$ are, respectively, the global structural mass, damping and stiffness matrices of the whole bridge with the dimensions of $6N \times 6N$. $\mathbf{F}^b(t)$ is the $6N \times 1$ buffeting force vector of the entire bridge under skew winds, which is determined by Equation 10.26.

Because the number of degrees of freedom (DOF) of the finite element model of a long-span bridge is large, the modal superposition scheme is commonly used to reduce computational efforts when solving the governing equation in the frequency domain. Traditionally, the complete quadratic combination (CQC) method and the square root of the sum of square (SRSS) method are employed for the solution [15,16]. However, the CQC method needs great computational efforts for high accuracy, while the SRSS method bears the loss of accuracy to some extent for its facility. In this connection, a so-called pseudo-excitation method developed by Lin *et al.* [17–19] is used to solve the governing equation (10.39) with less computational efforts and, at same time, sufficient accuracy.

Designate $\mathbf{\Phi}$ the $6N \times M_\Phi$ modal matrix containing the M_Φ modes of vibration of the entire bridge and introduce the following linear transformation for the buffeting displacement response of the bridge:

$$\mathbf{\Delta}(t) = \boldsymbol{\varphi}\boldsymbol{\eta}(t) = [\boldsymbol{\varphi}_1, \cdots, \boldsymbol{\varphi}_r, \cdots, \boldsymbol{\varphi}_{M_\Phi}]\boldsymbol{\eta}(t) \tag{10.41}$$

where:

$\boldsymbol{\varphi}_r$ is the $6N \times 1$ mode shape vector of the rth mode;
$\boldsymbol{\eta}(t)$ is the $M_\Phi \times 1$ vector of generalized displacement coordinates.

Then, Equation 10.41 can be reduced to:

$$\tilde{\mathbf{M}}^s\ddot{\boldsymbol{\eta}}(t) + (\tilde{\mathbf{C}}^s + \tilde{\mathbf{C}}^{se})\dot{\boldsymbol{\eta}}(t) + (\tilde{\mathbf{K}}^s + \tilde{\mathbf{K}}^{se})\boldsymbol{\eta}(t) = \tilde{\mathbf{F}}^b(t) \tag{10.42}$$

where $\tilde{\mathbf{F}}^b(t)$ is the $M_\Phi \times 1$ vector of the generalized buffeting force, determined with the following equations:

$$\tilde{\mathbf{F}}^b(t) = \mathbf{\Phi}^{\mathrm{T}}\mathbf{F}^b(t) = \mathbf{\Phi}^{\mathrm{T}}\mathbf{P}^b\mathbf{a}(t) \tag{10.43a}$$

where $\tilde{\mathbf{M}}^s$, $\tilde{\mathbf{K}}^s$, $\tilde{\mathbf{C}}^s$, $\tilde{\mathbf{K}}^{se}$ and $\tilde{\mathbf{C}}^{se}$ are, respectively, the diagonal matrices of the generalized structural stiffness and damping, and the generalized aerodynamic stiffness and damping matrices with dimensions of $M_\Phi \times M_\Phi$:

$$\tilde{\mathbf{M}}^s = \mathbf{\Phi}^{\mathrm{T}}\mathbf{M}^s\mathbf{\Phi} \tag{10.43b}$$

$$\tilde{\mathbf{C}}^s = \mathbf{\Phi}^{\mathrm{T}}\mathbf{C}^s\mathbf{\Phi} = \tilde{\mathbf{M}}^s\mathbf{Diag}(2\zeta_1\omega_1, \cdots, 2\zeta_{M_\Phi}\omega_{M_\Phi}), \tilde{\mathbf{C}}^{se} = \mathbf{\Phi}^{\mathrm{T}}\mathbf{C}^{se}\mathbf{\Phi} \tag{10.43c}$$

$$\tilde{\mathbf{K}}^s = \mathbf{\Phi}^{\mathrm{T}}\mathbf{K}^s\mathbf{\Phi} = \tilde{\mathbf{M}}^s\,\mathbf{Diag}(\omega_1^2, \cdots, \omega_{M_\Phi}^2), \tilde{\mathbf{K}}^{se} = \mathbf{\Phi}^{\mathrm{T}}\mathbf{K}^{se}\mathbf{\Phi} \tag{10.43d}$$

where ω_r and ζ_r ($r = 1, \ldots, M_\Phi$) are the natural circular frequency and damping ratio, respectively, of the rth mode of vibration of the bridge.

Provided that the fluctuating wind components in the vector $\mathbf{a}(t)$ are stationary random processes and the bridge vibration is linear, $\tilde{\mathbf{F}}^b(t)$, $\boldsymbol{\eta}(t)$ and $\mathbf{\Delta}(t)$ are also stationary random processes. In accordance with random vibration theory [20], the cross-spectral density matrices of $\boldsymbol{\eta}(t)$ and $\mathbf{\Delta}(t)$ can be found from the cross-spectral density matrix of the wind turbulence $\mathbf{S}_{\mathrm{aa}}(\omega)$ as follows:

$$\mathbf{S}_{\eta\eta}(\omega) = \tilde{\mathbf{H}}^*(\omega)\mathbf{S}_{\tilde{\mathrm{F}}\tilde{\mathrm{F}}}(\omega)\tilde{\mathbf{H}}^{\mathrm{T}}(\omega), \mathbf{S}_{\Delta\Delta}(\omega) = \mathbf{\Phi}\mathbf{S}_{\eta\eta}(\omega)\mathbf{\Phi}^{\mathrm{T}} \tag{10.44}$$

$$\mathbf{S}_{\tilde{\mathrm{F}}\tilde{\mathrm{F}}}(\omega) = \mathbf{\Phi}^{\mathrm{T}}\mathbf{P}^{b^*}\mathbf{S}_{\mathrm{aa}}(\omega)\mathbf{P}^{b^{\mathrm{T}}}\mathbf{\Phi} \tag{10.45}$$

where $\mathbf{S}^b_{\tilde{\mathrm{F}}\tilde{\mathrm{F}}}(\omega)$ is the $M_\Phi \times M_\Phi$ matrix of the generalized buffeting force spectra; and the generalized matrix of frequency response functions of the bridge is:

$$\tilde{\mathbf{H}}(\omega) = \left[(\tilde{\mathbf{K}}^s + \tilde{\mathbf{K}}^{se} - \omega^2\tilde{\mathbf{M}}^s) + \mathrm{i}\omega(\tilde{\mathbf{C}}^s + \tilde{\mathbf{C}}^{se})\right]^{-1} \tag{10.46}$$

A direct computation of Equation 10.44 is very time-consuming. However, because the cross-spectral matrix of wind turbulence, $\mathbf{S}_{\mathrm{aa}}(\omega)$, is a non-negative definite Hermitian matrix, it can be expressed as the sum of sub-spectral matrices using the $\mathbf{L}^*\mathbf{D}\mathbf{L}^{\mathrm{T}}$ decomposition as follows:

$$\mathbf{S}_{\mathrm{aa}}(\omega) = \mathbf{l}^*(\omega)\mathbf{d}(\omega)\mathbf{l}^{\mathrm{T}}(\omega) = \sum_{j=1}^{m_p}\mathbf{S}_{\mathrm{aa},j}(\omega), \mathbf{S}_{\mathrm{aa},j}(\omega) = d_j\mathbf{l}_j^*(\omega)\mathbf{l}_j^{\mathrm{T}}(\omega) \tag{10.47}$$

where:

$m_p \leq 3m$ is the rank of the spectral matrix $\mathbf{S}_{aa}(\omega)$;
$\mathbf{d}(\omega)$ is a $3m \times 3m$ real diagonal matrix;
$\mathbf{l}(\omega)$ is a $3m \times 3m$ lower triangular matrix with all its diagonal elements being unity;
$d_{ja}(\omega)$ is the jth non-zero diagonal element of $\mathbf{d}(\omega)$;
$\mathbf{l}_j(\omega)$ is the jth column of $\mathbf{l}(\omega)$.

As a result, Equation 10.45 becomes:

$$\mathbf{S}^b_{\tilde{F}\tilde{F}} = \sum_{j=1}^{m_p} d_j \tilde{\mathbf{l}}^*_j(\omega)\tilde{\mathbf{l}}^T_j(\omega), \tilde{\mathbf{l}}_j(\omega) = \mathbf{\Phi}^T\mathbf{P}^b(\omega)\mathbf{l}_j(\omega) \tag{10.48}$$

Obviously, the generalized spectral matrix $\mathbf{S}^b_{\tilde{F}\tilde{F}}(\omega)$ is also Hermitian, and it can also be decomposed with $\mathbf{L}^*\mathbf{D}\mathbf{L}^T$ as follows:

$$\mathbf{S}^b_{\tilde{F}\tilde{F}}(\omega) = \mathbf{L}^*(\omega)\mathbf{D}(\omega)\mathbf{L}^T(\omega) = \sum_{r=1}^{M_p} D_r(\omega)\mathbf{L}^*_r(\omega)\mathbf{L}^T_r(\omega) \tag{10.49}$$

where:

$M_p \leq M_\Phi$ is the rank of $\mathbf{S}^b_{\tilde{F}\tilde{F}}(\omega)$;
$\mathbf{D}(\omega)$ is a real $M_\Phi \times M_\Phi$ diagonal matrix;
$\mathbf{L}(\omega)$ is a $M_\Phi \times M_\Phi$ lower triangular matrix with all its diagonal elements being unity;
$D_r(\omega)$ is the rth non-zero diagonal element of $\mathbf{D}(\omega)$;
$\mathbf{L}_r(\omega)$ is the rth column of $\mathbf{L}(\omega)$.

Physically, $\sqrt{D_r}\mathbf{L}_r(\omega)e^{i\omega t}$ can be regarded as a harmonic generalized pseudo excitation of the generalized system governed by Equation 10.42. One can then get:

$$\tilde{\mathbf{M}}\ddot{\boldsymbol{\eta}}_{p,r}(\omega, t) + \tilde{\mathbf{C}}\dot{\boldsymbol{\eta}}_{p,r}(\omega, t) + \tilde{\mathbf{K}}\boldsymbol{\eta}_{p,r}(\omega, t) = \sqrt{D_r}\mathbf{L}_r(\omega)e^{i\omega t} \tag{10.50}$$

where $\boldsymbol{\eta}_{p,r}(\omega,t)$ is the rth harmonic generalized pseudo displacement response vector corresponding to the rth harmonic generalized pseudo wind excitation. Since this is a determinate dynamic problem, the generalized pseudo displacement response can be easily found as follows:

$$\boldsymbol{\eta}_{p,r}(\omega, t) = \tilde{\mathbf{H}}(\omega)\mathbf{L}_r(\omega)\sqrt{D_r(\omega)}\, e^{i\omega t} \tag{10.51}$$

Then, according to the principle of pseudo excitation method, the cross-spectral matrices of the generalized buffeting response can be found as follows:

$$\mathbf{S}_{\eta\eta}(\omega) = \sum_{r=1}^{M_p} \boldsymbol{\eta}^*_{p,r}(\omega, t)\boldsymbol{\eta}^T_{p,r}(\omega, t) = \sum_{r=1}^{M_p} D_r(\omega)\tilde{\mathbf{H}}^*(\omega)\mathbf{L}^*_r(\omega)\mathbf{L}^T_r(\omega)\tilde{\mathbf{H}}^T(\omega) \tag{10.52}$$

From Equation 10.44, one can then obtain the cross-spectral matrices of buffeting response as:

$$\mathbf{S}_{\Delta\Delta}(\omega) = \sum_{r=1}^{M_p} D_r(\omega)\tilde{\boldsymbol{\Delta}}^*_{p,r}(\omega)\tilde{\boldsymbol{\Delta}}^T_{p,r}(\omega), \mathbf{S}_{\dot{\Delta}\dot{\Delta}}(\omega) = \omega^2\mathbf{S}_{\Delta\Delta}(\omega), \mathbf{S}_{\ddot{\Delta}\ddot{\Delta}}(\omega) = \omega^4\mathbf{S}_{\Delta\Delta}(\omega) \tag{10.53}$$

$$\tilde{\boldsymbol{\Delta}}_{p,r}(\omega) = \mathbf{\Phi}\tilde{\mathbf{H}}(\omega)\mathbf{L}_r(\omega) \tag{10.54}$$

Furthermore, the RMS (root mean square) responses of the nodal displacement, velocity and acceleration can be calculated through the integral of corresponding spectrum in the frequency domain:

$$\sigma_{\Delta_i} = \sqrt{2\int_0^{+\infty} S_{\Delta_i}(\omega)d\,\omega} \tag{10.55}$$

where:

$\mathbf{\Delta}\mathrm{i} = (x_i, y_i, z_i, \theta_{xi}, \theta_{yi}, \theta_{zi})^{\mathrm{T}}$ and $\boldsymbol{\sigma}_{\mathbf{\Delta}_i}$ are the nodal displacement vector and RMS response vector of the nodal displacement at the *i*th node;

$\mathbf{S}_{\mathbf{\Delta}_i}$ is the auto-spectral vector at the *i*th node, extracted from the diagonal elements of the cross-spectral matrices $\mathbf{S}_{\Delta\Delta}(\omega)$.

10.3 Formulation in the Time Domain

The frequency domain approach for predicting the buffeting response of long-span cable-supported bridges under skew winds is now extended to the time domain [21]. The buffeting forces on an oblique strip of the bridge deck in the mean wind direction are derived in terms of aerodynamic coefficients measured under skew winds and equivalent fluctuating wind velocities, with aerodynamic impulse functions [22] included. The time histories of equivalent fluctuating wind velocities and then buffeting forces along the bridge deck are simulated using the spectral representation method [23,24]. The self-excited forces on an oblique strip of the bridge deck are represented by the convolution integrals involving aerodynamic impulse functions and structural motions [22]. The aerodynamic impulse functions of self-excited forces are derived from experimentally measured flutter derivatives under skew winds, using rational function approximations. The governing equation of motion of a long-span cable-supported bridge under skew winds is established using the finite element method and is solved by using the Newmark numerical method.

10.3.1 Buffeting Forces due to Skew Winds in the Time Domain

Based on the discussions in Section 10.2 and Equation 10.26, the buffeting forces due to skew winds in the global structural coordinate system in the time domain can be given as

$$\mathbf{F}^b(t) = \mathbf{P}^b\mathbf{a}(t) = \sum_{i=1}^{n} \mathbf{T}_i^b\overline{\mathbf{F}}_i^b(t) \tag{10.56}$$

where:

$\overline{\mathbf{F}}_i^b(t)$ is the 12×1 vector of buffeting forces at the nodes of the *i*th element with respect to the global structural coordinate *XYZ*-system;

$\mathbf{T}_i^b$ $(i = 1, \ldots, n)$ is the $N \times 12$ matrix, with its elements being either zero or unit to locate the vector $\overline{\mathbf{F}}_i^b(t)$ at the proper position in the global vector $\mathbf{F}^b(t)$;

n is the number of the total elements on which the buffeting forces need to be accounted.

With reference to Equation 10.25 and considering one element as a segment, the expression of $\overline{\mathbf{F}}_i^b(t)$ can be written as:

$$\overline{\mathbf{F}}_i^b(t) = \overline{\mathbf{T}}_{GsLs,i}\tilde{\mathbf{N}}_i^T\mathbf{T}_{LsL\overline{w},i}\overline{\mathbf{P}}_i^b(t) \tag{10.57}$$

where:

$\overline{\mathbf{P}}_i^b(t) = \overline{\mathbf{f}}^b(t) \approx \overline{\mathbf{A}}^b \overline{\mathbf{a}}(t)$ is the 6×1 vector of buffeting forces per unit length of the ith oblique strip (element) of the bridge deck with respect to the local mean wind coordinate $\overline{qph}$-system;

$\mathbf{T}_{LsL\overline{w}}$, is the 6×6 transformation matrix from the local wind coordinate system $\overline{qph}$ to the local structural coordinate system xyz for the ith element;

$\tilde{\mathbf{N}}_i$ is the 6×12 matrix of the displacement interpolation functions of the ith element as used in the conventional finite element method;

$\overline{\mathbf{T}}_{GsLs,i}$ is the 12×12 transformation matrix from the local xyz-system to the global XYZ-system for the ith element.

The vector $\overline{\mathbf{P}}_i^b(t)$ is the function of the air density ρ, the element width B_i, the mean wind speed at the center of the element $\overline{U}_i$, the fluctuations of wind velocity at the center of the element $u_i(t)$, $v_i(t)$ and $w_i(t)$, the aerodynamic impulse functions of the element under skew winds, the aerodynamic coefficients and their derivatives of the element under skew winds, the coordinate transformation matrix $\mathbf{T}_{LrGw,i}$ and others. The resulting expression is as follows:

$$\overline{\mathbf{P}}_i^b(t) = \begin{Bmatrix} C_{\overline{q},i}^b(t) \\ D_{\overline{p},i}^b(t) \\ L_{\overline{h},i}^b(t) \\ M_{\overline{\alpha},i}^b(t) \\ M_{\overline{\gamma},i}^b(t) \\ M_{\overline{\varphi},i}^b(t) \end{Bmatrix} = \frac{\rho \overline{U}_i B_i}{2} \begin{bmatrix} a_{11} u_{C_{\overline{q},i}^b,eq}(t) + a_{12} v_{C_{\overline{q},i}^b,eq}(t) + a_{13} w_{C_{\overline{q},i}^b,eq}(t) \\ a_{21} u_{D_{\overline{p},i}^b,eq}(t) + a_{22} v_{D_{\overline{p},i}^b,eq}(t) + a_{23} w_{D_{\overline{p},i}^b,eq}(t) \\ a_{31} u_{L_{\overline{h},i}^b,eq}(t) + a_{32} v_{L_{\overline{h},i}^b,eq}(t) + a_{33} w_{L_{\overline{h},i}^b,eq}(t) \\ a_{41} u_{M_{\overline{\alpha},i}^b,eq}(t) + a_{42} v_{M_{\overline{\alpha},i}^b,eq}(t) + a_{43} w_{M_{\overline{\alpha},i}^b,eq}(t) \\ a_{51} u_{M_{\overline{\gamma},i}^b,eq}(t) + a_{52} v_{M_{\overline{\gamma},i}^b,eq}(t) + a_{53} w_{M_{\overline{\gamma},i}^b,eq}(t) \\ a_{61} u_{M_{\overline{\varphi},i}^b,eq}(t) + a_{62} v_{M_{\overline{\varphi},i}^b,eq}(t) + a_{63} w_{M_{\overline{\varphi},i}^b,eq}(t) \end{bmatrix} \tag{10.58}$$

where:

$C_{\overline{q},i}^b(t)$, $D_{\overline{p},i}^b(t)$, $L_{\overline{h},i}^b(t)$, $M_{\overline{\alpha},i}^b(t)$, $M_{\overline{\gamma},i}^b(t)$ and $M_{\overline{\varphi},i}^b(t)$ are the buffeting crosswind force, drag, lift, pitching moment, rolling moment and yawing moment on the ith oblique element of the bridge deck with respect to the local wind coordinate $\overline{qph}$ system, respectively;

a_{jk} ($j = 1,2 \ldots, 6$; $k = 1,2 \ldots, 6$) are the coefficients which are the function of aerodynamic coefficients and their derivatives of the element under skew winds and the coordinate transformation matrix $\mathbf{T}_{LrGw,i}$[21];

$u_{\overline{P}_i^b,eq}(t)$, $v_{\overline{P}_i^b,eq}(t)$, $w_{\overline{P}_i^b,eq}(t)$ ($\overline{P}_i^b = C_{\overline{q},i}^b, D_{\overline{p},i}^b, L_{\overline{h},i}^b, M_{\overline{\alpha},i}^b, M_{\overline{\gamma},i}^b$ and $M_{\overline{\varphi},i}^b$) are six sets of equivalent fluctuating wind velocities defined as follows:

$$\begin{aligned} &\left\{ u_{C_{\overline{q},i}^b,eq}(t) \quad v_{C_{\overline{q},i}^b,eq}(t) \quad w_{C_{\overline{q},i}^b,eq}(t) \right\}^{\mathrm{T}} \\ &= \left\{ \int_{-\infty}^{t} I_{C_{\overline{q},i}^{b,u}}(t-\tau) u_i(\tau) d\tau \quad \int_{-\infty}^{t} I_{C_{\overline{q},i}^{b,v}}(t-\tau) v_i(\tau) d\tau \quad \int_{-\infty}^{t} I_{C_{\overline{q},i}^{b,w}}(t-\tau) w_i(\tau) d\tau \right\}^{\mathrm{T}} \end{aligned} \tag{10.59a}$$

$$\begin{aligned} &\left\{ u_{D_{\overline{p},i}^b,eq}(t) \quad v_{D_{\overline{p},i}^b,eq}(t) \quad w_{D_{\overline{p},i}^b,eq}(t) \right\}^{\mathrm{T}} \\ &= \left\{ \int_{-\infty}^{t} I_{D_{\overline{p},i}^{b,u}}(t-\tau) u_i(\tau) d\tau \quad \int_{-\infty}^{t} I_{D_{\overline{p},i}^{b,v}}(t-\tau) v_i(\tau) d\tau \quad \int_{-\infty}^{t} I_{D_{\overline{p},i}^{b,w}}(t-\tau) w_i(\tau) d\tau \right\}^{\mathrm{T}} \end{aligned} \tag{10.59b}$$

$$\left\{ u_{L^b_{\bar{h},i},eq}(t) \quad v_{L^b_{\bar{h},i},eq}(t) \quad w_{L^b_{\bar{h},i},eq}(t) \right\}^{\mathrm{T}} = \left\{ \int_{-\infty}^{t} I_{L^{b,u}_{\bar{h},i}}(t-\tau)u_i(\tau)d\tau \quad \int_{-\infty}^{t} I_{L^{b,v}_{\bar{h},i}}(t-\tau)v_i(\tau)d\tau \quad \int_{-\infty}^{t} I_{L^{b,w}_{\bar{h},i}}(t-\tau)w_i(\tau)d\tau \right\}^{\mathrm{T}} \tag{10.59c}$$

$$\left\{ u_{M^b_{\bar{\alpha},i},eq}(t) \quad v_{M^b_{\bar{\alpha},i},eq}(t) \quad w_{M^b_{\bar{\alpha},i},eq}(t) \right\}^{\mathrm{T}} = \left\{ \int_{-\infty}^{t} I_{M^{b,u}_{\bar{\alpha},i}}(t-\tau)u_i(\tau)d\tau \quad \int_{-\infty}^{t} I_{M^{b,v}_{\bar{\alpha},i}}(t-\tau)v_i(\tau)d\tau \quad \int_{-\infty}^{t} I_{M^{b,w}_{\bar{\alpha},i}}(t-\tau)w_i(\tau)d\tau \right\}^{\mathrm{T}} \tag{10.59d}$$

$$\left\{ u_{M^b_{\bar{\gamma},i},eq}(t) \quad v_{M^b_{\bar{\gamma},i},eq}(t) \quad w_{M^b_{\bar{\gamma},i},eq}(t) \right\}^{\mathrm{T}} = \left\{ \int_{-\infty}^{t} I_{M^{b,u}_{\bar{\gamma},i}}(t-\tau)u_i(\tau)d\tau \quad \int_{-\infty}^{t} I_{M^{b,v}_{\bar{\gamma},i}}(t-\tau)v_i(\tau)d\tau \quad \int_{-\infty}^{t} I_{M^{b,w}_{\bar{\gamma},i}}(t-\tau)w_i(\tau)d\tau \right\}^{\mathrm{T}} \tag{10.59e}$$

$$\left\{ u_{M^b_{\bar{\varphi},i},eq}(t) \quad v_{M^b_{\bar{\varphi},i},eq}(t) \quad w_{M^b_{\bar{\varphi},i},eq}(t) \right\}^{\mathrm{T}} = \left\{ \int_{-\infty}^{t} I_{M^{b,u}_{\bar{\varphi},i}}(t-\tau)u_i(\tau)d\tau \quad \int_{-\infty}^{t} I_{M^{b,v}_{\bar{\varphi},i}}(t-\tau)v_i(\tau)d\tau \quad \int_{-\infty}^{t} I_{M^{b,w}_{\bar{\varphi},i}}(t-\tau)w_i(\tau)d\tau \right\}^{\mathrm{T}} \tag{10.59f}$$

where:

the superscript T represents the matrix operation of transpose;

$I_{\bar{P}^{b,u}_i}(t-\tau)$, $I_{\bar{P}^{b,v}_i}(t-\tau)$ and $I_{\bar{P}^{b,w}_i}(t-\tau)$ ($\bar{P}^b_i = C^b_{\bar{q},i}, D^b_{\bar{p},i}, L^b_{\bar{h},i}, M^b_{\bar{\alpha},i}, M^b_{\bar{\gamma},i}$ and $M^b_{\bar{\varphi},i}$) are the aerodynamic impulse functions of the ith oblique element.

Equation 10.59 indicates that six sets of equivalent turbulent wind velocities are needed to describe the corresponding six-component buffeting forces.

Each set of equivalent turbulent wind velocities along the bridge deck can be simulated using the spectral representation method [23,24]. Let us take the equivalent turbulent wind velocities associated with the drag force as an example. The equivalent turbulent wind velocities along the bridge deck in the lateral direction can be expressed as a multivariate stochastic process:

$$\mathbf{f}_{D^b_{\bar{p}},eq}(t) = \left\{ u_{D^b_{\bar{p},1},eq}(t) \quad v_{D^b_{\bar{p},1},eq}(t) \quad w_{D^b_{\bar{p},1},eq}(t) \quad \ldots\ldots \quad u_{D^b_{\bar{p},n},eq}(t) \quad v_{D^b_{\bar{p},n},eq}(t) \quad w_{D^b_{\bar{p},n},eq}(t) \right\}^{\mathrm{T}} \tag{10.60}$$

The cross-spectral density matrix of $\mathbf{f}_{D^b_{\bar{p}},eq}(t)$ is a $3n \times 3n$ matrix $\mathbf{S}^0_{D^b_{\bar{p}}}(\omega)$ given by

$$\mathbf{S}^0_{D^b_{\bar{p}}}(\omega) = \begin{bmatrix} \mathbf{S}_{D^b_{\bar{p}},11}(\omega) & \cdots & \mathbf{S}_{D^b_{\bar{p}},1n}(\omega) \\ \cdots & \ddots & \cdots \\ \mathbf{S}_{D^b_{\bar{p}},n1}(\omega) & \cdots & \mathbf{S}_{D^b_{\bar{p}},nn}(\omega) \end{bmatrix} \tag{10.61}$$

$$\mathbf{S}_{D^b_{\bar{p}},jk}(\omega) = \begin{bmatrix} S_{D^b_{\bar{p}},eq,uu}(P_j,P_k,\omega) & S_{D^b_{\bar{p}},eq,uv}(P_j,P_k,\omega) & S_{D^b_{\bar{p}},eq,uw}(P_j,P_k,\omega) \\ S_{D^b_{\bar{p}},eq,vu}(P_j,P_k,\omega) & S_{D^b_{\bar{p}},eq,vv}(P_j,P_k,\omega) & S_{D^b_{\bar{p}},eq,vw}(P_j,P_k,\omega) \\ S_{D^b_{\bar{p}},eq,wu}(P_j,P_k,\omega) & S_{D^b_{\bar{p}},eq,wv}(P_j,P_k,\omega) & S_{D^b_{\bar{p}},eq,ww}(P_j,P_k,\omega) \end{bmatrix}$$

$$(j = 1,2,\ldots,n;\, k = 1,2,\ldots,n) \tag{10.62}$$

$$S_{D_{\bar{p}}^b,eq,ab}(P_j,P_k,\omega)=\bar{I}_{D_{\bar{p}}^{b,a}}(\omega)\bar{I}^*_{D_{\bar{p}}^{b,b}}(\omega)\sqrt{S_{ab}(P_j,\omega)S_{ab}(P_k,\omega)}R_{ab}(P_j,P_k,\omega) \tag{10.63}$$
$$(a=u,v,w;b=u,v,w)$$

where:

$\mathbf{S}_{D_{\bar{p}}^b,jk}(\omega)$ is the 3×3 matrix of the cross-spectral density functions between three equivalent turbulent wind velocities at the center point (point j) of the jth element and the center point (point k) of the kth element;
$R_{ab}(P_j,P_k,\omega)$ is the coherence function between turbulent wind components a and b at point j and point k;
$S_{ab}(P_j,\omega)$ is the cross-spectral density function between turbulent wind components a and b at the same point j;
the product $\bar{I}_{D_{\bar{p}}^{b,a}}(\omega)\bar{I}^*_{D_{\bar{p}}^{b,b}}(\omega)$ is defined as aerodynamic admittance functions;
$\bar{I}_{D_{\bar{p}}^{b,a}}(\omega)$ and $\bar{I}_{D_{\bar{p}}^{b,b}}(\omega)$ are, respectively, the Fourier transform of $I_{D_{\bar{p}}^{b,a}}(t)$ and $I_{D_{\bar{p}}^{b,b}}(t)$;
the superscript * denotes a complex conjugate operation.

The matrix $\mathbf{S}^0_{D_{\bar{p}}^b}(\omega)$ can be decomposed into the following product with Cholesky's method:

$$\mathbf{S}^0_{D_{\bar{p}}^b}(\omega)=\mathbf{H}_{Dp\bar{b}}(\omega)^{D_{\bar{p}}^b}\mathbf{H}^{\mathrm{T}*}_{D_{\bar{p}}^b}(\omega) \tag{10.64}$$

Based on the spectral representation method [23,24], the equivalent turbulent wind velocities along the bridge deck,$\mathbf{f}_{D_{\bar{p}}^b}(t)$, associated with the drag force, can be simulated by the following series as $\overline{N}\rightarrow\infty$:

$$f_{D_{\bar{p}}^b,j}(t)=\sqrt{2\Delta\omega}\sum_{m=1}^{j}\sum_{l=1}^{\overline{N}}\left|H_{D_{\bar{p}}^b,jk}(\omega_{ml})\right|\cos(\omega_{ml}t-\theta_{D_{\bar{p}}^b,jk}(\omega_{ml})+\Phi_{ml}),\ j=1,2,3,\ldots,3n \tag{10.65}$$

where:

$\Delta\omega=\omega_{up}/\overline{N}$ is the frequency increment;
ω_{up} is an upper cutoff frequency beyond which the elements of the cross-spectral density matrix $\mathbf{S}^0_{D_{\bar{p}}^b}(\omega)$ are assumed to be zero;
$H_{D_{\bar{p}}^b,jk}(\omega_{ml})$ is the element of $\mathbf{H}_{D_{\bar{p}}^b}(\omega_{ml})$ at the jth row and kth column;
$\theta_{D_{\bar{p}}^b,jk}(\omega_{ml})=\tan^{-1}\left\{\mathrm{Im}[H_{D_{\bar{p}}^b,jk}(\omega_{ml})]/\mathrm{Re}[H_{D_{\bar{p}}^b,jk}(\omega_{ml})]\right\}$, which is the complex phase angle of $H_{D_{\bar{p}}^b,jk}(\omega_{ml})$; $\mathrm{Im}[H_{jk}(\omega_{ml})]$ and $\mathrm{Re}[H_{D_{\bar{p}}^b,jk}(\omega_{ml})]$ represent the imaginary and real parts of $H_{D_{\bar{p}}^b,jk}(\omega_{ml})$ respectively;
$\Phi_{1l},\ldots,\Phi_{jl}$, $l=1,2,\ldots,\overline{N}$ are the sequences of independent random phase angles distributed uniformly over the interval $[0, 2\pi]$;
ω_{ml} is of the double-indexing of the frequencies:

$$\omega_{ml}=(l-1+m/3n)\Delta\omega,\ l=1,2,\ldots,\overline{N} \tag{10.66}$$

The application of the fast Fourier transform technique to the above algorithm can improve dramatically the computational efficiency for simulating the equivalent turbulent wind velocities $\mathbf{f}_{D_{\bar{p}}^b}(t)$. The detail procedure of the fast Fourier transform technique is given by Deodatis [24]. The simulated time histories of equivalent turbulent wind velocities along the bridge deck associated with the drag force include not only the characteristics of incoming turbulent wind but also the aerodynamic admittance functions of the bridge deck.

The same procedure can be applied to generate the time histories of equivalent turbulent wind velocities along the bridge deck associated with other forces. Finally, the time histories of buffeting forces along the bridge deck can be obtained through Equations 10.56 to 10.58.

10.3.2 Self-Excited Forces due to Skew Winds in the Time Domain

The self-excited forces acting on the bridge deck due to wind-structure interaction can be expressed as a function of mean wind speed, the aerodynamic impulse functions associated with the flutter derivatives measured under skew winds, and the bridge motion [22]. By performing a series of coordinate transformations, the vector of self-excited forces acting on the bridge deck due to skew winds in the global structural coordinate system can be found as:

$$\mathbf{F}^{se}(t) = \sum_{i=1}^{n} \mathbf{T}_i^{se} \overline{\mathbf{F}}_i^{se}(t) \tag{10.67}$$

where:

$\overline{\mathbf{F}}_i^{se}(t)$ is the 12×1 vector of self-excited forces at the nodes of the ith element of the bridge deck with respect to the global structural coordinate XYZ-system;
$\mathbf{T}_i^{se}$ $(i = 1, \ldots, n)$ is the $N \times 12$ matrix, with its elements being either zero or unit to locate the vector $\overline{\mathbf{F}}_i^{se}(t)$ at the proper position in the global vector $\mathbf{F}^{se}(t)$.

The expression of $\overline{\mathbf{F}}_i^{se}(t)$ can be written as:

$$\overline{\mathbf{F}}_i^{se}(t) = \overline{\mathbf{T}}_{GsLs,i} \tilde{\mathbf{N}}_i^T \mathbf{T}_{LsLr} \overline{\mathbf{P}}_i^{se}(t) \tag{10.68}$$

where:

$\overline{\mathbf{P}}_i^{se}(t)$ is the 6×1 vector of self-excited forces per unit length of the ith oblique element of the bridge deck with respect to the local reference coordinate qph-system;
$T_{Lslr,i}$ is the 6×6 transformation matrix from the local reference coordinate system qph to the local structural coordinate system xyz for the ith element.

The vector $\overline{\mathbf{P}}_i^{se}(t)$ includes six components:

$$\overline{\mathbf{P}}_i^{se}(t) = \left\{ C_{q,i}^{se}(t) \quad D_{p,i}^{se}(t) \quad L_{h,i}^{se}(t) \quad M_{\alpha,i}^{se}(t) \quad M_{\gamma,i}^{se}(t) \quad M_{\varphi,i}^{se}(t) \right\}^{\mathrm{T}} \tag{10.69}$$

where $C_{q,i}^{se}(t)$, $D_{p,i}^{se}(t)$, $L_{h,i}^{se}(t)$, $M_{\alpha,i}^{se}$, $M_{\gamma,i}^{se}(t)$ and $M_{\varphi,i}^{se}(t)$ are, respectively, the self-excited crosswind force, drag force, lift force, pitching moment, rolling moment and yawing moment on the ith oblique element of the bridge deck with respect to the local reference coordinate qph system.

Generally, only the self-excited drag force $D_{p,i}^{se}(t)$, the lift force $L_{h,i}^{se}(t)$ and the pitching moment $M_{\alpha,i}^{se}(t)$ are regarded to be important to the buffeting response prediction of the bridge [12,13]. Correspondingly, wind tunnel tests were performed to determine the flutter derivatives associated with these three forces under skew winds (see Section 10.5), and the other three forces, $C_{q,i}^{se}(t)$, $M_{\gamma,i}^{se}(t)$ and $M_{\varphi,i}^{se}(t)$ are omitted.

The vector $\overline{\mathbf{P}}_i^{se}(t)$ is the function of the air density ρ, the element width B_i, the mean wind speed at the center of the element $\overline{U}_i$, the aerodynamic impulse functions associated with the flutter derivatives of the element under skew winds, and the structural motion at the center of the element. It can be expressed as follows:

$$\overline{\mathbf{P}}_i^{se}(t) = \begin{Bmatrix} 0 \\ D_{p,i}^{se}(t) \\ L_{h,i}^{se}(t) \\ M_{\alpha,i}^{se}(t) \\ 0 \\ 0 \end{Bmatrix}$$

$$= \frac{\rho \overline{U}_i^2}{2} \begin{Bmatrix} 0 \\ \int_{-\infty}^{t} f_{D_{p,i}^{se}}(t-\tau)\delta_{p,i}^{ce}(\tau)d\tau + \int_{-\infty}^{t} f_{D_{h,i}^{se}}(t-\tau)\delta_{h,i}^{ce}(\tau)d\tau + B_i \int_{-\infty}^{t} f_{D_{\alpha,i}^{se}}(t-\tau)\delta_{\alpha,i}^{ce}(\tau)d\tau \\ \int_{-\infty}^{t} f_{L_{p,i}^{se}}(t-\tau)\delta_{p,i}^{ce}(\tau)d\tau + \int_{-\infty}^{t} f_{L_{h,i}^{se}}(t-\tau)\delta_{h,i}^{ce}(\tau)d\tau + B_i \int_{-\infty}^{t} f_{L_{\alpha,i}^{se}}(t-\tau)\delta_{\alpha,i}^{ce}(\tau)d\tau \\ B_i \int_{-\infty}^{t} f_{M_{p,i}^{se}}(t-\tau)\delta_{p,i}^{ce}(\tau)d\tau + B_i \int_{-\infty}^{t} f_{M_{h,i}^{se}}(t-\tau)\delta_{h,i}^{ce}(\tau)d\tau + B_i^2 \int_{-\infty}^{t} f_{M_{\alpha,i}^{se}}(t-\tau)\delta_{\alpha,i}^{ce}(\tau)d\tau \\ 0 \\ 0 \end{Bmatrix} \tag{10.70}$$

where $\delta_{p,i}^{ce}(t)$ and $\delta_{h,i}^{ce}(t)$ are the transverse displacements along the axis p and the axis h, respectively, and $\delta_{\alpha,i}^{ce}(t)$ is the torsional displacement around the axis q at the center of the ith element with respect to the local reference coordinate system qph. These displacements, with respect to the local reference coordinate system, can be related to those at the nodes of the element with respect to the local structural coordinate system through the coordinate transformation and displacement interpolation function.

$f_{D_{a,i}^{se}}(t)$, $f_{L_{a,i}^{se}}(t)$, and $f_{M_{a,i}^{se}}(t)$ $(a = p,\ h,\ \alpha)$ are the aerodynamic impulse functions, which can be obtained from the experimentally measured flutter derivatives of the oblique element of the bridge deck using the rational function approximation approach. For example, $f_{M_{\alpha,i}^{se}}(t)$ can be given by:

$$f_{M_{\alpha,i}^{se}}(t) = c_{M_{\alpha,i}^{se},1}\delta_{\alpha,i}^{ce}(t) + \frac{B_i}{\overline{U}_i} c_{M_{\alpha,i}^{se},2}\dot{\delta}_{\alpha,i}^{ce}(t) + \sum_{k=3}^{m_{M_{\alpha,i}^{se}}} c_{M_{\alpha,i}^{se},k} \int_{-\infty}^{t} e^{-\frac{d_{M_{\alpha,i}^{se},k}\overline{U}_i}{B_i}(t-\tau)} \dot{\delta}_{\alpha,i}^{ce}(\tau)d\tau \tag{10.71}$$

where:

the value of $m_{M_{\alpha,i}^{se}}$ determines the level of accuracy of the approximation;

$c_{M_{\alpha,i}^{se},i}$ and $d_{M_{\alpha,i}^{se},k}$ $(i = 1,2,\ \ldots, m_{M_{\alpha,i}^{se}};\ k = 3,\ \ldots, m_{M_{\alpha,i}^{se}})$ are the dimensionless coefficients, which can be determined by the non-linear least-squares fit of the flutter derivatives A_2^* and A_3^* simultaneously as follows [25]:

$$A_3^*(V_i) = \frac{V_i^2 c_{M_{\alpha,i}^{se},1}}{4\pi^2} + \sum_{k=3}^{m_{M_{\alpha,i}^{se}}} \frac{V_i^2 c_{M_{\alpha,i}^{se},k}}{V_i^2 d_{M_{\alpha,i}^{se},k}^2 + 4\pi^2};\ A_2^*(V_i) = \frac{V_i c_{M_{\alpha,i}^{se},2}}{2\pi} + \sum_{k=3}^{m_{M_{\alpha,i}^{se}}} \frac{V_i^3 c_{M_{\alpha,i}^{se},k} d_{M_{\alpha,i}^{se},k}}{V_i^2 d_{M_{\alpha,i}^{se},k}^2 + 8\pi^2} \tag{10.72}$$

in which $V_i = (2\pi \overline{U}_i)/(B_i\omega)$ is the reduced mean wind velocity. The substitution of Equation 10.71 to Equation 10.70 then yields:

$$\overline{\mathbf{P}}_i^{se}(t) = \begin{Bmatrix} 0 \\ D_{p,i}^{se}(t) \\ L_{h,i}^{se}(t) \\ M_{\alpha,i}^{se}(t) \\ 0 \\ 0 \end{Bmatrix} = \frac{\rho \overline{U}_i^2}{2} \mathbf{K}_i^{se} \begin{Bmatrix} 0 \\ \delta_{p,i}^{ce}(t) \\ \delta_{h,i}^{ce}(t) \\ \delta_{\alpha,i}^{ce}(t) \\ 0 \\ 0 \end{Bmatrix} + \frac{\rho \overline{U}_i B_i}{2} \mathbf{C}_i^{se} \begin{Bmatrix} 0 \\ \dot{\delta}_{p,i}^{ce}(t) \\ \dot{\delta}_{h,i}^{ce}(t) \\ \dot{\delta}_{\alpha,i}^{ce}(t) \\ 0 \\ 0 \end{Bmatrix} + \frac{\rho \overline{U}_i^2}{2} \overline{\mathbf{P}}_{lag,i}^{se}(t) \tag{10.73}$$

$$\mathbf{K}_i^{se} = \begin{bmatrix} 0 & 0 & 0 & 0 & 0 & 0 \\ 0 & c_{D_{p,i}^{se},1} & c_{D_{h,i}^{se},1} & B_i c_{D_{\alpha,i}^{se},1} & 0 & 0 \\ 0 & c_{L_{p,i}^{se},1} & c_{L_{h,i}^{se},1} & B_i c_{L_{\alpha,i}^{se},1} & 0 & 0 \\ 0 & B_i c_{M_{p,i}^{se},1} & B_i c_{M_{h,i}^{se},1} & B_i^2 c_{M_{\alpha,i}^{se},1} & 0 & 0 \\ 0 & 0 & 0 & 0 & 0 & 0 \\ 0 & 0 & 0 & 0 & 0 & 0 \end{bmatrix}_i ; \mathbf{C}_i^{se} = \begin{bmatrix} 0 & 0 & 0 & 0 & 0 & 0 \\ 0 & c_{D_{p,i}^{se},2} & c_{D_{h,i}^{se},2} & B_i c_{D_{\alpha,i}^{se},2} & 0 & 0 \\ 0 & c_{L_{p,i}^{se},2} & c_{L_{h,i}^{se},2} & B_i c_{L_{\alpha,i}^{se},2} & 0 & 0 \\ 0 & B_i c_{M_{p,i}^{se},2} & B_i c_{M_{h,i}^{se},2} & B_i^2 c_{M_{\alpha,i}^{se},2} & 0 & 0 \\ 0 & 0 & 0 & 0 & 0 & 0 \\ 0 & 0 & 0 & 0 & 0 & 0 \end{bmatrix}_i \tag{10.74a}$$

$$\overline{\mathbf{P}}_{lag,i}^{se}(t) = \begin{Bmatrix} 0 \\ \sum_{k=3}^{m_{D_{p,i}^{se}}} c_{D_{p,i}^{se},k} I_{D_{p,i}^{se},k}(t) + \sum_{k=3}^{m_{D_{h,i}^{se}}} c_{D_{h,i}^{se},k} I_{D_{h,i}^{se},k}(t) + B_i \sum_{k=3}^{m_{D_{\alpha,i}^{se}}} c_{D_{\alpha,i}^{se},k} I_{D_{\alpha,i}^{se},k}(t) \\ \sum_{k=3}^{m_{L_{p,i}^{se}}} c_{L_{p,i}^{se},k} I_{L_{p,i}^{se},k}(t) + \sum_{k=3}^{m_{L_{h,i}^{se}}} c_{L_{h,i}^{se},k} I_{L_{h,i}^{se},k}(t) + B_i \sum_{k=3}^{m_{L_{\alpha,i}^{se}}} c_{L_{\alpha,i}^{se},k} I_{L_{\alpha,i}^{se},k}(t) \\ B_i \sum_{k=3}^{m_{M_{p,i}^{se}}} c_{M_{p,i}^{se},k} I_{M_{p,i}^{se},k}(t) + B_i \sum_{k=3}^{m_{M_{h,i}^{se}}} c_{M_{h,i}^{se},k} I_{M_{h,i}^{se},k}(t) + B_i^2 \sum_{k=3}^{m_{M_{\alpha,i}^{se}}} c_{M_{\alpha,i}^{se},k} I_{M_{\alpha,i}^{se},k}(t) \\ 0 \\ 0 \end{Bmatrix} \tag{10.74b}$$

where $I_{P_{\alpha,i}^{se},k}(t)$ ($P_{\alpha,i}^{se} = D_{p,i}^{se}, D_{h,i}^{se}, D_{\alpha,i}^{se}, L_{p,i}^{se}, L_{h,i}^{se}, L_{\alpha,i}^{se}, M_{p,i}^{se}, M_{h,i}^{se}, M_{\alpha,i}^{se}$; $k = 3, \ldots, m_{P_{\alpha,i}^{se}}$) are the convolution integrations of the ith element, which can be calculated using a recursive algorithm. For instance, $I_{M_{\alpha,i}^{se},3}(t)$ can be computed by:

$$\begin{aligned} I_{M_{\alpha,i}^{se},3}(t) &= \int_{-\infty}^{t} e^{-\frac{d_{M_{\alpha,i}^{se},3}\overline{U}_i}{B_i}(t-\tau)} \dot{\delta}_{\alpha,i}^{ce}(\tau) d\tau \\ &\approx \frac{\Delta t}{2} \dot{\delta}_{\alpha,i}^{ce}(t) + e^{-\frac{d_{M_{\alpha,i}^{se},3}\overline{U}_i}{B_i}\Delta t} \left[I_{M_{\alpha,i}^{se},3}(t-\Delta t) + \frac{\Delta t}{2} \dot{\delta}_{\alpha,i}^{ce}(t-\Delta t) \right] \end{aligned} \tag{10.75}$$

10.3.3 Governing Equation and Solution in the Time Domain

The governing equation of motion of a long-span cable-supported bridge under skew winds in the frequency domain is expressed by Equation 10.39, in which the self-excited forces are shifted to the left side of the equation. In the time domain, the self-excited forces remain at the right side of the equation expressed as:

$$\mathbf{M}^s\ddot{\mathbf{\Delta}}(t) + \mathbf{C}^s\dot{\mathbf{\Delta}}(t) + \mathbf{K}^s\mathbf{\Delta}(t) = \mathbf{F}^{bu}(t) + \mathbf{F}^{se}(t) \tag{10.76}$$

where:

$\mathbf{M}^s$, C^s and $\mathbf{K}^s$ are, respectively, the $N \times N$ mass, damping, and stiffness matrices of the entire bridge;
$\mathbf{F}^{bu}(t)$ and $\mathbf{F}^{se}(t)$ are, respectively, the buffeting and self-excited force vectors of N dimensions due to skew winds;
$\mathbf{\Delta}(t)$ is the global nodal displacement vector of N-dimensions;
N is the number of the total degrees of freedom of the entire bridge.

Newmark's constant-average-acceleration scheme is used to find the step-by-step solution for the governing equation of motion of the bridge under skew winds. Rewrite the governing equation of motion for the time step $t + \Delta t$:

$$\mathbf{M}^s\ddot{\mathbf{\Delta}}(t+\Delta t) + \mathbf{C}^s\dot{\mathbf{\Delta}}(t+\Delta t) + \mathbf{K}^s\mathbf{\Delta}(t+\Delta t) = \mathbf{F}^{bu}(t+\Delta t) + \mathbf{F}^{se}(t+\Delta t) \tag{10.77}$$

where Δt is the time interval. The formulations for the nodal displacement, velocity, and acceleration at time $t + \Delta t$ can be obtained as:

$$\mathbf{\Delta}(t+\Delta t) = \tilde{\mathbf{K}}^{-1}\tilde{\mathbf{F}}(t+\Delta t) \tag{10.78a}$$

$$\ddot{\mathbf{\Delta}}(t+\Delta t) = a_0(\mathbf{\Delta}(t+\Delta t) - \mathbf{\Delta}(t)) - a_2\dot{\mathbf{\Delta}}(t) - a_3\ddot{\mathbf{\Delta}}(t) \tag{10.78b}$$

$$\dot{\mathbf{\Delta}}(t+\Delta t) = \dot{\mathbf{\Delta}}(t) + a_6\ddot{\mathbf{\Delta}}(t) + a_7\ddot{\mathbf{\Delta}}(t+\Delta t) \tag{10.78c}$$

where

$$\tilde{\mathbf{K}} = \mathbf{K}^s + a_0\mathbf{M}^s + a_1\mathbf{C}^s \tag{10.79a}$$

$$\begin{aligned}\tilde{\mathbf{F}}(t+\Delta t) = {} & \mathbf{F}^{bu}(t+\Delta t) + \mathbf{F}^{se}(t+\Delta t) + \mathbf{M}^s(a_0\mathbf{\Delta}(t) + a_2\dot{\mathbf{\Delta}}(t) + a_3\ddot{\mathbf{\Delta}}(t)) \\ & + \mathbf{C}^s(a_1\mathbf{\Delta}(t) + a_4\dot{\mathbf{\Delta}}(t) + a_5\ddot{\mathbf{\Delta}}(t))\end{aligned} \tag{10.79b}$$

in which a_i $(i = 0, 1 \ldots, 7)$ are the constant coefficients given by [26]:

$$a_0 = \frac{1}{\beta\Delta t^2};\, a_1 = \frac{\gamma}{\beta\Delta t};\, a_2 = \frac{1}{\beta\Delta t};\, a_3 = \frac{1}{2\beta} - 1 \tag{10.80a}$$

$$a_4 = \frac{\gamma}{\beta} - 1;\, a_5 = \left(\frac{\gamma}{\beta} - 2\right)\frac{\Delta t}{2};\, a_6 = (1-\gamma)\Delta t;\, a_7 = \gamma\Delta t \tag{10.80b}$$

where γ and β are taken as 0.5 and 0.25 in this study. Note that the self-excited force vector on the bridge, $\mathbf{F}^{se}(t + \Delta t)$ in Equation 10.77, is the function of bridge motion. Iterations should be performed in each time step. For instance, for the time step $t + \Delta t$, first use the self-excited force $\mathbf{F}^{se}(t)$ in the time step t to replace $\mathbf{F}^{se}(t + \Delta t)$ in Equation 10.77 to compute the motion of the bridge deck, then use the computed bridge motion to calculate the self-excited forces again. Repeat the above two steps until the bridge motion is converged to the prescribed criteria.

10.4 Aerodynamic Coefficients of Bridge Deck under Skew Winds

To perform a satisfactory comparison of buffeting response of the Tsing Ma suspension bridge between field measurement and analysis and to verify the proposed frameworks for buffeting analysis of long-span cable-supported bridges under skew winds, the aerodynamic coefficients and flutter derivatives of the bridge deck under skew winds have to be measured through wind tunnel tests. Thus, special test rigs and the models of a typical deck section of the Tsing Ma Bridge were designed. The tests were carried out in the TJ-2 Wind Tunnel of the State Key Laboratory for Disaster Reduction in Civil Engineering at Tongji University, Shanghai, China. Following is a brief description of the wind tunnel tests. Detailed information on the design of sectional models, the development of test rigs and measurement systems and the analysis of test results can be found in Zhu *et al.* [27,28].

Rather than traditional wind tunnel tests for measuring aerodynamic coefficients of a bridge deck under normal wind, a sectional model of parallelogram was used to measure aerodynamic coefficients of a bridge deck under skew winds. The oblique sectional model was installed vertically in the wind tunnel (see Figure 10.8).

To alleviate 3-D flow effects around the two ends of the oblique sectional model, two surrounding segments (upper and lower) were introduced and each was mounted close to one end of, but free from any contact with, the sectional model (the measured segment). To accommodate a few yaw angles in the test using one sectional model, two adjustable parts were designed and placed between the top part and the middle part, and between the bottom part and the middle part, of the model separately (see Figure 10.9).

Each adjustable part consisted of four small triangular wedges used to adjust the middle part to form a measured segment, and the top and bottom parts to form the upper and lower surrounding segments for five designated yaw angles ($\bar{\beta}$=0°, 5°, 13°, 20° and 31°). The geometric scale of the deck sectional model of the Tsing Ma Bridge was set as 1:100 after a careful consideration of many factors. These included the size of the working section of the wind tunnel, the requirement for the model to have a

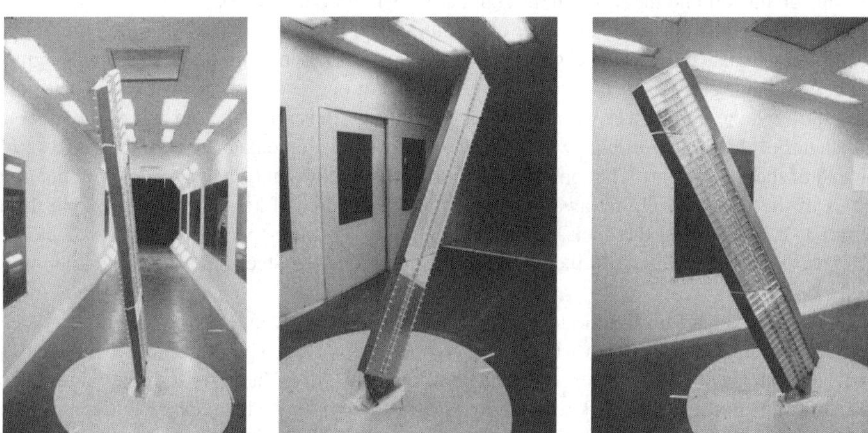

(a) Leeward view (b) View of deck top surface (c) View of deck bottom surface

Figure 10.8 Static sectional model of the Tsing Ma bridge deck ($\bar{\beta} = 31°$, $\bar{\theta} = -10°$) (from [27]) (Reproduced with permission from Elsevier).

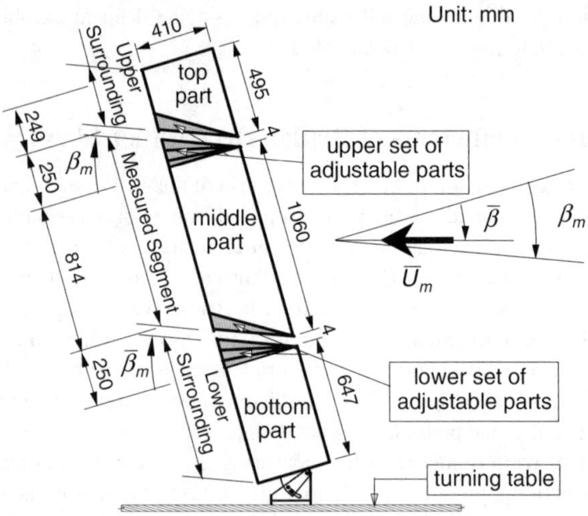

Figure 10.9 Configuration and constitution of static deck sectional model (from [27]) (Reproduced with permission from Elsevier).

sufficiently large aspect ratio (length-over-width), the high stiffness and lightweight requirement for the measured segment of the model, and the sensitivity and capacity of the force balance.

As a result, the width of sectional model is 410 mm. The total length of the test model is 2210 mm, in which the length of the oblique sectional model (the middle measured segment) is 1060 mm, the length of longer side of the upper surrounding segment is 495 mm and the length of shorter side of the lower segment is 647 mm. The gaps between the top part and the middle part, and between the bottom part and the middle, part, are each 4 mm.

The bottom part of the model was mounted rigidly on the turntable of the wind tunnel through a special apparatus called a yaw angle device, which was made of steel and designed in such a way that the model could be adjusted to meet the desired yaw angle. The desired angle of inclination was realized by rotating the turntable. There were two steel pipes installed on the top of the bottom part of the model – one on the left and the other on the right, as shown in Figure 10.10.

The left steel pipe had an outer diameter of 48 mm. Fixed at the top of the left steel pipe was a six-component force balance, shown in Figure 10.10 as a black cylinder. The middle measured segment of the model was then rigidly connected to the force balance through its strong diaphragm. The right steel pipe went through the middle part without any contact to rigidly support the top part of the model. The outer diameter of the right steel pipe was 42 mm. The top of the force balance was 508 mm away from the top end of the middle part. The longitudinal axis of the force balance was parallel with the longitudinal axis of the deck model, but away from the deck longitudinal axis at 97.5 mm along the deck width and at 2.2 mm along the deck height with reference to the upper surface of the deck model. Enough spaces were kept between the measured segment and the steel pipes to prevent any contact during the test.

The TJ-2 Wind Tunnel of Tongji University is a boundary layer tunnel of closed-circuit-type. The working section of the tunnel is 3 m wide, 2.5 m high and 15 m long. The achievable mean wind speed ranges from 0.5 m/sec to 68.0 m/sec, adjustable continuously. All of the tests for determining aerodynamic coefficients of the bridge deck were carried out in smooth flow. The thickness of boundary layers near the walls, floor and ceiling of the tunnel was less than 150 mm. The maximum non-uniformity of mean wind within the cross section, excluding the boundary areas, was less than 1% and the turbulence intensity within the same cross section was less than 0.5%. The horizontal yaw angle and vertical inclination angle of the mean wind in the tunnel were less than 0.5°.

Figure 10.10 Force balance and steel pipes in model (from [27]) (Reproduced with permission from Elsevier).

The tests were carried out under a mean wind of 15 m/sec. To enrich the database for generating curves of aerodynamic coefficients and to understand the effect of model end angle, it was decided that the model was to be tested under a wide combination of yaw angles, from 0° to 34° at an interval of 1° or 2°, and inclination angles from −10° to +10° at an interval of 1°. As a result, there were a total of 735 cases tested. For each test case, the measurement was repeated for at least three times. Each single measurement lasted for 10 seconds, with a sample frequency of 500 Hz. The trial tests showed that the test results in terms of aerodynamic coefficients were repeatable. The discrepancies among the three test results were small at the levels from 0.2% to 1%.

The six aerodynamic forces/moments applied on the test segment were measured by the six-component force balance. The measured forces/moments were then used to calculate the aerodynamic coefficients with respect to the local wind coordinates $\overline{qph}$ and the true width of the cross-section. The measurement data were selected and grouped to form two databases: one for lift force, drag force and pitching moment coefficients; the other for crosswind force, rolling moment and yawing moment coefficients. Interpolation and the curve-fitting technique were employed on each database to generate the curves of six aerodynamic coefficients as functions of both mean wind inclination and yaw angle.

Figure 10.11 depicts variations of six aerodynamic coefficients with the angle of inclination for a series of yaw angles. It can be seen that, with the increase of wind inclination, the lift coefficient grows from negative to positive, while the pitching moment coefficient goes down from positive to negative, according to the sign convention set in this investigation. The drag coefficient is always positive and attains the minimum value in the inclination range between 2.0° and 3.0° for all the concerned yaw angles. The slopes of drag coefficient curves are negative up to the inclination of 2.0° and 3.0° and then become positive.

Within the inclination range from −7.0° to +7.0°, the crosswind force coefficient decreases monotonically for all the concerned yaw angles. For yaw angles greater than 10°, the rolling moment coefficient decreases from positive value to negative value with the increase of inclination angle. When the yaw angle is less than 15°, the yawing moment coefficient becomes small as the inclination grows up. However, when the yaw angle ranges from 20° to 35°, the yawing moment coefficient curve has a maximum value around $\overline{\theta} = 0°$. Both the rolling and yawing moment coefficients are very small when the yaw angle is less than 5°.

Major conclusions drawn from the wind tunnel tests are as follows:

- The drag, lift, and pitching moment coefficients were much larger than the crosswind force, rolling moment and yawing moment coefficients, even at a large yaw angle of 31°.
- The variation of lift coefficient with yaw angle was small for an angle of inclination less than 5°. The value of drag coefficient, however, decreased with increasing yaw angle for all the concerned angles

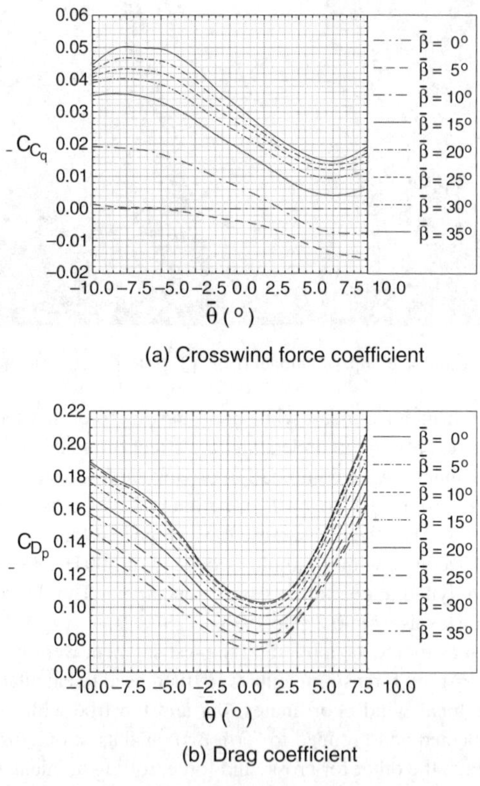

Figure 10.11 Variations of aerodynamic coefficients with angle of inclination (from [27]) (Reproduced with permission from Elsevier).

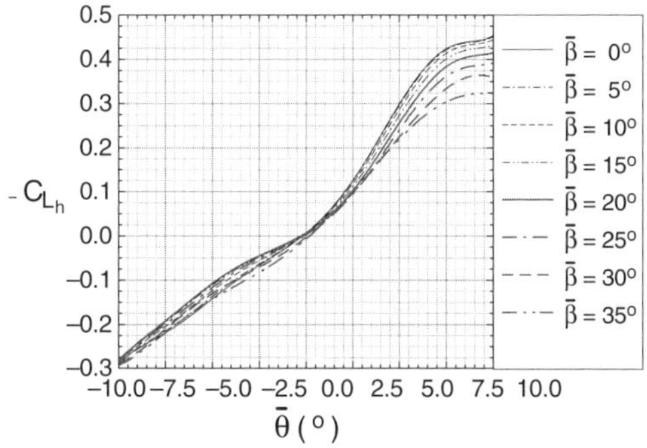

(c) Lift coefficient

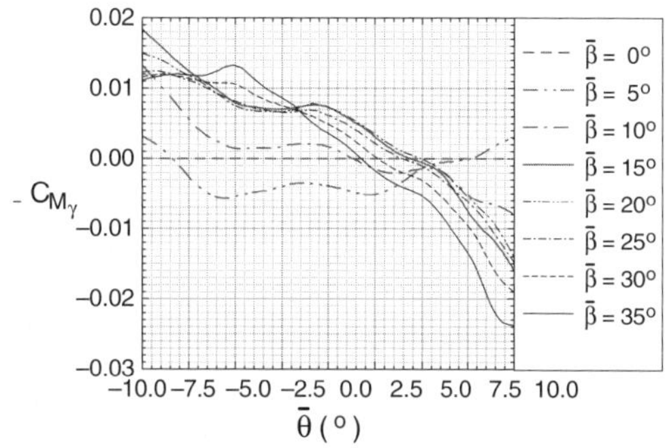

(e) Rolling moment coefficient

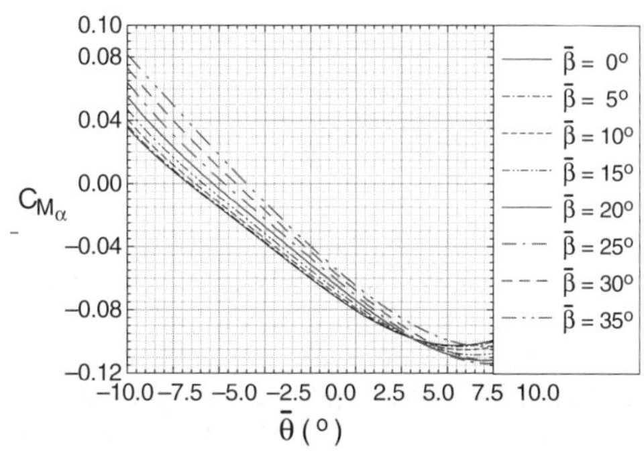

(d) Pitching moment coefficient

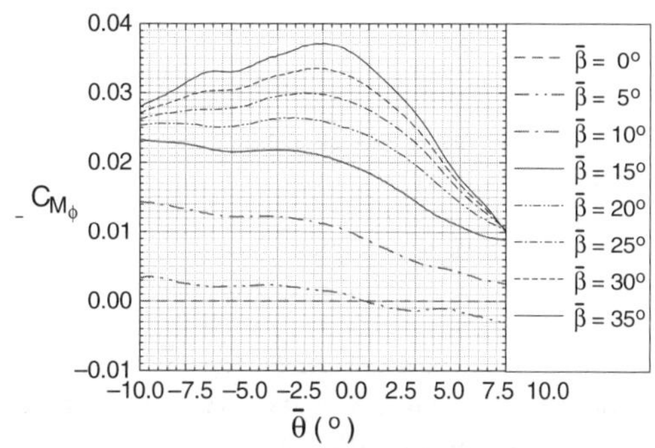

(f) Yawing moment coefficient

Figure 10.11 *(Continued)*

of inclination, whereas the value of pitching moment increased with increasing yaw angle for angles of inclination less than 10°.

- With increase of wind inclination, the lift coefficient changed from negative value to positive value, while the pitching moment coefficient went down from positive to negative. The drag coefficient was always positive and attained the minimum value in the inclination range between 2.0° and 3.0° for all the concerned yaw angles.
- The coefficients of the aerodynamic force along the model axis q, and the aerodynamic moments around the model axes p and h, were not zero for the non-zero yaw angle, as is assumed in the cosine rule. They became noticeably large for larger yaw angles.
- The coefficients of the aerodynamic forces along the model axes p and h, and the aerodynamic moment around the model q-axis, did not comply with the traditional cosine rule, particularly for large yaw angle cases.

10.5 Flutter Derivatives of Bridge Deck under Skew Winds

To facilitate the investigation of flutter derivatives of the bridge deck under skew winds, the dynamic sectional deck model was designed as an oblique strip composed of five major parts: one middle part, two end parts and two support arms between the middle and end parts (see Figure 10.12) [28].

The length of the middle part was 2.407 m for a 1/100 geometric scale and it remained unchanged in all the tests. The two end parts of the model were trapezoidal, and its length changed with designated

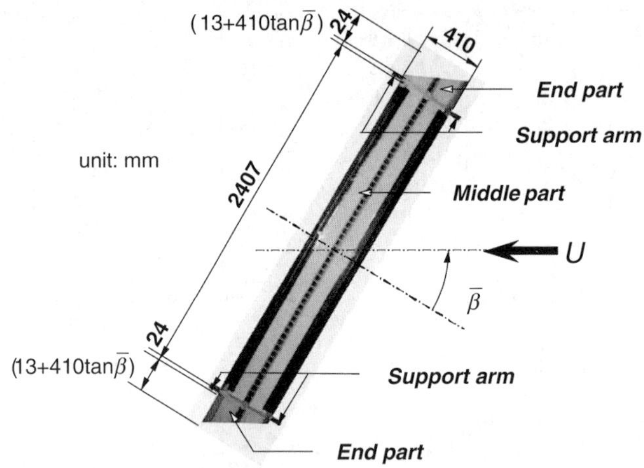

Figure 10.12 Dynamic sectional model of bridge deck under yaw wind (from [28]) (Reproduced with permission from Elsevier).

yaw angle. Five yaw angles – i.e. 0°, 5°, 13°, 20° and 31° – were concerned in this investigation. In correspondence with the five yaw angles, five pairs of the end parts were made. The short side length of the end part was 0.013 m for all the test cases, while the long side length varied according to the function of $0.013 + 0.41\tan\bar{\beta}$. The two support arms, each 0.024 m wide, were suspended by eight helical springs from the wind tunnel and kept horizontal in all the test cases when there was no wind in the tunnel (see Figure 10.13).

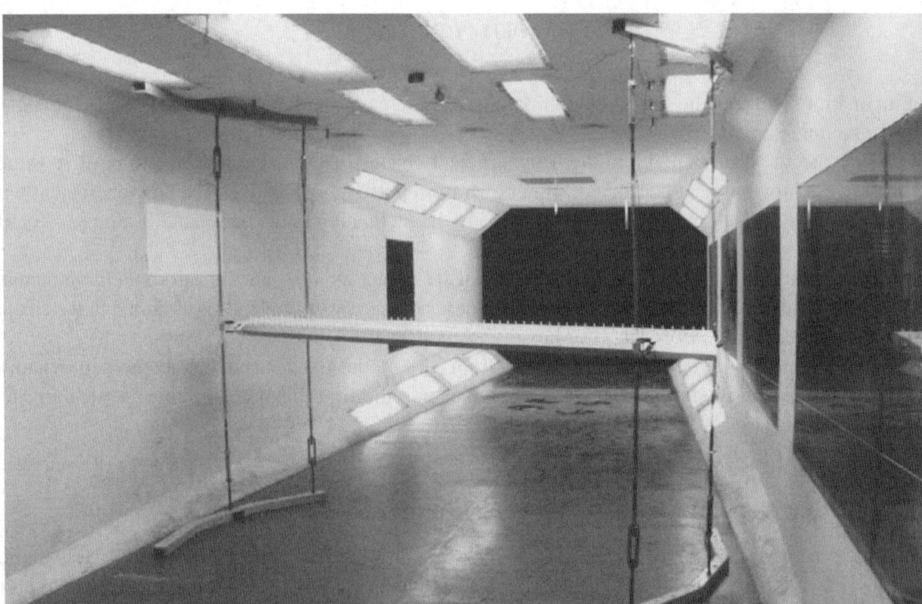

Figure 10.13 Installation of dynamic sectional model in TJ-2 wind tunnel ($\bar{\beta} = 31°$) (from [28]) (Reproduced with permission from Elsevier).

The middle part and the two end parts were then properly connected to the two support arms to form a complete test model. The two ends of the model were always kept in parallel to the incident wind in the wind tunnel. To facilitate adjustment of the sectional model to each of the five designated wind yaw angles, there were two broken-line tracks fixed on the ceiling and another two on the floor of the wind tunnel (see Figure 10.13). Five sets of screw eyes were bored and labeled on the tracks for the five designated yaw angles. There were two sets of holes on the aluminum support arm. Corresponding to the two sets of holes on the support arm, there were two sets of screw eyes drilled on the aluminum end plate of the middle part just next the big hole. These holes and screw eyes were used to adjust the inclination of the model. The inclination of the model concerned in this study ranged from 0° to ±5°. The length-to-width ratio of the model in all the test cases was greater than 6.0. Thus, the three-dimensional flow effect around the two ends of the model was regarded insignificant.

The test sectional model simulated the first symmetric vertical bending mode of vibration and the first symmetric torsional mode of vibration of the bridge [29]. The frequency ratio between the first torsional mode and the first vertical bending mode of similar shape is 0.1366 Hz. The stiffness of eight helical springs and the distance between each pair of springs across the model width were determined in accordance with the similarity law to satisfy the required frequency ratio. To have a better wind tunnel simulation using a 2DOF sectional deck model, the effects of motion of other bridge components, such as towers and cables, and the effects of other modal components can be taken into consideration indirectly. In this connection, the effective mass and the effective mass moment of inertia of the prototype bridge deck are used and targeted in the design, together with the frequency ratio.

The coupled 2DOF free decay vibration was performed to identify the eight flutter derivatives A_i^* and H_i^* ($i = 1,2,3,4$) in conjunction with the unifying least square (ULS) method [30]. Theoretically, the ULS method can be used to identify all of the 18 flutter derivatives using a 3DOF sectional bridge deck model. However, due to the lack of a more inclusive experimental setup to accommodate the 3DOF bridge deck model, only the eight flutter derivatives associated with the vertical and torsional motions, namely H_i^* and A_i^* ($i = 1,2,3,4$), were measured. The flutter derivatives are functions of reduced velocity/frequency, wind inclination, and wind yaw angle.

Seven model inclinations ($\bar{\theta} = 0°, \pm 2°, \pm 3°, \pm 5°$) and five yaw angles ($\bar{\beta} = 0°, 5°, 13°, 20°, 31°$) were considered, resulting in a total of 35 cases. To obtain the time-histories of free decay curves of the model at a given position and a given wind speed, two accelerometers were mounted on one support arm and a third was mounted on the other support arm. The vertical response of the model was obtained by averaging the outputs from the two accelerometers arranged in the diagonal. The torsional response was gained by multiplying the difference between the two outputs from the two accelerometers on the same support arm by a factor.

The acceleration signals from the accelerometers were sampled at 500 Hz. The digital acceleration data from the three channels were immediately analyzed using a computer program implementing the ULS method to extract the flutter derivatives. The test was repeated at least three times for each wind speed and each model position, and the average values of flutter derivatives were stored into the test database for further analysis.

The test data of flutter derivatives for a given model position were interpolated/extrapolated and fitted to generate the design curves of eight flutter derivatives H_i^* and A_i^* ($i = 1,2,3,4$). Figure 10.14 shows the three major flutter derivatives H_1^*, A_2^* and A_3^* measured at an inclination $\bar{\theta} = 0°$ and $\pm 3°$ for yaw angles $\bar{\beta} = 0°$, $\bar{\beta} = 5°$, $\bar{\beta} = 13°$, $\bar{\beta} = 20°$, and $\bar{\beta} = 31°$. Further details can be found in [28].

Major conclusions can be drawn from the test results as follows:

- The effect of wind yaw angle on the flutter derivatives H_1^*, H_2^*, H_3^*, A_2^* and A_3^* was, in general, not conspicuous in the range of lower reduced velocity, but became considerable in the range of higher reduced velocity. The flutter derivatives H_4^*, A_1^* and A_4^*, however, oscillated remarkably with reduced velocity and were affected by the wind yaw angle within the whole range of reduced velocity.

- The eight flutter derivatives were sensitive to wind inclination, particularly in the range of higher reduced velocity.
- The critical wind speed of the Tsing Ma Bridge may not increase with increasing wind yaw angle. The lowest critical wind speed could occur when the mean wind deviated from the normal of the bridge axis. The discrepancies between the two sets of flutter critical wind speeds obtained from the single DOF torsional flutter analysis and the two DOF coupled flutter analysis are very small (<8%) when wind inclination is not zero. However, the discrepancy becomes considerably larger (more than 15%) when wind inclination is zero.

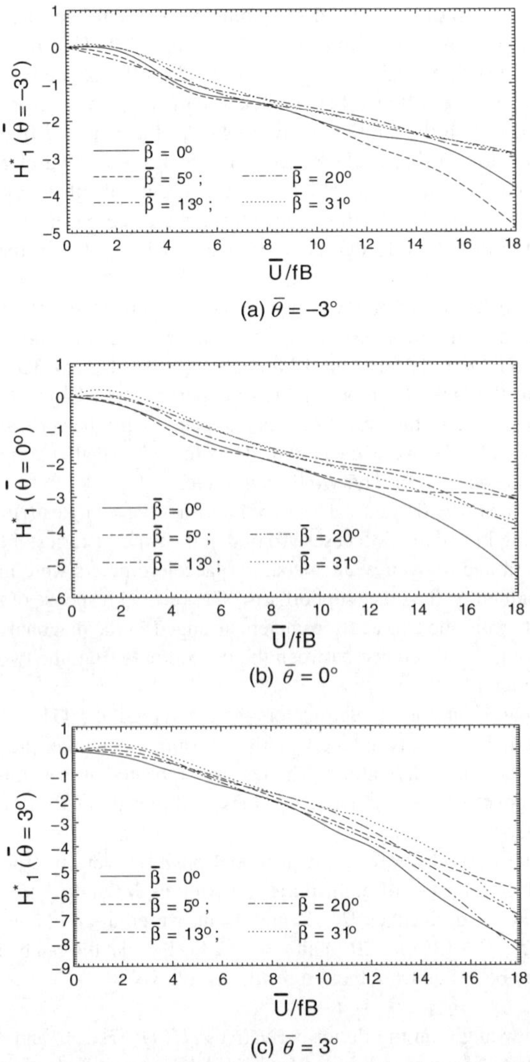

Figure 10.14 Flutter derivative curves: H_1^* vs. $\overline{U}/fB$. Flutter derivative curves: A_2^* vs. $\overline{U}/fB$. Flutter derivative curves: A_3^* vs. $\overline{U}/fB$ (from [28]) (Reproduced with permission from Elsevier).

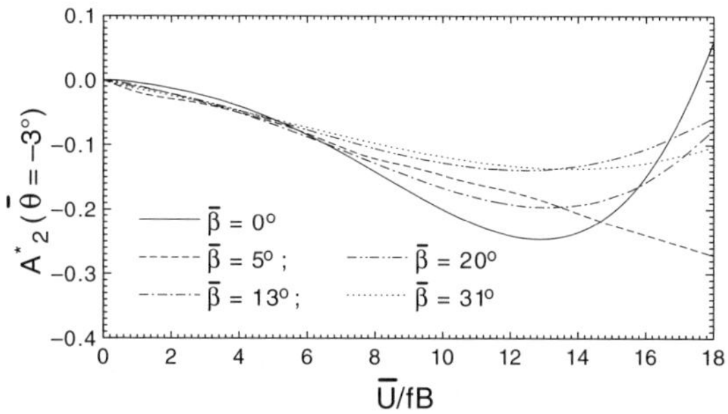

(d) $\bar{\theta} = -3°$

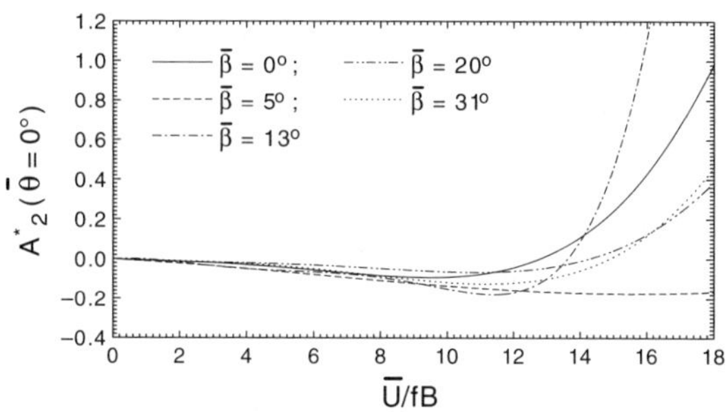

(e) $\bar{\theta} = 0°$

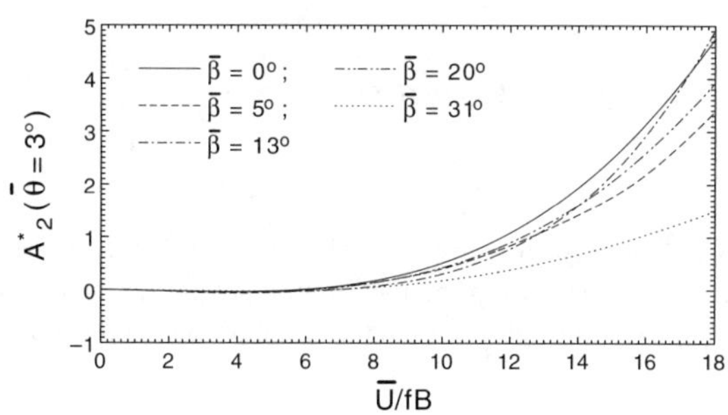

(f) $\bar{\theta} = 3°$

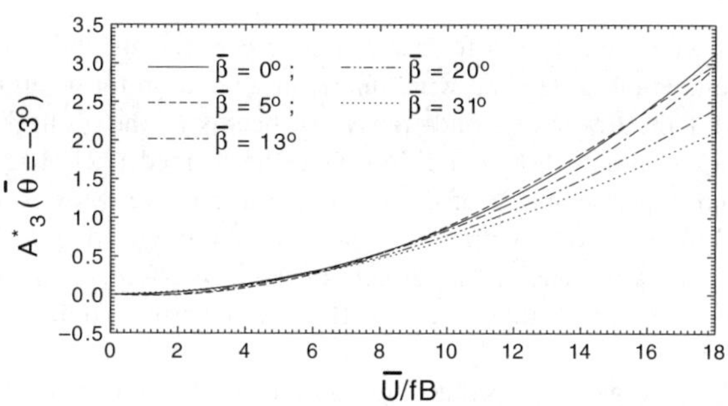

(g) $\bar{\theta} = -3°$

Figure 10.14 *(Continued)*

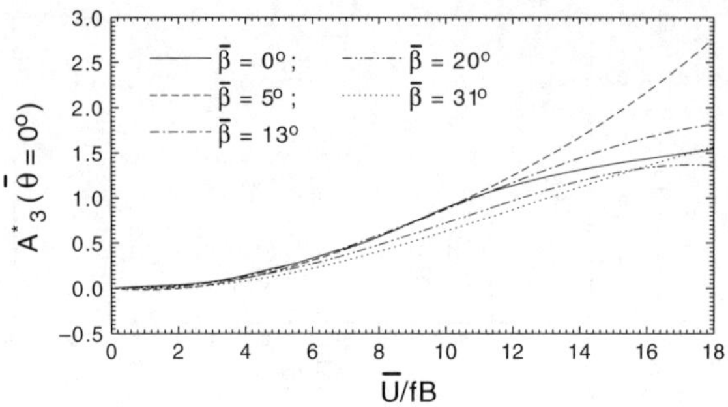

(h) $\bar{\theta} = 0°$

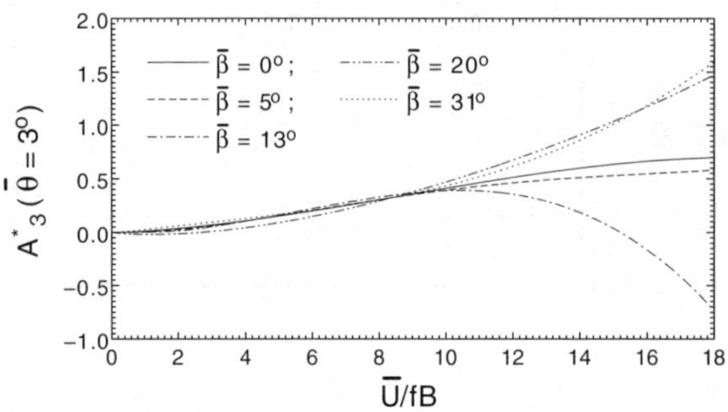

(i) $\bar{\theta} = 3°$

Figure 10.14 (*Continued*)

- In general, the major flutter derivatives estimated by the empirical formulas based on the skew wind theory [31] deviated considerably from the measured results. The major flutter derivatives could be estimated using the empirical formulae only in the range of lower reduced velocity.

10.6 Aerodynamic Coefficients of Bridge Tower Under Skew Winds

To be consistent with wind tunnel tests for aerodynamic coefficients and flutter derivatives of the bridge deck under skew winds, a detailed wind tunnel investigation for determining the mean aerodynamic coefficients of the tower legs and transverse beams of the Tsing Ma Bridge in various arrangements, and under smooth flow condition, were performed [32]. The aerodynamic coefficients of the lower and upper segments of the windward and leeward tower legs, and those of the transverse beams at different levels, with and without the dummy bridge deck model, were measured as a function of yaw wind angle. The effects of wind interference among the tower components, and the influence of the bridge deck on the tower aerodynamic coefficients, were also investigated.

The Tsing Ma Bridge has two H-shaped concrete towers, each is composed of two reinforced concrete legs and four deep transverse pre-stressed concrete beams. The two towers are almost identical, so only one tower was selected and tested. According to the wind tunnel size, the geometrical scale of the tower was selected as 1 : 100. The height of the tower model was 1935.1 mm.

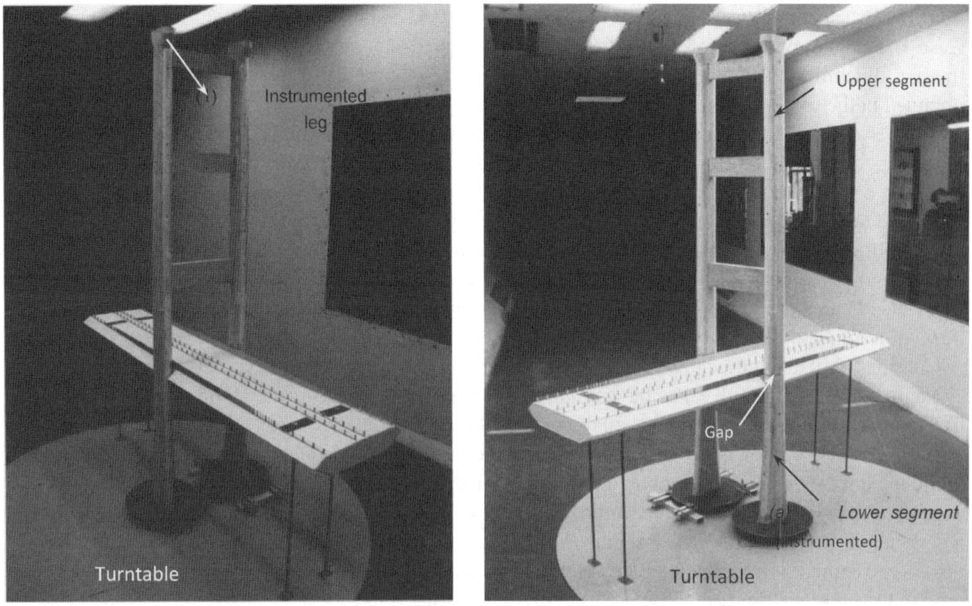

(a) Entire leg measurement with bridge deck (b) Lower leg segment measurement with bridge deck

Figure 10.15 Tower leg measurements with bridge deck (from [32]) (Reproduced with permission from Techno .Press).

The tower model was assembled from two tower legs and four transverse beams (see Figure 10.15). A circular steel plate was fixed at the base of each leg. The two legs and the four transverse beams could be disconnected and remounted in different ways, depending on the test cases. To consider the influence of the bridge deck on the wind flow around the bridge tower, a section equivalent to a 180.00 m long segment of the full-scale bridge deck was connected to the bridge tower. This section was considered to be a dummy deck model in the wind tunnel tests. It was made of foam plastic and mounted on the tower model symmetrically with respect to the tower plane. The transverse beam below the dummy deck model is defined as lower transverse beam, and all other beams as upper transverse beams.

A total of nine cases were considered in this wind tunnel investigation.

In Case 1, the aerodynamic forces acting on one single tower leg were measured with the dummy deck, the dummy tower leg and the transverse beams in place. The four transverse beams were connected to the dummy tower leg, which was fixed to the top plate of the turntable. The instrumented tower leg was centrally mounted on a five-component force balance through the aluminum link. The force balance was vertically connected to the balance support post, which was firmly mounted at the center of the base plate of the turntable. The dummy deck model was symmetrically installed with respect to the tower, but it was fixed directly to the turntable through four vertical thin bars (Figure 10.15). Small gaps of about 3 mm in width were left between the instrumented tower leg and the four transverse beams, and between the instrumented tower leg and the deck model.

In Cases 2 to 5, the aerodynamic forces acting on each transverse beam were measured with the dummy deck model in place. The measurement of aerodynamic forces on each transverse beam was indirect; for example, to determine the aerodynamic forces on the upper transverse beam, the element was disconnected from the dummy tower leg and rigidly connected to the instrumented tower leg at the original level. The total aerodynamic forces acting on the instrumented tower components, including both the instrumented tower leg and the upper transverse beam, were measured in a way similar to Case 1. The aerodynamic forces on the upper transverse beam were then calculated by subtraction of the aerodynamic forces on the instrumented tower leg obtained in Case 1 from the measurements obtained in the present case.

The aerodynamic forces acting on the instrumented tower leg without the dummy deck model in place were measured in Case 6. The comparison between the aerodynamic forces measured in Case 1 and Case 6 would reveal the influence of the presence of the dummy deck model.

The aerodynamic forces acting on a single tower leg without any dummy models were measured in Case 7. In this case, the dummy tower leg and the four transverse beams in Case 6 were removed, leaving only one tower leg on the force balance. The influence of the dummy tower leg and the four transverse beams on the aerodynamic forces acting on the instrumented tower leg could be evaluated through comparison between the aerodynamic forces measured in Case 6 and Case 7.

Considering the significant variation in width of the cross-section of a tower leg from the base to an elevation of 70.00 m in full scale, the instrumented tower leg was thus further divided into two parts at a height of 700.0 mm in Case 8 and a small horizontal gap of about 3 mm was left between the upper and lower segments, as shown in Figure 10.15. The upper segment was connected to the dummy tower leg, while the lower segment was mounted on the force balance. All other arrangements, including the four dummy transverse beams and the dummy deck model, were kept the same as those in Case 1. The aerodynamic forces on the lower segment of the tower leg were measured in this case. Those on the upper leg segment were calculated by subtracting the aerodynamic forces on the lower segment measured in the present case from those on the entire leg measured in Case 1.

In the last case (i.e. Case 9), the aerodynamic forces on the lower segment of the tower leg without the deck influence were measured. After the aerodynamic forces on the lower segment had been measured, those on the upper segment of the tower leg without the deck influence were taken as the difference between the aerodynamic forces on the entire tower leg (measured in Case 6) and those on the lower segment (measured in Case 9).

The five-component force balance used in these tests was a base-supported strain balance which met the requirements for high sensitivity and high stiffness, to avoid the model vibration and to ensure the measurement accuracy. The yaw wind angle β_0 is defined as zero when the wind is normal to the longitudinal alignment of the bridge (see Figure 10.16). The designated angle β_0 was then calculated from the angle β_T, which represents the clockwise angle rotation of the turntable (see Figure 10.16).

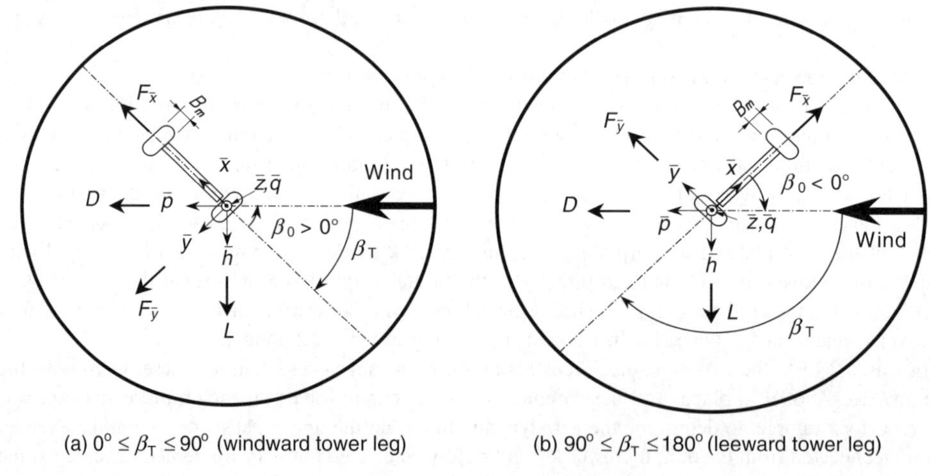

Figure 10.16 Positive directions of aerodynamic forces on tower legs (from [32]) (Reproduced with permission from Techno.Press).

For each test case, the measurements were repeated at least three times and the average values would be presented. Based on some trial tests, the test mean wind speed was finally selected as 15 m/sec. Each single measurement lasted for 10 seconds, with a sample frequency of 500 Hz. The time-averaged method was used to analyze the data to obtain the mean aerodynamic coefficients. Since the blockage ratio was less than 7.5%, no blockage corrections were applied to the results. The aerodynamic forces measured by the balance under designated wind directions were first analyzed to obtain the mean coefficients of the drag/lift/crosswind force of the tower members. These coefficients were then interpolated using spline functions, or fitted using polynomial functions, to obtain the aerodynamic coefficient curves. It should be noted that all the coefficients provided here are with reference to the wind coordinate $\overline{qph}$-system.

The drag and lift coefficients of the windward and leeward entire legs of the Tsing Ma bridge tower, with and without the presence of the deck segment (Case 1 and Case 6) are shown in Figure 10.17 as a function of yaw wind angle β_0 from 0° to 90°. The drag and lift coefficients of the single tower leg without any dummy models in place (Case 7) are also plotted in Figure 10.17.

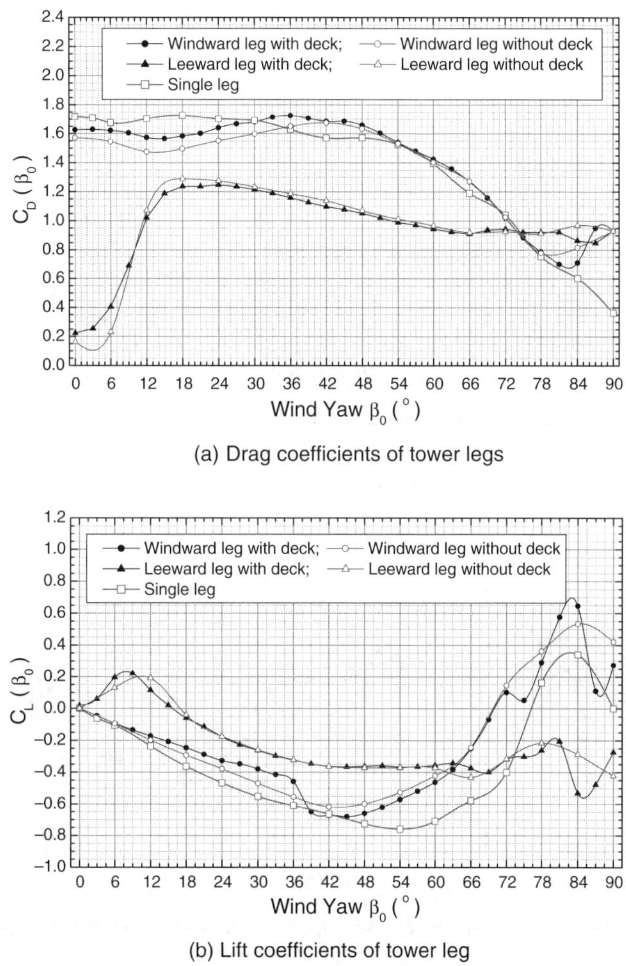

Figure 10.17 Aerodynamic coefficients of tower leg (from [32]) (Reproduced with permission from Techno.Press).

The measured drag and crosswind force coefficients of the four transverse beams of the Tsing Ma bridge tower with the dummy deck segment are plotted in Figure 10.18 as a function of yaw wind angle β_0.

Based on the measured results, the following observations and comments can be made:

1. The drag coefficients of the windward tower leg with the dummy leeward tower leg and four transverse beams had a slight difference from those of the single tower leg for most of the concerned yaw wind angles. The influence of the leeward leg and of the transverse beams on the lift coefficient of the windward leg was relatively small when $\beta_0 < 42°$, but became remarkable when $\beta_0 > 42°$. The interference effect with the deck segment on both the drag and lift coefficients of the windward tower leg was insignificant in the range of $0° \leq \beta_0 \leq 70°$.
2. The drag coefficients of the leeward tower leg with the dummy windward leg and transverse beams were significantly different from those of both the single tower leg and the windward tower leg,

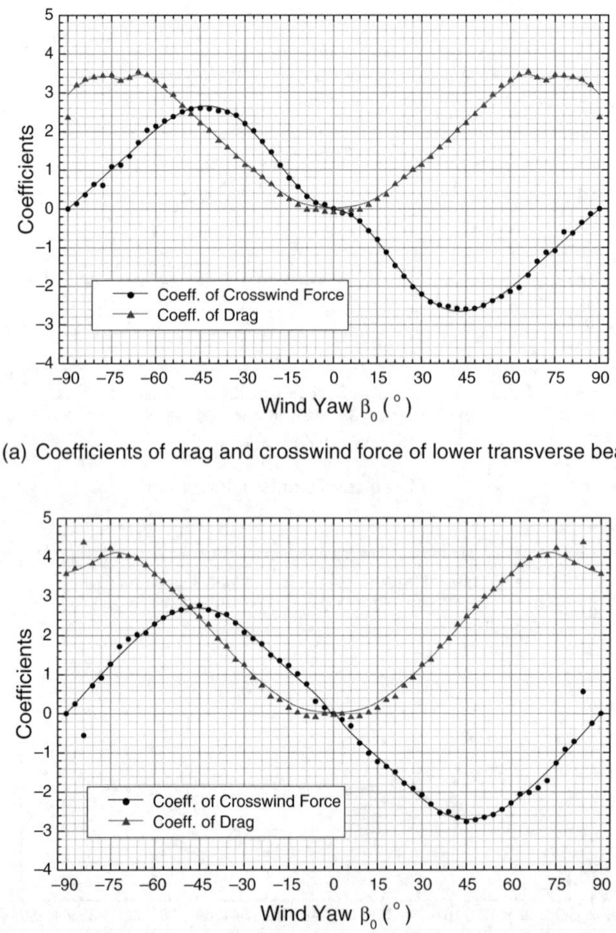

(a) Coefficients of drag and crosswind force of lower transverse beam

(b) Coefficients of drag and crosswind force of lower-middle transverse beam

Figure 10.18 Aerodynamic coefficients of transverse beams (from [32]) (Reproduced with permission from Techno.Press).

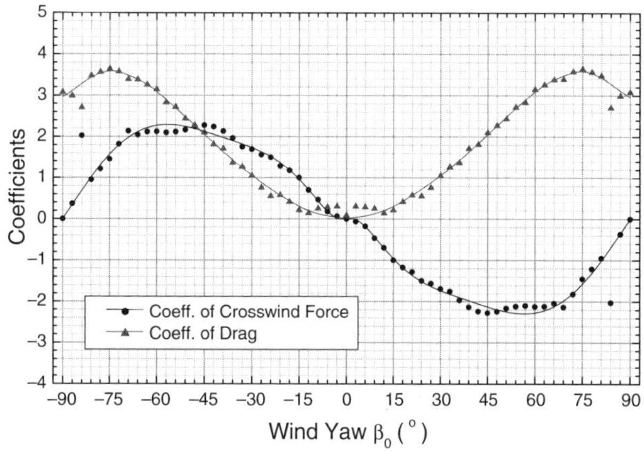

(c) Coefficients of drag and crosswind force of upper-middle transverse beam

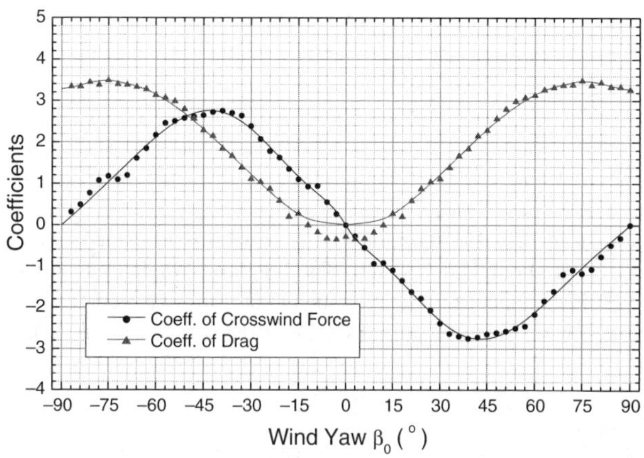

(d) Coefficients of drag and crosswind force of upper transverse beam

Figure 10.18 (*Continued*)

because of the sheltering effect from the windward tower leg and the transverse beams. The same observation was made for the lift coefficients of the leeward tower leg. The influence of the dummy deck segment on both the drag and lift coefficients of the leeward leg was also insignificant when $0° \leq \beta_0 \leq 80°$.

3. The variation of the drag coefficient of the upper windward leg segment with yaw wind angle was very similar to that of the drag coefficient of the entire windward leg. The pattern of the drag coefficient of the lower windward leg segment was, however, slightly different from that of the upper windward leg segment. The drag coefficient of the lower windward leg segment was larger than that of the upper windward leg segment when β_0 was smaller than 75°.
4. The variation of the drag coefficient of the upper segment of the leeward leg with β_0 was quite similar to that of the drag coefficient of the entire leeward leg. The variation of the drag coefficient of the lower segment of the leeward leg with β_0 was also similar to that of the drag coefficient of the entire leeward leg, except for when $\beta_0 > 70°$. The drag coefficients of both upper and lower

segments of the leeward leg were significantly different from those of the corresponding segments of the windward leg.

5. The lift coefficients of the upper segments of both the windward and leeward tower legs, with and without the dummy deck segment, were very similar to those of the entire windward and leeward tower legs, respectively. The lift coefficients of the lower segment of the windward and leeward tower legs, however, were to a certain extent different from those of the upper segments.
6. The variations of both drag and crosswind force coefficients with yaw wind angle were similar to each other for all the four transverse beams. The crosswind force coefficients varied with β_0 in a manner similar to a sine wave. The drag coefficient curves, however, seemed to have the shape of a cosine wave, but it was distorted in magnitude for β_0 ranging from 0° to 27° for the lower transverse beam, and from 0° to 15° for the other three transverse beams.

10.7 Comparison with Field Measurement Results of Tsing Ma Bridge

10.7.1 Typhoon Sam and Measured Wind Data

After developing at about 680 km east-northeast of Manila on 19 August 1999, the tropical depression Sam moved west-northwestwards over the Pacific and intensified into a tropical storm at that night. It then moved north-westerly towards the coast of Guangdong and became a typhoon during late morning of 22 August near Hong Kong. Typhoon Sam finally made landfall over the eastern part of Sai Kung in Hong Kong at around 6 pm on 22 August. Following landfall, Sam traversed the northeastern part of the New Territories at a speed of about 25 km/h and crossed into Shenzhen, before weakening gradually over inland Guangdong on 23 August (see Figure 10.19).

Hong Kong Observatory recorded a maximum hourly-mean wind speed of about 27 m/sec and a maximum gust wind speed of about 41 m/sec at a 75 m height at Waglan Island during the passage of Typhoon Sam. The lowest instantaneous pressure at mean sea level was recorded as 979.0 hPa [33]. The WASHMS timely recorded wind velocity and bridge buffeting responses. The sampling frequency for

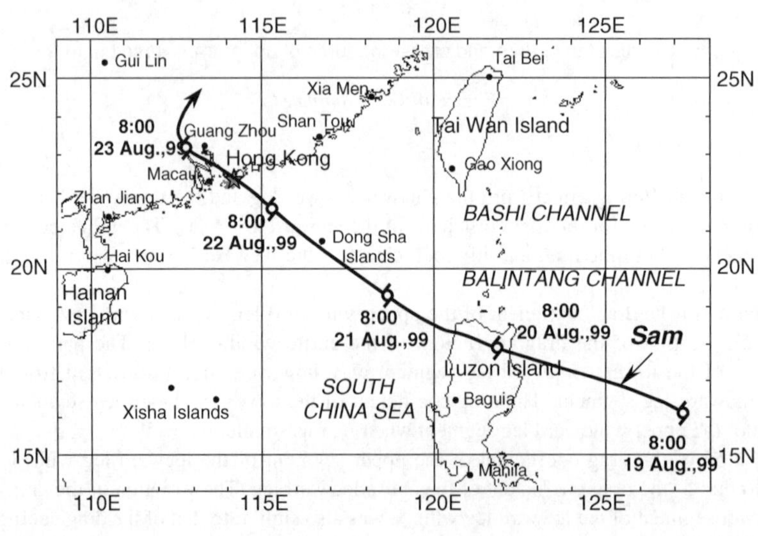

Figure 10.19 Moving tracks of Typhoon Sam (from [34]) (Reproduced with permission from Elsevier).

wind velocity was 2.56 Hz and the cut-off frequency was 1.28 Hz. The sampling frequency of acceleration response was 25.6 Hz and the cut-off frequency was 12.8 Hz.

After a careful examination of all the measured wind velocity time histories, one-hour record of wind velocity between 14:11 to 15:11 Hong Kong Time (HKT) on 22 August, 1999 was selected for the analysis. During this period, incident wind blew to the Tsing Ma Bridge from the direction near to the north. Therefore, the wind data recorded by the anemometers at the south side of the bridge deck were contaminated by the bridge deck itself and were not suitable for the analysis of natural wind structures.

Due to technical reasons, the mechanical anemometers installed at the deck and the top of the towers failed to record the wind azimuth. As a result, wind characteristics of Typhoon Sam surrounding the bridge could be extracted only from the wind speed time histories recorded by the three-component ultrasonic anemometer installed on the north side of the bridge deck at the mid-span. By analyzing the three components of the recorded wind velocity, it was found that the hourly-mean wind speed was about 17.1 m/sec and the mean wind blew from north-northeast.

The global hourly-mean wind yaw angle β_0 and inclination θ_0 were, respectively, $-29.15°$ and $2.25°$. The time histories of fluctuating wind speeds $u(t)$, $v(t)$ and $w(t)$ in the longitudinal (alongwind), lateral and upward directions were also extracted from the measured three components of wind velocity. It was found that the turbulence intensities were about 18.6%, 20.4% and 14.5% for $u(t)$, $v(t)$ and $w(t)$, respectively, and the corresponding integral scales of turbulence were 228 m, 116 m and 84 m, based on Taylor's hypothesis. Spectral analysis was performed to find the one-side normalized auto-spectra of fluctuating wind speeds $u(t)$, $v(t)$ and $w(t)$. The co-spectra and quadrature spectra between every two of the three fluctuating wind speeds were also analyzed. The curve fitting of non-linear least squares was carried out for all the measured spectra [34].

10.7.2 Measured Bridge Acceleration Responses

To be consistent with wind analysis, only the acceleration response data recorded from 14:11 to 15:11 HKT on 22 August 1999 were analyzed [34,35]. These included the lateral, vertical and torsional accelerations at the three deck sections in the main span (ATTJD, ATTID and ATTFD) and one deck section in the Ma Wan side span (ATTBD). They also included the lateral and vertical acceleration responses at the four cable sections (ABTQC, ABTLC, ATTGC and ABTCC) and the longitudinal acceleration response at ATTGC.

Note that when computing the buffeting response of the Tsing Ma Bridge in the frequency domain, a lower bound and an upper bound of frequency should be set. This depends on how many modes of vibration should be included in the computation and what is the valid frequency range for flutter derivatives obtained from wind tunnel tests. In this study, the upper bound of frequency used in the computation was 0.75 Hz, which is higher than the 45th natural frequency (0.7062 Hz) of the bridge. The lower bound of frequency used in the computation was set as 0.025 Hz, because the flutter derivatives measured from the wind tunnel are available only for the reduced velocity lower than 18 (i.e. only for the frequency higher than 0.0232 Hz when the mean wind speed is 17.1 m/sec and the deck width is 41 m). Therefore, to have a reasonable comparison between computed and measured buffeting responses, the measured response time histories were put through a digital bandpass filter with the same lower and upper bounds of frequency used in the computation.

The RMS acceleration responses were then calculated, using the filtered response time histories. The results are plotted in Figure 10.20 for the bridge deck and in Figure 10.21 for the bridge cable. The Hilbert-Huang transform (HHT) method was used to identify the modal damping ratios from the measured acceleration responses [36]. The first modal damping ratio of the bridge was found to be 1.0% in the lateral motion, 2.2% in the vertical motion and 0.44% in the torsion.

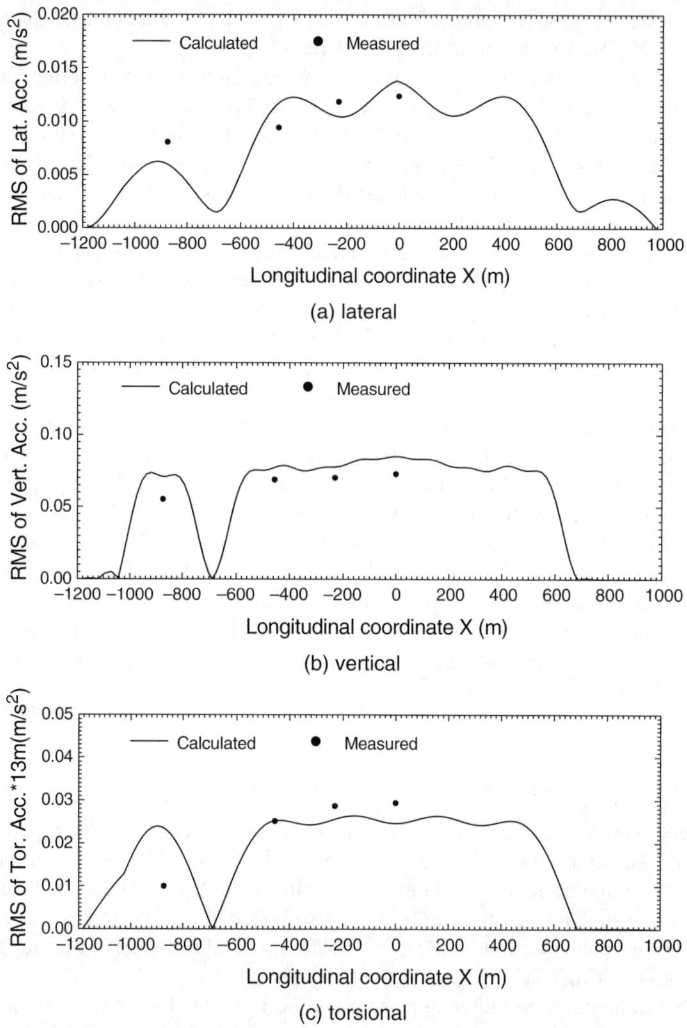

Figure 10.20 Measured and computed RMS responses of deck acceleration (from [34]) (Reproduced with permission from Elsevier).

10.7.3 Input Data to Computer Simulation

To perform a comparison between the measured and computed buffeting responses of the Tsing Ma Bridge, the buffeting response was computed using the analytical method proposed in this chapter. The first 45 modes of vibration of the bridge were included in the buffeting analysis. The modal damping ratios identified from the measured response time histories were used. The measured mean wind speed of 17.1 m/sec at an elevation of 75.314 m with global yaw angle of −29.15° and global inclination of 2.25° were used as input wind parameters. The mean wind speed and direction were considered to be uniform along the bridge deck in the computation.

According to the Hong Kong code of practice on wind effects, a power low with the exponent of 0.33 was adopted to describe the mean wind profile. The auto-spectra and the cross-spectra of three fluctuating wind speed components were taken as the fitted curves to the measured spectra. The friction

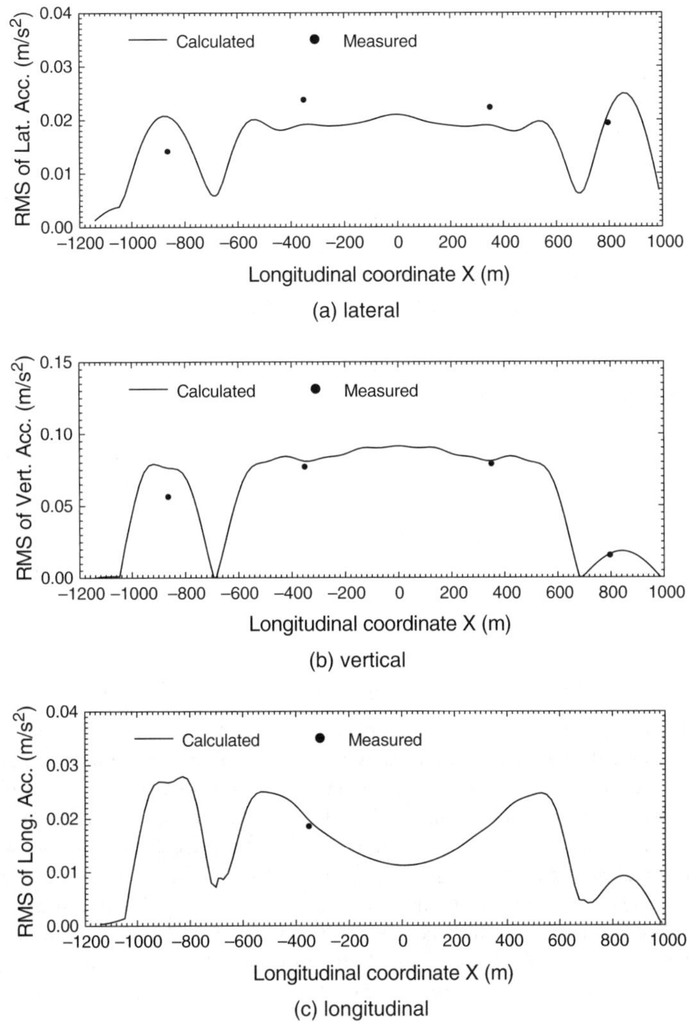

Figure 10.21 Measured and computed RMS responses of cable acceleration (from [34]) (Reproduced with permission from Elsevier).

velocity u_* was estimated as 1.69 m/sec from the measured horizontal shear stress. The exponential decay coefficients for the determination of the root-coherence functions could not be estimated from the field measurements, and were taken as those suggested by Simiu and Scanlan [8]. Phase spectra were also not available, and thus they were taken as zero.

The aerodynamic coefficients of the bridge deck, tower legs and tower transverse beams measured from the wind tunnel tests under yaw winds were used in the buffeting analysis. As for the coefficients of drag and crosswind forces of the main cables under skew winds, the formulae based on the traditional cosine rule were employed [31]. The aerodynamic forces on the bridge hangers (suspenders) were neglected because they were very small.

To include the effects of aeroelastic forces in the buffeting analysis, the eight flutter derivatives of the bridge deck measured under skew winds were used in the computation. The flutter derivatives of P_1^*, P_3^*, P_5^*, H_5^* and A_5^* were not available from the wind tunnel tests and, hence, formulae based on the quasi-steady theory were employed. The remaining five flutter derivatives were considered insignificant to the bridge buffeting response and neglected in the computation.

There were no measurement data available on the aerodynamic admittance functions of the Tsing Ma Bridge, so the empirical formula and other measures had to be used. For the bridge tower components, including tower legs and transverse beams, the aerodynamic admittance functions were set to unity. For the main cables, the formula suggested by Vickery [37] for the aerodynamic admittance function of a circular cylinder was used. For the bridge deck, the aerodynamic admittance functions proposed by Davenport [3] were employed for the nine aerodynamic admittance functions associated with the drag, crosswind force, and yawing moment of the bridge deck. The other nine admittance functions associated with the lift, pitching moment and rolling moment were set to unity.

The three-dimensional dynamic finite element model of the Tsing Ma suspension Bridge has been established and updated by Xu *et al.* [29], using the dynamic properties from the ambient vibration measurement. Three-dimensional Timoshenko beam elements, with rigid arms, were used to model the two bridge towers. The cables and suspenders were modeled by cable elements, accounting for geometric non-linearity due to cable tension. The hybrid steel bridge deck was represented by a single beam with equivalent cross-sectional properties, determined from a finite element analysis using detailed sectional modes. The connections between bridge components and the supports of the bridge were properly modeled. To constitute the damping matrix of the bridge, the Rayleigh damping assumption was used.

10.7.4 Comparison of Buffeting Response in the Frequency Domain

From the computed acceleration response spectra of the bridge deck, the RMS acceleration responses of the bridge deck could be obtained by integration of the spectra in the frequency domain. The frequency range for the integration was from 0.025–0.75 Hz, in order to have a fair comparison. The computed RMS acceleration responses of the bridge deck are plotted in Figure 10.20 for the lateral, vertical, and torsional vibrations, respectively. The RMS acceleration responses measured at the four specified deck sections are also plotted in Figure 10.20 for comparison. The measured RMS acceleration responses were directly obtained from the measured acceleration time histories. It can be seen from Figure 10.20 that, for the main span, the computed RMS acceleration responses of the bridge deck in the lateral, vertical and torsional directions are all close to the measured results; the relative differences are less than 25%. For the Ma Wan side span, the computed RMS acceleration responses in the lateral and vertical directions are also close to the measured ones but, for the torsional vibration, the relative difference between computed and measured RMS response is as high as 139%.

The comparison between the computed and measured RMS acceleration responses of the main cable is shown in Figure 10.21 for the lateral, vertical, and longitudinal vibration, respectively. A good agreement is seen between the computed and measured RMS acceleration responses for the vertical and longitudinal vibration of the cable in the main span and for the lateral and vertical vibration of the free cable in the Tsing Yi side span. The relative differences between the computed and measured RMS responses are less than 8.1%. For the lateral acceleration of the main span cable, the computed RMS acceleration responses also agree with the measured results, with the differences less than 20%. However, for the cable section ABTCC in the Ma Wan side span, the differences between the computed and measured RMS responses are significant for both the lateral and vertical vibrations, with relative differences of 46.3% and 35.1%, respectively.

To assess the effect of skew winds on the buffeting response of the Tsing Ma Bridge, the buffeting responses were computed for a wide combination of wind yaw angle and inclination by using the presented method in the frequency domain [38], and the results were compared with those of the bridge under normal wind. The comparative results, which are consistent in general with those of wind tunnel tests [39,40], show that the variations of buffeting responses are not monotonous with wind yaw angle and inclination, and that the normal wind case may not be the worst case.

10.7.5 Comparison of Buffeting Response in the Time Domain

Based on the concept of equivalent turbulent wind velocities described in Section 10.4, only two sets of equivalent turbulent wind velocities need to be simulated according to the aerodynamic admittance functions used for the Tsing Ma Bridge. One set of equivalent turbulent wind velocities is used for simulating the buffeting lift force, pitching moment and rolling moment acting on the bridge deck, in which the aerodynamic admittance functions are set to unity and the spectra of the equivalent turbulent wind velocities are actually equal to the measured wind spectra from Typhoon Sam for a given position. Another set of equivalent turbulent wind velocities is used for simulating the buffeting drag force, crosswind force and yawing moment on the bridge deck, in which the spectra of the equivalent turbulent wind velocities are the measured wind spectra from Typhoon Sam multiplied by Davenport's aerodynamic admittance functions for a given position. The measured auto-spectra and co-spectra are assumed to be the same for all points along the bridge deck and used in the simulation of buffeting forces. Thus, the two sets of equivalent turbulent wind velocities to be simulated are both multivariate stochastic processes.

The time interval and duration used in the simulation are 0.0625 sec and 3600 sec, respectively. The total number of points along the bridge deck in the simulation n is 119. The upper cutoff frequency ω_{up} is 4π and the frequency increment $\Delta\omega$ is $4\pi/1024$. The two sets of equivalent turbulent wind velocities are generated with the same random phase angles distributed uniformly over the interval $(0, 2\pi)$, so that all the buffeting forces can be regarded to be simulated simultaneously.

The time histories of the simulated three-dimensional turbulent velocity at point ATTJD are shown in Figure 10.22 for the second set of equivalent turbulent wind velocities. To examine the accuracy of the

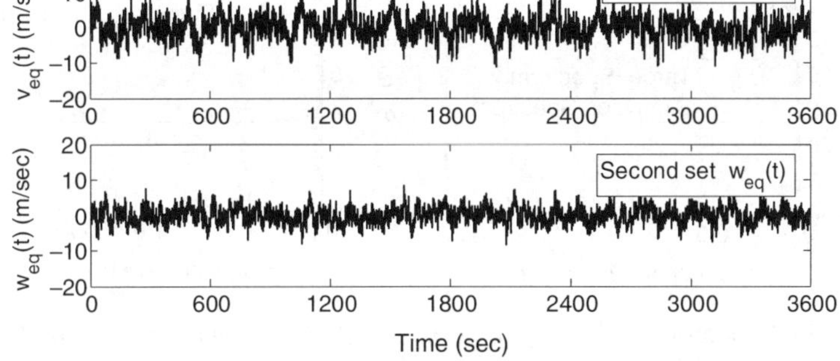

Figure 10.22 Equivalent fluctuating wind velocities at point ATTJD (from [21]) (Reproduced with permission from Techno.Press).

simulated results, the auto-spectra and cross-spectra of the simulated equivalent wind velocities at point ATTJD are computed using the time histories plotted in Figure 10.23 for the second set of equivalent turbulent wind velocities and compared with the target auto-spectra and cross-spectra. It is found that the simulated auto-spectra and cross-spectra match quite well to their respective target auto-spectra and cross-spectra.

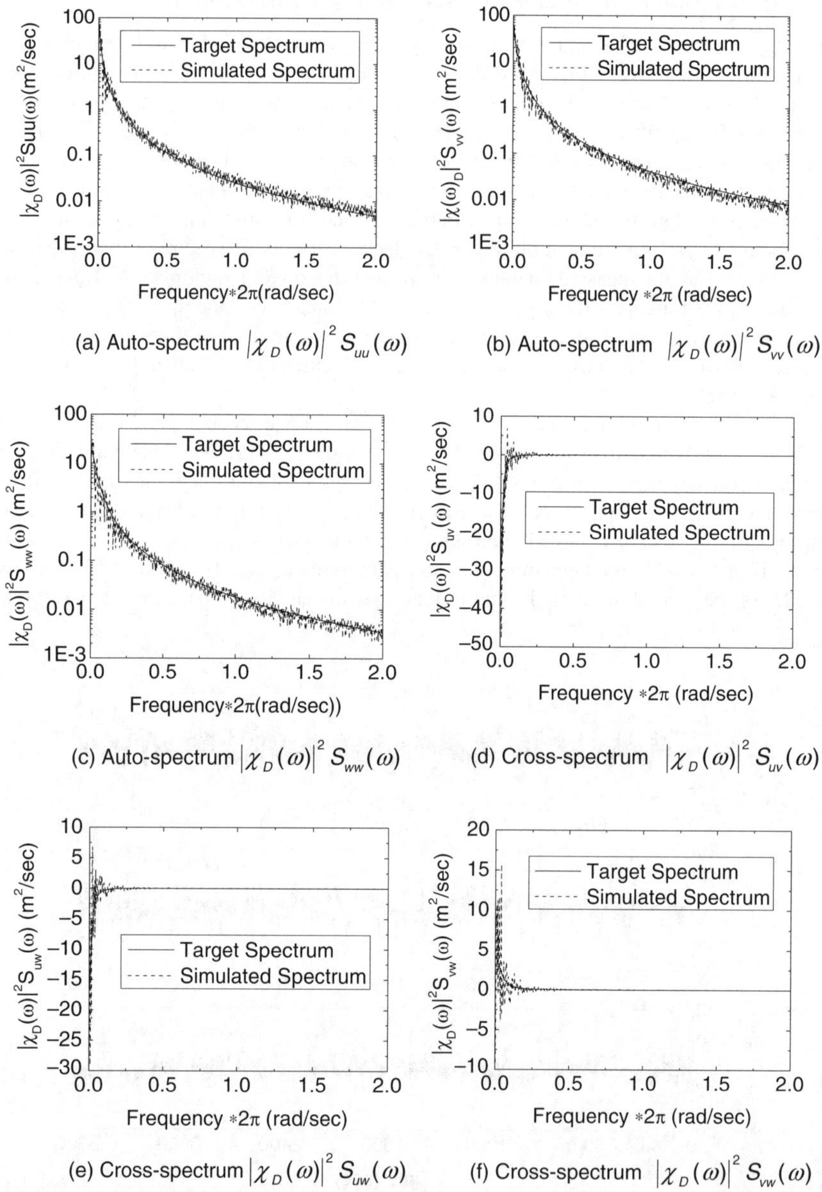

Figure 10.23 Auto/cross spectra of equivalent fluctuating wind velocities versus targeted spectra (from [21]) (Reproduced with permission from Techno.Press).

The time histories of lateral, vertical, and torsional acceleration responses of the Tsing Ma suspension bridge deck at the mid-main span (point ATTJD) under skew winds during Typhoon Sam from 14:11 to 15 : 11 HKT on 22 August 1999 were computed and are shown in Figure 10.24.

The root mean square (RMS) acceleration responses of the bridge deck computed from the response time histories are plotted in Figure 10.25, together with the measured RMS acceleration responses at the four specified deck sections in the lateral, vertical and torsional directions, respectively.

It can be seen that for the main span, the computed RMS acceleration responses of the bridge deck in the lateral, vertical and torsional directions are close to the measured results. The relative differences are less than 22% for the lateral response and less than 10% for the vertical and torsional responses. For the Ma Wan side span, the computed RMS acceleration response in the vertical direction is also close to

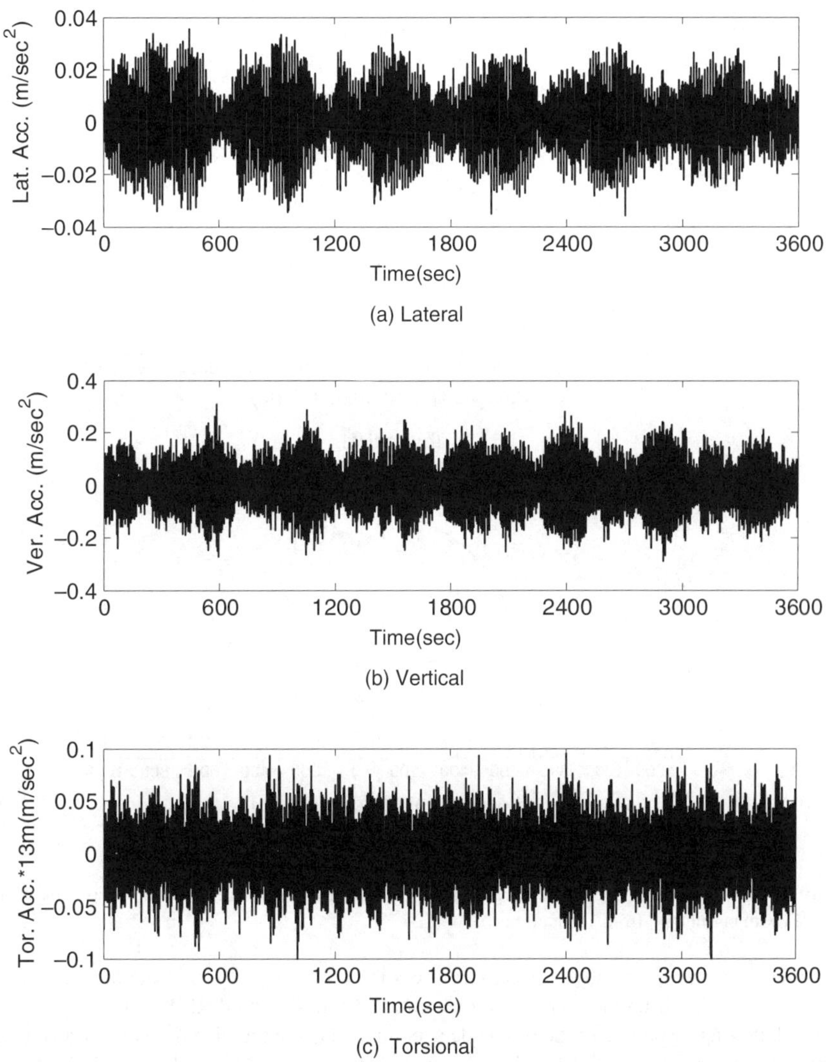

Figure 10.24 Time histories of acceleration responses of bridge deck at point ATTJD (from [21]) (Reproduced with permission from Techno.Press).

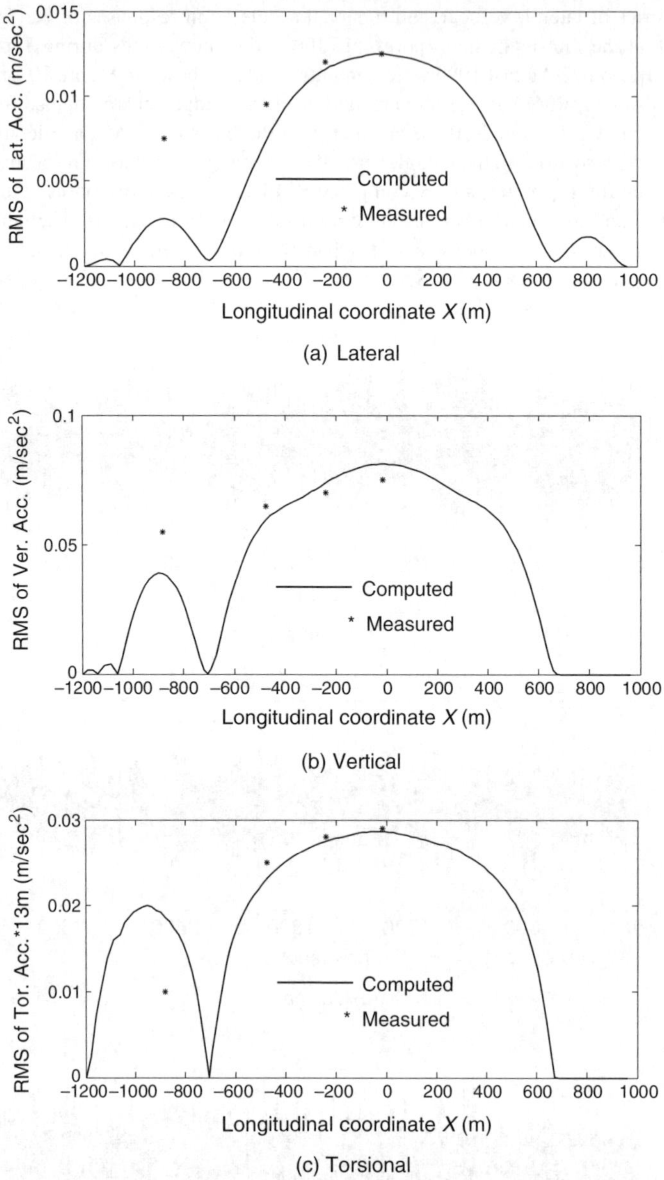

Figure 10.25 Comparison between measured and computed deck RMS acceleration responses (from [21]) (Reproduced with permission from Techno.Press).

the measured one, and the relative difference is less than 40%. However, in the lateral and torsional directions, the relative differences between the computed and measured RMS responses are as high as 130%. The buffeting responses of the bridge deck computed using the time domain approach were also compared with those computed, using the frequency domain approach (see Section 10.7.4). A reasonable agreement was achieved, but there still existed some differences. One of the reasons for the difference is that, in the time domain, the buffeting forces on the main cables and towers of the bridge were not included.

Nevertheless, the case study presented in this chapter is just one single event or demonstration. It is obviously insufficient to verify the proposed frameworks for buffeting analysis of long-span cable-supported bridges under skew winds, because some information regarding the modeling of aerodynamic forces, such as admittance functions, is not available for the Tsing Ma Bridge under skew winds, and the wind characteristics measured during Typhoon Sam are not comprehensive. Therefore, further investigation on these topics, and more case studies with comprehensive field and wind tunnel measurements and computation of buffeting responses, will be needed in the future before solid conclusions can be reached.

10.8 Notations

a	Subscript representing one of the turbulence components, u, v, w
$\mathbf{a}(t)$	$3m$-dimensional vector of turbulence components of entire bridge
$\bar{\mathbf{a}}(t)$	3D vector of turbulence components of a segment
$\bar{\mathbf{A}}^b$	6×3 aerodynamic coefficient matrix of buffeting forces per unit length with respect to $\overline{qph}$-system due to fluctuating winds
A_i^*	Flutter derivatives for self-excited pitching moment on bridge deck due to deck motion
B	Characteristic width of bridge component
$\mathbf{B}$	$\mathbf{Diag}(B, B, B, B^2, B^2, B^2)$
$\tilde{\mathbf{C}}$	6D vector of aerodynamic coefficients at $\tilde{\beta}$ and $\tilde{\theta}$
$\bar{\mathbf{C}}$	6D vector of aerodynamic coefficient at $\bar{\beta}$ and $\bar{\theta}$
$\mathbf{C}^s$	$6N \times 6N$ global structural damping matrix
$\tilde{\mathbf{C}}^s$	$M_\Phi \times M_\Phi$ generalized damping matrix of structure
$\mathbf{C}^{se}$	$6N \times 6N$ global aerodynamic damping matrix
$C_{C_{\bar{q}}}$	Aerodynamic coefficient of crosswind force along axis $\bar{q}$
$\tilde{\mathbf{C}}^{se}$	$M_\Phi \times M_\Phi$ generalized aerodynamic damping matrix of structure
$C_{D_{\bar{p}}}$	Aerodynamic coefficient of drag along axis $\bar{p}$
$C_{L_{\bar{h}}}$	Aerodynamic coefficient of lift along axis $\bar{h}$
$C_{M_{\bar{\alpha}}}$	Aerodynamic coefficient of pitching moment around axis $\bar{q}$
$C_{M_{\bar{\gamma}}}$	Aerodynamic coefficient of rolling moment around axis $\bar{p}$
$C_{M_{\bar{\varphi}}}$	Aerodynamic coefficient of yawing moment around axis $\bar{h}$
$C^b_{\bar{q}}(t)$	Buffeting crosswind force per unit length along axis $\bar{q}$
$C^{se}_{q,i}(t)$	Self-excited crosswind force on the ith oblique element of the bridge deck with respect to qph-system
$C^b_{\bar{q},i}(t)$	Buffeting crosswind force on the ith oblique element of the bridge deck with respect to $\overline{qph}$-system
$C^a_{X_u}$	Turbulence coherence decay coefficient of component a in X_u direction
$C^a_{Y_v}$	Turbulence coherence decay coefficient of component a in Y_v direction
$C^a_{Z_w}$	Turbulence coherence decay coefficient of component a in Z_w direction
$\mathbf{d}(\omega)$	$3m \times 3m$ real diagonal matrix of $\mathbf{L}^*\mathbf{D}\mathbf{L}^\mathrm{T}$ decomposition of $\mathbf{S}_{\mathrm{aa}}(\omega)$
$d_j(\omega)$	jth nonzero diagonal element of matrix $\mathbf{d}(\omega)$
$\mathbf{D}(\omega)$	$M_\Phi \times M_\Phi$ real diagonal matrix of $\mathbf{L}^*\mathbf{D}\mathbf{L}^\mathrm{T}$ decomposition of $\mathbf{S}^b_{\tilde{\mathrm{F}}\tilde{\mathrm{F}}}(\omega)$
$D^b_{\bar{p}}(t)$	Buffeting drag per unit length along axis $\bar{p}$
$D^b_{\bar{p},i}(t)$	Buffeting drag on the ith oblique element of the bridge deck with respect to $\overline{qph}$-system
D^{se}_p	Self-excited drag per unit length along axis p
$D^{se}_{p,i}(t)$	Self-excited drag on the ith oblique element of the bridge deck with respect to qph-system
$D_r(\omega)$	rth nonzero diagonal element of matrix $\mathbf{D}(\omega)$

$\mathbf{F}^b(t)$	6*N*-dimensional vector of buffeting forces in *XYZ*-system
$\overline{\mathbf{F}}_i^b(t)$	12×1 vector of buffeting forces at the nodes of the ith element with respect to *XYZ*-system
$\mathbf{F}_{e,k}^b(t)$	12D vector of nodal buffeting forces of *k*th element in *XYZ*-system
$\mathbf{F}^{se}(t)$	6*N*-dimensional vector of aeroelastic forces in *XYZ*-system
$\overline{\mathbf{F}}_i^{se}(t)$	12×1 vector of self-excited forces at the nodes of the ith element of the bridge deck with respect to XYZ-system
$\tilde{\mathbf{f}}^{ad}(t)$	6D vector of total wind forces per unit length in $\tilde{q}\tilde{p}\tilde{h}$-system
$\overline{\mathbf{f}}^b(t)$	6D vector of buffeting forces per unit length in $\overline{qph}$-system
$\mathbf{f}_{D_{\bar{p}}^b,eq}(t)$	$3n \times 1$ vector of equivalent turbulent wind velocities along the bridge deck in the lateral direction
$\mathbf{f}_{e,k}^b(t)$	12D nodal buffeting force vector of *k*th element in *xyz*-system
$f_{P_{a,i}^{se}}(t-\tau)$	Aerodynamic impulse functions
G_S	Used as a subscript, representing the global structural coordinate *XYZ*-system
G_W	Used as a subscript, representing the global wind coordinate $X_uY_vZ_w$-system
$\tilde{\mathbf{H}}(\omega)$	$M_\Phi \times M_\Phi$ matrix of generalized frequency response function
$\mathbf{H}_{D_{\bar{p}}^b}(\omega_{ml})$	Cholesky decomposition of the matrix $\mathbf{S}_{D_{\bar{p}}^b}^0(\omega)$
H_i^*	Flutter derivative for deck self-excited lift
$I_{\overline{P}_i^{b,u}}(t-\tau)$	Aerodynamic impulse functions of the ith oblique element
$I_{P_{a,i}^{se},k}(t)$	Convolution integrations of the *i*th element
K	Reduced frequency
$\tilde{\mathbf{K}}$	$M_\Phi \times M_\Phi$ matrix of total generalized stiffness of bridge
$\mathbf{K}^s$	$6N \times 6N$ global structural stiffness matrix
$\tilde{\mathbf{K}}^s$	$M_\Phi \times M_\Phi$ generalized stiffness matrix of structure
$\mathbf{K}^{se}$	$6N \times 6N$ global aerodynamic stiffness matrix
$\tilde{\mathbf{K}}^{se}$	$M_\Phi \times M_\Phi$ generalized aerodynamic stiffness matrix of structure
$\mathbf{l}(\omega)$	*j*th column of lower triangular matrix $\mathbf{l}(\omega)$
$\mathbf{l}_j(\omega)$	$3m \times 3m$ lower triangular matrix of $\mathbf{L}^*\mathbf{D}\mathbf{L}^\mathrm{T}$ decomposition of $\mathbf{S}_{\mathrm{aa}}(\omega)$
$\mathbf{L}(\omega)$	$M_\Phi \times M_\Phi$ lower triangular of $\mathbf{L}^*\mathbf{D}\mathbf{L}^\mathrm{T}$ decomposition of $\mathbf{S}_{\overline{\mathrm{F}}\overline{\mathrm{F}}}^b(\omega)$
$L_a^{x_u}$	Length scale of turbulence component *a* in the along wind direction
L_h^{se}	Aeroelastic lift per unit length along axis *h*
$L_{h,i}^{se}(t)$	Self-excited lift on the *i*th oblique element of the bridge deck with respect to *qph*-system
$L_{\bar{h}}^b(t)$	Buffeting lift per unit length along axis $\bar{h}$
$L_{\bar{h},i}^b(t)$	Buffeting lift on the *i*th oblique element of the bridge deck with respect to $\overline{qph}$ -system
$L_{i,k}$	Length of *i*th segment of *k*th element
L_k	Length of *k*th element
$\mathbf{L}_r(\omega)$	*r*th column of lower triangular matrix $\mathbf{L}(\omega)$
L_r	Used as a subscript, representing the local reference coordinate *qph*-system
L_s	Used as a subscript, representing the local structural coordinate *xyz*-system
$L_{\overline{w}}$	Used as a subscript, representing the local mean wind $\overline{qph}$ -system
$L_{\tilde{w}}$	Used as a subscript, representing the local instantaneous wind $\tilde{q}\tilde{p}\tilde{h}$-system
m	Total number of segments in bridge FE model
M	Number of elements of bridge FE model
$\mathbf{M}^s$	$6N \times 6N$ global structural mass matrix
$\tilde{\mathbf{M}}^s$	$M_\Phi \times M_\Phi$ generalized mass matrix of structure
M_Φ	Number of modes used in buffeting analysis
M_α^{se}	Self-excited pitching moment per unit length around axis *q*
$M_{\alpha,i}^{se}$	Self-excited pitching moment on the *i*th oblique element of the bridge deck with respect to *qph*-system
$M_{\bar{\alpha}}^b(t)$	Buffeting pitching moment per unit length around axis $\bar{q}$

$M^{b}_{\overline{\alpha},i}$ Buffeting pitching moment on the ith oblique element of the bridge deck with respect to $\overline{qph}$ -system

$M^{se}_{\gamma,i}(t)$ Self-excited rolling moment on the ith oblique element of the bridge deck with respect to qph-system

$M^{b}_{\overline{\gamma}}(t)$ Buffeting rolling moment per unit length around axis $\overline{p}$

$M^{b}_{\overline{\gamma},i}(t)$ Buffeting rolling moment on the ith oblique element of the bridge deck with respect to $\overline{qph}$ -system

$M^{se}_{\varphi,i}(t)$ Self-excited yawing moment on the ith oblique element of the bridge deck with respect to qph-system

$M^{b}_{\overline{\phi}}(t)$ Buffeting yawing moment per unit length around axis $\overline{h}$

$M^{b}_{\overline{\varphi},i}(t)$ Buffeting yawing moment on the ith oblique element of the bridge deck with respect to $\overline{qph}$-system

$\tilde{\mathbf{N}}_i$ 6×12 matrix of the displacement interpolation functions

$\overline{N}$ Number of frequency increments

n Frequency of turbulence

n_k Total number of segments of kth element

n_{xa} Modified turbulence frequency with turbulence scale effect

$\mathbf{N}_{\delta,k}(\xi)$ 6×12 displacement interpolation matrix of kth element in xyz-system

$\mathbf{P}^b$ $6N \times 3m$ coefficient matrix of buffeting forces

$\overline{\mathbf{P}}^{b}_{i}(t)$ 6×1 vector of buffeting forces per unit length of the ith oblique strip (element) of the bridge deck with respect to $\overline{qph}$-system

P^{*}_{i} Flutter derivatives for self-excited drag on bridge deck

$\overline{P}^{b}_{i}$ Subscript representing $C^{b}_{\overline{q},i}$, $D^{b}_{\overline{p},i}$, $L^{b}_{\overline{h},i}$, $M^{b}_{\overline{\alpha},i}$, $M^{b}_{\overline{\gamma},i}$ and $M^{b}_{\overline{\varphi},i}$

$\mathbf{P}^{b}_{i,k}(t)$ 12×3 coefficient matrix of nodal buffeting forces of kth element in XYZ-system due to wind turbulence on the ith segment

$\overline{\mathbf{P}}^{se}_{i}(t)$ 6×1 vector of self-excited forces per unit length of the ith oblique strip (element) of the bridge deck with respect to $\overline{qph}$-system

$P^{se}_{a,i}$ Subscript representing $D^{se}_{p,i}, D^{se}_{h,i}, D^{se}_{\alpha,i}, L^{se}_{p,i}, L^{se}_{h,i}, L^{se}_{\alpha,i}, M^{se}_{p,i}, M^{se}_{h,i}, M^{se}_{\alpha,i}$

qph Local reference coordinate system of an element

$\overline{qph}$ Local instantaneous wind coordinate system of an element

$\tilde{q}\tilde{p}\tilde{h}$ Local mean wind coordinate system of an element

$R_{a_{1P_1}a_{2P_2}}$ Root-coherence function between the fluctuating wind components a_1 at point P_1 and a_2 at point P_2

$\mathbf{S}_{\mathrm{aa}}(\omega)$ 3 m × 3 m spectral matrix of wind turbulence on the entire bridge

$\mathbf{S}^{0}_{D^{b}_{\overline{p}}}(\omega)$ 3 n × 3 n dimensional cross spectral density matrix of $\mathbf{f}_{D^{b}_{\overline{p}},eq}(t)$

$\mathbf{S}_{D^{b}_{\overline{p}},jk}(\omega)$ 3×3 matrix of the cross spectral density functions between three equivalent turbulent wind velocities at the center point (point j) of the jth element and the center point (point k) of the kth element

$\mathbf{S}^{b}_{\mathrm{FF}}(\omega)$ 6 N × 6 N spectral matrix of nodal buffeting force $\mathbf{F}^{b}(t)$ of the entire bridge

$\mathbf{S}_{\overline{a}_{i,k}\overline{a}_{j,l}}(\omega)$ 3×3 spectral matrix between one of the turbulence components at the ith segment center of the kth element and one of those at the jth segment center of the lth element

$\mathbf{S}^{b}_{\tilde{\mathrm{F}}\tilde{\mathrm{F}}}(\omega)$ $M_{\Phi} \times M_{\Phi}$ spectral matrix of generalized buffeting force $\tilde{\mathbf{F}}^{b}(t)$

$S_{a_{1P_1}a_{2P_2}}(\omega)$ Spectrum between the turbulence component a_1 at point P_1 and a_2 at point P_2

$\mathbf{S}_{a_{i,k}a_{j,l}}$ Cross-spectrum between a turbulence component at the center of the ith segment of the kth element and a turbulence component at the center of the jth segment of the lth element

$\mathbf{S}_{\Delta\Delta}(\omega)$ $M_{\Phi} \times M_{\Phi}$ response spectral matrix of bridge nodal displacements

$\mathbf{S}_{\eta\eta}(\omega)$ $6N \times 6N$ response spectral matrix of generalized displacements

$\mathbf{T}_i^b$	$N \times 12$ matrix with its elements being either zero or unit to locate the vector $\overline{\mathbf{F}}_i^b(t)$ at the proper position in the global vector $\mathbf{F}^b(t)$
$\mathbf{T}_i^{se}$	$N \times 12$ matrix with its elements being either zero or unit to locate the vector $\overline{\mathbf{F}}_i^{se}(t)$ at the proper position in the global vector $\mathbf{F}^{se}(t)$
$\mathbf{T}_{S_1S_2}$	3×3 transformation matrix from the system S_2 to the system S_1 (S_1 and S_2 ($S_1 \neq S_2$) can be any two of G_W, G_S, L_s, L_r, $L_{\overline{w}}$ and $L_{\tilde{w}}$)
$\overline{\mathbf{T}}_{S_1S_2}$	**Diag**($\mathbf{T}_{S_1S_2}$, $\mathbf{T}_{S_1S_2}$)
$\hat{\mathbf{T}}_{S_1S_2}$	**Diag**($\mathbf{T}_{S_1S_2}$, $\mathbf{T}_{S_1S_2}$, $\mathbf{T}_{S_1S_2}$, $\mathbf{T}_{S_1S_2}$)
t_{ij}	The element of the ith row and jth column of the 3×3 matrix $\mathbf{T}_{LrGw}$
$\overline{U}$	Mean wind velocity
$u(t)$	Fluctuating wind speed along mean wind
$u_{\overline{P}_i^b,eq}(t)$	Equivalent fluctuating wind speed along mean wind
V(t)	Instantaneous wind speed
$v(t)$	Lateral fluctuating wind speed in horizontal direction
$v_{\overline{P}_i^b,eq}(t)$	Equivalent fluctuating wind speed along mean wind
$w(t)$	Upward fluctuating wind speed
$w_{\overline{P}_i^b,eq}(t)$	Equivalent fluctuating wind speed along mean wind
XYZ	Global structural coordinate system
$X_uY_vZ_w$	Global wind coordinate system
xyz	Local structural coordinate system of an element
β_0	Global yaw angle of mean wind relative to bridge
$\overline{\beta}$	Local yaw angle of mean wind relative to an element
$\tilde{\beta}$	Local yaw angle of instantaneous wind relative to an element
$\boldsymbol{\Delta}(t)$	$6N$-dimensional vector of bridge nodal displacement in XYZ-system
$\boldsymbol{\Delta}_i$	6D vector of displacement response at ith node
Δt	Time step
$\Delta\omega$	Frequency increment
$\Delta\beta$	Increment of local wind inclination due to fluctuating wind speed
$\Delta\theta$	Increment of local wind yaw due to fluctuating wind speed
$\delta_h(t)$	Dynamic translational displacement of an element section along axis h
$\delta_{p,i}^{ce}(t)$	Transverse displacements along the axis p at the center of the ith element
$\delta_p(t)$	Dynamic translational displacement of an element section along axis p
$\delta_{h,i}^{ce}(t)$	Transverse displacements along the axis h at the center of the ith element
$\delta_\alpha(t)$	Dynamic angular displacement of an element section around axis q
$\delta_{\alpha,i}^{ce}(t)$	Torsional displacement around the axis q at the center of the ith element
Z_r	rth modal damping ratio of bridge in static air
Γ	Gamma function
$\boldsymbol{\eta}(t)$	M_Φ-dimensional vector of generalized displacement coordinates
$\boldsymbol{\eta}_{p,r}(\omega,t)$	rth harmonic generalized pseudo displacement response vector
θ_0	Global mean wind inclination angle
$\overline{\theta}$	Local inclination of mean wind relative to an element
$\tilde{\theta}$	Local inclination of instantaneous wind relative to an element
ξ	Reduced coordinate of element
ρ	Air density
$\boldsymbol{\sigma}_{\boldsymbol{\Delta}\,i}$	6D vector of RMS displacement response at ith node
$\boldsymbol{\varphi}_r$	$6N$-dimensional mode shape vector of rth mode
$\boldsymbol{\Phi}$	$6N \times M_\Phi$ modal matrix

$\chi_{\bar{f}a}$ Aerodynamic admittance functions between the buffeting force $\bar{f}$ and turbulence component a ($\bar{f} = C_{\bar{q}}, D_{\bar{p}}, L_{\bar{h}}, M_{\bar{\alpha}}, M_{\bar{\gamma}}, M_{\bar{\phi}}$; $a = u, v, w$)
ω Circular frequency of structural vibration or wind turbulence
ω_r rth modal circular frequency of bridge in static air
ω_{up} Upper cutoff frequency

References

1. Zhu, L.D. and Xu, Y.L. (2005) Buffeting response of long-span cable-supported bridges under skew winds. Part 1: theory. *Journal of Sound and Vibration*, **281**(3–5), 647–673.
2. Davenport, A.G. (1961) The application of statistical concepts to the wind loading of structures. *Proceedings of the Institution of Civil Engineers*, **19**(4), 449–472.
3. Davenport, A.G. (1962) Buffeting of a suspension bridge by storm winds. *Journal of the Structural Division*, **88**(3), 233–268.
4. Xie, J. and Tanaka, H. (1991) Buffeting analysis of long span bridges to turbulent wind with yaw angle. *Journal of Wind Engineering and Industrial Aerodynamics*, **37**(1), 65–77.
5. Kimura, K. and Tanaka, H. (1992) Bridge buffeting due to wind with yaw angles. *Journal of Wind Engineering and Industrial Aerodynamics*, **42**(1–3), 1309–1320.
6. Scanlan, R.H. (1993) Bridge buffeting by skew winds in erection stages. *Journal of Engineering Mechanics*, **11**(2), 251–269.
7. Roberts, J.B. and Surry, D. (1973) Coherence of grid-generated turbulence. *Journal of Engineering Mechanics Division*, **99**(6), 1227–1245.
8. Simiu, E. and Scanlan, R.H. (1996) *Wind Effects on Structures*, 3rd edn, John Wiley & Sons, Inc., New York.
9. Dyrbye, C. and Hansen, S.O. (1996) *Wind Loads on Structures*, John Wiley & Sons Ltd., Chichester, England, UK.
10. Katsuchi, H., Jones, J.P., and Scanlan, R.H. (1999) Multimode coupled flutter and buffeting analysis of the Akashi-Kaikyo bridge. *Journal of Structural Engineering*, **125**(1), 60–70.
11. Scanlan, R.H. and Gade, R.H. (1977) Motion of suspension bridge spans under gusty wind. *Journal of the Structural Division*, **103** (ST9), 1867–1883.
12. Scanlan, R.H. (1978) The action of flexible bridge under wind, II: buffeting theory. *Journal of Sound and Vibration*, **60**(2), 201–211.
13. Jain, A., Jones, N.P., and Scanlan, R.H. (1996) Coupled buffeting analysis of long-span bridges. *Journal of Structural Engineering*, **122**(7), 716–725.
14. Scanlan, R.H. and Jones, N.P. (1990) Aeroelastic analysis of cable-stayed bridges. *Journal of Structural Engineering*, **11**(2), 279–297.
15. Clough, R.W. and Penzien, J. (1975) *Dynamics of Structures*, McGraw-Hill Inc., New York, USA.
16. Wilson, E.L., Der Kiureghian, A., and Bayo, E.P. (1981) A replacement for the SRSS method in seismic analysis. *Earthquake Engineering and Structural Dynamics*, **9**(2), 187–192.
17. Xu, Y.L., Sun, D.K., Ko, J.M., and Lin, J.H. (1998) Buffeting analysis of long span bridges: a new algorithm. *Computers and Structures*, **68**(4), 303–313.
18. Lin, J.H. (1992) A fast CQC algorithm of PSD matrices for random seismic response. *Computer and Structures*, **44**(3), 683–687.
19. Lin, J.H., Zhang, W.S., and Li, J. (1994) Structural responses to arbitrarily coherent stationary random excitations. *Computer and Structures*, **50**(5), 629–633.
20. Newland, D.E. (1975) *An Introduction to Random Vibrations, Spectral and Wavelet Analysis*, 3rd edn, Longman Singapore Publishers Pte. Ltd., Singapore.
21. Liu, G., Xu, Y.L., and Zhu, L.D. (2004) Time domain buffeting analysis of long suspension bridges under skew winds. *Wind and Structures*, **7**(6), 421–447.
22. Chen, X.Z., Matsumoto, M., and Kareem, A. (2000) Time domain flutter and buffeting response analysis of bridges. *Journal of Engineering Mechanics*, **126**(1), 7–16.
23. Shinozuka, M. and Deodatis, G. (1991) Simulation of stochastic processes by spectral representation. *Apply Mechanics Review*, **44**(4), 191–203.
24. Deodatis, G. (1996) Simulation of ergodic multivariate stochastic process. *Journal of Engineering Mechanics*, **122**(8), 778–787.
25. Bucher, C.G. and Lin, Y.K. (1988) Stochastic stability of bridges considering coupled modes. *Journal of Engineering Mechanics*, **114**(12), 2055–2071.
26. Bathe, K.J. (1982) *Finite Element Procedures in Engineering Analysis*, Prentice-Hall, Englewood Cliffs, N.J.
27. Zhu, L.D., Xu, Y.L., Zhang, F., and Xiang, H.F. (2002) Tsing Ma bridge deck under skew winds—Part I: Aerodynamic coefficients. *Journal of Wind Engineering and Aerodynamics*, **90**(7), 781–805.
28. Zhu, L.D., Xu, Y.L., and Xiang, H.F. (2002) Tsing Ma bridge deck under skew winds—Part II: flutter derivatives. *Journal of Wind Engineering and Aerodynamics*, **90**(7), 807–837.

29. Xu, Y.L., Ko, J.M., and Zhang, W.H. (1997) Vibration studies of Tsing Ma suspension bridge. *Journal of Bridge Engineering*, **2**(4), 149–156.
30. Gu, M., Zhang, R.X., and Xiang, X.F. (2000) Identification of flutter derivatives of bridge decks. *Journal of Wind Engineering and Aerodynamics*, **84**(2), 151–162.
31. Scanlan, R.H. (1999) Estimates of skew wind speeds for bridge flutter. *Journal of Bridge Engineering*, **4**(2), 95–98.
32. Zhu, L.D., Xu, Y.L., Zhang, F., and Xiang, H.F. (2003) Measurement of aerodynamic coefficients of tower components of Tsing Ma Bridge under yaw winds. *Wind and Structures*, **6**(1), 53–70.
33. Hong Kong Observatory (1999) Typhoon Sam (9910): 19–23 August 1999, http://www.info.gov.hk/hko/informtc/sam/report.htm.
34. Xu, Y.L. and Zhu, L.D. (2005) Buffeting response of long-span cable-supported bridges under skew winds. Part 2: case study. *Journal of Sound and Vibration*, **281**(3–5), 675–697.
35. Xu, Y.L., Zhu, L.D., and Xiang, H.F. (2003) Buffeting response of long suspension bridges to skew winds. *Wind and Structures*, **6**(3), 179–196.
36. Chen, J. and Xu, Y.L. (2002) Identification of modal damping ratios of structures with closely spaced modal frequencies: HHT method. *Structural Engineering and Mechanics*, **14**(4), 417–434.
37. Vickery, B.J. (1966) Fluctuating lift and drag on a long cylinder of square cross-section in a smooth and in a turbulent stream. *Journal of Fluid Mechanics*, **25**(3), 481–494.
38. Zhu, L.D. (2002) Buffeting Response of Long Span Cable-supported Bridges under Skew Winds: Field Measurement and Analysis, Ph. D. Thesis, The Hong Kong Polytechnic University, Hong Kong.
39. Tanaka, H. and Davenport, A.G. (1982) Response of taut strip models to turbulent wind. *Journal of Engineering Mechanics Division*, **108** (EM1), 33–49.
40. Diana, G., Falco, M., Bruni, S. *et al.* (1995) Comparison between wind tunnel tests on a full aeroelastic model of the proposed bridge over Stretto di Messina and numerical results. *Journal of Wind Engineering and Industrial Aerodynamics*, **54–55**, 101–113.

11

Multiple Loading-Induced Fatigue Analysis

11.1 Preview

The term "fatigue", as it relates to materials and structural components, means damage and fracturing due to repeated cyclic stresses caused by dynamic loads. A significant feature of fatigue is that the load causing it is not large enough to cause immediate failure. Instead, failure occurs after a certain number of load fluctuations have been experienced, i.e. after the accumulated damage has reached a critical level.

Many long-span cable-supported bridges have been built around the world, and most of them are steel or steel-concrete composite structures. When these bridges are built in wind-prone regions, as discussed in Chapters 4 and 10, they will suffer considerable buffeting-induced vibration, which appears within a wide range of wind speeds and lasts for almost the whole design life of the bridge. As a result, the frequent occurrence of buffeting response of relatively large amplitude may cause fatigue damage to steel structural members and their connections.

Furthermore, as described in Chapter 6, in addition to fluctuating wind loading, long-span cable-supported bridges are also subject to other types of dynamic loads, such as highway loading and/or railway loading, which also affect the fatigue life of the bridge. Therefore, fatigue analysis and fatigue reliability of long-span cable-supported bridges under multiple dynamic loadings will be investigated here. To fulfill this task, the dynamic stress analysis of multi-load cable-supported bridges will first be performed. However, this is not a easy job, since it requires a complex dynamic finite element model of the bridge, including all important bridge components, various dynamic loading models for running trains, running road vehicles and high winds, and interactive models between the bridge and wind, bridge and trains, and bridge and road vehicles.

This chapter first presents a structural health monitoring (SHM)-oriented finite element model (FEM) of a long-span bridge, taking the Tsing Ma Bridge in Hong Kong as an example, in which stresses/strains in important bridge components can be directly computed and some can be directly compared with measured ones. A numerical procedure for buffeting-induced stress analysis based on the established FEM is then presented. Significant improvements of the numerical procedure are that the effects of the spatial distribution of both buffeting forces and self-excited forces on the bridge deck structure are taken into account, and the local structural behaviors linked to strain and stress that can cause local fatigue damage are estimated directly.

Wind Effects on Cable-Supported Bridges, First Edition. You-Lin Xu.

Field measurement data, including wind, acceleration and stress, recorded by the SHM system (as detailed in Chapter 9) during Typhoon York are analyzed and compared with the numerical results to verify, to some extent, the proposed stress analysis procedure. Once the proposed procedure is confirmed, it can be used to predict buffeting stresses and their distributions in the locations where no sensors are installed.

A framework for assessing long-term buffeting-induced fatigue damage to a long-span cable-supported bridge is then presented in this chapter by integrating wind and structure interaction with continuum damage mechanics (CDM)-based fatigue damage assessment method. By taking the Tsing Ma Bridge as an example, the joint probability density function of wind speed and direction(as detailed in Chapter 5) and the numerical procedure for buffeting-induced stress analysis of the bridge are used to identify stress characteristics at hot spots in critical steel members under different wind speeds and directions. The cumulative fatigue damage to the critical steel members at hot spots during the bridge's design life is evaluated using the CDM-based fatigue damage evolution model.

Another framework for dynamic stress analysis of long-span cable-supported bridges under combined wind, railway and highway loadings is then presented. The bridge model, train models and road vehicle models are respectively established using the finite element method, as discussed in Chapter 6. The connections between the bridge and trains, and between the bridge and road vehicles, are respectively considered in terms of wheel-rail and tire-road surface contact conditions. The spatial distributions of both buffeting forces and self-excited forces over the bridge deck surface are considered to facilitate dynamic stress analysis.

The Tsing Ma suspension bridge and the field measurement data recorded by the SHM system are utilized as a case study to examine the proposed framework. The information on the concerned loadings measured by the SHM system is taken as inputs for the computation simulation, and computed local stress responses are compared with the measured ones to verify the stress analysis procedure of the bridge under combined loadings.

Nevertheless, given that a long period is involved in fatigue damage accumulation in long-span bridges, and that the computation time for the dynamic stress responses to the combined action of multiple loading is very demanding, the proposed framework for dynamic stress analysis of long-span bridges under combined loadings will be simplified. Taking the Tsing Ma Bridge as an example again, a computationally efficient engineering approach is proposed for dynamic stress analysis of a long-span suspension bridge under combined railway, highway and wind loadings. The fatigue-critical locations are then determined for key bridge components, and databases of the dynamic stress responses at the critical locations are established; 120 years of time histories of the dynamic stresses induced by individual loading are generated based on the databases.

The corresponding stress time histories due to the combined action of multiple loading are compiled and fatigue analysis is then performed to compute the cumulative fatigue damage over the design life of 120 years. The cumulative fatigue damage induced by individual loading and the damage magnification due to multiple loading are finally investigated. The subsequent fatigue reliability analysis will be presented in Chapter 14.

11.2 SHM-oriented Finite Element Modeling

11.2.1 Background

Most of the existing buffeting analysis methods are based on the aerodynamic strip theory and the quasi-steady linear theory [1–3], and they are actually a combination of numerical, experimental, and analytical approaches. The finite element method is often used to model a bridge deck using 3-D equivalent beam finite elements. Wind tunnel tests of bridge section models provide flutter derivatives and aerodynamic coefficients. Buffeting response of the bridge deck is then determined, either in the frequency domain, based on random vibration theory, or in the time domain through simulation [4–7].

Nevertheless, the modeling of a complex bridge deck using equivalent beam finite elements may oversimplify the problem. Ignorance of spatial distribution of both buffeting forces and self-excited forces across the cross-section of the bridge deck may have a considerable impact on the accuracy of the buffeting-induced stress responses. There is also an increasing trend to install SHM systems in long-span cable-supported bridges to monitor their safety and functionality at both global and stress levels [8,9]. Currently used buffeting analysis methods should be improved to accommodate SHM-oriented stress analysis. In line with this task, this section presents a SHM-oriented finite element model (FEM) for the Tsing Ma suspension bridge, in which stresses/strains in important bridge components can be directly computed and compared with measured ones [10].

11.2.2 Main Features of Tsing Ma Bridge

The Tsing Ma Bridge, stretching from the Tsing Yi Island to the Ma Wan Island, is a suspension bridge with a main span of 1,377 m that carries a dual three-lane highway on the upper level of the bridge deck and two railway tracks and two carriageways on the lower level within the bridge deck, as shown in Figure 11.1 [11].

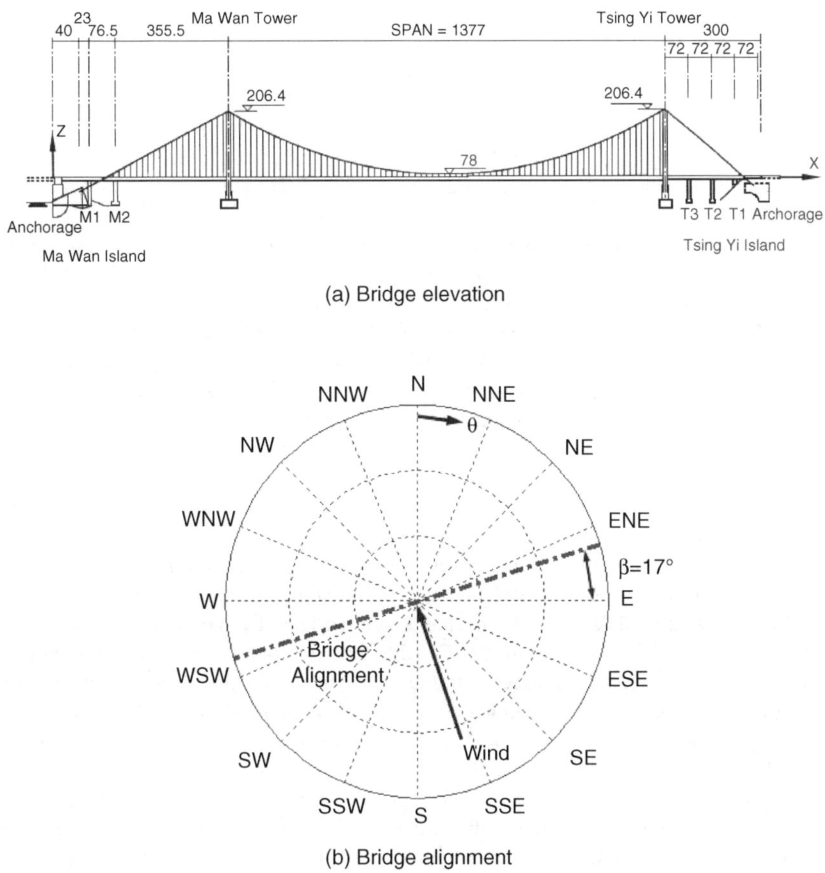

Figure 11.1 Configuration of the Tsing Ma Bridge (unit: m) (from [10]) (Reproduced with permission from Elsevier).

The height of the towers is 206 m, measured from the base level to the tower saddle. Two main cables, 36 m apart, in the north and south are accommodated by four saddles located at the tops of the tower legs in the main span. On the Tsing Yi side, the main cables are extended from the tower saddles to the main anchorage through splay saddles, forming a 300 m Tsing Yi side span. On the Ma Wan side, the main cables extended from the Ma Wan tower are held first by the saddles on Pier M2 at a horizontal distance of 355.5 m from the Ma Wan tower, and then by the main anchorage through splay saddles at the Ma Wan abutment. The bridge deck is a hybrid steel structure continuing between the two main anchorages.

11.2.3 Finite Element Modeling of Tsing Ma Bridge

Modeling work is executed using the commercial software packages MSC/PATRAN as model builder and MSC/NASTRAN as finite element solver. The modeling work is based on the previous model [12] with the following principles:

1. Model geometry should accurately represent actual geometry.
2. One analytical member should represent one real member.
3. Stiffness and mass should be simulated and quantified properly.
4. Boundary and continuity conditions should accurately represent reality.
5. The model should be sufficiently detailed at both global and local levels to facilitate subsequent model updating and buffeting-induced stress analysis.

The deck is a hybrid steel structure consisting of Vierendeel cross-frames supported on two longitudinal trusses, acting compositely with stiffened steel plates that carry the upper and lower highways. The bridge deck at the main span is a suspended deck and the structural configuration is typical for every 18 m segment.

Figure 11.2a illustrates a typical 18 m suspended deck module consisting of mainly longitudinal trusses, cross-frames, highway decks, railway tracks and bracings. The upper and lower chords of the longitudinal trusses are of box section, while the vertical and diagonal members of the longitudinal trusses are of I-section. They are all modeled as 12-DOF beam elements (CBAR), based on the principle of one element for one member. The upper and lower chords of the cross-frames are of T-section dominantly, except for some segments with I-section for the cross-bracing systems.

The inner struts, outer struts, and the upper and lower inclined edge members of the cross-frames all are of I-section. All of the members in the cross-frames are modeled as 12-DOF beam elements (CBAR) with actual section properties, except for the edge members, which are assigned large elastic modulus and significantly small density to reflect the real situation, where the joint is heavily stiffened for the connection with the suspender. All of the members in the cross-bracings are of box section, while all of the members in the sway bracings are of circular hollow section. These members all are modeled as 12-DOF beam elements (CBAR) with actual section properties.

Each railway track is modeled as an equivalent beam modeled by special 14-DOF beam elements (CBEAM), which are similar to the elements (CBAR), but with additional properties such as variable cross-section, shear centre offset from the neutral axis, wrap coefficient and others. The railway tracks are meshed every 4.5 m according to the interval of the adjacent cross-frames. The modulus of elasticity, the density and Poisson's ratio for all members, except for the edge members, are taken as $2.05 \times 10^{11}\ \mathrm{N/m^2}$, $8500\ \mathrm{kg/m^3}$ and 0.3, respectively.

Deck plates and deck troughs comprise orthotropic decks, and the accurate modeling of stiffened deck plates is complicated. To keep the problem manageable, two-dimensional anisotropic quadrilateral plate-bending elements (CQUAD4) are employed to model the stiffened deck plates. The equivalent section properties of the elements are estimated by a static analysis and the material properties of steel are used in the first instance, but it is updated subsequently.

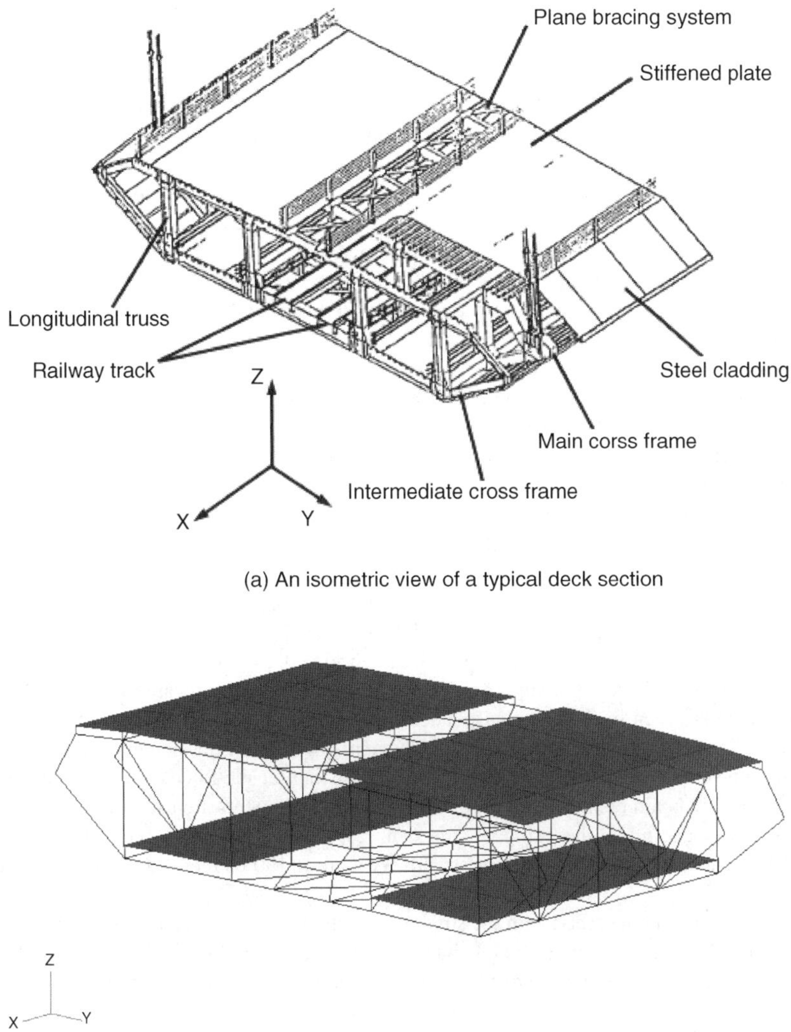

(a) An isometric view of a typical deck section

(b) FE model of an 18-m deck section

Figure 11.2 A typical 18 meters deck section at the main span (from [10]) (Reproduced with permission from Elsevier).

The connections between the deck plates and the chords of the cross-frames and the longitudinal trusses involve the use of MPC (Multi-Point Connection). Proper offsets of neutral axes for the connections between the components are considered to maintain the original configuration. In the modeling of the concerned typical 18 m deck module, a total of 130 nodes, with 172 CBAR elements, 16 CBEAM elements, 24 CQUAD4 elements and 50 MPCs are used. The skeleton view of the 3-D finite element model of the 18 m deck module is shown in Figure 11.2b.

The deck modules at the Ma Wan tower, at the Ma Wan approach span, at the Tsing Yi tower and at the Tsing Yi approach span are constructed using the same principle as the deck module at the main span, while considering the differences in the shape and size of cross-frames, longitudinal trusses and other members.

The Ma Wan tower and the Tsing Yi tower are reinforced concrete structures, and each tower consists of two reinforced concrete legs which are linked by four reinforced concrete portal beams. The bridge towers are represented by multilevel portal frames. The tower legs are modeled using 12-DOF beam elements (CBAR). The tower leg, from its foundation to the deck level, is meshed with an element of length 5 m. At the deck level, the tower leg is meshed according to the positions of the lateral bearings. Although the dimension of the cross-section of the tower leg varies from its bottom to its top, the geometric properties of the beam element are assumed to be constant along its axis, with an average value based on the design drawings.

The four portal beams of either tower are also modeled using CBAR elements, but with different section geometric properties. The deck-level portal beam of each tower is divided at the four particular positions, which correspond to the four vertical bearings between the bridge deck and the tower. The mass density, Poisson's ratio and the modulus of elasticity of reinforced concrete for the towers are estimated to be $2500\,\mathrm{kg/m^3}$, 0.2 and $3.4 \times 10^{10}\,\mathrm{N/m^2}$, respectively.

The two side spans on the Ma Wan side and the Tsing Yi side are supported by two and three piers, respectively. All supporting piers in the side spans are reinforced concrete structures. Piers M1, T2 and T3 are similarly modeled as a portal frame using 12-DOF beam elements (CBAR). Pier M2 is also modeled as a portal frame using 12 CBAR elements, in which the upper portal beam is meshed according to the four vertical bearing positions. The wall panel of pier T1 is represented by an equivalent portal frame with 25 CBAR elements. The mass density, Poisson's ratio and the modulus of elasticity of reinforced concrete for the piers are taken as $2500\,\mathrm{kg/m^3}$, 0.2 and $3.4 \times 10^{10}\,\mathrm{N/m^2}$, respectively.

The cable system is the major system supporting the bridge deck, and it consists of two main cables, 95 pairs of suspender units and 95 pairs of cable bands. CBEAM elements are used to model the main cables. The cable between the adjacent suspender units is modeled by one beam element of a circular cross-section. The DOFs for the rotational displacements of each beam element are released at both ends, because the cable is considered to be capable of resisting tensile force only. To model each cable in the main span, 77 beam elements are used, while 26 and 8 elements are used to model one cable on the Ma Wan side span and on the Tsing Yi side span respectively. Each suspender unit is modeled by one CBEAM element to represent the four strands. A total of 190 elements are used to model all the suspender units.

Since this study is not concerned with the stress distribution around the connection between the main cable and suspenders, the modeling of cable bands is ignored in the global bridge model. The connections between the main cables and suspenders are achieved simply by sharing their common nodes. To model the cable system, the geometry of cable profile should be determined. The geometric modeling of the two parallel main cables follows the profiles of the cables under the design dead load at a design temperature of 23 °C, based on the information from the design drawings.

The horizontal tension in the main cable from pier M2 to the Ma Wan anchorage is 400 013 kN, but it is 405 838 kN in other parts of the main cable. The tension forces in the suspenders on the Ma Wan side span are taken as 2610 kN, but they are 4060 kN in the other suspenders. The mass densities for both cables and suspenders are taken as $8200\,\mathrm{kg/m^3}$. The area of cross-section is $0.759\,\mathrm{m^2}$ for the main cables and $0.018\,\mathrm{m^2}$ for the suspenders. The modulus of elasticity is greatly influenced by the tension in the main cables and suspenders, which is estimated as $1.95 \times 10^{11}\,\mathrm{N/m^2}$ and $1.34 \times 10^{11}\,\mathrm{N/m^2}$ respectively at the design temperature of 23 °C and will be updated subsequently.

By integrating the bridge components with the proper modeling of the connections and boundary conditions, the entire global bridge model is established, as shown in Figure 11.3. This involves 12 898 nodes, 21 946 elements (2906 plate elements and 19 040 beam elements) and 4788 MPCs. Although the geometric features and supports of bridge deck have been modeled in a great detail in the established 3-D finite element model of the Tsing Ma Bridge, modeling discrepancies from the bridge as built still exist.

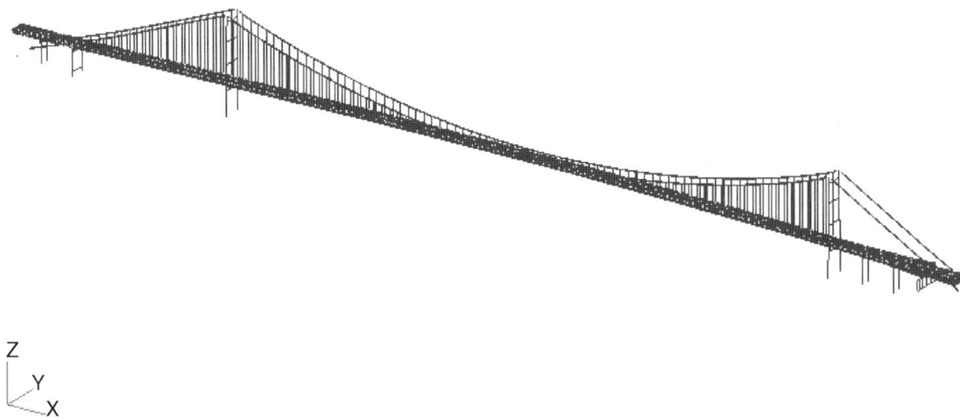

Figure 11.3 3-D Finite element model of the Tsing Ma Bridge (from [10]) (Reproduced with permission from Elsevier).

The modeling discrepancies mainly come from four sources:

1. the simplified modeling of stiffened plates;
2. the uncertainties in pavement mass and others;
3. the uncertainties in the stiffness of bearings;
4. the rigid connection assumption.

Updating the model is therefore necessary. In this regard, the objective function representing the differences between the analytical and measured natural frequencies is minimized. The measured first 18 natural frequencies and mode shapes of the bridge [11] are used in the updating process. The correlation of computed mode shapes with measured ones is evaluated using modal assurance criterion. It turns out that the updated complex FE model could provide comparable and credible structural dynamic modal characteristics. Further details can be found in Liu *et al.* [10].

11.3 Framework for Buffeting-Induced Stress Analysis

11.3.1 Equation of Motion

The governing equation of motion with respect to the static equilibrium position of the bridge, based on the SHM-oriented finite element model in the presence of buffeting forces and self-excited forces, can be expressed as [10]:

$$\mathbf{M}\ddot{\mathbf{X}}(t) + \mathbf{C}\dot{\mathbf{X}}(t) + \mathbf{K}\mathbf{X}(t) = \mathbf{F}^{bf} + \mathbf{F}^{se} \tag{11.1}$$

where:

M, **C** and **K** are the global structural mass, damping and stiffness matrices of the bridge with the dimensions of $6N \times 6N$, in which N is the total number of nodes in the FEM;

$\mathbf{X}(t) = \{\mathbf{X}_1(t), \mathbf{X}_2(t), \ldots, \mathbf{X}_N(t)\}^T$ is the nodal displacement vector of the bridge and $\mathbf{X}_j(t)$ is the *6x1* displacement vector of the *j*th node in the global coordinate system, with the first three being translational displacements and the last three being rational displacements;

each over-dot denotes one order of partial differentiation with respect to time;

$\mathbf{F}^{bf}$ and $\mathbf{F}^{se}$ are the corresponding nodal buffeting force vector and self-excited force vector, respectively.

11.3.2 Buffeting Forces

Most of the buffeting analysis methods are based on the aerodynamic strip theory and the quasi-steady theory. A bridge deck is often modeled using 3-D equivalent beam finite elements. Buffeting forces and self-excited forces act at the centre of elasticity of the deck section. By assuming no interaction between buffeting forces and self-excited forces, the equivalent buffeting forces acting at the centre of elasticity of the deck segment of unit length are expressed as:

$$\mathbf{F}_{ei}^{bf} = \mathbf{A}_{ei}^{bf}\mathbf{q}_i \tag{11.2}$$

in which:

$$\mathbf{F}_{ei}^{bf} = \begin{Bmatrix} D_{ei}^{bf} \\ L_{ei}^{bf} \\ M_{ei}^{bf} \end{Bmatrix}; \quad \mathbf{q}_i = \begin{Bmatrix} u_i \\ w_i \end{Bmatrix}; \mathbf{A}_{ei}^{bf} = \frac{1}{2}\rho U_i^2 B_i L_i \begin{bmatrix} \chi_{D_{bu}}\left(\dfrac{2C_{Di}}{U_i}\right) & \chi_{D_{bw}}\left(\dfrac{C'_{Di}}{U_i}\right) \\ \chi_{L_{bu}}\left(\dfrac{2C_{Li}}{U_i}\right) & \chi_{L_{bw}}\left(\dfrac{C'_{Li}+C_{Di}}{U_i}\right) \\ \chi_{M_{bu}}\left(\dfrac{2C_{Mi}}{U_i}\right)B_i & \chi_{M_{bw}}\left(\dfrac{C'_{Mi}}{U_i}\right)B_i \end{bmatrix} \tag{11.3}$$

where:

D_{ei}^{bf}, M_{ei}^{bf}, L_{ei}^{bf} are the equivalent buffeting drag, moment, and lift, respectively, on the ith node of the equivalent bridge deck beam (see Figure 11.4);
ρ is the air density;
U_i is the mean velocity of the incident wind perpendicular to the ith deck segment;
B_i and L_i are the width and length of the bridge deck segment at the ith node, respectively;

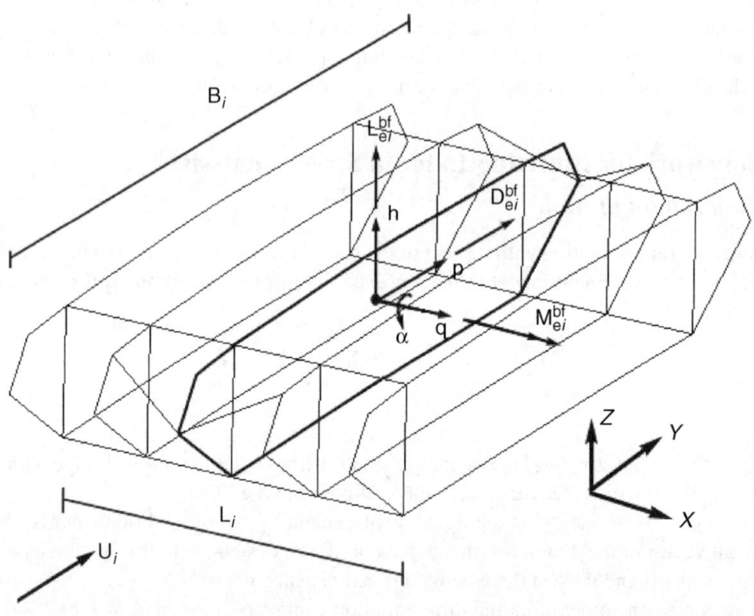

Figure 11.4 Buffeting forces at the center of elasticity of the ith deck section (from [10]) (Reproduced with permission from Elsevier).

C_{Di}, C_{Li}, C_{Mi} are the drag, lift and moment coefficients, respectively, of the *i*th bridge deck segment;
$C'_{Di} = dC_{Di}/d\alpha$; $C'_{Li} = dC_{Li}/d\alpha$; $C'_{Mi} = dC_{Mi}/d\alpha$;
α is the angle of attack of normal incident wind, referring to the horizontal plane of the deck;
$u_i(t)$ and $w_i(t)$ are the horizontal and vertical components of fluctuating wind, respectively; and
$\chi_{D_{bu}}$, $\chi_{D_{bw}}$, $\chi_{M_{bu}}$, $\chi_{M_{bw}}$, $\chi_{L_{bu}}$, and $\chi_{L_{bw}}$ are the aerodynamic transfer functions between fluctuating wind velocities and buffeting forces.

In reality, the equivalent buffeting forces are actually associated with the spatial distribution of wind pressures on the surface of the bridge deck, as shown in Figure 11.5.

Ignorance of spatial distribution and/or aerodynamic transfer function of buffeting forces across the cross-section of the bridge deck may have a considerable impact on the accuracy of buffeting response prediction. Furthermore, local structural behaviour of the bridge deck associated with local stress and strain, which can cause local damage, could not be predicted directly by the currently used approaches based on the equivalent buffeting forces. Therefore, the currently used buffeting analysis methods should be improved to accommodate SHM-oriented stress analysis.

Let us assume that wind pressure distribution on the surface of the *i*th section of the bridge deck can be measured at any given time through wind tunnel tests or field measurements. The buffeting force at the *k*th node, due to wind pressures acting on the *j*th element of the *i*th section of the bridge deck (see Figure 11.5), can be obtained by:

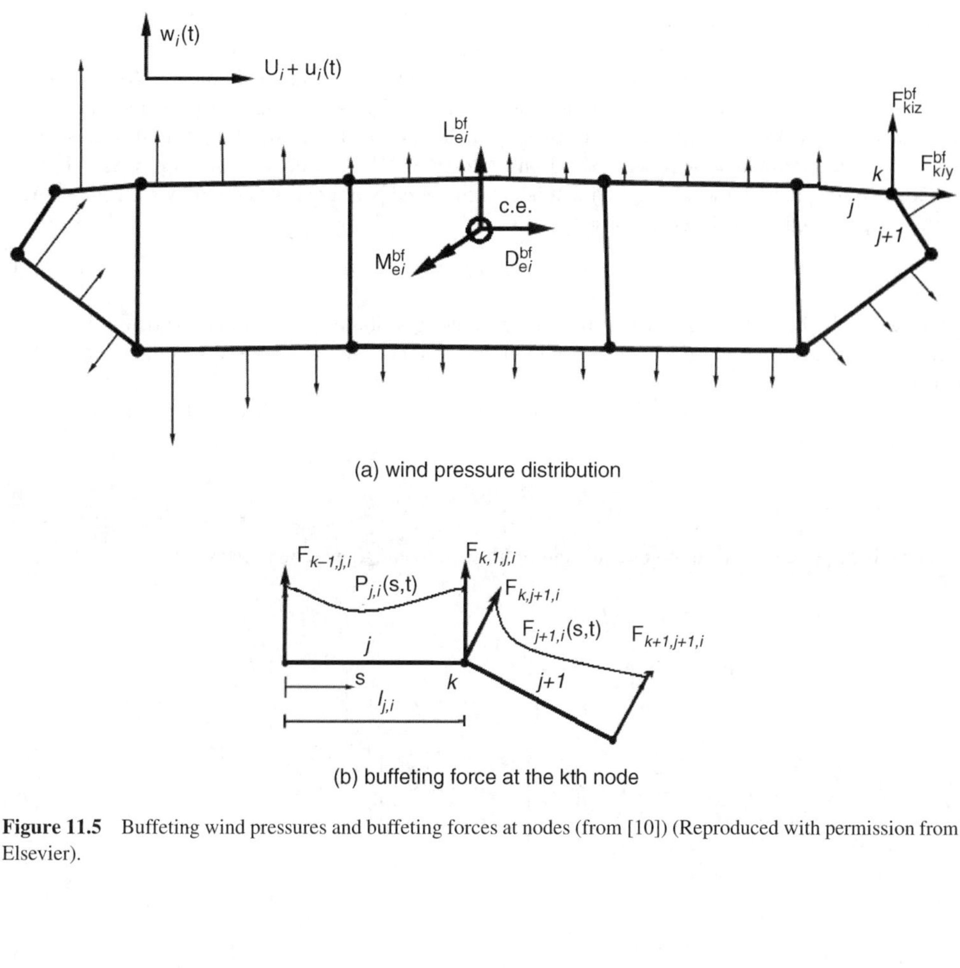

Figure 11.5 Buffeting wind pressures and buffeting forces at nodes (from [10]) (Reproduced with permission from Elsevier).

$$\{F_{k-1}(t), F_k(t)\}_{ji} = \left\{ \int_0^{l_{ji}} (1 - s/l_{ji}) p_{ji}(s,t) \mathrm{d}s, \int_0^{l_{ji}} (s/l_{ji}) p_{ji}(s,t) \mathrm{d}s \right\} \tag{11.4}$$

where:

l_{ji} is the length of jth element in the ith section ($j = 1, 2, \ldots, N_{si}$);
N_{si} is the number of elements used to model the ith deck section;
$p_{ji}(s,t)$ is wind pressure distribution over the jth element;
s is the local element coordinate;
$\{F_{k-1}(t), F_k(t)\}_{ji}$ are the buffeting forces at the (k–1)th and kth nodes, respectively, in the local coordinate of the jth element.

The buffeting force $\{F_k(t)\}_{ji}$ at the kth node of the jth element in the local coordinate can then be converted to $\{F_{ky}, F_{kz}\}_{ji}$ in the qph coordinate system or the xyz global coordinate system (see Figure 11.5). Finally, by adding the buffeting forces at the kth node from all the connecting elements together, the buffeting force vector at the kth node in the ith deck section in the global coordinate system can be obtained as $\mathbf{F}_{ki}^{\mathrm{bf}} = \left\{0, F_{kiy}^{\mathrm{bf}}, F_{kiz}^{\mathrm{bf}}, 0, 0, 0\right\}^{\mathrm{T}}$, in which $k = 1, 2, \ldots, N_{di}$ and N_{di} is the number of nodes used to model the ith deck section. After this procedure is applied to all the nodes in all the sections of the SHM-oriented finite element model of the bridge, the buffeting force vector in Equation 11.1 can be formed.

Nevertheless, it is almost impossible in practice to obtain fluctuating wind pressure distribution for the whole bridge deck as a function of time. Fluctuating wind pressure distribution on a typical bridge deck section may not be available in most cases, including the Tsing Ma Bridge used in this study. Therefore, an approximate approach based on the matrix $\mathbf{A}_{ei}^{bf}$ in Equation 11.3 is proposed here to tackle the problem, with the assumption that fluctuating wind pressure distribution over the jth element in the ith deck section can be decomposed as:

$$p_{ji}(s,t) = p_{ji}^{\mathrm{m}}(s) p_{ji}^{\mathrm{t}}(t) \tag{11.5}$$

where $p_{ji}^{\mathrm{m}}(s)$ is the time-invariant part of wind pressure distribution, which may be determined with reference to the mean wind pressure distribution from either wind tunnel tests or computational fluid dynamics (CFD).

The time-dependent part of wind pressure distribution, $p_{ji}^{\mathrm{t}}(t)$, can be related to the fluctuating wind speeds $u_i(t)$ and $w_i(t)$ as:

$$p_{ji}^{\mathrm{t}}(t) = \begin{bmatrix} B_{uji} & B_{wji} \end{bmatrix} \begin{Bmatrix} u_i(t) \\ w_i(t) \end{Bmatrix} \tag{11.6}$$

The wind pressure distribution over the entire ith section can then be written as:

$$\mathbf{p}_i(s,t) = \begin{bmatrix} p_{1i}^{m}(s) & & 0 \\ & p_{2i}^{m}(s) & & \\ & & \cdots & \\ 0 & & & p_{N_s i}^{m}(s) \end{bmatrix} \mathbf{B}_i \begin{Bmatrix} u_i(t) \\ w_i(t) \end{Bmatrix} \tag{11.7a}$$

$$\mathbf{B}_i = \begin{bmatrix} B_{u1i} & B_{w1i} \\ B_{u2i} & B_{w2i} \\ \cdots & \cdots \\ B_{uN_{si}i} & B_{wN_{si}i} \end{bmatrix} \tag{11.7b}$$

By introducing only three independent unknown variables in Equation 11.7a and comparing with Equation 11.3, the three variables can be determined using the following relationships for either the u_i-component or the w_i-component:

$$\sum_{k=1}^{N_d} F_{kiy}^{bf} = D_{ei}^{bf} \tag{11.8a}$$

$$\sum_{k=1}^{N_d} F_{kiz}^{bf} = L_{ei}^{bf} \tag{11.8b}$$

$$\sum_{k=1}^{N_d} (F_{kiz}^{bf} h_{ki} - F_{kiy}^{bf} p_{ki}) = M_{ei}^{bf} \tag{11.8c}$$

in which p_{ki} and h_{ki} are the lateral and vertical coordinates, respectively, of the kth node with respect to the centre of elasticity of the ith deck section.

As shown in Figure 11.4, the p-axis, h-axis and q-axis originate at the centre of elasticity and are parallel to the directions of y, z and x in the global coordinate system, respectively. The positive angle α is about the q-axis anticlockwise. Note from Equation 11.8 that the resultant forces of all the nodal buffeting forces in the ith section are actually equal to the equivalent buffeting forces acting at the centre of elasticity at any given time. This approach therefore ensures that the global bridge responses computed by this approach remain the same as those predicted by the currently used methods. After the wind pressure distributions over all the sections of the bridge deck are determined by the proposed approach, the buffeting force vector in Equation 11.1 can be formed.

11.3.3 Self-Excited Forces

The self-excited forces on a bridge deck are due to the interaction between wind and bridge motion. In the time domain, they are often expressed in terms of convolution integrals with impulse response functions [5]. The self-excited forces acting at the centre of elasticity of the ith deck section can be expressed in the matrix form [13].

$$\mathbf{F}_{ei}^{se} = \mathbf{E}_{ei}\mathbf{X}_{ei}(t) + \mathbf{G}_{ei}\dot{\mathbf{X}}_{ei}(t) + \mathbf{H}_{ei}\ddot{\mathbf{X}}_{ei}(t) + \hat{\mathbf{F}}_{ei}^{se} \tag{11.9}$$

where:

$\mathbf{F}_{ei}^{se} = \{0, D_{ei}^{se}, L_{ei}^{se}, M_{ei}^{se}, 0, 0\}^T$ is the equivalent self-excited force acting at the centre of elasticity of the ith deck section (see Figure 11.6);

$\mathbf{X}_{ei}(t)$, $\dot{\mathbf{X}}_{ei}(t)$, $\ddot{\mathbf{X}}_{ei}(t)$ are the displacement, velocity, acceleration vectors, respectively, with respect to the centre of elasticity of the ith section in the qph coordinate system;

$\hat{\mathbf{F}}_{ei}^{se}$ is the part of the self-excited forces reflecting aerodynamic phase lag;

$\mathbf{E}_{ei}$, $\mathbf{G}_{ei}$ and $\mathbf{H}_{ei}$ are the aeroelastic stiffness, aeroelastic damping and aeroelastic mass matrices, respectively, of the ith deck section with respect to the centre of elasticity. The details of these matrices can be found in the literature [13].

Similar to buffeting forces, actual information on the spatial distribution of self-excited forces over the surface of the ith deck section is not presently available for the purpose of research here. Consequently, to simulate the spatial distribution, the self-excited forces expressed by Equation 11.9 with respect to the centre of elasticity of the ith deck section are distributed to the nodal lines of the ith section of the health monitoring-oriented finite element model of the bridge deck. The distributions are

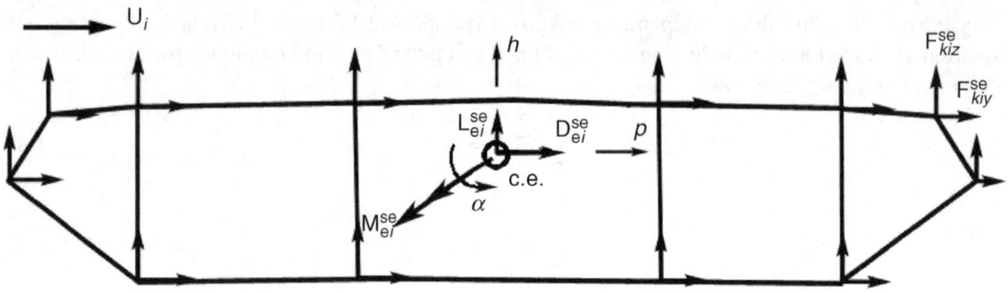

Figure 11.6 Self-excited forces at the center of elasticity and at the nodes in the *i*th deck section (from [10]) (Reproduced with permission from Elsevier).

based on the rigid body motion relationships between the motions at the nodal lines and those at the centre of elasticity of the *i*th deck section [14]. Based on the finite element model of the given *i*th deck section, the position of the centroid of the *i*th deck section can be determined in terms of the geometry of the section. The displacement relationship between the nodal lines and the centre of elasticity of the *i*th section can then be given as follows:

$$\mathbf{X}_{ei} = \mathbf{N}_i^{se}\mathbf{X}_i \tag{11.10}$$

where:

$\mathbf{X}_i = \{\mathbf{X}_{1i}, \mathbf{X}_{2i}, \cdots, \mathbf{X}_{N_{di}}\}^{\mathrm{T}}$ is the displacement vector of all the nodes in the *i*th deck section in the *qph* coordinate system;

$\mathbf{N}_i^{se}$ is the displacement transformation matrix, which can be expressed as:

$$\mathbf{N}_i^{se} = \begin{bmatrix} \mathbf{0} \\ \mathbf{N}_{Di}^{se} \\ \mathbf{N}_{Li}^{se} \\ \mathbf{N}_{Mi}^{se} \\ \mathbf{0} \\ \mathbf{0} \end{bmatrix} \tag{11.11a}$$

$$\mathbf{N}_{Di}^{se} = [0,\ a_{1i},\ h_{ci}b_{1i},\ 0,\ 0,\ 0, \cdots, 0,\ a_{N_{d}i},\ h_{ci}b_{N_{d}i},\ 0,\ 0,\ 0] \tag{11.11b}$$

$$\mathbf{N}_{Li}^{se} = [0,\ 0,\ a_{1i} - p_{ci}b_{1i},\ 0,\ 0,\ 0, \cdots, 0,\ 0,\ a_{N_{d}i} - p_{ci}b_{N_{d}i},\ 0,\ 0,\ 0] \tag{11.11c}$$

$$\mathbf{N}_{Mi}^{se} = [0,\ 0,\ b_{1i},\ 0,\ 0,\ 0, \cdots, 0,\ 0,\ b_{N_{d}i},\ 0,\ 0,\ 0] \tag{11.11d}$$

where:

h_{ci} and p_{ci} are the vertical and lateral coordinates, respectively, of the centroid of the *i*th deck section with respect to the *qph* coordinate system;

$a_{ki} = \dfrac{\sum_j l_{jki}}{2L_{ci}}$, in which $j = 1, 2, \ldots, N_{si}$ and $k = 1, 2, \ldots, N_{di}$, and L_{ci} = the summation of the lengths of all the elements in the *i*th deck section;

$\sum_j l_{jki}$ = the summation of the lengths of all the elements connected at the *k*th node;

$b_{ki} = \dfrac{1}{N_{di}p_{ki}}$, in which p_{ki} is the lateral coordinate of the *k*th node in the *qph* coordinate system.

The form and the size of the displacement transformation matrix depend on the geometry and discretization of the deck cross-section.

The substitution of Equation 11.10 into Equation 11.9 yields the self-excited forces expressed in terms of the nodal displacements vector for the ith deck section:

$$\mathbf{F}_{ei}^{se} = \mathbf{E}_{ei}\mathbf{N}_i^{se}\mathbf{X}_i + \mathbf{G}_{ei}\mathbf{N}_i^{se}\dot{\mathbf{X}}_i + \mathbf{H}_{ei}\mathbf{N}_i^{se}\ddot{\mathbf{X}} + \hat{\mathbf{F}}_{ei}^{se} \tag{11.12}$$

By applying the virtual work principle, the self-excited forces at the centre of elasticity of the ith section can be distributed to all of the nodes in the ith section by:

$$\mathbf{F}_i^{se} = (\mathbf{N}_i^{se})^{\mathrm{T}}\mathbf{F}_{ei}^{se} = \mathbf{E}_i\mathbf{X}_i + \mathbf{G}\dot{\mathbf{X}}_i + \mathbf{H}\ddot{\mathbf{X}}_i + (\mathbf{N}_i^{se})^{\mathrm{T}}\hat{\mathbf{F}}_{ei}^{se} \tag{11.13}$$

in which $\mathbf{F}_i^{se} = \left\{\mathbf{F}_{1i}^{se}, \mathbf{F}_{2i}^{se}, \cdots, \mathbf{F}_{N_d i}^{se}\right\}^{\mathrm{T}}$ is the nodal self-excited force vector and $\mathbf{F}_{ki}^{se} = \left\{0, F_{kiy}^{se}, F_{kiz}^{se}, 0, 0, 0\right\}^{\mathrm{T}}$; $\mathbf{E}_i = (\mathbf{N}_i^{se})^{\mathrm{T}}\mathbf{E}_{ei}\mathbf{N}_i^{se}$, $\mathbf{G}_i = (\mathbf{N}_i^{se})^{\mathrm{T}}\mathbf{G}_{ei}\mathbf{N}_i^{se}$ and $\mathbf{H}_i = (\mathbf{N}_i^{se})^{\mathrm{T}}\mathbf{H}_{ei}\mathbf{N}_i^{se}$ are the aeroelastic stiffness, aeroelastic damping and aeroelastic mass matrices, respectively, of the ith section related to the nodal self-excited forces. By applying the same assembling procedure to all the deck sections, the self-excited force vector in Equation 11.1 can be formed accordingly:

$$\mathbf{F}^{se} = \mathbf{EX} + \mathbf{G}\dot{\mathbf{X}} + \mathbf{H}\ddot{\mathbf{X}} + (\mathbf{N}^{se})^{\mathrm{T}}\hat{\mathbf{F}}^{se} \tag{11.14}$$

where **E**, **G**, and **H** are the aeroelastic stiffness, aeroelastic damping and aeroelastic mass matrices, respectively, of the bridge related to the nodal self-excited forces in the global coordinate system.

11.3.4 Determination of Bridge Responses

The mode superposition technique is adopted in this study to solve Equation 11.1 and compute buffeting-induced bridge responses [10]. The nodal displacement vector in Equation 11.1 can be expressed by the mode shape matrix and the generalized displacement vector:

$$\mathbf{X}(t) = \mathbf{\Phi q}(t) \tag{11.15}$$

where:

$\mathbf{q}(t) = \{q_1(t), q_2(t), \cdots, q_{N_m}(t)\}^{\mathrm{T}}$ is the generalized displacement vector;
N_m is the number of the interested modes involved in the computation;
$\mathbf{\Phi} = [\mathbf{\Phi}_1, \mathbf{\Phi}_2, \cdots, \mathbf{\Phi}_{N_m}]$ is the mode shape matrix with dimensions $6N \times N_m$.

The equation of motion of the wind-excited bridge, i.e. Equation 11.1, can then be rewritten as:

$$\overline{\mathbf{M}}\ddot{\mathbf{q}} + \overline{\mathbf{C}}\dot{\mathbf{q}} + \overline{\mathbf{K}}\mathbf{q} = \mathbf{Q}^{bf} + \mathbf{Q}^{se} \tag{11.16}$$

where:

$\overline{\mathbf{M}}$, $\overline{\mathbf{C}}$ and $\overline{\mathbf{K}}$ are the generalized mass, damping and stiffness matrices, respectively, of the bridge with the dimensions $N_m \times N_m$;
$\mathbf{Q}^{bf} = \mathbf{\Phi}^{\mathrm{T}}\mathbf{F}^{bf}$ and $\mathbf{Q}^{se} = \mathbf{\Phi}^{\mathrm{T}}\mathbf{F}^{se}$ are the generalized buffeting and self-excited force vectors, respectively.

The substitution of Equation 11.14 to Equation 11.16 yields:

$$(\overline{\mathbf{M}} - \mathbf{\Phi}^{\mathrm{T}}\mathbf{H\Phi})\ddot{\mathbf{q}} + (\overline{\mathbf{C}} - \mathbf{\Phi}^{\mathrm{T}}\mathbf{G\Phi})\dot{\mathbf{q}} + (\overline{\mathbf{K}} - \mathbf{\Phi}^{\mathrm{T}}\mathbf{E\Phi})\mathbf{q} = \mathbf{Q}^{bf} + \hat{\mathbf{Q}}^{se} \tag{11.17}$$

in which $\hat{\mathbf{Q}}^{\text{se}} = \mathbf{\Phi}^{\text{T}}(\mathbf{N}^{\text{se}})^{\text{T}}\hat{\mathbf{F}}^{\text{se}}$. It is convenient to normalize the modal matrix to satisfy the following orthogonal condition:

$$\mathbf{\Phi}^{\text{T}}\mathbf{M}\mathbf{\Phi} = \mathbf{I}, \mathbf{\Phi}^{\text{T}}\mathbf{K}\mathbf{\Phi} = \text{diag}\left[\omega_1^2, \omega_2^2, \cdots, \omega_{N_{\text{m}}}^2\right] \tag{11.18}$$

The generalized mass matrix is then the unit matrix $\overline{\mathbf{M}} = \mathbf{I}$. The generalized stiffness matrix is obtained by the natural frequencies only, namely, $\overline{\mathbf{K}} = \text{diag}\left[\omega_1^2, \omega_2^2, \cdots, \omega_{N_{\text{m}}}^2\right]$ where ω_i is the ith circular natural frequency of the bridge structure.

The generalized damping matrix is expressed in the form of modal viscous damping ratios, i.e. $\overline{\mathbf{C}} = \text{diag}\left[2\zeta_1\omega_1, 2\zeta_2\omega_2, \cdots, 2\zeta_{N_{\text{m}}}\omega_{N_{\text{m}}}\right]$ where ζ_i is the ith modal damping ratio of the bridge structure. The generalized displacement vector $\mathbf{q}(t)$ in Equation 11.17 can be solved using the Newmark implicit integral algorithm. The nodal displacement, velocity, and acceleration vectors can then be determined based on Equation 11.15.

Once the nodal displacement vector is determined, the element stress induced by the elastic deformation of the jth element without considering initial strains and stresses can be computed by:

$$\boldsymbol{\sigma}_j = \mathbf{D}_j\mathbf{L}_j\mathbf{N}_j\mathbf{X}_j \tag{11.19}$$

where:

$\boldsymbol{\sigma}_j$ is the jth element stress vector, namely, $\boldsymbol{\sigma}_j = \{\sigma_x, \sigma_y, \sigma_z, \tau_{xy}, \tau_{yz}, \tau_{zx}\}_j^T$;
$\mathbf{X}_j$ is the nodal displacement vector of the jth element;
$\mathbf{N}_j$ is the shape function of the jth element;
$\mathbf{L}_j$ is the differential operator that can transfer the displacement field to the strain field;
$\mathbf{D}_j$ is the elastic matrix which establishes the relationship between the stress and the strain of the jth element.

The modal stresses of all the elements at their end sections in the global coordinate system can be derived from the following relationship:

$$\mathbf{\Gamma} = \mathbf{DLN\Phi} \tag{11.20}$$

where **D**, **L** and **N** are the elastic matrix, the differential operator and the shape function, respectively, of the bridge structure in the global coordinate system. With the introduction of the modal stresses, the stresses of the element can be obtained from the superposition of the modal stresses with the generalized displacements.

11.4 Comparison with Field Measurement Results of Tsing Ma Bridge

The framework for buffeting-induced stress analysis using the SHM-oriented finite element model of the Tsing Ma Bridge established here should be verified before it can be used in practice. The verification of the framework is based on the field measurement data recorded by the WASHMS [8].

It has been more than ten years since the Tsing Ma Bridge was opened to the public in 1997. However, only on September 16, 1999, during Typhoon York (which was the strongest typhoon since 1983 and that of the longest duration on record in Hong Kong), were all vehicles except trains prohibited from running on the bridge. This event provides a distinctive opportunity to examine the proposed procedure for buffeting-induced stress analysis of long suspension bridges. Wind data from the anemometers, bridge deck acceleration responses from the accelerometers, and bridge deck stress responses from the strain gauges recorded by the WASHMS during this event, as shown in Figure 11.7, are therefore analyzed subsequently.

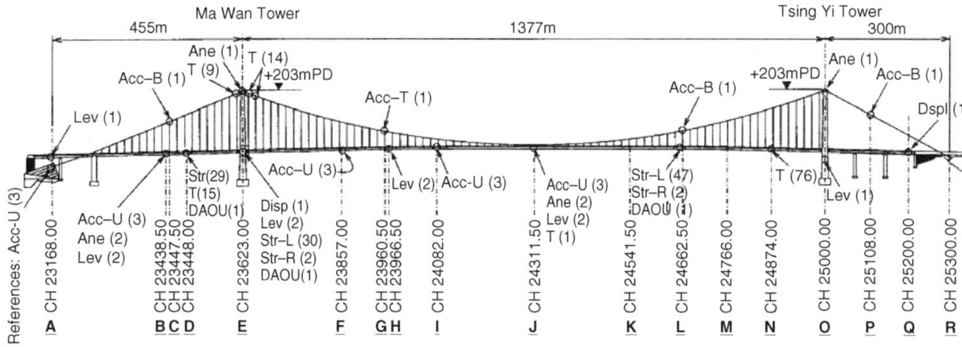

Figure 11.7 The layout of sensory system in the Tsing Ma Bridge (from [10]) (Reproduced with permission from Elsevier).

11.4.1 Wind Characteristics

On September 12, 1999, tropical depression York developed at about 430 km northeast of Manila and intensified into a tropical storm on the next day over the South China Sea [15]. After moving northwest for almost two days, Typhoon York passed Hong Kong on the early morning of September 16. Signal No.10 was forced to hoist for 11 hours – the longest on record in Hong Kong. All vehicles except trains were prohibited from running on the Tsing Ma Bridge for two and a half hours from 3:30 pm to 6:00 pm on September 16.

Wind data of 2.5 hours duration, recorded by the anemometer at the top of the Tsing Yi tower, were analyzed first. These data were evenly divided into segments of duration three minutes each, with a 1.5-minute overlap between neighboring segments. The duration of three minutes was chosen based on the fact that it takes about 1.5–2.0 minutes for a train to pass through the whole bridge completely, and the fact that some anemometers at the deck level were out of order during Typhoon York.

For each data segment, the mean wind direction was determined, by which wind data recorded by the two ultrasonic anemometers (WI-TJN-01 and WI-TJS-01) at the deck level in the main span were selected with the principle that the anemometer selected should directly face the incident wind and the wind data selected were not contaminated due to the bridge deck itself. For all wind data segments selected, the mean wind direction at the deck level was calculated. The data segments with wind direction perpendicular to the bridge alignment (with a tolerance of $\pm 14^\circ$) were then taken as qualified data segments. The qualified data segments during which there were no trains running on the bridge were finally selected.

Displayed in Figure 11.8 are the three-minute time histories of fluctuating wind components in alongwind and upward directions for the case where there was no train passing through the bridge and

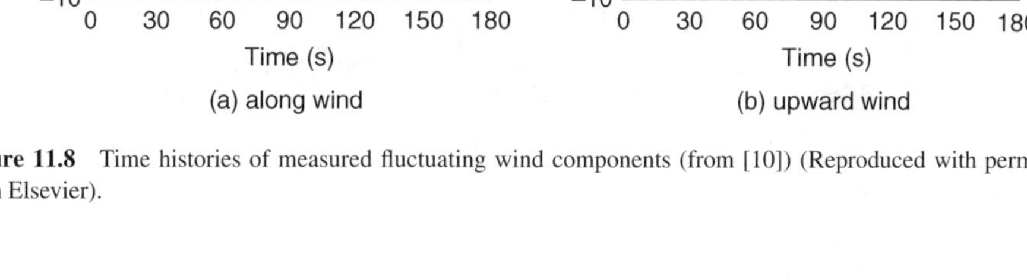

Figure 11.8 Time histories of measured fluctuating wind components (from [10]) (Reproduced with permission from Elsevier).

the bridge was subjected to mainly high cross-winds. The mean wind speed, mean wind direction, mean wind incidence, alongwind turbulence intensity and upward turbulence intensity corresponding to the wind time histories shown in Figure 11.8 are 18.8 m/sec, −1.4°, 3.5°, 10.3% and 7.3%, respectively. Since the mean wind direction refers to the axis perpendicular to the bridge alignment, the mean wind direction of the selected data segment is almost perpendicular to the bridge alignment, so as to facilitate comparison between the measured and computed bridge responses.

To obtain wind auto spectrum, the Fast Fourier Transform (FFT) technique was used, and each three-minute time history was re-sampled at 51.2 Hz and then divided into four sub-segments with a properly overlapped length,so that each segment contained 4096 data points. The frequency resolution in the spectral analysis was 0.0125 Hz. The piecewise smoothing method and the hamming window were adopted in the spectral analysis. The wind auto-spectra were fitted using the non-linear least squares method, with the following objective function:

$$n_f S_j(n_f) = \frac{n_f a}{\left(1 + b n_f^{1/m}\right)^{cm}} \,(\mathrm{m}^2/\mathrm{s}) \tag{11.21}$$

where:

the subscript j of S can be u or w (along wind and upward wind);
c is a constant exponent using 5/3;
n_f is the frequency of wind turbulence in Hz;
a, b and m are the parameters to be fitted.

For the concerned two time histories, the fitted parameters a, b and m are, respectively, 81.11, 7.29 and 1.43 for the alongwind spectrum and 10.17, 60.69, and 0.54 for the upward spectrum. The fitted auto-spectra, together with other wind characteristics, will be used for the numerical simulation of the stochastic wind speed field for the whole bridge deck.

11.4.2 Measured Acceleration Responses of Bridge Deck

The first 80 modes of vibration of the bridge are included in the numerical computation of bridge deck acceleration and stress responses. The highest frequency in the computed acceleration response is about 1.1 Hz. To have a reasonable comparison between the measured and computed buffeting responses, the measured acceleration time histories should go through a digital low pass filter with an upper bound of frequency 1.1 Hz. Time histories of one lateral and two vertical acceleration responses of the bridge deck at each of the four sections (Section B, F, I., J in Figure 11.7), corresponding to the three-minute wind time histories, were processed.

Figure 11.9 shows the positions of two vertical accelerometers, horizontally separated by 26 m, measuring accelerations in the vertical direction, and one lateral accelerometer measuring acceleration

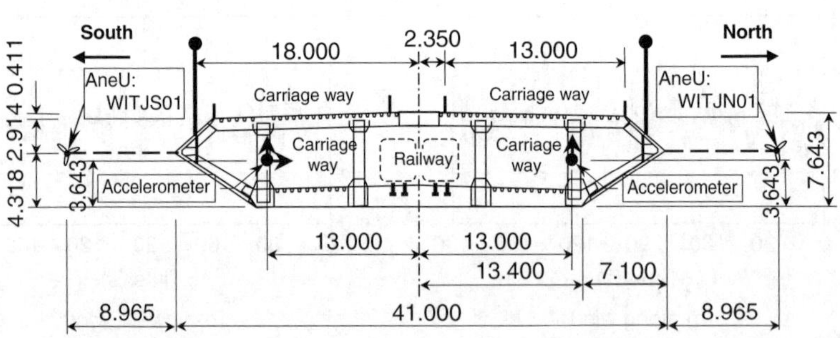

Figure 11.9 Positions of three accelerometers in the deck cross-section (from [10]) (Reproduced with permission from Elsevier).

Table 11.1 Measured and computed acceleration responses of bridge deck (from [10]) (Reproduced with permission from Elsevier)

Lateral acceleration							
Cross-section	Tag no. of accelerometer	RMS (m m/sec^2)			Maximum (m m/sec^2)		
		Measured	Computed	RD*(%)	Measured	Computed	RD (%)
B	AT-TBS-01	17.47	16.56	5.20	49.88	52.38	−5.01
F	AT-TFS-01	9.16	10.43	−13.93	24.49	38.57	−57.49
I	AT-TIS-01	10.03	10.41	−3.80	30.76	35.32	−14.81
J	AT-TJS-01	10.56	11.67	11.68	31.91	41.53	−30.16
Vertical acceleration (RD: Relative Difference)							
Cross-section	Tag no. of accelerometer	RMS (m m/sec^2)			Maximum (m m/sec^2)		
		Measured	Computed	RD (%)	Measured	Computed	RD (%)
B	AT-TBS-01	70.77	51.56	27.14	224.73	194.64	13.39
	AT-TBN-01	69.46	50.12	27.84	193.76	162.90	15.93
F	AT-TFS-01	61.84	62.86	−1.66	205.01	206.63	−0.79
	AT-TFN-01	66.22	64.31	2.88	206.41	254.50	−23.30
I	AT-TIS-01	63.67	71.39	−12.13	182.64	258.55	−41.56
	AT-TIN-01	65.78	67.07	−1.96	211.41	260.95	−23.43
J	AT-TJS-01	69.35	72.33	−4.29	216.76	229.70	−5.97
	AT-TJN-01	69.84	76.31	−9.27	250.94	227.94	9.17

in the lateral direction. The maximum and root mean square (RMS) acceleration responses were calculated and the results are listed in Table 11.1 to compare with the computed ones. In Table 11.1, Section B is located in the Ma Wan side span, whereas Sections F, I and J are located at 1/6, 1/3 and 1/2 of the distance along the main span, respectively.

11.4.3 Measured Stresses of Bridge Deck

For the purpose of comparison with the computed results, ten strain gauges, arranged on the longitudinal trusses of the bridge deck at Section L, were selected. Section L is located at almost 3/4 of the main span form the Ma Wan tower side. The tag number and location detail of each strain gauge can be found in Table 11.2 and Figure 11.10.

The strain gauges with the tag numbers "SS-TLN-xx" and "SS-TLS-xx" are the single linear strain gauge arranged on the north and south longitudinal trusses of Section L, respectively. The tag numbers "SP-TLN-xx" and "SP-TLS-xx" represent a pair of linear strain gauges arranged on the north and south longitudinal trusses of Section L, respectively, with one single output. Note that the selected strains are all located at the middle of the elements to avoid the effect of stress concentration and to facilitate the comparison.

The measured strain was converted to the measured stress by multiplying the modulus of elasticity for steel material $E = 2.05 \times 10^{11}\ \mathrm{N/m^2}$. The stress time histories then went through a digital low pass filter with an upper bound of frequency 1.1 Hz, to be compatible with the highest frequency involved in the computation results. The time histories of the ten stress responses of the bridge deck at Section L corresponding to the three-minute wind time histories were further processed to remove the mean

Table 11.2 Tag number and location details of ten strain gauges used in this study (from [10]) (Reproduced with permission from Elsevier)

Tag no. of strain gauge	Location details	Element number
SP-TLN-01	North truss top chord	31 341
SS-TLN-01	North truss top chord	
SP-TLN-05	North truss bottom chord	32 341
SS-TLN-03	North truss bottom chord	
SP-TLS-12	South truss bottom chord	36 341
SS-TLS-09	South truss bottom chord	
SP-TLN-02	North truss diagonal	32 341
SP-TLN-03	North truss diagonal	32 342
SP-TLS-02	South truss diagonal	36 341
SP-TLS-03	South truss diagonal	36 342

stresses. The maximum and root mean square (RMS) stress responses were then calculated, and the results are listed in Table 11.3 for comparison with the computed ones.

It is necessary to verify the proposed stress analysis procedure through comparison with field measurement data. The wind characteristics obtained from the measured wind data, together with other information, are used for the digital simulation of stochastic wind velocity field around the bridge. The simulated wind velocity field is then used to generate buffeting forces at the nodes as input parameters, together with self-excited forces at the nodes, to the SHM-oriented finite element model of the Tsing Ma Bridge to predict bridge responses The computed acceleration and stress responses of the bridge deck are finally compared with the measured acceleration and stress responses in order to verify the proposed approach to some extent.

11.4.4 Wind Field Simulation

Since fluctuating wind pressure distribution over the deck surface of the Tsing Ma Bridge is not available, the approximate approach proposed in Section 11.3 is used to estimate buffeting forces at the nodes

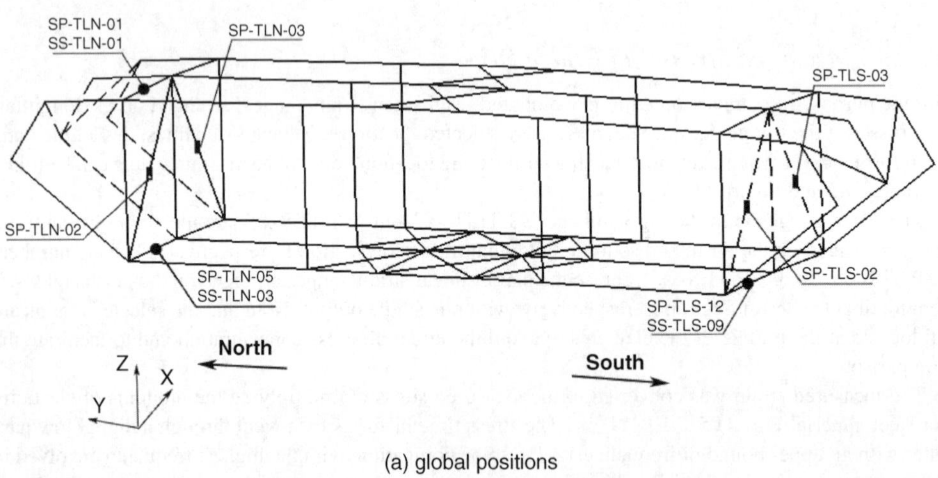

Figure 11.10 Locations of ten strain gauges used in this study (from [10]) (Reproduced with permission from Elsevier).

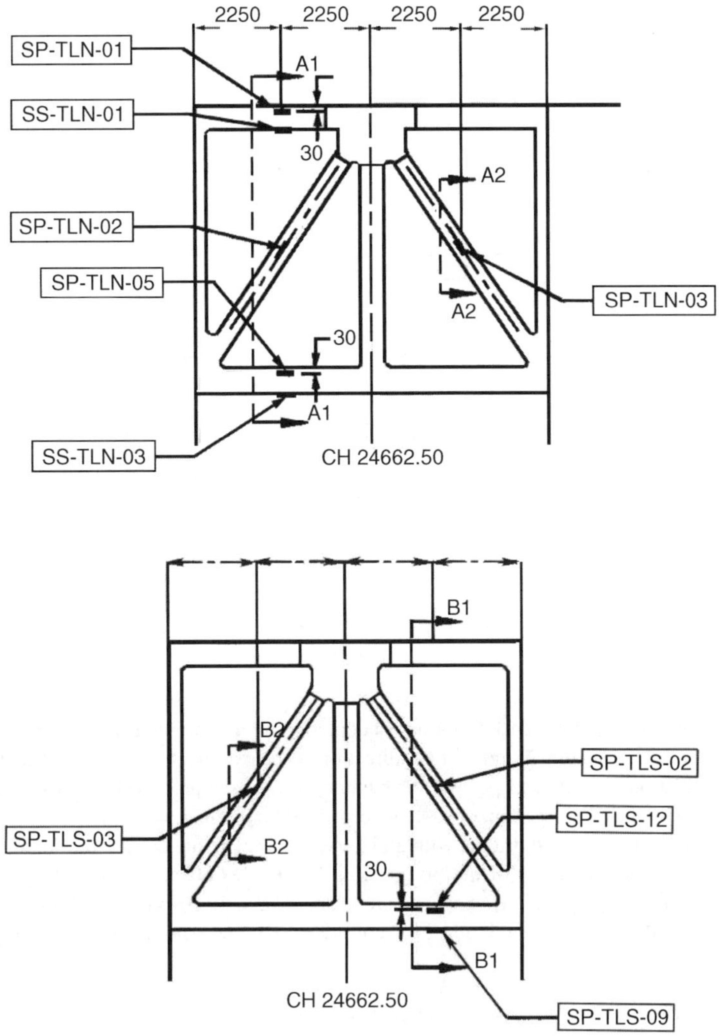

(b) positions in longitudinal trusses

Figure 11.10 *(Continued)*

of the FEM of the bridge. In this regard, a series of time histories of fluctuating wind velocity in horizontal and vertical directions at various points along the bridge deck is essential. However, there were only two anemometers installed along the bridge deck on each side. This situation was further worsened for the bridge subjected to Typhoon York because, during the typhoon, the anemometers installed at deck level were out of order for a certain period. Therefore, the horizontal and vertical wind auto-spectra, the mean wind speed, the mean wind incidence and the turbulent intensities obtained from the measured wind velocities at the mid-main span of the bridge deck are assumed to be constant along the bridge deck.

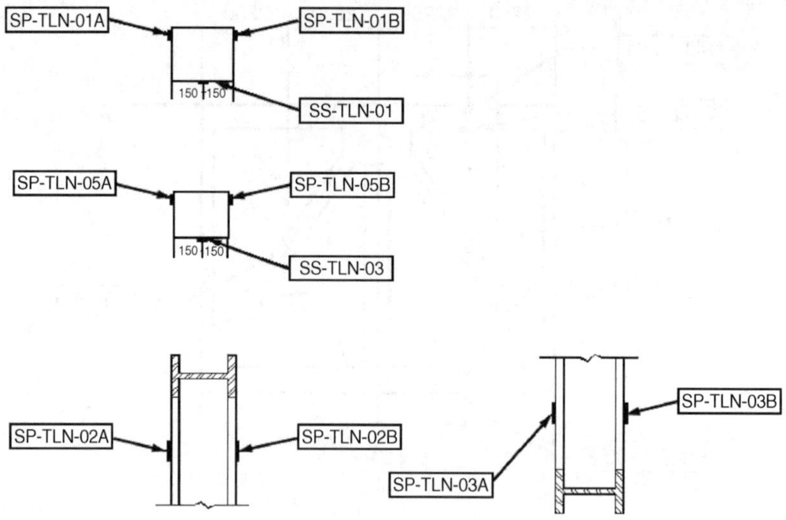

(c) local positions

Figure 11.10 *(Continued)*

The exponential form of coherence function is adopted to reflect turbulent wind correlation along the bridge deck in both the horizontal and vertical directions. The exponential decay coefficient is selected as 16 in the simulation [16]. A fast spectral representation approach proposed by Cao *et al.* [17] is then adopted here for the digital simulation of stochastic wind velocity field.

A total of 120 points along the bridge, with an interval of 18.0 m, are considered in the simulation of wind field. The average elevation of the bridge deck is taken as 60 m. The sampling frequency and duration used in the simulation of wind speeds are, respectively, 50 Hz and ten minutes (the use of ten minutes duration rather than three minutes is to consider the proper simulation of dynamic interaction

Table 11.3 Measured and computed stress responses of bridge deck (from [10]) (Reproduced with permission from Elsevier)

Tag no. of strain gauge	RMS (MPa)			Maximum (MPa)		
	Measured	Computed	RD (%)	Measured	Computed	RD (%)
SP-TLN-01	1.4824	0.9909	33.15	5.1765	3.7662	27.24
SS-TLN-01	1.4246	0.8895	37.56	4.7470	3.3586	29.25
SP-TLN-05	1.7117	1.1068	35.34	4.8009	3.4796	27.52
SS-TLN-03	1.9424	1.4428	25.72	5.2498	4.0755	22.37
SP-TLS-12	1.6625	1.2157	26.88	5.8009	4.7426	18.24
SS-TLS-09	1.8363	1.5844	13.72	6.3699	5.8040	8.88
SP-TLN-02	1.4489	1.4354	0.93	4.0886	4.4693	−9.31
SP-TLN-03	1.4083	1.4270	−1.33	3.9745	4.4464	−11.87
SP-TLS-02	1.3889	1.3107	5.63	4.1517	4.2061	−1.31
SP-TLS-03	1.4190	1.3080	7.82	4.3249	4.0949	5.32

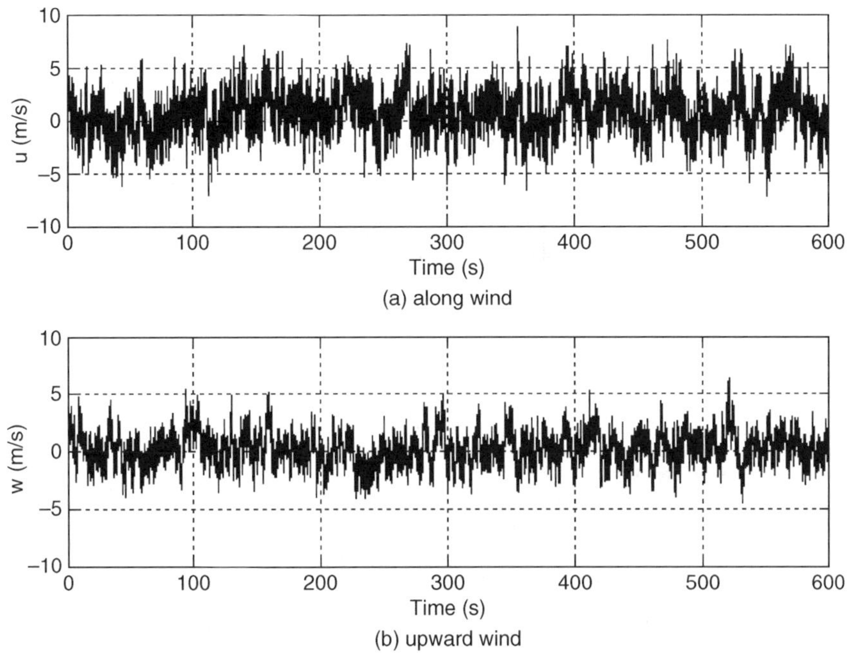

Figure 11.11 Time histories of simulated fluctuating wind components at mid-main span (from [10]) (Reproduced with permission from Elsevier).

between wind and bridge). The corresponding frequency interval and the time interval of wind velocity are 0.0015 Hz and 0.02 sec respectively. Figure 11.11 illustrates the simulated turbulent wind velocity time-histories of ten minutes duration in the alongwind (horizontal) and upward (vertical) directions at the mid-main span of the bridge deck.

11.4.5 Buffeting Forces and Self-Excited Forces

With reference to the mean wind pressure distributions on the typical deck sections of the Tsing Ma Bridge, the time-invariant part of wind pressure distribution over the three typical deck sections is decided and shown in Figure 11.12. The three typical deck sections have 12, 10 and 8 nodes, respectively.

It can be seen from Figure 11.12 that the time-invariant part of wind pressure distribution over each deck section contains three independent variables p_1, p_2 and p_3. These variables can be determined using Equation 11.8 for the u_t-component and the w_t-component separately. In the determination of the three variables, the drag, lift, and moment coefficients of the bridge deck of 12-node section, measured from wind tunnel tests, are 0.104, 0.210 and 0.082, respectively, at the wind angle of attack of 3.5° with respect to the deck width of 41 m [6]. The first derivatives of the drag, lift, and moment coefficients with respect to the same wind angle of attack are −0.172, 1.719 and 0.344, respectively. The aerodynamic coefficients for the other two deck sections are not available, so the aerodynamic coefficients for the 12-node section are applied to the other two deck sections. The aerodynamic transfer functions between fluctuating wind velocities and buffeting forces in Equation 11.3 are assumed to be one. After the wind pressure distributions over all of the sections of the bridge deck are determined, the buffeting force vector in Equation 11.1 can be formed, according to the procedure mentioned in Section 11.3 of this chapter.

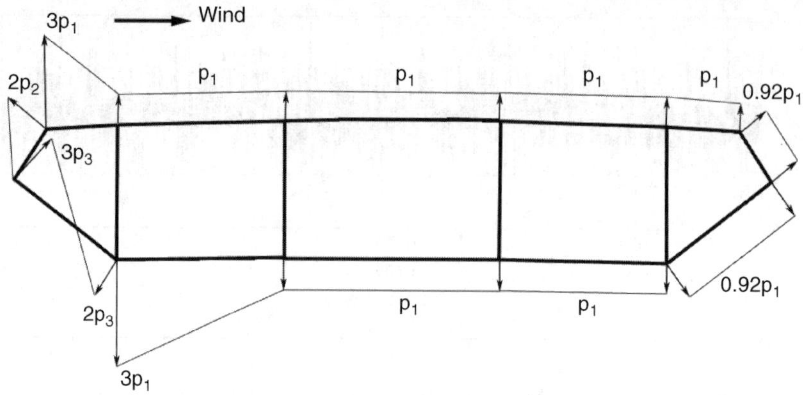

(a) 12-node section

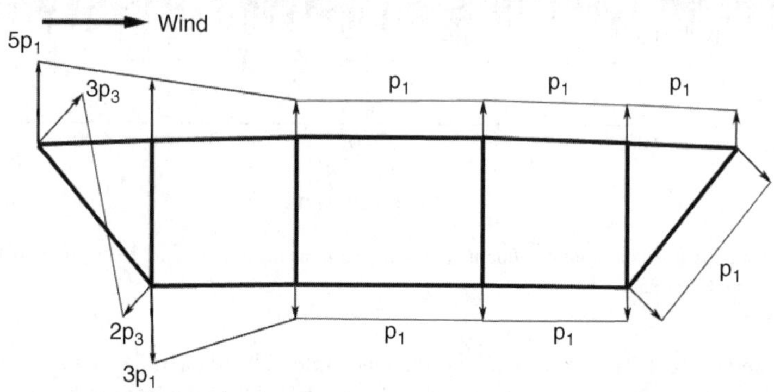

(b) 10-node section

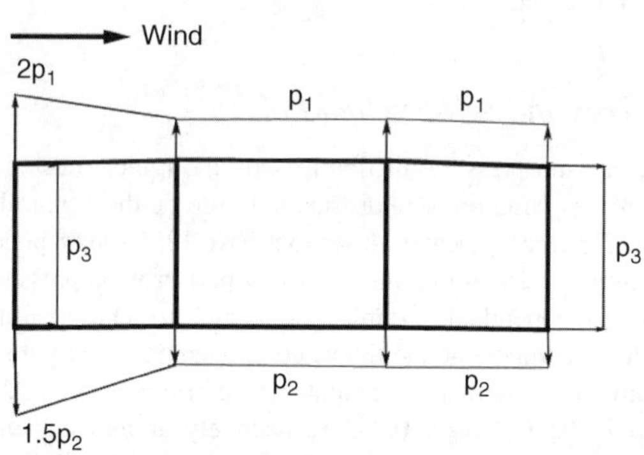

(c) 8-node section

Figure 11.12 Time-invariant part of wind pressure distribution over three typical deck sections (from [10]) (Reproduced with permission from Elsevier).

To determine the self-excited forces at the nodes of the FEM of the Tsing Ma Bridge, the aeroelastic stiffness matrix and the aeroelastic damping matrix of the typical deck section with respect to the centre of elasticity in Equation 11.9 will be determined first. The aerodynamic mass matrix in Equation 11.9 is normally neglected in practice. The number of terms used in $\hat{\mathbf{F}}_{ei}^{se}$ of Equation 11.9 is normally set as 2.

Due to the lack of wind tunnel test results on lateral flutter derivatives, only the vertical and rotational motions of the bridge deck are taken into account in the simulation of self-excited forces. Moreover, the coupled terms are neglected, since they have smaller effects on the self-excited forces. As a result, a total of 12 frequency independent coefficients are determined by using the measured flutter derivatives and the least squares fitting method, and they are used to determine the matrices $\mathbf{E}_{ei}$ and $\mathbf{G}_{ei}$ and the coefficients in the vector $\hat{\mathbf{F}}_{ei}^{se}$ for the ith typical deck section.

Because of the geometrical symmetry with respect to the mid-vertical axis of the bridge deck section, the centre of elasticity and the centroid of the deck cross-section are both in the vertical axis. By taking this geometric feature into account, the displacement transformation matrix $\mathbf{N}_i^{se}$ in Equation 11.10 for the ith typical deck section can be determined easily. The aeroelastic stiffness matrix $\mathbf{E}$, the aeroelastic damping matrix $\mathbf{G}$, the coefficients in the vector $\hat{\mathbf{F}}^{se}$, and the self-excited forces at the nodes of the FEM of the Tsing Ma Bridge in the global coordinate system, can then be determined using Equations 11.13 and 11.14.

11.4.6 Comparison of Bridge Acceleration Responses

The buffeting-induced acceleration and stress responses of the Tsing Ma Bridge are computed using the mode superposition technique, as discussed in Section 11.13. and the first 80 modes of vibration of the bridge are considered in the computation, with the highest frequency being 1.1 Hz. The damping ratios for all the modes of vibration are taken as 0.5%. The generalized displacement vector $\mathbf{q}(t)$ in Equation 11.17 is solved using the Newmark implicit integral algorithm with $\beta = 0.25$. The nodal displacement, velocity, and acceleration vectors are determined based on Equation 11.15. The number of time steps used in the computation is 2^{15}, with a time interval of 0.02 seconds. This is the same as that used in the simulation of wind velocity field.

The acceleration responses are computed for all the nodes of the FEM of the bridge. Some are then converted to the acceleration responses at the locations of accelerometers through the shape function of the element for comparison. Figure 11.13 illustrates the computed and measured acceleration response time histories of three minutes duration for Section J of the bridge deck in the vertical direction and the lateral direction on south side. The computed and measured maximum accelerations and RMS accelerations are listed in Table 11.1, together with the relative differences (RD: the measured one minus the computed one, then divided by the measured one) for the two vertical points and one lateral point, in each of the four sections (Sections B, F, I and J) where the accelerometers were installed.

It can be seen from Figure 11.13 that the computed vertical and lateral acceleration time histories are similar in both pattern and magnitude with the measured ones. Since the time period of the computed time histories does not coincide with that of the measured ones, an exact comparison is impossible. It can be seen from Table 11.1 that, for lateral accelerations, the computed RMS accelerations are closer to the measured ones than the computed maximum accelerations for the four deck sections concerned. The relative differences in the RMS lateral accelerations range from 5.2% to 13.93%. The relative differences in the maximum lateral accelerations range from 5.01% to 57.49%.

Similar observations are made for the vertical acceleration responses. The relative differences in the RMS vertical accelerations range from 1.66% to 27.84%, whereas those in the maximum vertical accelerations range from 0.79% to 41.56%. The relatively large difference in the maximum acceleration responses is due to the relatively short time period of the measured data available. Further comparison using the long period of measured data is desirable in the future.

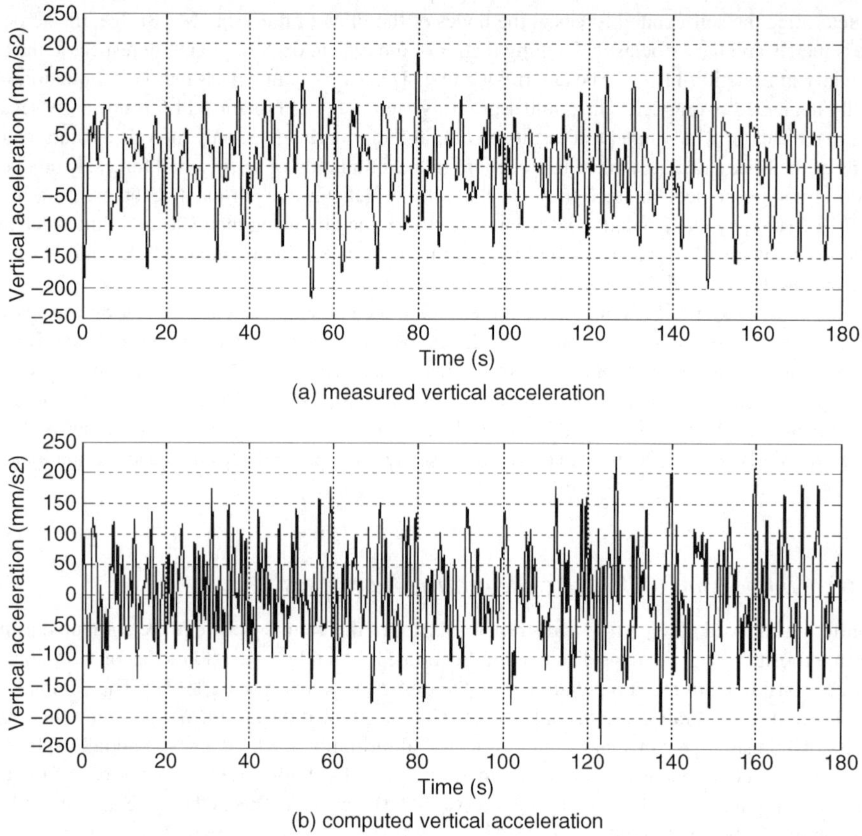

Figure 11.13 Computed and measured acceleration response time histories of three minutes duration (Section J, South) (from [10]) (Reproduced with permission from Elsevier).

11.4.7 Comparison of Bridge Stress Responses

The ten strain gauges concerned in this study are actually arranged in seven elements, as listed in Table 11.2. Four strain gauges (SP-TLN-02, SP-TLN-03, SP-TLS-02 and SP-TLS-03) are stuck on and along the neutral axis of the diagonal elements, as shown in Figure 11.10, and therefore these stresses are caused by the axial forces only. The other six strain gauges are stuck on either the top or bottom surface of the truss chords. These stresses are caused by both bending moments and axial forces.

For comparison with the measured stresses at the locations of nine strain gauges, the modal stresses of the relevant seven elements at their end sections are computed according to Equation 11.20, using the commercial computer program. The modal stresses are then multiplied by the generalized displacement vector to yield the stress time histories at five points of the end section of each element. Of these five points, four are located at each corner of the end section and one point is situated at the centroid of the end section. For the four diagonal elements (see Figure 11.10), the stress time history at the centroid of the end section is taken as the stress time history at the location of the strain gauge. For the truss

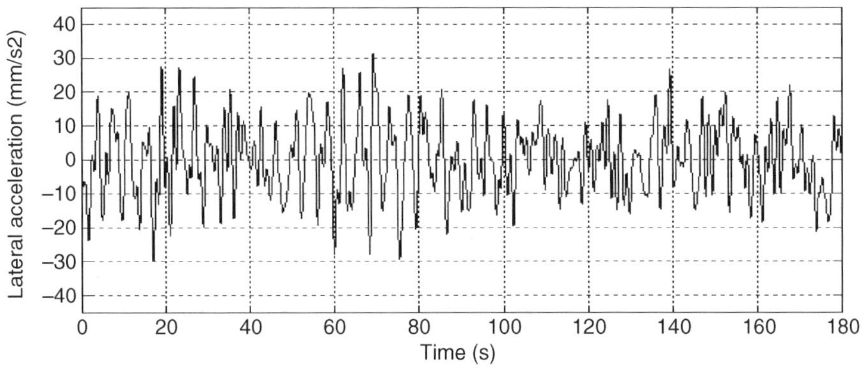

(c) measured lateral acceleration

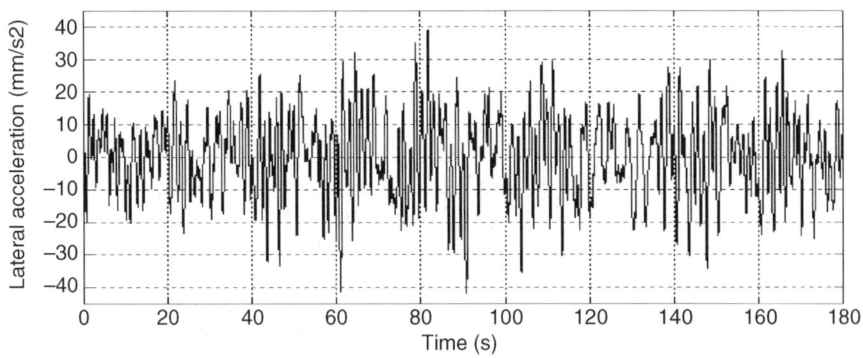

(d) computed lateral acceleration

Figure 11.13 *(Continued)*

chords, the stress time histories at the corners of the two end sections are properly combined to give the stress time history at the location of the strain gauge.

Figure 11.14 illustrates the computed and measured stress response time histories of three minutes duration at the location of strain gauge SP-TLN-01. The computed and measured maximum stresses and RMS stresses are listed in Table 11.3, together with the relative differences (RD) for the ten strain gauges concerned. It can be seen from Figure 11.14 that the computed stress time history is similar to the measured one. Since the time period of the computed time history does not coincide with that of the measured time history, exact comparison is again impossible.

It can be seen from Table 11.3 that the computed RMS and maximum stresses on the four diagonal elements (SP-TLN-02, SP-TLN-03, SP-TLS-02 and SP-TLS-03) are in good agreement with the measured ones. The relative differences in the RMS stress range from 0.93% to 7.82%. The relative differences in the maximum stress range from 1.31% to 11.87%. Nevertheless, the computed RMS and maximum stresses on the truss chords do not compare so well with the measured ones. The relative differences in the RMS stress range from 13.72% to 37.56%, whereas those in the maximum stress range from 8.88% to 29.25%. The relatively large difference in the chord stress comparison is attributed to the complex connections between the chords of the longitudinal trusses and the orthotropic deck plates. The accurate modeling of stiffened deck plates is very complicated and not fulfilled in this study.

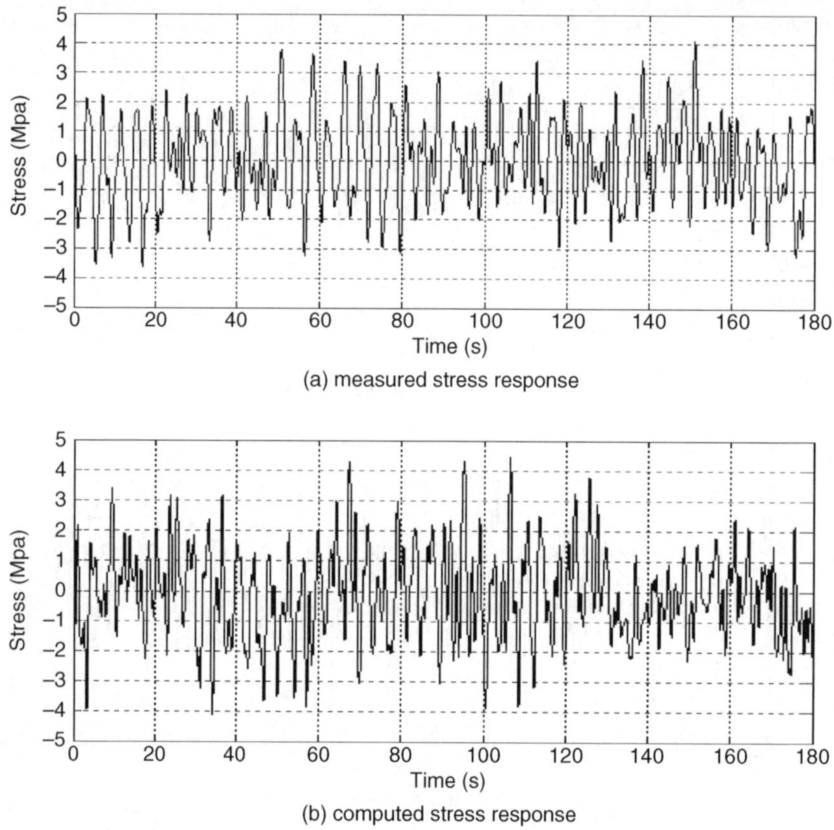

Figure 11.14 Computed and measured stress response time histories of three minutes duration (SP-TLN-01) (from [10]) (Reproduced with permission from Elsevier).

11.5 Buffeting-Induced Fatigue Damage Assessment

11.5.1 Background

Although much work has been conducted on traffic-induced fatigue damage of steel bridges [18], there has only been very limited research on buffeting-induced fatigue damage of long suspension bridges. Virlogeux [19] analyzed fatigue life of the Normandy cable-stayed bridge in France due to buffeting, in which the background component of buffeting response and the effect of wind direction were not taken into consideration. Gu *et al.* [20] considered both the background component and wind direction effects when they estimated buffeting-induced fatigue damage of the steel girders of the Yangpu cable-stayed bridge in Shanghai in the frequency-time domain. They found that the effects of wind direction on the fatigue damage of the Yangpu Bridge are significant, but the predicted fatigue life due to buffeting is much longer than the design life of the bridge.

The buffeting-induced fatigue analyses mentioned above were based on Miner's Law, which is widely used in the fatigue design of steel structures for its simplicity. However, Miner's Law does not associate fatigue damage with its physical mechanism, such as fatigue crack initiation and growth. It does not consider load sequence effects and load cycles below the fatigue limit which can actually propagate micro-cracks if the cracks have already been initiated. On the other hand, fatigue crack initiation and growth at the micro-scale in the vicinity of welds can be well described by continuum damage mechanics (CDM) [21].

Li *et al.* [22] recently applied a CDM-based fatigue model to evaluate the effect of one typhoon on fatigue damage to the steel deck of the Tsing Ma Bridge. They found that the increment of fatigue damage generated by hourly stress spectrum for the typhoon loading could be much greater than that by daily stress spectrum for normal traffic loading. However, their analysis was based on a single particular typhoon event. The long-term effects of buffeting forces on fatigue damage, associated with the joint probability distribution of wind speed and direction, were not considered.

Furthermore, many key issues remain unsolved as to how to take full advantage of the real-time data from the SHM system for effective and reliable health assessment of the bridges. Li *et al.* [22] used the strain data recorded by the WASHMS installed in the Tsing Ma Bridge during Typhoon York to evaluate single typhoon-induced fatigue damage of the bridge at the strain gauge points.

Note that the number of sensors is always limited for a long-span suspension bridge, and the locations of structural defects or degradation may not be at the same positions as the sensors, so it is possible that the worst structural conditions may not be monitored directly by sensors. In this regard, a numerical procedure for buffeting-induced stress analysis of a long-span suspension bridge using the SHM-oriented finite element model are introduced in Sections 11.2 and 11.3 and verified through comparison with the field measurement data in Section 11.4. This numerical procedure will be used to find the most critical locations of stress of the bridge and to identify stress characteristics at hot spots of critical steel members at different levels of wind speed and direction in this section [23]. The accumulative fatigue damage to critical members at their hot spots during the bridge design life will be evaluated by using a CMD-based fatigue damage model, taking into consideration the long-term effects of buffeting forces [23].

11.5.2 Joint Probability Density Function of Wind Speed and Direction

For estimation of wind-induced fatigue damage, it is necessary to have information on the distribution of the complete population of wind speeds at a bridge site [23]. Various probability density functions (PDF) have been proposed to model a complete population of wind speeds [24], but the convenience of the two-parameter Weibull distribution has encouraged its greater use than the other distributions [25]. With the lower bound being zero, the cumulative distribution function (CDF) and PDF of the Weibull form are, respectively:

$$P_u(U) = 1 - \exp\left[-\left(\frac{U}{c}\right)^k\right] \tag{11.22}$$

$$f_u(U) = \frac{k}{c}\left(\frac{U}{c}\right)^{k-1}\exp\left[-\left(\frac{U}{c}\right)^k\right] \tag{11.23}$$

where:

U is the wind speed;
c (>0) is the scale parameter with the same unit as wind speed;
k (>0) is the shape parameter without dimension.

One possible weakness in the Weibull distribution is that it neglects the effect of wind direction [26]. Since wind-induced fatigue damage to a bridge is closely related to wind direction, a joint probability distribution of wind speed and wind direction has to be used for this study. To this end, a practical joint probability distribution function, introduced in Chapter 5, is adopted in this chapter for a complete population of wind speed and wind direction, based on two assumptions:

1. The distribution of the component of wind speed for any given wind direction follows the Weibull distribution.
2. The interdependence of wind distribution in different wind directions can be reflected by the relative frequency of occurrence of wind.

$$P_{u,\theta}(U,\theta) = P_\theta(\theta)\left(1-\exp\left[-\left(\frac{U}{c(\theta)}\right)^{k(\theta)}\right]\right) = \iint f_\theta(\theta) f_{u,\theta}(U, k(\theta), c(\theta))\, du\, d\theta \qquad (11.24)$$

$$f_{u,\theta}(U,k(\theta),c(\theta)) = \frac{k(\theta)}{c(\theta)}\left(\frac{U}{c(\theta)}\right)^{k(\theta)-1}\exp\left[-\left(\frac{U}{c(\theta)}\right)^{k(\theta)}\right] \qquad (11.25)$$

$$P_\theta(\theta) = \int_0^\theta f_\theta(\theta)\, d\theta \qquad (11.26)$$

where:

$0 \le \theta < 2\pi$;
$P_\theta(\theta)$ is the relative frequency of occurrence of wind in wind direction θ.

The occurrence frequency $P_\theta(\theta)$, as well as the distribution parameters $k(\theta)$ and $c(\theta)$, can be estimated using wind data recorded at the bridge site.

In Chapter 5, wind records of hourly mean wind speed and direction within the period between 1 January 2000 and 31 December 2005 from the anemometer installed on the top of the Ma Wan tower of the Tsing Ma Bridge were used to find the joint probability density function of hourly mean wind speed and direction for monsoon winds. The relative frequency of wind direction and the scale and shape parameters of the Weibull function given in Chapter 5 are presented in polar plot in Figures 11.15a to (11.15)c, respectively.

It can be seen that the dominant monsoon direction is the east, and the scale and shape parameters do not vary significantly with wind direction. For the convenience of subsequent calculation, the data given in Table 11.4 and Figure 11.15 regarding the relative frequency of wind direction $f_\theta(\theta)$, the scale parameter $c(\theta)$ and the shape parameter $k(\theta)$, are fitted by the following harmonic functions [27]:

$$f_\theta(\theta) = a^f + \sum_{m=1}^{n_f} b_m^f \cos\left(m\theta - c_m^f\right) \qquad (11.27a)$$

$$c(\theta) = a^c + \sum_{m=1}^{n_c} b_m^c \cos\left(m\theta - c_m^c\right) \qquad (11.27b)$$

$$k(\theta) = a^k + \sum_{m=1}^{n_k} b_m^k \cos\left(m\theta - c_m^k\right) \qquad (11.27c)$$

where:

a, b_m and c_m are the coefficients to be determined, whose superscripts f, c and k denote the relative frequency, the scale and shape parameters respectively;
n_f, n_c and n_k are the order of harmonic functions.

When the above joint probability density function is used to estimate fatigue damage to the bridge, the maximum wind speed U_{max} for a given wind direction should be determined for a designated

Table 11.4 Identified parameters in different wind direction sectors (from [23]) (Reproduced with permission from Elsevier)

Direction	Record no.	$f_\theta(\theta)$	$c(\theta)$	$k(\theta)$	Coefficient of determination
N	258	0.013	8.501	1.394	0.6943
NNE	730	0.037	9.100	1.823	0.9165
NE	550	0.028	6.931	1.940	0.8914
ENE	2712	0.137	7.270	1.913	0.9826
E	4042	0.204	7.222	2.146	0.9840
ESE	2888	0.146	6.764	2.344	0.9929
SE	2252	0.114	6.384	2.215	0.9916
SSE	1597	0.081	6.819	2.242	0.9887
S	1653	0.084	7.266	2.221	0.9753
SSW	880	0.045	7.460	1.949	0.9723
SW	515	0.026	6.465	2.521	0.9987
WSW	253	0.013	5.904	2.295	0.9823
W	439	0.022	6.080	2.462	0.9771
WNW	275	0.014	5.392	2.438	0.9840
NW	160	0.008	5.454	1.992	0.8268
NNW	571	0.029	5.745	1.649	0.9248
Total	19 775	1.000	6.995	2.042	0.9869

fatigue damage evolution period. The probability of the wind speed lower than the maximum wind speed within the jth wind direction range Ω_j, i.e. P_j^{E}, can be determined by the probability distribution function of wind speed:

$$P_j^{\mathrm{E}} = \int_{\Omega_j} \left(\int_0^{U_{\max,j}} f_{u,\theta}(U, k(\theta), c(\theta))dU \right) d\theta = \int_{\Omega_j} \left(1 - \exp\left[-\left(\frac{U_{\max,j}}{c(\theta)} \right)^{k(\theta)} \right] \right) d\theta \tag{11.28}$$

The exceedance probability of the maximum wind speed within the jth wind direction range can be expressed as:

$$1 - P_j^{\mathrm{E}} = \frac{1}{N_j + 1} \tag{11.29}$$

in which N_j is the total numbers of wind records within the jth wind direction range, which can be given by:

$$N_j = T_0 n_0 \int_{\Omega_j} f_\theta(\theta) d\theta \tag{11.30}$$

where:

T_0 is the fatigue damage evolution period;
n_0 is the number of wind records per year for all directions.

By using Equations 11.28 to 11.30, the maximum wind speed $U_{\max,j}$ for the jth wind direction range can be obtained, and the results are listed in Table 11.5 for a wind return period of 120 years, which is actually the design life of the bridge.

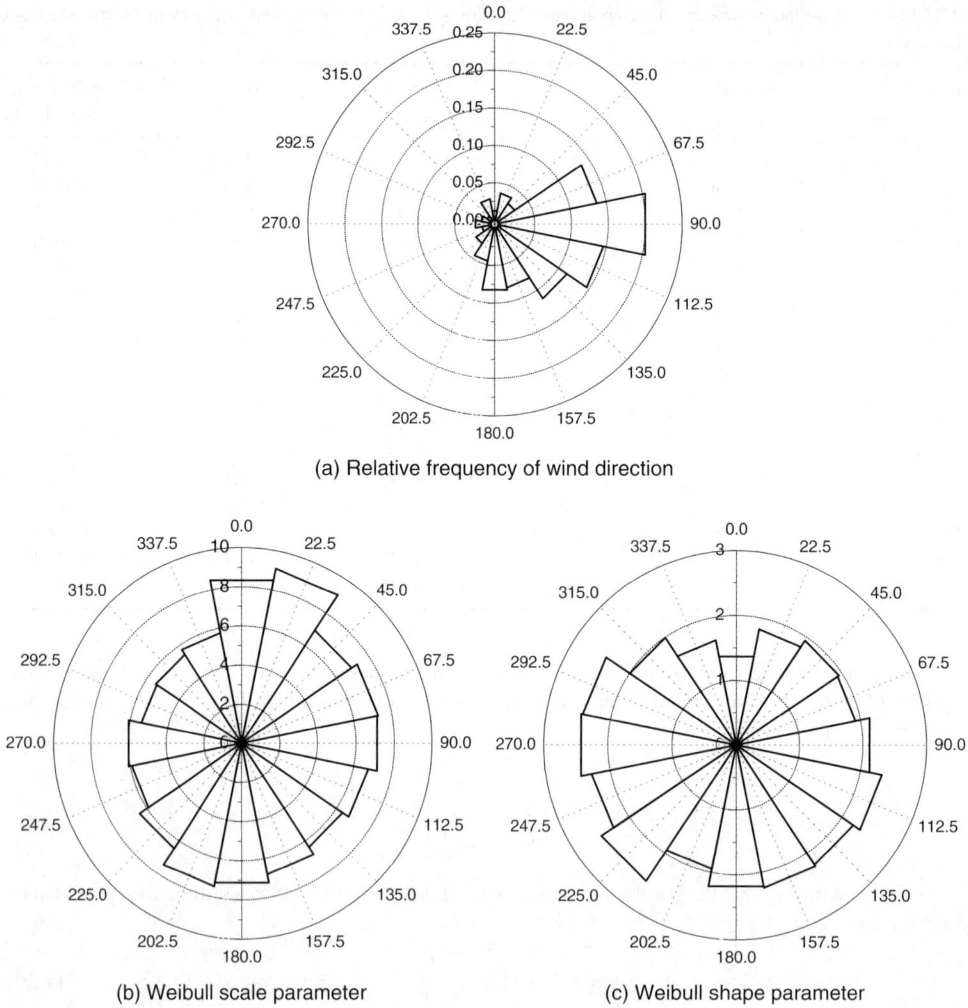

Figure 11.15 Relative frequency of wind direction and Weibull scale and shape parameters (from [23]) (Reproduced with permission from Elsevier).

The maximum wind speed obtained at the top of the tower is converted to the average deck level of the Tsing Ma Bridge for buffeting-induced stress analysis using the following equation:

$$U(z_{\mathrm{d}}) = \left(\frac{z_{\mathrm{d}}}{z_{\mathrm{t}}}\right)^{\alpha} U(z_{\mathrm{t}}) \tag{11.31}$$

where:

$U(z_{\mathrm{d}})$ is the mean wind speed at the average deck level z_{d};
$U(z_{\mathrm{t}})$ is the mean wind speed at the top of the tower z_{t};
α is the exponent of wind profile.

In this study, the average deck level is taken as 60 m. The exponent α is taken to be 0.30 for winds over the over-land fetch (see Figure 11.1b, clockwise from the SW to SE) and 0.2 for

Table 11.5 Maximum hourly mean wind speeds of 120-year return period (from [23]) (Reproduced with permission from Elsevier)

Direction	$U_{\max}$ (m/sec)		
	at tower top	at deck level	perpendicular to the deck
N	39.63	27.06	−25.89
NNE	31.45	21.47	−16.57
NE	21.89	14.94	−7.02
ENE	25.34	17.31	−1.66
E	22.36	15.27	4.46
ESE	18.79	12.83	8.16
SE	18.63	14.44	12.75
SSE	19.36	15.01	14.94
S	20.86	16.18	15.47
SSW	24.03	18.63	14.38
SW	15.62	10.66	5.01
WSW	15.02	10.26	0.98
W	14.90	10.17	−2.97
WNW	13.04	8.91	−5.66
NW	15.56	10.62	−9.38
NNW	22.28	15.21	−15.14

winds over the open-sea fetch (clockwise from the SE to SW). The converted maximum wind speeds at the average deck level, and their components perpendicular to the bridge deck, are also listed in Table 11.5, in which the positive sign indicates wind from the south and the negative sign denotes wind from the north.

11.5.3 Critical Stresses and Hot Spot Stresses

In the SHM-oriented FE model of the bridge, there are a total of 15 904 beam elements used to model the bridge deck [10]. To find the most critical beam elements and the corresponding most critical stresses in the bridge deck, a buffeting-induced stress analysis is carried out by considering a 15 m/sec mean wind perpendicular to the bridge axis from the south for an hour. The von Karman spectra, together with other wind characteristics, are first used for numerical simulation of the stochastic wind velocity field for the entire bridge deck. The stochastic wind velocity field comprises a series of time histories of fluctuating wind velocity in the horizontal and vertical directions at various points along the bridge deck.

In the simulation, the turbulence intensity is taken 24% in the horizontal direction and 17 % in the vertical direction by considering the most turbulent cases in the field. The integral length scale is taken as 251 m in the horizontal direction and 56 m in the vertical direction. The wind incidence is assumed to be zero. The exponential form of coherence function is adopted to reflect turbulent wind correlation along the bridge deck in both horizontal and vertical directions. The exponential decay coefficient is selected as 16 in the simulation [16]. A fast spectral representation approach, as proposed by Cao *et al.* [17], is adopted for the digital simulation of stochastic wind velocity field. A total of 120 points along the bridge with an interval 18.0 m are considered in the simulation of wind field. The average elevation of the bridge deck is taken as 60 m. Since this study concerns buffeting-induced stresses other than traffic-induced stresses, the sampling frequency and duration used in the simulation of wind speeds are, respectively, 16 Hz and 3600 sec.

The numerical method introduced in Section 11.3 is then used to estimate the buffeting forces and self-excited forces at the nodes of the FEM of the bridge. In determining the buffeting forces at the nodes, the drag, lift, and moment coefficients of the bridge deck are taken as 0.135, 0.090 and 0.063, respectively, at a zero wind angle of attack with respect to the deck width of 41 m, based on the wind tunnel test results. The first derivatives of the drag, lift, and moment coefficients with respect to the same wind angle of attack are −0.253, 1.324 and 0.278, respectively. The aerodynamic transfer functions between the simulated fluctuating wind velocities and the buffeting forces are assumed to be one.

Due to the lack of wind tunnel test results on lateral flutter derivatives, only the vertical and rotational motions of the bridge deck are taken into account in the simulation of self-excited forces. A total of 12 frequency independent coefficients are determined by using the measured flutter derivatives and the least squares fitting method. They are used to determine the aeroelastic stiffness matrix and the aeroelastic damping matrix, by which the self-excited forces at the nodes of the FEM of the bridge can be determined.

The buffeting-induced stress responses of the bridge are finally computed using the mode superposition technique. The first 80 modes of vibration of the bridge are considered in the computation, with the highest frequency being 1.1 Hz. The damping ratios for all the modes of vibration are taken as 0.5%. The modal stresses multiplied by the generalized displacement vector yield the stress time histories at five points of the end section of each element, with four of these points located at each corner of the end section and one point situated at the centroid of the end section.

By comparing the maximum values and the standard deviations of all stress time histories, the cross-section of the bridge deck at the Ma Wan tower is identified as the most critical section, in which six elements – nos. 34 111 and 38 111, 40 881 and 48 611, 58 111 and 59 111 are identified as the most critical elements. Nos. 34 111 and 38 111 are the bottom chords of the outer north and south longitudinal trusses, respectively, on the main span side (see Figure 11.16). Elements 40 881 and 48 611 are the bottom chords of the inner north and south longitudinal trusses on the main span side, respectively, and nos. 58 111 and 59 111 are the bottom chords in the middle of the cross-frame close to the north and south inner longitudinal trusses, respectively.

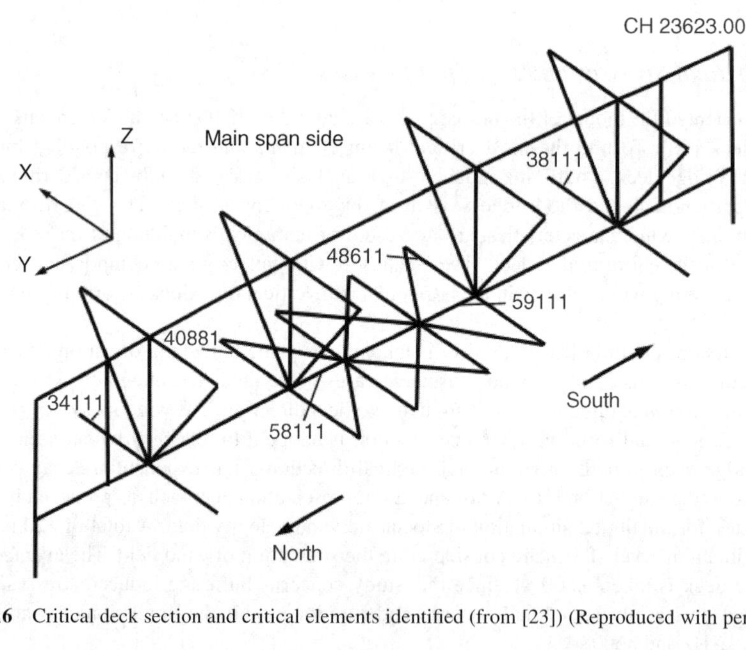

Figure 11.16 Critical deck section and critical elements identified (from [23]) (Reproduced with permission from Elsevier).

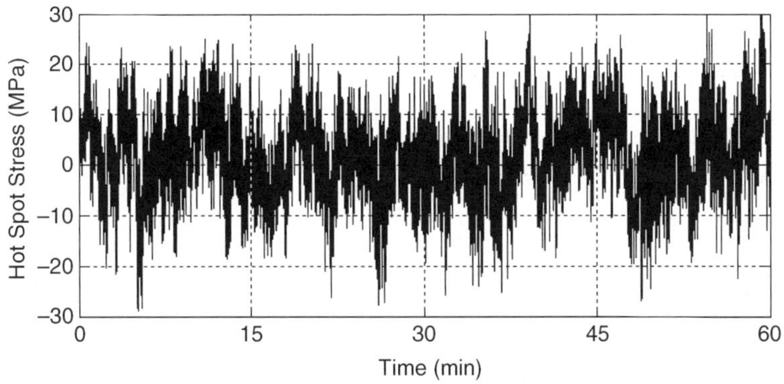

Figure 11.17 Time history of hot spot stress of element 40 881 (from [23]) (Reproduced with permission from Elsevier).

The hot spot stress approach, which considers the stress concentration at welded joints, has been widely used in hollow steel tubular structure fatigue design and analysis [28]. To apply this approach to the fatigue damage assessment of the bridge under traffic loading, Chan *et al.* [9] computed the stress concentration factors (SCF) of typical welded joints of the bridge deck based on both global and local finite element models. The hot spot stress block cycles were then determined by multiplying the nominal stress block cycles by the SCF. This approach, and the SCF obtained by Chan *et al.* [9] are used in this study.

In this regard, the maximum stress at the middle section of each critical element is computed based on the stresses at the two ends of the element and is taken as the nominal stress of the element. The hot spot stress is then determined by multiplying the nominal stress by the corresponding SCF. The type of welded connection for elements 34 111, 38 111, 40 881 and 48 611 in this study is classified as F2 according to BS5400 [29], and the SCF is taken as 1.95 [9]. The type of welded connection for elements 58 111 and 59 111 is classified as F, and the SCF is taken as 1.44. Depicted in Figure 11.17 is the one-hour time history of the hot spot stress for element 40 881.

11.5.4 Hot Spot Stress Characteristics

For the subsequent buffeting-induced fatigue damage assessment of the bridge deck at the six hot spot stress locations for a wind return period of 120 years, the preceding exercise has to be repeated for mean wind speeds from 5–30 m/sec, with an interval of 5 m/sec for winds over the overland fetch and from 5–20 m/sec at an interval of 5 m/sec for winds over the open-sea fetch, respectively. This yields a total of 60 one-hour time histories of the hot spot stresses for the bridge deck. For each of these one-hour time histories, the rainflow counting method [30] is applied to obtain the hot spot stress characteristics within one hour for different wind speeds and directions. The hot spot stress characteristics include the total number of stress cycles (N_r), the stress range (σ_r), the mean value (σ_m) of each stress cycle and the maximum stress range ($\sigma_{r,max}$). The mean value of each stress cycle includes the mean stress caused by the mean wind speed. As a result, a total of 60 data sets of hot spot stress characteristics are produced.

Figure 11.18 displays the total number of stress cycles and the maximum stress range against mean wind speed and wind terrain for the hot spot stress at element 40 881. It can be seen that variations in the total number of stress cycles with mean wind speed and wind terrain are not considerable within the range from 2200 to 2300 cycles. The variation of the maximum stress range with wind terrain is very small, but the effect of mean wind speeds on the maximum stress range is significant.

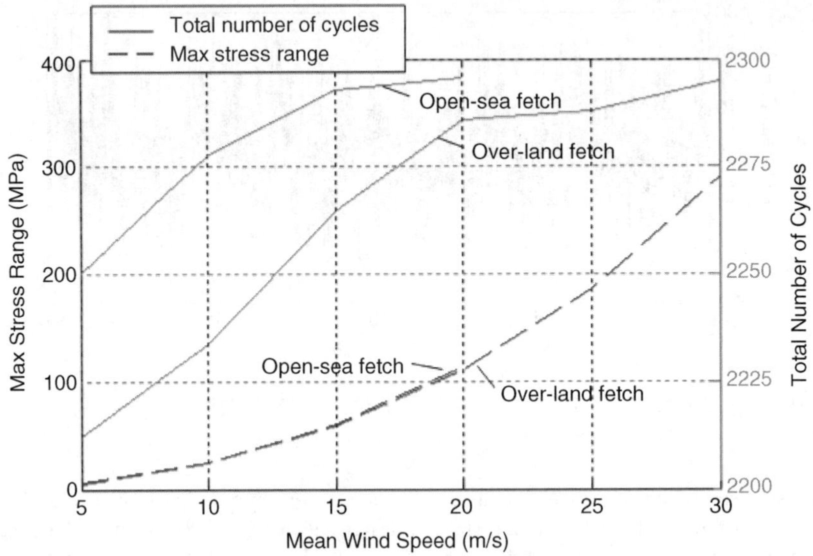

Figure 11.18 Variations of total number of cycles and maximum stress range of hot spot stress at element 40 881 (from [23]) (Reproduced with permission from Elsevier).

11.5.5 Damage Evolution Model

The growth in damage of the material is considered as a progressive internal deterioration, which induces a loss of the effective cross-sectional area that carries loads. The damage index D for the isotropic damage in the continuum damage mechanics (CDM) is often defined as:

$$D = \frac{A_n - A_e}{A_n} \tag{11.32}$$

where:

A_n is the nominal cross-section area;
A_e is the effective cross-section area considering area loss due to damage.

Based on thermodynamics and potential of dissipation, the rate of damage for high cycle fatigue can be expressed as a function of the accumulated micro-plastic strain, the strain energy density release rate and the current state of damage in CDM [31]. The micro-plastic strain (often neglected in a low-cycle fatigue problem) and its accumulation must be considered for high-cycle fatigue damage, even if macro-plastic strain does not exist [32]. In the one-dimensional situation, the equation for the rate of fatigue damage, $\dot{D}$, can be written as:

$$\dot{D} = \frac{\sigma^2 |\sigma - \overline{\sigma}|^{\beta} \langle \dot{\sigma} \rangle}{(1 - D)^{\alpha} B} \tag{11.33}$$

where:

$\overline{\sigma} = \sigma_m$ is the mean stress;
the symbol $\langle\,\rangle$ denotes the McCauley brackets, where $\langle x \rangle = x$ for $x > 0$ and $\langle x \rangle = 0$ for $x < 0$;
α, β and B are the material properties.

The material properties can be determined using the Woehler curves obtained through uniaxial periodic fatigue tests under strain-controlled condition and direct damage measurements of the material

[31,32]. Equation 11.33 is a general constitutive model for high cycle fatigue, and it can be integrated over time for the cycles with different mean stresses and stress ranges. For example, when considering $\sigma_m = 0$, and neglecting the variation of $(1 - D)^\alpha$ in one stress cycle, integrating Equation 11.32 over the cycle yields:

$$\frac{\delta D}{\delta N} = \frac{\sigma_a^{\beta+3}}{B(\beta+3)(1-D)^\alpha} \tag{11.34}$$

where σ_a is the stress amplitude of the cycle and N is the number of cycles.

In this study, the rainflow counting method has been applied to obtain the hot spot stress characteristics within one hour for different wind speeds and directions. The hot spot stress characteristics include the total number of stress cycles (N_r), the stress range (σ_r), the mean value (σ_m) of each stress cycle and the maximum stress range ($\sigma_{r,m}$) within one hour. Since a normal fatigue life of a bridge may be over 100 years, the effects of load sequence on damage accumulation within one hour can be neglected. Nevertheless, the effects of load sequence are taken into consideration in this study at one hour intervals. Thus, by considering one hour as one block, the fatigue damage rate generated by one block of stress cycles with zero mean stress can be expressed as:

$$\frac{\delta D}{\delta N_b} = \sum_{j=1}^{m_b} \frac{\sigma_{a,j}^{\beta+3}}{B(1-D)^{\alpha_j}(\beta+3)} \tag{11.35}$$

where:

m_b is the total number of stress cycles in the block;
N_b is the number of blocks.

If mean stress is not equal to zero, the fatigue damage rate generated by one block of stress cycles can be determined by the following expression [32]:

$$\frac{\delta D}{\delta N_b} = \sum_{j=1}^{m_b} \frac{\left[\left(\sigma_{r,j} + 2\sigma_{m,j}\right)\sigma_{r,j}\right]^{\frac{\beta+3}{2}}}{B(1-D)^{\alpha_j}(\beta+3)} \tag{11.36}$$

where:

$\sigma_{r,j} = 2\,\sigma_{a,j}$ is the jth stress range;
α_j depends on the jth stress range, i.e. $\alpha_j = f(\sigma_{r,j})$.

Within one hour block, the maximum stress range affects the damage increment most. Equation 11.36 can be rewritten as:

$$(1-D)^{\alpha_e}\delta D = \sum_{j=1}^{m_b} \frac{\left[\left(\sigma_{r,j} + 2\sigma_{m,j}\right)\sigma_{r,j}\right]^{\frac{\beta+3}{2}}}{B(1-D)^{\alpha_j-\alpha_e}(\beta+3)} \delta N_b \tag{11.37}$$

where α_e is determined by the maximum stress range $\sigma_{r,max}$ through the function $\alpha_j = f(\sigma_{r,j})$.

Let us consider the damage accumulation in the kth block. Integrating Equation 11.37 over the kth block yields:

$$\int_{D_{k-1}}^{D_k} (1-D)^{\alpha_{e,k}}\delta D = \int_{N_{b,k-1}}^{N_{b,k}} \left(\sum_{j=1}^{m_{b,k}} \frac{\left[\left(\sigma_{r,jk} + 2\sigma_{m,jk}\right)\sigma_{r,jk}\right]^{\frac{\beta+3}{2}}}{B(\beta+3)} (1-D)^{\alpha_{e,k}-\alpha_{j,k}} \right) \delta N_b \tag{11.38}$$

Making the approximation $(1-D)^{\alpha_{\mathrm{e,k}}-\alpha_{j,k}} = (1-D_{k-1})^{\alpha_{\mathrm{e,k}}-\alpha_{j,k}}$ [31], the damage evolution model for fatigue damage assessment of a long suspension bridge in this study is given as:

$$D_k = 1 - \left\{ (1-D_{k-1})^{\alpha_{\mathrm{e},k}+1} - \frac{(\alpha_{\mathrm{e},k}+1)}{B(\beta+3)} \left(\sum_{j=1}^{m_{b,k}} \left[(\sigma_{\mathrm{r},jk} + 2\sigma_{\mathrm{m},jk})\sigma_{\mathrm{r},jk} \right]^{\frac{\beta+3}{2}} (1-D_{k-1})^{\alpha_{\mathrm{e},k}-\alpha_{j,k}} \right) \right\}^{\frac{1}{1+\alpha_{\mathrm{e},k}}} \tag{11.39}$$

Buffeting-induced fatigue in the welded joints of a bridge is a cumulative process over many years. Buffeting-induced fatigue damage accumulation, including both fatigue crack initiation and growth in micro-scale, can be well estimated by using Equation 11.39 evolutionally, as shown in the next section. For the bridge concerned in this study, $\alpha_j = k_\alpha \sigma_{r,j} + \alpha_0$ is used. The parameters k_α and α_0 are taken as -0.135 MPa and 101.4 MPa, respectively, for both of the connection types F2 and F. $(\beta+3) = 3.01$ is used for both the connection types. $[B(\beta+3)] = 3.41 \times 10^{13}$ is used for connection type F2 and $[B(\beta+3)] = 5.11 \times 10^{13}$ is used for connection type F. According to BS5400 [29], if the stress range $\sigma_{r,j}$ is less than σ_{c}, it should be reduced in the proportion $\left(\frac{\sigma_{\mathrm{r,j}}}{\sigma_{\mathrm{c}}}\right)^2$, where σ_{c} is the stress limit to fatigue. The stress limit to fatigue used in this study is 35 MPa for connection type F2 and 40 MPa for connection type F.

11.5.6 *Buffeting-Induced Fatigue Damage Assessment*

To assess the buffeting-induced fatigue damage to the bridge deck at the identified six hot spot stress locations for a wind return period of 120 years, the occurrence sequences of 1 051 200 (= 24 × 365 × 120) blocks in one hour duration should be determined in consideration of different wind speeds and directions before Equation 11.39 can be applied [23]. The number of hourly wind records in the *j*th wind direction range Ω_j in one year can be determined by:

$$n_{o,j} = n_o \int_{\Omega_j} f_\theta(\theta) d\theta \tag{11.40}$$

where:

$n_0 = 8760$ is the total number of hourly wind records in one year;
$f_\theta(\theta)$ is the relative frequency of wind direction.

After the number of wind records in the *j*th wind direction range is determined, the number of hourly wind records in the *i*th wind speed range within the *j*th wind direction range can then be determined, by considering a normalized cumulative distribution function as follows:

$$\overline{P}_u(U_i, \theta_j) = \frac{1 - \exp\left[-\left(\frac{U_{i,j}}{c(\theta_j)} \right)^{k(\theta_j)} \right]}{1 - \exp\left[-\left(\frac{U_{\max,j}}{c(\theta_j)} \right)^{k(\theta_j)} \right]} \tag{11.41}$$

If the hourly mean wind speeds in the *j*th wind direction range are divided into several wind speed ranges from zero to $U_{\max,j}$ at an interval of 5 m/sec, the number of hourly wind records in the *i*th wind speed range within the *j*th wind direction range is then given as:

$$n_o(i,j) = n_{o,j} \left[P_u(5(i), \theta_j) - P_u(5(i-1), \theta_j) \right] \quad (i = 1, 2, \ldots) \tag{11.42}$$

Table 11.6 Distribution of the number of hourly wind records in one year (from [23]) (Reproduced with permission from Elsevier)

Direction	0–5	5–10	10–15	15–20	20–25	25–	SUM
N	98	91	46	17	5	1	258
NNE	66	84	41	11	1	0	203
NE	165	206	73	1	0	0	445
ENE	455	527	139	11	0	0	1132
E	602	802	189	4	0	0	1597
ESE	526	720	118	0	0	0	1364
SE	404	476	62	0	0	0	942
SSE	320	399	66	1	0	0	786
S	235	334	90	4	0	0	663
SSW	152	212	54	4	0	0	422
SW	100	121	7	0	0	0	228
WSW	74	79	2	0	0	0	155
W	73	75	1	0	0	0	149
WNW	78	59	0	0	0	0	137
NW	68	44	1	0	0	0	113
NNW	83	64	18	1	0	0	166
SUM	3499	4293	907	54	6	1	8760

By Equations 11.40 and 11.42, the distribution of the number of hourly wind records in one year within the designated wind direction and wind speed ranges can be obtained. The results from this study are listed in Table 11.6.

Because only monsoon wind effect on the bridge is considered in this study, and the monsoon wind in Hong Kong is normally southerly (from 90° to 270° in Figure 11.15) in summer and northerly (from 270° to 90° in Figure 11.15) in winter, it is assumed that the monsoon wind blows over the open-sea fetch in summer and over the overland fetch in winter. Two random permutation sequences of uniform distribution are then generated according to the total number of wind records in summer and in winter, respectively, in a particular year. The first random permutation sequence brings out the occurrence sequence of wind records over the open-sea fetch, while the second random permutation sequence leads to the occurrence sequence of wind records over the over-land fetch for that particular year.

Each wind record is then converted to a wind block according to its wind direction and wind speed. The hot spot stress characteristics corresponding to each wind block can be best found from one of the 60 data sets of the hot spot stress characteristics obtained in advance. The fatigue damage accumulation of the bridge deck at each hot spot stress location can finally be processed using the damage evolution model one year after another, up to a total of 120 years, by assuming zero damage at the beginning of fatigue damage accumulation.

Figure 11.19 shows the damage evolution of the bridge deck at the six hot spot stress locations during the 120-year period. It can be seen that the damage index increases with time. A slight non-linear relationship between the damage index and time can be observed for the hot spot stress of element 48 611, which indicates the non-linear nature of fatigue initiation and growth and the capability of the damage evolution model used in this study.

Listed in Table 11.7 are the fatigue damage indices of the bridge deck at the six hot spot stress locations at the end of the 120 years. It can be seen that monsoon wind-induced fatigue damage to the bridge deck is not significant. Note that this study does not take into account typhoon effects and traffic effects. Nevertheless, the procedure proposed in this study can also be applied to typhoon wind induced fatigue damage, and this will be done when long-term field measurement data on typhoons are available.

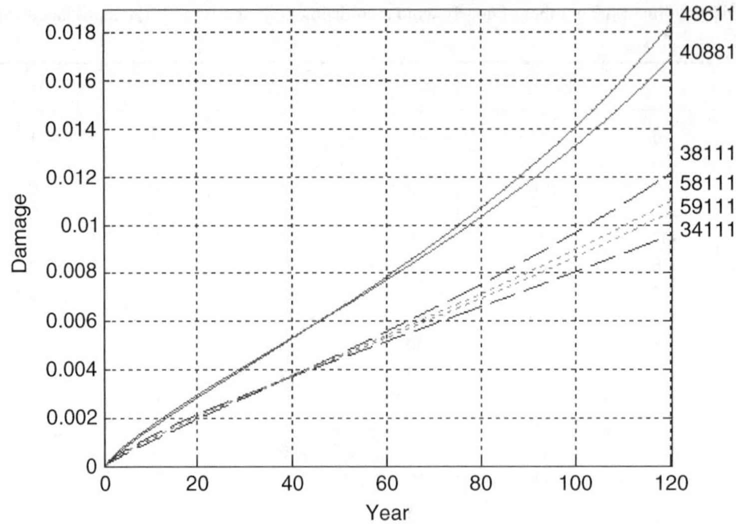

Figure 11.19 Damage evolution during the 120-year return period (from [23]) (Reproduced with permission from Elsevier).

Table 11.7 Fatigue damage at the end of 120 years (from [23]) (Reproduced with permission from Elsevier)

Locations	38 111	34 111	40 881	48 611	58 111	59 111
Damage index	0.0121	0.0096	0.0169	0.0183	0.0110	0.0105

11.6 Framework for Multiple Loading-Induced Stress Analysis

11.6.1 Equation of Motion

To establish a framework for multiple loading-induced stress analysis [33], a long-span cable-supported bridge, trains and road vehicles are regarded as three subsystems and represented by three finite element models, using a commercial software package. A long-span cable-supported bridge can be modeled using beam elements, plate elements, shell elements and others. A train usually consists of several vehicles, and each vehicle is in turn is composed of car body, bogies, wheel-sets, primary suspension systems connecting the wheel-sets to the bogies, and secondary suspension systems connecting the bogies to the car body. The car body, bogies and wheel-sets can be modeled by either beam elements, shell elements or rigid bodies. The primary or secondary suspension systems can be modeled by spring elements and dashpot elements. There is a variety of configurations of road vehicles in reality, such as a tractor with trailers with different axle spacings. Each tractor or trailer is composed of a car body, axle sets and the suspension systems connecting the two components. Similar to a train, the car body and axle sets can be modeled by either beam elements, shell elements or rigid bodies. The suspension systems can be modeled by spring elements and dashpot elements.

Except for linear spring elements in the suspension units, the spring element non-linear parts and all dashpot elements are treated as pseudo forces in the train and road vehicle subsystems in this study. The subsystems are coupled through contacts between bridge and trains and between bridge and road vehicles. Wind forces may act on all three subsystems. The equations of motion of bridge, train and road vehicle subsystems can then be expressed as [33]:

$$\begin{cases} \mathbf{M}_b\ddot{\mathbf{X}}_b + \mathbf{C}_b\dot{\mathbf{X}}_b + \mathbf{K}_b\mathbf{X}_b = \mathbf{F}_b \\ \mathbf{M}_r\ddot{\mathbf{X}}_r + \mathbf{C}_r\dot{\mathbf{X}}_r + \mathbf{K}_r\mathbf{X}_r = \mathbf{F}_r \\ \mathbf{M}_h\ddot{\mathbf{X}}_h + \mathbf{C}_h\dot{\mathbf{X}}_h + \mathbf{K}_h\mathbf{X}_h = \mathbf{F}_h \end{cases} \tag{11.43}$$

where:

$\mathbf{M}_b$, $\mathbf{K}_b$, $\mathbf{M}_r$, $\mathbf{K}_r$, $\mathbf{M}_h$ and $\mathbf{K}_h$ are the mass and stiffness matrices of the bridge, train and road vehicle subsystems, respectively;
$\mathbf{C}_b$, $\mathbf{C}_r$ and $\mathbf{C}_h$ are the damping matrices of the bridge, train and road vehicle subsystems, respectively;
$\mathbf{X}_b$, $\dot{\mathbf{X}}_b$, $\ddot{\mathbf{X}}_b$, $\mathbf{X}_r$, $\dot{\mathbf{X}}_r$, $\ddot{\mathbf{X}}_r$, $\mathbf{X}_h$, $\dot{\mathbf{X}}_h$ and $\ddot{\mathbf{X}}_h$ are respectively the displacement, velocity and acceleration vectors of the bridge, train and road vehicle subsystems;
$\mathbf{F}_b$, $\mathbf{F}_r$ and $\mathbf{F}_h$ are the force vectors of the bridge, train and road vehicle subsystems, which can be expressed as:

$$\begin{cases} \mathbf{F}_b = \mathbf{F}_{rb} + \mathbf{F}_{hb} + \mathbf{F}_{wb} \\ \mathbf{F}_r = \mathbf{F}_{pr} + \mathbf{F}_{br} + \mathbf{F}_{wr} \\ \mathbf{F}_h = \mathbf{F}_{ph} + \mathbf{F}_{bh} + \mathbf{F}_{wh} \end{cases} \tag{11.44}$$

In Equation 11.44, the forces acting on the bridge subsystem $\mathbf{F}_b$ include contact forces $\mathbf{F}_{rb}$ due to train-bridge interaction, contact forces $\mathbf{F}_{hb}$ due to road vehicle-bridge interaction, and wind forces $\mathbf{F}_{wb}$. The forces acting on the train subsystem $\mathbf{F}_r$ include pseudo forces $\mathbf{F}_{pr}$ produced by non-linear parts of all springs, as well as all dashpots in the train subsystem, contact forces $\mathbf{F}_{br}$ due to train-bridge interaction, and wind forces $\mathbf{F}_{wr}$. The forces acting on the road vehicle subsystem $\mathbf{F}_h$ include pseudo forces $\mathbf{F}_{ph}$ produced by non-linear parts of all springs, as well as all dashpots in the road vehicle subsystem, contact forces $\mathbf{F}_{bh}$ due to road vehicle-bridge interaction, and wind forces $\mathbf{F}_{wh}$.

These force vectors will be discussed in detail subsequently. In consideration that a large number of DOFs are involved in the finite element models of the three subsystems, the mode superposition method is adopted to make the dynamic stress analysis of the bridge under multi-loadings manageable. Using this method, Equation 11.43 can be rewritten as:

$$\begin{cases} \ddot{\mathbf{q}}_b + 2\boldsymbol{\xi}_b\boldsymbol{\omega}_b\dot{\mathbf{q}}_b + \boldsymbol{\omega}_b^2\mathbf{q}_b = \boldsymbol{\Phi}_b^{\mathrm{T}}\mathbf{F}_b \\ \ddot{\mathbf{q}}_r + 2\boldsymbol{\xi}_r\boldsymbol{\omega}_r\dot{\mathbf{q}}_r + \boldsymbol{\omega}_r^2\mathbf{q}_r = \boldsymbol{\Phi}_r^{\mathrm{T}}\mathbf{F}_r \\ \ddot{\mathbf{q}}_h + 2\boldsymbol{\xi}_h\boldsymbol{\omega}_h\dot{\mathbf{q}}_h + \boldsymbol{\omega}_h^2\mathbf{q}_h = \boldsymbol{\Phi}_h^{\mathrm{T}}\mathbf{F}_h \end{cases} \tag{11.45}$$

where:

$q_b = \left[q_{b1}, q_{b2}, \cdots, q_{bN_{bm}}\right]^{\mathrm{T}}$, $q_r = \left[q_{r1}, q_{r2}, \cdots, q_{rN_{vm}}\right]^{\mathrm{T}}$ and $q_h = \left[q_{h1}, q_{h2}, \cdots, q_{hN_{hm}}\right]^{\mathrm{T}}$ are the modal coordinate vector of the bridge, train, and road vehicle subsystems, respectively;
N_{bm}, N_{rm} and N_{hm} are the number of interested modes of vibration of the bridge, train and road vehicle subsystems, respectively;
$\boldsymbol{\omega}_b$, $\boldsymbol{\xi}_b$, $\boldsymbol{\Phi}_b$, $\boldsymbol{\omega}_r$, $\boldsymbol{\xi}_r$, $\boldsymbol{\Phi}_r$, $\boldsymbol{\omega}_h$, $\boldsymbol{\xi}_h$ and $\boldsymbol{\Phi}_h$ are respectively the modal frequency matrix, modal damping ratio matrix and mode shape matrix related to the interested modes of vibration of the bridge, train and road vehicle subsystems.

Note that both the modal frequency matrix and the modal damping ratio matrix are a diagonal matrix. The terms on the right side of Equation 11.45 are actually the functions of the relevant subsystem's responses, so numerical iterations are required to find the solutions. Once the generalized coordinates of the bridge are found, the element stress vector $\boldsymbol{\sigma}_{b,j}$ of the jth element of the bridge without considering initial strains and stresses can be obtained as follows:

$$\boldsymbol{\sigma}_{b,j} = \boldsymbol{\Gamma}_{b,j}\mathbf{q}_b \tag{11.46}$$

where $\boldsymbol{\Gamma}_{b,j}$ denotes the modal stress matrix, which can be determined by:

$$\boldsymbol{\Gamma}_{b,j} = \mathbf{D}_{b,j}\mathbf{L}_{b,j}\mathbf{N}_{b,j}\mathbf{T}_{b,j}\boldsymbol{\Phi}_{b,j} \tag{11.47}$$

in which:

$\boldsymbol{\Phi}_{b,j}$ is the modal shape matrix with respect to nodes of the jth element;
$\mathbf{T}_{b,j}$ is the coordinate transfer matrix from global coordinate to local coordinate with respect to nodes of the jth element;
$\mathbf{N}_{b,j}$ is the shape function transforming local node displacements to element displacement field;
$\mathbf{L}_{b,j}$ is the differential operator transforming the element displacement field to the element strain field;
$\mathbf{D}_{b,j}$ denotes the elastic matrix representing stress-strain relationship.

11.6.2 Pseudo Forces in Trains and Road Vehicles

The springs and dashpots in the suspension units of a vehicle are non-linear in some cases. In this study, the linear part of spring stiffness is included in the finite element model of the vehicle, while all the damping forces and non-linear restoring forces of the suspension units are treated as pseudo forces to facilitate non-linear analysis [33]. For example, the pseudo forces produced by a dashpot with a saturation property in a local coordinate can be expressed as:

$$\begin{cases} f^{\mathrm{L}}_{\mathrm{vn,i}} = \begin{cases} -F_{\mathrm{maxij}}(\dot{\delta}^{\mathrm{L}}_{\mathrm{v,i}} - \dot{\delta}^{\mathrm{L}}_{\mathrm{v,j}})/v_{\mathrm{0ij}} & \left|\dot{\delta}^{\mathrm{L}}_{\mathrm{v,i}} - \dot{\delta}^{\mathrm{L}}_{\mathrm{v,j}}\right| < v_{\mathrm{0ij}} \\ -F_{\mathrm{maxij}}\mathrm{sign}(\dot{\delta}^{\mathrm{L}}_{\mathrm{v,i}} - \dot{\delta}^{\mathrm{L}}_{\mathrm{v,j}}) & \left|\dot{\delta}^{\mathrm{L}}_{\mathrm{v,i}} - \dot{\delta}^{\mathrm{L}}_{\mathrm{v,j}}\right| \geq v_{\mathrm{0ij}} \end{cases} \\ f^{\mathrm{L}}_{\mathrm{vn,j}} = -f^{\mathrm{L}}_{\mathrm{vn,i}} \end{cases} \tag{11.48}$$

where:

$f^{\mathrm{L}}_{\mathrm{vn,i}}$ and $f^{\mathrm{L}}_{\mathrm{vn,j}}$ are the pseudo forces acting on the nodes i and j, respectively, along the $\overline{x}$-axis of the local coordinates as shown in Figure 11.20;
$\dot{\delta}^{\mathrm{L}}_{\mathrm{v,i}}$ and $\dot{\delta}^{\mathrm{L}}_{\mathrm{v,j}}$ are the velocities of the nodes i and j, respectively, along the $\overline{x}$-axis of the local coordinates;

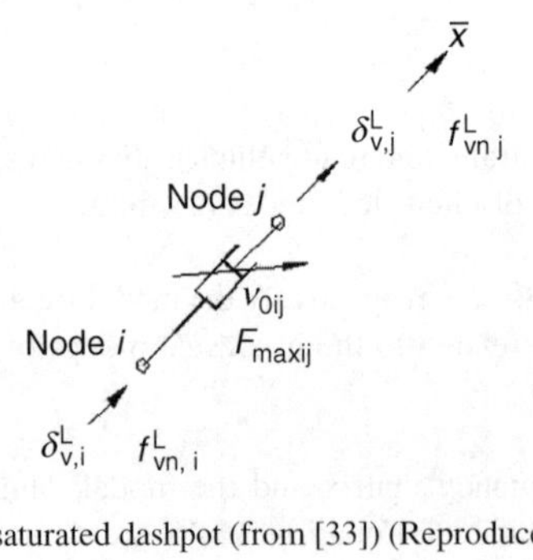

Figure 11.20 Pseudo force of saturated dashpot (from [33]) (Reproduced with permission from ASCE).

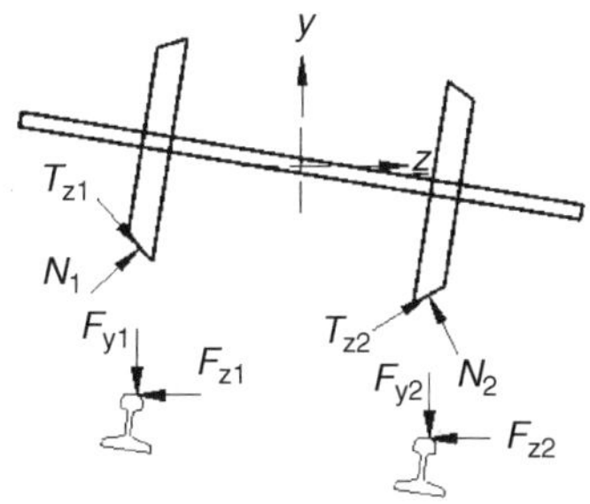

Figure 11.21 Contact forces acting on wheels and rails (from [33]) (Reproduced with permission from ASCE).

v_{0ij} is a saturated velocity, which is an intrinsic property of a dashpot;
F_{maxij} is the saturated damping force when the dashpot velocity exceeds v_{0ij}.

Equation 11.48 can be used to simulate dry friction when v_{0ij} is of very small value. Using the transformation matrix between the local coordinates and the global coordinates and by considering all the damping forces and non-linear restoring forces of the suspension units in either the train subsystem or the road vehicle subsystem, the pseudo force vectors $\mathbf{F}_{pr}$ and $\mathbf{F}_{ph}$ in Equation 11.44 can be obtained.

11.6.3 Contact Forces between Train and Bridge

The train subsystem and the bridge subsystem are coupled through contact between the wheels and rails (see Figure 11.21). Wheel-rail contact geometry computation is the basis for determining contact forces, so it is often carried out first to find the contact geometry parameters as functions of relative lateral displacement and yawing angle of the wheel-set to the rails. Two hypotheses are adopted for wheel-rail contact geometry computation: wheel and rail are regarded as rigid bodies; and the wheel is not allowed to jump away from the rail.

As plotted in Figure 11.21, F_{y1} and F_{y2} are the contact forces on the rails in the vertical direction; F_{z1} and F_{z2} denote the contact forces on the rails in the lateral direction; N_1 and N_2 are the normal contact forces on the wheels; T_{z1} and T_{z2} are the lateral creeping forces on the wheels; and the subscripts 1 and 2 stand for the left and right rail/wheel, respectively.

Eight equations are required to find the aforementioned eight contact forces between wheels and rails [33]. Four equations are established, based on the relationships between the contact forces on the rails and the contact forces on the wheels. Two equations are established by considering the equilibrium conditions of the wheel set. The last two equations are formed from the relationship between the lateral creep forces and normal contact forces according to the Kalker creepage theory. Once the contact forces on the rails are determined for all the wheel sets of the train subsystem, the force vector $\mathbf{F}_{rb}$ due to the train-bridge interaction in Equation 11.44 can be established for the bridge subsystem. Note that the positions of contact forces on the rails vary with time. Therefore, the force vector $\mathbf{F}_{rb}$ also varies with time and must be updated for every time interval.

Wheel non-jump condition is considered in this study. Therefore, vertical displacement of a wheel is dependent on the rail under the wheel, and it is not necessary to assign vertical degrees of freedom to the wheels of the train subsystem. As a result, the vertical forces produced by the motions of the wheel sets on the bogies of the train subsystem are found first, based on the equilibrium condition of the wheel-set. All of the vertical forces are then used to form the force vector $\mathbf{F}_{br}$ due to train-bridge interaction in Equation 11.44 for the train subsystem (excluding the vertical motion of the wheels). The force vector $\mathbf{F}_{br}$ is a function of time, depending on the position of the train on the bridge.

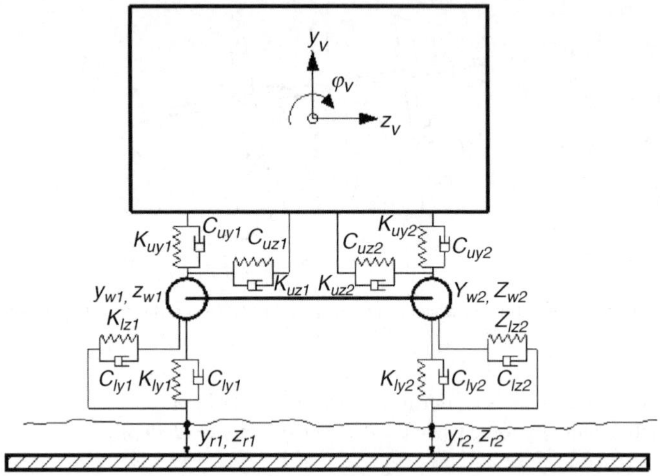

Figure 11.22 Schematic model of a road vehicle (from [33]) (Reproduced with permission from ASCE).

11.6.4 Contact Forces between Road Vehicles and Bridge

This study considers wind-vehicle-bridge interaction, and the lateral motion of vehicle body should be considered. Therefore, the axle set is connected to the vehicle body through two suspension units. One of these is the parallel combination of a spring and a dashpot in the vertical direction (y-direction), while the other is the parallel combination of a spring and a dashpot in the lateral direction (z-direction), as shown in Figure 11.22.

The connection of the axle set to the bridge deck is also realized through the two units representing the dynamic characteristics of the tire: one is the parallel combination of a spring and a dashpot in the vertical direction, while the other is the parallel combination of a spring and a dashpot in the lateral direction. Some common assumptions are adopted in this study to calculate the contact forces between road surface and vehicle tires, including that the vertical surface profile of the bridge deck pavement is not too rough to make the tires jump or leave the riding surface, so that a road vehicle's tires can be assumed to remain in contact with the bridge deck all times.

The contact between the road surface and the moving tire of the vehicle is further assumed to be a point contact. There is no sliding of the vehicle tires in the lateral direction. As a result, the interaction forces between the road vehicle and the bridge deck surface, which act on the axle set and the road surface respectively, can be determined from the relative motions of the axle set of the vehicle and the road surface of the bridge deck by the following equations:

$$\begin{cases} F_{yq} = K_{lyq}(y_{wq} - y_{rq}) + C_{lyq}(\dot{y}_{wq} - \dot{y}_{rq}) \\ F_{zq} = K_{lzq}(z_{wq} - z_{wq}) + C_{lzq}(\dot{z}_{wq} - \dot{z}_{rq}) \end{cases} \quad (q = 1, 2) \tag{11.49}$$

where:

F_{yq} and F_{zq} ($q = 1, 2$) are the interaction forces in the y-direction and z-direction respectively, and the subscripts 1 and 2 stand for the left and right tire respectively;

K_{lyq} and K_{lzq} are the stiffness of linear elastic springs for a tire in the y-direction and z-direction respectively;

C_{lyq} and C_{lzq} are the damping coefficients of viscous dashpots for a tire in the y-direction and z-direction respectively;

y_{wq} and $\dot{y}_{wq}$ are respectively the displacement and velocity of the axle set in the y-direction;
y_{rq} and $\dot{y}_{rq}$ are respectively the displacement and velocity of the bridge deck surface in the y-direction at the contact point;
z_{wq} and $\dot{z}_{wq}$ are respectively the displacement and velocity of the axle set in the z-direction;
z_{rq} and $\dot{z}_{rq}$ are respectively the displacement and velocity of the bridge deck surface in the z-direction at the contact point.

Note that the displacement and velocity of the bridge deck surface should consider the road roughness. Once the interaction forces on the bridge deck surface are determined for all the axles of the road vehicle subsystem, the force vector $\mathbf{F}_{hb}$ due to the road vehicle-bridge interaction in Equation 11.44 can be established for the bridge subsystem.

Note also that the positions of interaction forces on the deck surface vary with time. Therefore, the force vector $\mathbf{F}_{hb}$ also varies with time and must be updated for every time interval. The interaction forces on the axles of a road vehicle are the counter-forces acting on the deck surface, so the force vector $\mathbf{F}_{bh}$ due to the road vehicle-bridge interaction in Equation 11.44 can be established for the road vehicle subsystem.

11.6.5 Wind Forces on Bridge

Wind forces acting on a long-span cable-supported bridge are mainly the static wind forces due to mean wind, the buffeting forces due to turbulent wind and the self-excited forces due to interaction between wind and bridge motion. The mean stresses of the bridge caused by the static wind forces can be readily determined separately, and only the buffeting forces $\mathbf{F}_{bf}$ and self-excited force $\mathbf{F}_{se}$ are considered in this study:

$$\mathbf{F}_{wb} = \mathbf{F}_{bf} + \mathbf{F}_{se} \tag{11.50}$$

This study focuses on dynamic stress analysis of a long-span cable-supported bridge under multi-types of loadings. To this end, the SHM-oriented finite element model of a long-span bridge with significant modeling features at a stress level is used. The numerical procedure for wind-induced stress analysis based on this model was introduced in Section 11.4 and verified in Section 11.5, in which the buffeting forces acting at the center of elasticity of the deck cross section are distributed to the nodes of the deck section in terms of wind pressure distribution around the deck section.

After such a distribution is applied to all the sections of the bridge, the buffeting force vector $\mathbf{F}_{bf}$ in Equation 11.50 can be formed. Furthermore, by applying the virtual work principle, the self-excited forces at the centre of elasticity of the bridge deck section can also be distributed to the nodes of the bridge deck to form the self-excited force vector $\mathbf{F}_{se}$ in Equation 11.50.

$$\mathbf{F}_{se} = \mathbf{E}\mathbf{X}_b + \mathbf{G}\dot{\mathbf{X}}_b + \mathbf{H}\ddot{\mathbf{X}}_b + (\mathbf{N}_{se})^{\mathrm{T}}\hat{\mathbf{F}}_{se} \tag{11.51}$$

where:

E, **G**, and **H** are the aeroelastic stiffness, aeroelastic damping and aeroelastic mass matrices, respectively, of the bridge related to the nodal self-excited forces in the global coordinate system;
$\mathbf{N}_{se}$ is the displacement transformation matrix from the local coordinate system to the global coordinate system;
$\hat{F}_{se}$ is the part of the self-excited forces reflecting aerodynamic phase lag.

11.6.6 Wind Forces on Vehicles

Trains and road vehicles are assumed to travel along the bridge deck at constant velocities in this study. The aerodynamic wind forces acting on a car body of either train or road vehicle are determined using the quasi-steady approach. As this study focuses on dynamic stresses of bridge rather than safety of vehicles, only the three components of aerodynamic forces (i.e. lift force F_L, side force F_S and rolling moment M_R) acting on the center of gravity of the car body are considered [34].

$$\begin{cases} F_L = \frac{1}{2}\rho A_f C_L(\varphi)\overline{V}_R^2 + \frac{1}{2}\rho A_f \overline{V}_R^2 \left[\chi_{F_{Lu}}(n) C_L(\varphi) \frac{2\overline{u}u}{\overline{V}_R^2} + \chi_{F_{Lw}}(n) C_L'(\varphi) \frac{w}{\overline{u}} \right] \\ F_S = \frac{1}{2}\rho A_f C_S(\varphi)\overline{V}_R^2 + \frac{1}{2}\rho A_f \overline{V}_R^2 \left[\chi_{F_{Su}}(n) C_S(\varphi) \frac{2\overline{u}u}{\overline{V}_R^2} + \chi_{F_{Sw}}(n) C_S'(\varphi) \frac{w}{\overline{u}} \right] \\ M_R = \frac{1}{2}\rho A_f H C_R(\varphi)\overline{V}_R^2 + \frac{1}{2}\rho A_f H \overline{V}_R^2 \left[\chi_{M_{Ru}}(n) C_R(\varphi) \frac{2\overline{u}u}{\overline{V}_R^2} + \chi_{M_{Rw}}(n) C_R'(\varphi) \frac{w}{\overline{u}} \right] \end{cases} \tag{11.52}$$

where:

$\overline{u}$ is the mean wind speed component;
u and w are, respectively, the longitudinal and vertical fluctuating wind speed;
$\overline{V}_R$ is the mean wind velocity relative to the vehicle;
A_f is the reference area, which is normally taken as the frontal area of the car body;
ρ is the air density;
H is the reference height, which is normally taken as the height of the center of gravity of the car body above the ground;
$C_i(\varphi)$ and $C'_i(\varphi)$ ($i = L, S, R$) are the lift force, side force and rolling moment coefficients and their derivatives at $\alpha = 0$;
α is the incidence angle;
$\chi_{F_{Lu}}(n)$, $\chi_{F_{Lw}}(n)$, $\chi_{F_{Su}}(n)$, $\chi_{F_{Sw}}(n)$, $\chi_{M_{Ru}}(n)$ and $\chi_{M_{Rw}}(n)$ are the aerodynamic transfer functions between the fluctuating wind velocities and aerodynamic forces;
n is the frequency in Hz.

By considering all the car bodies in the train subsystem, the force vector $\mathbf{F}_{wr}$ in Equation 11.44 for the train subsystem can be formed. Also by considering all the car bodies in the road vehicle system, the force vector $\mathbf{F}_{wh}$ in Equation 11.44 for the road vehicle subsystem can be obtained. Note that the longitudinal and vertical fluctuating wind speeds in Equation 11.52 are compatible with the wind speeds used for determining wind forces acting on the bridge deck, and therefore the correlation of wind velocity along the bridge deck is taken into consideration. Nevertheless, the aerodynamic transfer functions in Equation 11.52 are taken as unit in this study for simplification. Furthermore, for road vehicles, the derivatives of the aerodynamic force coefficients are taken as zero because of lack of data.

11.6.7 Numerical Solution

The springs and dashpots in the suspension units of a vehicle may be non-linear, and therefore the pseudo forces produced by these elements are non-linear. Furthermore, the contact forces between the train and bridge and between the road vehicles and bridge are functions of dynamic responses of both the bridge and vehicles, as are the self-excited wind forces. Therefore, the equations of motion of the wind-vehicle-bridge system expressed by Equation 11.45 can be generally written as [33]:

$$\begin{cases} \ddot{\mathbf{q}}_b + 2\boldsymbol{\xi}_b\boldsymbol{\omega}_b\dot{\mathbf{q}}_b + \boldsymbol{\omega}_b^2\mathbf{q}_b = \boldsymbol{\Phi}_b^{\mathrm{T}}\mathbf{F}_b(\mathbf{q}_b, \dot{\mathbf{q}}_b, \ddot{\mathbf{q}}_b, \mathbf{q}_r, \dot{\mathbf{q}}_r, \ddot{\mathbf{q}}_r, \mathbf{q}_h, \dot{\mathbf{q}}_h, \ddot{\mathbf{q}}_h) \\ \ddot{\mathbf{q}}_r + 2\boldsymbol{\xi}_r\boldsymbol{\omega}_r\dot{\mathbf{q}}_r + \boldsymbol{\omega}_r^2\mathbf{q}_r = \boldsymbol{\Phi}_r^{\mathrm{T}}\mathbf{F}_r(\mathbf{q}_b, \dot{\mathbf{q}}_b, \ddot{\mathbf{q}}_b, \mathbf{q}_r, \dot{\mathbf{q}}_r, \ddot{\mathbf{q}}_r) \\ \ddot{\mathbf{q}}_h + 2\boldsymbol{\xi}_h\boldsymbol{\omega}_h\dot{\mathbf{q}}_h + \boldsymbol{\omega}_h^2\mathbf{q}_h = \boldsymbol{\Phi}_h^{\mathrm{T}}\mathbf{F}_h(\mathbf{q}_b, \dot{\mathbf{q}}_b, \ddot{\mathbf{q}}_b, \mathbf{q}_h, \dot{\mathbf{q}}_h, \ddot{\mathbf{q}}_h) \end{cases} \tag{11.53}$$

Iterative schemes are required to find numerical solutions of the above equation. In this study, the step-by-step explicit integration method [35] is adopted and the iterative scheme is summarized as follows:

1. Decide the time duration and time interval, Δt, for the dynamic response time histories required. Generate or provide wind speed time histories along the bridge deck. Take the responses of bridge, train and road vehicle subsystems at the last time step t as the initial values of the iteration for the current time step $t+\Delta t$.
2. Use the responses of the bridge, train and road vehicle subsystems at the last iteration step k to estimate the pseudo forces acting on the trains and road vehicles for the current iteration step $k+1$.
3. Determine the contact forces and wind forces, and solve the equation of motion of the train and road vehicle subsystems in Equation 11.53 using the step-by-step explicit integration method to find the responses of trains and road vehicles at the iteration step $k+1$.
4. Use the train and road vehicle responses at the iteration step $k+1$ and the bridge responses at the iteration step k to estimate the pseudo forces acting on the bridge for the current iteration step $k+1$.
5. Determine the contact forces and wind forces, and solve the equation of motion of the bridge subsystem in Equation 11.53, using the step-by-step explicit integration method to find the bridge responses at the iteration step $k+1$.
6. Use the train and road vehicle responses obtained in step 3 and the bridge responses obtained in step 5 for the next iteration step until the convergence is reached for all of three equations in Equation 11.53. At this point, a new step can be started from step 1.

11.7 Verification by Case Study: Tsing Ma Bridge

To examine the proposed framework for multiple loading-induced stress analysis, a case study is performed in this section [33], for which the Tsing Ma Bridge in Hong Kong is selected. In consideration of the nature of the case study, a time period of 140 seconds from 08:24:00 pm to 08:26:20 pm on 19th November 2005 is selected, during which one train and 29 heavy road vehicles of over four tons were identified running on the bridge. During this time period, bridge stress response time histories recorded by dynamic strain gauges (see Figure 11.23) were used.

Information on the train, such as the arriving instant, running speed and heading direction, was converted from the strain response time history recorded by a special set of strain gauges arranged under the railway beams at Section L (see Figure 11.23). Information on 29 heavy road vehicles was recorded by dynamic weigh-in-motion (WIM) stations, including the lane number, arriving instant, running speed and heading direction. Wind data were collected by the anemometers installed at both bridge deck and towers.

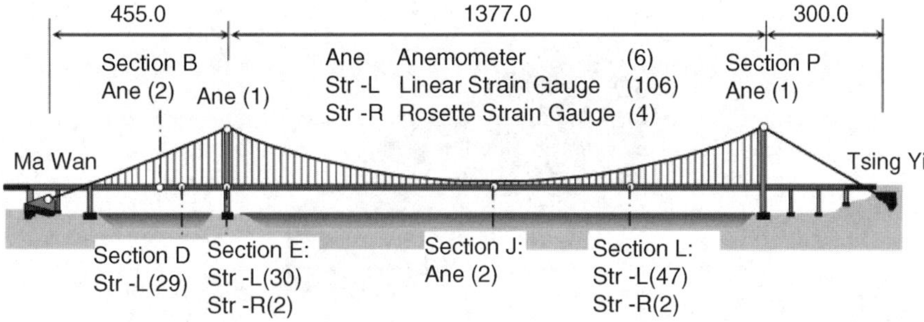

Figure 11.23 Distribution of dynamic strain gauges and anemometers in the Tsing Ma Bridge (from [33]) (Reproduced with permission from ASCE).

Figure 11.24 FEM of a railway vehicle (from [33]) (Reproduced with permission from ASCE).

11.7.1 Finite Element Models of Bridge, Train and Road Vehicles

Considering the requirement of stress analysis of local bridge components, the SHM-oriented FEM of the Tsing Ma Bridge established in Section 11.2 and shown in Figure 11.3 is used. The train running on the bridge is an eight-carriage (vehicle) train of a total length of approximately 182 m.

Figure 11.24 shows the FEM of a four-axle railway vehicle, which is composed of one car body, two bogies and four wheel sets. The car body, bogies and wheel-sets are regarded as rigid components. The primary and secondary suspension systems are characterized by two springs and dashpots modeled by spring elements and dashpot elements in either the horizontal direction or vertical direction. In the modeling of railway vehicles, the actual bogie loads of the typical railway vehicle running over the bridge and recorded by the WASHMS in the concern time duration are used; other train parameters, such as damping ratio and springs, were provided by MTR Corporation of Hong Kong, Limited [36]. The total degrees of freedom of the railway vehicle are 23. The 23 modal frequencies and modal shapes of the vehicle model within a frequency range up to about 30 Hz are used in the subsequent dynamic analysis.

The road vehicles of over four tons running on the bridge are regarded as typical high-sided road vehicles. Figure 11.25 shows the FEM of a two-axle high-sided road vehicle, which is composed of a car body, two axle-sets and the connections between the two components. The car body and axle-sets in each vehicle model are regarded as rigid components. Springs and dashpots are respectively modeled by spring elements and dashpot elements to connect the vehicle bodies and axle sets. Axle sets and contact points at bridge deck surfaces are connected by spring elements and dashpot elements representing the dynamic characteristics of the tires. The actual axle loads recorded by the WASHMS are adopted, together with other parameters used by Xu and Guo [37] for a two-axle high side road vehicle. The total degrees of freedom of one road vehicle are 11. The 11 modal frequencies and modal shapes of the road vehicle within a frequency range up to about 12 Hz are used in the subsequent dynamic analysis.

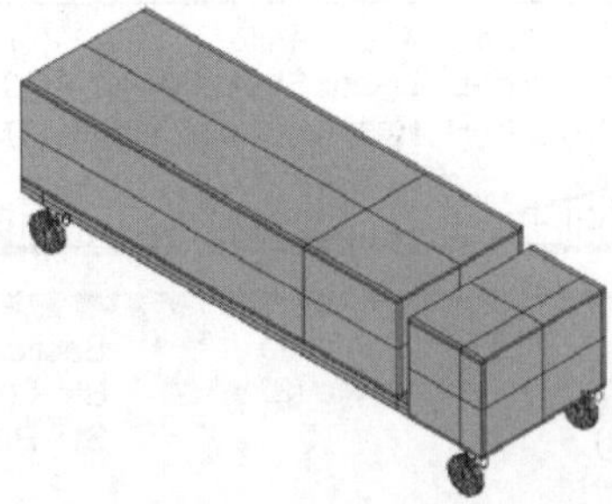

Figure 11.25 FEM of a road vehicle (from [33]) (Reproduced with permission from ASCE).

11.7.2 Rail Irregularities and Road Roughness

The rail irregularities and the roughness of the bridge deck surface are important factors that determine the contact forces between train and bridge and between road vehicles and bridge. The rail irregularities and road surface roughness are both considered in this study. They are assumed to be a zero-mean stationary Gaussian random process and expressed through the inverse Fourier transformation on a power spectral density function:

$$y_v(x) = \sum_{k=1}^{N} \sqrt{2S(f_k)\Delta f}\cos(2\pi f_k x + \theta_k) \tag{11.54}$$

where:

$S(f)$ is the power spectrum density (PSD) function;
$f_k = f_l + (n - 1/2)\Delta f$;
$\Delta f = (f_u - f_l)/N$;
f_u and f_l are, respectively, the lower and upper cutoff frequencies;
θ_k is the random phase angle uniformly distributed between 0 and 2π.

Vertical and lateral rail irregularities are considered for right and left rails of the railway track. The rail irregularity in railway engineering is often represented by a one-sided PSD function. The PSD functions of rail irregularities developed by the Research Institute of China Railway Administration [35] are used in the case study. All rail irregularities are expressed using a unified rational formula as:

$$S(f) = \frac{A(f^2 + Bf + C)}{f^4 + Df^3 + Ef^2 + Ff + G} \tag{11.55}$$

where:

$f = 1/\lambda$ (m^{-1}) is the spatial frequency in cycle/m (λ is the wavelength);
A to G are the specific parameters, but these are different for vertical and lateral rail irregularities. The values of these parameters can be found in [35].

The road surface roughness is also described by a PSD function in both the vertical and lateral directions. The following PSD functions are used for simulating road surface roughness [38,39]:

$$S(f) = A_r\left(\frac{f}{f_0}\right)^{-2} \tag{11.56}$$

where:

f is the spatial frequency (cycle/m);
f_0 is the discontinuity frequency of $(1/2\pi)$ (cycle/m);
A_r is the roughness coefficient (m^3/cycle), depending on the road condition.

11.7.3 Wind Force Simulation

In this case study, wind forces act on both the bridge and the road vehicles running on the bridge decks, but not on the train. The first step of wind force simulation is the numerical simulation of the stochastic wind velocity field at the entire bridge deck. The wind information recorded by the anemometers at

Section J of the bridge deck, as well as some empirical formulae, are utilized for stochastic wind velocity field simulation.

In the selected time period, the measured mean wind speed was 11.91 m/sec and its direction was from north to south. The standard deviation of wind turbulence was 1.310 m/sec in the horizontal direction and 0.679 m/sec in the vertical direction. The integral length scales were estimated at 256.7 m in the horizontal direction and 40.8 m in the vertical direction. A fast spectral representation approach [17] is adopted for the digital simulation of stochastic wind velocity field in the horizontal and vertical directions at points along the bridge deck. The exponential form of coherence function is adopted to reflect turbulent wind correlation along the bridge deck in both horizontal and vertical directions. A total of 120 simulation points along the bridge with an interval of 18.0 m are considered in the simulation of wind field. The average elevation of the bridge deck is taken as 60 m. To keep the spatial correlation of turbulent wind along the bridge deck for road vehicles, the wind speeds at the centre of gravity of a road vehicle are interpolated from the wind velocities at two adjacent simulation points of the bridge deck.

Buffeting forces and self-excited forces on the bridge deck are both considered in the case study. The drag, lift and moment coefficients of the bridge deck at the zero wind angle of attack with respect to the deck width of 41 m were 0.135, 0.090, and 0.063, and their first derivatives were −0.253, 1.324, and 0.278. The buffeting forces acting at the center of elasticity of the deck cross section were then distributed to the nodes of the deck section in terms of wind pressure distribution around the deck section. The self-excited forces at the bridge deck were determined by Equation 11.51. Further information on the flutter derivatives from wind tunnel tests of the Tsing Ma Bridge could be found in [40,41].

In summary, due to the lack of wind tunnel results on lateral flutter derivatives, only the vertical and rotational motion flutter derivatives derived from wind tunnel test results were taken into account in the simulation of self-exciting forces. By using the non-linear least-squares method to fit the measured flutter derivatives at different reduced frequencies, a total of 12 frequency-independent coefficients were determined and used in the buffeting analyses.

Since the train ran on the lower level within the bridge deck, no wind forces were acting on the train. The road vehicles ran on the upper level of the bridge deck, so that wind forces did act on the road vehicles. For road vehicles, the derivatives of the aerodynamic force coefficients in Equation 11.52 are taken as zero because of lack of measurement data.

11.7.4 Selected Results

The Tsing Ma bridge deck is a hybrid steel structure consisting of Vierendeel cross-frames supported on two longitudinal trusses acting compositely with stiffened steel plates. Based on the aforementioned information, dynamic stresses of the bridge are computed for all bridge components. The dynamic stresses computed at the locations of two strain gauges (SS-TLS-12 and SP-TLS-02) installed in Figure 11.10 are selected to compare with the measured stresses for verification to some extent [33]. Strain gauges SS-TLS-12 and SP-TLS-02 are actually located at the most critical section (Section L in Figure 11.23), with relatively high stress fluctuations. In addition, strain gauge SS-TLS-12 is installed on the railway beam between the two neighboring cross-frames, and a large bending moment is produced in the beam when the bogies of a train are close to this location. Strain gauge SS-TLS-12 is therefore sensitive to local stress responses induced by train bogies.

Strain gauge SP-TLS-02 is installed on the diagonal member of the outer longitudinal truss, which mainly experiences tension or compression and is not sensitive to local stresses induced by train bogies. For the local member with the strain gauge SS-TLS-12, the railway loads are transferred almost directly to the member; however, highway loads are not, as these are transferred to the member through the stiffened plates and the cross-frames. For the local member with the strain gauge SS-TLS-02, both the highway and railway loads are transferred to the member through the stiffened plates, the railway beams, the cross-frames and the longitudinal trusses.

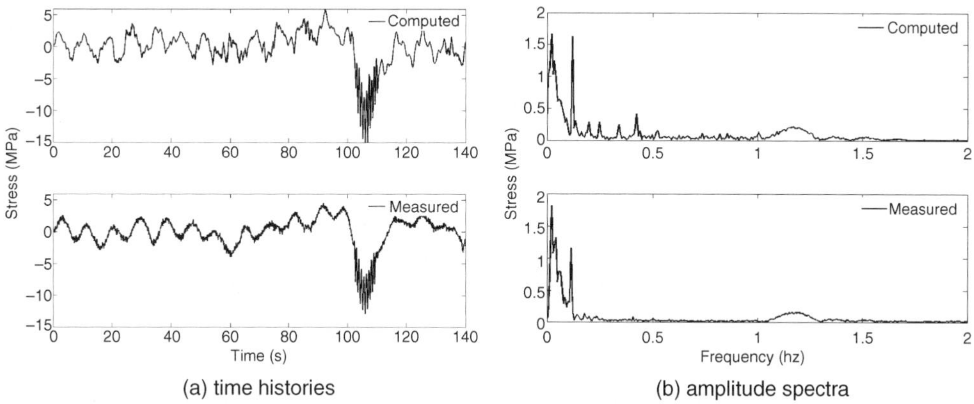

Figure 11.26 Comparison of measured and computed stress responses (SS-TLS-12) to wind, train and road vehicles (from [33]) (Reproduced with permission from ASCE).

Displayed in Figure 11.26a and Figure 11.27a are the computed and measured 140-second stress time histories at the locations concerned. In addition, the computed and measured stress amplitude spectra computed from the corresponding time histories are plotted in Figure 11.26b and Figure 11.27b for a frequency range less than 2 Hz. The sample frequency for measured stress responses is 51.2 Hz, and the number of points involved in the FFT analysis is 7168.

Although the duration of time history used is not long enough to eliminate potential statistical bias, it is considered to be acceptable because the FFT analysis results are used for comparison only. It can be seen from the figures that both time histories and spectra of computed stresses are close to those measured by the corresponding strain gauges, which indicates to some extent that the proposed framework could effectively predict the dynamic stress responses at local components of a long suspension bridge under railway, highway and wind loading.

From Figures 11.26a and 11.27a it can also be seen that the largest stress fluctuations appear between the first 100–115 sec of the stress time histories. This is because, at this instant, the train actually ran close to the locations of the concerned strain gauges. These results, together with others, show that running railway vehicles actually play a predominating role in bridge dynamic stress responses,

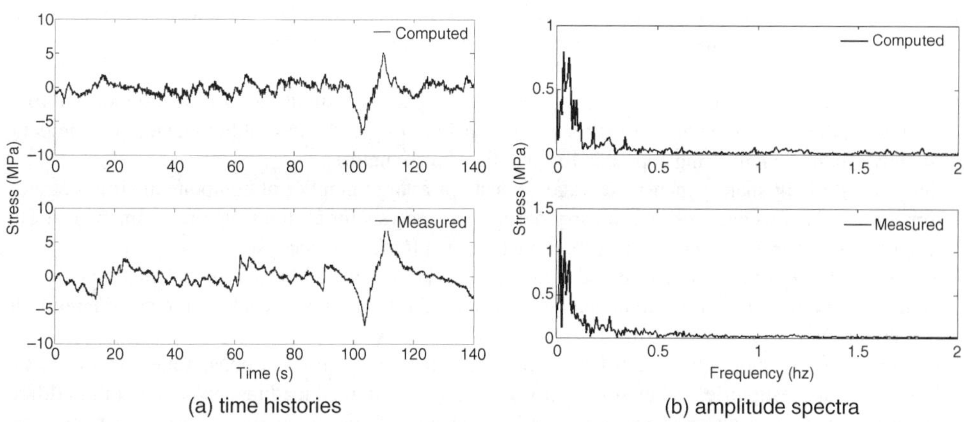

Figure 11.27 Comparison of measured and computed stress responses (SP-TLS-02) to wind, train and road vehicles (from [33]) (Reproduced with permission from ASCE).

compared to running road vehicles and fluctuating wind loading. Nevertheless, the measured stress time histories used not only verify the framework for the case with a combination of railway, highway and wind loadings, but also verify the framework for the case without railway loading.

By taking the stress time histories shown in Figure 11.26a as an example, the stress time histories between 0–80 seconds are actually due to highway and wind loadings only. Since the computed stress time history is found to be similar to the measured one during the period from 0–80 sec, this indicates that the method is also effective in predicting stress response due to road vehicles and wind.

The comparative results demonstrate that the proposed framework could effectively predict the dynamic stress responses of local components of the bridge under railway, highway and wind loading. The results also show that the running train played a predominant role in bridge stress responses, compared to running road vehicles and fluctuating wind loading [33].

11.8 Fatigue Analysis of Long-Span Suspension Bridges under Multiple Loading

Given that the long-span period is involved in fatigue damage accumulation in long-span bridges and the computation time for the dynamic stress responses to the combined action of multiple loading is very demanding, the proposed framework in Sections 11.6 and 11.7 for dynamic stress analysis of long-span bridges under combined loadings will be simplified. This section presents a general framework for fatigue analysis of a long-span suspension bridge under multiple loading by integrating computer simulation with measurement data from a SHM system [42].

Taking the Tsing Ma Bridge as an example, a computationally efficient engineering approach is first proposed for dynamic stress analysis of the bridge under railway, highway and wind loading. The fatigue-critical locations are then determined for key bridge components, and databases of the dynamic stress responses at the critical locations are established. Based on the databases, 120 years of time histories of the dynamic stresses induced by individual loading are generated. The corresponding stress time histories due to the combined action of multiple loading are compiled. Fatigue analysis is then performed to compute the cumulative fatigue damage over the design life of 120 years. The cumulative fatigue damage induced by individual loading and the damage magnification due to multiple loading are finally investigated.

11.8.1 Establishment of Framework

To establish a framework for fatigue analysis of long-span suspension bridges under combined action of railway, highway and wind loading, some key issues need to be considered.

- First, given that a great number of stress time histories caused by multiple loading are required for a complete fatigue assessment of a long-span suspension bridge, it is desirable to develop a computationally efficient engineering approach for dynamic stress analysis.
- Second, as a long-span suspension bridge consists of a large number of components, it is not only impossible, but also unnecessary, to carry out fatigue analysis for all the structural components. The fatigue-critical locations should be properly determined for fatigue analysis.
- Third, the design life of the concerned bridge should be designated before the calculation of wind-induced stress responses for fatigue analysis, because the wind intensity taken into consideration in the fatigue analysis is related to the design life.
- Fourth, databases should be established in order to generate the stress response time histories of the bridge over its design life. Databases of railway, highway and wind loading need to be built in different ways because of different properties of loading type. Wind-induced stress responses are computed in one hour to build a database for fatigue analysis. As urban passenger trains often follow a regular timetable that is similar on different days, railway-induced stress time histories are computed

in one day, and daily time histories are used to compose the database. The database of highway stress time histories is also composed of daily time histories, as highway traffic conditions among different days are found to be similar.

- Fifth, multiple loading-induced fatigue damage should be calculated based on the stress responses induced by multiple types of loading, rather than the summation of damage induced by individual loading. Fatigue analysis should therefore be applied directly to the multiple loading-induced stress time histories, which is the superposition of stress responses induced by three individual loadings.
- Finally, it is recommended that measured data be adopted in the computation of fatigue damage as far as possible, to represent better the real conditions of the bridge.

Taking the above issues into consideration, a framework for the fatigue analysis of a long-span suspension bridge under multiple types of loading within the design life is proposed and summarized as follows:

1. Develop a computationally efficient engineering approach for dynamic stress analysis.
2. Designate the design life of the concerned suspension bridge.
3. Determine the fatigue-critical locations of key structural components of the bridge.
4. Establish databases of the dynamic stress responses at the fatigue-critical locations induced by railway, highway and wind loading, respectively, using an engineering approach.
5. Generate the multiple loading-induced dynamic stress time histories at the fatigue-critical locations within the design life based on the databases established in step (4).
6. Set the initial damage $D_0 = 0$ and time step Δt.
7. Count the number of stress cycles at different stress range levels from the multiple load-induced stress time history in the kth time step using the rainflow counting method [30].
8. Compute the increase in the level of fatigue damage ΔD_k in the kth time step for a given fatigue-critical location.
9. Compute the cumulative fatigue damage $D_k = D_{k-1} + \Delta D_k$ and the cumulative service time $t_k = t_{k-1} + \Delta t$ in the kth time step; and
10. Move to the next time step and go from step (7) to the end of the design life.

Fatigue damage accumulated in the time step can be calculated using the Palmgren-Miner rule based on the two-slope S-N curves in [29] for simplicity. The fatigue assessment of a long-span bridge under multiple loadings by using the continuum damage mechanics needs further investigation.

$$\Delta D = \Delta D_H + \Delta D_L \tag{11.57}$$

where:

$$\Delta D_H = \sum_{i=1}^{N_1} \frac{n_i (\sigma_{r,i})^m}{K_2} \quad \text{if } \sigma_{r,i} \geq \sigma_{r,0} \tag{11.58a}$$

$$\Delta D_L = \sum_{i=1}^{N_2} \frac{n_i (\sigma_{r,i})^{m+2}}{K_2 (\sigma_{r,0})^2} \quad \text{if } \sigma_{r,i} < \sigma_{r,0} \tag{11.58b}$$

where:

K_2 and m are constants relevant to the fatigue detail;
K_2 is determined from constant amplitude experiments corresponding to a probability of failure of 2.3%;
n_i is the applied number of stress cycles at the stress range level $\sigma_{r,i}$;

N_1 and N_2 are the number of stress range levels $\sigma_{r,i}$ in the stress time histories above and below $\sigma_{r,0}$, respectively;
$\sigma_{r,0}$ is the constant amplitude fatigue limit, which is defined as $N = 10^7$.

11.8.2 Simplifications used in Engineering Approach

Although the coupled dynamic approach presented in Section 11.6 provides an accurate estimation of bridge dynamic stresses, the complexity of the framework makes computation very time-consuming. It is impractical to apply the coupled dynamic approach to fatigue analysis of a long-span suspension bridge. In this regard, two major simplifications are adopted here to simplify the coupled dynamic approach and lead to the engineering approach, based on the features and properties of long-span suspension bridges under normal operation condition, with a trade-off between computational accuracy and efficiency.

The first major simplification is to neglect coupled effects of multiple load-induced dynamic stresses, because wind speed is normally not too high when vehicles are running on the bridge. Under extreme wind conditions, such as when a strong typhoon is blowing, bridge traffic management systems come into effect and the bridge is closed to traffic. Therefore, it is reasonable to assume that the coupled effects of dynamic stress responses of the bridge induced by railway, highway and wind loading are insignificant under normal operation conditions, and that the bridge motions induced by railway, highway and wind loading are considered to be independent of each other. As a result, the bridge stress response at a given point induced by the combined effects of the three types of loading can be approximately obtained by the synchronous superposition of stress responses induced by individual loadings:

$$\sigma_b = \sigma_{rb} + \sigma_{hb} + \sigma_{wb} \tag{11.59}$$

where σ_{rb}, σ_{hb}, and σ_{wb} are the bridge stress responses induced by railway, highway, and wind loading, respectively.

Another major simplification is to neglect the dynamic magnification related to vehicle dynamics. This is because the self-weight of a long-span suspension bridge carrying both trains and road vehicles is much larger than the weight of a train and/or a series of road vehicles. Furthermore, this study concerns the dynamic stress response of the bridge rather than the safety of vehicles. As a result, trains and road vehicles can be simplified as a series of moving forces on the bridge deck. Moreover, through the analysis of three resonance conditions, it is found that the impact factor of a long-span cable-supported bridge under a series of moving forces is often small [43].

Based on the above, the bridge stress responses induced by trains and road vehicles can be calculated on the basis of a series of static forces and stress influence lines in this study, i.e. without considering dynamic magnification. Wind-induced stress responses of a long-span suspension bridge are, however, computed based on the aerodynamic analysis.

11.8.3 Dynamic Stress Analysis using Engineering Approach

On the basis of the two major simplifications proposed in the preceding section, the engineering approach for dynamic stress analysis of long-span suspension bridges under multiple loading can be implemented by the following four steps:

1. Analysis of railway-induced bridge dynamic stress based on stress influence lines.
2. Analysis of highway-induced bridge dynamic stress based on stress influence lines.
3. Analysis of wind-induced dynamic stress using buffeting theory.
4. Combination of the stress responses induced by multiple types of loading by the superposition method in the time domain.

The procedures for the first three steps are presented as follows.

To determine railway-induced dynamic stress responses of a long-span suspension bridge, the stress influence lines should be established. To derive the stress influence lines for a given fatigue-critical location, the stress response at the designated location due to a unit vertical force moving along the railway tracks from one end of the bridge to the other end must be computed. The abscissa of the stress influence line denotes the position of the unit vertical force in the longitudinal direction of the bridge, and the ordinate of the stress influence line, the so-called stress influence coefficient, $\mathbf{\Phi}$, is the stress response induced by the unit vertical force at the corresponding position.

Railway loading is then determined in terms of a series of moving vertical forces. For example, the railway loading for an eight-car train with 32 wheel-sets can be represented by 32 vertical forces, with each force coming from one wheel-set. The railway loading information is used to determine the railway loading for a given train and to simulate the railway vehicle flow running along the bridge. This information can be obtained from the train data recorded at the bridge site and includes at least the number of trains, the number and types of railway vehicles in a train, arrival instant, running speed, heading direction, railway track in use, number of bogies, bogie weight and bogie spacing. Underlying assumptions include a constant speed for a typical train running across the bridge on a given railway track. The computational procedure of the stress time history under railway loading is summarized as follows:

1. Establish the database of railway loading stress influence lines for a given stress output point.
2. Update the train information at the instant t, which includes the number of railway vehicles comprising the train and wheel-set locations.
3. Determine the vertical loading $f_{\mathrm{k,ij}}$ due to the ith wheel-set in the jth railway vehicle on the kth railway track using the train information obtained.
4. Determine the stress influence coefficient $\Phi_{\mathrm{k,ij}}$ due to the ith wheel-set in the jth railway vehicle on the kth railway track using the stress influence line database.
5. Calculate the railway load-induced stress σ_{rb} by the triple summation of the product of the stress influence coefficient $\Phi_{\mathrm{k,ij}}$ and axle load $f_{\mathrm{k,ij}}$.
6. Move to the next time instant and go from step (2) to the end of the given duration of stress responses.

To consider the dynamic stresses induced by road vehicles running along the bridge on different traffic lanes, highway loading stress influence line for each traffic lane should be established, and highway loading should be determined based on the measured road vehicle data. For example, the highway loading of a typical four-axle road vehicle can be represented by four vertical forces, with each load coming from one axle. To determine not only the highway loading due to a given road vehicle, but also the road vehicle flow running along the bridge, the highway loading information should include at least the number and types of road vehicles, traffic lane in use, arrival time, heading direction, running speed, axle number, axle weight and axle spacing. Underlying assumptions include a constant speed and no switching of the traffic lane for a given road vehicle running along the bridge. The computational procedure of the stress time history under highway loading can be derived in a similar way to that under railway loading.

Long-span suspension bridges that are built in wind-prone regions suffer considerable buffeting-induced vibration. Therefore, wind-induced dynamic stress responses should also be considered. Wind-induced dynamic stress response time histories can be computed using a step-by-step procedure. In the first step, wind characteristics in a given time period, e.g. one hour or ten minutes, are identified from wind data collected by anemometers installed at the bridge site. In the second step, the stochastic wind velocity at the simulation points along the bridge deck, and the normal mean wind speed in the time period of concern, are generated on the basis of the wind characteristics acquired from the measured wind data. The buffeting and self-exciting forces on the surface of the bridge deck are then computed. In the third step, the wind-induced stress responses in the time period of concern are computed at the given stress output points, using an integration method. The procedure then moves to the next time period and goes from the first step to the end of the given duration of stress responses.

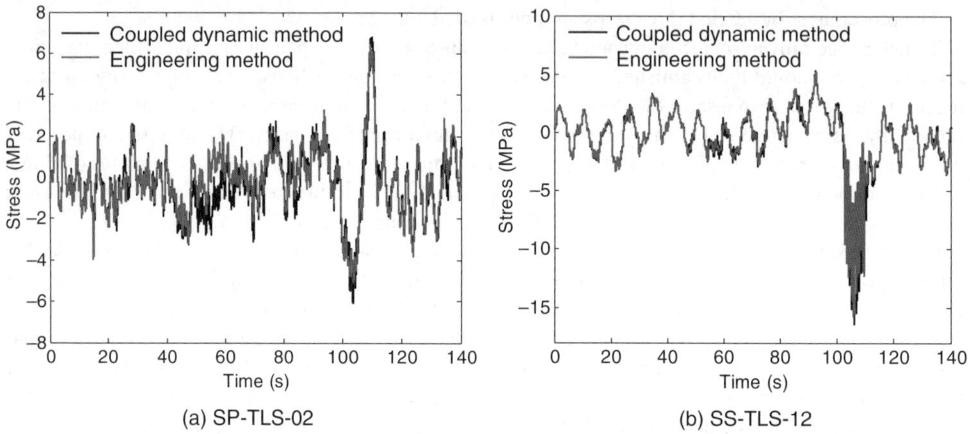

Figure 11.28 Stress time histories under railway, highway, and wind loading (from [42]) (Reproduced with permission from Elsevier).

11.8.4 Verification of Engineering Approach

To validate the computational accuracy of the engineering approach, the stress responses induced by multiple types of loading computed using the engineering approach are compared with those calculated, using the coupled dynamic approach. A 140-second dynamic stress time history was computed by the coupled dynamic method, as shown in Figures 11.26 and 11.27 and re-plotted in Figure 11.28. It is used as a reference for comparison.

During this period, there was one train running on the north track heading toward Hong Kong Island and 29 road vehicles weighing over four tons running along the bridge. The normal mean wind speed is 11.91 m/sec. The standard deviation of turbulent wind is 1.310 m/sec in the horizontal direction and 0.679 m/sec in the vertical direction. The integral length scales are 256.7 m in the horizontal direction and 40.8 m in the vertical direction.

To use the engineering approach to compute 140-second dynamic stress time history, the railway and highway loading stress influence lines of the bridge should be established. For each of the two strain gauges specified, two stress influence lines corresponding to the two railway tracks are established. Six stress influence lines corresponding to the six highway traffic lanes are also established. All stress influence lines are generated by structural analyses based on the finite element model of the bridge.

In the computation of stress response, the train and road vehicle information is updated at each time step, and the length of the time step Δt is 0.02 sec. The acquired wind characteristics are adopted to generate the stochastic wind velocity field of the entire bridge deck, then the buffeting and self-excited forces on the bridge deck are estimated. The stress time histories under wind loading at the concerned locations are computed. Based on the stress responses induced by railway, highway and wind loading, respectively, the multiple load-induced stress responses are computed using the superposition method.

The result, computed by the engineering method, is also plotted in Figure 11.28 for comparison. The figure shows that the stress time histories computed using the engineering approach match well with those from the coupled dynamic approach. The relative differences in the peak-to-peak stress responses (the response obtained by the coupled dynamic method minus that by the engineering approach, divided by one predicted by the coupled dynamic method) at the location of strain gauges SP-TLS-02 and SS-TLS-12 are 16.1% and 5.4%, respectively. The 16.1% error is the worst case, but this error will not exaggerate the final fatigue damage because fatigue damage depends on a large number of stress ranges rather than peak stresses. The results demonstrate that the level of computational accuracy of the engineering approach is acceptable.

In addition to computational accuracy, computational efficiency is also an important factor for the engineering approach. Most of the trains running across the bridge follow a timetable on a daily basis and, thus, the cycle of railway loading is close to one day. As hundreds of trains and thousands of road vehicles run across the bridge every day, the computational efficiency of the engineering approach is tested for one day only. The measured train, road vehicle, wind and strain data collected on 19 November 2005 are used for dynamic stress computation and comparison. This day was chosen as the wind was relatively strong and the traffic was heavy.

The gross train weight (GTW) ranged from 282.7 and 402.2 tons, with running speed ranging from 25–38 m/sec. On the same day, 8225 and 8623 heavy road vehicles weighing over 30 kN ran across the bridge, using the north and south three-lane carriageway, respectively. The gross vehicle weight (GVW) ranged from 4–54 tons. The mean wind speed and direction were obtained from wind data recorded by the anemometers installed at the middle of the bridge deck. The hourly mean wind speed perpendicular to the bridge axis ranged from 2–13 m/sec on that day. Turbulence intensities were taken as 24% and 17% in the horizontal and vertical directions by considering the most turbulent cases in the field. The integral length scales were taken as 251 m and 56 m in the horizontal and vertical directions, respectively.

As wind-induced dynamic stress responses are dominated by vibration modes of a relatively low frequency, only the first 153 modes of vibration up to 2 Hz were included in this case for the stress response computation. 24-hour time periods of the railway-, highway- and wind-induced stress responses were calculated using the engineering approach. Based on the daily stress responses induced by the three individual loadings, the multiple load-induced stress response was obtained by superposition.

The daily multiple load-induced stress time histories computed using the engineering approach at the location of strain gauges SS-TLS-12 and SP-TLS-02 are shown in Figure 11.29a and Figure 11.30a, respectively. The measured ones are shown in Figure 11.29b and Figure 11.30b for comparison.

It can be seen that the computed stress time histories agree well with the measured ones. The relative differences in the root mean square (RMS) of the stress responses are calculated to determine the relative differences (the measured value minus the computed one, divided by the measured one) at the two typical locations. The relative differences in the RMSs of the stress responses are 12.9% and 8.4% for strain gauges SS-TLS-12 and SP-TLS-02, respectively.

The coupled dynamic approach is actually not applicable for computation of the daily dynamic stress responses, as it takes an intolerably long time, whereas only several minutes are required for the engineering approach if the stress influence lines are available. The small relative differences between the computed and measured time histories and a short computation time for the engineering approach demonstrate a high level of computational efficiency and an acceptable level of computational accuracy.

11.8.5 Determination of Fatigue-Critical Locations

The above section proposes an engineering approach for dynamic stress analysis. In the next step, the fatigue-critical locations will be determined for the key structural components of a long-span suspension bridge. Given that the main structural components of and loadings on one long-span suspension bridge can be quite different from another, the determination of fatigue-critical locations is case-dependent. The Tsing Ma suspension bridge is taken as an example for illustration. The key structural components of the bridge can be classified into 55 components in 15 groups. The details of the classification of the components in each group are given in Table 11.8 [44].

The fatigue-critical locations are determined through the stress analysis of each component. To make sure that the size of the FEM is not too large to be used for dynamic analysis, some types of bridge components cannot be modeled exactly. For instance, if the orthotropic deck of the bridge were modeled using shell elements, the size of the FEM would be too large to be used. Therefore, the orthotropic deck between the two adjacent cross frames at an interval of 4.5 m was simply modeled by a plate element that was fixed to the two cross-frames at its two ends in the longitudinal direction, and free on the other two sides in the lateral direction. Such a model makes it impossible to obtain actual stresses of

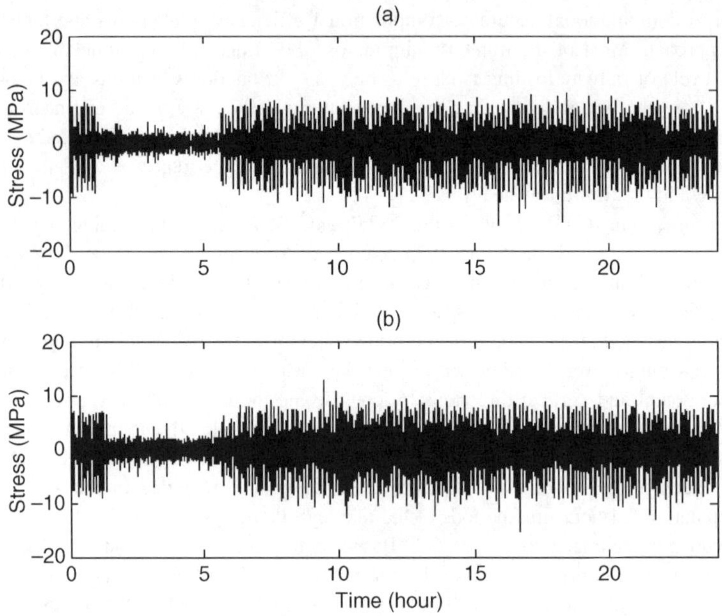

Figure 11.29 Daily stress time histories under multiple types of loading at SP-TLS-02: (a) calculated; (b) measured (from [42]) (Reproduced with permission from Elsevier).

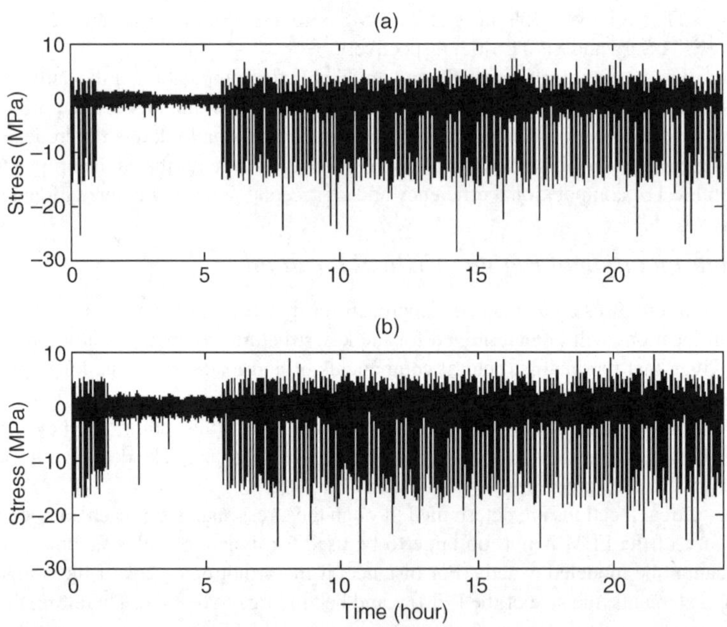

Figure 11.30 Daily stress time histories under multiple types of loading at SS-TLS-12: (a) calculated; (b) measured (from [42]) (Reproduced with permission from Elsevier).

Table 11.8 Classification of the structural components of Tsing Ma Bridge (from [42]) (Reproduced with permission from Elsevier)

Name of group	Name of component	Group no.	Component no.	Serial no.
Suspension cables	Main cables	1	(a)	1
	Strand shoes		(b)	2
	Shoe anchor rods		(c)	3
	Anchor bolts		(d)	4
	Cable clamps & bands		(e)	5
Suspenders	Hangers	2	(a)	6
	Hanger connections: stiffeners		(b)	7
	Hanger connections: bearing plates		(c)	8
Towers	Legs	3	(a)	9
	Portals		(b)	10
	Saddles		(c)	11
Anchorages	Chambers	4	(a)	12
	Pre-stressing anchors		(b)	13
	Saddles		(c)	14
(Piers: M1, M2, T1, T2, T3)	Legs	5	(a)	15
	Cross-beams		(b)	16
Outer longitudinal trusses	Top chord	6	(a)	17
	Diagonal		(b)	18
	Vertical post		(c)	19
	Bottom chord		(d)	20
Inner longitudinal trusses	Top chord	7	(a)	21
	Diagonal		(b)	22
	Vertical Post		(c)	23
	Bottom chord		(d)	24
Main cross-frames	Top web	8	(a)	25
	Sloping web		(b)	26
	Bottom web		(c)	27
	Bottom chord		(d)	28
Intermediate cross-frames	Top web	9	(a)	29
	Sloping web		(b)	30
	Bottom web		(c)	31
	Bottom chord		(d)	32
Plan bracings	Upper deck	10	(a)	33
	Lower deck		(b)	34
Deck	Troughs	11	(a)	35
	Plates		(b)	36
Railway beams	T-sections	12	(a)	37
	Top flanges		(b)	38
	Connections		(c)	39

(continued)

Table 11.8 *(Continued)*

Name of group	Name of component	Group no.	Component no.	Serial no.
Bearings	Rocker bearings at Ma Wan Tower	13	(a)	40
	PTFE bearings at Tsing Yi Tower		(b)	41
	PTFE bearings at Pier T1		(c)	42
	PTFE bearings at Pier T2		(d)	43
	PTFE bearings at Pier T3		(e)	44
	PTFE bearings at Tsing Yi anchorage		(f)	45
	Rocker bearings at M2		(g)	46
	PTFE bearings at M1		(h)	47
	Hinge bearing at Lantau anchorage		(i)	48
Movement joints	Highway movement joint	14	(a)	49
	Railway movement joint		(b)	50
Tsing Yi approach deck	Top chord	15	(a)	51
	Diagonal		(b)	52
	Vertical post		(c)	53
	Bottom chord		(d)	54
	Diagonals (K-bracings)		(e)	55

the orthotropic deck. Apart from these components, some other types are neglected because they are not critical to fatigue in practice. The bridge components taken into consideration for fatigue analysis in this study are highlighted in grey in Table 11.8.

As the fatigue damage of the Tsing Ma Bridge is induced by the combined effect of railway, highway and wind loading, the fatigue-critical locations should be determined on the basis of the multiple types of loading. However, this is very difficult, because so many stress analyses are required for a great number of structural components under a large number of loading combinations in which different intensities of the three loadings must be considered. Some simplifications are therefore necessary.

The fatigue damage induced by railway and highway loading was separately investigated and it was found that, for most bridge components except for the upper deck, the fatigue damage was mainly caused by moving trains and that the contribution of moving road vehicles was small. In addition, wind-induced fatigue damage to the bridge was not significant [45].

Therefore, railway loading is a dominant factor for the fatigue damage of the bridge. Given that almost all of the trains running across the Tsing Ma Bridge since November 2005 have been 16-bogie trains, a standard train is defined to represent all 16-bogie trains by taking the weight of each bogie as the mean weight of the relevant bogies of all 16-bogie trains in November 2005. The standard train has a fixed configuration, and the railway loading of the train is represented by 32 vertical forces. The standard train is then adopted to compute the railway-induced dynamic stress responses of members in a given type of bridge component, and then the responses are compared to each other to determine the fatigue-critical members and locations.

Equations 11.58a and 11.58b indicate that fatigue damage is the function of the stress range level σ_r and number of stress cycles n. To simplify the criteria for determining fatigue-critical location, an assumption is adopted that the number of stress cycles induced by a standard train is almost the same for all components of the same type, and difference only exists in the stress range level. This assumption is acceptable because the stress fluctuations at all components of the same type induced by the standard train running across are found to have a similar pattern. In addition, the equations demonstrate that the damage is most sensitive to the maximum stress range $\Delta\sigma_{max}$, because fatigue damage is a function of $(\sigma_r)^m$ or $(\sigma_r)^{m+2}$.

$\Delta\sigma_{max}$ is therefore selected as the index for determining the fatigue-critical locations of bridge components of the same type. To make the problem manageable, $\Delta\sigma_{max}$ is approximately decided by using the

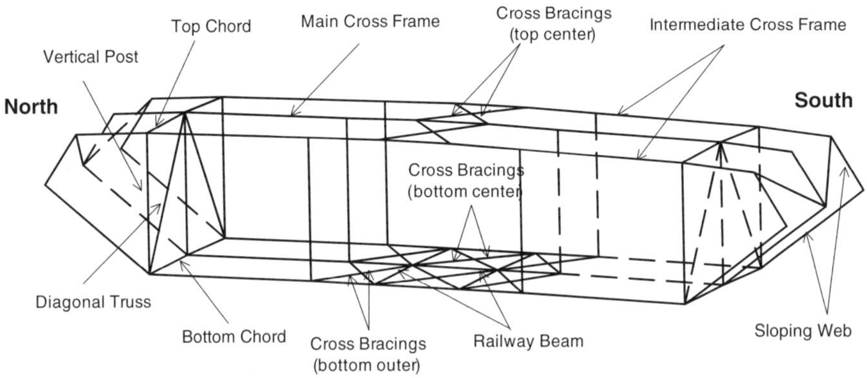

Figure 11.31 Major structural components of bridge deck (from [42]) (Reproduced with permission from Elsevier).

difference of the maximum and minimum stress in the stress time history, based on the principle of the level crossing method. As fatigue is critical to the tension and reversal members, additional structural analysis should be performed to check the net stress in the member due to the dead and superposed dead loads, plus an extreme live load. If it is positive, then the member is finally defined as a fatigue-sensitive member.

Let us take the diagonal members of outer longitudinal trusses as an example to illustrate the determination of fatigue-critical locations (see Figure 11.31). Given the symmetry of the cross-sections of the bridge, the standard train is supposed to run on the north railway track and, accordingly, only the outer longitudinal truss on the north needs to be considered. The stress time histories at the stress output points of all of the diagonal members of the north outer longitudinal truss are computed based on the standard train running across the bridge on the north railway track. The maximum stress ranges are subsequently estimated.

Figure 11.32 shows the maximum stress ranges of the diagonal members of the north outer longitudinal truss due to the standard train running on the north side of the bridge deck. The potential fatigue-critical locations in the diagonal members of the truss can be determined from the figure – the diagonal

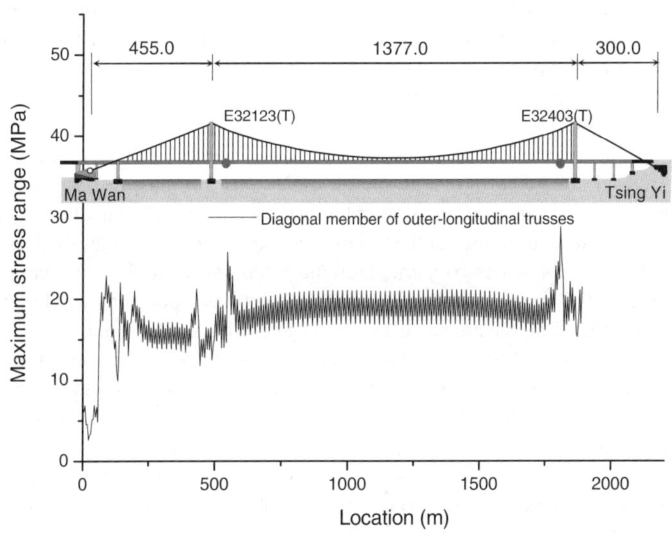

Figure 11.32 Maximum stress ranges of diagonal members in north outer longitudinal trusses (from [42]) (Reproduced with permission from Elsevier).

member E32123 (T) close to the Ma Wan tower in the main span, and the diagonal member E32403 (T) close to the Tsing Yi tower in the main span. "T" or "B" in brackets denotes that the potential fatigue-critical location is at the top or bottom flange of the cross-section of the member at the two ends. Similar procedures are applied to the other bridge components to determine the potential fatigue-critical locations.

Net stresses at the potential fatigue-critical locations are also checked for determining the final fatigue-critical locations. The results demonstrate that the fatigue-critical sections of the bridge deck are around the bridge towers. Within these fatigue-critical sections, six of the strain points are chosen for fatigue analysis, i.e. the following elements:

- E32123 (T) at the top flange of the outer longitudinal diagonal member close to the Ma Wan Tower;
- E34415 (B) at the bottom flange of the outer longitudinal bottom chord of the Tsing Yi Tower;
- E40056 (T) at the top flange of the inner longitudinal top chord of the Tsing Yi Tower;
- E40906 (B), at the bottom flange of the inner longitudinal bottom chord of the Tsing Yi Tower;
- E39417 (B), at the bottom flange of the T-section of the railway beam of the Tsing Yi Tower; and
- E55406 (T) at the top flange of the bottom web of the cross-frame close to the Tsing Yi Tower.

11.8.6 Databases of Dynamic Stress Responses to Different Loadings

In this section, the databases of dynamic stress responses induced by railway, highway, and wind loading at the critical locations of the Tsing Ma Bridge are established based on the loading information recorded by the WASHMS.

Long-span suspension bridges built in wind-prone regions suffer from considerable wind-induced vibration, which appears within a wide range of wind speeds and lasts for almost the whole design life of the bridge. A joint probability distribution function of the mean wind speed and direction is utilized to describe wind intensity at the bridge site [23]. The distribution of wind speed for any given wind direction is assumed to follow Weibull distribution. The parameters in the distribution are determined from monsoon wind records of hourly mean wind speed and direction during the period from 1 January 2000 to 31 December 2005, which were collected by an anemometer installed on the top of the Ma Wan tower. Given that the measured typhoon wind records are not enough to establish a reliable joint probability distribution, only monsoon wind is of concern in this study.

The maximum wind speed at the top of the tower in each wind direction within the 120-year design life is then obtained from the joint probability distribution. The maximum wind speed obtained at the top of the tower is converted to the average deck level of the bridge. The maximum hourly mean wind speed at the deck level is 25.89 m/sec in the north direction for winds over the overland fetch, and 15.47 m/sec in the south direction for winds over the open-sea fetch [23]. Finally, a database of hourly wind-induced dynamic stress responses at the fatigue-critical locations is established: from 5–26 m/sec at an interval of 1 m/sec for winds over the overland fetch and from 5–16 m/sec at an interval of 1 m/sec for winds over the open-sea fetch. The stress fluctuations induced by wind of a normal hourly mean wind speed less than 5 m/sec are neglected, as they contribute little to fatigue.

The database includes a total of 34 one-hour time histories for each fatigue-critical location. The nominal stress of each fatigue-critical member is computed based on the stresses at five points of the two ends of the member. The hot spot stresses, which reflect the stress concentration at welded joints, should be considered in fatigue analysis [30]. The hot spot stresses at the fatigue-critical locations are determined by multiplying the nominal stresses by the stress concentration factor (SCF).

Note that the value of SCF depends largely on the local geometry of the connection details. However, there are numerous fatigue-critical locations in this bridge, and the local geometry of the connection details at these locations is quite different. In consideration that an SCF of 1.4 was used in the design of the concerned bridge for almost all connections, this number is also used for the six identified fatigue-critical locations in this study. The fatigue damage at fatigue-critical locations refers hereafter to fatigue damage at these hot spots.

Since it is almost impossible to predict railway traffic volume in distant future for the Tsing Ma Bridge, one month of railway traffic that is close to the current traffic conditions is adopted here to establish the database of railway-induced stress responses at the fatigue-critical locations for fatigue analysis. Monthly railway traffic for November 2005 is selected to establish the database, and more than 90% of trains in this period were of the 16-bogie type. In addition, the daily average number of trains in this month reached the maximum in the time period of concern.

The daily time histories of railway-induced stress responses at the fatigue-critical locations are computed using the stress influence lines for railway loading and the railway loading information measured in each day during November 2005. No large difference can be found in the stress time histories for these days. The railway loading information in each day of November 2005 is adopted to compute 30 daily railway-induced stress time histories at the fatigue-critical locations, to compile the database of railway-induced dynamic stress responses.

The highway traffic on the Tsing Ma Bridge has been monitored through seven dynamic weigh-in-motion (WIM) stations installed near the Lantau Administration Building since August 1998. The road vehicle data for November 2005 are adopted to build a database of highway-induced stress responses, because this month reached a maximum number of monthly vehicles and other months had slightly smaller vehicle numbers. Highway-induced stress time histories are also computed in one-day units. The daily time histories of highway-induced stress responses at the fatigue-critical locations are computed using the stress influence lines for highway loading and the highway loading information measured in each day of November 2005. No large differences can be found in the stress time histories. Finally, 30 daily stress response time histories at the fatigue-critical locations are computed to create the database of highway-induced dynamic stress response.

11.8.7 Multiple Load-Induced Dynamic Stress Time Histories in Design Life

Reflecting the 120-year design life of the Tsing Ma Bridge, 120 years of time histories of the dynamic stresses induced by railway, highway, and wind loading need to be generated for fatigue analysis. To generate them, the hourly mean wind speeds and directions for 120 years should be first obtained. A two-step Monte Carlo simulation (MCS) method is adopted to draw out 1 051 200 ($120 \times 365 \times 24$) pairs of hourly mean wind speed and direction for 120 years. In the first step, the mean wind direction is extracted through MCS according to the relative frequency of the mean wind direction without considering wind speed. In the second step, the mean wind speed at the top of the tower is drawn out through MCS according to the probability distribution of the mean wind speed at the given mean wind direction, under the condition that it is not larger than the maximum wind speed in this direction.

The mean wind speed and direction are paired after two steps of MCS, then converted into the hourly normal mean wind speed to generate a sequence of 120 years. As the monsoon wind in Hong Kong normally is southerly (from 90° to 270°) in summer and northerly (from 270° to 90°) in winter, the sequence of hourly normal mean wind speeds in each year is adjusted to consider this. For each hourly normal mean wind speed in the sequence, the corresponding wind-induced dynamic stress response can be found in the database established in the previous section. As wind blowing in two directions is of concern, the mean wind direction in each hour of the sequence is adopted to determine whether the wind is blowing in the direction of the overland fetch or the open-sea fetch. As the database is established for different levels of mean wind speed at an interval of 1 m/sec, rounding towards infinity is adopted to handle the hourly normal mean wind speeds in the sequence. Finally, 1 051 200 hours of wind-induced dynamic stress time histories are generated at each fatigue-critical location to compose a time history of 120 years.

In addition, 120 years of railway-induced stress time histories are generated at the fatigue-critical locations based on the database of 30 daily time histories. An integer between 1 and 30 is randomly

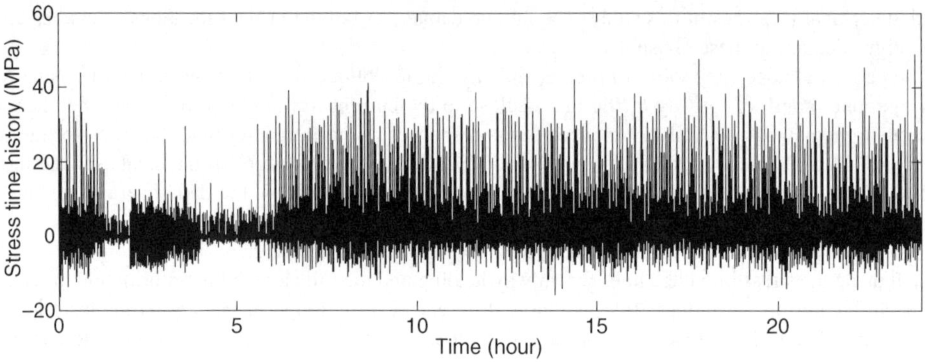

Figure 11.33 A sample stress time history due to multiple types of loading (from [42]) (Reproduced with permission from Elsevier).

drawn out for each day to generate a random number sequence of 120 years. For each item in the sequence, the corresponding daily railway-induced dynamic stress responses can be found in the database established in the aforementioned section. Finally, 43,800 daily time histories at each fatigue-critical location are used to compose 120 years of railway-induced dynamic stresses. As the database of highway-induced stress responses is also based on 30 daily time histories, a similar processing method is applied to obtain 120 years of highway-induced dynamic stresses at the fatigue-critical locations.

Based on the engineering approach proposed in the previous section, stress responses at the critical locations induced by the combined effect of railway, highway and wind loading can be approximately obtained from the three responses induced by individual loadings by superposition. Therefore, a 120-year time history of the stress induced by multiple types of loading is determined from those induced by railway, highway and wind loading individually. It should be noted that the bridge is closed to traffic when the mean wind speed recorded on site is very high; therefore, the bridge stress responses under this condition are induced by wind loading only. Figure 11.33 shows a sample daily multiple load-induced hot-spot stress time history at the critical location E32123.

11.8.8 Fatigue Analysis at Fatigue-Critical Locations

The rainflow counting method is applied to the 120-year multiple load-induced stress time history, and the number of stress cycles in different stress range levels can be obtained. A stress range spectrum is defined as the percentage of the number of stress ranges in each stress range set to the total number in all sets. Figure 11.34a shows the stress range spectrum at E32123. It demonstrates that most of the stress ranges are at the low levels, as 92.0% are less than 8 MPa.

Because fatigue damage is much more sensitive to high level stress rather than low level stress, a stress range of large amplitude may make a great contribution to fatigue damage, even though it occurs less frequently. Fatigue damage in each stress range set is computed using Equation 11.58. The type of welded connection at the six fatigue-critical locations in this study is classified as F according to British Standard [28], with $\sigma_{r,0} = 40$ MPa, $K_2 = 6.3 \times 10^{11}$ and $m = 3$. A fatigue damage spectrum is defined as the percentage of fatigue damage in each stress range set to the total damage in all sets. Figure 11.34b displays the fatigue damage spectrum at E32123. The figure shows that the contribution of stress ranges in the low levels (less than 8 MPa) to fatigue damage is small, and that the greatest fatigue damage is in the stress range of 36–44 MPa.

Based on the multiple load-induced stress time histories over the period of 120 years and the time step $\Delta t = 1/365$ year, the curves of cumulative fatigue damage within 120 years at the fatigue-critical

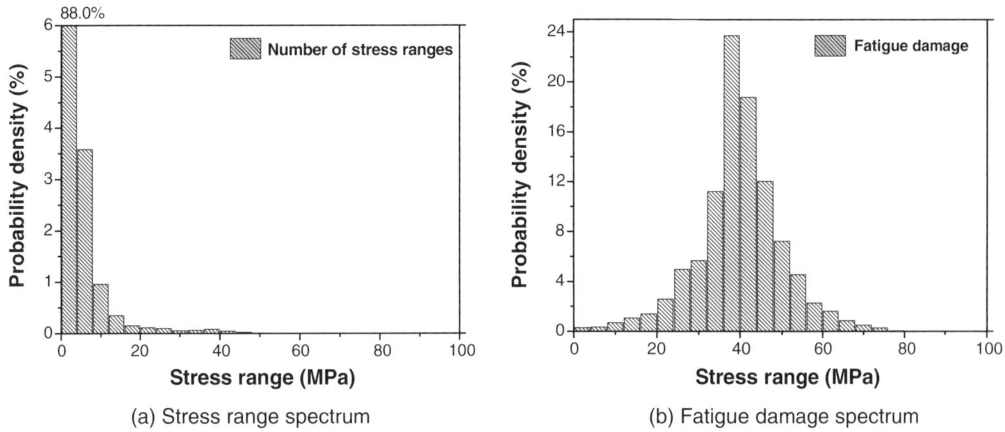

Figure 11.34 Multiple load-related spectra (from [42]) (Reproduced with permission from Elsevier).

locations can be computed. The cumulative fatigue damage ΔD_k on the kth day is calculated based on the daily stress time history using Equations 11.57 and 11.58, and the cumulative fatigue damage D_k is updated by adding the new damage on this day.

Figure 11.35 shows the cumulative fatigue damage curves at the fatigue-critical locations within a design life of 120 years. Note that the structure is in danger when the cumulative fatigue damage is > 1. The maximum of the 120 years of cumulative fatigue damage at the fatigue-critical locations of the Tsing Ma Bridge is very close to one, which implies that the health condition of the bridge is satisfactory. In addition, the cumulative damage curves seem to be very linear. This is because Miner's model is a linear damage model, and traffic loading is assumed to remain stable over the design life.

In addition to the fatigue damage induced by multiple types of loading, the fatigue damage induced by each individual loading type is also investigated. The 120 years of cumulative fatigue damage induced by railway, highway and wind loading are respectively computed, based on the three stress

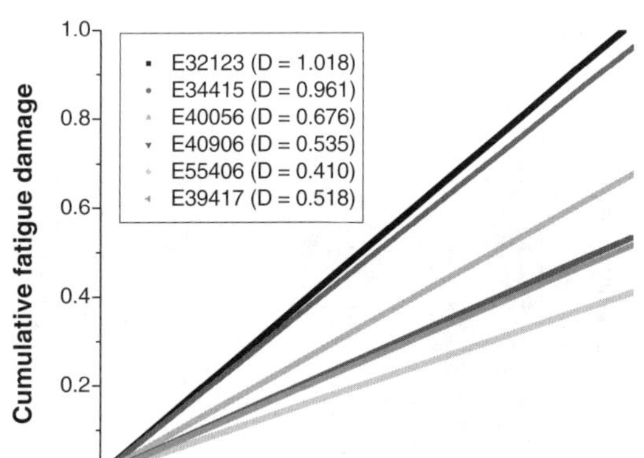

Figure 11.35 Cumulative fatigue damage curves at fatigue-critical locations (from [42]) (Reproduced with permission from Elsevier).

Table 11.9 120 years of cumulative fatigue damage under different loading types (from [42]) (Reproduced with permission from Elsevier)

Fatigue-critical locations	Loading types			
	Railway (R)	Highway (H)	Wind (W)	R + H + W
E32123 (T)	0.70	0.048	0.011	1.02
E34415 (B)	0.66	0.044	0.0092	0.96
E40056 (T)	0.52	0.0022	0.0057	0.68
E40906 (B)	0.42	0.0025	0.0052	0.54
E55406 (T)	0.34	0.0037	0.0016	0.41
E39417 (B)	0.48	0.0020	0.0074	0.52

responses under the different loadings. The results of the damage at different fatigue-critical locations are listed in Table 11.9.

It is found that railway loading plays a dominant role in the fatigue damage of the Tsing Ma Bridge, and that the damage induced by highway loading is greater than that due to wind loading at some locations, whereas the reverse is true in other locations. It is also found that fatigue damage due to

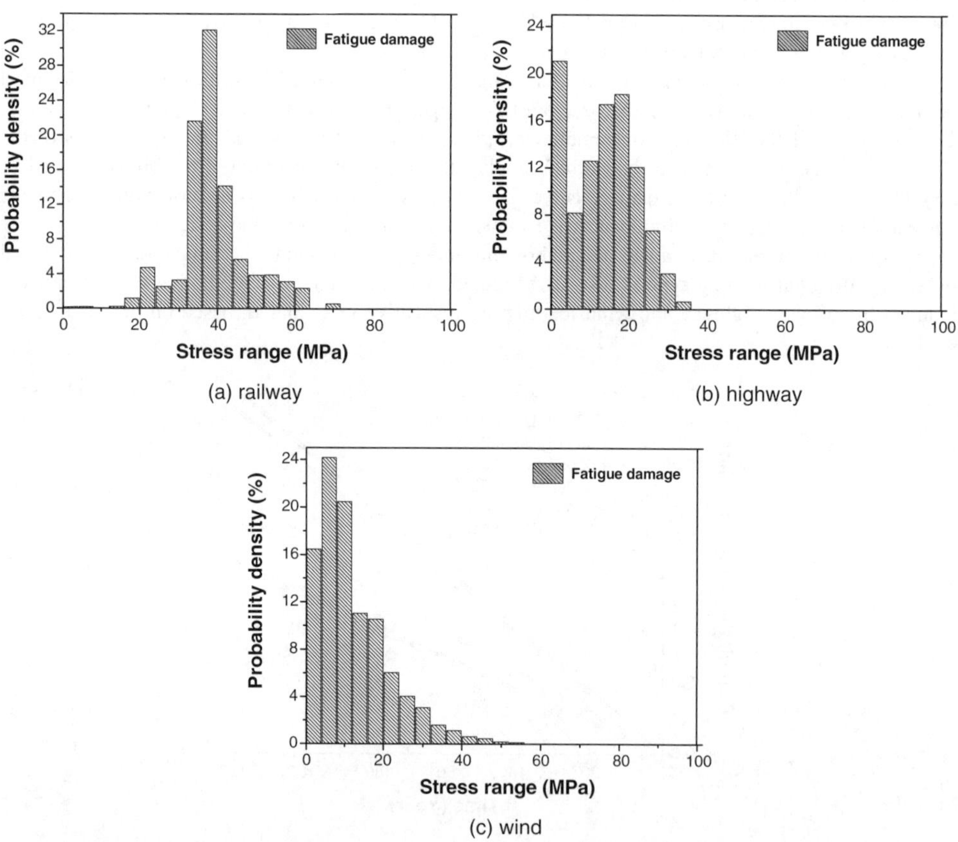

Figure 11.36 Fatigue damage spectra (from [42]) (Reproduced with permission from Elsevier).

combined effects of railway, highway and wind loading is larger than the sum of fatigue damage due to each of individual loadings, as fatigue damage is the function of m-power stress range (non-linear relationship), and stress ranges induced by multiple loading are larger than those caused by individual loading. In addition, the fatigue damage spectra of railway, highway and wind loading are investigated on the basis of the 120-year time histories, and the results are shown in Figures 11.36a to 11.36c.

The figure shows that the spectra are quite different. For example, the greatest fatigue damage induced by railway loading is in the stress range of 32–40 MPa, while that induced by highway loading is in the range of 0–4 and 8–24 MPa, and that induced by wind loading in the range of 0–12 MPa. To study the combined effect of multiple types of loading on fatigue damage, a multiple load magnification factor is defined as the ratio of the fatigue damage due to the combined effect of the three loadings to the sum of the damage due to each individual loading. The factors at the six fatigue-critical locations are computed and range from 1.06–1.35. The maximum factor is at critical locations E32123 and E34415, at which the fatigue damage induced by highway and wind loading is much closer to that induced by railway loading than at the other critical locations. The results indicate that the combined effect of multiple loads must be considered in a bridge subject to multiple types of loading, especially in the case in which the contributions of different loadings to fatigue damage are close.

11.9 Notations

A_e	Effective cross-section area
A_n	Nominal cross-section area
A_f	Reference area
A_r	Roughness coefficient depended on road condition
a	Fitted parameter for wind spectrum (Equation 11.21)
a^c	Fitted coefficient in harmonic function for scale parameter
a^f	Fitted coefficient in harmonic function for relative frequency
a^k	Fitted coefficient in harmonic function for shape parameter
B	(i) Width of bridge deck segment (Equation 11.3)
	(ii) Material property (Equation 11.33)
b	Fitted parameter for wind spectrum (Equation 11.21)
b_m^c	Fitted coefficient in harmonic function for scale parameter
b_m^f	Fitted coefficient in harmonic function for relative frequency
b_m^k	Fitted coefficient in harmonic function for shape parameter
$\mathbf{C}$	Global structural damping matrices of bridge
$\overline{\mathbf{C}}$	Generalized damping matrix of bridge
C_D	Drag coefficient
C_L	Lift coefficient
C_M	Moment coefficient
C_R	Rolling moment coefficient
c	(i) Constant exponent (Equation 11.21)
	(ii) Scale parameter for Weibull form (Equation 11.22, Equation 11.23)
c_m^c	Fitted coefficient in harmonic function for scale parameter
c_m^f	Fitted coefficient in harmonic function for relative frequency
c_m^k	Fitted coefficient in harmonic function for shape parameter
D	(i) Elastic matrix (Equation 11.20)
	(ii) Damage index (Equation 11.32)
$\dot{D}$	Rate of fatigue damage
D_{ei}^{bf}	Equivalent buffeting drag on the ith node of the equivalent bridge deck beam

$\mathbf{D}_{b,j}$	Elastic matrix representing stress-strain relationship
$\mathbf{D}_j$	Elastic matrix establishing the relationship between the stress and the strain of the *j*th element
$\mathbf{E}$	Aeroelastic stiffness matrix of the bridge
$\mathbf{E}_{ei}$	Aeroelastic stiffness matrix of the *i*th deck section with respect to the center of elasticity
$\mathbf{E}_i$	Aeroelastic stiffness matrix of the *i*th section related to nodal self-excited force
F_L	Lift force
F_S	Side force
$\mathbf{F}^{\text{bf}}$	Nodal buffeting force vector
$\mathbf{F}^{\text{se}}$	Nodal self-excited force vector
$\mathbf{F}_{br}$	Contact vector due to train-bridge interaction
$\mathbf{F}_{hb}$	Force vector due to road vehicle-bridge interaction
$\mathbf{F}_{bh}$	Contact force due to road vehicle-bridge interaction
$\mathbf{F}_{ei}^{\text{se}}$	Equivalent self-excited force of the *i*th deck section
$\mathbf{F}_{ki}^{\text{bf}}$	Buffeting force vector at the kth node in the *i*th deck section in global coordinate system
$\hat{\mathbf{F}}^{\text{se}}$	Self-excited force part reflecting aerodynamic phase lag
F_{maxij}	Saturated damping force
$\mathbf{F}_p$	Pseudo force vector
F_q	Interaction force between the road vehicle and the bridge deck surface
$\mathbf{F}_{rb}$	Force vector due to train-bridge interaction
$\{F_k(t)\}_{ji}$	Buffeting force at the *k*th node of the jth element in the local coordinate
f	Spatial frequency
f_0	Discontinuity frequency
f_{u}	(i) Probability density function of wind speed (ii) Upper cutoff frequency (Equation 11.54)
f_l	Lower cutoff frequency
f_θ	Relative frequency of wind direction
$f_{u,\theta}$	Joint density function of wind speed and wind direction
f_{vn}^{L}	Pseudo forces along the $\overline{x}$-axis of the local coordinates
$\mathbf{G}$	Aeroelastic damping matrix of the bridge
$\mathbf{G}_{ei}$	Aeroelastic damping matrix of the *i*th deck section with respect to the center of elasticity
$\mathbf{G}_i$	Aeroelastic damping matrix of the *i*th deck section related to the nodal self-excited force
$\mathbf{H}$	Aeroelastic mass matrix of the bridge
H	Reference height
$\mathbf{H}_{ei}$	Aeroelastic mass matrix of the *i*th deck section with respect to the center of elasticity
$\mathbf{H}_i$	Aeroelastic mass matrix of the *i*th deck section related to the nodal self-excited force
h_{ci}	Vertical coordinate of the centroid of the *i*th deck section
h_{ki}	Vertical coordinate of the *k*th node
$\mathbf{K}$	Global structural stiffness matrices of the bridge
$\overline{\mathbf{K}}$	Generalized stiffness matrices of the bridge
k	Shape parameter for Weibull distribution
k_α	Parameter related to material property α
L_{ei}^{bf}	Equivalent buffeting lift on the ith node of the equivalent bridge deck beam
L	(i) Length of the bridge deck segment (Equation 11.3) (ii) Differential operator (Equation 11.20)
$\mathbf{L}_{b,j}$	Differential operator transforming the element displacement field to the element strain field
L_{ci}	Summation of the lengths of all the elements in the *i*th deck section
$\mathbf{L}_j$	Differential operator transferring the displacement field to the strain field

K_2	Constant relevant to the fatigue detail
K_{lyq}	Stiffness of linear elastic springs for a tire in the y-direction
K_{lzq}	Stiffness of linear elastic springs for a tire in the z-direction
l_{ji}	Length of jth element in the ith section
M_R	Rolling moment
M_{ei}^{bf}	Equivalent buffeting moment on the ith node of the equivalent bridge deck beam
$\mathbf{M}$	Global structural mass matrix of the bridge
$\overline{\mathbf{M}}$	Generalized mass matrix of the bridge
m	(i) Fitted parameter for wind spectrum (Equation 11.21) (ii) Constant relevant to the fatigue detail (Equation 11.58)
m_{b}	Total number of stress cycles in one block
N	(i) Total number of nodes (Equation 11.1) (ii) Shape function (Equation 11.20) (iii) Number of stress cycles (Equation 11.34)
N_b	Number of blocks
$\mathbf{N}_{b,j}$	Shape function transforming local node displacements to element displacement field
N_{m}	Number of the interested modes involved in the computation
N_j	(i) Shape function of the jth element (Equation 11.19) (ii) Total numbers of wind records within the jth wind direction range (Equation 11.29)
N_{r}	Total number of stress cycles
N_{si}	Element number used to model the ith deck section
N_{di}	Node number used to model the ith deck section
$\mathbf{N}_i^{\mathrm{se}}$	Displacement transformation matrix
n	Frequency
n_i	Applied number of stress cycles at the stress range level $\sigma_{r,i}$
n_0	Number of wind records per year for all wind directions
n_f	(i) Frequency of turbulence wind (Equation 11.21) (ii) Order of harmonic function for relative frequency (Equation 11.27)
n_c	Order of harmonic function for scale parameter
n_k	Order of harmonic function for shape parameter
P_u	Cumulative distribution function of wind speed
$P_{u,\theta}$	Joint cumulative distribution function of wind speed and wind direction
P_θ	Occurrence frequency of wind direction
P_j^{E}	Probability of wind speed lower than the maximum wind speed within the jth wind direction range
$\overline{P}_u$	Normalized cumulative distribution function of wind speed
p_{ci}	Lateral coordinate of the centroid of the ith deck section
p_{ki}	Lateral coordinate of the kth node
$p_{ji}^{\mathrm{m}}(s)$	Time-invariant part of wind pressure distribution
$p_{ji}^{\mathrm{t}}(t)$	Time-dependent part of wind pressure distribution
$P_{ji}(s,t)$	Wind pressure distribution over the jth element
$P_\theta(\theta)$	Relative frequency of occurrence of wind in wind direction θ
$\mathbf{Q}^{\mathrm{bf}}$	Generalized buffeting force vector
$\mathbf{Q}^{\mathrm{se}}$	Generalized self-excited force vector
$\mathbf{q}(t)$	Generalized displacement vector
$S(f)$	Power spectrum density function
s	Local element coordinate
T_0	Fatigue damage evolution period
$\mathbf{T}_{b,j}$	Coordinate transfer matrix from global coordinate to local coordinate
U	Mean wind speed

U_{max}	Maximum wind speed
$\overline{u}$	Mean wind speed component
$u(t)$	Horizontal fluctuating wind component
$\overline{V}_R$	Mean wind velocity relative to the vehicle
v_{0ij}	Saturated velocity
$w(t)$	Vertical fluctuating wind component
$\boldsymbol{X}(t)$	Nodal displacement vector of bridge
y_{wq}	Displacement of axle set in the y-direction
y_{rq}	Displacement of bridge deck surface in the y-direction at the contact point
z_{d}	Height of the average deck level
z_{t}	Height of the top of tower
z_{wq}	Displacement of the axle set in z-direction
z_{rq}	Displacement of bridge deck surface in the y-direction at the contact point
$\Delta\sigma_{\mathrm{max}}$	Maximum stress range
ΔD_{k}	Cumulative fatigue damage
ΔU	Interval of wind speed
Δt	Time interval
$\Delta\theta$	Interval of wind direction
$\boldsymbol{\Phi}$	(i) Mode shape matrix (ii) Stress influence coefficient
Ω_j	jth wind direction range
$\boldsymbol{\Gamma}_b$	Modal stress matrix
θ	(i) Wind direction (ii) Random phase angle (Equation 11.54)
ρ	Air density
α	(i) Attack angle of normal incident wind (Equation 11.3) (ii) Positive angle about the q-axis anticlockwise (Equation 11.8) (iii) Exponent of wind profile (Equation 11.31) (iv) Material properties (Equation 11.33)
α_0	Parameter related to material property α
α_j	Material property in jth stress range
β	Material property
ω_i	ith circular natural frequency of the bridge structure
$\boldsymbol{\omega}$	Modal frequency matrix
σ_{b}	Bridge stress response
σ_{rb}	Bridge stress responses induced by railway loading
σ_{hb}	Bridge stress responses induced by highway loading
σ_{wb}	Bridge stress responses induced by wind loading
σ_a	Stress amplitude
σ_c	Stress limit to fatigue
$\boldsymbol{\sigma}_j$	jth element stress vector
σ_{m}	Mean value of each stress cycle
σ_{r}	Stress range of each stress cycle
$\sigma_{r,j}$	jth stress range
$\sigma_{\mathrm{r,max}}$	Maximum stress range
$\overline{\sigma}$	Mean value of stress cycle
$\boldsymbol{\xi}$	Modal damping ratio matrix
ζ_i	ith modal damping ratio of the bridge structure
λ	Wavelength
χ	Aerodynamic transfer function

References

1. Davenport, A.G. (1962) Buffeting of a suspension bridge by storm winds. *Journal of the Structural Division, ASCE*, **88**(3), 233–270.
2. Scanlan, R.H. and Gade, R.H. (1977) Motion of suspended bridge spans under gusty wind. *Journal of the Structural Division, ASCE*, **103**(9), 1867–1883.
3. Lin, Y.K. and Yang, J.N. (1983) Multimode bridge response to wind excitations. *Journal of Engineering Mechanics, ASCE*, **109**(2), 586–603.
4. Jain, A. Jones, N.P., and Scanlan, R.H. (1996) Coupled flutter and buffeting analysis of long-span bridges. *Journal of Structural Engineering, ASCE*, **122**(7), 716–725.
5. Chen, X.Z. Matsumoto, M., and Kareem, A. (2000) Time domain flutter and buffeting response analysis of bridges. *Journal of Structural Engineering, ASCE*, **126**(1), 7–16.
6. Xu, Y.L., Sun, D.K., Ko, J.M., and Lin, J.H. (2000) Fully coupled buffeting analysis of Tsing Ma suspension bridge. *Journal of Wind Engineering and Industrial Aerodynamics*, **85**(1), 97–117.
7. Liu, G. Xu, Y.L., and Zhu, L.D. (2004) Time domain buffeting analysis of long suspension bridges under skew winds. *Wind and Structures-An International Journal*, **7**(6), 421–447.
8. Wong, K.Y. (2004) Instrumentation and health monitoring of cable-supported bridges. *Journal of Structural Control and Health Monitoring*, **11**(2), 91–124.
9. Chan, T.H.T., Zhou, T.Q., Li, Z.X., and Guo, L. (2005) Hot spot stress approach for Tsing Ma Bridge fatigue evaluation under traffic using finite element method. *Structural Engineering and Mechanics*, **19**(3), 261–279.
10. Liu, T.T., Xu, Y.L., Zhang, W.S. *et al.* (2009) Buffeting-induced stresses in a long suspension bridge: structural health monitoring oriented stress analysis. *Wind and Structures-An International Journal*, **12**(6), 479–504.
11. Xu, Y.L. Ko, J.M., and Zhang, W.S. (1997) Vibration studies of Tsing Ma long suspension bridge. *Journal of Bridge Engineering, ASCE*, **2**(4), 149–156.
12. Wong, K.Y. (2002) Structural identification of Tsing Ma Bridge. *The Transactions of Hong Kong Institution of Engineers*, **10**(1), 38–47.
13. Xu, Y.L. Xia, H., and Yan, Q. S. (2003) Dynamic response of suspension bridge to high wind and running train. *Journal of Bridge Engineering, ASCE*, **8**(1), 46–55.
14. Lau, D.T. Cheung, M.S., and Cheng, S.H. (2000) 3D flutter analysis of bridges by spline finite-strip method. *Journal of Structural Engineering, ASCE*, **126**(10), 1246–1254.
15. HKO (1999) Typhoon York (9915): 12–17 September 1999, Hong Kong Observatory, (http://www.info.gov.hk/hko/informtc/tork/report.htm).
16. Simiu, E. and Scanlan, R.H. (1996) *Wind Effects on Structures*, Wiley, New York.
17. Cao, Y.D., Xiang, H.F., and Zhou, Y. (2000) Simulation of stochastic wind velocity field on long-span bridges. *Journal of Engineering Mechanics, ASCE*, **126**(1), 1–6.
18. Fisher, J.W. (1984) *Fatigue and Fracture in Steel Bridges – Case Studies*, John Wiley & Sons.
19. Virlogeux, M. (1992) Wind design and analysis for the Normandy Bridge, in *Aerodynamics of Large Bridges* (ed A. Larsen), Balkema, Rotterdam, the Netherlands, pp. 183–216.
20. Gu, M., Xu, Y.L., Chen, L.Z., and Xiang, H.F. (1999) Fatigue life estimation of steel girder of Yangpu cable-stayed Bridge due to buffeting. *Journal of Wind Engineering and Industrial Aerodynamics*, **80**(3), 383–400.
21. Lemaitre, J. (1996) *A Course on Damage Mechanics*, Springer, Verlag.
22. Li, X.Z. Chan, T.H.T., and Ko, J.M. (2002) Evaluation of typhoon induced damage for Tsing Ma Bridge. *Engineering Structures*, **24**(8), 1035–1047.
23. Xu Y.L., Liu T.T., Zhang W.S. (2009) Buffeting-induced fatigue damage assessment of a long suspension bridge. *International Journal of Fatigue*, **31**(3), 575–586.
24. Corotis, R.B. Sigl, A.B., and Klein, J. (1978) Probability models of wind velocity magnitude and persistence. *Solar Energy*, **20**(6), 483–493.
25. Morgan, V.T. (1995) Statistical distributions of wind parameters at Sydney, *Australia. Journal of Renewable Energy*, **6**(1), 39–47.
26. Dixon, J.C. and Swift, R.H. (1984) The directional variation of wind probability and Weibull speed parameters. *Atmospheric Environment*, **18**(10), 2041–2047.
27. Coles, S.G. and Walshaw, D. (1995) Directional modeling of extreme wind speeds. *Applied Statistics-Journal of the Royal Statistical Society Series C*, **43**(1), 139–157.
28. [7608BS] (1993) *Code of Practice for Fatigue Design and Assessment of Steel Structures*, British Standards Institution.
29. [5400BS] (1993) *Part 10, Code of Practice for Fatigue*, British Standards Institution.
30. Downing, S.D. and Socie, D.F. (1982) Simple rainflow counting algorithms. *International Journal of Fatigue*, **4**(1), 31–40.
31. Krajcinovic, D. and Lemaitre, J. (1987) *Continuum Damage Mechanics: Theory and Applications*, Springer, *Vienna*.
32. Li, Z.X. Chan, T.H.T., and Ko, J.M. (2001) Fatigue damage model for bridge under traffic loading: application made to Tsing Ma Bridge. *Theoretical and Applied Fracture Mechanics*, **35**(1), 81–91.
33. Chen Z.W., Xu Y.L., Li Q., and Wu D.J. (2011) Dynamic stress analysis of long suspension bridges under wind, railway and highway loading. *Journal of Bridge Engineering, ASCE*, **16**(3), 383–391.

34. Xu, Y.L. and Ding, Q.S. (2006) Interaction of railway vehicles with track in cross-winds. *Journal of Fluid and Structures*, **22**(3), 295–314.
35. Zhai, W.M. (2007) *Vehicle-Track Coupling Dynamics*, 3rd edn, Chinese Railway Press, Beijing, China, (in Chinese).
36. Xia, H. Xu, Y.L., and Chan, T.H.T. (2000) Dynamic interaction of long suspension bridges with running trains. *Journal of Sound and Vibration*, **237**(2), 263–280.
37. Xu, Y.L. and Guo, W.H. (2003) Dynamic analysis of coupled highway vehicle and cable-stayed bridge systems under turbulent wind. *Engineering Structures*, **25**(4), 473–486.
38. Dodds, C.J. and Robson, J.D. (1973) The description of road surface roughness. *Journal of Sound Vibration*, **31**(2), 175–183.
39. Wang, T.L. and Huang, D.Z. (1992) Cable-stayed bridge vibration due to road surface roughness. *Journal of Structural Engineering*, **118**(5), 1354–1374.
40. Ding, Q. and Lee, P.K.K. (2000) Computer simulation of buffeting actions of suspension bridges under turbulent wind. *Computers and Structures*, **76**(6), 787–797.
41. Guo, W.W., Xu, Y.L., Xia, H. *et al.* (2007) Dynamic response of suspension bridge to typhoon and trains. II: numerical results. *Journal of Structural Engineering*, **133**(1), 12–21.
42. Chen, Z.W., Xu, Y.L., Xia, Y., Li, Q. and Wong, K.Y. (2011) Fatigue analysis of long-span suspension bridges under multiple loading: Case study. *Engineering Structures*, **33**, 3246–3256.
43. Xu, Y.L., Li, Q., Wu, D.J., and Chen, Z.W. (2010) Stress and acceleration response analysis of vehicle and long-span bridge systems using the mode superposition method. *Engineering Structures*, **32**(5), 1356–1368.
44. Wong, K.Y. (2007) Stress and traffic loads monitoring of Tsing Ma Bridge. Proceedings of China Bridge Congress, (CD ROM).
45. Xu, Y.L., Liu, T.T., and Zhang, W.S. (2009) Buffeting-induced fatigue damage assessment of a long suspension bridge. *International Journal of Fatigue*, **31**(3), 575–586.

12

Wind-Induced Vibration Control

12.1 Preview

It has been discussed in previous chapters that long-span cable-supported bridges are susceptible to flutter instability, vortex-induced vibration, buffeting-induced vibration, vehicle-wind-bridge interaction, multiple loading-induced fatigue and rain-wind-induced cable vibration. To ensure the safety and functionality of bridges and vehicles under wind loading, it is necessary in some cases to find efficient and effective measures for mitigating excessive wind-induced vibration to a acceptable level, and for improving bridge stability to avoid collapse. In general, there are three ways to mitigate wind-induced vibration of a long-span bridge: structural modification, aerodynamic measures and mechanical measures. The realization of these mitigation technologies may also make the construction of super-long-span bridges come true and elevate bridge design concepts from a static and passive level to one with dynamism and adaptability.

A brief introduction on wind-induced vibration control methods will be first given in this chapter. Various aerodynamic measures are then introduced for flutter control, vortex-induced vibration mitigation and cable vibration mitigation. Different mechanical measures – including passive control systems, active control systems, and semi-active control systems – are also addressed for wind-induced vibration control. Damping stay cables of a cable-stayed bridge using adjustable fluid dampers is finally presented as a case study.

12.2 Control Methods for Wind-Induced Vibration

Various possibilities exist to reduce wind-induced vibration of long-span bridges [1,2]. Control methods for reducing wind-induced structural vibration can be divided into three major categories:

1. structural modification;
2. aerodynamic measures; and
3. mechanical measures.

Structural modification is applied to the three important dynamic properties of a bridge structure, i.e. stiffness, damping and mass, through the change of structural materials and/or the rearrangement of structural systems. It can be applied to the entire bridge or bridge components, such as bridge deck, towers, and cables. It is found that an increase in the damping capacity is always beneficial for

Wind Effects on Cable-Supported Bridges, First Edition. You-Lin Xu.

mitigating vortex-induced vibration, buffeting-induced vibration and galloping instability. Nevertheless, the effect of additional damping upon flutter-related critical wind speed is low.

In reality, concrete bridges tend to have more damping than steel bridges. Composite steel-concrete bridges also have considerably higher values of damping compared to those of steel bridges. An estimation of the structural damping is crucial but it is the parameter which is least analytically tractable, because its estimation relies totally on limited measured full-scale data.

The common aim of stiffness modification for a long-span cable-supported bridge is to increase the torsional stiffness of the bridge deck so as to improve the bridge stability under wind loading. For example, deep truss girders are more torsionally rigid compared with flat box girders, but they exhibit higher wind forces [3]. To increase the torsional stiffness of a super long-span suspension bridge, cable system modifications may be considered by using more than two main cable systems[4], as shown in Figure 12.1. The deployment of eccentric mass was also applied during construction of the Humber Bridge in England [5], and for the bridge in service [6] if it has relatively close natural frequencies of vertical and torsional motions. The drawback of mass modification is an enlarged dead load and static torsional deformation. To reduce vibration of stay cables in a cable-stayed bridge, cross-ties have been used to increase the in-plane stiffness of stay cables by connecting them together, but this may affect the aesthetic view of stay cables due to a complex cable network and connections.

Aerodynamic measures alter the flow pattern around bridge components (such as bridge deck, towers and cables) to reduce directly the aerodynamic force or moment acting on it by selecting better cross-sectional shapes or by installing some aerodynamic devices on the bridge component. A typical example is to install wedge-shaped fairings to the two longitudinal edges of the bridge deck (see Figure 12.2) to make the bridge deck section streamlined so as to reduce vortex shedding, abate buffeting forces and improve flutter instability [7].

Since wind-bridge interaction is so complicated, no analytical methods are available at present to facilitate the optimal selection of cross-sectional shapes. Wind tunnel tests and computational simulations, as discussed in Chapter 7 and Chapter 8 respectively, have to be used to perform aerodynamic modifications. Although many types of aerodynamic devices, such as vanes and stabilizers, have been found to be effective in the suppression of wind-induced response and the improvement of motion stability, the alterations to the shape of the bridge deck and the substantial nature of the alterations may be impractical and aesthetically undesirable in modern long-span bridges. In particular, the addition of aerodynamic devices to a bridge deck often increases buffeting forces.

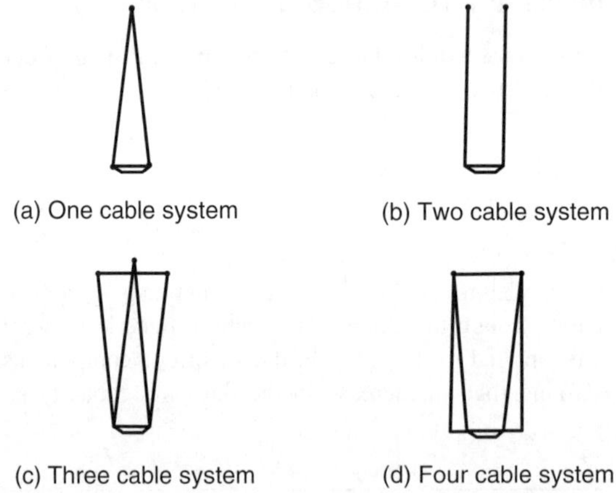

Figure 12.1 Cross-section of alternative suspension cable systems.

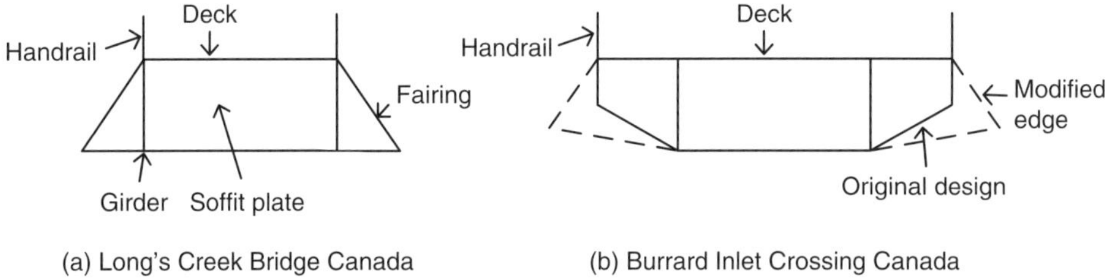

Figure 12.2 Aerodynamic measures to reduce wind-induced vibration.

Mechanical measures, which include passive control technology, active control technology, semi-active technology and hybrid control technology, modify overall structural characteristics, leading to a reduction of wind-induced vibration [8]. They can be applied to the bridge as a whole or to any individual bridge component. Passive control systems operate without requiring an external power source and utilize the motion of a bridge structure to develop control forces. Passive control devices, such as metallic dampers, friction dampers, viscoelastic dampers, tuned mass dampers and tuned liquid column dampers, do not increase vibrational energy and are thus inherently stable [9]. The passive control devices mainly enhance the overall damping capacity of a bridge structure, so as to absorb the energy of wind-induced vibration and thereby reduce it. The basic configurations of a conventional structure and a structure with a passive control system are shown schematically in Figure 12.3a and 12.3b, in which PED stands for passive energy dissipation [10].

Compared with passive control systems, active control systems require external power supply to apply forces to a bridge structure in a prescribed manner. Control forces are developed on the basis of feedback from sensors which measure the excitation and/or the response of the bridge structure. These forces can be used to add or dissipate energy in the bridge structure. An active control system usually consists of:

(a) sensors installed at suitable locations of the bridge structure to measure either the external excitations, or the structural response quantities, or both;
(b) devices to process the measured information and to compute the necessary control forces based on a given control algorithm; and
(c) actuators, usually powered by external energy sources, to produce the required forces.

The basic configuration of a structure with an active control system is shown schematically in Figure 12.3c [10,11]. An active control system mainly changes the overall structural characteristics to reduce wind-induced vibration, but it needs an external power supply and may cause instability of the bridge structure because of many uncertainties involved in the bridge structure, wind loading and control system. Active control systems are also proposed to control some aerodynamic devices to change flow patterns around and reduce wind forces on the bridge deck.

Semi-active control is a compromise between passive and active control systems [12–14]. Semi-active control systems require much less input energy in comparison with the active type, and the input energy is used to modify the damper properties, leading to the optimal control of structure vibration. Control forces are also developed based on feedback from sensors which measure the excitation and/or the response of the structure. Examples include variable orifice dampers, variable friction damper system and magnetorheological (MR) dampers. Figure 12.3d shows the basic configuration of a structure with a semi-active control system schematically.

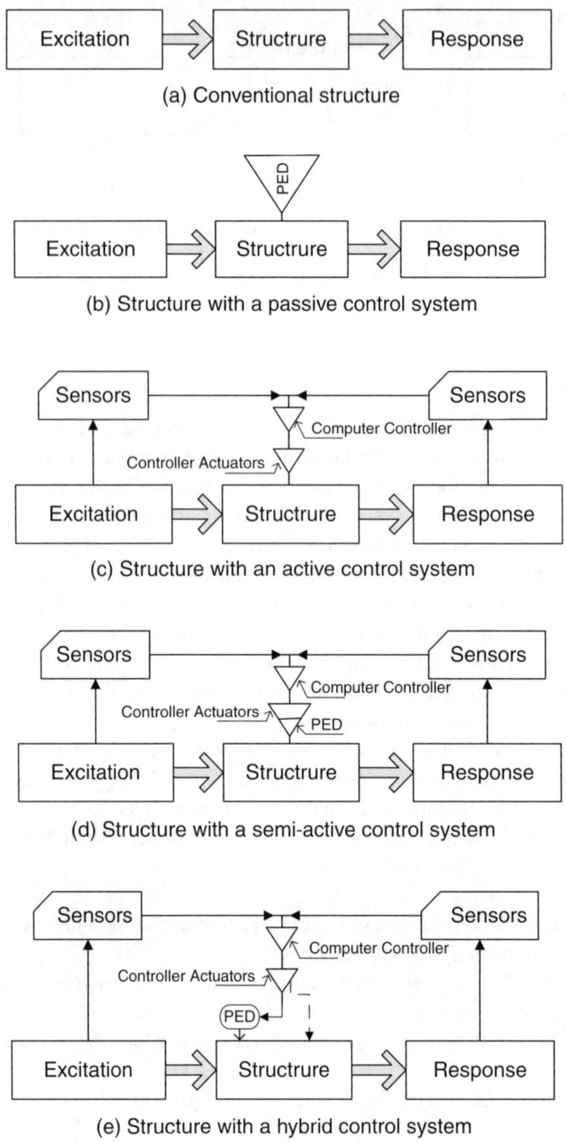

Figure 12.3 Structure with various control systems.

A hybrid system consists of an active system and a passive control system and thus increases the performance and robustness of the control system (see Figure 12.3e). A hybrid mass damper (HMD) is a combination of a passive tuned mass damper and an active control actuator. This has been employed in full-scale building structures, but not yet in bridge structures.

12.3 Aerodynamic Measures for Flutter Control

Flutter instability will lead to the structural collapse of a long-span cable-supported bridge, and hence the critical wind speed associated with flutter instability of the bridge needs to be higher than the maximum design wind speed at the bridge site. When a long-span bridge is predicted to have its intrinsic limit in reaching the required critical wind speed, it is necessary to adopt some aerodynamic or mechanical measures to improve its aerodynamic performance.

12.3.1 Passive Aerodynamic Measures

Aerodynamic forces arise when air flows around bridge deck sections and interacts with them. Passive aerodynamic measures thus aim to directly reduce the aerodynamic forces through improving the configuration of cross-sections of the bridge deck. Passive aerodynamic measures are the most common approach among all the flutter control methods for long-span cable-supported bridges. Some commonly-used measures are introduced below.

Wedge-shaped fairings added at both ends of the bridge deck cross section (see Figure 12.2) can make the cross-section tend to streamline and, hence, improves the air flow pattern around the deck, reduces drag forces and vortex shedding and enhances flutter stability. Wind tunnel studies and flutter analysis have shown that such measures can effectively improve the flutter stability of cable-supported bridges, and the smaller the tip angle of the edge fairings, the greater the improvement of flutter stability [7]. With the same tip angle of the edge fairings, the increase of tip length will enhance the aerodynamic stability of the bridge.

The aerodynamic performance of a bridge deck can be further enhanced by longitudinal open slots in the deck [15–22]. This is a unique feature observed from recently constructed suspension bridges. The adoption of a "slotted" deck solution not only enhances aerodynamic stability, but also prevents the periodic formation and shedding of large vortices in the wake of the bridge deck.

The aerodynamic performance of the slotted box girder can be further improved if it is used together with some aerodynamic devices. Ueda *et al.* [23] investigated the flutter stability of a super-long-span suspension bridge with a central span length of 2500 m by considering three types of central slots: central slots only, central slots with central stabilizer, central slots with both central stabilizer and horizontal stabilizer (see Figure 12.4).

The wind tunnel test results showed that the critical flutter wind speed increased by about 35% and reached 62 m/sec after the central stabilizer was added to the central slots. If a horizontal stabilizer was further added to the central slots with the central stabilizer, the critical flutter wind speed was improved again by 33% and reached 82.2 m/sec. The aerodynamic advantages of this solution have also been explored during the design of the proposed 3300 m long suspension bridge for the crossing of the Messina Strait [24], and the relevant wind tunnel studies were performed rigorously by Diana [25].

Other alternative passive aerodynamic measures involve modifying wind loading acting on a bridge by adding aerodynamic appendages to the deck section [26]. The idea is to locate a winglet above (or below) the bridge deck, running parallel to the bridge axis, far enough from the deck that it lies in the undisturbed flow field. The presence of the winglets modifies the aeroelastic loads acting on the bridge and increases the flutter associated with the critical wind speed. Arco and Aparicio [27] investigated four types of aerodynamic appendages added to the bridge cross-section according to the position of the winglets, as shown in Figure 12.5.

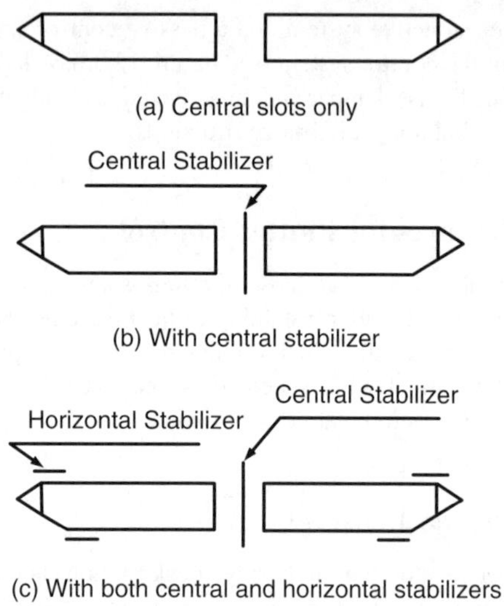

Figure 12.4 Slotted box girder with central stabilizer and horizontal stabilizer.

In Case A, a pair of winglets are symmetrically fixed to the bridge cross-section. In Case B, a pair of winglets are fixed to the hangers. In Case C, only one winglet is fixed to the bridge cross-section in the leeward position. In Case D, a pair of winglets are fixed to both the windard and leedward positions. These results demonstrated that, compared with Cases A and B, Case C gave the best aerodynamic performence and the highest critical wind speed. When this finding was applied to the Great Belt Bridge, using aerodynamic appendages arranged like Case C, it was found that the critical wind speed was clearly enhanced. The increase in critical wind speed in Case C acounted to more than 35%, compared with the normal case without any appendages. The increase of critical wind speed in Case B was somewhat modest, at 21%, and was even less (11%) in Case A. For Case D, care must be given to the selection of the phase angle, for a wrong selection of the angle may cause the instability of the bridge structure.

All of these passive aerodynamic measures are viewed as economical and reliable methods that do not require special routine maintenance. They have been validated through wind tunnel tests, computational simulations, and field measurements.

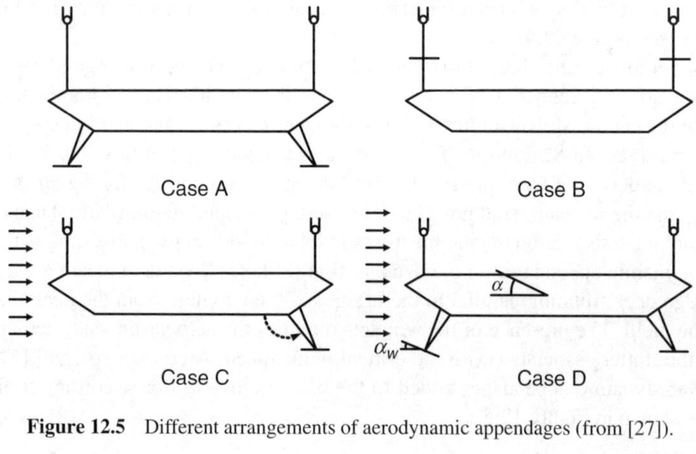

Figure 12.5 Different arrangements of aerodynamic appendages (from [27]).

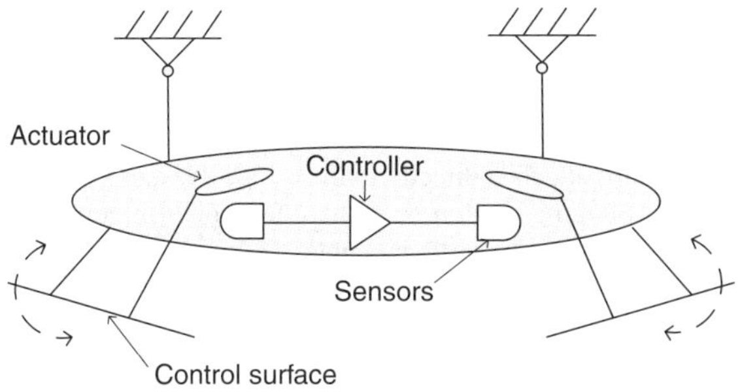

Figure 12.6 Active aerodynamic control of bridge deck flutter by control surface.

12.3.2 Active Aerodynamic Control

Active aerodynamic control methods for long-span bridges are inspired by those employed in the aerospace industry for aircraft control. Ostenfeld and Larsen [3] presented an active aerodynamic control system by using actively controlled surfaces attached beneath the two edges of the bridge deck cross-section (see Figure 12.6).

The rotations of the surfaces are actively controlled according to a feedback algorithm in such a way that the aerodynamic forces acting on the bridge deck provide a stabilizing action. These stabilizing aerodynamic forces are not produced directly by the actuators in this method, and they actually result from the change of air flow around the bridge deck via the rotation of the surfaces controlled by the actuators.

Kobayashi and Nagaoka conducted wind tunnel experiments on the sectional model of a bridge and obtained an increase of flutter critical wind speed of factor 2 [28]. The control algorithm used in the experiments was based on the principle that the rotation of the control surfaces is proportional to the rotation of the bridge deck. Wilde and Fujino [29] carried out theoretical analysis of such a control surface system. They applied a rational function approximation to model unsteady aerodynamics and, under the assumption of no flow interaction between the control surfaces and the deck, they derived a time domain equation of motion for the control system. Accuracy of the approximation was achieved by multilevel linear and non-linear optimization. Their numerical simulation results show that the aerodynamic control of flutter by actively controlled surfaces can stabilize the bridge at any wind speed. For the control to be effective, it is necessary to locate control surfaces outside the turbulent boundary layer and as far away from the local flow pattern around the deck as is practically possible.

Active aerodynamic control may also be achieved by additional flaps attached directly to the edges of the bridge deck [30]. In this system, the flow pattern around the deck is affected by the motion of the flaps (see Figure 12.7). Therefore, the stabilizing action not only comes from the aerodynamic forces generated on the flaps, but also from modification of the aerodynamic forces induced on the bridge deck.

Wilde *et al.* [31] referred to this control system as the "deck-flaps system" and presented the modeling procedure and the control algorithm for controlling flutter of long-span bridges. Three optimal configurations of the deck-flaps system, with different flap hinge locations, were proposed and investigated. For each hinge location, four flap sizes were considered: 1.5 m, 3.0 m, 4.5 m and 6.0 m. All of

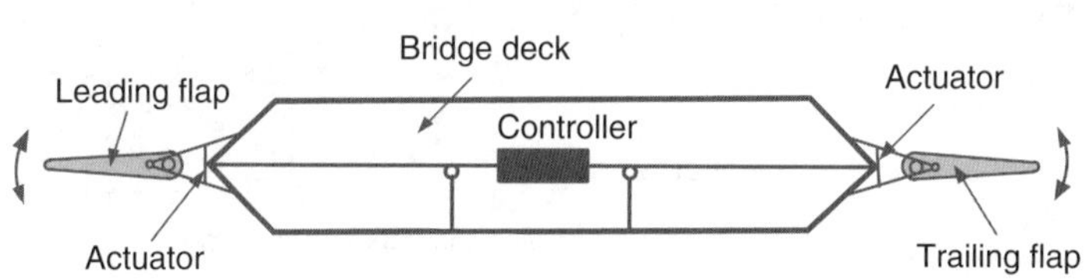

Figure 12.7 Cross-section of bridge deck with active flaps for flutter control.

these considered configurations of the control system could stabilize the bridge against flutter and divergence in the selected wind speed range. The best performance was obtained by the system with the flaps located on the edges of the deck and the hinges located in the middle of the flaps.

Omenzetter *et al.* [32–35], on the other hand, proposed similar control systems but of the passive type, consisting of auxiliary flaps attached directly to the bridge deck, as shown in Figure 12.8.

When the deck undergoes pitching motion or relative horizontal motion with respect to the main cables, the flap rotation is governed by additional cables spanned between the flaps and an auxiliary transverse beam supported by the main cables of the bridge. Since the cables can only pull the flaps,

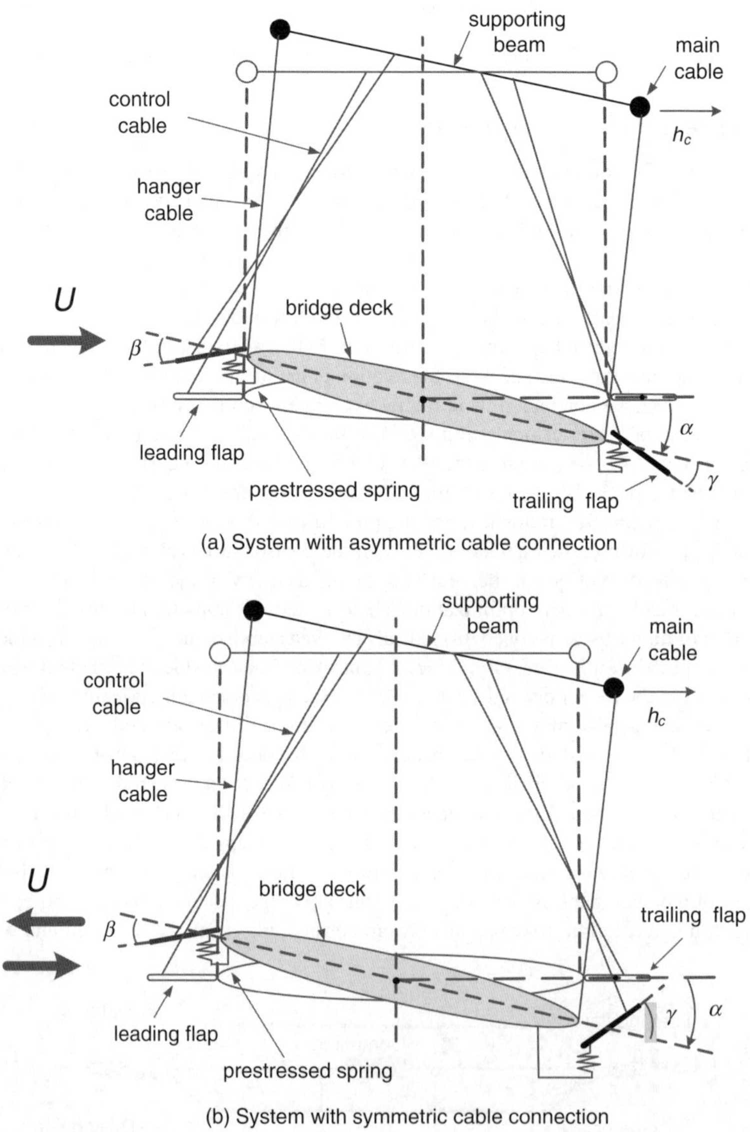

Figure 12.8 Aerodynamic control of bridge deck flutter with passive flaps (Figure courtesy of Prof. Yozo Fujino).

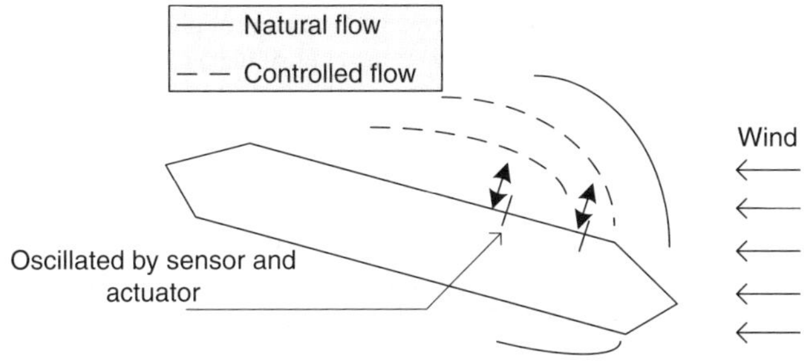

Figure 12.9 Active aerodynamic control by changing air flow.

but not push them, additional pre-stressed springs are used to reverse the motion of the control surfaces. The control system shown in Figure 12.8a is an asymmetric cable connecting system, which can work properly for wind coming from only one direction and requires alteration of its configuration as wind direction changes. The stabilizing action of the system shown in Figure 12.8b comes from a symmetric cable connection system, which is, in contrast, independent of wind direction.

For these two systems, Omenzetter *et al.* [32–35] derived the time-domain mathematical model of self-excited forces on the sectional model of the deck-flaps system and conducted numerical simulations of a suspension bridge using a three-dimensional FEM model.

Although both systems employ the same idea of flutter control, their dynamic characteristics and performance differ significantly. The system with asymmetric cable connection offers high increase in critical wind speed. The action of control flaps strongly modifies the properties of the flutter mode and its stiffness and damping are markedly increased. However, this system requires changes in its configuration as wind direction changes. To achieve its best performance, it requires large flaps and significant stiffness of the supporting system.

The improvement in critical wind speed for the system with symmetric cable connection is limited. Properties of the flutter mode are not strongly altered and the degree of stability is lower. The gust response of the control flaps is large. Nevertheless, this system can work properly for the wind coming from any direction. It uses flaps of small size, and the stiffness of the supporting system is much less compared to the system with asymmetric cable connection.

There was an active aerodynamic control idea for flutter in aeronautics that utilized a fence-type spoiler to change airflow around the airfoil surface, as shown in Figure 12.9 [36]. Kwon *et al.* [36] applied similar idea to a bridge deck and designed a new aerodynamic controller to control plates properly according to the motion of the bridge deck without requiring external power supply.

Figure 12.10 schematically shows the mechanism of the proposed aerodynamic control method. The key idea of this system is to properly activate the control plates according to the bridge deck motion. A constant phase angle between the control plate and deck motion should be kept for successful control.

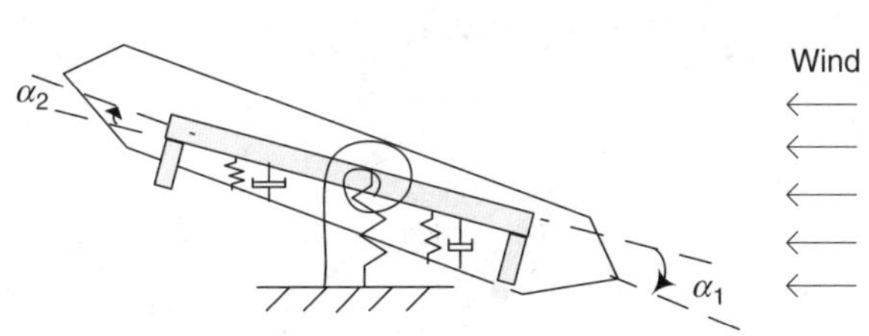

Figure 12.10 Control mechanism of active aerodynamic control system.

There are two slits on the bottom flange of the deck, and the control plates oscillate through these slits. When the bridge deck maintains its horizontal state, the control plates do not move and the deck keeps its equilibrium position. However, when the deck is oscillated by the wind, the control plates move up or down, according to the motion of the decks, and thus the air flow is changed.

A seesaw-like rod inside the deck activates the control plates. Its mechanism is similar to a tuned mass damper (TMD). The phase angle between the deck and control plate can be adjusted by the mass and stiffness of the rod. Because the inertial force of a TMD is used for moving control plates for aerodynamic control and not for direct vibration control, the mass ratio of this system can be greatly reduced in comparison with a traditional TMD for structural control. This is one of the great benefits of the proposed system.

A spring-supported section model test was conducted in order to verify the proposed control method. The wind tunnel test results showed that the increasing ratio of the critical flutter wind speed was 64.6. For studying the applicability of the proposed system to a long-span bridge, a long-span suspension bridge with a central span length of 3000 m was used as a numerical example. It showed that the flutter onset velocity increased by 51% when 60% of the center span equipped the controller.

12.4 Aerodynamic Measures for Vortex-Induced Vibration Control

Since the selection of a bridge deck configuration depends on many factors, such as structural and economical advantages, the basic deck shape does not necessarily have optimal aerodynamic efficiency. As a consequence, long-span bridges may be subjected to vortex-induced vibration. However, compared with the flutter control of long-span bridges, the studies on vortex-induced vibration control are relatively weak. The following will give a brief introduction on aerodynamic measures for vortex-induced vibration control of long-span cable-supported bridges.

During the final phase of deck erection and road surfacing works of the Storebælt suspension bridge in Denmark in 1998, vertical deck oscillations of low frequency were observed [37]. The results from both field measurements and wind tunnel tests indicated that the observed oscillations were due to periodic von Karman type vortex shedding in the wake of the mono box girder of the bridge. With the aid of wind tunnel tests, Larsen *et al.* [37] then designed and installed guide vanes to the main suspension girder for suppression of vortex-induced vibration of the bridge (see Figure 12.11).

The guide vanes were composed of curved steel plates running along the bottom/lower side panel joints and were tested on a 1: 60 scale section model of the bridge girder. This guide vane design was proved very efficient in eliminating vortex-induced oscillations of the bridge by both field measurements and wind tunnel tests. In addition, the guide vanes improved the flutter stability of the bridge deck marginally, with the critical wind speed for onset of flutter rising from 69 m/sec to 73 m/sec.

The Stonecutters cable-stayed bridge in Hong Kong has a main span of 1018 m, and the bridge deck is composed of two wedge-shaped box girders with an open gap between them (see Figure 12.12). Vortex shedding and lock-in with the lowest vertical mode of the bridge was expected at low wind

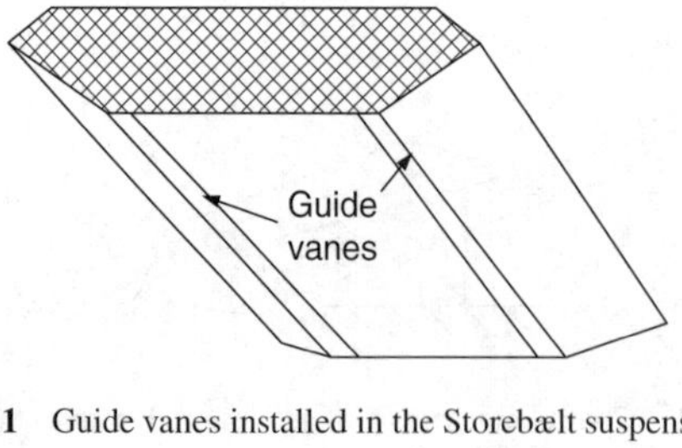

Figure 12.11 Guide vanes installed in the Storebælt suspension bridge.

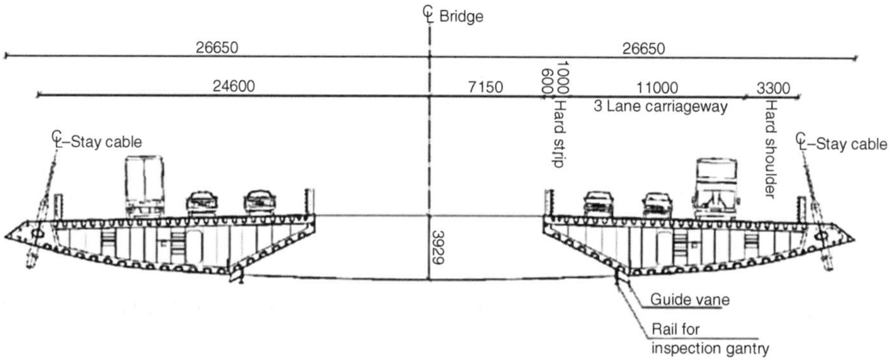

Figure 12.12 Guide vanes installed in the Stonecutters cable-stayed bridge.

speeds in the range from 5–8 m/sec. Since there is open sea in the south-west wind direction, low turbulence cannot in general be relied upon for mitigation of vortex shedding excitation, and von Karman type vortices were expected to form and to be shed at regular time intervals in the near wake of the upwind girder. This vortex shedding process may lock in with a vertical mode of vibration and thus excite vertical oscillation of the bridge deck. Furthermore, this action may be amplified as the shed vortices drift across the gap and impinge on the downwind girder (see Figure 12.13a). Therefore, the mitigation of vortex-induced vibration was considered as an important issue for the bridge [38].

Based on the success of using guide vanes to mitigate vortex-induced oscillations of the Storebælt Bridge in 1998, similar guide vanes were proposed to be mounted at the knuckle lines between the individual box bottom plates and the inclined side panels facing the gap (Figure 12.13b). It was hoped to create flows which would prevent or diminish rhythmic vortex formation at the upwind knuckle line and, in turn, reduce vortex-induced forces on the upwind box as well as vortex impingement on the downwind box. However, wind tunnel tests on a spring-suspended section model of 1: 80 scale did not demonstrate the effectiveness of the guide vanes.

At first glance, the results were difficult to understand, but the calculation of the Reynolds number based on the chord length of the individual box sections yielded a Reynolds number around 18 000, indicating laminar flow conditions and, thus, a thick boundary layer at the position of the upwind guide vane. Under such conditions, the upwind guide vane would be well immersed in the boundary layer growing on the curved soffit plate, thus preventing sufficient flow through the guide vane to inhibit vortex formation in the near wake of the upwind box section. Large 1: 20 scale section model tests, with a higher Reynolds number, were then devised and performed. The guide vane design which had failed in the 1: 80 scale tests proved to be very efficient in the 1: 20 scale tests, completely eliminating vortex shedding response [38].

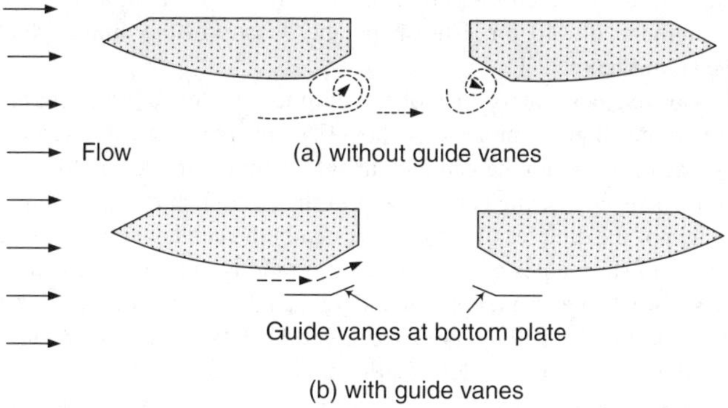

Figure 12.13 Anticipated flow patterns.

The Xihoumen Bridge in China is a long-span suspension bridge with a 1650 m main span. The bridge deck is a twin box girder. Ge *et al.* [39] carried out both conventional sectional model tests and large-scale sectional model tests of the bridge and found vortex-induced vibration with large amplitude. Several aerodynamic measures, including guide vane, grid plate and adjustable wind barrier, were proposed and tested to find the most effective and feasible measure that could be applied to the real bridge. Later, vertical vortex-induced vibration did occur in the prototype bridge [40]; an adjustable wind barrier was then selected and installed on the girder deck to suppress the vortex-induced vibration of the bridge.

12.5 Aerodynamic Measures for Rain-Wind-Induced Cable Vibration Control

With widespread popularity of cable-stayed bridges around the world, longer spans are being constructed by employing increasingly longer stay cables. These cables are laterally flexible structural members with very low fundamental frequency and very little inherent damping and, for this reason, they have been known to be susceptible to wind-induced vibration, as well as rain-wind-induced vibration [41,42]. Recognition of this susceptibility of stay cables leads to the use of some aerodynamic measures. The following is a brief introduction to the aerodynamic measures used for rain-wind-induced vibration control of stay cables on the Stonecutters Bridge. The aerodynamic measures considered were helical fillets and dimples added on the cable surface. Wind tunnel investigation was carried out to determine drag coefficients of the 1: 1 scale section models of stay cables manufactured with a diameter of 140 mm, 160 mm and 180 mm, which are representative of medium, long and very long prototype cables of the Stonecutters Bridge [43]. Rain-wind tunnel investigation was also conducted to assess the efficiency of helical fillets and dimples for mitigating rain-wind-induced cable vibration [44].

12.5.1 Wind Tunnel Investigation and Cable Drag Coefficients

Cable section models were designed and made of 1: 1 scale stainless steel cylinders covered with high density polyethylene tube, which is the same material as used in the prototype stay cables, to ensure Reynolds number similarity and the same surface conditions for cable model and prototype. The wind tunnel tests were conducted in the high-speed working section of the CLP Wind Tunnel at the Hong Kong University of Science and Technology. Figure 12.14 shows some of the cable section models used in the wind tunnel investigation, including cables with smooth surface, cables with helical fillets and cables with dimples. Detailed information on the wind tunnel and the test apparatus can be found in Section 7.10 in Chapter 7.

Helical fillets of a rectangular cross-section were stuck on the smooth surface of the cable model, with an effective step of 0.3 m, to simulate the prototype cable with helical fillets. The helical fillet was made of light wood with a square cross-section of 2×2 mm. A cable model of dimpled surface was manufactured of a diameter of 140 mm. The shape and density of the dimples on the surface of the cable model are shown in Figure 12.15.

The lengths of major axis and minor axis of the elliptic dimple were 8.5 mm and 6.5 mm, respectively. The periphery of the elliptic dimples was about 0.5 mm above the cable surface, while the central part of the dimples was pressed into the cable by about 1.5 mm. Dimples of elliptical shape and 2 mm depth were made on the surface of the cable model in the pattern shown in Figure 12.15. The density ratio of the surface area of the dimples to the surface area of the cable was approximately 0.14.

The drag coefficients of stay cables of different configurations were investigated through wind tunnel tests with respect to cable diameter, inclination and yaw angle. Figure 12.16 exhibits the test rig used and the experimental set-up. The aerodynamic force coefficients of the stay cables of smooth surface were compared with those of the stay cables of either helical fillets or dimples, so that the difference in static wind load on the stay cables due to surface change could be estimated. The major results obtained

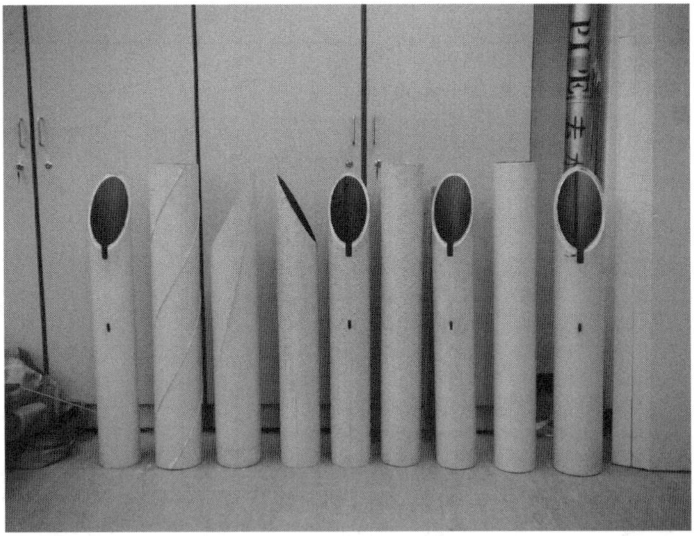

Figure 12.14 Selected cable section models used in wind tunnel investigation.

from the wind tunnel investigation could be summarized in the following, and the full results can be found in the reference [43].

The drag coefficients of the stay cables with a smooth surface were larger when wind was perpendicular to the bridge deck axis than when the wind was along the bridge deck axis. When the wind was perpendicular to the bridge deck axis, the drag coefficients of the stay cables decreased, with increasing mean wind speed ranging from 10–20 m/sec. The drag coefficients of the stay cables with the helical fillets were only slightly larger than those of the same cables with smooth surface when the wind was perpendicular to the bridge deck axis. When the wind was along the bridge deck axis, the drag coefficients of the stay cables with helical fillets may not be larger than those of the same cables with smooth surface. The drag coefficients of the cable with dimpled surface were smaller than those of the cable with smooth surface in all the wind directions concerned. The drag coefficient of the dimpled cable

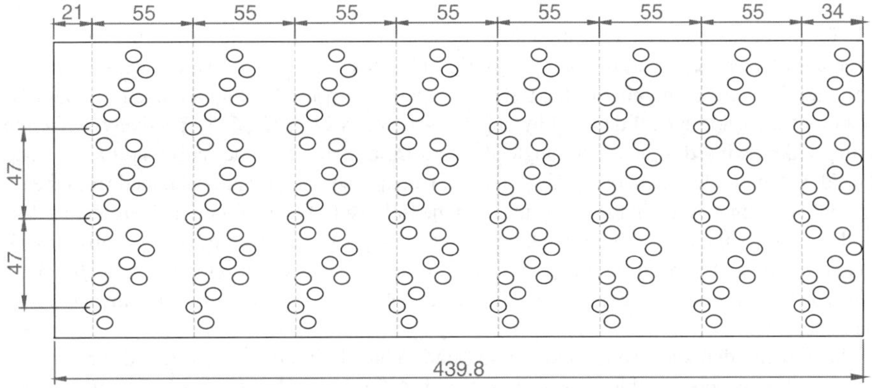

Figure 12.15 Dimple pattern on the cable surface (unit in mm).

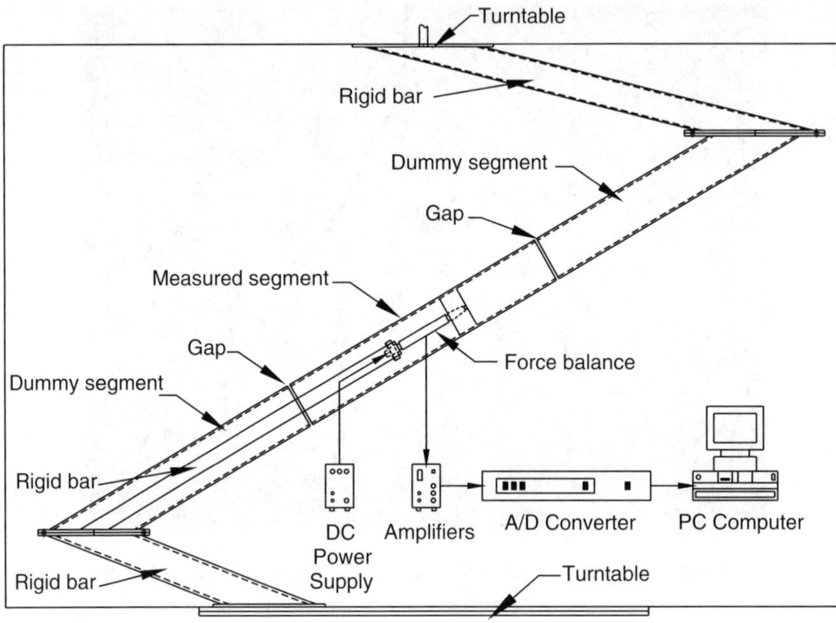

Figure 12.16 Wind tunnel test arrangement for inclined cables (from [63]) (Reproduced with permission from Multi-Science Publishing Co Ltd).

was also much less influenced by wind speed, compared with the smooth cable. In summary, static wind load on the cable with dimpled surface is the smallest among the three types of cable surface.

12.5.2 Rain-Wind Tunnel Investigation of Stay Cables of Different Surfaces

Cable section models were also designed and made of 1: 1 scale stainless steel cylinders covered with high density polyethylene tube, which is the same material as used in the prototype stay cables. The diameters of two cable models tested were 139 mm (140) and 169 mm (170) respectively. Three cable surfaces were considered: cables with smooth surface, cables with helical fillets, and cables with dimples.

Figure 12.17 shows the schematic diagram of the test arrangement. The natural frequency was 0.63 Hz for the cable model of 170 mm diameter and 0.91 Hz for the cable model of 140 mm diameter. With the aid of an oil damper, the damping ratio of the test cable model varied within a range from 0.1% to 1%. The Scruton number of the test cable model also varied within a range from 0.9 to 8. The inclination of the cable model denoted by angle α was set as 21° and 36° respectively, while the yaw angle of incident wind designated by angle β varied from 0° to 65°. The rain-wind tunnel tests were conducted by the Southwest Jiaotong University in China. The major results obtained from the investigation could be summarized in the following, and the full results can be found in the reference [44].

For the 170 mm diameter cable model, rain-wind-induced vibrations were clearly observed for the yaw angle of incident wind from 30° to 55°. The strongest cable vibration occurred at $\beta = 30°$ and a wind speed of 12 m/sec. The maximum amplitude of cable vibration was up to 440 mm, about 2.6 times the cable diameter.

For the 140 mm diameter cable model, rain-wind-induced vibrations were found for all the tested yaw angles. The strongest cable vibration occurred at $\beta = 35°$ and a wind speed of 10.5 m/sec. The maximum amplitude of cable vibration was more than 138 mm, almost equal to the cable diameter.

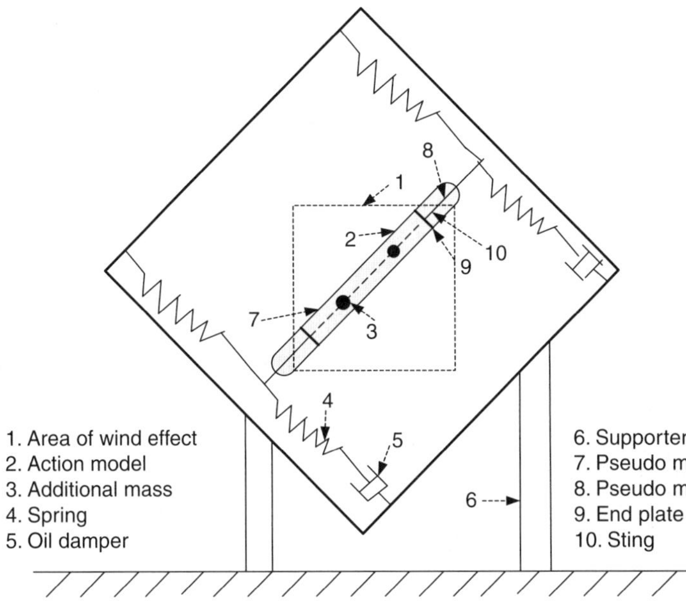

Figure 12.17 Rain-wind tunnel test arrangement for inclined cables.

As for the effects of rain intensity, it was found that the strongest cable vibration emerged at a rain intensity of 75 mm/h for the cable model of 170 mm diameter. The maximum amplitude of cable vibration was about 482 mm, about 2.85 times the cable diameter. For the cable model of 140 mm diameter, the strongest vibration emerged at a rain intensity of 85 mm/h. The maximum amplitude of cable vibration was about 153 mm, which was 1.1 times the cable diameter.

For the cable model with either helical fillets or dimples, no significant cable vibration was observed for most of the test cases. The vibration amplitude of the cable model with helical fillet increased slightly with the increasing height of the fillet. Relatively strong vibration was only observed for the cable model of 170 mm diameter with dimple at $\beta = 40^\circ$ and a rain intensity of 80 mm/h. The maximum vibration amplitude was about 0.95 times the diameter of the cable model.

It could be concluded that the helical fillets could reduce rain-wind-induced cable vibration significantly but amplify static wind load. The dimples could reduce both rain-wind-induced cable vibration and static wind load, but rain-wind-induced cable vibration might still emerge for particular wind directions and cable diameters.

12.6 Mechanical Measures for Vortex-Induced Vibration Control

Sometimes, particularly after a bridge is completed, it is difficult to find aerodynamic measures that are effective for vortex-induced vibration control. When satisfactory aerodynamic solutions cannot be found, the control of vortex-induced vibration relies on mechanical measures. Most popular mechanical measures for vortex-induced vibration control are tuned mass dampers (TMD). A TMD is a device consisting of a mass attached to a structure in such a way that it oscillates at the same frequency of the bridge structure, but with a phase shift. The mass is attached to the structure via a spring-dashpot system, and the energy is dissipated by the dashpot as relative motion develops between the mass and the structure. The schematic diagram of a TMD attached to a flexible beam is shown in Figure 12.18.

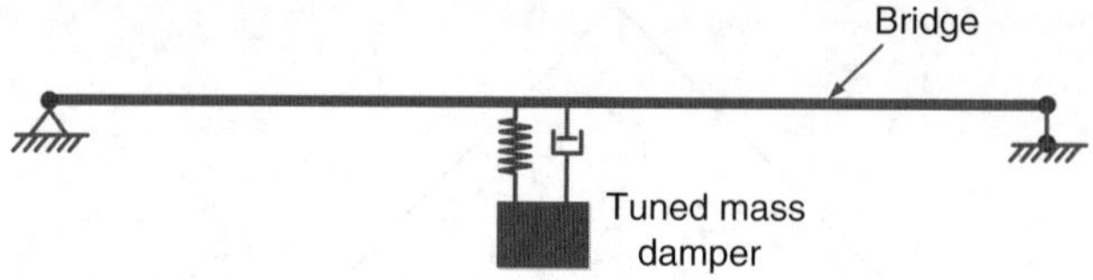

Figure 12.18 Schematic diagram of a TMD attached to a bridge.

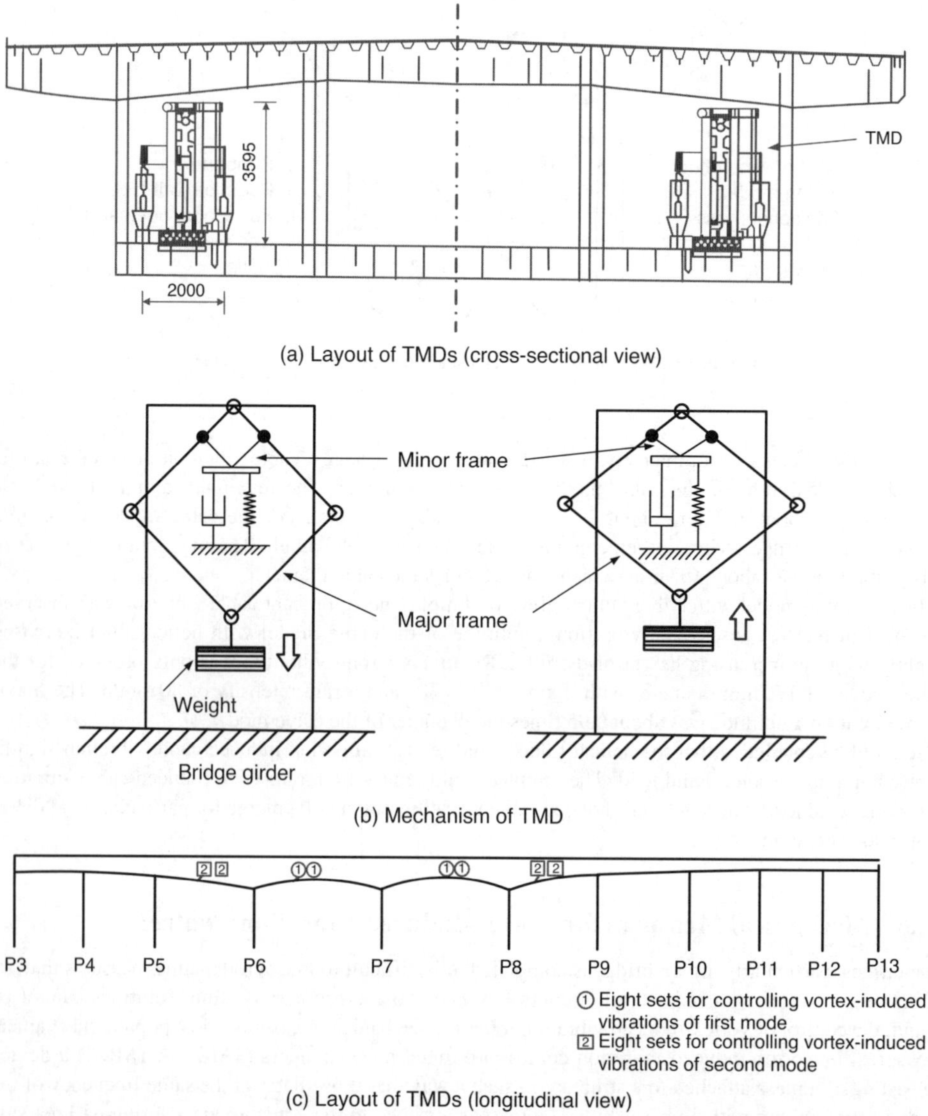

Figure 12.19 Tuned mass dampers and their locations of Trans-Tokyo Crossing Bridge (Figure courtesy of Prof. Yozo Fujino).

The Trans-Tokyo Bay Crossing Bridge in Japan, completed in 1997, is a ten-span continuous steel box girder bridge with a total length of 1630 m [45]. Although this bridge is not a cable-supported bridge, the experience of vortex-induced vibration control obtained from the bridge is invaluable for long-span cable-supported bridges. In the design phase of the bridge, various wind tunnel tests were carried out on two-dimensional sectional models as well as a complete three-dimensional model [46]. These tests revealed that vortex-induced vertical flexural vibration would develop in the bridge under the wind normal to the bridge axis. A variety of aerodynamic mitigation strategies were tested, but complete suppression of vibration was not achieved. As a result, consideration was given to adding TMDs to the bridge if excessive vibration did happen after completion of the bridge.

Significant vortex-induced vibration was, indeed, observed after the girder erection was completed in 1994. The vortex-induced first-mode of vibration peaked at a wind speed of around 16–17 m/sec with maximum amplitude over 50 cm [45]. It was decided to install 16 TMDs in the girders to control its first and second modes of vibration. Vibration in the higher modes was controlled aerodynamically by attaching vertical plates to the posts of crash barriers in the girders.

The TMDs for the bridge were installed inside the box girder, as shown in Figure 12.19a. The structure of a TMD consisted of a major frame and a minor frame in the shape of a pantograph (see Figure 12.19b). The weight of the TMD was mounted on the lower end of the major frame via coil springs and oil dampers, and the TMD moved up and down with girder vibration. The use of two frames made it possible to reduce the TMDs' size considerably. Eight TMDs were installed for the first mode and eight for the second mode, as shown in Figure 12.19c.

A two degrees of freedom model was adopted in the determination of the parameters of the TMDs, and those for the first mode were designed independently of those for the second mode. The natural frequency of the TMDs was 0.33 Hz for the first mode and 0.47 for the second mode, and the mass ratio was 1.16% for both modes of vibration. The logarithmic decrement damping of the bridge-TMD system was targeted to be 0.22 for both modes of vibration. The maximum stroke of the TMD was $\pm$600 mm for the first mode and $\pm$800 mm for the second mode. The field measurements were made before and after installation of the TMDs, to access their performance. The measurement results showed that the performance of the TMDs was satisfactory: under the almost same wind conditions, the amplitude of girder vibration without TMDs was over 40 cm, but the amplitude of girder vibration with the TMDs was just 5–6 cm.

12.7 Mechanical Measures for Flutter Control

A number of studies have been conducted by utilizing auxiliary damping devices for controlling bridge flutter, although most of them have not yet been implemented in real long-span cable-supported bridges. These devices include passive control systems, active control systems and semi-active control systems.

12.7.1 Passive Control Systems for Flutter Control

Various passive control systems, such as TMDs, tuned liquid dampers (TLD), and tuned liquid column dampers (TLCD) have been proposed and investigated for controlling bridge flutter. Nobuto *et al.* [47] carried out both numerical analyses and wind tunnel tests of a bridge section model to investigate the effectiveness of TMDs for flutter control. Their results showed that the critical flutter wind speed was increased by about 14%. However, the performance of the TMD system was very sensitive to the tuning condition. Gu *et al.* [48] also conducted both numerical analyses and wind tunnel tests of a box section model to confirm the performance of TMDs for controlling flutter. Two TMDs were placed on the leading and trailing edges of the deck, and their mechanical model is shown in Figure 12.20.

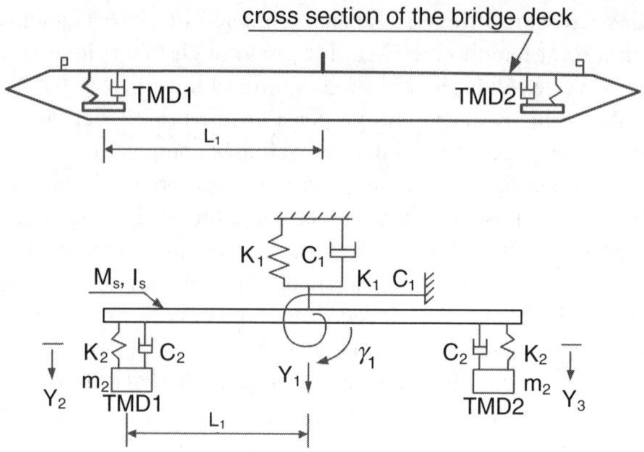

Figure 12.20 Mechanical model of TMDs and bridge deck for flutter control (from [48]).

Their results showed that TMDs could increase the critical flutter wind speed significantly. The TMDs with more than 5.6% inertia mass moment ratio could increase the critical flutter wind speed of the Tiger-Gate Bridge section model with wind screens by more than 40%. Another TMD study was carried out by Pourzeynali and Datta [49], in which a combined vertical and torsional TMD system was applied to increase the critical flutter wind speed. The proposed system had two degrees of freedom, tuned close to the frequencies corresponding to the vertical and torsional symmetric modes of the bridge for coupled flutter control. The maximum improvement of the critical flutter wind speed was about 2.03 times the uncontrolled critical wind speed for a 5% TMD damping ratio.

However, all of the above studies indicated that the performance of a TMD is very sensitive to frequency tuning: a slight shift of the frequency tuning may lead to a significant deterioration of control performance of the TMD. Kwon and Park [50] therefore proposed to use multiple-tuned mass dampers (MTMD) to overcome the problem. They investigated two types of MTMD – some with frequencies that were equally spaced and some that were not equally spaced. They found that irregular multiple-tuned mass dampers that have unequal frequency interval and different damping ratios offered more flexible designs and provided better performance for controlling bridge flutter.

While the effectiveness of TMDs in controlling bridge flutter has been demonstrated in these studies for specific bridges, their limitations and the dependence of control performance on the dynamic and aerodynamic characteristics of bridge structures were addressed by Chen and Kareem [51]. Their study showed that the performance of TMDs strongly depends on the bridge's aerodynamic characteristics. For a hard-type flutter characterized by negative damping that grows rapidly with increasing wind speed beyond the flutter onset, the influence of structural modal damping on the critical flutter speed was insignificant. Accordingly, the effectiveness of TMDs in controlling this type of flutter was rather marginal. However, for a soft-type flutter, in which the negative damping of bridges grows slowly with increasing wind speed, addition of auxiliary damping may result in significantly higher critical flutter wind speed. For this type of flutter, a reliable estimate of structural modal damping was critical for accurate estimation, and auxiliary dampers such as TMDs would be relatively effective in controlling such type of flutter.

Another class of passive control systems is TLD, in which liquids are used to provide all of necessary characteristics of the secondary system (see Figure 12.21a). In this system, the liquid not only supplies the required secondary mass but also damping through viscous action, primarily in the boundary layers. Meanwhile, the gravity force of the liquid provides the necessary restoring mechanism. As a result, the secondary system has characteristic frequencies that can be used for optimal performance, similar to TMDs, for flutter control.

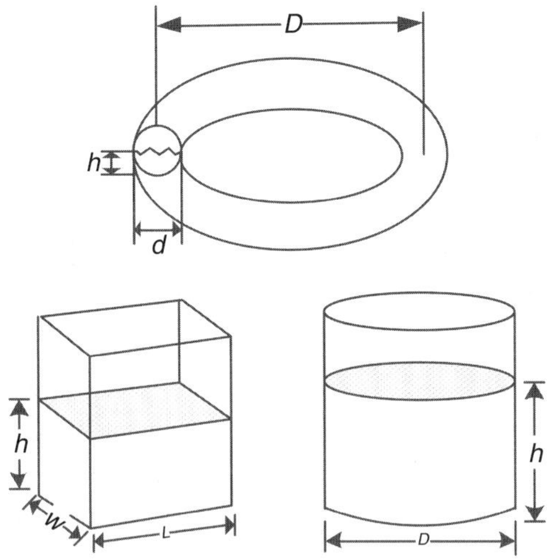

(a) Tuned liquid damper

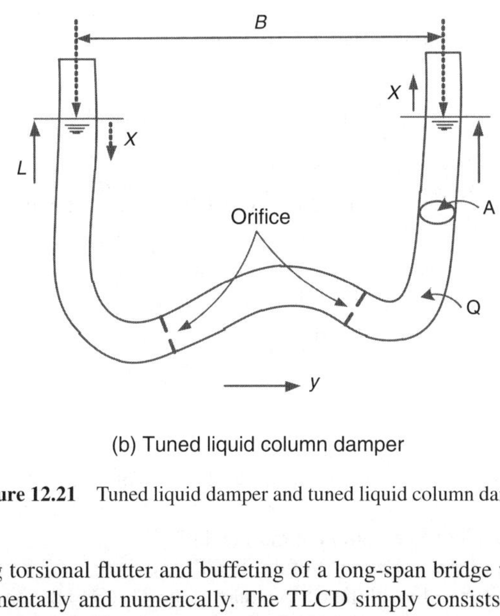

(b) Tuned liquid column damper

Figure 12.21 Tuned liquid damper and tuned liquid column damper.

TLCD for controlling torsional flutter and buffeting of a long-span bridge were also investigated by Xue *et al.* [52], experimentally and numerically. The TLCD simply consists of a U-shaped container filled with water, and structural vibration energy is dissipated as the water passes through an orifice with inherent head-loss characteristics (see Figure 12.21b). For controlling torsional flutter of a bridge deck, Figure 12.22 displays a schematic diagram of the experimental setup and instrumentation, in which a TLCD model is installed inside a bridge deck unit model [52].

The bridge deck unit model is supported by a pivot at the middle point of its horizontal frame so that the bridge deck could rotate around the pivot. Two springs are installed at the two ends of the horizontal frame to provide the deck with torsional stiffness. The test platform is rigidly connected to the horizontal frame, using four vertical steel members. The whole deck unit could thus be seen as a single degree of freedom system rotating around the pivot. The container of a TLCD is fixed on the test platform and

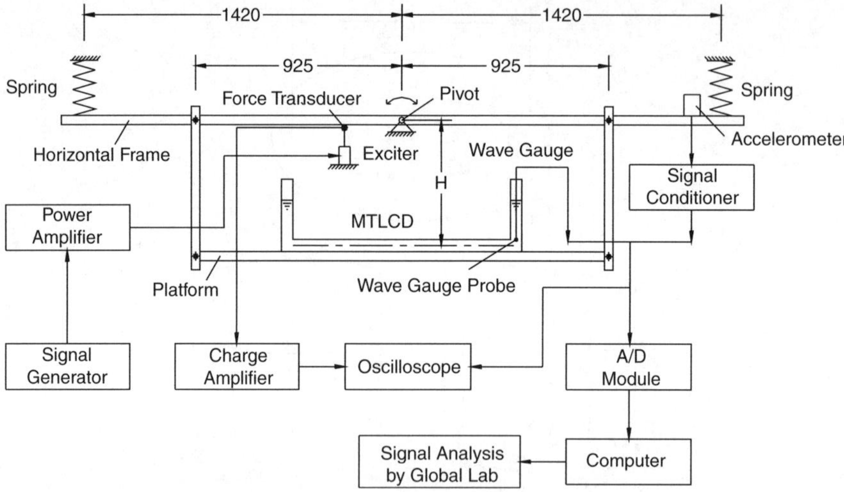

Figure 12.22 Experimental set-up and instrumentation of TLCD for flutter control.

the weight of the container without water is regarded as part of the deck weight. The coupled equations of motion of the bridge girder equipped with a TLCD for torsional vibration control can be expressed as:

$$(I_s + I_d)\ddot{\theta} + \rho AB\left(H + \frac{L-B}{2}\right)\ddot{W} + C_s\dot{\theta} + (K_s + \rho ALgH)\theta + \rho AgBW = M_s \tag{12.1a}$$

$$\rho AL\ddot{W} + \frac{\rho A}{2}\delta|\dot{W}|\dot{W} + 2\rho AgW + \rho AB\left(H + \frac{L-B}{2}\right)\ddot{\theta} + \rho AgB\theta = 0 \tag{12.1b}$$

with the condition:

$$W \leq \frac{L-B}{2} - \frac{d}{2} \tag{12.1c}$$

where:

I_s is the mass moment of inertia of the bridge girder with respect to the pivot;
I_d is the mass moment of inertia of water column of the TLCD with respect to the pivot;
ρ is the density of water;
A is the cross-sectional area of water column of the TLCD;
B is the horizontal width of the water column of the TLCD;
H is the distance from the centre line of the bottom tube of the TLCD to the pivot;
L is the total length of water column;
C_s is the damping coefficient of the structure;
K_s is the torsional stiffness of the structure;
g is the acceleration of gravity;
δ is the head loss coefficient of the TLCD governed by the opening ratio of orifice;
θ is the torsional displacement of the structure;
W is the relative motion of water to the container;
d is the thickness of water column;
M_s is the external moment excitation.

The experimental and numerical results show that TLCD is also an effective device for increasing critical wind speed and for reducing buffeting response of a long-span suspension bridge [52].

12.7.2 Active Control Systems for Flutter Control

With increasing span of cable-supported bridges, a few studies have been made on active and semi-active control systems for flutter control. One of the active control systems being considered is active mass dampers (AMD), which change the dynamic properties of the bridge structure to enhance flutter stability in terms of active control forces or moments generated by the actuators that are, in turn, regulated by control algorithms according to the feedback from the structural responses and/or external excitations. AMDs can be assembled and operate completely inside the bridge girder.

Two types of AMDs were investigated by Korlin and Starossek [53]: the rotational mass damper (RMD) and the movable eccentric mass damper (MEMD). The equations of motion were established on the basis of a two degrees of freedom model of the bridge structure.

Figure 12.23a shows the two degrees of freedom structure with a RMD: the vertical displacement and the rotational displacement. An additional rotating mass was implemented in the center of the bridge girder for active control. The control variable was the rotational acceleration of damper mass. Figure 12.23b displays the two degrees of freedom model of the bridge structure with MEMD. In this case, the active control was exerted by a moment due to gravity force of the additional mass. The eccentricity was the control variable. An optimal linear static feedback controller was used for flutter suppression in both the analysis and the experiment. The motion-induced aerodynamic forces considered in the analysis were obtained experimentally in terms of frequency dependent flutter derivatives. The flutter analysis of the uncontrolled and the controlled bridge structure was carried out using the Hurwitz criterion.

Various control scenarios under different flow regimes were simulated, and the analytical and experimental results were found in good agreement. A deterioration of control performance and flutter stability was found as the effect of saturations in the control. The large energy demand of the proposed AMDs in combination with linear control was a potential challenging issue. Alternative devices and/or control laws should be a topic of further research.

Achkire *et al.* [54] presented another active control strategy for flutter control of cable-stayed bridges. As shown in Figure 12.24, the flutter control of the bridge is realized through active tendon control of stay cables. The control law adopted was decentralized integral force feedback, which guaranteed control stability. The efficiency of the active control strategy was demonstrated by experiments on small laboratory models using piezoelectric actuators, which laid on a foundation for further investigation.

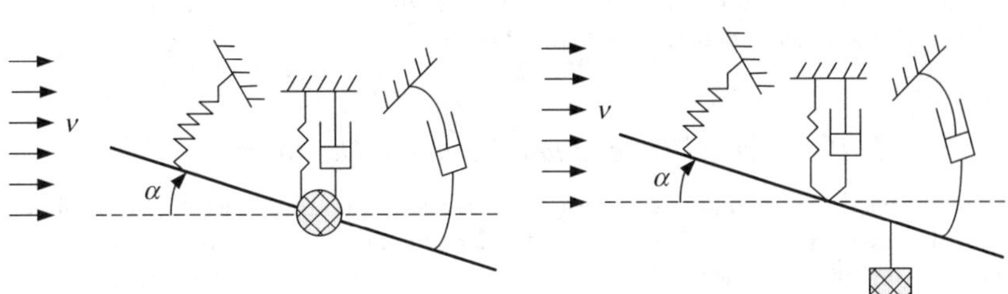

(a) Rotating mass damper (b) Movable eccentric mass damper

Figure 12.23 Active mass dampers for flutter control of bridge deck.

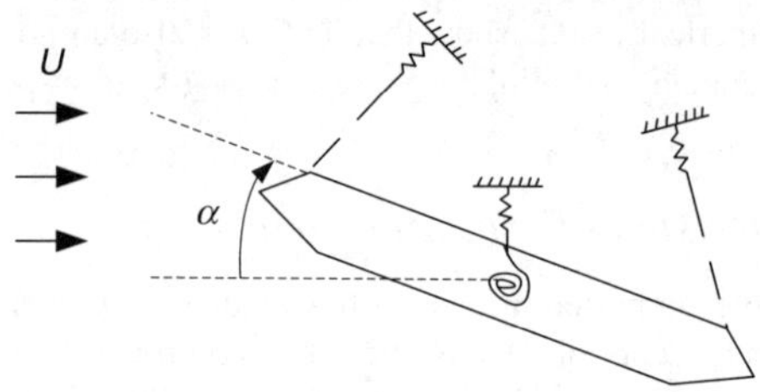

Figure 12.24 Active tendons for flutter control of bridge dec.

12.7.3 Semi-Active Control Systems for Flutter Control

A combined vertical and torsional semi-active TMD (STMD) system was proposed by Pourzeynali and Datta for flutter control of long-span suspension bridges [55]. The semi-active TMDs had adjustable damping which could be actively controlled. A fuzzy logic feedback control algorithm was used to obtain the optimal damping of TMDs for optimal reduction of the bridge response at the flutter wind speed. The numerical results showed the superior performance of semi-active control over passive control.

For the uncontrolled case, the critical flutter wind speed was about 55.52 m/sec. With a passive TMD system having a 20% damping ratio, the critical flutter wind speed increased to 98 m/sec. With nearly the same value of the maximum variable damping (21.6%) of the STMD system, the bridge underwent a decaying oscillation at a wind speed of 110 m/sec. The maximum torsional oscillation was reduced from 0.02 rad to 0.0063 rad – much less than that for sustained flutter oscillation of the bridge alone. Therefore, the STMD control system with variable damping is a promising way to control the flutter conditions of long-span suspension bridges.

12.8 Mechanical Measures for Buffeting Control

When the aerodynamic count measures are not able to reduce wind-induced buffeting response sufficiently, mechanical measures may be considered. These can be passive, active, semi-active or hybrid control systems, as discussed for flutter control, and these control systems can also be applied to bridge decks, cables and towers. The use of tuned mass dampers for mitigating wind-induced buffeting responses of long-span cable-supported bridges has been discussed in detail in the references [56,57]. For a wide coverage of the concerned topic, this section introduces multiple pressurized tuned liquid column dampers [58] and semi-active tuned liquid column dampers [59] for mitigating buffeting responses in either the construction or completion stages.

12.8.1 Multiple Pressurized Tuned Liquid Column Dampers

The pressurized tuned liquid column damper (PTLCD) originates from tuned liquid column dampers (TLCD), as discussed in Section 12.7.1. It is a U-shaped container of uniform cross-sectional area, with liquid filled into the container and two chambers at its two ends filled with compressed air of static pressure P_0, as shown in Figure 12.25.

When the damper experiences a vibration due to structural motion, the volume of the two end chambers varies due to liquid motion inside the PTLCD. The change in the volume of an air chamber leads to

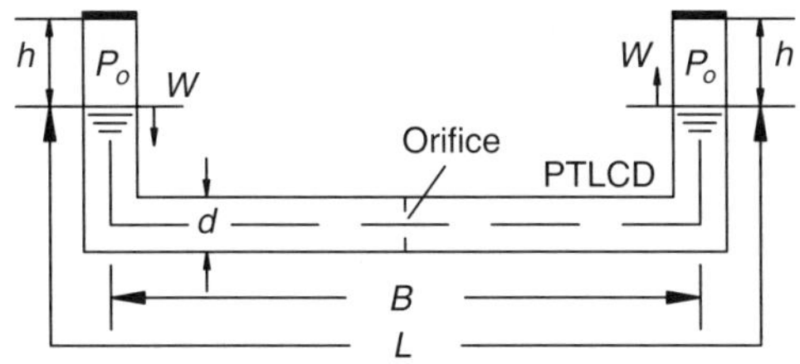

Figure 12.25 Schematic diagram of pressurized tuned liquid column damper (from [58]) (Reproduced with permission from Elsevier).

a variation of air chamber pressure and, thus, a pressure difference between the two chambers. The restoring force F_p, due to the pressure difference between the two chambers, can be expressed in the following form:

$$F_p = P \times A \tag{12.2}$$

where:

P is the net pressure between the two air chambers;
A is the cross-sectional area of container.

With an assumption that the variation of pressure and volume inside the air chamber is an isothermal process, the pressure and volume change can be described by Boyle's law, i.e.:

$$P_0 V_0 = P_R(V_0 - AW) = P_L(V_0 + AW) \tag{12.3}$$

where:

P_L and P_R are the pressure in the left and right air chamber of PTLCD respectively;
V_0 is the volume of air inside the chamber at the static position of liquid;
W is the liquid displacement of the PTLCD.

The restoring force F_p acting on the water at time t is then determined by:

$$PA = (P_R - P_L)A = P_o A\left[\left(1 - \frac{W}{h}\right)^{-1} - \left(1 + \frac{W}{h}\right)^{-1}\right] \tag{12.4}$$

where h is the air chamber height.

Equation 12.4 shows that the restoring force due to the pressure acting on the liquid inside the PTLCD is a non-linear function of water displacement. Iterations are generally required to solve Equation 12.4 in the time domain. However, if the ratio W/h is small, the restoring force in Equation 12.4 can be approximately expressed as a linear function of water displacement, i.e.:

$$PA \approx P_o A\left[\left(1 + \frac{W}{h}\right) - \left(1 - \frac{W}{h}\right)\right] = \frac{2P_o AW}{h} \tag{12.5}$$

Equation 12.5 shows that the restoring force due to the pressure inside the air chamber of the PTLCD is dependent on liquid displacement. For the case of a liquid column damper without air pressure, it becomes a traditional TLCD and the natural frequency of liquid in the TLCD is given by[51]:

$$\omega^2 = \frac{2\,\rho_w A g}{\rho_w A L} = \frac{2g}{L} \tag{12.6}$$

where ρ_w is the density of liquid inside TLCD.

Note that the natural frequency of liquid motion in the TLCD depends solely on the liquid column length L of the TLCD. By applying a static pressure inside the two sealed air chambers at the two ends of TLCD, the natural frequency of liquid motion in the PTLCD becomes:

$$\omega^2 \approx \frac{2\,\rho_w A g + \dfrac{2P_0 A}{h}}{\rho_w A L} = \frac{2g}{L}\left(1 + \frac{P_0}{\rho_w g h}\right) \tag{12.7}$$

Equation 12.7 shows that the natural frequency of liquid motion inside the PTLCD is determined not only by the length of liquid column but also the static pressure P_0. For a given liquid column length, the natural frequency of the liquid motion inside the PTLCD can be increased by the factor $(1 + P_0/\rho g h)$, comparing with the traditional TLCD, which greatly facilitates the frequency tuning requirement. By re-arranging Equation 12.7, the static pressure P_0 inside the PTLCD is given by:

$$P_0 = \frac{\rho_w L h}{2}\left[\omega^2 - \frac{2g}{L}\right] \tag{12.8}$$

Equation 12.8 provides a way of determining the required static pressure P_0 after the frequency of the PTLCD and its liquid column length L are selected.

To consider the interaction between the liquid dampers and a long-span cable-supported bridge under wind excitation, it is expedient to derive the finite element model of the PTLCD or MPTLCD, which is then incorporated into the finite element model of the bridge to form a coupled bridge-MPTLCD system. Let us consider the MPTLCD, which consists of n_T small PTLCD units. Each small PTLCD unit is installed below the torsional centre of the bridge deck and at the locations where vibration amplitudes of the bridge in the lateral and torsional directions are the largest. The MPTLCDs are connected to the transverse beams of the bridge deck by roller supports and simply supports, as shown in Figure 12.26.

Two additional nodes, named node 1 and node 2, are generated at the positions where the MPTLCD is connected to the bridge. These nodes reflect the motion of the MPTLCD interacted with the motion of the bridge. From a view point of practical use, the distance between the vertical columns B is the same for all PTLCD units. The axial deformation of the transverse beam between the two supports is assumed to be negligible and, hence, the lateral displacement of the PTLCD units is taken as x and the torsional displacement of the PTLCD unit is then determined by:

$$\theta = \frac{y_2 - y_1}{B} \tag{12.9}$$

where y_1 and y_2 are the vertical displacements of node 1 and node 2 respectively, as shown in Figure 12.26.

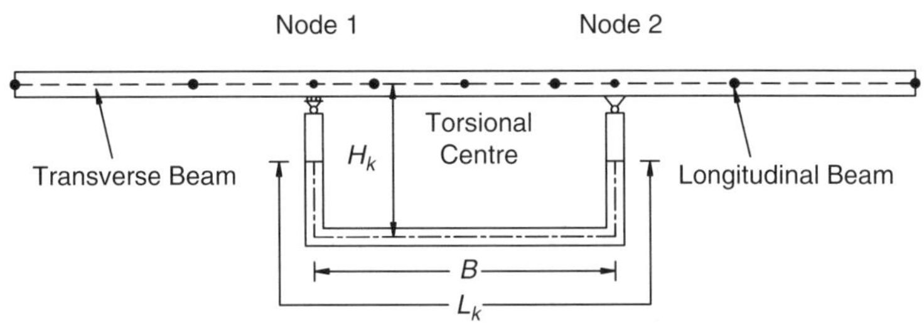

Figure 12.26 Connections between bridge deck and liquid damper (from [58]) (Reproduced with permission from Elsevier).

x shown in Figure 12.26 is the lateral displacement of node 2, which is not involved in the determination of the torsional displacement. The mass matrix $[M]$ and the stiffness matrix $[K]$ of the damper element can be obtained by formulating the Lagrangian of all PTLCDs used.

A real triple-tower cable-stayed bridge with an overall length of 1177 m, two main spans of 448 m and 475 m and two side spans of 127 m each, was used for a case study [58] to assess numerically the performance of MPTLCD for suppressing combined lateral and torsional vibration of the bridge (see Figure 12.27).

From the dynamic characteristics of this bridge, the dynamic lateral displacement response of the bridge deck was dominated by the first five lateral modes of vibrations, but the dynamic torsional displacement response was dominated by the first torsional mode of vibration only. Therefore, it was decided that for each main span, five MPTLCDs would be installed at the middle section (point A or point B) and would be tuned to the first five lateral frequencies of the bridge respectively, while one MPTLCD would be installed at the middle section (point A or point B) and would be tuned to the first torsional frequency of the bridge. The positions of point A and Point B are also shown in Figure 12.27. The standard deviation displacement and acceleration responses of the bridge deck are listed in Tables 12.1 and 12.2, respectively, for mean wind speeds of 20 m/sec and 50 m/sec at points A and B, with and without control.

It can be seen from Tables 12.1 and 12.2 that both the standard deviation displacement and acceleration responses in either lateral or torsional directions are reduced by the MPTLCDs effectively. The standard deviation displacement reduction in the lateral direction at point A reaches the level of 25% at a mean wind speed of 20 m/sec, and the level of 29% at a mean wind speed of 50 m/sec. The reduction of standard deviation torsional displacement response can reach the level of 21% at a mean wind speed

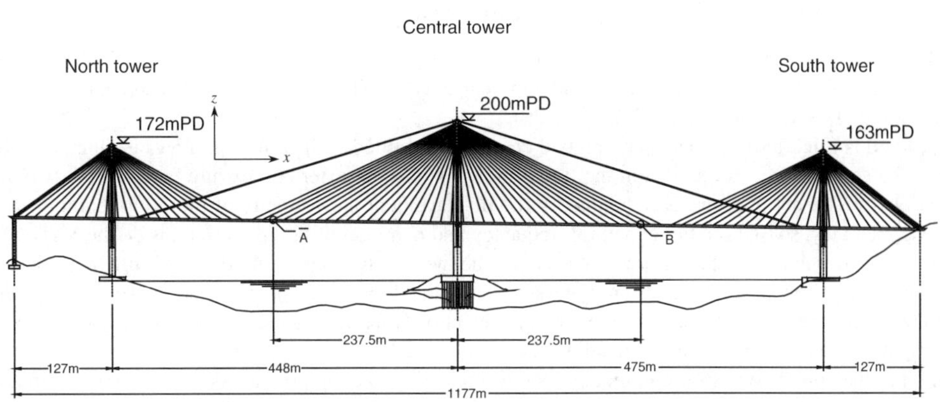

Figure 12.27 Configuration of long-span cable-stayed bridge (from [58]) (Reproduced with permission from Elsevier).

Table 12.1 Standard deviation of deck displacement with and without control (PTLCD) (from [58]) (Reproduced with permission from Elsevier)

	Lateral (m)		Torsional (rad)	
Location	A	B	A	B
Without control ($U_m = 20$ m/sec)	0.0028	0.0032	0.00 041	0.00 043
With control ($U_m = 20$ m/sec)	0.0021	0.0023	0.00 032	0.00 034
	(−25.0%)	(−28.1%)	(−22.0%)	(−20.9%)
Without control ($U_m = 50$ m/sec)	0.0287	0.0323	0.00 386	0.00 402
With control ($U_m = 50$ m/sec)	0.0204	0.0228	0.00 287	0.00 303
	(−28.9%)	(−29.4%)	(−25.7%)	(−24.6%)

Table 12.2 Standard deviation of deck acceleration with and without control (PTLCD) (from [58]) (Reproduced with permission from Elsevier)

	Lateral (m/sec^2)		Torsional (rad/sec^2)	
Location	A	B	A	B
Without control ($U_m = 20$ m/sec)	0.0137	0.0134	0.0048	0.0047
With control ($U_m = 20$ m/sec)	0.0107	0.0108	0.0037	0.0037
	(−21.9%)	(−19.4%)	(−22.9%)	(−21.3%)
Without control ($U_m = 50$ m/sec)	0.1375	0.1341	0.0467	0.0463
With control ($U_m = 50$ m/sec)	0.1058	0.1075	0.0357	0.0354
	(−23.1%)	(−19.8%)	(−23.6%)	(−23.5%)

of 20 m/sec and the level of 25% at a mean wind speed of 50 m/sec at point B. The standard deviations of lateral and torsional displacement responses of the bridge deck along the bridge axis at the mean wind speed of 50 m/sec are plotted in Figure 12.28, which shows that the maximum standard deviation of the displacement response of the bridge occurs near the midpoints of the two main spans and is reduced significantly by the MPTLCDs. However, the displacement responses of the deck near the towers are hardly reduced by the MPTLCDs. Further details on analytical models and numerical results can be found in the reference [58].

12.8.2 Semi-Active Tuned Liquid Column Dampers

PTLCD can be further developed to be a more robust PTLCD called semi-active tuned liquid column dampers (SATLCD) with frequency adaptability capacity. SATLCD is also a U-shaped container with uniform cross-sectional area (see Figure 12.29).

Liquid is filled into its container, and two chambers are filled with compressed air of static pressure P_0. The control force is applied on the liquid in terms of a net external pressure between the two air chambers. The net pressure is regulated by the displacement and velocity of the liquid column in a prescribed way, so that the target natural frequency and damping of liquid motion inside the SATLCD can be easily achieved. The net pressure between the two air chambers, sensed by pressure transducers, is forced to follow or track the desired pressure determined by a computer in accordance with a given control algorithm and a targeted frequency. Any deviation from the desired pressure is fed back into the computer to take corrective action to adjust the servo valves.

Thus, the control system is continually monitoring and correcting pressure deviation to maintain the desired pressure acting on the liquid column and the targeted frequency. In practice, a number of SATLCDs are required for reducing vibration of a bridge. The control force is composed of two parts:

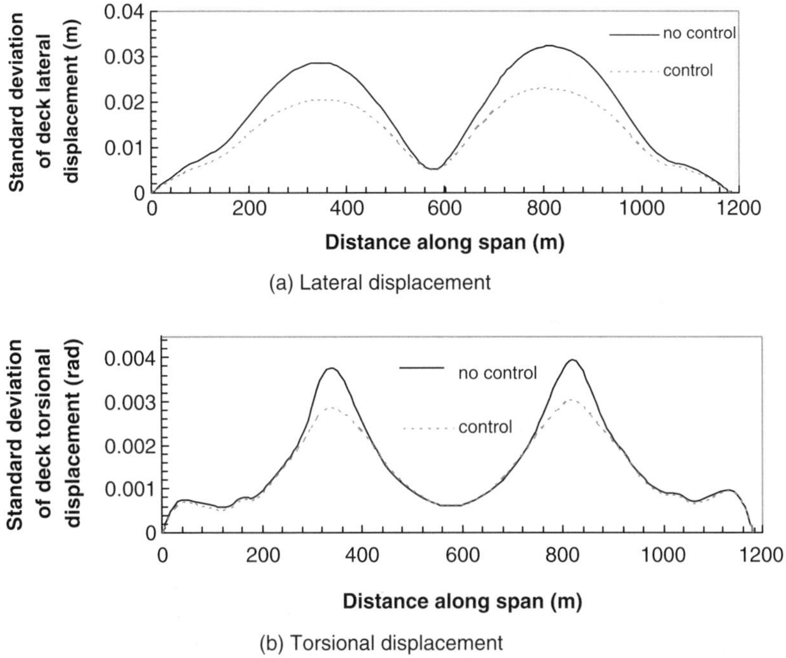

Figure 12.28 Comparisons of standard deviations of bridge deck displacements (from [58]) (Reproduced with permission from Elsevier).

one is displacement feedback force for the control of liquid oscillation frequency, while the other is velocity feedback force for the control of liquid damping. The displacement feedback control force $u_{1k}(t)$ in the kth SATLCD is given by:

$$u_{1k}(t) = S_k \times W_k(t) \tag{12.10}$$

where:

S_k is the constant displacement feedback gain of the kth SATLCD;
$W_k(t)$ is the displacement of the liquid column.

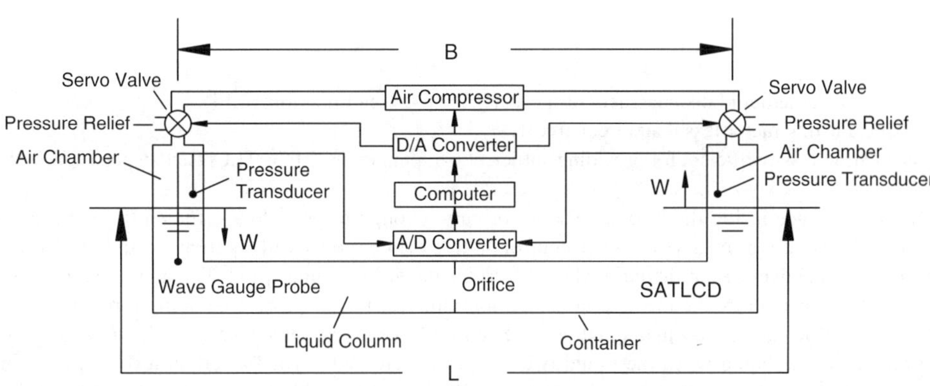

Figure 12.29 Schematic diagram of semi-active tuned liquid column damper (from [59]) (Reproduced with permission from Techno.Press).

The direction of the control force $u_{1k}(t)$ is in the same (opposite) direction as the liquid displacement $W_k(t)$ when the constant displacement feedback gain is positive (negative). With the control force expressed by Equation 12.10, the circular natural frequency, ω_k, of liquid motion in the kth SATLCD, can be determined by:

$$\omega_k^2 = \frac{2g}{L_k} + \frac{S_k}{\rho_w A_k L_k} \tag{12.11}$$

where:

L_k is the length of the kth SATLCD;
A_k is the cross-sectional area of the kth SATLCD;
ρ_w is the density of liquid inside the kth SATLCD.

For a targeted frequency of the liquid damper, it can easily be seen from Equation 12.11 that the liquid column length of the kth SATLCD is given by:

$$L_k = \frac{2g}{\omega_k^2} + \frac{S_k}{\rho_w A_k \omega_k^2} \tag{12.12}$$

Clearly, the liquid column length can be increased or decreased by adjusting the constant displacement feedback gain while keeping its frequency unchanged. The SATLCD is therefore more flexible than the traditional TLCD, in which S_k is equal to zero and there is no way for changing the liquid column length if the frequency is given. Once the frequency and length of the liquid column are decided, the required constant displacement feedback gain of the kth SATLCD can be determined by:

$$S_k = m_k \left[\omega_k^2 - \frac{2g}{L_k} \right] \tag{12.13}$$

where m_k is the mass of the kth SATLCD.

Malfunction of a SATLCD may result from excessive liquid motion when the bridge is subjected to high winds. An on-off control algorithm is therefore employed to make sure that liquid motion inside the SATLCD is within the tolerable limit. This velocity feedback control force is regulated by manipulating the pressure in accordance with the on-off control strategy as follows:

$$u_{2k}(t) = \begin{cases} \frac{1}{2}\rho_w A_k \delta_m \left|\dot{W}_k\right| \dot{W}_k & \textit{if } (W_k > K_p \times W_a) \text{ and } (\dot{W}_k \times W_k > 0) \\ 0 & \text{otherwise} \end{cases} \tag{12.14}$$

where:

W_a is the tolerable liquid displacement of the kth SATLCD (see Equation 12.14);
K_p is an adjusting factor less than or equal to 1;
δ_m is the head loss coefficient for providing sufficient damping to the kth SATLCD and depends on $K_p W_a$.

Since the space inside the bridge deck is not large enough to provide a sufficiently long vertical column, the adoption of this particular control strategy (Equation 12.14) is to avoid the overflow of liquid inside the damper, particularly when the bridge under high wind speed. The non-linear damping force selected is to correlate this additional damping force with the passive damping force in term of head loss coefficient for easy manipulation. The control force is so selected that the additional damping is provided to the liquid when the liquid is continually increased beyond a certain level ($K_p W_a$). The factor K_p decides how the liquid displacement is close to the tolerable liquid displacement and when the additional damping control force should be added to the damper.

In other words, a larger value of the factor K_p would require a large value of δ_m to suppress the liquid motion as it gets closer to the tolerable liquid displacement than when there is a smaller value of K_p. The total control force acting on the kth SATLCD is then the sum of the control force, u_{1k}, based on the feedback of liquid displacement and the control force, u_{2k}, based on the feedback of liquid velocity:

$$u_k = \begin{cases} S_k W_k + \frac{1}{2}\rho_w A_k \delta_m |\dot{W}_k| \dot{W}_k & \text{if } (W_k > K_p \times W_a) \text{ and } (\dot{W}_k \times W_k > 0) \\ S_k W_k & \text{otherwise} \end{cases} \tag{12.15}$$

The desired control force acting on the liquid column can be provided by regulating the air pressure in the right chamber with respect to the air pressure in the left chamber, to obtain a net pressure $P_k(t)$. The relation between the net pressure and the control force can be expressed as:

$$u_k = P_k(t) \times A_k \tag{12.16}$$

The net pressure $P_k(t)$ in the kth SATLCD can be obtained from Equation 12.15 as:

$$P_k(t) = \begin{cases} \frac{S_k W_k}{A_k} + \frac{1}{2}\rho_w \delta_m |\dot{W}_k| \dot{W}_k & \text{if } (W_k > K_p \times W_a) \text{ and } (\dot{W}_k \times W_k > 0) \\ \frac{S_k W_k}{A_k} & \text{otherwise} \end{cases} \tag{12.17}$$

Inside the kth SATLCD, the air pressure in the left chamber P_L and in the right chamber P_R is then determined, respectively, by:

$$P_L = P_0 - \frac{P_k(t)}{2} \tag{12.18a}$$

$$P_R = P_0 + \frac{P_k(t)}{2} \tag{12.18b}$$

To consider the interaction between semi-active liquid column dampers and a long-span cable-stayed bridge under wind excitation, it is expedient to derive the finite element model of the liquid dampers so that it can be incorporated into the finite element model of the bridge. This procedure is similar to that for MPTLCD. Let us consider a total of n_T units of SATLCD installed below the torsional centre of the bridge deck and at the locations where vibration amplitudes of the bridge in the lateral and torsional directions are the largest. The SATLCD units are connected to the transverse beams of the bridge deck by roller supports and simple supports in a similar way to the PTLCD, as shown in Figure 12.26. Two additional nodes, node 1 and node 2, are generated at the positions where the SATLCD units are connected to the bridge. These two additional nodes reflect the motion of the SATLCD units interacted with the motion of the bridge. From a view point of practical use, the distance between the two vertical columns B is the same for all SATLCD units. The axial deformation of the transverse beam between the two supports is assumed to be negligible and, hence, the lateral displacement of the SATLCD units is taken as x and the torsional displacement of the SATLCD units is then determined by Equation 12.9. The Lagrangian of all SATLCD units can be expressed as follows:

$$L_d = \sum_{k=1}^{n} \begin{bmatrix} \frac{1}{2}m_k \dot{W}_k^2 + \frac{1}{2}m_k \dot{x}^2 + \frac{1}{2}I_k \dot{\theta}^2 + m_k \alpha_k \dot{W}_k \dot{x} + G_k \dot{W}_k \dot{\theta} + m_k \overline{H}_k \dot{x}\dot{\theta} \\ +m_k g \overline{H}_k \cos\theta - m_k g \alpha_k W_k \sin\theta - \frac{m_k g}{L_k} W_k^2 \cos\theta - \frac{S_k}{2} W_k^2 \end{bmatrix} \tag{12.19}$$

where:

I_k is the second moment of the liquid mass inside the kth SATLCD;
α_k is the liquid length ratio $= B_k/L_k$;
$\overline{H}_k$ is the distance from the mass center of liquid inside the kth SATLCD to the torsional (elastic) center of the bridge deck;
G_k is the first moment of the liquid mass in the kth SATLCD;
and the last term in Equation 12.19 is the potential energy of liquid due to the control force u_{1k} in the kth SATLCD.

In Equation 12.19, the effect of vertical acceleration of the bridge deck on the liquid motion is neglected because it is relatively small compared with the acceleration due to gravity. The expressions for I_k, $\overline{H}_k$ and G_k are given as follows:

$$I_k = m_k\left(\alpha_k\left[H_k^2 + \frac{B_k^2}{12}\right] + (1-\alpha_k)\left[H_k^2 + \frac{B_k^2}{4} - \frac{H_k(L_k - B_k)}{2} + \frac{(L_k - B_k)^2}{12}\right]\right) \tag{12.20a}$$

$$\overline{H}_k = H_k - \frac{(L_k - B_k)^2}{4L_k} \tag{12.20b}$$

$$G_k = m_k\alpha_k\left(H_k + \frac{L_k - B_k}{2}\right) \tag{12.20c}$$

where H_k is the vertical distance between the centerline of the horizontal part of the container and the torsional center of the bridge deck.

Equation 12.19 is subjected to the condition that the liquid should be fully retained in the horizontal part of the SATLCD and, thus, the following equation should always be satisfied:

$$W_k \le \frac{L_k - B_k}{2} - \frac{d_k}{2} = W_a \tag{12.21}$$

where d_k is the thickness of the liquid column in the kth SATLCD.

The entries of mass matrix m_{ij} and stiffness matrix k_{ij} of the SATLCD (damper element) can be determined by:

$$m_{ij} = \frac{\partial}{\partial \dot{q}_j}\left(\frac{\partial L_d}{\partial \dot{q}_i}\right) \tag{12.22a}$$

$$k_{ij} = -\frac{\partial}{\partial q_j}\left(\frac{\partial L_d}{\partial q_i}\right) \tag{12.22b}$$

$$[q]^T = [x \quad y_1 \quad y_2 \quad W_1 \quad \ldots\ldots \quad W_N] \tag{12.22c}$$

After some manipulations, the mass matrix of the damper element $[M]$ and the stiffness matrix of the damper element $[K]$ can be written as:

$$[M] = \begin{bmatrix} M_{11} & M_{12} \\ M_{12}^T & M_{22} \end{bmatrix} \tag{12.23a}$$

$$[K] = \begin{bmatrix} K_{11} & K_{12} \\ K_{12}^T & K_{22} \end{bmatrix} \tag{12.23b}$$

where:

$$M_{11} = \frac{1}{B^2}\begin{bmatrix} m_d B^2 & -mB & -mB \\ -mB & I_d & -I_d \\ -mB & -I_d & I_d \end{bmatrix} M_{12} = \begin{bmatrix} m_1\alpha_1 & m_2\alpha_2 & \cdots & m_N\alpha_N \\ -\frac{G_1}{B} & -\frac{G_2}{B} & \cdots & -\frac{G_N}{B} \\ \frac{G_1}{B} & \frac{G_2}{B} & \cdots & \frac{G_N}{B} \end{bmatrix} \tag{12.24a}$$

$$M_{22} = diag(m_1, m_2, \ldots, m_N) \tag{12.24b}$$

$$K_{11} = \frac{mg}{B^2}\begin{bmatrix} 0 & 0 & 0 \\ 0 & 1 & -1 \\ 0 & -1 & 1 \end{bmatrix} K_{12} = \frac{g}{B}\begin{bmatrix} 0 & 0 & \cdots & 0 \\ -m_1\alpha_1 & -m_2\alpha_2 & \cdots & -m_N\alpha_N \\ m_1\alpha_1 & m_2\alpha_2 & \cdots & m_N\alpha_N \end{bmatrix} \tag{12.24c}$$

$$K_{22} = diag(m_1\omega_1^2, m_2\omega_2^2, \ldots, m_N\omega_N^2) \tag{12.24d}$$

where:

m_d is the total liquid mass of all SATLCD units;

I_d is the total second moment of the liquid mass of all SATLCD units with respect to the torsional center of the bridge deck;

$m = \sum_{k=1}^{N} m_k \overline{H}_k$.

The vertical motion of liquid inside the liquid column is taken into consideration by including the restoring force of liquid due to gravitational force in the vertical direction. The vertical inertia effect of liquid mass in the global finite element model is also considered by modeling it as lumped masses at the corresponding node.

In consideration of the features of a long-span cable-stayed bridge during construction, SATLCDs with frequency adaptability were investigated to suppress combined lateral and torsional vibration of a long-span cable-stayed bridge under different stages of cantilever construction [59]. A real long-span cable-stayed bridge, as shown in Figure 12.27, was selected as a case study. Five different construction stages of the concerned bridge were considered (see Figure 12.30).

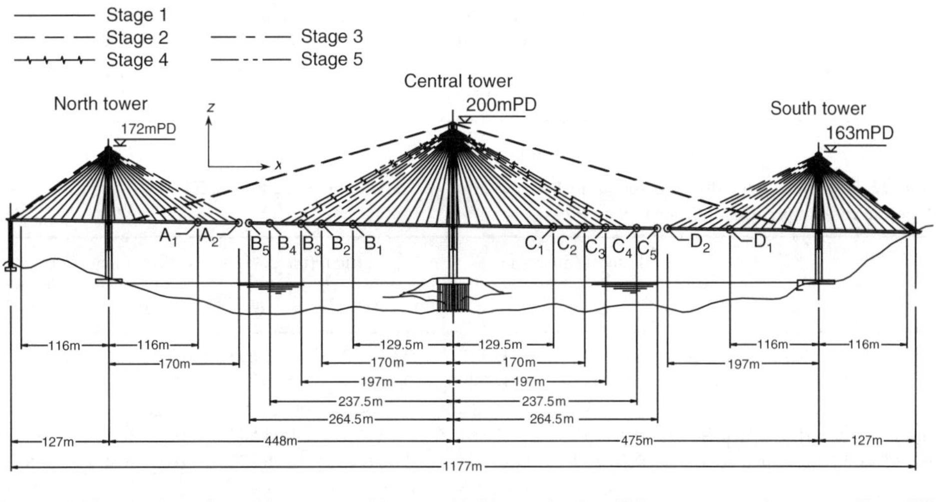

Figure 12.30 Configuration of long-span cable-stayed bridge under five different construction stages (from [59]) (Reproduced with permission from Techno.Press).

Figure 12.30 shows that the bridge under construction stage 1 is divided into three parts, which are erected simultaneously. Points A and D represent the locations at the tips of the cantilevers of the two side bridge decks. Points B and C represent the locations at the tips of the double cantilever of the bridge deck at the central tower part. Each part of the bridge deck is free at its two ends, and its transverse restraint is provided by the tower only. The deck-to-tower connections offer longitudinal and lateral displacement restraints, with essentially completely free rotation about all three axes, together with free vertical displacement. The span length of all incomplete bridge decks becomes longer at stage 2 and the central pylon is stabilized by the longitudinal stabilizing cables. The incomplete bridge deck, at two side spans, is fixed transversely and vertically at one end. The span of the incomplete bridge deck at the central tower is increased gradually from stage 3 to stage 5, at which point the bridge is almost completed.

The performance of SATLCD was examined by studying the double deck displacement and acceleration at points B and C. The results for the mean wind speeds at 20 m/sec and 50 m/sec are tabulated in Tables 12.3 and 12.4.

It can be seen from Tables 12.3 and 12.4 that both the standard deviation displacement and acceleration responses in either lateral or torsional direction are reduced by the SATLCD effectively. The standard deviation displacement reduction in the lateral direction reaches the level of 43% at a mean wind speed of 20 m/sec and the level of 29% at a mean wind speed of 50 m/sec. The reduction in standard deviation torsional displacement can reach 15% at a mean wind speed of 20 m/sec and 18% at a mean wind speed of 50 m/sec. The standard deviations of lateral, vertical and torsional displacement responses of the bridge deck along the bridge longitudinal axis at a mean wind speed of 20 m/sec are plotted in Figure 12.31.

It can be seen that the maximum standard deviation of the displacement response of the bridge deck occurs at the tips of the two cantilevers. The lateral displacement response of the bridge deck in the

Table 12.3 Standard deviation of deck displacement with and without control (SATLCD) (from [59]) (Reproduced with permission from Techno.Press)

	Lateral (m)		Vertical (m)		Torsional (rad)	
Location	B	C	B	C	B	C
Without control ($U_m = 20$ m/sec)	0.29 746	0.29 755	0.37 318	0.37 437	0.004 041	0.003 652
With control ($U_m = 20$ m/sec)	0.16 782	0.16 739	0.37 243	0.37 459	0.003 176	0.003 094
	(−43.6%)	(−43.7%)	(−0.2%)	(0.06%)	(−21.4%)	(−15.3%)
Without control ($U_m = 50$ m/sec)	1.87 074	1.87 170	2.45 735	2.46 240	0.031 593	0.029 666
With control ($U_m = 50$ m/sec)	1.32 582	1.32 442	2.44 805	2.45 722	0.023 934	0.024 204
	(−29.1%)	(−29.2%)	(−0.38%)	(−0.21%)	(−24.2%)	(−18.4%)

Table 12.4 Control standard deviation of deck acceleration with and without control (SATLCD) (from [59]) (Reproduced with permission from Techno.Press)

	Lateral (m/sec^2)		Vertical (m/sec^2)		Torsional (rad/sec^2)	
Location	B	C	B	C	B	C
Without control ($U_m = 20$ m/sec)	0.05 392	0.05 410	0.42 710	0.41 693	0.012 370	0.011 126
With control ($U_m = 20$ m/sec)	0.04 027	0.04 049	0.41 139	0.40 715	0.009 073	0.008 924
	(−25.3%)	(−25.2%)	(−3.68%)	(−2.35%)	(−26.6%)	(−19.8%)
Without control ($U_m = 50$ m/sec)	0.42 718	0.42 861	3.12 393	3.05 544	0.101 623	0.095 820
With control ($U_m = 50$ m/sec)	0.34 159	0.34 373	3.00 655	2.96 235	0.077 574	0.078 167
	(−20.0%)	(−19.8%)	(−3.76%)	(−3.05%)	(−23.7%)	(−18.4%)

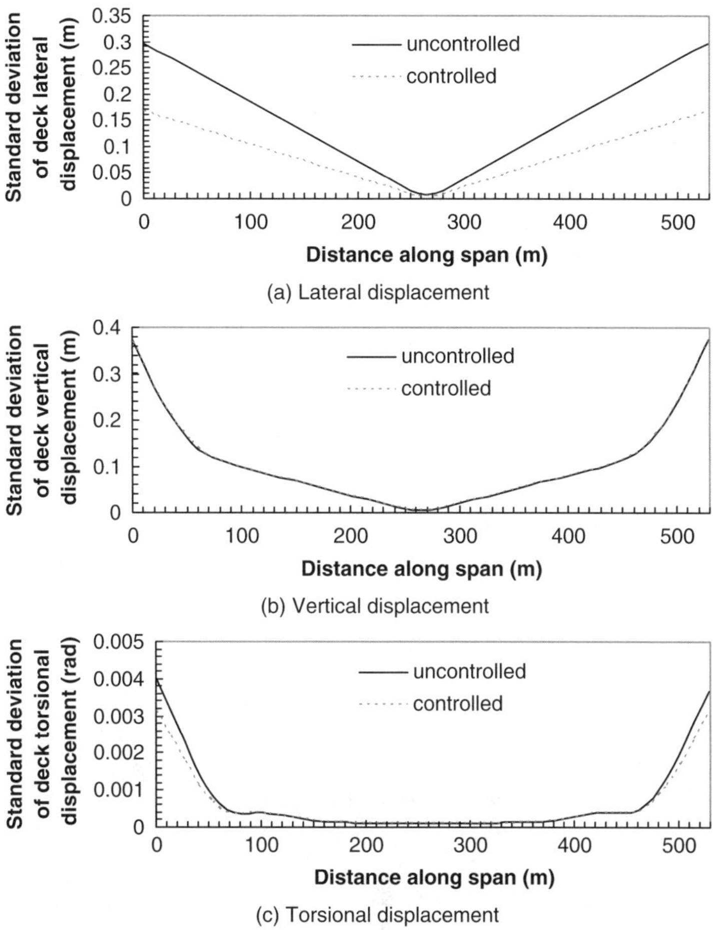

Figure 12.31 Standard deviations of deck displacements of the bridge under construction stage 5 (from [59]) (Reproduced with permission from Techno.Press).

uncontrolled case increases linearly with increasing distance from the central tower, but the torsional displacement in the uncontrolled case is fairly small and increases suddenly near the tip of the cantilever. This may imply that the torsional stiffness of the deck near the tip of the cantilever is smaller. The lateral displacement response of the whole bridge deck is reduced effectively by the SATLCD. However, for reduction of torsional displacement, only the part with significant torsional vibration can be reduced.

12.9 Mechanical Measures for Rain-Wind-Induced Cable Vibration Control

Similar to flutter and buffeting control, mechanical measures for rain-wind-induced cable vibration control can also be classified as passive control systems, active control systems, semi-active control systems and hybrid control systems, among which passive control systems (particularly viscous (oil) dampers) are widely used in practice throughout the world. In this connection, this session only introduces viscous dampers for mitigating rain-wind-induced cable vibration. The knowledge acquired from Chapter 5 will be useful in understanding the materials presented in this section.

The analytical model for predicting rain-wind-induced response of a stay cable with a viscous damper installed near the cable anchorage is presented based on the work of Zhou and Xu [60]. Most stay cables in a cable-stayed bridge are of small sag, in the order of 1% sag-to-length ratio, but with a high tension-to-weight ratio [61]. As a preliminary analytical study, the effects of sag and bending stiffness on cable vibration are neglected and only a taut inclined cable is considered, as shown in Figure 12.32a.

The inclination of the taut cable is denoted by angle α, and the yaw angle of incident wind is designated by angle β. To reduce possible cable vibration, discrete viscous dampers may be attached to the cable near its anchorage. The in-plane transverse motion $w(x,t)$ of a taut cable with

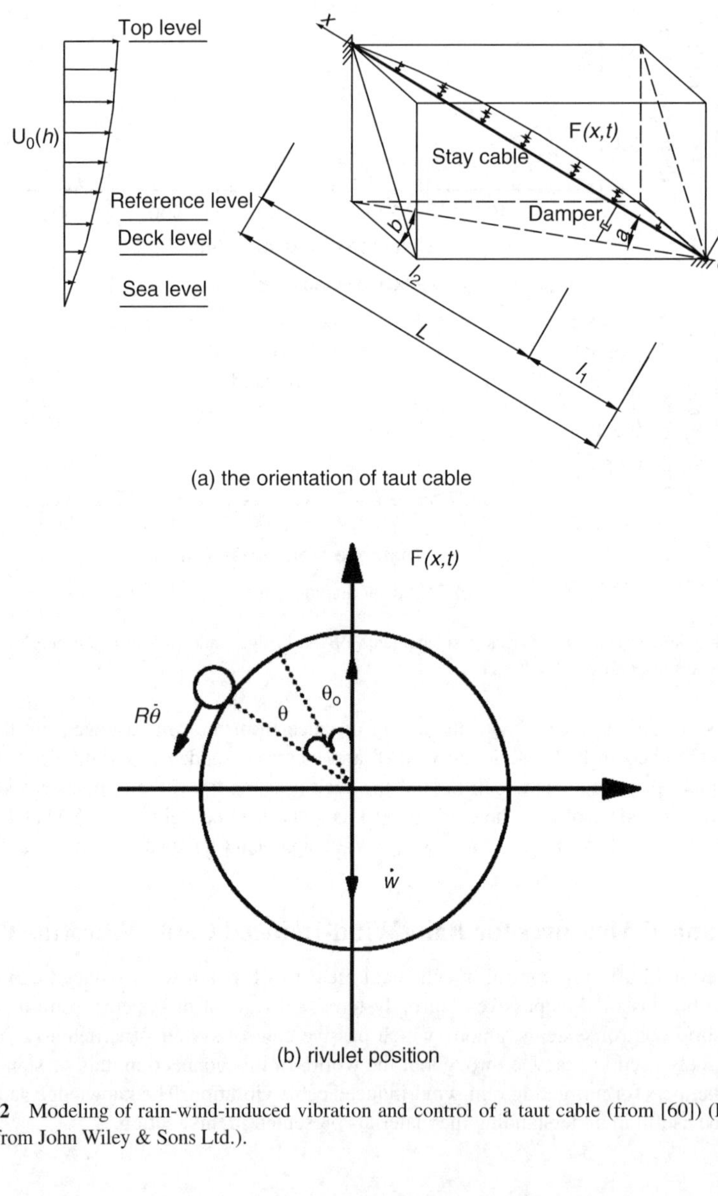

Figure 12.32 Modeling of rain-wind-induced vibration and control of a taut cable (from [60]) (Reproduced with permission from John Wiley & Sons Ltd.).

a viscous damper attached near its anchorage under rain-wind excitation can be described by the following partial differential equation of motion:

$$m\frac{\partial^2 w}{\partial t^2} - T\frac{\partial^2 w}{\partial x^2} + c\frac{\partial w}{\partial t} + F(x,t) + F_d(t)\delta(x - l_1) = 0, \quad 0 \le x \le L \tag{12.25}$$

with the boundary conditions $w(0,t) = w(L,t) = 0$,
where:

$w(x,t)$ is the in-plane transverse displacement normal to the cable axis at position x;
m is the cable mass per unit length;
T is the cable tension force;
c is the internal cable damping coefficient;
$F(x,t)$ is the rain-wind-induced force on the cable;
$\delta(\bullet)$ is the Delta function;
l_1 is the length between the cable anchorage to the damper location;
L is the length of the cable;
$F_d(t)$ is the damper force on the cable.

When $F_d(t) = 0$, Equation 12.25 is reduced to the equation of motion of a taut cable under rain-wind excitation without control.

The fundamental assumption used in determining rain-wind-induced force on a stay cable is the quasi-steady approximation in consideration that rain-wind-induced cable oscillations are often very slow, as discussed in Chapter 5. By assuming that the upper rivulet vibrates circumferentially over the surface of the cable, rain-wind-induced force normal to the cable axis per unit length in the vertical direction, without considering turbulence and axis flow effects, can be expressed as:

$$F(x,t) = \frac{\rho D U_{rel}^2(x)}{2}\left[C_L(\phi)\cos(\phi + \theta + \theta_o) + C_D(\phi)\sin(\phi + \theta + \theta_o)\right] \tag{12.26}$$

where:

$U_{rel}(x)$ is the relative mean wind speed with respect to the cable motion;
D is the cable diameter;
ρ is the air density;
$\theta(x,t)$ is the dynamic angular displacement of upper rivulet with respect to the static position of upper rivulet $\theta_0(x)$ as shown in Figure 12.32b;
C_L is the lift coefficient of the cable with rivulet;
C_D is the drag coefficient of the cable with rivulet;
ϕ is the wind angle of attack.

Expansion of both drag and lift coefficients using the first three terms of Taylor's series leads to:

$$C_L(\phi) = A_0 + A_1\phi + \frac{A_2}{2}\phi^2 + \frac{A_3}{6}\phi^3 \tag{12.27a}$$

$$C_D(\phi) = B_0 + B_1\phi + \frac{B_2}{2}\phi^2 + \frac{B_3}{6}\phi^3 \tag{12.27b}$$

The substitution of Equations 12.27a and 12.27b, as well as the relationship between the relative mean wind speed $U_{rel}(x)$ and the incident mean wind speed $U_0(x)$ into Equation 12.26 and then the reservation of the linear terms of $\dot{w}$, θ, θ_o only yield:

$$F(x,t) = \frac{\rho D U_0(x)}{2}\left[R\dot{\theta}(x,t)\Gamma_1(x,t) + \dot{w}(x,t)\Gamma_2(x,t) + U_0(x)\theta(x,t)\Gamma_3(x,t)\right] \tag{12.28}$$

where:

R is the cable radius;

$\Gamma_1(\bullet)$, $\Gamma_2(\bullet)$ and $\Gamma_3(\bullet)$ are the aerodynamic functions which are not only the function of cable inclination angle, wind yaw angle, wind stagnation point and the mean wind speed via the static position of rivulet, but also the function of cable motion and rivulet motion, which may change the values of drag and lift coefficients.

Detailed information on the aerodynamic functions can be found in Section 5.4.2 in Chapter 5, in which ε is defined as the stagnation point influence factor. The stagnation point influence factor is introduced because, for rain-wind-induced vibration of a stay cable, wind direction is often not perpendicular to the cable axis, and the existence of an upper rivulet may change wind flow surrounding the cable. The stagnation point of incident wind on the surface of the cable is therefore not the same as that determined by the Bernoulli equation. No experimental results are available at present with respect to the position of stagnation point on a stay cable with upper rivulet. Nevertheless, numerical studies carried out by taking the stagnation point as a variable have indicated that the stagnation point influence factor could be taken as 0.4 [62].

In rain-wind-induced cable vibration, the motion of upper rivulet is actually coupled with the motion of cable. Therefore, in principle, the motion of upper rivulet should be taken as a variable and predicted together with the motion of cable. This, however, requires a very complicated analytical model for rain-wind-induced cable vibration. In this study, based on the observations from either field measurements or simulated rain-wind tunnel tests, the motion of upper rivulet, $\theta(x,t)$, is assumed to be harmonic and to follow the excited mode shape of cable as long as a steady-state cable vibration is concerned:

$$\theta(x,t) = a \cdot W(x)\sin(\omega_r t) \tag{12.29}$$

where:

a is the maximum amplitude of rivulet motion and is assumed to be a constant in this study;

$W(x)$ is the motion profile of rivulet along the cable axis and is assumed to be the same as the excited mode shape of cable;

ω_r is the frequency of rivulet motion and it is assumed to be the same as the natural frequency of cable corresponding to the excited mode shape.

For a taut cable without damper, the motion profile and frequency of rivulet corresponding to the ith mode of cable vibration are given by:

$$W(x) = \sin\left(\frac{\pi i}{L}x\right) \tag{12.30a}$$

$$\omega_i = i\omega_{o1} \tag{12.30b}$$

where $\omega_{o1} = \frac{\pi}{L}\sqrt{\frac{T}{m}}$ is the first circular natural frequency of the taut cable without damper.

For a taut cable with a viscous damper installed near its anchorage, its mode shapes are complex in general. Based on the asymptotic approximation [63], the motion profile of rivulet along the cable axis corresponding to the ith mode of vibration of the cable with a viscous damper near its anchorage can be approximately given by:

$$W(x) \cong \pm\sqrt{(1-\Theta_{ci})\sin^2\left(\pi ix/L\right) + H(x-l_1)\Theta_{ci}\sin^2(\pi i(x-l_1)/l_2)} \tag{12.31}$$

where:

$l_2 = L - l_1$ is the length from damper location to the top end of cable;
H(•) is the heaviside function.

$$\mathrm{H}(x - l_1) = \begin{cases} 0 & 0 \le x < l_1 \\ 1 & l_1 \le x \le L \end{cases} \tag{12.32}$$

Θ_{ci} is called the "clamping ratio" for the *i*th mode of vibration, which can be expressed as:

$$\Theta_{ci} \approx \frac{(\pi^2 \kappa)^2}{1 + (\pi^2 \kappa)^2} \tag{12.33}$$

where the non-dimensional damper damping parameter κ for the *i*th mode of vibration is defined as:

$$\kappa = \frac{c_L}{mL\omega_{o1}} i \frac{l_1}{L} \tag{12.34}$$

in which c_L is the damping coefficient of viscous damper.

The selection of positive or negative sign in Equation 12.31 depends on the phase angle ϕ_i:

$$\phi_i = \arctan\left(\mathrm{H}(x - l_1)\sqrt{\Theta_{ci}} \sin\left(\frac{\pi i (x - l_1)/l_2)}{\sqrt{(1 - \Theta_{ci})} \sin(\pi i x / L} \right) \right) \tag{12.35}$$

Also based on the asymptotic approximation [63], the motion frequency of rivulet corresponding to the *i*th circular natural frequency of the cable with a viscous damper near its anchorage can be approximately given by:

$$\omega_r \cong i\left(1 + \Theta_{ci} \frac{l_1}{l_2}\right)\omega_{o1} \tag{12.36}$$

For a stay cable in a cable-stayed bridge, mean wind speed actually varies along with cable axis. Mean wind speed is zero at the sea level and increases with height above the sea level in the atmospheric boundary layer, as shown in Figure 12.32a. The mean wind speed profile could be approximately expressed by a power-law function:

$$U_0(h) = U_r \left(\frac{h}{h_r}\right)^{\alpha_1} \tag{12.37}$$

where:

$U_0(h)$ is the mean wind speed at height h above the sea level;
U_r is the mean wind speed at a chosen reference height h_r;
α_1 is the exponent depending on the surface roughness.

Since the height h can be expressed as a function of x along the cable axis, Equation 12.37 can be rewritten as:

$$U_0(x) = U_0(h(x)) = U_r \left(\frac{h_0}{h_r} + \frac{x\sin\alpha}{h_r}\right)^{\alpha_1} \tag{12.38}$$

where h_0 is the height of lower anchorage of the cable above the sea level.

Linear viscous damper is considered in this study. Thus, the damper force acting perpendicularly on the cable at the position $x = l_l$ is given by:

$$F_d(t) = c_L \dot{w}(l_1, t) \tag{12.39}$$

For a taut cable with a linear viscous damper near its anchorage, Pacheco *et al.* [64] obtained the so-called universal design curve for the modal damping ratio in the cable through numerical complex-eigenvalue analysis. The mode shapes of the taut cable without damper were used as shape functions, and several hundred terms were required for adequate convergence in the solution, creating a computational burden. Main and Jones [65] recently used the clamped and unclamped mode shapes as shape functions in developing the numerical formulation of the eigenvalue problem of a taut cable with a linear or non-linear damper, which significantly reduces computational effort. This study extends their method to the case of rain-wind-induced vibration and control of a taut cable with a viscous damper near its anchorage.

The in-plane transverse displacement $w(x,t)$ of a taut cable with a linear viscous damper in Equation 12.25 can be assumed to be of the form:

$$w(x, \tau) = L[\mathbf{\Phi}(x)^T \mathbf{z}(\tau)] = L[\mathbf{\Phi_o}(x)^{\mathbf{T}} \mathbf{z_o}(\tau)^T + \mathbf{\Phi_c}(x)^T \mathbf{z_c}(\tau)^T] \tag{12.40}$$

where:

$\tau = \omega_{01} t$ is the non-dimensional time;
$\mathbf{\Phi}(x) = [\mathbf{\Phi_0}(x)^T \mathbf{\Phi_c}(x)^T]^T$ is a vector of shape functions;
$\mathbf{z}(\tau) = [\mathbf{z_0}(\tau)^T \mathbf{z_c}(\tau)^T]^T$ is a vector of non-dimensional time-varying coefficients;
$\mathbf{\Phi_0}(x)$ is a vector of unclamped mode shapes of cable, which is actually the vector of mode shapes of a taut cable without damper;
$\mathbf{\Phi_c}(x)$ is a vector of clamped mode shapes of cable, which is the same as the vector of mode shapes of the taut cable when the damper becomes a clamp;
$\mathbf{z_0}(\tau)$ is the vector of the time-varying coefficient corresponding to the vector of unclamped mode shapes;
$\mathbf{z_c}(\tau)$ is the vector of the time-varying coefficient corresponding to the vector of clamped mode shapes.

The vector of the unclamped mode shapes of the taut cable is given by:

$$\mathbf{\Phi_o}(x) = [\Phi_{o1}(x) \quad \Phi_{o2}(x) \quad \Phi_{o3}(x) \quad \ldots]^T \tag{12.41a}$$

$$\Phi_{oi}(x) = \sin\left(\frac{\pi i x}{L}\right) \tag{12.41b}$$

The vector of the clamped mode shapes of the taut cable is expressed as:

$$\mathbf{\Phi}_c(x) = [\Phi_{c1}(x) \quad \Phi_{c2}(x) \quad \Phi_{c3}(x) \quad \ldots]^T \tag{12.42a}$$

$$\Phi_{ci}(x) = \mathrm{H}(x - l_1)\sin\left[\frac{\pi i(x - l_1)}{L - l_1}\right] \tag{12.42b}$$

Substituting Equation 12.40 into Equation 12.25 and using Galerkin's method lead to the following equation of motion for rain-wind-induced vibration of a taut stay cable with a viscous damper:

$$\mathbf{M}\ddot{\mathbf{z}} + \mathbf{C}\dot{\mathbf{z}} + \mathbf{K}\mathbf{z} + \tilde{\mathbf{F}} = 0 \tag{12.43}$$

The mass matrix **M** and the stiffness matrix **K** in Equation 12.43 are given by:

$$\mathbf{M} = \begin{bmatrix} \mathbf{\Lambda_o} & \mathbf{\Pi_{oc}} \\ \mathbf{\Pi_{oc}^T} & \mathbf{\Lambda_c} \end{bmatrix} \tag{12.44a}$$

$$\mathbf{K} = \begin{bmatrix} \mathbf{\Lambda_o'} & \mathbf{\Pi_{oc}'} \\ \mathbf{\Pi_{oc}'^T} & \mathbf{\Lambda_c'} \end{bmatrix} \tag{12.44b}$$

The expression of each sub-matrix in Equations 12.44a and 12.44b can be found in the literature [63]. The damping matrix **C** in Equation 12.43 can be expressed as:

$$\mathbf{C} = 2\zeta_s\mathbf{M} + \mathbf{C_0} + \mathbf{C}_a \tag{12.45}$$

in which ζ_s is the first structural modal damping ratio of the cable:

$$\mathbf{C_0} = \frac{2c_L}{m(L\omega_{o1})}\begin{bmatrix} \mathbf{\Phi_o}(l_1)\mathbf{\Phi_o}(l_1)^T & 0 \\ 0 & 0 \end{bmatrix} \tag{12.46}$$

is the damping matrix due to the viscous damper; and

$$\mathbf{C_a} = \frac{\rho D}{L(m\omega_{o1})}\int_0^L \mathbf{\Phi}(x)U_0(x)\Gamma_2(x,\tau)\mathbf{\Phi}(x)^T dx \tag{12.47}$$

is the aeroelastic damping matrix due to the rain-wind-cable interaction. The normalized wind-rain-induced force in Equation 12.43 is given by:

$$\tilde{\mathbf{F}} = \frac{\rho D}{\pi^2 T}\int_0^L \Phi(x)U_0(x)\left[R\omega_{o1}\dot{\theta}(x,\tau)\Gamma_1(x,\tau) + U_0(x)\theta(x,\tau)\Gamma_3(x,\tau)\right]dx \tag{12.48}$$

The advantage of the numerical formulation presented above is that a good approximate solution could be obtained by using a small number of shape functions. As pointed out in [63], for a particular mode *i* of vibration of a taut cable with a damper near its anchorage, an accurate approximation could be obtained using just two shape functions: the *i*th unclamped mode shape $\Phi_{Oi}(x)$ and the *i*th clamped mode shape $\Phi_{ci}(x)$. This feature greatly facilitates the numerical analysis of rain-wind-induced vibration and control of stay cable in a cable-stayed bridge.

Nevertheless, Equation 12.43 is a strongly non-linear equation, due to the physical interaction between cable, rivulet and wind. The rain-wind-induced force expressed in terms of Equation 12.48 is also an implicit function of cable motion and rivulet motion. The Runge-Kutta-Fehlberg method is therefore employed to find the numerical solutions [66]. This method is adaptive and easily implemented for the concerned numerical problem, in which a stay cable is divided into a number of sections along the cable axis and the numerical solutions are found at these sections for a series of time intervals, using the relevant unclamped and clamped shape functions. By using the Runge-Kutta-Fehlberg method, only six evaluations are required per step and local truncation error can be kept below a prescribed value by varying the step length.

Note that rain-wind-induced force depends on static rivulet position and cable motion, while the static rivulet position varies with wind speed. Rain-wind-induced force on the cable needs to be computed section-by-section, starting from the static rivulet position and cable motion at the previous time step.

Stay cable No.14 on the Meikonishi West Bridge, as investigated by Hikami and Shiraishi [42], is taken as an example in this study. The length of the cable is 75 m; the diameter of the cable is 0.14 m; the mass per unit length is 51 kg/m; the tension force in the cable is 1.147×10^6 N; the first structural damping ratio of the cable is estimated at 0.0011; the inclination of the cable is about 45°; and the height of lower anchorage of the cable above the sea level h_0 is 45 m.

The field measurements carried out by Hikami and Shiraishi [42] recorded that when cable No. 14 was down-inclined along wind direction, with a wind yawed angle about 45°, the simultaneous occurrence of wind and rain caused excessive vibration of the cable in the first mode of vibration, with a natural frequency about 1 Hz. They also conducted a series of wind tunnel tests with rain conditions simulated, and reproduced rain-wind-induced cable vibration in their model tests. They observed from the model tests that the rivulet oscillated in a circumferential direction at the same period as the cable motion, and the amplitude of rivulet motion was around 10°. The drag and lift coefficient curves of the cylinder with artificial upper rivulet, and the rivulet static position curve as the function of mean wind speed, were also given by Yamaguchi [67] and Hikami and Shiraishi [42], respectively, and these are reproduced in Figures 12.33a and 12.33b, respectively.

In addition to the information provided above, the power exponent α_1 in the wind profile is selected as 0.16 for an open fetch. The stagnation point influence factor ε is selected as 0.4. The wind reference height h_r is 47 m above the sea level and the mean wind speed at the reference height is taken as a variable. The cable is then divided into 50 sections along the cable axis. The basic parameters used in the simulation of rain-wind-induced cable vibration are listed in Table 12.5. The analytical model and the numerical method proposed are finally applied to the cable with proper initial conditions, and an error tolerance of 10^{-4}, to compute rain-wind-induced displacement response of the cable mode by mode.

To investigate whether viscous dampers can mitigate rain-wind-induced cable vibration, a linear viscous damper is installed perpendicularly to the cable No.14 at the location of 2% of the cable length from the low cable anchorage. The non-dimensional damper damping parameter κ is taken as a variable, and the mean wind speed at the reference level is selected as 8.0, 9.0 and 10.0 m/sec, respectively. All other parameters of the cable and the rivulet remain the same as those mentioned above. Only the first mode of vibration of the cable is considered here.

Depicted in Figure 12.34 are rain-wind-induced maximum displacement responses of cable No. 14 against the non-dimensional damper damping parameter $\kappa \times \pi^2$ for the three mean wind speeds. It can be seen that for a given mean wind speed, there is an optimal value of $\kappa \times \pi^2$, by which the displacement response of the cable reaches its minimum. This optimal value is the same for three mean wind speeds and is around 1.0, i.e. the optimal non-dimensional damper damping parameter κ is equal to $1/\pi^2$. By using this optimal value, rain-wind-induced maximum displacement responses of the cable with the optimal damper are computed and compared with those of the cable without any damper for a series of mean wind speeds. The computed results are plotted in Figure 12.35, from which it can be seen that the damper with the optimal damping coefficient can effectively mitigate rain-wind-induced cable vibration. The global maximum displacement response of the cable is reduced from 0.22 m in the case without damper, to 0.03 m with the optimal damper.

Another parameter considered for evaluation of damper effectiveness is damper location. While all the other parameters of the cable and the rivulet are kept the same as those used above, the damper location is taken as a variable with the location at 2% and 4% of cable length from the low cable anchorage, respectively. The corresponding optimum non-dimensional damper damping parameter κ is $1/\pi^2$ for both locations. The computed rain-wind-induced maximum displacement responses of the cable are plotted in Figure 12.36 for a series of mean wind speeds at the reference level. Clearly, the further the damper is located away from the low anchorage, the more effective it is in reducing rain-wind-induced displacement response.

On the one hand, the above study shows that there is optimal damper damping coefficient by which rain-wind-induced displacement response of the cable can be reduced to its minimum, and that this optimal value is about $1/\pi^2$ in terms of the non-dimensional damper damping parameter κ. On the other

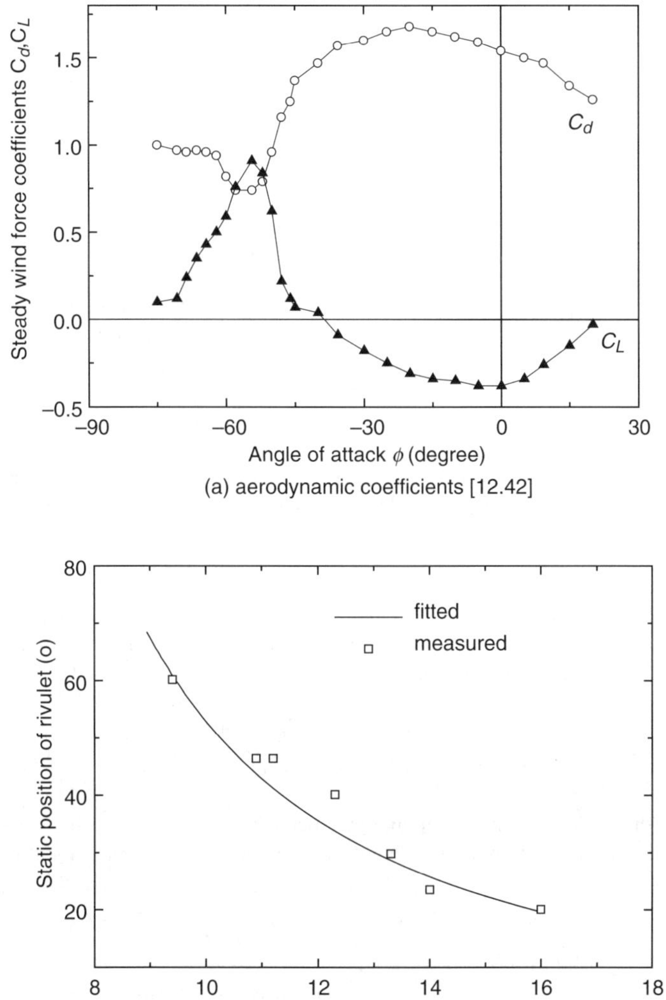

(a) aerodynamic coefficients [12.42]

(b) static position of upper rivulet [12.67]

Figure 12.33 Aerodynamic properties of rain-wind-induced cylinder vibration (from [60]) (Reproduced with permission from John Wiley & Sons Ltd.).

hand, in current practice, the optimal damper damping parameter for a taut cable can be estimated using the so-called universal design curve obtained from a complex eigenvalue analysis [63,64,68], and the corresponding optimal modal damping ratio in the cable is then used to estimate the rain-wind-induced displacement response of the cable. It is thus interesting to know whether the optimal values and the rain-wind-induced displacement responses obtained from these two ways are the same.

Table 12.5 Basic parameters used in the simulation of rain-wind-induced cable vibration (from [60]) (Reproduced with permission from John Wiley & Sons Ltd)

Parameters	L (m)	D (m)	T ($\times 10^6$N)	m (kg/m)	ζ_s (%)	α (°)	β (°)	h_0 (m)	h_r (m)	E	α_1
Values	75	0.14	1.147	51	0.1	45	45	45	47	0.4	0.16

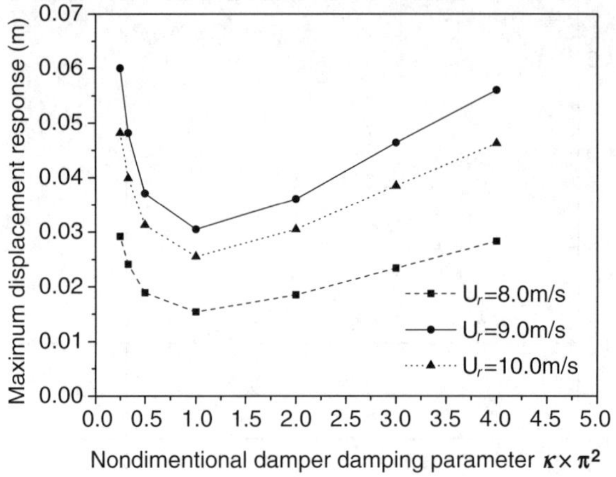

Figure 12.34 Rain-wind-induced maximum displacement responses of the cable with the non-dimensional damper damping parameter $\kappa \times \pi^2$ (from [60]) (Reproduced with permission from John Wiley & Sons Ltd.).

From the complex eigenvalue analysis of a taut cable with a linear viscous damper at the location l_l from the low cable anchorage, the relationship between the ith modal damping ratio ζ_i in the cable and the non-dimensional damper damping parameter κ can be derived and expressed as:

$$\frac{\zeta_i}{(l_1/L)} = \frac{\pi^2\kappa}{1 + (\pi^2\kappa)^2} \tag{12.49}$$

When the non-dimensional damper damping parameter κ is taken as $1/\pi^2$, the modal damping ratio in the cable reaches its maximum value. Therefore, the optimal non-dimensional damper damping

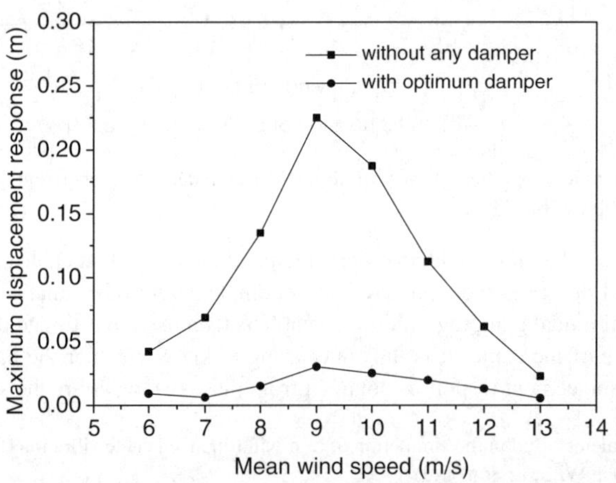

Figure 12.35 Comparison of rain-wind-induced maximum displacement responses between the cable with optimal damper and without any damper (from [60]) (Reproduced with permission from John Wiley & Sons Ltd.).

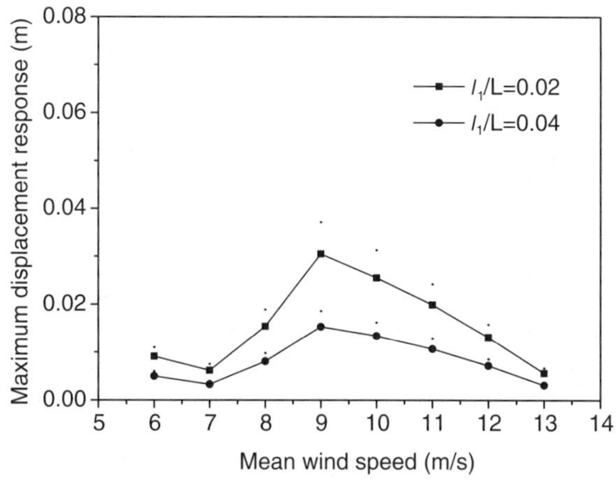

Figure 12.36 Effect of damper location on cable response (from [60]) (Reproduced with permission from John Wiley & Sons Ltd.).

parameters obtained from the two methods are actually the same. Furthermore, rain-wind-induced maximum displacement responses of cable No. 14 are computed for four cases:

- Case 1 is the cable with the optimal damper at the 2% location;
- Case 2 is the cable without damper, but with the additional optimal modal damping ratio obtained from the eigenvalue analysis of the cable with the optimal damper at the 2% location;
- Case 3 is the cable with the optimal damper at the 4% location;
- Case 4 is the cable without damper, but with the additional optimal modal damping ratio obtained from the eigenvalue analysis of the cable with the optimal damper at the 4% location.

The computed rain-wind-induced maximum displacement responses of cable No.14 are plotted in Figure 12.37 against different mean wind speeds at the reference level. It can be seen that the

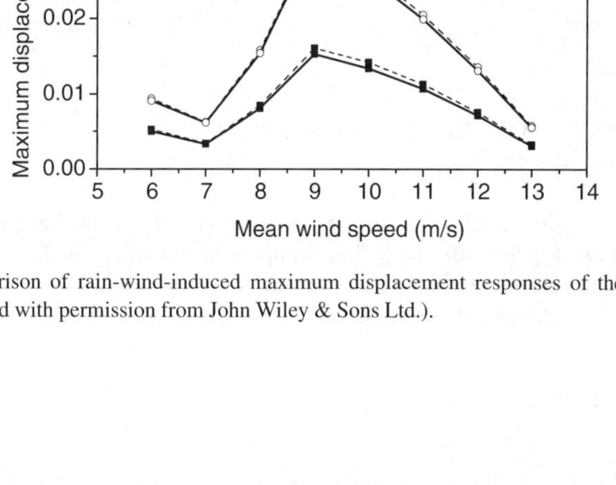

Figure 12.37 Comparison of rain-wind-induced maximum displacement responses of the cable for four cases (from [60]) (Reproduced with permission from John Wiley & Sons Ltd.).

rain-wind-induced maximum displacement responses of the cable computed from the two ways for a given damper location are almost the same, indicating that the current practice is acceptable.

12.10 Case Study: Damping Stay Cables in a Cable-Stayed Bridge

There are often hundreds of stay cables in a cable-stayed bridge, each of which features unique dynamic characteristics and requires a specific viscous damper to achieve the best control performance. If the viscous coefficient of the damper deviates from its optimum value, the modal damping ratio of the stay cable decreases rapidly. This requirement engenders many problems in the manufacture, implementation and maintenance of fluid dampers for a cable-stayed bridge. The use of MR with a semi-active control algorithm [69] may be an alternative, but some practical issues need to be solved before it can be accepted by the engineering professions. A new approach for damping vibration of stay cables in a cable-stayed bridge by using adjustable fluid dampers has been recently proposed by Xu and Zhou [70].

Different from a common passive fluid damper with a fixed number of orifices in its piston head, an adjustable fluid damper with shape memory alloy (SMA) actuators are installed inside the piston head in an adjustable fluid damper to control the number of orifices, in order to change damper parameters for the best control of a group of stay cables. After the optimum viscous coefficient of the damper of an adjustable fluid damper is achieved for a given cable, the damper then works as a passive fluid damper to maintain the practical merit of passive fluid dampers. A SMA actuator possesses at least two main characteristics:

1. It could control a small mechanical valve inside the damper piston to close or open an orifice.
2. the mechanical valve could be firmly locked at a prescribed position within a working temperature range.

A schematic diagram of a SMA actuator is shown in Figure 12.38a, and the SMA actuators installed in the piston of a prototype adjustable fluid damper are depicted in Figure 12.38b. There are two SMA wires, each of which is connected to one end of the valve block and fixed on the piston via two pulleys. By using an impulse current to heat one wire each time, the SMA wire will pull the valve block to open or close an orifice through the positioning plate and locating ball.

Two types of prototype adjustable fluid dampers were designed and manufactured: Type 1, with 1.5 mm diameter orifices for large viscous coefficient of the damper; and Type 2, with 1.8 mm diameter orifices for relatively small viscous coefficient of the damper. The two prototype dampers were extensively calibrated and tested, and the experimental results demonstrated that the damper performance could be well described by the following Maxwell model [71]:

$$F_d + \lambda_d \frac{dF_d}{dt} = c_d v \tag{12.50}$$

where:

F_d is the damper force;
λ_d is the relaxation time constant;
c_d is the viscous coefficient of the damper at zero frequency;
v is the velocity of the piston head.

Table 12.6 lists the experimental results of the viscous coefficient of the dampers and the relaxation time constants of the two types of adjustable fluid dampers developed. It can be seen that the relaxation time constant increases with increasing viscous coefficient of the damper at zero frequency. Both the viscous coefficient of the damper at zero frequency and the relaxation time constant become larger as

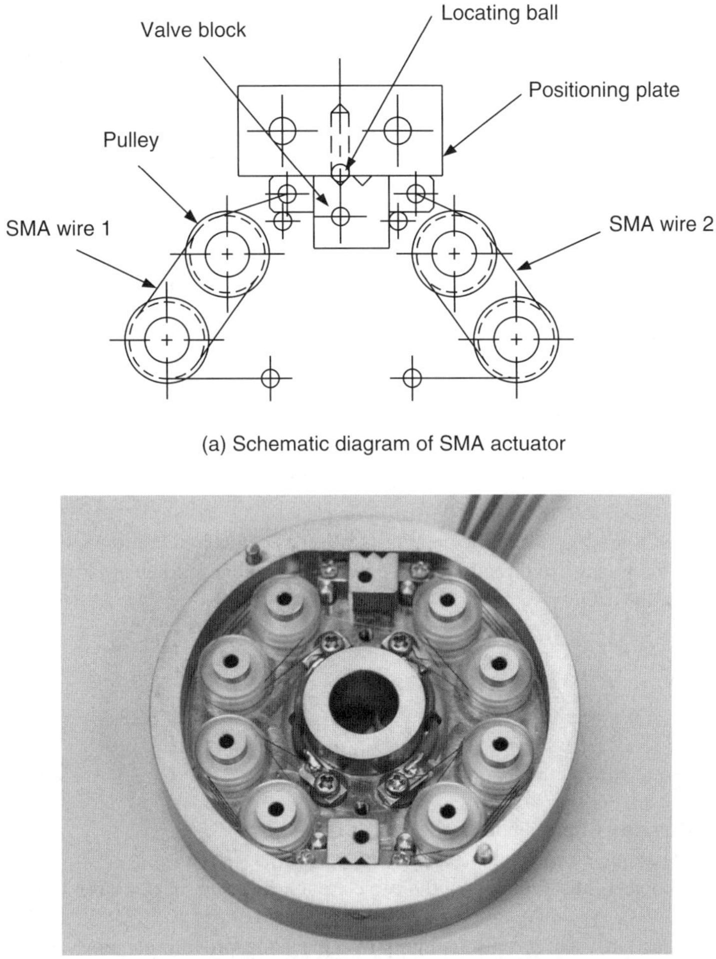

(a) Schematic diagram of SMA actuator

(b) Prototype of SMA actuator within a piston

Figure 12.38 Principle and prototype of SMA actuator (from [70]) (Reproduced with permission from Elsevier).

the opened orifice number n becomes smaller. More details on adjustable fluid dampers and the test results could be found in Li *et al.* [72].

In consideration that most stay cables in a cable-stayed bridge are of small sag (in the order of 1% sag-to-length ratio) but with a high tension-to-weight ratio, only a taut cable with an adjustable fluid damper installed near one of cable anchorage is considered, as shown in Figure 12.39.

The adjustable fluid damper described by the Maxwell model can be represented by a dashpot and a spring connected in series. The effect of the damper support is also considered in terms of a spring connected to the damper in series. If the length between the left cable anchorage and the damper is denoted as l_1, then the length between the right cable anchorage and the damper is $l_2 = L - l_1$, in which L is the total length of the cable. In practice, the length l_1 is much smaller than the length l_2.

The internal structural damping of a stay cable is very small compared with the damping provided by a properly-designed fluid damper, and it is thus neglected here. By considering the cable-damper

Table 12.6 The viscous coefficient of the dampers c_d and relaxation time constants λ_d of two types of adjustable fluid dampers against the opened orifice number n (from [70]) (Reproduced with permission from Elsevier)

n	Damper Type 1 (1.5 mm orifices)		Damper Type 2 (1.8 mm orifices)	
	c_d (Ns/m)	λ_d (sec.)	c_d (Ns/m)	λ_d (sec.)
10	137 000	0.0038	89 000	0
9	149 000	0.0053	101 000	0
8	173 000	0.0059	112 000	0.0013
7	198 000	0.0066	129 000	0.0021
6	228 000	0.0102	146 000	0.0056
5	267 000	0.0128	169 000	0.0069
4	332 000	0.0164	213 000	0.0098
3	446 000	0.0212	278 000	0.0158
2	621 000	0.0338	427 000	0.0228

system in two parts, x_1 and x_2, using the damper position as a division, the free vibration of the cable-damper system in the transverse direction can be described by the following partial differential equation for each part of the cable:

$$m\frac{\partial^2 w_k(x_k,t)}{\partial t^2} = T\frac{\partial^2 w_k(x_k,t)}{\partial x_k^2} \qquad (k = 1,2) \tag{12.51}$$

where:

$w_k(x_k, t)$ is the transverse displacement of the cable at point x_k;
x_k is the coordinate along the cable chord axis in the kth part;
the boundary conditions of the two parts of the cable are $w_1(0, t) = w_2(0, t) = 0$ for all t.

At the damper location, there is a discontinuity in cable slope, providing a transverse force matching the damper force F_d:

$$T\left[-\left.\frac{\partial w_2}{\partial x_2}\right|_{x_2=l_2} - \left.\frac{\partial w_1}{\partial x_1}\right|_{x_1=l_1}\right] = F_d \tag{12.52}$$

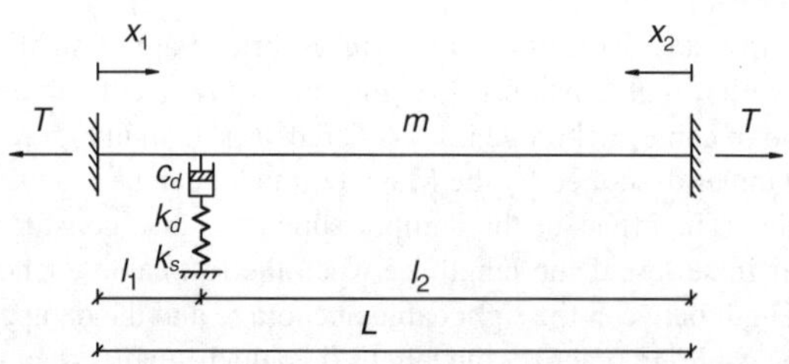

Figure 12.39 A taut cable with a fluid damper near its anchorage (from [70]) (Reproduced with permission from Elsevier).

Since the velocity of the damper piston is the same as the velocity of the cable at damper location, Equation 12.50 could be rewritten as:

$$F_d + \lambda \frac{dF_d}{dt} = c_d \frac{\partial w_1}{\partial t}\bigg|_{x_1=l_1} \tag{12.53}$$

where:

$$\lambda = \lambda_d + \lambda_s = c_d/k_d + c_d/k_s \tag{12.54}$$

in which:

λ is the total relaxation time constant considering both damper stiffness and damper support stiffness;
λ_d is the relaxation time constant of damper itself;
λ_s is the equivalent relaxation time constant of damper support;
k_d is the damper stiffness;
k_s is the damper support stiffness.

Although the Maxwell damper was not considered previously [65,73], the frequency equation of the cable with an adjustable fluid damper can be obtained by simply substituting the complex mechanical impedance $c_d/(1+\lambda\omega_{01}\chi)$ in place of the viscous coefficient c:

$$(1+\lambda\omega_{o1}\chi)[coth(\pi\chi l_1/L) + coth(\pi\chi l_2/L)] + \frac{c_d}{\sqrt{Tm}} = 0 \tag{12.55}$$

where χ is the dimensionless eigenvalue that is complex in general.

For specific values of $c_d/\sqrt{Tm}$, l_1/L and $\lambda\omega_{o1}$, Equation 12.55 can be directly solved numerically to obtain a series of complex eigenvalues. In practice, the damper is often installed near the cable anchorage and l_1/L is rather small. Cable vibration mitigation also focuses on the first few modes of vibration only. Based on Equation 12.55 and the work of Main and Jones [65], the approximate (asymptotic) solution of the ith non-dimensional modal damping ratio of the cable-damper system can be found as:

$$\frac{\zeta_i}{l_1/L} \cong \frac{\pi^2\kappa_i}{1+(\pi^2\kappa_i + i\lambda\omega_{o1})^2} \tag{12.56}$$

where:

$$\kappa_i \equiv \frac{c_d}{mL\omega_{01}} i(l_1/L) \tag{12.57}$$

in which κ_i is termed the non-dimensional viscous coefficient of the damper for the ith mode of vibration.

It can also be found from Equation 12.56 that the maximum attainable damping ratio is $\zeta_{i,\max} \cong \frac{1}{2}\frac{l_1}{L}\left[\sqrt{(i\lambda\,\omega_{o1})^2+1} - i\lambda\,\omega_{o1}\right]$ when $\kappa_{i,opt} = \sqrt{1+(i\lambda\omega_{o1})^2}/\pi^2$, where $\kappa_{i,opt}$ is the optimal non-dimensional viscous coefficient of the damper to achieve the maximum attainable modal damping ratio $\zeta_{i,\max}$.

Figure 12.40a displays the asymptotic solutions for $\frac{\zeta_i}{l_1/L}$ and κ_i for the first five modes of vibration, obtained by using Equation 12.56, for the cable-damper system with $l_1/L = 0.02$, $\lambda = 0.01$ (sec),

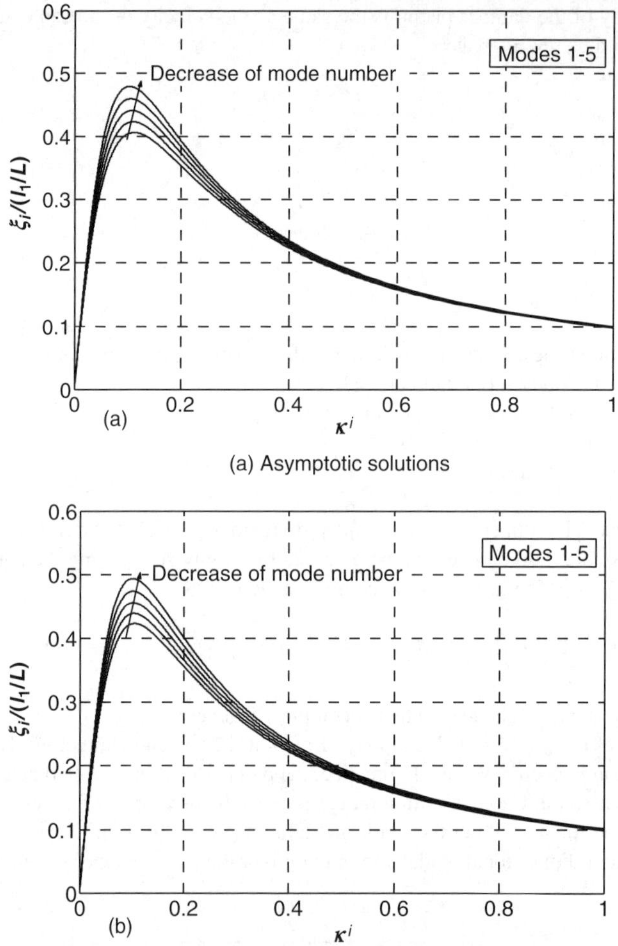

Figure 12.40 Cable modal damping ratio vs. non-dimensional viscous coefficient of the damper ($l_1/L = 0.02$, $\lambda = 0.01$ (sec), $\omega_{01} = 4.152$ (rad/sec), $i = 1$–5) (from [70]) (Reproduced with permission from Elsevier).

$\omega_{01} = 4.152$ (rad/sec). Figure 12.40b depicts the exact solutions. It turns out that the attainable modal damping ratios obtained by Equation 12.56 are only slightly smaller than the exact solutions. However, both the asymptotic solutions and the exact solutions show that the attainable modal damping ratios are smaller for higher modes of vibration. This is due to the relaxation time constant, which is different from the case of the cable with a linear viscous damper, where the relaxation time constant is zero and all the curves overlap with each other.

The effects of relaxation time constant on attainable modal damping ratio can be found in Figure 12.41 for $\lambda = 0.01$, 0.04, 0.06 (sec) and for $l_1/L = 0.02$, $i = 1$, $\omega_{01} = 4.152$ (rad/sec), in which the results in Figure 12.41a are obtained from the asymptotic solution while those in Figure 12.41b are computed using the exact solution. The use of λ rather than the non-dimensional quantity $\lambda\omega_{01}$ is to be consistent with the experimental results provided in [72]. Again, the asymptotic results are very close to the accurate results

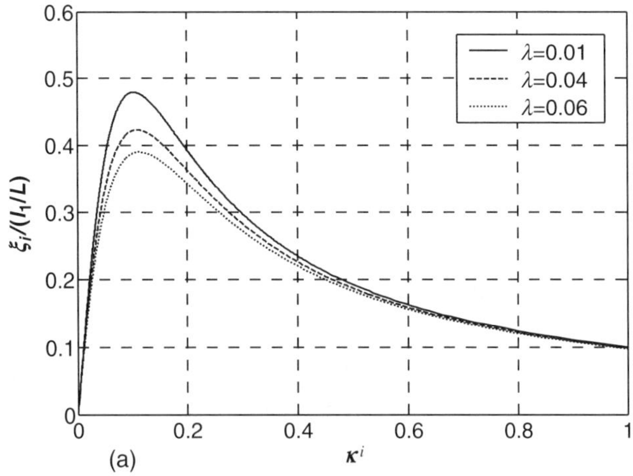

(a) Asymptotic solutions

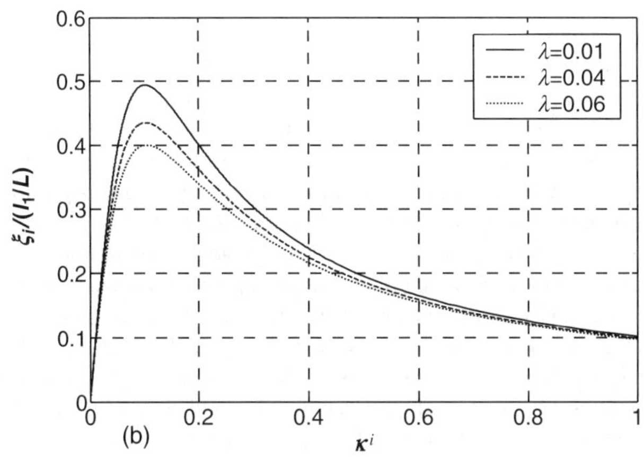

(b) Exact solutions

Figure 12.41 Effects of relaxation time constant ($l_1/L = 0.02$, $\lambda = 0.01$–0.06 (sec), $\omega_{01} = 4.152$ (rad/sec), $i = 1$) (from [70]) (Reproduced with permission from Elsevier).

for different relaxation time constants λ. The maximum attainable non-dimensional modal damping ratio decreases as the relaxation time constant λ increases; such a reduction accounts for about 20% when the relaxation time constant $\lambda = 0.06$ (sec), comparing to $\lambda = 0$.

As the relaxation time constant is associated with damper stiffness and damper support stiffness, one may conclude that both the damper stiffness and the damper support stiffness will reduce the control effectiveness of the damper. Thus, it may be necessary to include the frequency dependence property in the analysis if a damper exhibits Maxwell characteristics and/or the damper support is not stiff enough.

From Equations 12.56 and 12.57, it can be seen that when a fluid damper is installed at a given location of a cable, the damper will have an optimum viscous coefficient of the damper that can achieve the maximum modal damping ratio in the cable. The optimum viscous coefficient of the damper,

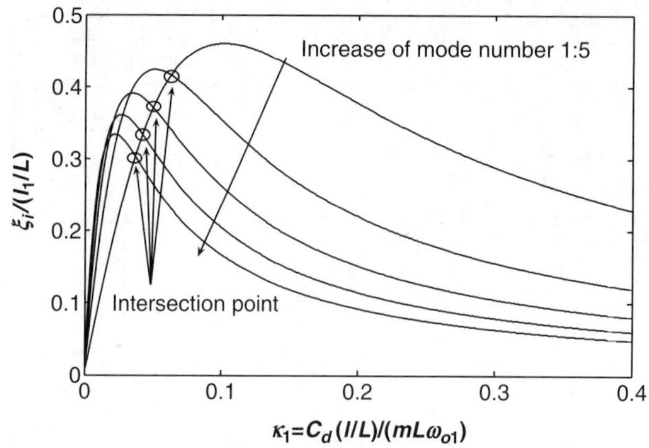

Figure 12.42 Intersection points for non-dimensional viscous coefficient of the damper ($l_1/L = 0.02$, $\lambda = 0.02$ (sec), $\omega_{01} = 4.152$ (rad/sec), $i = 1{:}\ 5$) (from [70]) (Reproduced with permission from Elsevier).

however, depends on not only cable properties but also cable vibration mode. Let us consider a cable-damper system with $l_1/L = 0.02$, $\lambda = 0.02$ (sec), $\omega_{01} = 4.152$ (rad/sec).

Figure 12.42 plots the normalized modal damping ratios against the first non-dimensional viscous coefficient of the damper $\kappa_1 = c_d(l_1/L)(mL\omega_{01})$ for the first five modes of vibration based on Equation 12.56. It can be seen that the lower mode of vibration has the higher maximum modal damping ratio. The corresponding optimum viscous coefficient of the damper c_d is also significantly different; the lower mode of vibration needs the larger optimum viscous coefficient of the damper.

Clearly, it is not possible for one fluid damper to achieve the maximum modal damping ratio for all modes of vibration of a given cable. Furthermore, as the optimum viscous coefficient of the damper depends on the properties and vibration modes of the cable and the location of the damper, it will be different for different cables. This implies that, ideally, every cable needs its own damper. However, this would cause many problems in the process of manufacture, installation and maintenance. Therefore, some compromise must be made and the grouping process is unavoidable in practice. This study proposes a most favorable design principle to select the favorable viscous coefficient of the damper for each cable and to group these favorable viscous coefficients of the damper according to the adjustable levels of one or two adjustable fluid dampers.

The most favorable design principle is to select the number of modes of vibration required to be damped for all stay cables in a cable-stayed bridge and at the same time, to comply with the requirement that the modal damping ratios in the concerned vibration modes for all stay cables should be greater than the least modal damping ratio limit ζ_{limit}. The determination of ζ_{limit} should consider many practical factors. For example, it may be selected as 0.5% damping ratio or 3% logarithmic decrement

Adopting this favorable design principle, one should plot the ζ_i-ζ_d curves similar to those in Figure 12.42 for each cable. Then, select the initial number of modes of vibration required to be damped, for instance *i*. Find the intersection point of the *ith* modal damping ratio curve and the first modal damping ratio curve for each cable from their ζ_i-ζ_d curves. The viscous coefficient of the damper corresponding to the intersection point can then be regarded as the favorable viscous coefficient of the damper, and the corresponding modal damping ratio is regarded as the favorable modal damping ratio. If not all of the favorable modal damping ratios are greater than ζ_{limit}, the number of modes of vibration required to be damped should be reduced until all the favorable modal damping ratios are greater than ζ_{limit}. The favorable viscous coefficient of the damper for each cable can be found using Equation 12.56.

$$\text{Let } \frac{\zeta_1}{l_1/L} = \frac{\zeta_i}{l_1/L} \Rightarrow \frac{\pi^2 \dfrac{c_d}{mL\omega_{01}}(l_1/L)}{1+\left[\pi^2 \dfrac{c_d}{mL\omega_{01}}(l_1/L)+\lambda\omega_{o1}\right]^2} = \frac{\pi^2 \dfrac{c_d}{mL\omega_{01}}i(l_1/L)}{1+\left[\pi^2 \dfrac{c_d}{mL\omega_{01}}i(l_1/L)+i\lambda\omega_{o1}\right]^2} \tag{12.58}$$

then

$$c_{d,opt} = \frac{mL\omega_{01}}{\pi^2(l_1/L)}\left(\frac{1}{\sqrt{i}} - \lambda\omega_{o1}\right) \tag{12.59}$$

where $c_{d,opt}$ is the favorable viscous coefficient of the damper.

The long-span cable-stayed bridge taken as a case study has a total length of 1596 m and a main span of 1018 m. The height of the two towers is nearly 300 m, measured from the base level to the top of the towers. The bridge tower is of a single composite column, and the stay cables are of the parallel wire strand type, made up of 7 mm wires. There are a total of 224 stay cables, and the length of the longest stay cable is about 540 m. For the sake of clear presentation, only one quarter of the stay cables of the bridge are considered, including 28 stay cables in the main span and 28 in the side span.

Figure 12.43 displays the computed first natural frequency, $f_1 = \omega_{01}/(2\pi)$, of each cable without any damper. The lowest first natural frequency is $0.244Hz$ for cable No. 228 in the main span and the highest first natural frequency is $1.159Hz$ for cable No. 102 in the side span. Figure 12.44 shows the ratio of the damper location to the original cable length, l_1/L, for each cable. The ratio ranges from 0.018 to 0.031. Let us assume that only one adjustable fluid damper is installed, perpendicular to the cable axis in the vertical plane near the low cable anchorage for each cable. As discussed before, the effect of the damper support stiffness may be considered. Figure 12.45 displays the damper support stiffness for each cable. It can be seen that the damper support stiffness increases as cable length becomes shorter.

The favorable design principle is used in this case study and the number of vibration modes is taken as two for all the stay cables concerned. The favorable design of adjustable fluid dampers is, in fact, an

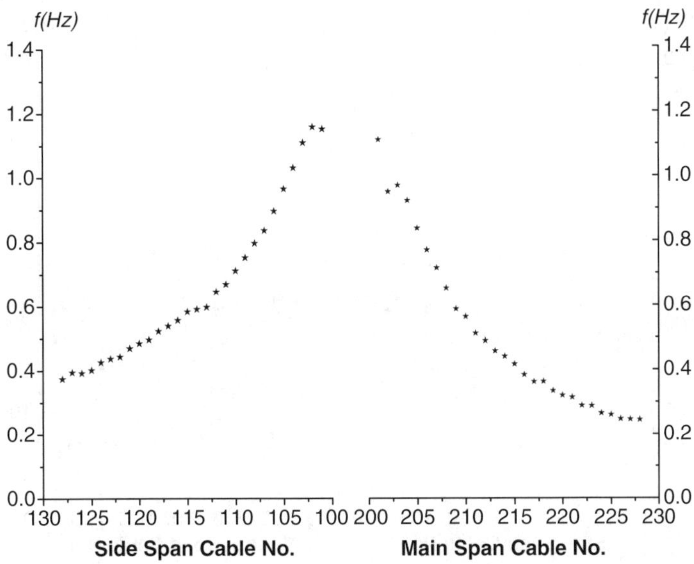

Figure 12.43 First natural frequencies of stay cables (from [70]) (Reproduced with permission from Elsevier).

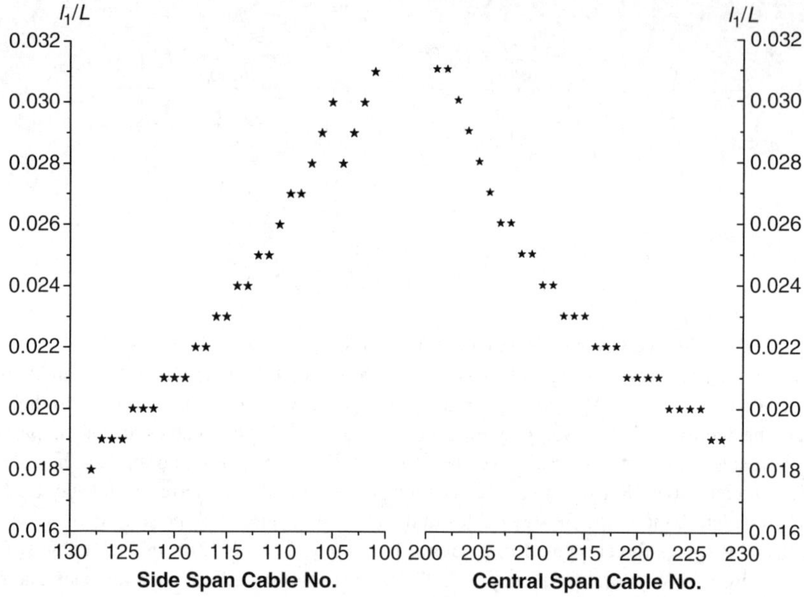

Figure 12.44 Ratios of damper location to cable length (from [70]) (Reproduced with permission from Elsevier).

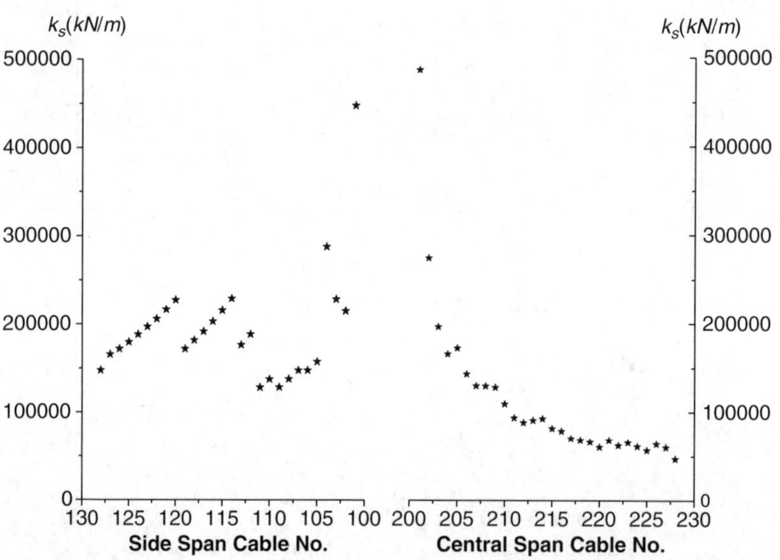

Figure 12.45 Damper support stiffness for each cable (from [70]) (Reproduced with permission from Elsevier).

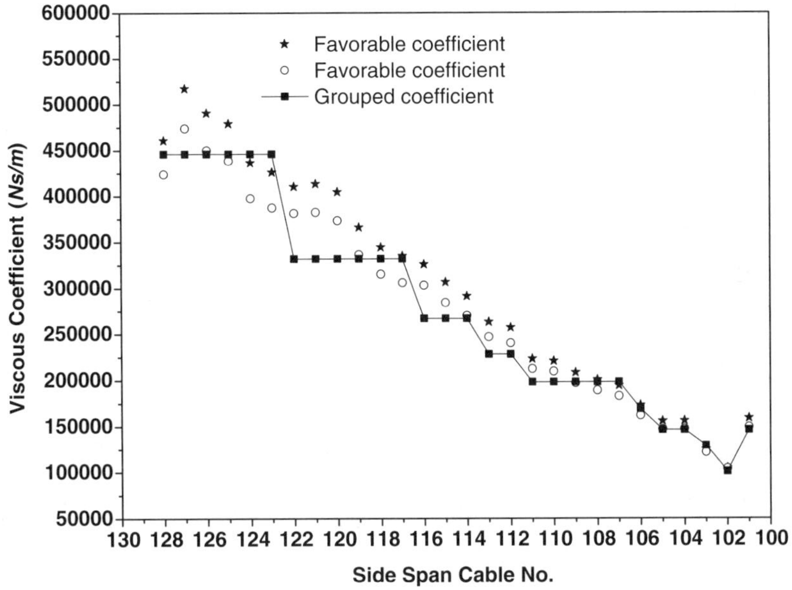

Figure 12.46 Viscous coefficient of the damper for stay cables in side span (from [70]) (Reproduced with permission from Elsevier).

iterative process. First, assume the damper is an ideally linear viscous damper and the damper support is perfectly rigid, so that the relaxation time constant is equal to zero. Then use Equation 12.59 to find the initial favorable viscous coefficient of the damper for each cable, as shown in Figure 12.46 for stay cables in the side span and in Figure 12.47 for stay cables in the main span.

Based on such information, adjustable fluid dampers are designed with enough adjustable levels to cover a wide range of viscous coefficient of the damper. The adjustable fluid dampers are manufactured and calibrated to find the viscous coefficient of the damper and relaxation time constants. In this case study, two types of prototype adjustable fluid dampers, which have ten orifices in the piston head – eight of them controllable, for achieving changes in damping at nine levels – were designed and manufactured: Type 1, with 1.5 mm diameter orifices for large viscous coefficient of the damper; and Type 2, with 1.8 mm diameter orifices for relatively small viscous coefficient of the damper. The two prototype dampers were extensively calibrated and tested.

The viscous coefficient of the damper and the relaxation time constants obtained from the tests are listed in Table 12.6. According to this, and the required initial favorable viscous coefficient of the damper as shown in Figures 12.46 and 12.47, it can be determined that the damper Type 2 should used for those cables requiring a favorable viscous coefficient of the damper below 173 000 *Ns/m* and the damper Type 1 should be used for all other cables. One may also decide the adjustable levels of each damper, using the number of opened orifices in the damper to "best fit" the initial favorable viscous coefficient of the damper.

Figure 12.48 shows the grouping results of damper type and adjustable level. The next step is to calculate the total relaxation time constant for each cable, which includes both damper stiffness and damper support stiffness, as shown in Figure 12.49. The results indicate that, in this case study, the damper support stiffness is much larger than the damper stiffness.

Equation 12.59 is then used again to determine the favorable viscous coefficient of the damper, but with the relaxation time constants included. These results are plotted in Figures 12.46 and 12.47, and compared with the grouping results. If the comparison is not satisfactory, the grouping can be finely

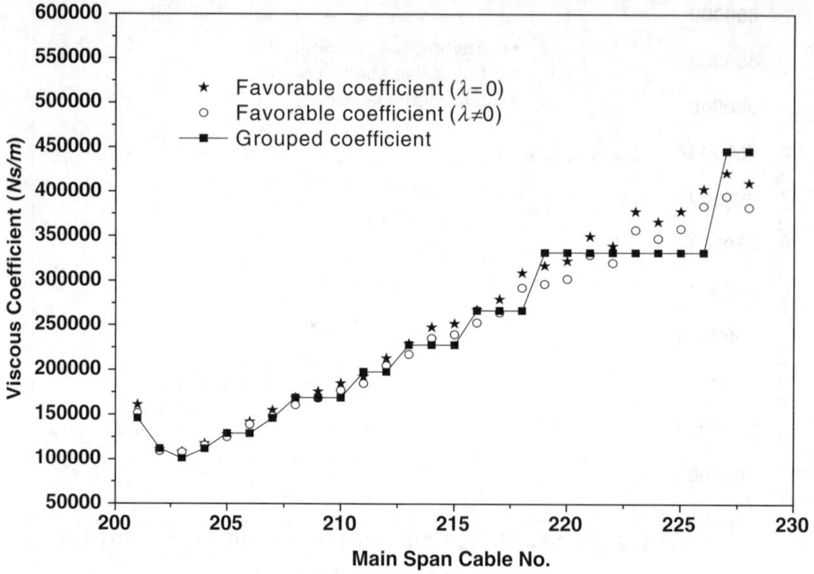

Figure 12.47 Viscous coefficient of the damper for stay cables in main span (from [70]) (Reproduced with permission from Elsevier).

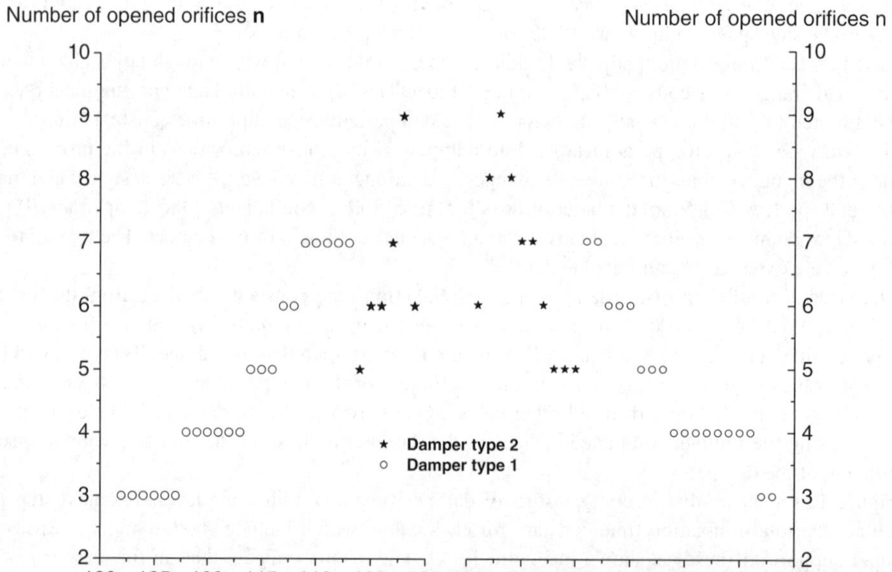

Figure 12.48 Grouping results of damper type and adjustable level (from [70]) (Reproduced with permission from Elsevier).

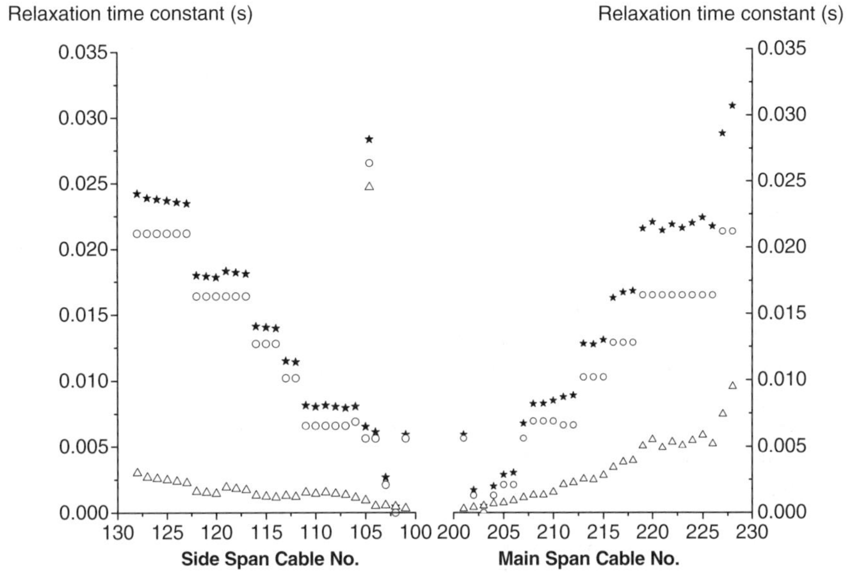

Figure 12.49 Relaxation time constants λ, λ_d, λ_s (from [70]) (Reproduced with permission from Elsevier).

tuned until both are close to each other. It can be seen from Figures 12.46 and 12.47 that some differences exist between the initial and final favorable viscous coefficient of the damper, but these are not very significant, because only the first two modes of vibration are considered in this case study.

Finally, based on the final grouping results, the favorable modal damping ratios in the first two modes of vibration for each cable can be recalculated and checked to see if they are greater than ζ_{limit}. The favorable modal damping logarithmic decrements obtained and shown in Figure 12.50 for this case study demonstrate that the modal damping logarithmic decrements in the first two modes of all the stay

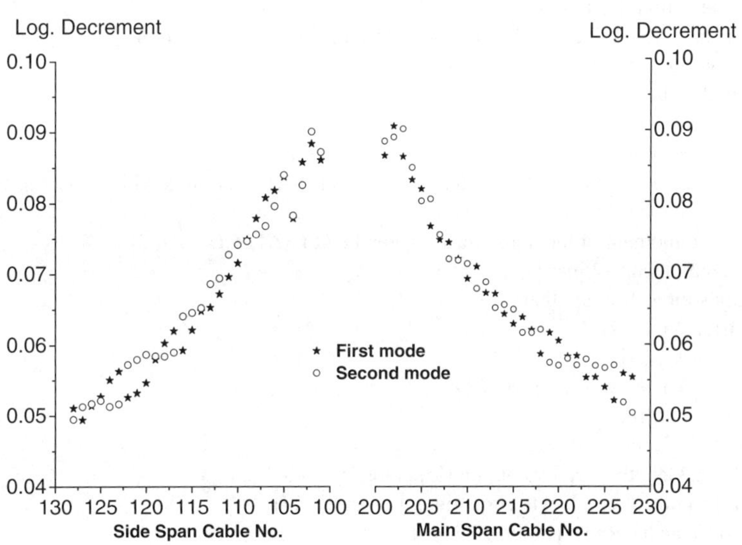

Figure 12.50 Modal damping logarithmic decrements in stay cables (from [70]) (Reproduced with permission from Elsevier).

cables concerned are greater than 4%. It should be pointed out, however, that the favorable modal damping logarithmic decrements shown in Figure 12.50 are obtained on the basis of the cable being assumed to be taut. For the first mode of vibration of the longest stay cable, sag effect may reduce the modal damping logarithmic decrement to some extent.

12.11 Notations

A	Cross-sectional area of container
A_k	Cross-sectional area of the kth SATLCD
a	Maximum amplitude of rivulet motion
B	Horizontal width of the water column of the TLCD
$\mathbf{C}$	Damping matrix
$\mathbf{C_a}$	Aeroelastic damping matrix due to the wind-rain-cable interaction
$\mathbf{C_0}$	Damping matrix due to the viscous damper
C_D	Drag coefficient of the cable with rivulet
C_L	Lift coefficient of the cable with rivulet
c	Internal cable damping coefficient
c_d	Viscous coefficient of the damper at zero frequency
$c_{d,opt}$	Favorable viscous coefficient of the damper
c_L	Damping coefficient of viscous damper
D	Cable diameter
d_k	Thickness of the liquid column in the kth SATLCD
$F(x,t)$	Rain-wind-induced force on the cable
$F_d(t)$	Damper force on the cable
F_p	Restoring force
$\tilde{\boldsymbol{F}}$	Normalized rain-wind-induced force
f_1	First natural frequency
G_k	First moment of the liquid mass in the kth SATLCD
g	Acceleration due to gravity
H	Distance from the center line of the bottom tube of the TLCD to the pivot
$H(\bullet)$	Heaviside function
h	Air chamber height
h_r	Wind reference height
h_0	Lower anchorage height of the cable above the sea level
I_d	Total second moment of the liquid mass of all SATLCD units with respect to the torsional center of the bridge deck
I_k	Second moment of the liquid mass inside the kth SATLCD
I	Mass moments of inertia
K_p	Adjusting factor less than or equal to 1
$[K]$	Stiffness matrix
$\boldsymbol{K}$	Stiffness matrix
k_{ij}	Stiffness matrix of the SATLCD
k_d	Damper stiffness
k_s	Damper support stiffness
L	(i) Total length of water column (Equations (12.1a–12.1c) (ii) Liquid column length (Equations 12.6 and 12.7) (iii) Cable length (Equation 12.25)
L_k	Length of the kth SATLCD

l_1	Length between the cable anchorage to the damper location
l_2	Length from damper location to the top end of cable
m	Cable mass per unit length (Equations 12.25 and 12.51)
m_d	Total liquid mass of all SATLCD units
m_k	Mass of the kth SATLCD
m_{ij}	Mass matrix of the SATLCD
$[M]$	Mass matrix
$\boldsymbol{M}$	Mass matrix
P	Net pressure between the two air chambers
P_L	Pressure in the left air chamber
P_R	Pressure in the right air chamber
P_0	Static pressure
R	Cable radius
S_k	Constant displacement feedback gain of the kth SATLCD
T	Cable tension force
t	Time
$U_{rel}(x)$	Relative mean wind speed with respect to the cable motion
$U_0(x)$	Incident mean wind speed
$U_0(h)$	Mean wind speed at height h above the sea level
U_r	Mean wind speed at a chosen reference height h_r
$u_{1k}(t)$	Displacement feedback control force in the kth SATLCD
u_{2k}	Control force based on the feedback of liquid velocity
V_0	Volume of air inside the chamber at the static position of liquid
W	(i) Relative motion of water to the container (Equations (12.1a–12.1c) (ii) Liquid displacement (Equation 12.3)
W_a	Tolerable liquid displacement of the kth SATLCD
$W(x)$	Motion profile of rivulet along the cable axis
$w(x,t)$	In-plane transverse displacement normal to the cable axis at position x
x	Lateral displacement of node 2
y_1	Vertical displacement of node 1
y_2	Vertical displacement of node 2
$\mathbf{z}(\tau)$	Vector of non-dimensional time-varying coefficient
$\mathbf{z_c}(\tau)$	Vector of time-varying coefficient corresponding to the vector of clamped mode shapes
$\mathbf{z_0}(\tau)$	Vector of time-varying coefficient corresponding to the vector of unclamped mode shapes
$\overline{H}_k$	Distance from the mass center of liquid inside the kth SATLCD to the torsional (elastic) center of the bridge deck
ρ_w	Liquid density
ρ	Air density
α	Inclination angle of the taut cable
α_1	Exponent depending on surface roughness
α_k	Liquid length ratio
β	Yaw angle of incident wind
ε	Stagnation point influence factor
ω_k	Circular natural frequency
ω_r	Frequency of rivulet motion
ω_{01}	First circular natural frequency of the taut cable without damper
θ	Torsional displacement
λ	Total relaxation time constant considering both damper stiffness and damper support stiffness
λ_d	Relaxation time constant of damper itself
λ_s	Equivalent relaxation time constant of damper support

τ	Non-dimensional time
$\theta(x,t)$	Dynamic angular displacement of upper rivulet with respect to the static position of upper rivulet $\theta_0(x)$
$\zeta_{i,\max}$	Maximum attainable modal damping ratio
ζ_{limit}	Least modal damping ratio
χ	Dimensionless eigenvalue
ν	Velocity of piston head
κ	Non-dimensional damper damping parameter
κ_i	Non-dimensional viscous coefficient of the damper for the ith mode of vibration
$\kappa_{i,opt}$	Optimal non-dimensional viscous coefficient of the damper
ϕ	(i) Wind angle of attack (Equations 12.26 and 12.27a, 12.27b) (ii) Phase angle (Equation 12.35)
δ_m	Head loss coefficient for providing sufficient damping to the kth SATLCD
$\delta(\bullet)$	Delta function
$\mathbf{\Phi}(x)$	Vector of shape function
$\mathbf{\Phi_c}(x)$	Vector of clamped mode shapes of cable
$\mathbf{\Phi_0}(x)$	Vector of unclamped mode shapes of cable
Θ_{ci}	"Clamping ratio" for the ith mode of vibration
$\Gamma(\bullet)$	Aerodynamic function

References

1. Fujino, Y., Kimura, K., and Tanaka, H. (2012) *Wind Resistant Design of Bridges in Japan: Development and Practices*, Springer Verlag, Japan.
2. Fujino, Y. (2002) Vibration, control and monitoring of long-span bridges- recent research, developments and practice in Japan. *Journal of Constructional Steel Research*, **58** (1), 71–97.
3. Ostenfeld, K. and Larsen, A. (1992) Bridge engineering and aerodynamics, in *Aerodynamics of Large Bridges: Proceedings of the First International Symposium on Aerodynamics of Large Bridges* (ed. A. Larsen), Balkema, Rotterdam, The Netherlands, pp. 3–22.
4. Astiz, M.A. (1998) Flutter stability of very long suspension bridges. *Journal of Bridge Engineering, ASCE*, **3** (3), 132–139.
5. Branceleoni, F. (1992) The construction phase and its aerodynamic issues, in *Aerodynamics of Large Bridges: Proceedings of the First International Symposium on Aerodynamics of Large Bridges* (ed. A. Larsen), Balkema, Rotterdam, The Netherlands, pp. 147–158.
6. Wilde, K., Fujino, Y., and Prabis, V. (1996) Effects of eccentric mass on flutter of long span bridge. Proceedings of 2nd International Workshop on Structural Control, Hong Kong, pp. 564–574.
7. Xiang, H.F. (2005) *Modern Theory and Practice on Bridge Wind Resistance*, China Communications Press, Beijing (in Chinese).
8. Housner, G.W., Bergman, L.A., Caughey, T.K. *et al.* (1997) Structural control: Past, present and future. *Journal of Engineering Mechanics, ASCE*, **123** (9), 897–971.
9. Soong, T.T. and Dargush, G.F. (1997) *Passive Energy Dissipation Systems in Structural Engineering*, John Wiley & Sons, Inc., New York, N.Y.
10. Spencer, B.F. Jr. and Soong, T.T. (1999) New applications and development of active, semi-active and hybrid control techniques for seismic and non-seismic vibration in the USA. Proceedings of International Post-SMiRT Conference on Seismic Isolation, Passive Energy Dissipation and Active Control of Vibration of Structures, Cheju, Korea, August, pp. 23–25.
11. Soong, T.T. (1990) *Active Structural Control: Theory and Practice*, Longman Scientific, Essex, UK.
12. Spencer, B.F. Jr., Dyke, S.J., Sain, M.K., and Carlson, J.D. (1997) Phenomenological model of a magnetorheological damper. *Journal of Engineering Mechanics, ASCE*, **123** (3), 230–238.
13. Spencer, B.F. Jr. and Nagarajaiah, S. (2003) State of the art of structural control. *Journal of Structural Engineering, ASCE*, **129** (7), 845–856.
14. Xu, Y.L., Qu, W.L., and Ko, J.M. (2000) Seismic response control of frame structures using magnetorheological/electrorheological dampers. *Earthquake Engineering and Structural Dynamics*, **29** (5), 557–575.
15. Richardson, J.R. (1981) The Development of the Concept of the Twin Suspension Bridge, National Maritime Institute, NMIR125.
16. Sato, H., Toriumi, R., and Kusakabe, T. (1995) Aerodynamic characteristics of slotted box girders. Proceedings of Bridges into the 21st Century, Hong Kong.

17. Larsen, A. and Astiz, M.A. (1998) Aeroelastic consideration for the Gibraltar bridge feasibility study, in *Bridge Aerodynamics* (eds Larsen, A. and Esdahl, S.), Balkema, Rotterdam, pp. 165–173.
18. Matsumoto, M., Yoshzumi, F., Yabutani, T. *et al.* (1999) Flutter stabilization and heaving-branch flutter. *Journal of Wind Engineering and Industrial Aerodynamics*, **83** (1–3), 289–299.
19. Sato, H., Kusuhara, S., Ogi, K., and Matsufuji, H. (2000) Aerodynamic characteristics of super long-span bridges with slotted box girder. *Journal of Wind Engineering and Industrial Aerodynamics*, **88** (2–3), 297–306.
20. Sato, H., Hirahara, N., Fumoto, K. *et al.* (2002) Full aeroelastic model test of a super long-span bridge with slotted box girder. *Journal of Wind Engineering and Industrial Aerodynamics*, **90** (12–15), 2023–2032.
21. Xiang, H.F. and Ge, Y.J. (2003) On aerodynamic limit to suspension bridges. Proceedings of the 11th International Conference on Wind Engineering, Texas, USA, pp. 65–80.
22. Simiu, E. and Miyata, T. (2006) *Design of Buildings and Bridges for Wind: A Practical Guide for ASCE 7 Users and Designers of Special Structures*, John Wiley and Sons, New York.
23. Ueda, T., Tanaka, T., and Matsushita, Y. (1998) Aerodynamic stabilization for super long-span suspension bridges. IABSE Symposium on Long Span and High Rise Structures, Japan.
24. Brown, W.C. (1996) Development of the deck for the 3300 m span Messina Crossing. 15th IABSE Congr. Rep., IABSE, Zurich, pp. 1019–1030.
25. Diana, G. (1993) Analytical and wind tunnel simulations for the aeroelastic design of the Messina Straits Bridge. Proceedings of International Seminar on Utilization of Large Boundary Wind Tunnel, Tsukuba, Japan, pp. 183–202.
26. Raggett, J.D. (1987) Stabilizing pair of winglets for slender bridge decks. Bridges and Transmission Line Structures, Sixth Structural Congress of ASCE, New York, pp. 292–302.
27. Arco, D.C. and Aparicio, Á.C. (1999) Improving suspension bridge wind stability with aerodynamic appendages. *Journal of Structural Engineering*, **125** (12), 1367–1375.
28. Kobayashi, H. and Nagaoka, H. (1992) Active control of flutter of a suspension bridge. *Journal of Wind Engineering and Industrial Aerodynamics*, **41** (3), 143–151.
29. Wilde, K. and Fujino, Y. (1998) Aerodynamic control of bridge deck flutter by active surfaces. *Journal of Engineering Mechanics, ASCE*, **124** (7), 718–727.
30. Kobayashi, H., Ogawa, R., and Taniguchi, S. (1998) Active flutter control of a bridge deck by ailerons. Proceedings of 2nd World Conference on Structural Control, Kobori ed., Kyoto, Japan, Vol. 3, pp. 1841–1848.
31. Wilde, K., Omenzetter, P., and Fujino, Y. (2001) Suspension of bridge flutter by active deck-flaps control system. *Journal of Engineering Mechanics*, **127** (1), 80–89.
32. Omenzetter, P., Wilde, K., and Fujino, Y. (2000) Suppression of wind-induced instabilities of a long span bridge by a passive deck-flaps control system Part I: formulation. *Journal of Wind Engineering and Industrial Aerodynamics*, **87** (1), 61–79.
33. Omenzetter, P., Wilde, K., and Fujino, Y. (2000) Suppression of wind-induced instabilities of a long span bridge by a passive deck-flaps control system Part II: Numerical simulations. *Journal of Wind Engineering and Industrial Aerodynamics*, **87** (1), 81–91.
34. Omenzetter, P., Wilde, K., and Fujino, Y. (2002) Study of passive deck flaps flutter control system on full bridge model. I: theory. *Journal of Engineering Mechanics*, **128** (3), 264–279.
35. Omenzetter, P., Wilde, K., and Fujino, Y. (2002) Study of passive deck flaps flutter control system on full bridge model. II: results. *Journal of Engineering Mechanics*, **128** (3), 280–286.
36. Kwon, S.D., Jung, M.S.S., and Chang, S.P. (2000) A new passive aerodynamic control method for bridge flutter. *Journal of Wind Engineering and Industrial Aerodynamics*, **86** (2–3), 187–202.
37. Larsen, A., Esdahl, S., Andersen, J.E., and Vejrum, T. (2000) Storebaelt suspension bridge-vortex shedding excitation and mitigation by guide vanes. *Journal of Wind Engineering and Industrial Aerodynamics*, **88** (2–3), 283–296.
38. Larsen, A., Savage, M., Lafrenière, A. *et al.* (2008) Investigation of vortex response of a twin box bridge section at high and low Reynolds numbers. *Journal of Wind Engineering and Industrial Aerodynamics*, **96** (6–7), 934–944.
39. Ge, Y.J., Yang, Y.X., and Cao, F.C. (2011) VIV sectional model testing and field measurement of Xihoumen suspension bridge with twin box girder. 13th International Conference on Wind Engineering, July 10–15, Netherlands.
40. Li, H., Laima, S., Ou., J. *et al.* (2011) Investigation of vortex-induced vibration of a suspension bridge with two separated steel box girders based on field measurements. *Journal of Wind Engineering and Industrial Aerodynamics*, **33** (6), 1894–1907.
41. Federal Highway Administration (2007) Wind Induced Vibration of Stay Cables, FHWA-HRT-05-084.
42. Hikami, Y. and Shiraishi, N. (1988) Rain-wind induced vibrations of cables in cable stayed bridges. *Journal of Wind Engineering and Industrial Aerodynamics*, **29** (1–3), 409–418.
43. Xu, Y.L., Li, Y.L., and Shum, K.M. (2004) Stonecutters Bridge: Investigation of Aerodynamic Forces on Stay Cables - Wind Tunnel Investigation, Report to Highways Department of Hong Kong, October.
44. Li, M.S., Liao, H.L., Chen, X., and He, X.D. (2003) Stonecutters Bridge: Rain-Wind-Induced Vibration Tests of Stay Cables, Report to Ove Arup & Partners Hong Kong Ltd, March.
45. Fujino, Y. and Yoshida, Y. (2002) Wind-induced vibration and control of Trans-Tokyo Bay Crossing Bridge. *Journal of Structural Engineering*, **128** (8), 1012–1025.
46. Hirai, S., Honda, A., Kato, H. *et al.* (1993) Aerodynamic stability of Trans-Tokyo Bay Highway Bridge. *Journal of Wind Engineering and Industrial Aerodynamics*, **49** (1–3), 487–496.

47. Nobuto, J., Fujino, Y., and Ito, M. (1988) A study on the effectiveness of TMD to suppress a coupled flutter of bridge deck. *Proceedings of Japan Society of Civil Engineering, Hiroshima, Japan*, **5** (5), 413–416.
48. Gu, M., Chang, C.C., Wu, W., and Xiang, H.F. (1998) Increase of critical flutter wind speed of long-span bridges using tuned mass dampers. *Journal of Wind Engineering and Industrial Aerodynamics*, **73** (2), 111–123.
49. Pourzeynali, S. and Datta, T.K. (2002) Control of flutter of suspension bridge deck using TMD. *Journal of Wind and Structures*, **5** (5), 407–422.
50. Kwon, S.D. and Park, K.S. (2004) Suppression of bridge flutter using tuned mass dampers based on robust performance design. *Journal of Wind Engineering and Industrial Aerodynamics*, **92** (11), 919–934.
51. Chen, X.Z. and Kareem, A. (2003) Efficacy of tuned mass dampers for bridge flutter control. *Journal of Structural Engineering, ASCE*, **129** (10), 1291–1300.
52. Xue, S.D., Ko, J.M., and Xu, Y.L. (2000) Control of torsional flutter and buffeting of long span bridges using liquid column damper, in Proceedings of. International Conference on Advances in Structural Dynamics (eds Ko, J.M and Xu, Y.L.), Elsevier, Oxford, pp. 1413–1420.
53. Korlin, R. and Starossek, U. (2007) Wind tunnel test of an active mass damper for bridge decks. *Journal of Wind Engineering and Industrial Aerodynamics*, **95** (4), 267–277.
54. Achkire, Y., Bossens, F., and Preumont, A. (1998) Active damping and flutter control of cable-stayed bridges. *Journal of Wind Engineering and Industrial Aerodynamics*, **74–76**, 913–921.
55. Pourzeynali, S. and Datta, T.K. (2005) Semiactive fuzzy logic control of suspension bridge flutter. *Journal of Structural Engineering*, **131** (6), 900–912.
56. Gu, M., Chen, S.R., and Chang, C.C. (2002) Parametric study on multiple tuned mass dampers for buffeting control of Yangpu Bridge. *Journal of Wind Engineering and Industrial Aerodynamics*, **89** (11–12), 987–1000.
57. Lin, Y.Y., Cheng, C.M., and Lee, C.H. (2000) A tuned mass damper for suppressing the coupled flexural and torsional buffeting response of long-span bridges. *Engineering Structures*, **22** (9), 1195–1204.
58. Shum, K.M., Xu, Y.L. and Guo, W.H. (2008) Wind-induced vibration control of long span cable-stayed bridges using multiple pressurized tuned liquid column dampers, *Journal of Wind Engineering and Industrial Aerodynamics*, **96** (2), 166–192.
59. Shum, K.M., Xu, Y.L., and Guo, W.H. (2006) Buffeting response control of a long span cable-stayed bridge during construction using semi-active tuned liquid column dampers. *Wind & Structures-An International Journal*, **9** (4), 271–296.
60. Zhou, H.J. and Xu, Y.L. (2007) Wind-rain-induced vibration and control of stay cables in a cable-stayed bridge. *Journal of Structural Control and Health Monitoring*, **14** (7), 1013–1033.
61. Irvine, H.M. (1981) *Cable Structures*, MIT Press, Cambridge, Massachusetts.
62. Xu, Y.L. and Wang, L.Y. (2003) Analytical study of wind-rain-induced cable vibration: SDOF model. *Journal of Wind Engineering and Industrial Aerodynamics*, **91** (1–2), 27–40.
63. Main, J.A. (2002) Modeling the Vibrations of a Stay Cable with Attached Damper, PhD thesis, The Johns Hopkins University.
64. Pacheco, B.M., Fujino, Y., and Sulekh, A. (1993) Estimation curve for modal damping in stay cables with viscous damper. *Journal of Structural Engineering*, **119** (6), 1961–1979.
65. Main, J.A. and Jones, N.P. (2002) Free vibrations of cable with attached damper I: Linear viscous damper. *Journal of Engineering Mechanics*, **128** (10), 1062–1071.
66. Butcher, J.C. (1987) *The Numerical Analysis of Ordinary Differential Equations-Runge-Kutta and General Linear Methods*, John Wiley & Sons, Inc., NY, USA.
67. Yamaguchi, H. (1990) Analytical study on growth mechanism of rain vibration of cables. *Journal of Wind Engineering and Industrial Aerodynamics*, **33** (1–2), 73–80.
68. Krenk, S. (2000) Vibrations of a taut cable with an external damper. *Journal of Applied Mechanics*, **67** (4), 772–776.
69. Christenson, R.E., Spencer, B.F., and Johnson, E.A. (2006) Experimental verification of smart cable damping. *Journal of Engineering Mechanics ASCE*, **132** (3), 268–278.
70. Xu, Y.L. and Zhou, H.J. (2007) Damping cable vibration for a cable-stayed bridge using adjustable fluid dampers. *Journal of Sound and Vibration*, **306** (1–2), 349–360.
71. Singh, M.P., Verma, N.P., and Moreschi, L.M. (2003) Seismic analysis and design with Maxwell dampers. *Journal of Engineering Mechanics*, **129** (3), 273–282.
72. Li, Z.Q., Xu, Y.L., and Zhou, L.M. (2006) Adjustable fluid damper with SMA actuators. *Smart Materials and Structures*, **15** (5), 1483–1492.
73. Krenk, S. and Høgsberg, J.R. (2005) Damping of cables by a transverse force. *Journal of Engineering Mechanics*, **131** (4), 340–348.

13

Typhoon Wind Field Simulation

13.1 Preview

As discussed in Chapter 1, tropical cyclones are intense cyclonic storms that occur over the tropical oceans, mainly in late summer and early autumn. The strongest tropical cyclones have occurred in the Caribbean, where they are called hurricanes, off the northwest coast of Australia, where they are known as cyclones, and in the South China Sea, where they are called typhoons. A mature typhoon has a three-dimensional vortex structure, with a horizontal dimension for several hundred kilometers and a vertical dimension of the order of ten kilometers. The central zone of a typhoon, known as the eye, has a diameter of a few tens of kilometers and consists of relatively cloudless and quiescent air. The strongest winds occur just outside the eye wall, up to a distance of about 200 km from the typhoon center. Typhoons normally travel as whole entities at speeds of 5–50 km per hour. They generally rage for several days and cause extensive damage due to their high wind speeds.

Long-span cable-supported bridges built in typhoon regions are inevitably affected by typhoon winds. They must be designed to withstand typhoon winds during their design lives. The determination of typhoon wind speed, direction and profile, among other characteristics, in the atmospheric boundary layer (ABL) for a bridge within a given design life is an important task. Long-term field measurements are desirable to obtain typhoon wind characteristics, but in many cases there are no, or very few, field measurement data available at a bridge site, so that the reliable estimation of typhoon wind characteristics at the bridge site could not be achieved. The use of the Monte Carlo simulation method, in conjunction with a typhoon wind field model and some statistical data, is a universal approach for typhoon wind field simulation to determine typhoon wind speed, direction and profile.

After a brief review of the existing typhoon wind field models used in engineering community, this chapter presents a refined typhoon wind field model based on some simplifications of the three-dimensional Navier-Stokes equations, but including the effect of temperature and the variation of central pressure difference with height. The refined typhoon wind field model is solved by using a decomposition method. The typhoon wind speed and direction predicted by the refined model for Typhoon York are compared with those measured at Waglan Island in Hong Kong during Typhoon York. Mean wind speed profiles in the vertical direction at different distances from the typhoon center are also predicted by the refined model, and the results are compared with the field observation data from GPS-based dropsondes and the power law profiles.

The Monte Carlo simulation method, together with the refined model and statistical distributions of typhoon parameters, is then used to predict non-directional and directional design typhoon wind speeds

Wind Effects on Cable-Supported Bridges, First Edition. You-Lin Xu.

at Waglan Island. The predicted results are compared with those directly from wind measurement data recorded by the anemometers installed on the island.

Since long-span cable-supported bridges are often not only located in a typhoon-prone region, but also surrounded by complex terrain, topographic effect on typhoon winds and interaction between winds and terrain surface roughness need to be considered. Therefore, a numerical simulation procedure for predicting directional design wind speeds and profiles for a bridge site surrounded by complex terrain, and a case study by applying this simulation procedure to the Stonecutters bridge site, are finally presented. The results obtained in this chapter will be used in Chapter 15 for buffeting response analysis of long-span cable-supported bridges to non-stationary typhoon wind.

13.2 Refined Typhoon Wind Field Model

13.2.1 Background

The modeling of typhoon wind field has been investigated by many researchers [1–9]. The models developed for civil structures may be classified into two categories: numerical models and analytical models. Numerical models have often been developed by simplifying the three-dimensional Navier-Stokes equations through introduction of some empirical formulas based on the observation data of typhoons in the field. The solutions are then sought by using numerical methods.

For example, Shapiro solved the momentum equations for a slab boundary layer of constant depth under an imposed symmetric pressure distribution [5]. Georgiou used the Shapiro model to determine wind field at a gradient height, and the surface wind speeds were then estimated using an empirical ratio of surface to gradient wind speeds based on observation data [6]. The model proposed by Vickery and Twisdale was also based on the Shapiro model, which considered the equation of horizontal motion, vertically averaged over the height of the ABL, and the rate of decay of typhoon after landfall [7]. A finite-difference scheme was used to solve the equation for the steady-state wind field over a set of nested rectangular grid. The model was then improved by taking into account the effects of sea surface roughness change and the air-sea temperature difference on the estimated surface-level wind speeds [10,11].

Instead of numerical models, Yoshizumi derived an analytical solution from the Navier-Stokes equation by assuming that the gradient wind velocity could be obtained by the summation of the typhoon translation velocity and the wind velocity relative to the typhoon center [12]. The surface wind velocity was then obtained by using the approximate solution of Rosenthal [13]. Meng *et al.* started from the Navier-Stokes equation, but envisaged typhoon-induced mean wind velocity as the combination of a gradient wind in the free atmosphere and a surface wind caused by friction on the ground surface [8,14]. The perturbation analysis was then performed to obtain the tangential and radial boundary layer velocity in the friction region.

All of the existing typhoon wind field models consider the physical features of the typhoon boundary layer caused by friction at the earth's surface, but most neglect the influence of temperature and assume that the central pressure difference of a typhoon does not vary with height above the ground. The numerical simulation results obtained by meteorologists have suggested that it is necessary to include temperature, due to dissipative heating from the surface friction in forecasting typhoon structure and intensity [15].

Field observations made from the data recorded by the GPS dropsonde [16] showed that the central pressure difference of a typhoon actually decreases with increasing height above the ground. The assumption that the central pressure difference remains constant with height may affect the prediction of design typhoon wind speed and wind profile. A refined typhoon wind field model, including the influence of temperature and the variation of the central pressure difference with height, was therefore proposed [17] and is introduced in this chapter.

13.2.2 Refined Typhoon Wind Field Model

The deduction of the refined typhoon wind field model is based on the three dimensional Navier-Stokes equations. Equations 13.1a and 13.1b are the continuity and momentum equations, respectively, whereas Equations 13.1c and 13.1d are respectively the state equation and thermodynamic equation [18].

$$\frac{d\rho}{dt} = -\rho\nabla \cdot \boldsymbol{v} \tag{13.1a}$$

$$\frac{d\boldsymbol{v}}{dt} = -\frac{1}{\rho}\nabla p - g\boldsymbol{k} - 2\boldsymbol{\Omega} \times \boldsymbol{v} + \boldsymbol{F} \tag{13.1b}$$

$$p = \rho RT \tag{13.1c}$$

$$Jdt = c_v dT + pd\alpha \tag{13.1d}$$

where:

$\boldsymbol{v} = (u, v, w)$ is the wind velocity vector, in which u, v and w are wind components in the x, y and z directions;
ρ is the air density;
p is the atmospheric pressure;
∇ is the three-dimensional del operator in the rectangular coordinate system;
g is the gravitational constant;
$\boldsymbol{k}$ is the unit vector in the z-direction;
$\boldsymbol{\Omega}$ is the angular velocity of the earth;
$\boldsymbol{F}$ is the friction force;
R is the ideal gas constant;
T is the temperature;
J is the adiabatic heat source;
c_v is the heat capacity at constant volume;
α is equal to $1/\rho$.

Equation 13.1b can be decomposed into the horizontal momentum and the vertical momentum equations. Because the Coriolis force in the vertical direction is much smaller than that in the horizontal direction [18], it can be neglected in the vertical direction. As a result, Equation 13.1 can be re-written as

$$\frac{d\rho}{dt} = -\rho\nabla \cdot \boldsymbol{v} \tag{13.2a}$$

$$\frac{d\boldsymbol{v}_h}{dt} = -\frac{1}{\rho}\nabla_h p - f\boldsymbol{k}_h \times \boldsymbol{v}_h + \boldsymbol{F}_h \tag{13.2b}$$

$$\frac{dw}{dt} = -\frac{1}{\rho}\frac{\partial p}{\partial z} - g + F_v \tag{13.2c}$$

$$p = \rho RT \tag{13.2d}$$

$$Jdt = c_v dT + pd\alpha \tag{13.2e}$$

where:

the subscript h means the horizontal;
$\boldsymbol{v}_h$ is the wind velocity in the horizontal plane;
f is the Coriolis parameter;
$\boldsymbol{k}_h$ is the unit vector in horizontal plane;
$\boldsymbol{F}_h$ and F_v are the friction forces in the horizontal and vertical direction, respectively.

It is common to replace temperature with potential temperature θ, which is defined as:

$$\theta = T\left(\frac{p_0}{p}\right)^{R/c_p} \tag{13.3}$$

where:

p_0 is the pressure at zero-plane;
$c_p = c_v + R$ is the heat capacity at constant pressure.

It is difficult to directly use Equation 13.2 to predict design typhoon wind speeds for bridges, and some simplifications have to be made to obtain a typhoon wind field model. The hydrostatic approximation is adopted because the vertical scale of a typhoon is much smaller than its horizontal scale. As a result, Equation 13.2c can be simplified as:

$$-\frac{1}{\rho}\frac{\partial p}{\partial z} - g = 0 \tag{13.4}$$

By using Equations 13.2d, 13.2e, 13.3 and 13.4, one may have:

$$\frac{\partial \Pi}{\partial z} = -\frac{g}{\theta} \tag{13.5a}$$

$$\Pi = c_p\left(\frac{p}{p_0}\right)^{R/c_p} \tag{13.5b}$$

where:

Π is the Exner function.

A geometric height-based terrain-following coordinate transformation can be used to deal with the difficulties at the low boundary that arises from using the conventional z-coordinate. In this regard, the following coordinate transformation was proposed [19]:

$$z^* = \frac{z_t(z - z_s)}{z_t - z_s} \tag{13.6}$$

where:

z^* is the terrain-following vertical coordinate;
z_t is the height of the wind model top;
z_s is the height of the lower boundary.

The terrain-following coordinate transformation can be done by substituting Equation 13.6 into Equation 13.5a. Equation 13.5a then becomes:

$$\frac{\partial \Pi}{\partial z^*} = -\left(1 - \frac{z_s}{z_t}\right)\frac{g}{\theta} \tag{13.7}$$

For the boundary condition $p = p_0$ at $z = 0$, and by using Equations 13.5b, 13.6 and 13.7, the atmospheric pressure p, which varies with height, can be expressed by:

$$p = p_0(1 - gz/\theta c_p)^{c_p/R} \tag{13.8}$$

Since wind flow in a typhoon boundary layer is treated as incompressible atmosphere, the air density is regarded as constant and the continuity Equation 13.2a can be neglected. As a result, Equation 13.2 can be reduced to the following two equations:

$$\frac{d\boldsymbol{v}_h}{dt} = -\frac{1}{\rho}\nabla_h p - f\boldsymbol{k}_h \times \boldsymbol{v}_h + \boldsymbol{F}_h \tag{13.9a}$$

$$p = p_0(1 - gz/\theta c_p)^{c_p/R} \tag{13.9b}$$

It can be seen that Equation 13.9 includes the effect of temperature and reflects the variation of pressure with height. However, Equation 13.9 cannot be solved directly because the continuity Equation 13.2a is not included, so that the number of unknown quantities is greater than the number of equations. Nevertheless, the following analytical model for the radial profiles of sea level pressure in a typhoon was proposed by Holland, based on field measurement data [3]:

$$p_0 = p_{c0} + \Delta p_0 \exp[-(r_m/r)^B] \tag{13.10}$$

where:

p_{c0} is the central pressure of a typhoon at zero-plane;
Δp_0 is the central pressure difference at zero plane equal to $p_{m0} - p_{c0}$;
p_{m0} is the ambient pressure (theoretically at infinite radius) at zero-plane;
r_m is the radius to maximum winds;
r is the radial distance from the typhoon center;
B is Holland's radial pressure profile parameter, taking on values between 0.5 and 2.5.

Equations 13.9 and 13.10 become the basic equations of the refined model for typhoon wind field over sea. The typhoon wind field over land can also be obtained by using these equations in conjunction with the typhoon wind decay model described in the next section.

13.2.3 Typhoon Wind Decay Model

Once a typhoon makes landfall, it weakens as the central pressure rises. Proper modeling of the decay of a typhoon is important for the prediction of typhoon winds after landfall. The form of the decay model used in this chapter is one proposed by Vickery and Twisdale [20] for North America, but the parameters in the model are different. The decay model is defined as:

$$\Delta p_0(t) = \Delta p_{0n} \exp(-at + b) \tag{13.11a}$$

where:

$\Delta p_0(t)$ is the central pressure difference at zero plane of a typhoon at t hours after landfall;
Δp_{0n} is the central pressure difference at zero plane at the time when the typhoon makes landfall;
a is the decay constant;
b is the random variable with a normal distribution.

The decay constant a is given as:

$$a = a_0 + a_1 \Delta p_{0n} + \varepsilon \tag{13.11b}$$

where:

a_0 and a_1 are the constants determined by a standard linear regression analysis;
ε is the error term with a zero-mean normal distribution.

13.2.4 Remarks

Equations 13.9, 13.10 and 13.11a are the complete basic equations of the refined typhoon wind field model. When the typhoon is over sea, the refined typhoon wind field model contains Equations 13.9 and 13.10 only. Equations 13.9b and 13.10 can be substituted into Equation 13.9a, and the basic equations of the refined typhoon wind field model can then be regarded as only one horizontal motion equation. When a typhoon makes landfall, the refined typhoon wind field contains all the three equations. Equations 13.9b,13.10 and 13.11 can be substituted into Equation 13.9a and the basic equations of the refined typhoon wind field model can also be regarded as only one horizontal motion equation.

Therefore, how to find the solution for the refined typhoon wind field model becomes how to solve Equation 13.9a. This horizontal motion equation can be solved by using the method proposed by Meng *et al.* [8] and it is given briefly in the following section for the sake of understanding and completion.

13.3 Model Solutions

13.3.1 Decomposition Method

The horizontal typhoon-induced wind velocity $\boldsymbol{v}_h$ is decomposed as the friction free wind velocity $\boldsymbol{v}_g$ and the friction induced wind velocity $\boldsymbol{v}_f$:

$$\boldsymbol{v}_h = \boldsymbol{v}_g + \boldsymbol{v}_f \tag{13.12}$$

Correspondingly, the horizontal motion equation (Equation 13.9a) is divided into the following two equations:

$$\frac{\partial \boldsymbol{v}_g}{\partial t} + \boldsymbol{v}_g \cdot \nabla_h \boldsymbol{v}_g = -\frac{1}{\rho}\nabla_h p - f\boldsymbol{k}_h \times \boldsymbol{v}_g \tag{13.13a}$$

$$\frac{\partial \boldsymbol{v}_f}{\partial t} + \boldsymbol{v}_f \cdot \nabla_h \boldsymbol{v}_f + \boldsymbol{v}_f \cdot \nabla_h \boldsymbol{v}_g + \boldsymbol{v}_g \cdot \nabla_h \boldsymbol{v}_f = -f\boldsymbol{k}_h \times \boldsymbol{v}_f + \boldsymbol{F}_{hf} \tag{13.13b}$$

In the free atmosphere, the friction free wind velocity $\boldsymbol{v}_g$ in Equation 13.13a is equal to the translation velocity of the typhoon $\boldsymbol{c}$. The first term in Equation 13.13b is considered smaller than the viscosity term and the inertia term in typhoon boundary layer, and thus it can be disregarded. As a result, Equation 13.13 becomes:

$$(\boldsymbol{v}_g - \boldsymbol{c}) \cdot \nabla_h \boldsymbol{v}_g = -\frac{1}{\rho}\nabla_h p - f\boldsymbol{k} \times \boldsymbol{v}_g \tag{13.14a}$$

$$\boldsymbol{v}_f \cdot \nabla_h \boldsymbol{v}_f + \boldsymbol{v}_f \cdot \nabla_h \boldsymbol{v}_g + \boldsymbol{v}_g \cdot \nabla_h \boldsymbol{v}_f = -f\boldsymbol{k} \times \boldsymbol{v}_f + \boldsymbol{F}_{hf} \tag{13.14b}$$

13.3.2 Friction-Free Wind Velocity

For Equation 13.14a, by using a moving cylindrical coordinate system whose origin coincides with the typhoon center and by considering that the radial speed v_{rg} is smaller than the tangential speed $v_{\theta g}$ in the free atmosphere, the tangential wind speed $v_{\theta g}$ is found approximately as:

$$v_{\theta g} = \frac{1}{2}(c \sin\theta_r - fr) + \left[\left(\frac{c \sin\theta_r - fr}{2}\right)^2 + \frac{r}{\rho}\frac{\partial p}{\partial r}\right]^{1/2} \tag{13.15}$$

where θ_r is the angle between the typhoon translation direction and the vector from the center of pressure field to the site of interest.

The radial wind speed v_{rg} is found as:

$$v_{rg} = -\frac{1}{r}\int_0^r \frac{\partial v_{\theta g}}{\partial r} dr \tag{13.16}$$

Different from the work of Meng *et al.* [8], the tangential wind speed $v_{\theta g}$ varies with the height from the ground because the variation of central pressure difference with the height (Equation 13.9b) is considered.

13.3.3 Friction-Induced Wind Velocity

For Equation 13.14b, two boundary conditions – one at the upper atmosphere and one above the ground surface – must be considered. The perturbation analysis is performed to obtain the distribution of friction-induced wind velocity $\boldsymbol{v}_f$. The friction velocity components $v_{\theta f}$ and v_{rf} are assumed to be small compared with the friction-free wind velocity, and the first derivatives of the friction velocity components are also presumably small compared with the friction-free velocity components. Equation 13.14b can then be linearized as:

$$-\left(2\frac{v_{\theta g}}{r} + f\right) v_{\theta f} = k_m \frac{\partial^2 v_{rf}}{\partial z^2} \tag{13.17a}$$

$$\left(\frac{\partial v_{\theta g}}{\partial r} + \frac{v_{\theta g}}{r} + f\right) v_{rf} = k_m \frac{\partial^2 v_{\theta f}}{\partial z^2} \tag{13.17b}$$

where k_m is the vertical coefficient of eddy viscosity.

Note that the tangential speed $v_{\theta g}$ in the free atmosphere can be determined by Equation 13.15, and the following two abbreviations can be introduced into Equation 13.17:

$$\xi = \left(\frac{\partial v_{\theta g}}{\partial r} + \frac{v_{\theta g}}{r} + f\right)^{1/2} \Big/ \left(2\frac{v_{\theta g}}{r} + f\right)^{1/2} \tag{13.18a}$$

$$\lambda = \left(\frac{\partial v_{\theta g}}{\partial r} + \frac{v_{\theta g}}{r} + f\right)^{1/4} \left(2\frac{v_{\theta g}}{r} + f\right)^{1/4} / (2k_m)^{1/2} \tag{13.18b}$$

After some manipulations, and by using two boundary conditions, the friction-induced wind velocity components can be obtained as:

$$v_{\theta f} = e^{-\lambda z^*}[D_1 \cos(\lambda z^*) + D_2 \sin(\lambda z^*)] \tag{13.19a}$$

$$v_{rf} = -\xi e^{-\lambda z^*}[D_2 \cos(\lambda z^*) - D_1 \sin(\lambda z^*)] \tag{13.19b}$$

where:

$$D_1 = -\frac{\chi(\chi+1)v_{\theta g} - \chi v_{rg}/\xi}{1+(\chi+1)^2} \tag{13.20a}$$

$$D_2 = \frac{\chi v_{\theta g} + \chi(\chi+1)v_{rg}/\xi}{1+(\chi+1)^2} \tag{13.20b}$$

$$\chi = \frac{C_d}{k_m\lambda}|\boldsymbol{v}_s| = \frac{C_d}{k_m\lambda}\sqrt{v_{\theta s}^2 + v_{rs}^2} \tag{13.20c}$$

where:

C_d is the drag coefficient on the ground surface, which can be estimated according to the formula proposed by Meng *et al.* [8];

$\boldsymbol{v}_s$ is the horizontal wind velocity on the ground surface.

The value of k_m is taken as 100 m^2/sec, according to the data obtained from hurricane observation [12]. Since $\boldsymbol{v}_s$ is included in the formula of χ, the values of $v_{\theta f}$ and v_{rf} must be found from iterative computation. The friction-free wind speeds of $v_{\theta g}$ and v_{rg} on the ground surface can be used for the initial values of $v_{\theta s}$ and v_{rs}.

13.3.4 Procedure of Typhoon Wind Field Simulation

To simulate typhoon wind field over sea, substitute Equation 13.10 into Equation 13.9b and calculate the pressure gradient $\nabla_h p$ in Equation 13.9a. The pressure gradient, in conjunction with typhoon key parameters, including typhoon position, translation direction and translation speed, are substituted into Equations 13.15 and 13.16 to find the friction-free wind velocity $\boldsymbol{v}_g$. Then, by taking $\boldsymbol{v}_g$ as the initial value of the friction-induced wind velocity $\boldsymbol{v}_f$, $\boldsymbol{v}_f$ can be calculated from Equation 13.19 through iterative computation. Finally, based on the above results, the typhoon-induced wind velocity $\boldsymbol{v}_h$ at any height can be determined through the vector summation of $\boldsymbol{v}_g$ and $\boldsymbol{v}_f$ by using the formula shown in Equation 13.12. The same procedure can be applied to simulate typhoon wind field over land, but $\nabla_h p$ is calculated by substituting Equation 13.11 into Equation 13.10 and then into Equation 13.9b.

13.4 Model Validation

The refined typhoon wind field model needs to be verified through comparisons with full-scale typhoon data as many times as possible. Typhoon York is used for validation in this chapter as an example [17].

13.4.1 Typhoon York

Typhoon York was the strongest typhoon passing by Hong Kong since 1983. The tropical depression York developed at about 430 km northeast of Manila on 12 September 1999 and intensified into a tropical storm on the next day over the South China Sea [21]. The movement of the storm was erratic, heading north at first on 14 September, then turning to the northwest. After having moved northwesterly for almost two days, it passed by Hong Kong on the morning of 16 September. The eye of Typhoon York was closest to the Hong Kong Observatory (HKO) Headquarters around 10 am on 16 September, when it was about 20 km to the southwest. The typhoon signal No.10 was forced to hoist for 11 hours – the longest on record in Hong Kong. Typhoon York finally made landfall near Zhuhai, a city in the

Guangdong Province of China, and then it weakened gradually further inland on 17 September (see Figures 13.1a and 13.1b).

During its passage over Hong Kong and Guangdong Province, Typhoon York caused widespread flooding, severe traffic disruption and serious damage to buildings, signboards, trees and others. HKO

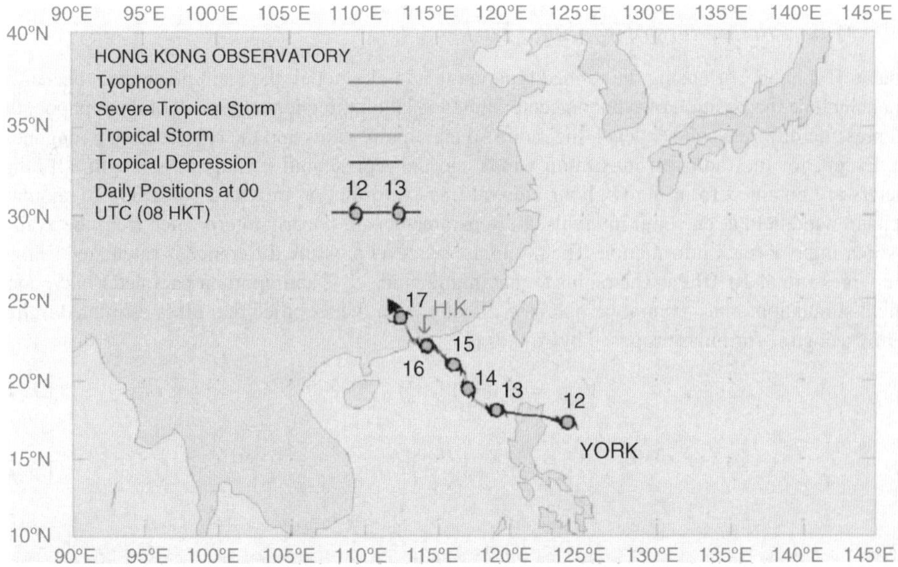

(a) Track of typhoon York: 12-17 September 1999

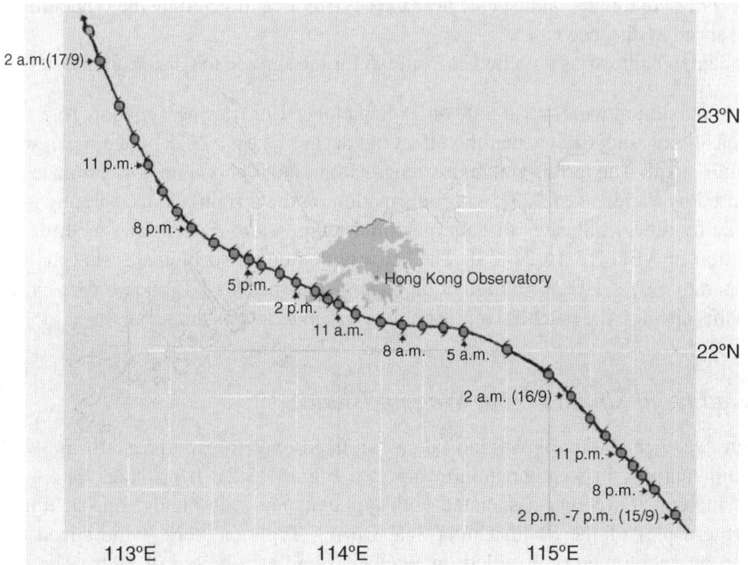

(b) Track of typhoon York over Hong Kong on 16 September 1999

Figure 13.1 Track of Typhoon York (from [17]) (Reproduced with permission from Multi-Science Publishing Co Ltd).

recorded a maximum hourly wind speed of 42 m/sec and a maximum gust of 65 m/sec at a height of 90 m on Waglan Island during the passage of the typhoon. This gust was the highest recorded at Waglan Island since the station was established. The lowest instantaneous mean sea-level pressure was recorded as 970.7 hPa.

13.4.2 Main Parameters of Typhoon York

To simulate Typhoon York using the refined typhoon wind field model, the main parameters needed as input data include the position of typhoon center, the translation direction and speed of the typhoon, the central pressure difference at sea level, the radius to maximum winds and the representative roughness length. Except for the radius to maximum winds and the representative roughness length, all other parameters are obtained from the six-hour interval track information provided by HKO. To simulate the typhoon wind field at one hour intervals, these parameters are linearly interpolated from the available six-hour interval track information. The minimum sea level pressure difference is calculated using a periphery pressure of 1010 hPa. The radius to maximum winds r_m is an important parameter in typhoon wind field simulation, but it cannot be obtained directly from the recorded data. It is estimated in this chapter by using the formulas proposed by Anthes [22].

$$V_{\max} = 6.3(1013 - p_{c0})^{0.5} \tag{13.21}$$

$$V_r = V_{\max}\left(\frac{r_m}{r}\right)^{\alpha_1} \tag{13.22}$$

where:

$V_{\max}$ is the maximum wind speed in m/sec;
$r_m < r < r_0$, and r_0 is a large radial distance near the edge of area affected by the typhoon;
V_r is the wind speed at distance r;
α_1 is the exponent, which varies with height from 0.5 for the surface to 0.7 for higher elevations.

The radius to maximum winds r_m at 6:00 on 15 September 1999 during Typhoon York is estimated as 58 km. The sensitivity study shows that the effect of varying r_m by $\pm 20\%$ on the mean wind speeds of Typhoon York is small. The terrain roughness length along the typhoon track is estimated by using the method proposed by Pande *et al.* [23], with information on the terrain and topography of Hong Kong. The work done by others indicates that the potential temperature θ changes very little in radial and vertical direction in ABL [15,16,24–26]. The sensitivity study also indicates that the change in the potential temperature by ± 10 k would affect the mean wind speeds of Typhoon York by less than 5%. Therefore, in this chapter, the potential temperature is taken as a constant value.

13.4.3 Wind Field Simulation at Waglan Island

The automatic weather station at Waglan Island, situated over the southeastern tip of Hong Kong, is HKO's major outpost in weather monitoring (see Figure 13.2). It provides data critical for the early alert of inclement weather associated with typhoons and rain bands approaching Hong Kong from the northern part of the South China Sea. During Typhoon York, wind speed and direction were recorded by anemometers installed on Waglan Island at a height of 90 m. The data recorded at Waglan Island are of interest because the wind flows over an open sea for a long fetch and, if corrected for the topographic error, the data can provide relatively accurate information on the characteristics of typhoons over sea.

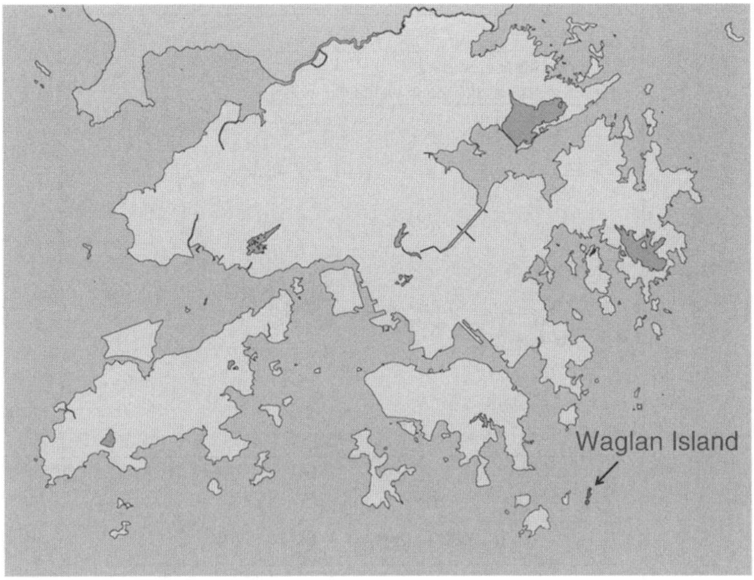

Figure 13.2 Location of Waglan Island in Hong Kong (from [38]) (Reproduced with permission from The Hong Kong Institute of Steel Construction).

The measured wind speed and direction are compared with those predicted by the refined typhoon wind field model, and the comparison results are plotted in Figure 13.3. The horizontal coordinate is time in hours, and the concerned 24-hour duration of wind data starts from 6:00 HKT (Hong Kong Time), 15 September and ends at 6:00 HKT, 16 September. It can be seen that the refined typhoon wind field model can predict the wind speed satisfactorily within the entire duration of 24 hours. The refined typhoon wind field model underestimates the maximum wind speed by about 7% compared with the measured one. The tendency of wind direction predicted by the refined typhoon wind field model agrees well with the measured one. The difference between the predicted wind direction and the measured wind direction is within 30° in the first 22 hours and becomes relatively large in the following three hours.

13.4.4 Spatial Distribution of Typhoon Wind Field

Further validation of the refined typhoon wind field model is done by comparing the spatial distribution of simulated typhoon wind field at upper atmosphere with the visible imagery of the typhoon wind field [27]. The visible imagery of the wind field of Typhoon York at 11:00 am on 16 September is shown in Figure 13.4 [21]. The spatial distribution of simulated typhoon wind field can be obtained through an isotach analysis.

Figure 13.5a shows the isotach distribution of hourly mean wind speed of Typhoon York at a height of 400 m at 11:00 am on 16 September. The mean wind speeds are expressed in a coordinate system fixed on the earth. The relevant typhoon parameters include the central pressure difference Δp of 32 hPa, the radius to maximum winds r_m of 58 km, the translation speed c of 10 m/sec, and the translation direction β of 158°. The visible imagery of the wind field indicates the maximum wind speed located approximately in the right front quadrant with reference to the typhoon center (see Figure 13.4). The simulated wind field produces the maximum wind speed greater than 27 m/sec, which is also roughly situated in the right front quadrant (see Figure 13.5a).

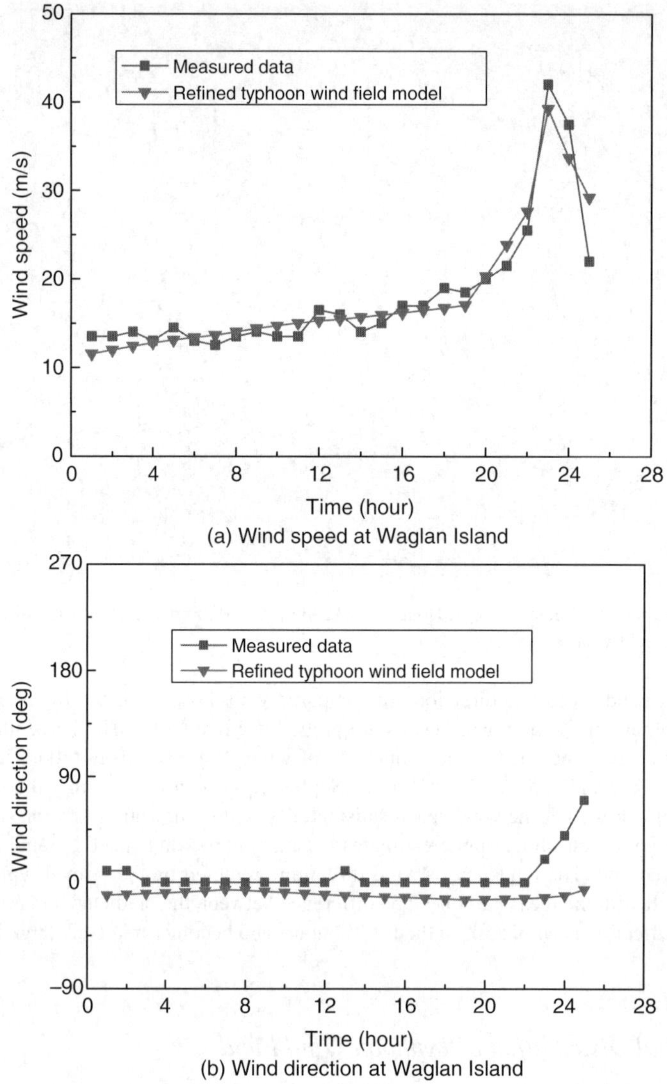

Figure 13.3 Comparison of predicted and measured wind velocity at Waglan Island.

An outer secondary maximum wind speed of 20 m/sec occurs at about 45 km west-southwest from the typhoon center. The position of the secondary maximum wind speed is associated with a convection band. This feature can also be identified from the visible imagery of Typhoon York. Furthermore, Figure 13.5a shows that the winds are stronger to the right of the direction of movement than to the left, relative to the earth. The difference in the maximum wind velocity between the right and left hand side of the typhoon movement direction is almost equivalent to the magnitude of the typhoon translation velocity.

Figure 13.4 Visible imagery of Typhoon York at around 11 a.m. on 16 September 1999 (from [17]) (Reproduced with permission from Multi-Science Publishing Co Ltd).

The isotach distribution of one-hour mean wind speed of Typhoon York is also analyzed at a height of 10 m, representing the near ground surface wind, and the results are plotted in Figure 13.5b. It can be seen that the isotach becomes different from that at 400 m height. This is attributed to the heterogeneity of surface roughness and topography of the concerned location. Furthermore, the ratio of the maximum wind speed at 10 m to that at 400 m is about 0.6, which is similar to the value of 0.62 suggested by Vickery and Twisdale [7]. This ratio is also within the 0.44–0.78 range, as evaluated by Powell [27].

13.4.5 Wind Speed Profiles in Vertical Direction

One of the major factors affecting the magnitude of wind loading on bridges is the mean wind speed profile in the vertical direction, as discussed in Chapter 2. The mean wind speed profiles of Typhoon York are generated by the refined model and compared with field data and power law profiles. The typhoon parameters used to generate typhoon wind profiles have been given in Section 13.4.4 for spatial distribution of the typhoon wind field. Figure 13.6 shows the pressure and central pressure difference variation with height at 135° counterclockwise from the East of typhoon York. It can be seen that central pressure, ambient pressure and central pressure difference all decrease with height at different rates. The central pressure difference decreases from 31.6 hPa near the ground surface to 21.6 hPa at about 800 m height, which affects wind speed profiles as discussed below.

The development of the global positioning system (GPS)-based dropsondes [28] has made it possible to obtain wind speed profiles within nearly all positions of a typhoon. By using the profiles of lowest 3000 m obtained by GPS dropsondes at Atlantic, Eastern and Central Pacific during years of 1997–1999, Powell *et al.* [29] and Franklin *et al.* [30] obtained averaged eye-wall wind speed profiles and outer-vortex wind speed profiles. To compare with these profiles, the wind speed profiles at a distance of 60 km, 100 km, 150 km from the typhoon center and at a 135° counter-clockwise from the East are predicted by the refined typhoon wind field model. In addition, the power law mean wind speed profiles for normal wind conditions are also computed.

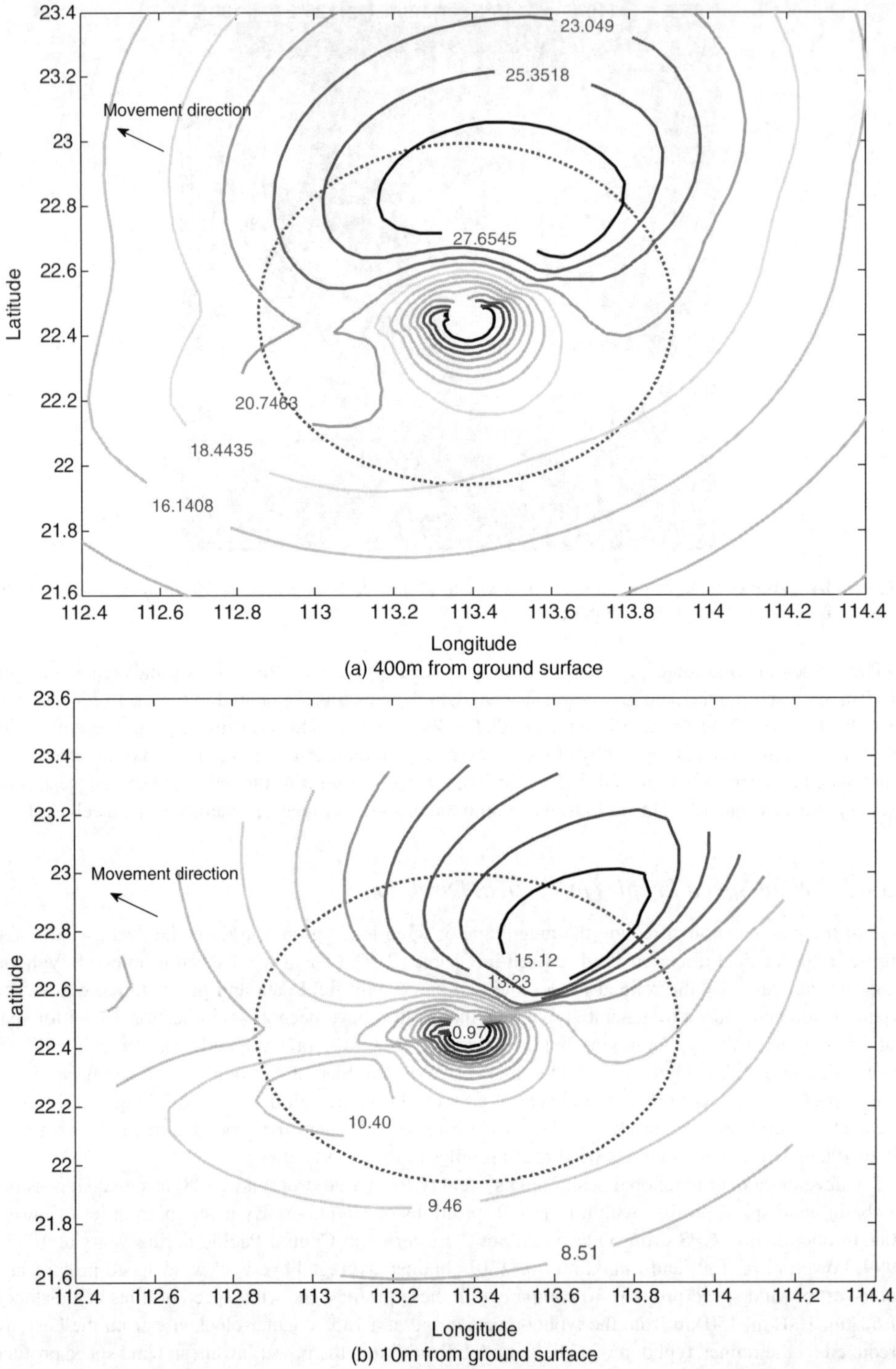

Figure 13.5 Isotaches of wind velocity from the refined typhoon wind field model (from [17]) (Reproduced with permission from Multi-Science Publishing Co Ltd).

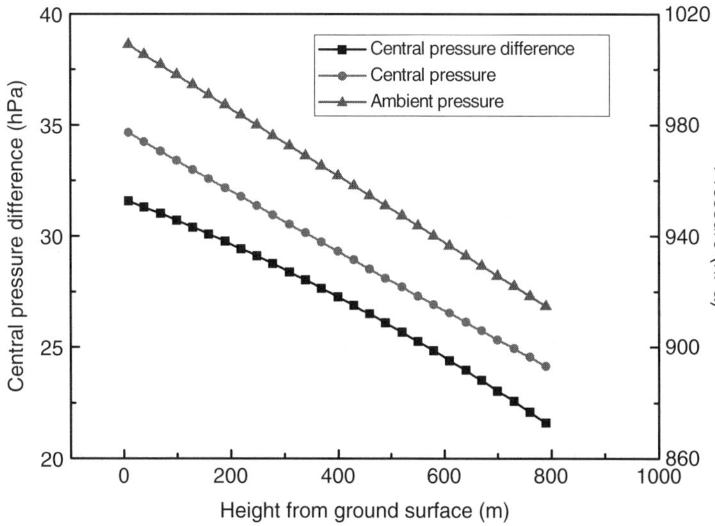

Figure 13.6 Variation of pressure and central pressure difference with height (from [17]) (Reproduced with permission from Multi-Science Publishing Co Ltd).

Figure 13.7 shows the mean wind speed profiles predicted by the refined model for the eye-wall region (at a distance of 60 km from the typhoon center) as well as those obtained by Powell *et al.* [29] and Franklin *et al.* [30]. The two power law wind speed profiles with an exponent of 0.06 and 0.10 are also plotted in Figure 13.7. It can be seen that the wind speed profile predicted by the refined model is close to those given by Powell *et al.* [29] and Franklin *et al.* [30]. It is also interesting to note that the wind speed profiles predicted by the refined model and those given by Powell *et al.* [29] and Franklin

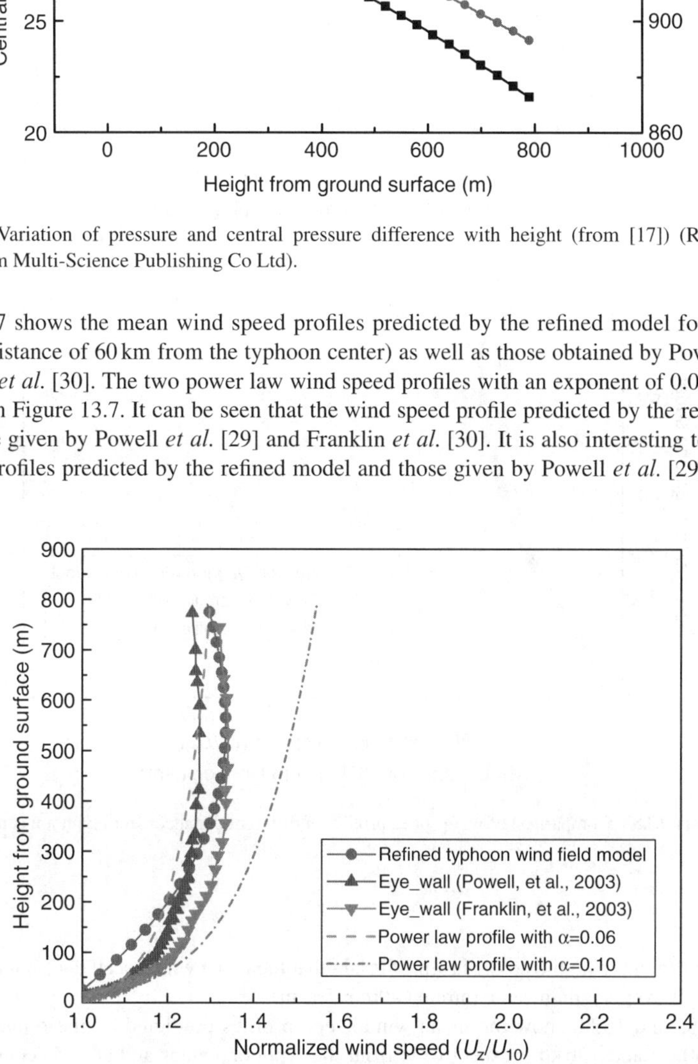

Figure 13.7 Comparison of wind speed profiles in eye-wall region (from [17]) (Reproduced with permission from Multi-Science Publishing Co Ltd).

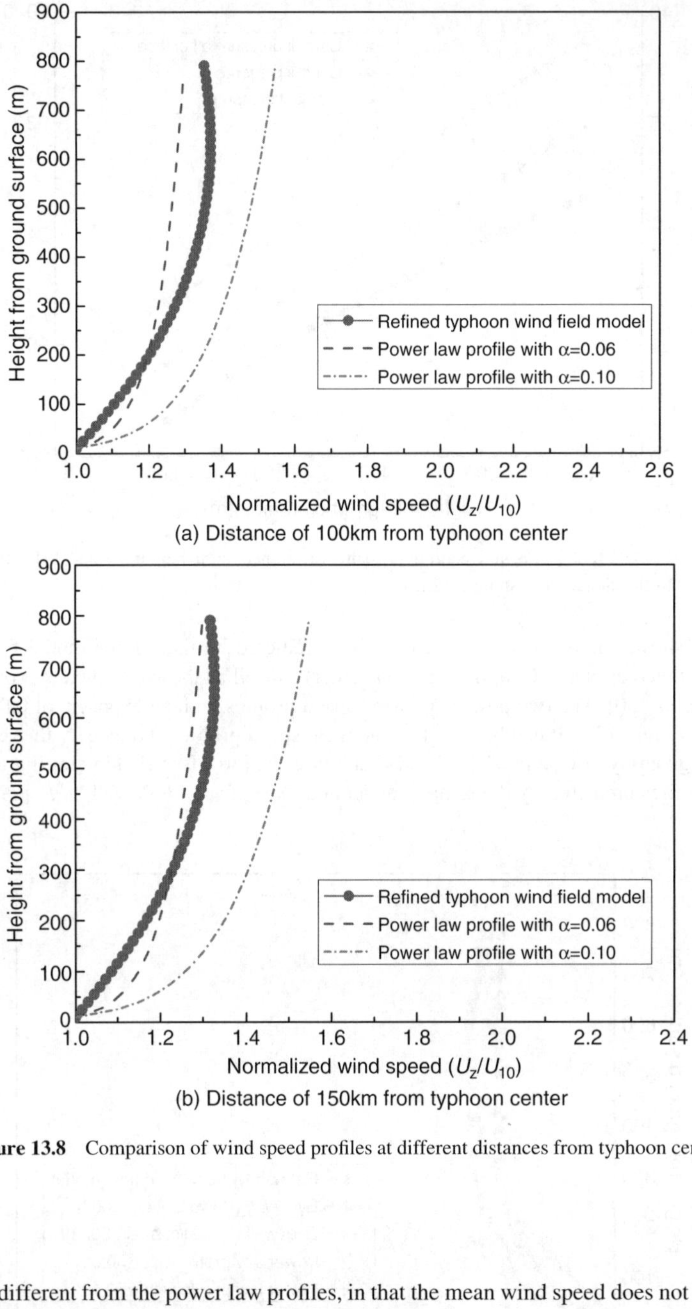

Figure 13.8 Comparison of wind speed profiles at different distances from typhoon center.

et al. [30] are different from the power law profiles, in that the mean wind speed does not increase when reaching a height over a certain value (around 400 or 500 m).

Figures 13.8a and 13.8b show the mean wind speed profiles predicted by the refined model at a distance of 100 km and 150 km, respectively, from the typhoon center and at 135° counterclockwise from the east. The two power law profiles, of an exponent of 0.06 and 0.10, are also plotted in Figure 13.8. Also, the wind speed profiles predicted by the refined model at different distances in the

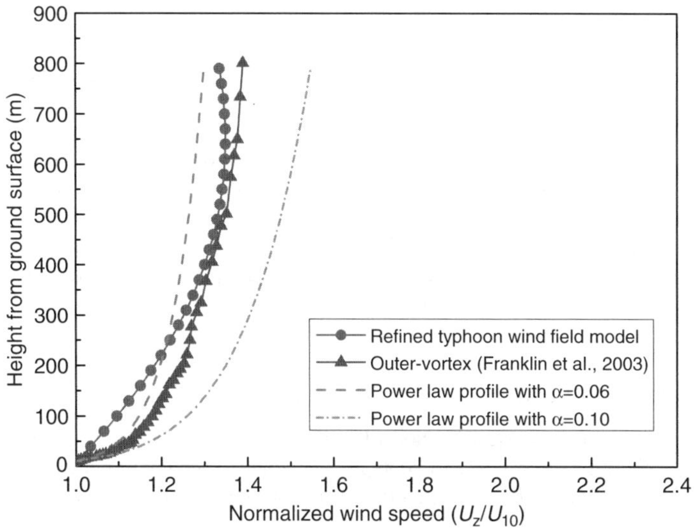

Figure 13.9 Comparison of wind speed profiles in outer-vortex region.

outer-vortex region, are averaged and compared with that given by Franklin *et al.* [30], and the results are shown in Figure 13.9. Once again, the averaged mean wind profile predicted by the refined model for the outer-vortex region is close to the measured one. It can be seen that the refined typhoon wind field model can give a reasonable agreement with the observed mean wind profile, both in the eye-wall region and the outer-vortex region.

13.5 Monte Carlo Simulation

13.5.1 Background

The design wind speed for a given return period can be determined by using the observation data of annual maximum wind speeds recorded at the concerned bridge site. However, the observation data are often not enough for a reliable estimation of long return period design wind speed. For example, the annual maximum wind speeds of at least several thousand years are required to estimate the design wind speed with a return period of 50 years.

Typhoon simulation is an alternative approach to evaluate extreme wind speeds over long periods of time by the so-called Monte Carlo simulation. The use of this method to determine design typhoon wind speeds for bridges and structures was probably first suggested by Russell [31,32]. Within the simulation, probability distributions of typhoon parameters, such as annual occurrence rate, paths, central pressure difference, translation velocity, approach angle, minimum of closest distance and radius to maximum winds, are first determined from the observation data. The typhoon wind field model, with the typhoon decay model after landfall, are then used to generate a series of typhoons by using the Monte Carlo simulation. The design wind speed for a given return period and location is finally decided through extreme wind analysis.

The method suggested by Russell was then expanded and improved by many researchers [1–10,20,33–37]. The study by Batts, *et al.* [4] is a milestone, being the first study to examine the entire United States coastline, and it provides a rational means to estimate typhoon wind speeds along the Gulf and Atlantic coasts of the United States. Vickery and Twisdale [20] developed an improved prediction methodology for typhoon wind speeds with an emphasis placed on the importance of typhoon wind field model and decay model. Vickery *et al.* [10] further developed the simulation approach, where the typhoon full track and the central pressure varying with the sea surface temperature were modeled.

A series of typhoons is generated in this chapter using the Monte Carlo simulation method in conjunction with the refined typhoon wind field model [38]. The statistical distributions of six typhoon key parameters, i.e. the central pressure difference Δp, the translation velocity c, the approach angle θ, the minimum of closest distance $d_{\min}$, the radius to maximum winds r_{m} and the annual occurrence rate λ, are established on the basis of typhoon wind field data recorded in the area concerned, which in this chapter is around Waglan Island in Hong Kong, due to its location and the quality and quantity of wind data available [38].

13.5.2 Typhoon Wind Data

Typhoon wind field data for the Western North Pacific region are recorded every six hours by HKO. The typhoon wind field data from 1947 to 2006 are available to the author, but only typhoons of their tracks within a radius of 250 km from Waglan Island are considered in this chapter.

13.5.3 Probability Distributions of Key Parameters

The central pressure difference Δp_0 is the difference between the central pressure and the ambient pressure at zero plane, and it plays an important role in estimating design typhoon wind speeds. Both lognormal distribution and three-parameter Weibull distribution are used to fit the observed data. The observed and fitted probability distribution function (PDF) and cumulative distribution function (CDF) of Δp_0 are shown in Figure 13.10, and the results of statistical parameter estimation are listed in Table 13.1, in which $\mu_{\ln}$ and $\sigma_{\ln}$ are the mean value and standard deviation of the lognormal distribution, and γ, β, η are the location parameter, shape parameter and scale parameter, respectively, of the Weibull distribution.

The translation speed c of a typhoon is determined by using the distance of the typhoon center between the two recorded points divided by 6 hours from the typhoon wind field database. The results show that the translation velocity ranged from 2 km/h to 65 km/h. Both normal distribution and lognormal distribution are used to fit the observed data. The observed and fitted PDF and CDF of the translation velocity c are shown in Figure 13.11, and the results of statistical parameter estimation are listed in Table 13.1, in which μ and σ are the mean value and standard deviation, respectively, of the normal distribution.

The approach angle θ of a typhoon indicates the typhoon moving direction and it is also estimated based on the adjacent positions of the typhoon center in a six-hour interval. It is expressed to the nearest ten degrees with respect to the north clockwise. The observed data are fitted using both normal distribution and bi-normal distribution. The observed and fitted PDF and CDF of θ are shown in Figure 13.12, and the results of statistical parameter estimation are listed in Table 13.1, in which μ_1 and μ_2, σ_1 and σ_2 are the mean values and standard deviations of the two random variables in the bi-normal distribution, and a_1 is the weighting factor.

The minimum of closest distance $d_{\min}$ is the minimum perpendicular distance from the simulation point to the typhoon translation direction, and it can be estimated according to the positions of both the typhoon center and the location of interest, as well as the moving direction of the typhoon. It is defined as positive if the concerned location is located to the right of the typhoon direction, otherwise it is negative. The observed minimum of closest distance is fitted by both uniform distribution and trapezoidal distribution. The observed and fitted PDF and CDF of the minimum of closest distance $d_{\min}$ are shown in Figure 13.13, and the results of statistical

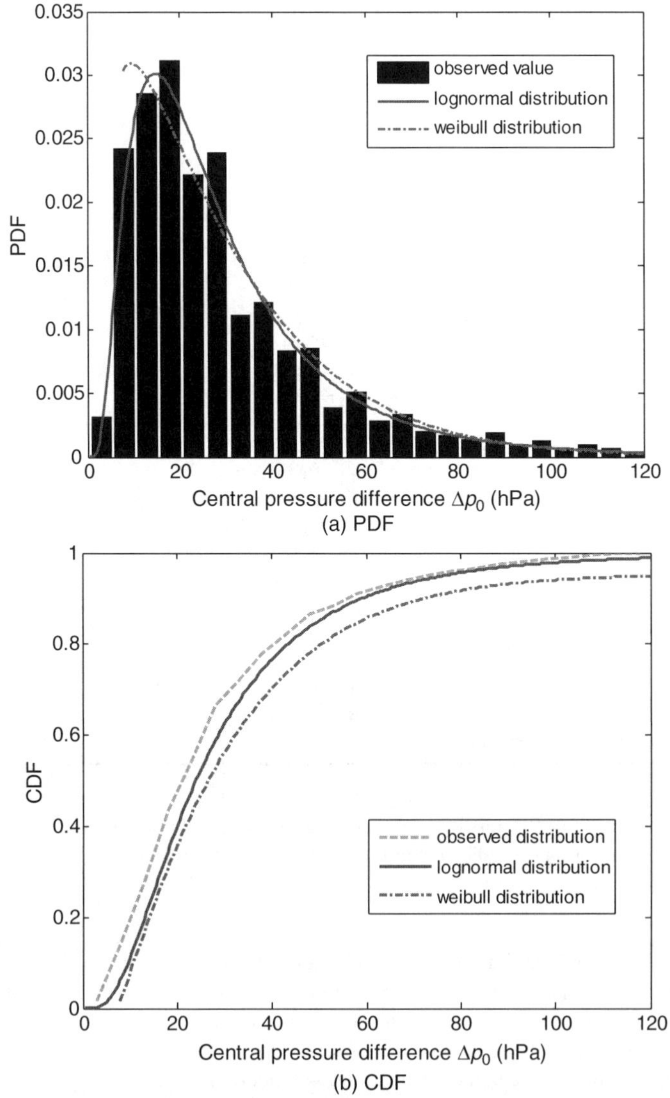

Figure 13.10 Observed and fitted distributions of central pressure difference Δp_0 (from [38]) (Reproduced with permission from The Hong Kong Institute of Steel Construction).

parameter estimation are listed in Table 13.1, in which ν is the value in the uniform distribution, and A and B are the minimum and maximum values, respectively, of trapezoidal probability density function.

The radius to maximum wind r_m describes the range of the most intensive typhoon wind speed, and it can be estimated according to the method proposed by Anthes [22]. The observed data is fitted by using the lognormal distribution only. The observed and fitted PDF and CDF of r_m are shown in Figure 13.14, and the results of statistical parameter estimation are listed in Table 13.1.

Table 13.1 Key parameters and probability distributions of typhoons (from [38]) (Reproduced with permission from The Hong Kong Institute of Steel Construction)

Parameter	Distribution	Parameters in distribution	D	$d_{F,n}$	*K-S* Test
Translation velocity c (m/s)	Normal	$\mu = 19.1$ $\sigma = 9.04$	0.065	0.019	Lognormal
	Lognormal	$\mu_{\ln} = 2.83$ $\sigma_{\ln} = 0.53$	0.027		
Approach angle β (degree)	Normal	$\mu = -46.4$ $\sigma = 58.9$	0.194	0.029	bi-normal
	bi-normal	$\mu_1 = -69.2$ $\sigma_1 = 26.4$ $a_1 = 0.828$ $\mu_2 = -69.2$ $\sigma_2 = 26.4$	0.027		
central pressure difference Δp_0 (hPa)	Lognormal	$\mu_{\ln} = -3.19$ $\sigma_{\ln} = 0.70$	0.035	0.026	Lognormal
	Weibull	$\gamma = 6.07$ $\beta = 1.12$ $\eta = 25.56$	0.195		
radius to maximum wind r_m (km)	Lognormal	$\mu_{\ln} = 4.19$ $\sigma_{\ln} = 1.01$	0.027	0.031	Lognormal
minimum of closest distance $d_{\min}$ (km)	Uniform	$v = 0.002$	0.046	0.027	Trapezoidal
	Trapezoidal	$A = 0.0017$ $B = 0.0023$	0.021		
annual occurrence rate λ	Poisson	$\lambda = 2.9$			

The annual occurrence rate λ refers to the occurrence times of typhoons per year and it is modeled by using the Poisson distribution only. The result is also shown in Table 13.1.

13.5.4 K-S *Test*

Since the *K-S* test can avoid measurement bias and can be effectively applied to five, but preferably ten or more observations, it is performed here to determine the distribution by providing the best fit to the historical typhoon wind field data, except for the annual occurrence rate λ.

The *K-S* test is based upon the maximum distance D between the probability cumulative distribution function $F(x)$ and the empirical cumulative distribution function $F_n(x)$.

$$D = \max|F(x) - F_n(x)| \tag{13.23}$$

The distribution function $F(x)$ is rejected if D is much greater than the specified value $d_{F,n}$. By choosing the cutting off upper-tail area of 0.05, the value of $d_{F,n}$ is defined as:

$$d_{F,n} = \frac{1.36}{\sqrt{n}} \tag{13.24}$$

where n is the number of observations.

Based on the *K-S* tests, the distributions of key parameters discussed in Section 13.5.3 are finally decided and shown in Table 13.1, together with their respective D values.

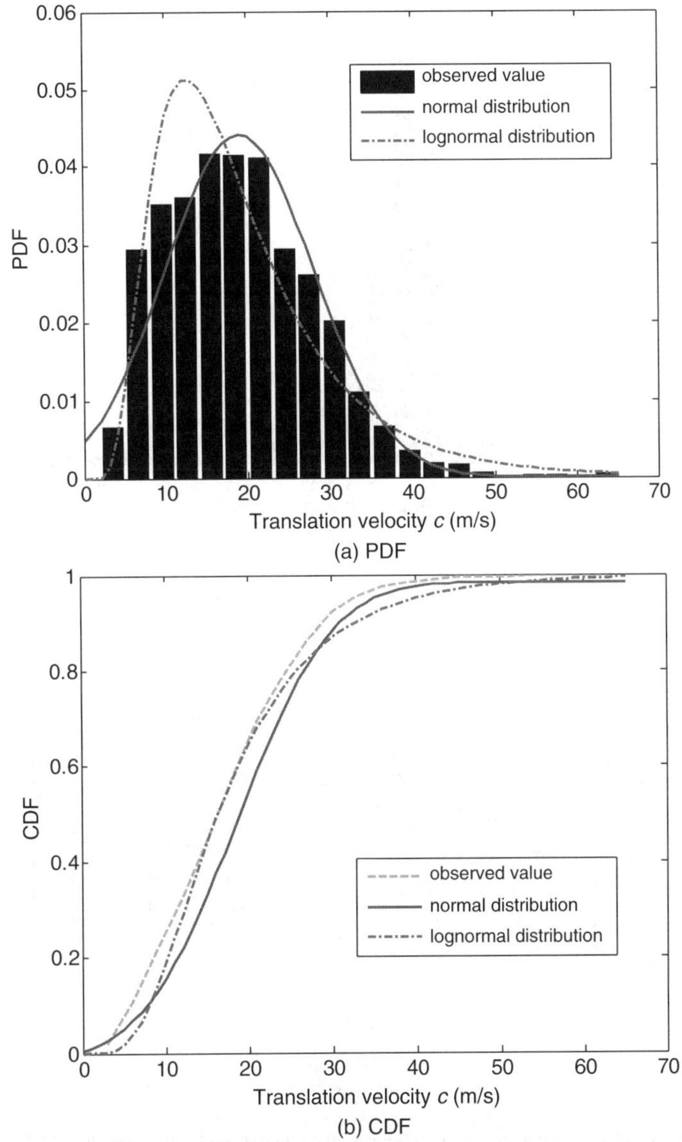

Figure 13.11 Observed and fitted distributions of translation velocity c (from [38]) (Reproduced with permission from The Hong Kong Institute of Steel Construction).

13.5.5 Typhoon Wind Decay Model Parameters

The parameters in the typhoon wind decay model can also be obtained through linear regression analysis of the selected typhoon wind field data. The results from the regression analysis indicate that the two constants a_0 and a_1 are 0.0237 and 0.00012, respectively. The mean value and standard deviation are 0.0 and 0.0193 for the random variable ε and -0.011 and 0.162 for the random variable b. Note that

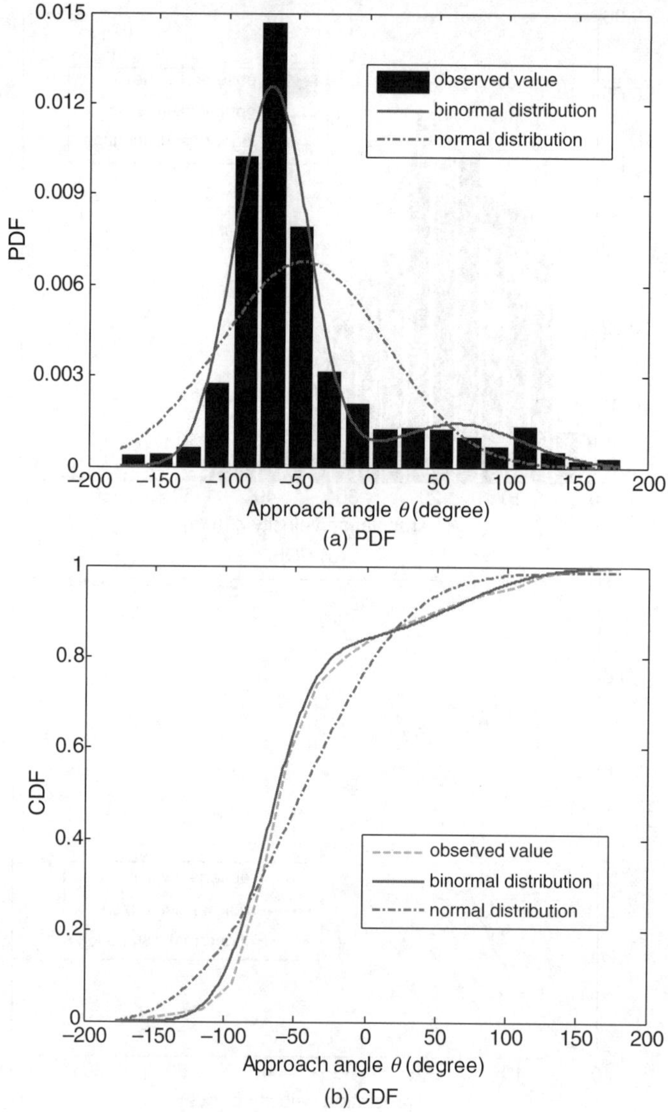

Figure 13.12 Observed and fitted distributions of approach angle θ (from [38]) (Reproduced with permission from The Hong Kong Institute of Steel Construction).

the correlations between the above typhoon parameters are not considered in this chapter. How to generate typhoons with the correlations between typhoon parameters deserves further investigation.

13.5.6 Procedure for Estimating Extreme Wind Speeds and Averaged Wind Profiles

The basic steps of using the Monte Carlo simulation method to estimate extreme wind speeds for the area concerned can be summarized as follows:

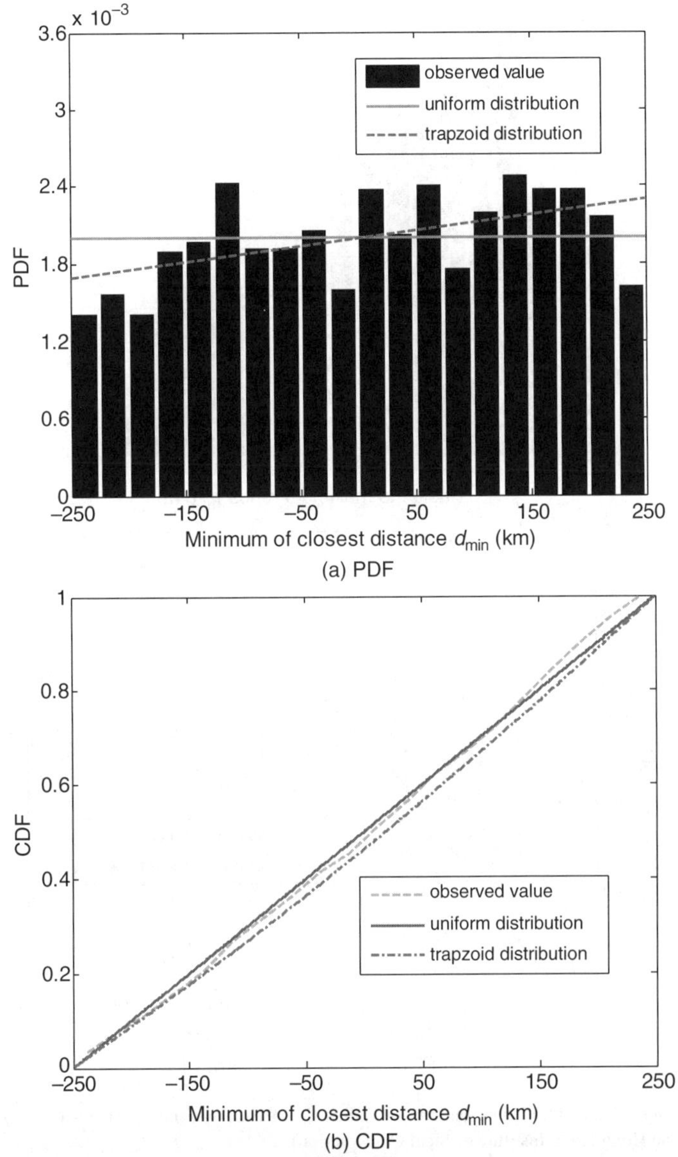

Figure 13.13 Observed and fitted distributions of minimum of closest distance $d_{\min}$ (from [38]) (Reproduced with permission from The Hong Kong Institute of Steel Construction).

(a) Generate a set of typhoon parameters based on the established probability distributions.

(b) Generate a typhoon wind field based on the generated typhoon parameters and using the refined typhoon wind field model, in which the central pressure difference $\Delta p_0(t)$ is held constant until landfall, and afterwards the central pressure difference is determined by the typhoon wind decay model.

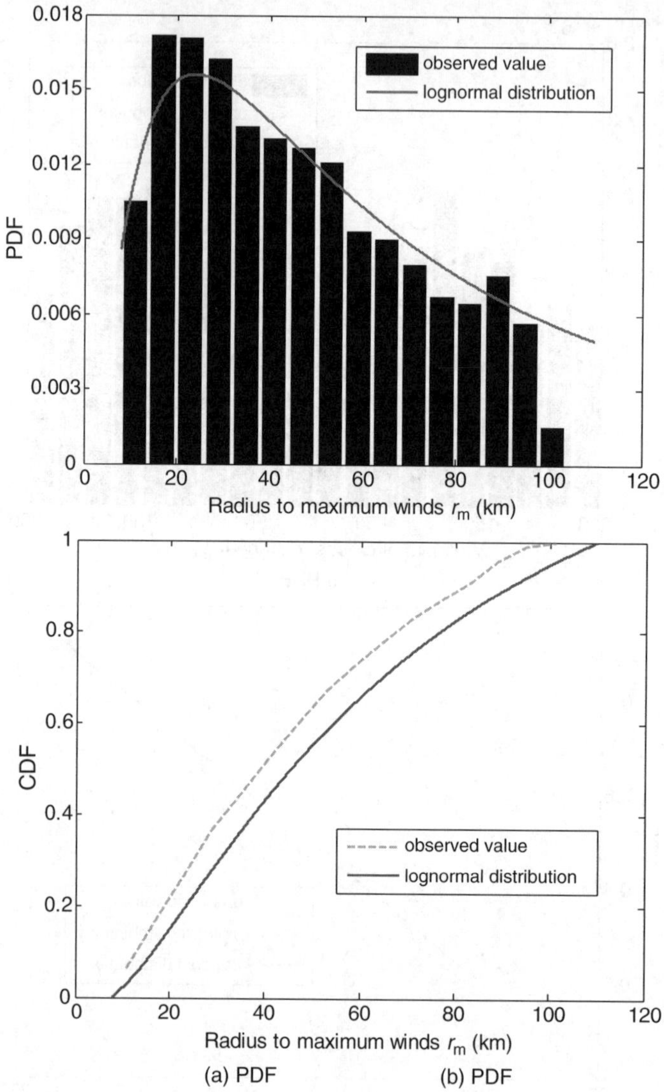

Figure 13.14 Observed and fitted distributions of radius to maximum wind r_m (from [38]) (Reproduced with permission from The Hong Kong Institute of Steel Construction).

(c) Compute the mean wind speed and direction at the concerned area due to the moving typhoon.
(d) Repeat the above steps for thousands of typhoons.
(e) Carry out the extreme wind analysis to determine design wind speeds in terms of a given return period.

In the simulation process in this chapter, the typhoon translation velocity c is held constant for each typhoon, while each simulated typhoon travels along a straight line path. To simulate the typhoon wind field at hourly intervals, all the parameters are interpolated linearly from the available six-hourly track information provided by HKO.

Steps a) to d) of the Monte Carlo method provide not only thousands of hourly mean wind speeds and directions, but also thousands of mean wind speed profiles at Waglan Island. Since typhoon mean wind speed profiles vary with the distance from the typhoon center, the obtained profiles need to be classified according to the distance from the typhoon center. In this chapter, only the two regions (i.e. eye-wall region and outer-vortex region) are considered. All of the mean wind speed profiles belonging to these regions are averaged to obtain the averaged typhoon wind profile (also called the design typhoon wind profile) for each region.

13.6 Extreme Wind Analysis

13.6.1 Basic Theory

The basic theory of extreme wind speed analysis based on the Monte Carlo simulation method can be found in Simiu and Scanlan [39], and is actually an extension of the extreme wind speed analysis introduced in Chapter 2. Denote F_v as the probability that the wind speed in any one typhoon simulated is less than some value v. The probability that the highest wind U in n typhoons simulated is less than v can be written as:

$$F(U < v|n) = (F_v)^n = F_v^n \tag{13.25}$$

According to the total probability theorem, the probability that $U < v$ in τ years is:

$$F(U < v, \tau) = \sum_{n=0}^{\infty} F(U < v|n)p(n, \tau) \tag{13.26}$$

where $p(n, \tau)$ is the probability that n typhoons will occur in τ years.

Assuming that $p(n, \tau)$ follows a Poisson distribution, Equation 13.26 becomes:

$$F(U < v, \tau) = \sum_{n=0}^{\infty} F_v^n \frac{(\lambda\tau)^n e^{-\lambda\tau}}{n!} = e^{-\lambda\tau} \sum_{n=0}^{\infty} \frac{(\lambda\tau F_v)^n}{n!} = e^{-\lambda\tau(1-F_v)} \tag{13.27}$$

where λ is the annual occurrence rate of typhoons in the area of interest for the site being considered.

When $\tau = 1$, $F(U < v, \tau)$ becomes the probability of occurrence of wind speeds less than v in any one year:

$$F(U < v) = \sum_{n=0}^{\infty} F_v^n \frac{(\lambda)^n e^{-\lambda}}{n!} = e^{-\lambda} \sum_{n=0}^{\infty} \frac{(\lambda F_v)^n}{n!} = e^{-\lambda(1-F_v)} \tag{13.28}$$

Suppose that m typhoons are simulated using the Monte Carlo simulation method and the maximum wind speed induced by each typhoon for the site can be arranged in an ascending order as $v_1, v_2, \ldots, v_m$. By considering the wind speed v_i, the probability that $U < v_i$ in any one typhoon can be determined by:

$$F_{v_i} = \frac{i}{m+1} \tag{13.29}$$

The probability of occurrence of wind speeds less than v_i in any one year is:

$$F(U < v_i) = e^{-\lambda\left(1-\frac{i}{m+1}\right)} \tag{13.30}$$

Extreme wind speeds in typhoons are usually fitted with one of the family of the Generalized Extreme Value (GEV) distributions. The Type I Extreme Value (Gumbel) distribution is used in the Hong Kong wind code. The Gumbel distribution function can be written in the form:

$$F(v) = \exp[-\exp(-\alpha(v-\mu))] \tag{13.31}$$

where:

v is the annual maximum wind speed;
α is the dispersion;
μ is the mode.

Using Equation 13.29, the dispersion and the mode in Equation 13.31 can be estimated.

The design wind speed for a given return period T_R (in year) and a given structure life N (in year) can be finally determined by:

$$[F(v)]^N = \left[1 - \frac{1}{T_R}\right]^N = 1 - R \tag{13.32}$$

where R is the associated risk of having a wind speed higher than v in N years.

13.6.2 Extreme Wind Speed Analysis using the Refined Typhoon Wind Field Model

The Waglan Island meteorological station in Hong Kong is chosen as a reference location for extreme wind speed analysis. The surrounding condition around the station can be regarded as open terrain, and the adjusted measurement height of the station is 90 m above sea level. By using the Monte Carlo simulation method, more than 9000 typhoons are generated and the hourly mean wind speed and direction of each typhoon at the reference height of 90 m of the station are calculated. The Gumbel distribution is employed for extreme wind speed value analysis. Both non-directional and directional design wind speeds of 50-year return period at the Waglan Island anemometer height are computed (Note: in Hong Kong, the return period of design wind speeds at present is 50 years for buildings and 120 years for long-span bridges). The results show that the non-directional design wind speed of a 50-year return period at the station is 48.2 m/sec. The directional design wind speeds of a 50-year return period in 36 direction sectors at Waglan Island, obtained from the refined typhoon wind field model, are shown in Figure 13.15. It can be seen that the asymmetrical directional distribution of design wind speeds is produced by using the refined typhoon wind field model.

13.6.3 Extreme Wind Speed Analysis based on Wind Measurement Data

Hourly mean wind speeds and directions at Waglan Island are recorded by HKO, and the wind measurement data during a period from 1970 to 1999 are available to the authors. The data are allocated according to wind direction at a 10° interval. The annual maximum wind speed in each of 36 direction sectors, and the annual maximum wind speed among all these directions, are obtained for each year, from which extreme wind speed analysis is performed. By using the Gumbel distribution, the non-directional design wind speed of a 50-year return period at Waglan Island is 46.9 m/sec, which is also provided by the Hong Kong wind code. The directional design wind speed of a 50-year return period in each direction at Waglan Island obtained from the wind measurement data is shown in Figure 13.15.

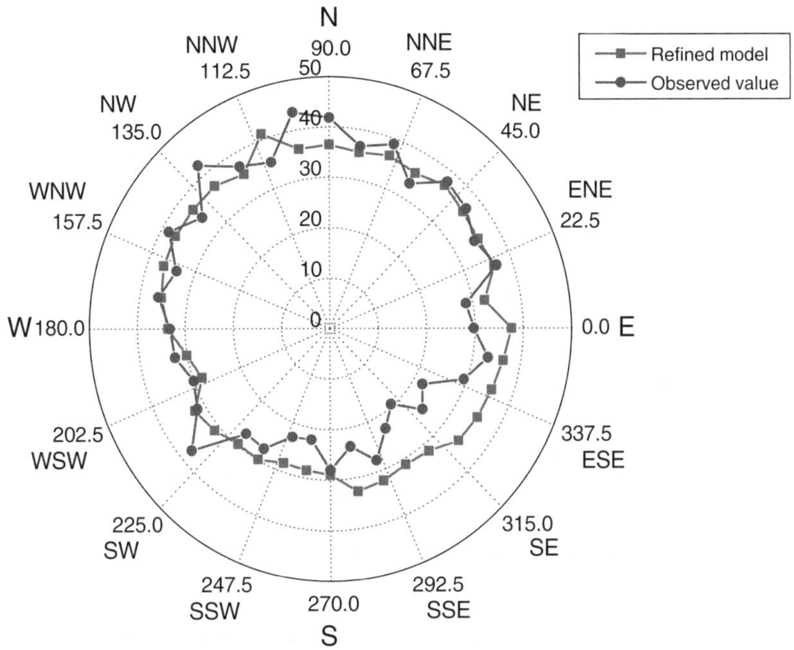

Figure 13.15 Directional design wind speeds of a 50 year return period at Waglan Island (from [38]) (Reproduced with permission from The Hong Kong Institute of Steel Construction).

13.6.4 Comparison of Results and Discussion

The design wind speed of a 50-year return period predicted by the refined typhoon wind field model for Waglan Island is 48.2 m/sec. This value is slightly higher than the design wind speed of 46.9 m/sec predicted based on the wind measurement data. The error between the refined typhoon wind field model and the wind measurement data may be because the number of typhoons used in the Monte Carlo simulation method is much larger than that actually recorded in Waglan Island during the concerned period.

In considering the effects of wind direction on design wind speeds, the largest design wind speeds of a 50-year return period predicted based on the wind measurement data, and by using the refined typhoon wind field model, are 43.6 m/sec and 41.1 m/sec and occur in the direction of 90° and 110° degree, respectively. The smallest are 19.5 m/sec and 28.2 m/sec and happen in the direction of 300° and 200° or 250°, respectively. The simulation errors between the refined typhoon wind field model and the wind measurement data are shown in Figure 13.16 for 36 directions. The mean error produced by the refined typhoon wind field model is 4.14 m/sec. The standard deviation is 3.37 m/sec.

13.6.5 Mean Wind Speed Profile Analysis

By using the Monte Carlo simulation method described in Section 13.5.6, hourly mean wind speed profiles from thousands of typhoons are generated by using the refined typhoon wind field model. Only the mean wind speed profiles with the mean wind speed over 5 m/sec at 10 m height and the distance from the typhoon center within the range from r_m to $5r_m$ are counted and analyzed. By averaging all the mean wind speed profiles within the two regions generated by the model, the averaged eye-wall and outer-vortex wind profiles are obtained and compared with the measured wind profiles.

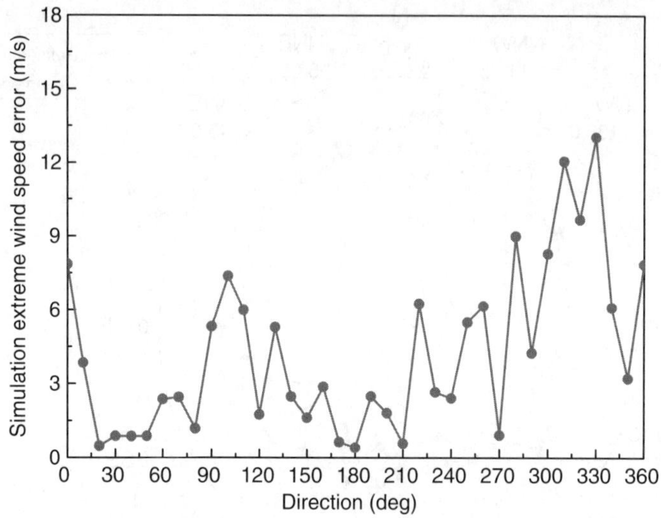

Figure 13.16 Simulation errors in the refined model compared with measurement data.

Figure 13.17 displays the averaged wind profiles predicted by the refined model for the eye-wall region (from r_m to $2r_m$) and those obtained by Powell *et al.* [29] and Franklin *et al.* [30] from the measured data. It can be seen that the averaged wind profile predicted by the refined typhoon wind field model is close to the measured ones. The relative difference of the averaged wind speed profile from the refined typhoon wind field model to that given by Powell *et al.* [29] is 6% at 700 m high.

Figure 13.18 displays the averaged wind profiles predicted by the refined model for the outer-vortex region (from $2r_m$ to $5r_m$) and that obtained by Franklin *et al.* [30] from the measured data. In lower positions (less than 300 m above the ground), the averaged wind speeds predicted by the model are

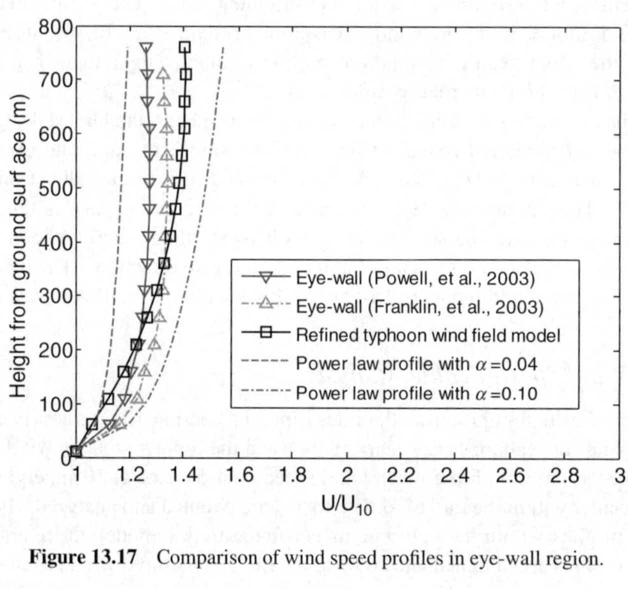

Figure 13.17 Comparison of wind speed profiles in eye-wall region.

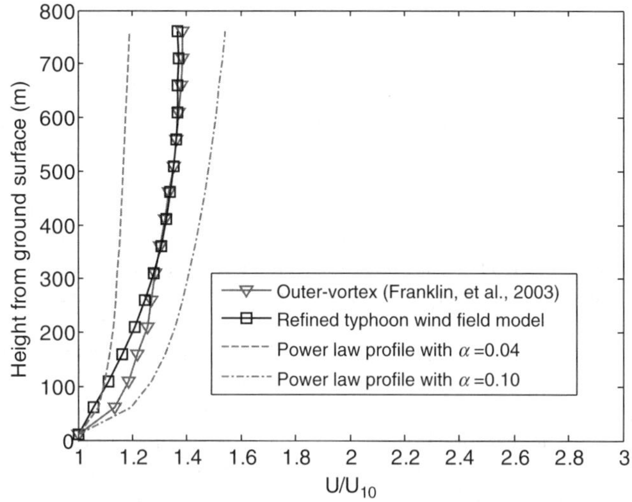

Figure 13.18 Comparison of wind speed profiles in outer-vortex region.

relatively smaller compared with the measured ones. In higher positions (greater than 300 m above the ground), the model gives a better estimation on wind speeds, compared with the measured mean wind profile. The relative difference of the averaged wind speed profile from the model to the measured profile is 1% at 700 m high.

13.7 Simulation of Typhoon Wind Field over Complex Terrain

13.7.1 Background

Since long-span cable-supported bridges are often not only located in a typhoon prone region but also surrounded by complex terrain, topographic effect on typhoon winds and interaction between winds and terrain surface roughness must be considered [40–42]. Therefore, a numerical simulation procedure for predicting directional design typhoon wind speeds and profiles for a bridge site surrounded by complex terrain is presented in this section, with a case study by applying this simulation approach [43].

In this procedure, directional typhoon wind speeds and profiles at the upstream of complex terrain are first generated by using the refined typhoon wind field model, together with the Monte Carlo simulation method, and by considering the interaction between typhoon wind and terrain surface roughness. The representative directional upstream typhoon wind speeds and profiles are then used as inputs and the computational fluid dynamics (CFD) simulation is used as a tool, together with the topographic model, to obtain representative directional typhoon wind speeds and profiles at the site.

Artificial neural networks (ANN) for predicting directional design typhoon wind speeds and profiles at the site are trained using the representative directional upstream typhoon wind speeds and profiles, as well as those at the site obtained from the CFD simulation. Based on the trained ANN, thousands of directional wind speeds and profiles for the site can be generated. The extreme wind speed analysis is performed to determine the directional design wind speeds, and statistical analysis is conducted to obtain the directional averaged mean wind profiles. Finally, the Stonecutters bridge site, surrounded by complex terrain in Hong Kong, is chosen as a case study to examine the feasibility of the proposed numerical simulation procedure.

13.7.2 Directional Upstream Typhoon Wind Speeds and Profiles

Consider a structure site surrounded by complex terrain and select an appropriate radius to draw a circle to cover the complex terrain, so that topographic effects on typhoon wind field at the site can be sufficiently considered. Then, to simulate directional upstream typhoon wind speeds and profiles in consideration of the interaction between typhoon winds and terrain surface roughness, the average and directional surface roughness length z_0 over the complex terrain is determined by using information from the Digital Elevation Model (DEM) provided by the CGIAR (Consultative Group on International Agricultural Research) consortium, together with local land use information. The probability distributions of key typhoon parameters are established based on typhoon data collected for the area concerned.

Based on the established probability distributions, the Monte Carlo simulation method is used to generate a set of key typhoon parameters for a typhoon at its initial position. This set of parameters, and the computed average surface roughness length z_0, are used as inputs to the typhoon wind field model to generate a typhoon wind field, for which the wind speed and direction can be identified for the site. If the directional surface roughness length identified is not the same as z_0, this set of key typhoon parameters and the directional surface roughness length are used to re-generate a typhoon wind field until the typhoon wind direction is consistent with the directional surface roughness length used. The wind speed, direction and profile obtained by this method are regarded as one of the directional upstream typhoon wind speeds and profiles for the site over the complex terrain.

The model typhoon is then moved forward at one-hour time intervals, and the above procedure is repeated to obtain another directional upstream wind speed and profile for the site at each time step, until the typhoon disappears. Repeating the above steps for thousands of typhoons, a complete assembly of directional upstream typhoon wind speeds and profiles can be obtained.

13.7.3 Representative Directional Typhoon Wind Speeds and Profiles at Site

It is very time-consuming to use the CFD simulation together with the topographic model of the site to directly simulate thousands of directional typhoon wind speeds and profiles at the site by using thousands of directional upstream wind speeds and profiles as inputs for the purpose of predicting directional design wind speeds and profiles at the site. To make the computation manageable, for this purpose, ANN is trained by using the representative directional upstream typhoon wind speeds and profiles, as well as those at the site obtained from the CFD simulation. The trained ANN is then used to generate thousands of directional wind speeds and profiles at the site from thousands of directional upstream wind speeds and profiles.

For the CFD simulation, the topography of the complex terrain surrounding the site is modeled in consideration of appropriate computational domains, meshes and boundary conditions. For a given wind direction, representative directional upstream typhoon wind speeds and profiles are selected from all of the generated upstream typhoon wind speeds and profiles described in Section 13.7.2 to cover a certain range of wind speeds and profiles.

Taking a representative upstream typhoon wind speed and profile as an inlet to the topographic model, the CFD simulation, with the Reynolds-averaged Navier-Stokes (RANS) method, is then performed [44]. Details of the RANS method used in the CFD simulation for steady, incompressible flow can be found in Chapter 8. The representative typhoon wind speed and profile at the site surrounded by the complex terrain is finally obtained in the designated wind direction. The above procedure continues until there are enough representative directional typhoon wind speeds and profiles at the upstream and the site.

13.7.4 Training ANN Model for Predicting Directional Typhoon Wind Speeds and Profiles

The representative upstream and site typhoon wind speeds and profiles described in Section 13.7.3 are used as input and output data to train an ANN model for predicting thousands of

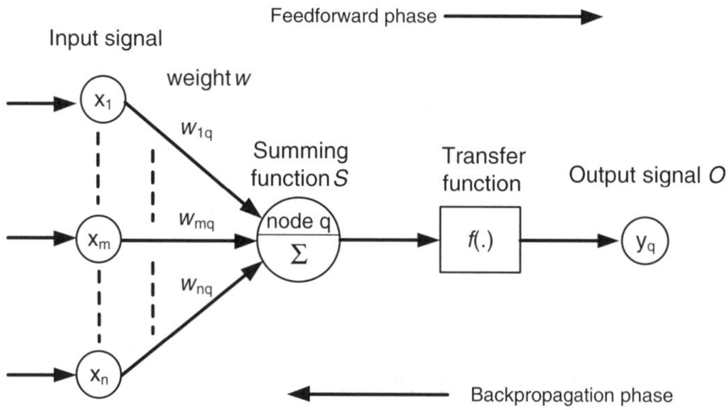

Figure 13.19 A simple ANN using BPN method.

directional typhoon wind speeds and profiles at the site over complex terrain. During the training, the ANN model learns how to map the input parameters from the representative directional upstream typhoon wind speeds and profiles into the output parameters from the representative directional typhoon wind speeds and profiles at the site, through an iterative process that involves optimization and some mathematical operations.

The back propagation network (BPN) method [45] is selected to train the ANN model. As a simple example, an elementary BPN neuron with n inputs and weighted connections is shown in Figure 13.19.

The training of this simple ANN involves two phases: feed forward phase and back propagation phase. In the feed forward phase, each input signal x_i $(i = 1, \ldots, n)$ is weighted with an approximate weight w, and the products are then summed up at the neuron of the network. This summation S of the weighted inputs at the neuron is modified by a transfer function f, thereby generating an output signal O. In practice, a continuous, non-linear logistic or sigmoid transfer function is commonly used, because it meets the differentiability requirement of the back propagation algorithm given by:

$$f(S) = \frac{1}{1 + e^{-aS}} \tag{13.33}$$

where:

a is the slope parameter of the sigmoid function;
$f()$ is the transfer function.

The output signal O is then compared with the designated output, from which an error signal is obtained and the back propagation phase is activated. The error signal is transmitted back to the neuron of the network. Based on the error signal received, the connection weights w for the network are updated. A new output signal O is then obtained by using the new weights, and a new error signal is calculated. The above procedure is repeated until the error signal is small enough to satisfy the pre-set criterion. The training is discontinued when all the training data are used up.

Once the ANN model is trained, it is then validated by using a few new sets of input and output data to make sure it has proper internal relationships. The validated ANN model and thousands of the directional upstream typhoon wind speeds and profiles can then be used to obtain thousands of directional typhoon wind speeds and profiles at the site surrounded by the complex terrain, for the estimation of directional design wind speeds for a given return period and averaged wind profiles for design.

13.7.5 Directional Design Typhoon Wind Speeds and Profiles at Site

Based on thousands of directional typhoon wind speeds and profiles at the site and the annual occurrence rate of typhoons in the area of interest, a series of directional annual maximum wind speeds at the site over the complex terrain can be found for a given height above the sea level. By performing an extreme wind speed analysis, the directional design wind speeds at the site can be determined for a given return period. By averaging mean wind speed profiles for a given wind direction, the directional design mean wind profile can be produced. The flow chart of the entire numerical simulation procedure for predicting directional design wind speeds and profiles for the site over the complex terrain can be seen in Figure 13.20.

13.8 Case Study: Stonecutters Bridge Site

13.8.1 Topographical Conditions

The site of the Stonecutters Bridge in Hong Kong is chosen as a case study (see Figure 13.21). This is because this bridge is an important structure (one of the longest cable-stayed bridges in the world) which is located in a typhoon-prone region, and the bridge site is surrounded by complex terrain.

The major topographical conditions that may affect wind characteristics are described in the following using eight cardinal directions [46]:

(a) Wind from the north to northeast of the bridge site: the near field effect may be arising from buildings in Kwai Chung and container port terminals, whereas the far field effect may be

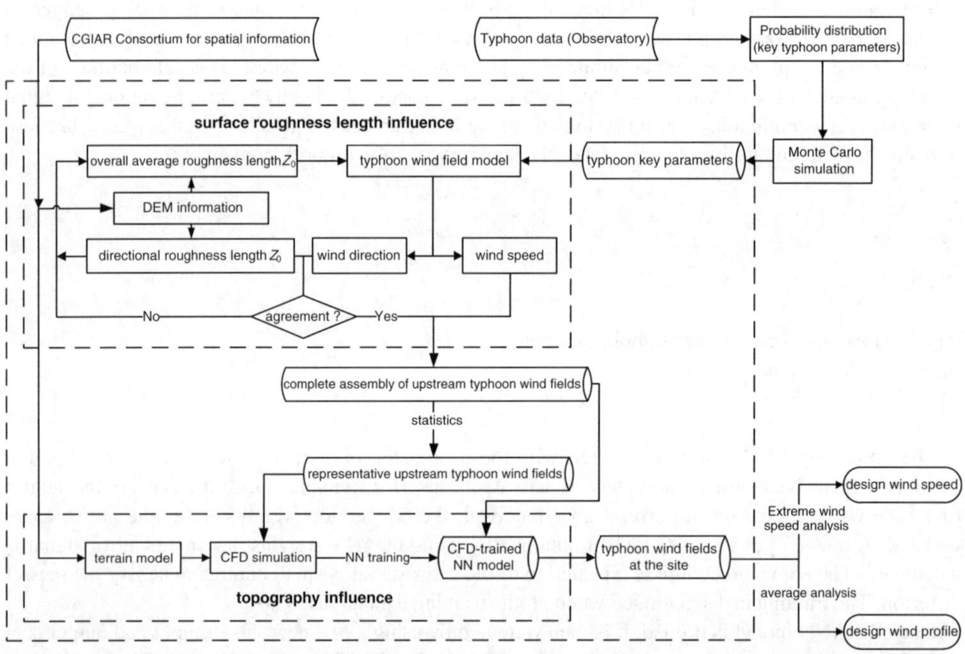

Figure 13.20 Numerical simulation procedure for predicting directional design wind speeds and profiles for a site over complex terrain.

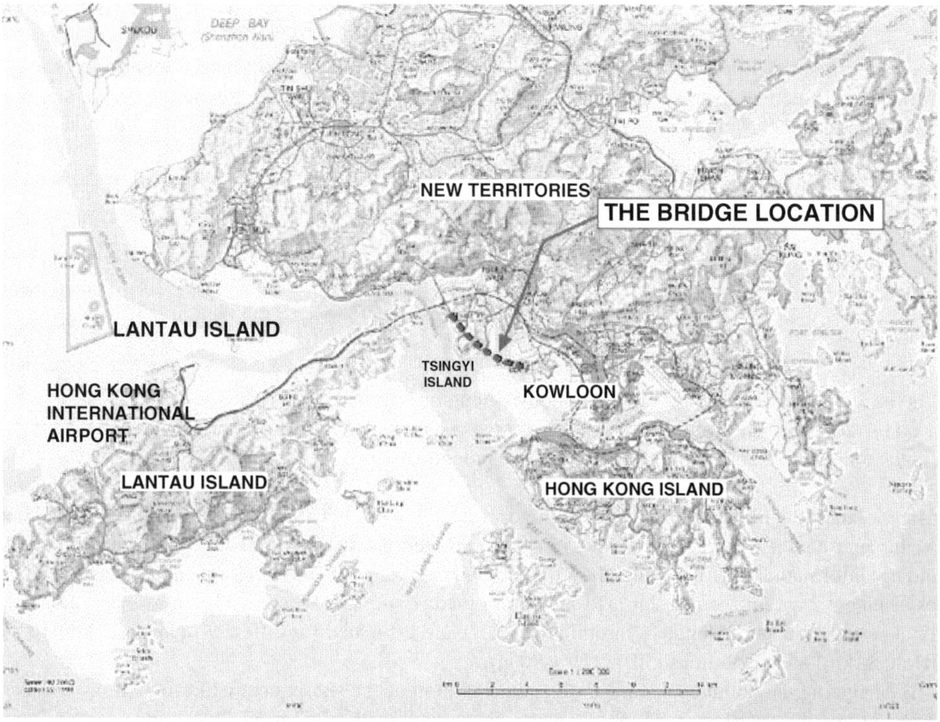

Figure 13.21 Hong Kong map and location of the Stonecutters Bridge.

caused by the mountains of Tai Mo Shan and Grassy Hill at the respective heights of 957 m and 647 m.

(b) Wind from the northeast to east of the bridge site: the near field effect may be arising from buildings in Cheung Sha Wan and North Kowloon and container port terminals whereas the far field effect may be induced by the mountains of Beacon Hill, Lion Rock, Tate's Cairn and Kowloon Peak at the respective heights of 457 m, 495 m, 577 m and 602 m.

(c) Wind from the east to southeast of the bridge site: the near field effect may be arising from container port terminals and Stonecutter Island, whereas the far field effect may be induced by the mountains of Victoria Peak at a height of 552 m.

(d) Wind from the southeast to south of the bridge site: the near field effect may be arising from the open sea, whereas the far field effect may be induced by the mountains of Mount Davis and Victoria Peak at a height of 269 m and 552 m respectively.

(e) Wind from the south to southwest of the bridge site: the effect of open sea.

(f) Wind from the southwest to west of the bridge site: the near field may be arising from the open sea, whereas the far field effect may be induced by Lantau Island.

(g) Wind from the west to northwest of the bridge site: the near field effect may be arising from the mountain of Tsing Yi, at a height of 334 m, whereas the far field effect may be generated by the mountains of Fa Peng northeast of Lantau Island, at a height of 273 m.

(h) Wind from the northwest to north of the bridge site: the near field effect may be arising from the mountain of Tsing Yi, whereas the far field effect may be generated by the Tai Mo Shan mountain.

13.8.2 Directional Upstream Typhoon Wind Speeds and Profiles

To obtain directional upstream typhoon wind speeds and profiles for the Stonecutters bridge site, the average and directional surface roughness lengths for the complex terrain around the bridge site must be determined. The average and directional surface roughness lengths can be estimated by using the equation below:

$$z_0 = \frac{\sum_{i=1}^{N} z_{0i} A_i}{A} \tag{13.34}$$

where:

z_0 is the average or directional surface roughness length;
z_{oi} and A_i are the surface roughness length and area occupied by the surface roughness element i;
A is the total area of complex terrain that the N elements occupy.

In this study, a circular area taking the bridge site as a center with an appropriate radius is considered. The circular area is further divided into 16 cardinal directions. By using the DEM information and the land use information in Hong Kong, the surface roughness length z_{oi} and area A_i for each surface roughness element i within the circular area are determined. Using Equation 13.34, the average and directional surface roughness length z_0 around the bridge are estimated for different radiuses (5 km, 10 km, 20 km, 30 km) and resolutions (30 m, 60 m, 90 m).

By analyzing the obtained results and in consideration of the subsequent CFD simulation, the average and directional surface roughness lengths for the area with a radius of 30 km and a resolution of 30 m are used for the typhoon simulation and given in Figure 13.22 and Table 13.2. It can be seen that the surface roughness length varies significantly with cardinal direction. The maximum directional surface roughness length z_0 is about 2.9 m in the ESE direction, mainly caused by container port terminals

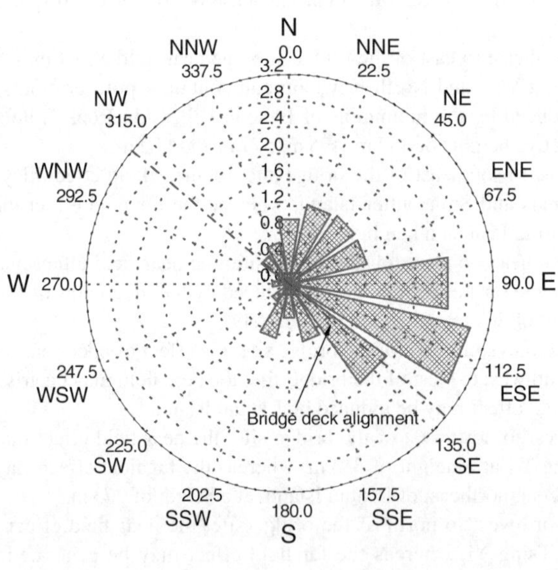

Figure 13.22 Directional surface roughness length around the bridge site.

Table 13.2 Directional surface roughness length z_0 for the complex terrain around the bridge site

Sector	True azimuth (degrees)	z_0 (m)	Sector	True azimuth (degrees)	z_0 (m)
N	0	0.992	S	180	0.533
NNE	22.5	1.245	SSW	202.5	0.861
NE	45	1.282	SW	225	0.334
ENE	67.5	1.262	WSW	247.5	0.087
E	90	2.515	W	270	0.258
ESE	112.5	2.913	WNW	292.5	0.192
SE	135	1.882	NW	315	0.532
SSE	157.5	0.909	NNW	337.5	0.649

and Stonecutter Island. The minimum directional surface roughness length z_0 is about 0.09 m in the WSW direction, which faces the open sea. The average surface roughness length z_0 for the entire area is 1.027 m.

Using the Monte Carlo simulation method, the refined typhoon wind field model described in Section 13.2.2, the probability distributions of key typhoon parameters listed in Table 13.1, and the average and directional surface roughness lengths, thousands of directional upstream typhoon wind speeds and profiles for the bridge site are obtained.

13.8.3 Representative Typhoon Wind Speeds and Profiles

The topographic model used for CFD simulation is set up based on the DEM information provided by the CGIAR Consortium. The area of computational domain, taking the bridge site as a center, is more than 30 km in width. Because the highest mountain within the computational domain is 957 m high, the computational domain in the vertical direction is set as 4 km, so that the flow field can be fully developed. Figure 13.23 shows the top view of the topographic model around the bridge site, and Figure 13.24 displays the mesh grid for the topographic model.

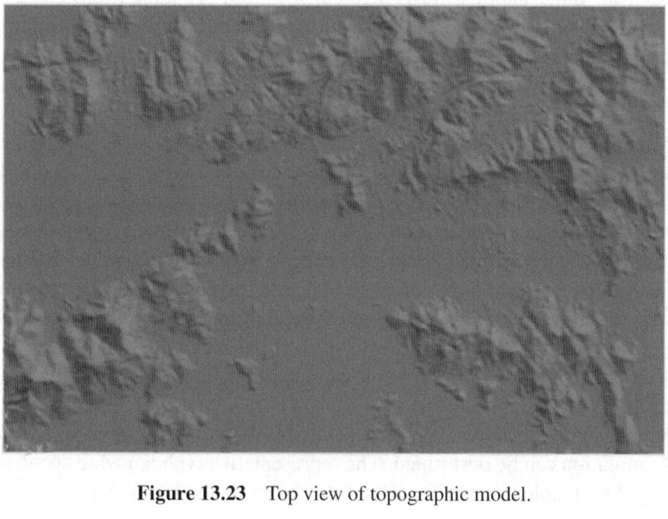

Figure 13.23 Top view of topographic model.

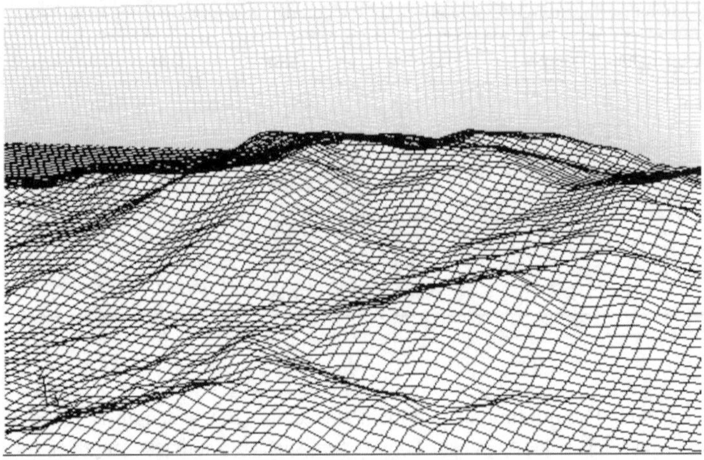

Figure 13.24 Mesh grid of topographic model.

In the horizontal plan, the topographic model is meshed by 200 × 100 grids. The vertical level is divided into 80 grids, leading to a total of 1.65 million nodes for CFD simulation, with an expansion factor of 1.05. The surface of the topographic model is modeled as a non-slip wall boundary. The flow inlet boundary is set as wind inlet profile, with a representative directional upstream wind speed and profile, while the outlet boundary is specified as fully developed outflow boundary conditions. The side and top boundaries are all defined in such a way that the gradients of flow variables (including velocity and pressure) normal to those boundary faces are zero. The RANS method is applied to perform CFD simulation to obtain a representative typhoon wind speed and profile at the bridge site, in which the finite volume method and the first order upwind scheme for spatial discretization are used and the SIMPLEC method is adopted to solve velocity and pressure simultaneously. More information on RANS can be found in Chapter 8. The computation is discontinued when residuals for all the variables are smaller than 10^{-3}.

For a given wind direction, representative directional upstream typhoon wind speeds and profiles are selected from thousands of the generated upstream typhoon wind speeds and profiles to cover a certain range of mean wind speeds at a reference height and a certain range of power law exponents used in the power law wind profile. In this regard, all the upstream typhoon wind speed profiles obtained in Section 13.8.2 are fitted by power law profiles, for the sake of simplification, by taking the 10 m height as a reference height. As a result, a series of mean wind speeds v_{10} at the 10 m height and a series of power law exponents α can be determined from all the upstream typhoon wind speed profiles.

By assuming that the power law exponents α in a given wind direction complies with a normal distribution, the range of α with a confidence level of 0.95 for all the upstream typhoon wind profiles in the given wind direction can be determined. Listed in Table 13.3 are the mean value, standard deviation and confidence range of α in each wind direction. It can be seen that the mean value of power law exponent varies significantly with direction. The variation of standard deviation of the power law exponent is relatively small. Based on the mean wind speeds at the 10 m height and the power law exponents in Table 13.3, the representative upstream typhoon wind speeds and profiles in different wind directions are determined and are shown in Table 13.4.

By taking a representative upstream typhoon wind speed and profile as an inlet to the topographic model, the CFD simulation can be performed. The representative typhoon wind speed and profile at the site surrounded by the complex terrain is obtained. For example, by taking representative upstream

Table 13.3 Statistical results of power law exponent α for upstream typhoon wind profiles

Sector	Mean value	Standard deviation	Confidence range	Sector	Mean value	Standard deviation	Confidence range
N	0.142	0.039	[0.07 0.22]	S	0.114	0.039	[0.04 0.19]
NNE	0.149	0.040	[0.07 0.23]	SSW	0.129	0.039	[0.05 0.21]
NE	0.149	0.038	[0.07 0.22]	SW	0.098	0.032	[0.04 0.16]
ENE	0.154	0.045	[0.07 0.24]	WSW	0.071	0.027	[0.02 0.12]
E	0.217	0.048	[0.12 0.31]	W	0.075	0.026	[0.02 0.13]
ESE	0.218	0.059	[0.10 0.33]	WNW	0.084	0.026	[0.03 0.13]
SE	0.161	0.054	[0.06 0.27]	NW	0.107	0.036	[0.04 0.18]
SSE	0.134	0.052	[0.03 0.24]	NNW	0.128	0.036	[0.06 0.20]

Table 13.4 Representative directional upstream typhoon wind speeds and profiles

Sector	Power law exponent α	Wind speed (m/sec)	Sector	Power law exponent α	Wind speed (m/sec)
N	0.07,0.16,0.22	10,30,50	S	0.04,0.12,0.19	10,30,50
NNE	0.07,0.16,0.23	10,30,50	SSW	0.05,0.12,0.21	10,30,50
NE	0.07,0.16,0.22	10,30,50	SW	0.04,0.12,0.16	10,30,50
ENE	0.12,0.16,0.24	10,30,50	WSW	0.02,0.08,0.13	10,30,50
E	0.12,0.22,0.31	10,30,50	W	0.02,0.08,0.13	10,30,50
ESE	0.10,0.22,0.33	10,30,50	WNW	0.03,0.08,0.13	10,30,50
SE	0.06,0.16,0.27	10,30,50	NW	0.04,0.12,0.18	10,30,50
SSE	0.03,0.16,0.24	10,30,50	NNW	0.06,0.12,0.20	10,30,50

typhoon wind speeds and profiles in the N direction with 10 m high wind speeds v_{10} of 10 m/sec, 30 m/sec, 50 m/sec and the power law exponent α of 0.22, the representative typhoon wind speeds profiles at the bridge site are obtained from the CFD simulation and are shown in Figure 13.25. Repeating the procedure for all the wind directions, a complete assembly of the representative upstream and site typhoon wind speeds and profiles are obtained and are subsequently used as input and output data to train an ANN model for predicting directional design typhoon wind speeds and profiles for the site.

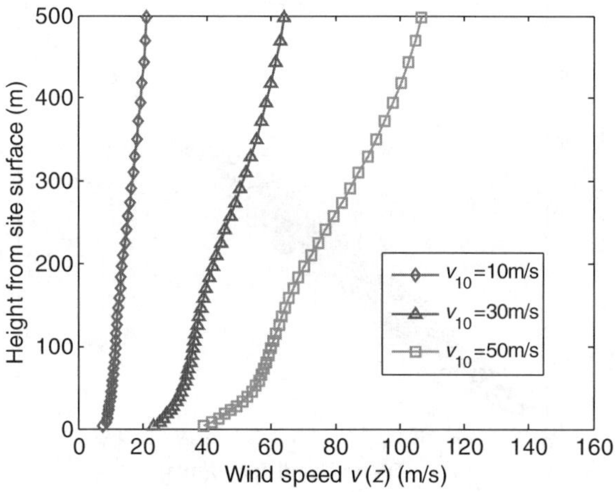

Figure 13.25 Representative wind profiles at bridge site from CFD simulation (N direction).

13.8.4 Establishment of ANN Model

The architecture of the ANN model used in this case study is shown in Figure 13.26 with the input and output variables. It consists of an input layer, a hidden layer and an output layer. The input variables for the ANN model are the directional upstream typhoon wind speeds $v(z)$ at a series of heights z above the upstream ground, as well as wind direction $\theta(z)$. The output parameters for the ANN model are wind speeds $v'(z)$ and wind directions $\theta'(z)$ at a series of heights z above the bridge site.

To train the ANN model, directional upstream and site typhoon wind speeds and profiles are selected pair by pair. Each is normalized by its maximum value so that they vary between 0 and 1. The normalized pairs are used as input and output parameters to train the ANN model. By trial and error, the neuron numbers in the hidden layer are set as 25. The learning rate parameter is kept constant with 0.001. The maximum training iterative step is set at 2000. The training of the network is carried out until the its root mean square error is smaller than 0.0001. Figure 13.27 shows the comparative results

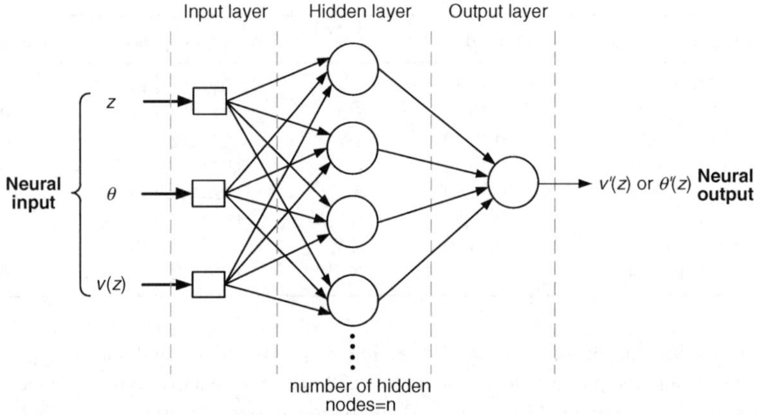

Figure 13.26 Architecture of ANN model used in case study.

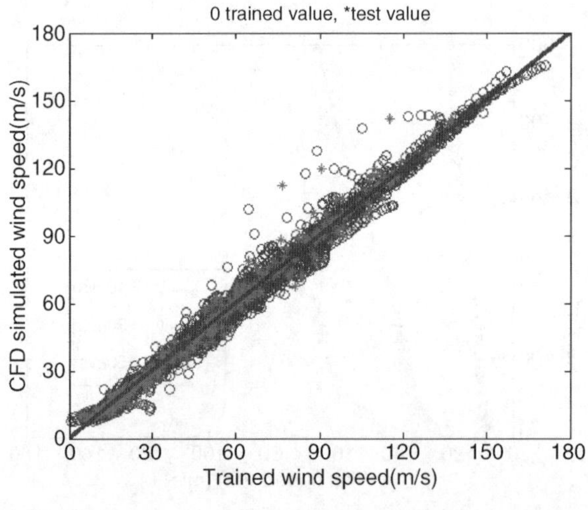

Figure 13.27 Comparison of wind speeds from CFD and ANN.

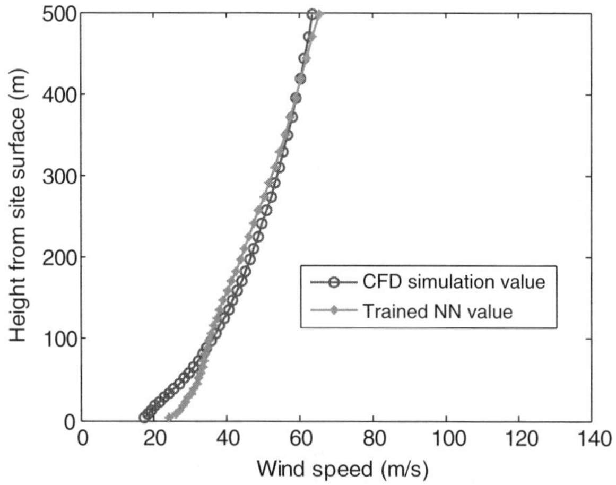

Figure 13.28 Comparison of wind profiles from CFD simulation and trained ANN model (ENE direction).

of wind speeds from both the CFD simulation and the trained ANN model. The results show that the mean difference of wind speeds is about 6%.

After the ANN model is trained, it needs to be checked by using a few new sets of data which are not used in the training. For this task, the upstream typhoon wind speeds and profiles in the ENE direction with a reference wind speed of 30 m/sec and a power law exponent of 0.22, and in the E direction with a reference wind speed of 30 m/sec and a power law exponent of 0.30, are used to confirm the accuracy of the ANN model for predicting typhoon wind speeds and profiles at the bridge site. The comparative results from the trained ANN model and the CFD simulation are shown in Figure 13.28 for the ENE direction and in Figure 13.29 for the E direction.

The results show that the mean difference in wind speeds is about 9% and 8% in the ENE direction and the E direction, respectively. It can be concluded that the trained ANN model can be used effectively to predict typhoon wind speed profiles at the bridge site.

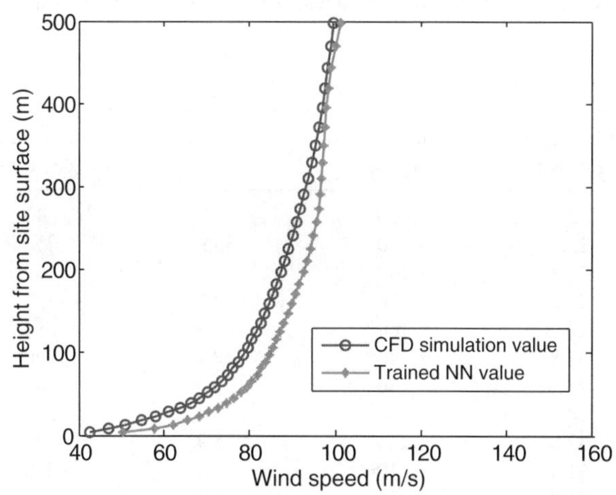

Figure 13.29 Comparison of wind profiles from CFD simulation and trained ANN model (E direction).

13.8.5 Directional Design Wind Speeds and Wind Profiles

Using the trained ANN model, the directional typhoon wind speeds and profiles at the bridge site are predicted on the basis of the directional upstream typhoon wind speeds and profiles obtained by generating 3000 typhoons for the bridge site. Taking 10 m as the reference height, series of annual maximum wind speeds at 10 m high in each of 36 direction sectors are obtained based on the Poisson distribution of annual occurrence rate of typhoons. Then, by using the Type I extreme distribution, the directional design wind speed of a 50-year return period is determined for the bridge site in each of 36 directional sectors, and the results are shown in Figure 13.30 and Table 13.5.

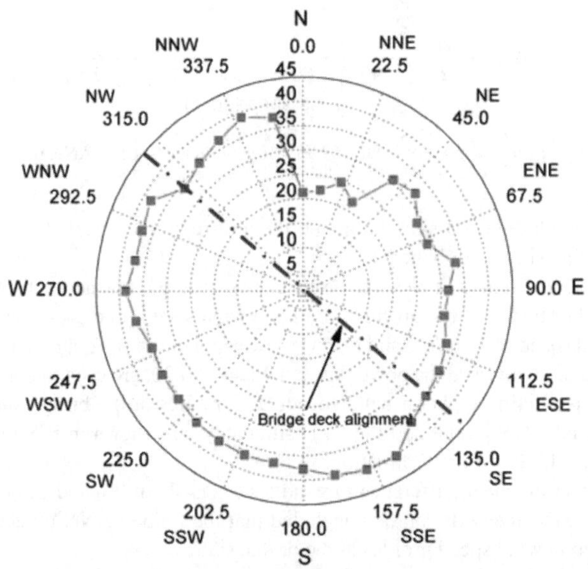

Figure 13.30 Directional design wind speeds at bridge site at 10 m high.

Table 13.5 Directional design wind speeds at bridge site at 10 m high

True azimuth (degrees)	Wind speed (m/sec)	True azimuth (degrees)	Wind speed (m/sec)	True azimuth (degrees)	Wind speed (m/sec)
0	22.8	120	33.7	240	34.4
10	22.5	130	33.9	250	34.3
20	24.8	140	35.4	260	36.4
30	22.2	150	39.1	270	37.1
40	29.9	160	38.9	280	35.7
50	31.8	170	38.4	290	36.6
60	28.9	180	36.9	300	37.3
70	28.4	190	35.9	310	31.2
80	32.8	200	35.9	320	33.1
90	31.4	210	35.6	330	34.6
100	31.8	220	35.4	340	36.9
110	33.0	230	34.4	350	35.4

It can be seen that the minimum design wind speed of about 23 m/sec at the 10 m height occurs mainly from the N direction to the ENE direction, due to the mountains. The maximum design wind speed comes from the S direction at about 38 m/sec, due to open sea. By averaging all the mean wind speed profiles for a given wind direction, the directional averaged mean wind profile at the bridge site surrounded by the complex terrain can be obtained as the directional design wind speed profile. The results are shown in Figure 13.31, and they clearly indicate that the design wind speed profile varies with wind direction due to the topographic effects.

The comparisons of the design wind speed profiles in the 40°, 180° and 270° directions at the bridge site, with the corresponding averaged upstream mean wind profiles in the same direction, are shown in Figure 13.32.

It can be seen that the effects of the complex terrain on the wind speed profile is very significant in the 40° direction, due to the shielding effects of the upstream mountains over 500 m height in this direction. When wind blows from the 180° direction, it is over open sea and there is a little speed-up effect generated by Hong Kong Island or Lantau Island. As a result, the design wind speed profile at the bridge site is similar to the averaged upstream mean wind profile in this direction.

When wind blows from the 270° direction, there is a relatively large speed-up effect caused by mountains in the west-east direction and Lantau Island. The design wind speed profile produces relatively larger mean wind speeds than the averaged upstream mean wind profile in this direction. In summary, the effects of the complex terrain around the Stonecutters bridge site on the design wind speed profile vary with direction and are significant in some wind directions.

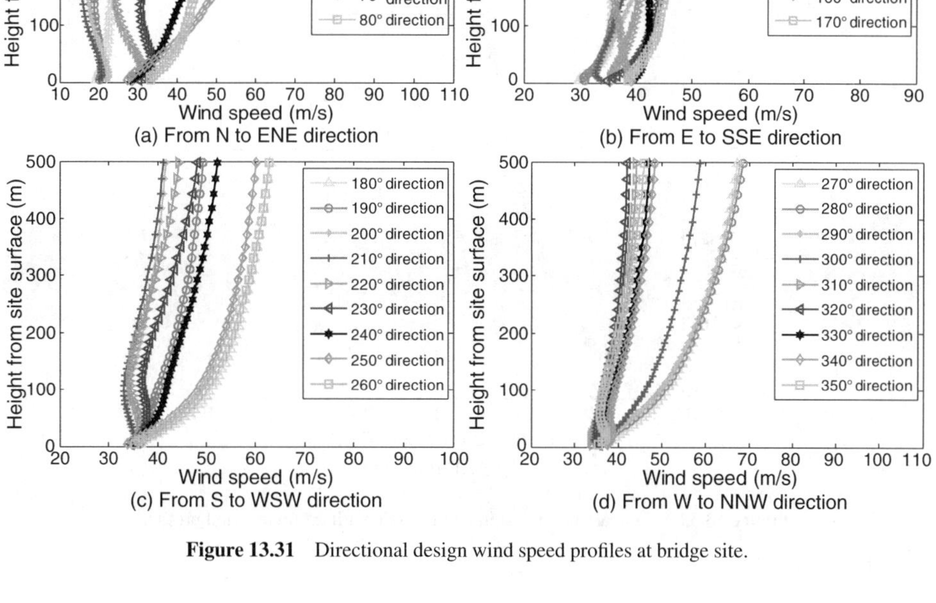

Figure 13.31 Directional design wind speed profiles at bridge site.

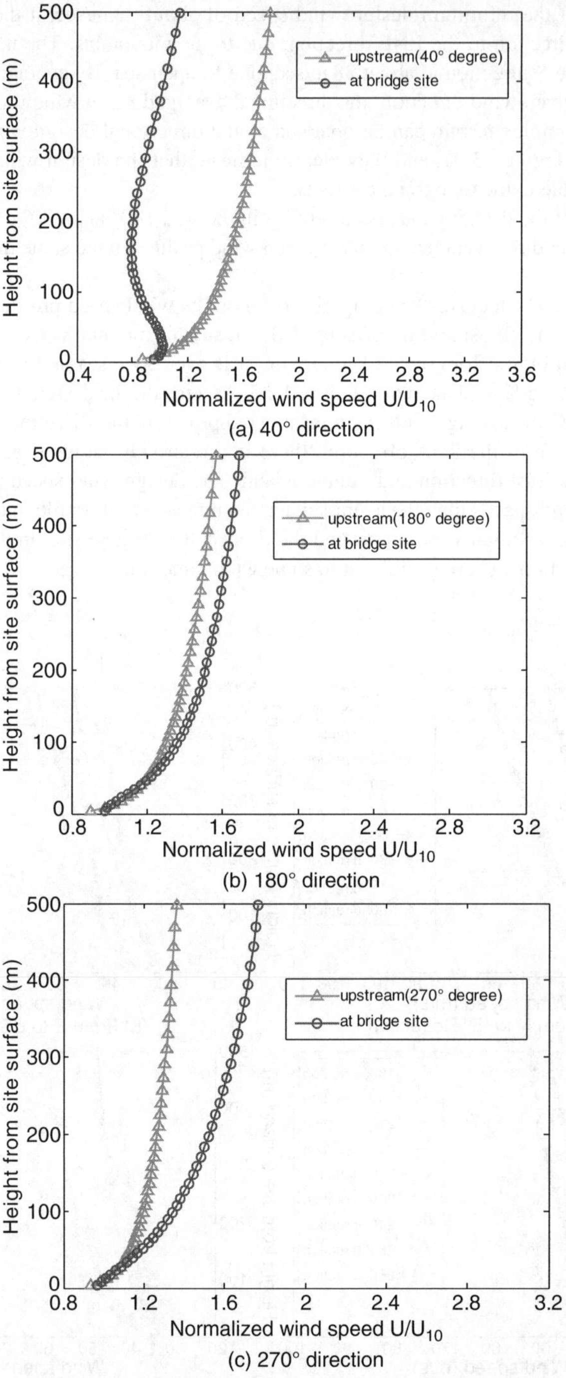

Figure 13.32 Comparison of design wind profile with upstream wind profile.

13.9 Notations

A	Decay constant
B	Holland's radial pressure profile parameter
$\boldsymbol{C}$	Translation velocity of the typhoon
c_p	Heat capacity at constant pressure
c_v	Heat capacity at constant volume
C_d	Drag coefficient
$d_{\min}$	Minimum of closest distance
F	Coriolis parameter
F	Friction force
F_h	Friction forces in the horizontal direction
F_v	Friction forces in the vertical direction
G	Gravitational constant
J	Adiabatic heat source
$\boldsymbol{K}$	Unit vector in the z-direction
$\boldsymbol{k_h}$	Unit vector in horizontal plane
k_m	Vertical coefficient of eddy viscosity
P	Atmospheric pressure
p_0	Pressure at zero plane
p_{c0}	Central pressure of a typhoon at zero plane
p_{m0}	Ambient pressure at zero plane
R	Radial distance from the typhoon center
R	Ideal gas constant
r_0	A large radial distance near the edge of area affected by the typhoon
r_m	Radius to maximum winds
T	Temperature
$\boldsymbol{v}_g$	Friction free wind velocity
$\boldsymbol{v}_h$	Wind velocity in the horizontal plane
v_{rg}	Radial wind speed
$\boldsymbol{v}_s$	Horizontal wind velocity on the ground surface
$v_{\theta g}$	Tangential wind speed
$V_{\max}$	Maximum wind speed
V_r	Wind speed at distance r
z_0	Surface roughness length
z_s	Height of the lower boundary
z_t	Height of the wind model top
α	$1/\rho$
β	Typhoon translation direction
Θ	Potential temperature; approach angle
θ_{r}	Angle between the typhoon translation direction and the vector from the center of pressure field to the site of interest
Λ	Annual occurrence rate
P	Air density
Δp	Central pressure difference
Δp_0	Central pressure difference at zero plane
Δp_{0n}	Central pressure difference at zero plane at the time when the typhoon makes landfall
$\prod$	Exner function

References

1. Chow S.H. (1970) A Study of the Wind Field in the Planetary Boundary Layer of a Moving Tropical Cyclone, Master Thesis, New York University, New York.
2. Gomes L. Vickery B.J. (1976) On the prediction of tropical cyclone gust speeds along the Northern Australian coast. *Institution of Engineers Australia, Civil Engineering Transaction* **CE18** (2) 197640–49
3. Holland, G.J. (1980) An analytical model of the wind and pressure profile in hurricanes. *Monthly Weather Review*, **108** (8), 1212–1218.
4. Batts, M.E., Simiu, E., and Russell, L.R. (1980) Hurricane wind speeds in the United States. *Journal of the Structural Division*, **106** (10), 2001–2016.
5. Shapiro, L.J. (1983) The asymmetric boundary layer flow under a translating hurricane. *Journal of the Atmospheric Sciences*, **40** (8), 1984–1998.
6. Georgiou, P.N. (1985) Design Wind Speeds in Tropical Cyclone-Prone Regions, PhD Thesis, University of Western Ontario, London, Ontario, Canada.
7. Vickery, P.J. and Twisdale, L.A. (1995) Wind-field and filling models for hurricane wind-speed predictions. *Journal of Structural Engineering*, **121** (11), 1700–1709.
8. Meng, Y., Matsui, M., and Hibi, K. (1995) An analytical model for simulation of the wind field in a typhoon boundary layer. *Journal of Wind Engineering and Industrial Aerodynamics*, **56** (2–3), 291–310.
9. Thompson, E.F. and Cardone, V.J. (1996) Practical modeling of hurricane surface wind fields. *Journal of Waterway, Port, Coastal, and Ocean Engineering*, **122** (4), 195–205.
10. Vickery, P.J., Skerlj, P.F., Steckley, A.C., and Twisdale, L.A. (2000) Hurricane wind field model for use in hurricane simulations. *Journal of Structural Engineering*, **126** (10), 1203–1221.
11. Vickery, P.J. (2005) Simple empirical models for estimating the increase in the central pressure of tropical cyclones after landfall along the coastline of the United States. *Journal of American Meteorological Society*, **44**, 1807–1826.
12. Yoshizumi, S. (1968) On the asymmetry of wind distribution in the lower layer of a typhoon. *Journal of the Meteorological Society of Japan*, **46** (3), 153–159.
13. Rosenthal, S.L. (1962) A Theoretical Analysis of the Field of Motion in the Hurricane Boundary Layer, NHRP Report, N56, U. S. Department of Commerce.
14. Meng, Y., Matsui, M., and Hibi, K. (1997) A numerical study of the wind field in a typhoon boundary layer. *Journal of Wind Engineering and Industrial Aerodynamics*, **67–68**437–448.
15. Jin, Y., Thompson, T., Wang, S., and Liu, C.S. (2007) A numerical study of the effect of dissipative heating on tropical cyclone intensity. *Weather and Forecasting*, **22**, 950–966.
16. Kepert, J.D. (2006) Observed boundary layer wind structure and balance in the hurricane core. Part I: Hurricane Georges. *Journal of the Atmospheric Sciences*, **63**, 2169–2193.
17. Huang, W.F. and Xu, Y.L. (2012) A refined model for typhoon wind field simulation in boundary layer. *Advances in Structural Engineering* **15** (1), 77–89.
18. Holton, J.R. (2004) *An Introduction to Dynamic Meteorology*, 4th edn, Elsevier Academic Press.
19. Galchen, T. and Somerville, R.C.J. (1975) On the use of a coordinate transformation for the solution of Navier-Stokes equations. *Journal of Computational Physics*, **17**, 209–228.
20. Vickery, P.J. and Twisdale, L.A. (1995) Prediction of hurricane wind speeds in the United States. *Journal of Structural Engineering*, **121** (11), 1691–1699.
21. Hong Kong Observatory (1999) Typhoon York (9915): 12–17 September 1999 http://www.hko.gov.hk/informtc/90s/york/york .htm.
22. Anthes, R.A. (1982) *Tropical Cyclones: Their Evolution, Structure, and Effects*, American Meteorological Society, Boston.
23. Pande, M., Ho, T.C.E., Mikitiuk, M. *et al.* (2002) Implications of typhoon York on the design wind speeds in Hong Kong. *Journal of Wind Engineering and Industrial Aerodynamics*, **90** (12–15), 1569–1583.
24. Charles, J.L. (1952) On the low-level structure of the typhoon eye. *Journal of Meteorology*, **9** (4), 285–290.
25. Wang, Y.Q. (2001) An explicit simulation of tropical cyclones with a triply nested movable mesh primitive equation model: TCM3. Part I: Model description and control experiment. *Journal of Monthly Weather Review*, **129**, 1370–1394.
26. Wang, Y.Q. (2002) An explicit simulation of tropical cyclones with a triply nested movable mesh primitive equation model: TCM3. Part II: Model refinements and sensitivity to cloud microphysics parameterization. *Journal of Monthly Weather Review*, **130**, 3022–3036.
27. Powell, M.D. (1982) The transition of the hurricane Frederic boundary layer wind field from the open Gulf of Mexico to landfall. *Journal of Monthly Weather Review*, **110**, 1912–1932.
28. Hock, T.F. and Franklin, J.L. (1999) The NCAR GPS dropwindsonde. *Bulletin of the American Meteorological Society*, **80** (3), 407–420.
29. Powell, M.D., Vickery, P.J., and Reinhold, T.A. (2003) Reduced drag coefficient for high wind speeds in tropical cyclones. *Nature*, **422**, 279–283.
30. Franklin, J.L., Black, M.L., and Valde, K. (2003) GPS dropwindsonde wind profiles in hurricanes and their operational implications. *Weather and Forecasting*, **18**, 32–44.

31. Russell, L.R. (1968) Probability Distributions for Texas Gulf Coast Hurricane Effects of Engineering Interest, PhD thesis, Stanford University, Stanford, California.
32. Russell, L.R. (1971) Probability distributions for hurricane effects. *Journal of the Waterways, Harbors and Coastal Engineering Division*, **97** (1), 139–154.
33. Martin, G.S. (1974) Probability distributions for hurricane wind gust speeds on the Australian Coast. Proceedings of Institution of Engineers Australia Conference on Applied Probability Theory to Structural Design, Melbourne.
34. Tryggvason, B.V., Davenport, A.G., and Surry, D. (1976) Predicting wind-induced response in hurricane zones. *Journal of the Structural Division*, **102** (12), 2333–2350.
35. Georgiou, P.N., Davenport, A.G., and Vickery, B.J. (1983) Design wind speeds in regions dominated by tropical cyclones. *Journal of Wind Engineering and Industrial Aerodynamics*, **13** (1–3), 139–152.
36. Fujii, T. and Mitsuta, Y. (1986) Simulation of winds in typhoons by a stochastic model. *Journal of Wind Engineering*, **28**, 1–12.
37. Vickery, P.J. (2009) U. S. Hurricane wind speed risk and uncertainty. *Journal of Structural Engineering*, **135** (3), 301–320.
38. Huang, W.F., Xu, Y.L., Li, C.W. and Liu, H.J. (2011). Prediction of design typhoon wind speed and profile using refined typhoon wind field model, *Advanced Steel Construction*, **7** (4), 387–402.
39. Simiu, E. and Scanlan, R.H. (1996) *Wind Effects on Structures*, 3rd edn, John Wiley & Sons, Inc., New York.
40. Bitsuamlak, G.T., Stathopoulos, T., and Bedard, C. (2004) Numerical simulation of wind flow over complex terrain: review. *Journal of Aerospace Engineering*, **17** (4), 135–145.
41. Chock, G.Y.K. and Cochran, L. (2005) Modeling of topographic wind speed effects in Hawaii. *Journal of Wind Engineering and Industrial Aerodynamics*, **93** (8), 623–638.
42. Wakes, S.J., Maegli, T., Dickinson, K.J., and Hilton, M.J. (2010) Numerical simulation of wind flow over a complex topography. *Environmental modelling & Software*, **25** (2), 237–247.
43. Xu, Y.L., Huang, W.F. and Liu, H.J. (2011) Simulation of directional typhoon wind speeds and profiles over complex terrain. *Proceedings of the 8th International Advanced School on Structural Wind Engineering*, Hong Kong, November 14–18, 34–51.
44. Cheng, Y., Lien, F.S., Yee, E., and Sinclair, R. (2003) A comparison of large eddy simulations with a standard k-ε Reynolds-averaged Navier-Stokes model for the prediction of a fully developed turbulent flow over a matrix of cubes. *Journal of Wind Engineering and Industrial Aerodynamics*, **91** (11), 1301–1328.
45. Rumelhart, D.E., Hinton, G.E., and Williams, R.J. (1986) Learning internal representation by error propagation, in *Parallel Distributed Processing* (eds D.E. Rumelhart*et al.*), MIT Press, Cambridge, MA, pp. 318–362.
46. Chen F J., Hui, C.H., and Xu, Y.L. (2007) A comparative study of stationary and nonstationary wind models using field measurements. *Boundary Layer Meteorology*, **22** (1), 105–121.

14

Reliability Analysis of Wind-Excited Bridges

14.1 Preview

The reliability analysis of a wind-excited long-span cable-supported bridge aims to predict the probability that the bridge will fulfill its prescribed functions under the action of stochastic wind loading within the stipulated time period. The fulfillment of the prescribed functions of a wind-excited bridge must ensure that the bridge keeps its serviceability, durability, stability and safety within its design life. Therefore, the reliability analysis should find either the reliability index or the failure probability of the bridge subject to aerostatic instability, buffeting, vortex shedding, flutter, galloping and rain-wind-induced vibration.

The occurrence probability of rain-wind-induced stay cable vibration has been investigated in Chapter 5, while the buffeting reliability analysis of long-span cable-supported bridges under non-stationary winds will be carried out in Chapter 15. This chapter therefore first provides a brief introduction to the fundamentals of reliability analysis, then it focuses on the reliability analysis of the bridge subject to aerostatic instability, flutter, buffeting and vortex shedding induced by stationary winds. The fatigue reliability analysis of long-span bridges under combined railway, highway and wind loadings is finally analyzed by two approaches based on Miner's rule and continuum damage mechanics (CDM) respectively. Since there are many challenging issues to be solved with respect to the reliability analysis and performance-based design of wind-excited bridges, readers should take this chapter as an elementary introduction.

14.2 Fundamentals of Reliability Analysis

14.2.1 Limit-States

Reliability is the probability that a system will perform its function over a specified period of time and under specified service conditions [1]. For a wind-excited bridge, this function is assessed mainly in terms of its wind-induced response, which involves many uncertainties that come from random wind loads, boundary conditions, physical properties, modeling, analytical methods and others.

The bridge response is considered satisfactory when the requirements imposed on the bridge's performance are met within an acceptable degree of certainty. Each of these requirements is termed as a limit-state. The reliability analysis is concerned with the calculation and prediction of the probability of

Wind Effects on Cable-Supported Bridges, First Edition. You-Lin Xu.

limit-state violations at any stage during the life of a bridge. For most structures, the limit-state can be divided into two categories:

(a) ultimate limit-states (ULSs) that are related to a collapse of part or all of the structure; and
(b) serviceability limit-states (SLSs) that are related to disruption of the normal use of the structures.

Examples of the most common ultimate limit-states are instability, fatigue, and collapse, with a relatively low probability of occurrence. Examples of serviceability limit-states are excessive deflection and excessive vibration, with a relatively high probability of occurrence.

Generally, the limit-state indicates the margin of safety between the resistance and the load of a structure. The limit-state function, $g(X)$, and the probability of failure, P_f, can be defined as:

$$g(X) = R(X) - S(X) \tag{14.1}$$

$$P_f = [g(X) < 0] \tag{14.2}$$

where R is the resistance and S is the loading of the structure.

Both $R(X)$ and $S(X)$ are functions of random variables $X = \{x_1, x_2, \ldots, x_n\}^T$. The notation $g(X) < 0$ denotes the failure region. Likewise, $g(X) = 0$ and $g(X) > 0$ indicate the failure surface and safe region, respectively. Clearly, the reliability P_r ($=1 - P_f$) means the probability of success which is equal to one minus the probability of failure P_f. The mean and standard deviation of the limit-state, $g(X)$, can be determined from the elementary definition of the mean and variance. The mean of $g(X)$ is:

$$\mu_g = \mu_R - \mu_S \tag{14.3}$$

where μ_R and μ_s are the means of R and S, respectively.

The standard deviation of $g(X)$ is:

$$\sigma_g = \sqrt{\sigma_R^2 + \sigma_S^2 - 2\rho_{RS}\sigma_R\sigma_S} \tag{14.4}$$

where:

ρ_{RS} is the correlation coefficient between R and S;
σ_R and σ_S are the standard deviations of R and S, respectively.

If the resistance and the loading are uncorrelated ($\rho_{RS} = 0$), the standard deviation becomes:

$$\sigma_g = \sqrt{\sigma_R^2 + \sigma_S^2} \tag{14.5}$$

The safety index or reliability index, β, is defined as:

$$\beta = \frac{\mu_g}{\sigma_g} \tag{14.6}$$

The safety index indicates the distance of the mean margin of safety from $g(X) = 0$. Figure 14.1 shows a geometrical illustration of the safety index in a one-dimensional case.

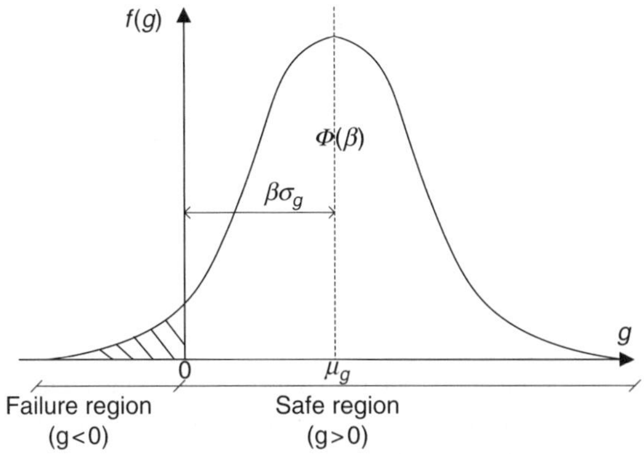

Figure 14.1 Probability density for limit-state.

The shaded area of Figure 14.1 identifies the probability of failure. If the probability density function of the limit-state function is a normal (Gaussian) distribution, the relationship between the safety index and the probability of failure can be expressed as:

$$P_f = 1 - \Phi(\beta) = \Phi(-\beta) = 1 - P_r \tag{14.7}$$

where $\Phi(\beta)$ is the normal cumulative distribution function.

14.2.2 First-Order Second Moment (FOSM) Method

The Taylor series expansion is often used to linearize the limit-state $g(X) = 0$. If the first-order Taylor series expansion is used to estimate reliability, this method is referred to as the first-order second moment (FOSM) method. The FOSM method is also called the mean value FOSM (MVFOSM) because it is a point expansion method at the mean point, the second moment is the highest-order statistical result used in the analysis and higher moments, which might describe skew and flatness of the distribution, are ignored. Assuming that the variables X are statistically independent, the approximate limit-state function at the mean is written as:

$$\tilde{g}(X) = g(\mu_X) + \nabla g(\mu_X)^T (X_i - \mu_{X_i}) \tag{14.8}$$

where:

$\mu_X = \{\mu_{X_1}, \mu_{X_2}, \ldots \mu_{X_n}\}^T$;

$\nabla g(\mu_X)^T$ is the gradient of g evaluated at μ_X.

$$\nabla g(\mu_X)^T = \left\{ \frac{\nabla g(\mu_X)}{\partial x_1}, \frac{\nabla g(\mu_X)}{\partial x_2}, \ldots, \frac{\nabla g(\mu_X)}{\partial x_n} \right\} \tag{14.9a}$$

$$(X_i - \mu_{X_i}) = \{X_1 - \mu_{X_i}, X_2 - \mu_{X_2}, \ldots, X_n - \mu_{X_n}\}^T \tag{14.9b}$$

The mean value of the approximate limit-state function $\tilde{g}(X)$ is:

$$\mu_{\tilde{g}} = E[g(\mu_X)] = g(\mu_X) \tag{14.10}$$

The standard deviation of the approximate limit-state function is:

$$\sigma_{\tilde{g}} = \sqrt{Var[\tilde{g}(X)]} = \sqrt{[\nabla g(\mu_X)^T]^2 Var(X)} = \left[\sum_{i=1}^{n}\left(\frac{\partial g(\mu_X)}{\partial x_i}\right)^2 \sigma_{X_i}^2\right] \tag{14.11}$$

The reliability index β is computed as:

$$\beta = \frac{\mu_{\tilde{g}}}{\sigma_{\tilde{g}}} \tag{14.12}$$

Equation 14.12 is the same as Equation 14.6 with Equation 14.5 if the limit-state function is linear. If the limit-state function is non-linear, the approximate limit-state surface $g(X) = 0$ is obtained by linearizing the original limit-state function at the mean value point. Therefore, this method is called the mean-value method, and the β given in Equation 14.12 is called a MVFOSM reliability index.

Although the implementation of FOSM or MVFOSM is simple, it has been shown that the accuracy is not acceptable for low probability of failure ($P_f < 10^{-5}$) or for highly non-linear responses [2]. The safety index approach to reliability analysis is actually a mathematical optimization problem for finding the point on the structural response surface (limit-state approximation) that has the shortest distance from the origin to the surface in the standard normal space. Hasofer and Lind [3] provided a geographic interpretation of the safety index and improved the FOSM method by introducing the Hasofer and Lind (HL) transformation.

14.2.3 Hasofer and Lind (HL) Method

The improvement of the HL method, compared with the MVFOSM, comes from changing the expansion point from the mean value point to the most probable point (MPP). Searching for the MPP on the limit-state surface is a key step in the method. Figure 14.1 shows how the reliability index could be interpreted as the measure of the distance from the origin to the failure surface. In the one-dimensional case, the standard deviation of the safety margin was conveniently used as the scale. To obtain a similar scale in the case of multiple variables, Hasofer and Lind [3] proposed a linear mapping of the basic variables into a set of normalized and independent variables, u_i.

Consider the fundamental case with the independent variables of resistance, R, and loading, S, which are both normally distributed. First, Hasofer and Lind introduced the standard normalized random variables:

$$\hat{R} = \frac{R - \mu_R}{\sigma_R};\ \hat{S} = \frac{S - \mu_S}{\sigma_S} \tag{14.13}$$

where:

μ_R and μ_s are the mean values of random variables R and S, respectively;
σ_R and σ_s are the standard deviations of R and S, respectively.

Next, transform the limit-state surface $g(R, S) = R - S = 0$ in the original (R, S) coordinate system into the limit-state surface in the standard normalized $(\hat{R}, \hat{S})$ coordinate system. Here, the shortest distance from the origin in the $(\hat{R}, \hat{S})$ coordinate system to the failure surface $\hat{g}(\hat{R}, \hat{S}) = 0$ is equal to the safety-index, $\beta = \hat{O}P^* = (\mu_R - \mu_S)/\sqrt{\sigma_R^2 + \sigma_S^2}$, as shown in Figure 14.2. The point $P^*\left(\hat{R}^*, \hat{S}^*\right)$ on $\hat{g}\left(\hat{R}, \hat{S}\right) = 0$, which corresponds to this shortest distance, is often referred to as the MPP.

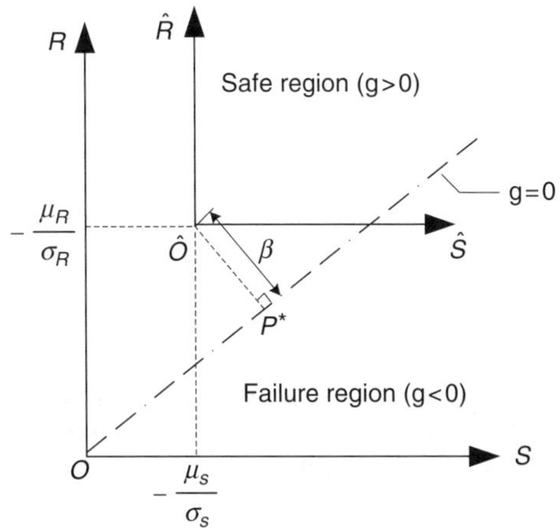

Figure 14.2 Geometrical illustration of safety-index.

Mathematically, the safety-index β, the shortest distance from the origin to the failure surface $g(U) = 0$, can be expressed as:

$$\beta = \min_{U \in g(U)=0} \left(U^T U\right)^{\frac{1}{2}} \tag{14.14}$$

The above safety-index is called the Hasofer and Lind (HL) safety-index, β_{HL}. The values of safety indices given in Equations 14.12 and 14.14 are the same when the failure surface is a hyperplane. The value of β_{HL} is the same for the true failure surface as well as for the approximate tangent hyperplane at the design point.

Equation 14.14 shows that safety-index, β, is the solution of a constrained optimization problem in the standard normal space:

$$\text{Minimize} : \beta = \left(U^T U\right)^{\frac{1}{2}} \tag{14.15a}$$

subject to:

$$g(U) = 0 \tag{14.15b}$$

There are many algorithms that can solve this problem, such as mathematical optimization schemes or other iteration algorithms. The HL iteration method proposed by Hasofer and Lind [3] is introduced here. Assume that the limit-state surface with n-dimensional normally distributed and independent random variables X is:

$$g(X) = g\left(\{x_1, x_2, \cdots, x_n\}^T\right) = 0 \tag{14.16}$$

This limit-state function can be linear or non-linear. Transform the variables into their standardized forms:

$$u_i = \frac{x_i - \mu_{x_i}}{\sigma_{x_i}} \tag{14.17}$$

where μ_{x_i} and σ_{x_i} represent the mean and the standard deviation of x_i, respectively. The limit-state function is transformed into:

$$g(X) = g\left(\left\{\sigma_{x_1}u_1 + \mu_{x_1}, \sigma_{x_2}u_2 + \mu_{x_2}, \cdots, \sigma_{x_n}u_n + \mu_{x_n}\right\}^T\right) \tag{14.18}$$

The normal vector from the origin $\hat{O}$ to the limit-state surface $g(U)$ generates an intersection point P^*, as shown in Figure 14.2. The distance from the origin to the MPP is the safety-index β. The first-order Taylor series of expansion of g(U) at the MPP, U^*, is:

$$\tilde{g}(U) \approx g(U^*) + \sum_{i=1}^{n} \frac{\partial g(U^*)}{\partial U_i}(u_i - u_i^*) \tag{14.19}$$

where n denotes the iteration number of the recursive algorithm. From Equation 14.17, we have:

$$\frac{\partial g(U)}{\partial u_i} = \frac{\partial g(X)}{\partial x_i}\sigma_{x_i} \tag{14.20}$$

The shortest distance from the origin to the above approximate failure surface given in Equation 14.19 is:

$$\hat{O}P^* = \beta = \frac{g(U^*) - \sum_{i=1}^{n} \frac{\partial g(U^*)}{\partial x_i}\sigma_{x_i}u_i^*}{\sqrt{\sum_{i=1}^{n}\left(\frac{\partial g(U^*)}{\partial x_i}\sigma_{x_i}\right)^2}} \tag{14.21}$$

The direction cosine of the unit outward normal vector is given as:

$$\cos\theta_{x_i} = \cos\theta_{u_i} = -\frac{\frac{\partial g(U^*)}{\partial u_i}}{|\nabla g(U^*)|} = -\frac{\frac{\partial g(X^*)}{\partial x_i}\sigma_{x_i}}{\left[\sum_{i=1}^{n}\left(\frac{\partial g(X^*)}{\partial x_i}\sigma_{x_i}\right)^2\right]^{1/2}} = \alpha_i \tag{14.22}$$

where α_i expresses the relative effect of the corresponding random variable on the total variation. Thus it is called the sensitivity factor. The coordinates of the point P^* are computed as:

$$u_i^* = \frac{x_i^* - \mu_{x_i}}{\sigma_{x_i}} = \hat{O}P^*\cos\theta_{x_i} = \beta\cos\theta_{x_i} \tag{14.23}$$

The coordinates corresponding to P^* in the original space are:

$$x_i^* = \mu_{x_i} + \beta\sigma_{x_i}\cos\theta_{x_i}(i = 1, 2, \cdots, n) \tag{14.24}$$

Since P^* is a point on the limit-state surface,

$$g\left(\{x_1^*, x_2^*, \cdots, x_n^*\}^T\right) = 0 \tag{14.25}$$

The main steps of the HL iteration method can be summarized as follows:

(a) Define the appropriate limit-state function of Equation 14.16.
(b) Set the mean value point as an initial design point and compute the gradients of the limit-state function at this point.
(c) Compute the initial β using the mean-value method, $\beta = \mu_{\tilde{g}}/\sigma_{\tilde{g}}$ and its direction cosine.
(d) Compute a new design point using Equations 14.23 and 14.24, function value and gradients at this new design point.
(e) Compute the safety-index β using Equation 14.21 and the direction cosine or sensitivity factor from Equation 14.22.
(f) Repeat steps (d) to (f) until the estimate of β converges.
(g) Compute the coordinates of the design point or most probable failure point (MPP).

It can be seen from Equations 14.12 and 14.21 that the difference between the MVFOSM method and the HL method is that the HL method approximates the limit-state function using the first-order Taylor expansion at the design point instead of the mean value point. Also, the MVFOSM method does not require iterations, while the HL method needs several iterations to converge for non-linear problems.

Different approximate response surfaces correspond to different methods for failure probability calculations. If the response surface is approached by a first-order approximation at the MPP, the method is called the first-order reliability method (FORM); if the response surface is approached by a second-order approximation at the MPP, the method is called the second-order reliability method (SORM). Historically, the HL transformation method is often referred to as FORM. The HL method usually provides better results than the mean-value method for non-linear problems, but there is no guarantee that the HL algorithm converges in all situations. Furthermore, the MVFOSM and HL methods only consider normally distributed random variables, so cannot be used for non-Gaussian random variables.

For non-Gaussian random variables, the Hasofer Lind-Rackwitz Fiessler (HL-RF) method can be used, which is actually based on the transformation of non-Gaussian variables into equivalent normal variables [4]. The HL method or FORM usually works well when the limit-state surface has only one minimal distance point and the function is nearly linear in the neighborhood of the design point. However, if the failure surface has large curvatures (high non-linearity), the failure probability estimated by FORM using the safety index may give unreasonable and inaccurate results. To resolve this problem, the second-order reliability method (SORM) can be considered [5,6].

14.2.4 Monte Carlo Simulation (MCS) and Response Surface Method (RSM)

In structural reliability analysis, the Monte Carlo Simulation (MCS) is the only known technique to estimate the probability of failure, P_f accurately, regardless of the complexity of the structure or the limit-state [7]. The sampling set of the corresponding random variables are first generated according to the probability density functions. The mathematical model of limit-station function $g(X)$ is then set, which can determine failures for the drawing samples of the random variables. After conducting simulations using the generated sampling set, the probabilistic characteristics of the response of the structure can be obtained. If $g(X)$ is violated, the structure or structural element has "failed."

The trial is repeated many times to guarantee convergence of the statistical results. If N trials are conducted, the probability of failure is given approximately by $P_f = N_f/N$, where N_f is the number of trials for which $g(X)$ is violated out of the N experiments conducted [1]. MCS becomes computationally intensive for the reliability analysis of complex structures with low failure probabilities, and the technique can be infeasible when the analysis requires a large number of computationally intensive simulations. In such cases, the first-order reliability method (FORM) and the second-order reliability method (SORM) are also difficult to apply, since the true limit-state usually cannot be easily expressed explicitly.

The response surface method (RSM) is another technique which can provide an efficient and accurate estimation of structural reliability, regardless of the complexity of the failure process [8]. The RSM approximates the true limit-state function, using simple and explicit mathematical functions (typically quadratic polynomials) of the random variables involved in the limit-state function. By fitting the response surface to a number of designated sample points of the true limit-state, an approximated limit-state function is constructed. As this function is explicit, FORM or SORM can then be applied to estimate the probability of failure directly.

14.2.5 Threshold Crossing

The above subsections concern time-invariant structural failures. When a stable structure is subject to random dynamic excitation, it may fail upon the occurrence of one of the following two events [9]:

(a) First-passage failure: the dynamic response $X(t)$ at a critical point of a structure reaches, for the first time, either an upper bound level or a lower bound level (threshold).
(b) Fatigue failure: failure occurs when the accumulated damage as $X(t)$ fluctuates reaches its threshold value.

For the study of both types of failure, it is useful to obtain the probabilistic characteristics of the following two items:

(a) within a given time interval the number of times that the random process $X(t)$ crossed a certain threshold; and
(b) within a given time interval the number of peaks in $X(t)$ which are above (or below) a certain level.

The studies of items (a) and (b) are known as the threshold crossings and the peak distribution.

For a stochastic process $X(t)$, the time of passing a given level within a given time interval is a random variable.

Figure 14.3 shows a sample function of a stochastic process and the situations that the sample time process passes the level $x(t) = a$ $(a > 0)$ with a positive and negative slope (upward and downward passage, respectively) [10]. The probability distribution density of this random variable per unit time is called the rate of passage (or threshold crossings) and is designated as $N(t)$. The condition that the stochastic process passes the threshold with a positive slope in the time interval $(t,\ t + dt)$ can be expressed by:

$$\begin{cases} x(t) < \alpha \\ x(t + dt) > \alpha \end{cases} \tag{14.26}$$

Considering that $x(t + dt) = x(t) + \dot{x}(t)dt$, the condition can be rewritten as:

$$\begin{cases} x(t) < \alpha \\ \dot{x}(t)dt > \alpha - x(t) \end{cases} \tag{14.27}$$

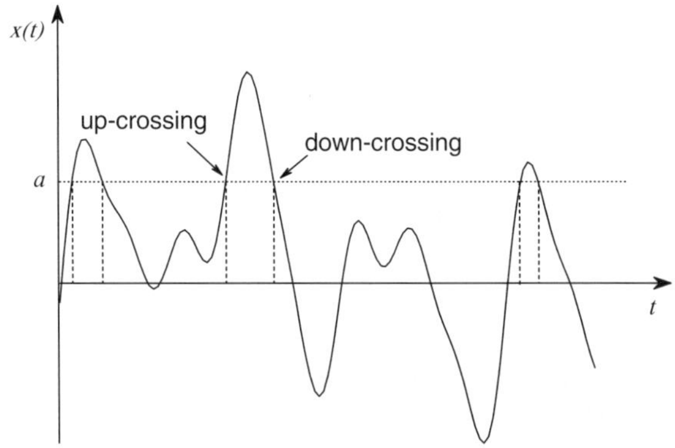

Figure 14.3 Threshold crossings in a sample function of a random process.

Assuming that the joint probabilistic density function (PDF) of $X(t)$ and $\dot{X}(t)$ is $f_{X\dot{X}}(x, \dot{x}, t)$, the probability that passages happen in the time interval $(t, t + dt)$ is:

$$N^{+}(\alpha, t)dt = \int_{0}^{\infty} \int_{\alpha - \dot{x}dt}^{\alpha} f_{X\dot{X}}(x, \dot{x}, t)dx d\dot{x} \tag{14.28}$$

Since the upper and lower limits of the integration of the second integral differ by an infinitesimal quantity, $\dot{x}dx$, we can use the mean value theorem to set:

$$\int_{\alpha - \dot{x}dt}^{\alpha} f_{X\dot{X}}(x, \dot{x}, t)dx = \dot{x}dt \cdot f_{X\dot{X}}(\alpha, \dot{x}, t) \tag{14.29}$$

Substituting Equation 14.29 into 14.28 and cancelling dt from both sides yields:

$$N^{+}(\alpha, t) = \int_{0}^{\infty} \dot{x} f_{X\dot{X}}(\alpha, \dot{x}, t)d\dot{x} \tag{14.30}$$

This is the rate of threshold crossings at time instant t with a positive slope. Similarly, the rate of threshold crossings of the level $x(t) = -\alpha$ at time instant t with a negative slope can be found as:

$$N^{-}(-\alpha, t) = \int_{-\infty}^{0} |\dot{x}| f_{X\dot{X}}(-\alpha, \dot{x}, t)d\dot{x} \tag{14.31}$$

The rate of crossings of both the upper and lower levels at time instant t can be finally written as:

$$N(\alpha, t) = \int_{-\infty}^{\infty} |\dot{x}| f_{X\dot{X}}(\alpha, \dot{x}, t)d\dot{x} \tag{14.32}$$

This result is known as Rice's formula [11]. For a stationary random process $X(t)$, $f_{X\dot{X}}(x, \dot{x}, t) = f_{X\dot{X}}(x, \dot{x})$ and Equation 14.32 reduces to:

$$N(\alpha) = \int_{-\infty}^{\infty} |\dot{x}| f_{X\dot{X}}(\alpha, \dot{x})d\dot{x} \tag{14.33}$$

For a zero-mean stationary Gaussian stochastic process, we have:

$$N^+(\alpha) = N^-(-\alpha) = \frac{1}{2\pi}\frac{\sigma_{\dot{x}}}{\sigma_x}\exp\left(-\frac{a^2}{2\sigma_x^2}\right) \tag{14.34}$$

$$N^+(0) = N^-(0) = \frac{1}{2\pi}\frac{\sigma_{\dot{x}}}{\sigma_x} \tag{14.35}$$

where:

σ_x and $\sigma_{\dot{x}}$ are the standard deviation of $X(t)$ and $\dot{X}(t)$respectively;
$N^+(0)$ and $N^-(0)$are the expected rate of zero crossings with positive and negative slope respectively.

14.2.6 Peak Distribution

A peak in a sample function $x(t)$ of a random process $X(t)$ occurs at time $t \in (t_1, t_1 + dt)$ above the level $x(t) = \alpha$ when:

(a) $x(t) > \alpha$,
(b) $\dot{X}(t) = 0$, and
(c) $\ddot{X}(t) < 0$.

The expected number of peaks per unit time above the level $x(t) = \alpha$ can be derived as [12]:

$$N_p(\alpha, t) = \int_\alpha^\infty dx \int_{-\infty}^0 |\ddot{x}| f_{X\dot{X}\ddot{X}}(x, 0, \ddot{x}, t) d\ddot{x} \tag{14.36}$$

The expected number of total peaks per unit time without any restriction on level, $M(t)$, is obtained by letting $\alpha \to -\infty$:

$$M_p(t) = N_p(-\infty, t) = \int_{-\infty}^\infty dx \int_{-\infty}^0 |\ddot{x}| f_{X\dot{X}\ddot{X}}(x, 0, \ddot{x}, t) d\ddot{x} \tag{14.37}$$

Using the relative frequency definition of probability, the probability distribution function of the peaks can be expressed as:

$$F_p(\alpha, t) = 1 - \frac{N_p(\alpha, t)}{M_p(t)} = 1 - \frac{1}{M_p(t)} \int_\alpha^\infty dx \int_{-\infty}^0 |\ddot{x}| f_{X\dot{X}\ddot{X}}(x, 0, \ddot{x}, t) d\ddot{x} \tag{14.38}$$

Then the probability density function of the peaks is:

$$f_p(\alpha, t) = \frac{\partial}{\partial \alpha} F_p(\alpha, t) = \frac{1}{M_p(t)} \int_{-\infty}^0 |\ddot{x}| f_{X\dot{X}\ddot{X}}(\alpha, 0, \ddot{x}, t) d\ddot{x} \tag{14.39}$$

For a stationary random process:

$$f_p(\alpha) = f_p(\alpha, t) \tag{14.40}$$

If $X(t)$ is a stationary normal random process with zero mean, we have:

$$f_{X\dot{X}\ddot{X}}(x,0,\ddot{x}) = \frac{1}{(2\pi)^{2/3}|\mathbf{K}|^{1/2}}\exp\left[-\frac{1}{2|\mathbf{K}|}\left(\sigma_{\dot{x}}^2\sigma_{\ddot{x}}^2x^2 + 2\sigma_{\dot{x}}^4x\ddot{x} + \sigma_x^2\sigma_{\dot{x}}^2\ddot{x}^2\right)\right] \tag{14.41}$$

where **K** is the covariance matrix given by:

$$\mathbf{K} = \begin{bmatrix} \sigma_x^2 & 0 & -\sigma_{\dot{x}}^2 \\ 0 & \sigma_{\dot{x}}^2 & 0 \\ -\sigma_{\dot{x}}^2 & 0 & \sigma_{\ddot{x}}^2 \end{bmatrix} \tag{14.42}$$

The probability density function of the peaks can be obtained by substituting Equations 14. 41 and 14. 42 into Equations 14.37 and 14.39:

$$f_p(\alpha) = \frac{\left(1-\xi^2\right)^{1/2}}{\left(2\pi\sigma_x^2\right)^{1/2}}\exp\left\{-\frac{\alpha^2}{2\sigma_x^2\left(1-\xi^2\right)}\right\} + \frac{\xi\alpha}{2\sigma_x^2}\left\{1+\operatorname{erf}\left[\frac{\xi\alpha}{\sqrt{2\sigma_x^2}\left(1-\xi^2\right)^{1/2}}\right]\exp\left(-\frac{\alpha^2}{2\sigma_x^2}\right)\right\} \tag{14.43a}$$

$$\xi = \frac{N^+(0)}{M_p} = \frac{\sigma_{\dot{x}}^2}{\sigma_x\sigma_{\ddot{x}}} = \frac{\lambda_2}{\sqrt{\lambda_0\lambda_4}} \tag{14.43b}$$

$$M_p = \frac{1}{2\pi}\frac{\sigma_{\ddot{x}}^2}{\sigma_{\dot{x}}^2} = \frac{1}{2\pi}\left(\frac{\lambda_4}{\lambda_2}\right)^{(1/2)} \tag{14.43c}$$

where:

ξ is the ratio of the rate of zero crossings with positive slope to the expected number of total peaks per unit time without any restriction on level;

$\lambda_o = \sigma_x^2, \quad \lambda_2 = \sigma_{\dot{x}}^2, \quad \lambda_4 = \sigma_{\ddot{x}}^2;$

erf(•) is the error function defined by:

$$\operatorname{erf}(y) = \frac{1}{\sqrt{2\pi}}\int_{-y}^{y}\exp\left(-\frac{u^2}{2}\right)du \tag{14.44}$$

Obviously, $0<\xi<1$. The upper limit ($\xi=1$) corresponds to a special case in which the expected number of total peaks is equal to the rate of zero crossings with a positive slope. For this special case, Equation 14.43a converges to the upper limit as:

$$f_p^u(\alpha) = \frac{\alpha}{\sigma_x^2}\left\{\exp\left(-\frac{\alpha^2}{2\sigma_x^2}\right)\right\} \tag{14.45}$$

which is a Rayleigh distribution. On the other hand, when ξ is very small, which means that the expected number of total peaks is much larger than the rate of zero crossings, Equation 14.43a approaches to the lower limit as:

$$f_p^l(\alpha) = \frac{1}{\left(2\pi\sigma_x^2\right)^{1/2}} \exp\left\{-\frac{\alpha^2}{2\sigma_x^2}\right\} \tag{14.46}$$

which represents a Gaussian distribution.

14.3 Reliability Analysis of Aerostatic Instability

The aerostatic instability of long-span bridges has been discussed in Chapter 3, where uncertainties of geometric parameters, material properties, force coefficients and other factors are not considered. Furthermore, the critical wind speed for the aerostatic instability of a long-span bridge cannot be expressed as an explicit function of random variables. The corresponding limited state function in the aerostatic reliability analysis will be implicit functions of these random variables.

Su *et al.* [13] used a combination of the response surface method (RSM) and the Monte Carlo Simulation method (MCS) to investigate aerostatic reliability of long-span bridges, in which non-linear effects due to geometric non-linearity and deformation-dependent aerostatic loads are taken into consideration, and the geometric parameters, material parameters and force coefficients of the bridge deck are regarded as random variables. Their approach is implemented in the following steps:

(a) Construct a limit-state function for reliability analysis of aerostatic instability of a long-span bridge.

$$g(x_1, x_2, \ldots, x_n) = R - S(x_1, x_2, \ldots, x_n) \tag{14.47}$$

where:

R is the threshold value of critical wind velocity for aerostatic instability;
S is the critical wind velocity for aerostatic instability of a given bridge;
$\{x_1, x_2, \ldots, x_n\}$ are random variables representing geometric parameters, material parameters and force coefficients of the bridge deck, respectively.

(b) Find the design point P^* by a two-level iteration: the outer loop iteration of the surface response function and the inner loop iteration of the HL method. During the outer loop iteration, FE analyses of aerostatic stability are used as numerical tests to determine the coefficients of the response surface function.
(c) Generate a new response surface function using the design point P^* obtained in step b) and the mean value point P_{m} as two centers for numerical tests, and let it be the final form of the response surface function.
(d) Calculate the failure probability regarding the aerostatic instability of the bridge by the MCS method, based on the final response surface function obtained in step c).

It should be pointed out that, in this chapter, it is not intended to consider the overall reliability of a long-span bridge under the action of wind loads. The reliability studied in this chapter for a wind-excited bridge is event-dependent.

14.4 Flutter Reliability Analysis

The flutter instability of long-span cable-supported bridges has been discussed in Chapter 4, where all structural and aerodynamic parameters are regarded as deterministic parameters. The probabilistic flutter analysis of long-span bridges has not been reported as adequately as in the case of deterministic analysis of long-span bridges against flutter. Ostenfeld-Rosenthal *et al.* [14] presented the reliability analysis of flutter response of cable-supported bridges by considering various uncertainties involved in actual extreme wind speeds, conversion of results from model tests to prototype structures, turbulence intensity, and structural damping. Ge *et al.* [15] used three approaches based on the first-order reliability method (FORM) to study the failure probability of bridges due to flutter. Pourzeynail and Datta [16] presented a reliability analysis of suspension bridges against flutter failure using the fundamental theory of reliability. Prenninger *et al.* [17] investigated the reliability of two types of long-span bridge structures – a suspension and a cable-stayed bridge under wind loading.

In these studies, the limit-state function is explicitly expressed in terms of the random variables; the uncertainties included in resistances and loads are simply considered in the function as continuous random variables by introducing some assumptions. However, in practical applications, the flutter response of long-span bridges is an implicit function of the random variables. Under such a condition, the first-order reliability method and basic theory of reliability cannot be attempted.

Cheng *et al.* [18] proposed a flutter reliability analysis method through a combination of the advantages of the response surface method (RSM), first-order reliability method (FORM), and the importance sampling updating method. The method is especially applicable for the reliability evaluation of complex structures, of which the limit-state surfaces are not known explicitly. The procedure of their proposed method is stated as:

(a) Determine the values of the random variables at the chosen sampling points.
(b) Conduct deterministic flutter analysis using these values of the random variables.
(c) Use the RSM to construct the approximate limit-state function.
(d) After the approximated limit-state function is determined, FORM with the Hasofer-Lind-Rackwitz-Fiessler algorithm is applied to obtain the reliability index.
(e) Apply the importance sampling updating method to improve the obtained reliability index.

Cheng *et al.* [18] presented a case study in which the following limit-state function was used for flutter reliability analysis of a long-span suspension bridge:

$$g(X) = G_u\bar{U} - X_{12}U_{cr}(X_1, X_2, \ldots, X_{11}) \tag{14.48}$$

where:

G_u is the gust speed factor;
$\bar{U}$ is the basic wind speed;
$X_1, X_2, \ldots, X_{11}$ are the random variables representing elastic modulus, cross-sectional area, cross-sectional moment of inertia, mass density, mass moment of inertia, model damping ratio and flutter derivatives.

G_u and $\bar{U}$ are also taken as random variables. U_{cr} is the critical flutter wind speed calculated by the method provided in Chapter 4, but using the generated values from the set of the random variables.

A second-order polynomial without cross-terms was used to approximate the aforementioned implicit limit-state function when the RSM is used:

$$\widehat{g}(X) = b_0 + \sum_{i=1}^{k} b_i X_i + \sum_{i=1}^{k} b_{ii} X_i^2 \tag{14.49}$$

where:

$\widehat{g}(X)$ is the approximate limit-state function;
X_i $(i=1,2,\ldots,k)$ is the ith random variable;
k is the total number of random variables;
b_0, b_i, b_{ii} are the coefficients to be determined by fitting the response surface to a number of designated sample points of the true limit-state and then solving a set of simultaneous equations.

14.5 Buffeting Reliability Analysis

The buffeting response analysis of a long-span cable-supported bridge has been addressed in Chapter 4. The failure of the bridge due to buffeting can be either first-passage failure or fatigue failure. In first-passage failure, which will be discussed in this subsection, the buffeting response $X(t)$ at a critical point in a bridge structure reaches, for the first time, either an upper bound level or a lower bound level. In the fatigue failure, which will be discussed in subsequent sections, failure occurs when the accumulated damage as $X(t)$ fluctuates reaches its limit value.

14.5.1 Failure Model by First Passage

In consideration that the total buffeting response is the sum of the mean wind response and the fluctuating wind response, the upper bound level, U, and the lower bound level, L, which will be reached for the first time by the fluctuating buffeting response at a critical point in a bridge structure, can be expressed in terms of the resistance R and the mean response $\bar{X}$[19]:

$$U = R - \bar{X} \tag{14.50a}$$

$$L = R + \bar{X} \tag{14.50b}$$

The mean response here is caused by dead load, superimposed dead load and static wind load. If the resistance R and the mean response $\bar{X}$are regarded as deterministic parameters, the upper and lower bound levels are also deterministic. If the resistance and the mean response are regarded as random variables, the upper and lower bound levels are random bound levels.

In this chapter, the bound levels are regarded as deterministic values. When both the upper and lower bound levels are taken into consideration in buffeting reliability analysis, the safety range can be expressed as $-L \leq X(t) \leq U$ and the dynamic reliability with the double-sided boundary condition within the time period T can be defined as:

$$P_r(U,-L,T) = P\{\max x(t) \leq U \cap \min x(t) \geq -L,\ 0 < t \leq T\} = 1 - P_f(U,-L,T) \tag{14.51}$$

where $P_f(U,-L,T)$ is the probability of the time that the maximum or minimum response reaches the respective bound level for the first time, i.e. the failure probability of the double-sided boundary problem. In bridge wind engineering, this often considers only the fluctuating buffeting response in reliability analysis, and the mean buffeting response is considered separately. Thus, the upper and lower bound levels are the same ($U=-L$) and it is called the symmetry boundary condition. The dynamic reliability can be defined in terms of the absolute maximum response $Z_m = \max|x(t)|$.

$$P_r(U,T) = P\{Z_m \leq U,\ 0 < t \leq T\} = 1 - P_f(U,T) \tag{14.52}$$

It means that the dynamic reliability $P_r(U,T)$ of the bridge within the time period T is the probability that the absolute maximum response is not greater than the bound level U. In other words, the dynamic failure probability $P_f(U,T)$of the bridge is the probability of time that the absolute maximum response reaches the bound level for the first time within the time period.

14.5.2 Reliability Analysis based on Threshold Crossings

The first passage problem in random vibration has been studied for more than 65 years since it was first formulated by Rice [11]. Over the years, many researchers have addressed the first passage theory with various applications. Some of the important contributions include works by Lin [9], Crandall [20] and Vanmarcke [21]. An earlier development of the theory was summarized by Vanmarcke [22]. Though an exact theoretical solution still remains to be found, it is possible to introduce some reasonable assumptions to develop approximate approaches to solve the first-passage problems in structural engineering. Two most representative approximate solution methods are the Poisson crossing method [23] and the Markovian crossing method [21,24].

The Poisson crossing method for the first passage problem is based on two assumptions:

(a) in a tiny time interval, the positive (negative) crossing occurs at most once; and
(b) the times of crossings happening in different time intervals are independent.

As a result, the crossing process can be regarded as a Poisson process. The structural dynamic reliability in the time interval $(0,T)$ for the double-sided boundary problem can be obtained as follows [10]:

$$P_r(U,-L,T) = \exp\left\{-\int_0^T [N^+(U,t)+N^-(-L,t)]dt\right\} = 1 - P_f(U,-L,T) \tag{14.53}$$

The fluctuating buffeting response of a long-span bridge is often regarded as a zero-mean stationary Gaussian stochastic process. The rate of crossings of both the upper and lower levels can be expressed as:

$$N^+(U) = \frac{1}{2\pi}\frac{\sigma_{\dot{x}}}{\sigma_x}\exp\left(-\frac{U^2}{2\sigma_x^2}\right), N^-(-L) = \frac{1}{2\pi}\frac{\sigma_{\dot{x}}}{\sigma_x}\exp\left(-\frac{L^2}{2\sigma_x^2}\right) \tag{14.54}$$

where:

σ_x^2 and $\sigma_{\dot{x}}^2$ are the variances of $X(t)$ and $\dot{X}(t)$, respectively;
$-L$ and U are the lower and upper failure bounds of the bridge structure, respectively.

By substituting Equation 14.54 into Equation 14.53, the dynamic reliability of the bridge structure at the critical point within the time period T based on the Poisson crossing method can be calculated by:

$$P_r^P(U,-L,T) = \exp\left\{-\frac{T}{2\pi}\frac{\sigma_{\dot{x}}}{\sigma_x}\left[\exp\left(-\frac{U^2}{2\sigma_x^2}\right)+\exp\left(-\frac{L^2}{2\sigma_x^2}\right)\right]\right\} = 1 - P_f^P(U,-L,T) \tag{14.55}$$

For the symmetry boundary condition, Equation 14.55 reduces to:

$$P_r^P(U,-U,T) = \exp\left\{-\frac{T}{\pi}\frac{\sigma_{\dot{x}}}{\sigma_x}\left[\exp\left(-\frac{U^2}{2\sigma_x^2}\right)\right]\right\} = 1 - P_f^P(U,-U,T) \tag{14.56}$$

Related studies demonstrate that, for stationary Gaussian structural responses, when the threshold approaches an infinitely high value, the above method may give an accurate solution of the dynamic reliability [25]. However, when the threshold is not so high, some errors occur in the results. In fact, for narrow-band processes, the dynamic reliability computed by the Poisson crossing method tends to be low (on the conservative side), because their crossing events are not independent but occur in cluster [21,25].

An improved estimate of the failure probability was thus proposed by Vanmarcke [21], considering the envelope process as defined by Cramer and Leadbetter [26]. It is assumed that the next crossing event is related to the present crossing event, whereas it is independent of past events. Therefore, the crossing process is a Markov process and the method is called the Markovian crossing method. By using this method, the dynamic reliability of a bridge structure at a critical location with a non-stationary Gaussian response and a symmetry boundary condition can be computed by:

$$P_r^M(U,-U,T) = \exp\left\{-\int_0^T \phi(t)dt\right\} = 1 - P_f^M(U,-U,T) \tag{14.57}$$

in which:

$$\phi(t) = \frac{1}{\pi}\frac{\sigma_{\dot{x}}(t)}{\sigma_x(t)}\exp\left(-\frac{U^2}{2\sigma_x^2(t)}\right)\frac{1-\exp\left[-\sqrt{(\pi/2)}q^{1.2}(t)\cdot\dfrac{U}{\sigma_x(t)}\right]}{1-\exp\left(-\dfrac{U^2}{2\sigma_x^2(t)}\right)} \tag{14.58}$$

where q is the spectral bandwidth factor, defined as:

$$q = \left(1 - \frac{\lambda_2^2}{\lambda_0\lambda_4}\right)^{1/2} \tag{14.59}$$

For a stationary Gaussian response within the time period T and a symmetry boundary condition, the dynamic reliability is given by:

$$P_r^M(U,-U,T) = \exp\left\{-\frac{T}{\pi}\frac{\sigma_{\dot{x}}}{\sigma_x}\exp\left(-\frac{U^2}{2\sigma_x^2}\right)\frac{1-\exp\left[-\sqrt{(\pi/2)}q^{1.2}\cdot\dfrac{U}{\sigma_x}\right]}{1-\exp\left(-\dfrac{U^2}{2\sigma_x^2}\right)}\right\} = 1 - P_f^M(U,-U,T) \tag{14.60}$$

14.5.3 Reliability Analysis based on Peak Distribution

Another approximate method for calculating reliability or failure probability of a structure due to the first passage is based on the probability distribution of structural response peaks. It is assumed that there are a total of N response peaks $X_m = \{X_1, X_2, \ldots, X_N\}$ within the time interval T and that these peaks are independent. $Z_m = \max\{X_m\}$ is the largest peak. The probability that Z_m does not exceed the bound level U within the time period T is given by [24]:

$$P_{Z_m}(U,T) = P_{prob}\{Z_m \le U,\ 0 < t \le T\} = [P_{X_m}(U,T)]^N \tag{14.61}$$

where $P_{X_m}(U,T)$ is the probability distribution function of peaks X_m ($m = 1,2, \ldots, N$).

Clearly, the dynamic reliability of a bridge structure at a critical location within the time period T can be estimated using the probability given by Equation 14.61:

$$P_r(U,T) = P_{Z_m}(U,T) = [P_{X_m}(U,T)]^N = 1 - P_f(U,T) \tag{14.62}$$

As discussed in Section 14.2.6 for a zero-mean stationary Gaussian random process, the upper limit of the probability density function of peaks is a Rayleigh distribution (Equation 14.45), whereas the lower limit is a Gaussian distribution (Equation 14.46). The upper limit and the lower limit of the probability distribution function of peaks can then be found as:

$$P^u_{X_m}(\alpha) = \int_{-\infty}^{\alpha} f^u_p(\alpha)d\alpha = \int_{-\infty}^{\alpha} \frac{\alpha}{\sigma_x^2}\exp\left(-\frac{\alpha^2}{2\sigma_x^2}\right)d\alpha = 1 - \exp\left(-\frac{\alpha^2}{2\sigma_x^2}\right) \tag{14.63}$$

$$P^l_{X_m}(\alpha) = \int_{-\infty}^{\alpha} f^l_p(\alpha)d\alpha = \int_{-\infty}^{\alpha} \frac{1}{\left(2\pi\sigma_x^2\right)^{1/2}}\exp\left(-\frac{\alpha^2}{2\sigma_x^2}\right)d\alpha = \Phi\left(\frac{\alpha}{\sigma_x}\right) \tag{14.64}$$

where $\Phi(\beta)$ is the standard Gauss distribution function.

For a zero-mean stationary Gaussian random process with the symmetry boundary condition, the upper limit and the lower limit of the dynamic reliability of the bridge structure at the critical point within the time period T can be found as:

$$P^u_r(U,-U,T) = \left[1 - \exp\left(-\frac{U^2}{2\sigma_x^2}\right)\right]^N = 1 - P^u_f(U,-U,T) \tag{14.65}$$

$$P^l_r(U,-U,T) = \left[\Phi\left(\frac{U}{\sigma_x}\right)\right]^N = 1 - P^l_f(U,-U,T) \tag{14.66}$$

where the expected number N of peaks within the time period T can be calculated by:

$$N = N^+(0)T = \frac{\sigma_{\dot{x}}T}{2\pi\sigma_x} \tag{14.67}$$

14.5.4 Notes on Buffeting Reliability Analysis

It is suggested in the literature [19] that both the upper and lower limits of the buffeting reliability based on peak distribution should be calculated, and that the buffeting reliability based on threshold crossings should also be calculated for both the Poisson distribution and the Markov distribution. Then, when $0 \leq q \leq 0.35$, the random dynamic response is a narrow band process and the dynamic reliability of the bridge structure with the symmetry boundary condition, $P^N_r(U,-U,T)$, should satisfy the following inequality:

$$P^u_r(U,-U,T) < P^N_r(U,-U,T) < P^M_r(U,-U,T) \tag{14.68}$$

where:

$P^u_r(U,-U,T)$ is the dynamic reliability calculated based on the Rayleigh distribution;
$P^M_r(U,-U,T)$ is the dynamic reliability calculated using the Markovian crossing method.

When $0.35 \leq q \leq 1$, the random dynamic response is a broad band process, and the dynamic reliability of the bridge structure with the symmetry boundary condition, $P_r^B(U, -U, T)$, should satisfy the following inequality:

$$P_r^P(U, -U, T) < P_r^B(U, -U, T) < P_r^I(U, -U, T) \quad (14.69)$$

where:

$P_r^P(U, -U, T)$ is the dynamic reliability calculated using the Poisson crossing method;
$P_r^I(U, -U, T)$ is the dynamic reliability calculated based on the Gaussian distribution.

14.6 Reliability Analysis of Vortex-Induced Vibration

Vortex-induced forces and vortex-induced responses of a long-span bridge have been discussed in Chapter 4. It has been recognized that considerable vibration amplitude may occur when the vortex-shedding frequency is close to one of the natural frequencies of the bridge structure. In practice, aerodynamic measures or mechanical measures should be taken to eliminate vortex-induced vibration, but sometimes this cannot be achieved. Vortex-induced response is of limited amplitude and does not cause structural collapse, but it can result in relatively large displacements and discomfort to the drivers. Moreover, vortex-induced vibration commonly occurs at low wind speeds, so that its occurrence probability is high, resulting in long-term fatigue damage.

For a vortex-induced discomfortability problem, the buffeting reliability analysis method described in Section 14.5 can be applied in principle, in which vortex-induced responses are often narrow band. For a vortex-induced fatigue problem, it is necessary to estimate the occurrence period of vortex-induced vibration within the bridge's design life [27]. In this connection, the joint probability distribution function of wind speed and direction must first be found for the designated bridge site. Wind tunnel tests are then required to find out the occurrence ranges of wind speed and direction in which vortex-induced vibration will occur. After the occurrence period is determined, the buffeting-induced fatigue reliability analysis method, which will be introduced in the next subsection, can be applied in principle to estimate the fatigue damage caused by vortex-induced vibration.

14.7 Fatigue Reliability Analysis based on Miner's Rule for Tsing Ma Bridge

In the last few decades, many long-span suspension bridges have been built throughout the world. Some of these bridges carry both trains and road vehicles and are located in wind-prone regions, and therefore they are subjected to the long-term combined action of railway, highway and wind loading. For such bridges, fatigue is one of the crucial structural safety issues with which bridge engineers are mostly concerned.

Fatigue damage of steel bridge structures under traffic loading has been investigated by many researchers [28,29], and fatigue analysis using Miner's rule is a well-known approach for its evaluation, in which the magnitude of cyclic stress ranges is the principal parameter. It has been found that a small error in stress ranges could lead to a significant discrepancy in the estimation of structural service life. However, uncertainties arising from loads as well as structural modeling are unavoidable, which may call the estimation of fatigue damage in question [30,31]. Furthermore, very little of the literature discusses fatigue reliability of long-span suspension bridges under multiple types of random loading.

Recently-developed structural health monitoring (SHM) technology provides a better solution for these problems. The SHM technology has been discussed and the comprehensive SHM system installed

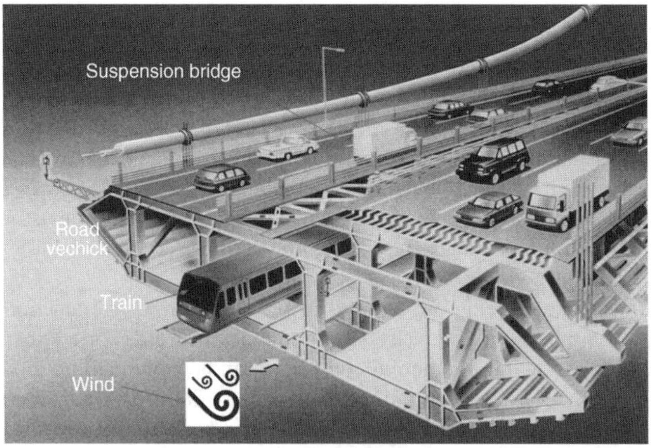

(a) Tsing Ma Bridge under railway, highway, and wind loading

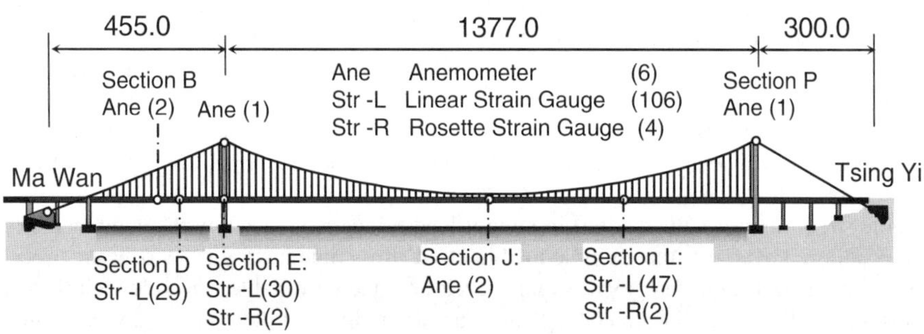

(b) Locations of strain gauges and anemometers

Figure 14.4 The Tsing Ma Bridge and locations of strain gauges and anemometers (from [52]) (Reproduced with permission from Elsevier).

in the Tsing Ma suspension bridge in Hong Kong has been demonstrated in Chapter 9. Figure 14.4 illustrates the Tsing Ma suspension bridge and the locations of strain gauges and anemometers.

An engineering approach for dynamic stress analysis of long-span suspension bridges under multiple types of dynamic loading has been presented and verified in Chapter 11 using the measurement data acquired from the SHM system of the Tsing Ma Bridge [32,33]. These works make it possible to propose a framework in this section for the fatigue reliability analysis of long-span suspension bridges under the long-term combined action of railway, highway and wind loading.

A limit-state function in terms of the daily sum of m-power stress ranges is first defined. Probabilistic models of railway, highway, and wind loading are established, based on the measurement data acquired from the SHM system of the Tsing Ma Bridge. The daily stochastic stress responses induced by the multiple types of random loading are simulated at the fatigue-critical locations of the bridge deck by using the finite element method and the Monte Carlo Simulation (MCS), together with the loading probabilistic models established. The probability distribution of the daily sum of m-power stress ranges is estimated, based on the daily stochastic stress responses. The probability distribution of the sum of m-power stress ranges for a given time period is then evaluated in consideration of future traffic growth patterns. Finally, the fatigue failure probabilities of the bridge at the fatigue-critical locations are calculated for different time periods.

14.7.1 Framework for Fatigue Reliability Analysis

This section sets out the framework for the fatigue reliability analysis of long suspension bridges under railway, highway and wind loading [34]. Given the particularity of such bridges, some key issues need to be considered. First, there are uncertainties in the different types of fatigue loading (railway, highway and wind loading). Thus, probabilistic models must be established for each individual loading. Second, as uncertainties exist in the random combination of the multiple loadings, it is necessary to calculate the multiple load-induced stochastic stress responses. Finally, uncertainties also exist in the prediction of future traffic loading, so future loadings and traffic growth trends are assumed to estimate the fatigue damage accumulation in the period concerned.

The first step of the structural reliability analysis is to define a limit-state function which adequately describes the relationship between the resistance and the applied load for a structural member. The cumulative fatigue damage within the period concerned can be computed based on the British S-N curves and Palmgren-Miner's rule. The fatigue limit-state is defined as cumulative damage that is greater than or equal to the damage accumulation index Δ, which is typically taken as one when the deterministic method is adopted.

Distinct from the deterministic fatigue analysis, the randomness in both the fatigue loading and the fatigue resistance are considered in the reliability analysis. Δ is a random variable in terms of resistance, and it is assumed to be lognormal, with a mean value of 1.0 and a coefficient of variation (CoV) of 0.3 for metallic materials [35]. The fatigue detail coefficient K is a random variable in terms of resistance. It is also assumed to be lognormal, with a CoV of 0.3, and its mean value for different fatigue detail classes can be obtained in British Standard [36]. Given that the uncertainty in bridge models can be largely reduced by modeling the bridge details and model updating, it is not treated as a random variable in this study.

Dynamic stress (or stress fluctuation) in a multi-load bridge is induced by the combined effects of multiple loadings. Given that the multiple load-induced stress time history is a stochastic process, the numbers of stress cycle at different stress range levels are random variables. In addition, urban passenger trains often follow a regular timetable that is similar on different days, and highway traffic conditions on certain days are also similar. Thus, the cycle of railway and highway traffic is close to one day. As a result, the daily sum of m-power stress ranges $S_{\mathrm{mr,j}}$ ($\mathrm{j} = 1, \ldots, N_{\mathrm{b}}$, where N_{b} is the total number of days in the period concerned), is treated as a random variable in terms of fatigue loading.

Taking these random variables into account, the limit-state function for fatigue reliability analysis is defined as follows:

$$g(\mathbf{X}) = g\left(K, \Delta, \sum_{\mathrm{j}=1}^{N_{\mathrm{b}}} S_{\mathrm{mr,j}}\right) = \Delta - \frac{1}{K}\sum_{\mathrm{j}=1}^{N_{\mathrm{b}}} S_{\mathrm{mr,j}} \tag{14.70}$$

$$S_{\mathrm{mr,j}} = \sum_{\mathrm{i}=1}^{N_1} n_{\mathrm{i}}\left(\sigma_{\mathrm{r,i}}\right)^{\mathrm{m}} + \frac{1}{\left(\sigma_{\mathrm{r,0}}\right)^2}\sum_{i=1}^{N_2} n_{\mathrm{i}}\left(\sigma_{\mathrm{r,i}}\right)^{\mathrm{m+2}} \tag{14.71}$$

where n_i is the applied number of stress cycles at the stress range level $\sigma_{r,i}$, which is counted from the multi-load-induced daily stochastic stress response time history using the rainflow counting method. According to the two-slope S-N curves defined in British Standard [36], if the stress range level $\sigma_{r,i}$ is less than the fatigue limit $\sigma_{r,0}$, then it will be reduced in proportion. N_1 and N_2 are the number of stress range levels above $\sigma_{r,i}$ and below $\sigma_{r,0}$, respectively.

The object of the structural reliability analysis of a structural member or system is to estimate its failure probability. The fatigue failure probability P_{f} can be evaluated and related to the reliability index using the following relationship:

$$P_f = P(g(\mathbf{X}) < 0) = \Phi(-\beta) \tag{14.72}$$

The reliability index β is estimated based on the limit-state function using the FORM method. As the variables Δ and K are not normally distributed, a simple approximate transformation method is applied to transform the non-normal distribution into a normal distribution for use in the FORM method. One of most commonly used recursive algorithms, the HL-RF method [37], is adopted to solve the reliability index β.

Given that the limit-state function in Equation 14.72 is non-linear, several iterations are required to obtain convergence to β. If convergence is not achieved after the first iteration, then the process is repeated until β_{n-1} and β_n in the step $n-1$ and n satisfy the stopping criterion $|\beta_n - \beta_{n-1}|/\beta_{n-1} \leq \varepsilon_r$, where $\varepsilon_r = 0.001$, used in this study.

The estimation of the probability distribution of $\sum_{j=1}^{N_b} S_{mr,j}$ and its distribution parameters is the main concern, and can be achieved through the following steps:

(a) Establish probabilistic models of the railway, highway and wind loads, based on the measured load data.
(b) Generate the multiple load-induced daily stochastic stress responses at the fatigue-critical locations through the combination of the stress responses induced by the individual loadings generated by the established probabilistic models.
(c) Estimate the probability distribution of S_{mr} from the samples, which are computed from the generated daily stochastic stress responses.
(d) Estimate the probability distribution of $\sum_{j=1}^{N_b} S_{mr,j}$ from the samples of S_{mr} in the period concerned, based on assumed future loadings and traffic growth patterns.

The establishment of probabilistic loading models is a critical step in describing the randomness in railway, highway and wind loadings. Given that the loading conditions of a long suspension bridge can be quite different from those of other bridges, the establishment of these models is case-dependent. The Tsing Ma suspension bridge, which is located in a wind-prone region and carries both trains and road vehicles, is taken as the case study here.

14.7.2 Probabilistic Model of Railway Loading

Information on the trains running across the Tsing Ma Bridge is converted from the typical strain data recorded under the two rail tracks on the bridge. As the railway traffic in November 2005 is close to the average traffic conditions, data on the trains in this month are adopted to build a database of railway loading parameters. The main parameters include arrival instant, running speed, heading direction, number of bogies (two wheel-sets in each bogie and two bogies in each vehicle), bogie weight and bogie spacing.

The measured train data analysis indicates that some of the railway loading parameters can be considered as variables, whereas others can be regarded as constants. As almost all trains running across the bridge since 2005 have been eight-car trains, trains are all assumed to have a configuration equivalent to that of a standard eight-car train. The running speeds of the trains are assumed to be constant and equal to the mean train speed recorded in the database. The weight distribution of the 16 bogie loads in an eight-car train is also assumed to be the same for all trains.

The gross train weight (GTW) and train arrival time are treated as random variables. The random nature of the GTW is mainly due to uncertainties in the number of passengers, whereas the random nature of the train arrival time is due to many reasons, such as variability in the running speed and unexpected events. Given that most passing trains follow a scheduled timetable, except for the occasional unexpected event leading to delay or cancellation, the scheduled arrival time of each train is assumed to be a constant, and the difference between the actual and scheduled arrival times is assumed to be a random variable. As a result, the actual arrival time of the *i*th train is the sum of the scheduled arrival time and a random deviation.

The histogram of the GTW is shown in Figure 14.5 in terms of the established database for the GTW. The histogram cannot be fitted by a single conventional probability distribution function. Thus, a mixture model distribution [38] is used to describe the probability distribution of the GTW, and this is also applied to fit the probability distribution of the random variables in the highway loading model. Suppose that the random variable X can be described as a mixture of n components of the random variables Y_i. The probability density function of X, which is denoted as $f_X(x)$, is then a weighted sum of the probability density functions of its components$f_{Y_i}(x)$.

$$f_X(x) = \sum_{i=1}^{n} \alpha_i f_{Y_i}(x) \tag{14.73}$$

where the weighted ratio $0 < \alpha_i < 1$ and $\alpha_i + \ldots + \alpha_n = 1$. The parameters α_i and f_{Y_i} are determined using the expectation-maximization (EM) algorithm, which is a two-step method for finding the maximum likelihood estimates of the parameters in statistical models [39].

The theoretical probability density function is fitted by a mixture of two weighted normal distributions using the EM algorithm. The mean value, standard deviation and weighted ratio of the first normal function are 328.1 tons, 34.2 tons and 38.2%, and that of the second normal function are 338.1 tons, 12.3 tons and 61.8%, respectively. Figure 14.5 shows both the measured and fitted distributions. It can be seen that the latter matches the former quite well.

Figure 14.6 gives the scheduled arrival time of trains heading in both directions, which is estimated from the mean arrival time based on the train arrival time database. The random variable, which represents the deviation of the actual arrival time from the scheduled arriving time, is modeled by a single zero-mean normal distribution with a standard deviation of 59.4 seconds (see Figure 14.7).

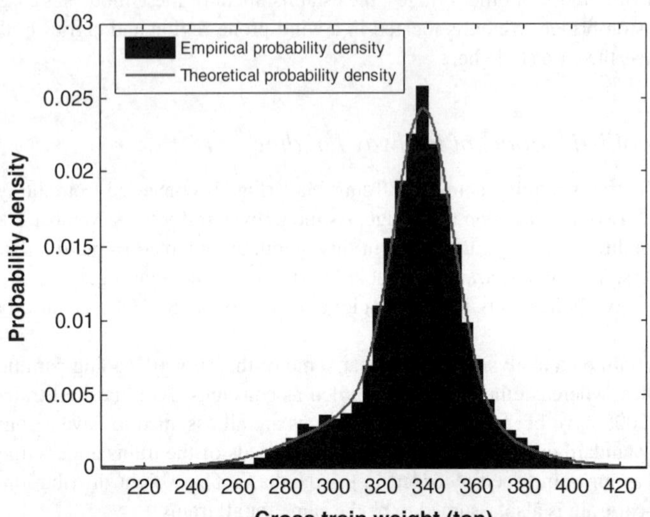

Figure 14.5 Fitting the distribution of gross train weight (from [34]) (Reproduced with permission from ASCE).

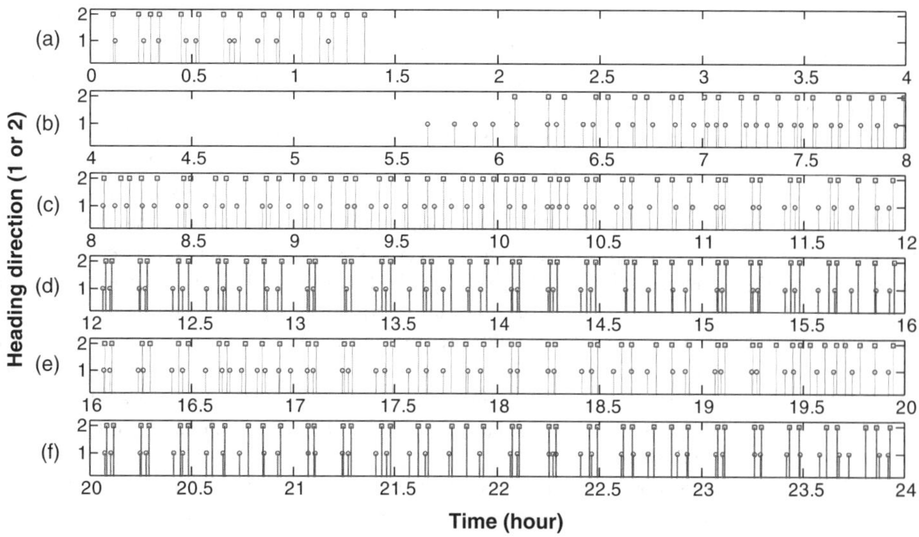

Figure 14.6 Daily scheduled arrival time of trains: (a) 0 : 00 ~ 4 : 00; (b) 4 : 00 ~ 8 : 00; (c) 8 : 00 ~ 12 : 00; (d) 12 : 00 ~ 16 : 00; (e) 16 : 00 ~ 20 : 00; (f) 20 : 00 ~ 24 : 00.

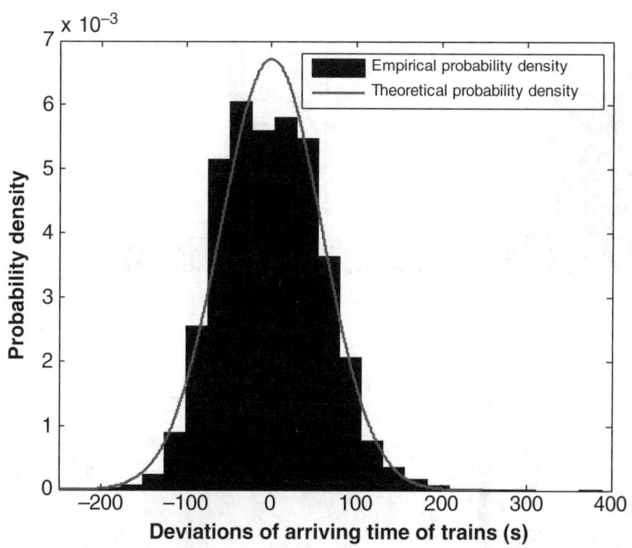

Figure 14.7 Fitting the distribution of the deviation in arrival time (from [34]) (Reproduced with permission from ASCE).

14.7.3 Probabilistic Model of Highway Loading

Information on the road vehicles running across the bridge is recorded by dynamic weigh-in-motion (WIM) stations at the approach to the Lautau Toll Plaza near the bridge. Highway traffic in November 2005 is close to the average traffic conditions, and information on traffic lanes is available for the measured data. Thus, the road vehicle data in this month are adopted to build a database of highway loading parameters. However, only heavy road vehicles with a gross vehicle weight (GVW) of over 30 kN are included in the database, as road vehicles with a lower GVW contribute little to the fatigue damage. The main road vehicle parameters include the vehicle type, arrival time, running speed, heading direction, traffic lane used, axle number, axle weight and axle spacing.

Given that the length of a road vehicle is very small compared with the total length of the suspension bridge, a road vehicle is simplified into a force concentrated on the vehicle center, rather than the forces on its axles. Thus, the axle number, axle weight and axle space of a typical road vehicle are not considered. The running speed of a heavy road vehicle is assumed to be a constant, and is determined as the mean value of all of the road vehicles in the database. The GVW and time interval between successive vehicles are treated as random variables. The first variable describes the loading intensity of highway vehicles, while the second is related to the frequency of occurrence of highway vehicles.

Analysis of the measured road vehicle data indicates that most heavy road vehicles use the slow lane preferentially, and then the middle lane, but they seldomly use the fast lane. The road vehicles running on the bridge can be classified into eight categories, as detailed in Chapter 6. As the road traffic conditions differ among the slow, middle and fast lanes, the probability distributions of the GVW on the different lanes must be established.

Figure 14.8 displays the histograms of the GVW on the slow, middle and fast lanes in both directions, which are estimated based on the GVW database. There is more than one peak in the distribution of the GVW on the slow and middle lanes, which may be due to road vehicles in different categories running

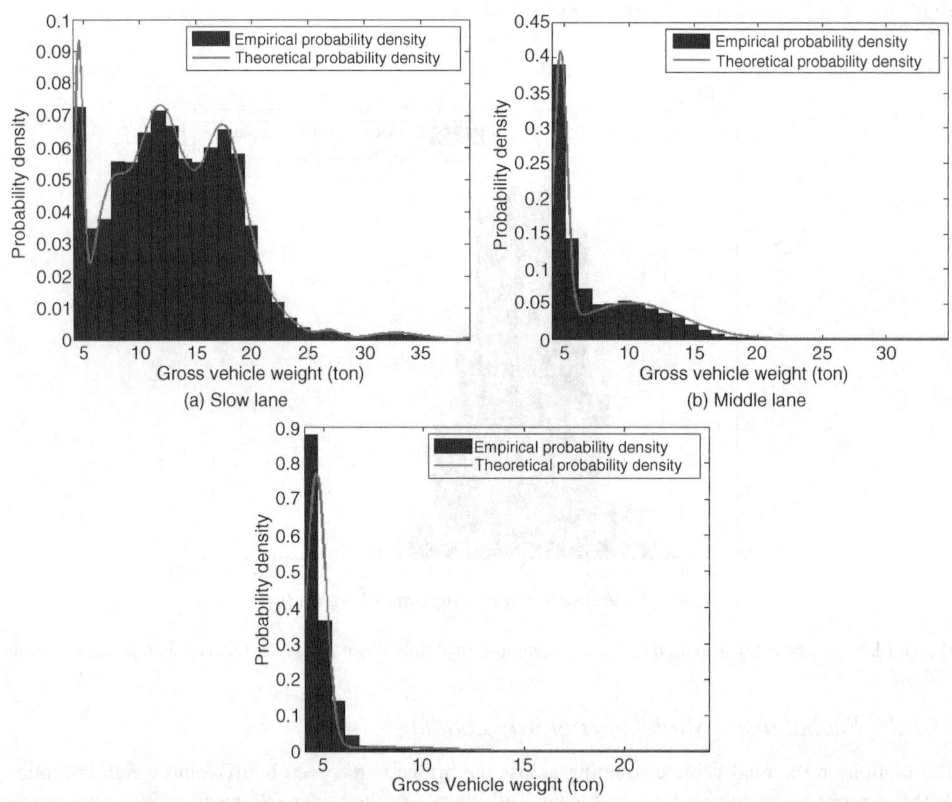

Figure 14.8 Fitting the distribution of gross vehicle weight.

across lanes. In this regard, a mixture of multiple normal distributions is used to describe the theoretical density function of the GVW on the slow, middle and fast lanes, as shown in Figure 14.8. The theoretical density function fits the measured histogram quite well.

As the time interval of successive vehicles during "rush hour" and "normal hour" may be different, these are separately fitted using different probability distributions. The time period from 23:00 to 8:00 is defined as "normal hour", during which fewer road vehicles pass over the bridge, whereas the period from 8:00 to 23:00 is called "rush hour", during which more road vehicles run on the bridge. The time interval between successive vehicles on each traffic lane can be derived from the vehicle arrival time database.

The histograms of the time interval of successive vehicles on the slow, middle and fast lanes during "rush hour" and "normal hour" are shown in Figure 14.9, which shows that the occurrence probability decreases as the time interval increases on all traffic lanes and in all periods. The mean time interval during "normal hour" is larger than that during "rush hour." Of the three lanes, the smallest mean time interval is that for the slow lane, followed by the middle lane, and then the fast lane. The probability density functions of the time interval of successive vehicles are modeled as an exponential distribution, as shown in Figure 14.9. The theoretical density function matches the histogram well.

14.7.4 Probabilistic Model of Wind Loading

The mean wind speed and direction are two random variables considered in the probabilistic model of wind loading. A joint probability distribution function of the mean wind speed and direction is utilized to describe the wind intensity at the bridge site. The distribution of the wind speed for any given wind direction follows the Weibull distribution. The distribution parameters are determined from wind records of the hourly mean wind speed and direction in the period 1 January 2000 to 31 December 2005, which were collected by an anemometer installed on the top of the Ma Wan tower of the Tsing Ma Bridge (see Chapter 2). In addition, the maximum hourly normal mean wind speed at the deck level is obtained through the following steps:

(a) The maximum wind speed at the top of the tower in each wind direction within the 120-year design life is obtained from the joint probability distribution.
(b) The maximum wind speed obtained at the top of the tower is then converted to the average deck level of the bridge.
(c) Finally, the different wind speed directions are converted to normal mean wind speeds perpendicular to the deck.

Based on the normal mean wind speed at the bridge deck and the other wind characteristics, the buffeting and self-excited forces over the bridge decks can be computed. More details can be found in Chapters 2, 4 and 11.

14.7.5 Multiple Load-Induced Daily Stochastic Stress Response

In addition to the randomness in each loading type, the random combination of the various loadings must also be considered. In this section, the daily stochastic stress responses induced by multiple types of loading are computed to consider the randomness due to multiple loading effects. In the first step, the dominant parameters are determined from probabilistic loading models using Monte Carlo Simulation (MCS). The stress responses respectively induced by the railway, highway, and wind loadings are then computed separately by the finite element-based stress analysis methods and superposed.

Because the train arrival time is scheduled on a daily basis, the one-day time history of the stress responses induced by railway loading is computed as one sample. Given the differences in the highway traffic conditions between "normal hour" and "rush hour," the hourly stress response due to highway loading is computed and then extended to the one-day stress response time history. As the hourly mean

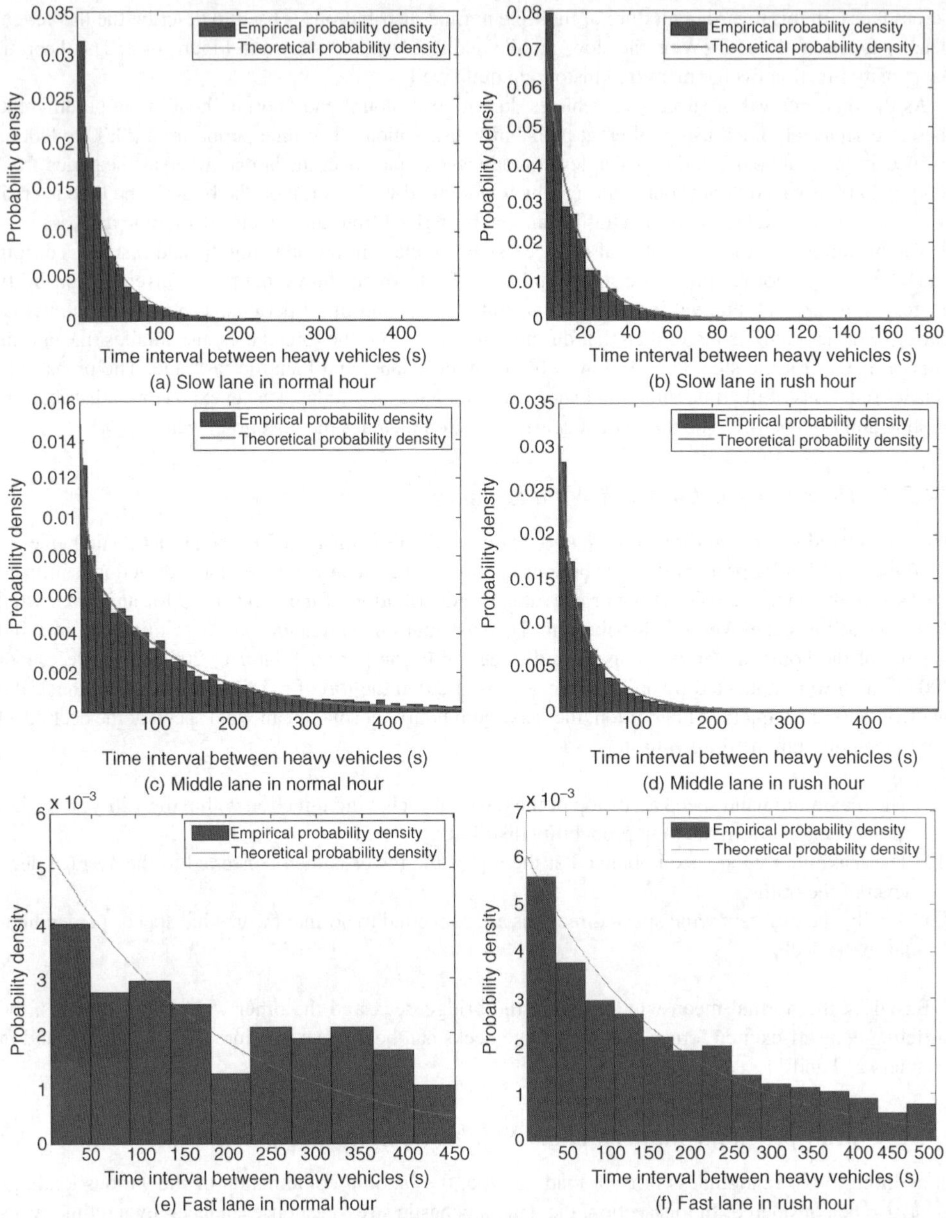

Figure 14.9 Fitting the distribution of the time interval between heavy road vehicles.

wind speed is considered, the hourly stress response due to wind loading is computed and then extended to the one-day stress response time history. Because the bridge is closed to traffic when the mean wind speed recorded on site is very high, stress responses under this condition are assumed to be induced by wind loading only.

Figure 14.10a presents a flow chart of the computation of the daily stochastic stress response due to railway loading. To calculate this stress response, the GTW of the *i*th train on the *j*th track is selected

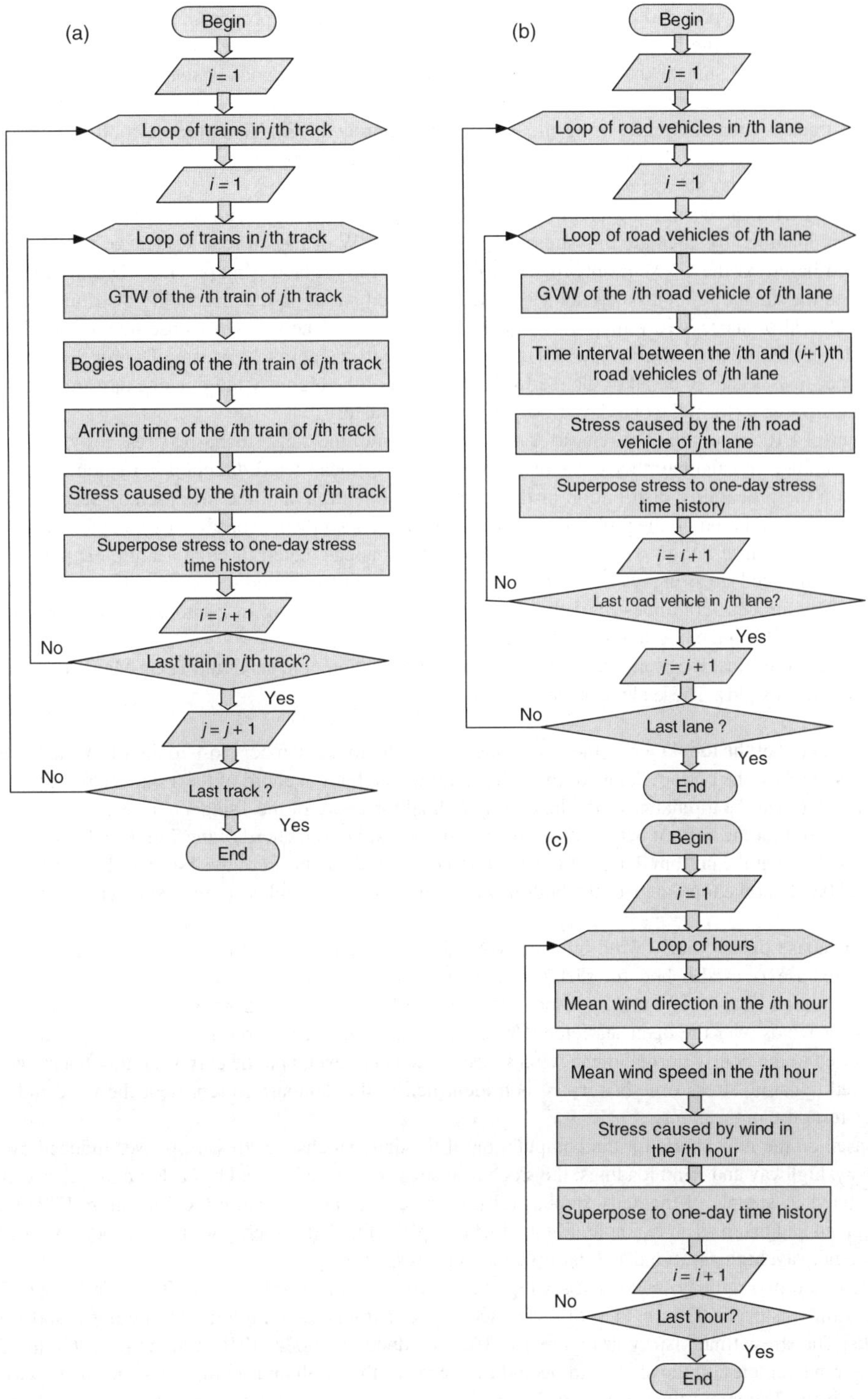

Figure 14.10 Flow charts for computing the daily stochastic stress responses: (a) railway; (b) highway; and (c) wind loading.

based on the GTW probability distribution (Figure 14.5) and is then proportionally distributed to each bogie according to the bogie loading distribution. The arrival time of the *i*th train on the *j*th track is determined by the sum of the scheduled arrival time (Figure 14.6) and a random deviation which is obtained based on the normal distribution (Figure 14.7). The stress response caused by the *i*th train on the *j*th track is then computed using the engineering approach [33]. This process continues until the stresses caused by all of the trains traveling in both directions in one day are computed and superposed to obtain the one-day stress response time history.

Figure 14.10b shows the flow chart for the computation of the daily stochastic stress response due to highway loading. To calculate this stress response, the GVW of the *i*th road vehicle of the *j*th lane is selected based on the GVW probability distribution for the *j*th lane (Figure 14.8). The time interval between the *i*th and $(i+1)$th road vehicles is then selected based on the probability distribution of the time interval of successive vehicles, as shown in Figure 14.9. The stress response due to the *i*th road vehicle on the *j*th lane is then computed using the engineering approach. Again, this process continues until the stresses caused by all of the road vehicles on all of the highway lanes in one day are computed and superposed to form the one-day stress response time history.

Figure 14.10c shows the flow chart for the computation of the daily stochastic stress response due to wind loading. To calculate this stress response, the hourly mean wind direction is randomly selected based on the probability distribution of the mean wind direction, and the mean wind speed is subsequently selected based on the probability distribution of the mean wind speed in a given wind direction. The stress response due to wind loading in the *i*th hour is calculated through a finite element-based buffeting-induced stress analysis that considers both buffeting forces and self-excited forces (see Chapter 4). This process continues until the stresses due to wind loading in 24 hours are computed and superposed to form the one-day stress time history.

Six fatigue-critical locations are determined from the key components of the Tsing Ma Bridge for the fatigue analysis [34]. These elements are:

- E32123 (T) at the top flange of the outer longitudinal diagonal member close to the Ma Wan Tower;
- E34415 (B) at the bottom flange of the outer longitudinal bottom chord of the Tsing Yi Tower;
- E40056 (T) at the top flange of the inner longitudinal top chord of the Tsing Yi Tower;
- E40906 (B) at the bottom flange of the inner longitudinal bottom chord of the Tsing Yi Tower;
- E39417 (B) at the bottom flange of the T-section of the railway beam of the Tsing Yi Tower; and
- E55406 (T) at the top flange of the bottom web of the cross-frame close to the Tsing Yi Tower.

A database of the hourly wind-induced dynamic stress responses at these locations is established. The database is established for different hourly normal mean wind speeds and directions, from 5–26 m/s at an interval of 1 m/s for winds over the overland fetch, and from 5–16 m/s at an interval of 1 m/s for winds over the open sea fetch. Here, the mean wind speed and direction in a given hour are converted to the hourly normal mean wind speed at the deck level, and the corresponding hourly wind-induced dynamic stress time history is then identified in the database to represent the wind-induced stress response in this hour.

Based on the flow charts for the computation of the daily stochastic stress responses induced by the railway, highway and wind loadings, the stochastic stress response induced by each loading type can be calculated. A sample of the daily stochastic stress responses at a fatigue-critical location, E32123, is computed and taken as an example. Figure 14.11 displays the daily stochastic stress responses induced by the railway, highway, wind and multiple loadings, respectively.

There are two railway tracks and six highway traffic lanes available on the Tsing Ma Bridge. The stress time history in Figure 14.11(a) is induced by 222 trains traveling on both the north and south tracks. The stress time history in Figure 14.11(b) is induced by 5082, 1716, 524, 5146, 1760 and 526 heavy road vehicles on the slow, middle and fast lanes of the north and south three-lane carriageways, respectively. The stress time history in Figure 14.11(c) is induced by wind at an hourly normal mean

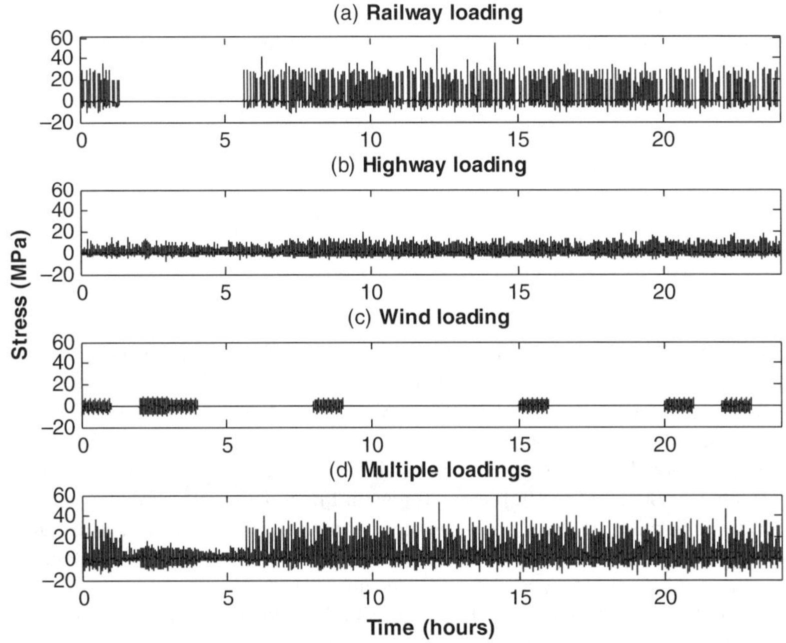

Figure 14.11 Sample of the daily stochastic stress time history (from [34]) (Reproduced with permission from ASCE).

wind speed that ranges from 2–10 m/s. To obtain the multiple load-induced stress time history, the superposition method is applied to these three time histories using the engineering approach [33]. Figure 14.11(d) gives a sample of the daily stochastic stress time history induced by the combination of railway, highway and wind loadings.

14.7.6 Probability Distribution of the Daily Sum of m-power Stress Ranges

Based on the multi-load-induced daily stochastic stress time histories generated at the fatigue-critical locations, the daily sum of m-power stress ranges, S_{mr}, can be calculated using Equation 14.71. S_{mr} is a random variable because uncertainties exist in the stochastic stress time history induced by the simultaneous action of the railway, highway and wind loading. As a sufficient number of samples must be considered to estimate the probability distribution of S_{mr} and its distribution parameters exactly, the sample size must first be determined.

This can be achieved by discussing the accuracy of the parameter estimates for the mean value and variance based on statistical error formula, as several probability distribution parameters are related to them. The random error of the estimated parameters is the main concern, because it can be reduced by considering more samples. The normalized random error of the variable x is given by:

$$\varepsilon_r = \frac{\sqrt{\mathrm{Var}[\hat{x}]}}{x} = \frac{\sqrt{E[\hat{x}^2] - E[\hat{x}]^2}}{x} \tag{14.74}$$

where $\hat{x}$ is an estimator for the parameter x.

Based on the assumption that samples are from a normal distribution with a mean value $\mu_x \neq 0$ and a variance σ_x^2, the normalized random error of the mean estimate can be given by Julius and Allan [40]:

$$\varepsilon_r[\hat{\mu}_x] \approx \frac{1}{\sqrt{N}}\left(\frac{\sigma_x}{\mu_x}\right) \approx \frac{1}{\sqrt{N}}\left(\frac{\hat{\sigma}_x}{\hat{\mu}_x}\right) \tag{14.75}$$

and the normalized random error of the variance estimate can be given by:

$$\varepsilon_r\left[\hat{\sigma}_x^2\right] \approx \sqrt{\frac{2}{N}} \tag{14.76}$$

Equations 14.75 and 14.76 indicate that the normalized random errors of the mean and variance estimate decrease with the increasing sample size N. The normalized random error of the variance estimate decreases to 0.1 when $N = 200$, and the convergence of the mean estimate is better than that of the variance estimate if $\hat{\mu}_x$ is larger than $\hat{\sigma}_x$.

The probabilistic models of railway and highway loading are established based on loading information from November 2005, and the traffic conditions in this month are regarded as analogous to the current loading conditions. 200 daily stochastic stress time histories under current loading conditions are computed at a given fatigue-critical location, and 200 samples of S_{mr} are then obtained from these to estimate the probability distribution of S_{mr} and its parameters.

As an example, Figure 14.12 shows the histogram and theoretical distributions of S_{mr} at the fatigue-critical location E32123. The histogram is estimated based on the 200 generated samples, and the theoretical density function is modeled as a normal distribution. The theoretical density function matches the histogram quite well. The mean value and STD of the fitted normal distribution are 1.15×10^{12} and 3.22×10^{8}.

Given that both the railway and highway loading may increase in future, possible future loadings are assumed and the fatigue damage induced by them is computed. The first possible future loading assumes a 30% increase in both railway and highway loading, compared with the current traffic

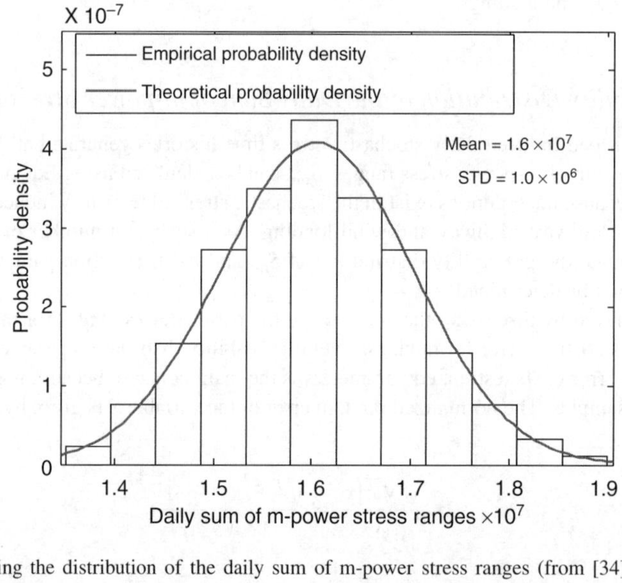

Figure 14.12 Fitting the distribution of the daily sum of m-power stress ranges (from [34]) (Reproduced with permission from ASCE).

Table 14.1 Mean value/STD of the daily sum of m-power stress ranges

Load case	Fatigue-critical location					
	E32123	E34415	E40056	E40906	E55406	E39417
CL: R + H + W	1.6×10^7/ 1.0×10^6	1.5×10^7/ 1.0×10^6	9.5×10^6/ 7.5×10^5	9.2×10^6/ 7.6×10^5	6.1×10^6/ 3.0×10^5	6.7×10^6/ 4.4×10^5
FL1: 1.3R + 1.3H + W	3.8×10^7/ 1.8×10^6	3.8×10^7/ 1.5×10^6	2.5×10^7/ 1.3×10^6	2.5×10^7/ 1.3×10^6	1.9×10^7/ 5.8×10^5	2.0×10^7/ 8.5×10^5
FL2: 1.1R + 2.0H + W	3.8×10^7/ 1.6×10^6	3.2×10^7/ 1.4×10^6	1.6×10^7/ 1.0×10^6	1.7×10^7/ 9.9×10^5	1.2×10^7/ 4.5×10^5	1.1×10^7/ 6.2×10^5

Note: CL – Current loading, FL1 – Future loading 1, FL2 – Future loading 2; R – Railway loading, H – Highway loading, W – Wind loading.

loading. Another possible future loading assumes a 10% increase in railway loading and a 100% increase in highway loading. The latter assumption is based on the finding that the current railway loading is close to the design railway loading of the Tsing Ma Bridge, whereas the current highway loading is much less than the design loading. That is to say, there is more room for an increase in the highway loading than in the railway loading.

The growth in traffic loading is simulated by increasing the GTW in the probabilistic model of railway loading and the GVW in the probabilistic model of highway loading, respectively. The probabilistic model of wind loading remains constant, because it is established on the basis of the 120-year design life of the bridge. Following the aforementioned procedure, the random variable S_{mr} at the six fatigue-critical locations under three types of loading is fitted with a normal distribution. The mean values and STDs of these distributions are listed in Table 14.1.

A comparison of the current loading and the first future loading scenario indicates that fatigue damage is very sensitive to traffic growth, because a mere 30% increase in traffic loading leads to more than 100% increase in S_{mr} (fatigue damage). Although a linear relationship can be found between the traffic growth and the traffic induced stress range, the relationship between traffic growth and fatigue damage is non-linear, because the fatigue damage is a function of m-power stress range (m = 3 or 5 for most of detail classes in British Standard). A comparison of the two future loading scenarios indicates that there is more room for increasing the highway loading than the railway loading, because S_{mr} in the first future scenario (30% increase in both railway and highway loading) is more dangerous than in the second scenario (10% increase in railway loading and 100% increase in highway loading).

14.7.7 Probability Distribution of the Sum of m-power Stress Ranges within the Period

To estimate the probability distribution of the variable $\sum_{\mathrm{j}=1}^{N_{\mathrm{b}}} S_{\mathrm{mr,j}}$, some assumptions are made about future traffic growth patterns, so that $S_{\mathrm{mr,j}}$ can be generated for the whole period. The period concerned in this study is designated as 120 years, which is equal to the design life of the Tsing Ma Bridge. First, no traffic growth in 120 years is assumed – that is to say, the traffic conditions in the whole period are the same as the current loading conditions. S_{mr} under the current loading conditions is found to follow a normal distribution, and the distribution parameters are estimated in Table 14.1. Thus, the random variables of $S_{\mathrm{mr},i}$ ($i = 1, \ldots, N$, $N = 120 \times 365$) are assumed to follow an identical normal distribution, with a mean value of μ and an STD of σ. 120 years of S_{mr} are generated from the normal distribution using the MCS method, and then summed to obtain a sample of $\sum_{\mathrm{j}=1}^{N_{\mathrm{b}}} S_{\mathrm{mr,j}}$.

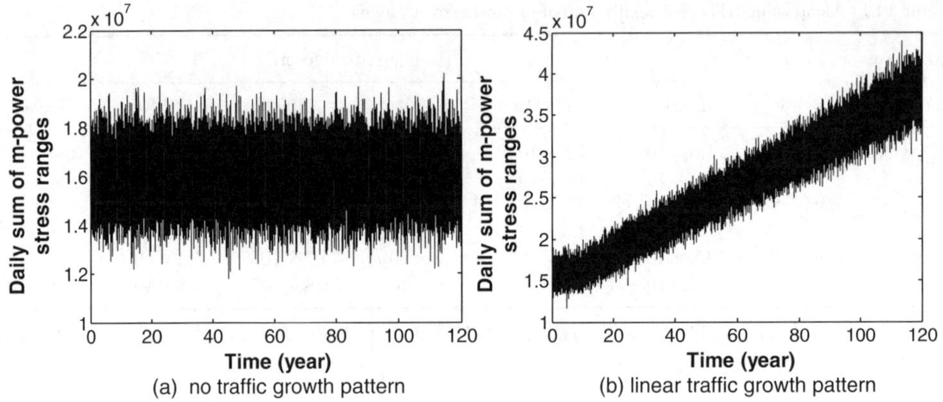

Figure 14.13 Samples of daily m-power stress ranges over 120 years (from [34]) (Reproduced with permission from ASCE).

Figure 14.13a shows a sample of 120 years of S_{mr} at the fatigue-critical location of E32123 under the condition of no traffic growth. The histogram of $\sum_{j=1}^{N_b} S_{mr,j}$ over the 120 years is estimated based on 200 samples, and the theoretical density function is modeled as a normal distribution. Figure 14.14a demonstrates that the theoretical distribution matches the histogram quite well.

Actually, according to the central limit theorem, $\sum_{j=1}^{N_b} S_{mr,j}$ follows a normal distribution with a mean value of $N_b\mu$ and an STD of $\sqrt{N_b}\sigma$ when N_b is sufficiently large [41]. The mean value and STD estimated using the central limit theorem are consistent with those fitted from the normal distribution in Figure 14.14a.

Then, $S_{mr,j}$ in the period concerned are generated based on the assumptions of different traffic growth patterns. The variable S_{mr} in November 2005 follows the distribution under the current loading conditions, as it is determined based on the railway and highway loading data in that

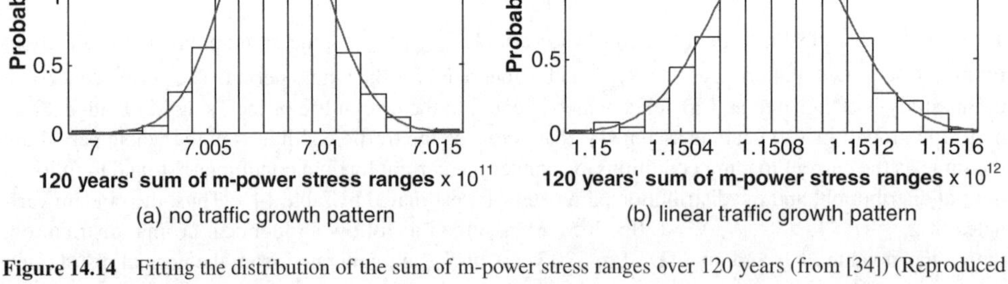

Figure 14.14 Fitting the distribution of the sum of m-power stress ranges over 120 years (from [34]) (Reproduced with permission from ASCE).

month. S_{mr} in the last month of the 120-year period is assumed to follow the distribution under the future loading condition, and its distribution parameters are listed in Table 14.1. Another important assumption is that the variable S_{mr} in a given month follows a constant normal distribution, and that the distribution parameters of the mean value and STD among different months in the 120 years change following the traffic growth patterns defined in the following.

The first growth pattern is assumed to take a linear form, in which no growth takes place in the first T_r years but does take place from T_r to T_t in a linear fashion at a constant growth rate α_g. The growth function is given as:

$$X = X_0 \quad t \le T_r, \quad X = X_0(1+\alpha_g t) \quad t > T_r \quad \text{(linear)} \tag{14.77}$$

Another two growth patterns are also assumed, both of which are exponential types [42]. Simple algebraic considerations lead to the following expressions:

$$X = X_0 \quad t \le T_r, \quad X = X_0(1+n\alpha_g)^{t/(nT_r)} \quad t > T_r \quad \text{(Exp-1)} \tag{14.78}$$

and:

$$X = X_0 \quad t \le T_r, \quad X = X_0\left\{1+(1+n\alpha_g)\left[1-\left(\frac{1}{1+n\alpha_g}\right)^{t/(nT_r)}\right]\right\} \quad t > T_r \quad \text{(Exp-2)} \tag{14.79}$$

To facilitate the comparison among the different growth patterns, the parameters of the exponential types are determined by assuming that when $t = T_t = nT_r$, the variable $X = X_0(1+\alpha_g T_t)$. The distribution parameters of S_{mr} remain constant in the first eight years of the bridge's life (from 1998 to 2005) and are equal to the parameters under the current loading conditions. They increase in the months between 2006 and 2117, following three growth patterns. The mean value of S_{mr} at E32123 is taken as an example. The mean value under the current loading is $X_0 = 1.6 \times 10^7$, and that under the first future loading scenario is $X_{Tt} = 3.8 \times 10^7$.

Figure 14.15 shows the evolution of the normalized mean value X/X_0 over time for different traffic growth patterns – linear, Exp-1 and Exp-2, respectively. The figure shows the difference in mean values over 120 years among the different patterns. The largest value is that of the Exp-2 pattern, followed by the linear pattern and then the Exp-1 pattern. Based on this finding, the mean value of the normal distribution in each month of the 120 years can be determined. The same procedure is repeated to obtain the STD for each month of the 120 years. The normal distribution of S_{mr} in each month of the 120 years is then determined once the pairs of mean values and STDs are known. Finally, the S_{mr} in each month is determined on the basis of the corresponding distribution using the MCS method, to compose a sample of 120 years of S_{mr}.

Figure 14.13b shows a sample of 120 years of S_{mr} at E32123 for the linear traffic growth pattern from the current loading to the first future loading scenario. Compared with the no growth pattern in Figure 14.13a, a notable growth in S_{mr} occurs over the 120 years. Similarly, the histogram is estimated from 200 samples of $\sum_{j=1}^{N_b} S_{mr,j}$ for 120 years, and the theoretical density function is modeled as a normal distribution (see Figure 14.14b). The theoretical density function fits the histogram quite well. The same procedure is applied to determine the mean value and STD of $\sum_{j=1}^{N_b} S_{mr,j}$ over 120 years, at different fatigue-critical locations and under different traffic growth patterns.

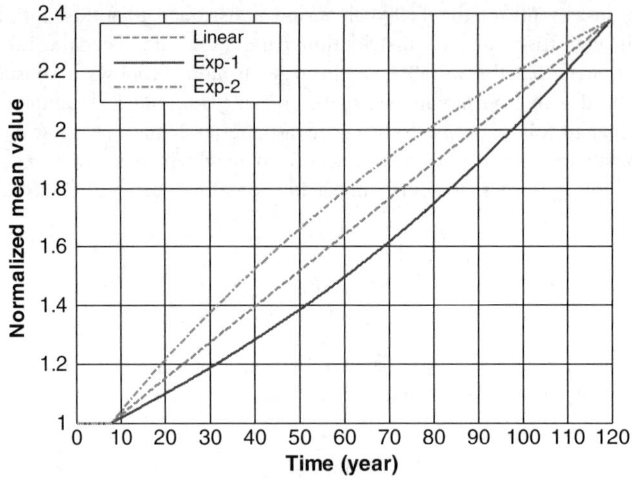

Figure 14.15 Evolution of the normalized mean value X/X_0 over time ($T_r = 8$ years, $T_t = 120$ years) (from [34]) (Reproduced with permission from ASCE).

14.7.8 Reliability Analysis Results

Based on the limit-state function in Equation 14.70 and the distribution parameters of the variables K, Δ and S_{mr}, the fatigue reliability index β can be solved using Equation 14.72. According to British Standard [36], the type of welded connection for the six fatigue-critical locations concerned is classified as F. The mean value and STD of K for detail F are $\mu_K = 1.73 \times 10^{12}$ and $\sigma_K = 0.52 \times 10^{12}$. The mean value and STD of Δ are $\mu_\Delta = 1.0$ and $\sigma_\Delta = 0.3$. The distribution parameters of $\sum_{j=1}^{N_b} S_{mr,j}$ across 120 years at the various fatigue-critical locations under different traffic growth patterns are determined in the foregoing part.

The fatigue damage at the fatigue-critical location of E32123 under the different traffic growth patterns is taken as an example to study the evolution of the fatigue reliability index over time (see Figure 14.16). The first pattern is no traffic growth, while the other three patterns are growth from the current loading to the first future loading scenario of a linear and exponential type, respectively. The probabilistic model of wind loading remains constant, as it is established on the basis of the 120-year bridge design life.

To compute the fatigue reliability indices at different time epochs at intervals of ten years, the distribution parameters of $\sum_{j=1}^{N_b} S_{mr,j}$ within the period from the bridge being opened to traffic to each subsequent time epoch is estimated from the generated 200 samples. Figure 14.16 shows that the fatigue reliability indices decrease with time, and the decreasing velocity in the curve without traffic growth (constant) is much slower than that in the other three curves with traffic growth (linear, Exp-1 and Exp-2). Among the three traffic growth patterns, the fastest decreasing velocity occurs for the Exp-2 pattern, followed by the linear pattern and then the Exp-1 pattern. The fatigue failure probabilities are then estimated from the fatigue reliability indices using Equation 14.72.

The evolution of the fatigue failure reliability over time at E32123 is shown in Figure 14.17. The figure indicates that the fatigue failure probabilities increase with time, and that the failure probability without traffic growth is smaller than that of the three patterns with traffic growth. Among the three growth patterns, the failure probability is largest for the Exp-2 pattern, followed by the linear pattern and then the Exp-1 pattern.

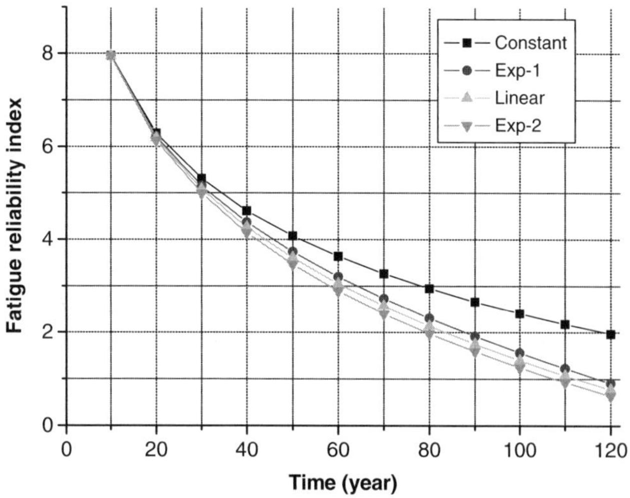

Figure 14.16 Evolution of the fatigue reliability index over time.

The failure probabilities at the fatigue-critical locations at the end of 120 years are computed, and the results are listed in Table 14.2. The highest fatigue failure probability is at E32123 and the lowest is at E55406. A failure probability of 2.3% is recommended in British Standard [36], above which the concerned structural components are regarded as being in danger. The highest failure probabilities at the end of 120 years at the six fatigue-critical locations under current traffic conditions without growth are close to the reference failure probability, which implies that the health condition of the bridge is satisfactory in terms of fatigue. This conclusion is consistent with that based on the deterministic fatigue analysis [33]. Special attention should be paid to future traffic growth, because it may lead to a failure

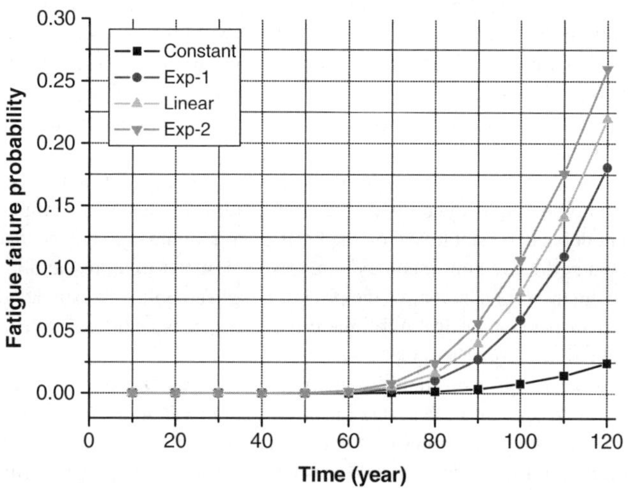

Figure 14.17 Evolution of the fatigue failure probability over time (from [34]) (Reproduced with permission from ASCE).

Table 14.2 Fatigue failure probabilities at the end of 120 years at the fatigue-critical locations (Miner's rule)

Load case	Growth pattern	Fatigue-critical location					
		E32123	E34415	E40056	E40906	E55406	E39417
CL	Constant	0.024	0.017	6×10^{-3}	5×10^{-3}	8×10^{-6}	2×10^{-5}
CL → FL1	Linear	0.22	0.20	0.03	0.03	4×10^{-3}	6×10^{-3}
	Exp-1	0.18	0.16	0.02	0.02	2×10^{-3}	3×10^{-3}
	Exp-2	0.26	0.25	0.05	0.04	8×10^{-3}	0.01
CL → FL2	Linear	0.22	0.13	5×10^{-3}	6×10^{-3}	3×10^{-4}	3×10^{-4}
	Exp-1	0.18	0.11	5×10^{-3}	5×10^{-3}	3×10^{-4}	3×10^{-4}
	Exp-2	0.26	0.16	6×10^{-3}	8×10^{-3}	5×10^{-4}	4×10^{-4}

probability at the end of 120 years that is much greater than the reference level. More details can be found in the literature [34].

14.8 Fatigue Reliability Analysis based on Continuum Damage Mechanics

The fatigue assessment of long-span steel bridges under long-term combined action of railway, highway and wind loading represents a necessary but challenging task. This is because fatigue damage accumulation involving fatigue crack initiation and growth is actually a non-linear process during the service life of a bridge [43,44]. The fatigue damage accumulation also involves many uncertainties in both fatigue loading and fatigue resistance. For a bridge subjected to multiple types of loading, uncertainties become more complicated, making the fatigue assessment of such a bridge difficult to perform.

Miner's rule is widely used in civil engineering for the fatigue damage assessment and reliability analysis of steel structures. However, Miner's rule is a linear damage model and does not address the actual physical mechanism of fatigue crack initiation and growth. It also does not consider the fatigue loading sequence effect, leading to either overly optimistic or pessimistic results [45]. Fatigue damage models based on continuum damage mechanics have been recently proposed in the field of engineering mechanics to deal with the mechanical behavior of a deteriorating medium at the continuum scale [46–48]. These models are highly non-linear in terms of damage evolution, and some of them were also tried by some researchers to estimate fatigue damage of long-span suspension bridges under a single type of loading [49–51]. However, none of these investigations refer to fatigue damage assessment of suspension bridges under multiple types of loading and fatigue reliability analysis with uncertainties.

In this connection, this section presents a framework for fatigue assessment of a long-span suspension bridge under combined highway, railway and wind loadings by using a continuum damage model (CDM) [52]. This model is first established on the basis of continuum damage mechanics with an effective stress range and an effective non-linear accumulative parameter to represent all of the stress ranges within a daily block of stress time history of the bridge. It is then applied to estimate damage accumulation of the Tsing Ma suspension bridge at fatigue-critical locations, and the results are compared with those estimated by the linear Miner's model.

A limit-state function for fatigue reliability analysis based on CDM is also defined by introducing appropriate random variables into CDM. One of these random variables is the daily sum of m-power stress ranges, and its probability distribution is determined based on the measurement data recorded by the structural health monitoring system installed in the bridge. The Monte Carlo Simulation (MCS) is then adopted to generate the random variables and to calculate failure probability. Finally, the failure probabilities of the Tsing Ma Bridge at the end of 120 years are estimated for different loading scenarios.

14.8.1 Basic Theory of Continuum Damage Mechanics

The damage growth of a material is considered to consist of a progressive internal deterioration that causes some loss in the effective cross-section area which carries loads. In the continuum damage mechanics, the damage index D for isotropic damage is often defined as:

$$D = \frac{S - \tilde{S}}{S} \tag{14.80}$$

where:

S is the nominal cross-section area;
$\tilde{S}$ is the effective cross-section area when loss due to damage is taken into account.

Based on thermodynamics and dissipation potential, the rate of damage for high-cycle fatigue can be expressed as a function of the accumulated micro-plastic strain, the strain energy density release rate and the current state of damage [47]. The micro-plastic strain (which is often neglected in a low-cycle fatigue problems) and its accumulation must be considered for high-cycle fatigue damage, even if macro-plastic strain is not present [49]. In a one-dimensional situation, the equation for the rate of fatigue damage $\dot{D}$ can be given as:

$$\dot{D} = \frac{\sigma^2 |\sigma - \bar{\sigma}|^{\beta'}}{B'(1-D)^{\alpha}} \langle \dot{\sigma} \rangle \tag{14.81}$$

where:

$\bar{\sigma} = \sigma_m$ is the mean stress over the stress cycle;
the symbol $\langle\ \rangle$ denotes the McCauley brackets, where $\langle x \rangle = x$ for $x > 0$ and $\langle x \rangle = 0$ for $x < 0$;
α, β' and B' are the material parameters.

Equation 14.81 is a general constitutive model for high-cycle fatigue and can be integrated over time for cycles with different mean stresses and stress ranges. For instance, if $\bar{\sigma} = 0$ and the variation of $(1-D)^{\alpha}$ in a single stress cycle is neglected, then integrating Equation 14.81 over the cycle yields:

$$\frac{\delta D}{\delta N} = \frac{2\sigma_a^{\beta'+3}}{B'(\beta+3)(1-D)^{\alpha}} \tag{14.82}$$

where σ_a is the amplitude of the stress cycle.

By considering the mean stress effect [53], Equation 14.82 can be rewritten as [49]:

$$\frac{\delta D}{\delta N} = \frac{[(\sigma_r + 2\sigma_m)\sigma_r]^{\frac{(\beta'+3)}{2}}}{B(\beta'+3)(1-D)^{\alpha}} \tag{14.83}$$

where $B = 2^{\beta'+2}B'$.

Integrating this equation over N stress cycles, in which $\sigma_m = 0$ and σ_r = constant, yields the damage accumulation:

$$D = 1 - \left[1 - \frac{(\alpha+1)}{B(\beta'+3)}(\sigma_r)^{(\beta'+3)}N\right]^{\frac{1}{(\alpha+1)}} \tag{14.84}$$

Many experiments have been conducted on the fatigue failure of structural details under different constant stress ranges, and S-N curves have been established based on these experimental results. Given that far too few experiments have been conducted to determine the parameters B, β' and α in Equation 14.84, they can be expressed by using parameters of the S-N curves. The number of stress cycles to failure N_f under the stress range σ_r ($\sigma_r \geq \sigma_{r,0}$, $\sigma_{r,0}$ is the fatigue limit) can be determined based on the British S-N curves [36]:

$$N_f = K_2(\sigma_r)^{-m} \tag{14.85}$$

Equation 14.84 can be written as:

$$N_f = \left[1-(1-D_f)^{(\alpha+1)}\right]\frac{B(\beta'+3)}{(\alpha+1)}\sigma_r^{-(\beta'+3)} \tag{14.86}$$

where the damage at failure, D_f, is an intrinsic material property that is dependent on the durability of the material [43,54].

As the values of α adopted for different amplitudes of stress ranges occurring on a bridge are deemed to be sufficiently large to make $\left[1-(1-D_f)^{(\alpha+1)}\right]$ very close to 1, the parameters B, β' and α can be expressed by m and K_2 by comparing Equations 14.85 with 14.86:

$$\begin{cases} \beta'+3 = m \\ \dfrac{B(\beta'+3)}{\alpha+1} = K_2 \end{cases} \tag{14.87}$$

Substituting this equation into Equation 14.84 gives:

$$D = 1-\left[1-\frac{1}{K_2}(\sigma_r)^m N\right]^{\frac{1}{\alpha+1}} \tag{14.88}$$

To extend the approach to fatigue analysis under variable-amplitude loading, Equation 14.88 is expressed as:

$$D(k) = 1-\left[(1-D(k-1))^{(\alpha_k+1)}-(\sigma_{r,k})^m/K_2\right]^{\frac{1}{\alpha_k+1}} \tag{14.89}$$

where:

$\sigma_{r,k}$ and α_k are the stress range and non-linear accumulative parameter for the kth stress cycle, respectively;
$D(k)$ is the cumulative fatigue damage after the kth stress cycle.

The equation can also be derived using stepwise iteration from the initial damage $D(0)=0$:

$$\begin{cases} D(1) = 1-\left[1-(\sigma_{r,1})^m/K_2\right]^{\frac{1}{\alpha_1+1}} \\ D(2) = 1-\left[(1-D(1))^{(\alpha_2+1)}-(\sigma_{r,2})^m/K_2\right]^{\frac{1}{\alpha_2+1}} \\ \qquad\cdots \\ D(k) = 1-\left[(1-D(k-1))^{(\alpha_k+1)}-(\sigma_{r,k})^m/K_2\right]^{\frac{1}{\alpha_k+1}} \end{cases} \tag{14.90}$$

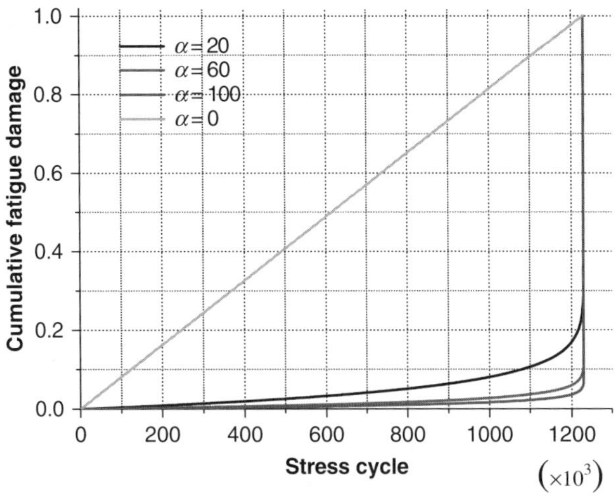

Figure 14.18 Fatigue damage accumulation for different α (from [52]) (Reproduced with permission from Elsevier).

14.8.2 Non-Linear Properties of Fatigue Damage Accumulation

Fatigue damage accumulation is non-linear because it is induced by fatigue crack initiation at a very slow pace, then fatigue crack growth at a relatively fast pace, then finally and suddenly, fatigue failure. In the continuum damage mechanics-based fatigue damage model, the non-linear accumulative parameter α is the parameter that controls the non-linearity. Numerical simulations are carried out at this point to study the sensitivity of α to fatigue damage accumulation. In the simulations, the constant stress range $\sigma_r = 80$ MPa is used, $K_2 = 6.3 \times 10^{11}$ and m = 3 are adopted for detail class F in British Standard [36], and different values of α ($\alpha = 0$, 20, 60 and 100) are tried to compute the fatigue damage accumulation over a certain number of cycles until failure by using Equation 14.88.

Figure 14.18 shows the fatigue damage accumulation curves for different values of α. It demonstrates that the cumulative damage increases with the number of stress cycles applied. The cumulative rate of damage is small at the initial stage, but increases at late stages, showing a strong non-linear trend. The non-linearity of the cumulative damage curve increases with the growth of α. Note that when $\alpha = 0$ is adopted, the non-linear damage model simply becomes the linear Miner's model. Therefore, a low α represents a conservative estimate of damage accumulation. Figure 14.18 also shows that the number of cycles to failure is the same for different values of α. This is because the parameters in Equation 14.88 are determined by the S-N curves.

14.8.3 Continuum Damage Model used in This Study

Long-span suspension bridges carrying railway and highway in wind prone regions undergo variable-amplitude loading induced by the combined effect of the multiple loads. Hence, cycle-by-cycle computation may be required to estimate the damage accumulation by using Equation 14.90. However, given that the design life of a long-span suspension bridge is normally a hundred years or more, it is difficult to compute so many stress cycles occurring over such a long period using the cycle-by-cycle approach.

To simplify the computation and make it feasible for practical application, the effective parameters must be derived for a given block of stress time history, and the cumulative fatigue damage in the block can then be computed using these effective parameters, rather than the complete set of parameters for all of the stress cycles in the block. For example, urban passenger trains often follow a regular timetable that is similar on different days, and highway traffic conditions are also similar across different days.

Thus, railway and highway traffic operates almost on a one-day cycle, and the daily stress time history can thus be considered as one block. The effective stress range σ_{re} can be defined to represent all of the stress ranges in the daily block according to the two-slope S-N curves defined in British Standard [36], in which, if the stress range level $\sigma_{r,i}$ is less than the fatigue limit $\sigma_{r,0}$, it will diminish proportionally. The effective stress range in the kth block can thus be expressed as:

$$\sigma_{re,k} = \left\{ \left[\sum_{i=1}^{N_1} n_i \left(\sigma_{r,i}\right)^m + \frac{1}{\left(\sigma_{r,0}\right)^2} \sum_{i=1}^{N_2} n_i \left(\sigma_{r,i}\right)^{m+2} \right] \Big/ N_{r,k} \right\}^{\frac{1}{m}} \tag{14.91}$$

where:

n_i is the number of stress cycles applied at the stress range level $\sigma_{r,i}$, which is determined from the stress response time history in the kth block using the rainflow counting method;
N_1 and N_2 are the number of stress range levels above $\sigma_{r,i}$ and below $\sigma_{r,0}$, respectively;
$N_{r,k}$ is the total number of cycles in the kth block;
m is a parameter in the S-N curve and is taken as 3.0.

The non-linear accumulative parameter α_e is another effective parameter that is also determined from the stress cycles in the daily block. As the daily block contains a large number of cycles in different stress range amplitudes, the value of α varies for different cycles. Given that the damage estimation would tend to conservative side if a low parameter α is adopted, the effective accumulative parameter $\alpha_{e,k}$ is thus adopted as the minimum α in the kth block for simplicity, and it can be determined by the maximum stress range in the block according to the function given in [49]. By introducing the effective parameters $\sigma_{re,k}$ and $\alpha_{e,k}$ into Equation 14.89, the damage accumulation in the kth block can be estimated by the following continuum damage model:

$$D(k) = 1 - \left[(1 - D(k-1))^{(\alpha_{e,k}+1)} - N_{r,k} \left(\sigma_{re,k}\right)^m \Big/ K_2 \right]^{\frac{1}{\alpha_{e,k}+1}} \tag{14.92}$$

14.8.4 Verification of Continuum Damage Model

Before the continuum damage model (CDM) expressed by Equation 14.92 can be applied for fatigue assessment of long-span suspension bridge under multiple loadings, some verification is necessary. Since a wind and structural health monitoring system (WASHMS) has been installed in the Tsing Ma Bridge since 1997, it has recorded many stress time histories.

Figure 14.19 shows a typical daily multiple load-induced hot spot stress time history recorded at one of the fatigue-critical locations of the bridge. The fatigue damage accumulation within this daily block is computed using the cycle-by-cycle method and the effective parameter method, respectively, based on Equation 14.92. The obtained results are compared with each other to see if the effective parameter method can yield reasonable damage estimation.

Figure 14.20 shows the stress range sequence determined from the daily stress time history (see Figure 14.19) using the rainflow counting method. There are a total of 72,333 stress cycles in the sequence in this time history, and the effective stress range $\sigma_{re} = 6.0$ MPa is calculated from these using Equation 14.91. The maximum stress range in the block is 69.1 MPa, and the corresponding effective non-linear accumulative parameter $\alpha_e = 92.1$. The fatigue damage accumulation within the block is calculated by using Equation 14.90 with the step-by-step method and the effective parameter method.

The results are shown in Figure 14.21, and they demonstrate that the final cumulative damage values calculated with the two methods are very close, although the result from the effective parameter method is slightly more conservative than that from the step-by-step method. This indicates that the effective

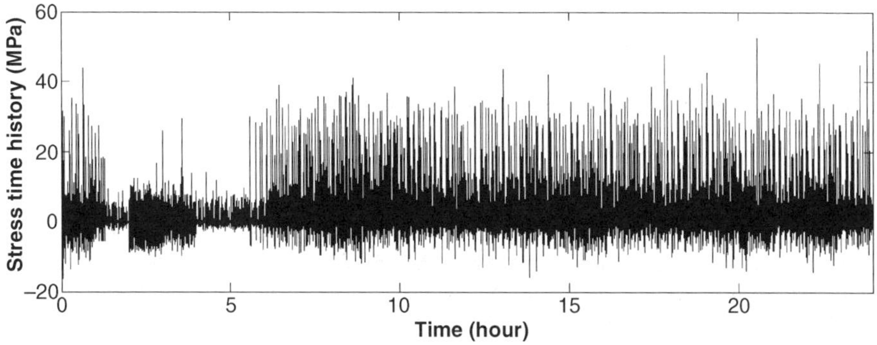

Figure 14.19 A sample stress time history under multiple types of loading (from [52]) (Reproduced with permission from Elsevier).

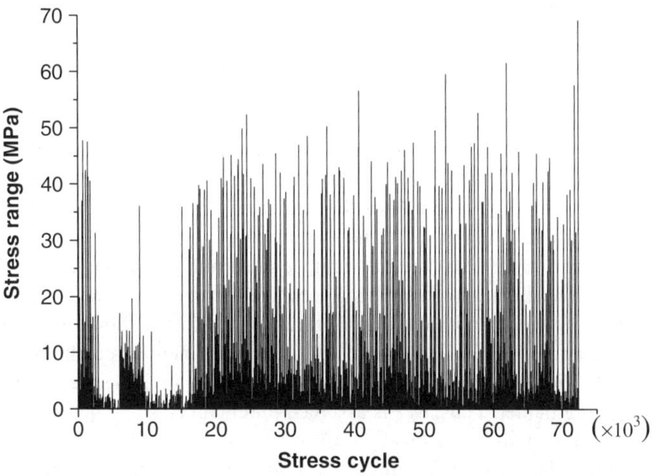

Figure 14.20 Sample daily stress range sequence (from [52]) (Reproduced with permission from Elsevier).

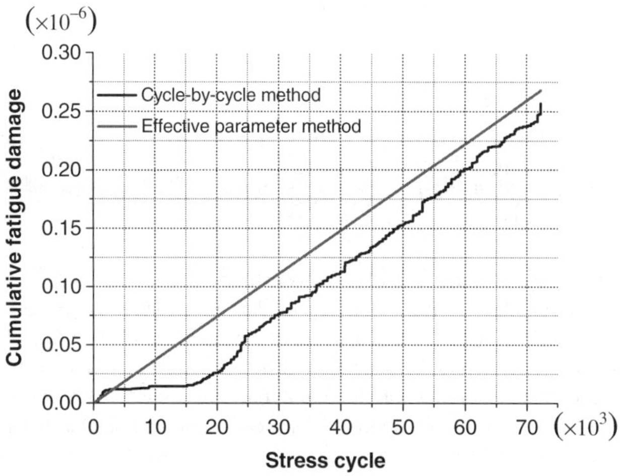

Figure 14.21 Fatigue damage accumulation in a daily block (from [52]) (Reproduced with permission from Elsevier).

parameter method, neglecting the load sequence effect on damage accumulation within a daily block, is reasonable and effective for estimating damage accumulation for a daily stress cycle interval. The effective parameter method or the continuum damage model (CDM) makes it possible to estimate the fatigue damage accumulation of a long-span suspension bridge under multiple loading.

14.8.5 Framework of Fatigue Reliability Analysis

To consider the uncertainties that exist in the non-linear process of damage accumulation, this section proposes a framework for fatigue reliability analysis based on the CDM model. The first step of fatigue reliability analysis is to define a limit-state function that adequately describes the relationship between fatigue resistance and fatigue loading for a fatigue-sensitive structural member. As indicated by Equation 14.92, the fatigue limit-state can be defined as the point at which the cumulative fatigue damage stops increasing when more loads are applied, such as $D(k) \leq D(k-1)$. In other words, it is the point at which the structure cannot suffer any further damage.

For computational simplicity, the continuous process of damage accumulation is reduced to a discrete process and updated at daily intervals, as expressed in Equation 14.92. Distinct from deterministic fatigue analysis, the randomness in both the fatigue loading and the fatigue resistance need to be considered in fatigue reliability analysis. In the fatigue loading aspect, given that the multiple loading-induced stress time history is a stochastic process, the number of stress cycles at a given stress range level is random. Furthermore, as the operation of urban passenger trains often follows a daily timetable, the cycle of railway and highway traffic is close to one day. Consequently, the daily sum of m-power stress ranges $S_{mr,k}$ ($k=1, \ldots, N_b$, where N_b is the total number of days in the time period concerned) is treated as a random variable in terms of fatigue loading.

In the fatigue resistance aspect, the fatigue damage accumulation index Δ and the fatigue detail coefficient K are regarded as random variables to replace the parameter $K_2 = K\Delta$. According to the findings of the previous numerical simulations, different values of α_e should be used to consider the non-linear damage accumulation. The loading condition is assumed to remain constant over a given period, and daily stochastic stress time histories from this period are generated on the basis of a constant probabilistic loading model. Thus, $\alpha_{e,k}$ is assumed to remain constant during the same period, but to change across different periods.

Taking these random variables and constants into account, the limit-state function for fatigue reliability analysis based on the CDM model is defined as:

$$g(\mathbf{X}) = D(k) - D(k-1) \tag{14.93}$$

$$D(k) = 1 - \left[(1 - D(k-1))^{\alpha_{e,k}+1} - S_{mr,k}/(K\Delta)\right]^{\frac{1}{\alpha_{e,k}+1}} \tag{14.94}$$

$$S_{mr,k} = \sum_{i=1}^{N_1} n_i (\sigma_{r,i})^m + \frac{1}{(\sigma_{r,0})^2} \sum_{i=1}^{N_2} n_i (\sigma_{r,i})^{m+2} = N_{r,k}(\sigma_{re,k})^m \tag{14.95}$$

where $S_{mr,k} = N_{r,k}(\sigma_{re,k})^m$ is the daily sum of m-power stress ranges.

The fatigue damage accumulation index, Δ, is regarded as a random variable modeled by a lognormal distribution with a mean value μ_Δ of 1.0 and a standard deviation σ_Δ of 0.3 [35]. This random variable actually accounts for the uncertainty associated with the damage to failure when fatigue model is applied to deal with problems involving variable-amplitude stress ranges. The fatigue detail coefficient K is assumed to be a lognormal distribution, and its mean value and standard deviation for different fatigue detail classes are obtained from the relevant information provided in British Standard [36].

Based on the defined limit-state function, the failure probability P_f of a structural member or system can be estimated. In this connection, the Monte Carlo Simulation (MCS) is adopted, which involves the repeated drawing of random samples of all of the random variables involved in the limit-state function and simply checks whether a new "failure" or "non-failure" has resulted (or, equivalently, whether the limit-state function has a value of less than zero). Base on N simulations, P_f can be estimated using the following equation:

$$P_f \approx \frac{1}{N}\sum_{i=1}^{N} I[g(\mathbf{X}_i) < 0] \tag{14.96}$$

where:

$g(\mathbf{X}_i)$ is the value of the limit-state function in the ith simulation. $I[g(\mathbf{X}_i) < 0] = 1$ if $g(\mathbf{X}_i) < 0$, and $I[g(\mathbf{X}_i) < 0] = 0$ otherwise;
$\mathbf{X}_i = \{X_{i,1}, X_{i,2}, \ldots, X_{i,n}\}$ is a vector of n random variables;
$X_{i,1}$ and $X_{i,2}$ are the random variables Δ and K;
$X_{i,3} \sim X_{i,n}$ are the random variables $S_{mr,k}$ ($k = 1, \ldots, N_b$, N_b is the number of daily blocks).

14.8.6 Reliability Analysis Results

In this section, the fatigue reliability (or failure probability) at the six fatigue-critical locations of the Tsing Ma Bridge is solved. First, the probability distributions of the random variables are determined for the reliability analysis. The mean value and STD of K for detail F are $\mu_K = 1.73 \times 10^{12}$ and $\sigma_K = 0.52 \times 10^{12}$. The mean value and STD of Δ are $\mu_\Delta = 1.0$ and $\sigma_\Delta = 0.3$.

The probability distributions of S_{mr} remain constant within a given month, but change across different months for the three traffic growth patterns. The distribution parameters of $S_{mr,k}$ ($k = 1, \ldots, 120 \times 365$) in different months across 120 years at the various fatigue-critical locations under different traffic growth patterns were determined in the previous subsection. Correspondingly, a constant α_e is adopted here for the no traffic growth pattern, where α_e changes across different months for the three growth patterns.

Second, the random variables of K, Δ, and $S_{mr,k}$ are randomly sampled from the corresponding distributions to yield a random variable vector. Third, the vector is substituted into the limit-state function to calculate the value of $g(\mathbf{X}_i)$ in Equation 14.96. Fourth, N_f is increased by 1 if $g(\mathbf{X}_i) < 0$. Finally, the steps are repeated N times and the probability of failure P_f is computed by N_f/N. Note that N should be sufficiently large for a good estimate of P_f. $N_f \geq 10$ is adopted for this estimation.

Using this procedure, the fatigue failure probabilities at the end of 120 years are estimated at the six fatigue-critical locations and the results are listed in Table 14.3. The results are then compared with

Table 14.3 Fatigue failure probabilities at the end of 120 years at the fatigue-critical locations. (CDM model)

Load Case	Growth Pattern	Fatigue-Critical Location					
		E32123	E34415	E40056	E40906	E55406	E39417
CL	Constant	0.024	0.017	6×10^{-3}	5×10^{-3}	8×10^{-6}	2×10^{-5}
CL → FL1	Linear	0.21	0.19	0.03	0.03	4×10^{-3}	6×10^{-3}
	Exp-1	0.17	0.15	0.02	0.02	2×10^{-3}	3×10^{-3}
	Exp-2	0.24	0.23	0.05	0.04	8×10^{-3}	0.01
CL → FL2	Linear	0.21	0.12	5×10^{-3}	6×10^{-3}	3×10^{-4}	3×10^{-4}
	Exp-1	0.17	0.10	5×10^{-3}	5×10^{-3}	3×10^{-4}	3×10^{-4}
	Exp-2	0.24	0.15	6×10^{-3}	8×10^{-3}	5×10^{-4}	4×10^{-4}

those estimated by Miner's model (see Table 14.2). If there is no growth in traffic over the bridge's design life, then there is no difference between the failure probabilities estimated by the two models, because a constant α_e is assumed for the entire 120 years and, thus, there is no non-linear damage accumulation. If, however, traffic growth occurs in certain typical patterns, then the failure probabilities estimated by the CDM model are smaller than those estimated by Miner's model. This can be explained by the previous finding that the cumulative fatigue damage under an increasing loading sequence is smaller than that under a decreasing loading sequence [52].

14.9 Notations

D	The damage index
erf(•)	The error function
$F_p(\alpha, t)$	The probability distribution function of the peaks
$f_p(\alpha, t)$	The probability density function of the peaks
$f_{X\dot{X}}(x, \dot{x}, t)$	The joint probabilistic density function PDF of X(t) and $\dot{X}(t)$
$g(X)$	The limit-state function
$\tilde{g}(X)$	The approximate limit-state function
$\mathbf{K}$	Covariance matrix
L	The lower bound level
$M_p(t)$	The expected number of total peaks per unit time without any restriction on level
$N(t)$	The rate of passage
$N_p(\alpha, t)$	The expected number of peaks per unit time above the level $x(t) = \alpha$
N_1	The number of stress range levels above $\sigma_{r,i}$
N_2	The number of stress range levels above $\sigma_{r,0}$
n	The iteration number of the recursive algorithm
n_i	The applied number of stress cycles at the stress range level $\sigma_{r,i}$
P_f	The probability of failure
$P_f(U, -L, T)$	The probability of the time that the maximum or minimum response reaches the respective bound level for the first time
P_r	The probability of success
$P_{X_m}(U, T)$	The probability distribution function of peaks
$P_{X_m}^u$	The upper limit of the probability distribution function of peaks
$P_{X_m}^l$	The lower limit of the probability distribution function of peaks
$P_{Z_m}(U, T)$	The probability that Z_m does not exceed the bound level U within the time period T
q	The spectral bandwidth factor
R	The resistance of the structure
S	The loading of the structure
U	The upper bound level
X	The random variable vector
Z_m	The absolute maximum response
α_i	The sensitivity factor
β	The safety index or reliability index
β_{HL}	The Hasofer and Lind (HL) safety-index
$\nabla g()$	The gradient of g
ε_r	The normalized random error of the variable x
μ_g	Mean of $g(X)$
μ_R	Mean of R
μ_S	Mean of S

μ_X	The mean of random variables
ξ	The ratio of the rate of zero crossings with positive slope to the expected number of total peaks per unit time without any restriction on level
ρ_{RS}	The correlation coefficient between R and S
σ_g	The standard deviation of $g(X)$
$\sigma\tilde{g}$	The standard deviation of the approximate limit-state function
σ_r	The standard deviation of R
$\sigma_{r,i}$	The stress range level
σ_S	The standard deviation of S
$\Phi(\beta)$	The normal cumulative distribution function

References

1. Choi, S.K., Grandhi, R.V., and Canfield, R.A. (2007) *Reliability-Based Structural Design*, Springer-Verlag, London.
2. SAE (1997) Integration of Probabilistic Methods into the Design Process, Aerospace Information Report 5080, Society of Automotive Engineers.
3. Hasofer, A.M. and Lind, N.C. (1974) Exact and invariant second-moment code format. *Journal of Engineering Mechanics Division, ASCE*, **100** (EM1), 111–121.
4. Hohenbichler, M. and Rackwitz, R. (1981) Non-normal dependent vectors in structural safety. *Journal of Engineering Mechanics Division, ASCE*, **107** (EM6), 1227–1238.
5. Cai, G.Q. and Elishakoff, I. (1994) Refined second-order approximations analysis. *Structural Safety*, **14**(4), 267–276.
6. Der Kiureghian, A., Lin, H.Z., and Hwang, S.J. (1987) Second order reliability approximations. *Journal of Engineering Mechanics, ASCE*, **113**, 1208–1225.
7. Melchers, R.E. (1999) *Structural Reliability Analysis and Prediction*, 2nd edn, John Wiley & Sons.
8. Faravelli, L. (1989) Response surface approach for reliability analysis. *Journal of Engineering Mechanics, ASCE*, **115**(12), 2763–2781.
9. Lin, Y.K. (1967) *Probabilistic Theory of Structural Dynamics*, McGraw-Hill, New York.
10. Li, J. and Chen, J.B. (2009) *Stochastic Dynamics of Structures*, Wiley, Singapore.
11. Rice, S.O. (1944) Mathematical analysis of random noise. *Bell System Technical Journal*, **23**, 282–332.
12. Nigam, N.C. (1983) *Introduction to Random Vibrations*, MIT, London.
13. Su, C., Luo, X., and Yun, T. (2010) Aerostatic reliability analysis of long span bridges. *Journal of Bridge Engineering, ASCE*, **15**(3), 260–268.
14. Ostenfeld-Rosenthal, P., Madsen, H.O., and Larsen, A. (1992) Probabilistic flutter criteria for long span bridges. *Journal of Wind Engineering and Industrial Aerodynamics*, **42**(1–3), 1265–1276.
15. Ge, Y.J., Xiang, H.F., and Tanaka, H. (2000) Application of a reliability analysis model to bridge flutter under extreme winds. *Journal of Wind Engineering and Industrial Aerodynamics*, **86**(2–3), 155–167.
16. Pourzeynali, S. and Datta, T.K. (2002) Reliability analysis of suspension bridges against flutter. *Journal of Sound and Vibration*, **254**(1), 143–162.
17. Prenninger, P.H.W., Matsumoto, M., Shiraishi, N. *et al.* (1990) Reliability of bridge structures under wind loading: Consideration of uncertainties of wind load parameters. *Journal of Wind Engineering and Industrial Aerodynamics*, **33**(1–2), 385–394.
18. Cheng, J., Cai, C.S., Xiao, R.C., and Chen, S.R. (2005) Flutter reliability analysis of suspension bridges. *Journal of Wind Engineering and Industrial Aerodynamics*, **93**(10), 757–775.
19. Xiang, H.F. (2005) *Modern Theory and Practice on Bridge Wind Resistance*, China Communications Press, Beijing, (in Chinese).
20. Crandall, S.H. (1970) First-crossing probabilities of the linear oscillator. *Journal Sound and Vibration*, **12**(3), 285–299.
21. Vanmarcke, E.H. (1975) On the distribution of the first-passage time for normal stationary processes. *Journal Applied Mechanics, ASME*, **42**, 215–220.
22. Vanmarcke, E.H. (1983) *Random Field: Analysis and Synthesis*, MIT Press, Cambridge, MA.
23. Coleman, J.J. (1959) Reliability of aircraft structures in resisting chance failure. *Operations Research*, **7** (5), 639–645.
24. Crandall, S.H., Chandiramani, K.L., and Cook, R.G. (1966) Some first passage problems in random vibrations. *Journal of Applied Mechanics, Transactions of ASME*, **33** (3), 532–538.
25. Cramer, H. (1966) On the intersection between the trajectories of a normal stationary stochastic process. *Arkiv fur Mathematik*, **6**, 337–349.
26. Cramer, H. and Leadbetter, M.R. (1967) *Stationary and Related Stochastic Process*, Wiley, New York.
27. Ge, Y.J. and Xiang, H.F. (2008) Recent development of bridge aerodynamics in China. *Journal of Wind Engineering and Industrial Aerodynamics*, **96**(6–7), 736–768.

28. Chan, T.H.T., Li, Z.X., and Ko, J.M. (2001) Fatigue analysis and life prediction of bridges with structural health monitoring data -Part II: Application. *International Journal of Fatigue*, **23**(1), 55–64.
29. Zhou, Y. (2006) Assessment of bridge remaining fatigue life through field strain measurement. *Journal of Bridge Engineering, ASCE*, **11** (6), 737–744.
30. Chung, H.C. (2004) Fatigue Reliability and Optimal Inspection Strategies for Steel Bridges, PhD Thesis, The Faculty of the Graduate School, The University of Texas at Austin, USA.
31. Kwon, K. and Frangopol, D.M. (2010) Bridge fatigue reliability assessment using probability density functions of equivalent stress range based on field monitoring data. *International Journal of Fatigue*, **32**(8), 1221–1232.
32. Chen, Z.W., Xu, Y.L., Li, Q., and Wu, D.J. (2011) Dynamic stress analysis of long suspension bridges under wind, railway and highway loadings. *Journal of Bridge Engineering, ASCE*, **16**(3), 383–391.
33. Chen, Z.W., Xu, Y.L., and Wong, K.Y. (2012) Fatigue analysis of long-span suspension bridges under multiple loading: Case study. *Engineering Structures*, **33**, 3246–3256.
34. Chen, Z.W., Xu, Y.L., and Wang, X.M. (2012) SHMS-based fatigue reliability analysis of multi-loading suspension bridges. *Journal of Structural Engineering, ASCE*, **138**(3), 299–307.
35. Wirsching, P.H. (1984) Fatigue reliability for offshore structures. *Journal of Structural Engineering, ASCE*, **110**(10), 2340–2356.
36. BS (1980) *BS5400: Part 10, Code of Practice for Fatigue*, British Standards Institution, London.
37. Rackwitz, R. and Fiessler, B. (1978) Structural reliability under combined random load sequence. *Computers and Structures*, **9**, 489–494.
38. McLachlan, G. and Peel, D. (2000) *Finite Mixture Models*, Wiley, New York.
39. Bilmes, J. (1998) A gentle tutorial of the EM algorithm and its application to parameter estimation for Gaussian mixture and hidden Markov models. Proceedings of International Computer Science Institute, ICSI Technical Report TR-97-02.
40. Julius, S. and Allan, G. (2000) *Random Data: Analysis and Measurement Procedures*, 3rd edn, Wiley, New York.
41. Roussas, G.G. (2007) *Introduction to Probability*, Academic Press, Boston.
42. Righiniotis, T.D. (2006) Effects of increasing traffic loads on the fatigue reliability of a typical welded bridge detail. *International Journal of Fatigue*, **28**(8), 873–880.
43. ASCE (1982) Committee on fatigue and fracture reliability of the committee on structural safety and reliability of the structural division, fatigue reliability: 1–4. *Journal of Structural Engineering, ASCE*, **108**, 3–88.
44. BS (1993) *BS7608: Code of Practice for Fatigue Design and Assessment of Steel Structures*, British Standards Institution, London.
45. Lemaitre, J. (1996) *A Course of Damage Mechanics*, Springer, Berlin.
46. Kachanov, L.M. (1986) *Introduction to Continuum Damage Mechanics*, Martinus Nijhoff, Dordrecht.
47. Krajcinovic, D. and Lemaitre, J. (1987) *Continuum Damage Mechanics: Theory and Applications*, Springer, Vienna.
48. Chaboche, J.L. and Lesne, P.M. (1988) Non-linear continuous fatigue damage model. *Fatigue and Fracture of Engineering Materials and Structures*, **11** (1), 1–17.
49. Li, Z.X., Chan, T.H.T., and Ko, J.M. (2001) Fatigue damage model for bridge under traffic loading: Application made to Tsing Ma Bridge. *Theoretical and Applied Fracture Mechanics*, **35**(1), 81–91.
50. Liu, T.T., Xu, Y.L., Zhang, W.S. *et al.* (2009) Buffeting-Induced stresses in a long suspension bridge: structural health monitoring orientated stress analysis. *Wind and Structures*, **12**(6), 479–504.
51. Xu, Y.L., Liu, T.T., Zhang, W.S. *et al.* (2009) Buffeting-induced fatigue damage assessment of a long suspension bridge. *International Journal of Fatigue*, **31** (3), 575–586.
52. Xu, Y.L., Chen, Z.W., and Xia, Y. (2012) Fatigue assessment of multi-loading suspension bridges using continuum damage model. *International Journal of Fatigue*, **40**, 27–35.
53. Smith, K.N., Watson, P., and Topper, T.H. (1970) A stress–strain function for the fatigue of metals. *Journal of Materials, ASTM*, **5**(4), 767–778.
54. Bhattacharya, B. and Ellingwood, B. (1998) Continuum damage mechanics analysis of fatigue crack initiation. *International Journal of Fatigue*, **20** (9), 631–639.

15

Non-Stationary and Non-Linear Buffeting Response

15.1 Preview

The buffeting analysis of long-span cable-supported bridges has been addressed in Chapter 4 for strong winds perpendicular to the bridge deck, and in Chapter 10 for strong skew winds to the bridge deck. Strong winds in both chapters are assumed to be stationary random processes that represent well monsoons and gales (extra-tropical cyclones) over flat terrains. However, some extreme wind events, such as typhoons, downbursts and tornados, actually exhibit different characteristics from monsoons and gales. The assumption of stationary processes may not be valid for those extreme wind events because of their vortex or convective origins.

For example, if a bridge site is close to the external eye wall of a strong typhoon over complex terrain, changes in wind direction and convective turbulences will be significant and, consequently, the statistical characteristics of typhoon winds over the bridge site will evolve with time considerably. For such cases, it is more realistic to model typhoon winds as non-stationary random processes rather than stationary random processes, and to perform non-stationary buffeting analysis of long-span bridges accordingly.

Furthermore, the excessive buffeting responses of long-span cable-supported bridges under extreme wind events may be accompanied by both structural and aerodynamic non-linearities. "Structural non-linearities" refers to geometric and material non-linearities; aerodynamic non-linearities may origin from flow separation near deck edges, higher order fluid memory and bridge motions [1]. A non-linear phenomenon clearly observed from experiments is that aerodynamic forces acting on a moving bridge deck in turbulence flow depend on the deck motions with hysteretic characteristics.

Nevertheless, non-stationary and non-linear buffeting analysis of long-span cable-supported bridges is a challenging issue and needs much research. This chapter provides only an elementary introduction to the topic. Two non-stationary probabilistic models for typhoon winds are first presented. Model I regards mean wind as time-varying, while the remaining fluctuating wind is stationary. In he model II, in addition to time-varying mean wind speed, the fluctuating wind is regarded as a non-stationary process characterized by an evolutionary power spectral density (EPSD) function.

With the non-stationary wind model, the method for predicting non-stationary buffeting response of long-span cable-supported bridge is proposed in the time-frequency domain, and the buffeting response is estimated in terms of the time-varying mean response and the standard deviation response from the EPSD. A method for determining the extreme value of non-stationary buffeting response for the design of purpose is also put forward. For performing non-stationary and non-linear buffeting responses in the

Wind Effects on Cable-Supported Bridges, First Edition. You-Lin Xu.

time domain, both unconditional and conditional simulations of non-stationary fluctuating winds are necessary and are addressed in this chapter. Non-linear buffeting analysis is finally discussed, including observations from wind tunnel tests and approaches for modeling aerodynamic non-linearities.

15.2 Non-Stationary Wind Model I

There are several models proposed for non-stationary winds but most of them refer to downbursts. Chen and Letchford [2] suggested a three-layer model for describing non-stationary winds resulting from downbursts, where the three layers represent time-varying mean wind speed, time-varying variance and time-varying wind spectrum. The fluctuating wind speed components of downbursts are characterized by either evolutionary power spectral density (EPSD) functions [2,3] or time-varying auto-regressive model [4] or wavelets [5]. Chen and Letchford's model has been used to predict downburst-induced responses of building structures and transmission towers [6–8].

For typhoons, Xu and Chen [9] characterized typhoon wind as a time-varying mean wind speed plus stationary fluctuating components. This concept was also implied in the modeling of Hurricane Lili by Wang and Kareem [10]. Hu *et al.* [11] further characterized typhoon wind as a time-varying mean wind speed plus non-stationary fluctuating components featured by EPSD. This section will introduce the non-stationary model I for typhoon winds.

15.2.1 Non-Stationary Wind Model I

A preliminary study of non-stationary wind data recorded in the field during a nearby typhoon to the Tsing Ma Bridge reveals that the mean wind speed in the longitudinal wind direction over one hour is often subject to a significant temporal trend with slowly time-varying feature [9]. Therefore, the non-stationary wind speed in the longitudinal wind direction is modeled as a deterministic time-varying mean wind speed plus a zero-mean stationary random process for fluctuating wind speed.

$$U(t) = \overline{U}(t) + u(t) \tag{15.1}$$

where:

$U(t)$ is the total wind speed time history within a specified time range $[0, T]$ in the longitudinal wind direction;
$\overline{U}(t)$ is the deterministic time-varying mean wind speed;
$u(t)$ is the fluctuating wind component in the longitudinal wind direction, which can be modeled as a zero-mean stationary Gaussian process.

This model is actually the basic "trend plus stationary residual" model for non-stationary random process as discussed by Bendat and Piersol [12] and Priestley [13]. It simply allows mean wind speed to vary over time in an arbitrary fashion, providing a great flexibility for considering the variation of mean wind speed with time. Clearly, if the wind speed $U(t)$ is a strictly stationary random process, the time-varying mean wind speed $\overline{U}(t)$ in Equation 15.1 becomes a constant mean wind speed $\overline{U}$. Consequently, the non-stationary wind speed model expressed by Equation 15.1 can be also used for characterizing stationary wind speeds in a traditional way.

15.2.2 Empirical Mode Decomposition

The key issue of using the aforementioned model to characterize non-stationary wind speeds is how to extract a time-varying mean wind speed from a wind speed time history recorded in the field. To circumvent this difficulty, a data processing method termed the empirical model decomposition (EMD), developed by Huang *et al.* [14], is employed. With the EMD, any complicated data set can be decomposed into a finite but often small number of intrinsic mode functions (IMF). This decomposition is adaptive and applicable to non-linear and non-stationary processes because it is based on the local characteristic time scale of the data. The following is the brief description of the procedure involved in the EMD.

Let $U(t)$ represent a non-stationary wind speed time history. The upper and lower envelopes of $U(t)$ are constructed by connecting its local maxima and minima, respectively, using a cubic spline line. The mean of the two envelopes is then computed. The difference between the original time history and the mean value is called the first IMF, $c_1(t)$, if it satisfies the two conditions:

1. within the data range, the number of extrema and the number of zero-crossings are equal or differ by one only;
2. at any point, the mean value of the envelope defined by the local maxima and the envelope defined by the local minima is zero.

This process is called a sifting process, and the resulting IMF represents a simple oscillatory mode as a counterpart to the simple harmonic function, but it is much more general and admits a well-behaved Hilbert transform. The difference between $U(t)$ and $c_1(t)$ is then treated as a new time history and subjected to the same sifting process, giving the second IMF, $c_2(t)$. The EMD procedure continues until the residue becomes so small that it is less than a predetermined value of consequence, or the residue becomes a monotonic function. The original time history $U(t)$ is finally expressed as the sum of the IMFs plus the final residue:

$$U(t) = \sum_{j=1}^{N} c_j(t) + r(t)_N \tag{15.2}$$

where:

N is the number of IMF components;
$r(t)_N$ is the final residue.

The final residue $r(t)_N$ of a wind speed time history $U(t)$ of one hour duration can be defined as the time-varying mean wind speed $\overline{U}_{1/3600}(t)$ with frequency contents less than 1/3600 Hz. Clearly, the EMD naturally extracts the time-varying mean wind speed from the recorded wind speed time history without any prior information.

The concept of the aforementioned time-varying mean wind speed can be extended to more general case in which a time-varying mean wind speed is taken as the sum of the last a few IMFs plus the final residue. The highest frequency "n" (the intermittency frequency) involved in such a time-varying mean wind speed $\overline{U}_n(t)$ may be decided according to the first natural frequency of the concerned bridge structure, so that the dynamic effect of the time-varying mean wind speed on the bridge structure can be completely neglected. Such a time-varying mean wind speed can be extracted from the measured wind speed time history through an intermittency check [15] implemented during the sifting process of the wind data.

The implementation procedure of the intermittency check is to process the original wind speed time history $U(t)$ through EMD with a targeted intermittency frequency, to obtain all IMFs that have frequencies greater than this frequency and then to remove all obtained IMFs from $U(t)$ to yield the time-varying mean wind speed which contains frequencies lower than the targeted intermittency frequency. In such a process, the intermittency frequency, in conjunction with the sampling frequency of the time history, is converted to the maximum intermittency to be implemented. This maximum specifies the greatest number of points between the consecutive extrema of an allowable IMF. It will be demonstrated in the following section that such a continuously time-varying mean wind speed is more natural than the traditional time-averaged mean wind speed.

15.2.3 Non-Stationary Wind Characteristics

After a time-varying mean wind speed $\overline{U}_n(t)$ is identified, the subtraction of $\overline{U}_n(t)$ with frequency contents less than n Hz from the measured wind speed $U(t)$ then produces the fluctuating wind speed $u(t)$. Because $u(t)$ is assumed to be a zero-mean stationary Gaussian process, the standard deviation and the probability density function (PDF) of the fluctuating wind speed can be obtained by:

$$\sigma_u^2 = \frac{1}{T}\int_0^T u^2(t)\mathrm{d}t \tag{15.3}$$

$$p(u) = \frac{1}{\sqrt{2\pi}\sigma_u}\mathrm{e}^{-u^2/2\sigma_u^2} \tag{15.4}$$

The wind spectrum of fluctuating wind speed can be obtained using traditional spectral analysis. However, the mean wind speed concerned is time-varying for non-stationary wind, and the turbulence intensity is also time-dependent over the time interval T. To be consistent with the turbulence intensity defined by the traditional approach, that obtained by the EMD is defined as the mean value of the time-varying turbulence intensity over the time interval T:

$$I_u = \frac{1}{T}\int_0^T \frac{\sigma_{u,T}}{\overline{U}_{1/T}(t)}\mathrm{d}t \tag{15.5}$$

where $\sigma_{u,T}$ is the standard deviation of fluctuating wind speed over the time interval T.

The definition of gust factor is correspondingly changed to the maximum ratio of t_1 time-varying mean wind speed to the corresponding part of the hourly time-varying mean wind speed:

$$G(t_1) = \max\left[\frac{\overline{U}_{1/t_1}(t)}{\overline{U}_{1/3600}(t)}\right] \tag{15.6}$$

15.2.4 Case Study: Typhoon Victor

On 2 August 1997, about three months after the opening of the Tsing Ma Bridge to the public, Typhoon Victor crossed over the bridge and made landfall over the western part of the New Territories of Hong Kong. The WASHMS installed in the bridge timely recorded wind speed time-histories of seven hours duration. The wind data recorded by the six anemometers are used in the following analysis [9]. More information on Typhoon Victor and the WASHMS can be found in Chapter 9.

The evaluation of stationarity of wind data is performed on wind samples of one-hour duration. Seven wind samples can be extracted from a seven-hour duration record of the six anemometers, so a total of 42 wind samples are obtained for evaluation. Each one-hour sample is further divided into 30 segments. The run test, as described by Bendat and Piersol [12], is applied to each sample to evaluate its stationarity in terms of the mean square value.

Listed in Table 15.1 are the test results for all wind samples with a significance level of 0.05. It can be seen that only 13 out of the 42 samples pass the stationarity test. In particular, wind samples recorded between 20:00 to 21:00 Hong Kong Time (HKT) from all six anemometers are non-stationary.

Table 15.1 Stationarity test results from traditional approach (from [9]) (Reproduced with permission from ASCE)

Anemometer	17:00–18:00	18:00–19:00	19:00–20:00	20:00–21:00	21:00–22:00	22:00–23:00	23:00–24:00
WITBN01	•	□	□	□	•	□	□
WITBS01	□	□	□	□	•	□	□
WITET01	•	•	□	□	□	□	□
WITJN01	•	□	•	□	•	•	•
WITJS01	□	□	•	□	□	□	□
WITPT01	□	•	•	□	□	□	□

•: Stationary □: Non-stationary.

To understand the stationarity test results, the ten-minute averaged mean wind speed and mean wind direction at the two anemometers (WITJN01 and WITJS01) are computed using the traditional approach and plotted against time in Figures 15.1a and 15.1b, respectively. It can be seen that there is a sudden change of wind direction from north-east to south-west, and that the mean wind speed becomes very small around 20:00 HKT. This indicates that the eye of Typhoon Victor had just crossed over the bridge at that time. As a result, wind direction became quite unstable in the next hour (from 20:00 and 21:00 HKT), and the corresponding wind samples thus fail to pass the stationarity test.

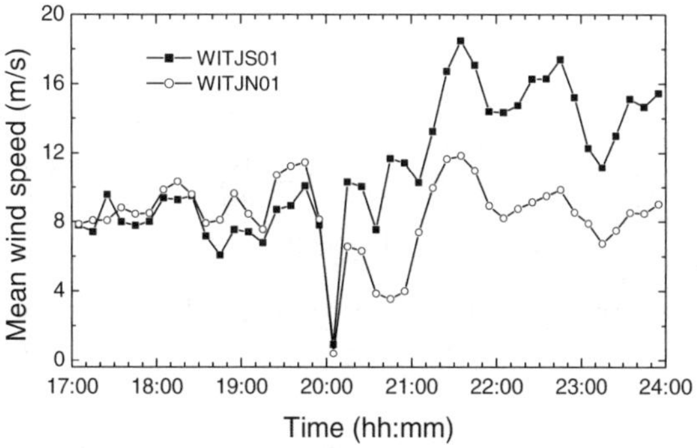

(a) Mean Wind Speed

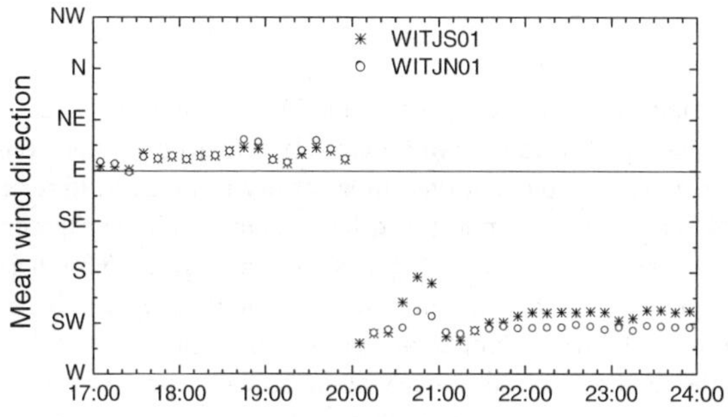

(b) Mean Wind Direction

Figure 15.1 Variations of ten-minute mean wind speed and direction (from [9]) (Reproduced with permission from ASCE).

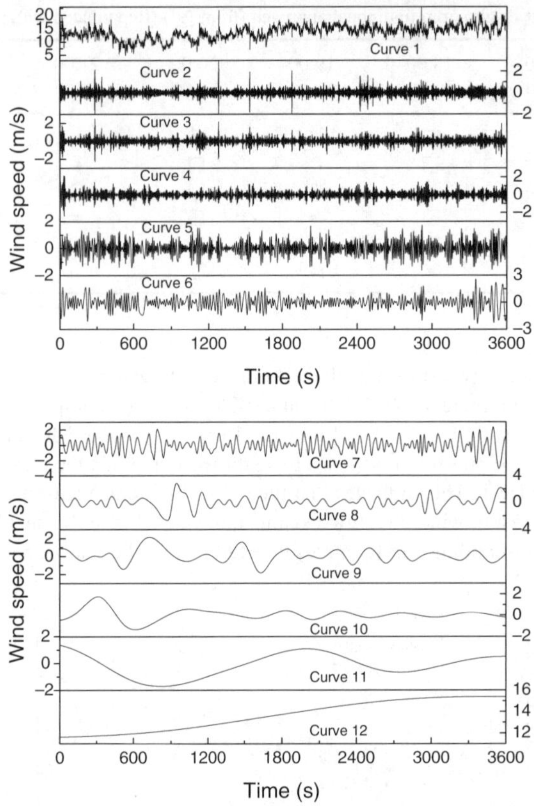

Figure 15.2 Empirical mode decomposition components (from [9]) (Reproduced with permission from ASCE).

The EMD method is then applied to the wind samples that fail to pass the stationarity test. The results obtained from the wind speed time history of one-hour duration recorded between 23:00 and 24:00 HKT are shown in Figure 15.2.

The first curve (curve 1) displayed in Figure 15.2 is the original longitudinal wind speed time history. After applying the EMD to this time history, a total of ten IMF components (curves 2–11) and one final residue (curve 12) are obtained. It can be seen that the IMF components start from the shortest to the longest periods in the sequence. The last curve (curve 12) is actually a monotonic function of time. As pointed out by Huang *et al.* [14], if the analyzed time history has a trend, this must be the final residue obtained from the EMD. This can be demonstrated here again by plotting the original wind speed time history together with the final residue (curve 12), as shown in Figure 15.2b. It is interesting to see that curve 12 has a lower frequency (less than 1/3600) but larger amplitude than that of any IMF component.

The hourly mean value of the wind sample, as computed by the traditional approach, is 13.5 m/sec, while the amplitude of the final residue (time-varying mean wind speed) varies between 11.7 m/sec and 15.4 m/sec, with a mean value of 13.6 m/sec. Such a continuously time-varying mean wind speed cannot be obtained by the traditional approach, but the wind sample failing to pass the stationarity test is actually attributed to the existence of such a time-varying mean wind speed. Since the highest frequency in the time-varying mean wind speed is less than 1/3600 Hz, it will not cause any dynamic effect on long-span cable-supported bridges.

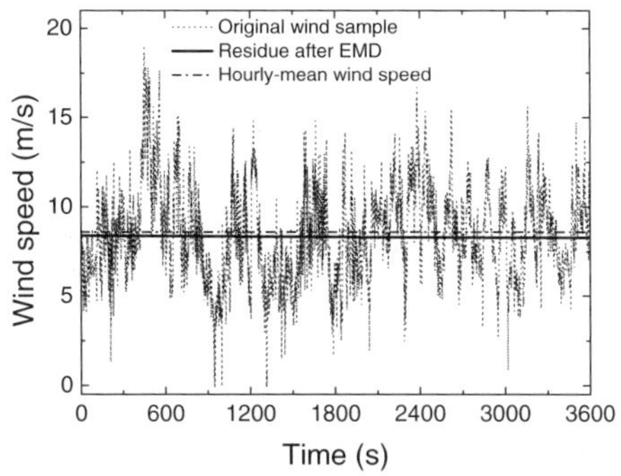

(a) Wind Sample from WITJS01 (19:00-20:00HKT)

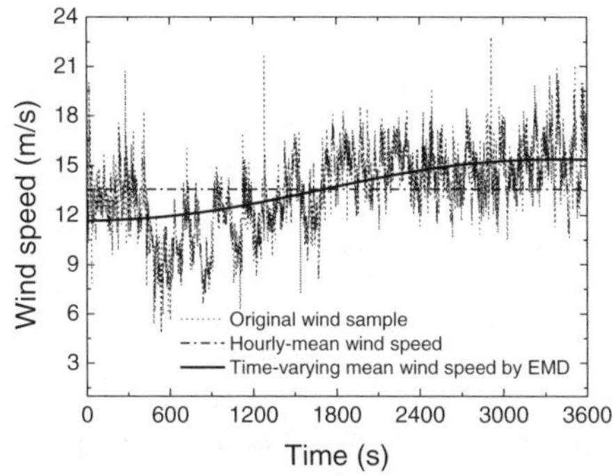

(b) Wind Sample from WITJS01 (23:00-24:00HKT)

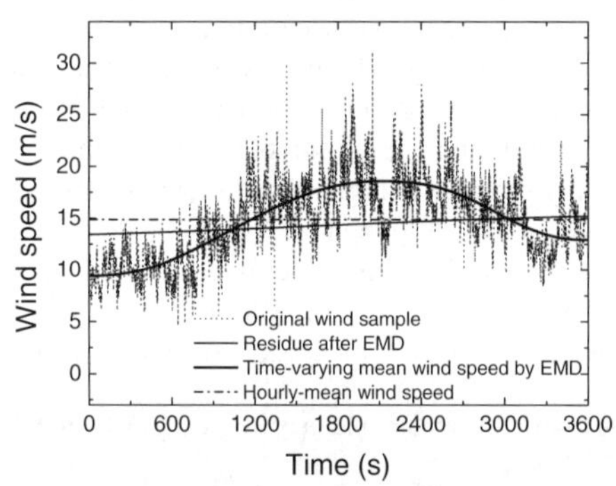

(c) Wind Sample from WITJS01 (21:00-22:00 HKT)

Figure 15.3 Time-varying and hourly-mean wind speeds (from [9]) (Reproduced with permission from ASCE).

To further understand the final residue, the EMD is also applied to the wind speed time history recorded at the anemometer WITJS01 between 19:00 and 20:00 HKT, which passes the stationarity test. The final residue identified by the EMD is plotted in Figure 15.3a, together with the original wind speed time history and the hourly-mean wind speed obtained from the traditional approach. Clearly, the final residue identified by the EMD is almost constant and very close to the hourly mean wind speed.

Table 15.2 Stationarity test results from EMD at intermittency frequency of 1/3600 Hz (from [9]) (Reproduced with permission from ASCE)

Anemometer	17:00–18:00	18:00–19:00	19:00–20:00	20:00–21:00	21:00–22:00	22:00–23:00	23:00–24:00
WITBN01	•	•	□	□	•	□	□
WITBS01	□	□	•	□	•	•	□
WITET01	•	•	•	□	•	•	□
WITJN01	•	•	•	□	•	•	•
WITJS01	□	□	•	□	□	•	•
WITPT01	•	•	•	□	•	□	□

•: Stationary □: Non-stationary.

For the wind samples that fail to pass the stationarity test, the final residue identified by the EMD is subtracted from the original time history to form a new sample. The stationarity test is then applied to each new sample, with the same number of segments and significance level as used in the first stationarity test. The second stationarity test results are listed in Table 15.2 and demonstrates that, among 29 non-stationary samples identified from the first stationarity test, 11 wind samples now pass the second stationarity test, including the wind sample recorded at one anemometer between 23:00 and 24:00 HKT. However, the six samples recorded during the sudden change of wind direction (between 20:00 and 21:00 HKT) still cannot pass the second stationarity test. If these six samples are not counted, the number of stationary samples is increased from 36% to 67% of 36 wind samples.

The highest frequency (the intermittency frequency) involved in the aforementioned time-varying mean wind speed (the final residue) identified by the EMD is less than 1/3600 Hz. Now, let us increase the highest frequency involved in the time-varying mean wind speed from 1/3600 Hz to 1/1200 Hz and use the EMD, together with the intermittence check, to find the corresponding time-varying mean wind speed from wind samples that fail to pass the second stationarity test.

Figure 15.3c shows the wind sample recorded between 21:00 and 22:00 HKT, which fails to pass both the first and second stationarity tests, and the time-varying mean wind speed with the highest frequency less than 1/1200 Hz. It can be seen that such a time-varying mean wind speed, rather than the final residue, reflects the trend of the original wind speed. The time-varying mean wind speed is then subtracted from the original wind sample to form a new sample that is, in turn, subjected to the third stationarity test. It is found that six more wind samples pass the third stationarity test, as listed in Table 15.3, leading to the stationary samples being 83% of 36 wind samples. This result indicates that most of the wind samples measured during Typhoon Victor can be described by the proposed non-stationary wind model, except for those recorded during the sudden change of wind direction.

The time-varying mean wind speed obtained by the EMD at a designated frequency level may be related to the traditional time-averaged mean wind speed over the corresponding time interval. Take

Table 15.3 Stationarity test results from EMD at intermittency frequency of 1/1200 Hz (from [9]) (Reproduced with permission from ASCE)

Anemometer	17:00–18:00	18:00–19:00	19:00–20:00	20:00–21:00	21:00–22:00	22:00–23:00	23:00–24:00
WITBN01	•	•	□	□	•	□	□
WITBS01	□	•	•	□	•	•	•
WITET01	•	•	•	□	•	•	□
WITJN01	•	•	•	□	•	•	•
WITJS01	□	•	•	□	•	•	•
WITPT01	•	•	•	□	•	•	•

•: Stationary □: Non-stationary.

the wind speed time history recorded by the anemometer WITJS01 between 23:00 and 24:00 HKT as an example (see Figure 15.3b).

Figures 15.4a to 15.4c show the time-varying mean wind speed obtained by the EMD at the intermittency frequency level of 1/600, 1/300 and 1/200 Hz, respectively. Figures 15.4a to 5.4c also show the traditional time-averaged mean wind speed over the time interval 600, 300 and 200 seconds. It can be

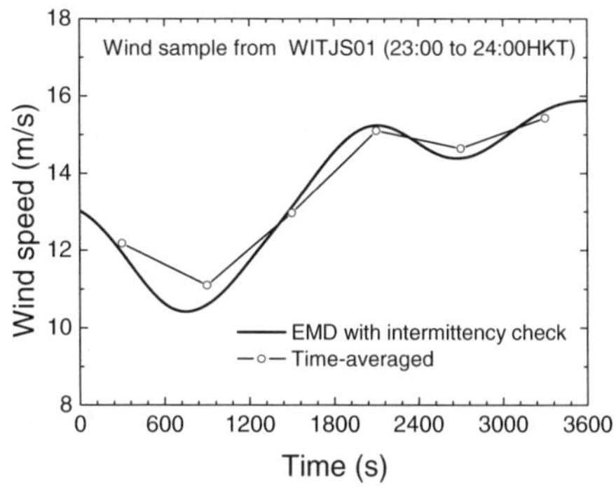

(a) 600 Seconds Time Interval

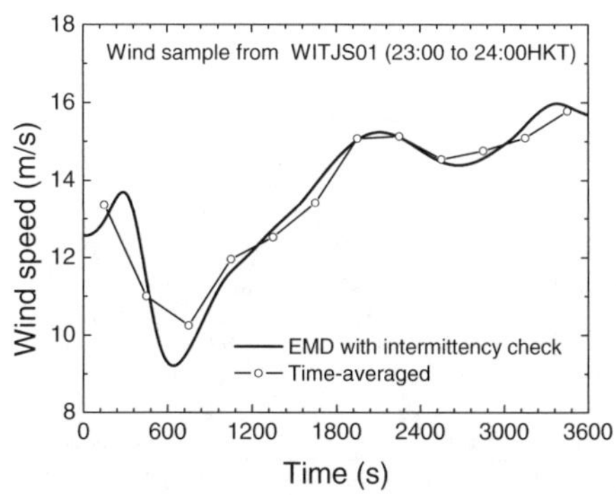

(b) 300 Seconds Time Interval

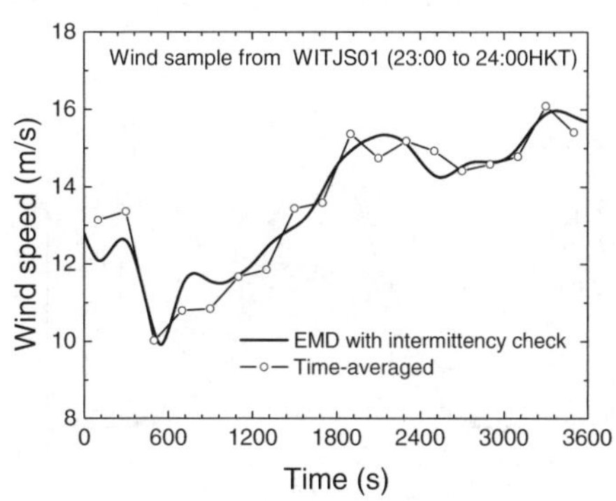

(c) 200 Seconds Time Interval

Figure 15.4 Comparison between time-varying and time-averaged mean wind speeds (from [9]) (Reproduced with permission from ASCE).

seen that the time-varying mean wind speed identified by the EMD is similar to the corresponding time-averaged mean wind speed obtained by the traditional approach. That identified by the EMD is, however, a continuous function of time with a designated frequency level, which is more natural than the traditional time-averaged mean wind speed with the certain time interval.

Shown in Figure 15.5a are the probability distributions of fluctuating wind speed recorded between 19:00 and 20:00 HKT (Figure 15.3a), together with a Gaussian density function. One probability

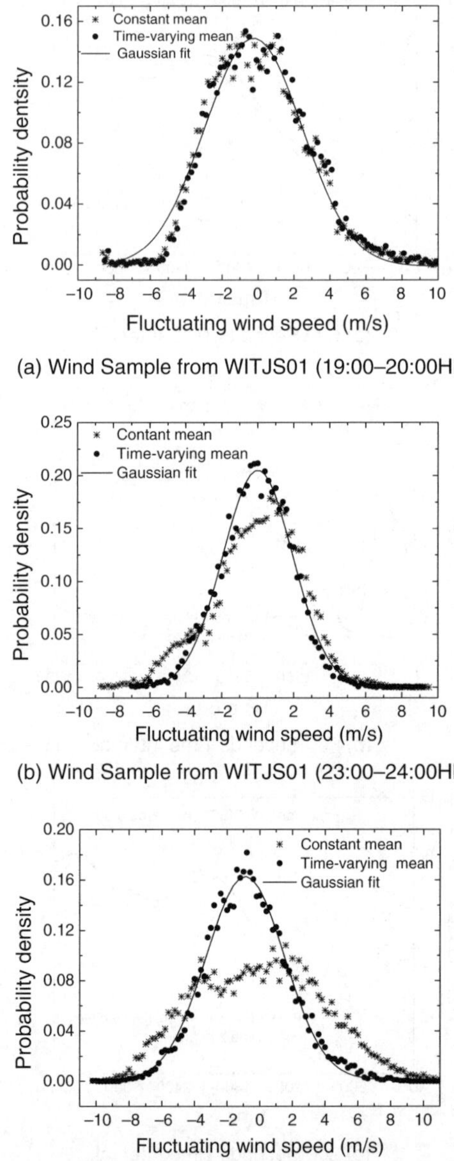

(a) Wind Sample from WITJS01 (19:00–20:00HKT)

(b) Wind Sample from WITJS01 (23:00–24:00HKT)

(c) Wind Sample from WITJS01 (21:00–22:00HKT)

Figure 15.5 Probability distributions of wind speed (from [9]) (Reproduced with permission from ASCE).

distribution is computed from the fluctuating wind speed obtained by subtracting the constant mean wind speed from the original wind sample. The other is obtained from the fluctuating wind speed obtained by subtracting the time-varying mean wind speed at a frequency level less than 1/3600 Hz from the original wind sample. It can be seen that the probability distributions obtained by the two approaches are close to each other and both comply with the Gaussian distribution.

Displayed in Figure 15.5b are the two probability distributions of fluctuating wind speed recorded between 23:00 and 24:00 HKT (Figure 15.3b), together with a Gaussian density function, obtained using the same approaches as for the first wind sample. It can be seen that for this sample, only the fluctuating wind speed obtained using the time-varying mean wind speed complies with the Gaussian distribution. The probability distribution of the fluctuating wind speed obtained using the constant mean wind speed deviates from the Gaussian distribution significantly. This indicates that the subtraction of the time-varying mean wind speed from the original non-stationary wind sample leads to a stationary Gaussian-distributed fluctuating wind speed.

Further demonstration is made on the fluctuating wind speed recorded between 21:00 and 22:00 HKT (Figure 15.3c), as shown in Figure 15.5c. One probability distribution is computed based on the constant mean wind speed, but the other is obtained based on the time-varying mean wind speed at a frequency level less than 1/1200 Hz. It can be seen that the latter fits the Gaussian distribution well, while the fluctuating wind speed obtained from the constant mean wind speed departs from the Gaussian distribution significantly. It may thus be concluded that the proposed non-stationary wind model, together with the EMD method, does provide a means of characterizing non-stationary wind speeds.

The three wind samples used in the probability analysis are taken to investigate the wind spectra of stationary and non-stationary wind samples. The spectral analysis is performed on each wind sample in two ways: one is performed on the fluctuating wind speed obtained from the constant mean wind speed (the traditional approach); the other is on the stationary fluctuating wind speed gained from the time-varying mean wind speed (EMD).

Figure 15.6a demonstrates the spectral density functions of fluctuating wind speed recorded between 19:00 and 20:00 HKT. The two spectral density functions are almost overlapping within the entire frequency range concerned. This is because the time-varying mean wind speed at a frequency level less than 1/3600 is almost the same as the hourly mean wind speed as shown in Figure 15.3a. Figure 15.6b shows the spectral density functions of fluctuating wind speed recorded between 23:00 and 24:00 HKT. It can be seen that the amplitude of the spectral density function obtained using the time-varying mean wind speed is slightly lower than that obtained using the hourly mean wind speed in the frequency range lower than 0.01 Hz. In the latter, the two spectral density functions are very similar. The same trend can be found in the spectral density functions of fluctuating wind speed recorded between 21:00 and 22:00 (Figure 15.6c). The reduction of spectral amplitude in the low frequency range may affect wind-induced dynamic response of super-long suspension bridges.

In order to compare the turbulence intensity obtained by the stationary and non-stationary approaches, turbulence intensities of ten minutes duration are computed using the wind samples between 21:00 and 24:00 HKT.

Shown in Figure 15.7 are the ten-minute turbulence intensities obtained by the EMD and the traditional approach. It can be seen that the two approaches give almost the same value of turbulence intensity over the entire three-hour duration, except for the two points at 21:15 HKT and 23:05 HKT. After a careful examination of the wind speed time history, it is found that the trend of wind speed varies rapidly within the ten-minute period centered at either 21:15 HKT or 23:05 HKT. This leads to overestimation of turbulence intensity by the traditional approach. The turbulence intensity estimated by the EMD seems to be more reasonable than the traditional approach.

A comparison of gust factors obtained using the two approaches is made in the following. Figure 15.8 shows the variations of gust factor with gust duration computed by stationary and non-stationary approaches in terms of the three one-hour wind samples from the anemometer WITJS01 between 21:00 and 22:00 HKT, 22:00 and 23:00 HKT, and 23:00 and 24:00 HKT. The gust factors obtained by the

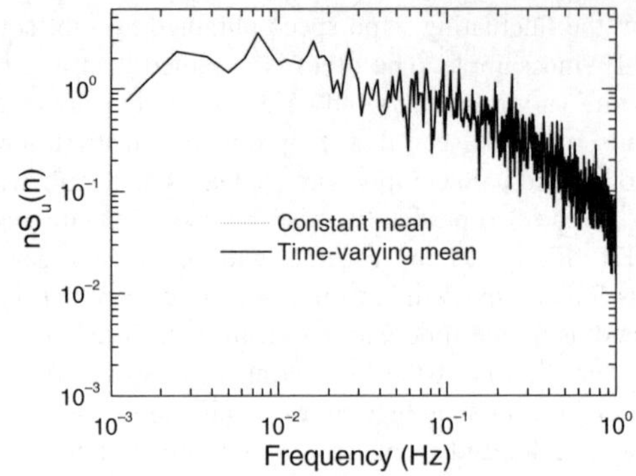

(a) Wind Sample from WITJS01 (19:00–20:00HKT)

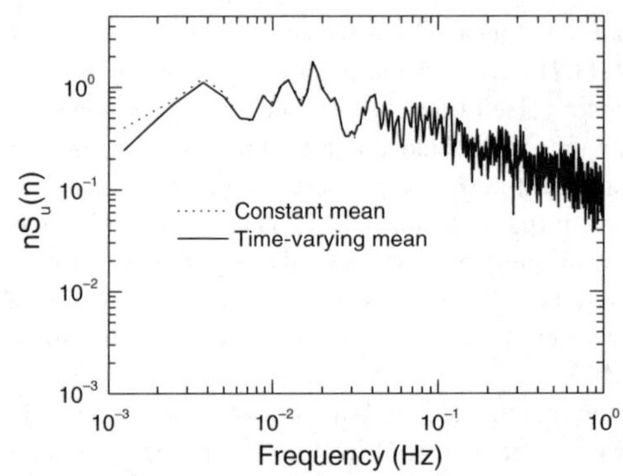

(b) Wind Sample from WITJS01 (23:00–24:00HKT)

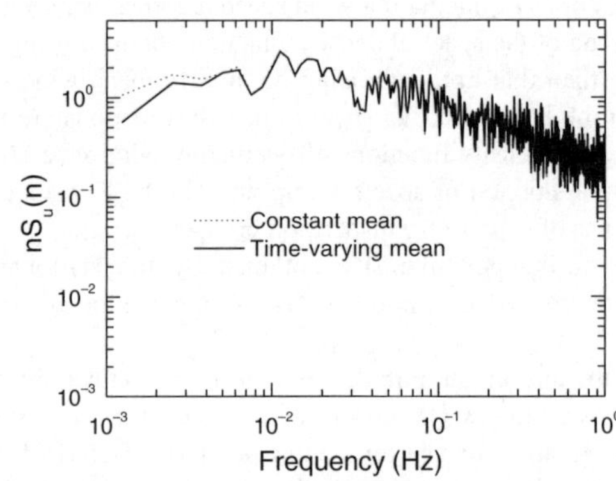

(c) Wind Sample from WITJS01 (21:00–22:00 HKT)

Figure 15.6 Wind spectra (from [9]) (Reproduced with permission from ASCE).

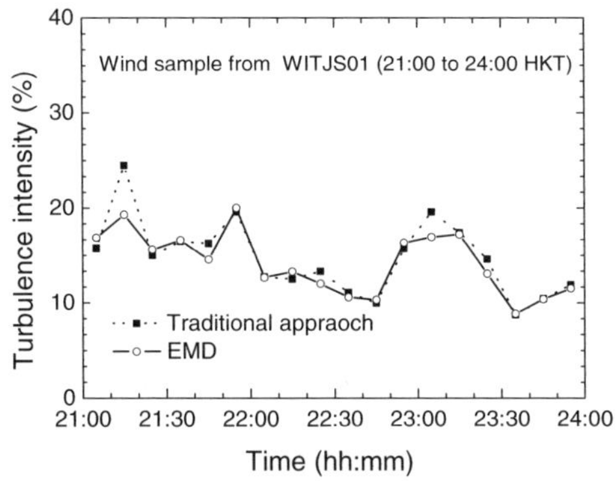

Figure 15.7 Turbulence intensities (from [9]) (Reproduced with permission from ASCE).

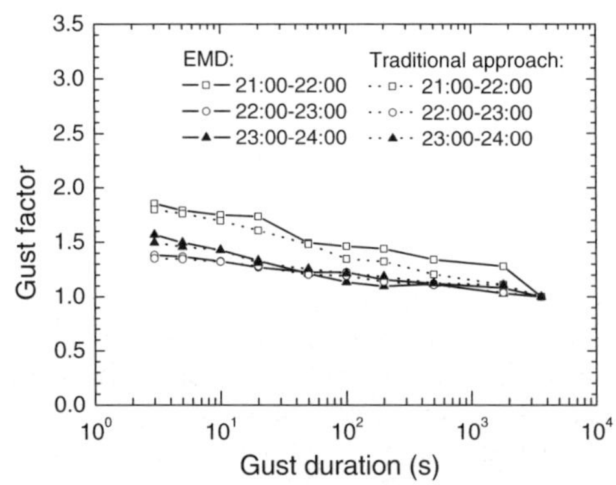

Figure 15.8 Gust factors (from [9]) (Reproduced with permission from ASCE).

EMD are almost the same as those obtained by the traditional approach. The variation of gust factor with gust duration obtained by either approach confirms approximately to a straight line if the curve is plotted in semi-logarithmic scale.

15.3 Non-Stationary Wind Model II

As demonstrated in the last section, for some wind samples, subtraction of the time-varying mean wind speed from the original non-stationary wind sample still cannot make the wind sample pass the stationarity test. A second non-stationary wind model is therefore needed. In model II, the fluctuating wind is further regarded as a non-stationary process characterized by an evolutionary power spectral density (EPSD) function in addition to the time-varying mean wind speed [11]. The EPSD function can be established by extending the stationary model with instantaneous statistical properties (note: the term "instantaneous" here means that, for each time instant, the corresponding statistical properties are estimated over a specified short time interval whose center is the current time instant). In this connection, the time-varying mean wind speed needs to be obtained using a way different from the EMD.

15.3.1 Time-Varying Mean Wind Speed and Mean Wind Profile

In the non-stationary typhoon wind model II, the longitudinal component $U(t)$ is still regarded as the sum of time-varying mean wind speed $\overline{U}(t)$ and zero-mean fluctuating wind speed component $u(t)$, but $\overline{U}(t)$ is determined by short-time averaging over a short time interval T:

$$\overline{U}(t) = \frac{1}{T}\int_{t-T/2}^{t+T/2} U(t)\mathrm{d}t \tag{15.7}$$

Clearly, this requires the length of wind speed time history to be at least twice the time interval T. Equation 15.7 implies that $\overline{U}(t)$ contains frequency contents lower than $1/T$ and actually varies with time very slowly compared with the fluctuating wind speed component $u(t)$. To consider the mean wind speed profile in the vertical plane, the time-varying mean wind speed can be further expressed as:

$$\overline{U}(t) = \overline{U}_0(z) \cdot \eta_0(t) \tag{15.8}$$

where:

$\overline{U}_0(z)$ is the mean value of the time-varying mean wind speed $\overline{U}(t)$, which is almost the same as the conventional time-invariant mean wind speed depending on the height above the ground z;
$\eta_0(t)$ is the time-varying function of mean wind speed, which is assumed to be independent of the height above the ground.

In such a way, the mean wind speed profile can be readily incorporated into the non-stationary wind model. Furthermore, it is assumed that the vertical wind component $W(t)$ has a zero mean, which is similar to the assumption used in the stationary buffeting analysis. As a result, the vertical component $W(t)$ is regarded as the fluctuating wind speed $w(t) = W(t)$ only.

During the passage of a typhoon, the mean wind speed profile at a bridge site surrounded by a complex terrain generally does not conform to the traditional logarithmic law or power law. A numerical method for predicting directional typhoon mean wind speed profiles over a complex terrain, $\overline{U}_0(z)$, has been recently proposed by Xu *et al.* [16] and is presented in Chapter 13. The method they proposed involves a refined typhoon wind field model, Monte Carlo Simulation, computational fluid dynamics (CFD) simulation and artificial neural networks (ANN).

15.3.2 Evolutionary Spectra

After extracting the time-varying mean wind speed from $U(t)$, the remained fluctuating wind speed component $u(t)$ may still possess time-varying characteristics of wind turbulence. Therefore, $u(t)$ is represented by a zero-mean oscillatory process that admits the representation [17]:

$$u(t) = \int_0^{+\infty} A(\omega, t)\mathrm{e}^{\mathrm{i}\omega t}\mathrm{d}\xi(\omega) \tag{15.9}$$

where:

$i = \sqrt{-1}$;
$A(\omega, t)$ is a slowly varying function with time;

$\xi(\omega)$ is a zero-mean Gaussian orthogonal increment process having the properties:

$$\mathrm{E}[\mathrm{d}\xi(\omega)\mathrm{d}\xi(\omega')^*] = \begin{cases} 0 & \omega \neq \omega' \\ \mu(\omega)\mathrm{d}\omega & \omega = \omega' \end{cases} \tag{15.10}$$

where:

the superscript "*" denotes the complex conjugate;
$\mu(\omega)\mathrm{d}\omega$ is variance of the increment process.

The EPSD of $u(t)$ at time t can be then written as:

$$S_{uu}(\omega, t) = |A(\omega, t)|^2 \mu(\omega) \tag{15.11}$$

Based on observations on measured typhoon wind data, a closed-form formula for $S_{uu}(\omega,t)$ can be proposed by incorporating time-varying parameters into stationary power spectral density functions. In this study, the von Karman spectrum presented in Chapter 2 is selected and extended to the evolutionary von Karman spectrum with time-varying variance, mean wind speed and integral length scale:

$$\frac{S_{uu}(\omega, t)}{\sigma_u^2(t)} = \frac{1}{2\pi} \frac{\dfrac{4L_u(t)}{\overline{U}(t)}}{\left[1 + 70.8\left(\dfrac{4\omega L_u(t)}{2\pi\overline{U}(t)}\right)^2\right]^{\frac{5}{6}}} \tag{15.12}$$

where $L_u(t)$ is the time-varying integral length scale; and $\sigma_u^2(t)$ is the time-varying variance of $u(t)$.

$$\sigma_u^2(t) = \frac{1}{T_U}\int_{t-T_U/2}^{t+T_U/2} u^2(t)\mathrm{d}t \tag{15.13}$$

The time interval T_U in Equation 15.13 should be less than T in order to involve frequency contents higher than $1/T$. The time-varying turbulence intensity can be expressed as:

$$I_u(t) = \sigma_u(t)/\overline{U}(t) \tag{15.14}$$

Similarly, the vertical wind speed component $w(t)$ can also be defined as an oscillatory process with the evolutionary von Karman spectrum as:

$$\frac{S_{ww}(\omega, t)}{\sigma_w^2(t)} = \frac{1}{2\pi} \frac{\dfrac{4L_w(t)}{\overline{U}(t)}\left[1 + 755\left(\dfrac{L_w(t)}{\overline{U}(t)}\right)^2\right]}{\left[1 + 283\left(\dfrac{L_w(t)}{\overline{U}(t)}\right)^2\right]^{\frac{11}{6}}} \tag{15.15}$$

where $\sigma_w^2(t)$ and $L_w(t)$ are the vertical time-varying variance and integral length scale, respectively.

15.3.3 Coherence Function

For the non-stationary buffeting analysis of long-span cable-supported bridges, the exponential coherence function used in the stationary wind model as described in Chapter 2 can be directly adopted.

$$\gamma_{pp,ij}(\omega) = \exp\left\{ -\frac{\omega}{2\pi} \frac{\left[C_{pz}{}^2 (z_i - z_j)^2 + C_{py}{}^2 (y_i - y_j)^2 \right]^{\frac{1}{2}}}{\frac{1}{2}\left[\overline{U}_0(z_i) + \overline{U}_0(z_j) \right]} \right\} \tag{15.16}$$

where the symbol p represents u or w; z_i and y_i denote the vertical and lateral coordinates; C_{pz} and C_{py} are the decay coefficients. This coherence function is time-invariant, which meets the requirement of Priestley's EPSD theory [18]. The coherence function between the vertical and longitudinal wind components is disregarded. Based on coherence functions, the cross-EPSD can be constructed by:

$$S_{pp,ij}(\omega, t) = S_{pp,ii}(\omega, t) S_{pp,jj}(\omega, t) \gamma_{pp,ij}(\omega) \tag{15.17}$$

15.3.4 Case Study: Typhoon Dujuan

Figure 15.9 shows the Stonecutters Bridge, with a main span of 1018 m, in Hong Kong. A field measurement system was installed near the bridge site in 2002 to measure strong winds and to ascertain wind turbulence parameters for the design purpose of the bridge, and the system has recorded several strong typhoons passing by Hong Kong [19]. Among all the available records, Typhoon Dujuan, which affected the site on 29 August, 2003, is selected because its eye has passed nearby the site at a close distance, although the Stonecutters Bridge was completed at the end of 2009 [11].

The wind speed time histories recorded at a height of 30 m above the ground during Typhoon Dujuan are used to establish the non-stationary wind field model. The measured longitudinal wind speed time history $U(t)$ of one-hour duration is shown in Figure 15.10a.

Figure 15.9 Stonecutters Bridge in Hong Kong.

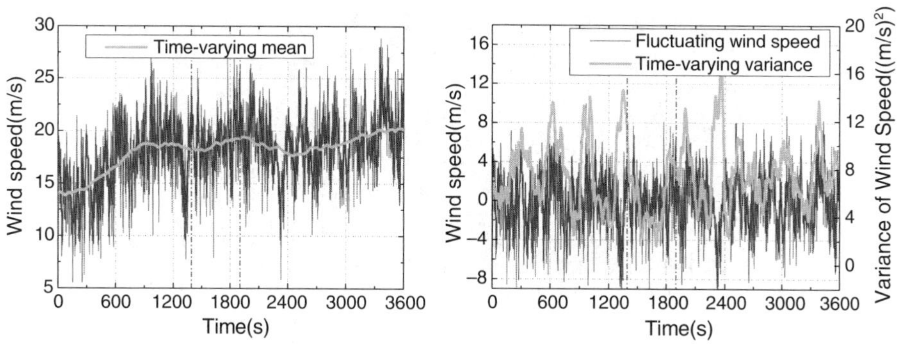

(a) Original time history with time-varying mean speed

(b) Fluctuating wind time history with time-varying variance

Figure 15.10 Longitudinal wind speed time histories recorded during Typhoon Dujuan.

The time-varying mean wind speed $\overline{U}(t)$ is estimated by using the short-time averaging over $T = 3600$ sec, and the result is shown in Figure 15.10a. The fluctuating wind speed time history $u(t) = U(t) - \overline{U}(t)$ is then computed and displayed in Figure 15.10b. The time-varying variance is also estimated by using the short-time averaging with a time window $T_U = 60$ sec and the resulting $\sigma_u^2(t)$ is shown in Figure 15.10b. The time-varying turbulence intensity $I_u(t)$ is determined by using Equation 15.14 and shown in Figure 15.11a.

The EPSD function $S_{uu}(\omega,t)$ of fluctuating wind speed $u(t)$ can be estimated by using the method proposed by Priestley [13,17,20], and the results are shown in Figure 15.12a. $L_u(t)$ is then obtained by using Equation 15.12 to fit the estimated EPSD and shown in Figure 15.11b. The evolutionary von Karman spectrum within the specified time period can be finally found using Equation 15.12 in terms of the estimated $\overline{U}(t)$, $I_u(t)$, and $L_u(t)$, and it is shown in Figure 15.12b to compare with the measured one shown in Figure 15.12a. To avoid end effects, $\overline{U}(t)$, $\sigma_u^2(t)$ and $S_{uu}(\omega,t)$ within the specified one-hour duration are actually computed by using the measured wind speed time history of at least two hours duration.

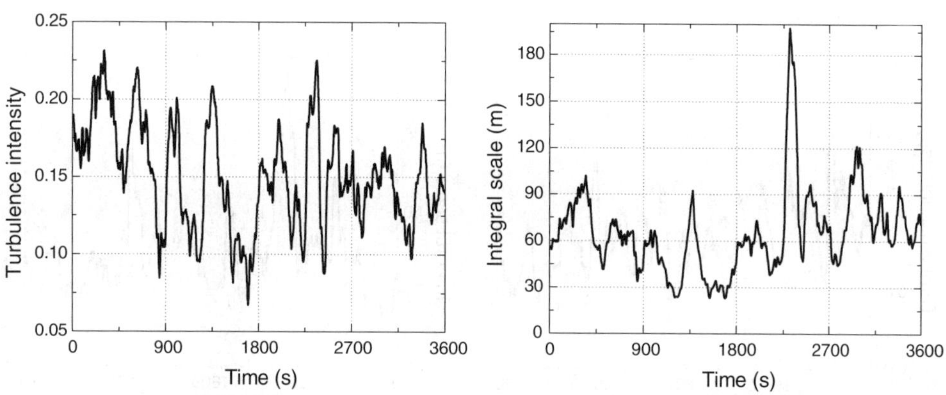

(a) Turbulence intensity

(b) Integral length scale

Figure 15.11 Longitudinal time-varying turbulence characteristics.

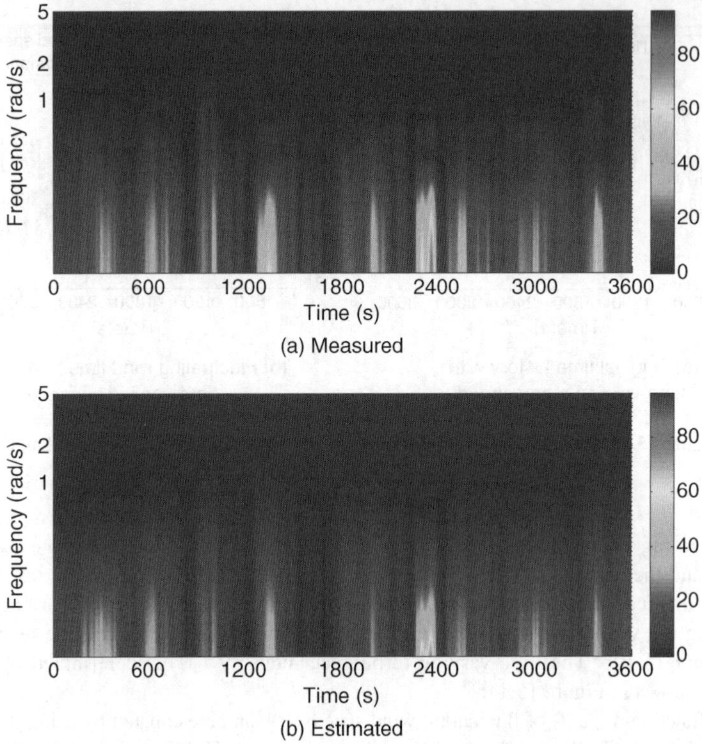

Figure 15.12 EPSD of longitudinal fluctuating wind speed.

Similar procedure is applied to the measured vertical turbulence component $w(t)$. The time-varying mean vertical wind speed estimated from measured time histories is close to zero.

Figures 15.13a and 15.13b show the turbulence intensity and the turbulence integral scale of vertical fluctuating wind speed, while the measured and estimated vertical evolutionary von Karman spectra are displayed in Figures 15.14a and 15.14b respectively.

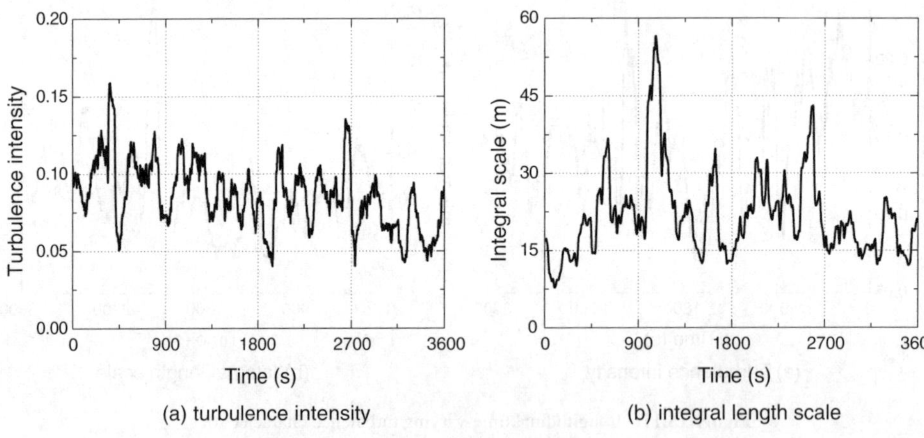

Figure 15.13 Vertical time-varying turbulence characteristics.

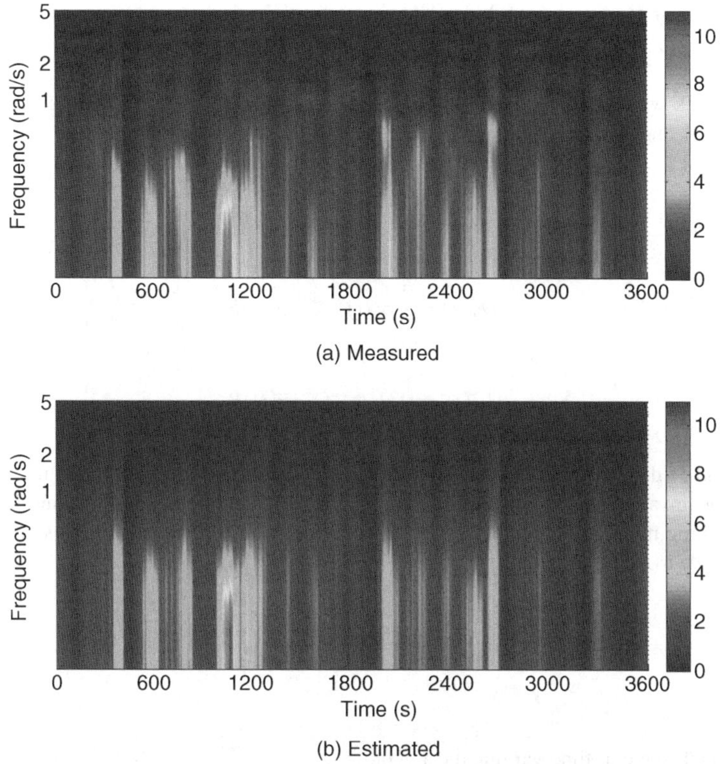

Figure 15.14 EPSD of vertical fluctuating wind speed.

The mean wind profile and the coherence function cannot be obtained from the measurement data. The mean wind speed profile shown in Figure 15.15 is used for the case study in the next section, which is obtained by a CFD simulation with the topographic model around the bridge site and Typhoon Dujuan wind data [16].

For the coherence function expressed by Equation 15.16, the decay coefficients C_{pz} and C_{py} are taken as 8.3 and 6.4 for $p = u$ and 3.8 and 4.8 for $p = w$.

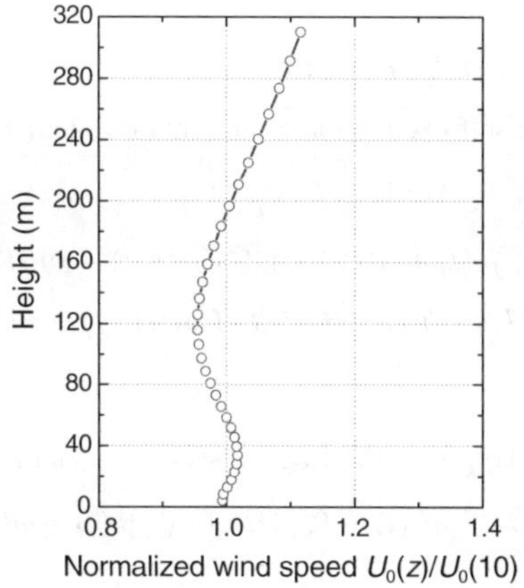

Figure 15.15 Typhoon mean wind speed profile over complex terrain.

15.4 Buffeting Response to Non-Stationary Wind

15.4.1 Time-Varying Mean Wind Forces

For a bridge deck with unit length and width B, the mean wind forces acting on its center due to the non-stationary mean wind speeds described above can be expressed as:

$$L_m(t) = \frac{1}{2}\rho \overline{U}(t)^2 C_L(\overline{\alpha}(t))B \tag{15.18a}$$

$$D_m(t) = \frac{1}{2}\rho \overline{U}(t)^2 C_D(\overline{\alpha}(t))B \tag{15.18b}$$

$$M_m(t) = \frac{1}{2}\rho \overline{U}(t)^2 C_M(\overline{\alpha}(t))B^2 \tag{15.18c}$$

where L, D and M denote the lift force, drag force, and pitching moment, in which the subscript "m" denotes the mean value; ρ is the air density; C_L, C_D and C_M denote the aerodynamic lift force, drag force and pitching moment coefficients. These coefficients are functions of the instantaneous mean wind attack angle $\overline{\alpha}(t)$.

$$\overline{\alpha}(t) = \overline{\alpha}_0(t) + \overline{\alpha}_S(t) \tag{15.19}$$

where:

$\overline{\alpha}_0(t)$ is the attack angle of time-varying mean wind;
$\overline{\alpha}_S(t)$ is the torsional response of the bridge deck due to the time-varying mean wind forces and moment.

As a result, $\overline{\alpha}$ is time-varying, and the force coefficients, which depend on $\overline{\alpha}(t)$, are also time-dependent. The mean wind forces and moment evolve with time and should be determined by iteration. Nevertheless, the mean wind forces vary with time very slowly, as the time-varying mean wind speed contains frequency contents lower than $1/T$.

15.4.2 Non-Stationary Self-Excited Forces

Under non-stationary winds, the self-excited forces and moment acting on the bridge deck unit can be expressed as

$$L_{se}(t) = \rho B^2 \omega \begin{bmatrix} H_1^*(\overline{\alpha}(t), \overline{U}_R(t))\dot{h}(t) + BH_2^*(\overline{\alpha}(t), \overline{U}_R(t))\dot{\alpha}(t) + \omega BH_3^*(\overline{\alpha}(t), \overline{U}_R(t))\alpha(t) \\ + \omega H_4^*(\overline{\alpha}(t), \overline{U}_R(t))h(t) + H_5^*(\overline{\alpha}(t), \overline{U}_R(t))\dot{p}(t) + \omega H_6^*(\overline{\alpha}(t), \overline{U}_R(t))p(t) \end{bmatrix} \tag{15.20a}$$

$$D_{se}(t) = \rho B^2 \omega \begin{bmatrix} P_1^*(\overline{\alpha}(t), \overline{U}_R(t))\dot{p}_k(t) + BP_2^*(\overline{\alpha}(t), \overline{U}_R(t))\dot{\alpha}(t) + \omega BP_3^*(\overline{\alpha}(t), \overline{U}_R(t))\alpha(t) \\ + \omega P_4^*(\overline{\alpha}(t), \overline{U}_R(t))p_k(t) + P_5^*(\overline{\alpha}(t), \overline{U}_R(t))\dot{h}(t) + \omega P_6^*(\overline{\alpha}(t), \overline{U}_R(t))h(t) \end{bmatrix} \tag{15.20b}$$

$$M_{se}(t) = \rho B^3\omega \begin{bmatrix} A_1^*(\overline{\alpha}(t),\overline{U}_R(t))\dot{h}(t) + BA_2^*(\overline{\alpha}(t),\overline{U}_R(t))\dot{\alpha}(t) + \omega BA_3^*(\overline{\alpha}(t),\overline{U}_R(t))\alpha(t) \\ +\omega A_4^*(\overline{\alpha}(t),\overline{U}_R(t))h(t) + A_5^*(\overline{\alpha}(t),\overline{U}_R(t))\dot{p}(t) + \omega A_6^*(\overline{\alpha}(t),\overline{U}_R(t))p(t) \end{bmatrix} \quad (15.20c)$$

where:

the subscript "*se*" denotes the self-excited forces;
P_i^*, H_i^*, and A_i^* ($i = 1, \ldots, 6$) are the flutter derivatives which are functions of the instantaneous mean wind attack angle and the reduced mean wind speed $\overline{U}_R(t) = 2\pi\overline{U}(t)/\omega B$;
$h_k(t)$, $p_k(t)$ and $\alpha_k(t)$ are the vertical and lateral dynamic displacement responses and the dynamic torsional response of the bridge deck unit;
each over-dot denotes the differentiation with respect to time.

Note that the flutter derivatives are the functions of time due to indirect effects of the time-varying mean wind speed, and they vary with time very slowly.

15.4.3 Non-Stationary Buffeting Forces

The non-stationary buffeting forces on the bridge deck unit can be expressed by:

$$L_b(t) = \frac{1}{2}\rho\overline{U}(t)B\{2C_L(\overline{\alpha}(t))u(t) + [C_L'(\overline{\alpha}(t)) + C_D(\overline{\alpha}(t))]w(t)\} \quad (15.21a)$$

$$D_b(t) = \frac{1}{2}\rho\overline{U}(t)B[2C_D(\overline{\alpha}(t))u(t) + C_D'(\overline{\alpha}(t))w(t)] \quad (15.21b)$$

$$M_b(t) = \frac{1}{2}\rho\overline{U}(t)B^2[2C_M(\overline{\alpha}(t))u(t) + C_M'(\overline{\alpha}(t))w(t)] \quad (15.21c)$$

where:

the subscript "*b*" denotes the buffeting forces;
the superscript "$\circ'$" means the differentiation with respect to the angle of attack.

The quasi-steady assumption is adopted and the unit aerodynamic admittance is used in Equations 15.21a to 15.21c. Apart from the time-varying mean wind speed, the aerodynamic coefficients and their derivatives are also time-dependent, due to the time-dependent angle of attack. Furthermore, the fluctuating wind speed components $u(t)$ and $w(t)$ are non-stationary and have time-varying statistical properties characterized by the EPSD.

15.4.4 Governing Equations of Motion

The governing equations of motion for buffeting response of a long-span bridge under non-stationary winds can be expressed as:

$$\mathbf{M}\ddot{\mathbf{X}}(t) + \mathbf{C}\dot{\mathbf{X}}(t) + \mathbf{K}\mathbf{X}(t) = \mathbf{F}_m(t) + \mathbf{F}_{se}(t) + \mathbf{F}_b(t) \quad (15.22)$$

where:

$\mathbf{M}$, $\mathbf{C}$ and $\mathbf{K}$ are the $N \times N$ mass, damping and stiffness matrices of the finite element model of the entire bridge, respectively;
N is the total number of degree of freedoms;
$\mathbf{X}(t)$ is the N-dimensional buffeting response vector of the whole bridge in the global coordinate system.

The buffeting response vector is a non-stationary vector process which can be written as:

$$\mathbf{X}(t) = \overline{\mathbf{X}}(t) + \mathbf{x}(t) \tag{15.23}$$

where:

$\overline{\mathbf{X}}(t)$ and $\mathbf{x}(t)$ denote the time-varying mean wind displacement response and the fluctuating buffeting displacement response, respectively;
$\dot{\mathbf{X}}(t)$ and $\ddot{\mathbf{X}}(t)$ are the non-stationary velocity and acceleration vectors;
$\mathbf{F}_m$, $\mathbf{F}_{se}$ and $\mathbf{F}_b$ are the loading vectors of mean wind forces, self-excited forces and buffeting forces, respectively.

Suppose that the entire bridge deck is divided into M elements for considering wind loading. The wind loading vectors can then be expressed as:

$$\mathbf{F}_m(t) = \mathbf{T}_s(t)\mathbf{C}_{bu}(t)\overline{\mathbf{U}}(t) \tag{15.24a}$$

$$\mathbf{F}_{se}(t) = \mathbf{A}_s(t)\mathbf{x}(t) + \mathbf{A}_d(t)\dot{\mathbf{x}}(t) \tag{15.24b}$$

$$\mathbf{F}_b(t) = \mathbf{T}_s(t)\mathbf{C}_b(t)\mathbf{\Theta}(t) \tag{15.24c}$$

where:

$\overline{\mathbf{U}}(t) = \left[\overline{U}_1(t)\cdots\overline{U}_M(t)\right]$ and $\mathbf{\Theta}(t) = \left[\mathbf{u}(t)^{\mathrm{T}}\ \mathbf{w}(t)^{\mathrm{T}}\right]^{\mathrm{T}}$ are the vectors of time-varying mean wind speeds and non-stationary fluctuating wind speeds;
$\mathbf{u}(t) = [u_1(t)\cdots u_M(t)]^{\mathrm{T}}$ and $\mathbf{w}(t) = [w_1(t)\cdots w_M(t)]^{\mathrm{T}}$;
$\mathbf{C}_b(t) = [\mathbf{C}_{bu}(t)\mathbf{C}_{bw}(t)]$ is the time-dependent aerodynamic force coefficients matrix;
$\mathbf{C}_{bu}(t) = \mathrm{diag}\left[\mathbf{C}_{bu}^1(t)\cdots\mathbf{C}_{bu}^M(t)\right]$; $\mathbf{C}_{bw}(t) = \mathrm{diag}\left[\mathbf{C}_{bw}^1(t)\cdots\mathbf{C}_{bw}^M(t)\right]$;

and:

$$\mathbf{C}_{bu}^k(t) = \frac{1}{2}\rho\overline{U}_k(t)Bl_k\left[2C_{L,k}(t)\quad 2C_{D,k}(t)\quad 2BC_{M,k}(t)\right]^{\mathrm{T}} \tag{15.25a}$$

$$\mathbf{C}_{bw}^k(t) = \frac{1}{2}\rho\overline{U}_k(t)Bl_k\left[C'_{L,k}(t) + C_{D,k}(t)\quad C'_{D,k}(t)\quad BC'_{M,k}(t)\right]^{\mathrm{T}} \tag{15.25b}$$

are the element force coefficient matrices; l_k is the length of the kth deck element; $\mathbf{A}_s(t) = \mathbf{T}_s(t)\{\text{diag}[\mathbf{A}_s^1(t)\cdots\mathbf{A}_s^M(t)]\}\mathbf{T}_s(t)^{\mathrm{T}}$ and $\mathbf{A}_d(t) = \mathbf{T}_s(t)\{\text{diag}[\mathbf{A}_d^1(t)\cdots\mathbf{A}_d^M(t)]\}\mathbf{T}_s(t)^{\mathrm{T}}$ are the time-dependent aeroelastic stiffness and damping matrices, in which:

$$\mathbf{A}_s^k = \rho B^2\omega^2 l_k \begin{bmatrix} H_{4,k}^*(t) & H_{6,k}^*(t) & BH_{3,k}^*(t) \\ P_{6,k}^*(t) & P_{4,k}^*(t) & BP_{3,k}^*(t) \\ BA_{4,k}^*(t) & BA_{6,k}^*(t) & B^2A_{3,k}^*(t) \end{bmatrix}, \quad \mathbf{A}_d^k = \rho B^2\omega l_k \begin{bmatrix} H_{1,k}^*(t) & H_{5,k}^*(t) & BH_{2,k}^*(t) \\ P_{5,k}^*(t) & P_{1,k}^*(t) & BP_{2,k}^*(t) \\ BA_{1,k}^*(t) & BA_{5,k}^*(t) & B^2A_{2,k}^*(t) \end{bmatrix} \tag{15.26}$$

In the above, $\mathbf{T}_s(t)$ is the $N \times M$ coordinate transformation matrix, which is also time-dependent as it depends on the instantaneous mean wind angle of attack.

15.4.5 Time-Varying Mean Wind Response

The mean wind force vector $\mathbf{F}_m(t)$ evolves with time very slowly. Therefore, although the mean wind response can be predicted for each time instant, the prediction can be performed by static analysis without considering dynamic effects, because the fundamental frequency of a long-span bridge is much larger than $1/T$. The basic equation for static analysis can be derived from Equation 15.22 at the rth time instant:

$$\mathbf{K}\overline{\mathbf{X}}(t_r) = \mathbf{F}_m(\overline{\mathbf{U}}(t_r), \overline{\boldsymbol{\alpha}}(t_r)) \tag{15.27}$$

Nevertheless, the above equation is a non-linear equation because the mean wind attack angle vector $\overline{\boldsymbol{\alpha}}(t_r)$ depends on the mean wind response vector $\overline{\mathbf{X}}(t_r)$. An iterative procedure is introduced to find both $\overline{\boldsymbol{\alpha}}(t_r)$ and $\overline{\mathbf{X}}(t_r)$. The obtained time-varying mean wind attack angle $\overline{\boldsymbol{\alpha}}(t)$ within the concerned time period is used for the dynamic response analysis.

15.4.6 Modal Equations for Non-Stationary Buffeting Response

By eliminating the time-varying mean wind force and response and by using the modal superposition method, Equation 15.22 can be transformed into:

$$\ddot{\mathbf{q}}(t) + \tilde{\mathbf{C}}(t)\dot{\mathbf{q}}(t) + \tilde{\mathbf{K}}(t)\mathbf{q}(t) = \mathbf{Q}_b(t) \tag{15.28}$$

where:

$\tilde{\mathbf{C}}(t) = \mathbf{\Phi}^{\mathrm{T}}\mathbf{C}\mathbf{\Phi} - \mathbf{\Phi}^{\mathrm{T}}\mathbf{A}_d(t)\mathbf{\Phi}$ and $\tilde{\mathbf{K}}(t) = \mathbf{\Lambda} - \mathbf{\Phi}^{\mathrm{T}}\mathbf{A}_s(t)\mathbf{\Phi}$ are the generalized modal damping and stiffness matrices, and they are time-dependent;
$\mathbf{Q}_b(t) = \mathbf{\Phi}^{\mathrm{T}}\mathbf{F}_b(t)$ is the generalized buffeting force vector;
$\mathbf{\Phi}$ is the modal matrix with only the first N_s order of structural modes, which is normalized as:

$$\mathbf{\Phi}^{\mathrm{T}}\mathbf{M}\mathbf{\Phi} = \mathbf{I}_{N_S\times N_S}, \ \mathbf{\Phi}^{\mathrm{T}}\mathbf{K}\mathbf{\Phi} = \mathbf{\Lambda} = \text{diag}(\omega_j^2), \quad j = 1, 2, \ldots, N_S \tag{15.29}$$

in which:

ω_j is the jth natural frequency of the bridge;
$\mathbf{q}(t)$ denotes the generalized modal coordinate vector such that:

$$\mathbf{x}(t) = \mathbf{\Phi}\mathbf{q}(t) \tag{15.30}$$

It is noted that $\mathbf{Q}_b(t)$ in Equation 15.18a is a N_S-dimensional non-stationary Gaussian vector process with zero mean, referring to Equation 15.24c and characterized by the EPSD matrix rather than the PSD matrix:

$$\mathbf{S}_{\mathbf{Q}_b\mathbf{Q}_b}(\omega, t) = \mathbf{\Phi}^{\mathrm{T}}\mathbf{S}_{\mathbf{F}_b\mathbf{F}_b}(\omega, t)\mathbf{\Phi} = \mathbf{\Phi}^{\mathrm{T}}\mathbf{T}_s(t)\mathbf{C}_b(t)\mathbf{S}_{\mathbf{\Theta\Theta}}(\omega, t)\mathbf{C}_b(t)^{\mathrm{T}}\mathbf{T}_s(t)^{\mathrm{T}}\mathbf{\Phi} \tag{15.31}$$

15.4.7 Pseudo Excitation Method for Solving Modal Equations

The pseudo excitation method (PEM) is a highly efficient and accurate method for random vibration analysis, and it has been used for stationary buffeting analysis of the bridge [21]. The PEM is extended in this study to solve Equation 15.28 in the time-frequency domain. First, the pseudo excitation vector is defined as:

$$\tilde{\mathbf{Q}}_{b,j}(\omega, t) = \mathbf{A}_{\mathbf{Q}_b,j}(\omega, t)\mathrm{e}^{\mathrm{i}\omega t} \tag{15.32}$$

where the matrix $\mathbf{A}_{\mathbf{Q}_b}(\omega, t)$ can be obtained by the Cholesky decomposition of the EPSD matrix $\mathbf{S}_{\mathbf{Q}_b\mathbf{Q}_b}(\omega, t)$, which satisfies:

$$\mathbf{S}_{\mathbf{Q}_b\mathbf{Q}_b}(\omega, t) = \mathbf{A}_{\mathbf{Q}_b}(t, \omega)\mathbf{A}_{\mathbf{Q}_b}(t, \omega)^{\mathrm{T}} \tag{15.33}$$

By replacing the loading term $\mathbf{Q}_b(t)$ in Equation 15.28 with the pseudo excitation $\tilde{\mathbf{Q}}_{b,j}(\omega, t)$, the pseudo response $\mathbf{y}_j(\omega,t)$ can be then obtained by solving the dynamic equation as follows:

$$\ddot{\mathbf{y}}_j(\omega, t) + \tilde{\mathbf{C}}(t)\dot{\mathbf{y}}_j(\omega, t) + \tilde{\mathbf{K}}(t)\mathbf{y}_j(\omega, t) = \tilde{\mathbf{Q}}_{b,j}(\omega, t) \tag{15.34}$$

The Newmark step-by-step integration method can be used to find $\mathbf{y}_j(\omega,t)$ numerically. Consequently, the EPSD matrix of buffeting responses $\mathbf{x}(t)$ can be obtained as:

$$\mathbf{S}_{\mathbf{xx}}(\omega, t) = \sum_{j=1}^{N_S} \left[\mathbf{\Phi}\mathbf{y}_j(\omega, t)\right]^* \left[\mathbf{\Phi}\mathbf{y}_j(\omega, t)\right]^{\mathrm{T}} \tag{15.35}$$

Once the EPSD matrix is obtained, other statistics, such as time-varying standard deviations of the non-stationary buffeting responses, can be computed readily.

15.4.8 Case Study: Stonecutters Bridge

In this section, the proposed framework is applied to the Stonecutters Bridge to demonstrate how to predict its non-stationary buffeting response. By using the updated FE model of the Stonecutters Bridge, the first 20 natural frequencies and modal shapes are computed and used for the non-stationary buffeting analysis. A modal damping ratio of 2% is adopted for each mode of vibration. Wind forces are supposed to act on the bridge deck, bridge tower, and bridge piers at the nodes shown in Figure 15.16. The aerodynamic coefficients and flutter derivatives of the bridge deck were provided by the Hong Kong Highways Department obtained from wind tunnel sectional model tests.

The time-varying mean wind responses of the bridge are displayed in Figures 15.17a to 15.17c for the deck unit at the middle point of the main span of the bridge (location S1 in Figure 15.16); the three

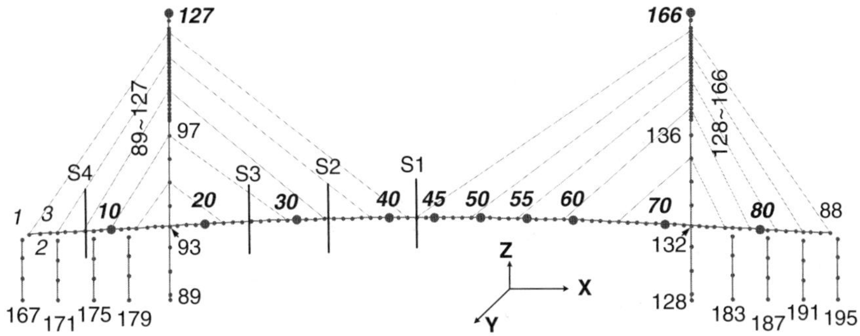

Figure 15.16 Distribution of nodes with wind forces applied.

plots correspond to vertical and horizontal displacements and torsional angle, respectively. It can be observed that the variation amplitudes of mean wind responses are all about 33%, indicating that the degree of influence of time-varying mean wind speed on bridge mean responses is obvious, although it is not very significant. The computed time-varying mean torsional angle $\overline{\alpha}_S(t)$ of the bridge deck (Figure 15.17c) will be used as the instantaneous mean wind attack angle $\overline{\alpha}(t)$ in the following analysis in consideration of $\overline{\alpha}_0 = 0$ in the wind field model.

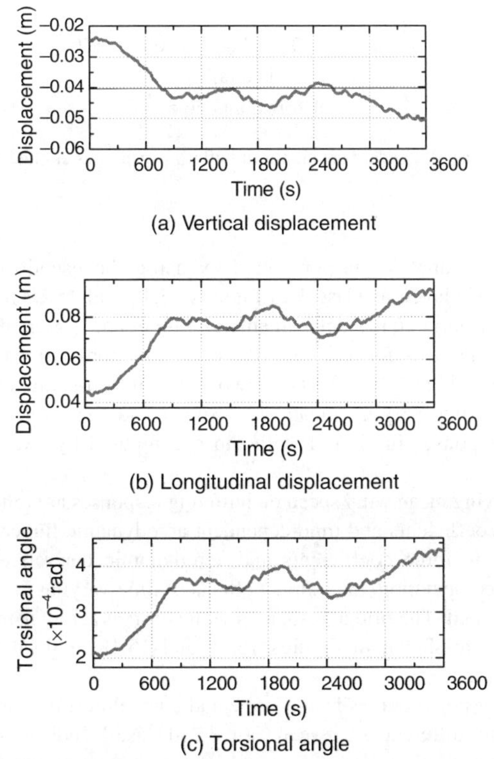

Figure 15.17 Mean wind responses at the middle point of main span.

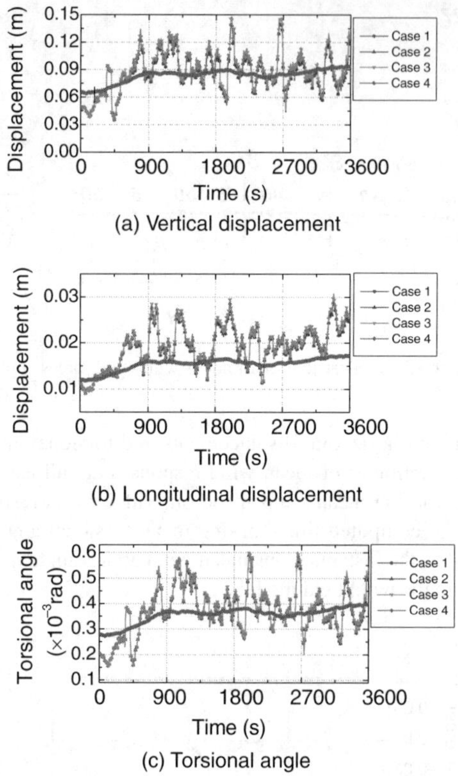

Figure 15.18 Time-varying standard deviations of buffeting responses at middle point of main span.

Non-stationary buffeting analysis is performed by using the pseudo excitation method. The computed responses are displayed as Case 1 in Figures 15.18 and 15.19 for time-varying standard deviations of vertical and horizontal displacements and torsional angles of the bridge at the location S1, which serve as the basis for assessing the effects of non-stationary typhoon wind on the buffeting response of the bridge. The effects of two types of non-stationary factors on the bridge responses, including the time-varying mean wind speed and the EPSD, are then estimated by comparing the bridge responses in Case 1 with those computed by excluding one of the above factors.

The effects of time-varying mean wind speed on buffeting responses are reflected by the time-dependent aerodynamic force coefficients and time-dependent aerodynamic flutter derivatives. By keeping the matrix of aerodynamic force coefficients and aerodynamic derivatives at the first time step unchanged for the entire computation, the indirect effects of time-varying mean wind speed on buffeting responses can be excluded. The bridge responses with constant aerodynamic force coefficients and aerodynamic derivatives are shown in Figures 15.18 and 15.19, defined as Case 2 and Case 3, respectively.

It can be seen that the response curves from Case 2 and 3 are almost the same as those from Case 1, except for a very few minor differences between Case 3 and Case 1, indicating that the indirect effects of time-varying mean wind speed on non-stationary buffeting responses can be ignored in this special case study. This can be confirmed by Figures 15.20 and 15.21, which show that the time-dependent

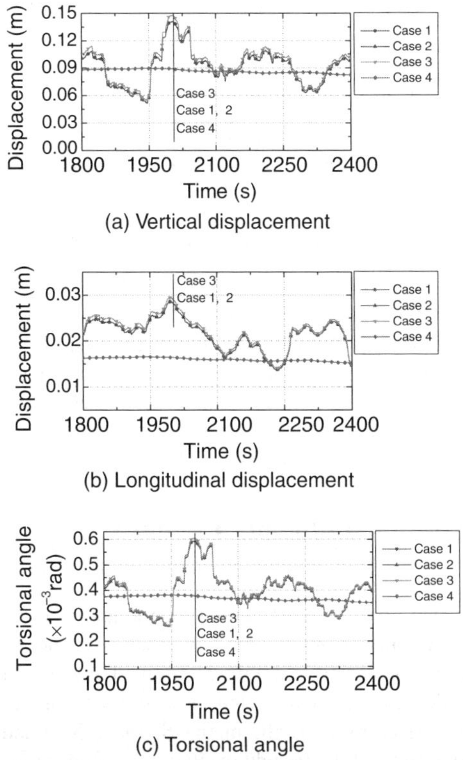

Figure 15.19 Detailed view of time-varying standard deviations of buffeting responses at middle point of main span.

aerodynamic force coefficients and derivatives do not change considerably during the selected time period. This is consistent with relatively small changes of time-varying mean wind speed and wind attack angle.

The effects of EPSD can be investigated by considering the fluctuating components of wind velocity to be stationary processes. The time-invariant PSD functions are estimated from the longitudinal wind speed time history after eliminating the time-varying mean wind speed and the vertical wind speed time history. These functions are used to form a time-invariant PSD matrix. The bridge responses are then computed and displayed as Case 4 in Figures 15.18 and 15.19. In comparing Case 4 with Case 1, a relatively large difference can be observed from Figures 15.19b and 15.19c in detail. The relative

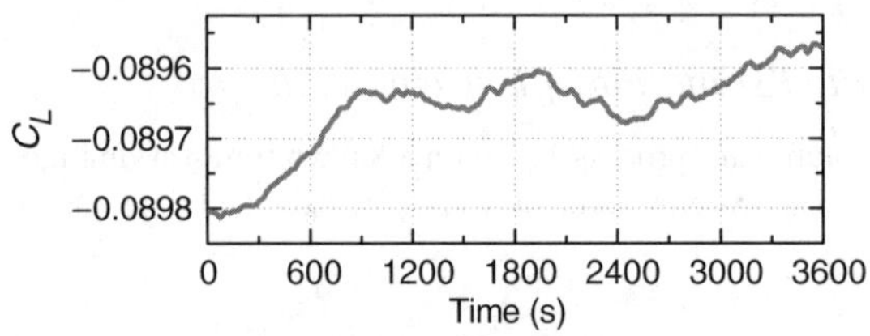

Figure 15.20 Time-dependent aerodynamic coefficients C_L.

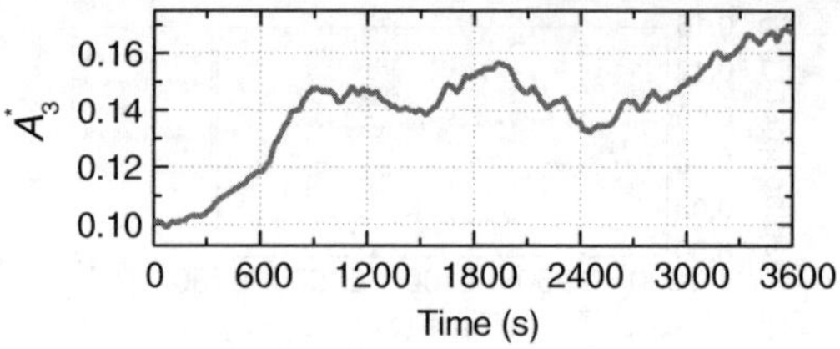

Figure 15.21 Time-dependent aerodynamic derivatives A_3^* at $\omega = 1$.

differences between the two cases are 7%, 27% and 6% for the mean values of standard deviations of vertical, lateral and torsional responses, respectively. Therefore, it is necessary to consider the non-stationary buffeting analysis of the bridge under typhoon winds with time-varying mean wind speed and EPSD.

15.5 Extreme Value of Non-Stationary Response

15.5.1 Background

As previously mentioned, when non-stationary wind speed is modeled as the sum of a time-varying mean wind speed and an evolutionary Gaussian random process for the fluctuating part, the buffeting response $\mathbf{X}(t)$ of the bridge can be correspondingly determined as the sum of the time-varying mean response $\overline{\mathbf{X}}(t)$ and the evolutionary Gaussian random process $\mathbf{x}(t)$ of the bridge, where $\mathbf{x}(t)$ is a zero-mean random process characterized by its EPSD matrix $\mathbf{S_{xx}}(\omega,t)$. $\overline{\mathbf{X}}(t)$ and $\mathbf{S_{xx}}(\omega,t)$ can be evaluated separately by using static analysis and the pseudo excitation method, respectively, as demonstrated in previous sections.

In general, the exact solution for the extreme value of a random process is not available, while the Monte Carlo solution is often computationally inefficient [22]. Thus, it is desirable to develop approximate solutions for the extreme value of non-stationary buffeting response. Probabilistically, the extreme value problem can be described to predict the extreme value of an evolutionary random process with a time-varying mean over a specified time interval. For a zero-mean Gaussian random process, a few approximations for the cumulative distribution function (CDF) and statistical moments of its extreme value have been developed [8,23–30].

On the other hand, the extreme value of a Gaussian process with non-zero time-varying mean can be paraphrased as two-sided time-dependent barrier problems [31] or curve crossing problems [32,33]. This section will therefore derive approximate probabilistic properties of the extreme value of non-stationary buffeting response, based on the curve-crossing problem and Microlev's approximation, and the derived formulae will be applied to assess the extreme value of non-stationary buffeting response of a real long-span cable-supported bridge.

15.5.2 Approximate Estimation of Extreme Value

Consider a Gaussian evolutionary process $X(t)$ with a known time-varying mean $m(t)$ and single-sided EPSD function $S_{XX}(\omega,t)$. The extreme value of $X(t)$ over a specified time interval $[0, T]$ is also a random variable:

$$X_m = \max_{t\in[0,T]} |X(t)| \tag{15.36}$$

The problem at hand is to find approximate formulae for probabilistic characteristics of X_m, including the CDF, $F_{X_m}(x,T)$ of the random variable X_m, the mean value μ_{X_m} and the standard deviation σ_{X_m} of X_m. To this end, $F_{X_m}(x,T)$ is first expressed as a function of the up-crossing rate, including the effects of mean value. Time-dependent spectral characteristics of the evolutionary process $X(t)$ proposed by Michaelov *et al.* [34] are then utilized to derive two types of approximation of the up-crossing rate by incorporating the formulas derived by [22] with the effects of time-varying mean. The CDF of extreme value of $X(t)$ is finally estimated by both the Possion and Vanmarcke approximation, from which μ_{X_m} and σ_{X_m} can be evaluated.

Generally, the extreme value CDF of a Gaussian non-stationary random process $X(t)$ with $X(0)=0$ can be represented by:

$$F_{X_m}(x,T) = \exp[-N_X(x,T)] = \exp\left[-\int_0^T \eta_X(x,t)\mathrm{d}t\right] \tag{15.37}$$

where $N_X(x,t)$ is the mean number of crossings over level $x(x>0)$; and $\eta_X(x,t)$ is the unconditional transient crossing rate:

$$\eta_X(x,t) = \eta_X^+(x,t) + \eta_X^-(-x,t) \tag{15.38}$$

where η_X^+ and η_X^- denote respectively the up-crossing and down-crossing rate of the process $X(t)$. Because $X(t)$ is embedded with a non-zero time-varying mean $m(t)$, η_X^+ and η_X^- are not equal and should be computed separately. However, the down-crossing rate can be transferred into an equivalent up-crossing rate by taking advantage of the symmetry of the fluctuating part $X(t)-m(t)$, so that the down-crossing rate of $X(t)$ over $-x$ is equal to the up-crossing rate of $X(t)$ over $x+m(t)$.

$$\eta_X^-(-x,m;t) = \eta_X^+(x,-m;t) \tag{15.39}$$

Jointly utilizing Equations 15.37 to 15.39 leads to:

$$F_{X_m}(x,T) = \exp\left[-\int_0^T \left[\eta_X^+(x,-m;t) + \eta_X^+(x,m;t)\right]\mathrm{d}t\right] \tag{15.40}$$

Equation 15.40 indicates that the determination of $F_{X_m}(x,T)$ requires computation of the time-varying crossing rate $\eta_X^+(x,m;t)$. Two types of approximation of $\eta_X^+(x,m;t)$ will be derived subsequently. Before that, the time-varying spectral characteristics will be introduced, laying down a foundation for the formulas of $\eta_X^+(x,m;t)$.

To estimate the extreme value of $X(t)$, the four types of spectral characteristics are of importance and can be evaluated from $\mathbf{S_{xx}}(\omega,t)$. The standard deviations of $X(t)$ and the derivative $\dot{X}(t)$ with respect to time can be computed by:

$$\sigma_X(t)^2 = \int_0^{+\infty} S_{XX}(\omega,t)\mathrm{d}\omega \tag{15.41a}$$

$$\sigma_{\dot{X}}(t)^2 = \int_0^{+\infty} S_{\dot{X}\dot{X}}(\omega,t)\mathrm{d}\omega = \int_0^{+\infty} \left[\dot{A}(\omega,t)^2 + \omega^2 A(\omega,t)^2\right]\mathrm{d}\omega \tag{15.41b}$$

where $A(\omega,t) = \sqrt{S_{XX}(\omega,t)}$ is the time-frequency modulation function.

The correlation coefficient $\rho_{X\dot{X}}(t)$ between $X(t)$ and $\dot{X}(t)$ can be computed by:

$$\rho_{X\dot{X}}(t) = \frac{\mathrm{cov}\left[X(t), \dot{X}(t)\right]}{\sigma_X(t)\sigma_{\dot{X}}(t)} = \frac{\int_0^{+\infty} A(\omega,t)\dot{A}(\omega,t)\mathrm{d}\omega}{\sigma_X(t)\sigma_{\dot{X}}(t)} \tag{15.42}$$

Equation 15.42 indicates that, unlike in the stationary situation, where $X(t)$ is independent of $\dot{X}(t)$, the non-stationary process $X(t)$ makes the correlation coefficient $\rho_{X\dot{X}}(t)$ non-zero. However, $\rho_{X\dot{X}}(t)$ may not deviate from zero significantly if the modulation function $A(\omega,t)$ varies with time slowly.

Another time-dependent spectral characteristic is the bandwidth factor $q_X(t)$:

$$q_X(t) = \sqrt{1 - \left\{\frac{\int_0^{+\infty} \omega A(\omega,t)^2\mathrm{d}\omega}{\sigma_X(t)\sigma_{\dot{X}}(t)}\right\}^2} \tag{15.43}$$

where $0 \leq q_X(t) \leq 1$ denotes the instantaneous bandwidth of $X(t)$ in the frequency domain.

Low values of $q_X(t)$ represent that $X(t)$ is a narrow-band process, and vice versa. For band-limited white noises, the value of this factor is 0.5.

15.5.3 Possion Approximation

The two types of approximation considered for the extreme value of an evolutionary process are the Possion and Vanmarcke approximations modified by Michaelov [22], which will be further revised to accommodate the time-varying mean. The Possion approximation is based on the jointly Gaussian PDF $p(x, \dot{x}; t)$ of the process $X(t)$ and its derivative $\dot{X}(t)$:

$$\begin{aligned} p(x,\dot{x};t) = {} & \frac{1}{\sqrt{2\pi}\sigma_X(t)\sigma_{\dot{X}}(t)\sqrt{1-\rho_{X\dot{X}}^2(t)}} \\ & \times \exp\left\{-\frac{1}{2[1-\rho_{X\dot{X}}^2]}\left[\left(\frac{x-m(t)}{\sigma_X(t)}\right)^2 - 2\rho_{X\dot{X}}(t)\left(\frac{x-m(t)}{\sigma_X(t)}\right)\left(\frac{\dot{x}-\dot{m}(t)}{\sigma_{\dot{X}}(t)}\right) + \left(\frac{\dot{x}-\dot{m}(t)}{\sigma_{\dot{X}}(t)}\right)^2\right]\right\} \end{aligned} \tag{15.44}$$

Consequently, under the critical assumption that the crossings are independent, Equation 15.44 leads to the up-crossing rate over level $x > m(t)$ as [32]:

$$\begin{aligned} \eta_X^{P+}(x,m;t) &= \int_{\dot{m}(t)}^{+\infty} [\dot{x} - \dot{m}(t)]p(x,\dot{x};t)\mathrm{d}\dot{x} \\ &= \frac{\sigma_{\dot{X}}(t)}{\sigma_X(t)}[1-\rho_{X\dot{X}}(t)]^{1/2}\phi\left[\frac{x-m(t)}{\sigma_X(t)}\right][\{\phi[\xi^P(t)] + \xi^P(t)\Phi[\xi^P(t)]\}] \end{aligned} \tag{15.45}$$

where:

the superscript "P" denotes the Possion approximation;

$\phi(x)$ and $\Phi(x)$ denote standard Gaussian PDF and CDF, respectively; and $\xi^P(t)$ denotes the revised crossing level with consideration of the effects of both time-dependent mean value and its derivatives:

$$\xi^{P}(t)=\frac{\dot{m}(t)}{\sigma_{\dot{X}}(t)[1-\rho_{X\dot{X}}(t)]^{\frac{1}{2}}}-\frac{\rho_{X\dot{X}}(t)[x-m(t)]}{\sigma_{X}(t)[1-\rho_{X\dot{X}}(t)]^{\frac{1}{2}}} \tag{15.46}$$

The results expressed by Equation 15.45 have been also given for computing the curve crossing rate of non-stationary processes [32].

15.5.4 Vanmarcke Approximation

The Vanmarcke approximation is an alternative of the Possion approximation. In the stationary situation, the Possion approximation generally leads to very conservative results, mainly attributed to the independent assumption of crossings. The Vanmarcke approximation is therefore proposed on the two state Markov assumption of crossings of the envelope of $X(t)$, $V(t)$. The Vanmarke approximation is expressed by the crossing rate of $V(t)$, of which the mean value is also $m(t)$. The jointly PDF of the envelope function $V(t)$ and its derivative $\dot{V}(t)$ is given as:

$$\begin{aligned}p(v,\dot{v};t)=&\frac{1}{\sqrt{2\pi}\sigma_{X}(t)\sigma_{\dot{X}}(t)q_{X}(t)\sqrt{1-\alpha_{X}^{2}(t)}}\\&\times\exp\left\{-\frac{1}{2[1-\alpha_{X}^{2}(t)]}\left[\left(\frac{v-m(t)}{\sigma_{X}(t)}\right)^{2}-2\alpha_{X}(t)\left(\frac{v-m(t)}{\sigma_{X}(t)}\right)\left(\frac{\dot{v}-\dot{m}(t)}{\sigma_{\dot{X}}(t)q_{X}(t)}\right)+\left(\frac{\dot{v}-\dot{m}(t)}{\sigma_{\dot{X}}(t)q_{X}(t)}\right)^{2}\right]\right\}\end{aligned} \tag{15.47}$$

where:

$$\alpha_{X}(t)=\frac{\rho_{X\dot{X}}(t)}{q_{X}(t)} \tag{15.48}$$

Equation 15.48 renders the up-crossing rate of $V(t)$ as:

$$\begin{aligned}\eta_{X}^{V+}(x,m;t)=&\frac{1}{2\pi}\frac{\sigma_{\dot{X}}(t)}{\sigma_{X}(t)}[1-\rho_{X\dot{X}}(t)]^{\frac{1}{2}}\frac{1}{\exp\left[\frac{1}{2}\left(\frac{x-m(t)}{\sigma_{X}(t)}\right)^{2}\right]-1}\\&\times\left\{1-\exp\left\{-\pi\left(\frac{x-m(t)}{\sigma_{X}(t)}\right)\sqrt{\frac{q_{X}^{2}(t)-\rho_{X\dot{X}}^{2}(t)}{1-\rho_{X\dot{X}}^{2}(t)}}\{\phi[\xi^{V}(t)]+\xi^{V}(t)\Phi[\xi^{V}(t)]\}\right\}\right\}\end{aligned} \tag{15.49}$$

where:

$$\xi^{V}(t)=\frac{\dfrac{\dot{m}(t)}{\sigma_{\dot{X}}(t)q_{X}(t)}-\dfrac{\alpha_{X}(t)[m(t)-x]}{\sigma_{X}(t)}}{\sqrt{1-\alpha_{X}^{2}(t)}} \tag{15.50}$$

The superscript "V" denotes the Vanmarcke approximation. Equation 15.50 involves the effect of the bandwidth factor $q_X(t)$.

15.5.5 Statistical Moment of Extreme Value

Substituting the obtained up-crossing rate (Equation 15.45 or Equation 15.49) into Equation 15.40, the CDF of extreme value of the Possion or Vanmarcke approximation can be obtained. Accordingly, $\mu_{X_m}(T)$ and $\sigma_{X_m}(T)$ of $X_m(t)$ can be computed in terms of the CDF:

$$\mu_{X_m}(T) = \int_{\max|m(t)|}^{+\infty} F_{X_m}(x, T)\mathrm{d}x \tag{15.51}$$

$$\sigma_{X_m}^2(T) = \int_{\max|m(t)|}^{+\infty} x^2 F_{X_m}(x, T)\mathrm{d}x - \mu_{X_m}^2(T) \tag{15.52}$$

Equations 15.51 and 15.52 imply that the probability of X_m being larger than the absolute maximum of time-varying mean value $m(t)$ is ignorable. Traditionally, μ_{X_m} is used as the representative value of the extreme of $X(t)$ as discussed in Chapter 14.

15.5.6 Case Study: Stonecutters Bridge

This sub-section assesses the extreme value of typhoon-induced non-stationary buffeting displacement response of the Stonecutters Bridge using the above-described approximations. The non-stationary buffeting responses were computed as Case 1 in Section 15.4.8. As a demonstration, the time-varying mean response, EPSD, time-varying RMS value and temporal mean value of EPSD of the vertical displacement at the middle point of the main span of the bridge (location S1 in Figure 15.16) are displayed in Figures 15.22 to 15.25, respectively.

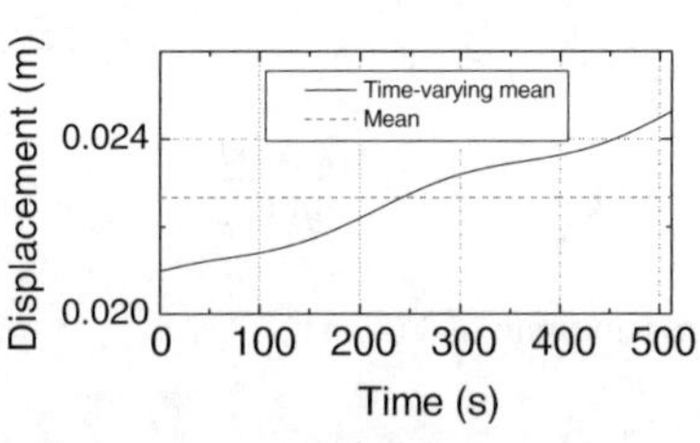

Figure 15.22 Time-varying mean of vertical displacement at S1.

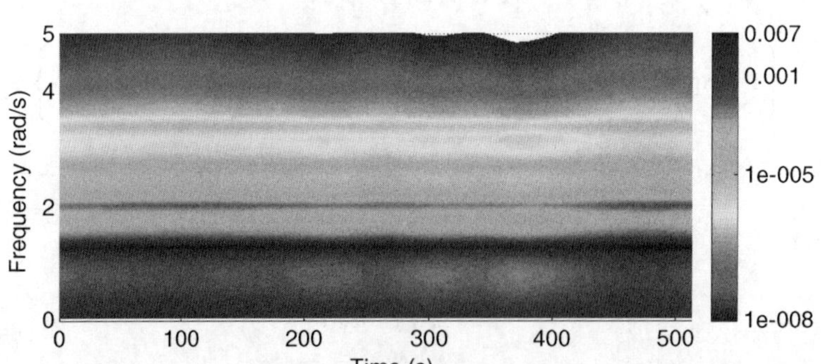

Figure 15.23 EPSD of vertical displacement at S1.

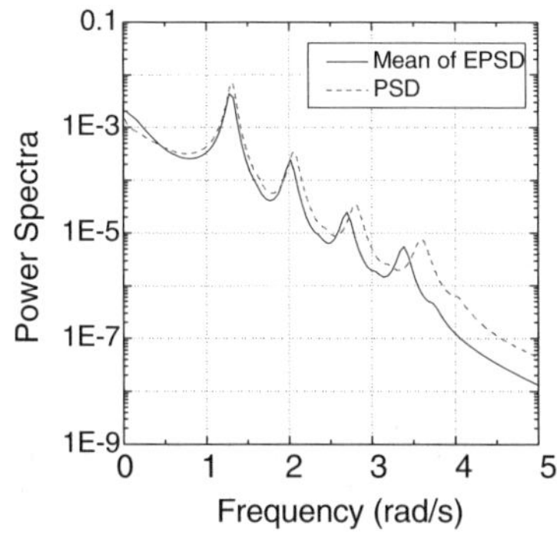

Figure 15.24 Temporal mean of EPSD and PSD vertical displacement at S1.

The formulas derived above for the extreme value of a non-stationary response shall be examined through the comparison of the extreme displacement obtained by the Possion approximation and the Vanmarcke approximation with that from the Monte Carlo Simulation. Let us take the vertical displacement at the location S1 during (18:00-23:12 HKT) as an example. The three types of time-varying spectral characteristics are computed: the standard deviation $\sigma_{\dot{X}}(t)$ of $\dot{X}(t)$, the correlation coefficient $\rho_{X\dot{X}}(t)$ between $X(t)$ and $\dot{X}(t)$, and the bandwidth factor $q_X(t)$. The results are displayed in Figures 15.25 and Figure 15.26.

Note that the values of $\rho_{X\dot{X}}(t)$ are small, so its effect on the extreme value may be ignored [22]. In the specified time interval, the bandwidth factor fluctuates around 0.53, which is close to the bandwidth factor of white noise. With these spectral characteristics, the CDF of extreme value of the vertical displacement computed by the Possion approximation and the Vanmarcke approximation are shown in Figures 15.27 and 15.28.

It is observed that the CDFs from the two approximations are similar. Using the estimated CDF, the mean value and standard deviation of the extreme value are computed by using Equations 15.51 and 15.52. On the other hand, the Monte Carlo Simulation of mean value and standard deviation of the

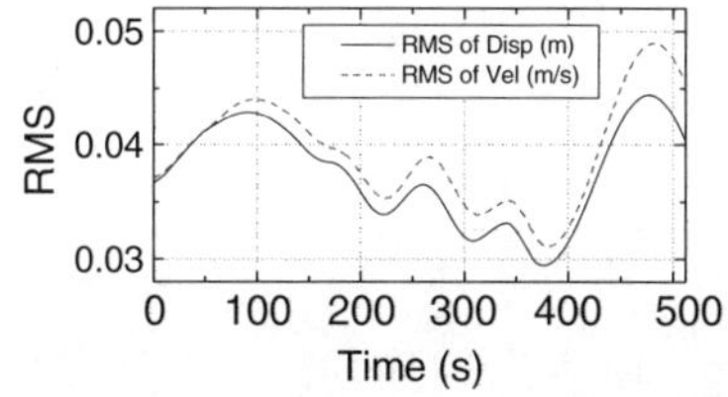

Figure 15.25 Time-varying standard deviations of vertical displacement and velocity at S1.

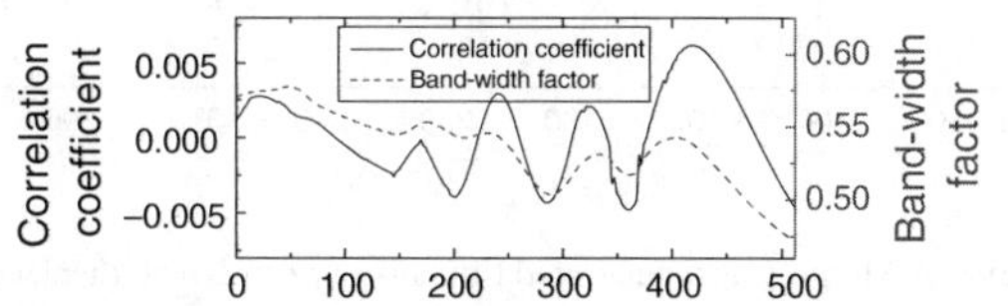

Figure 15.26 Time-dependent correlation coefficient $\rho_{X\dot{X}}(t)$ and bandwidth factor $q_X(t)$ of vertical displacement at S1.

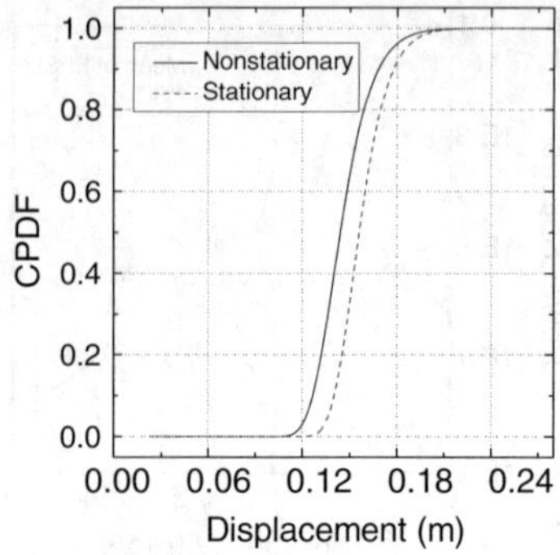

Figure 15.27 Possion approximation of CDF of extremes of vertical displacement at S1.

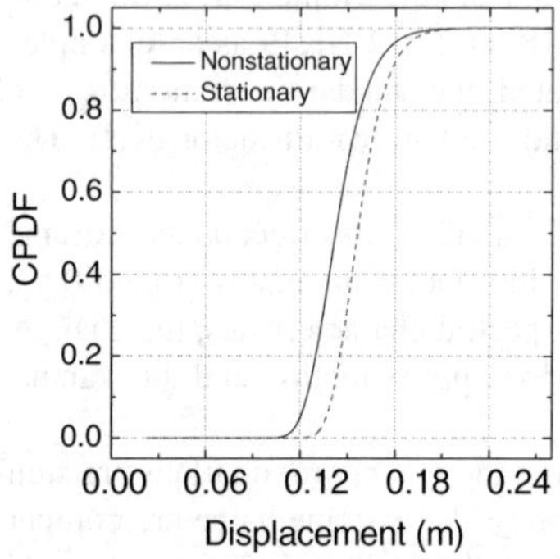

Figure 15.28 Vanmarcke approximation of CDF of extremes of vertical displacement at S1.

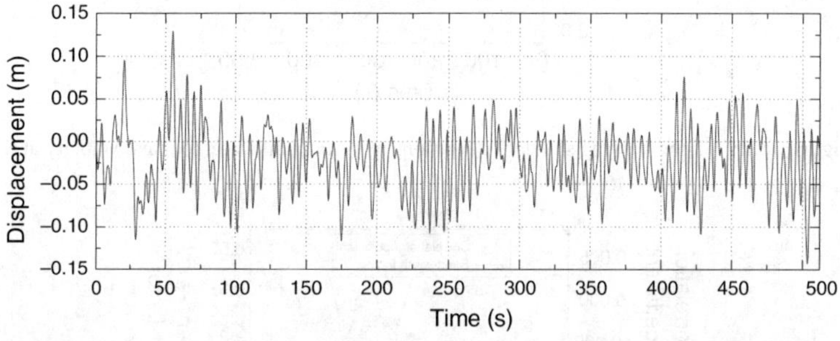

Figure 15.29 A Monte-Carlo generated time history of vertical displacement at S1.

extreme value is performed by considering a total of 100 000 time histories of the vertical displacement.

Figure 15.29 shows one generated sample time history of the vertical displacement by the spectral representation method [35], which conforms to the time-varying mean and EPSD in the ensemble meaning. Table 15.4 lists the mean values and standard deviations of extreme values of non-stationary vertical, lateral and torsional buffeting displacements obtained by the two types of approximations as well as the Monte Carlo Simulation. It can be seen that the mean values and standard deviations from both approximations well match those from the Monte Carlo Simulation with only slight deviations. Furthermore, the Possion approximation always yields an extreme mean value larger than that from the Monte Carlo Simulation. The mean value from the Vanmarcke approximation may be larger or smaller than that from the Monte Carlo Simulation, but it is always closer to the Monte Carlo solution than the Possion approximation.

In addition, the mean values of extreme values are also computed by using the existing Possion and Vanmarcke approximations [22] and listed in Table 15.5. The existing approximations do not include the effect of time-varying mean value, and thus the final results are the sum of the mean of time-varying mean value plus the results from the existing approximations. In comparison with the results from the revised approximations and Monte Carlo Simulation, it can be seen that the existing approximations overestimate the mean values of extremes of non-stationary buffeting responses, and they deviate from the Monte Carlo solutions more than the revised approximations.

Using the revised approximations, the extreme mean values of non-stationary buffeting displacement responses of the whole bridge deck are computed and shown in Figures 15.30 and 15.31 for the Possion and Vanmarcke approximations, respectively.

The corresponding extreme values of stationary buffeting displacements are also shown in the two figures. Note that the extreme values of non-stationary vertical buffeting displacements of the bridge deck are smaller than those of stationary buffeting displacements, while the extreme values of the

Table 15.4 Mean value and standard deviations of extreme values

LOC	Mean value					Standard deviation		
	MC	P	V	P-WTM	V-WTM	MC	P	V
	Vertical Displacement (m)							
S1	0.14 471	0.14 605	0.14 408	0.499	0.327	0.01 665	0.097	0.01 709
S2	0.08 340	0.08 375	0.08 290	0.08 872	0.08 798	0.00 925	0.00 888	0.00 945
S3	0.03 467	0.03 495	0.03 472	0.03 693	0.03 673	0.00 363	0.00 350	0.00 368
S4	1.339E-04	1.347E-04	1.334E-04	1.427E-04	1.4E-04	1.544E-05	1.472E-05	1.573E-05
	Lateral Displacement (m)							
S1	0.06 587	0.06 628	0.06 585	0.06 747	0.06 711	0.00 370	0.00 354	0.00 380
S2	0.04 960	0.04 993	0.04 965	0.05 075	0.05 051	0.00 271	0.00 260	0.00 277
S3	0.02 641	0.02 659	0.02 649	0.02 692	0.02 683	0.00 134	0.00 129	0.00 136
S4	3.848E-03	3.910E-03	3.883E-03	4.007E-03	3.985E-03	2.500E-04	2.4E-04	2.563E-04
	Torsional Angle (rad)							
S1	7.542E-04	7.641E-04	7.608E-04	8.052E-04	8.024E-04	7.220E-05	7.047E-05	7.296E-05
S2	5.488E-04	5.532E-04	5.512E-04	5.767E-04	5.750E-04	4.360E-05	4.221E-05	4.405E-05
S3	3.601E-04	3.633E-04	3.617E-04	3.727E-04	3.714E-04	2.309E-05	2.213E-05	2.343E-05
S4	9.675E-05	9.8E-05	9.751E-05	1.005E-04	9.996E-05	6.8E-06	5.968E-06	6.306E-06

MC: Monte Carlo Simulation; P: Possion approximation proposed; V: Vanmarcke approximation proposed; P-WTM: Possion approximation of Microlev without effects of mean value; V-WTM: Vanmarcke approximation of Microlev without effects of mean value.

Table 15.5 Mean values of extreme displacements

LOC	Possion approximation			Vanmarcke approximation		
	Non-stationary	Stationary	Diff.(%)	Non-stationary	Stationary	Diff.(%)
	Vertical Displacement (m)					
S1	0.14 605	0.15 781	7.45	0.14 408	0.15 565	7.44
S2	0.08 375	0.09 048	7.43	0.08 290	0.08 956	7.44
S3	0.03 495	0.03 754	6.90	0.03 472	0.03 729	6.92
S4	1.347E-04	1.441E-04	6.54	1.334E-04	1.427E-04	6.54
	Lateral Displacement (m)					
S1	0.06 628	0.06 540	−1.35	0.06 585	0.06 526	−0.90
S2	0.04 993	0.04 953	−0.81	0.04 965	0.04 943	−0.43
S3	0.02 659	0.02 681	0.83	0.02 649	0.02 677	1.05
S4	3.910E-03	3.790E-03	−3.	3.883E-03	3.781E-03	−2.70
	Torsional Angle (rad)					
S1	7.641E-04	7.772E-04	1.68	7.608E-04	7.742E-04	1.72
S2	5.532E-04	5.656E-04	2.21	5.512E-04	5.641E-04	2.29
S3	3.633E-04	3.636E-04	0.10	3.617E-04	3.630E-04	0.35
S4	9.8E-05	9.597E-05	−2.26	9.751E-05	9.574E-05	−1.84

lateral and torsional displacements are almost the same for both non-stationary and stationary buffeting responses. The extreme values of non-stationary and stationary buffeting displacements of the locations S1-S4 are listed in Table 15.5, together with the relative differences between these buffeting responses. It can be seen that the differences in the extreme vertical displacements between non-stationary and stationary buffeting responses are larger than 6.5%. In contrast, the differences in the extreme lateral and torsional displacements between non-stationary and stationary buffeting responses are small.

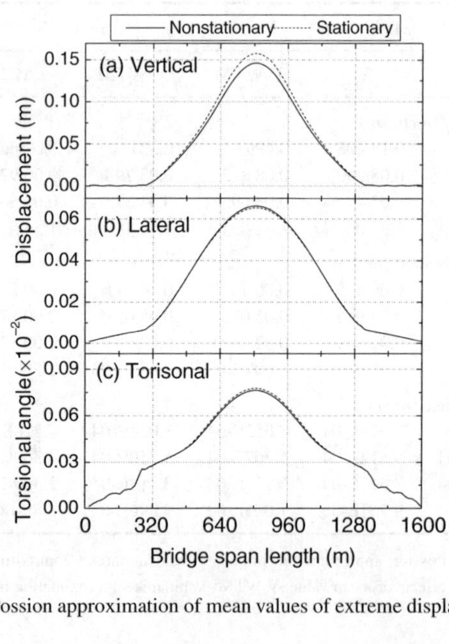

Figure 15.30 Possion approximation of mean values of extreme displacements of deck.

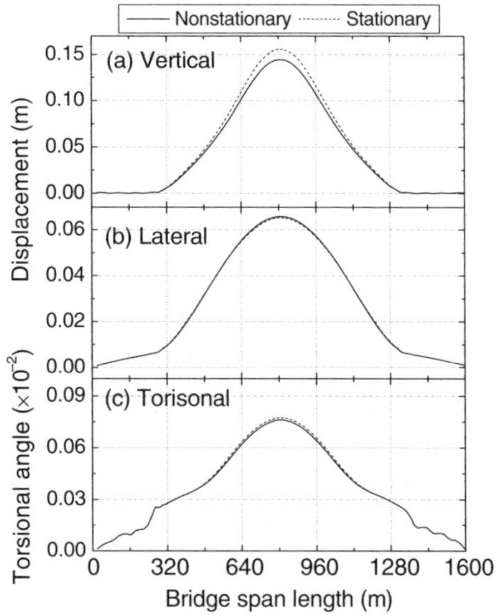

Figure 15.31 Vanmarcke approximation of mean values of extreme displacements of deck.

15.6 Unconditional Simulation of Non-Stationary Wind

15.6.1 Background

Non-linear buffeting analysis is often performed in the time domain. Time domain buffeting analysis requires a complete set of fluctuating wind speed time histories at a number of locations over the bridge deck, towers, and piers. However, the health monitoring system can usually only provide wind speed records at a limited number of locations, because of the limited number of anemometers. As a result, it is necessary to expand the limited measurements to the complete set of wind speed time histories by unconditional or conditional simulation techniques. The unconditional simulation generates wind speed time histories at all the locations that conform to the probabilistic model, with properties estimated from the measured time histories.

On the other hand, the conditional simulation technique generates wind speed time histories at unmeasured locations that not only conform to the probabilistic model, but also are conditioned by the measured time histories. The generated time histories are therefore compatible with the measured time histories, so that the required complete set consists of the generated time histories at the unmeasured locations and the measured time histories.

The conditional simulation technique can make use of more information embedding in the measurements than the unconditional technique. This section will introduce the unconditional simulation of zero-mean non-stationary fluctuating wind characterized as an evolutionary process by the spectral representation method, and the next section will present the Kriging-based conditional simulation method for non-stationary fluctuating winds based on limited measurements.

15.6.2 Unconditional Simulation

The spectral representation method (SRM) introduced in Chapter 4 for the unconditional simulation of stationary fluctuating wind speed is extended in this section to the unconditional simulation of non-stationary fluctuating wind speed. The SRM for stationary processes can be revised to accommodate the evolutionary power spectral density (EPSD) matrix of non-stationary processes [35,36].

Consider a zero-mean multivariate real Gaussian oscillatory vector process $\mathbf{X}(t)$ ($t \in [0, T]$), where $0 < T < \infty$, which is used to model the zero-mean non-stationary fluctuating wind speeds. $\mathbf{X}(t)$ contains M component processes, whose jth component $X_j(t)$ admits the spectral representation as in Equation 15.9. The evolutionary power spectral density (EPSD) at time t can be written as:

$$S_{jk}(\omega, t) = a_j(t, \omega)a_k(t, \omega)^* \mu_{jk}(\omega) = \sqrt{S_{jj}(\omega, t)S_{kk}(\omega, t)}\gamma_{jk}(\omega) \tag{15.53}$$

where:

$S_{jj}(\omega,t)$ and $S_{jk}(\omega,t)$ denote the auto and cross EPSDs, respectively;
$\gamma_{jk}(\omega)$ is the coherence function, which is time-invariant [13,18].

The EPSD matrix of $\mathbf{X}(t)$ can be then formed by $S_{jj}(\omega,t)$ and $S_{jk}(\omega,t)$ as:

$$\mathbf{S}(\omega, t) = \begin{bmatrix} S_{11}(\omega, t) & S_{12}(\omega, t) & \cdots & S_{1M}(\omega, t) \\ S_{21}(\omega, t) & S_{22}(\omega, t) & \cdots & S_{21}(\omega, t) \\ \vdots & \vdots & \ddots & \vdots \\ S_{M1}(\omega, t) & S_{M2}(\omega, t) & \cdots & S_{MM}(\omega, t) \end{bmatrix} \tag{15.54}$$

Corresponding to the EPSD matrix, the time-varying correlation matrix of $\mathbf{X}(t)$, $\mathbf{R}(t_1,t_2) = \mathrm{E}[\mathbf{X}(t_1)\mathbf{X}(t_2)^{\mathrm{T}}]$, can be also formed by auto- and cross-correlation functions $R_{jj}(t_1,t_2)$ and $R_{jk}(t_1,t_2)$. The relationship between time-varying correlation functions and corresponding EPSDs can be derived as:

$$R_{jk}(t_1, t_2) = \int_{-\infty}^{+\infty} \sqrt{S_{jj}(\omega, t_1)S_{kk}(\omega, t_2)}\gamma_{jk}(\omega)\exp[\mathrm{i}\omega(t_1 - t_2)]\mathrm{d}\omega \tag{15.55}$$

Equation 15.55 can be used to compute the correlation function from the EPSD. However, it is not easy to compute the accurate EPSD from the corresponding time-varying correlation function because the correlation function can only provide a "blurred" observation of the EPSD [37].

The SRM proposed by [35] aims to generate sample time histories of $\mathbf{X}(t)$ in terms of the prescribed EPSD matrix $\mathbf{S}(\omega,t)$. The unconditional simulation $\mathbf{X}^U(t)$ can be readily obtained by:

$$\mathbf{X}^U(t) = \sum_{j=1}^{M}\sum_{p=1}^{N} \mathfrak{P}j(\omega_p, t)\cos(\omega_p t + \psi_{pj}) \tag{15.56}$$

where:

ψ_{pj} is the random phase angle distributed uniformly over $[0, 2\pi]$; $\mathfrak{P}_j(\omega_p, t)$ is the jth column of the matrix $\mathfrak{P}(\omega_p, t)$ obtained by the Cholesky decomposition of the EPSD matrix $\mathbf{S}(\omega,t)$ satisfying $\mathbf{S}(\omega, t) = \mathfrak{P}(\omega_p, t)\mathfrak{P}(\omega_p, t)^T$. It can be derived that the time-varying correlation $\mathbf{R}^U(t_1,t_2) = \mathrm{E}[\mathbf{X}^U(t_1)\mathbf{X}^U(t_2)^{\mathrm{T}}]$ of the simulation $\mathbf{X}^U(t)$ can approach the target time-varying correlation matrix $\mathbf{R}(t_1,t_2)$ in ensemble.

15.7 Conditional Simulation of Non-Stationary Wind

15.7.1 Background

The conditional simulation of typhoon-induced wind speeds aims at generating non-stationary fluctuating wind speed time histories that not only conform to the prescribed probabilistic model but also are compatible with the measured time histories probabilistically. The conditional simulation of non-stationary fluctuating wind speeds can be carried out in different ways. This part of the chapter is aimed at proposing a conditional simulation method of non-stationary fluctuating wind speeds for long-span cable-supported bridges with low computational demands.

First, the problem of conditional simulation of fluctuating wind speeds is described in terms of the evolutionary processes. In order to deal with this problem, the conditional simulation method is proposed by applying the Kriging method to the random vector of the Fourier coefficients of evolutionary processes. To alleviate the numerical difficulties when dealing with the conditional simulation of fluctuating wind speed time histories at several hundreds of locations for non-stationary buffeting analysis of a long-span bridge, properties of the covariance matrix of Fourier coefficients are investigated and an alternative computational formula of the matrix is thus derived. Based on that derived formula, a fast algorithm for the simulation method with less computational demands is proposed. The procedure for the resulting fast algorithm-based conditional simulation method is also presented. The fast algorithm is finally applied to the typhoon-induced evolutionary fluctuating wind speeds on a long-span bridge to demonstrate its capabilities.

15.7.2 Problem Statement

Assume that during a typhoon wind event, non-stationary fluctuating wind speed time histories are measured at n locations of a long-span bridge and denoted as $\mathbf{x}_\alpha^0(t)$ $(\alpha=1\ldots n)$, but the time-domain buffeting analysis of the bridge requires wind speed time histories at M $(M>n)$ locations. Therefore, the fluctuating wind speed time histories at m $(m=M=n\ )$ unmeasured locations, denoted as $\mathbf{x}_\beta^S(t)$ $(\beta=n+1\ldots n+m)$, must be generated. With this understanding, the problem of conditional simulation in this study can be described in the following. The fluctuating wind speeds are regarded as a zero-mean non-stationary Gaussian vector process $\mathbf{X}(t)$ of $M=n+m$ components characterized by a known EPSD matrix $\mathbf{S}(\omega,t)$.

We want to simulate unknown components $\mathbf{X}_\beta^S(t)$ $(\beta=n+1\ldots n+m)$ by taking the measured components $\mathbf{x}_\alpha^0(t)$ $(\alpha=1\ldots n)$ as conditions so as to generate the complete set of time histories $\mathbf{x}^S(t)=\left[\mathbf{x}_\alpha^0(t),\mathbf{x}_\beta^S(t)\right]^{\mathrm{T}}$ that conform to $\mathbf{X}(t)$. The conformation requires at least that:

1. the simulated components $\mathbf{X}_\beta^S(t)$will satisfy the prescribed covariance matrix; and
2. the covariance between the simulated components $\mathbf{X}_\beta^S(t)$ and the measured components $\mathbf{X}_\alpha^0(t)$ will also satisfy the prescribed covariance matrix.

For the sake of clarification, the notations in this section use the following conventions:

- there is no superscript for the original stochastic processes/variables;
- use the superscript “0” for the measured stochastic processes/variables;
- use the superscript “S” for the simulated stochastic processes/variables;
- use uppercase letters to represent stochastic processes/variables;
- use lowercase letters to represent the sample time histories/realizations of the stochastic processes/variables.

15.7.3 Conditional Simulation Method

The conditional simulation method proposed is to apply the Kriging method to the zero-mean Gaussian random vector $\mathbf{F_{AB}}=\left[(\mathbf{F}_{\mathbf{A}_1\mathbf{B}_1})^{\mathrm{T}}\cdots(\mathbf{F}_{\mathbf{A}_M\mathbf{B}_M})^{\mathrm{T}}\right]^{\mathrm{T}}$. This random vector comprises the random Fourier coefficients of the evolutionary process $\mathbf{X}(t)$, i.e. $\mathbf{F}_{\mathbf{A}_j\mathbf{B}_j} = \left[A_{1_j} \cdots A_{N_j}, B_{2_j} \cdots B_{N_j}\right]^{\mathrm{T}} = \left[\mathbf{A}_j^{\mathrm{T}}\ \mathbf{B}_j^{\mathrm{T}}\right]^{\mathrm{T}}$, and A_{l_j} and B_{l_j} are the random Fourier coefficients of the component $X_j(t)$.

$$A_{l_j} = \frac{2}{T}\int_0^T X_j(t)\cos\omega_l t\,\mathrm{d}t \qquad l = 1, 2, \ldots, N \tag{15.57a}$$

$$B_{l_j} = \frac{2}{T}\int_0^T X_j(t)\sin\omega_l t\,\mathrm{d}t \qquad l = 2, 3, \ldots, N \tag{15.57b}$$

in which $\omega_l = 2\pi(l-1)/T$. The covariance matrix $\mathbf{C_{AB}}$ of $\mathbf{F_{AB}}$ consists of $(M \times M)$ sub-matrices $\mathbf{C}_{\mathbf{F}_j\mathbf{F}_k}$.

$$\mathbf{C}_{\mathbf{F}_j\mathbf{F}_k} = \begin{bmatrix} \mathbf{C}_{\mathbf{A}_j\mathbf{A}_k} & \mathbf{C}_{\mathbf{A}_j\mathbf{B}_k} \\ \mathbf{C}_{\mathbf{B}_j\mathbf{A}_k} & \mathbf{C}_{\mathbf{B}_j\mathbf{B}_k} \end{bmatrix} = \begin{bmatrix} \mathrm{E}\left[\mathbf{A}_j\mathbf{A}_k^{\mathrm{T}}\right] & \mathrm{E}\left[\mathbf{A}_j\mathbf{B}_k^{\mathrm{T}}\right] \\ \mathrm{E}\left[\mathbf{B}_j\mathbf{A}_k^{\mathrm{T}}\right] & \mathrm{E}\left[\mathbf{B}_j\mathbf{A}_k^{\mathrm{T}}\right] \end{bmatrix} \tag{15.58}$$

where the entries of sub-matrices $\mathbf{C}_{\mathbf{A}_j\mathbf{A}_k}$, $\mathbf{C}_{\mathbf{A}_j\mathbf{B}_k}$, $\mathbf{C}_{\mathbf{B}_j\mathbf{A}_k}$ and $\mathbf{C}_{\mathbf{B}_j\mathbf{B}_k}$ are $\mathrm{E}(A_{p_j}A_{q_k})$, $\mathrm{E}(A_{p_j}B_{q_k})$, $\mathrm{E}(B_{p_j}A_{q_k})$ and $\mathrm{E}(B_{p_j}B_{q_k})$ computed by:

$$\mathrm{E}(A_{p_j}A_{q_k}) = \frac{4}{T^2}\int_0^T\int_0^T R_{jk}(t_1, t_2)\cos\omega_p t_1 \cos\omega_q t_2\,\mathrm{d}t_1\,\mathrm{d}t_2 \tag{15.59a}$$

$$\mathrm{E}(A_{p_j}B_{q_k}) = \frac{4}{T^2}\int_0^T\int_0^T R_{jk}(t_1, t_2)\cos\omega_p t_1 \sin\omega_q t_2\,\mathrm{d}t_1\,\mathrm{d}t_2 \tag{15.59b}$$

$$\mathrm{E}(B_{p_j}A_{q_k}) = \frac{4}{T^2}\int_0^T\int_0^T R_{jk}(t_1, t_2)\sin\omega_p t_1 \cos\omega_q t_2\,\mathrm{d}t_1\,\mathrm{d}t_2 \tag{15.59c}$$

$$\mathrm{E}(B_{p_j}B_{q_k}) = \frac{4}{T^2}\int_0^T\int_0^T R_{jk}(t_1, t_2)\sin\omega_p t_1 \sin\omega_q t_2\,\mathrm{d}t_1\,\mathrm{d}t_2 \tag{15.59d}$$

According to the known (measured) and unknown (to be simulated) components of $\mathbf{X}(t)$, the covariance matrix $\mathbf{C_{AB}}$ can be partitioned into four sub matrices:

$$\mathbf{C_{AB}} = \underset{\qquad n\cdot(2N-1)\quad m\cdot(2N-1)}{\begin{bmatrix} \mathbf{C}_{\mathbf{AB},\alpha\alpha} & \mathbf{C}_{\mathbf{AB},\alpha\beta} \\ \mathbf{C}_{\mathbf{AB},\beta\alpha} & \mathbf{C}_{\mathbf{AB},\beta\beta} \end{bmatrix}} \begin{matrix} n\cdot(2N-1) \\ m\cdot(2N-1) \end{matrix} \tag{15.60}$$

Write the Fourier coefficient vector of the known components $\mathbf{F}_\alpha^0$ as:

$$\mathbf{F}_\alpha^0 = \left[(\mathbf{A}_1^0)^{\mathrm{T}}(\mathbf{B}_1^0)^{\mathrm{T}} \cdots (\mathbf{A}_n^0)^{\mathrm{T}}(\mathbf{B}_n^0)^{\mathrm{T}}\right]^{\mathrm{T}} \tag{15.61}$$

and the unknown Fourier coefficients vector $\mathbf{F}_\beta^S$ as:

$$\mathbf{F}_\beta^S = \left[(\mathbf{A}_{n+1}^S)^{\mathrm{T}}(\mathbf{B}_{n+1}^S)^{\mathrm{T}} \cdots (\mathbf{A}_M^S)^{\mathrm{T}}(\mathbf{B}_M^S)^{\mathrm{T}}\right]^{\mathrm{T}} \tag{15.62}$$

The Kriging method will give

$$\mathbf{F}_{\beta}^{S} = \left[\mathbf{C}_{\mathbf{AB},\alpha\beta}\right]^{\mathrm{T}}\left[\mathbf{C}_{\mathbf{AB},\alpha\alpha}\right]^{-1}(\mathbf{F}_{\alpha}^{0} - \mathbf{F}_{\alpha}^{S*}) + \mathbf{F}_{\beta}^{S*} \tag{15.63}$$

where $\mathbf{F}_{\alpha}^{S^*}$ and $\mathbf{F}_{\beta}^{S*}$ are the two parts of the unconditionally simulated Fourier coefficients vector $\mathbf{F}_{\mathbf{AB}}^{S^*} = \left[\left(\mathbf{F}_{\alpha}^{S^*}\right)^{\mathrm{T}}, \left(\mathbf{F}_{\beta}^{S^*}\right)^{\mathrm{T}}\right]^{\mathrm{T}}$, satisfying that:

$$\mathrm{E}\left[\mathbf{F}_{\mathbf{AB}}^{S^*}\left(\mathbf{F}_{\mathbf{AB}}^{S^*}\right)^{\mathrm{T}}\right] = \mathbf{C}_{\mathbf{AB}} \tag{15.64}$$

$\mathbf{F}_{\mathbf{AB}}^{S^*}$ can be obtained by [38]:

$$\mathbf{F}_{\mathbf{AB}}^{S^*} = \mathbf{Q}\mathbf{U}^{S} \tag{15.65}$$

where $\mathbf{U}^{S}$ is a vector of uncorrelated standard Gaussian random variables of the size $M(2N-1)\times 1$; and $\mathbf{Q}$ is the lower triangular matrix given by the Cholesky decomposition:

$$\mathbf{C}_{\mathbf{AB}} = \mathbf{Q}\mathbf{Q}^{\mathrm{T}} \tag{15.66}$$

By taking the inverse Fourier transform on the conditionally simulated Fourier coefficient vector $\mathbf{F}_{\beta}^{S}$, the conditionally simulated components $\mathbf{X}_{\beta}^{S}(t)$ can be produced:

$$X_{j}^{S}(t) = \frac{A_{1_j}}{2} + \sum_{l=2}^{N}\left(A_{l_j}\cos\omega_l t + B_{l_j}\sin\omega_l t\right) \qquad j = n+1,\ldots,M \tag{15.67}$$

As a result, the entire conditionally simulated evolutionary vector process can be written as $\mathbf{X}^{S}(t) = \left[\mathbf{X}_{\alpha}^{0}(t), \mathbf{X}_{\beta}^{S}(t)\right]$.

To ensure that the conditional simulation method described above is correct, it is necessary to demonstrate that the conditional simulation $\mathbf{X}^{S}(t)$ is a version of $\mathbf{X}(t)$. It can be proved that the simulated vector of Fourier coefficients $\mathbf{F}_{\mathbf{AB}}^{S} = \left[\left(\mathbf{F}_{\alpha}^{0}\right)^{\mathrm{T}}\left(\mathbf{F}_{\beta}^{S}\right)^{\mathrm{T}}\right]^{\mathrm{T}}$ is zero-mean and conforms to the prescribed covariance matrix. It highlights that when the order of Fourier coefficients approaches infinity, $\mathbf{X}^{S}(t)$ is indeed a version of $\mathbf{X}(t)$. This implies that the simulation $\mathbf{X}^{S}(t)$ possesses the time-varying correlation functions matrix:

$$\mathbf{R}_{\beta\beta}^{S}(t_1,t_2) = \mathbf{R}_{\beta\beta}(t_1,t_2);\ \mathbf{R}_{\alpha\beta}^{S}(t_1,t_2) = \mathbf{R}_{\alpha\beta}(t_1,t_2) \tag{15.68}$$

The first part of Equation 15.68 indicates that the simulation $\mathbf{X}_{\beta}^{S}(t)$ has an unbiased correlation matrix, and the second part indicates the simulation $\mathbf{X}_{\beta}^{S}(t)$ is compatible with the measurement $\mathbf{X}_{\alpha}^{0}(t)$.

15.7.4 Computational Difficulties in Conditional Simulation

The buffeting analysis of a long-span bridge often requires the conditional simulation of non-stationary fluctuating wind speed time histories at hundreds of locations (say, M). Meanwhile, to ensure the quality of the conditional simulation, the time steps involved in the simulated time histories must be large enough (say, N). This may cause computational difficulty in the Cholesky decomposition of the covariance matrix $\mathbf{C_{AB}}$, as shown in Equations 15.65 and 15.66. It is well known that the Cholesky decomposition of a d-dimensional matrix requires $[(d^3/6)+(d^2/2)-(2d/3)]$ multiplications [39].

For large numbers of M and N, the Cholesky decomposition of $\mathbf{C_{AB}}$ of the size $M(2N-1)\times M(2N-1)$ requires an extremely long computational time, so that it is almost computationally infeasible for a personnel computer. For instance, if $M=100$ and $N=2048$, the size of $\mathbf{C_{AB}}$ will be as large as $204\,800\times 204\,800$ and the Cholesky decomposition of $\mathbf{C_{AB}}$ requires more than 1.4×10^{15} multiplications. Therefore, it is necessary to find a fast algorithm for the Cholesky decomposition so that the proposed conditional simulation method can be applied for buffeting analysis of long-span bridges in practice.

15.7.5 Alternative Formulas for Decomposition

To alleviate high computational demand for the Cholesky decomposition of matrix $\mathbf{C_{AB}}$, alternative formulas that can help to avoid the direct computation of the Cholesky decomposition of $\mathbf{C_{AB}}$ are derived. A fast algorithm is accordingly developed, and the implementation procedure for the complete conditional simulation is presented in this section.

The entries of $\mathbf{C_{AB}}$ are computed by the integration of the time-varying correlation functions according to Equation 15.59. The correlation function $R_{jk}(t_1,t_2)$ is computed from the EPSD in terms of Equation 15.55, which can be written as:

$$R_{jk}(t_1,t_2)=2\int_0^{+\infty}\sqrt{S_{jj}(\omega,t_1)S_{kk}(\omega,t_2)}\gamma_{jk}(\omega)\cos\omega(t_1-t_2)\mathrm{d}\omega \tag{15.69}$$

This is because the EPSD $S_{jj}(\omega,t)$ is an even function with respect to frequency and the coherent function $\gamma_{jk}(\omega)$ of fluctuating wind speeds is real if the effect of the airwave passage delay is neglected [40]. Substituting Equation 15.69 into Equation 15.59 and separating the integral variables t_1 and t_2 yields:

$$\begin{aligned}\mathrm{E}(A_{p_j}A_{q_k})=\frac{8}{T^2}\int_0^{+\infty}\gamma_{jk}(\omega)\bigg[&\int_0^T\sqrt{S_{jj}(\omega,t_1)}\cos\omega t_1\cos\omega_p t_1\,\mathrm{d}t_1\int_0^T\sqrt{S_{kk}(\omega,t_2)}\cos\omega t_2\cos\omega_q t_2\,\mathrm{d}t_2\\&+\int_0^T\sqrt{S_{jj}(\omega,t_1)}\sin\omega t_1\cos\omega_p t_1\,\mathrm{d}t_1\int_0^T\sqrt{S_{kk}(\omega,t_2)}\sin\omega t_2\cos\omega_q t_2\,\mathrm{d}t_2\bigg]\mathrm{d}\omega\end{aligned} \tag{15.70}$$

Define the integrals as follows:

$$I_{pj}=I_j(\omega,\omega_p)=\int_0^T\sqrt{S_{jj}(\omega,t)}\cos\omega t\cos\omega_p t\,\mathrm{d}t \tag{15.71a}$$

$$J_{pj}=J_j(\omega,\omega_p)=\int_0^T\sqrt{S_{jj}(\omega,t)}\sin\omega t\cos\omega_p t\,\mathrm{d}t \tag{15.71b}$$

where $p=(1,2,\ldots,N)$ and $j=(1,2,\ldots,M)$. By using Equation 15.71, Equation 15.70 can be written in a simple form:

$$\mathrm{E}(A_{p_j}A_{q_k})=\frac{8}{T^2}\int_0^{+\infty}\gamma_{jk}(\omega)\left[I_{jp}I_{kq}+J_{jp}J_{kq}\right]\mathrm{d}\omega \tag{15.72}$$

Similarly, define the two more integrals K_{pj} and L_{pj} with $p = (2,3, \ldots, N)$:

$$K_{pj} = \int_0^T \sqrt{S_{jj}(\omega, t)} \cos \omega t \sin \omega_p t \, \mathrm{d}t \tag{15.73a}$$

$$L_{pj} = \int_0^T \sqrt{S_{jj}(\omega, t)} \sin \omega t \sin \omega_p t \, \mathrm{d}t \tag{15.73b}$$

Equations 15.59b to 15.59d can be simplified as follows:

$$\mathrm{E}(A_{p_j} B_{q_k}) = \frac{8}{T^2} \int_0^{+\infty} \gamma_{jk}(\omega) \left[I_{jp} K_{kq} + J_{jp} L_{kq} \right] \mathrm{d}\omega \tag{15.74a}$$

$$\mathrm{E}(B_{p_j} A_{q_k}) = \frac{8}{T^2} \int_0^{+\infty} \gamma_{jk}(\omega) \left[K_{jp} I_{kq} + L_{jp} J_{kq} \right] \mathrm{d}\omega \tag{15.74b}$$

$$\mathrm{E}(B_{p_j} B_{q_k}) = \frac{8}{T^2} \int_0^{+\infty} \gamma_{ij}(\omega) \left[K_{jp} K_{kq} + L_{jp} L_{kq} \right] \mathrm{d}\omega \tag{15.74c}$$

By Equation 15.72 and Equation 15.74, $\mathbf{C_{AB}}$ can be computed by:

$$\mathbf{C_{AB}} = \frac{8}{T^2} \int_0^{+\infty} \left[\mathbf{C}^1_{\mathbf{AB}}(\omega) \cdot \mathbf{C}^1_{\mathbf{AB}}(\omega)^{\mathrm{T}} + \mathbf{C}^2_{\mathbf{AB}}(\omega) \cdot \mathbf{C}^2_{\mathbf{AB}}(\omega)^{\mathrm{T}} \right] \mathrm{d}\omega \tag{15.75}$$

where $\mathbf{C}^1_{\mathbf{AB}}(\omega)$ and $\mathbf{C}^2_{\mathbf{AB}}(\omega)$ are $M(2N-1) \times M$-dimensional matrices:

$$\mathbf{C}^1_{\mathbf{AB}}(\omega) = [\mathbf{P}(\omega) \otimes \mathbf{e}_1] * \left[\mathbf{e}_2^{\mathrm{T}} \otimes \mathbf{D}_1(\omega) \right] \tag{15.76a}$$

$$\mathbf{C}^2_{\mathbf{AB}}(\omega) = [\mathbf{P}(\omega) \otimes \mathbf{e}_1] * \left[\mathbf{e}_2^{\mathrm{T}} \otimes \mathbf{D}_2(\omega) \right] \tag{15.76b}$$

where:

"$\otimes$" denotes the Kronecker tensor product;
"$*$" denotes the array multiplication (i.e. $\mathbf{A}^*\mathbf{B}$ is the element-by-element product);
$\mathbf{e}_1$ and $\mathbf{e}_2$ are the $(2N-1)$ -dimensional and M-dimensional vectors of all ones, respectively.

$\mathbf{P}(\omega)$ is the lower triangular matrix obtained by the Cholesky decomposition of the $M \times M$-dimensional coherence matrix $\boldsymbol{\gamma}(\omega)$:

$$\boldsymbol{\gamma}(\omega) = \mathbf{P}(\omega)\mathbf{P}(\omega)^{\mathrm{T}} \tag{15.77}$$

$\mathbf{D}_1(\omega)$ and $\mathbf{D}_2(\omega)$ both are $M(2N-1) \times 1$-dimensional vectors:

$$\mathbf{D}_1(\omega) = \left[\mathbf{D}_{1,1}(\omega)^{\mathrm{T}} \cdots \mathbf{D}_{1,M}(\omega)^{\mathrm{T}} \right]^{\mathrm{T}}, \; \mathbf{D}_2(\omega) = \left[\mathbf{D}_{2,1}(\omega)^{\mathrm{T}} \cdots \mathbf{D}_{2,M}(\omega)^{\mathrm{T}} \right]^{\mathrm{T}} \tag{15.78}$$

$$\mathbf{D}_{1,j}(\omega) = \begin{bmatrix} \mathbf{I}_j \\ \mathbf{K}_j \end{bmatrix}, \; \mathbf{D}_{2,j}(\omega) = \begin{bmatrix} \mathbf{J}_j \\ \mathbf{L}_j \end{bmatrix} \tag{15.79}$$

where:

$\mathbf{I}_j = [I_{1j}, \ldots, I_{Nj}]^{\mathrm{T}}$ is of the size $N \times 1$;
$\mathbf{K}_j = [K_{2j}, \ldots, K_{Nj}]^{\mathrm{T}}$ is of the size $(N-1) \times 1$;
$\mathbf{J}_j = [J_{1j}, \ldots, J_{Nj}]^{\mathrm{T}}$;
$\mathbf{L}_j = [L_{2j}, \ldots, L_{Nj}]^{\mathrm{T}}$.

Equation 15.75 indicates that the covariance matrix $\mathbf{C}_{\mathbf{AB}}$ can be alternatively computed by an integration of the sum of two frequency-dependent sub-matrices, and each sub-matrix is actually expressed by its Cholesky decomposition in terms of $\mathbf{C}^1_{\mathbf{AB}}(\omega)$ and $\mathbf{C}^2_{\mathbf{AB}}(\omega)$.

15.7.6 Fast Algorithm for Conditional Simulation

By Equation 15.75, the Fourier coefficient vector $\mathbf{F}^{S^*}_{\mathbf{AB}}$, which is originally computed by Equations 15.65 and 15.66, can be computed by the following algorithm:

$$\mathbf{F}^{S^*}_{\mathbf{AB}} = \frac{2\sqrt{2\Delta\omega'}}{T}\sum_{l'=1}^{N'}\left[\mathbf{F}^1(\omega_{l'}) + \mathbf{F}^2(\omega_{l'})\right] \tag{15.80}$$

where:

$\omega_{l'} = (l'-1)\Delta\omega'$ ($l' = 1, \ldots, N'$) is the discretized frequency;
$\Delta\omega'$ is the frequency interval; and

$$\mathbf{F}^1(\omega_{l'}) = \mathbf{C}^1_{\mathbf{AB}}(\omega_{l'})\mathbf{U}^1_{l'},\ \mathbf{F}^2(\omega_{l'}) = \mathbf{C}^2_{\mathbf{AB}}(\omega_{l'})\mathbf{U}^2_{l'} \tag{15.81}$$

where $\mathbf{U}^1_{l'}$ and $\mathbf{U}^2_{l'}$ both are $(2N-1)$-dimensional vectors consisting of independent standard Gaussian random variables. It can be observed from Equation 15.76 that the computation of $\mathbf{C}^1_{\mathbf{AB}}(\omega)$ and $\mathbf{C}^2_{\mathbf{AB}}(\omega)$ only involves the Cholesky decomposition of the matrix $\boldsymbol{\gamma}(\omega)$ of the size $M \times M$. It thus requires much less computational time, compared with the Cholesky decomposition of the matrix $\mathbf{C}_{\mathbf{AB}}$ of the size $M(2N-1) \times M(2N-1)$.

Let us prove that the covariance of the Fourier coefficient vector $\mathbf{F}^{S^*}_{\mathbf{AB}}$ simulated by Equation 15.80 is equal to the matrix $\mathbf{C}_{\mathbf{AB}}$, as required by Equation 15.64. From Equation 15.80, one may have

$$\mathrm{E}\left[\mathbf{F}^{S^*}_{\mathbf{AB}}\cdot(\mathbf{F}^{S^*}_{\mathbf{AB}})^{\mathrm{T}}\right] = \frac{8\Delta\omega}{T^2}\sum_{l'=1}^{N'}\sum_{s'=1}^{N'}\left\{\begin{array}{l}\mathrm{E}\left[\mathbf{F}^1(\omega_{l'})\mathbf{F}^1(\omega_{s'})^{\mathrm{T}}\right] + \mathrm{E}\left[\mathbf{F}^2(\omega_{l'})\mathbf{F}^1(\omega_{s'})^{\mathrm{T}}\right] + \\ \mathrm{E}\left[\mathbf{F}^1(\omega_{l'})\mathbf{F}^2(\omega_{s'})^{\mathrm{T}}\right] + \mathrm{E}\left[\mathbf{F}^2(\omega_{l'})\mathbf{F}^2(\omega_{s'})^{\mathrm{T}}\right]\end{array}\right\} \tag{15.82}$$

From Equation 15.81, it can be derived that:

$$\mathrm{E}\left[\mathbf{F}^m(\omega_{l'})\mathbf{F}^n(\omega_{s'})^{\mathrm{T}}\right] = \mathbf{C}^m_{\mathbf{AB}}(\omega_{l'})\cdot\mathrm{E}\left[\mathbf{U}^m_{l'}(\mathbf{U}^n_{s'})^{\mathrm{T}}\right]\cdot\mathbf{C}^n_{\mathbf{AB}}(\omega_{s'})^{\mathrm{T}} \tag{15.83}$$

in which $\mathbf{U}^m_{l'}$ is independent of $\mathbf{U}^n_{s'}$ and $\mathbf{U}^n_{l'}$ for all $m \neq n$ and $l' \neq s'$. Thus:

$$\mathrm{E}\left[\mathbf{U}^m_{l'}(\mathbf{U}^n_{s'})^{\mathrm{T}}\right] = \begin{cases}\boldsymbol{\Xi}_{(2N-1)\times(2N-1)} & m = n \text{ and } l' = s' \\ 0 & \text{otherwise}\end{cases} \tag{15.84}$$

where $\boldsymbol{\Xi}$ is the unit matrix.

Substituting Equations 15.83 and 15.84 into Equation 15.82 finally yields:

$$\mathrm{E}\left[\mathbf{F}_{\mathbf{AB}}^{S^*}\cdot(\mathbf{F}_{\mathbf{AB}}^{S^*})^{\mathrm{T}}\right] = \frac{8\Delta\omega'}{T^2}\left\{\sum_{k'=1}^{N'}\left[\mathbf{C}_{\mathbf{AB}}^{1}(\omega_{l'})\cdot\mathbf{C}_{\mathbf{AB}}^{1}(\omega_{l'})^{\mathrm{T}} + \mathbf{C}_{\mathbf{AB}}^{2}(\omega_{l'})\cdot\mathbf{C}_{\mathbf{AB}}^{2}(\omega_{l'})^{\mathrm{T}}\right]\right\} \tag{15.85}$$

The right hand side of Equation 15.85 is the discretized version of the integral in Equation 15.75 according to the rectangular convention. Therefore, it can be concluded that:

$$\mathrm{E}\left[\mathbf{F}_{\mathbf{AB}}^{S^*}\cdot(\mathbf{F}_{\mathbf{AB}}^{S^*})^{\mathrm{T}}\right] \doteq \mathbf{C}_{\mathbf{AB}} \tag{15.86}$$

which indicates that the simulated $\mathbf{F}_{\mathbf{AB}}^{S^*}$ has a unbiased covariance matrix.

Furthermore, in the computation of $\mathbf{C}_{\mathbf{AB}}^{1}(\omega)$, the vector $\mathbf{I}_j$ is computed from the EPSD using Equation 15.71a, which can be rewritten as:

$$I_{pj}^{l'} = I_j(\omega_{l'}, \omega_p) = \sum_{r=1}^{N_T}\sqrt{S_{jj}(\omega_{l'}, t_r)}\cos\omega_{l'}t_r\cos\omega_p t_r\,\Delta t \tag{15.87}$$

where $t_r = (r-1)\Delta t$.

If the number of frequency intervals in the discretized frequency series ω_p is chosen as a power of 2, $I_{pj}^{l'}$ can be computed efficiently by using the FFT technique:

$$\mathbf{I}_j^{l'} = \Delta t\mathrm{Re}\left\{\mathrm{FFT}_{N_T}\circ\left[\sqrt{S_{jj}(\omega_{l'}, t_r)}\cos\omega_{l'}t_r\right]\right\} \tag{15.88}$$

where $\mathrm{FFT}_{N_T}\circ[\,]$ denotes the N_T-points FFT operator.

Similarly, the vectors $\mathbf{J}_j$, $\mathbf{K}_j$ and $\mathbf{L}_j$ can also be computed by using FFT:

$$\mathbf{J}_j^{l'} = \Delta t\mathrm{Re}\left\{\mathrm{FFT}_{N_T}\circ\left[\sqrt{S_{jj}(\omega_{l'}, t_r)}\sin\omega_{l'}t_r\right]\right\} \tag{15.89a}$$

$$\mathbf{K}_j^{l'} = \Delta t\mathrm{Im}\left\{\mathrm{FFT}_{N_T}\circ\left[\sqrt{S_{jj}(\omega_{l'}, t_r)}\cos\omega_{l'}t_r\right]\right\} \tag{15.89b}$$

$$\mathbf{L}_j^{l'} = \Delta t\mathrm{Im}\left\{\mathrm{FFT}_{N_T}\circ\left[\sqrt{S_{jj}(\omega_{l'}, t_r)}\sin\omega_{l'}t_r\right]\right\} \tag{15.89c}$$

By taking advantage of the FFT technique and using the simplified computation method for computing $\mathbf{F}_{\mathbf{AB}}^{S^*}$ expressed by Equation 15.80, a fast algorithm can be yielded. The ratio of computational time of the matrix decomposition between the original conditional simulation method and the method with the fast algorithm is approximately $1/(8N^2)$. If the number $N = 2048$, this ratio is as small as 3×10^{-8}, which indicates that the fast algorithm substantially reduces the computational time of the proposed conditional simulation method. As a result, the conditional simulation of non-stationary fluctuating wind speeds for buffeting analysis of a long-span bridge can now be performed readily using a personal computer.

15.7.7 Implementation Procedure

To implement the fast algorithm-based conditional simulation method for non-stationary fluctuating wind speeds, the *a priori* known information is listed in the following. Simulation parameters include the time interval Δt, the number of time intervals N_T, the frequency interval $\Delta\omega' = \Delta\omega$, and the number of frequency intervals $N = N'$.

By virtue of the sampling theorem, one may take $N = N_T/2$ and $\Delta\omega = 2\pi/(N_T\Delta t)$. To invoke FFT, N_T should be a power of 2. The EPSDs, coherence functions and the measured time histories are known as $S_{jj}(\omega_{l'}, t_r)$, $\gamma(\omega_{l'})$ and $\mathbf{x}_\alpha^{0(I)}(t_r)$, where $\omega_{l'} = (l' - 1)\Delta\omega'$ and $t_r = (r - 1)\Delta t$. Based on the above given information, the conditional simulation procedure can be implemented by the following steps.:

1. Compute vectors $\mathbf{I}_j^{l'}$, $\mathbf{J}_j^{l'}$, $\mathbf{K}_j^{l'}$ and $\mathbf{L}_j^{l'}$ using Equations 15.88 and 15.89.
2. Form vectors $\mathbf{D}_1(\omega_{l'})$ and $\mathbf{D}_2(\omega_{l'})$ using Equations 15.78 and 15.79.
3. Compute the Cholesky decomposition of $\boldsymbol{\gamma}(\omega_{l'})$ using Equation 15.77.
4. Form matrices $\mathbf{C}_{\mathbf{AB}}^1(\omega_{l'})$ and $\mathbf{C}_{\mathbf{AB}}^2(\omega_{l'})$ by Equation 15.76.
5. Generate two $(2N-1)$-dimensional sample vectors $\mathbf{u}_{l'}^1$ and $\mathbf{u}_{l'}^2$ using standard Gaussian random numbers.
6. Compute sample vectors $\mathbf{f}^1(\omega_{l'})$ and $\mathbf{f}^2(\omega_{l'})$ using Equation 15.81.
7. Repeat steps 1–6 to get $\mathbf{f}^1(\omega_{l'})$, $\mathbf{f}^2(\omega_{l'})$, $\mathbf{C}_{\mathbf{AB}}^1(\omega_{l'})$ and $\mathbf{C}_{\mathbf{AB}}^2(\omega_{l'})$ for all frequencies $\omega_{l'}$.
8. Compute sample matrix $\mathbf{f}_{\mathbf{AB}}^{S^*}$ using Equation 15.80.
9. Separate $\mathbf{f}_{\mathbf{AB}}^{S^*}$ into two sub vectors $\mathbf{f}_\alpha^{S^*}$ and $\mathbf{f}_\beta^{S^*}$.
10. Compute the covariance matrix $\mathbf{C}_{\mathbf{AB}}$ using Equation 15.75.
11. Partition $\mathbf{C}_{\mathbf{AB}}$ into the four sub-matrices using Equation 15.60.
12. Compute the Fourier coefficient vector $\mathbf{f}_\alpha^0$ by applying FFT to measured time histories according to Equations 15.57 and 15.61.
13. Compute the simulated Fourier coefficient vector $\mathbf{f}_\beta^S$ using Equation 15.63.
14. Apply the inverse FFT to the simulated Fourier coefficient vector $\mathbf{f}_\beta^S$ by Equations 15.62 and 15.67 to yield the simulated time histories $\mathbf{x}_\beta^{S(I)}(t_r)$.
15. The complete set of the *i*th sample of the simulated non-stationary fluctuating wind speed time histories can be produced as $\mathbf{x}^{S(I)}(t_r) = \left[\mathbf{x}_\alpha^{0(I)}(t_r),\ \mathbf{x}_\beta^{S(I)}(t_r)\right]$.
16. More samples of non-stationary fluctuating wind speed time histories can be generated by repeating the above steps.

15.7.8 Validation and Application

To demonstrate its feasibility and capability, the proposed fast algorithm-based conditional simulation method is applied to simulate a complete set of fluctuating wind speed time histories based on part of wind speed time histories measured during the strong typhoon Dujuan passing by the Stonecutters Bridge. The proposed conditional simulation method is then validated through the simulation of fluctuating wind speed time histories at the first three locations of the bridge. The complete set of fluctuating wind speed time histories is finally simulated.

To carry out the non-stationary buffeting analysis of the bridge in the time domain, a total of 198 locations (points) are selected as the action points of buffeting forces (see Figure 15.16). This includes points 1–88 on the bridge deck, points 89–127 on the left tower, points 128–166 on the right tower and points 167–198 on the eight piers. The longitudinal and vertical non-stationary fluctuating wind speeds along the bridge deck are modeled as a 198-variate evolutionary vector process $\mathbf{U}(t)$ and an 88-variate evolutionary vector process $\mathbf{W}(t)$, respectively. The EPSD matrices of $\mathbf{U}(t)$ and $\mathbf{W}(t)$ are constructed by assuming the auto-EPSDs are uniform for all the points.

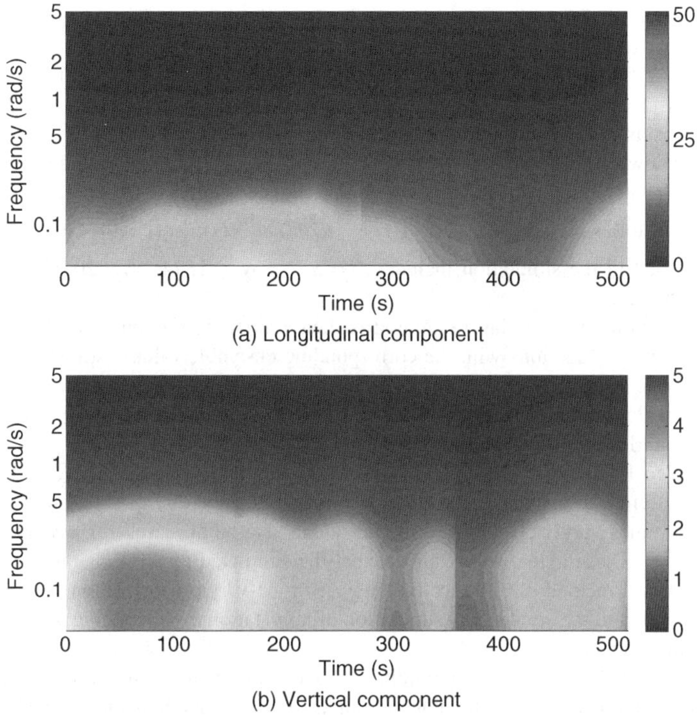

Figure 15.32 Auto EPSDs of turbulent winds.

The auto-EPSDs adopted for longitudinal and vertical components are shown in Figure 15.32, which are estimated from the wind speed time histories recorded by the measurement system during Typhoon Dujuan on 29 August, 2003. The empirical exponential coherence function in Equation 15.16 is adopted in the probabilistic model of $\mathbf{U}(t)$ and $\mathbf{W}(t)$, while the coherence between the vertical and longitudinal components is disregarded.

Using the prescribed probabilistic model of turbulence winds established in terms of the limited measured non-stationary fluctuating wind speed time histories, the complete set of time histories can be conditionally simulated by the proposed fast algorithm-based conditional simulation method. However, before performing this large-sized simulation problem, let us first validate the method by applying it for the conditional simulation of a small part of the non-stationary wind field.

To validate the proposed method, the longitudinal components at points 1, 2 and 3 of the bridge are conditionally simulated first. This non-stationary fluctuating wind field is modeled as a tri-variate zero-mean vector process $\mathbf{U}(t) = [U_1(t)U_2(t)U_3(t)]$. The auto-EPSD of $S_{U1}(\omega,t)$ uses the measurement-based function as shown in Figure 15.32, whereas the auto EPSDs of $U_2(t)$ and $U_3(t)$ are assumed to have the following forms:

$$S_{U2}(\omega, t) = S_{U1}(1.1\omega, 1.1t);\ S_{U3}(\omega, t) = S_{U1}(1.2\omega, 1.2t) \tag{15.90}$$

Jointly using these auto EPSDs and the coherence function, the EPSD matrix of $\mathbf{U}(t)$ can be determined. The fluctuating wind speed time histories at the point 1 are simulated unconditionally for 10,000 times by using the spectral representation method. The generated sample time histories are then used as the measured fluctuating wind speed time histories $u_1^{0(I)}(t)$ ($I=1\ldots 10\ 000$). Accordingly, the conditional simulation of fluctuating wind speed time histories for all the three points is also performed for 10,000 times ($I=1\ldots 10\ 000$) using the proposed fast algorithm-based simulation method. For each time, it is assumed that only the measured time history at the point 1 $u_1^{0(I)}(t)$ is known. That is, the *a priori* known information involves $u_1^{0(I)}(t)$ and the EPSD matrix, while the time histories $u_2^{S(I)}(t)$ and $u_3^{S(I)}(t)$ are unknown and to be simulated. The simulated complete set of fluctuating wind speed time histories are denoted as $\mathbf{u}^{S(I)}(t) = \left[u_1^{0(I)}(t)u_2^{S(I)}(t)u_3^{S(I)}(t)\right]$ for $I=(1, 2, \ldots, 10\ 000)$. The parameters used in the simulation include: $M=3$, $N=N'=1024$, $N_T=2048$, $\Delta t=0.25$ sec and $\Delta\omega=\Delta\omega'=2\pi/512$ rad/sec.

The proposed conditional simulation method is then validated by comparing the target values of time-varying correlation functions with the corresponding ensemble values estimated from the conditional simulated time histories. The theoretical values are computed from the EPSD matrix. The ensemble correlation functions are estimated by taking ensemble average over the 10 000 sets of conditionally simulated time histories $\mathbf{u}^{S(I)}(t)$.

Figures 15.33a to 15.33f show the comparison of correlations $R_{jk}(t_1,t_2)$, ($j=1\ldots 3$), ($k=1\ldots j$) associated with different time instants. From each subplot, it can be observed that the ensemble values match the target values very well. Good agreements can also be observed for any other time instant $t_1 \in (0.25, 512)$ sec. These agreements indicate that the simulated results generated by the proposed conditional simulation method can reserve the unbiased time-varying correlation functions. Thus, this method can be applied to the conditional simulation of non-stationary fluctuating wind speed time histories for the entire bridge.

Based on the given EPSD matrix of longitudinal components, the measured (reference) time histories at all the 198 points are generated by the unconditional simulation method and denoted as $\mathbf{u}^{0(1)}(t)$. It is then assumed that the measured time histories are only available at 12 points: 10, 20, 30, 40, 45, 50, 55, 60, 70, 80, 127 and 166 (see Figure 15.16).

To perform the non-stationary buffeting analysis of the bridge, a complete set of fluctuating wind speed time histories $\mathbf{u}^{S(1)}(t)$ are generated under the conditions of the measured time histories at the 12 points and the given EPSD matrix. This implementation only takes about two hours on a personal computer equipped with an Intel i5 750 CPU but, without the fast algorithm, this simulation is not feasible by such a computer.

Similarly, the vertical fluctuating wind speed time histories $\mathbf{w}^{S(1)}(t)$ are also simulated conditionally, but only the 88 points along the bridge deck are considered under the conditions of the measured time histories at ten points (10, 20, 30, 40, 45, 50, 55, 60, 70, and 80). The conditional simulation uses the same parameters as the example presented above, except for $M=198$ for $\mathbf{U}(t)$ and $M=88$ for $\mathbf{W}(t)$. In addition, another set of fluctuating wind speed time histories are also generated by using the unconditional simulation method based on the given EPSD matrices only, and these are denoted the unconditionally simulated time histories $\mathbf{u}^{US(1)}(t)$ and $\mathbf{w}^{US(1)}(t)$.

The feasibility and capability of the conditional simulation method are then demonstrated through the comparison of the measured, conditionally simulated and unconditionally simulated fluctuating wind speed time histories.

Figures 15.34 to 15.36 show the longitudinal fluctuating wind speed time histories at points 42 and 25 and the vertical fluctuating wind speed time histories at point 11, with subplots (a), (b), and (c) showing the measured, conditionally simulated and unconditionally simulated fluctuating wind speed time histories, respectively. It can be seen that the conditional simulated time histories are similar to the corresponding measured time histories, but they are not similar to the unconditionally simulated ones. This observation suggests that the conditional simulation can represent the measured time histories more closely than the unconditional simulation.

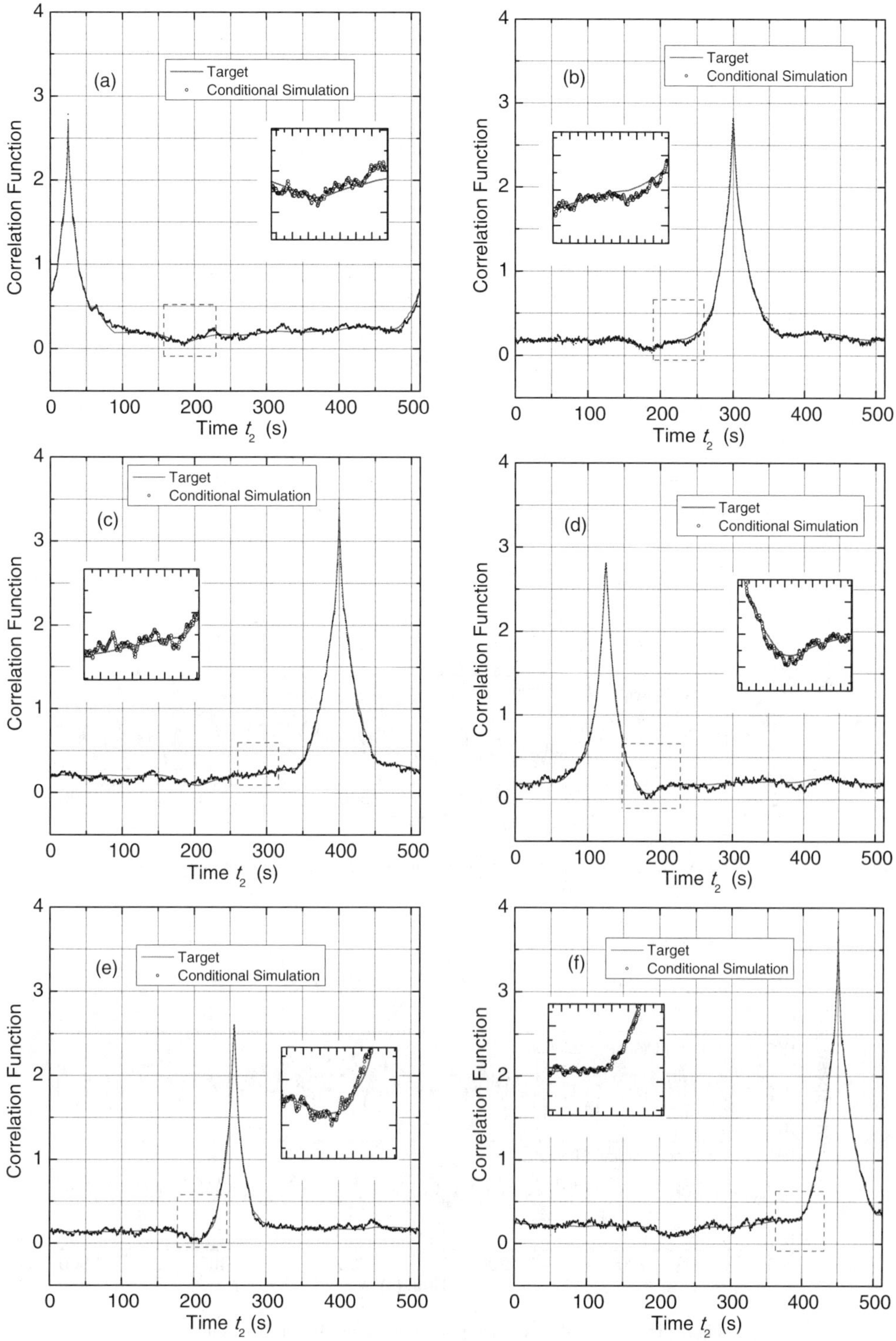

Figure 15.33 Ensemble and target time-varying correlation functions. (a) $R_{11}(t_1, t_2)$ at $t_1 = 25$ sec; (b) $R_{21}(t_1, t_2)$, $t_1 = 300$ sec; (c) $R_{22}(t_1, t_2)$, $t_1 = 400$ sec; (d) $R_{31}(t_1, t_2)$, $t_1 = 125$ sec; (e) $R_{32}(t_1, t_2)$, $t_1 = 256$ sec; (f) $R_{33}(t_1, t_2)$, $t_1 = 450$ sec.

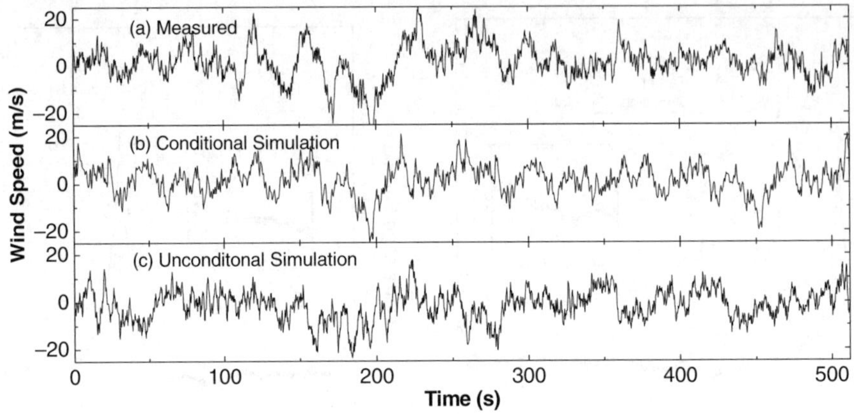

Figure 15.34 Longitudinal fluctuating wind speed time histories at point 42.

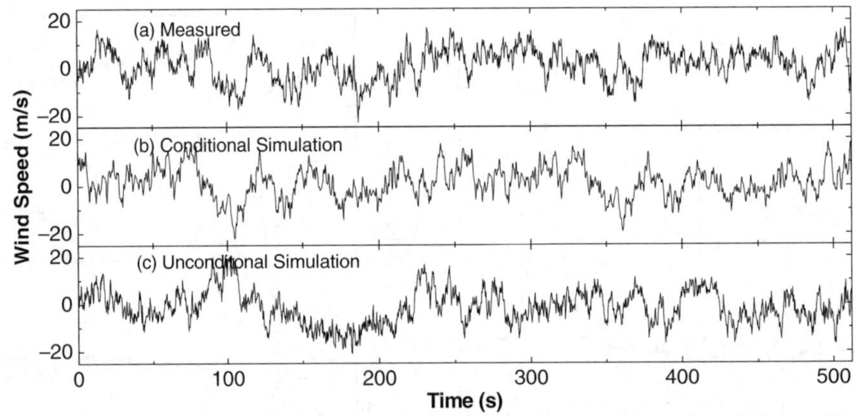

Figure 15.35 Longitudinal fluctuating wind speed time histories at point 25.

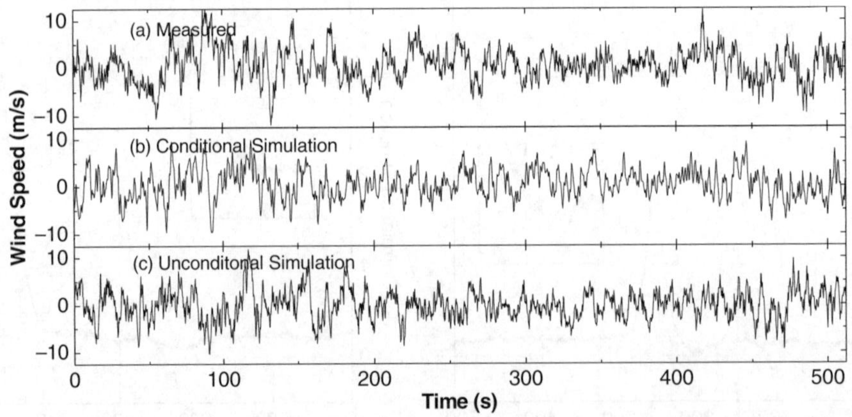

Figure 15.36 Vertical fluctuating wind speed time histories at point 11.

15.8 Non-Linear Buffeting Response

15.8.1 Introduction

Non-linearities should be accounted for predicting excessive responses of long-span cable-supported bridges induced by high wind speeds. The non-linearities involved in predicted buffeting responses stem from both structural and aerodynamic origins. The structural origin engenders geometric and material non-linearities of long-span bridge structures [1]. The aerodynamic non-linearities of bridge decks are due to fluid memory effects, vortices at the deck edges and characteristics generated by bridge deck motions, and they may display apparent hysteretic behavior that is significantly different from linear manner [41].

Under turbulent wind with high wind speeds, large amplitudes of bridge deck motions can be expected, in which large fluctuations of instantaneous angle of attack may make the effects of aerodynamic non-linearities notable and of critical importance. A typical situation in which aerodynamic non-linearities are indispensible is the bridge response under turbulence wind speeds near the flutter critical point, which is critical for the safety of a long-span cable-supported bridge.

With respect to traditional frameworks for prediction of buffeting responses, structural non-linearities can be readily included in the time-domain buffeting analysis based on non-linear finite element method, but the effects of aerodynamic non-linearities can be captured only partially. In traditional frameworks, aerodynamic forces due to turbulent winds are always divided into static forces due to mean wind speeds, buffeting forces due to fluctuating winds, and self-excited forces due to bridge deck motions.

In modeling buffeting and self-excited forces, the quasi-steady assumption implies the effects of instantaneous angle of attack only, by linearizing the relationship of aerodynamic forces and the angle of attack via Taylor series expansion around the angle of attack due to static forces. The time domain buffeting analysis framework developed by Chen *et al.* [42] further includes unsteady effects by rational function approximation of flutter derivatives, admittance functions, joint acceptance functions and convolutional integral forms of both buffeting and self-excited forces. It also includes the effects of instantaneous angles of attack under static forces. Non-linear effects of the second order term of fluctuating wind speed on buffeting responses were also assessed by Gurley *et al.* [43], considering the accompanying non-Gaussian properties. Nevertheless, the division of aerodynamic forces into three separate parts is questionable in the non-linear aerodynamics of a bridge deck, and a feasible framework for predicting non-linear aerodynamics responses of long-span bridges under high winds is required.

Some attempts in modeling non-linear aerodynamics have been made recently, among which, three types of approaches are introduced in this section. The first model [44–47] incorporates the unsteady time domain buffeting analysis with time-dependent aerodynamic coefficients and flutter derivatives in terms of time-varying effective instantaneous angle of incidence due to low frequency turbulence components and attendant deck rotations. This model offers a linear approximation of non-linear aerodynamics by expanding the buffeting and self-excited forces via Taylor series around the time-varying instantaneous angle of incidence.

The second and third models treat the aerodynamic forces as a whole, based on the hysteretic aerodynamic behavior, and avoid the division of aerodynamic forces into several parts. The second model assumes the hysteretic relationship between the aerodynamic forces and the angle of incidence observed from wind tunnel tests. The hysteretic relationship is modeled by a rheological model [48], a polynomial model [48] or a joint rheological and polynomial model [49]. The third type of model [41] uses a general non-linear system to describe the complex non-linear relationship by specifying all components of turbulent wind speeds and bridge deck motions as inputs and outputs of the system. A cellular automata nested artificial neural network (ANN) is then applied to model this non-linear aerodynamic system.

15.8.2 Linearization Model for Non-Linear Aerodynamic Forces

This model is the basis of the unsteady time domain buffeting analysis framework proposed by Chen *et al.* [42]. The framework accounts for unsteady aerodynamic forces and aerodynamic non-linearities due to initial wind angle of attack of the bridge deck. Furthermore, the instantaneous angle of incidence is time-dependent and, thus, the corresponding instantaneous aerodynamic coefficients are time-dependent, while the flutter derivatives and admittance functions that vary with the angle of incidence are dependent on both frequency and time. To consider the non-linear effects due to the angle of incidence, Chen and Kareem [44] decomposed the instantaneous angle of attack $\alpha_e(t)$ into low frequency and high frequency components:|

$$\alpha_e(t) = \alpha_e^l(t) + \alpha_e^h(t) \tag{15.91}$$

and linearized the aerodynamic forces **F** around the low frequency part of angle of incidence:

$$\mathbf{F} = \mathbf{F}(\alpha_e(t)) \approx \mathbf{F}(\alpha_e^l(t)) + \frac{\mathrm{d}\mathbf{F}}{\mathrm{d}\alpha}\bigg|\alpha_e^l(t) \cdot \alpha_e^h(t) = \mathbf{F}^l + \mathbf{F}_{se}^h + \mathbf{F}_b^h \tag{15.92}$$

where:

the superscripts l and h denote the low-frequency and high-frequency components;
the subscripts se and b denote self-excited and buffeting forces.

The quasi-steady approach is then used to determine the low-frequency part of aerodynamic forces and the attendant responses. The high-frequency part of aerodynamic forces is computed by the unsteady aerodynamic theory in terms of the time-dependent aerodynamic coefficients, flutter derivatives and admittance functions associated with the low-frequency angle of incidence. With resorting to the rational function approximation, buffeting responses due to the high-frequency aerodynamic buffeting and self-excited forces are computed in the time domain. The non-linear buffeting responses in total are the sum of low- and high-frequency buffeting responses.

The procedure for performing non-linear buffeting analysis using this framework is described briefly here. First, the turbulence components $u(t)$ and $w(t)$ are decomposed into low- and high-frequency components u^l, u^h and w^l, w^h by band-limited digital filtering, respectively. Although a reasonable criterion for the demarcation between low and high frequencies remains unresolved, the first order of natural frequency of the bridge structure can be adopted temporarily.

Second, the quasi-steady low-frequency aerodynamic forces are determined as:

$$L^l = F_L^l \cos\phi^l + F_D^l \sin\phi^l,\ D^l = F_D^l \cos\phi^l - F_L^l \sin\phi^l,\ M^l = F_M^l \tag{15.93a}$$

$$F_L^l = \frac{1}{2}\rho U_r^2 BlC_L(\alpha_e^l(t)),\ F_D^l = \frac{1}{2}\rho U_r^2 BlC_D(\alpha_e^l(t)),\ F_M^l = \frac{1}{2}\rho U_r^2 B^2 lC_M(\alpha_e^l(t)) \tag{15.93b}$$

$$U_r(t) = \sqrt{(\overline{U} + u^l - \dot{p}^l)^2 + (w^l - \dot{h}^l + m_l b\dot{\alpha}^l)^2} \tag{15.93c}$$

$$\alpha_e^l(t) = \alpha_s + \phi^l(t) + \alpha^l(t),\ \phi^l(t) = \arctan\left(\frac{w^l - \dot{h}^l + m_l b\dot{\alpha}^l}{\overline{U} + u^l - \dot{p}^l}\right) \tag{15.93d}$$

where:

$m_1 = 0.5$;
$B = 2b$ is the bridge deck width;
α_s is the static wind angle of attack;
U_r is the instantaneous relative velocity between deck and wind;
$\phi^l(t)$ is the low-frequency part of the angle of incidence of the relative velocity;
L^l, D^l, and M^l are the low-frequency components of aerodynamic lift force, drag force and pitching moments along with the coordinate system defined with respect to mean wind speed $\overline{U}$;
F_L^l, F_D^l, and F_M^l are the low-frequency components of aerodynamic lift force, drag force and pitching moments along with the coordinate system defined with respect to the instantaneous relative velocity U_r;
h^l, p^l, and α^l are the low-frequency components of dynamic vertical, horizontal, and torsional displacements of the bridge deck;
the over dot denotes the derivative with respect to time.

The aerodynamic forces are dependent on the angle of incidence and thus on the torsional motion of the bridge deck $\alpha^l(t)$.

Third, for each time instant, the aerodynamic coefficients, flutter derivatives, admittance functions and jointly acceptance functions are interpolated with respect to $\alpha_e^l(t)$ and are then fitted by rational functions. The rational function approximation then gives the unsteady high-frequency force components via convolutional integration. For example, the high-frequency components of buffeting and self-excited lift forces are expressed as:

$$L_{se}^h(t) = \frac{1}{2}\rho\overline{U}^2 \int_{-\infty}^{t} (I_{Lh}(\alpha_e^l, t-\tau)h^h(\tau) + I_{Lp}(\alpha_e^l, t-\tau)p^h(\tau) + I_{L\alpha}(\alpha_e^l, t-\tau)\alpha^h(\tau))\mathrm{d}\tau \tag{15.94a}$$

$$\begin{aligned} L_b^h(t) = -\frac{1}{2}\rho\overline{U}^2 l \int_{-\infty}^{t}\int_{-\infty}^{\tau_2} \Bigg(& (J_{Lu}(\alpha_e^l, t-\tau_2)I_{Lu}(\alpha_e^l, \tau_2-\tau_1)\frac{u^h(\tau_1)}{\overline{U}} \\ & + J_{Lw}(\alpha_e^l, t-\tau_2)I_{Lw}(\alpha_e^l, \tau_2-\tau_1)\frac{w^h(\tau_1)}{\overline{U}}\Bigg)\mathrm{d}\tau_1\mathrm{d}\tau_2 \end{aligned} \tag{15.94b}$$

where:

I_{Lj} denotes the aerodynamic impulse functions between the lift force and bridge deck motion ($j = h$, p and α) given by the rational approximation of flutter derivatives, or between lift force and turbulence components ($j=u$ and w) given by the rational approximation of admittance functions;
J_{Lj} ($j = u$ and w) is the impulse response function given by rational approximation of jointly acceptance functions.

Finally, the buffeting responses associated with those non-linear buffeting and self-excited forces are obtained by the Newmark Beta step-by-step integration method, with non-linear iterations at each time instant.

The linear approximation model of non-linear aerodynamics proposed by Chen and Kareem has been applied to buffeting analysis of a long-span suspension bridge with a main span of 2000 m and the same aerodynamic characteristics as those of the Messina Straits bridge deck section. In the non-linear buffeting analysis, the spatiotemporal characteristics of aerodynamic forces are apparent, but the associated effects on buffeting responses are not notable.

By comparison with the results from the linear unsteady buffeting analysis, it can be seen that the buffeting responses from the non-linear analysis are close to those from the linear analysis, with aerodynamic characteristics interpolated with respect to the static wind angles of attack. The characteristics of buffeting responses under wind speeds close to critical flutter wind speed display a gradual fluttering boundary, which is different from the sharp edge without effects of non-linear aerodynamics. The gradual fluttering boundary is a phenomenon similar to that observed from wind tunnel tests.

15.8.3 Hysteretic Behavior of Non-Linear Aerodynamic Forces

Wind tunnel tests on the forces and motions of the deck section of the Messina Strait Bridge indicate that the relationship between the total aerodynamic forces and the effective angle of attack forms a hysteretic loop. It seems more reasonable to represent this hysteretic behavior of aerodynamic forces by modeling the force components along each direction as a whole, rather than dividing the aerodynamic forces into three parts of static, buffeting and self-excited forces. From this viewpoint, Diana *et al.* [49–51] first investigated the aerodynamic hysteretic behavior using a newly designed wind tunnel test rig. The test rig can be used to study basic characteristics of hysteretic behavior, provide basic data for modeling hysteretic loops of specific deck sections, and validate the possible models of hysteretic behavior. Three types of models were presented, including the rheological model, the polynomial model and the hybrid model, to represent the relationship between aerodynamic forces and angle of attack. These models, their identification methods and the simulation results in comparison with wind tunnel test results, are introduced briefly in the following.

Let us consider the aerodynamic forces $F_i(t)$ ($I = L, D, M$) and the instantaneous angle of attack angle ψ:

$$F_L = \frac{1}{2}\rho U_r^2 BlC_L,\ F_D = \frac{1}{2}\rho U_r^2 BlC_D,\ F_M = \frac{1}{2}\rho U_r^2 B^2 lC_M \tag{15.95a}$$

And:

$$\psi = \alpha + \arctan\left(\frac{w - \dot{h} + B^*\dot{\alpha}}{U_r}\right),\ U_r = \overline{U} + u - \dot{p} \tag{15.95b}$$

where:

B^* is an effective bridge deck width;
α is the torsional angle of deck motion (see Figure 15.37b).

The non-linear relationship between $F_i(t)$ and ψ can be represented as $F_i = F_i(\psi, \overline{U}_R)$ or, more conveniently, as the force coefficients $C_i = C_i(\psi, \overline{U}_R)$ (i.e. normalized forces), where $\overline{U}_R = \overline{U}/(fB)$ is the reduced velocity. The effective wind angle of attack is involved in this non-linear relationship, which implies a basic hypothesis that the effects on the variation of ψ due to the deck motion $\dot{h} + B^*\dot{\alpha}$ and due to the turbulence w weigh equivalently.

A plot of relationship between the lift force C_L and motion $\dot{h} + B^*\dot{\alpha}$ is shown in Figure 15.38 with a typical hysteretic loop. Essentially, the loop emerges because there is a phase delay between the force and the angle of attack, which may dissipate energy or pump energy into the system depending on the rotational direction of the loop. To reveal characteristics of the hysteretic loop, Diana *et al.* [49–51] performed extensive wind tunnel studies, from which they built mathematical models for the hysteretic loops.

The wind tunnel tests were performed on two bridge deck sections. The first one (Test A, Figure 15.37b) used the deck of the Messina Strait Bridge with a length scale of 1 : 60, on which the aerodynamic forces are sensitive to the variation of wind angle of attack. The second one (Test B) modeled a single-box deck section from an actual highway bridge.

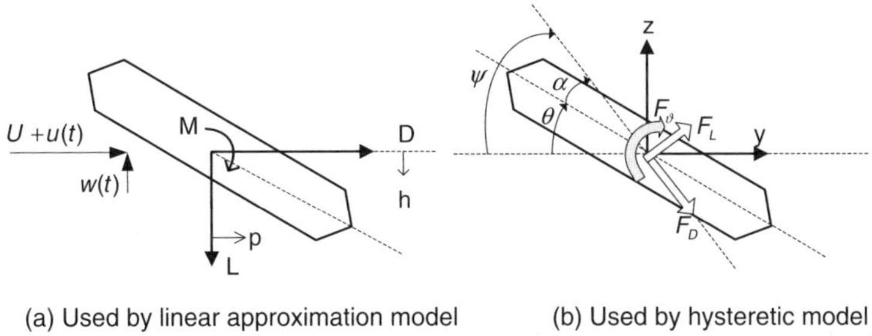

Figure 15.37 Sectional model of bridge deck.

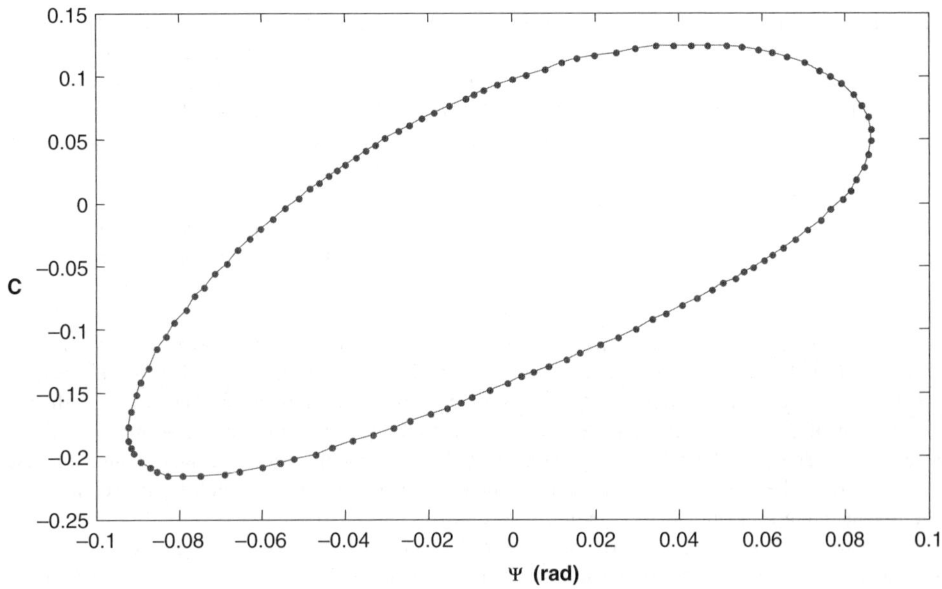

Figure 15.38 A typical hysteretic loop.

The newly established test rig [51] could perform both forced vibration tests and free vibration tests on deck sectional models. For forced vibration tests, three computer-controlled hydraulic actuators were employed to drive harmonic motions of vertical and torsional motions of the sectional model with specified average angles of attack and amplitudes under the smooth flow conditions with turbulence intensities of less than 2%. For free vibration tests, the sectional models were suspended by steel cables elastically to simulate the sectional model movement with specified vertical, horizontal and torsional modal properties. An active turbulence generator was employed to generate the required harmonic wind flows with various frequencies and amplitudes, but with negligible residual turbulence.

Aerodynamic forces on the sectional models, the motion of the models, and incoming wind velocities at the leading edge, were measured simultaneously in the wind tunnel tests. Test A used a dynamometric balance to measure the forces acting on the central dynamometric part of the sectional model, plus three accelerometers and two laser displacement transducers to measure motions and two pilot tubes and two hot wire anemometers to measure wind speeds.

In Test B, the aerodynamic forces were measured via a pressure measurement system consisting of 78 pressure taps surrounding a middle section of the sectional model, thus avoiding the subtraction problem of inertia forces in forced vibration tests. Test B also utilized two laser transducers and three infrared cameras to measure displacements of the model in forced vibration tests and in free vibration tests, respectively, and employed a four-hole probe for measuring instantaneous wind velocities.

Three categories of test cases were performed on both deck section models, defined as "forced vertical harmonic vibration", "forced torsional harmonic vibration" and "admittance test". The vertical test was used only to measure the aerodynamic derivatives, while the torsional test and the admittance test were conducted with various parameters to capture their associated effects on non-linear aerodynamics. The varied parameters include reduced wind velocities U^* (via mean wind velocities and motion frequencies), the mean angle of torsional oscillation $\overline{\alpha}$ and the amplitude of torsional oscillation $\Delta\alpha$. For more details of the experimental setup, please refer to the literature [50,51] for Test A and the reference [49] for Test B.

Based on the different types of measurements, the wind tunnel tests on the two deck section models provided force characteristics such as: aerodynamic coefficients, aerodynamic derivatives and admittance functions of the bridge deck; motion characteristics, including the non-linear vibration properties and instability on-sets; and, more importantly, they gave the non-linear relationship between aerodynamic forces and deck motions. The experimental data also provided possibilities for analyzing the effects of various factors on aerodynamic non-linearities. The following observations on the hysteretic behavior of aerodynamic forces have been summarized from the experiments.

In both Test A and Test B, the results from torsional tests provided comparisons between the hysteretic loops versus the torsional angle α, and those versus the instantaneous angle of attack $\psi = \alpha + B^*\dot{\alpha}/U_r$. It is found that the loops associated with ψ are more regular, clear and reasonable, implying the dependency of the loops on the angle of attack.

In the torsional test and the admittance test, the instantaneous angles of attack were $\psi_\alpha = \alpha + \arctan(B^*\dot{\alpha}/U_r)$ and $\psi_w = \alpha + \arctan(w/U_r)$, respectively. It can be observed from the results of both Test A and Test B that if a proper value of equivalent bridge deck width B^* were selected, the loop plot in terms of ψ_a from the torsional test was similar to the loop plot in terms of ψ_w from the admittance test with an equal reduced velocity. As a result, the components of the equivalence angle of attack ψ, generated by the torsional motion and by turbulence, could be regarded with equal weight, which supports the hypothesis implied in Equation 15.96.

The lift force hysteretic loops corresponding to different reduced velocities from Test A indicated that high reduced velocity corresponded to small hysteretic loop. For very high reduced velocity representing low frequency of deck motion and high mean wind speed, the non-linear effects of aerodynamic forces were insignificant and, thus, the quasi-steady theory may be applicable. The lift force hysteretic loops corresponding to different mean angles of torsional oscillation from Test A suggested that the hysteretic loops are around the mean angles with different tracks but similar areas.

The moment coefficient hysteretic loop with varying amplitudes of torsional oscillation from Test B indicated that the increase of the amplitude not only enlarges the cycles but also changes the shape of the loop, especially for the large amplitudes. Test B also discovered that if the amplitude from $w/U_r = \tan(0.5^\circ)$ is doubled to $\tan(1^\circ)$ while other conditions remain unchanged ($\overline{\alpha} = 3.5^\circ$ and $\overline{U}_R = 10$), the displacement responses can be ten times greater than in the original case, and the variation of instantaneous angle of attack can drive the deck sectional model in and out of the instability threshold.

Figure 15.39 shows the frequency responses of global displacements from Test B excited by harmonic turbulence with amplitude $w/U_r = \tan(3^\circ)$ at a frequency of 0.37 Hz. A distinct super-harmonic resonance phenomenon can be observed: the harmonic excitations at 0.37 Hz generated oscillations at not only 0.37 Hz but also 2*0.37 Hz and 3∗0.37 Hz. This is a typical characteristic of non-linear vibration, which can be also observed for aerodynamic forces from Test B and for both motions and forces from Test A.

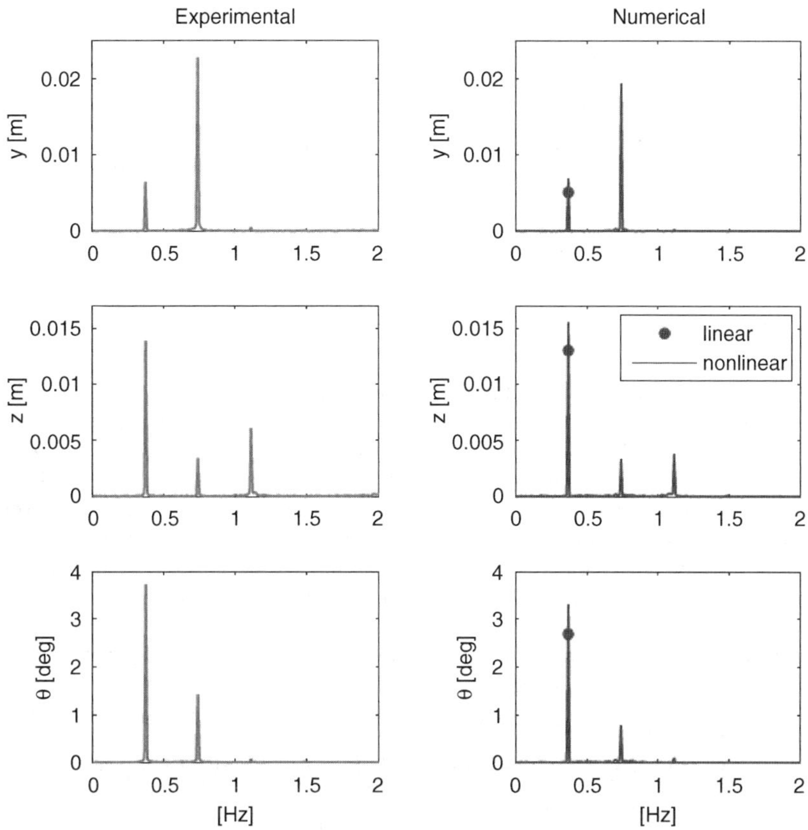

Figure 15.39 Global displacements with super harmonics (Figure courtesy of Prof. Giorgio Diana).

15.8.4 Hysteretic Models for Non-Linear Aerodynamic Forces

To describe the hysteretic loops between the aerodynamic forces and motion angle mathematically, both parametric and non-parametric approaches have been employed by Diana *et al.* [52].

The rheological model is a non-parametric model which utilizes the analogy between aerodynamic non-linear system and mechanical non-linear system, aiming to establish the model in terms of familiar non-linear mechanical components such as springs, dampers, bump stops and others. In the rheological model, four blocks are defined and used. These are shown as blocks A, C, D and E in Figures 15.40a to 15.40d.

Block A is a non-linear spring whose stiffness is non-linear. It can be used to reproduce the non-linear trend of the hysteretic loops between the forces F_i and ψ, which can be regarded as the relationship of mean forces with ψ.

Block B connects a series of Kevin-Voight elements (a spring and a damper in parallel) in parallel with a Coulomb friction skate at each extremity. Each Kevin-Voight element can generate a hysteretic loop in relation to both ψ and $\dot{\psi}$, and thus this block can generate the sum of several hysteretic loops. Availability of each Kevin-Voight element depends on the adhesion condition of a friction skate controlled by threshold of reduced velocities, which models the effects such that, as the reduced velocity increases, the hysteretic loops are dropped gradually by halting the corresponding Kevin-Voight elements.

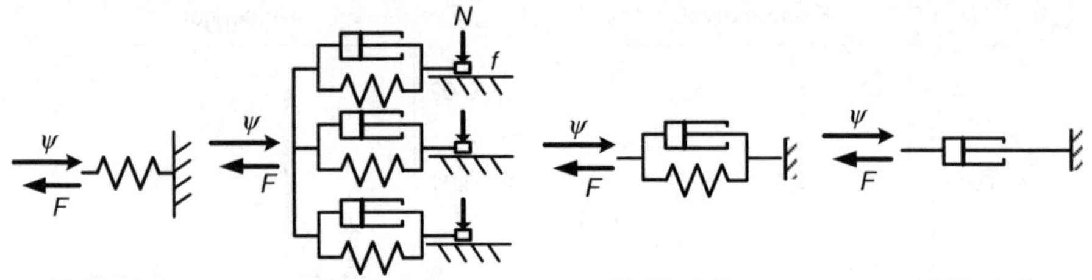

Figure 15.40 Rheological blocks.

Block D is a Kevin-Voight element with a bump stop, which is used to model a hysteretic loop that is effective only at a specific range of ψ. Block E is a single damper that can reproduce part of the force related to $\dot{\psi}$ only, which is also effective only in one direction of the motion.

Identification of the model should focus on the hysteretic loops within the range of low reduced velocity and on the deformation of the hysteretic loops with the changes of the reduced velocity. A simple model including only Block A and Block C without skates can represent the behavior at the high reduced velocity satisfactorily. However, to consider the dependence of the loop shape on the reduced velocity and the complicated shape, a full model should be used, which may include all of the four blocks.

The polynomial model in the time domain describing the hysteretic relationship between force coefficients $C_i\,(i = L, D, M)$ and ψ can be expressed as [48]:

$$C_i = C_i^{ST}(\psi) + \beta_1 + \beta_2\psi + \beta_3\dot{\psi} + \beta_4\psi^2 + \beta_5\psi\dot{\psi} + \beta_6\psi^3 + \beta_7\psi^2\dot{\psi} \tag{15.96}$$

where:

C^{ST} is the static aerodynamic force coefficient;

β_1 to β_7 are the parameters that can be identified by a weighted least square curve fitting procedure through minimizing errors of Equation 15.96 with the experimental hysteretic loops.

Note that the fitting process should take into account the hysteretic loops at various reduced velocities, and that higher weights are given to the loops associated with low reduced velocities. Figure 15.41 compares the loops obtained experimentally from Test A and numerically by Equation 15.96 at different low reduced velocities, indicating the capability of Equation 15.96 as a good representation of the non-linear aerodynamic forces. This model was applied to determine flutter critical wind speed of a full scale long-span bridge, yielding critical wind speed values lower than the linear flutter analysis procedure [48].

The hybrid model jointly uses the polynomial model as shown in Equation 15.96 with a rheological element as shown in Figure 15.42.

The rheological element is active under the conditions [49]:

1. $\psi > \psi_{\text{threshold}}$ when $\dot{\psi}$ becomes negative; and
2. $\dot{\psi} < 0$.

The rheological element is introduced for simulating the relationship near the instability on-set with the amplitude of the angle of attack crossing in and out of the instability threshold. The model is

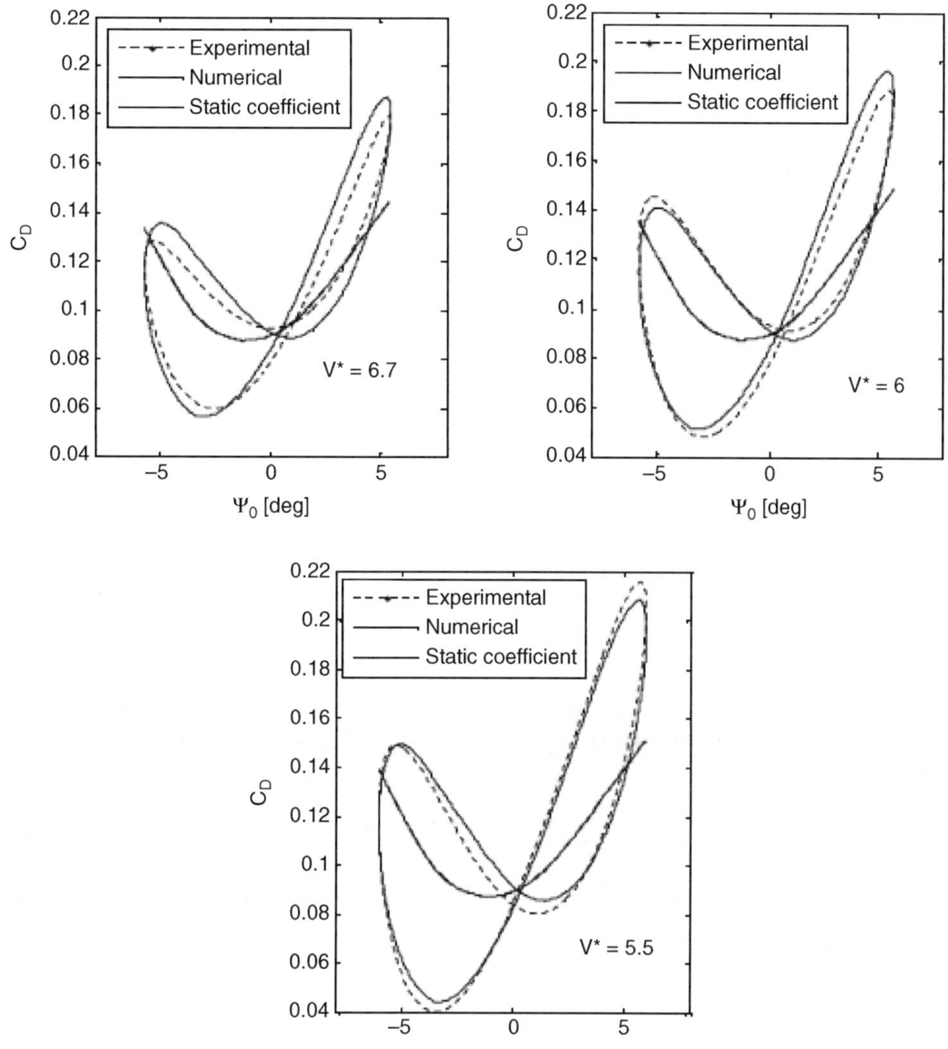

Figure 15.41 Polynomial model (Figure courtesy of Prof. Giorgio Diana).

validated by Figure 15.41. The results shown in Figure 15.41 indicate that the displacements of a sectional model of a suspended bridge computed by this hybrid model can reproduce the non-linear characteristics embedded in experimental results (super harmonic resonance) that cannot be obtained using linear aerodynamic force models.

15.8.5 ANN-Based Hysteretic Model of Non-Linear Buffeting Response

This model is proposed by Wu and Kareem [41]. It is suggested that the relationship between the responses of bridge deck and the turbulence wind speeds can be represented by a general

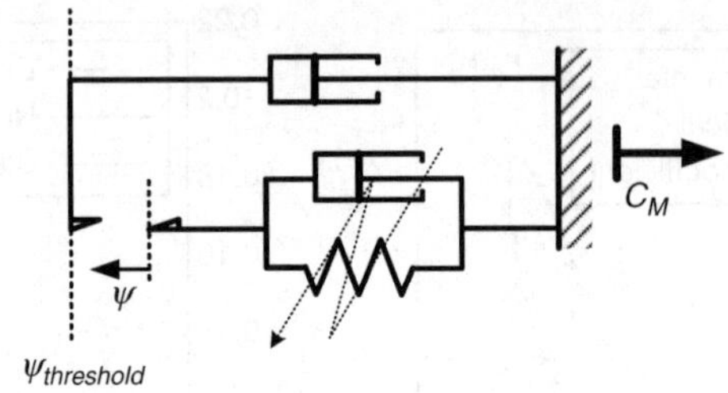

Figure 15.42 Rheological element in the hybrid model.

non-linear system:

$$\ddot{h}^n = \Re_L^n + f_L^n(h, \dot{h}, \ddot{h}, \alpha, \dot{\alpha}, \ddot{\alpha}, \overline{U}, u, w) \tag{15.97a}$$

$$\ddot{\alpha}^n = \Re_M^n + f_M^n(h, \dot{h}, \ddot{h}, \alpha, \dot{\alpha}, \ddot{\alpha}, \overline{U}, u, w) \tag{15.97b}$$

where:

f_L^n and f_M^n are the non-linear excitation force functions;
$\Re_L^n$ and $\Re_M^n$ are the restoring force functions having the forms of non-linear differential equation:

$$\dot{\Re}_L^n = \mathbb{R}_L^n(\Re_L^n, h, \dot{h}, \alpha, \dot{\alpha}),\ \dot{\Re}_M^n = \mathbb{R}_M^n(\Re_M^n, h, \dot{h}, \alpha, \dot{\alpha}) \tag{15.98}$$

where $\mathbb{R}$ denotes unknown non-linear functions.

Using the first-order difference equation between the time step t_{n+1} and t_n, the acceleration responses can be expressed as:

$$\ddot{h}^{n+1} = \mathbb{Q}_L(h^n, \dot{h}^n, \ddot{h}^n, \alpha^n, \dot{\alpha}^n, \ddot{\alpha}^n, \overline{U}^n, u^n, w^n, \overline{U}^{n+1}, u^{n+1}, w^{n+1}) \tag{15.99a}$$

$$\ddot{\alpha}^{n+1} = \mathbb{Q}_M(h^n, \dot{h}^n, \ddot{h}^n, \alpha^n, \dot{\alpha}^n, \ddot{\alpha}^n, \overline{U}^n, u^n, w^n, \overline{U}^{n+1}, u^{n+1}, w^{n+1}) \tag{15.99b}$$

where $\mathbb{Q}$ represents unknown non-linear functions.

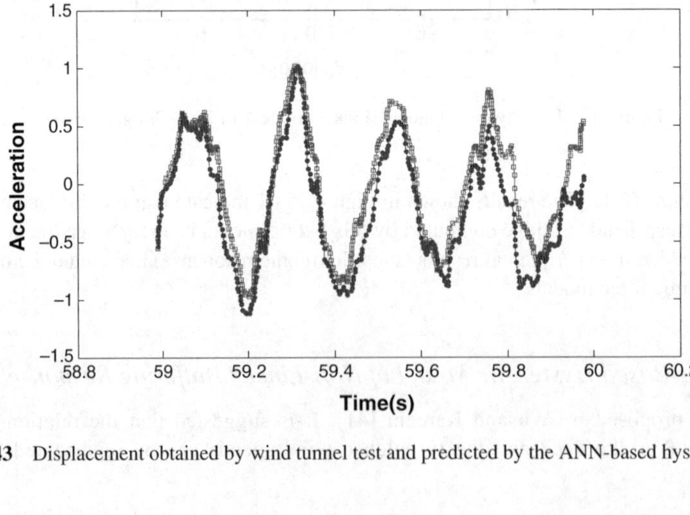

Figure 15.43 Displacement obtained by wind tunnel test and predicted by the ANN-based hysteretic model.

The ANN is suitable to represent those non-linear functions, since the inputs and outputs can be obtained to reveal the complex relationship with different weights of each input component. The cellular automata (CA) based-ANN is then used to model Q_L and Q_M simultaneously, since the vertical and torsional displacements are coupled. The CA scheme is used to find out the unnecessary connections between the input and hidden layers of the ANN, yielding that the torsional acceleration and vertical turbulence are mostly connected with neurons in the ANN. The inputs can include only the terms in Equation 15.99 or the associated higher order terms.

The ANN-based model has been used to represent the hysteretic behavior of two-dimensional responses of a bridge with single-box section, which shows that the higher-order model is appropriate for this example. Figure 15.43 shows one of the tested non-linear torsional acceleration time series, along with the one predicted by the ANN, indicating acceptable accuracy.

15.9 Notations

$\otimes$	Kronecker tensor product
$*$	Array multiplication, i.e. $\mathbf{A}^*\mathbf{B}$ is the element-by-element product
$A(\omega,t)$	Time-frequency modulating function slowly varying with time
$\mathbf{A}_d(t)$	Time-dependent aeroelastic damping matrix
A_i^*	Aerodynamic derivatives corresponding to dynamic torsional angle
A_{l_j}	Cosine Fourier coefficients of component $X_j(t)$
$\mathbf{A}_{\mathbf{Q}_b}(\omega, t)$	Cholesky decomposition evolutionary power spectral density matrix of generalized buffeting force
$\mathbf{A}_s(t)$	Time-dependent aeroelastic stiffness matrix
B	Bridge deck width
B^*	Effective bridge deck width
B_{l_j}	Sinusoid Fourier coefficients of component $X_j(t)$
$c_j(t)$	The jth intrinsic mode function of empirical model decomposition
$\mathbf{C}$	$N \times N$ damping matrix
$\tilde{\mathbf{C}}(t)$	Generalized modal damping matrix
$\mathbf{C}_{\mathbf{AB}}$	Covariance matrix of vector of random Fourier coefficients of evolutionary vector process $\mathbf{X}(t)$
$\mathbf{C}_b(t)$	Time-dependent aerodynamic force coefficients matrix
$\mathbf{C}_{bu}(t)$	Time-dependent aerodynamic force coefficients matrix associated with longitudinal fluctuating wind speeds
$\mathbf{C}_{bw}(t)$	Time-dependent aerodynamic force coefficients matrix associated with vertical fluctuating wind speeds
C_D	Aerodynamic drag force coefficients
$\mathbf{C}_{\mathbf{F}_j\mathbf{F}_k}$	Covariance matrix of vector of random Fourier coefficients of component process $X_j(t)$
C_i	Aerodynamic force coefficients defined with respect to deck coordinate system ($i = L, D, M$)
C_i^{ST}	Static aerodynamic force coefficient ($i = L, D, M$)
C_L	Aerodynamic lift force coefficients
C_M	Aerodynamic pitching moment coefficients
C_{py}	Decay coefficients of coherence function of turbulence component $p = u, w$ with respect to the lateral coordinate y
C_{pz}	Decay coefficients of coherence function of turbulence component $p = u, w$ with respect to the vertical coordinate z
$D_b(t)$	Buffeting drag force

D^l	Low-frequency component of aerodynamic drag force along with the coordinate system defined with respect to mean wind speed $\overline{U}$
$D_m(t)$	Time-varying mean drag force
$D_{se}(t)$	Self-excited drag force
$\mathbf{e}_1$	$(2N-1)$-Dimensional vectors of all ones
$\mathbf{e}_2$	M-Dimensional vector of all ones
$\mathbf{F}$	Total aerodynamic forces
$\mathbf{F_{AB}}$	Vector of random Fourier coefficients of evolutionary vector process $\mathbf{X}(t)$
$\mathbf{F^S_{AB}}$	Conditional simulation of vector of Fourier coefficients $\mathbf{F_{AB}}$
$\mathbf{F^{S^*}_{AB}}$	Unconditional simulation of vector of Fourier coefficients $\mathbf{F_{AB}}$
$\mathbf{F}_{\mathbf{A}_j\mathbf{B}_j}$	Vector of random Fourier coefficients of component process $X_j(t)$
$\mathbf{F}_b$	Loading vectors of buffeting forces
$\mathbf{F}_b^h$	High-frequency part of buffeting forces
F_D^l	Low-frequency component of aerodynamic drag force along with the coordinate system defined with respect to the instantaneous relative velocity
$F_i(t)$	Aerodynamic forces defined with respect to deck coordinate system ($i = L, D, M$)
$\mathbf{F}^l$	Low-frequency part of total aerodynamic forces
F_L^l	Low-frequency component of aerodynamic lift force along with the coordinate system defined with respect to the instantaneous relative velocity
$\mathbf{F}_m$	Loading vectors of mean wind forces
F_M^l	Low-frequency component of aerodynamic pitching moment along with the coordinate system defined with respect to the instantaneous relative velocity
$\mathbf{F}_{se}$	Loading vectors of self-excited forces
$\mathbf{F}_{se}^h$	High-frequency part of self-excited forces
$F_{X_m}(x,T)$	Cumulative distribution function of extreme value of $X(t)$ over specified time interval $[0, T]$
$\mathbf{F}_\alpha^0$	Vector of random Fourier coefficients of known fluctuating wind speed components $\mathbf{X}_\alpha^0(t)$
$\mathbf{F}_\beta^{S^*}$	Unconditional simulation of vector of Fourier coefficients corresponding to unknown components $\mathbf{F}_\beta^S$
$\mathbf{F}_\alpha^{S^*}$	Unconditional simulation of vector of Fourier coefficients corresponding to known components $\mathbf{F}_\alpha^0$
$\mathbf{F}_\beta^S$	Vector of random Fourier coefficients of conditional simulation of fluctuating wind speed components $\mathbf{X}_\beta^S(t)$
$\mathbf{f^S_{AB}}$	Conditionally simulated Fourier coefficients of $\mathbf{x}^S(t)$
f_L^n and f_M^n	Non-linear excitation force functions
$\mathbf{f}_\alpha^0$	Fourier coefficients of known fluctuating wind speed time series $\mathbf{x}_\alpha^0(t)$
$\mathbf{f}_\alpha^{S^*}$	Unconditionally simulated Fourier coefficients corresponding to known components $\mathbf{X}_\alpha^0(t)$
$\mathbf{f}_\beta^S$	Fourier coefficients of conditional simulated fluctuating wind speed time histories $\mathbf{x}_\beta^S(t)$
$\mathbf{f}_\beta^{S^*}$	Unconditionally simulated Fourier coefficients corresponding to unknown components $\mathbf{X}_\beta^S(t)$
$G(t_1)$	Gust factor defined as the maximum ratio of t_1 time-varying mean wind speed to the corresponding part of the hourly time-varying mean wind speed
H_i^*	Aerodynamic derivatives corresponding to vertical dynamic displacement
$h_k(t)$	Vertical dynamic displacement response of bridge deck unit
h^l	Low-frequency component of dynamic vertical displacement of bridge deck
$\ddot{h}n, \ddot{\alpha}n$	Vertical and torsional motion acceleration of bridge deck at time instant t_n

I_{Lj}	Aerodynamic impulse functions between the lift force and bridge deck motion ($j = h$, p and α) given by the rational approximation of flutter derivatives, or between lift force and turbulence components ($j = u$ and w) given by rational approximation of admittance functions
I_u	Turbulence intensity of longitudinal fluctuating wind speed
$I_u(t)$	Longitudinal time-varying turbulence intensity
$I_w(t)$	Vertical time-varying turbulence intensity
J_{Lj}	Impulse response function given by rational approximation of jointly acceptance functions ($j = u$ and w)
$\mathbf{K}$	$N \times N$ stiffness matrix
$\tilde{\mathbf{K}}(t)$	Generalized modal stiffness matrix
l	Index of frequency series for random Fourier coefficients A_{l_j} and B_{l_j} in Section 15.7
L^l	Low-frequency component of aerodynamic lift force along with the coordinate system defined with respect to mean wind speed $\overline{U}$
$L_b(t)$	Buffeting lift force
$L_b^h(t)$	High-frequency component of buffeting lift force
l_k	Length of the kth bridge deck element
l'	Index of frequency series used to compute integration of evolutionary power spectral density function to obtain time-varying correlation function and to obtain $\mathbf{F}_{\mathbf{AB}}^{S^*}$
$L_m(t)$	Time-varying mean lift force
$L_{se}(t)$	Self-excited lift force
$L_{se}^h(t)$	High-frequency component of self-excited lift force
$L_u(t)$	Longitudinal time-varying integral length scale
$L_w(t)$	Vertical time-varying integral length scale
$\mathbf{M}$	$N \times N$ mass matrix
M	Number of elements of bridge deck in Section 15.3; number of components of evolutionary vector process $\mathbf{X}(t)$ in Section 15.6
$M_b(t)$	Buffeting pitching moment
M^l	Low-frequency component of aerodynamic pitching moment with the coordinate system defined with respect to mean wind speed $\overline{U}$
$M_m(t)$	Time-varying mean pitching moment
$M_{se}(t)$	Self-excited pitching moment
$m(t)$	Time-varying mean value of $X(t)$
N	Total number of degree of freedoms in Section 15.4
N'	Number of frequency intervals used to compute integration of evolutionary power spectral density function to obtain time-varying correlation function and to obtain $\mathbf{F}_{\mathbf{AB}}^{S^*}$
N_S	Number of structural modes
N_T	Number of time intervals of fluctuating wind speed time histories
$N_X(x,t)$	Mean number of crossings over level $x(x > 0)$ of $X(t)$ over specified time interval $[0, T]$
$\mathbf{P}(\omega)$	Cholesky decomposition of the $M \times M$-dimensional coherence matrix $\boldsymbol{\gamma}(\omega)$:
P_i*	Aerodynamic derivatives corresponding to lateral dynamic displacement
$p_k(t)$	Lateral dynamic displacement response of bridge deck unit
p^l	Low-frequency component of dynamic lateral displacement of bridge deck
$p(u)$	Probability density function of longitudinal fluctuating wind speed
$\mathbf{Q}$	Lower triangular matrix given by Cholesky decomposition of $\mathbf{C_{AB}}$
$\mathbf{Q}_b(t)$	Generalized buffeting force vector
$\tilde{\mathbf{Q}}_{b,j}(\omega, t)$	Pseudo excitation vector
$\mathbf{q}(t)$	Generalized modal coordinate vector
$q_X(t)$	Time-dependent band-width factor of $X(t)$
$r(t)_N$	Final residue of empirical model decomposition after N intrinsic mode functions

$\mathbf{R}(t_1,t_2)$	Time-varying correlation matrix of $\mathbf{X}(t)$
$R_{jj}(t_1,t_2)$	Auto time-varying correlation of $X_j(t)$
$R_{jk}(t_1,t_2)$	Cross time-varying correlation between $X_j(t)$ and $X_k(t)$
$\Re_L^n$ and $\Re_M^n$	Restoring force functions having the forms of non-linear differential equation
$\mathbb{R},\mathbb{Q}$	Unknown non-linear functions to be modeled by artificial neutral network
$\mathbf{R}^U(t_1,t_2)$	Time-varying correlation matrix of unconditional simulation $\mathbf{X}^U(t)$
$\mathbf{R}_{ab}(t_1,t_2)$	Target time-varying correlation matrix of $\mathbf{X}_\beta^S(t)$
$\mathbf{R}_{\alpha\beta}^S(t_1,t_2)$	Time-varying cross correlation matrix between conditional simulation of unknown components $\mathbf{X}_\beta^S(t)$ and known components $\mathbf{X}_\alpha^0(t)$
$\mathbf{R}_{\beta\beta}(t_1,t_2)$	Target time-varying correlation matrix between $\mathbf{X}_\beta^S(t)$ and $\mathbf{X}_\alpha^0(t)$
$\mathbf{R}_{\beta\beta}^S(t_1,t_2)$	Time-varying correlation matrix of conditional simulation of unknown components $\mathbf{X}_\beta^S(t)$
$\mathbf{S}(\omega,t)$	Evolutionary power density matrix of $\mathbf{X}(t)$
$S_{pp,ij}(\omega,t)$	Cross evolutionary spectral density function between turbulence components $p=u$ or w at point i and point j
$\mathbf{S}_{\mathbf{Q}^b\,\mathbf{Q}^b}(\omega,t)$	Evolutionary power spectral density matrix of generalized buffeting force
$S_{uu}(\omega,t)$	Auto evolutionary power spectral density function of longitudinal fluctuating wind speed $u(t)$
$S_{ww}(\omega,t)$	Auto evolutionary power spectral density function of vertical fluctuating wind speed $w(t)$
$\mathbf{S}_{\mathbf{xx}}(\omega,t)$	Evolutionary power spectral density matrix of fluctuating buffeting response vector
$\mathbf{S}_{\boldsymbol{\Theta\Theta}}(\omega,t)$	Evolutionary power spectral density matrix of non-stationary fluctuating wind speed vector
T	Time interval for computing short-time averaged time-varying mean wind speed
$\mathbf{T}_s(t)$	$N\times M$ coordinate transformation matrix
T_U	Time interval for computing short-time averaged time-varying standard deviation of fluctuating wind speed
t_r	Time series corresponding to simulated fluctuating wind speed time histories
U_r	Instantaneous relative velocity between deck and wind
$\mathbf{U}^S$	Vector of uncorrelated standard Gaussian random variables of the size $M(2N-1)\times 1$
$U(t)$	Total longitudinal wind speed time history
$\overline{\mathbf{U}}(t)$	Vector of time-varying mean wind speeds
$\mathbf{U}(t)$	Evolutionary process of longitudinal fluctuating wind speeds in Section 15.6
$\mathbf{U}_{l'}^1, \mathbf{U}_{l'}^2$	$(2N-1)$-dimensional vectors consisting of independent standard Gaussian random variables
$\overline{U}$	Constant mean wind speed
$\overline{U}_0(z)$	Mean value of the time-varying mean wind speed $\overline{U}(t)$
$\overline{U}_{1/3600}(t)$	Time-varying mean wind speed with frequency contents less than 1/3600 Hz
$\overline{U}_n(t)$	Time-varying mean wind speed with frequency contents less than n
$\overline{U}_R$	Reduced mean wind speed
$\overline{U}_R(t)$	Reduced mean wind speed that is time-varying corresponding to time-varying mean wind speed
$\overline{U}(t)$	Time-varying mean wind speed
$u(t)$	Fluctuating wind component in the longitudinal wind direction
$\mathbf{u}(t)$	Vector of longitudinal non-stationary fluctuating wind speeds
$\mathbf{u}^{0(l)}(t)$	ith sample of measured (reference) time histories of longitudinal fluctuating wind speed, which is generated by unconditional simulation procedure
u^l, u^h	Low- and high-frequency components of longitudinal fluctuating wind speed
$\mathbf{u}_{l'}^1, \mathbf{u}_{l'}^2$	Sample realization of $(2N-1)$-dimensional vectors filling with independent standard Gaussian random numbers

$\mathbf{u}^{S(I)}(t)$	ith sample of conditionally simulated time histories of longitudinal fluctuating wind speeds
$\mathbf{u}^{US(I)}(t)$	ith sample of unconditionally simulated time histories of longitudinal fluctuating wind speed
$V(t)$	Envelope of $X(t)$
$W(t)$	Total vertical wind speed time history
$\mathbf{W}(t)$	Evolutionary process of vertical fluctuating wind speeds in Section 15.6
$w(t)$	Fluctuating wind component in the longitudinal wind direction
$\mathbf{w}(t)$	Vector of vertical non-stationary fluctuating wind speeds
$\mathbf{w}^{0(I)}(t)$	ith sample of measured (reference) time histories of vertical fluctuating wind speed, which is generated by unconditional simulation procedure
w^l, w^h	Low- and high-frequency components of vertical fluctuating wind speed
$\mathbf{w}^{S(I)}(t)$	ith sample of conditionally simulated time histories of vertical fluctuating wind speeds
$\mathbf{w}^{US(I)}(t)$	ith sample of unconditionally simulated time histories of vertical fluctuating wind speed
$X(t)$	A Gaussian evolutionary process in Section 15.4 with a known time-varying mean $m(t)$ and single-sided EPSD function $S_{XX}(\omega,t)$
$\mathbf{X}(t)$	N-dimensional buffeting displacement response vector of the whole bridge in the global coordinate system in Section 15.3; zero-mean multivariate real Gaussian oscillatory vector process with M components in Section 15.6
$\dot{X}(t)$	Derivative of $X(t)$ with respect to time
$\dot{\mathbf{X}}(t)$	Buffeting velocity response vector
$\ddot{\mathbf{X}}(t)$	Buffeting acceleration response vector
$\overline{\mathbf{X}}(t)$	Time-varying mean wind response vector
$X_j(t)$	jth component of $\mathbf{X}(t)$
X_m	Extreme value of $X(t)$ over specified time interval $[0, T]$
$\mathbf{X}^S(t)$	Evolutionary processes of conditional simulation of fluctuating wind speed components at all of M measured locations
$\mathbf{X}^U(t)$	Unconditional simulation of $\mathbf{X}(t)$
$\mathbf{X}^0_\alpha(t)$	Evolutionary processes of known fluctuating wind speed components at n measured locations ($\alpha = 1 \ldots n$)
$\mathbf{X}^S_\beta(t)$	Evolutionary processes of conditional simulation of unknown fluctuating wind speed components at m unmeasured locations ($\beta = n+1 \ldots n+m$)
$\mathbf{x}(t)$	Fluctuating buffeting response vector in Section 15.4;
$\mathbf{x}^S(t)$	Conditionally simulated fluctuating wind speed time histories at all of M measured locations
$\mathbf{x}^{S(I)}(t_r)$	ith sample time histories of whole set of conditionally simulated fluctuating wind speed components
$\mathbf{x}^0_\alpha(t)$	Known fluctuating wind speed time histories at n measured locations ($\alpha = 1 \ldots n$)
$\mathbf{x}^{0(I)}_\alpha(t_r)$	ith sample time histories of known fluctuating wind speed components
$\mathbf{x}^S_\beta(t)$	Conditionally simulated fluctuating wind speed time histories at m unmeasured locations ($\beta = n+1 \ldots n+m$)
$\mathbf{x}^{S(I)}_\beta(t_r)$	ith sample time histories of conditionally simulated unknown fluctuating wind speed components
$\mathbf{y}_j(\omega,t)$	Pseudo response vector
z	Height above the ground
$\overline{\alpha}(t)$	Instantaneous mean wind attack angle
$\overline{\alpha}_0(t)$	Attack angle of time-varying mean wind

$\overline{\alpha}$	Mean value of torsional angle of bridge deck
$\alpha_e(t)$	Instantaneous angle of attack
$\alpha_e^l(t), \alpha_e^h(t)$	Low- and high-frequency components of instantaneous angle of attack
$\alpha_k(t)$	Dynamic torsional response of bridge deck unit
α^l	Low-frequency component of dynamic torsional angle of bridge deck
α_s	Static wind angle of attack
$\overline{\alpha}_S(t)$	Torsional response of the bridge deck due to the time-varying mean wind forces and moment
β_1 to β_7	Parameters in polynomial hysteretic aerodynamic force model identified by weighted least square curve fitting procedure
$\mathfrak{P}(\omega_p, t)$	Cholesky decomposition of evolutionary power spectral density matrix $\mathbf{S}(\omega,t)$
$\mathfrak{P}_j(\omega_p, t)$	jth column of the matrix $\mathfrak{P}(\omega_p, t)$
Δt	Time interval of simulated fluctuating wind speed time histories
$\Delta\omega'$	Frequency interval used to compute integration of evolutionary power spectral density function to obtain time-varying correlation function and to obtain $\mathbf{F}_{\mathbf{AB}}^{S^*}$
$\boldsymbol{\gamma}(\omega)$	Coherence function matrix of evolutionary vector process $\mathbf{X}(t)$
$\gamma_{pp,ij}(\omega)$	Coherence function of between turbulence components $p = u$ or w at point i and point j
$\xi(\omega)$	Zero-mean Gaussian orthogonal increment process
$\xi^{\mathrm{P}}(t)$	Revised crossing level with consideration of the effects of both time-dependent mean value and its derivatives in Possion approximation
$\xi^{\mathrm{V}}(t)$	Revised crossing level with consideration of the effects of both time-dependent mean value and its derivatives in Vanmarcke approximation
$\mu(\omega)\mathrm{d}\omega$	Variance of the increment process.
μ_{X_m}	Mean value of extreme value of $X(t)$ over specified time interval $[0, T]$
ψ	Instantaneous angle of attack angle
$\mathbf{\Phi}$	Modal matrix with only the first N_S order of structural modes
$\Phi(x)$	Standard Gaussian cumulative distribution function
$\phi(x)$	Standard Gaussian probabilistic density function
$\phi^l(t)$	Low-frequency part of the angle of incidence of the relative velocity
ψ_{pj}	Random phase angle distributed uniformly over $[0, 2\pi]$
ψ_w	Instantaneous angle of attack angle induced by actively generated turbulence
ψ_α	Instantaneous angle of attack angle induced by forced torsional motions
ρ	Air density
$\rho_{X\dot{X}}(t)$	Time-dependent correlation coefficient $\rho_{X\dot{X}}(t)$ between $X(t)$ and
σ_u	Standard deviation of longitudinal fluctuating wind speed
$\sigma_u(t)$	Time-varying standard deviation of longitudinal fluctuating wind speed $u(t)$
$\sigma_w(t)$	Time-varying standard deviation of vertical fluctuating wind speed $w(t)$
σ_{X_m}	Standard deviation of extreme value of $X(t)$ over specified time interval $[0, T]$
$\mathbf{\Xi}$	Unit matrix
$\mathbf{\Theta}(t)$	Vector of non-stationary fluctuating wind speeds
ω_l	Frequency series for random Fourier coefficients A_{l_j} and B_{l_j}
$\omega_{l'}$	Frequency series used to compute integration of evolutionary power spectral density function to obtain time-varying correlation function and to obtain $\mathbf{F}_{\mathbf{AB}}^{S^*}$
$\eta_0(t)$	Time-varying function of mean wind speed independent of the height above the ground
$\eta_X(x,t)$	Unconditional transient crossing rate of $X(t)$ over level $x (x > 0)$
$\eta_X^+(x, t)$	Up-crossing rate of $X(t)$ over level x
$\eta_X^+(x, -m; t)$	Up-crossing rate of $X(t)$ (with time-varying mean value $m(t)$) over level x
$\eta_X^-(-x, t)$	Down-crossing rate of $X(t)$ over level $-x$
$\eta_X^-(-x, m; t)$	Down-crossing rate of $X(t)$ (with time-varying mean value $-m(t)$) over level $-x$

References

1. Kareem, A. (2009) The changing dynamics of aerodynamics: New frontiers. *APCWE-VII,* Taipei.
2. Chen, L. and Letchford, C.W. (2007) Numerical simulation of extreme winds from thunderstorm downbursts. *Journal of Wind Engineering and Industrial Aerodynamics*, **95**(9–11), 977–990.
3. Huang, G. and Chen, X. (2009) Wavelets-based estimation of multivariate evolutionary spectra and its application to nonstationary downburst winds. *Engineering Structures*, **31**(4), 976–989.
4. Chen, L. (2006) Vector time-varying autoregressive (TVAR) models and their application to downburst wind speeds. Ph. D. Thesis, Texas Tech University, Texas, United States.
5. Wang, L. (2007) Stochastic modelling and simulation of transient events. Ph.D. Thesis, University of Notre Dame, Notre Dame, Indiana, United States.
6. Chen, L. and Letchford, C.W. (2004) A deterministic-stochastic hybrid model of downbursts and its impact on a cantilevered structure. *Engineering Structures*, **26**(5), 619–629.
7. Chen, X. (2008) Analysis of alongwind tall building response to transient nonstationary winds. *Journal of Structural Engineering, ASCE*, **134**(5), 782–791.
8. Kwon, D.K. and Kareem, A. (2009) Gust-front factor: A new framework for wind load effects on structures. *Journal of Structural Engineering, ASCE*, **135**(6), 717–732.
9. Xu, Y.L. and Chen, J. (2004) Characterizing nonstationary wind speed using empirical mode decomposition. *Journal of Structural Engineering, ASCE*, **130**(6), 912–920.
10. Wang, L. and Kareem, A. (2004) Modeling of non-stationary winds in gust-fronts. CD-ROM, Proceeding of 9th ASCE Joint Specialty Conference on Probabilistic Mechanics and Structural Reliability.
11. Hu, L., Xu, Y. L., Huang, W.F. and Liu, H. J. (2011) Typhoon-induced nonstationary buffeting response of long-span bridges. Proceedings of 14th Asia Pacific Vibration Conference, Hong Kong, December 5–8, 173–182.
12. Bendat, J.S. and Piersol, A.G. (1986) *Random Data: Analysis and Measurement Procedures*, 3rd edn, John Wiley and Sons, New York.
13. Priestley, M.B. (1988) *Non-linear and Nonstationary Time Series Analysis*, Academic Press Limited, London.
14. Huang, N.E., Shen, Z., Long, S.R. *et al.* (1998) The empirical mode decomposition and the Hilbert spectrum for nonlinear and non-stationary time series analysis. *Proceedings of the Royal Society of London Series A-Mathematical Physical and Engineering Sciences*, **454**(1971), 903–995.
15. Huang, N.E., Shen, Z., and Long, S.R. (1999) A new view of nonlinear water waves: The Hilbert spectrum. *Annual Review of Fluid Mechanics*, **31**,417–457.
16. Xu, Y.L., Huang, W.F. and Liu, H.J. (2011) Simulation of directional typhoon wind speeds and profiles over complex terrain. Proceedings of the 8th International Advanced School on Structural Wind Engineering, Hong Kong, November 14–18, 34–51.
17. Priestley, M.B. (1965) Evolutionary spectra and non-stationary processes. *Journal of the Royal Statistical Society Series B-Statistical Methodology*, **27**(2), 204–237.
18. Mélard, G. and Schutter, A.H.D. (1989) Contributions to evolutionary spectral theory. *Journal of Time Series Analysis*, **10**(1), 41–63.
19. Chen, J., Hui, M.C.H., and Xu, Y.L. (2007) A comparative study of stationary and non-stationary wind models using field measurements. *Boundary-Layer Meteorology*, **122**(1), 105–121.
20. Priestley, M.B. (1966) Design relations for non-stationary processes. *Journal of the Royal Statistical Society Series B-Statistical Methodology*, **28**(1), 228–240.
21. Lin, J.H. (2004) *Pseudo Excitation Method in Random Vibration*, Science Press, Beijing (in Chinese).
22. Michaelov, G., Lutes, L.D., and Sarkani, S. (2001) Extreme value of response to nonstationary excitation. *Journal of Engineering Mechanics*, **127**(4), 352–363.
23. Corotis, R.B., Vanmarcke, E.H., and Cornell, C.A. (1972) First passage of nonstationary random processes. *Journal of Engineering Mechanics*, **98**(EM2) 401–414.
24. He, J. (2009) A reliability approximation for structures subjected to non-stationary random excitation. *Structural Safety*, **31**(4), 268–274.
25. Howell, L.J. and Lin, Y.K. (1971) Response of flight vehicles to nonstationary atmospheric turbulence. *AIAA Journal*, **9**(11), 2201–2207.
26. Roberts, J.B. (1976) First passage time for the envelope of a randomly excited linear oscillator. *Journal of Sound and Vibration*, **46**(1), 1–14.
27. Solomos, G.P. and Spanos, P.T.D. (1983) Structural reliability under evolutionary seismic excitation. *International Journal of Soil Dynamics and Earthquake Engineering*, **2**(2), 110–114.
28. Solomos, G.P. and Spanos, P.T.D. (1984) Probability of response to evolutionary process. *Journal of Applied Mechanics, Transactions ASME*, **51**(4), 907–912.
29. Spanos, P.T.D. and Lutes, L.D. (1980) Oscillator response to nonstationary excitation. *Journal of Engineering Mechanics*, **106**(2), 213–224.
30. Yang, J.-N. (1972) Nonstationary envelope process and first excursion probability. *Journal of Structural Mechanics*, **1**(2), 231–248.
31. Shinozuka, M. and Yao, J.T.P. (1967) On the two-sided time-dependent barrier problem. *Journal of Sound and Vibration*, **6**(1), 98–104.

32. Leadbetter, M.R. and Cryer, J.D. (1965) Curve crossing by normal processes and reliability implications. *SIAM Review*, **7**(2), 241–250.
33. Leadbetter, M.R. and Cryer, J.D. (1965) On the mean number of curve crossings by non-stationary normal processes. *The Annals of Mathematical Statistics*, **36**(2), 509–516.
34. Michaelov, G., Sarkani, S., and Lutes, L.D. (1999) Spectral characteristics of nonstationary random processes: A critical review. *Structural Safety*, **21**(3), 223–244.
35. Deodatis, G. (1996) Non-stationary stochastic vector processes: Seismic ground motion applications. *Probabilistic Engineering Mechanics*, **11**(3), 149–168.
36. Li, Y. and Kareem, A. (1997) Simulation of multivariate nonstationary random processes: Hybrid DFT and digital filtering approach. *Journal of Engineering Mechanics, ASCE*, **123**(12), 1302–1310.
37. Emresoy, M.K. and El-Jaroudi, A. (1999) Evolutionary spectrum estimation by positivity constrained deconvolution. *IEEE Transactions on Signal Processing*, **47**(3), 889–893.
38. Yamazaki, F. and Shinozuka, M. (1990) Simulation of stochastic fields by statistical preconditioning. *Journal of Engineering Mechanics*, **116**(2), 268–287.
39. Burden, R.L. and Faires, J.D. (2010) *Numerical Analysis*, 9th edn, Brooks/Cole/Cengage Learning, Mass, Boston, United States.
40. Di Paola, M. (1998) Digital simulation of wind field velocity. *Journal of Wind Engineering and Industrial Aerodynamics*, **74–76**(1), 91–109.
41. Wu, T. and Kareem, A. (2011) Modeling hysteretic nonlinear behavior of bridge aerodynamics via cellular automata nested neural network. *Journal of Wind Engineering and Industrial Aerodynamics*, **99**(4), 378–388.
42. Chen, X., Matsumoto, M., and Kareem, A. (2000) Time domain flutter and buffeting response analysis of bridges. *Journal of Engineering Mechanics, ASCE*, **126**(1), 7–16.
43. Gurley, K.R., Tognarelli, M.A., and Kareem, A. (1997) Analysis and simulation tools for wind engineering. *Probabilistic Engineering Mechanics*, **12**(1), 9–31.
44. Chen, X. and Kareem, A. (2001) Nonlinear response analysis of long-span bridges under turbulent winds. *Journal of Wind Engineering and Industrial Aerodynamics*, **89**(14–15), 1335–1350.
45. Chen, X. and Kareem, A. (2002) Advances in modeling of aerodynamic forces on bridge decks. *Journal of Engineering Mechanics, ASCE*, **128**(11), 1193–1205.
46. Chen, X. and Kareem, A. (2003) Aeroelastic analysis of bridges: Effects of turbulence and aerodynamic nonlinearities. *Journal of Engineering Mechanics, ASCE*, **129**(8), 885–895.
47. Diana, G., Cheli, F., Zasso, A., and Bocciolone, M. (1999) Suspension bridge response to turbulent wind: Comparison of new numerical simulation method results with full scale data, in *Wind Engineering into the 21 Century*, Balkema, Rotterdam, pp. 871–878.
48. Diana, G., Resta, F., Rocchi, D., and Argentini, T. (2008) Aerodynamic hysteresis: Wind tunnel tests and numerical implementation of a fully non linear model for the bridge aeroelastic forces. Proceedings of the AWAS 08, Jeju, Korea, pp. 944–960.
49. Diana, G., Rocchi, D., Argentini, T., and Muggiasca, S. (2010) Aerodynamic instability of a bridge deck section model: Linear and nonlinear approach to force modeling. *Journal of Wind Engineering and Industrial Aerodynamics*, **98**(6–7), 363–374.
50. Diana, G., Resta, F., and Rocchi, D. (2007) A new approach to model the aeroelastic response of bridges in time domain by means of a rheological model. Proceedings of the ICWE 12, Cairns, Australia, pp. 207–214.
51. Diana, G., Resta, F., Zasso, A. *et al.* (2004) Forced motion and free motion aeroelastic tests on a new concept dynamometric section model of the messina suspension bridge. *Journal of Wind Engineering and Industrial Aerodynamics*, **92**(6), 441–462.
52. Diana, G., Resta, F., and Rocchi, D. (2008) A new numerical approach to reproduce bridge aerodynamic non-linearities in time domain. *Journal of Wind Engineering and Industrial Aerodynamics*, **96**(10–11), 1871–1884.

16

Epilogue: Challenges and Prospects

16.1 Challenges

The previous chapters in this book have presented the basic elements of bridge wind engineering, which include wind characteristics in atmospheric boundary layer, mean wind load and aerostatic instability, wind-induced vibration and aerodynamic instability. They have also detailed the methods and techniques currently used for investigating wind effects on long-span cable-supported bridges, namely analytical methods, wind tunnel test techniques, CFD-based computational wind engineering techniques, wind and structural health monitoring techniques. Also outlined have been recent developments in modern bridge wind engineering: wind-induced vibration of stay cables; wind-vehicle-bridge interaction; buffeting response to skew winds; multiple loading-induced fatigue analysis; wind-induced vibration control; typhoon wind field simulation; uncertainty and reliability analysis; non-stationary and non-linear buffeting response analysis.

Although a considerable amount of information has been covered by the previous chapters, there are still some important issues that have not been fully discussed, and some challenging issues to be solved in the near future.

16.1.1 Typhoon Wind Characteristics and Topography Effects

Long-span cable-supported bridges built in typhoon regions are inevitably affected by typhoon winds, and they must be designed to withstand typhoon winds during their design lives. The determination of typhoon wind speed, direction and profile, among other characteristics, in the atmospheric boundary layer (ABL) for a long-span cable-supported bridge within a given design life, is an important task.

Great efforts have been made in the past to investigate the wind characteristics of typhoons. The limited field measurement results show that, although they have similar boundary layers to monsoons and gales on their outer edges, typhoons exhibit different wind profiles near the region of strongest winds. If a long-span cable-supported bridge is located in an area with complex topography, the effect of topography on typhoon structures will make typhoon wind structures more complex. This effect is much less clearly understood than those in the well-developed boundary layers of monsoons and gales. Also, there are virtually no measurement data on how typhoon wind structures change as they cross the coast. More field measurements with advanced measurement instruments and computational

Wind Effects on Cable-Supported Bridges, First Edition. You-Lin Xu.

simulations need to be carried out to reveal wind characteristics of typhoons near their eye walls, as well as the topographic and inland effects.

Also, a few long-span cable-supported bridges have been built to cross valleys in mountain areas. Wind conditions and topography effects on wind characteristics in mountain areas need to be carefully investigated, and wind effects on long-span cable-supported bridges in mountainous areas need to be closely monitored.

16.1.2 Effects of Non-Stationary and Non-Gaussian Winds

Because turbulence in gales and monsoons is mainly mechanical, strong winds caused by gales and monsoons are often assumed to be stationary and Gaussian random processes, and bridge responses are also assumed to be stationary, with a Gaussian distribution. However, some extreme wind events, such as typhoons, downbursts and tornados, actually exhibit different characteristics from monsoons and gales. The assumption of stationary and Gaussian processes may not, therefore, be valid for those extreme wind events, because of their vortex or convective origins. One may therefore question the appropriateness of current bridge aerodynamics analysis and bridge design based on the assumption of stationarity and Gaussianity of the extreme wind events.

It has also been noted that road vehicles running over a long-span bridge will be shielded briefly from the wind by the bridge towers but, when they pass out of this shelter, they enter a sharp-edged cross-wind gust. The sharp changes in gusts cause transient aerodynamic forces on the vehicles, and these forces are non-stationary and non-Gaussian, carrying with them an obvious danger of turning the vehicles over.

Therefore, non-stationary and non-Gaussian winds and their effects on long span bridges need to be investigated. The major challenges involved in this topic include the following:

- The nature of non-stationary winds and transient winds needs to be better quantified.
- Modelling and analysis tools to capture these features need to be established.
- Aerodynamic and aeroelastic analysis methods to predict non-stationary and transient responses of long-span bridges must be developed.

16.1.3 Effects of Aerodynamic Non-Linearity

Three types of non-linearities may be experienced by long-span cable-supported bridges: geometrical, material and aerodynamic. Geometrical non-linearities are due to stay cables or suspended cables, and they may also be initiated by large deformations and axial forces of other components of the bridge as well. Material non-linearities may arise from materials of bridge components that do not follow linear constitutive relationships. The effects of both geometrical and material non-linearities can be adequately modeled and incorporated in the analysis of a long-span bridge using finite element methods.

For many bridge deck sections, even at low levels of turbulence, the effective angle of incidence due to bridge motion and incoming wind fluctuations may vary to a level such that non-linearities in the aerodynamic forces may no longer be neglected. This is particularly true for super-long-span cable-supported bridges, due to their extreme slenderness. For such bridges, the quasi-steady assumption and the strip assumption currently used in bridge aerodynamic and aeroelastic analysis needs to be revisited. The effect of non-linear relationship between wind speed and wind force, and the difference between span-wise coherence of turbulent wind speed and span-wise coherence of aerodynamic forces, must be investigated.

The non-linear response to vortex shedding also remains an area that needs attention. Many bridge structures have exhibited unsatisfactory oscillations that are largely unpredicted, while others show the opposite trend. The effect of turbulence on vortex shedding and the ensuing oscillations need to be addressed in a systematic fashion.

Furthermore, flutter analysis is currently performed on the basis of the assumption of small amplitude. Flutter is actually a self-feeding and potentially destructive vibration to a long-span cable-supported bridge, where aerodynamic forces on the bridge deck couple with its motion. If the energy input by the aerodynamic forces due to strong winds in a cycle is larger than that dissipated by the damping in the bridge structure system, the amplitude of vibration of the bridge deck will increase. This increasing vibration will then amplify the aerodynamic forces, resulting in self-excited forces and self-exciting oscillations.

The vibration amplitude of the bridge deck can build up until the collapse of the bridge, and the build up to collapse, as well as the associated aeroelastic phenomenon, is really not yet understood. Non-linear buffeting and flutter analysis is essential to explore the mechanisms behind these phenomena and this task will require a new analysis framework in the time-frequency domain, with time-frequency signal representation such as wavelet transformation. It may also require a thorough understanding of the dynamic pressure field that envelops a bridge deck and gives rise to the fundamental forces. Note that most suction pressures on a bridge deck are non-stationary and non-Gaussian random processes.

16.1.4 Wind Effects on Coupled Vehicle-Bridge Systems

Road vehicles running on a long-span cable-supported bridge are more vulnerable to crosswind gust than on ground. Road vehicles may also be briefly shielded from the wind by the bridge towers but, when they pass out of the shelter, they enter a sharp-edged crosswind gust with a real danger of them turning over. Little information is publicly available on the safety of road vehicles running on a long-span cable-supported bridge, so vehicle restrictions in high winds have tended to be somewhat *ad hoc*.

The major challenges on this topic include at least:

(a) How to determine aerodynamic forces on vehicles when they are moving on a bridge deck?
(b) What are the transient characteristics of aerodynamic forces acting on a road vehicle when the vehicle is passing through the wake of a bridge tower?
(c) How to quantify accident risk of vehicles running on a bridge?

To protect vehicles of various types from adverse wind conditions, and to ensure the bridge can remain open to traffic, the present practice is to install wind fences on the bridge. However, there is no consensus on the optimum design of wind fences in terms of accident susceptibility and risk. Also, the use of wind fences may have an adverse effect on overall bridge stability. A full understanding of wind-vehicle-bridge interactions, and an optimum solution for the functionality and safety of both the vehicles and bridge, are needed.

16.1.5 Rain-Wind-Induced Vibration of Stay Cables

Stay cables in long-span cable-stayed bridges are laterally flexible structural members with very low fundamental frequency and very little inherent damping. These features make them susceptible to rain-wind-induced vibrations of large amplitude. A significant correlation has been observed between the occurrence of these large-amplitude vibrations and that of rain combined with wind. A substantial amount of research on the subject has already been conducted by researchers around the world. It is believed that water rivulet formation and its interaction with wind flow are the root cause of rain-wind-induced vibrations. With this understanding, various surface modifications have been proposed and tested, with the aim being the disruption of water rivulet formation. Traditional measures, such as external dampers, have also been applied to many newer bridges to mitigate rain-wind-induced cable vibration.

Although Chapters 5 and 12 shed some lights on this topic, it is not sufficient. The lack of a more in-depth understanding of the underlying mechanism of rain-wind-induced stay cable vibration has made the practical and consistent application of known mitigation methods difficult. This is because a wide range of different cable lengths (and thus a wide range of frequencies) on cable-stayed bridges actually produce a practical continuum of fundamental and higher mode frequencies. Any excitation mechanism with any arbitrary frequency is likely to find one or more cables with either a fundamental or higher mode frequency sympathetic to the excitation. Therefore, it would be of great assistance to have a model with the capability to predict, for an arbitrary stay cable, the major characteristics of rain-wind-induced cable vibration.

16.1.6 Uncertainty and Reliability Analysis

Prediction of wind-induced response and instability of long-span cable-supported bridges is associated with a number of uncertainties. These stem primarily from the random nature of wind, but also from physical modeling (e.g. wind load model and structure model), the methodology used (e.g. assumptions in buffeting analysis and flutter analysis), and the measurements involved (e.g. wind tunnel tests and field measurements). As indicated in Chapter 14, it is quite difficult to define uncertainty explicitly and to evaluate it accurately for wind-excited long-span cable-supported bridges. New probabilistic-based methodologies need to be developed to quantify these uncertainties, and the stochastic approaches we use now for prediction must be refined. The data obtained from field observations and measurements will be used to validate the stochastic approaches so that the reliability-based design philosophy can be applied eventually to wind-excited long-span cable-supported bridges.

The current design of wind-excited long-span cable-supported bridges is mainly a deterministic design – that is, a factor of safety-based design. Deterministic design may lead to a greater probability of failure than reliability-based design that accounts for uncertainties. This is because the design requirements are precisely satisfied in the deterministic approach, and any variation of the parameters could potentially violate the system constraints. Reliability-based design, on the other hand, facilitates robust designs that provide the designer with a guarantee of satisfaction in the presence of a given amount of uncertainty.

Nevertheless, the materials presented in Chapter 14 on bridge structure reliability or failure probability are very preliminary and case-dependent, and no systematic framework has been established yet. To realize the reliability-based design for wind-excited long-span cable-supported bridges, contemporary methods of reliability analysis must be used, together with uncertainty analysis, to provide an accurate estimation of safety index or failure probability of the bridges.

16.1.7 Advancing Computational Wind Engineering and Wind Tunnel Test Techniques

The computational wind engineering technique can now solve simple 2-D wind engineering problems with the features of low Reynolds number and low turbulence based on commercially available computational fluid dynamics (CFD) tools. However, most bridge wind engineering problems are 3-D problems with the features of high Reynolds number and high turbulence. The current CFD tools seem not to be able to handle these problems. In the development of boundary layer wind tunnels for testing bridge structures other than using aeronautical wind tunnels, the wind engineering community may have to develop their own CFD codes to solve their unique problems. This can be done in parallel to the advance of current wind tunnel test techniques that can provide better quality data for validation of the CFD codes developed. Broadly speaking, computational wind engineering needs also to cover other computation-oriented applications, such as developments of E-codes and standards for bridge designers and managers.

16.1.8 Application of Wind and Structural Health Monitoring Technique

Wind and structural health monitoring (SHM) is cutting-edge technology, as discussed in Chapter 9. The application and further development of long-term wind and SHM technique for long-span cable-supported bridges is vital for solving the aforementioned challenging issues in bridge wind engineering. Long-term measurements of winds are essential for understanding typhoon wind characteristics and topography effects and for determining failure probability or safety index of the bridge. Field measurement results of winds, wind pressures and wind-induced responses of the bridge can facilitate the refinement of current stochastic approaches and the validation of new methods that can account for non-stationarity, non-Gaussianity and non-linearity. Field observations of wind-vehicle-bridge interaction and rain-wind-induced stay cable vibrations will help finding excitation mechanisms behind them and developing corresponding mathematical models.

Field measurement results are also needed to validate the results obtained from either wind tunnel tests or CFD. Since this is long-term monitoring, SHM is a vital tool for promoting the sustainability of wind-excited long-span cable-supported bridges.

16.2 Prospects

The history and latest development of bridge wind engineering both demonstrate that any progress made in this subject stems from the synergy between research and practice. Mutual respect and effectively coordinated collaboration among professional engineers, academic researchers, government agencies, contractors, managers and owners are essential for advancing the field. All sectors will benefit from advanced knowledge for improving bridge structural reliability and operational efficiency, while reducing life-cycle maintenance and operating costs.

With the advance of bridge wind engineering, the aforementioned challenging issues will be eventually solved. This process should not be too long, considering that there is strong motivation from the need to construct super long-span cable-supported bridges to cross straits around the world. In addition to functionality and safety requirements, the demand on sustainability of long-span cable-supported bridges will become a new challenging issue. Wind-resistant bridges and wind energy harvest on bridge sites, representing both negative and positive impacts of wind, may be combined to form a self-sustainable system.

Another important aspect making bridge wind engineering sustainable as a subject is the education of future generations. It is most likely that some viewpoints introduced in this book will be overruled, and some methodologies will be improved, by our young generation in the near future. This was the main purpose that compelled the author to write this book.

Index

Wind Effects on Cable-Supported Bridges, First Edition. You-Lin Xu.